Materials and Mechanisms of Superconductivity
High Temperature Superconductors

On the cover: Scanning SQUID microscope image of a blanket coverage of Tl2201 on a π-ring geometry tetra-crystal sample, cooled in nominal zero field and SSM imaged at 4.2 K.
(From 'Phase-sensitive tests of pairing symmetry in cuprate superconductors' by C.C. Tsuei and J.R. Kirtley, pp. 4-11)

Invited Papers of the Fifth International Conference on

Materials and Mechanisms of Superconductivity High Temperature Superconductors

Beijing, China, February 28 – March 4, 1997

Editors:

Yu-Sheng He

Cryogenic Laboratory, Chinese Academy of Sciences, Beijing, China

Pei-Heng Wu

Department of Electronic Science and Engineering, University of Nanjing, China

Li-Fang Xu

Cryogenic Laboratory, Chinese Academy of Sciences, Beijing, China

Zhong-Xian Zhao

National Laboratory for Superconductivity and Institute of Physics, Chinese Academy of Sciences, Beijing, China

1997

ELSEVIER

AMSTERDAM-LAUSANNE-NEW YORK-OXFORD-SHANNON-SINGAPORE-TOKYO

ELSEVIER SCIENCE B.V.
Sara Burgerhartstraat 25
P.O. Box 211, 1000 AE Amsterdam, The Netherlands

REPRINTED FROM PHYSICA C VOLUMES 282-287 PART 1
(Parts II-IV not available in book form)

This volume contains the complete Table of Contents plus Author and Subject Indices of the proceedings (Parts I-IV).

ISBN: 0 444 82835 4

This book is printed on acid-free paper.

Transferred to digital printing 2006

Preface

Following the previous four meetings in the series (Interlaken, Switzerland, 1988; Stanford, USA, 1989; Kanazawa, Japan, 1991; and Grenoble, France, 1994), the International Conference on Materials and Mechanisms of Superconductivity – High Temperature Superconductors V (M^2S-HTSC V) was held in Beijing, China, February 28–March 4, 1997. These volumes are the Proceedings of the Beijing Conference.

Some 1100 scientists from 46 different countries or regions participated in the Beijing Conference contributing more than 1100 papers. In addition to these, 116 papers were invited. Covering quite completely the topics of recent interest in the field, these papers described the major achievements gained over the period since the last conference in Grenoble.

Due to the size limit of the journal in which the Proceedings would be published, we were forced to allot 2 pages to each contributed paper. To ensure the high academic qualities required by Physica C, each contributed paper was carefully read by at least one referee before it was accepted for publication in the Proceedings.

The papers in the Proceedings are arranged in such a way that the conference summaries come first, followed by the plenary talks and tutorial lectures. Then there are the other invited talks after which come all the contributed papers, both appearing according to the classifications.

As editors of the Proceedings, we regret that there were too many referees to list their names here. Nevertheless, we would like to take this opportunity to express our sincere gratitude to all of them for their hard work and high efficiency without which it would have been impossible for us to complete our job. Our special thanks go to Prof. Dr. P.F. de Châtel who made all the necessary arrangements for the publication of the Proceedings, and to Dr. M.B. Brodsky who carefully read many manuscripts and made valuable comments when he was busy attending the conference. The competence and enthusiasm of all the members of the Publication Committee and Local Scientific Committee is highly appreciated. We enjoyed working together with the staff of Elsevier and Physica C, in particular with Ms. J. Happée whose expertise and devotion contributed so much to the Proceedings.

Yu-Sheng He
Pei-Heng Wu
Li-Fang Xu
Zhong-Xian Zhao
Editors

Opening Address

Ladies and Gentlemen,

On behalf of the Organizing Committee, I am very pleased to announce the opening of the 5th International Conference on Materials and Mechanisms of Superconductivity – High Temperature Superconductors. I am also pleased to extend my warmest welcome to all the distinguished guests who are here at this opening ceremony, in particular to Professor Zhou Guang-Zhao, President and Academician of the Chinese Academy of Sciences, and President of the Association of Science and Technology of China; and Professor Zhang Cun-Hao, President of the National Nature Science Foundation of China.

Exactly 9 years back, we had the first meeting in Interlaken (Switzerland) organized by Professor K. Alex Müller and his colleagues. Since then meetings were held in Stanford (United States), Kanazawa (Japan), and Grenoble (France) subsequently.

We feel very much honored to get the chance to organize this 5th conference in Beijing. As the host, I heartily thank all of you for your support already given and to be given to us. We are very glad to have more than 1100 participants from 46 countries and regions gathering here, bringing with them more than 1200 papers to present at the meeting.

Now, let's welcome Professor Zhou Gunag-Zhao to give the opening speech.

Zhong-Xian Zhao
Professor and Academician
National Laboratory for Superconductivity
and Institute of Physics
Chinese Academy of Sciences

Opening Address

Ladies and Gentlemen,

First of all, on behalf of the Chinese Academy of Sciences, and the Association of Science and Technology of China, I would like to express my cordial congratulations on the opening of the International Conference on Materials and Mechanisms of Superconductivity – High Temperature Superconductors V (M^2S-HTSC V) in Beijing, and to warmly welcome all of you, scientists from all over the world, to this important event.

The discovery of high temperature superconductivity not only has opened many possibilities for potential technical applications, but also has provided a unique, challenging research subject for condensed matter physics and material sciences. As we all know, high temperature superconductivity appears in systems with strong electron correlation and constitutes one of the key issues in condensed matter physics. We believe that the understanding of its mechanism will greatly promote the future developments of this branch of science. The last ten years have witnessed great progress in the research on the physical properties, on the mechanisms, and on the technical applications. The very timely convening of this International Conference certainly will have profound impacts on these subjects.

For the last 20 years since the opening to the outside world, China has made remarkable advances in developing the socialist market economy with national specifics. The passing away of Mr. Deng Xiao-Ping is an enormous loss to the entire nation. As the chief architect of the policies on reform and openness, he was a great friend of the intellectuals and a warm advocate of respect for knowledge and respect for talents. The great cause initiated by him will be continued with further success.

The Chinese Government has established the policy of developing the national economy with particular emphasis on science and education, providing the Chinese scientists and engineers with great opportunities and challenges. We are also deepening reforms in science and technology systems to meet the ever increasing demands of social progress and economical development. For this purpose, we have to learn and to share experience with scientists all over the world.

The research on high temperature superconductivity has been very favorably supported by the Chinese Government, enabling the Chinese scientific and technological workers in this field to make remarkable progress. I am sure that this conference will further promote the international exchanges and collaborations between the Chinese scientists and their foreign colleagues.

I also believe that, via fruitful exchanges of new ideas and new results, this conference will contribute significantly to the research of high temperature superconductivity worldwide, advancing it to a new level.

I wish you a very successful conference and a very enjoyable stay in Beijing. Thank you all.

Guang-Zhao Zhou
President and Academician
The Chinese Academy of Sciences
President, The Association of Science
and Technology of China

Report from the Organizers

The 5th International Conference on Materials and Mechanisms of Superconductivity – High Temperature Superconductors V (M^2S-HTSC V) was held in Beijing, China, from February 28 to March 4, 1997. Some 1100 researchers from 46 countries or regions participated in the Conference, contributing more than 1100 papers in addition to 116 papers invited.

In organizing the conference our primary concern was to have all the major developments in the field properly covered, paying particular attention to those made since the last conference in Grenoble in July 1994. We also did our best to include as many different viewpoints as possible. The members of the International Advisory Committee, the Program Committee, and the Local Scientific Committee did an excellent job by recommending what they thought to be the most important achievements. With prominent scientists from (alphabetically) Argentina, Australia, Belgium, Brazil, Canada, China, France, Germany, India, Israel, Italy, Japan, Korea, New Zealand, Norway, Russia, Spain, Sweden, Switzerland, Tajikistan, The Netherlands, Ukraine, UK, and USA as the members of these committees, coordination of the effort on the global level was guaranteed. Relying on these recommendations, the members of the Program Committee made final decisions on whom to invite to talk on which topics so that both the speakers and the topics could represent the main trends witnessed in the period since the last conference. The program of the conference was worked out accordingly.

Right after the official opening of the conference on the first day, a plenary session was held to review the progress in areas of common interest such as the recent achievements of high T_c superconductivity, phase-sensitive tests of pairing symmetry in cuprate superconductors, and *c*-axis charge dynamics of high T_c cuprates. Then the conference was split into parallel sessions, oral (with invited and contributed talks) or poster. Almost every aspect of the field was covered. Nine microsymposia were arranged to focus attention respectively on pseudo-gaps, Sr_2RuO_4, two subsystem superconductivity, phase-sensitive experiments, stripe phase and phase separation, isotope effects, ladder compounds, single crystals growth and inhomogeneity, and high frequency electronic devices. And three tutorial lectures were given, especially to help the young participants. On the final day before the closing of the conference, a plenary session was held under the title of 'Reviews, Prospects and Concluding Remarks' to discuss the open questions, experiments, and materials needed for applications.

Two prizes were awarded at this conference. The Bardeen Prize sponsored by his friends and former colleagues went to P.W. Anderson for his contributions to the understanding of broken symmetry, the order parameter in the A and B phases of superfluid helium three and the role of impurities in metallic superconductors, while the Matthias Prize sponsored also by his friends and former colleagues was awarded to R.J. Cava and B. Batlogg for their leading work on a variety of superconductors through the successful combination of creative materials chemistry and physics. In addition to these two prizes, the International Advisory Committee decided that a third one, the Kamerlingh Onnes Prize, would be created especially for experimentalists in the field. The members of the editorial board of Physica C would look for the sponsors of this new prize. It was also decided that three sub-committees would be established accordingly.

Three satellite conferences, i.e., International Workshop on Critical Currents in Superconductors for Practical Applications, International Workshop on Crystal Chemistry of High T_c Superconductors, and International Symposium on the Intrinsic Josephson Effect and Terahertz Plasma Oscillations in High T_c Superconductors were arranged by independent organizing committees and held in Xi'an (China), Hefei (China), and Sendai (Japan) respectively.

The conference could not have been run without the support of all the researchers in the international superconductivity community. The effort of the members of the International Advisory Committee, Program Committee, Publication Committee, Local Scientific Committee, and Local Committee is gratefully acknowledged. Our special thanks go to the generous sponsors whose names are listed in the Proceedings.

It was decided in Beijing that the 6th conference in this series on Materials and Mechanisms of Superconductivity – High Temperature Superconductors (M^2S-HTSC VI) will be held in Houston (USA). We wish the organizers of the next conference a great success.

Zhong-Xian Zhao — *Co-chairman*
Zi-Zhao Gan — *Co-chairman*
Guo-Zhen Yang — *Co-chairman*
Duo Jin — *Secretary and Chairman of Local Committee*
Yu-Sheng He — *Chairman of Publication Committee*

The John Bardeen and Bernd Matthias Prizes

John Bardeen Prize

The award committee of M²S
is privileged to present the John Bardeen Prize to

Phillip W. Anderson

for

his contributions to the understanding of broken symmetry, the order parameter in the A and B phases of superfluid helium three and the role of impurities in metallic superconductors.

February 28, 1997 Chairman, award committee

The John Bardeen Prize supported by contributions of friends and former colleagues of John Bardeen

Bernd Matthias Prize

The award committee of M²S
is privileged to present the Bernd Matthias Prize to

Bertram Batlogg and Robert J. Cava

for

their leading work on a variety of superconductors through the successful combination of creative materials chemistry and physics.

February 28, 1997 Chairman, award committee

The Bernd Matthias Prize supported by contributions of friends and former colleagues of Bernd Matthias

Committee Members

Co-chairmen Zhong-Xian Zhao
Zi-Zhao Gan
Guo-Zhen Yang

Scientific Program Committee

Z.-Z. Gan, *Chairman*	*Beijing*	Y.-S. He	*Beijing*
T. Claeson	*Göteborg*	K. Kitazawa	*Tokyo*
J.R. Clem	*Ames*	T.M. Rice	*Zürich*
H. Fukuyama	*Tokyo*	P.-H. Wu	*Nanjing*
D.M. Ginsberg	*Urbana*	L. Yu	*Trieste / Beijing*

Local Scientific Committee

Guo-Zhen Yang, *Chairman*
Zhao-Jia Chen
Yuan-Dong Dai
Jun-Ren Fang
Sun-Qi Feng
Chang-De Gong
Ru-Shan Han
Hong-Cheng Li
Man-Yuan Li
Liang-Zhen Lin
Wei-Chuan Qiao
Li-Qin Shao
Jue-Lian Shen
Zhao-Bin Su
Hong-Jie Tao
Si-Shen Xie
Ding-Yu Xing
Shou-Sheng Yan
Qian-Sheng Yang
Xi-Xian Yao
Dao-Le Yin
Guan-Sen Yuan
Qi-Rui Zhang
Qi-Shao Zhang
Yu-Heng Zhang
Lian Zhou
Lu-Wei Zhou
Dao-Ben Zhu
Tie-Yong Zuo

Local Committee

Duo Jin, *Chairman*
Ying-Fei Chen
Lian-Hong Ma
Lai-Feng Li
Yi-Ping Liu
Li-Fang Xu
Xin-Min Xu
Yan Wang
Zhong Zhang
De-Juan Zheng

Publication Committee

Yu-Sheng He, *Chairman*
Jun Chen
Cheng Dong
Shi-Ping Feng
Zheng-Xiang Gao
Jian-Dong Guo
Chang-Qing Jin
Yang Li
Li Lu
Jia-Cai Nie
Han-Ting Wang
Shi-Guang Wang
Xiao-Ping Wang
Yu-Peng Wang
Hai-Hu Wen
Guang-Cheng Xiong
Ming-Liang Zhang
Xing Zhu

Acknowledgements

The organizing committee is indebted to our French colleagues, Organizer of M^2S-HTSC IV, Grenoble, 1994, for providing essential start-up funds.

We are grateful for the partial support from the following organizations:

The State Commission of Science and Technology
The State Commission of Education
The National Natural Science Foundation of China
The Chinese Academy of Sciences (CAS)
The Association of Science and Technology of China
South–South Foundation, CAS
The National Center for R&D on Superconductivity of China
China National Non-ferrous Metals Industry Corporation
China Sciences Group
Legend Group
Founder Group

ELSEVIER

Physica C 282–287 (1997) xix–xxiii

PHYSICA C

Conference summary

Reviews, prospects and concluding remarks. High-T_c superconductivity — Where next?

T.M. Rice

Theoretische Physik, ETH-Hönggerberg, CH-8093 Zürich, Switzerland

When the summary talks were discussed in the program committee, there was a strong wish that we should close with a more general outlook on high-temperature superconductivity and not just a summary of the theoretical results presented at the conference. After ten years of high-T_c superconductors and five conferences in this series, it seemed appropriate to take stock and look for new inspirations and thus I will attempt to look beyond the trenches, so to speak. The goal has always been clear, also in the series of conferences on the Materials and Mechanisms of Superconductivity that preceded the high-T_c era, and it can simply be summarized by the motto:

T_c: *Take me higher*!

This is our inspiration and it has been so through the feasts when T_c rose rapidly and the famines, when the highest T_c stood still for years on end.

The discovery of the high-T_c cuprate superconductors by Bednorz and Müller in 1986 was a watershed in this search. In the era prior to 1986 the conferences were dominated by classic BCS superconductors. The criteria to get a 'high T_c' was clear within the BCS theory, namely one should look for good metals with large densities of states and strong electron–phonon interactions. This is not to say that the theorists were successful in predicting materials, but at least the criteria necessary for a good superconductor were clear, thanks to the success of the BCS theory.

The new era that began in 1986 dominated by the high-T_c cuprate superconductors, is the era of unconventional superconductors. The definitive confirmation of the d-wave character of the superconductivity in the cuprates is a great advance since the last conference in Grenoble. This is due to the series of elegant phase sensitive experiments by various groups in different countries. At this conference these experiments were described in detail in the lectures by C.C. Tsuei and D. van Harlingen, so I will not repeat these here. Rather let me emphasize the obvious conclusion from this clear microscopic difference between the classic BCS superconductors and the high-T_c superconductors, namely that they have a different origin and that different criteria are required for high-T_c superconductivity than for the classic BCS superconductors. The task facing us then is much more difficult than previously. The pre-1986 era of classic BCS superconductors were the good old days when, at least one knew the criteria for good superconductors, even if the prediction of specific materials was not especially successful.

Another way of looking at the breakthrough of Bednorz and Müller, is that they succeeded in finding high temperature superconductivity in the transition metal oxides, a class of materials known for many years for a wide range of electronic phase transitions driven by strong electron–electron interactions and strong electron–phonon interactions. Although strictly speaking Cu is a noble metal, the partly filled 3d-shell in the copper oxides places them in the class of transition metal oxides. In Fig. 1 this diversity of electronic phase transitions is illus-

PII S0921-4534(97)00185-8

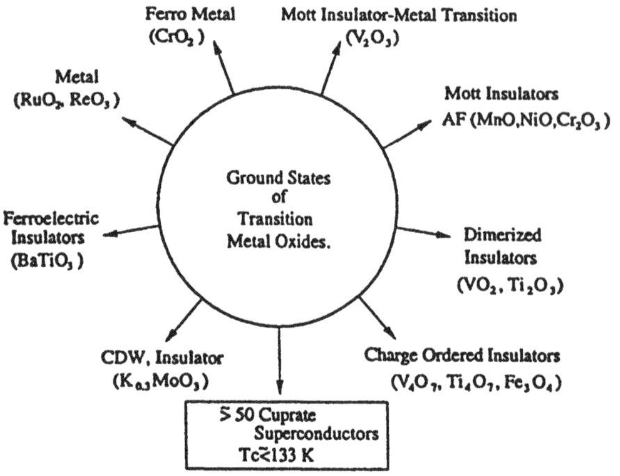

Fig. 1. Phases of transition metal oxides.

trated. It ranges from antiferromagnetic insulators (Cr_2O_3, NiO) to ferromagnetic metals (CrO_2) and metal–insulator transitions involving antiferromagnetism (V_2O_3), spin pairing (Ti_2O_3) or charge ordering (V_4O_7, Ti_4O_7). The group of high-T_c superconductors encompasses now over fifty materials listed by R.J. Cava in his talk. Viewed from the perspective of transition metal oxides, the highest transition temperature, T_c, of the cuprate superconductors ($T_c \simeq 133$ K at atmospheric pressure or $T_c \simeq 160$ K at higher pressures) is not that high at all and is exceeded by most of the transition temperatures encountered in the other phase transitions. This is an optimistic observation which suggests possibilities to raise T_c. But the fact remains that after ten years this group of 50 cuprate superconductors are all variations on a single theme, namely lightly doped CuO_2-planes, and there has not been a general revolution in the transition metal oxides nor a broad class of new superconductors.

If one approaches the problem in a conservative way and many of us, in physics are naturally conservative in our thinking, then one looks for the minimal generalization of the well established BCS theory to unconventional forms of superconductivity. The basis for BCS theory is the formation first of a Landau–Fermi liquid i.e. a quantum coherent state. Then one looks for strong interactions which crossover to an attractive pairing interaction in some higher angular momentum channel. Does this description work to give us a high-T_c superconductor? The answer to this question must be an overwhelming No! At this conference, finally among all the transition metal oxides which form Landau–Fermi liquids, a superconductor has been reported, namely Sr_2RuO_4. This new superconductor was the subject of a symposium and as Y. Maeno discussed, the evidence for unconventional superconductivity is considerable. All the ingredients are present: clear T^2-law in the resistivity and strong interactions causing large Landau parameters (e.g. a mass enhancement ~ 3 or 4), but the value of T_c is low, ~ 1 K! Especially since Ru is a 4d-transition metal a large energy scale might be expected. Further, there are many other metals with Landau–Fermi liquid states obeying a T^2-law in the resistivity. Some even, such as V_2O_3 at a pressure of 26 Kbar, are in close proximity to an antiferromagnetic insulator but have even lower values of T_c, that is if they have any T_c at all. The conclusion that follows is that the formation of a strongly quantum coherent Landau–Fermi liquid does not lead to a high value of T_c even if the residual interactions are strong. So we must look elsewhere — a point we will return to later.

If we look more closely at the cuprate superconductors then there is a lot of evidence in favour of the t–J model, or some variant thereof, as the basic underlying microscopic model. This model has a simple chemical interpretation. The parent stoichiometric insulators have an oxidation state Cu^{2+}, which is an $s=\frac{1}{2}$ ion, and the magnetic interactions are well described by a simple n.n. AF Heisenberg model with a large value of the exchange interaction, $J \simeq 1500$ K. Hole doping introduces the oxidation state Cu^{3+} in the low spin configuration, $s=0$ (a.k.a. the Zhang–Rice singlet). The Cu^{2+} and Cu^{3+} can interchange places with a hopping matrix element, t, that describes the electron transfer. The t–J model is then the simplest model that incorporates these two key processes. Of course, it may not accurately describe all interactions at low energies in the cuprates but if we ask for the simplest model which can describe a quantum fluid of $s=\frac{1}{2}$ Cu^{2+}, and $s=0$, Cu^{3+}-oxidation states, then this is it.

If we want to break out from purely cuprates then the nearest members in the periodic table are the nickelates, leading one to ask, are there nickelate analogs of the cuprates which would also be described by a t–J model? The first step would be to achieve a stoichiometric insulator with a $s = \frac{1}{2}$ Ni^{3+}-oxidation state. Since Cu^{2+} is formally $3d^9$ with 1 hole in the $d_{x^2-y^2}$-state of the e_g-manifold, a $s = \frac{1}{2}$ Ni^{3+}-oxidation state with $3d^7$ and 1 electron in the $d_{3z^2-r^2}$-state of the e_g-manifold, should be a direct analog. Already however this is a problem. Some years ago, Cava et al. [1] studied the alloy series $La_{2-x}Sr_xNiO_4$ which, at $x = 1$, has Ni in the oxidation state Ni^{3+}, and in a crystal structure familiar from the cuprates. By the reasoning quoted above one should look for a Mott insulator which would be directly analogous to the parent insulating cuprates. Instead Cava et al. [1] report that $LaSrNiO_4$ is a metal with an enhanced value of the susceptibility which does not undergo any obvious phase transition. Certainly it is neither a Mott insulator nor a high-T_c superconducting metal.

So what went wrong? Apparently we have not really achieved a $3d^7$-configuration. Rather it seems that the hybridization between Ni-3d and the O-2p orbitals is so strong that one has predominantly a $3d^8$L configuration, where L stands for a ligand hole in the O-2p orbitals. This interpretation implies that $LaSrNiO_4$ should be a 3d-analog of the heavy fermion metals made from rare earth or actinide elements (e.g. Ce or U). Of course it would be interesting in its own right to examine this material in more detail but it is not the analog of the cuprates that we are looking for.

Perhaps there are further tricks that one can play. For example, one might consider trying to replace the apical O with Cl to force the $3d^7$ configuration that one wants. Then if one can succeed in making a Mott insulator with a Ni^{3+}-oxidation state, one could try to dope with Ni^{2+} — again like Cu^{3+} in a low spin $s = 0$ configuration. Can this be achieved? I don't know. Will it give us a new family of high-T_c superconductors? Well, at this point I can do no better than quote from the lecture of M. Takano, who has much more experience in this business than me. He coined the memorable sentence, "*Diamonds or Garbage? – Hard to predict*".

I would like to make some remarks about the microscopic theories discussed here. There seems to be a consensus emerging about some points. One point is that the generic phase diagram of the high-T_c cuprates sketched in Fig. 2, is characterized by two critical points. One is in the underdoped regime and

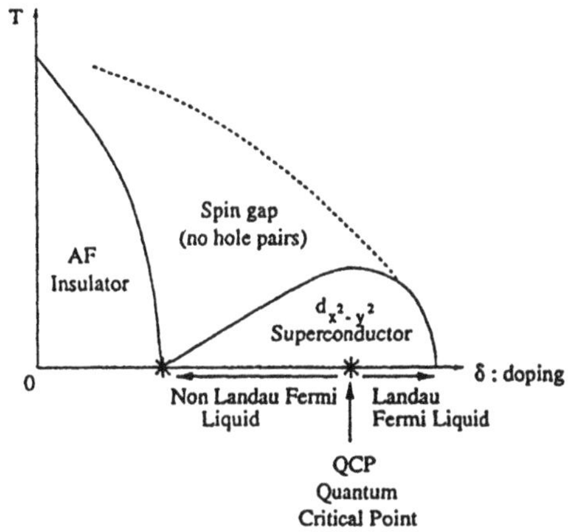

Fig. 2. Schematic phase diagram of high-temperature superconductors.

characterizes the transition from the AF ordered state to the superconducting state. This transition has not been discussed so much but at this conference S.C. Zhang presented some interesting new ideas based on a rotation of an order parameter of constant length in a five-dimensional space (i.e. SO(5) symmetry) which should stimulate more investigations of this transition. The other critical point is a quantum critical point, which should lie hidden under the superconducting transition, and describes a transition between Landau and non-Landau–Fermi liquids in the extrapolated normal phase, as shown in Fig. 2. The arguments for this anomalous behaviour at hole doping around and below optimal doping go back to the earliest theory papers on the high-T_c superconductors by P.W. Anderson. They have been made many times, e.g. from the form of the resistivity. At this conference, the lectures by S. Uchida and Y. Ando discussed how the resistivity perpendicular to the CuO_2-planes, $\rho_c(T) \to \infty$ as $T \to 0$ K. In the plane both behaviours $\rho_{ab}(T) \to \infty$ or to a constant as $T \to 0$ K were found when the superconductivity transition was suppressed in different systems. The latter implies 2D confinement and is incompatible with Landau–Fermi liquid behaviour as Anderson has stressed many times. The former is also highly anomalous. However, overdoped cuprates behave in a much more standard way supporting Landau–Fermi liquid behaviour. If we accept this reasoning then there must be a zero temperature quantum critical point which signals the breakdown of the Landau–Fermi liquid.

There is also considerable agreement that the way to achieve such a breakdown is through the approach to a new phase transition. But there is no agreement as to the nature of this phase transition. For example, the suggestions range from a lattice instability (C. Castellani), or a combination of stripes and frustrated phase separation (V. Emery and J. Zaanen) or a multi-orbital instability (C. Varma) to a magnetic instability such as AF order or a RVB state proposed by many speakers (e.g. H. Fukuyama, P. Lee, D. Pines, D. Scalpino). A feature of all these different proposals is that the instability is approached but not actually crossed, so that no symmetry is explicitly broken. For this reason, I personally favour the latter proposals since the strong quantum fluctuations of isotropic $s=\frac{1}{2}$ spin systems with AF interactions allow incipient AF order to be converted to RVB behaviour. Viewed in this way, it is this highly quantum nature that opens up a window allowing superconductivity to appear in ways not possible at a more classical phase transition.

One of the big open problems is to better understand the nature of this non-Landau–Fermi liquid in 2D. The anomalous behaviour of such a liquid is best characterized at low temperature and under clean conditions. The proposal of P. Lee to find the cleanest underdoped superconductor with a low value of T_c which can be suppressed in a laboratory magnetic field, is the most promising way to go, it seems to me. The characteristics that we know to date, such as 2D confinement mentioned earlier, and a truncated Fermi surface reported here by Z. Shen, answer some but not all questions we have about this intriguing state.

One related question concerns the static AF order and stripe behaviour reported here by J. Tranquada in (214) alloys with the LTT structure. Is this also a characteristic of the generic non-Landau–Fermi liquid or is it rather a fingerprint of a competing fixed point, competing that is with fixed point that is associated with superconductivity? For what it is worth, I tend to the latter point of view.

A new and promising direction since the Grenoble meeting, is to modify the cuprates. For example, as we heard in the ladder symposium from M. Takano, one can replace the infinite CuO_2-planes by Cu_2O_3 or Cu_3O_5-planes which are characterized by weakly coupled 2- and 3-leg ladder structures. This can be achieved keeping all Cu-ions in a square planar coordination with O, but modifying the intersquare coordination from all corner sharing (CuO_2 plane) to partly edge sharing. The latter causes much weaker ferromagnetic bonds, which in turn allows one to effectively break-up the cuprate plane into weakly coupled ladders. A key question that interested D. Scalapino, me and other theorists, was whether hole doping could be realized and superconductivity obtained. The answer reported here by J. Aktemitsu is yes. He and his collaborators found a superconducting transition with a $T_c \simeq 10$ K in a ladder cuprate $Ca_{13.4}Sr_{0.6}Cu_{14}O_{41}$ at a high pressure (~ 3 GPa) but only after an insulating phase which normally occurs below ~ 50 K, is suppressed. While the interpretation that the increasing Ca-content leads to hole doping of the ladders may not be definitively proven, it is certainly very reasonable. The nature of the insulating phase is not yet determined but if the predicted hole pairing, discussed for example here by D. Scalapino, occurs, it should be a CDW phase of hole pairs. This will be a crucial test of our understanding of the ladder cuprates.

A surprising feature is how close the similarity to the organic superconductors is. The 1D organics based on TMTSF molecules have a very similar phase diagram, and again superconductivity, as D. Jerome stresses, is only found in a narrow pressure window as an insulating phase (in this case a SDW phase) is suppressed. There are also 2D organic superconductors for example the $(ET)_2X$ compounds discussed here by K. Kanoda. In this case there are also strong similarities to the cuprates with a superconducting phase which most probably is 'd-wave' abutting a Mott insulating AF-order phase. Even more striking is the ratio of the values of T_c in the 2D and 1D systems. In both cases it is an order of magnitude, 12 K vs. 1.2 K in the organics and ~ 100 K vs. 10 K in the cuprates. Clearly, we have many interesting and unexpected puzzles to unravel.

Finally let me broaden the horizon to include the two other material systems with substantial values of T_c ($T_c \sim 30$ K) that were discovered since 1986, namely the A_3C_{60} and $Ba_{1-x}K_xBiO_3$ systems. Are there general features in common with the cuprates? I think there are. None of the above have clearly established quantum coherent Landau–Fermi liquids in the normal phase. Instead in all cases we have a small number of charge carriers coupled strongly to a large number of electronic states at infrared energy scales. Only a small fraction of the low energy optical weight goes into the superconducting δ-function peak and most remains in a mid-infrared peak. This suggests a strong interaction of the carriers through low lying electronic degrees of freedom. Further, it is tempting to speculate on a common origin for all high temperature superconductivity, as a state which generates quantum coherence for these charge carries that are somehow prevented from forming the usual Landau–Fermi liquid. Is this the general recipe for high-T_c superconductivity that we are seeking? I do not think one can answer yes definitively but at least this seems to be part of the story.

To conclude, an inspirational talk needs a slogan for our next meeting, which probably will be in the year 2000 in Texas. Since everything is bigger in Texas we could aim for *2000* in *2000*! But I live in Switzerland, a much smaller country, so I and also I suspect most of you, would be very happy with the more modest goal, *200* in *2000*. Thank you.

References

[1] R.J. Cava, B. Batlogg, T.T. Palstra, J.J. Krajewski, W.F. Peck, A.P. Ramirez, L.W. Rupp, Phys. Rev. B 43 (1991) 129.

ELSEVIER

Physica C 282–287 (1997) xxiv–xxx

PHYSICA C

Conference summary

Reviews, prospects and concluding remarks. Comments on experiments

Bertram Batlogg

Bell Laboratories, Lucent Technologies, Murray Hill, NJ 07974, USA

1. Cuprates

By far the largest fraction of work reported at this conference dealt with cuprate superconductors, with a focus on the elucidation of the electronic phase diagram as function of hole doping, particularly the pseudogap formation, on the superconducting state with the order parameter symmetry and the ensuing consequences, and the stacking of superconducting layers that are intrinsically Josephson coupled, and the large area of vortex physics with the rich phenomena in the static and dynamic limits.

2. The superconducting order parameter symmetry

A major milestone has now been reached in identifying the superconducting order parameter in (most) cuprates. An overwhelming majority of experiments indicate a predominant $d_{x^2-y^2}$ symmetry. This consensus has emerged from results that reflect the quasi-particle density of states and from experiments that are sensitive to the phase of the order parameter as it expresses itself in macroscopic quantum coherence effects. Impressive improvements in measurement techniques and in single-crystal preparation, together with cleverly conceived experiments have now led to quite a clear picture of the superconducting state. Those experiments include (in no particular order): 'corner'-junctions, 'tri-crystal' and 'tetra-crystal' junctions exhibiting spontaneously forming $\phi_0/2$ vortices, Andreev-scattering, *c*-axis tunneling, angle-resolved photo-electron spectroscopy (ARPES), magnetic torque measurements, NMR relaxation rate, magnetic penetration length parallel and perpendicular to the CuO_2-planes, electronic specific heat, electronic Raman scattering, quasi-particle tunneling, and others.

Questions remain about the admixture of non-d wave symmetry due to distortions of the CuO_2 plaquette (rectangular rather than square). While $d_{x^2-y^2}$ appears to be supported by experiments on hole-doped cuprates with one or more closely coupled CuO_2 layers, further work will be needed for the electron-doped Nd-cuprate.

These results on the pair symmetry pose some constraints on the models of the microscopic pairing mechanism, but it appears that almost all models under serious consideration will produce d-wave pairing.

3. Consequences of $d_{x^2-y^2}$ pairing

The node and sign change of the $d_{x^2-y^2}$ gap function have several far-reaching consequences. They influence, among others, the dissipation of moving vortices, the dissipation in the electromagnetic response and the current-carrying capacity across grain-boundaries, and junctions in general.

Whenever two grains touch with the slightest crystallographic misalignment, the order parameters

PII S0921-4534(97)00187-1

will also be misaligned. Given the d-wave symmetry, this leads to a suppression of the order parameter at the interface even when the two grains meet in an atomically clean way, i.e. without any 'dead' layer due to compositional modifications. Thus, an intrinsic Josephson junction is formed that affects the capacity of the grain boundary to carry supercurrent. The boundary between two thin film sections with different orientation on a macroscopic scale provides a particularly interesting example when looked at on a microscopic scale. Such a boundary is not smooth, but faceted, and follows the interface between small crystallites with varying degrees of misalignment. Thus, the critical current density varies along the boundary. This spatial variation of J_c can include even a local sign reversal, resulting in the formation of local current loops whose magnetic field has indeed been imaged with a scanning SQUID microscope.

Another consequence of d-wave pairing can be seen in the non-linear electrodynamic response, such as the power-dependent absorption or transmission of microwaves or terahertz radiation. The node in the gap function gives rise to quasi-particle creation of arbitrary small energies and a superfluid density that decreases linearly with the condensate velocity. The intrinsic pair breaking by photon absorption would have to be of concern in passive microwave devices, such as filters with very high Q-factor elements. At this stage, however, it appears that non-linearities and other unwanted properties of microwave filters are mainly caused by weak links in the superconducting thin films. As the film quality keeps improving, one will face the ultimate non-linear response due to the d-wave order parameter.

4. The electronic phase diagram

Given the variety of crystal structures and the chemical methods used to introduce holes into the CuO_2 layers, it is substantial progress that a 'generic' electronic phase diagram can be sketched for all compounds. Terms such as 'underdoped', 'optimally doped' and 'overdoped' have become common language to describe the degree of hole doping, which is the key parameter determining the electronic properties.

The study of the normal state properties has always been a central theme in the past ten years, and the unusual charge response, particularly at optimal doping, has been intensely studied and well documented. This has led to the recognition that the normal state of cuprates is most unusual and requires a description different from the Landau–Fermi-liquid concept, which so far had been most successful in describing metals. At this conference the underdoped region of the phase diagram has been on center stage. The discussion of the underdoped cuprates started with some intensity at the Grenoble conference with the recognition of a hole density dependent scaling temperature T^* in transport data, a suppression of the static and dynamic spin response, and a reduction of the electronic heat capacity. These data had been interpreted in terms of a pseudogap that develops well above T_c.

The phenomenology of the pseudogap has been much advanced since then. Many more measurements of spin and charge response have been performed, including in-plane and out-of-plane static and dynamic conductivity, Hall effect, uniform susceptibility, NMR relaxation rate and isomer shift, Raman phonon frequencies, as well as electronic Raman scattering. High resolution specific heat measurements over wide doping levels provide the thermodynamic signature of the pseudogap and allow for a quantitative estimate of the pseudogap (and the superconducting gap) as function of hole doping. It is observed in both single and double layer compounds. Accordingly, the magnitude of the pseudogap increases as the hole concentration is reduced away from the optimal level. The pseudogap is also seen directly in tunneling spectroscopy and in angle resolved photo-electron spectroscopy (ARPES). These latter techniques add an important aspect to the discussion because they probe the single particle density of states. ARPES found the pseudogap to have the same momentum (angle) independence as the superconducting gap and to increase slightly in the underdoped region. Similarly, the tunneling conductance spectra reflect the persistence of a gap of the same magnitude as the superconducting gap to temperatures well above T_c. In each case, the difference between underdoped and optimally (or overdoped) samples in striking. In the latter ones, the gap features disappear near T_c.

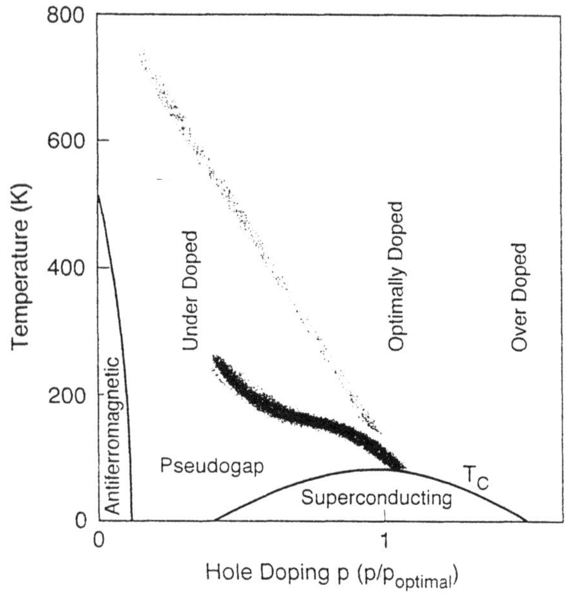

Fig. 1. Electronic phase diagram of layered cuprates (schematic). The transitions into the antiferromagnetically ordered and the superconducting states are well defined. Crossover regions in the underdoped part are indicated by shaded bands.

A generalized phase diagram is shown in Fig. 1. It captures the present situation and may need to be revised as our understanding develops further. While the superconducting T_c and the Neel points represent phase transitions, the two other shaded areas indicate crossover regions. The upper crossover may be associated with the development of antiferromagnetic correlations (manifested e.g. in the uniform static susceptibility). The crossover temperatures are not the same for all compounds. The first systematic studies to locate them for various compounds have been reported. (We note parenthetically that characteristic features in transport properties such as the in-plane resistivity appear to be connected also with the upper crossover. This is somewhat unexpected as transport tends to be dominated by processes with large momentum transfer). The lower crossover may be associated with the pseudogap development. The details of the lower crossover are not clear at the moment and require more studies. It is desirable to measure the magnitude of the pseudogap for various hole concentrations to high temperatures. The existence of pseudogaps certainly adds one more intriguing aspect to the physics of cuprate metals.

A new portion of the electronic phase diagram could be accessed recently through the application of very high (60 T) pulsed magnetic fields. These fields are sufficiently strong to suppress superconductivity when T_c is moderately high, such as in LSCO. When following the resistivity to the lowest temperatures one finds what appears to be a critical hole composition that separates non-metallic conductivity in the underdoped region from (anisotropic) metallic conductivity in the overdoped region. The crossover lies near the optimal concentration, at least for LSCO. It would be most desirable to expand these studies to several more compound families to clarify whether there indeed exists a $T = 0$ critical point for all of the doped layered cuprates.

Significant progress has also been reported from magnetic response measurements using inelastic neutron scattering. The susceptibility of LSCO has been determined on a absolute scale up to $\sim 1/4$ eV and in underdoped YBCO up to $\sim 1/8$ eV. In the slightly underdoped compound ($Sr_{0.14}$) a pronounced maximum at ~ 20 meV dominates the imaginary part of the susceptibility. New excitations at low energies with an instrumentally limited long correlation length have been discovered below T_c, whose origin remains to be explained. In YBCO the '41 meV resonance' has been followed into the underdoped region and its energy is found to decrease linearly with T_c. Again, the origin of this sharp excitation is not clear. Remarkably, its intensity disappears near T_c in the optimally doped sample, but persists to well above T_c in the underdoped sample. Thus, it will be interesting to clarify any possible connection to the occurrence of the pseudogap, and ultimately to the Neel state of the insulating parent compound.

5. Vortex physics

Understanding of vortex behavior is of great practical interest and much progress has been made in the past years. The basic picture of a vortex line is that of a stack of Josephson-coupled pancake vortices. In this highly successful approach only two intrinsic parameters dominate the physics: crystal structure and T_c. (Here we consider optimally doped compounds). The crystal structure determines the distance between CuO_2 layers and thus the effective electronic anisotropy and coupling strength between layers. T_c essentially determines the in-plane coherence length and therefore a basic energy scale. A complex vortex behavior as function of field and temperature results from the interaction between vortices, vortex pinning and coupling between the layers, because all these energy scales are comparable to the thermal energy.

The equilibrium phase diagram in the temperature-field plane is well mapped out by now. The occurrence of several phases and transitions (or crossovers) between them has been studied. Particularly impressive are the most recent thermodynamic measurements of the lattice melting transitions and the associated entropy changes, and the consistency with magnetic data. The richness of the phase diagram is comparable to that of ordinary molecular or atomic matter; hence, the term 'vortex matter'.

A broad range of experimental probes have been employed to study the dynamics of vortices. They provide both spatial and temporal resolution and give insight into the dynamics in response to variable drive forces. In Lorentz microscopy, 'vortex rivers' can be directly imaged as they flow around obstacles and as they settle in new patterns when the flow stops. Magnetic relaxation and I–V characteristics are employed to study the crossover from 2D to 3D behavior. Magnetico-optic microscopy reveals the local current density on a micrometer length scale. Nuclear magnetic resonance and muon spin resonance probe on an atomic level the longitudinal and transverse field fluctuations. Together with other techniques, such as STM, scanning SQUID microscopy and miniature Hall probe arrays, quite a complete picture of the vortex dynamics has emerged.

These insights are important for materials processing techniques to enhance the current carrying capability wires and cables. In addition, 'vortex matter' has also become a model system to study general topics such as non-linear dynamics or phase transition in the presence of disorder.

6. Atomic mass (isotope) effects

The influence of the atomic mass on the T_c of cuprates has been known in great detail for several years. Near optimal doping, the isotope effect is very small or zero, but increases even beyond $\frac{1}{2}$ in underdoped compounds with strongly suppressed T_c. More recently, a close correlation have been found between the pressure effect on T_c and the isotope effect. A microscopic understanding of these observations is missing at this point.

A new type of isotope effect has been discovered most recently in LSCO: the in-plane magnetic penetration length depends on the oxygen mass. When the hole concentration is kept constant during isotope exchange this effect indicates a change in the effective mass of the charge carriers that form Cooper pairs. These carriers are therefore interpreted as being of polaronic nature, at least in the underdoped region. The effect decreases towards optimal doping.

These studies raise some important questions about the coupling of atomic motion to the motion of charge in transition metal compounds with formally mixed valent metal ions. One would like to come to a quantitative understanding of the role of (isotropic) volume changes due to different valence states and of the Jahn–Teller type distortions. Those quantities are expected to vary on going across the 3d series, at least due to the varying degree of p–d hybridization.

7. Local charge inhomogeneities

It is convenient and customary to use the average hole density as the primary parameter in describing the cuprates, but there is experimental evidence suggesting the charge is not uniformly distributed in at least a few cases.

Best documented is the stage formation in oxygen-overdoped La_2CuO_{4+y}. Diffraction experiments show miscibility gaps and ordered arrangements of the extra oxygen ions, resembling the intercalation patterns in other transition metal oxides and chalcogenides. Microscopic probes will have to tell us how the charge is spatially distributed when the oxygen-overdoped samples become metallic and superconducting.

Static charge ordering in the form of stripes is well documented for nickelates and manganates. Also in underdoped LSCO/LBCO, there is experimental evidence suggesting stripe formation. This includes diffraction data and, more indirectly, low temperature thermal conductivity data. An intense discussion is now underway whether charge stripes are a generic feature and essential aspect of cuprate physics, or whether stripes are more specific to the LSCO and LBCO compounds with their uniquely complex rotation and tilting patterns of the CuO_6 octahedra. (For some of the researchers in this field this discussion goes back to the earliest years of cuprates. "We are still confused, but now on a more advanced level.")

Perhaps an even more subtle aspect of charge inhomogeneity occurs when oxygen ions can occupy several crystallographically inequivalent sites. Two studies, looking in great detail at Hg-1201 and YBCO with varying oxygen content, have revealed an apparent connection between site occupation and the 'generalize' electronic phase diagram. In both compounds a change in site-occupation numbers occurs near the optimal hole concentration, leading in over-oxygenated YBCO to two distinct superconducting phases with slightly different T_c.

While much of the discussion of cuprates can indeed focus on generic aspects common to all compounds, one cannot entirely forget that holes need to be introduced through chemical modifications of the charge reservoir layers, which may or may not cause significant perturbations of the local electronic structure. Such local inhomogeneities are expected to be more important in the underdoped than in the optimally doped region of the phase diagram, but further studies would have to confirm this speculation.

8. Cuprate chains and ladders

In an attempt to find a new way towards an understanding of cuprate superconductivity, an entirely new category of cuprates has been studied during the past few years. Stimulated by predictions from model calculations of doped chains and coupled doped chains (ladders), experimentalists indeed synthesized compounds with structural building blocks that realize to a high degree the model systems. This in itself is remarkable, particularly if one considers that those ladder compounds are made of the same constituent elements as the other cuprates of interest so far. Again, the elements Cu, O, Sr, La, Ca are combined, but now in different ratios, to form the ladder and chain compounds. In yet one more success of creative crystal chemistry one often needs to resort to stringent synthetic conditions, such as high pressure or very narrow temperature windows to stabilize these compounds.

In the undoped ladders, the most outstanding result is the alternation of magnetic and non-magnetic ground states for ladders of odd and even numbers of coupled chains. When a single chain is doped with holes, each hole renders one CuO_2 unit non-magnetic, in close analogy to the Zhang–Rice singlet formation in the layered cuprates. When holes are introduced in a (two leg) ladder, which has a gap in the spin excitation spectrum, the spin gap is reduced and the static susceptibility develops in a very similar way as, e.g. in LSCO with increasing Sr concentration. Ultimately, these doped ladders go superconducting as recently observed in $A_{14}Cu_{24}O_{41}$ ($A_{14} = Ca_{13.6}Sr_{0.4}$). The T_c of $\sim$ 12 K is reduced in comparison to the 2D layer compounds. The rapid development of the chain and ladder cuprates with the discovery of superconductivity is an outstanding chapter as it reflects the very close and productive synergy between theory and experiment.

9. Borocarbides

Discovered just shortly before the Grenoble meeting, these superconductors have been studied extensively during the intervening period. Their maximum T_c values are still near 24 K, and there is good evidence that superconductivity is mediated by electron–phonon coupling. The borocarbides provide a fruitful ground to study the interaction between magnetism and superconductivity. Through appropriate choice of the rare earth ions, the superconducting T_c and the magnetic ordering temperature can be adjusted over a wide range. The vortex lattice has been studied by decoration, neutron scattering and STM techniques and a transition from a square lattice to a hexagonal lattice was found upon increasing the applied field. This effect is rooted in the four-fold anisotropy of the electronic band structure and is now quantitatively understood. The mixed state exhibits some similarity to the one in cuprates: a vortex lattice melting transition has been seen.

10. Ruthenate superconductor Sr_2RuO_4

This compound has attracted much interest recently, not through the modest T_c of ~ 1 K, but through the discussion about the origin of superconducting pairing. A very detailed picture of the electronic structure near the Fermi-level has emerged from classical magneto-oscillation measurements. Three bands cross the Fermi-level and their mass is enhanced by a factor of ~ 4 compared to the values from band structure calculations. A Wilson ratio of ~ 1.8 indicates modest enhancements of the spin susceptibility. Remarkable is the superconducting state. Below the NMR spin relaxation rate drops only by about 50%. Al impurities very effectively depress T_c. The discussions now consider the possibility of an unconventional pairing state mediated by ferromagnetic spin fluctuations. One may well expect a conclusive answer to this speculation by the time of the next M^2S conference.

11. Towards applications

While part of the community is occupied with unravelling the puzzles of cuprate superconductivity, another part works diligently and with great technical success towards the practical application of the 'new' superconductors. Broadly speaking, the main application areas include magnetic sensors, active and passive electronic devices and high-power applications. Great strides have been made in overcoming major technical obstacles in all these fields.

HTS SQUID sensors have now been developed that are at 77 K as 'quiet' and sensitive as traditional SQUIDs at 4 K. Through a newly developed understanding of the origin of flux noise and the subsequent modifications in the design, one can now operate these SQUIDs even in the earth magnetic field. It appears, therefore, that the most daunting technical challenges have met with success.

Passive microwave devices have been fabricated, such as filters with unprecedentedly sharp skirts and low insertion loss. Thin film technology is the preferred implementation, as it benefits from the high critical current densities that the best films can support. In addition to developing the microwave filters themselves, much effort and engineering skills had to go into the systems development, including packaging and reliable, compact cooling. Prototype systems are now being tested at several locations in the US. They are part of cellular communication base stations and carry 'live' telephone traffic. The goal of these extended field tests is to assess quantitatively the overall systems benefits derived from superconducting filters.

Progress in manufacturing HTS wires underlies the development of a variety of prototypes that have been, or are now being, deployed for testing in realistic environments. They include: current limiters of different designs using HTS elements in different configurations, a 50 m machine-stranded cable conductor, a magnet to switch ion beams, a 5 kJ SMES system, a 200 hp AC synchronous motor, a 160 hp homopolar motor, and most recently, a three-phase 630 kVA output transformer converting 18.7 kV to 420 V. Further increases of the effective current densities and the manufacturing processes will bring the BSCCO based wires rapidly closer to early commercialization in many electric power applications. The long-term challenges include the translation of laboratory demonstration of bi-axially textured YBCO films into a viable manufacturing process. The intrinsic superiority of YBCO is a big motivation for work in this area.

The challenges for a commercially successful HTS application are manifold. They start with technical obstacles in harnessing the ultimate performance of superconductors, such as thin film deposition techniques, gentle patterning of films into useful devices, wire fabrication with optimal grain orientation and connectivity, and many more. As in previous conferences, many researchers have reported significant advances on many fronts, e.g. the large area (20×20 cm^2) deposition of highest quality films or the micro-slicing of SQUID pick-up coils for noise suppression.

Often under-appreciated, particularly by physicists, is the enormous amount of engineering necessary to incorporate a superconducting element into a system of sufficient ruggedness and functionality so it can be tested in real-life environments.

It appears that most potential HTS applications are at the stage of prototype evaluation. Technical improvements will emerge as a result of these tests, but they will not necessarily guarantee ultimate commercial success, because most of the envisioned applications are replacing traditional technologies. (SQUID sensors may well be an exception). The decisive factor then is a cost/performance analysis where new HTS technologies bring with them a performance advantage, but also usually a higher price tag. Therefore, it is part of reality that technical success is a necessary, but not always sufficient ingredient for success in the marketplace.

12. Outlook

The field of high temperature superconductivity has presented us with challenges on the conceptual and practical level that are much more fundamental, but also much richer, than we could anticipate ten years ago. As the experimental exploration of the key physical properties of cuprates has come to a stage where one talks about a 'generic' behavior and a 'generic' phase diagram, the groundwork has been laid for the construction of an all-encompassing understanding of these materials. One can easily formulate a list of short-term and long-term goals worth pursuing, but they would all be more or less bold and creative extrapolations of present issues. Doubtless, those studies will be pursued in the years to come. The most fascinating and invigorating aspect of working in this materials-based field is the possibility of yet more revolutionary breakthroughs. The discovery of cuprate superconductors ten years ago was such a momentous event, and we would like to believe that more surprises of this magnitude are awaiting us.

A note of thanks

The fifth conference in this series has been a great experience and we return from Beijing with the best of memories. We are grateful to the Co-Chairmen, the Scientific Program Committee, the Local Scientific Committee, the Publication Committee and the numerous other helpful people who made this meeting possible. The well-selected scientific program and the many cultural events provided the participants with five days of broad enrichment. May all who contributed to the success of this international conference accept the sincere thanks of all of us who were guests in Beijing.

Acknowledgements

I would like to thank all friends and colleagues who have provided me with preprints of their latest work and who discussed their results with me before and during the conference. As this comment on the experimental situation was not intended to be an all-encompassing conference summary, several interesting areas had to be omitted. For this, I ask your indulgence.

ELSEVIER

Physica C 282-287 (1997) xxxi-xxxix

PHYSICA C

Conference summary

Reviews, prospects and concluding remarks. Materials needs for applications

Shoji Tanaka

Superconductivity Research Laboratory / ISTEC 1-10-13 Shinonme, Koto-ku, Tokyo 135, Japan

Abstract

It is expected that applications of high-T_c superconductors will be made in a very wide range of industries. After ten years of very intensive studies on these materials, the directions of material researches are becoming clear. They are classified into four categories: (1) Finding new materials having high-T_c, e.g. 150 K. (2) Development of materials and processing of bulk materials showing a high magnetic levitation force and greater trapped magnetic field: e.g. trapped magnetic field = 3 T at 77 K. (3) Development of superconducting tapes or wires showing a high critical current J_c: e.g. $J_c = 10^5$ A/cm^2 at 77 K, 5 T. (4) Development of new superconducting devices having a very high-speed response and a very small power consumption: e.g. speed = 10 ps; power = 1 nW. At present, it is most important to choose the best materials and the best processing in each item. In this paper the prospects for those problems will be shown and the expected future applications will be summarized. © 1997 Elsevier Science B.V.

1. Introduction

In the past ten years, the research on high-T_c superconductors has been very intensive, and many kinds of applications have been tried. Only a few applications, however, have come onto the market: high-T_c SQUID, microwave devices and Bi-compound tapes. The reason is quite clear; as the high-T_c superconducting materials have complex structures and are quite peculiar, it has been rather difficult to control these materials artificially. Furthermore, superconductivity itself is quite different from that of low-T_c materials. For instance, the superconductivity is strongly two-dimensional and so it is very sensitive to the structure of crystals. Once the crystallinity of the sample is damaged, its superconductivity characteristic disappears.

From my view point, however, we now understand all the features of these materials, as a result of the intensive research in the past. And we are entering a new era of developments from the era of fundamental research, as is shown in Fig. 1. In the era of development, we must choose what kinds of

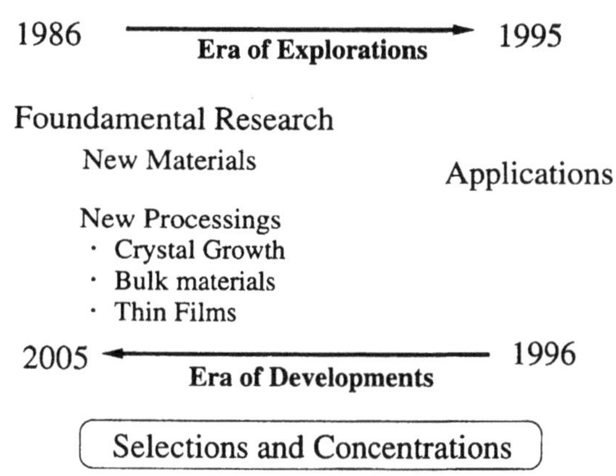

Fig. 1. A view on the progress of R&D.

PII S0921-4534(97)00189-5

Semiconductors and Superconductors

Historical Survey

Semiconductors	Superconductors
Se - Rectifier Cu_2O - Rectifier (ZnO -N - type (Cu_2O - P - type	$La_{2-x}Ba_xCuO_4$ (214)
Ge	$Y_1Ba_2Cu_3O_7$
Si	$Nd_1Ba_2Cu_3O_7$, $SmBa_2Cu_3O_7$ $Y_1Ba_2Cu_4O_8$
Compound Semicondutors III-V (GaAs) II - IV (ZnSe)	Bi - Compound Tl - Compound Hg- Compound

Fig. 2. Historical survey on the evolution of semiconductors and that of high-T_c superconductors.

materials are most plausible for future applications, and after selecting materials for development we must concentrate our efforts on choosing the most suitable means of processing the materials for each application.

We may compare the history of evolution of semiconductors with that of high-T_c superconductors (Fig. 2). Even though the semiconducting characteristics of GaAs (III–V compound) is superior to that of Si, most applications at present are dominated by Si. This is due to the fact that Si is much simpler than GaAs in its crystal structure and also much more stable. These are essentially vital in the course of miniaturizing devices.

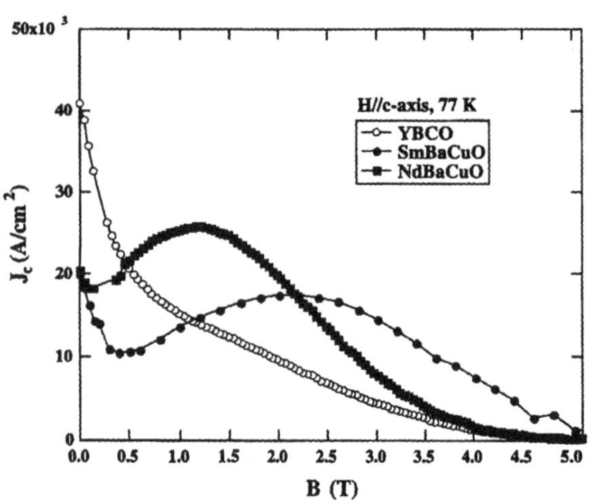

Fig. 3. Magnetic field dependencies of critical currents of YBCO, SmBCO and NdBCO.

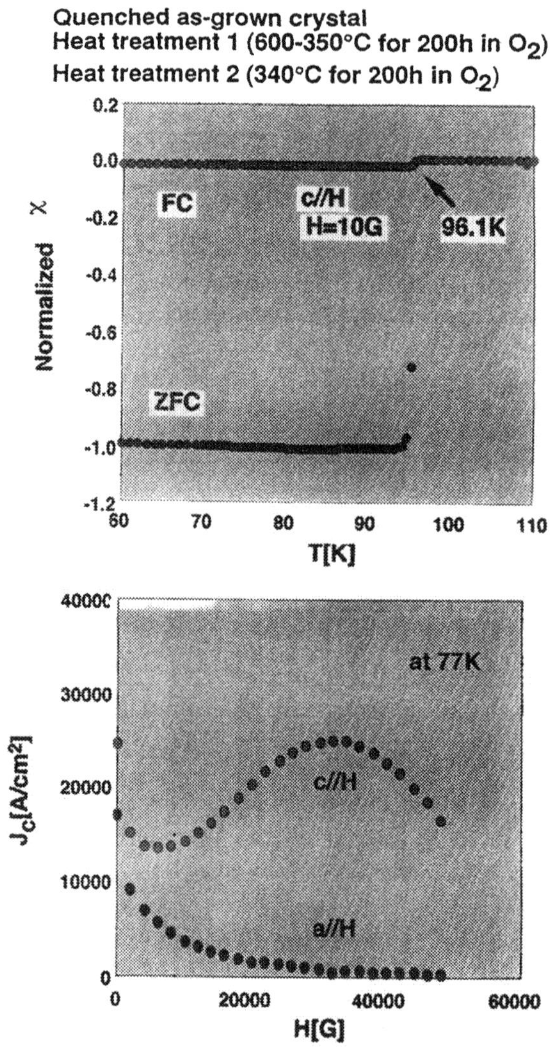

Fig. 4. Superconductivity and magnetic field dependence of critical current in a single crystal of NdBCO.

In the case of high-T_c superconductors, this situation may be similar when we think about its future applications. In the past, more than thirty different kinds of high-T_c materials have been discovered, but 123 materials and Bi-, Tl- and Hg-compounds could be candidates when we set the operating temperature beyond 77 K.

Of those materials, from my personal opinion, the 123 compounds may be the most plausible for future applications since the crystal structure is simple and it is easy to obtain single phase specimens, and finally it is easier to introduce strong pinning centers, as compared to Bi-, Tl- and Hg-compounds.

In this paper, the special features of the 123 compound, especially NdBCO and SmBCO are explained in Section 2, and possible applications of these materials will be described in Section 3.

2. Special features of NdBCO and SmBCO

As was mentioned above, 123 materials may be the most plausible materials in future applications of high-T_c superconductors. In 1993, we found that in 123 materials YBCO and NdBCO have different characteristics [1]. At present it is not certain which material is best for future applications. However, since the characteristics of YBCO are already well-known, the characteristics of NdBCO and SmBCO will be explained in this section.

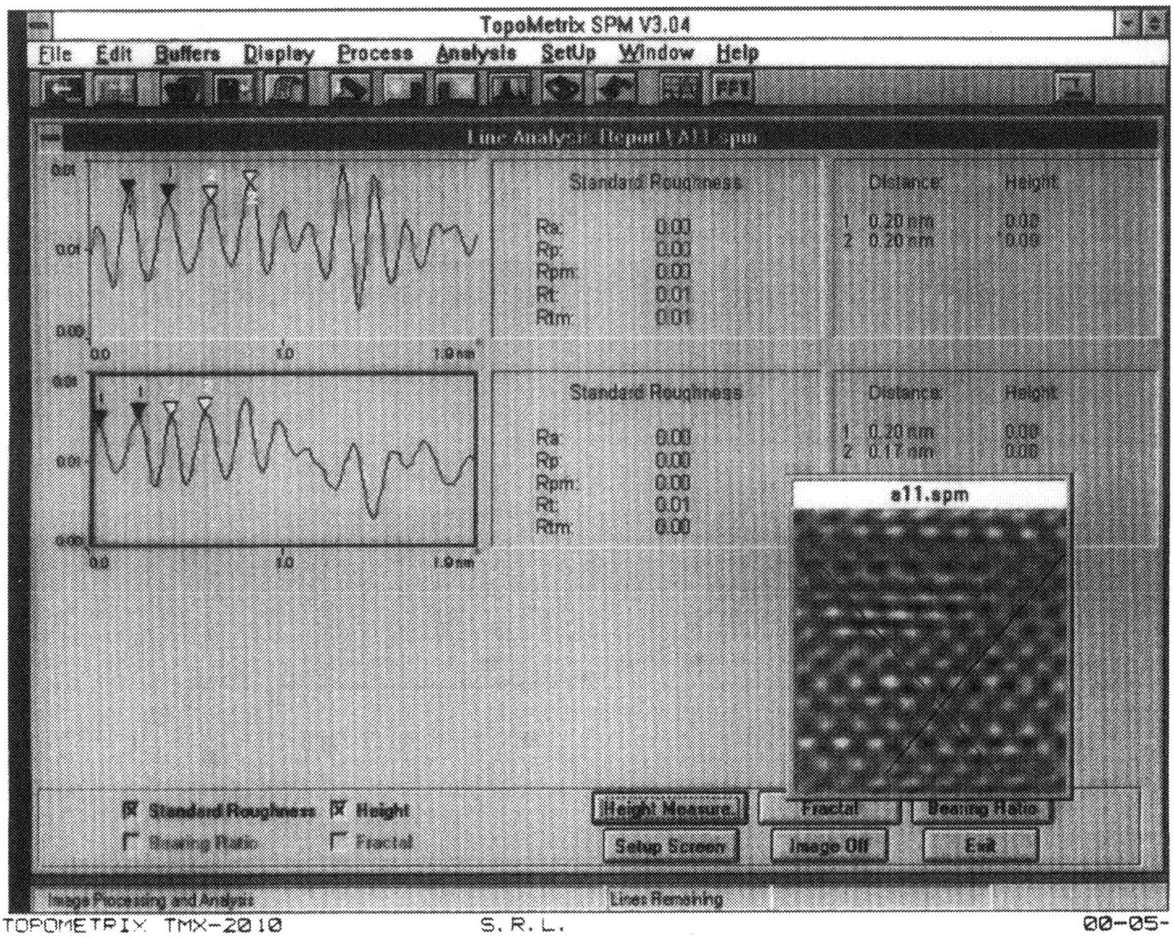

Fig. 5. Structure and roughness of surface of NdBCO thin film.

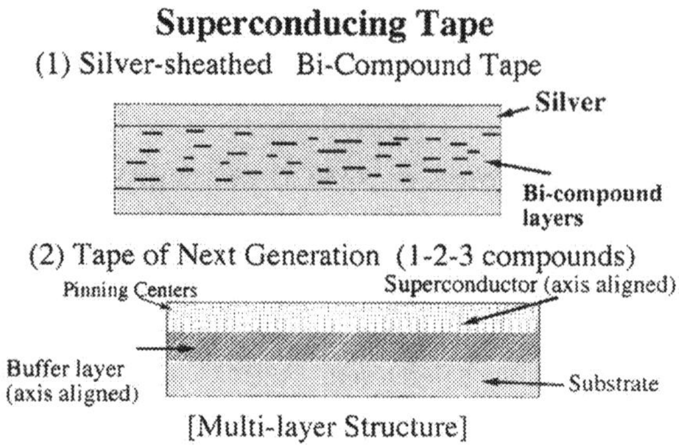

Fig. 6. Structure of superconducting tapes.

Trials for Superconducting Tape or Fiber at 77K

Substrate	Buffer Layer	Superconductors
Ni Alloy	· IBAD : YSZ (Fujikura, Los Alamos, Geneve) · Sputtering : YSZ (Sumitomo)	YBCO
Pure Ni	RABITS : YSZ (Oak Ridge)	YBCO
Pure Ag	(Hitachi, Toshiba)	Tl compound YBCO
Oxide Fiber (STO, YSZ)	Sputtering MOCVD:YBCO (ISTEC)	YBCO

Fig. 7. Trials for superconducting tape or fiber at 77 K.

In the series of rare earth elements, Y belongs to a heavy element group and Nd and Sm belong to a light element group. Thus, the ionic radius of Nd^{3+} is much larger than that of Y^{3+} and further it is almost the same as that of Ba^{2+}.

In 1993, we found that substitution of Ba^{2+} for Nd^{3+} or Sm^{3+} in the lattice sites sometimes occurs. The fraction of this substitution depends on the crystal growth process and the annealing process [1]. Even though the fraction is small, the effect of this substitution on the superconductivity is significant and the critical current increases drastically in a high magnetic field region.

In Fig. 3, the magnetic field dependencies of NdBCO and SmBCO which are made by the OCMG method, are shown [1]. In these materials, critical currents show a remarkable peak effect in a high magnetic field region and this is quite different from that of YBCO. Even in the single crystal of NdBCO, a similar peak effect can be observed after an appropriate annealing process, and it must be mentioned that the critical current in a high magnetic field region shows strong anisotropy (Fig. 4) [2].

The thin films of NdBCO show very peculiar characteristics. As shown in Fig. 5, the surface of thin films made by the laser ablation method is very

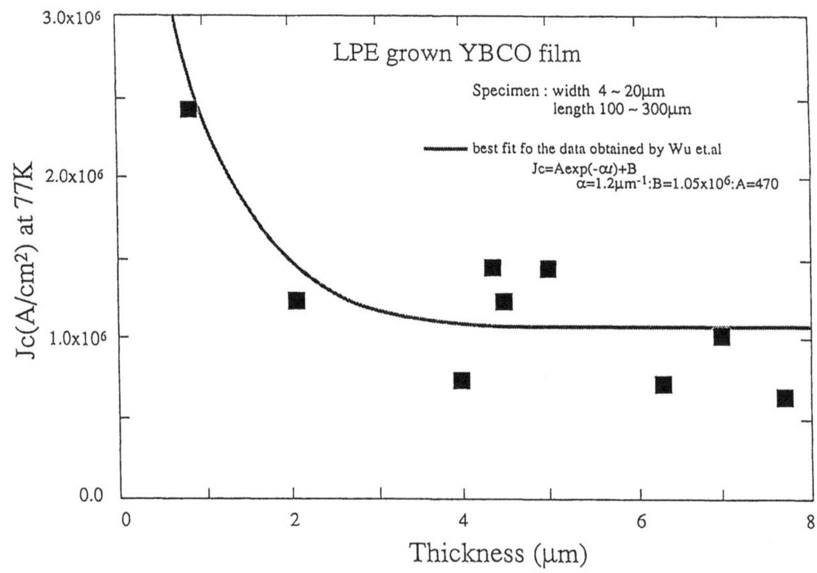

Fig. 8. Critical current of LPE grown YBCO film at 77 K.

flat and its roughness is nearly 10 Å. Furthermore, the atomic structure of the surface is clearly observed by AFM even after exposure to the ambient atmosphere [3]. The reason for this peculiar phenomenon is not clear, but it must be very useful in the development of future electronic devices.

3. Materials problems for practical applications

Looking at the very wide range of future applications, the problems to be solved urgently are classified into four subjects.

3.1. Superconducting tapes and wires

The superconducting tapes made by silver-sheathed Bi-compounds have been developed in many places in the temperature range of 20–30 K. However, at the liquid nitrogen temperature the critical current decreases remarkably with increasing the magnetic field even below 1 T. We call the Bi-compound tape a tape of the first generation. If we want to have a new kind of tapes or wires which can be used at 77 K and 5 T, we have to develop new tapes or wires; we call them those of the second generation. The materials for the second generation tapes and wires must be 123 materials.

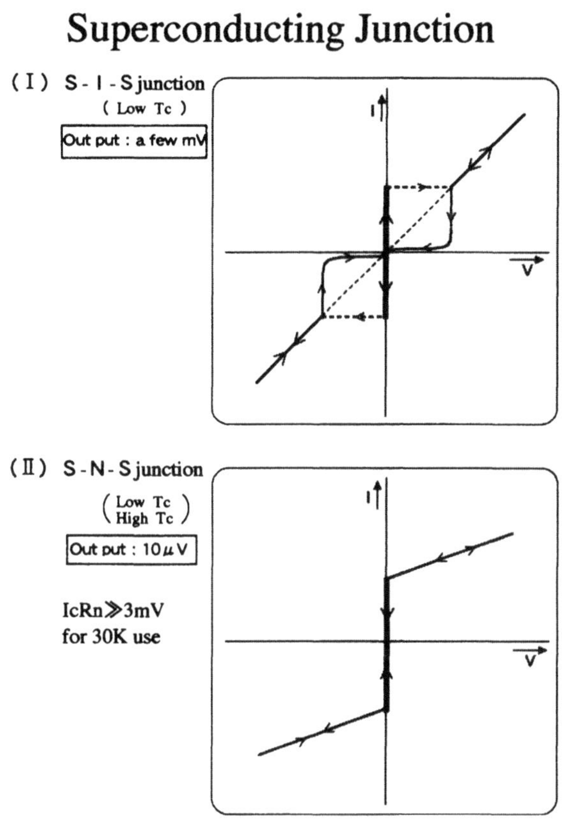

Fig. 10. Characteristics of superconducting junctions.

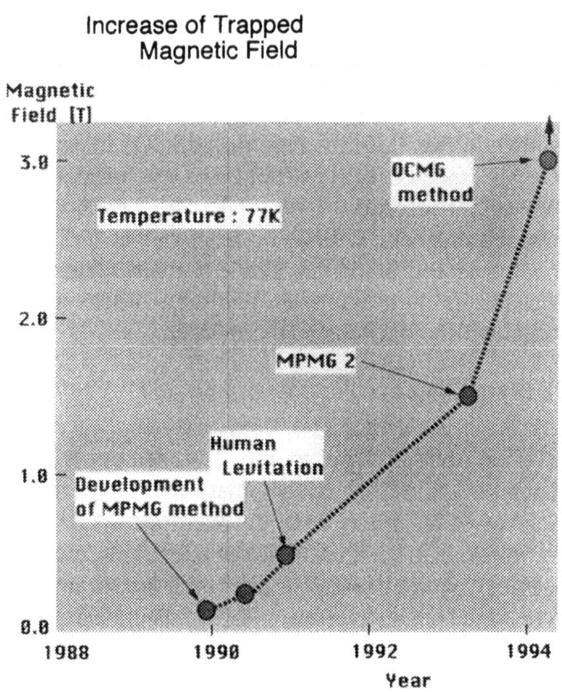

Fig. 9. Increase of trapped magnetic field.

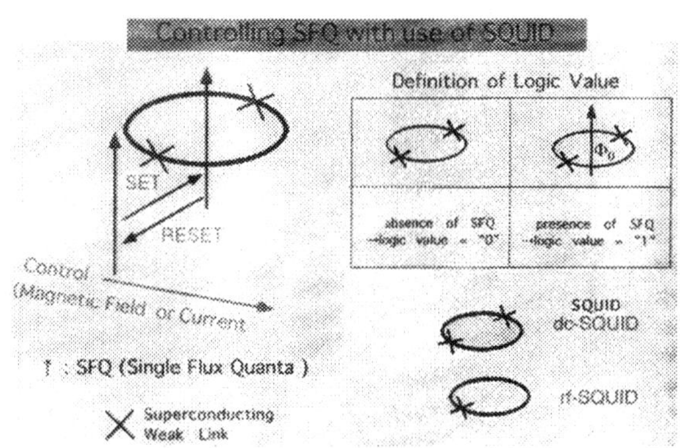

Fig. 11. Principle of single flux quantum device.

In order to make long superconducting tapes or wires, we must solve the grain boundary problem, where the weak links appear among grain boundaries. Fortunately, in Bi-compound tape, this grain boundary problem is not serious. However, it is almost impossible to introduce strong pinning centers, which are still effective at 77 K and 5 T, into the Bi-compound, as the two-dimensionality in this material is so large that the structure of magnetic vortex is very complicated.

As grain boundary problem is quite serious in the 123 compound, we must minimize the density of the weak links. A comparison between the structure of the Bi-compound tape and the 123 tape is shown in Fig. 6. Here the structure of the 123 tape is tentative and many trials have been made in the past few years, as is shown in Fig. 7.

Of these trials, the liquid phase epitaxy method, which was developed in ISTEC, is unique [4]; the critical current density in this rather thick film is almost independent of the thickness over 10 μm in thickness and is about 10^6 A/cm^2 as is shown in Fig. 8.

In the course of the development of the second generation tapes or wires, various kinds of combinations of substrates, buffer layers and superconductors should be examined in the near future and the best combination should be decided in the next five years.

3.2. Superconducting bulk materials

A special feature of the superconducting bulk materials is that the materials are limited to the 123 compounds and they are made by the melt and

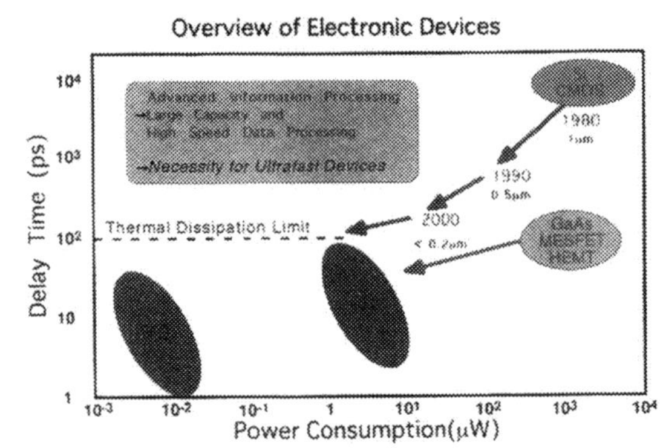

Fig. 12. Overview of electronic devices.

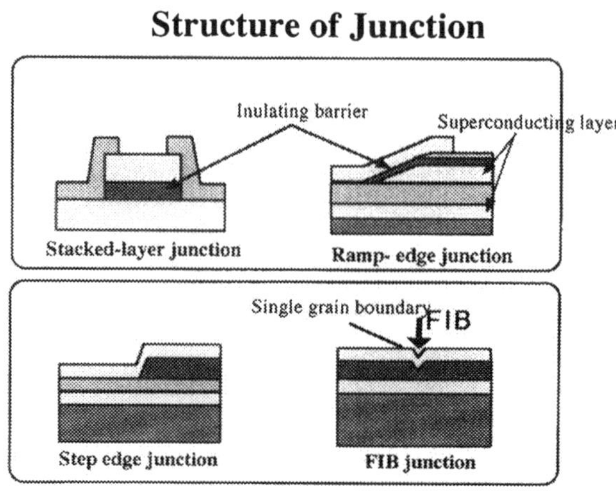

Fig. 13. Structure of junction.

solidification processing. Using these materials, we can obtain a strong magnetic levitation force and a strong trapped magnetic field. Both of them come from the large critical current flowing in the bulk which is determined by the grain boundary effect and the density and the strength of the introduced pinning centers, as happens in the case of superconducting tapes and wires.

Thus the problems in bulk materials are: (1) how to grow a large single domained bulk, and (2) how to introduce strong pinning centers into the bulk. In the case of YBCO, the so-called MPMG method is very suitable in making a high performance bulk as is shown in Fig. 9. In the case of NdBCO, the so-called OCMG method is successful and the trapped magnetic field reached 3 T at 77 K. We expect that the trapped magnetic field could reach 5 T at 77 K, if we could succeed in making single domained bulk materials of 10 cm in diameter, and then various applications will be opened for the market.

3.3. Development of electronic devices

It is thought that in the 21st century a highly sophisticated information society will be constructed in the world and in order to make the society more efficient it is necessary to develop new electronic devices, which should be very high speed and have very low power consumption. A candidate of such new devices is superconducting electronic devices.

In the superconducting devices, the most plausible device may be the one where the Josephson junc-

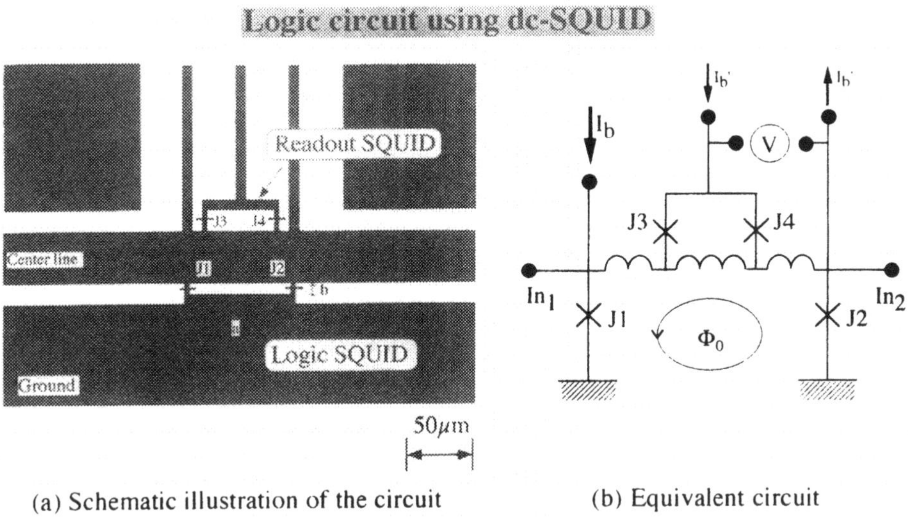

Fig. 14. Logic circuit using DC-SQUID.

tions are efficiently utilized. There are two types of Josephson junctions, one is an S–I–S junction and the other is an S–N–S junction; characteristics of both types of junctions are shown in Fig. 10. In high-T_c superconductors, only the S–N–S junction can be made artificially. Thus the single quantum flux device is the only solution.

The principle and the structure of the single flux quantum (SFQ) device is shown in Fig. 11, and the expected characteristics of the SFQ device are shown in Fig. 12 in comparison with those of semiconductors devices. In this figure, we can expect that the theoretical characteristic of the SFQ device is much better than the expected C-MOS device by two orders of magnitude in both delay time and power consumption.

Material problems in making stable Josephson junctions of high-T_c superconductor are not easy to solve. Until now, four types of S–N–S junctions have been tried as is shown in Fig. 13, and it is not yet obvious what kind of junctions is the most plausible for future integrated circuits.

In this paper, a logic circuit using DC-SQUID, which was developed in our laboratory [5,6], is

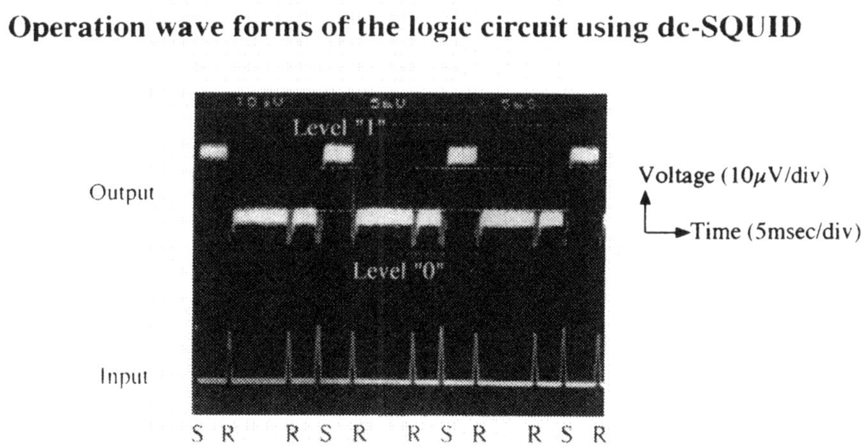

Fig. 15. Operation wave forms of the logic circuit using DC-SQUID.

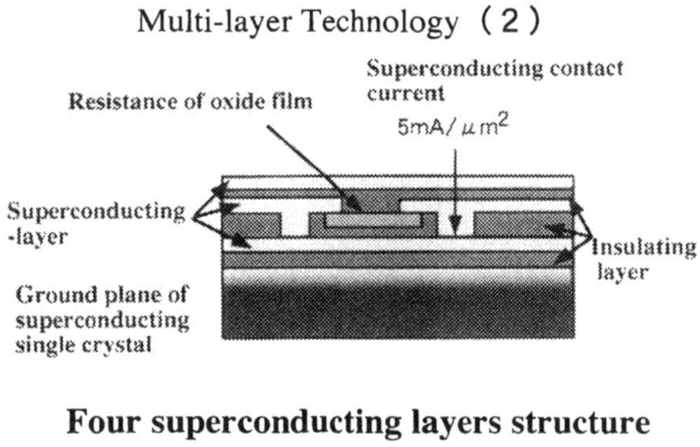

Fig. 16. Multi-layer structure in future SFQ circuit.

shown as an example in Fig. 14. This device consists of a logic SQUID and a read-out SQUID, and the current through the logic SQUID can be measured by the read-out SQUID. The result is shown in Fig. 15. As can be seen in this figure, the logic device shows a beautiful flip-flop operation corresponding to the set and the reset signals.

However, we must expect many difficult problems in the future integration of these devices on one chip. For instance, the expected layer structure as shown in Fig. 16 must be constructed. In such a multilayer structure, it is very difficult to keep good superconductivity in each superconducting layers. One solution might be using a single crystal of the high-T_c material as ground plane, as is shown in this figure. Homo-epitaxial or quasi-homo-epitaxial growth of 123 films on the surface of the single crystal of the 123 material seems to be successful in making such multi-layer structure. In Fig. 17 an example of the homo-epitaxial structures is shown [7]. In Fig. 16 the boundary between the surface of the YBCO single crystal and the thin film can be hardly observed. This means that the homo-epitaxial growth is almost perfect.

Even though we may expect many difficulties in making future SFQ circuits, we must show great patience to overcome those technical barriers by developing new processing methods.

3.4. New materials

It is very important scientifically to discover new high-T_c superconductors. Since consensus on the mechanism of high-T_c superconductivity is not yet established, the new materials must accelerate the understanding of the mechanism of high-T_c superconductivity.

From the practical point of view, new materials having a critical temperature of beyond 150 K are very important. As new devices operating beyond

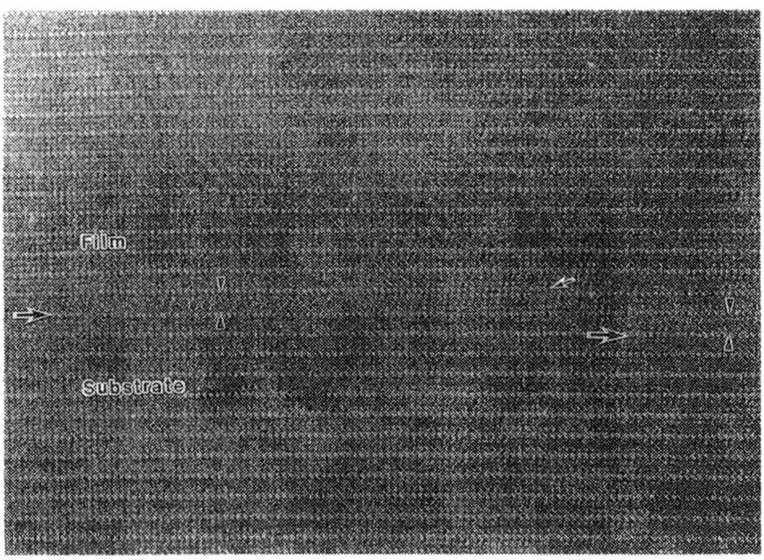

Fig. 17. A c-YBCO film grown on a c-YBCO single crystal by MOCVD.

liquid natural gas (LNG) temperature (110 K) will become possible. We have to keep endeavours in this direction in future.

4. Summary

As mentioned in Section 1, we are getting into new era of the developments of practical applications of high-T_c superconductors and the targets of the developments are becoming increasingly clear. Here it must be stressed that we must keep close contact with the fundamental problems in materials research; crystal growth, high quality thin films and so on. In these mysterious materials, there must be many fundamental problems left unsolved. It is believed that most of the materials problems for practical applications mentioned in Section 3 will be solved in the coming several years, and the market will be opened wide to high-T_c superconductors.

References

[1] S.I. Yoo, N. Sakai, T. Higuchi, M. Murakami, Appl. Phys. Lett. 65 (1994) 633.
[2] N. Nakamura, Y. Yamada, T. Hirabayashi, Y. Ikuhara, Y. Shiohara, T. Tanaka, Physica C 259 (1996) 295.
[3] M. Badaye, W. Ting, K. Fukushima, N. Koshizuka, T. Morishita, S. Tanaka, Appl. Phys. Lett. 67 (1995) 2155.
[4] S. Miura, K. Hashimoto, F. Wang, Y. Enomoto, T. Morishita, Physica C, in press.
[5] H. Fuke, K. Saitoh, T. Utagawa, Y. Enomoto, Appl. Phys. Lett. 69 (19) (1996) 2921–2923.
[6] K. Saitoh, H. Fuke, T. Utagawa, Y. Enomoto, Jpn. J. Appl. Phys. 35 (11A) (1996) L1415–L1418.
[7] J.G. Wen, T. Zama, T. Morishita, N. Koshizuka, to be submitted.

Listing of sections

Part I

Plenary talks 1
Tutorial lectures 17
Invited papers 35

Part II

Contributed papers 435

Section A. Materials, synthesis, structure and physical chemistry

A1. Phase diagrams 437
A2. Bulk material synthesis, single crystal growth 457
A3. Thin films 549
A4. Multilayers and superlattices 681
A5. Non copper oxides 715
A6. Alkali metal fullerenes 733
A7. Organic superconductors (no contributions) 741
A8. Conventional superconductors 741
A9. Rare earth cuprates: 123, 214, ... 743
A10. Layer cuprates: Tl, Bi, Hg, ... 811
A11. New materials 929

Section B. Physical properties and theory I

Ba. Spectroscopy, electronic and structural properties

Ba1. Spectroscopy (photoemission, EELS, ...) 961
Ba2. Formation of gap states and photoinduced doping 1003
Ba3. Optical spectroscopy 1007
Ba4. Phonon spectroscopy 1027
Ba5. Structural and phonon anomalies 1051
Ba6. Defects ordering and structural coherence 1089
Ba7. Metal–non-metal transitions 1099
Ba8. Properties of the ladder compounds 1113

Part III

Bb. Transport, magnetic and thermodynamic properties

Bb1. Resistivity, Hall effect, thermoelectric power, optical conductivity, thermal conductivity, ... 1123
Bb2. Magnetic phases and static properties 1289
Bb3. Dynamics of spin excitations, neutrons, NMR, μSR, ... 1331
Bb4. Specific heat 1395
Bb5. Others 1427

Bc. Superconductivity properties

Bc1. Magnetic penetration depths 1459
Bc2. Superconducting gap, symmetry, Andreev reflection and tunneling 1471
Bc3. Superconducting phase transitions and fluctuations 1533
Bc4. Nonequilibrium superconductivity (no contributions) 1573
Bc5. Microwave absorption, acoustic properties 1573
Bc6. Others 1605

Bd. Theory

Bd1. Band structure: ab initio and cluster calculations 1619
Bd2. Weak coupling theory: BCS, van Hove scenario, magnetic coupling 1641
Bd3. Strong correlation aspects: Hubbard model, t–J, multiband model, spin–charge separation, ... 1667
Bd4. Non-Fermi-liquid behavior 1797
Bd5. Local pair superconductivity 1811
Bd6. Strong electron–phonon coupling 1823
Bd7. Symmetry of the superconducting order parameter 1833
Bd8. Numerical simulation 1861
Bd9. Phase separation 1871

Be. Non-cuprate superconductors

Be1. Heavy fermions 1881
Be2. Organic superconductors 1887
Be3. Alkali metal fullerenes 1913
Be4. Conventional superconductors (no contributions) 1929
Be5. Other oxide superconductors 1929
Be6. Others 1933

Part IV

Section C. Theory and physical properties II (phenomenology of the superconducting state)

Ca. Flux lattice: structure and phase diagram

Ca1. Structure of the vortex lattice, effects of anisotropy 1943
Ca2. Effects of the layered structure, 3D versus 2D behavior 1983
Ca3. Thermal fluctuations and melting of the vortex lattice 2015
Ca4. Disorder and the vortex glass 2067
Ca5. Kosterlitz–Thouless transition 2077
Ca6. Imaging of the vortex lattice 2083

Cb. Pinning and flux dynamics

Cb1. Pinning mechanisms, irreversibility line 2091
Cb2. Magnetization and relaxation in the critical state 2195
Cb3. Quantum dynamics of vortices 2265
Cb4. Critical currents 2269
Cb5. Flux dynamics, magnetoresistance in the superconducting state 2317
Cb6. AC and vibrational measurements 2375

Cc. Josephson effects and networks

Cc1. Physical properties of single Josephson junctions 2399
Cc2. Josephson properties in layered structures 2421
Cc3. Proximity effects 2441
Cc4. Josephson and superconducting networks 2447

Section D. Applications

D1. General overview of applications (no contributions) 2455
D2. Fabrication and properties of Josephson junctions 2455
D3. SQUIDs and magnetometry 2477
D4. Josephson circuitry 2491
D5. Superconductive transistors and hybrids 2495
D6. Passive and active microwave devices 2511
D7. Detectors (IR, X-ray, particles, ...) 2545
D8. Processing of ceramics, wires and tapes 2553
D9. Strong current applications (bearings, magnets, machines, transmissions, ...) 2623

Physica C 282–287 (1997) xliii–xciv
North-Holland

Contents

Preface . . . vii
Opening address by Z.X. Zhao . . . ix
Opening address by G.Z. Zhou . . . xi
Report from the Organizers . . . xiii
Awards . . . xv
Committee Members . . . xvii
Acknowledgements . . . xviii
Conference Summaries
Reviews, prospects and concluding remarks: High-T_c superconductivity – where next?, T.M. Rice . . . xix
Reviews, prospects and concluding remarks: Comments on experiments, B. Batlogg . . . xxiv
Reviews, prospects and concluding remarks: Materials needs for applications, S. Tanaka . . . xxxi
Listing of Sections . . . xli

PART I

PLENARY TALKS

Recent achievements in high-temperature superconductivity
K.A. Müller . . . 3
Phase-sensitive tests of pairing symmetry in cuprate superconductors
C.C. Tsuei and J.R. Kirtley . . . 4
Spin gap effects on the c-axis and in-plane charge dynamics of high-T_c cuprates
S. Uchida . . . 12

TUTORIAL LECTURES

Vortex matter
G. Blatter . . . 19
Introduction to the structure and chemistry of superconducting materials
R.J. Cava . . . 27

INVITED PAPERS

Novel two-dimensional perovskites
J.G. Bednorz . . . 37
Strategy of the synthesis of new layered cuprates and oxycarbonates: from the ceramics to the thin films
B. Raveau, M. Hervieu and C. Michel . . . 41
Superconducting properties of homologous layered cuprates synthesized under high pressure
K. Tanabe, T. Tatsuki, A. Tokiwa-Yamamoto, T. Tamura, X.-J. Wu and S. Adachi . . . 45
The high-pressure synthesis and superconductivity of $Pr_{0.5}Ca_{0.5}Ba_2Cu_3O_z$
Y.S. Yao, Y.F. Xiong, D. Jin, J.W. Li, F. Wu, J.L. Luo and Z.X. Zhao . . . 49
Higher T_c's by chemical substitutions in layered cuprates containing Hg in the reservoir blocks
M. Marezio and F. Licci . . . 53

Search for novel high temperature superconducting materials: a new 126 K system, $Ba_2Ca_{n-1+x}Cu_{n+y}O_z$
C.W. Chu, Y. Cao, Z.L. Du, L. Gao, K. Ross, I. Rusakova, Y.Y. Sun, N.L. Wu and Y.Y. Xue 57

Structural features, oxygen and fluorine doping in Cu-based superconductors
E.V. Antipov, S.N. Putilin, R.V. Shpanchenko, V.A. Alyoshin, M.G. Rozova, A.M. Abakumov, D.A. Mikhailova, A.M. Balagurov, O. Lebedev and G. Van Tendeloo 61

High pressure synthesis and structure of the superconducting mercury cuprates $(Hg_{1-x}M_x)Ba_2Ca_{n-1}Cu_nO_{2+2n+\delta}$ with $M = C, S$
P. Bordet, S. Loureiro, S. LeFloch, J.J. Capponi, C. Chaillout, J. Chenavas, J.L. Tholence, E.V. Antipov, P.G. Radaelli and M. Marezio 65

Quaternary borocarbides – exotic superconducting and magnetic materials
R. Nagarajan 69

Superconductivity in Ru-based double perovskite – the possible existence of a new superconducting pairing state
D.Y. Chen, F.Z. Chien, D.C. Ling, J.L. Tseng, S.R. Sheen, M.J. Wang and M.K. Wu 73

High gas pressure crystal growth and properties of $HgBa_2Ca_{n-1}Cu_nO_{2n+2+\delta}$, $Sr_{0.73}CuO_2$ and $Ca_xA_{1-x}CuO_2$
J. Karpinski, H. Schwer, E. Kopnin, R. Molinski, G.I. Meijer and K. Conder 77

Single crystal growth of RE-123 superconductor and its superconductive properties
Y. Shiohara, M. Nakamura, X. Yao, M. Kambara and T. Umeda 81

Low-frequency spin fluctuations in the superconducting $La_{2-x}Sr_xCuO_4$
K. Yamada, C.H. Lee, Y. Endoh, G. Shirane, R.J. Birgeneau and M.A. Kastner 85

Improvements in crystal growth and crystal homogeneity and its impact on physics
A. Erb, E. Walker, J.-Y. Genoud and R. Flükiger 89

Microscopic X-ray characterization of inhomogeneities in YBCO crystals with sharp superconducting transitions
S.B. Qadri, E.F. Skelton, M.S. Osofsky, V.M. Browning, J.Z. Hu, P.R. Broussard, M.E. Reeves and W. Prusseit 93

Defects that control the properties of Tl- and Hg-based superconductors
J.D. Jorgensen, O. Chmaissem, J.L. Wagner, W.R. Jensen, B. Dabrowski, D.G. Hinks and J.F. Mitchell 97

Electron crystallographic study of superconducting oxides and related compounds
F.H. Li, H.F. Fan and Z.X. Zhao 101

Scanning tunneling microscopy and spectroscopy study of substitution- and defect-doped films of the infinite layer compound
R. Feenstra, T. Matsumoto and T. Kawai 105

YBCO film deposition on very large areas up to $20 \times 20\,cm^2$
H. Kinder, P. Berberich, W. Prusseit, S. Rieder-Zecha, R. Semerad and B. Utz 107

Epitaxial ferroelectric/superconductor heterostructures
Yu.A. Boikov, Z.G. Ivanov, E. Olsson and T. Claeson 111

Studies of high temperature superconducting thin films by transmission electron microscopy
L.P. Guo, Y. Yang, Y.Q. Zhou, Y. Wu, Y.L. Qin, O. Jin and L. Li 115

Pseudo spin gap in high-T_c's manifested by the NMR experiments
H. Yasuoka 119

Recent photoemission data from underdoped Bi2212 single crystals
Z.-X. Shen 123

The spin gap and extended t–J model
H. Fukuyama and H. Kohno 124

Probing the symmetry of the pairing state of unconventional superconductors by SQUID interferometer measurements
D.J. Van Harlingen 128

Symmetry of the order parameter: implications for the transport properties of grain boundaries
J. Mannhart, H. Hilgenkamp and Ch. Gerber 132

Pairing symmetry in single-layer tetragonal $Tl_2Ba_2CuO_{6+\delta}$ from in-plane torque anisotropy
C. Rossel, M. Willemin, J. Hofer, H. Keller, Z.F. Ren and J.H. Wang 136

Andreev reflections from $La_{2-x}Sr_xCuO_4$ single crystals
G. Deutscher, N. Achsaf, D. Goldschmidt, A. Revcolevschi and A. Vietkine 140

Anisotropic penetration depth measurements of high-T_c superconductors
C. Panagopoulos, J.R. Cooper, T. Xiang, G.B. Peacock, I. Gameson, P.P. Edwards, W. Schmidbauer and J.W. Hodby . . . 145

Impurity effects on a 2-leg spin ladder compound $SrCu_2O_3$
M. Takano, M. Azuma, Y. Fujishiro, M. Nohara, H. Takagi, M. Fujiwara, H. Yasuoka, S. Ohsugi, Y. Kitaoka and R.S. Eccleston 149

Superconductivity in the ladder compound $Sr_{2.5}Ca_{11.5}Cu_{24}O_{41}$ (single crystal)
T. Nagata, M. Uehara, J. Goto, N. Komiya, J. Akimitsu, N. Motoyama, H. Eisaki, S. Uchida, H. Takahashi, T. Nakanishi and N. Môri . . . 153

Numerical studies of the 2-leg Hubbard ladder
D.J. Scalapino . . . 157

Transport properties of MBE grown cuprate spin ladders
M. Laguës, J.Y. Laval, C.F. Beuran, C. Deville Cavellin, B. Eustache, J.B. Moussy, C. Partiot and X.Z. Xu . . . 162

Charge stripes and spin correlations in copper-oxide superconductors
J.M. Tranquada . . . 166

Phase separation, staging and spin fluctuations in La_2CuO_{4+y}
Y. Endoh, R.J. Birgeneau, M.A. Kastner, Y.S. Lee, G. Shirane, S. Wakimoto, B.O. Wells and K. Yamada . . . 170

Pairing and phase coherence in high temperature superconductors
V.J. Emery, S.A. Kivelson and O. Zachar . . . 174

Dynamical stripe correlations and the spin fluctuations in cuprate superconductors
J. Zaanen and W. van Saarloos . . . 178

Local structure study of the underdoped–overdoped transition in $YBa_2Cu_3O_x$ ($x = 6.806$–6.984) by yttrium EXAFS
J. Röhler, P.W. Loeffen, K. Conder and E. Kaldis . . . 182

Evidence for two-component superconductivity in the femtosecond optical and transient photoconducting response of $YBa_2Cu_3O_{7-\delta}$
D. Mihailovic, C.J. Stevens, B. Podobnik, J. Demsar, M. Zavrtanik, D. Smith and J.F. Ryan . . . 186

EPR evidence for Jahn–Teller polarons in $La_{1-x}Ca_xMnO_{3+y}$
A. Shengelaya, G.-m. Zhao, H. Keller and K.A. Müller . . . 190

Bose–Einstein to BCS crossover picture for high-T_c cuprates
Y.J. Uemura . . . 194

Experimental studies of the isotope effect in high T_c superconductors
J.P. Franck . . . 198

Oxygen isotope effects in the manganites and cuprates: evidence for polaronic charge carriers
G.-m. Zhao, M.B. Hunt, K. Conder, H. Keller and K.A. Müller . . . 202

Electronic states of the superconductor Sr_2RuO_4
Y. Maeno . . . 206

Spin fluctuation and unconventional pairing in Sr_2RuO_4
Y. Kitaoka, K. Ishida, K. Asayama, S. Ikeda, S. Nishizaki, Y. Maeno, K. Yoshida and T. Fujita . . . 210

Sr_2RuO_4 – a realization of spin triplet pairing?
M. Sigrist, D. Agterberg, T.M. Rice and M.E. Zhitomirsky . . . 214

Angle-resolved photoemission study of Sr_2RuO_4; an extended van-Hove singularity in non-cuprate superconductor
T. Takahashi, T. Yokoya, A. Chainani, H. Ding, J.C. Campuzano, M. Kasai and Y. Tokura . . . 218

Raman scattering by electronic excitations in high T_c superconductors
M. Cardona, T. Strohm and X. Zhou . . . 222

NMR studies of the original magnetic properties of the cuprates: influence of impurities and defects
H. Alloul, J. Bobroff, A. Mahajan, P. Mendels, Y. Yoshinari, G. Collin and J.F. Marucco . . . 226

Cross-over temperatures and spin-gap in high T_c cuprate superconductors: an NMR approach
C. Berthier, M.-H. Julien, O. Bakharev, M. Horvatić and P. Ségransan . . . 227

Magnetic coherence in the superconducting and normal states of $La_{2-x}Sr_xCuO_4$
G. Aeppli, T.E. Mason, H.A. Mook, A. Schroeder and S.M. Hayden . . . 231

Doping dependence of the magnetic resonance peak in $YBa_2Cu_3O_{6+x}$
B. Keimer, H.F. Fong, S.H. Lee, D.L. Milius and I.A. Aksay . . . 232

Phase separation, pseudogap and impurity scattering in the HTS cuprates
J.L. Tallon, G.V.M. Williams, N.E. Flower and C. Bernhard . . . 236

Low-temperature normal-state resistivity of high-T_c cuprates
Y. Ando, G.S. Boebinger, A. Passner, N.L. Wang, C. Geibel, F. Steglich, T. Kimura, M. Okuya, J. Shimoyama, K. Kishio, K. Tamasaku, N. Ichikawa and S. Uchida . . . 240

Unusual field dependence of thermal conductivity in $La_{2-x}Sr_xCuO_4$: suppression of quasi-particle coherence
N.P. Ong, K. Krishana and T. Kimura . . . 244

The role of density of states fluctuations in the normal state anomalies of high T_c superconductors
A.A. Varlamov 248
Nonlinear electrodynamics in cuprate superconductors
J. Orenstein, J. Bokor, E. Budiarto, J. Corson, R. Mallozzi, I. Bozovic and J.N. Eckstein 252
Evidence for multi-component superconducting order parameter from microwave measurements of $YBa_2Cu_3O_{7-\delta}$ single crystals
S. Sridhar, H. Srikanth, Z. Zhai, B.A. Willemsen, T. Jacobs, A. Erb, E. Walker and R. Flükiger 256
The charge-density-wave quantum-critical-point scenario
C. Castellani, C. Di Castro and M. Grilli 260
Marginal Fermi-liquid state and pairing instability in a general model of copper-oxides
C.M. Varma 264
The SO(5) theory of high T_c superconductivity
S.-C. Zhang 265
Bipolaron kinetics of high-T_c cuprates
A.S. Alexandrov 269
Understanding high temperature superconductors: progress and prospects
D. Pines 273
The unusual normal and superconducting states of underdoped cuprates
P.A. Lee 279
Effects of impurities in spin gap state of cuprates
N. Nagaosa 283
Electronic excitations in strongly correlated systems; 1D vs. 2D
S. Maekawa and T. Tohyama 286
Phenomenological aspects of phase fluctuation in high T_c superconductors: Cooper pair droplets
B.K. Chakraverty and T.V. Ramakrishnan 290
High T_c superconductors: the van Hove scenario, a review
J. Bok and J. Bouvier 294
Electron correlation, metal–insulator transition and superconductivity in quasi-2D organic systems, $(ET)_2X$
K. Kanoda 299
The normal and superconducting states of $CeIn_3$ near the border of antiferromagnetic order
I.R. Walker, F.M. Grosche, D.M. Freye and G.G. Lonzarich 303
Current issues in fulleride superconductivity
K. Prassides, K. Tanigaki and Y. Iwasa 307
The role of 2D pancake vortices in high-temperature superconductors
J.R. Clem, T. Pe and M. Benkraouda 311
Tunneling spectroscopy and STS observation of vortices on high temperature superconductors
Ø. Fischer, Ch. Renner, I. Maggio-Aprile, A. Erb, E. Walker, B. Revaz and J.-Y. Genoud 315
Van der Waals attraction of vortices in superconductors
G. Blatter and V.B. Geshkenbein 319
Vortex matter phase transitions in $Bi_2Sr_2CaCu_2O_8$
R.A. Doyle, B. Khaykovich, M. Konczykowski, E. Zeldov, N. Morozov, D. Majer, P.H. Kes and V. Vinokur 323
Calorimetric evidence for the occurrence of a first-order transition in the mixed state of untwinned $YBa_2Cu_3O_{7-\delta}$
A. Schilling, R.A. Fisher, N.E. Phillips, U. Welp, W.K. Kwok and G.W. Crabtree 327
Properties of the vortex liquid in the classical and quantum regime
P.H. Kes, M.H. Theunissen and B. Becker 331
The universal scaling law in magnetic phase diagram of high temperature superconductors
K. Kitazawa, J. Shimoyama, H. Ikuta, T. Sasagawa and K. Kishio 335
Real-time observation of vortex interaction with pinning centers using electron waves
A. Tonomura 339
Thick superconductors in a perpendicular magnetic field
E.H. Brandt 343
Negative-μ vortex dynamics in high-T_c superconducting films
R. Griessen, A.F.Th. Hoekstra, H.H. Wen, G. Doornbos and H.G. Schnack 347

Non-solitary determination of vortex-glass transition temperatures in $YBa_2Cu_3O_{7-\delta}$ superconducting thin films
H.-h. Wen, X.-x. Yao, R.-l. Wang, H.-c. Li, S.-q. Guo and Z.-x. Zhao 351

The superconducting energy gap and vortex lattice structure in $LuNi_2B_2C$
Y. De Wilde, M. Iavarone, U. Welp, V. Metlushko, A.E. Koshelev, I. Aranson, G.W. Crabtree, P.L. Gammel, D.J. Bishop and P.C. Canfield 355

Thermal fluctuations of flux lines and nuclear relaxation
F. Borsa, P. Carretta, M. Corti, A. Rigamonti and B.J. Suh 359

Bragg- and moving-glasses: a theory of disordered vortex lattices
T. Giamarchi and P. Le Doussal 363

Vortex dynamics within the BCS theory
P. Ao and X.-M. Zhu 367

Flux pinning in melt processed RE–Ba–Cu–O
M. Murakami, N. Sakai, N. Chikumoto, H. Kojo, T. Higuchi and S.I. Yoo 371

A path to high-field performance of cuprate superconductors above 100 K with fission induced defects
L. Krusin-Elbaum, D. López, J.R. Thompson, R. Wheeler, J. Ullmann, C.C. Tsuei, C.W. Chu and Q.M. Lin 375

Novel flux phases in superconductors with a periodic pinning array
V.V. Moshchalkov, M. Baert, E. Rosseel, V.V. Metlushko, M.J. Van Bael and Y. Bruynseraede 379

Josephson plasma in cuprate high T_c superconductors
M. Tachiki 383

Intrinsic Josephson effects on BSCCO 2212 single crystal whiskers
Yu.I. Latyshev, P. Monceau and V.N. Pavlenko 387

Interlayer phase coherence in the vortex state of $Bi_2Sr_2CaCu_2O_{8+\delta}$ proved by Josephson plasma resonance
Y. Matsuda, M. Kosugi, M.B. Gaifullin, K. Kumagai, K. Hirata, N. Chikumoto, M. Konczykowski, S. Watauchi, J. Shimoyama and K. Kishio 391

High T_c coplanar resonators for microwave applications and scientific studies
C.E. Gough, A. Porch, M.J. Lancaster, J.R. Powell, B. Avenhaus, J.J. Wingfield, D. Hung and R.G. Humphreys 395

Harmonic frequency mixing at millimeter waveband using high T_c Josephson junction and its applications
P.H. Wu, H.B. Wang, Q.H. Cheng and W.W. Xu 399

Single flux quantum logic and its non-latching operation
Y. Okabe and H. Kodaka 403

Eddy current nondestructive material evaluation based on HTS SQUIDs
M. Mück, M. v. Kreutzbruck, U. Baby, J. Tröll and C. Heiden 407

Application of high temperature superconducting quantum interference devices in transient electromagnetic method for magnetotelluric soundings
S.G. Wang, L.H. Zhang, C.J. Wang and Y.D. Dai 411

Novel high-T_c superconducting three terminal devices
Z.G. Ivanov 415

HTS SNS Josephson junctions: interfaces and mechanisms
K. Char 419

Dynamics of stacked Josephson junctions
N.F. Pedersen 423

Progress in BSCCO-2223 tape technology
A.P. Malozemoff, Q. Li and S. Fleshler 424

The combinatorial synthesis and evaluation of functional materials
X.-D. Xiang and P.G. Schultz 428

Preparation and characterization of melt-textured $Nd_{1+x}Ba_{2-x}Cu_3O_y$ made in air
A.M. Hu, Z.X. Zhao, W. Gawalek, M. Wu, T. Strasser and P. Goernert 431

List of Contributors xcv

Analytic Subject Index cxv

PART II

CONTRIBUTED PAPERS

Section A. MATERIALS, SYNTHESIS, STRUCTURE AND PHYSICAL CHEMISTRY

A1. Phase diagrams

Crystallisation in Bi_2O_3–SrO–CaO–CuO system, and the influence of stoichiometry, and oxygen content on the T_c of Bi-2212 crystals
M. Ionescu, V. Murashov and S.X. Dou 437

Evaluation of thermodynamic data and phase diagram determination in the partially open Bi–Sr–Cu–(O) system
D. Sedmidubský, M. Nevriva and J. Leitner 439

Thermodynamic and nonstoichiometric behavior of the lead-doped and lead-free Bi-2212 systems
M. Tetenbaum, M. Hash, B.S. Tani, J.S. Luo and V.A. Maroni 441

Discrete compounds of homologous series $Ba_{m+n}Bi_mO_y$–$Ba_mBi_{m+n}O_y$ and disordered materials in the Ba–Bi–O system
V.I. Nikolaichik and L.A. Klinkova 443

Phase diagram and crystal growth of La_2CuO_4
C. Chen, I. Kosseva, T. Hargreave, F. Wondre, J.W. Hodby and B.M. Wanklyn 445

X-ray absorption study of rare-earth doped molten Ba–Cu–O system in various oxygen partial pressures
M. Kambara, H.S. Chauhan, A. Endo, Y. Shiohara and T. Umeda 447

The introduction of palladium into superconducting cuprate materials
R.J. Crooks, D.B. Currie, A.L. Hector and M.T. Weller 449

Magnetic properties and metal–non-metal transition in the $YBa_2Cu_{3-x}Li_xO_z$ system
M. Nicolas-Francillon, F. Maury and R. Ollitrault 451

Classification of superconducting oxide structures
J. Hauck and K. Mika 453

A2. Bulk material synthesis, single crystal growth

Growth of high purity $YBa_2Cu_4O_8$ and $Y_2Ba_4Cu_7O_{15-\delta}$ single crystals in $BaZrO_3$ crucibles under high oxygen pressure, and absence of magnetic "fishtail" effect
J.-Y. Genoud, A. Erb, B. Revaz and A. Junod 457

10 years of crystal growth of the 123- and 124- high T_c superconductors: from Al_2O_3 to $BaZrO_3$. Progress in crystal growth and sample quality and its impact on physics
A. Erb, E. Walker, J.-Y. Genoud and R. Flükiger 459

Growth kinetic of top-seeded YBCO single domains
P. Gautier-Picard, X. Chaud, E. Beaugnon and R. Tournier 461

Growth shape analysis of round $YBa_2Cu_3O_{7-x}$ single crystal fabricated by the pulling method
M. Egami and Y. Shiohara 463

An opposite behavior of semiconducting and superconducting YBCO single crystal
X.C. Qin, H. Zhang, Q.R. Feng and Y. Zhao 465

Spiral growth and (110) twins in $Bi_2Sr_2CaCu_2O_y$ crystals
X.L. Wang, S.-T. Yau, H.K. Liu and S.X. Dou 467

Fabrication of NdBaCuO single crystal by a flux method
S. Gong, D. Shi, G. Zhou, C. Cai, Y. Fu and H. Zhang 469

Determination of the *a* and *b* crystalline axes in $YBa_2Cu_3O_7$ single crystal by a polarizing light set up
R. Desmicht, A. Sacuto and J. Jegoudez 471

Effect of preparation conditions on the growth of Bi-based superconducting single crystals
H. Imao, K. Toyoda, S. Kishida and H. Tokutaka 473

Crystallinity improvement of $Bi_2Sr_2CaCu_2O_{8+\delta}$ single crystal by TSFZ method
T. Mochiku, K. Hirata and K. Kadowaki 475

Growth of $Bi_2Sr_2Ca_{n-1}Cu_nO_y$ single crystals in oxygen-reduced atmospheres
S. Kishida, E. Hosokawa, T. Tanikawa, H. Tokutaka and K. Fujimura 477

Crystal growth of $PrBa_2Cu_3O_{7-y}$
K. Oka, Z. Zou and T. Ito . . . 479

A comparative characteristic study of a flux-grown and a travelling solvent floating-zone-grown crystal of $PrBa_2Cu_3O_x$
Z. Zou, K. Oka, T. Ito and Y. Nishihara . . . 481

$Ca_{1-x}R_xCuO_2$ (R = Sr, La) single crystals with infinite-layer structure: high Ar gas pressure synthesis and properties
E.M. Kopnin, H. Schwer, J. Jun, G.I. Meijer, R. Molinski, K. Conder and J. Karpinski . . . 483

The fabrication and properties of a single domain YBCO disk
H.T. Ren, L. Xiao, Y.L. Jiao, S.A. Chang, X.H. Wang and Y.L. Wang . . . 485

Seeded directional growth of superconducting bulk Y123
M. Boffa, M. Boutet, A. Di Trolio, S. Pace, N. Sparvieri, A.M. Testa and A. Vecchione . . . 487

Optimization of $YBa_2Cu_3O_7$ superconducting thick layers produced by plasma spray technique
A.G. Cunha, M.T.D. Orlando, F.G. Emmerich, C. Larica and E.M. Baggio-Saitovitch . . . 489

Growth rate estimation of $Ln_{2-x}R_xCuO_4$ single crystals formed in solution
C.T. Lin, E. Schönherr and K. Peters . . . 491

Crystal growth and characterization of chain cuprates $SrCuO_2$, Sr_2CuO_3 and spin-ladder $Sr_{14}Cu_{24}O_{41}$
A. Revcolevschi, A. Vietkine and H. Moudden . . . 493

Effects of Ag on preparation and J_c of textured YBCO superconductor
Z.G. Fan, Y.Q. Shan, W.H. Wang, X.Y. Wang, D.W. Soh and Z.X. Zhao . . . 495

Steam corrosion dynamics and protection for YBCO superconductor
Y.Q. Shan, Z.G. Fan, J.H. Du, Q.L. Gong, Y.M. Cai, W.H. Wang, L.L. Chen and Z.X. Zhao . . . 497

Effect of Pt with SnO_2 (PbO) addition on the microstructure and magnetization of melt-powder-melt-growth YBaCuO
C.-B. Cai, S.-M. Gong, Y.-X. Fu and H. Zhang . . . 499

Processing of superconductive YBCO single-grain monoliths by top-seeding-melt-texturing
J. Wang, I. Monot, S. Marinel, M.P. Delamare, J. Provost and G. Desgardin . . . 501

Application of oxygen controlled melt growth (OCMG) in ternary RE123 systems
M. Muralidhar, H.S. Chauhan, T. Saitoh, K. Segawa, K. Kamada and M. Murakami . . . 503

Peculiarities of texturing of Y-123 single domain samples by the use of Sm-123 seed and Y-211 substrate
V.A. Murashov, S.X. Dou, H.K. Liu, J. Horvat, P. Schätzle, G. Krabbes, J. Klosowski and T. Wolf . . . 505

Influence of the precursor and dopants on the chemistry and the texture formation of melt processed $YBa_2Cu_3O_{7-\delta}$ superconductors
I. Monot, J. Wang, M.P. Delamare, S. Marinel, M. Hervieu, J. Provost and G. Desgardin . . . 507

High-pressure synthesis of $(M,C)(Ba,Sr)_2Ca_2Cu_3O_9$ (M = Al and Ga)
A. Iyo, T. Saya, K. Tokiwa, Y. Tanaka, N. Terada, M. Tokumoto and H. Ihara . . . 509

A comparative study on the kinetics and phases formation of the (Bi,Pb)-2223 between sol–gel and solid state methods
V. Rouessac, S. Nhien, J. Wang and G. Desgardin . . . 511

High pressure synthesis and superconducting properties of $Sr_2Ca_{n-1}Cu_nO_y$ (n = 1–4) and $Sr_2Ca_{n-1}Cu_nO_{2n+\delta}F_{2\pm y}$ (n = 2–5)
T. Kawashima, Y. Matsui and E. Takayama-Muromachi . . . 513

Synthesis of $YSr_2Cu_3O_y$ under ambient pressure
T. Motohashi, A. Yoshikawa, J.-i. Shimoyama and K. Kishio . . . 515

Structural and physical properties of Al substituted $Y_{1-x}Al_xBa_2Cu_3O_y$
Y. Yu, X.N. Xu, W.M. Chen, A.M. Sun, S.Y. Ding, X. Jin and X.X. Yao . . . 517

Modelling of RE–Ba–Cu–O growth using transparent organic analogues and numerical simulations
G.J. Schmitz, M. Seeßelberg, B. Nestler and R. Terborg . . . 519

Characterizations of grain-structure in shocked YBCO particles
H. Kezuka, M. Kikuchi, E. Ohshima, H. Hıkosaka and Y. Syono . . . 521

TEM-studies of nano-structures of ultrafine YBCO particles
H. Kezuka, Z. Xi and Q. Zhang . . . 523

Magnetic susceptibility during the peritectic recombination of YBaCuO
X. Chaud, E. Beaugnon and R. Tournier . . . 525

Thermodynamics and kinetics of the oxygen exchange in single chains $YBa_2Cu_3O_{7-x}$ (123) and $Y_2Ba_4Cu_7O_{15-x}$ (247) superconductors
K. Conder, Ch. Krüger and J. Karpinski . . . 527

The effect of processing condition on the morphology of bulk YBCO superconductors by melt slowly cooling process
W.M. Yang, L. Zhou, J.R. Wang, C.P. Zhang, Y. Feng, P.X. Zhang and X.Z. Wu . . . 529

Effect of additives (Pt or CeO_2) on $Nd_4Ba_2Cu_2O_{10}$ particles size in melt processed $NdBa_2Cu_3O_x$ superconductor
S. Matsuoka, M. Sumida, T. Umeda and Y. Shiohara . . . 531

The effect of N_2 annealing on the structure and superconductivity of $La_{2-x}Sr_xCuO_4$
C. Dong, L. Lin, H. Chen, F. Wu, S.L. Jia, Y. Yang and Z.X. Zhao . . . 533

Superconductivity of La-deficient $La_{2-x}CuO_{4\pm\delta}$ treated under various oxygen partial pressures
K. Sakai and R. Aoki . . . 535

Crystal growth and superconductivity of $(La,Sr)_2CuO_4$ with SrF_2 as additive
C. Chen, J.W. Hodby, M. Spears, F. Wondre, B.M. Wanklyn, M. Slaski and C. Gough . . . 537

Ambient pressure synthesis of tetragonal Sr_2CuO_{3+x}
M. Bonvalot, E. Beaugnon, D. Bourgault, M. Núñez-Regueiro and R. Tournier . . . 539

On the properties of $Pr_2CuO_{4\pm\delta}$ and $RE_{2-x}Pr_xCuO_{4\pm\delta}$ synthesized under elevated oxygen pressure conditions
P.W. Klamut, A. Sikora, Z. Bukowski, B. Dabrowski and J. Klamut . . . 541

A new superconductor in lead cuprate $(Pb_{0.5}Ba_{0.5})Sr_2(Y_{0.5}Ca_{0.5})Cu_2O_y$
Q.R. Feng, H. Zhang, D.P. Yu and J.D. Guo . . . 543

The role of temperature gradient in the growth of Bi-system whiskers
Y.Q. Zhou, H. Jin, K.Q. Ruan, P. Zheng, Z.J. Chen and L.Z. Cao . . . 545

Introduction of the ultrafine MgO to the high T_c melt-texture $YBa_2Cu_3O_{7-\delta}$
Z.H. He, X.M. Xiong, Y.S. Lai, C.H. Liu, Y.Y. Luo and J.X. Zhang . . . 547

A3. Thin films

The growth mechanism of pulsed laser deposited and DC magnetron sputtered high T_c $YBa_2Cu_3O_7$ thin films
J.Z. Liu and L. Li . . . 549

Microstructure and growth of the epitaxial superconducting $YBa_2Cu_3O_7$ films
G.C. Xiong, G.J. Lian, Y.J. Li and Z.Z. Gan . . . 551

Effect of outgrowths on the surface microwave resistance of $YBa_2Cu_3O_7$ thin films
L.P. Guo and L. Li . . . 553

The surface structure of $YBa_2Cu_3O_{7-x}$ single crystals at high temperatures
F. Wang, H. Zama, H. Ohtsuka, M. Sato and T. Morishita . . . 555

Nanoscale engineering: tailored transport properties by self-organization in $YBa_2Cu_3O_{7-\delta}$ thin films
T. Haage, J. Zegenhagen, Ch. Jooss, R. Warthmann, J.Q. Li, H.-U. Habermeier and M. Cardona . . . 557

Spiral growth in pulsed laser deposited $YBa_2Cu_3O_{7-\delta}$ films
B. Dam, J.H. Rector, J. Huijbregtse and R. Griessen . . . 559

AFM study of growth of $Bi_2Sr_{2-x}La_xCuO_6$ thin films
H. Yang, H. Tao, Y. Zhang, D. Yang, L. Li and Z. Zhao . . . 561

Morphology of $Bi_2Sr_{2-x}La_xCu_1O_{6+\delta}$ thin films grown on untilted and tilted substrates
Y.Z. Zhang, L. Li, D.G. Yang, H.J. Tao, H.T. Yang, Y.L. Qin, W. Liu, B.R. Zhao and Z.X. Zhao . . . 563

Synthesis and characterization of $La_xSr_{1-x}CuO_{2-\delta}$ thin films
J.G. Hou and A.W. Sleight . . . 565

Highly oriented YBCO thin films deposited on cold rolling silver substrates and the study on recrystallization textures in silver
H.-L. Suo, M.-L. Zhou, L.-H. Zhai, Z. Sun, X.-F. Xu, W. Han and T.-Y. Zuo . . . 567

Comparative analysis of YBCO thin films prepared by spray-pyrolysis with different precursors
S.E. Jacobo, C.M.C. Vera, A.C. Razzitte, S. Duhalde, G. Paissan and G. Quintana . . . 569

Crystal structure of YBCO thin films grown at substrate temperatures of about 500°C
A. Köhler, S. Linzen, J. Kräußlich, P. Seidel, B. Freitag and W. Mader . . . 571

In situ film deposition of superconducting borocarbides by pulsed laser ablation technique
M.R. Cimberle, C. Ferdeghini, P. Guasconi, D. Marrè, M. Putti, A.S. Siri, F. Canepa, P. Manfrinetti and A. Palenzona . . . 573

Morphology evolution and crack formation of $YBa_2Cu_3O_7$ on (110)$SrTiO_3$ substrates
K.H. Wu, S.P. Chen, J.Y. Juang, T.M. Uen and Y.S. Gou . . . 575

Preparation of thin epitaxial YBCO films by ablation using fourth harmonics of Nd:YAG laser
T. Kusumori and H. Muto . . . 577

Crystal orientation control of $YBa_2Cu_3O_{7-x}$ thin films prepared by pulsed laser deposition
J.D. Suh and G.Y. Sung . . . 579

Properties of $Nd_{1+x}Ba_{2-x}Cu_3O_{7-\delta}$ films on YSZ and $SrTiO_3$ synthesized by RF flash plasma evaporation
Y. Minagawa and Y. Shiohara 581

XRD and micro Raman characterization of epitaxial Bi-2201, Bi-2212 and Bi-2223 thin films
M. Holiastou, N. Poulakis, D. Palles, E. Liarokapis, D. Niarchos, U. Frey and H. Adrian 583

Preparation of $YBa_2Cu_3O_{7-\delta}$ thin films by hot-wall type MOCVD
Y. Ito, S. Taniguchi, Y. Yoshida, M. Iwata, Y. Takai and I. Hirabayashi 585

Characterization of homoepitaxial $YBa_2Cu_3O_x$ films grown by metalorganic chemical vapor deposition
H. Zama, N. Tanaka, S. Koyama and T. Morishita 587

Formation of thin superconducting YBaCuO layers by oxygen implantation
V.V. Bazarov, V.Yu. Petukhov, I.A. Faizrakhmanov, G.G. Gumarov, V.A. Shustov, E.P. Zheglov, V.A. Zhikharev and I.B. Khaibullin 589

Correlation between surface resistance of $YBa_2Cu_3O_{7-\delta}$ thin films and oxygen pressure during deposition process
R.X. Wu, S.Z. Yang, Y.L. Zhou, B.B. Jin, M. Qian, P.H. Wu, J.H. Gu and D.S. Ding 591

Direct measurement of internal strains in $YBa_2Cu_3O_{7-x}$ thin films on $SrTiO_3$ substrate
M. Li, X. Jin and J. Xu 593

Preparation of *a*-axis oriented $YBa_2Cu_3O_{7-y}$ films by metalorganic chemical vapor deposition using liquid sources
Y. Yoshida, Y. Ito, H. Nagai, Y. Takai and I. Hirabayashi 595

An atomically smooth (110) surface obtained by (110) YBCO homoepitaxy on (110) YBCO single crystal substrates
T. Usagawa, Y. Ishimaru, J. Wen, S. Koyama and Y. Enomoto 597

Growth of (110) and (103)/(013) oriented epitaxial $SmBa_2Cu_3O_{7-\delta}$ thin films on $SrTiO_3$ and characterization with Raman spectroscopy
J. Brunen, T. Strach, J. Zegenhagen and M. Cardona 599

The preparation of smooth and particle-free $YBa_2Cu_3O_{7-x}$ superconducting films by laser ablated plasma plume collision deposition
Y.-C. Fan, A.G. Fitzgerald, B.E. Storey, H.-C. Xu and F.H. Hubbard 601

Growth mechanism studies of $YBa_2Cu_3O_{7-x}$ superconducting films by atomic force and scanning tunneling microscopes
A.G. Fitzgerald, Y.-C. Fan, B.E. Storey and H.-C. Xu 603

Growth and characterization of *a*-axis oriented $YBa_2Cu_3O_{7-x}$ thin films on $SrTiO_3$ substrates
P.J. Wu, G.R. Liu, L. Chen, T. Yang, H. Chen, W. Liu, M.Q. Huang, J.C. Nie and L. Li 605

Synthesis of in-plane aligned, *a*-axis oriented $RBa_2Cu_3O_{7-\delta}$ (R = Y, Nd) films on $LaSrGaO_4$ substrates with $Gd_2CuO_4/PrBa_2Cu_3O_{7-\delta}$ buffer layers
A. Sawa, S. Hayashi, H. Obara, H. Yamasaki and S. Kosaka 607

Fabrication of a cylindrically textured YBCO film on a YSZ cylinder and its critical currents
D. Goo, J. Kim and D. Youm 609

Ion beam assisted deposition of biaxially textured YSZ thin films as buffer layers for YBCO superconducting films
Y. Mao, C. Ren, J. Yuan, F. Zhang, X. Liu, S. Zou, E. Zhou and H. Zhang 611

Improvement of in-plane alignment of $YBa_2Cu_3O_{7-x}$ films on polycrystalline alumina substrates using biaxially aligned CeO_2/YSZ buffer layers
K.Y. Chen, S. Afonso, Q. Xiong, G.J. Salamo and F.T. Chan 613

Fabrication of growth of textured YBCO thin film on very thin metallic substrate
D. Goo, D. Youm, J. Yoo, K. Ahn and J. Yoon 615

High J_c YBCO film on NiCr alloy with aligned YSZ buffer
E.Y. Zhou, S.M. Gong, Y.J. Mao, M.P. Yuan, H. Zhang, J. Yuan, G.Q. Yang, X.H. Liu and S.C. Zou 617

Preparation and characterization of epitaxial $La_{0.5}Sr_{0.5}CoO_3$ films on (001) $LaAlO_3$ by pulsed laser deposition
W. Wu, P.W. Chan, K.H. Wong, J.T. Cheung, X.-G. Li and Y.H. Zhang 619

Chemical diffusion of oxygen in a *c*-axis-oriented $YBa_2Cu_3O_{7-\delta}$ film
X.-J. Jin, M. Li, Y.-M. Cao and L. Xu 621

Differences in the nucleation rate of $YBa_2Cu_3O_{7-\delta}$ on patterned (001) $LaAlO_3$ substrates
M. Gustafsson, E. Olsson, H.R. Yi, D. Winkler and T. Claeson 623

Optimization of T_c of epitaxial GdBaCuO growth superconducting thin film by inverted cylindrical sputtering
Y. Li, J. Geerk, R. Smithy, X. Liu and J. Xu 625

Growth and characterization of "infinite layer" films doped with p-type carriers
Y. Sun, P. Sun, J.S. Niu, Y.G. Sun and B.K. Ma 627

Morphology of Bi2201 thin films fabricated on the thermal and chemical treated MgO(100)
F. Ahmed, Y. Qi, M. Kawakami, K. Sakai and R. Aoki 629

Molecular beam epitaxy of a single atomic layer: experimental results for cuprates and lattice-gas simulations
F.C. Beuran, J-L. Domange, X.Z. Xu, C. Deville-Cavellin, V. Mairet and M. Lagues 631

Epitaxial growth of $Bi_2Sr_{2-x}La_xCuO_{6+\delta}$ thin films by in-situ DC- and RF- magnetron sputtering
Y.Z. Zhang, L. Li, D.G. Yang, Y.L. Qin, B.R. Zhao, S.L. Jia, H. Chen, F. Wu, H.J. Tao, H.T. Yang, J.W. Li, W. Liu and Z.X. Zhao 633

Thin film growth of high-T_c superconductor by infrared-ray irradiation
M. Akinaga and L. Rinderer 635

Influence of thallination conditions upon properties of TBCCO films deposited from an aerosol
A. Morales, A. Conde-Gallardo, M. Jergel, C. Falcony, S. Jiménez-Sandoval, E. Chavira and J.G. Cabañas Moreno . . . 637

Properties of Tl-based thin films and thin film microstrips deposited by evaporation of BaF_2, CaF_2 and Cu components
S. Chromik, V. Štrbík, M. Jergel, F. Hanic, A. Conde-Gallardo, C. Falcony, E. Chavira and Š. Beňačka 639

Enhancement of critical current density in $Tl_2Ba_2CaCu_2O_y$ superconducting thin film on polycrystalline Al_2O_3 substrates using textured YSZ buffer layers
Q. Xiong, S. Afonso, K.Y. Chen, G. Salamo and F.T. Chan 641

Highly *c*-axis oriented $HgBa_2Ca_2Cu_3O_y$ thick and thin films with $T_c > 130$ K
Y. Moriwaki, T. Sugano, C. Gasser, K. Nakanishi, S. Adachi and K. Tanabe 643

Preparation of superconducting $HgBa_2CaCu_2O_x$ and $Hg_{0.8}Bi_{0.2}Ba_2CaCu_2O_x$ films by means of annealing of mercury-free precursor films
J.D. Guo, G.C. Xiong, D.P. Yu, Q.R. Feng, X.L. Xu, G.J. Lian, K. Xiu and Z.H. Hu 645

High T_c superconducting oxycarbonate thin films grown by laser ablation: promising new materials
W. Prellier, A. Tebano, J.L. Allen, B. Mercey, J.F. Hamet, M. Hervieu and B. Raveau 647

Homogeneity measurement of J_c of large-scale high T_c thin films
J.-W. Qiu, B. Wu, X.-b. Yang, Z.-M. Tang, L.-W. Zhou, C.-X. Ren, G.-L. Chen, R.-L. Wang and H.-C. Li 649

Structure and electrical properties of metastable $Sr_{1-x}CuO_{2-\delta}$ thin films
J.G. Hou, P.M. Woodwardand and A.W. Sleight 651

On the role of Ag in improved $YBa_2Cu_3O_{7-\delta}$ films
Y. Li, J.R. Liu, X.T. Cui, Y. Cao, Z.J. Qu, Q.Y. Chen, C.W. Chu and W.K. Chu 653

Transversal thermovoltages of (119) $Bi_2Sr_2CaCu_2O_{8+\delta}$ thin films on vicinal (110) $SrTiO_3$ substrates
P. Haibach, U. Frey, H. Adrian, F. Lankes and K.F. Renk 655

Enhancement of superconductivity by low energy Ar ion milling in epitaxial $YBa_2Cu_3O_7$ thin films
L. Chen, T. Yang, J.C. Nie, P.J. Wu, M.Q. Huang, G.R. Liu and Z.X. Zhao 657

Determination of the *c*-axis thermal diffusivity in $YBa_2Cu_3O_{7-\delta}$ thin film by 'Mirage' technique
B.M. Wu, Y.L. Du, X.P. Wang, J.W. Fang, S.Y. Zhang and S. Sheng 659

T_c reduction and related phases at early stages of $YBaCuO_{7-\delta}$ film growth on $LaSrAlO_4$
H.-U. Habermeier, P.X. Zhang, T. Haage and J.Q. Li 661

Effect of laser beam profile on the ablated HTS target surface
Y.S. Jeong, S.Y. Lee, M.K. Choi, J.Y. Kim, B.-D. Chung, I.-S. Yang, B. Oh and S.-H. Moon 663

The *ab*-anisotropy of twinfree $YBa_2Cu_3O_{7-\delta}$ films above and below T_c
B. Dam, J. Rector, R. Surdeanu, R.J. Wijngaarden, G. Koster, F. Peerdeman, J. van Berkum, D.G. de Groot and R. Griessen . 665

Superconducting properties of doped-YBCO ultra thin films
G.Y. Sung and J.D. Suh 667

Finite range scattering of Ni and Zn impurities in Y-123 thin films
T. Kluge, G. Jakob, M. Speckmann and H. Adrian 669

Analysis of persistent photoconductivity on oxygen deficient $YBa_2Cu_3O_{7-x}$ thin films
D. Giratá, B. Arenas, R. Hoyos, J. Osorio, M.E. Gómez, J. Heiras and P. Prieto 671

Transport properties of layered cuprate films with multiple fluorite-type blocks (Pb-32*n*2 phase) and of Pb-3212 phase films
S. Ikegawa and Y. Motoi 673

Growth-induced in plane anisotropy of flux-line pinning in $YBa_2Cu_3O_7$ thin films
H.-U. Habermeier, T. Haage, J.Q. Li, J. Zegenhagen and A. Forkl 675

$NdBa_2Cu_3O_{7-\delta}$ superconducting thin films made by the PLD method
S.-H. Moon and B. Oh 677

RHEED/AFM analysis of YBCO epitaxial thin films grown by laser MBE
M.R. Cimberle, A. Diaspro, C. Ferdeghini, E. Giannini, G. Grassano, D. Marrè, I. Pallecchi, M. Putti, R. Rolandi and A.S. Siri 679

A4. Multilayers and superlattices

The study of the $YBa_2Cu_3O_7/PBa_2Cu_3O_7$ multilayer films by high resolution electron microscopy
L. Wei, Z. Yun, S. Jijun, Z. Bairu and L. Lin . . . 681

Fabrication and characterization of vias for contacting $YBa_2Cu_3O_{7-x}$ multilayers
J.W. Cooksey, S. Afonso, W.D. Brown, L.W. Schaper, S.S. Ang, R.K. Ulrich, G.J. Salamo and F.T. Chan . . . 683

Fabrication techniques and electrical properties of $YBa_2Cu_3O_{7-x}$ multilayers with rf sputtered amorphous SiO_2 interlayers
S. Afonso, K.Y. Chen, Q. Xiong, F.T. Chan, G.J. Salamo, J.W. Cooksey, S. Scott, S. Ang, W.D. Brown and L.W. Schaper . . . 685

Superlattice fabrication of perovskite oxide with different B-site cations
I. Tsukada, M. Nose and K. Uchinokura . . . 687

Microstructure and strain relaxation in $YBa_2Cu_3O_7$ / $PrBa_2Cu_{3-x}Ga_xO_7$ superlattices
J.-P. Contour, M. Drouet, O. Durand, J.-L. Maurice and A. Gauzzi . . . 689

Superconducting and magnetic properties of $(La,Ca)MnO_3/YBa_2Cu_3O_{7-\delta}$ multilayer films
M. Isa, T. Nishizaki, M. Fujiwara, T. Naito and N. Kobayashi . . . 691

Properties of multilayer samples with interface of superconducting $YBa_2Cu_3O_7$ and ferromagnetic $Pr_{0.7}Sr_{0.3}MnO_3$
G.C. Xiong, G.J. Lian, J.F. Kang, Y.F. Hu, Y. Zhang and Z.Z. Gan . . . 693

High quality $YBa_2Cu_3O_{7-x}/SrTiO_3$ bilayers for superconducting electric field effect transistor
J.D. Suh, G.Y. Sung, S.G. Lee and D.K. Kim . . . 695

Epitaxial growth of infinite layer $MCuO_2$ (M = Ba, Sr, Ca) thin films and their superlattices by laser molecular beam epitaxy
H.S. Wang, Z.Q. Peng, O. Jin, Z.H. Chen, H.B. Lu, Y.L. Zhou, L. Li and G.Z. Yang . . . 697

Structural and ferroelectric properties of $BaTiO_3/YBa_2Cu_3O_7$ heterostructures prepared by laser molecular beam epitaxy
H.S. Wang, Y.W. Liu, K. Ma, Z.Q. Peng, D.F. Cui, H.B. Lu, Y.L. Zhou, Z.H. Chen, L. Li and G.Z. Yang . . . 699

Particulate-free a-$YBa_2Cu_3O_{7-x}/La_{0.5}Sr_{0.5}CoO_3/a$-$YBa_2Cu_3O_{7-x}$ sandwiched thin film on (100)$SrTiO_3$ fabricated by pulsed laser deposition with a shadow mask
W. Wu, K.H. Wong, P.W. Chan, J.T. Cheung, X.-G. Li and Y.H. Zhang . . . 701

YBCO and YSZ buffer layer growth on metallic tape using sputtering method
D.Q. Shi, J. Wang, J.M. Hao, X.P. Wang, Z.Q. Hua and G.S. Yuan . . . 703

Study of structures and dielectric dispersion behavior of $BaTiO_3/YBa_2Cu_3O_{7-x}$ bilayer films
A. Zhu, C. Li, K. Ma, Z. Wang, X. Xiong, W. Wu, C. Wang, L. Wu, J. Li, L. Li, Z. Chen and Z. Mai . . . 705

Transport properties of $Sr_xCuO_{2-\delta}$ film in YBCO/$Sr_xCuO_{2-\delta}$/YBCO system
S. Koba, R. Nagata, M. Onishi, Y. Ozono, K. Shibata, S. Higo, I. Kawano, T. Kushida, Y.W. Kim and T. Ogushi . . . 707

Epitaxial growth and tunneling properties of $YBa_2Cu_3O_{7-x}/PrBa_2Cu_3O_{7-x}/YBa_2Cu_3O_{7-x}$ trilayer structures
E. Baca, W. Lopera, M.E. Gómez, J. Heiras, P. Prieto, R. Di Leo, F. Bobba, A. Nigro, P. Romano and A.M. Cucolo . . . 709

Tunneling characteristics of epitaxial $YBa_2Cu_3O_{7-x}/Y_2O_3/YBa_2Cu_3O_{7-x}$ planar type junctions
M. Chacón, G. Bolaños, W. Lopera, A. Pulzara, M.E. Gómez, J. Heiras and P. Prieto . . . 711

Physical properties of $Pb(Zr,Ti)O_3/YBa_2Cu_3O_7$ integrated films
B.R. Zhao, L.X. Cao, J.H. Jiang, Y. Xu, K. Ma, L. Li, Y. Zhu, D.F. Zhang, J.H. Zhao, X.J. Li, B. Xu, Y.Z. Zhang, T. Yang and Z.X. Zhao . . . 713

A5. Non copper oxides

Heat capacity measurement on single crystalline YNi_2B_2C grown by floating zone method
H. Takeya, S. Miyamoto, K. Yamada, N. Nonose and K. Kadowaki . . . 715

Systematic variation of superconductivity and phase stability for the pseudo-quaternary borocarbide system $Y(Pd_{1-x}Rh_x)_2B_2C$
Y.Y. Hsu, T.I. Hsu and H.C. Ku . . . 717

$BaZrO_3$: Skull-melting, characterisation and interactions with YBCO
S. Gerhold, M. Aigner and W. Assmus . . . 719

Electrochemical syntheses of Ba–K–Bi–O crystals in hot alkaline solution and water rich melts
L.Z. Zhao, B. Yin, J.B. Zhang, J.W. Li, C.Y. Xu, S.H. Liu, M. Xie, Y. Li, X.A. Chen and B.M. Chen . . . 721

Synthesis of Ba–K–Bi–O superconductor at 260°C by direct precipitation from KOH melts
L.Z. Zhao, B. Yin, J.B. Zhang, J.W. Li, C.Y. Xu, M. Xie and S.H. Liu . . . 723

Study of local structure of YNi_2B_2C superconductor by EXAFS
X. Ju, Y.X. Su, Y.B. Wang, Y. Tao and L. Wei . . . 725

Search for new superconducting oxides in Sr–Nb–O system
N. Peng and J.T.S. Irvine . . . 727

New layered perovskite ruthenates: Ca_2RuO_4
S. Nakatsuji, S.-i. Ikeda and Y. Maeno 729

New insulating materials for high-T_c superconductor device applications
R. Hojczyk, C.-L. Jia, U. Poppe and K. Urban 731

A6. Alkali metal fullerenes

Dynamic of transformation under the heat treatment of a new carbon phase carbolite studied by calorimetry and X-ray powder diffraction
A.V. Palnichenko, A.F. Gurov, V.N. Kopylov, K. Kusano and S. Tanuma 733

Structure and conductivity of C_{60}/Pd multilayer films
B. Wang, H. Wang, Y. Li, B. Xie and J. Hou 735

Study of microstructure and interfacial interaction of Al-C_{60} co-evaporated films
Y. Li, Y. Wang, W. Xu, J.G. Hou and Y.H. Zhang 737

Growth of high-quality $C_{60}(111)$ single crystalline film on $Ni_3Co(111)$ surface
J. Zeng, Y. Wang, Y. Li and J.G. Hou 739

A7. Organic superconductors (no contributions)

A8. Conventional superconductors

A new tin niobium selenide superconductor
A. Soreau-Leblanc, P. Molinié and J.C. Jumas 741

A9. Rare earth cuprates

Synthesis and characterisation of $La_{2-x}Ba_{1+x}Cu_2O_{6+\delta}$
E. Lee, W. Zhou and V. Rives 743

Relating structural properties and oxygen content to the electronic and magnetic states of $(Eu_{1.5-x}Pr_xCe_{0.5})Sr_2Cu_2NbO_{10-\delta}$
T.J. Goodwin, R.N. Shelton and H.B. Radousky 745

Correlation of lattice parameter, oxygen content and charge carrier density in $Nd_{2-x-y}La_yCe_xCuO_{4-\delta}$ superconductors
C. Wang, C. Dong, Z. Qiao, G. Che and Z. Zhao 747

Structural minals and generic relationship between complex copper-containing oxides
L. Leonyuk, G.-J. Babonas and V. Maltsev 749

Synthesis and structural characterisation of quintupled layered perovskites
M.J. Pack and M.T. Weller 751

Structural transition of La.Ba.Cu.O doped with iron
W.M. Chen, J. Chen, N.L. Chen, X. Zhang, X.S. Wu, X. Jin and S.S. Jiang 753

Superconducting properties of the oxidized $La_{2-x}Ba_xCuO_4$ system
G.D. Liu, Z.X. Zhao, G.C. Che, J.W. Li and F. Wu 755

Suppression of superconductivity in the $Y_{1-x}Pr_xBa_2Cu_{2.95}Li_{0.05}O_{7-\delta}$ system
H. Song, G. Cao, Y. Yu, Z. Jiao and Q. Zhang 757

X-ray Rietveld analysis of $Y_{0.4}Pr_{0.6}BaSrCu_3O_{7-\delta}$ and $Y_{0.4}Pr_{0.6}Ba_2Cu_3O_{7-\delta}$
H. Song, G. Cao, Y. Yu, G. Lu, Z. Jiao and Q. Zhang 759

Origin of the ionic size effect on Pr-induced T_c suppression in $R_{1-x}Pr_xBa_{2-y}Sr_yCu_3O_{7-\delta}$
G. Cao, H. Song, Z. Jiao and Q. Zhang 761

Structures and superconductivity of $Nd_{0.7}Sr_{1.3}Cu(O,F)_{4-\delta}$ with $T_c = 44$ K and its precursor
J.L. Yang, J.K. Liang, Z.F. Dong, G.H. Rao, Y. Shi and W.H. Tang 763

Substitution effects of Pr for La on structure and properties in the T*-type $La_{1-x}Pr_xGd_{0.8}Sr_{0.2}CuO_z$
T. Moriga, T. Takahashi, T. Takeuchi, Y. Haruyama, S. Takemoto and I. Nakabayashi 765

Superconductivity, hole concentration and peritectic transition of $La_{1.5}Ca_xBa_{1.5-x}Cu_3O_y$ compounds
W.N. Huang, C.M. Wang and H.-C.I. Kao 767

Synthesis and superconductivity of fluorides of $Pr_{2-x}Sr_xCuO_{4-y}$ ($x = 0.0, 0.4, 1.0$)
J.L. Yang, J.K. Liang, D. Jin, Y. Shi, W.H. Tang and G.H. Rao 769

Influence of oxygen post-annealing on superconducting properties of $NdBa_2Cu_3O_y$
C.-B. Cai, Y. Sun, S.-M. Gong, Y.-X. Fu and H. Zhang . . . 771

Superconductivity at 80 K in single crystals of structural type $A_{14}Cu_{24}O_{41}$
L. Leonyuk, G.-J. Babonas, V. Maltsev, A. Rėza, R. Szymczak and S. Kapelnitsky . . . 773

Synthesis and superconductivity of a new type of 1212 phase cuprate $(B,Cd)(Sr,Ba)_2YCu_2O_y$
Z. Xu, X. Zhang, Y. Ren, Z. Jiao and Q. Zhang . . . 775

Structural, transport and magnetic properties of $PrBa_{2-x}Sr_xCu_3O_{7-\delta}$
Y.G. Zhao, Y.P. Li, B. Zhang, S.Y. Xiong, B. Yin, J.W. Li, S.Q. Guo, J.W. Xiong, D.J. Dong, B.S. Cao and B.L. Gu . . . 777

Superconductivity, synthesis and magnetism of $CaLaBaCu_{3-x}(Zn,Fe)_xO_7$ tetragonal system
V.P.S. Awana, R. Singh, D.A. Landinez Tellez, J.M. Ferreira, J. Albino Aguiar and A.V. Narlikar . . . 779

Structural and superconducting properties of $La_{2-x}Nd_xCuO_{4+y}$ $(0 \leqslant x \leqslant 0.5)$ prepared by room temperature chemical oxidation
C. Rial, E. Morán, M.A. Alario-Franco, U. Amador and N.H. Andersen . . . 781

Study of synthesis of $Pr_{1-x}Ce_xBa_2Cu_3O_{7-\delta}$ and its transport and magnetic properties
S.Y. Xiong, Y.G. Zhao, D.J. Dong, S.Q. Guo, P.C. Song, L.W. Zhang, M.H. Zhu, B.S. Cao and B.L. Gu . . . 783

The vanadium effects on properties of $YBa_2Cu_{3-x}V_xO_{7+y}$
I.V. Hong, N.H. Dan and D.T. Dat . . . 785

X-ray diffraction studies on $La_{1.85}Sr_{0.15}CuO_4$ doped with Mg in high doping level
X.S. Wu, Z.Q. Mao, J. Lin, W.M. Chen, X. Jin, G.J. Xu, Y.H. Zhang, F.M. Pan and S.S. Jiang . . . 787

The high-pressure synthesis and characterization of some praseodymium-substituted rare-earth based $R_2Ba_4Cu_7O_{14+\delta}$ (R = Nd, Sm, Eu, Gd, Ho, Tm)
T.-M. Chen, Y.-L. Lai and F.S. Kao . . . 789

A role of the Cu–O chain for the pressure-induced change in the carrier number in $Y_{1-x}Ca_xBa_2Cu_3O_y$ system
T. Honma, N. Môri and M. Tanimoto . . . 791

Influence of oxygen content on T_c of $R_{1-x}Ca_xBa_2Cu_3O_{6+\delta}$ (R = Y, Gd: $x = 0$, 0.2)
K. Hatada and H. Shimizu . . . 793

The effect of excess strontium $(x \geqslant 2)$ on the phase and electric properties of $LnSr_xCu_{2.7}M_{0.3}O_z$ (M = Cr, Mo, W, Ln = rare earth element)
Z.Y. Chen, H.M. Luo, Y. Han, Y.T. Qian, Z.W. Mao and G.E. Zhou . . . 795

Rare-earth ionic size effects on T_c in the series $RBa_2Cu_3O_{7-\delta}$
X. Chen, J. Sha, Z. Xu, Z. Jiao and Q. Zhang . . . 797

Current dependence of superconducting transition in sintered polycrystalline $YBa_2Cu_3O_{7-x}$
S.B. Ota, M. Bose, B. Sahoo and T.V. Chandrasekhar Rao . . . 799

Structural modulations in superconducting $La_{2-x}K_xCuO_4$
M.A. Señaris, M.A. Alario-Franco, S. Stoll and A.M. Stacy . . . 801

On atoms charge states and orbital occupancy in $RBa_2Cu_3O_x$ high-temperature superconductors
V.E. Gusakov . . . 803

Synthesis of new praseodymium-based cuprates $(Pr,Cd)Sr_2YCu_2O_y$ and $(Pr,M)Sr_2(Gd,Ce)_2Cu_2O_z$ (M = Cu, Zn)
Z.Y. Chen, H.M. Luo, Y.T. Qian, L. Yang, L. Shi and G.E. Zhou . . . 805

Superconductivity and ordering of Pr in $CaLa_{1-x}Pr_xBaCu_3O_7$ compound
V.P.S. Awana, D.A. Landinez Tellez, J.M. Ferreira, J. Albino Aguiar, S.K. Malik, A.V. Narlikar, W.B. Yelon, R. Singh, S. Uma and E. Gmelin . . . 807

Synthesis and physical properties of $(Pr_{1-x}RE_x)Ba_2Cu_4O_8$ oxides (RE = Y, Yb, Nd, Gd)
S. Horii, Y. Yamada, N. Yamada, I. Hirabayashi and U. Mizutani . . . 809

A10. Layer cuprates

Crystal structure and physical properties of the spin-1/2 two leg ladder system, $Sr_{14-x}Ca_xCu_{24}O_{41+\delta}$
M. Isobe, J.Q. Li, Y. Matsui, T. Ohta, M. Onoda, F. Izumi, S. Nakano, T. Matsumoto, H. Hayakawa and E. Takayama-Muromachi 811

Order/disorder of tetrahedral-chains in $AlSr_2YCu_2O_y$ and related oxide superconductors examined by HRTEM
J. Ramírez-Castellanos, Y. Matsui, M. Isobe, F. Izumi, N.R. Khasanova and E. Takayama-Muromachi . . . 813

Effects of oxygen content upon phase stability of $(Pb,Cu)Sr_2(Y,Ca)Cu_2O_{7+\delta}$
B.J. Mitchell and J.T.S. Irvine . . . 815

Structural and physical properties of the $(Cu,C,B)Ba_2Ca_{n-1}Cu_nO_{2n+2+\delta}$ superconductors with T_c up to 130 K under pressure
C. Chaillout, E. Gautier, S. Le Floch, C. Acha, P. Bordet, M. Nunez-Regueiro and J.L. Tholence . . . 817

Synthesis of the $(Ga,Al)Sr_2Ca_{n-1}Cu_nO_{2n+3}$ ($n=2$–5) systems at ambient pressure
E. Chavira, E. Aguila, L. Baños and O. Navarro 819

Superconductivity of $(Pb,M)Sr_2(Y,Ca)Cu_2O_y$ ($M=Cu, Cd$) in relevance to the valence state of lead
M. Kikuchi, K. Kanehashi, E. Ohshima and Y. Syono 821

New cuprates with the 1212 structure $(Cd,M)(SrLa)_2YCu_2O_\delta$ ($M=Bi, In$)
S. Wang, Y. Qian and H. Xin 823

Degradation mechanism of $Bi_2Sr_2CaCu_2O_y$ single crystal in humid atmosphere
L. Yang, X.-F. Sun, W.-B. Wu, G.-E. Zhou and Y.-H. Zhang 825

Stabilization of Tl–Sr 1223 phase by high-pressure synthesis and Mo- and Re-substitution for Tl
E. Ohshima, T. Atou, M. Kikuchi and Y. Syono 827

Effect of dynamical scattering on crystal structure determination of $Bi_4(Sr_{0.75}La_{0.25})_8Cu_5O_y$
B. Lu and F.H. Li 829

Study of mismatch between two different blocks in Bi-system superconductors
H. Zhang, X.C. Qin, Y. Zhao, Y.Y. He and Q.R. Feng 831

Neutron and X-ray diffraction studies on $YBa_2(Cu_{1-y}Co_y)_3O_x$ superconductors with $0.02 \leqslant y \leqslant 0.04$
X.S. Wu, C. Gou, C.C. Lam, W.M. Chen, D.F. Chen, K. Sun, W. Ji, W.J. Liu, X. Jin and S.S. Jiang 833

Short-range-order state in the $Sr_2Nd_{0.5}Ca_{0.5}Cu_2O_{5+y}F_{1+\delta}$ superconductor
J. Li, M. Isobe, E. Takayama-Muromachi and Y. Matsui 835

A new series of layered pure perovskites $(ACuO_{2.5})_2(ATiO_3)_m$
K.D. Otzschi, K.R. Poeppelmeier, P.A. Salvador, T.O. Mason, W. Sinkler, H. Zhang and L.D. Marks 837

Superconductivity and structure of $Bi_2Sr_2CaCu_2O_y$ single crystals with various oxygen content
X. Sun, W. Wu, L. Wang, G. Zhou, X.-G. Li and Y. Zhang 839

Effect of structural homogeneity on the diamagnetic transition width in $Bi_2Si_2CaCu_2O_y$ single crystals
X. Zhao, W. Wu, X. Sun, G. Zhou, X.-G. Li and Y. Zhang 841

Correlation of grain size and ac susceptibility in the $Bi_2Sr_2CaCu_2O_8$ system
H. Jin, W.C. Zhao, F.M. Lou, K. Zhang, K.Q. Ruan, C.Y. Wang, Q.M. Liu, L.Z. Cao, Y.H. Zhang, Z.J. Chen and Z.Q. Hu . 843

T_c enhancement in silver substituted $Bi_{1.7}Pb_{0.3}Sr_{2-x}Ag_xCa_2Cu_3O_y$
Y. Yu, X.N. Xu, X. Jin, S.Y. Ding, Z.Y. Zheng and X.X. Yao 845

Model of the structural phase transition for $Bi_2Sr_2CaCu_2O_8$
E.E. Tornau, S. Lapinskas and A. Rosengren 847

Multiple layer structures and superconductivity
M.T. Weller, M.J. Pack, C.S. Knee, D.M. Ogborne and A. Gormezano 849

Formation of 2223 phase in the cation substituted Bi-based system
D. Sýkorová, O. Smrčková, K. Nováková and P. Vašek 851

Influence of Cu content on the formation and microstructures of (Bi,Pb)-2223 phase in the partial-melting and sintering process
X.Y. Lu, A. Nagata, M. Yasuda, K. Sugawara and S. Kamada 853

Controllable preparation of Bi-2223 phase through directed modification of precursors
V.V. Ischenko, O.A. Shlyakhtin, A.L. Vinokurov, H. Altenburg and N.N. Oleinikov 855

Superconducting properties of the mercury and Cu/C phases
J.L. Tholence, P. Toulemonde, A. Sulpice, C. Acha, M. Nunez-Regueiro, J.J. Capponi, C. Chaillout, M. Alario-Franco, E. Gautier, S. Le Floch, M. Marezio, P. Bordet, S.M. Loureiro, S.N. Putilin and E.V. Antipov 857

Dynamic study of oxygen in Cu–O plane in Bi-system superconductors
Y. Zhao, H. Zhang, X. Du, Y.Y. He, Q.R. Feng and S.Q. Feng 859

Superconductivity of nanostructured Bi–Sr–Ca–Cu–O materials
W.C. Zhao, H. Jin, M.L. Tian, K. Zhang, F.M. Lou, Y.C. Tian, K.Q. Ruan, C.Y. Wang, L.Z. Cao and Y.H. Zhang 861

Crystallization and phase evolution of the Bi–Sr–Ca–Cu–O nanostructured materials
W.C. Zhao, H. Jin, L.B. Wang, K. Zhang, M.L. Tian, L. Yang, F.Q. Li, K.Q. Ruan, C.Y. Wang, L.Z. Cao and Y.H. Zhang . 863

Effect of growth conditions and compositions on the morphology and stability of the solid–liquid interface of Bi–Sr–Ca–Cu–O grown by a floating zone method
G.D. Gu, G.J. Russell and N. Koshizuka 865

On the formation and properties of Bi-2212 phase with different cation composition
D.I. Grigorashev, K. Wissing, M. Mertens, M.S. Kuznetsov, N.N. Oleynikov and Yu.D. Tretyakov 867

Electronic structure and superconducting properties of doped BPSCC (2223) system
M.N. Khan, A.N. Kayani and A. ul Haq . . . 869

Synthesis and electric properties of new cuprates $(Pb,Cu,Cd)_3Sr_2(Nd,Ce)_2Cu_2O_z$
H.M. Luo, Z.Y. Chen, H.X. Wang, Y.T. Qian and W.Q. Zhang . . . 871

Effects of doping hole concentration in bismuth-contained superconducting cuprates
N.H. Sinh . . . 873

Superstructure defects in BSCCO (2212) whiskers
V.N. Timofeev and I.G. Gorlova . . . 875

Two-peak structure of AC susceptibility of $Bi_2Sr_2CaCu_2O_x$ tapes
P.N. Mikheenko, M. Ionescu, J. Horvat and S.X. Dou . . . 877

Change in crystal structure related with two different blocks in $Bi_{1.7}Pb_{0.3}Sr_2Ca_2Cu_3O_x$
H. Zhang, Y. Zhao, Q.R. Feng, S.Q. Feng, F. Ritter and W. Assmus . . . 879

Synthesis and properties of $(HgPb)_1(BaSr)_2Ca_2Cu_3O_z$ superconductors
S. Lee, N. Kiryakov, O. Plesenkova, D. Emelyanov and Y. Tretyakov . . . 881

The structures of the 126 K $Ba_2Ca_2Cu_3(Ca_xCu_y)O_{8+\delta}$, the 116 K $Ba_2Ca_3Cu_4(Ca_xCu_y)O_{10+\delta}$ and their transformed products
Y.Y. Xue, Y.Y. Sun, Z.L. Du, I. Rusakova, K. Ross, N.L. Wu, Y. Cao, L. Gao and C.W. Chu . . . 883

Pressure dependence of T_c in $HgBa_2CuO_{4+\delta}$ up to 18 GPa with $0.17 < \delta < 0.3$
X.D. Qiu, Q. Xiong, L. Gao, Y. Cao, Y.Y. Xue and C.W. Chu . . . 885

A new Hg-based homologous series of cuprate superconductors; (Hg,Tl)-22($n-1$)n ($n = 1$–5)
A. Tokiwa-Yamamoto, T. Tatsuki, X.-J. Wu, S. Adachi and K. Tanabe . . . 887

Structural study on the (Hg,Tl)–Ba–(Ca,Y)–Cu–O system by XRD and TEM
X.-J. Wu, A. Tokiwa-Yamamoto, T. Tatsuki, S. Adachi and K. Tanabe . . . 889

Partial melting of F-doped $Hg_xBa_2Ca_2Cu_3Re_{0.2}O_y$ filament
T. Goto . . . 891

The first step on the synthesis route to Bi and Hg superconductors
O. Smrčková, D. Sýkorová and K. Petrýdes . . . 893

Electron microscopy study of the Ba and Sr mercury-based superconductors $Hg_mM_2(Y,Ca)_{n-1}Cu_nO_y$, with M = Ba or Sr and $m = 1, 2$
C. Chaillout, S.M. Loureiro, P. Toulemonde, E. Gautier, J.J. Capponi and M. Marezio . . . 895

New phase of Hg-2435 in the Hg–Ba–Ca–Cu–O oxide system
Y.S. Yao, W. Liu, Y.J. Su, Y.F. Xiong, W.J. Ma, G.H. Cao, G.T. Zou and Z.X. Zhao . . . 897

New phases in the double-Hg layer system $Hg_2Ba_2LnCu_2O_{8-\delta}$ (Ln = Nd–Gd, Dy–Lu)
S.M. Loureiro, P. Toulemonde, C. Chaillout, E. Gautier, E.V. Antipov, J.J. Capponi and M. Marezio . . . 899

X-ray single crystal structure analysis of $Sr_{0.73}CuO_2$ and $Hg_{1-x}Pb_xBa_2Ca_{n-1}Cu_nO_{2n+2+\delta}$ compounds ($n = 1$–5, $x = 0$–0.5)
H. Schwer, K. Conder, E. Kopnin, G.I. Meijer, R. Molinski and J. Karpinski . . . 901

Anisotropic compressibility and pressure dependence of the superconducting transition temperature in $HgBa_2Ca_{n-1}Cu_nO_{2n+2+\delta}$ ($n = 1$, 2, and 3) series as a function of oxygen content
X.-j. Chen, Z.-a. Xu, Z.-k. Jiao and Q.-r. Zhang . . . 903

Dependence of superconducting transition temperature on oxygen content and number of CuO_2 layers in mercury-based cuprates
Q.-r. Zhang and X.-j. Chen . . . 905

Intergrowth structures and screw dislocations in Pb-doped Hg-1234 superconductor
K.C. Hung, C.C. Lam, J.Q. Li and H.M. Shao . . . 907

The surface chemistry of Hg-containing superconductors by HREM
W. Zhou, C. Panagopoulos, E.T.C. Lee, J.P. Hodges, G.B. Peacock, A. Asab, I. Gameson and P.P. Edwards . . . 909

Crystal structure of $(Hg_{0.7}Tl_{0.3})_2Ba_2(Y_{0.8}Ca_{0.2})Cu_2O_{8-\delta}$ synthesized at high pressure
T. Ohta, A. Tokiwa-Yamamoto, F. Izumi, A.W. Hewat and K. Tanabe . . . 911

Study of local structure changes of HBCCO superconductor with EXAFS
X. Ju, Y.X. Su, Y.B. Wang, Y. Tao, L. Wei and H.M. Shao . . . 913

Synthesis and characterization of $(Tl_{1-x}M_x)(Ba,Sr)_2Ca_2Cu_3O_z$ (M = Pb, K, Bi) cuprates: the chemical control of high-temperature superconductivity
T.-M. Chen and J.S. Ho . . . 915

Preparation and structure refinement of the Ti, Mo co-doped 123 compound $NdBa_2Cu_2(Ti,Mo)O_8$
R.K. Li, Z.S. Lou, Y.T. Qian and Z.Y. Chen . . . 917

Synthesis and thermal stabilization of nearly single phase $Hg(Tl)Ba_2Ca_2Cu_3O_{8+\delta}$ superconductors
Q.L. Xu, L.Z. Cao, C.Y. Xu, Y.H. Zhang, F. Foong and S.H. Liou . . . 919

Synthesis of Hg-based 1222 cuprates using sol–gel process combined with the sealed quartz-tube method
Y. Qian, K. Tang, X. Xu, L. Yang and Y. Zhang . . . 921

Phase transition of the $HgBa_2Ca_2Cu_3O_8$ superconducting phase by in-situ heating in transmission electron microscope
D.P. Yu, J.D. Guo, X.L. Xu, S.Q. Feng and Z. Zhang . . . 923

TEM observation of moiré fringe in the $HgBa_2Ca_2Cu_3O_{8+\delta}$ superconductor
H.M. Shao, C.C. Lam, S.D. Wang and X.X. Yao . . . 925

Superconducting properties of mercury cuprates single crystals
G. Villard, V. Hardy, A. Maignan, D. Pelloquin and B. Raveau . . . 927

A11. New materials

Calcium ladder cuprate films grown by Molecular Beam Epitaxy
C. Deville Cavellin, B. Eustache, C.J. Hatterer, J.-P. Lauriat, X.Z. Xu, V. Mairet, F. Beuran and M. Laguës . . . 929

Studies on a possible superconducting system Y–Cu–Se
Y. Huang, C. Dong, J. Liu, Y. Ni, Z. Zhao and Chen Liquan . . . 931

$La_{1.6}Sr_{0.4}CaCu_2O_{4+\delta}Cl_{2-y}$ superconductor synthesized at ambient pressure
G.C. Che, G.D. Liu, Z.X. Zhao, J.W. Li, H. Chen, S.Q. Guo, F. Wu, C. Dong and S.L. Jia . . . 933

Study on crystal structure and physical properties of $(Sc_{1-x}In_x)Ba_2CuO_y$
T. Shimada, S. Kambe, S. Ohshima, N. Ohnishi and K. Hiraga . . . 935

Depression of superconductivity by tilting of the 6-coordinated CuO_2 plane
G. Samukawa, S. Kambe, C. Yoshida, S. Ohshima and O. Ishii . . . 937

Synthesis and characterization of a new Pb-3212 phase: $Pb_{2-x}Cd_{2x}Cu_{1-x}Sr_2(Dy,Ca)Cu_2O_z$
W.H. Zhang, Y.T. Qian, K.B. Tang, A.W. Zhang and Z.Y. Chen . . . 939

Synthesis and properties of $(Bi,Cd)Sr_2(Gd_{2-x}Ce_x)Cu_2O_z$
K.B. Tang, Y.T. Qian, Y.G. Zhang, H. Cang and Y.H. Zhang . . . 941

Cerium-based layered cuprates $(Ce,Ni)Sr_2YCu_2O_y$ and $(Ce,M)Sr_2(Ln,Ce)_2Cu_2O_z$
H.M. Luo, Z.Y. Chen, Y.T. Qian and L. Yang . . . 943

Systematic studies on substitutional effects in layered manganocuprate $Eu_3Ba_2Mn_2Cu_2O_{12}$
I. Matsubara, R. Funahashi, N. Kida, K. Ueno and H. Ishikawa . . . 945

Crystal structure of a new series of 1212 type cuprate $RuSr_2LnCu_2O_z$
K.B. Tang, Y.T. Qian, L. Yang, Y.D. Zhao and Y.H. Zhang . . . 947

High-pressure syntheses of series of high T_c-superconductors $(Cu,X)Sr_2Ca_{n-1}Cu_nO_y$ (X = Ge, P, C, S)
E. Takayama-Muromachi, A.T. Matveev, M. Isobe, T. Kawashima and Y. Matsui . . . 949

Au-1212: a new superconducting gold cuprate with T_c above 80 K
P. Bordet, C. Chaillout, E. Gautier, M.F. Gorius, S. Le Floch, M. Perroux, J.J. Capponi, P. Toulemonde and J.L. Tholence . 951

Crystal growth of pure and Al, Mg, Ni, Si or Zn-substituted $CuGeO_3$ single crystals
G. Dhalenne, J.C. Rouchaud, A. Revcolevschi and M. Fedoroff . . . 953

Superconductivity in the charge density wave state of the quasi-two-dimensional monophosphate tungsten bronze $P_4W_{14}O_{50}$
C. Hess, C. Schlenker, G. Bonfait, J. Marcus, T. Ohm, C. Paulsen, J. Dumas, J.-L. Tholence, M. Greenblatt and M. Almeida 955

$Cu_{1-x}Tl_xBa_2Ca_3Cu_4O_{12-y}$ ($Cu_{1-x}Tl_x$-1234) superconductor with $T_c = 126$ K
H. Ihara, K. Tokiwa, K. Tanaka, T. Tsukamoto, T. Watanabe, H. Yamamoto, A. Iyo, M. Tokumoto and M. Umeda . . . 957

Genesis of superconductivity in the layered cuprates $(ACuO_{2.5})_2(ATiO_3)_m$; $2 \leqslant m \leqslant 4$
P.A. Salvador, K.D. Otzschi, H. Zhang, B. Dabrowski, J.D. Jorrensen, T.O. Mason, L.D. Marks and K.R. Poeppelmeier . . 959

Section B. PHYSICAL PROPERTIES AND THEORY I

Ba. Spectroscopy, electronic and structural properties

Ba1. Spectroscopy

The study of electronic structure in new Hg-based Sr-containing 1212-typed $(Hg_{0.5}Pb_{0.5})Sr_2(Ca_{1-x}Y_x)Cu_2O_7$ superconductors
R.S. Liu, J.M. Chen and S.F. Hu . . . 961

Polarization analysis of X-ray absorption at the O 1*s* absorption edge in $PrBa_2Cu_3O_7$: unusual electronic structure in CuO_2 plane
N. Yamada, H. Yamazaki, M. Watanabe, N. Takane, T. Iri, K.-I. Gondaira, J.Z. Liu, R.N. Shelton, M. Fujisawa, S. Shin, S.-I. Shamoto, M. Sato and K. Kohn . . . 963

Surface and interface study on MBE-grown $Nd_{1.85}Ce_{0.15}CuO_4$ thin films by photoemission spectroscopy and tunneling spectroscopy
M. Naito, H. Yamamoto and H. Sato . . . 965

Oxidation state of mercury and lead in $Hg_{0.66}Pb_{0.34}Ba_2Ca_2Cu_3O_{8+\delta}$ studied by X-ray photoelectron spectroscopy
H.M. Shao, C.C. Lam, X.S. Wu, L.W. Shao and X.X. Yao . . . 967

Effect of electronic weak localization of the charge transfer for oxygen-stoichiometric Y1:2:3 systems
Zhang Jincang, Zhang Ying, Xing Huaimin, Wei Shoutao, Cui Yujian, Dai Xianqi, Cheng Guosheng, Li Xigui and Wang Xizhong 969

Charge density distribution in alkaline-earth cuprates with one- or two- dimensional arrangements of Cu–O
K. Nakanishi, S. Adachi, A. Tsukamoto, K. Tanabe, W.-Z. Hu, T. Ida and M. Izumi . . . 971

Evidence of weakening of the Pr4f-O2p hybridization in $Y_{0.4}Pr_{0.6}Ba_{2-x}Sr_xCu_3O_{7-\delta}$: a photoemission spectrum study
G. Cao, Y. Qian, X. Liu, Y. Zhang, Z. Jiao and Q. Zhang . . . 973

Study of Fe-doping effect of Tl-system superconductor
Y.-B. Wang, Y. Li, X. Ju, L. Wei, W.-F. Wu and W.-H. Wang . . . 975

Comparative study of interatomic charge distribution in copper–oxygen linkage in $YBa_2Cu_3O_{6.9}$ and $CuGeO_3$
T. Ida, W.-Z. Hu, M. Izumi, K. Nakanishi, S. Adachi, K. Tanabe, S. Tanaka, Y. Sasago and K. Uchinokura . . . 977

Optical phonon contribution to the high T_c Cooper-pairing investigated by tunneling spectrum
R. Aoki, H. Murakami, M. Shirai, V.M. Svistunov and D.N. Afanassyev . . . 979

Hole distribution in the underdoped, optimally doped, and overdoped superconductors $(Tl_{0.5}Pb_{0.5})Sr_2(Ca_{1-x}Y_x)Cu_2O_7$
R.S. Liu, J.M. Chen and W.Y. Liang . . . 981

The surface characterization and stability of Pb-doped Hg-based HTSC
J. Feng, C.C. Lam, L.W. Shao, X. Jin, H.M. Shao and K.C. Hung . . . 983

Studies on $Ba_{1-x}K_xBiO_3$ single crystals by tunneling spectroscopy
M. Taira, M. Suzuki, S. Tanaka and X.G. Zheng . . . 985

A theoretical analysis of the ARPES of copper oxides
M.D. Núñez Regueiro . . . 987

Shift of core-level binding energy of strontium due to gamma irradiation in $Bi_2Sr_2CaCu_2O_y$ crystals
L.-B. Wang, W.-B. Wu, J.-S. Zhu, X.-F. Sun, X.-R. Zhao, G.-E. Zhou, X.-G. Li and Y.-H. Zhang . . . 989

Electronic states and excitation spectra of $A_{14-x}Ca_xCu_{24}O_{41}$ (A = Sr, La)
Y. Mizuno, T. Tohyama and S. Maekawa . . . 991

Photoemission and inverse-photoemission study on $Bi_2(Sr_{2-x}M_x)CuO_y$ (M = Nd and Gd) oxides
N. Sanada, Y. Suzuki, M. Shimomura, Y. Fukuda, H. Nameki and Y. Syono . . . 993

Electronic structure of $SrNbO_{3.45}$: angle-resolved photoemission results
D.H. Lu, C.S. Gopinath, M. Schmidt, T.R. Cummins, N. Nücker, S. Schuppler and F. Lichtenberg . . . 995

High-resolution angle-resolved photoemission study of $Sr_{14-x}Ca_xCu_{24}O_{41}$ ($x = 0$ and 6)
T. Yokoya, H.-D. Kim, A. Ashihara, H. Kumigashira, H. Fujisawa, T. Takahashi, M. Uehara, T. Nagata and J. Akimitsu . . . 997

ARPES studies of Pb substituted Bi2201 compounds
T. Takeuchi, H. Ding, J.C. Campuzano, T. Yokoya, T. Takahashi, I. Chong, T. Terashima and M. Takano . . . 999

Mössbauer study of $LnBaCuFeO_{5+\delta}$ (Ln = La, Pr, Y) phases: disappearance of the spectrum at low temperature
M.J. Ruiz-Aragón, E. Morán, U. Amador, N. Menéndez and J.D. Tornero . . . 1001

Ba2. Formation of gap states and photoinduced doping

Normal state spin gap of Bi-based superconductors
Z.-Q. Mao, G.-J. Xu, R.-P. Wang, K.-Q. Wang, M.-L. Tian and Y.-H. Zhang . . . 1003

Crystal-field and size effects in $(\text{rare-earth})_{2-z}Ce_zCuO_4$ superconductivity
H.A. Blackstead and J.D. Dow . . . 1005

Ba3. Optical spectroscopy

Vibrational and electronic Raman scattering of $HgBa_2Ca_2Cu_3O_{8+\delta}$ superconductor
X. Zhou, M. Cardona, D. Colson and V. Viallet . . . 1007

Electronic Raman scattering of overdoped $Tl_2Ba_2CuO_{6+\delta}$ in high magnetic fields
G. Blumberg, M. Kang and M.V. Klein . . . 1009

Far-infrared reflectance and Raman scattering study of infinite-layer $SrCuO_2$
X. Zhou, M. Cardona, W. König, J. Zegenhagen and Z.Y. Zhao 1011

Symmetry of the superconducting gap in $HgBa_2CaCu_2O_{6+\delta}$ single crystals from electronic Raman scattering
A. Sacuto, R. Combescot, N. Bontemps, P. Monod, V. Viallet and D. Colson 1013

Related Fermi level position and non-equilibrium electron relaxation in $Y_{1-x-y}Pr_xCa_yBa_2Cu_3O_7$ thin films
G.C. Xiong, Y.H. Zou, Z.J. Xia, P. Yuan, G.J. Lian and Z.Z. Gan 1015

Evolution of the electronic Raman scattering of high-T_c superconductors with doping
G. Blumberg, M. Kang and M.V. Klein 1017

Infrared properties of single crystal $SrTiO_3$, a substrate for high temperature superconducting films
T.R. Yang, C.C. Lu and H.C. Yang 1019

Study on the Pr-doped and Ce-doped $YBa_2Cu_3O_y$ system by XPS and Raman spectrum
L. Shi, Y. Huang, W. Pang, X. Liu, L. Wang, X.-G. Li, G. Zhou and Y. Zhang 1021

c-Axis infrared properties of $Sm_{2-x}Ce_xCuO_{4-\delta}$
J.H. Kim, B.-h. O, J.T. Markert and D. van der Marel 1023

Study of the birefringence by optical reflectivity in $YBa_2Cu_3O_7$ and its variation with a superconducting current
J. Durrell, G. Hauchecorne, J. Bok, J. Bouvier, A.C. Boccara, J.P. Contour and J.P. Roger 1025

Ba4. Phonon spectroscopy

Electronic Raman scattering in single and double CuO_2-layer Tl–Ba–(Ca)–Cu–O superconductors
M. Kang, G. Blumberg and M.V. Klein 1027

Raman study of untwinned $YBa_2Cu_3O_7$ single crystals: new evidence for the *x*–*y* anisotropy of the superconducting gap
M.F. Limonov, A.I. Rykov and S. Tajima 1029

Far infrared optical absorption study and characterization of $Y_1Ba_2Cu_3O_{7-\delta}/Pr_{0.5}Y_{0.5}Ba_2Cu_3O_{7-\delta}$ superlattices
T.R. Yang, H.C. Yang, H.E. Horng and Y.K. Chyan 1031

Study on the phonon spectroscopy and transport properties of $Y_{1-x}Nd_xSr_2Cu_{2.7}Mo_{0.3}O_{7-\delta}$
P. Luo, D. Jin, C. Xu, J. Zuo, K. Hu, X.-G. Li and Y. Zhang 1033

Thin films of high-T_c superconducting cuprates by multi-target laser ablation
S. Pessaud, M. Licheron, F. Gervais, C. Champeaux, P. Marchet and A. Catherinot 1035

Phonon dispersion in $Ba(Pb_{0.75}Bi_{0.25})O_3$
M. Braden and W. Reichardt 1037

Raman study of the $Hg_{0.7}Cr_{0.3}Sr_2CuO_{4+\delta}$ superconductors
S.-Y. Lee, B.-Y. Chang, I.-S. Yang, J.-H. Gwak, S.-J. Kim, J.-H. Choi, S.-I. Lee, J.V. Yakhmi, J.B. Mandal, B. Bandyopadhyay and B. Ghosh 1039

A peculiar temperature dependence of infrared absorption spectra of semiconducting $BaBiO_3$ and $Ba_{0.93}K_{0.07}BiO_3$ crystals
X. Ji, M.E. Kozlov, H. Minami and H. Uwe 1041

Improvement in the current density and cristallinity of superconducting YBCO single crystal by O^+ ion implantation
H.-Y. Nan, X. Yao, M. Murakami and S. Tanaka 1043

Brillouin light scattering, X-ray diffraction and neutron scattering spectroscopy of collective soundlike plane mode in HTSC cuprates
D.V. Fil and A.L. Zazunov 1045

Raman characterization of $YBa_2Cu_3O_{7-\delta}$ thin films
P.X. Zhang, T. Sekinger, U. Sticher, B. Leibold, H.-U. Habermeier and M. Cardona 1047

Raman and FT-ir studies on $(Hg,V)(La,Sr)_2CuO_{4+\delta}$, $(Hg,Cr)Sr_2CuO_{4+\delta}$ and $(Hg,V)Ba_2CuO_{4+\delta}$ superconductors
B. Bandyopadhyay, J.B. Mandal, A.J. Pal, P. Choudhury, P. Mandal and B. Ghosh 1049

Ba5. Structural and phonon anomalies

Low temperature Raman study of the $YBa_2Cu_3O_x$ ($6.44 \leqslant x \leqslant 6.98$) superconductor
D. Palles, D. Lampakis, N. Poulakis, K. Conder, E. Kaldis and E. Liarokapis 1051

Anomalies of mechanical and transport properies above T_c in underdoped $Gd_{1-x}Pr_xBa_2Cu_3O_{6+\delta}$
Q.M. Zhang, Y.N. Wang, H.M. Shen, X.S. Xu, W. Tian and Y.N. Huang 1053

New phonon peak in superconducting state of $YNi_2{}^{11}B_2C$
H. Kawano, H. Yoshizawa, H. Takeya and K. Kadowaki 1055

Anomalous behaviors in high temperature superconductors: lattice effect and d-electron delocalization effect of the normal state on the Bi, Tl, and Y-system superconductors
Y. Li, G. Cao, Y. Wang, R. Ma and Z. Zhao . . . 1057

Structural aspects of the phase transition in $(Nd_{1-x}Tb_x)_{1.85}Ce_{0.15}CuO_4$
M. Braden, P. Adelmann and P. Schweiss . . . 1059

Lattice instability due to carrier–phonon coupling in high-temperature superconductors
Q.M. Zhang, Y.N. Wang, H.M. Shen and Y.N. Huang . . . 1061

Possibility of strong electron–phonon interaction in double layered cuprates
K. Motida . . . 1063

Anomalous thermal expansion of high-T_c superconducting oxide system
N.V. Anshukova, A.I. Golovashkin, L.I. Ivanova, K.V. Kraiskaya, I.B. Krinetskii, L.I. Leonjuk and A.P. Rusakov . . . 1065

High-T_c superconductors with apical halogens
A.I. Golovashkin, N.V. Anshukova, L.I. Ivanova and A.P. Rusakov . . . 1067

Rietveld analysis and superconductivity of $La_{1.5}Ca_xBa_{1.5-x}Cu_3O_y$ compounds
H.-C.I. Kao, W.N. Huang, K.F. Lee and C.M. Wang . . . 1069

Microstructure and T_c suppression in La-cuprates
Y. Horibe, Y. Inoue and Y. Koyama . . . 1071

Superconductivity, X-ray structure and non-stoichiometry of $HgBa_2CuO_{4+\delta}$
V. Viallet, A. Bertinotti, J.-F. Marucco, D. Colson, L. Fruchter, G. Le Bras, V. Vulcanescu and J. Hammann . . . 1073

Properties of Tl-2201 thin films
Q.-H. Hu, L.-G. Johansson, H.Q. Chen, Z.G. Ivanov, D. Erts, H. Olin, T. Claeson and E.A. Stepantsov . . . 1075

Acoustic evidence against BCS behavior in monocrystal $YBa_2Cu_3O_7$
M. Lei, A. Migliori and H. Ledbetter . . . 1077

Charge order in oxygen deficient perovskite $La_{8-x}Sr_xCu_8O_{20}$
H. Yamaguchi, H. Matsuhata, T. Ito and K. Oka . . . 1079

Structural instability around T_c observed in Hg-1201 by neutron powder diffraction and EXAFS
P. Bordet, F. Duc, P.G. Radaelli, A. Lanzara, N. Saini, A. Bianconi and E.V. Antipov . . . 1081

Electronic and structural instabilities of single-crystalline $La_{2-x-y}Nd_ySr_xCuO_4$ near 1/8 doping
S. Sakita, F. Nakamura, J. Takase, T. Suzuki and T. Fujita . . . 1083

Structural instability in (Pb,Cu) 1212 superconducting cuprates
S.G. Titova, J.T.S. Irvine, B.J. Mitchell and V.F. Balakirev . . . 1085

Study of phonon anomalies in Tl-system superconductor near T_c
Y.-B. Wang, G. Cao, Y. Li, X. Ju, L. Wei and W. Wu . . . 1087

Ba6. Defects ordering and structural coherence

Dynamics of oxygen ordering in $YBa_2Cu_3O_{6+x}$ studied by neutron and high-energy synchrotron X-ray diffraction
T. Frello, N.H. Andersen, J. Madsen, M. Käll, M. von Zimmermann, O. Schmidt, H.F. Poulsen, J.R. Schneider and Th. Wolf . 1089

Lead diffusion and segregation in Pb-doped Bi-2223 superconductor
L.W. Zhang, Z.Y. Wang, S.M. Gong and T.G. Chen . . . 1091

Monoclinic 2212-BSCCO single crystals: phase transformation, twinning, evolution, superconductivity and doping studies
G. Yang, P. Shang, I.P. Jones, J.S. Abell and C.E. Gough . . . 1093

Investigation on the surface morphology of single crystal $Ba_{1-x}K_xBiO_3$
X.G. Zheng, M. Taira, M. Suzuki, C. Xu and K. Hirakawa . . . 1095

The excess oxygen model for Fe-doped superconductors
G. Chen, Y. Li, Y. Wang, G. Cao, R. Ma and S. Pan . . . 1097

Ba7. Metal–non-metal transitions

Study of electronic states of copper, bismuth, lead and oxygen atoms in some superconducting and related bismuth containing oxide phases
V.E. Fedorov, N.F. Zakharchuk, N.G. Naumov and U. Paek . . . 1099

The evolution of high-T_c superconductor electronic structure with doping
K.V. Mitsen and O.M. Ivanenko . . . 1101

$^{47/49}Ti$ and ^{139}La NMR studies on Mott transition in $La_{1-x}Sr_xTiO_3$
Y. Furukawa, I. Okamura, K. Kumagai, T. Goto, T. Fukase, Y. Taguchi and Y. Tokura 1103

NMR study on a metal–insulator transition in oxysulfides $(La_{1-x}Ca_xO)Cu_{1-x}Ni_xS$
Y. Furukawa, S. Ikeda, H. Watanabe, K. Kumagai, K. Mori, Y. Takano and K. Sekizawa 1105

Metal–insulator transition in the spinel $CuIr_2(S_{1-x}Se_x)_4$ system studied by NMR
S. Tsuji, K. Kumagai, N. Matsumoto and S. Nagata 1107

Phonon effects on superconducting phase in two-chain electron systems
K. Yonemitsu 1109

Features of crystal structures in the correlated-electron system $Sr_{1-x}La_xTiO_3$
M. Arao, F. Munakata and Y. Koyama 1111

Ba8. Properties of the ladder compounds

Magnetic properties of the ladder compound $LaCuO_{2.5}$
B. Normand and T.M. Rice 1113

^{63}Cu NMR study on spin ladder compounds $Sr_{14-x}Ca_xCu_{24}O_{41}$ ($x=9$, 11.5)
K. Magishi, S. Matsumoto, K. Ishida, Y. Kitaoka, K. Asayama, M. Uehara, T. Nagata and J. Akimitsu 1115

Phase diagram of the $S=\frac{1}{2}$ frustrated double ladder system
B. Normand, K. Penc, M. Albrecht and F. Mila 1117

Transport properties of $Sr_{14-x}Ca_xCu_{24}O_{41}$ with doped two-leg Cu–O ladders
N. Motoyama, H. Eisaki, T. Osafune, A. Yoshida, K. Yotsui and S. Uchida 1119

Electronic structure of $Sr_{14-x}Ca_xCu_{24}O_{41}$ investigated by optical study
T. Osafune, N. Motoyama, H. Eisaki and S. Uchida 1121

List of Contributors xcv

PART III

Bb. Transport, magnetic and thermodynamic properties

Bb1. Resistivity, conductivity, Hall effect, thermoelectric power, etc.

Transport properties of oxygen-deficient perovskite $La_{8-x}Sr_xCu_8O_{20}$
T. Ito, H. Yamaguchi, K. Oka, K. Nozawa and H. Takagi 1123

Out-of-plane microwave conductivity of the cuprates in the superconducting state
H. Kitano, T. Hanaguri and A. Maeda 1125

In-plane and out-of-plane resistivities of $PrBa_2Cu_3O_{7-\delta}$ and $Y_{0.47}Pr_{0.53}Ba_2Cu_3O_{7-\delta}$ single crystals
G.A. Levin, T. Stein, C.N. Jiang, C.C. Almasan, D.A. Gajewski, S.H. Han and M.B. Maple 1127

Normal-state out-of-plane and in-plane resistivities of $Y_{1-x}Pr_xBa_2Cu_3O_{7-\delta}$ $(0<x\leqslant 1)$ single crystals
C.C. Almasan, G.A. Levin, C.N. Jiang, T. Stein, D.A. Gajewski, S.H. Han and M.B. Maple 1129

The modulated investigations of the $YBa_2Cu_3O_{7-x}$ single crystals I–V characteristics nonlinearity near T_c
N.D. Kuzmichev and M.A. Vasyutin 1131

Anisotropic transport properties of $PrBa_2Cu_3O_7$
M. Goto, K. Takenaka, H. Eisaki and S. Uchida 1133

Temperature dependence of the resistivity anisotropy $\rho_c(T)/\rho_{ab}(T)$ of single crystals $PrBa_2Cu_3O_{6.6}$
A.I. Romanenko and L.P. Kozeeva 1135

Valence states of atoms and localization of conductivity electrons in $PrBa_2Cu_3O_{6+x}$
A.I. Romanenko, N.F. Zakharchuk, N.G. Naumov, V.E. Fedorov and U. Paek 1137

Reentrant transition in tetragonal superconductor $CaLaBaCu_3O_y$
M. Akinaga 1139

Crossover from London to Mattis–Bardeen behavior evidenced by fitting c-axis conductivity spectra of $YBa_2Cu_3O_{7-\delta}$
R.P.S.M. Lobo and F. Gervais 1141

Mixed-state properties of $YBa_2Cu_3O_y/Nd_{1.85}Ce_{0.15}CuO_4$ and $YBa_2Cu_3O_y/Y_{0.5}Pr_{0.5}Ba_2Cu_3O_y$ multilayers
H.C. Yang, H.W. Yu, H.E. Horng and F.F. Lin . . . 1143

Temperature dependence of the velocity of the wave of reflection and conduction in NdCeCuO, YBaCuO and LaSrCuO
E.M. Kudriavtsev, S.D. Zotov, V.V. Krivov and M. Autric . . . 1145

An improved analysis of resistivity from flux transformer geometry
G.A. Levin, T. Stein, C.N. Jiang, C.C. Almasan, D.A. Gajewski, S.H. Han and M.B. Maple . . . 1147

Properties of the mixed and normal states induced by magnetic field in a $Nd_{1.82}Ce_{0.18}CuO_{4-\delta}$ film
A.I. Ponomarev, G.I. Harus, A.N. Ignatenkov, N.K. Lerinman, L.D. Sabirzyanova, N.G. Shelushinina, N.A. Babushkina and L.M. Belova . . . 1149

Nonmetallic behavior of out-of-plane conduction in $La_{2-x}Sr_xCuO_4$
T. Goko, F. Nakamura, N. Kikugawa, Y. Uno, S. Sakita and T. Fujita . . . 1151

Evidence of the spin gap in $Hg_{0.9}Tl_{0.1}Ba_2Ca_2Cu_3O_{8+\delta}$
Y. Feng, K.Q. Ruan, K.Q. Wang, C.Y. Wang, L.Z. Cao, Q.L. Xu, F. Foong, B. Bedard and S.H. Liou . . . 1153

Model of c-axis resistivity of high-T_c layered superconductors
L. Zheng, X.-G. Li and Y. Zhang . . . 1155

Phase transition at $T \approx 600$ K and metastable states of single crystals $Bi_2Sr_2CaCu_2O_{8+x}$
A.I. Romanenko, O.B. Anikeeva, L.P. Kozeeva and U. Paek . . . 1157

Anisotropic resistivity and Cu^{2+} localized spin moment in $Bi_2Sr_2CaCu_2O_{8+x}$ crystals
X.-G. Li, X. Sun, W. Wu, L. Wang, Q. Chen, L. Shi and Y. Zhang . . . 1159

Alternative modeling of resistance versus temperature curves along the c direction of Bi-2212 crystals, suggesting a possible metal/insulator transition below T_c
C. Boulesteix, Y. Marietti, T. Badeche, O. Monnereau and V. Grachev . . . 1161

Role of the magnetic scattering in the quadratic temperature dependence of $Nd_{2-x}Ce_xCuO_{4-\delta}$ films in normal state
N.A. Babushkina, L.M. Belova and A.P. Zhernov . . . 1163

Evidence of the spin gap in $Bi_2Sr_2Ca_{1-x}Y_xCu_2O_y$ single crystals
K.Q. Ruan, Y. Feng, C.Y. Wang, X.H. Chen and L.Z. Cao . . . 1165

Large enhancement of T_c (50 K) by applying high pressure in the Hg-2212 superconductor
C. Acha, S.M. Loureiro, C. Chaillout, J.L. Tholence, J.J. Capponi, M. Marezio and M. Núñez-Regueiro . . . 1167

Anisotropic transport properties of impurity (Co) doped and oxygen controlled single-crystal $Bi_2Sr_2CaCu_2O_{8+\delta}$: evidence of temperature-dependent interlayer coupling and a pseudogap
T. Watanabe, T. Fujii and A. Matsuda . . . 1169

The study of microcurrent path of highly c-axis-oriented Bi(2223)/Ag tapes
Y. Sun, G. Xu, L. Jiang, J. Du and Y. Zhang . . . 1171

Critical currents in polycrystalline $(Bi,Pb)_2Sr_2Ca_2Cu_3O_{10}$ systems: a comparative magnetisation study
M. Dhallé, F. Marti, G. Grasso, A. Perin, J.C. Grivel, E. Walker and R. Flükiger . . . 1173

The dissipation of superconducting BSCCO films on the $I-T$ plane
L. Romanò and C. Paracchini . . . 1175

Observation of anisotropy of in-plane resistivities in $PrBa_2Cu_3O_7$ single crystals
X. Zhang, G. Cao, J. He, W. Yu, H. Song, Z. Jiao and Q. Zhang . . . 1177

The mechanism of hole carrier relaxation in high-T_c superconductors
O.M. Ivanenko and K.V. Mitsen . . . 1179

Current redistributions in superconductors with non-uniformly distributed T_c-inhomogeneities
Th. Siebold, C. Carballeira, J. Mosqueira, M.V. Ramallo and F. Vidal . . . 1181

Failure of scaling in material with low Γ value
M.R. Cimberle, C. Ferdeghini, G. Grassano, D. Marrè, M. Putti and A.S. Siri . . . 1183

Hall effect of single layer, tetragonal $Tl_2Ba_2CuO_{6+\delta}$ near optimal doping
A.W. Tyler and A.P. Mackenzie . . . 1185

In-plane resistivity of La-doped $Bi_2Sr_2CuO_{6+\delta}$ single crystals
W. Si, J. Xiong and Z. Zhao . . . 1187

Great enhancement of the critical temperature of the oxide superconductor $Tl_{1.8}Ba_{2.0}Ca_{2.6}Cu_{3.0}O_{10+\delta}$ under high pressure
C.Y. Han, W. Lin, Y.S. Wu, B. Yin and D.S. Tang . . . 1189

Hall effect measurements on ion bombarded YBaCuO films
E. Schäfer, R. Durner, A. Feucht, Th. Theilig and P. Ziemann . . . 1191

Mobility of positive holes in $YBa_2Cu_3O_{7-y}$ – effects of oxygen deficiency and light irradiation
H. Minami and H. Uwe . . . 1193

About the Hall effect temperature anomalies in the normal state of high-T_c superconductors
E.A. Pashitskii and A.V. Semenov . . . 1195

The Ba-site Pr substitution effects in $YBa_{2-x}Pr_xCu_3O_{7-\delta}$ system
Z.-a. Xu, W.-h. Yu, X.-j. Zhang, Y.-h. Ren, Z.-k. Jiao and Q.-r. Zhang . . . 1197

Electron irradiation effects in $YBa_2Cu_3O_{6+x}$ and $HgBa_2Ca_2Cu_3O_{8+\delta}$ crystals
F. Rullier-Albenque, P.A. Vieillefond, J.F. Marucco, D. Colson, V. Viallet and P. Lejay . . . 1199

Anomalous Hall effect in $YBa_2Cu_3O_{7-\delta}$ in low magnetic field
Y.-p. Wang, D.-l. Zhang and B.R. Zhao . . . 1201

Real-time Hall effect measurements of high critical currents epitaxial $YBa_2Cu_3O_{7-x}$ thin films
S.G. Zybtsev, I. Nizhankovskii and S.F. Kim . . . 1203

Effect of Zn impurities on the mixed state Hall effect in Bi-2212 single crystal
L.-X. Chen, H. Ikeda and R. Yoshizaki . . . 1205

The anomalous Hall effect of $(Bi,Pb)_2Sr_2Ca_2Cu_3O_x$ silver-sheathed tape
L. Jiang, Y. Sun, G. Xu and J. Du . . . 1207

Hall conductivity of Bi-based superconducting cuprates close to T_c
R. Jin and H.R. Ott . . . 1209

Study on physical properties of Bi-2201, Bi-2212 and Bi-2223 superconductors
K. Koike, S. Kambe, S. Ohshima, Y. Takasugi, T. Furusawa and T. Shiomi . . . 1211

Hall effect studies of the mixed state of *c*-axis oriented $Hg_{0.9}Tl_{0.1}Ba_2Ca_2Cu_3O_{8+\delta}$ superconducting films with $T_c > 124$ K
K.Q. Ruan, Y. Feng, C.Y. Wang, X.H. Chen, L.Z. Cao, Q.L. Xu, F. Foong, B. Bedard and S.H. Liou . . . 1213

Hall effect as a probe of the mechanisms of superconductivity
V.B. Geshkenbein, L.B. Ioffe and A.I. Larkin . . . 1215

Transport and optical properties of the striped spin and charge ordered state in Nd-doped La214 system
N. Ichikawa, S. Uchida, S. Tajima, Y. Ando and G.S. Boebinger . . . 1217

Out-of-plane magnetoresistance in $Bi_2(Sr,La)_2CuO_{6+\delta}$ single crystals
R. Yoshizaki, I. Tomitsuka and H. Ikeda . . . 1219

Magnetoresistivity of untwinned single crystals of $YBa_2Cu_3O_{7-\delta}$
J. Axnäs, Yu. Eltsev, W. Holm, B. Lundqvist and Ö. Rapp . . . 1221

Superconductivity effects in hopping conductivity of single-crystal La_2CuO_4 with excess oxygen
B.I. Belevtsev, N.V. Dalakova and A.S. Panfilov . . . 1223

Peak shifts in magnetization rotation – effects of anisotropy and pinning
S.K. Hasanain, S. Manzoor and M. Aftab . . . 1225

Physical and geometrical origins of the normal-state magnetoresistance in $YBa_2Cu_3O_7$
W. Göb, W. Lang and R. Sobolewski . . . 1227

The out-of-plane magnetoresistance of Sr_2RuO_4
N.E. Hussey, A.P. Mackenzie, J.R. Cooper, Y. Maeno, S. Nishizaki and T. Fujita . . . 1229

Enhanced magnetoresistance in La–Ca–Mn–O films on Si substrates using YBaCuO/CeO_2 heterostructures
W. Zhang, I.W. Boyd, M. Elliott and W. Herrenden-Harkerand . . . 1231

Possibility of the pinning of the stripe pattern of holes and spins by Zn in $La_{2-x}Sr_xCu_{1-y}Zn_yO_4$ with $x \approx 0.115$
Y. Koike, S. Takeuchi, Y. Hama, H. Sato, T. Adachi and M. Kato . . . 1233

Coexistence of bound pairs and independent carriers in the normal state of high T_c superconductors: thermoelectric power
Y.S. Ha, J.H. Lee and Y.W. Park . . . 1235

Thermopower of magnetically aligned $YBa_2Cu_3O_{7-\delta}$ ceramics: an approach for measuring anisotropic transport properties of ceramics
C.-J. Liu, H.J. Trodahl, J. Downes and M.P. Staines . . . 1237

Anomalies induced by calcium doping in NdCeCuO system
X.H. Chen, T.F. Li, R.P. Wang, Y.M. Qiu, K.Q. Ruan, L.Z. Cao, L. Shi, Y.B. Jia, G.E. Zhou and Y.H. Zhang . . . 1239

Thermoelectric power in normal state of Hg-1223 doped with Pb, Tl and Bi
T. Zhang, J. Guo, G. Liu, X. Xu, H. Dong and S. Yan . . . 1241

Annealing and thermoelectric power of (Cu,C)-1223, (Cu,C)-1234 and $Ba_2Ca_2Cu_3(Ca_uCu_v)O_{8+\delta}$
Y. Cao, F.Y. Lin, Z.L. Du, F. Chen, Y.Y. Xue and C.W. Chu . . . 1243

Thermopower of $HgBa_2Ca_{m-1}Cu_mO_{2m+2+\delta}$ under pressure
F. Chen, Q.M. Lin, Y.Y. Xue and C.W. Chu 1245

Investigation of the off-diagonal Seebeck effect and Peltier effect on textured $YBa_2Cu_3O_{7-\delta}$
Z.H. He, F.Q. Zhong, Y.Y. Luo, M. Wu, W. Gavalek, K.F. Liang and P.C.W. Fung 1247

Resistivity and thermoelectric power of Cu-substituted $YNi_{2-x}Cu_xB_2C$ superconductor
C.K. Choi, E.S. Choi, J.H. Lee, Y.S. Ha, Y.S. Song and Y.W. Park 1249

Comparison of thermopower behaviour in different superconductors
A.B. Kaiser 1251

Electron–phonon enhancement of thermopower in silver-doped superconducting Bi compounds
J.E. Rodríguez, A. Mariño and J. Giraldo 1253

The negative thermopower and the suppression of superconductivity in Nd-doped $La_{2-x}Ba_xCuO_4$ with x around 1/8
X. Bao, T. Fujita, Y. Maeno and J. Nyhus 1255

Transport phenomena in $(Bi_{1.6}Pb_{0.4})(Sr_{1.8}Ba_{0.2})Ca_2(Cu_{1-x}Fe_x)_3O_y$ thin films
G. Ilonca, A.V. Pop, D. Ciurchea, C. Corega, M. Ilonca and R. Deltour 1257

Thermoelectric power in $Bi_2Sr_2Ca_{1-x}Ce_xCu_2O_{8+y}$ single crystals and polycrystals based on NL model
R.P. Wang, H. Jin and H. Sekine 1259

Thermoelectric power anomalies induced by the lattice instability in doped Bi-2201 phase
Z.-Q. Mao, G.-J. Xu, L. Wan, L. Shi, R.-P. Wang, C.-G. Fan, G.-E. Zhou and Y.-H. Zhang 1261

Study on the transport properties in $Bi_{2-x}Pb_xSr_{1.65}La_{0.35}CuO_y$ system
G. Xu, Z. Mao, C. Xu, X. Liu, G. Zhou and Y. Zhang 1263

Superconductivity in the halogen-intercalated $Bi_2Sr_2(Gd_{0.82}Ce_{0.18})_2Cu_2O_{10+\delta}$ of the Bi-2222 phase
Y. Koike, T. Hisaki, M. Akoshima, M. Kato and T. Noji 1265

Effect of substitution for Cu in $Bi_2Sr_2Ca(Cu_{1-x}M_x)_2O_y$ (M = Fe, Co) on thermopower and band spectrum in the normal state
V.E. Gasumyants, M. Ye, E.V. Vladimirskaya, L.Yu. Stolypina and R. Deltour 1267

Thermoelectric power of Re-doped Hg1223 compounds
K. Komori, H. Sumioka, A. Yoshikawa, J. Shimoyama, H. Ikuta and K. Kishio 1269

Thermopower of overdoped Ba–K–Bi–O and Ni-substituted 214 cuprates
C.-J. Liu, H.Y. Tang, C.K. Subramaniam and A.B. Kaiser 1271

Investigation of laser-induced thermoelectric response in normal state HTS films
H.S. Huang, L.F. Zhang, B. Tang, H.Y. Zhang, M.L. Liu, Y.L. Zhou, G.Z. Yang and R.L. Geng 1273

The structure and laser-induced thermoelectric effect on tilt orientation TlBaCaCuO films
H. Zhang, H. Huang, B. Tang, L. Zhang, Z. Zhang and M. Liu 1275

Electrical resistivity and thermal conductivity of $NdBa_2(Cu_{1-z}Zn_z)_3O_{7-y}$ and the phonon transport model
S. Cao, K. Mori and K. Nishimura 1277

Normal-state Nernst effect of $YBa_2Cu_3O_x$ (x = 6.3–6.9): Experiment and analysis
V.E. Gasumyants, N.V. Ageev, I.E. Goldberg and V.I. Kaydanov 1279

Peak observed in the secondary voltage of a flux transformer configuration in $Bi_2Sr_2CaCu_2O_{8-x}$
M. Chung, H.-S. Chang, Y.J. Doh and H. Lee 1281

Study of photoirradiation for $YBa_2Cu_3O_{6+x}$ compounds and the electron structure by positron experiment
Cheng Guosheng, Shang Jiaxiang, Li Xigui, Dai Xianqi, Wang Xizhong and Zhang Jincang 1283

Heat transport in HTSC cuprate crystals in *ab*-plane and along *c*-direction
V.B. Efimov and L.P. Mezhov-Deglin 1285

Thermal conductivity of single crystals of $YBa_2(Cu_{1-x}Zn_x)_3O_7$
V.W. Wittorff, N.E. Hussey, J.R. Cooper, Chen Changkang and J.W. Hodby 1287

Bb2. Magnetic phases and static properties

Model for the magnetic phases observed at low temperatures in reduced $YBa_2Cu_3O_{6+x}$ materials with Al substitution for Cu or rare earth ions for Ba
N.H. Andersen and G. Uimin 1289

H_{c2} and electron characteristics of superconductors with radiation defects
S.I. Krasnosvobodtsev, N.P. Shabanova, V.S. Nozdrin, E.V. Pechen, S.V. Antonenko, G.I. Zhabrev and A.I. Golovashkin 1291

The effect of Pr ion concentration on superconducting parameters in $Y_{1-x}Pr_xBa_2Cu_3O_{7-\delta}$ single crystals
Y. Zhuo, J.-H. Choi, M.-S. Kim, J.-N. Park, M.-K. Bae, Z. Xianyu and S.-I. Lee 1293

Anomalies in the $M(T)$ curves of the ceramic superconductor $Y_{0.8}Pr_{0.2}Ba_2Cu_{2.90}Li_{0.10}O_{7-\delta}$
H. Tang, Z.Q. Yang, Y.Z. Wang, B.G. Shen, C. Zhang, X.D. Su, F.R. de Boer and G.W. Qiao 1295

Magnetization measurements on $Nd_{1-x}Ba_{2+x}Cu_3O_7$ thin films
X.G. Qiu, N. Kanda, J. Nishino, M. Kawasaki and H. Koinuma 1297

Vortex fluctuations in $YBa_2Cu_3O_{6.5}$ single crystal: evidence for 2D → 3D crossover
A. Poddar, R. Prozorov, Y. Wolfus, M. Ghinovker, B.Ya. Shapiro, A. Shaulov and Y. Yeshurun 1299

Temperature dependence of the lower critical field H_{c1} in $Bi_2Sr_2CaCu_2O_{8+\delta}$ single crystals
M. Nideröst, R. Frassanito, P. Visani, A.C. Mota and G. Blatter 1301

The static and dynamic properties of the vortices in $Bi_2Sr_2CaCu_2O_y$
T. Tsuboi, T. Hanaguri and A. Maeda 1303

The effect of splayed columnar defects on the magnetisation of $Bi_2Sr_2Ca_2Cu_3O_{10}$/Ag tapes
M. Dhallé, F. Marti, G. Grasso, B. Hensel, E. Paschoud, M. Victoria and R. Flükiger 1305

Anomalous normal-state transport property in Bi-2201 single crystals
G. Xu, Z. Mao, X. Xu, W. Wu and Y. Zhang 1307

Experimental study of the lower critical field in high-T_c-single crystals
G. Brandstätter, C. Böhmer, H.W. Weber and B.W. Veal 1309

Preparation and magnetic measurements of Ba–K–Pb–O and Ba–K–M–Pb–O (M = Tl, Sb) crystals
L.Z. Zhao, B. Yin, J.B. Zhang, J.W. Li and C.Y. Xu 1311

Magnetic phase transitions of the $DyNi_2B_2C$ single crystal
W.C. Lee, E.J. Joo, H.S. Cho, H.S. Lee and H.B. Kim 1313

Interplay between magnetism and superconductivity in $HoNi_2{}^{11}B_2C$
H. Yoshizawa, H. Kawano, H. Takeya and K. Kadowaki 1315

Thc effect of Co on the superconducting and magnetic ordering temperatures in $ErNi_2B_2C$
I. Felner, C. Godart and E. Alleno 1317

Magnetic properties of La_2CuO_4 with slight amounts of excess oxygen
Y. Okajima, M. Yamada and K. Yamaya 1319

Observation of ferromagnetic order in electrochemically oxidated $La_2CuO_{4+\delta}$
N.N. Garif'yanov and E.L. Vavilova 1321

Spin magnetic susceptibility of Cu–O chains in Sr_2CuO_3 and $SrCuO_2$
H. Eisaki, N. Motoyama and S. Uchida 1323

Magnetic properties and weak ferromagnetism of $Eu_2CuO_{4+\delta}$
Y.F. Xiong, Y.S. Yao, D. Jin, S.Q. Guo, S.M. Hong, X.L. Jia and Z.X. Zhao 1325

Crystal electric field splitting of R^{3+}-ions in RNi_2B_2C (R = rare earth)
U. Gasser, P. Allenspach, J. Mesot and A. Furrer 1327

Anomalous magnetic properties of the 2212-type rare earth cuprates $Bi_2Sr_2RCu_2O_{8+\delta}$ (R = Pr, Gd, Y)
T.I. Hsu, Y.Y. Hsu, Y.H. Toh, J.U. Lin and H.C. Ku 1329

Bb3. Dynamics of spin excitations, neutrons, NMR, μSR, etc.

NMR study of $YBa_2Cu_3O_xF_y$ ($x = 6, 6.7$, and $0 < y < 1.5$)
S.D. Goren, C. Korn, L. Frenkel Ben-Yakar, A. Shames, B. Bandyopadhyay, H. Shaked, P. Massiot and C. Perrin 1331

Small defects in YBCO single crystals: T_c after neutron irradiation and annealing
F.M. Sauerzopf, M. Werner, H.W. Weber, R.A. Suris, D.V. Kulikov, V.S. Kharlamov and Yu.V. Trushin 1333

Neutron spectroscopic study of crystalline electric-field in infinite-layer $Sr_{1-x}Nd_xCuO_2$
A. Mirmelstein, A. Podlesnyak, V. Bobrovskii, E. Mitberg, B. Goshchitskii, A. Muzychka, I. Sashin, R. Eccleston, J. Mesot, M. Zolliker, W. Henggeler and A. Furrer 1335

^{63}Cu and ^{89}Y NMR and spin susceptibility of $YBa_2Cu_3O_{6.9}$ disordered by neutron irradiation
A. Ananyev, A. Gerashenko, K. Mikhalev, Yu. Zhdanov, S. Verkhovskii, Eu. Medvedev and B. Goshchitskii 1337

Preparation and NMR study of fluorinated Pb-BSCCO-2223
L. Frenkel Ben-Yakar, A. Shames, B. Bandyopadhyay, S.D. Goren, C. Korn, H. Shaked, H. Selig, L. Ben-Dor and J.H. Greenberg 1339

ESR study on Cu^{2+} local magnetic moment of $Bi_2Sr_2CaCu_2O_y$ crystals
D. Jin, P. Luo, W. Wu, X. Sun, Y. Xu, R. Liang, X.-G. Li and Y. Zhang 1341

Hole-doping effect on spin gaps in $Sr_{14-x}A_xCu_{24}O_{41}$ (A = Ca, Y and La) by NMR studies
K. Kumagai, S. Tsuji, K. Maki, M. Kato and Y. Koike 1343

Strong evidence for an electronic crossover around 180 K in $YBa_2Cu_4O_8$, an NMR/NQR study
A. Suter, M. Mali, J. Roos, D. Brinkmann and J. Karpinski . . . 1345

Spin fluctuations parameters in $Tl_2Ba_2CaCu_2O_{8-\delta}$ as seen by ^{63}Cu and ^{17}O NMR
A. Gerashenko, K. Mikhalev, S. Verkhovskii, Yu. Piskunov and A. Yakubovskii . . . 1347

Er^{3+}-EPR study of Er-substituted $YBa_2Cu_3O_y$
H. Shimizu, K. Fujiwara and K. Hatada . . . 1349

Cu NQR study of spin fluctuation in Ni substituted $YBa_2Cu_3O_7$
Y. Tokunaga, K. Ishida, Y. Kitaoka and K. Asayama . . . 1351

Phase separation in La_2CuO_{4+y} single crystals studied by μSR and neutron diffraction
V.Yu. Pomjakushin, A. Amato, A.M. Balagurov, A.I. Beskrovny, V.N. Duginov, F.N. Gygax, D. Herlach, A.N. Ponomarev, A. Schenck, V.G. Simkin and A.A. Zakharov . . . 1353

Local suppression of superconductivity in Zn-substituted high T_c cuprates
B. Nachumi, Y. Fudamoto, A. Keren, K.M. Kojima, M. Larkin, G.M. Luke, J. Merrin, O. Tchernyshöv, Y.J. Uemura, N. Ichikawa, M. Goto and S. Uchida . . . 1355

Magnetic properties of $(Tl_{1-x}Pb_x)Sr_2CaCu_2O_7$ high-T_c oxide studied by ^{17}O NMR and SQUID: from overdoped superconductor to strongly overdoped metal
P.-V. Bellot, A. Trokiner, Y. Zhdanov, A. Yakubovskii, L. Shustov, S. Verkhovskii, S. Zagoulaev and P. Monod . . . 1357

Intrinsic localized magnetic moments in overdoped high-T_c cuprate $(Tl_{1-x}Pb_x)Sr_2CaCu_2O_7$ probed by ^{17}O NMR
P.-V. Bellot, A. Trokiner, A. Yakubovskii and L. Shustov . . . 1359

Electric field gradient and hole concentration in inequivalent CuO_2 layers of $Tl_2Ba_2Ca_2Cu_3O_{10-\delta}$ with different T_c
Yu. Piskunov, A. Geraschenko, K. Mikhalev, Yu. Zhdanov, S. Verkhovskii, A. Yakubovskii, A. Trokiner and P.V. Bellot . . . 1361

NMR/NQR studies of hyperfine field and electric field gradient in $RSr_2Cu_2GaO_7$, $RBa_2Cu_2NbO_8$ and $(R_{2-x}Ce_x)Sr_2Cu_2NbO_{10}$
A. Yamanaka, K. Kumagai, Y. Piskunov and A. Yakubovsky . . . 1363

NMR study in single-layered high-T_c compounds near the antiferromagnetic insulator phase boundary
T. Mito, K. Ishida, Y. Kitaoka, K. Asayama, H. Eisaki and S. Uchida . . . 1365

Cu NQR study in Zn- and Ni-doped $La_{1.825}Sr_{0.175}CuO_4$
K. Ishida, T. Tanaka, H. Yamanaka, T. Mito, Y. Tokunaga, K. Yoshida, K. Yamazoe, S. Ohsugi, Y. Kitaoka and K. Asayama . 1367

Anisotropic pairing in superconducting Sr_2RuO_4 ($T_c \sim 0.7\,K$)
K. Ishida, Y. Kitaoka, K. Asayama, S. Ikeda, S. Nishizaki, Y. Maeno, K. Yoshida and T. Fujita . . . 1369

Electronic spin dynamics in $Ca_{0.85}Sr_{0.15}CuO_2$ – an NMR relaxation study
F. Raffa, M. Mali, J. Roos, D. Brinkmann, M. Matsumura and K. Conder . . . 1371

Temperature dependence of spin susceptibility of $La_{2-x}Sr_xCuO_4$ – Cu Knight shift measurement
S. Ohsugi, Y. Kitaoka and K. Asayama . . . 1373

Neutron study of the spin dynamics in superconducting $La_{2-x}Sr_xCuO_4$
S. Petit, A.H. Moudden, B. Hennion, A. Vietkin and A. Revcolevschi . . . 1375

Temperature dependence of crystal-field transitions in underdoped superconductors: evidence for a pseudogap in the normal state
J. Mesot, G. Böttger, P. Berastegui, H. Mutka and A. Furrer . . . 1377

Influence of Ce-substitution on exchange field seen by Gd^{3+}-spin in $Gd_{2-x}Ce_xCuO_4$ compounds
H. Shimizu, S. Suzuki and K. Hatada . . . 1379

Magnetic properties and NMR investigations of the Hg-based cuprate Hg-1201
W. Hoffmann, M. Baenitz, M. Heinze, K. Lüders, A.A. Gippius and E.V. Antipov . . . 1381

Localization of positive muons in antiferromagnetic transition metal oxides
W.K. Dawson, K. Nishiyama, S. Ohira and W. Higemoto . . . 1383

NMR study of disorder effect on the superconductivity in $Tl_2Ba_2Ca_2Cu_3O_{10}$
G.-q. Zhen, Y. Kitaoka, K. Asayama, K. Hamada, H. Yamauchi and S. Tanaka . . . 1385

Temperature and doping dependences of spin susceptibility $La_{2-x}Sr_xCuO_4$
M. Eremin and S. Solovjanov . . . 1387

^{17}O NMR comparison of zinc and nickel substituted $YBa_2Cu_3O_{6.6}$
J. Bobroff, H. Alloul, Y. Yoshinari, P. Mendels, N. Blanchard, G. Collin and J.-F. Marucco . . . 1389

Charge fluctuations and magnetic inelastic neutron scattering in copper-oxide high-T_c superconductors
A.S. Moskvin, A.S. Ovchinnikov and O.S. Kovalev . . . 1391

Study of charge and spin order in $La_2NiO_{4+\delta}$ using ^{139}La-NMR
O.O. Bernal, H.B. Brom, M.L. de Kok, J. Witteveen and A.A. Menovsky . . . 1393

Bb4. Specific heat

Kondo effect induced suppression of superconductivity in $Y_{1-x}Pr_xBa_2Cu_3O_{7-\delta}$
D. Jin, J. Luo, Y. Wang, H. Jin, G. Rao and S. Yan 1395

Specific heat of antiferromagnetic spin fluctuations in cuprate superconductors
S.P. Kruchinin and S.K. Patapis 1397

Specific heat of high temperature superconductors in high magnetic fields
A. Junod, M. Roulin, B. Revaz, A. Mirmelstein, J.-Y. Genoud, E. Walker and A. Erb 1399

High resolution specific heat experiments on the melting line of the vortex glass in $YBa_2Cu_3O_x$: observation of second-order transitions up to 14 teslas
M. Roulin, A. Junod, E. Walker and A. Erb 1401

Magnetic field dependence of low temperature heat capacities of $GdBa_2Cu_4O_8$
J.C. Ho, Y.Y. Chen, Y.D. Yao, W.S. Huang, S.R. Sheen, J.C. Huang and M.K. Wu 1403

Superconducting and normal state energy gaps in $Y_{0.8}Ca_{0.2}Ba_2Cu_3O_{7-\delta}$ from the electronic specific heat
J.W. Loram, K.A. Mirza, J.R. Cooper and J.L. Tallon 1405

Thermal properties of $PrBa_2Cu_4O_8$ prepared at ambient oxygen pressure
H.D. Yang, C.W. Lin, J.-Y. Lin, T.H. Meen, H.L. Tsay, J.C. Huang, S.R. Sheen and M.K. Wu 1407

Heat capicity of $HoNi_2B_2C$
S. Miyamoto, H. Takeya and K. Kadowaki 1409

Specific heat evidence on d-wave superconductivity in $La_{1.85}Sr_{0.15}Cu_{1-y}Zn_yO_4$
K.A. Mirza, J.W. Loram and J.R. Cooper 1411

Pairing symmetry of superconducting Sr_2RuO_4 from specific heat measurements
S. Nishizaki, Y. Maeno, S. Farner, S. Ikeda and T. Fujita 1413

The relations of specific heat anomaly with variation of oxygen content in BiSrCaCuO(2212)
K.Q. Wang, Z.Q. Mao, Y. Feng, L.Z. Cao and Y.Q. Zhou 1415

The specific heat anomaly of HgBaCaCuO(1223) near T_c
K.Q. Wang, Y. Feng, L.Z. Cao and Q.L. Xu 1417

Anomalous electronic specific heat of $Bi_{1.8}Pb_{0.3}Sr_{1.9}CuO_y$
H. Mayama, Y. Okajima and K. Yamaya 1419

Comparative study of Tl-2201 single crystals with different temperatures of superconducting transition by means of specific heat method
V.N. Kopylov, N.N. Kolesnikov and A.V. Palnichenko 1421

Specific heat and electron photoemission of Pb- and Bi-doped Tl–(Ba,Sr)–Ca–Cu–O (1223) high-temperature superconductors
A. Kolodziejczyk, R. Zalecki, A. Kozłowski, G. Gritzner, W. König and M. Mair 1423

Specific heat anomalies on the vortex melting line in $YBa_2Cu_3O_{7-\delta}$ up to $B=23$ teslas: observation of first- and second-order transitions, effect of oxygen doping, field orientation, crystal purity and detwinning
A. Junod, M. Roulin, J.-Y. Genoud, B. Revaz, E. Walker, A. Erb, C. Marcenat, R. Calemczuk and F. Bouquet 1425

Bb5. Others

Pinning aspects of magnetostriction effects in superconductors
V.V. Eremenko, I.N. Nechiporenko and V.A. Sirenko 1427

New low activation energy processes in $La_2CuO_{4+\delta}$ by elastic energy loss experiments
G. Cannelli, R. Cantelli, F. Cordero, F. Trequattrini and M. Ferretti 1429

1/f Noise in several Hg-based superconductors
H. Dong, J. Guo, X. Xu, T. Zhang and S. Yan 1431

Temperature dependent noise in $YBa_2Cu_3O_y$ thin film
H.E. Horng, J.M. Wu, S.Y. Yang, H.C. Yang and J.D. Chern 1433

An accurate method of iodometric titration for determining oxygen content of superconducting cuprates
W.M. Chen, N.L. Chen, C.C. Lam, J. Chen, X.S. Wu, X. Zhang, D.B. Chen and X. Jin 1435

Anomalous behavior of Bi-system superconductors in the normal state and their correlation with superconductivity
H. Zhang, Q.R. Feng, Y. Zhao, F. Ritter and W. Assmus 1437

Size effect of 1/f noise in normal state of $YBa_2Cu_3O_7$
H. Dong, X. Hu, Y. Bei, R. Nie, T. Zhang and S. Yan 1439

Anomaly of thermal expansion in Bi2212 oxide near T_c
N. Yamada, M. Okaji, K. Nara, H. Kato, K. Michishita and Y. Kubo . . . 1441

Pressure effect on T_c in $RBa_2Cu_4O_8$ (R = Sm, Er, Dy) superconductors
J. Tang, Y. Okada, Y. Yamada, S. Horii, A. Matsushita, T. Kosaka and T. Matsumoto . . . 1443

Magnetic energy loss in polycrystalline $YBa_2Cu_3O_{7-\delta}$ with varying oxygen content
A.A. Pramana Dn., Kh.A. Ziq and N.M. Hamdan . . . 1445

Superconducting properties of $Bi_{1.8}Pb_{0.4}Sr_2Ca_2Cu_{3.2-x}Ge_xO_y$ compounds
Y.-Z. Wang, Z.-R. Li, Z.-Q. Yang, H. Tang, D.-C. Zeng, C. Zhang, L. Hua, X.-D. Su, W.-S. Yu, F.R. de Boer and G.-W. Qiao 1447

The experimental study of dynamic thermal expansion of Bi-based high T_c superconductors
J.D. Guo, W.L. Zhao, R.S. Qin, L. Hua, H. Tang, G.H. He, Y.Z. Wang, G.W. Qiao and B.L. Zhou . . . 1449

Effects of intercalation of lithium ion on superconductivity of YNi_2B_2C
Z. Wang, X. Huang and L. Chen . . . 1451

Possible observation of polaron pairs in highly doped $YBa_2Cu_3O_{6+x}$ by elastic energy loss
G. Cannelli, R. Cantelli, F. Cordero and F. Trequattrini . . . 1453

Grain size effect on magnetic levitation of YBCO superconducting samples
W.C. Chan, C.Y. Wang and J.J. Lee . . . 1455

Thermally activated dynamics in $La_2CuO_{4+\delta}$: tilts of the CuO_6 octahedra and interstitial O
G. Cannelli, R. Cantelli, F. Cordero, F. Trequattrini, C.R. Grandini and M. Ferretti . . . 1457

Bc. Superconductivity properties

Bc1. Magnetic penetration depths

Temperature dependence of superconducting carrier density and effective mass in $HgBa_2Ca_{n-1}Cu_nO_y$
R. Puźniak . . . 1459

Inductive response of double sided superconductor film
J.Y. Lee, T.-S. Hahn and S.S. Choi . . . 1461

Penetration depth in magnetically oriented, ceramic, Ni- and Zn-doped $La_{2-x}Sr_xCuO_4$
A.J. Zaleski and J. Klamut . . . 1463

Magnetic field penetration depth in single crystal $La_{1.85}Sr_{0.15}CuO_4$ and $Nd_{1.85}Ce_{0.15}CuO_4$
G.M. Luke, Y. Fudamoto, K. Kojima, M. Larkin, J. Merrin, B. Nachumi, Y.J. Uemura, J.E. Sonier, T. Ito, K. Oka, M. de Andrade, M.B. Maple and S. Uchida . . . 1465

ab-Plane surface impedance of the single-layer cuprate $Tl_2Ba_2CuO_{6+\delta}$
D.M. Broun, D.C. Morgan, R.J. Ormeno, S.F. Lee, A.P. Mackenzie, A.W. Tyler and J.R. Waldram . . . 1467

Flux pinning and grain coupling studied by magneto-optical visualization in $Bi_2Sr_2CaCu_2O_{8+\delta}$ ceramics
M.R. Koblischka, S.-L. Huang, T.H. Johansen, H. Bratsberg and K. Fossheim . . . 1469

Bc2. Superconducting gap, symmetry, Andreev reflection and tunneling

Quantum tunneling cyclic loop in high temperature superconductor probably is the microscopic origin of high temperature superconductivity
D. Zhou . . . 1471

Tunneling spectroscopy in $Bi_2Sr_2CaCu_2O_{8+x}$ single-crystal break junctions: reproducible determination of the Eliashberg function and role of the breakdown of Migdal's theorem
R.S. Gonnelli, G.A. Ummarino and V.A. Stepanov . . . 1473

Optical evidence of the superconducting gap in LNi_2B_2C (L = Lu and Y)
F. Bommeli, L. Degiorgi, P. Wachter, B.K. Cho, P.C. Canfield, R. Chau and M.B. Maple . . . 1475

Tunneling spectroscopy and symmetries in YBCO and NCCO
S. Kashiwaya, L. Alff, H. Takashima, N. Terada, T. Ito, K. Oka, Y. Tanaka, M. Koyanagi, S. Ueno and K. Kajimura . . . 1477

Local variation of electronic structure in BiSrCaCuO investigated by cryogenic STM/STS
K. Kitazawa, T. Hasegawa, T. Taneda, S. Matsuura, T. Endo, H. Sugawara, J. Shimoyama and K. Kishio . . . 1479

Peculiarities of 'tunneling' characteristics observed in HTS/metal point contact junctions
P. Seidel, A. Plecenik, M. Grajcar, M. Belogolovskii, K.-U. Barholz and A. Matthes . . . 1481

C-axis quasiparticle tunneling properties of planar $NdBa_2Cu_3O_{7-\delta}$ /$PrBa_2Cu_3O_{7-\delta}$ /$NdBa_2Cu_3O_{7-\delta}$ quasi-homostructures
G.A. Alvarez, T. Utagawa and Y. Enomoto . . . 1483

Orientation dependence of tunneling spectra in YBCO and NCCO
L. Alff, H. Takashima, S. Kashiwaya, N. Terada, T. Ito, K. Oka, Y. Tanaka and M. Koyanagi 1485
Tunneling in pure and substituted single crystals of $Bi_2Sr_2Ca_1Cu_2O_{8+x}$: s+d-wave pairing and gap enhancement
H. Hancotte, R. Deltour, D.N. Davydov, A.G.M. Jansen and P. Wyder 1487
Tunneling study of Co-substituted $Bi_2Sr_2CaCu_2O_8$
N. Tsuda, T. Arao, T. Hosokawa, Y. Shiina, N. Matsuda and D. Shimada 1489
Experimental study of quasiparticle tunneling in high T_c junctions
S.G. Lee, Y. Huh, Y.S. Hwang, I. Kim, Y.K. Park and G.Y. Sung 1491
Zero-bias anomaly in the tunneling spectroscopy of high-T_c superconducting films
R.T. Kao, S.J. Wang, J.Y. Juang, K.H. Wu, T.M. Uen and Y.S. Gou 1493
Superconducting coupling along the c-axis of [001] twist grain-boundaries in $Bi_2Sr_2CaCu_2O_{8+\delta}$ bicrystals
Q. Li, Y.N. Tsay, M. Suenaga, G.D. Gu and N. Koshizuka 1495
Tunneling studies on charge of electronic state accompanied with hole doping in $Bi_2Sr_2CaCu_2O_{8-x}$
S. Matsuo, M. Suzuki, S. Tanaka and X.G. Zheng 1497
STM/STS studies for doping effects on the symmetry and magnitude of superconducting gap in $Bi_2Sr_2CaCu_2O_{8+\delta}$
M. Oda, K. Hoya, R. Kubota, C. Manabe, N. Momono, T. Nakano and M. Ido 1499
Energy dependence of the normal density of states and of the electron–phonon spectral function in the quasiparticle tunneling of high-T_c superconductors
G.A. Ummarino, R.S. Gonnelli and V.A. Stepanov 1501
Scanning tunneling microscopy and spectroscopy study of $Nd_{2-x}Ce_xCuO_{4-y}$
B. Susła, R. Czajka, W. Sadowski and T. Klimczuk 1503
In-plane symmetry of the superconducting gap in YBCO probed by the angular dependence of thermal magnetoresistance
H. Aubin, K. Behnia, M. Ribault, R. Gagnon and L. Taillefer 1505
Observation of pseudogap in $Bi_2Sr_2CaCu_2O_{8+\delta}$ single crystals with electron tunneling spectroscopy
H.J. Tao, F. Lu and E.L. Wolf 1507
Fluctuation conductivity behavior in thin films of YBaCuO indicates s-wave pairing
S.K. Patapis and S.P. Kruchinin 1509
Temperature dependence of the paramagnetic Meissner effect in BSCCO small particles
Y. He 1511
Are all high-temperature superconductors p-type?
H.A. Blackstead and J.D. Dow 1513
Possible evidence for $d+s$ pairing symmetry in YBCO using tricrystal microbridges
J.H. Miller Jr., Z.G. Zou, J. Liu, Z.-S. Zheng and W.K. Chu 1515
Temperature dependence of the spectral function of the electron–phonon interaction for $Bi_2Sr_2CaCu_2O_8$
D. Shimada, N. Miyakawa and N. Tsuda 1517
Electron–phonon spectral function $\alpha^2F(\omega)$ determined by quasiparticle tunneling spectroscopy for $Bi_2Sr_2CaCu_2O_8$/Au junctions
N. Miyakawa, A. Nakamura, Y. Fujino, T. Kaneko, D. Shimada, Y. Shiina and N. Tsuda 1519
Theory of the quasi-particle spectra around a vortex in the two-dimensional t–J model
A. Himeda, M. Ogata, Y. Tanaka and S. Kashiwaya 1521
Internal friction related to the superconducting condensate in BiSrCaCuO (2223)
Y.N. Huang, X. Li, Q.M. Zhang, Y.N. Wang, P.C.W. Fung and P. Esquinazi 1523
Zero bias anomaly of a normal metal–d-wave superconductor junction in the presence of a Kondo-like magnetic impurity
Z.D. Wang and J.-X. Zhu 1525
Ferromagnetic scattering effects on quantum transport in a normal metal–superconductor junction
J.-X. Zhu and Z.D. Wang 1527
Fitting experimental current–voltage data using a simplified model for SNS junctions
L.A.A. Pereira, A.M. Luiz and R. Nicolsky 1529
Single electron tunneling and superconducting energy gap in $YBa_2Cu_3O_{7-\delta}$ and $TlBa_2Ca_2Cu_3O_{9\pm\delta}$ films
R.T. Kao, S.J. Wang, S.P. Chen, J.Y. Juang, K.H. Wu, T.M. Uen and Y.S. Gou 1531

Bc3. Superconducting phase transitions and fluctuations

Fluctuation conductivity of BSCCO (2212) whiskers
I.G. Gorlova, S.G. Zybtsev, V.Ya. Pokrovskii and V.N. Timofeev 1533

Density of states fluctuations in the magnetoresistivity of $YBa_2Cu_3O_{7-\delta}$
J. Axnäs, Yu. Eltsev, W. Holm, B. Lundqvist and Ö. Rapp 1535
Superconducting fluctuation of the YBCO films
S. Wang, K. Xiu, J. Wang, R. Nie, X. Hu, G. Lian, G. Xiong, S. Wang, Z. Liu, H. Chen and Y. Dai 1537
Fluctuation effects on the magnetization above and below the superconducting transition in $Tl_2Ba_2Ca_2Cu_3O_{10}$ crystals
J. Mosqueira, A. Maignan, Ch. Simon, C. Torrón, A. Wahl and F. Vidal 1539
Levanyuk–Ginzburg criterion for full-critical fluctuations in layered superconductors
M.V. Ramallo and F. Vidal 1541
Fluctuation effects and critical fields in $Tl_2Ba_2CuO_{6+x}$
V.N. Kopylov, T.G. Togonidze, P.H. Kes, N.N. Kolesnikov and T.W. Li 1543
Critical fluctuations at the superconducting transition of $TmBa_2Cu_3O_{7-\delta}$ thin films in transverse and longitudinal magnetic fields
P. Gehringer, W. Lang, E. Stangl and S. Proyer 1545
Fluctuational Ettingshausen effect in single crystals of $YBa_2Cu_3O_{7-\delta}$
L. Lu, Y.X. Jia, M. Fuhrer and A. Zettl 1547
Fluctuation-induced resistivity in YBCO across T_c from angular measurements in magnetic field
M. Giura, R. Fastampa, S. Sarti and E. Silva 1549
Definitive test of spin-fluctuation pairing: pair-breaking
H.A. Blackstead and J.D. Dow 1551
Effect of Zn impurity on the pressure dependence of superconducting transition temperature T_c in $YBa_2Cu_{3-x}Zn_xO_{7-\delta}$ single crystals
X. Li, L. Lu, S. Li, X.N. Jing, D.-l. Zhang and Z.Z. Wang 1553
A comparative study of the superconducting transition in Tl-2212 polycrystals and granular thin films
V. Ta Phuoc, J.C. Soret, L. Ammor, A. Ruyter, B. Martinie, A. Wahl, J. Lecomte and Ch. Simon 1555
The intrinsic upper critical field of high-T_c superconductors
X.-J. Xu, L. Fu and Y.-H. Zhang 1557
Optimization and enhancement of transition temperature for $HgBa_2Ca_2Cu_3O_{8+\delta}$ ceramic superconductors
H.M. Shao, C.C. Lam, K.C. Hung, X.S. Wu and X.X. Yao 1559
Out-of-plane magnetoresistance of $Bi_2Sr_2CaCu_2O_{8+x}$ single crystals: probing the fluctuation-induced pseudogap in the density of states
G. Heine, W. Lang, X.L. Wang and X.Z. Wang 1561
Effects of an isoelectric alloying element Ni–Zn on T_c in $YBa_2(Cu_{1-2z}Ni_zZn_z)_3O_{7-x}$ thin films
K. Hasegawa and K. Ogawa 1563
Probe of the Josephson vortex lattice under high magnetic field
V.N. Zavaritsky and W.Y. Liang 1565
Upper critical field of Pb films with an antidot lattice
E. Rosseel, T. Puig, M. Baert, M.J. Van Bael, V.V. Moshchalkov and Y. Bruynseraede 1567
Paraconductivity and dimensionality along two different directions in a plane of a thin film of YBaCuO with *c*-axis in it
S.K. Patapis, D. Yarmis, Ch. Katsoura, G. Wagner, R. Herzog, R.E. Somekh and S. Kruchinin 1569
Excess conductivity in $Y_{1-2x}Ca_xTb_xBa_2Cu_3O_{7-\delta}$
S.-H. Han, P. Lundqvist and Ö. Rapp 1571

Bc4. Nonequilibrium superconductivity (no contributions)

Bc5. Microwave absorption, acoustic properties

Microwave propertics of $YBa_2Cu_3O_{7-x}$ deposited on MgO substrates
M.C. Hsieh, T.Y. Tseng, K.H. Wu, J.Y. Juang, T.M. Uen and Y.S. Gou 1573
Internal friction studies of relaxation processes in $La_2CuO_{4+\delta}$
B. Kusz, M. Gazda and W. Sadowski 1575
Variation in microwave losses induced by dc and rf magnetic fields in HTS thin films
J.C. Gallop, A. Cowie and L.F. Cohen 1577
Spatially resolved measurements of HTS microwave surface impedance
J. Gallop, L. Hao and F. Abbas 1579
Phenomenological model of microwave electrodynamics of high-T_c superconductor metal oxides
A.L. Karuzskii 1581

Microwave dielectric resonance in the $YBa_2Cu_3O_7$ single crystal
V.N. Apakina, V.A. Dravin, Yu.F. Eltsev, A.L. Karuzskii, V.N. Murzin, A.V. Perestoronin and B.G. Zhurkin 1583

Experimental evidence for spatial dispersion effects and the change of sign of dielectric permittivity in superconductors from microwave measurements
V.N. Apakina, V.A. Dravin, Yu.F. Eltsev, A.L. Karuzskii, V.N. Murzin, A.V. Perestoronin, N.A. Volchkov and B.G. Zhurkin . 1585

Microwave surface resistance peaks in BSCCO thin films
E. Silva, M. Lanucara and R. Marcon 1587

Periodic microwave spectra due to quantised-flux penetration into intra-crystal Josephson links in Bi_{2212} single crystals detected by ESR spectrometer
H. Muto, T. Kusumori and N. Murayama 1589

Low field modulated microwave absorption and electrical properties in degraded and annealed granular $YBa_2Cu_3O_x$
T. Endo, S. Nagase, S. Sugiura, N. Hirate, M. Horie and S. Yamada 1591

Ultrasonic behavior in $La_{2-x}Ba_xCuO_{4-y}$ polycrystalline superconductor
C. Zhu, J. Su, X. Luo, S. Wang, W. Liu and L. Cao 1593

Plasma waves in anisotropic superconducting films below and above the plasma frequency
M.M. Doria, G. Hollauer, F. Parage and O. Buisson 1595

Field-dependent surface impedance of YBCO single crystals and evidence for a non-zero Hall term
J.J. Wingfield, J.R. Powell, A. Porch and C.E. Gough 1597

Sample size dependence of the Josephson plasma resonance in $Bi_2Sr_2CaCu_2O_{8+\delta}$
I. Kakeya, K. Kindo, T. Mochiku and K. Kadowaki 1599

Observation of coherent Josephson response in the non-linear *ab*-plane microwave impedance of $YBa_2Cu_3O_{6.95}$ single crystals
Z. Zhai, H. Srikanth, S. Sridhar, A. Erb, E. Walker and R. Flükiger 1601

Surface resistance of YBa_2Cu_3O thin films at low dc magnetic fields
C.K. Ong, G.C. Han, L.F. Chen, S.Y. Xu and B.T.G. Tan 1603

Bc6. Others

Superconductivity in the $YBa_{2-y}M_yCu_3O_x$ (M = La, Pr) material
D.H. Ha, Y.-I. Kim, I.-S. Kim, Y.K. Park, J.-C. Park and Y. Nishihara 1605

Effects of fast neutron irradiation on J_c and microstructure of PMP-processed YBCO
P.X. Zhang, L. Zhou, W.M. Bian, P. Ji, K.G. Wang, X.Z. Wu, R. Puzniak, A. Wisniewski, M. Baran and H. Szymczak . . 1607

Magnetically dependent adsorption of oxygen on superconducting adsorbent
L.L. Makarshin, D.V. Andreev and V.N. Parmon 1609

Effect of hydrogenation on positron annihilation in the superconductor of $YBa_2Cu_3O_{7-\delta}$
J.Q. Li, C.C. Lam, H.Q. Xiong, W.M. Chen, L.W. Shao, X. Jin, J. Feng, S.R. Qi and K.C. Hung 1611

Plasma treatment of high-temperature superconducting materials
Z.-Q. Yang, A. Ulyashin, N. Shlopak, H. Tang, X.-D. Su, C. Zhang, Y.-Z. Wang and G.-W. Qiao 1613

The influence of nano-Cu addition on the superconductivity of YBCO
X.D. Su, Z.Q. Yang, Y.Z. Wang, H. Tang, C. Zhang, L. Hua and G.W. Qiao 1615

T_c curves for $La_{2-x}M_xCuO_4$ from chemical equilibrium
A. Kallio, K.A. Gernoth, J. Hissa and T. Häyrynen 1617

Bd. Theory

Bd1. Band structure: ab initio and cluster calculations

Pressure dependence of the electric-field gradients in $YBa_2Cu_4O_8$
C.O. Rodriguez, G. Fabricius, M.G. Stachiotti and N.E. Christensen 1619

The Local Density Approximation (LDA) bandstructure and Fermi-surface of $YBa_2Cu_3O_7$ (Y123)
C.O. Rodriguez, R. Weht and N.E. Christensen 1621

Theoretical analysis of the Jahn–Teller effect in the La–Ba–Cu–O superconductor
A. Calles, J.R. Soto, J.J. Castro and E. Yépez 1623

Electronic structure of $RBa_2Cu_3O_7$ (R = Y, Pr, Nd, Sm, Eu, Gd, Dy, Ho, Tm, Yb and La): a LSDA and LSDA+U study
W.Y. Hu, M.C. Qian and Q.Q. Zheng 1625

Electronic structure of a multilayer system: $La_{2-1/6}Sr_{1/6}CuO_4/La_{2/3}Ca_{1/3}MnO_3$
W.Y. Hu and Q.Q. Zheng 1627

A mechanism of high T_c superconductivity and its relation with Bose–Einstein condensation
X. Dai and P. Hor . . . 1629

Normal state quasiparticle structure in the high-T_c superconductors
J. Costa-Quintana, F. López-Aguilar, M.M. Sánchez-López and I.S. Sandalov . . . 1631

Spin and magnetism of rare-earth intermetallic nickel borocarbides: $RENi_2B_2C$
Z. Zeng, D. Guenzburger, E.M. Baggio-Saitovitch and D.E. Ellis . . . 1633

Theoretical investigations of structural and vibronic properties of $YBa_2Cu_3O_7$ by first-principles atomic-force calculations
R. Kouba and C. Ambrosch-Draxl . . . 1635

Fermi surface-analysis of $YBa_2Cu_3O_7$ by full-potential LAPW calculations
B. Zangger and C. Ambrosch-Draxl . . . 1637

A theoretical study of the local electronic structure of two adjacent Cu–O-planes in $YBa_2Cu_3O_7$
H.U. Suter, E.P. Stoll, P. Hüsser, S. Schafroth and P.F. Meier . . . 1639

Bd2. Weak coupling theory: BCS, van Hove scenario, magnetic coupling

Spin gap and charge transport in underdoped high-T_c cuprates
J.-X. Li and C.-D. Gong . . . 1641

The cause of formation of the local gap in the high-T_c cuprates
F.-S. Liu and P. Ma . . . 1643

Local dynamical d-wave pairing in layered quantum paraelectric superconductors
S.R. Shenoy, V. Subrahmanyam and A.R. Bishop . . . 1645

Theory of negative magnetoresistance in the c-axis resistivity of high-T_c cuprates
D.-D. Yan, Y.-S. Yi, Z.-B. Su and L. Yu . . . 1647

Mixed $s+d$ symmetry states induced by orthorhombic distortion effect
M. Liu, H.Y. Teng, D.Y. Xing and J.M. Dong . . . 1649

Number parity effects in a mesoscopic superconducting sample
Z. Wang and C. Gong . . . 1651

Surface state effects in layered superconductors
H. Zhao and J.-L. Shen . . . 1653

Phase transition to mixed state in unconventional superconductors
E.A. Shapoval . . . 1655

Roles of LO-phonon and SO-phonon induced potentials for a bipolaron in the superconductor
C.M. Lee, S.W. Gu and C.C. Lam . . . 1657

Energy gap dispersion in bilayered cuprates
M.V. Eremin and I.A. Larionov . . . 1659

Renormalization group analysis of quasi-one-dimensional quantum antiferromagnets
Z. Wang . . . 1661

Analytical solutions of Ginzburg–Landau equations for high-T_c superconductors
J. Zhu, X. Li and Y. Zhang . . . 1663

The van Hove scenario of high T_c superconductors: the effect of doping
J. Bouvier and J. Bok . . . 1665

Bd3. Strong correlation aspects: Hubbard model, t–J, multiband model, spin–charge separation, etc.

Non-phonon superconductivity and the phase diagram in the boron–nickel compound
R.O. Zaitsev and S.A. Karamov . . . 1667

The Fermi surface of underdoped high-T_c superconducting cuprates
X. Dai, Z.B. Su and L. Yu . . . 1669

Results of relaxation function theory for the NMR measurements in La_2CuO_4
Y.-J. Wang, M.-R. Li and C.-D. Gong . . . 1671

Effects of the strong correlation on the photoemission spectroscopy
D. Shi, M. Xiao and Z. Li . . . 1673

Green's function theory of the two-dimensional anisotropic Heisenberg model on a square lattice
Y. Song . . . 1675

Non-phonon superconductivity of Sr_2RuO_4 compound
R.O. Zaitsev and Ju.V. Mikhailova 1677
Microscopic deduction of the Ginzburg–Landau equations on the basis of extended Hubbard model
R.O. Zaitsev 1679
Electronic theory for bilayer-effects in high-T_c superconductors
S. Grabowski, J. Schmalian and K.H. Bennemann 1681
g-on Mean field theory of the t–J model: uniform RVB state
F. Basco, H. Kohno, H. Fukuyama and G. Baskaran 1683
π Excitation and resonant neutron scattering: a mean-field study
H. Kohno, B. Normand and H. Fukuyama 1685
In-plane resistivity from the copper oxide material
S. Feng and Z. Huang 1687
Quasiparticle bands and superconductivity in bilayer cuprates
W.-G. Yin and C.-D. Gong 1689
The effect of Andreev reflection on transport currents and persistent currents in normal-metallic loops
Z. Cao, Y. Xie and R. Han 1691
Main factors determine T_c in superconducting copper-oxides
M.-L. Zhang 1693
Perturbation expansion for the p–d model around the atomic limit: a study on spin magnetic susceptibility
R. Citro and M. Marinaro 1695
Antiferromagnetic phase diagram based on Hubbard model under external magnetic field
S.-P. Hong, H. Doh and S.-H.S. Salk 1697
Bilayer coupling in cuprates based on t–J model
T.-H. Gimm, S.-P. Hong and S.-H.S. Salk 1699
Symmetry of the order parameter in correlated electron–phonon superconductor
K.I. Wysokiński 1701
Anisotropic s- or d-wave pairing in the Hubbard model
Y.M. Malozovsky and J.D. Fan 1703
Thermodynamical properties of the Hubbard model on finite-size clusters
C. Noce, M. Cuoco and A. Romano 1705
Functional Schrödinger representation of slave-boson and slave-fermion approaches for doped and undoped antiferromagnetic 2-D systems
S.-A. Ahn, I.E. Dikstein and S.H.S. Salk 1707
Zinc valency and local moments in impurity doped cuprates
B.C. den Hertog and M.P. Das 1709
Orthogonality instability in a planar antiferromagnet with one impurity
A. Ramšak 1711
A tight-binding model for Sr_2RuO_4
C. Noce and T. Xiang 1713
Problem on single hole in the two-dimensional Hubbard model with infinite repulsion
Ju.V. Mikhailova 1715
New diagram technique for periodic Anderson model
V.A. Moskalenko, D.F. Digor, M.I. Vladimir, L.A. Dogotaru and I.G. Porcescu 1717
Electron–phonon coupling in high-T_c materials
V.A. Moskalenko, L.A. Dogotaru and I.G. Porcescu 1719
A perturbative approach to high-T_c superconductivity
J.D. Fan and Y.M. Malozovsky 1721
Optical absorption for the integrable chain of electrons coupled with phonons
V.A. Ivanov, M.Ye. Zhuravlev and E.A. Ugolkova 1723
Towards the unified theory of the mechanism of superconductivity
J.D. Fan and Y.M. Malozovsky 1725
Localization induced by impurity doping in high-T_c superconductors
Y.M. Malozovsky and J.D. Fan 1727

Real-space pairing theories of superconductivity: influence of hole doping
E. Dagotto, D. Duffy, A. Nazarenko, S. Haas, A. Moreo and J. Riera . . . 1729
Quasiparticle momentum distributions in the t–J model
S. Nishimoto, R. Eder and Y. Ohta . . . 1731
Excitation spectra of an effective d-wave model for cuprate superconductivity
M. Yamaguchi, Y. Ohta and R. Eder . . . 1733
Non-rigid shell model and correlational mechanism of the local pairing
A.S. Moskvin, V.A. Korotaev, Yu.D. Panov and M.A. Sidorov . . . 1735
Superconductivity in the two-band singlet-hole model for copper-oxide plane
N.M. Plakida . . . 1737
Quasiparticles and superconductivity in the t–t'–J model
N.M. Plakida, V.S. Oudovenko, P. Horsch and A. Liechtenstein . . . 1739
Instabilities towards superconductivity and incommensurate bond-order waves in the t–J model
R. Zeyher and A. Greco . . . 1741
Analysis of optical conductivities of cuprates via doped Hubbard model
J. Hong and H.-Y. Kee . . . 1743
Hole pairing symmetry in attractive Hubbard model
C. Wang, O. Navarro, R. Oviedo-Roa and L.A. Pérez . . . 1745
Charge–spin separation and non-Fermi liquid physics in the normal state of quasi-1D charge density wave systems?
J. Voit . . . 1747
Impurity effects in spin liquids
G. Khaliullin, S. Krivenko, R. Kilian and P. Fulde . . . 1749
Density fluctuations and phonon renormalization in the t–J model
G. Khaliullin and P. Horsch . . . 1751
Model of 2D electron system accompanied by modulated structure and the resistivity anomaly in layered high-T_c superconductors
W. Kinase, S. Kuwata, K. Harada and Y. Nishi . . . 1753
Specific heat of the two-dimensional Hubbard model
F. Mancini, D. Villani and H. Matsumoto . . . 1755
The superconducting gap in the two-dimensional Hubbard model
A. Avella, F. Mancini, D. Villani and H. Matsumoto . . . 1757
Fermi surface and density of states in the two-dimensional t–t'–U model
A. Avella, F. Mancini, D. Villani and H. Matsumoto . . . 1759
The proximity effect and upper critical field in the Hubbard model with repulsion
Ju.V. Mikhailova and R.O. Zaitsev . . . 1761
The normal state in high-T_c cuprates and the gauge invariant carrieron model
I. Kanazawa . . . 1763
Variational Monte Carlo study of the two-chain Hubbard model
S. Koike, K. Yamaji and T. Yanagisawa . . . 1765
Strong indications of d-wave superconductivity in the two-dimensional Hubbard model obtained by the variational Monte Carlo calculations
K. Yamaji, T. Nakanishi and T. Yanagisawa . . . 1767
Effect of electron–lattice interaction on the bound states in the Peierls d–p model
A. Mishima . . . 1769
Spin fluctuations in magnetically coupled bilayer cuprates
J. Kishine . . . 1771
Electronic states in the t–J model
H. Matsumoto, T. Saikawa, F. Mancini and D. Villani . . . 1773
Doping dependence of the superconducting state of the cuprates
S. Grabowski, J. Schmalian and K.H. Bennemann . . . 1775
Analytic theory of the spin bipolaron in the t–J model
H. Barentzen . . . 1777
Magnetic properties of the 2D t–t' Hubbard model
U. Trapper, H. Fehske and D. Ihle . . . 1779

Spin susceptibility and magnetic short-range order in the 2D Heisenberg model
S. Winterfeldt and D. Ihle . . . 1781

The electronic structure of CuO_2 plane
X. Yang, T. Li and Z.-Z. Gan . . . 1783

Spectral properties of one-dimensional copper oxides
T. Tohyama, Y. Mizuno and S. Maekawa . . . 1785

ODLRO and η-pairing in high-T_c cuprate superconductors and the ground state of the extended Hubbard model
L.J. Dunne . . . 1787

Specific heat in the three-band Hubbard model
M. Marinaro and R. Citro . . . 1789

Site occupancy functional theory for a strongly correlated model
Y.-N. Xie, Z.-L. Cao and R.-S. Han . . . 1791

Dispersion relation and spectral functions of the t–J model and the antiferromagnetic insulator $Sr_2CuO_2Cl_2$
P.W. Leung, B.O. Wells and R.J. Gooding . . . 1793

Hubbard model predictions for the Knight shift in cuprates
J.M. Singer and P.F. Meier . . . 1795

Bd4. Non-Fermi-liquid behavior

Interacting fermions – coexistence of local and bound fermion pairs and breakdown of Fermi liquid behavior
R. Chaudhury . . . 1797

Tunneling effect and non-Fermi liquid behavior induced by an impurity in the Hubbard chain
Y. Wang and D. Jin . . . 1799

Marginal Fermi liquid behaviour within a Fermi liquid theory
A. Ben Ali, S. Charfi-Kaddour, R. Louati, R. Bennaceur and M. Héritier . . . 1801

Spectral properties of the planar t–J model
P. Prelovšek and J. Jaklič . . . 1803

Wave function and size of spin-polarons in the t–J model
A. Ramšak and P. Horsch . . . 1805

Origin of unconventional physical behaviour of copper oxide high-T_c superconductors
A.S. Moskvin . . . 1807

Cooper pairs as low-energy excitations in the normal state
O. Tchernyshyov and Y.J. Uemura . . . 1809

Bd5. Local pair superconductivity

Comparison between competing interaction energies of AF correlation and long range Coulomb repulsion in phase separation
L. Ye and R. Han . . . 1811

Polar Jahn–Teller centers and isotope effect in copper oxide high-T_c superconductors
A.S. Moskvin, A.S. Ovchinnikov, Yu.D. Panov and M.A. Sidorov . . . 1813

A 2-band Monte Carlo study of the pseudogap in high T_c superconductors
S.T. Bromley and D.J. Newman . . . 1815

Analytic results for the crossover from BCS superconductivity to Bose–Einstein condensation
M. Marini, F. Pistolesi and G.C. Strinati . . . 1817

Study of the phase diagram and the T_c pressure dependence for $HgBa_2CuO_{4-\delta}$ superconductors
E.V.L. de Mello and C. Acha . . . 1819

Exact infrared behavior of superfluid interacting bosons at zero temperature
C. Castellani, C. Di Castro, F. Pistolesi and G.C. Strinati . . . 1821

Bd6. Strong electron–phonon coupling

Polaronic effects in systems with lattice instabilities
M. Zoli . . . 1823

Analysis of the electron–phonon coupling in the superconductor $Ba_{0.7}K_{0.3}BiO_3$
O. Navarro and E. Chavira . . . 1825

On the stability of polaronic superlattices in strongly coupled electron–phonon systems
G. Wellein, H. Fehske, H. Büttner and A.R. Bishop . . . 1827
Spectral properties of the 2D Holstein t–J model
H. Fehske, G. Wellein, B. Bäuml and R.N. Silver . . . 1829
Superconductivity in $HgBa_2CuO_{4+\delta}$: a layered electron gas model
D. Varshney and P. Dubey . . . 1831

Bd7. Symmetry of the superconducting order parameter

The path integral model of d-pairing in CuO_2 planes of HTSC and the collective modes
P. Brusov, N. Brusova, P. Brusov and N. Harabaev . . . 1833
Magnetic field dependence of the density of states for thin films of d-wave superconductors
A.S. Mel'nikov . . . 1835
Impurity scattering in vortex states of d-wave superconductor
H. Won and K. Maki . . . 1837
Superconducting fluctuation and pseudogap phenomena in high-T_c cuprates
K. Maki and H. Won . . . 1839
Ginzburg–Landau equation and vortex state in d + s superconductivity
M.T. Béal-Monod and K. Maki . . . 1841
On the nature of the anisotropic gap structure in the high temperature superconductors: competition between s and d symmetry types
E.A. Pashitskii, V.I. Pentegov and A.V. Semenov . . . 1843
Linear response and collective oscillations in superconductors with d-wave pairing
S.N. Artemenko and A.G. Kobelkov . . . 1845
Ginzburg–Landau expansion and the slope of the upper critical field in disordered superconductors with anisotropic pairing
M.V. Sadovskii and A.I. Posazhennikova . . . 1847
Disorder effects in superconductors with anisotropic pairing: from Cooper pairs to compact bosons
M.V. Sadovskii and A.I. Posazhennikova . . . 1849
Impurity effects on the T_c suppression in the pair tunneling mechanism for high-T_c superconductors
Y. Bang . . . 1851
Conductance oscillation in ferromagnet–insulator–ferromagnet–superconductor (FIFS) structure
J. Shi, J. Dong and D.Y. Xing . . . 1853
Theory of d.c. Josephson current in d-wave superconductor/normal-metal/d-wave superconductor junctions
Y. Tanaka and S. Kashiwaya . . . 1855
Effects of atomic-scale roughness at the surface of $d_{x^2-y^2}$-wave superconductor
Y. Tanuma, Y. Tanaka, M. Yamashiro and S. Kashiwaya . . . 1857
Restoration of superconductivity at high magnetic fields in layered high-T_c and organic superconductors
A.G. Lebed and K. Yamaji . . . 1859

Bd8. Numerical simulation

Numerical simulations for superconductors with strong randomness
T. Koyama and S. Odashima . . . 1861
Macroscopic numerical analysis of pulsed field magnetization of a bulk high T_c superconductor
K. Morikawa and M. Tsuchimoto . . . 1863
Vortex-driven phase transition in 2d classical XY-model
S.B. Ota and S. Ota . . . 1865
Evolution of the quasiparticle band with doping in models of strongly correlated electrons for the high-T_c cuprates
A. Moreo, D. Duffy, A. Nazarenko, S. Haas, J. Riera and E. Dagotto . . . 1867
Study of surface terrace effect on RHEED intensity oscillations in thin-film growth by laser molecular-beam epitaxy
K.-j. Jin, S.-h. Pan and G.-z. Yang . . . 1869

Bd9. Phase separation

Phase separation in the one-dimensional t–V model
Y.C. Chen and H.Q. Lin . . . 1871

Charge fluctuation properties of two-dimensional three-band Hubbard model
J. Yang and J.-L. Shen . . . 1873
Phase separation in the 1-D extended Hubbard model
H.Q. Lin, E. Gagliano and D.K. Campbell . . . 1875
Magnetization investigation of phase separation in $La_2CuO_{4+\delta}$ system
X.L. Dong, B.R. Zhao, Z.X. Zhao, B. Xu, Y.Z. Huang, S.Q. Guo, J.W. Li, B. Yin, F. Wu, H. Chen, W.W. Huang, G.C. Che, L.H. Zhao and Z.Y. Xu . . . 1877
Enhancement of phase separation in light Cu-doped $La_2CuO_{4+\delta}$ system: evidence from TEM and magnetization measurements
B.R. Zhao, X.L. Dong, Z.X. Zhao, W. Liu, Z.F. Dong, X.F. Duan, B. Xu, L. Zhou, G.C. Che, S.Q. Guo, B. Yin, H. Chen, F. Wu, L.H. Zhao and Z.Y. Xu . . . 1879

Be. Non-cuprate superconductors

Be1. Heavy fermions

The collective modes in HTSC and heavy fermion superconductors (HFSC) under d-pairing
P. Brusov, N. Brusova, P. Brusov and N. Harabaev . . . 1881
Probing the superconducting state of UPd_2Al_3 thin films by tunneling spectroscopy
M. Jourdan, M. Huth, S. Mouloud and H. Adrian . . . 1883
Magnetic ordering in the frustrated Kondo lattice compound CePdAl
M.D. Núñez-Regueiro, C. Lacroix and B. Canals . . . 1885

Be2. Organic superconductors

Normal state properties of ET organic conductors
R. Louati, S. Charfi-Kaddour, M. Héritier, R. Bennaceur and A. Ben Ali . . . 1887
The transport and magnetic resonance properties of the organic superconductor $NMe_4[Ni(dmit)_2]_2$
D.P. Tunstall, G.J. Stockham, J.T.S. Irvine and S. Doidge-Harrison . . . 1889
The relationship between the critical temperature T_c and the quasiparticle effective mass in quasi-two-dimensional organic superconductors
I. Kanazawa . . . 1891
A new organic superconductor with $T_c = 7$ K
H. Yu and D. Zhu . . . 1893
STM spectroscopy of $(BEDT\text{-}TTF)_2Cu(NCS)_2$
K. Ichimura, T. Arai, K. Nomura, S. Takasaki, J. Yamada, S. Nakatsuji and H. Anzai . . . 1895
Thermodynamic property of organic superconductor κ-$(BEDT\text{-}TTF)_2X$ [$X = Cu(NCS)_2$, $Cu[N(CN)_2]Br$]
Y. Nakazawa and K. Kanoda . . . 1897
Charge transfer complexes based on BEDT-TTF vinylogues
L. Yu and D. Zhu . . . 1899
Interlayer transport in the organic superconductor κ-$(BEDT\text{-}TTF)_2Cu[N(CN)_2]Br$
F. Zuo, X. Su, G.C. Alexandrakis, J.A. Schlueter, M.E. Kelly and J.M. Williams . . . 1901
Thermal conductivity of organic superconductors
V.B. Efimov, M.K. Makova and L.P. Mezhov-Deglin . . . 1903
Electronic structure and superconductivity in κ-$(BEDT\text{-}TTF)_2X$ salts
V.A. Ivanov, E.A. Ugolkova and M.Ye. Zhuravlev . . . 1905
Electronic structure and Mott transition in κ-$(BEDT\text{-}TTF)_2X$ salts
V.A. Ivanov and K. Yakushi . . . 1907
Two-chain Hubbard model for superconductivity in $TTF[Ni(dmit)_2]_2$ taking account of electron–phonon interactions
R. Ramakumar and K. Yamaji . . . 1909
Single-particle excitation spectra of κ-$(ET)_2X$: an exact-diagonalization study
S. Nishimoto and Y. Ohta . . . 1911

Be3. Alkali metal fullerenes

Theoretical analysis for the specific heat and thermal parameters of solid C_{60}
J.R. Soto, A. Calles and J.J. Castro . . . 1913

Generalized symmetry and vibrational states in halogene–fullerene compounds
Yu.S. Grushko, Yu.E. Kitaev, S.G. Kolesnik, S.N. Kolesnik, L.V. Laisheva, M.F. Limonov and V.P. Smirnov 1915

Carbon isotope effect in single-crystal Rb_3C_{60}
M.S. Fuhrer, K. Cherrey and A. Zettl 1917

Electron–electron and electron–phonon interactions in alkali metal doped fullerenes: implications for the resistivity
D. Varshney and R.K. Singh 1919

Thermal conductivity of compacted from powder and crystalline C_{60} samples
V.B. Efimov, L.P. Mezhov-Deglin and R.K. Nikolaev 1921

Coulomb interactions in the Eliashberg theory of fullerene superconductors
H.-Y. Choi and G.-Y. Oh 1923

The peculiarities of the electron spectrum of two-dimensional nanostructures
V.A. Ivanov, E.A. Ugolkova, M.Ye. Zhuravlev and V.S. Yarunin 1925

Self-consistent treatment for dynamical Jahn–Teller effect in C_{60}^+: in comparison with C_{60}^-
Y.-S. Yi, L. Tian, C.-L. Wang, Z.-B. Su and L. Yu 1927

Be4. Conventional superconductors (no contributions)

Be5. Other oxide superconductors

EPR observation of localized hole pairs in BKBO and BPBO systems
A. Yakubovskii, S. Gudenko, A. Rusakov, A. Golovashkin and S. Verkhovskii 1929

Electronic state study of $BaBiO_3$ and related superconductors
H. Minami 1931

Be6. Others

Superconductivity in $Y_{1-x}R_xNi_2B_2C$ (R = Ce, U)
J. Luo, Y. Wang, Q. Ran, X. Tong, N. Tang, F. Yang and D. Jin 1933

The photoemission study of intermetallic compound superconductor YNi_2B_2C
J. Wang, X. Zhang, Y. Ruan, Z. Jiao, Q. Zhang, C. Fan, X. Zhang, Y. Su and G. Cao 1935

Raman spectra of niobium chalcogenide superconductors: $NbSe_2$ and $SnNb_5Se_9$
A. Leblanc-Soreau, P. Molinié and E. Faulques 1937

Magnetoresistance, susceptibility and magnetization measurements on RNiBC compounds (R = Er, Ho, Dy, Tb, and Gd)
J.C. Tróchez, D.R. Sánchez, B. Giordanengo, M.B. Fontes, M. Continentino and E.M. Baggio-Saitovitch 1939

Magnetic phase transitions in $(RC)_m(NiB)_2$ intermetallic compounds observed by a local probe
D.R. Sánchez, J.C. Tróchez, H. Micklitz, M.B. Fontes, B. Giordanengo and E. Baggio-Saitovitch 1941

List of Contributors xcv

PART IV

Section C. THEORY AND PHYSICAL PROPERTIES II (PHENOMENOLOGY OF THE SUPERCONDUCTING STATE)

Ca. Flux lattice: structure and phase diagram

Ca1. Structure of the vortex lattice, effects of anisotropy

Intrinsic and disorder-induced anisotropic vortex dynamics of high-temperature superconductors
N.-C. Yeh, W. Jiang, C.C. Fu, M. Konczykowski and F. Holtzberg 1943

The critical magnetic field in the high-T_c superconductive thin films
Y.-H. Zhang, L. Fu, X.-G. Li and J.-S. Zhu 1945

Investigation of the current-induced nonlinear vortex dynamics in $YBa_2Cu_3O_7$ single crystals using transport and transmittivity measurements
W. Jiang, N.-C. Yeh, C.-C. Fu, M. Konczykowski and F. Holtzberg 1947
Flux-lattice wrecking in low-dimensional superconductors
I.L. Maksimov and V.N. Dutyshev 1949
Twin plane influence on irreversibility line anisotropy in the *ab* plane of YBaCuO
A. Sulpice, S. Sanfilippo, X. Chaud, P. Germi and R. Tournier 1951
Disappearance of the force-free current configuration at the first order vortex lattice phase transition in $YBa_2Cu_3O_{7-\delta}$ single crystals
C.J. van der Beek, M.V. Indenbom, V. Berseth, W. Benoit, A. Erb and R. Flükiger 1953
Magnetic anisotropy in single crystals of $Bi_{2.1}Sr_{1.9}Ca_1Cu_2O_x$
G.D. Gu, R. Puzniak, K. Nakao, G.J. Russell and N. Koshizuka 1955
Vortices in a $La_{1.9}Sr_{0.1}CuO_4$ crystal: influence of the layered structure on magnetisation
Yu.V. Bugoslavsky, A.A. Zhukov, G.K. Perkins, A.D. Caplin, H. Kojima and I. Tanaka 1957
Vortex dynamics in oxygen deficient single crystals of $YBa_2Cu_3O_{7-\delta}$
B. Lundqvist, A. Rydh, Ö. Rapp and M. Andersson 1959
Flux line orientation in $YBa_2Cu_3O_{7-\delta}$ crystals in low magnetic fields
S. Koleśnik, H.W. Weber, T. Skośkiewicz, M. Sawicki and W. Sadowski 1961
Glass phases and melting of flux lattices in layered superconductors
B. Horovitz and A. Golub 1963
Vortex lattice melting and the peak effect in oblique field in $Bi_2Sr_2CaCu_2O_{8+y}$
S. Ooi, T. Tamegai and T. Shibauchi 1965
Finite-element method to the *d*-wave time-dependent Ginzburg–Landau equations
Q.-H. Wang and Z.D. Wang 1967
The effect of random disorder on intrinsic pinning in $YBa_2Cu_3O_{7-x}$ single crystals
K. Deligiannis, M. Oussena, P.A.J. de Groot, R. Gagnon and L. Taillefer 1969
Influence of the in-plane anisotropy on the vortex phase diagram in $YBa_2Cu_3O_\delta$
K. Deligiannis, M. Oussena, P.A.J. de Groot, R. Gagnon and L. Taillefer 1971
How to make superconducting-anisotropy least in high-T_c cuprate superconductors
H. Ihara, A. Iyo, K. Tanaka, K. Tokiwa, K. Ishida, N. Terada, M. Tokumoto, Y. Sekita, T. Tsukamoto, T. Watanabe and M. Umeda 1973
Non-resonant rf absorption evidence for reentrant melting of vortex lattice in $Bi_2Sr_2CaCu_2O_8$ single crystals
Y.S. Sudershan, A. Rastogi, S.V. Bhat, A.K. Grover, Y. Yamaguchi, K. Oka and Y. Nishihara 1975
Vortex lattice in the superconductor–conductor transition region and the voltage noise in a highly textured thin film of $YBa_2Cu_3O_{7-\delta}$
K. Frikach, A. Taoufik, S. Senoussi and A. Tirbiyin 1977
Influence of twins in the vortex pinning in YBaCuO single crystals
M. Boudissa, K. Frikach, A. Taoufik, S. Senoussi, O. Sarrhini and R. Halimi 1979
Magnetization of Y- and Ni-substituted Bi-2212 single crystals
D.H. Ha, K.W. Lee, K. Oka and Y. Nishihara 1981

Ca2. Effects of the layered structure, 3D versus 2D behavior

Dimensional crossover vortex fluctuations in irreversibility line of high-quality $Tl_2Ba_2CaCu_2O_8$ thin film
L. Ammor, J.C. Soret, A. Smina, V. Ta Phuoc, A. Ruyter, A. Wahl, B. Martinie, J. Lecomte and Ch. Simon 1983
Microwave surface resistance in BSCCO crystal: magnetic field and angular measurement
E. Silva, R. Fastampa, M. Giura, R. Marcon, C. Possanzini, S. Sarti and E.L. Wolf 1985
Vortex fluctuation in $HgBa_2Ca_3Cu_4O_{10+\delta}$
M.-S. Kim, W.-S. Kim, S.-I. Lee, S.-C. Yu, E.S. Itskevich and I. Kuzemskaya 1987
$PrBa_2Cu_3O_7$ thickness dependence of current–voltage behaviors in $YBa_2Cu_3O_7/PrBa_2Cu_3O_7$ multilayers
F. Ichikawa, B. Zhao, J.-J. Sun, B. Xu, L. Li, T. Fukami, T. Aomine and L. Rinderer 1989
Reflection of two-dimensionality on the temperature dependence of AC susceptibility
P.N. Mikheenko, J. Horvat, M. Ionescu and S.X. Dou 1991
Properties of upper critical field and pinning potential in epitaxial $GdBa_2Cu_3O_{7-\delta}$ thin film
X.W. Cao, X.J. Xu, Z.H. Wang, J. Fang, R.L. Wang and H.C. Li 1993
Microstructure anisotropy and critical currents in silver-sheathed Bi2223 tapes
Q.Y. Hu, H.K. Liu, H.W. Weber and S.X. Dou 1995

Dimensionality of flux pinning in various superconducting YBCO/PYBCO superlattices
S.Y. Yang, H.E. Horng and H.C. Yang 1997
On the relation between anisotropic 3D-XY/Josephson-junction array model and layered high T_c cuprate superconductors
C. Kawabata 1999
Two dimensional superconducting behavior of $Hg_{0.7}Cr_{0.3}Sr_2CuO_{4+\delta}$
J.-H. Choi, M.-S. Kim, S.-I. Lee, S.-Y. Lee, I.-S. Yang, J.V. Yakhmi, J.B. Mandal, B. Bandyopadhyay and B. Ghosh . . . 2001
Equilibrium magnetic properties and vortex fluctuation behavior of grain-aligned $Hg_{0.8}Pb_{0.2}Ba_{1.5}Sr_{0.5}Ca_2Cu_3O_y$
Y. Zhuo, J.-H. Choi, M.-S. Kim, W.-S. Kim, S. Lee and S.-I. Lee 2003
3D versus 2D behavior in ceramic $Tl_2Ba_2Ca_2Cu_3O_{10-\delta}$ high-T_c superconductors
W.J. Yeh and Z.Q. Yu 2005
Effects of a geometrical length scale on the arrowhead magnetisation of $Bi_2Sr_2CaCu_2O_y$ single crystals
G. Yang, J.S. Abell and C.E. Gough 2007
Nernst effect in the mixed state of $YBa_2Cu_3O_{7-\delta}$ and $Bi_{2.1}Sr_{1.8}Ca_1Cu_2O_{8+\delta}$
T. Sasaki, K. Yamada, K. Watanabe, S. Watauchi, K. Kishio and N. Kobayashi 2009
Effect of vortex-pair fluctuations on zero-field current–voltage characteristics of YBCO films
C. Camerlingo, C. Nappi, M. Russo and E. Sarnelli 2011
Melting and interlayer coherence of vortices in the quasi-two-dimensional organic superconductor, α-(BEDT-TTF)$_2$NH$_4$Hg(SCN)$_4$
H. Taniguchi, Y. Nakazawa and K. Kanoda 2013

Ca3. Thermal fluctuations and melting of the vortex lattice

The vortex phase diagram of $Bi_2Sr_2CaCu_2O_y$ investigated by the resistivity measurements
H. Ikuta, S. Watauchi, H. Kobayashi, Y. Nakayama, J. Shimoyama, K. Kitazawa and K. Kishio 2015
Field induced suppression of the vortex lattice melting transition in twinned $YBa_2Cu_3O_{7-\delta}$
R.M. Langan, S.N. Gordeev, M. Oussena, S. Pinfold, P.A.J. de Groot, L. Jansen, R. Gagnon and L. Taillefer 2017
Scaling of $I-V$ curves and flux creep in high-T_c superconductors
A.N. Lykov, C. Attanasio, L. Maritato and S.L. Prischepa 2019
Spatially resolved resistivity near the vortex lattice phase transition in $Bi_2Sr_2CaCu_2O_{8+\delta}$ single crystals
V. Berseth, M.V. Indenbom, C.J. van der Beek, G. D'Anna and W. Benoit 2021
Sublimation and hysteretic transition of the vortex-lattice in $Bi_2Sr_2CaCu_2O_8$
D.T. Fuchs, R.A. Doyle, E. Zeldov, D. Majer, W.S. Seow, T. Tamegai, S. Ooi, R.J. Drost, M. Konczykowski and P.H. Kes . 2023
Enhancement of vortex fluctuation in $Bi_2Sr_2CaCu_2O_8$ single crystal by iodine intercalation
K.W. Lee, D.H. Ha, Y.K. Park, J.C. Park, M.-S. Kim, S.-C. Yu and S.-I. Lee 2025
Thermodynamic evidence for reentrant peak effect in a clean single crystal of 2H-NbSe$_2$ and the effect of disorder and thermomagnetic history on it
S.S. Banerjee, N.G. Patil, S. Saha, S. Ramakrishnan, A.K. Grover, S. Bhattacharya, T.V. Chandrasekhar Rao, P.K. Mishra, G. Ravikumar, V.C. Sahni, C.V. Tomy, G. Balakrishnan, D.McK. Paul and M.J. Higgins 2027
Effects of hydrostatic pressure and oxygen doping on the vortex phase diagram in $Bi_2Sr_2CaCu_2O_{8+y}$
T. Tamegai, R. Yamada, S. Ooi, T. Shibauchi, C. Murayama and N. Mori 2029
Current-induced resistivity hysteresis in detwinned $YBa_2Cu_3O_{7-\delta}$ single crystals
S.N. Gordeev, P.A.J. de Groot, M. Oussena, A.V. Volkozub, R.M. Langan, A.P. Rassau, R. Gagnon and L. Taillefer . . . 2031
Organisation of vortex motion in detwinned $YBa_2Cu_3O_{7-\delta}$ single crystals
S.N. Gordeev, P.A.J. de Groot, M. Oussena, R.M. Langan, A.P. Rassau, R. Gagnon and L. Taillefer 2033
Hysteretic and non-hysteretic a.c. losses in detwinned $YBa_2Cu_3O_{7-\delta}$ single crystals
D. Bracanovic, S. Gordeev, S. Pinfold, M. Oussena, P.A.J. de Groot, R. Gagnon and L. Taillefer 2035
a.c. susceptibility studies of the vortex lattice melting transition in clean, detwinned $YBa_2Cu_3O_{6.92}$ single crystals
D. Bracanovic, S. Pinfold, M. Oussena, S. Gordeev, R. Langan, P.A.J. de Groot, R. Gagnon and L. Taillefer 2037
The peak effect as a phase boundary in detwinned $YBa_2Cu_3O_{7-\delta}$
R.M. Langan, S.N. Gordeev, M. Oussena, S. Pinfold, P.A.J. de Groot, R. Gagnon and L. Taillefer 2039
Josephson vortex lattice melting in $Bi_2Sr_2CaCu_2O_8$
M.S. Fuhrer, K. Ino, K. Oka, Y. Nishihara and A. Zettl 2041
AC and DC magnetisation studies of peak effect in a clean crystal of CeRu$_2$
N.G. Patil, S.S. Banerjee, S. Saha, K. Ghosh, S. Ramakrishnan, S. Bhattacharya, A.K. Grover, T.V. Chandrasekhar Rao, G. Ravikumar, P.K. Mishra, V.C. Sahni, E. Yamamoto, Y. Haga, M. Hedo, Y. Inada and Y. Onuki 2043

Angular dependence of the first-order vortex-lattice phase transition in $Bi_2Sr_2CaCu_2O_8$
B. Schmidt, M. Konczykowski, N. Morozov and E. Zeldov 2045

The property of the vortex glass in the vicinity of the pinned vortex liquid for textured bulk $YBa_2Cu_3O_7$
S.Y. Ding, C. Ren, Y. Ge, X.X. Yao, Y.X. Fu and C.B. Cai 2047

Upper critical field in the $Ba_{1-x}K_xBiO_3$ superconductor
P. Samuely, P. Szabó, A.G.M. Jansen, P. Wyder, T. Klein, J. Marcus and C. Escribe-Filippini 2049

The irreversibility line of magnetically grain-aligned Hg-1212 sample – evidences of flux line lattice melting
R. de Andrade Jr., E.R. Leite, A.J.C. Lanfredi and W.A. Ortiz 2051

Bose–glass and entanglement–disentanglement transitions in $YBa_2Cu_3O_7$
A.V. Samoilov, M. Konczykowski, M.V. Feigel'man and F. Holtzberg 2053

The universal scaling of flux lattice melting line of Bi2212 under $H \parallel c$
J.-i. Shimoyama, K. Kishio, H. Ikuta, S. Watauchi, M. Okuya and K. Kitazawa 2055

Simulation for the first-order vortex-lattice melting transition in high-T_c superconductors
X. Hu, S. Miyashita and M. Tachiki 2057

Observation of the vortex melting line in $YBa_2Cu_3O_7$ by specific heat measurements in very high magnetic field: $H \leqslant 23$ Tesla
C. Marcenat, R. Calemczuk, A. Erb, E. Walker and A. Junod 2059

Various vortex phase transitions in a weakly disordered superconductor YNi_2B_2C single crystal
M.-O. Mun, S.-I. Lee and W.C. Lee 2061

^{13}C and ^{1}H NMR study of vortex dynamics in a layered organic superconductor at selective nuclear sites
K. Kanoda, K. Miyagawa and A. Kawamoto 2063

NMR spin echo experiments as probes of vortex dynamics in $YBa_2Cu_3O_7$
C.H. Recchia, J.A. Martindale, C.H. Pennington, W.L. Hults and J.L. Smith 2065

Ca4. Disorder and the vortex glass

Effects of correlated disorder on vortex-lattice melting in BSCCO
B. Khaykovich, M. Konczykowski, R.A. Doyle, E. Zeldov, P.H. Kes, D. Majer and T.W. Li 2067

μSR study of the grain-size dependence of Y1237 vortex states
C. Boekema, D.W. Cooke, M. Merton and O. Vera 2069

Thickness dependence of vortex-glass transition in $YBa_2Cu_3O_{7-\delta}$ thin films
A. Sawa, H. Yamasaki, Y. Mawatari, E. Obara, M. Umeda and S. Kosaka 2071

Influence of columnar defects on interlayer coherence in $Bi_2Sr_2CaCu_2O_{8+\delta}$ from Josephson plasma resonance
M. Kosugi, Y. Matsuda, M.B. Gaifullin, K. Kumagai, N. Chikumoto, J. Shimoyama, K. Kishio, K. Hirata and M. Konczykowski 2073

Vortex liquid and its transition to solid for $YBa_2Cu_3O_7$
C. Ren, S.Y. Ding, L. Qiu, Y. Ge, F.Y. Lin and X.X. Yao 2075

Ca5. Kosterlitz–Thouless transition

The Berezinskii–Kosterlitz–Thouless transition in BSCCO whiskers: non-linear conductivity and size effects
V.Ya. Pokrovskii, I.G. Gorlova and S.G. Zybtsev 2077

Vortex pairs generation by the flowing current in planar superconductors
L. Romanò and C. Paracchini 2079

Is there a Kosterlitz–Thouless transition in YBCO films?
J.M. Repaci, C. Kwon, Q. Li, X. Jiang, T. Venkatesan, R.E. Glover III, C.J. Lobb and R.S. Newrock 2081

Ca6. Imaging of the vortex lattice

Determination of the quenching temperature for the vortex lattice in field-cooling decoration experiments
M. Marchevsky, P.H. Kes and J. Aarts 2083

Double-sided decoration on $NbSe_2$
M. Marchevsky, A. Keurentjes, P.H. Kes and J. Aarts 2085

Direct observation of remnant fields in Y_2BaCuO_5 and platinum doped $YBa_2Cu_3O_7$ bulk superconductors
Z.X. Gao, X. Yang, S.G. Wang, L. Xiao, H.T. Ren, Y.L. Jiao, Y.D. Dai and Z.Z. Gan 2087

Neutron diffraction by the flux line lattice in $YBa_2Cu_3O_{7-\delta}$ single crystals
G. Brandstätter, A. Vostner, H.W. Weber, T. Chattopadhyay, R. Cubitt, H. Fischer and G.A. Emel'chenko 2089

Cb. Pinning and flux dynamics

Cb1. Pinning mechanisms, irreversibility line

The modulated research of $YBa_2Cu_3O_{7-x}$ single crystals and ceramics magnetization
N.D. Kuzmichev and V.V. Slavkin . . . 2091

Positron annihilation study on stress-field pinning in (Eu,Y)-123 superconductors
Y. Li, C. Tang, Q. Ma, Y. Wang, G. Cao, T. Wei, W. Wang and T. Zhang . . . 2093

Interplay between correlated and uncorrelated disorder in bulk HTSC materials with surface gold ion implantation
E. Mezzetti, R. Gerbaldo, G. Ghigo, L. Gozzelino, B. Minetti and R. Cherubini . . . 2095

Flux pinning in $YBa_2Cu_3O_{7-\delta}$ through nanometer-sized MgO particles
Q. Chen, M. Fang, Z. Jiao, Q. Zhang, H. Wen and Z. Zhao . . . 2097

Flux pinning and irreversibility in $NdBa_2Cu_3O_x$ single crystals
B. Chen, A.K. Pradhan, K. Kuroda and N. Koshizuka . . . 2099

The equilibrium vortex state in heavy-ion irradiated $Bi_2Sr_2CaCu_2O_{8+\delta}$
C.J. van der Beek, M. Konczykowski, T.W. Li and P.H. Kes . . . 2101

The relationship of the critical current density and the lattice deformation in $Y_{1-x}RE_xBa_2Cu_3O_{7-\delta}$ epitaxial films
Z.S. Peng, J.M. Hao, B. Yin, Z.X. Zhao, Z.Q. Hua and B.C. Yang . . . 2103

Suppression of vortex motion in $La_{1.85}Sr_{0.15}CuO_4$ with the splayed configuration of columnar defects
X. Fan, M. Terasawa, T. Mitamura, T. Kohara, K. Ueda, H. Tsubakino, T. Kambara, Y. Awaya, M. Matsuda, G. Tatara, K. Fukushima and N. Takezawa . . . 2105

Magnetism and flux pinning of $Bi_{2.1}Sr_{1.9}Ca_1(Cu_{1-y}Fe_y)_2O_{8+x}$ single crystals
G.D. Gu, G.J. Russell and N. Koshizuka . . . 2107

Flux pinning and critical currents in Bi-2223/Ag concentric tapes
M. Majoroš, J. Kvitkovič, M. Polák, L. Martini and S. Zannella . . . 2109

Irreversibility line of Tl-1223/Ag sheathed superconducting composite tape
H. Tang, Y.Z. Wang, Z.Q. Yang, C. Zhang, L. Hua, F.R. de Boer, G.W. Qiao, S.H. Zhou and H.T. Peng . . . 2111

Thickness dependence of magnetic force acting on a magnetic dipole over a type-II superconducting thin film
T.J. Yang, J.C. Wei, J.L. Chen and L. Horng . . . 2113

Superconducting properties of four-CuO_2-layer $TlBa_2Ca_3Cu_4O_{11-\delta}$ single crystals
L. Zhang, R.N. Shelton, J.Z. Liu and Q.S. Yang . . . 2115

Reentrant peak effect and irreversibility line in untwinned single crystals of $YBa_2Cu_3O_y$
T. Nishizaki, T. Naito and N. Kobayashi . . . 2117

Anisotropy of fishtail effects in $EuBa_2Cu_3O_y$ crystal prepared by melt-textured growth method
C.-D. Wei, Z.-X. Liu, H.-T. Ren and L. Xiao . . . 2119

Local vortex mobility below the irreversibility line of $Tl_2Ba_2CaCuO_8$. A ^{205}Tl NMR study of the transverse relaxation in single crystals
J. Witteveen, E.G. Nikolaev, H.B. Brom and M.L. de Kok . . . 2121

Flux pinning in melt-textured $YBa_2Cu_3O_{7-y}$ with different sizes of secondary phase particles
C.H. Choi, S.H. Han, C. Andrikidis and Y. Zhao . . . 2123

Enhanced flux pinning and fishtail effect in spiral grown $Bi_2Sr_2CaCu_2O_y$ single crystals
X.L. Wang, J. Horvat, H.K. Liu and S.X. Dou . . . 2125

Influence of Y_2BaCuO_5 precipitate size on flux pinning effect in melt-texture-growth $YBa_2Cu_3O_{7-y}$
M. Fang, Q. Chen, Z. Jiao, Q. Zhang, H. Wen, Z. Zhao and C. Cai . . . 2127

Flux pinning mechanism in $Ca_{1-x}Cd_xLaBaCu_3O_{7-y}$ superconductor
M. Fang, Q. Chen, Z. Jiao, Q. Zhang, H. Wen and Z. Zhao . . . 2129

The vortex-glass transition and irreversibility line in epitaxial $YBa_2Cu_3O_{7-\delta}$ thin film
Z.H. Wang, J. Fang, K.B. Li, Z.Y. Chen and X.W. Cao . . . 2131

Evidence of defect pinning in the superconducting Bi–(Pb)–Sr–Ca–Cu–O system
S.X. Dou, J. Horvat, M. Ionescu, H.K. Liu, I. Kusevic and E. Babic . . . 2133

Effect of high-energy heavy-ion irradiation on the critical current density of Bi-2212 single crystals
T. Terai, T. Kobayashi, K. Kishio, J. Shimoyama, S. Okayasu and Y. Kazumata . . . 2135

Flux pinning properties of $NdBa_2Cu_3O_{7-d}$ single crystals flux-grown in low oxygen partial pressures
T. Higuchi, S.-I. Yoo, K. Waki, H. Fujimoto and M. Murakami . . . 2137

Surface pinning effects in $YBa_2Cu_3O_7$ bulk sample
Z.Y. Zeng, A.M. Sun, X.N. Xu, Y. Yu, S.Y. Ding and X.X. Yao . . . 2139
Flux flow and creep induced by a pulsed magnetic field in bulk YBCO
M. Yamazaki, T. Takizawa, T. Watanabe, N. Yamaura and M. Murakami . . . 2141
Comparative study of the flux pinning by columnar defects in high-T_c superconductors
N. Chikumoto, M. Konczykowski, K. Kishio and M. Murakami . . . 2143
Reversible suppression of the so-called fishtail effect in ultra pure single crystals of $YBa_2Cu_3O_{7-\delta}$
A. Erb, J.-Y. Genoud, M. Dhalle, E. Walker and R. Flükiger . . . 2145
Flux pinning in oxygen deficient $YBa_2Cu_3O_{7-\delta}$ thin films
G. Samadi Hosseinali, H.W. Weber, E. Stangl, S. Proyer and D. Bäuerle . . . 2147
The effect of twin plane boundaries on the vortex dynamics in twinned YBCO crystals
G.K. Perkins, A.A. Zhukov, Yu.V. Bugoslavsky, L.F. Cohen, T. Wolf, M. Kläser and A.D. Caplin . . . 2149
Vector magnetization studies of high purity $YBa_2Cu_3O_y$ single crystal near the *ab*-plane
A.A. Zhukov, Yu. Bugoslavsky, G.K. Perkins, J.V. Thomas, A.D. Caplin, H. Küpfer and T. Wolf . . . 2151
Transport hysteresis of critical current density under lower voltage for a Bi-based superconducting tape
L. Horng and Y.J. Huang . . . 2153
Vortex lock-in by columnar defects in $YBa_2Cu_3O_y$ single crystals
A.A. Zhukov, G.K. Perkins, Yu. Bugoslavsky, J. Totty, L.F. Cohen, A.D. Caplin, H. Küpfer, T. Wolf and G. Wirth . . . 2155
Frequency dependence of susceptibility higher harmonics in YBCO samples by numerical solutions of the non linear magnetic flux diffusion equation
D. Di Gioacchino, P. Tripodi, F. Celani, A.M. Testa and S. Pace . . . 2157
Vortex phase diagram in $YBa_2Cu_3O_x$ and $Bi_2Sr_2CaCu_2O_x$ via the MMWA hysteresis
Yu. Talanov, T. Shaposhnikova, Yu. Vashakidze and R. Khasanov . . . 2159
Flux pinning in $HgBa_2CuO_{4+x}$ single crystals
A. Wiśniewski, R. Puźniak, M. Baran, J. Karpiński, E. Kopnin, R. Moliński and H. Schwer . . . 2161
Abrikosov vortex loop near the surface of superconductor
A.V. Samokhvalov . . . 2163
Observation of step structures in the I–V characteristics of YBCO thin films
J. Azoulay, A. Verdyan and I. Lapsker . . . 2165
Criterion dependence of the irreversibility field in BSCCO-2212
V.N. Zavaritsky . . . 2167
Monte Carlo study of vortex-line dynamics and I–V characteristics with correlated and uncorrelated disorders
R. Sugano and T. Onogi . . . 2169
Intergranular pinning properties in high-T_c superconductor by ac susceptibility response
K.C. Hung, C.C. Lam, X. Jin and J. Feng . . . 2171
Effective pinning potential in Pb-doped Hg-1234 superconductor by ac susceptibility response
K.C. Hung, C.C. Lam, X. Jin and H.M. Shao . . . 2173
The thermal-assisted-flux-flow effect in infinite length rectangle-section sample of Pb-doped $HgBa_2Ca_2Cu_3O_{8+\delta}$
C.C. Lam, L. Jei, J.R. Jin, X. Jin, K.C. Hung, H.M. Shao and J. Feng . . . 2175
Fishtail effect and its mechanism in PMP Y–Ho–Ba–Cu–O superconductors
Y. Feng, L. Zhou, J. Wen and N. Koshizuka . . . 2177
Vortex confinement by top and bottom channeling in ion irradiated YBCO melt textured bulk
R. Cherubini, R. Gerbaldo, G. Ghigo, L. Gozzelino, E. Mezzetti and B. Minetti . . . 2179
Irreversibility line in deoxygenated $YBa_2Cu_3O_{7-\delta}$ single crystal
I. Abrego Castillo, L. Ghivelder and S. Salem-Sugui Jr. . . . 2181
Collapse of the vortex lattice inductance in untwinned $YBa_2Cu_3O_7$ observed by complex resistivity technique
H. Wu, P. Matl, N.P. Ong, R. Gagnon and L. Taillefer . . . 2183
Scaling behavior of fishtail effects in $YBa_2Cu_3O_y$ crystal prepared by melt-textured growth method
C.-D. Wei, Z.-X. Liu, H.-T. Ren and L. Xiao . . . 2185
Effect of columnar defects on the vortex dynamics in BSCCO 2212 single crystals
J.T. Totty, L.F. Cohen, G.K. Perkins, K. Kadowaki, G. Wirth, D.McK. Paul and G. Balakrishnan . . . 2187
Local magnetic measurement of strong pinning by columnar defects
M. Konczykowski, E. Zeldov, D. Majer and S. Bouffard . . . 2189

Peak effect and its disappearance in YBCO superconducting crystals
X.S. Ling, J.I. Budnick and B.W. Veal 2191
Pinning forces and scaling in high-T_c superconductors
M.R. Koblischka 2193

Cb2. Magnetization and relaxation in the critical state

U–J relation determined from magnetic relaxation
X.X. Yao, Y. Yu and M.J. Qin 2195
Nonlogarithmic regime of the temporal magnetization relaxation in thin-film superconductors
G.M. Maksimova and I.L. Maksimov 2197
Temperature dependence of inter- and intragranular currents in $(Bi,Pb)_2Sr_2Ca_2Cu_3O_x$ superconductors
P. Kottman, J. Paasi, M. Majoroš and V. Plecháček 2199
Study of paramagnetic frozen states in superconducting Nb and Ta samples
O.F. de Lima, M.A. Avila and C.A. Cardoso 2201
Strong pinning in heat treated $YBa_2Cu_3O_{7-\delta}$
J.W. Farmer, M. Kornecki and D.L. Cowan 2203
Lower critical magnetic field H_{c1} and intergranular coupling of Bi-2223/Ag concentric tapes
M. Majoroš, L. Martini and S. Zannella 2205
Fishtail and flux creep phenomena in the cubic $(K,Ba)BiO_3$ high T_c superconductor
T. Klein, W. Harneit, L. Baril, J. Marcus and C. Escribe Filippini 2207
Local magnetic relaxation in $Nd_{1.85}Ce_{0.15}CuO_{4-\delta}$ crystals
D. Giller, Y. Abulafia, R. Prozorov, Y. Wolfus, A. Shaulov, Y. Yeshurun, D. Majer, E. Zeldov, J.L. Peng and R.L. Greene . . 2209
Temperature dependence of the material parameters in the Kim–Anderson critical-state model for a melt-processed $YBa_2Cu_3O_7$
H. Yasuoka, S. Tochihara, Y. Tanaka and H. Mazaki 2211
Magnetic properties of sintered and deformed Ag/BiSCCO tapes
A. Goldgirsh, V. Beilin, E. Yashchin, I. Felner and M. Schieber 2213
Thickness dependence of the magnetic properties in $YBa_2Cu_3O_{7-\delta}$ thin films
R. Prozorov, E. Sheriff, Y. Yeshurun and A. Shaulov 2215
$U(J, H, \alpha)$ dependence in epitaxial $YBa_2Cu_3O_{7-\delta}$ thin film
Z.H. Wang, K.B. Li, J. Fang, J.L. Chen and X.W. Cao 2217
Crossover from fractal to non-fractal flux penetration in vicinal $Tl_2Ba_2CuO_{6+\delta}$ thin films
R. Surdeanu, R.J. Wijngaarden, R. Griessen, C. Rossel, Z.F. Ren and J.H. Wang 2219
Interlayer Josephson coupling in the vortex solid state of $Bi_2Sr_2CaCu_2O_{8+\delta}$ from Josephson plasma resonance
M.B. Gaifullin, Y. Matsuda, K.-i. Kumagai, M. Kosugi, K. Kadowaki, T. Mochiku and K. Hirata 2221
Geometrical locking of the irreversible moment to the normal of a thin plate superconductor
A.A. Zhukov, G.K. Perkins, Y.V. Bugoslavsky and A.D. Caplin 2223
Extraction of current density distribution from local magnetic measurements
Y. Abulafia, Y. Wolfus, A. Shaulov, R. Prozorov, D. Giller and Y. Yeshurun 2225
Effect of columnar defects on the vortex dynamics in a twinned YBCO 123 crystal
L.F. Cohen, A.A. Zhukov, G.K. Perkins, J. Totty, A.D. Caplin, H. Kupfer, T.H. Wolf and G. Wirth 2227
Frequency dependence of ac susceptibility in high temperature superconductors
M.J. Qin and X.X. Yao 2229
Meissner holes and turbulent flux structures in high-T_c superconductors
V.K. Vlasko-Vlasov, U. Welp, G.W. Crabtree, V.V. Kabanov and V.I. Nikitenko 2231
Time dependence of the field cooled magnetization in $Bi_2Sr_2CaCu_2O_x$ single crystal
J.T. Wang, C.L. Lin, T. Mihalisin and T.P. Chen 2233
Anomalous behaviour of vortex creep in twinned YBCO crystals
G.K. Perkins, A.A. Zhukov, Yu.V. Bugoslavsky, L.F. Cohen, A.D. Caplin and T. Wolf 2235
Pulsed magnetization of melt-processed YBaCuO superconductors
H. Fujimoto, K. Kawano, T. Higuchi, K. Nagashima and M. Murakami 2237
The second peak effect observed in $(La_{1-x}Sr_x)_2CuO_{4-\delta}$ single crystals
M. Okuya, T. Sasagawa, H. Ikuta, J.-i. Shimoyama, K. Kitazawa and K. Kishio 2239

Bulk pinning by planar defects in Ti-doped Bi–2212 single crystals
R.J. Drost, T.W. Li, P.H. Kes, R. Surdeanu, R.J. Wijngaarden and M. Konczykowski 2241
AC response and current density relaxation
X. Jin, C.C. Lam, K.C. Hung, C.M. Lee and H.M. Shao 2243
Physical characteristics of imaginary ac magnetic susceptibility of granular superconductor, $Hg_{0.69}Pb_{0.31}Ba_2Ca_2Cu_3O_{8+\delta}$
C.M. Lee, C.C. Lam, K.C. Hung, X. Jin and H.M. Shao 2245
Study of the local magnetic relaxation in melting-textured-growth $YBa_2Cu_3O_{7-\delta}$ bulk sample with Bi Hall probe
T. Zhang, G. Liu, H. Dong, H. Ren, L. Xiao, Y. Jiao and S. Yan 2247
Investigation of flux pinning and magnetic relaxation behavior in PMP $YBa_2Cu_{2.95}Sn_{0.05}O_y$ sample
Y. Feng, L. Zhou, W. Yang, P. Zhang, X. Wu, B. Yi, J. Li and Z. Zhao 2249
Diamagnetic peaks in magnetization versus temperature curves caused by flux trapped relaxation observed in $YBa_2Cu_3O_{7-\delta}$ samples
O.F. de Lima and R.A. Ribeiro 2251
Magnetic relaxation and collective creep in a grain aligned sample of $YBa_2Cu_3O_{7-\delta}$
K. Frikach, A. Taoufik, M. Boudissa and S. Senoussi 2253
Field sweep rate dependence of magnetization and E–J characteristics in $YBa_2Cu_3O_7$ films
Y. Mawatari, A. Sawa, H. Obara, M. Umeda and H. Yamasaki 2255
Collective creep near the second peak of magnetization curve in $LaBa_2Cu_3O_7$ and $La_{2-x}Sr_xCuO_4$ single crystals
Y. Kodama, K. Oka, Y. Yamaguchi, Y. Nishihara and K. Kajimura 2257
Local magnetic relaxation close to the second peak in BSCCO crystals
S. Berry, M. Konczykowski, P.H. Kes and E. Zeldov 2259
Pinning enhanced anomalous magnetization in single crystal $Nd_{1.85}Ce_{0.15}CuO_{4-\delta}$ superconductors
F. Zuo, J.L. Peng and R.L. Greene 2261
Influence of the magnetic field sweep rate on flux creep in melt textured sample of $YBa_2Cu_3O_{7-d}$
K. Frikach, A. Taoufik, M. Boudissa, S. Senoussi and R. Halimi 2263

Cb3. Quantum dynamics of vortices

Quantum and thermal creep in $Pr_{1.85}Ce_{0.15}CuO_{4+\delta}$ single crystals
A.F.Th. Hoekstra, R. Griessen, M. Brinkmann, K. Westerholt and H.H. Wen 2265
Strong quantum fluctuations in 400 Å thick YBCO thin films
J. Chen, D.L. Yin, P. Zheng, J. Hammann, G.C. Xiong, Q. Jiang, K. Wu, Z.J. Chen and D. Jin 2267

Cb4. Critical currents

Are there Lorentz-force-independent critical currents in Bi(2223)/Ag tape?
S.Q. Guo and R. Zen 2269
A study on the critical current density of BPSCCO/Ag tapes with different thermo-mechanical treatments
H. Zhang, Y. Sun and Y.H. Zhang 2271
Effect of γ–irradiation on the critical current density of Tl/F superconductors
N.M. Hamdan, A.S. Al-Harthi and M.F. Choudhary 2273
New chemical pinning center from uranium compound in melt-textured YBCO
Y. Ren, R. Weinstein and R.-P. Sawh 2275
Superconducting properties as a function of substrate type in BSCCO thin films
A. Mariño, F. Ichikawa, H. Rodriguez and L. Rinderer 2277
Increased flux pinning in both pure and carbon-nanotube-embedded Bi-2212 superconductors
S.-L. Huang, M.R. Koblischka, K. Fossheim, T.W. Ebbesen and T.H. Johansen 2279
Magnetization current density and irreversibility line of $Tl_2Ba_2CaCu_2O_8$ superconducting film
X.N. Xu, A.M. Sun, M.J. Qin, S.Y. Ding, S.L. Yan, H.H. Wen, X. Jin and X.X. Yao 2281
Influence of thermomechanical treatment of Bi-2223/Ag tapes on I_c vs. H at elevated fields
J. Horvat, Y.C. Guo, S.X. Dou and H.K. Liu 2283
Change in superconducting properties of Bi-2212 single crystal due to fast neutron irradiation followed by thermal annealing
T. Terai, T. Kobayashi, Y. Ito, K. Kishio and J. Shimoyama 2285
Transport current density up to 10 T at 77 K in filamentary $YBa_2Cu_3O_7$ superconductor
M. Takahama, T. Goto, K. Watanabe and S. Awaji 2287

The study of current distribution in Ag-Bi2223 tape using a scanning Hall-probe magnetometry
K. Kawano, S. Isoda and A. Oota . . . 2289

Measurement of local currents in superconductors using an in-situ high field magneto-optical microscope
R.J. Wijngaarden, R. Surdeanu, R. Griessen, A.A. Menovsky, J. Fendrich and W.-K. Kwok . . . 2291

C-axis magnetoresistance of a few atomic surface layers of the Bi:2212 single crystals
A. Yurgens, D. Winkler, N.V. Zavaritsky and T. Claeson . . . 2293

Influence of microstructure on the pinning properties of Nd123
N. Chikumoto, J. Yoshioka, M. Otsuka, S.I. Yoo and M. Murakami . . . 2295

Flux pinning and J_c–B–T surface in high-J_c BSCCO/Ag tapes
I. Kušević, P. Šimundić, J. Ivkov, E. Babić, W.G. Wang, H.K. Liu and S.X. Dou . . . 2297

Dynamic resistance and critical current distribution in high-J_c BSCCO/Ag tape
Ž. Marohnić, E. Babić, I. Kušević, H.K. Liu and S.X. Dou . . . 2299

Isotropic short columnar pinning centers from fission fragment damage in bulk melt-textured YBCO
Y. Ren, R. Weinstein, R.-P. Sawh and J. Liu . . . 2301

Relation between micro-structure and transport properties of epitaxial $YBa_2Cu_3O_{7-\delta}$ thin films
G. Doornbos, B. Dam, J.C. Martínez, R. Surdeanu, U. Poppe and R. Griessen . . . 2303

Specific local electrical behaviour of grain boundaries related to their structure in $Y(Dy)Ba_2Cu_3O_{7-\alpha}$ ceramics
J.Y. Laval, E. Gradys and C. Nguyen van Huong . . . 2305

AC susceptibility response and critical current density of a nominal $Hg_{1-x}Pb_xBa_2Ca_2Cu_3O_{8+\delta}$ superconductor for different critical state models
C.C. Lam, K.C. Hung, J. Feng, X. Jin and H.M. Shao . . . 2307

Scaling behavior of magnetic hysteresis loops and critical current density in melt-textured $(Y_{1-x}Pr_x)Ba_2Cu_3O_{7-\delta}$
Y.G. Xiao, B. Yin, J.W. Li, Z.X. Zhao, X.K. Fu, H.T. Ren and L. Xiao . . . 2309

Anisotropic flux penetration and current distribution in (001) tilted $YBa_2Cu_3O_{7-\delta}$ thin films
Ch. Jooss, R. Warthmann, A. Forkl, T. Haage, J. Zegenhagen, H.-U. Habermeier and H. Kronmüller . . . 2311

c-axis transport critical current in melt textured YBCO: twin planes influence on the pinning of the flux lines
S. Sanfilippo, D. Bourgault, C. Villard, P. Gautier Picard, Th. Fournier, A. Sulpice and R. Tournier . . . 2313

Dependence of critical-state models parameters on the intergrain resistivity in granular $Y_1Ba_2Cu_3O_{7-\delta}$
A. Díaz, M.T. González, J. Maza and F. Vidal . . . 2315

Cb5. Flux dynamics, magnetoresistance in the superconducting state

Modified Arrhenius formula for the time decay of magnetic states in type II superconductors
T. Di Matteo, S. Pace, R. De Luca and M. Polichetti . . . 2317

Angle-resolved resistivity measurements in a Bi-2212 single crystal with inclined columnar defects
F. Warmont, V. Hardy, A. Ruyter, Ch. Goupil, G. Villard, J. Provost and Ch. Simon . . . 2319

Mixed-state Hall effect of $YBa_2Cu_3O_7$ thin films at high current densities
W. Liebich, P. Gehringer, W. Göb, W. Lang, K. Bierleutgeb and D. Bäuerle . . . 2321

RF response of along-plane vortices in $Bi_2Sr_2CaCu_2O_y$
H. Haneda, T. Ishiguro, S. Watauchi, J. Shimoyama and K. Kishio . . . 2323

A general materials equation for type-II superconductors
D.-L. Yin, K.-X. Chen, J. Chen, C.-Y. Li, J.-L. Zhang and H. Jiang . . . 2325

Flux motion in underdoped $Y_{1-x}Pr_xBa_2Cu_3O_{7-\delta}$ single crystals
T. Stein, G.A. Levin, C.N. Jiang, C.C. Almasan, D.A. Gajewski, S.H. Han and M.B. Maple . . . 2327

Frequency dependence of the higher harmonics susceptibilities of hydrogen loaded and unloaded melted YBCO samples
P. Tripodi, D. Di Gioacchino, F. Celani, A. Spallone and D. Shi . . . 2329

Vortex lattice melting and dimensionality crossover in $Bi_2Sr_2CaCu_2O_{8+x}$
S.H. Han, Y. Zhao, G.D. Gu, G.J. Russell, Y.Y. He and N. Koshizuka . . . 2331

Anisotropic flux creep for $YBa_2Cu_3O_7$ in millisecond time window
S.Y. Ding, C. Ren, Y. Ge, F.Y. Lin, L. Qiu, X.X. Yao, Y.X. Fu and C.B. Cai . . . 2333

Magnetic flux noise in Nb thin film
S.H. Chun, W. Song, G.H. Koh, H.C. Kim and Z.G. Khim . . . 2335

Anisotropy of magnetoresistance of cylindrically textured YBCO film on YSZ cylinder
S. Oh, D. Goo, S. Oh and D. Youm . . . 2337

Effect of reversible fluxoid motion in superconducting Bi-2223 multifilamentary wire
E.S. Otabe, H. Matsuoka, M. Kiuchi, T. Matsushita, J. Fujikami and K. Sato 2339

A new result of the E–j relationship of thermally activated model of high-T_c superconductor
L. Fu, X.-J. Xu, H. Jin and Y.-H. Zhang 2341

The geometrical edge barrier in type I superconductors
H. Castro, L. Rinderer and B. Dutoit 2343

Angular dependence of the resistively-determined irreversibility line in YBCO
M. Giura, R. Fastampa, S. Sarti and E. Silva 2345

Linear AC response in the high-T_c superconductor of cylinder sample
J.F. Geng, X. Jin, W.M. Chen, K. Shen and A.M. Sun 2347

The effect of Ag concentration on the vortex pinning at the grain boundaries in the Ag/BSSCO composites
E. Mogilko and Y. Schlesinger 2349

Vortex dynamics and vortex melting in YBCO thin films studied by SQUID
W.J. Yeh, L.K. Yu and Y.Z. Zhang 2351

Numerical simulation of the flux dynamics of a d-wave superconductor
Z.D. Wang and Q.-H. Wang 2353

Vortex dynamics in planar pinning centers and Hall effects
Y. Mawatari and H. Obara 2355

Low angle deviation from the anisotropic effective mass model in epitaxial $YBa_2(Cu_{1-x}Zn_x)_3O_{7-\delta}$ thin films
J.-F. de Marneffe, J. Schroeder, R. Deltour, A.G.M. Jansen and P. Wyder 2357

Flux jumps in Y-123 and La-214
M. Xu, K. Zhang, H.M. Jaeger, D.G. Hinks, G.W. Crabtree, K.C. Goretta, Y. Zhao and C.H. Choi 2359

Formation of multiple-quanta vortices at circular defects in a superconducting thin film
J.C. Wei, L. Horng and T.J. Yang 2361

High current driven vortex instability in $YBa_2Cu_3O_{7-\delta}$ thin films – analysis in terms of self organized criticality
Z.L. Xiao and P. Ziemann 2363

Superconducting properties in bulk $(Bi_{1.6}Pb_{0.4})(Sr_{1.8}Ba_{0.2})Ca_2Cu_3O_y$ system
A.V. Pop, D. Ciurchea, Gh. Ilonca, A. Harabor and R. Deltour 2365

Thermally activated flux creep velocity in $Hg_{0.69}Pb_{0.31}Ba_2Ca_2Cu_3O_{8+\delta}$
C.C. Lam, X. Jin, K.C. Hung and H.M. Shao 2367

Zero-field flux noise in granular $Bi_2Sr_2CaCu_2O_8$
J. Magnusson, P. Nordblad and P. Svedlindh 2369

Vortex dynamics in association with high T_c superconductivity
K. Bark and S.-H.S. Salk 2371

Activation energies for depinning of vortices from intrinsic and extrinsic pinning centers in high T_c superconductors
P. Vašek, I. Janeček and V. Plecháček 2373

Cb6. AC and vibrational measurements

Josephson plasma resonance in the mixed state of $Bi_2Sr_2CaCu_2O_y$ containing columnar defects
T. Hanaguri, Y. Tsuchiya, S. Sakamoto, A. Maeda and D.G. Steel 2375

Homogeneity measurements of transition temperature and critical current density of large area HTSC thin films
X.H. Jiang, Z.X. Zhao, H.C. Li, R.L. Wang and Y.L. Zhou 2377

Temperature dependence of the surface barrier on $YBa_2Cu_3O_{7-\delta}$ grains detected by AC susceptibility measurements
M. Polichetti and F. Gömöry 2379

Amplitude dependence of AC response in high T_c superconductors
K.X. Chen, G. Lu, L.X. Xue, W.P. Bai, Y.D. Dai, C.D. Wei, K. Wu, Q.R. Feng, H.T. Ren, L. Xiao and D.L. Yin 2381

Sample processing pressure dependence of weak links in polycrystalline $YBa_2Cu_3O_7$
J.T. Wang, C.L. Lin and T.P. Chen 2383

Magnetic response of $YBa_2Cu_3O_{7-\delta}$ thin film in a perpendicular magnetic field
M.J. Qin, X.N. Xu, A.M. Sun, X.X. Yao, J.P. Zuo and S.Z. Yang 2385

The vortex phase diagram obtained by vibrating-reed technique in Hg-system superconductors
W. Tian, J.-S. Zhu, H.-M. Shao, J. Li and Y.-N. Wang 2387

Harmonic ac susceptibilities for textured $YBa_2Cu_3O_7$
Y. Ge, S.Y. Ding, Q. Ding, C. Ren, L. Qiu, X.X. Yao, Y.X. Fu and C.B. Cai 2389
Mechanical measurement for phase transition of vortex at low field in high-T_c cuprates
W. Tian, J.-S. Zhu, H.-M. Shao, Q.-M. Zhang and Y.-N. Wang 2391
Glassy exponent of melt textured YBCO extracted from magnetic hysteresis loops with various field sweep rates
A.M. Sun, X.N. Xu, Y. Yu, Z.Y. Zeng, X. Jin, X.X. Yao and Z.G. Fan 2393
A study of analogy between the vortex glass and spin glass
C. Ren, S.Y. Ding, Y. Ge, L. Qiu, X.X. Yao, Y.X. Fu and C.B. Cai 2395
Flux creep determined ac magnetic response in type-II superconductors
Z.Y. Zeng, X.N. Xu, A.M. Sun, S.Y. Ding and X.X. Yao 2397

Cc. Josephson effects and networks

Cc1. Physical properties of single Josephson junctions

Saturation of $1/f$ noise level of superconductor–constrictions–superconductor mesoscopic devices
A. Saito, K. Hamasaki, S. Anezaki and Z. Wang 2399
Self-pumping mixing phenomenon in YBCO Josephson junction
H. Shimakage, Y. Uzawa and Z. Wang 2401
The role of pair symmetry and intrinsic surface depression of the order parameter on the critical current of high-T_c SIS and SIS′ Josephson junctions
R.S. Gonnelli, G.A. Ummarino, C. Bravi and V.A. Stepanov 2403
Quasiparticle tunneling in $d/I/d$ junctions
S. Kashiwaya, Y. Tanaka, M. Koyanagi, S. Ueno and K. Kajimura 2405
Nonlinear voltage response due to thermal noise of Josephson junction
D. Yin, J. Chen, S.G. Wang, Y.D. Dai, S.Z. Wang, G.C. Xiong, K.X. Chen, S. Luo and Y.S. He 2407
The supercurrent spatial distribution of grain boundary junctions
K. Xiu, Q. Lu, S. Wang, Y. Dai, G. Xiong, G. Lian, R. Nie, S. Wang and Z. Gan 2409
Simplified model for current–voltage characteristics of a superconductor–normal metal–superconductor junctions
L.A.A. Pereira and R. Nicolsky 2411
Epitaxial deposition and performance of the high T_c biepitaxial grain boundary junctions
L.F. Chen, J. Yang, H.W. Gu, P. Ma, G.S. Yuan, S.A. Chang, L. Li, L. Chen and T. Yang 2413
Properties of normal metal shunted YBCO bi-crystal junctions
L. Hao, J. Gallop, R. Reed, J. Macfarlane and E. Romans 2415
Bloch wave oscillations in the mesoscopic Josephson junction
B. Chen, Q. Zhang, Y. Ren, J. Sha and Z. Jiao 2417
Grain boundary junctions with Ag-doped $YBa_2Cu_3O_{7-x}$ epitaxial thin films
G. Bolaños, G. den Ouden, M. Chacón, W. Lopera, M.E. Gómez, A. Pulzara, J. Heiras and P. Prieto 2419

Cc2. Josephson properties in layered structures

Microwave absorption of longitudinal Josephson plasma in cuprate high T_c superconductors
M. Tachiki, S. Takahashi and K. Kadowaki 2421
Direct observation of the Nambu–Goldstone (NG) mode in $Bi_2Sr_2CaCu_2O_8$ by means of microwave excitation technique
K. Kadowaki, I. Kakeya, K. Kindo and T. Mochiku 2423
Interference of Josephson plasma waves in high-T_c cuprate films
S. Takahashi and M. Tachiki 2425
Josephson plasma resonance in $Bi_2Sr_2CaCu_2O_{8+y}$ with columnar defects
T. Shibauchi, M. Sato, S. Ooi, T. Tamegai and M. Konczykowski 2427
Collective electromagnetic wave excitation in $Bi_2Sr_2CaCu_2O_{8+\delta}$ in magnetic field nearly parallel to the CuO_2-planes
M.B. Gaifullin, Y. Matsuda, K. Kadowaki, T. Mochiku and K.-i. Kumagai 2429
C-axis tunneling between adjacent superconducting CuO_2 bi- or trilayers
K. Schlenga, G. Hechtfischer, W. Walkenhorst, H.L. Johnson, F. Forsthofer, Ch. Helm, Ch. Preis, J. Keller, R. Kleiner and P. Müller 2431
Intrinsic Josephson junctions of $Tl_2Ba_2CaCu_2O_8$ thin films
S.L. Yan, L. Fang, M.S. Si, J. Wang, J.M. Hao and Q.X. Song 2433

Supercurrents in *c*-axis $Pb/Bi_2Sr_2CaCu_2O_8$ tunnel junctions
R. Kleiner, M. Mößle, W. Walkenhorst, G. Hechtfischer, K. Schlenga and P. Müller 2435
Glass transition related Josephson junction network model for Bi:2212 single crystal
S. Luo, Y. He and C.E. Gough 2437
Resistive state of intrinsic Josephson junctions in $Bi_2Sr_2CaCu_2O_{8+\delta}$
J. Takeya, S. Akita, S. Watauchi, J. Shimoyama and K. Kishio 2439

Cc3. Proximity effects

Quantum mesoscopic conductance oscillations of the normal metal in SNS junctions
H.X. Tang and Z.D. Wang 2441
Scanning tunneling spectroscopy on $YBa_2Cu_3O_y/PrBa_2Cu_3O_y/YBa_2Cu_3O_y$ sandwiched junctions and $YBa_2Cu_3O_y/Y_{0.5}Pr_{0.5}Ba_2Cu_3O_y$ superlattices
H.C. Yang, H.W. Yu, C.C. Kuo, H.E. Horng and S.Y. Yang 2443
Tunneling studies of Nb/Al superconducting proximity-effect bilayers
S.P. Zhao, H. Du, R.F. Wang, G.H. Chen, F.Z. Xu and Q.S. Yang 2445

Cc4. Josephson and superconducting networks

The effect of nonmagnetic and magnetic impurities in the normal metal layer – N of network of weak S–N–S junctions in composites on high temperature superconductors basis
M.I. Petrov, D.A. Balaev, S.V. Ospishchev, B.P. Khrustalev and K.S. Aleksandrov 2447
Transport properties of composites high temperature superconductor + semiconductor with different carrier concentration
M.I. Petrov, D.A. Balaev, K.A. Shaihutdinov, B.P. Khrustalev and K.S. Aleksandrov 2449
The observation of non-integral Shapiro steps of a high-T_c superconducting dc SQUID
J.D. Chern, H.C. Yang, J.H. Lu and H.E. Horng 2451
Thermally activated phase slippage in composites HTSC + CuO
M.I. Petrov, D.A. Balaev, K.A. Shaihutdinov, B.P. Khrustalev and K.S. Aleksandrov 2453

Section D. APPLICATIONS

D1. General overview of applications (no contributions)

D2. Fabrication and properties of Josephson junctions

Microstructures and reproducibility for several high-T_c Josephson junctions
J.G. Wen and N. Koshizuka 2455
Properties of $Y_1Ba_2Cu_3O_{7-\delta}$ and $Ba_{1-x}K_xBiO_3$ step-edge grain boundary junctions
J. Zhou, Z.M. Ji, X.B. Wang, L. Kang, H.B. Wang, J.Y. Zhang, S.Z. Yang, W.W. Xu and P.H. Wu 2457
Current–voltage characteristics of $YBa_2Cu_3O_7$ step-edge junctions on MgO substrates
Y.J. Feng, Y.Q. Shen, J. Mygind, N.F. Pedersen and P.H. Wu 2459
A “lift-off” technique for $YBa_2Cu_3O_7$ step-edge Josephson junctions
J.C. Nie, L. Chen, L. Li, T. Yang, M.Q. Huang, P.J. Wu, G.R. Liu and Z.X. Zhao 2461
Fabrication and properties of $YBa_2Cu_3O_{7-y}/PrGaO_3/YBa_2Cu_3O_{7-y}$ tunnel junctions
Y.S. Jiang, X.Y. Cai, K. Usami, T. Kobayashi and T. Goto 2463
Epitaxial growth and crystal structures of NbN/AlN/NbN trilayers fabricated at ambient substrate temperature
Z. Wang, A. Kawakami and Y. Uzawa 2465
Josephson junction arrays on Si membrane using focused electron beam irradiation
S.-J. Kim, K. Nakajima, J. Chen, H. Myoren, T. Yamashita and M. Esashi 2467
Fabrication of capacitively shunted high T_c Josephson junctions
S.-G. Lee, G.Y. Sung, Y. Huh and J.D. Suh 2469
Fabrication and character of reproducible high quality steps and step-edge junctions on STO substrates
H.-Y. Zhai, G.-H. Chen, F.-Z. Xu, J. Wang, R.-F. Wang and Q.-S. Yang 2471
TEM study of microstructure of $YBa_2Cu_3O_y$ bicrystal grain boundary junctions grown on Y_2O_3 stabilized ZrO_2 bicrystal
D.-P. Yu, X.-D. Hu, Y.-Z. Wang, Y.-D. Dai, S.-G. Wang, L.-P. You, Z.-Q. Liu and Z. Zhang 2473

Properties of doped-YBCO bicrystal grain-boundary junctions for Josephson field effect transistor
G.Y. Sung, J.D. Suh and S.-G. Lee 2475

D3. SQUIDs and magnetometry

Sensitive dc-SQUIDs and magnetometers using step-edge junctions
J.C. Nie, L. Chen, T. Yang, M.Q. Huang, P.J. Wu, G.R. Liu and L. Li 2477

Ditch-typed step-edge grain boundary junction and rf SQUID
P. Ma, S.Z. Wang, R.F. Wang, S.G. Wang, J.Z. Li, R.J. Nie, X.D. Hu and Y.D. Dai 2479

First-order gradiometer of high T_c rf SQUID
D. He and Y. Zhang 2481

Influence of high frequency cut-off on application of high T_c SQUID in transient electromagnetic method
Y.D. Dai, X.L. Cai, S.Z. Mei, S.Z. Wang and S.G. Wang 2483

QRNT: a primary thermometer based on an HTS SQUID
J. Gallop, L. Hao and R. Reed 2485

Modified hairpin $\lambda/2$ resonator for high T_c SQUID
X.L. Cai, S.G. Wang, D.F. He, K. Xiu and Y.D. Dai 2487

High-T_c SQUID series arrays using step edge junctions
T. Yang, J.C. Nie, M.Q. Huang, P.J. Wu, L. Chen and Z.X. Zhao 2489

D4. Josephson circuitry

Heterodyne mixer using superconductor–normal metal–superconductor junction
Y.A. Gorelov, L.A.A. Pereira, A.M. Luiz and R. Nicolsky 2491

Operation of an HTS SFQ logic circuit
J. Kang 2493

D5. Superconductive transistors and hybrids

Proposal of coplanar-type high-T_c superconducting field-effect devices
S. Suzuki and S. Oda 2495

New type of superconducting field-effect transistor
A.V. Okomel'kov 2497

Intrinsic model for high T_c superconducting field effect transistor operating in the weak inversion region
Q.Y. Cai, J.F. Jiang, B. Shen and M.Z. Tong 2499

Electrical properties of a $Ba_{1-x}K_xBiO_3$/Nb-doped $SrTiO_3$ Schottky junction
S. Suzuki, T. Yamamoto, H. Suzuki, K. Kawaguchi, K. Takahashi and Y. Yoshisato 2501

Multi-terminal Abrikosov vortex flow structures at 77 K
A. Davidson and N.F. Pedersen 2503

Feasibility of a new type of single electron tunneling transistor (SETT) made of high T_c superconductor
J.F. Jiang, Q.Y. Cai, B. Shen and M.Z. Tong 2505

Numerical simulation of the HTSC-single electron transistor amplifier
B. Shen, J.-F. Jiang and Q.-Y. Cai 2507

A possible new device: the semiconductor quantum well–superconductor junction
G. Ma and X. Dai 2509

D6. Passive and active microwave devices

Comparative study of miniaturised HTS microstrip H-shaped antennas with and without enhanced capacitance
Z. Wu, D. Lacey, G. Drossos, L.E. Davis, T.W. Button and P. Smith 2511

Power dependence of radiation efficiency of circularly polarized $YBa_2Cu_3O_7$ microstrip antennas
M.H. Zhu, B.S. Cao, X.X. Zhang, N. Li, Y.J. Wang, H.J. Yuan, Y. Chen, Y.G. Zhao, D.J. Dong, B.L. Gu, Y.L. Zhou, Z.H. Chen, H.B. Lu, X.P. Wang and T.J. Liu 2513

Design and performance of a 4.5 GHz circularly polarized $YBa_2Cu_3O_7$ microstrip antenna
M.H. Zhu, B.S. Cao, X.X. Zhang, W.H. Li, H.J. Yuan, Y.J. Wang, L.W. Zhang, D.J. Dong, M.L. Liu, D.F. Cui, M. He, Y.L. Zhou and T.J. Liu 2515

Research and design of high T_c superconducting microstrip antenna array
X.X. Zhang, M.H. Zhu, Y.J. Wang, N. Li, B.S. Cao, D.J. Dong, Y.G. Zhao, L.W. Zhang, A.S. He, X.P. Wang and T.J. Liu . 2517

YBaCuO junctions applied to microwave phase shifters
K. Suzuki, S. Tokunaga and Y. Enomoto 2519

Performance of a surface acoustic wave filter with $YBa_2Cu_3O_{7-\delta}$ superconducting electrodes
Z.H. Gong, A. Rønnekleiv and J.K. Grepstad 2521

Microwave bandpass filters with closed loop resonators and coupled lines using YBCO thin films
Y.S. Ha, S.K. Han, K.Y. Kang, S.S. Lee, D. Ahn and Y.W. Park 2523

High-temperature superconducting microwave delay lines
D. Wu, H. Wang, L. Hu, C. Zhang, B. Yang and X. Wang 2525

Demonstration of HTS microwave sub-systems with a pulse tube cryocooler
H. Li, W. Zhu, J. Cai, P. Zhang, Y. Zhou, A. He, Z. Gong, Y. He, P.A. Smith, T.W. Button, J.E. Holmes, M.J. Lancaster, A. Porch, B. Avenhaus, P. Woodall, F. Wellhöfer, C.E. Gough, G. Kaiser, P. Seidel and M. Thürk 2527

Kinetic inductance of coupled superconducting microstrip lines
J.F. Kang, R.Q. Han, G.C. Xiong, X.Y. Liu and Y.Y. Wang 2529

Investigation of low phase noise oscillator of high-T_c superconductors
K. Yang, Q.-s. Zhang, Z.-x. Luo, J. Lu, Z.-x. Tang, Y.-l. Zhou, H.-b. Nu, D.-f. Cui and X.-p. Wang 2531

Measurements of surface resistance for Nb thin films in millimeter wavelength region using resonator coupled Josephson junctions
A. Kawakami and Z. Wang 2533

A quantum statistics approach for characterizing superconductors
F. Abbas, J.C. Gallop and L. Hao 2535

Measurement system for accurate characterization of microwave surface resistance of high-T_c superconductors thin films
Z.-x. Luo, K. Yang, J. Lu, Z.-x. Tang and Q.-s. Zhang 2537

Factors affecting power handling capability of microstrip resonator
R.X. Wu, B.B. Jin, L. Kang, Y.L. Zhou, M. Qian, P.H. Wu, J. Lu, X.Y. Ren and Q.S. Zhang 2539

High power microwave circuits and the demand to HTS materials
Z.-Y. Shen, C. Wilker, P. Pang, D.W. Face, C.F. Carter III and C.M. Harrington 2541

Microwave performance of high T_c superconducting double ring resonator
S.Y. Song, C.Y. Lee, M.S. Lee, Y. Huh, I. Kim, Y.K. Park, J.M. Chung and H.S. Chang 2543

D7. Detectors

Infrared detector fabricated by YBCO Josephson junctions in series
M.Q. Huang, L. Chen, T. Yang, J.C. Nie, P.J. Wu, G.R. Liu and Z.X. Zhao 2545

A study of the superconductivity of YBCO via C^+ implantation
X.T. Liu, H.X. Lu, X.H. Liu, Q.S. Chen and S.M. Pan 2547

Terahertz responses in NbN/AlN/NbN tunnel junctions
Y. Uzawa, Z. Wang, A. Kawakami, S.C. Shi and T. Noguchi 2549

Thickness dependence of the thermoelectric voltages in $YBaCuO_{7-\delta}$ thin films on tilted substrate of $SrTiO_3$
P.X. Zhang, U. Sticher, B. Leibold and H.-U. Habermeier 2551

D8. Processing of ceramics, wires and tapes

Fabrication and microstructure of Hg-1223 tape
R.L. Meng, Y.Q. Wang, K. Lewis, C. Garcia, L. Gao, Y.Y. Xue and C.W. Chu 2553

The local variation of Ca_2PbO_4 in the core of $(Pb,Bi)_2Sr_2Ca_2Cu_3O_x$ superconducting tapes during thermal processing with different heating rates
L. Law, Y. Yang and C. Beduz 2555

Homogeneous phase distribution in Ag-doped (Pb,Bi)2223 alloy sheathed tapes
S.J. Fisher, C. Beduz, Y. Yang and M. Penny 2557

A novel non-destructive method for the determination of transverse critical current distribution in Ag sheathed PbBi2223 tapes
Y. Yang, T. Hughes, L.N. Law and C. Beduz 2559

Application of Bi-2212 superconducting whiskers for thick films
R. Funahashi, I. Matsubara, M. Konishi, S. Tanaka, K. Ueno and H. Ishikawa 2561

High-J_c $(Bi,Pb)_2Sr_2Ca_2Cu_3O_{10+x}$ tapes fabricated by electrophoretic deposition
J.K.F. Yau and C.C. Sorrell 2563

Fabrication and characterization of Y124 wires
H. Matsui, X.G. Zheng, S. Tanaka, M. Suzuki and T. Hoshino 2565

Effect of oxygen partial pressure on grain boundaries in Bi-2212/Ag tapes
H. Fujii, H. Kitaguchi, H. Kumakura and K. Togano 2567

Fabrication and properties of superconducting compacted BiPbSrCaCuO ceramics
O. Vikhreva, O. Zhdanok and V. Khrustov 2569

Transmission electron microscopic study of grain boundaries in $(Bi,Pb)_2Sr_2Ca_2Cu_3O_x$ silver-sheathed tapes
D.P. Grindatto, B. Hensel, G. Grasso, H.-U. Nissen and R. Flükiger 2571

Influence of the sinter-forging conditions on the mechanical properties of textured bulk (Bi,Pb)-2223 ceramics
V. Rouessac, G. Desgardin and M. Gomina 2573

Microstructure and electrical properties of YBaCuO tapes prepared by alternating electrodeposition and heat treatments
P. Régnier, S. Poissonnet, G. Villars and C. Louchet 2575

Properties of superconducting Bi-2223 tubular current leads
V. Plecháček and J. Hejtmánek 2577

Control of partially melted microstructure and critical current density for Bi-2212 tapes
R. Funahashi, I. Matsubara, K. Ueno and H. Ishikawa 2579

A remarkable J_c enhancement in Y124 ceramics due to Ag-addition
X.G. Zheng, H. Matsui, S. Tanaka, M. Suzuki and C. Xu 2581

Critical current density on Bi(2212) samples with a needle-like grain microstructure
O. Cabeza, O. Barca and F. Miguélez 2583

Optimization of the Bi(Pb)-2223 bulk technology
E. Pollert, V. Plecháček, J. Hejtmánek, N. Hudáková, J. Chval, Z. Jirák, K. Knížek and V. Šíma 2585

TEM study of Ag-clad Bi-based (2223) multifilamentary tapes in various stages of processing
H.K. Liu, R.K. Wang, X.H. Wang, R. Bhasale and S.X. Dou 2587

Texturing of HTSC-rods by laser-zone-melting
T. Wilke, C. Baumgarten and W. Assmus 2589

Determining factors of the critical current density of silver-sheathed Bi2223 tapes
Q.Y. Hu, H.K. Liu, H.W. Weber and S.X. Dou 2591

The extended *E–J* characteristics of flux diffusion and their role in the self-field AC losses in Ag-sheathed PbBi2223 tapes
T. Hughes, Y. Yang, D.M. Spiller, S. Fisher and C. Beduz 2593

Improvement of transport properties in Bi-based superconducting tapes through the modification of precursor powder and heat treatment
Y.C. Guo, J. Horvat, H.K. Liu and S.X. Dou 2595

Microstructure and electromagnetic properties of Bi2223/Ag superconducting wires and tapes prepared by the "continuous tube forming/filling (CTFF)" technique
Y.C. Guo, H.K. Liu, S.X. Dou and E.W. Collings 2597

Microstructure and mechanical properties of MTG YBCO
L.F. Li, Z. Zhang, D. Jin, Y.Y. Li and S. Meriani 2599

The role of Bi-2212 in Bi-2223 Ag-sheathed tapes fabricated by a controlled melt process
L. Hua, Y.Z. Wang, C. Zhang, Z.Q. Yang, G.W. Qiao, Y.C. Chuang, J.M. Yoo, H.S. Chung and H.D. Kim 2601

AC losses of Bi2223/Ag tapes prepared by different process
J.-X. Wang, Z.-Z. Duan, C.-W. Yang, J.-N. Deng, P.-X. Zhang, Y.-F. He, X.-Z. Wu and L. Zhou 2603

An optimum heating program for fabricating multifilamentary Bi-2223/Ag tapes
Z.Z. Duan, X.Z. Wu, H.L. Zheng, S.H. Wu, S.G. Xiong, X.K. Teng, P.X. Zhang and L. Zhou 2605

Texture hardening in PIT method and its effect on texture formation during annealing
C. Zhang, G.-W. Qiao, Y.-Z. Wang, H. Tang, Z.-Q. Yang, X.-D. Su and J.-R. Fang 2607

Effect of precursor and annealing on the quality of $Bi_2Sr_2CaCu_2O_x$/Ag thick films
Y.G. Zhao, L. Du, C.Z. Jin, P.C. Song, A.S. He, S.Q. Guo, D.J. Dong, L.W. Zhang, B.S. Cao and B.L. Gu 2609

Fabrication of cold-rolled Ag-sheathed Bi(2223) tapes by a partial-melting process
C.S. Li, Z.P. Xi, P.X. Zhang, X.Z. Wu and L. Zhou 2611

Pseudobinary phase diagrams of Bi cuprates
J. Hauck, S. Chernyaev, J.L. MacManus-Driscoll and K. Mika . . . 2613
First critical current round robin test on the Ag-sheathed 2223 BSCCO tape at China
C.Y. Hua, Z.Z. Duan, S.Q. Guo, L.Z. Cao, C.Y. Wang, S.Y. Ding, Y. Sun, Q. Zhou and Q.S. Yang . . . 2615
Comparison of fully-reacted and partly-reacted $Bi_2Sr_2CaCu_2O_y$ precursor powders
S.-L. Huang, D. Dew-Hughes and K. Fossheim . . . 2617
To enhance J_c of Bi-2223 Ag-sheathed superconducting tapes by improving grain alignment with magnetic field
Y. Ma and Z. Wang . . . 2619
Two-peak structure of AC susceptibility of $Bi_2Sr_2CaCu_2O_x$ tapes
P.N. Mikheenko, M. Ionescu, J. Horvat and S.X. Dou . . . 2621

D9. Strong current applications

Abrupt S–N transition of melt textured YBCO at 77 K. Application to current limitation and transport of high currents
L. Porcar, D. Bourgault, J.G. Noudem, X. Chaud, O. Belmont, P. Tixador, J.M. Barbut, M. Barrault and R. Tournier . . . 2623
High pulsed current and voltage measurements on sintered BiPbSrCaCuO ceramics superconductors
J.G. Noudem, L. Porcar, O. Belmont, D. Bourgault, J.M. Barbut, P. Tixador, M. Barrault and R. Tournier . . . 2625
An electronic resonant circuit with a high T_c superconducting inductor for high voltage generation
J.X. Jin, S.X. Dou, C. Grantham and H.K. Liu . . . 2627
Bi-2223/Ag HTS coil magnetic field properties for magnet and bias winding
J.X. Jin, C. Grantham, H.K. Liu and S.X. Dou . . . 2629
Numerical evaluation of levitation force of a stack with high T_c superconducting thin films
M. Tsuchimoto, K. Matsuura, N. Homma and M. Matsuda . . . 2631
Studies on combination for high-T_c superconducting permanent magnets
M. Qiu, S. Han, L. Lin, H. Ren, L. Xiao and Y. Jiao . . . 2633
Fabrication and characterization of HTS (Bi:2223) shielding cylinders
A. He, P. Hua, J. Zhang, X. Fu and Y. Zhou . . . 2635
Research on enhanced cooling of superconducting AC magnet
T. Guo, T. Zhu and Y. Zhou . . . 2637
Fabrication and characterisation of superconducting rings for fault current limiter application
M. Chen, T. Baumann, P. Unternährer and W. Paul . . . 2639
Superconducting lossless resistor (LLR) and LLR-based fault current limiter
J.-y. Chen, Z. Chen and Q. Lu . . . 2641
Operating principle of a high T_c superconducting saturable magnetic core fault current limiter
J.X. Jin, S.X. Dou, C. Grantham, D. Sutanto and H.K. Liu . . . 2643
Preparation of MTG Y(Ho)BCO large bulk samples with Y_2O_3 excess for cryomagnetic applications
M. Wu, T. Strasser, D. Litzkendorf, K. Fischer, W. Gawalek and P. Görnert . . . 2645
AC losses of Ag-doped YBCO superconductors
C.-W. Yang, Y.-Q. Shan, J.-N. Deng, J.-X. Wang, Y.-F. He and Z.-G. Fan . . . 2647
Static and dynamic stability of levitated ball above ring in MTG HTS-magnet systems
E.V. Postrekhin, L.W. Zhou, G. Liu, C.B. Cai, S.M. Gong and Y.X. Fu . . . 2649
The effect of processing parameters on property of Bi(2223) current lead
X. Fu, Y. Zhou, W. Lin, P. Hua, R. Zeng, X. Wang, S. Chang, N. Song, P. Ye, G. Yuan and L. Lin . . . 2651
Influence of superconductor properties on magnetic levitation force
C. Navau and A. Sanchez . . . 2653

List of Contributors . . . 2655
Analytic Subject Index . . . 2675

Plenary Talks and Tutorial Lectures

ELSEVIER

Physica C 282–287 (1997) 3

PHYSICA C

Extended Abstract

Recent achievements in high-temperature superconductivity

K. A. Müller

Physics Institute, University of Zürich, CH--8057 Zürich, Switzerland, and
IBM Research Division, Zürich Research Laboratory, CH--8803 Rüschlikon, Switzerland.

Since the Congress in Grenoble, substantial results in the field of copper-oxide superconductors have been obtained in several areas. A selection of them have been reviewed regarding the fundamental aspects, the penetration of magnetic fields, and the applications.

Into the first category belong the increasing experimental signatures for the existence of Jahn-Teller type polarons, comprising structural, charge, and spin properties including their dynamics. The existence of two types of carriers in the CuO_2 planes is indicated by short-time optical and special magnetic resonance experiments. A dominance of *d*-wave symmetry of the condensate results from tunneling data; however, other data indicate that also an *s*-wave component is present.

In the second category, the melting of vortex lattices and the vortex entanglement leading to the determination of the existing phase diagrams have been obtained. A very rewarding development in the high-current area is the stationary operation of (i) a current limiter in a hydroelectric power station and (ii) that of a transformer for a town electric utility.

0921-4534/97/$17.00 Elsevier Science B.V.
PII S0921-4534(97)00191-3

ELSEVIER

Physica C 282–287 (1997) 4–11

PHYSICA C

Phase-sensitive tests of pairing symmetry in cuprate superconductors

C.C. Tsuei and J.R. Kirtley [a]

[a]IBM T.J. Watson Research Center, Yorktown Heights, NY 10598, USA

A new class of experiments, based on the macroscopic effects of pair tunneling and flux quantization, has been developed for determining pairing symmetry. These tests make use of the half-integer flux quantum effect in superconducting loops containing controlled orientation Josephson junctions as a probe of the internal symmetry of Cooper pairs. The novel phase-sensitive techniques include SQUID interferometry, tricrystal magnetometry, and single Josephson junction modulation. When combined with indirect evidence from experiments such as ARPES, penetration depth, NMR, and Raman measurements, the results of these phase sensitive experiments unambiguously establish the singlet $d_{x^2-y^2}$ pairing symmetry in high-T$_c$ hole doped cuprate superconductors such as $YBa_2Cu_3O_7$, $GdBa_2Cu_3O_7$, $Bi_2Sr_2CaCu_2O_8$, and $Tl_2Ba_2CuO_6$.

1. INTRODUCTION

Ten years have elapsed since the discovery of high temperature superconductivity (HTS) in copper oxides. There is still no general consensus on the mechanism responsible for HTS, despite tremendous theoretical and experimental effort [1,2]. From the study of flux quantization, Andreev reflection, and NMR, one can conclude with confidence that HTS in copper oxides arises from singlet Cooper pairing, just as in conventional low-T$_c$ superconductors [3–5]. To find out what mediates the pairing interaction, it is crucial to first experimentally determine the pairing symmetry. Although an unambiguous test of the order parameter symmetry does not necessarily pin down a specific high-T$_c$ mechanism, it does put strong constraints on possible microscopic mechanisms of HTS, and helps to elucidate the nature of the superconducting state in this novel class of superconducting materials.

In terms of the BCS formalism, the order parameter for the superconducting state can be identified as the internal part of the pair wavefunction $\varphi(r)$, where r is the relative coordinate for the paired electrons. $\varphi(\vec{k})$, the wavevector $\vec{k}$-space representation of $\varphi(r)$, is directly related to the $\vec{k}$-dependent energy gap $\Delta(\vec{k})$ by $\varphi(\vec{k}) = \Delta(\vec{k})/E(\vec{k})$, where $E(\vec{k})$ is the quasiparticle energy.

The symmetry of the pair wavefunction in a particular superconducting material should reflect the corresponding crystal lattice symmetry, although the possibility of an order parameter with a symmetry lower than that of the crystal point group cannot be entirely neglected. According to Ginzberg-Landau theory, the pair wavefunction for a superconductor with one T$_c$ and singlet, even-parity pairing should transform like the single irreducible representation of the relevant crystal point group. Thus, in principle, all pairing states for a given superconductor can be enumerated using standard group theory [3–5]. For example, in a tetragonal single-layer cuprate superconductor such as $Tl_2Ba_2CuO_{6+\delta}$ (Tl2201), which has a point group C_{4v} for the CuO_2 square lattice planes, there are four possible distinct singlet pair states (Table 1).

Exactly which form of pairing symmetry prevails in a given superconductor is of course dictated by the underlying pairing interaction. In Table 1, the identity irreducible representation (A_{1g}) corresponds to the nodeless s-wave (isotropic or anisotropic) pairing symmetry seen in conventional low-T$_c$ superconductors. On the other hand, in certain situations strong electron correlation may favor non-s-wave, high angular momentum pair states, as demonstrated in superfluid ^{3}He (triplet p-wave pairing l=1, S=1), and possibly in some heavy Fermion superconductors such as UPt_3[6]. It is well established that a strong on-site Coulomb repulsion is present in

0921-4534/97/$17.00
PII S0921-4534(97)00193-7

Group-theoretic notation	A_{1g}	A_{2g}	B_{1g}	B_{2g}
Order Parameter basis function	constant	$xy(x^2-y^2)$	x^2-y^2	xy
Wavefunction name	s-wave	g	$d_{x^2-y^2}$	d_{xy}
Schematic representation of $\Delta(k)$ in B.Z.	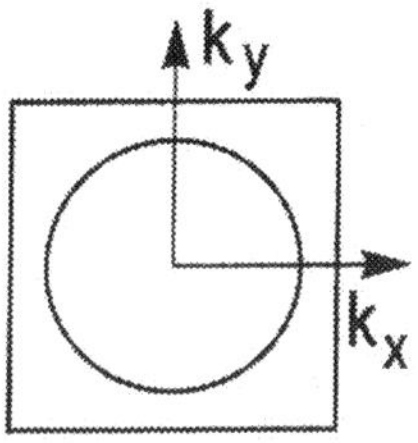		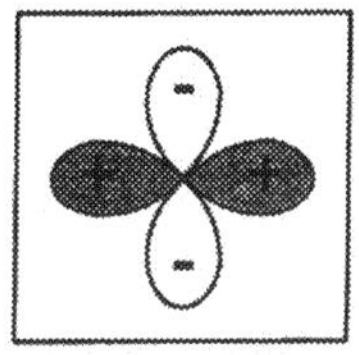	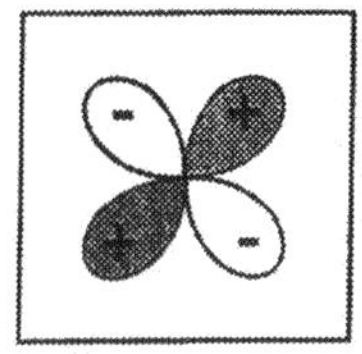

Table 1
Possible even-parity singlet pair states in a square lattice (point group C_{4v})

all cuprate superconductors. This rules out simple s-wave pairing and favors unconventional pair states such as d-wave (B_{1g}) pairing. To ascertain the existence of non s-wave pairing in cuprates, a phase-sensitive probe is required to detect the sign changes in the $\vec{k}$-dependent order parameter $\Delta(\vec{k})$. In this brief review, we describe a new class of phase-sensitive tests which are designed to probe the order parameter symmetry in cuprate superconductors. The implications of these phase-sensitive experiments for our understanding of HTS will also be discussed.

2. PHASE SENSITIVE TESTS

Before the recent advent of phase-sensitive tests of pairing symmetry in copper-oxide superconductors, the question of whether there is a node in $\Delta(\vec{k})$ on the Fermi surface, or more specifically whether there is d-wave or s-wave symmetry, was a topic of great controversy. This is mainly because conventional techniques, such as quasiparticle tunneling, NMR, angle-resolved photemissions spectroscopy (ARPES), and penetration depth measurements, can only obtain information about the magnitude of the order parameter, $|\Delta(\vec{k})|$, but not its sign. Furthermore, the interpretation of experimental data is always model-dependent, which makes indirect experiments inconclusive in determining pairing symmetry [3,4,6].

To unambiguously determine the order parameter symmetry in cuprates, a new class of phase-sensitive experiments have emerged. These tests of pairing symmetry make use of the half-integer flux quantum effect in a superconducting loop containing controlled orientation Josephson junctions as a probe of the internal symmetry of Cooper pairs. The half-integer flux quantum effect occurs when there is an odd number of Josephson junctions with "negative" supercurrents (I_s) in a superconducting loop.

A negative pair tunneling current can occur if the Josephson junction is made of an unconventional superconductor with nodes in the energy gap function $\Delta(\vec{k})$. The pair tunneling current of a Josephson junction between two superconductors i and j can be expressed by:

$$I_s^{ij} = I_c^{ij}(\theta_i, \theta_j) sin\Delta_{ij} \quad (1)$$

where θ_i and θ_j are the angles of its crystallographic axes with respect to the junction interface, such as a grain boundary, between i and j. As I_s^{ij} is determined by the product of the pro-

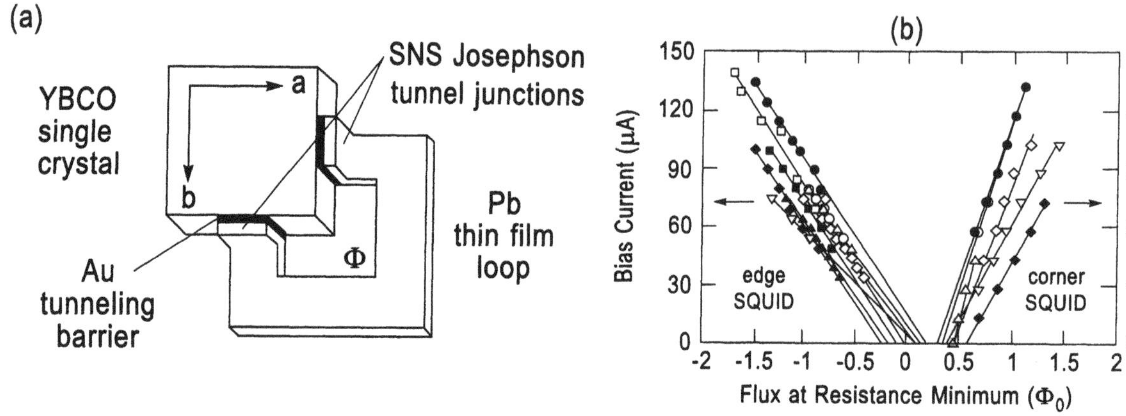

Figure 1. Single crystal YBCO SQUID geometry (a) and resistance minima vs. bias current (b) - after Ref. [9]

jections of the $\vec{k}$-dependent pair wavefunctions $\varphi(\vec{k})$ onto the normal to the junction interface, it can be negative if nodes are present in $\Delta(\vec{k})$, as is the case for $d_{x^2-y^2}$-wave superconductors[7] or the heavy Fermion system[8]. As seen from Eq. 1, a negative supercurrent can be translated as a phase shift of π at the junction. From consideration of free energy and flux quantization, the ground state of a superconducting loop with an odd number of sign changes in the circulating supercurrent (a π-loop) corresponds to a doubly degenerate spontaneous magnetization of $I_s L = \pm\Phi_0/2$, where L is the self-inductance of the loop, and $\Phi_0 = h/2e$ is the superconducting flux quantum, provided that the product of L and the minimum I_c^{ij} is much larger than Φ_0. The magnetic flux quantization of a π-loop can be described as

$$\Phi = \pm(n + 1/2)\Phi_0 \qquad n = 0, 1, 2, ... \qquad (2)$$

On the other hand, a superconducting loop with zero or an even-number of sign changes in I_s (a 0-loop) should show only standard flux quantization:

$$\Phi = \pm n\Phi_0 \qquad n = 0, 1, 2, ... \qquad (3)$$

By monitoring the presence or absence of the half-integer flux quantum effect in a variety of superconducting loop geometries and orientations, one can unambiguously detect the existence of nodes in $\Delta(k)$ via the effects arising from the sign change in pair tunneling current. The techniques used in the new phase-sensitive tests of pairing symmetry include: SQUID interferometry[9], scanning SQUID magnetometry[10,11] and Josephson junction modulation[12,13].

The first suggestion that vortices with non-integral multiples of the standard vorticity can result from an order parameter with nodes was by Volovik and Mineev[14], in a discussion of ^{3}He. Bulaevski et al.[15] were the first to suggest that superconducting rings could exhibit the half-integer flux quantum effect. They proposed that negative Josephson supercurrents could result from spin-flip scattering from magnetic impurities in the junctions. Geschkenbein, Barone, and Larkin [16] showed that half-integer flux quantization could also result from an order parameter with nodes, such as is expected for the heavy Fermion superconductors. Wohlleben and co-workers used the half-flux quantum effect to explain paramagnetic shielding effects often seen in granular samples of the cuprate superconductors, postulating spin-flip scattering effects in naturally occuring grain boundary rings in these samples[17]. Sigrist and Rice[7] argued that paramagnetic shielding effects in granular cuprate superconductors could arise instead from sign changes in an unconventional (d-wave) gap in these materials and proposed a SQUID interference experiment to test for such a sign change. The basic features and main results of order parameter symmetry experiments utilizing the half-integer flux quantum effect will be briefly described in the following:

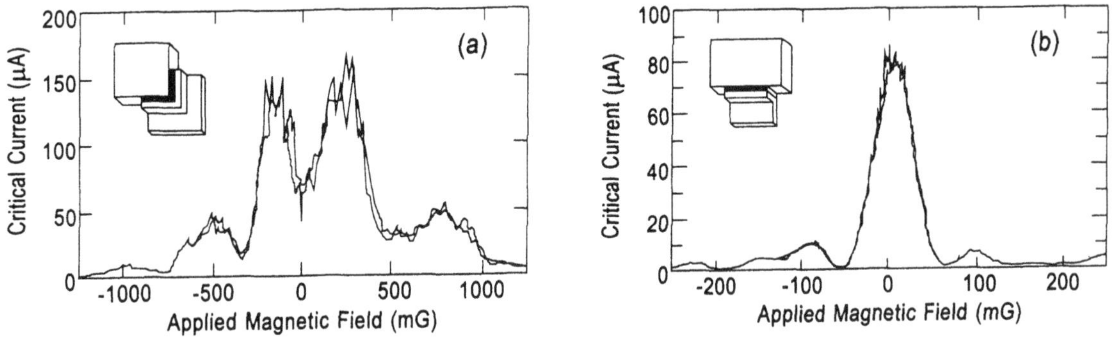

Figure 2. Magnetic field interference patterns for single crystal YBCO-Pb corner (a) and edge(b) single junctions - after Ref.[12]

3. SQUID INTERFEROMETRY

By measuring the interference effects in a YBCO-Pb dc-SQUID with the geometry proposed by Sigrist and Rice[7], Wollman et al.[9] made the first phase sensitive test of pairing symmetry in $YBa_2Cu_3O_7$ (YBCO). As shown in Fig. 1a, the SQUID consists of two SNS Josephson tunnel junctions located on the a and b faces of a single crystal of YBCO. Because of asymmetry of the critical currents of the two junctions making up the SQUID, the current used to bias the SQUID caused self-field effects. However, by extrapolating the magnetic field modulated dynamic resistance minimum of the SQUID to the zero-bias-current value, the phase shift of the order parameter $\Delta(\vec{k})$ between the orthogonal a and b directions in YBCO single crystals can be determined. The zero-bias-current data presented in Figure 1(b) for corner SQUIDs are scattered around $\Phi_0/2$, suggesting a phase-shift of π in accordance with d-wave pairing symmetry. Control experiments with SQUIDs fabricated only on one edge of the YBCO crystals showed no evidence for a sign change in the pair tunneling. Brawner and Ott[18] duplicated the SQUID interferometry experiment with Nb point contacts instead of Pb thin films. These experiments were controversial, because of concerns about the self-field effects, twinning, flux trapping, and corner demagnetization effects. By reversing the applied magnetic field and bias current used in thin-film YBCO-Pb SQUIDs, Mathai et al.[19] produced evidence for time-reversal invariant d-wave gap symmetry. This result is inconsistent with an $s + id$ mixed pair state in YBCO.

4. JOSEPHSON JUNCTION MODULATION

To address the problems of SQUID asymmetries and flux trapping in their original SQUID interference experiment, Wollman et al.[12] introduced, in 1995, the technique of single junction modulation as an alternative approach to determining the order parameter symmetry. They measured the magnetic field modulation of the critical current of YBCO-Au-Pb Josephson junctions fabricated on the a and b edges and straddling the $a - b$ corners of YBCO single crystals, and observed a dip in critical current near zero applied field in the corner junction (Fig. 2(a)). The edge junctions exhibited more conventional Fraunhofer diffraction patterns(Fig. 2(b)). These results again support d-wave pairing symmetry in YBCO. Similar conclusions were reached by Miller et al.[13] in a study of magnetic field modulation in YBCO tricrystal junctions operating in the short junction limit.

5. TRICRYSTAL MAGNETOMETRY

In tricrystal experiments[10,20–22], the ground state flux quantization of a three-junction cuprate ring, measured with a scanning SQUID microscope (SSM), is used as a general, systematic

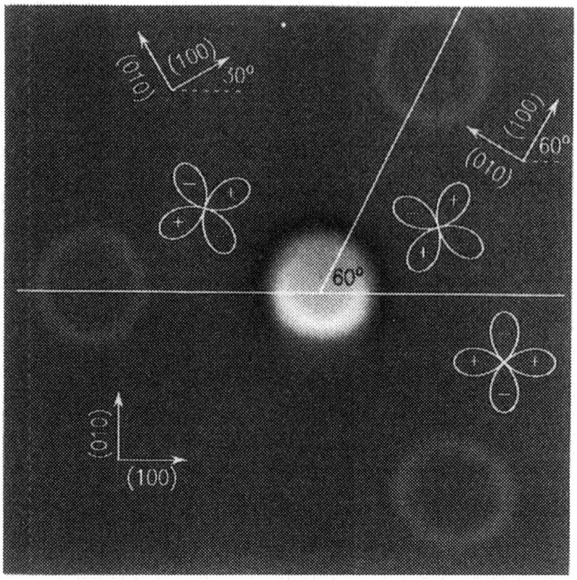

Figure 3. Scanning SQUID microscope image of four YBCO superconducting rings on a tricrystal substrate with a π-ring geometry. The solid white lines indicate the positions of the grain boundaries. The 4-leaf clovers indicate the orientations of assumed d-wave symmetry gaps aligned with the crystalline axes of the YBCO.

probe of pairing symmetry. For testing specifically the presumed $d_{x^2-y^2}$-wave pairing symmetry in cuprates, a tricrystal (100) $SrTiO_3$ substrate was designed for depositing an epitaxial film of YBCO, for example, and patterning a ring interrupted by three grain-boundary weak-link junctions. The grain misorientations and the angle between the grain-boundary planes were chosen so that a three-junction ring centered around the tricrystal point would have, by design, either an odd-number (π-ring) or an even-number (0-ring) of sign changes in the circulating supercurrent I_s (Eq. 1). In terms of d-wave pairing symmetry, the sign of I_s is determined in the perfectly smooth junction (clean) limit by the Sigrist-Rice formula[7]:

$$I_c^{ij}(\theta_i, \theta_j) \propto cos2\theta_i cos2\theta_j, \tag{4}$$

and in the opposite, maximum disorder limit[10], by:

$$I_c^{ij}(\theta_i, \theta_j) \propto cos2(\theta_i + \theta_j). \tag{5}$$

A ring of d-wave superconductor on the tricrystal geometry of Fig. 3 should show spontaneous magnetization of $\Phi_0/2$ in its ground state, regardless of whether the junction interface is in the clean or dirty limit. Also shown in Fig. 3, three rings with even-numbers of junctions (two located at the grain boundaries, one inside the grain) were fabricated as a control. These 0-rings should show only the conventional integer flux quantization. To measure directly the magnetic flux state of the rings, a high-resolution SSM was used. The design and operation of the SSM have been reported elsewhere[11]. The SSM images of the YBCO rings (taken at 4.2K and nominal zero field) show only flux of $\Phi_0/2$ threading through the 3-junction ring, while the control rings contain no flux at all. Our SSM was calibrated using 4 independent techniques[20,21]. The most accurate showed that the phase shift in the 3-junction ring was π to within $\pm 3\%$. The image shown in Fig. 3 represents the first direct observation of the half-integer flux quantum effect. This observation provides definitive evidence for d-wave pairing in the high-T_c cuprates such as YBCO. It is worthwhile to point out that, as a phase-sensitive test of pairing symmetry, the tricrystal magnetometry technique uses a scanning SQUID microscope for magnetic imaging, but since no external current is applied, it does not have the complications from self-field effects associated with the SQUID based phase-sensitive tests. As a result of the small coupling between the SSM's pickup loop and the cuprate ring, the ground state of the rings can be measured accurately and non-invasively. The small volume of the thin film rings used in the tricrystal experiments makes it unlikely that any significant flux trapping occurs in the sample. Furthermore, with our SSM, any flux trapped in the sample can be detected down to a level of $10^{-3}\Phi_0$.

We have directly observed the half-integer flux quantum effect in tricrystal YBCO disks, blanket films, and rings at 4.2K and a near-zero magnetic field, demonstrating that it is indeed a manifestation of the thermodynamic ground state arising from a π phase shift due to the existence of nodes in the energy gap[10,21,24]. We have demonstrated that the half-integer flux quantum

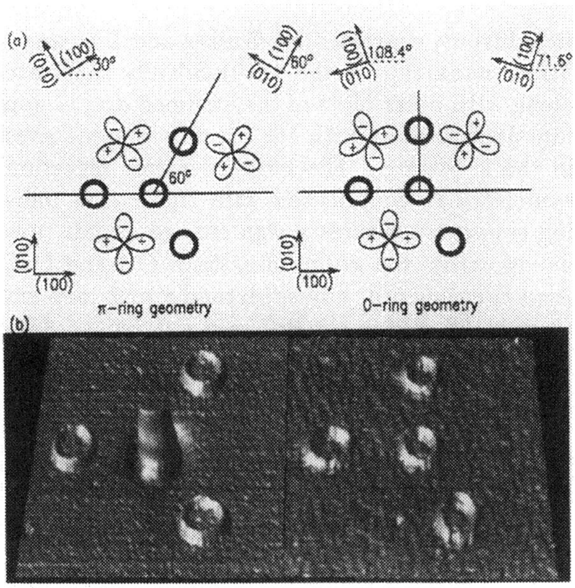

Figure 4. Tricrystal geometries (a) and scanning SQUID microscope images (b) of two YBCO tricrystal ring samples, one with a π-ring geometry (left) and a second with a 0-ring geometry, cooled in a field of less than 5mG and SSM imaged at 4.2K

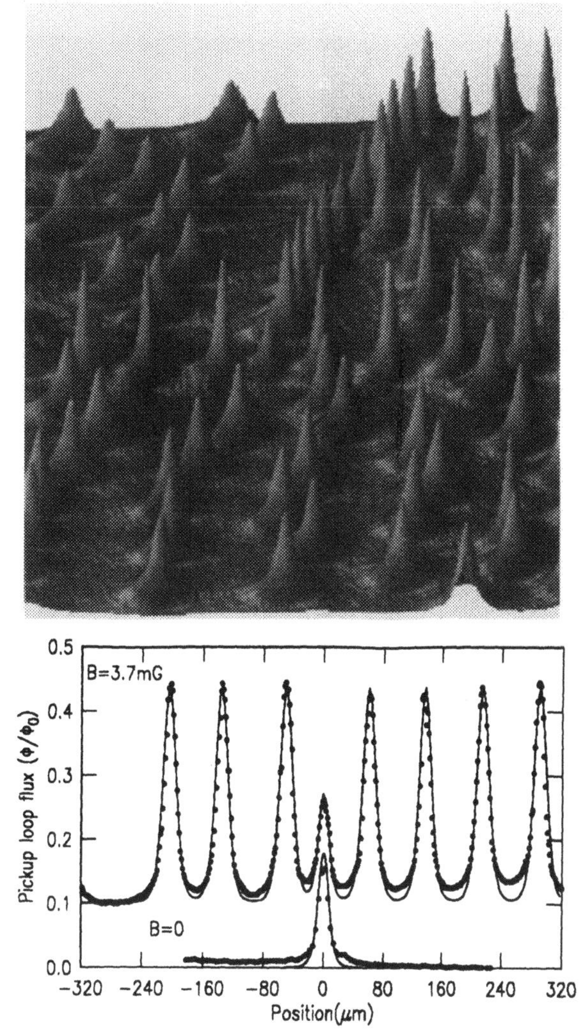

Figure 5. Scanning SQUID microscope image of a blanket coverage of Bi2212 on a π-ring geometry tricrystal substrate, cooled in a field of about 4mG and SSM imaged at 4.2K. Josephson vortices are trapped in the horizontal and diagonal grain boundaries, Abrikosov vortices are trapped in the grains away from the grain boundaries, and a half-flux quantum is spontaneously generated at the tricrystal point. A cross-section through the horizontal grain boundary (dots) is modelled well (solid line) with h/2e flux in the Josephson vortices away from the tricrystal point, and h/4e flux at the tricrystal point. An image taken after cooling in zero field shows only the half-flux quantum at the tricrystal point.

effect can be turned on and off by varying the tricrystal geometry in accordance with the d-wave formula form pairing tunneling (i.e. Eqs. (4) and (5)). As shown in Fig. 4, the presence of the spontaneously generated half-flux quantum in the experiments with the tricrystal geometrical configuration (a) but not (b) clearly rules out any symmetry independent mechanisms such as magnetic impurities at the grain boundaries as the cause of the π phase shift[20]. In addition, the extended s-wave (including g-wave) pairing, and the effects of twinning and CuO_2 bilayers on pairing symmetry have been ruled out by a series of tricrystal experiments[21,22]. The fact that the half-integer flux quantum effect has been observed in various tricrystal experiments with YBCO, Tl2201, Bi2212[23] (see Fig. 5), and GdBCO establishes the universality of the d-wave pair state in high-T_c hole-doped cuprate superconductors. No phase-sensitive test of pairing symmetry has been reported for the presumed s-

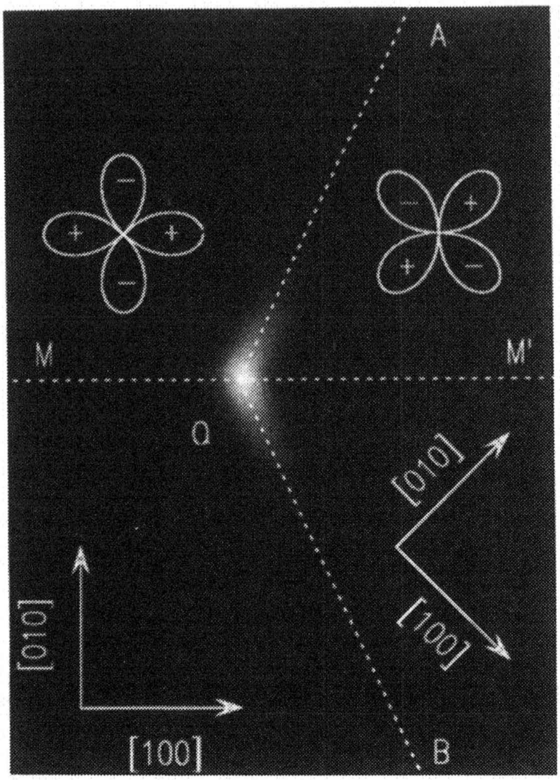

Figure 6. Scanning SQUID microscope image of a blanket coverage of Tl2201 on a π-ring geometry tetra-crystal sample, cooled in nominal zero field and SSM imaged at 4.2K.

wave electron-doped NdCeCuO system yet. The original tricrystal experiment, however, can not pinpoint the node lines ($\Delta(\vec{k})$=0) on the Fermi surface in the case of mixed $s+d$ pairing, unless there is a detailed model describing pair tunneling through a grain boundary weak-link. In this regard, the phase-sensitive test using a Pb-YBCO corner SQUID or a single Josephson junction also can not distinguish pure d-wave from mixed $s+d$ pair states as long as $d/s > 1$. This is because these interference experiments rely on a sign change in the order parameter between the a and b faces of YBCO, which is always the case if d-wave is dominant. We have recently carried out a $\pi/4$-rotated wedge tetracrystal experiment[25] to prove the existence of **pure** d-wave pairing in a cuprate superconductor based on symmetry considerations alone. Shown in Fig. 6 is the geometrical layout, suggested by Walker and Luettmer-Strathmann[26], of the (100) $SiTiO_3$ substrate along with polar plots of the assumed $d_{x^2-y^2}$ gap functions aligned with the crystallographic axes in the substrate. The built-in mirror reflection symmetry in conjunction with the $d_{x^2-y^2}$ pairing symmetry assures a sign change in pair tunneling across the grain boundaries OA and OB. As a consequence, any superconducting loop enclosing the wedge tip will give rise to the half-integer flux quantum effect. The geometrical layout in Fig. 6 is superimposed on the SSM image, taken at 4.2K and a field less than 1mG, of the Josephson vortex generated at the wedge tip in a tetragonal Tl2201 blanket film deposited on the tetracrystal substrate. A detailed modelling of the SSM data indicates that the total amount of flux threading through the wedge tip is indeed $\Phi_0/2$. Furthermore, the flux distribution for both wings of the vortex is very symmetric, as designed. Since in a tetragonal cuprate system with point group C_{4v} such as Tl2201, s-wave and d-wave pair states correspond to two distinctly different irreducible representations A_{1g} and B_{1g} respectively, an admixture of s and d pair states is not allowed. This is based on a rigorous group theoretic argument. Therefore, the observation of the half-integer flux quantum effect in this π-rotated wedge experiment represents definitive, model-independent pair tunneling evidence for pure $d_{x^2-y^2}$ pairing symmetry.

6. CONCLUDING REMARKS

Combined with the indirect evidence from experiments such as ARPES, penetration depth, NMR, and Raman measurements, the results of the phase-sensitive pairing symmetry tests unambiguously establish that the dominant order parameter symmetry in high-T_c hole-doped cuprate superconductors such as YBCO, GdBCO, Bi2212, and Tl2201 is that of the $d_{x^2-y^2}$ pair state. Based on c-axis tunneling[27] and the microwave induced step measurements in YBCO[28], a mixed $s+d$ pairing is possible, as expected from the orthorhombic crystal structure of YBCO. On the other hand, a recent $\pi/4$-rotated wedge tetracrystal phase-sensitive test in tetragonal single-layer Tl2201 demonstrates the existence of pure $d_{x^2-y^2}$-wave pairing symmetry. The settling of issues related to the universality of d-wave pairing, the effects of impurity scattering, band structure, etc., on pairing symmetry awaits more phase-sensitive experiments. Of particular interest are studies of pairing symmetry in the electron-doped NdCeCuO system and also in Tl2201, for example, as a function of oxygen content covering both the under- and over-doped regimes. The tricrystal magnetometry technique is especially well adapted for testing pairing symmetry in various superconducting materials. This phase-sensitive test should be extended to the study of order parameter symmetry in other unconventional superconductors such as the heavy Fermion systems.

The establishment of $d_{x^2-y^2}$ pairing symmetry in cuprates is important to our understanding of the origin of HTS, by providing well-defined constraints on possible models of the high-T_c mechanism. It can also help us in obtaining a better understanding of the phenomonology, particuarly in the mixed state, of this unconventional class of superconducting materials. The existence of nodes in the gap function allows low energy excitations and enhances dissipative effects. These properties should be studied with the ramifications of *d*-wave pairing in mind. The implications of *d*-wave pairing for understanding interface, grain boundary physics, Josephson junction characteristics, etc., should also be emphasized.

The authors wish to thank M. Bhushan, C.C. Chi, A. Gupta, Z.Z. Li, M.B. Ketchen, K.A. Moler, D.M. Newns, H. Raffy, Z.F. Ren, J.Z. Sun, G. Trafas, and J.H. Wang for invaluable assistance in the course of our tricrystal experiments.

REFERENCES

1. see reviews in Donald M. Ginsburg (ed.), Physical Properties of High Temperature Superconductors, vols I-IV, World Scientific, New York.
2. N.M. Plakida, "High T_c Superconductivity", Springer Verlag, Berlin, 1995.
3. J.F. Annett, N. Goldenfeld, and A.J. Leggett, J. Low Temp. Phys. 105 (1996) 473.
4. J.F. Annett, N. Goldenfeld, and S.R. Renn, in Ref. 1, Vol II, page 573.
5. D.J. Scalapino, Phys. Reports 250 (1995) 329.
6. M. Sigrist and K. Ueda, Rev. Mod. Phys. 63 (1991) 239.
7. M. Sigrist and T.M. Rice, J. Phys. Soc. Japan, 61 (1992) 4283.
8. V.B. Geshkenbein, A.I. Larkin, and A. Barone, Phys. Rev B 36 (1987) 235.
9. D.A. Wollman *et al*, Phys. Rev. Lett. 71 (1993) 2134.
10. C.C. Tsuei *et al*, Phys. Rev. Lett. 73 (1994) 593.
11. J.R. Kirtley *et al.* Appl. Phys. Lett. 66 (1995) 1138.
12. D.A. Wollman *et al.* Phys. Rev. Lett. 74 (1995) 795 .
13. J.H. Miller *et al.* Phys. Rev. Lett. 74 (1995) 2347.
14. G.E. Volovik and V.P. Mineev, JETP Lett. 24 (1976) 605.
15. L.N. Bulaevski, V.V. Kuzii, and A.A. Sobyanin, JETP Lett. 25 (1977) 290.
16. V.B. Geschkenbein, A.I. Larkin, and A. Barone, Phys. Rev. B 36 (1987) 235.
17. W. Braunisch *et al.*, Phys. Rev. Lett. 68 (1992) 1908.
18. D.A. Brawner and H.R. Ott, Phys. Rev. B 50 (1994) 6530.
19. A. Mathai *et al.* Phys. Rev. Lett. 74 (1995) 4523.
20. J.R. Kirtley *et al.* Nature 373 (1995) 225.
21. C.C. Tsuei *et al.* J. Phys. Chem. Solids 56 (1995) 1787.
22. C.C. Tsuei *et al.* Science 271 (1996) 329.
23. J.R. Kirtley *et al.* Europhys. Lett. 36 (1996) 707.
24. J.R. Kirtley *et al.* Phys. Rev. Lett. 76 (1996) 1336.
25. C.C. Tsuei, J.R. Kirtley, Z.F. Ren, J.H. Wang, H. Raffy, and Z.Z. Li (to be published in Nature, 1997).
26. M.B. Walker and J. Luettmer-Strathmann, Phys. Rev. B 54 (1996) 598.
27. A.G. Sun, D.A. Gajewski, M.B. Maple, and R.C. Dynes, Phys. Rev. Lett. 72 (1994) 2267.
28. R. Kleiner *et al.*, Phys. Rev. Lett. 76 (1996) 2161.

ELSEVIER
Physica C 282–287 (1997) 12–18
PHYSICA C

Spin gap effects on the c-axis and in-plane charge dynamics of high-T_c cuprates

S. Uchida

Department of Superconductivity and Department of Applied Physics, University of Tokyo
Hongo 7-3-1, Bunkyo-ku, Tokyo 113, Japan

The spin gap in the normal state is a phenomenon characteristic of high-T_c cuprates in the underdoped regime. The spin gap is apparently connected to the normal-state pseudogap seen in the c-axis optical conductivity and photoemission spectra, and the fundamental features of the cuprates, two-dimensional charge dynamics and reduced Drude spectral weight, result from the spin gap.

1. INTRODUCTION

It is an anomalous and unique phenomenon of high-T_c cuprates that a gap opens in the spin excitation spectrum at temperature T_s well above T_c [1,2]. It is also suggested that the decreasing magnetic susceptibility, which starts up at temperature T_0 higher than T_s, might be connected to a spin gap (formation of spin singlets). The spin gap is a universal property of the underdoped high-T_c cuprates, not restricted to bilayer systems. Recently, it has been confirmed by NMR for a single-layer $La_{2-x}Sr_xCuO_4$ (La214) [3] and a tri-layer Hg-based cuprate [4].

Closely connected to the spin gap, a pseudogap opens below temperature T^* in the angle-resolved-photoemission (ARPES) [5] and in the c-axis optical conductivity spectrum ($\sigma_c(\omega)$) [6]. The pseudogap deepens with lowering temperature and evolves, without changing its width, into the superconducting (SC) gap below T_c. The difference between pseudogap and SC gap is that the depressed spectral weight goes to a δ-function at $\omega = 0$ in $\sigma_c(\omega)$, giving rise to the so-called Josephson plasma [7], or forms a pile-up peak near the gap edge in ARPES below T_c, whereas it spreads over a wide energy range in the normal state.

It is found that associated with the pseudogap (or spin gap) the in-plane resistivity deviates from the T-linear behavior [8] and the T coefficient of the c-axis resistivity changes sign, showing semiconducting behavior at low temperatures [9]. The pseudogap and the anomalies in resistivity start at T^* which is near T_0 and become substantial at $T \sim T_s$.

In this paper, I review and summarize the effects of 'spin gap' on the c-axis and in-plane charge dynamics of the underdoped cuprates. It is illustrated based on the Zn-doping experiments that the fundamental features of the electronic state of the high-T_c cuprates result from the spin gap in the normal state.

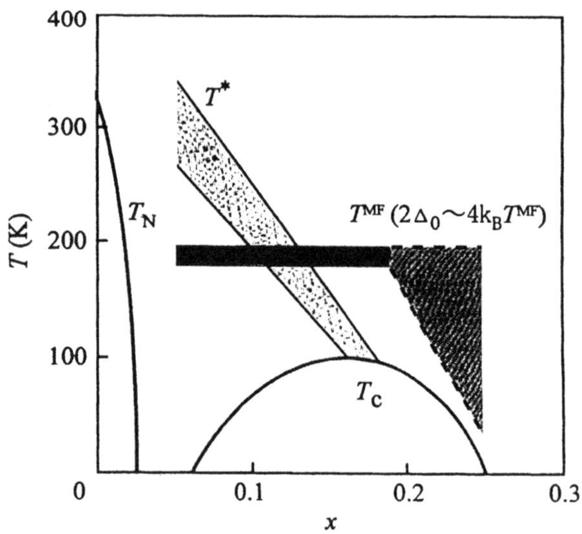

Figure 1. Doping dependences of various temperature (energy) scales of high-T_c cuprates. T_{MF} is defined by $2\Delta_0 = 4k_BT_{MF}$ with Δ_0 being the maximum of the d-wave SC gap.

PII S0921-4534(97)00195-0

2. SPIN GAP EFFECTS ON CHARGE DYNAMICS

The $\sigma_c(\omega)$ and ARPES experiments so far performed display that the width of the pseudogap and SC gap is almost independent of doping at least in the underdoped regime [10, 11]. The $\sigma_c(\omega)$ results suggest that the gap width is even material independent ($2\Delta_0 \sim$ 50-60 meV, Δ_0 is the maximum of the d-wave gap). This is in contrast to the strongly doping dependent characteristic temperatures T^* (or T_0 and T_s) and T_c (Fig. 1), indicating that T^* and T_c do not represent any energy scale, and T^* is not directly connected to the formation of preformed pairs (or the development of the amplitude of the SC order parameter).

2.1. c-axis dynamics

The electronic contribution to $\sigma_c(\omega)$ in the normal state of the underdoped cuprates is nearly ω-independent and does not form a Drude peak at $\omega = 0$, indicating incoherent nature of charge dynamics along the c-axis [12]. A pseudogap develops below T^* with σ_c depressed below $\sim$ 50 meV with lowering temperature. The semiconducting T dependence of the c-axis resistivity ρ_c is associated with the deepening of the pseudogap (Fig. 2). The connection to the spin gap is discussed in terms of interlayer charge hopping assisted by the spin fluctuations [13] or by spinon-holon recombination [14]. In this regard, the spin gap presents a barrier to c-axis charge transport[15].

A basic question that arises is whether ρ_c diverges or not if the normal state of the high-T_c cuprates lasts to $T = 0$ or whether $\sigma_c(\omega)$ is suppressed to zero without opening an SC gap. This question will be addressed by the Zn doping experiment.

2.2. In-plane dynamics

It should be emphasized that the pseudogap appears only in $\sigma_c(\omega)$, while the in-plane spectrum $\sigma_c(\omega)$ is dominated by a Drude peak at $\omega = 0$. The effect of spin gap or a change in the in-plane charge dynamics at T^* is seen in the change of the T dependence of the in-plane resistivity (ρ_{ab}). ρ_{ab} is linear in T above T^* and decreases more steeply (following $\rho_{ab} \sim T^{\alpha}$ with

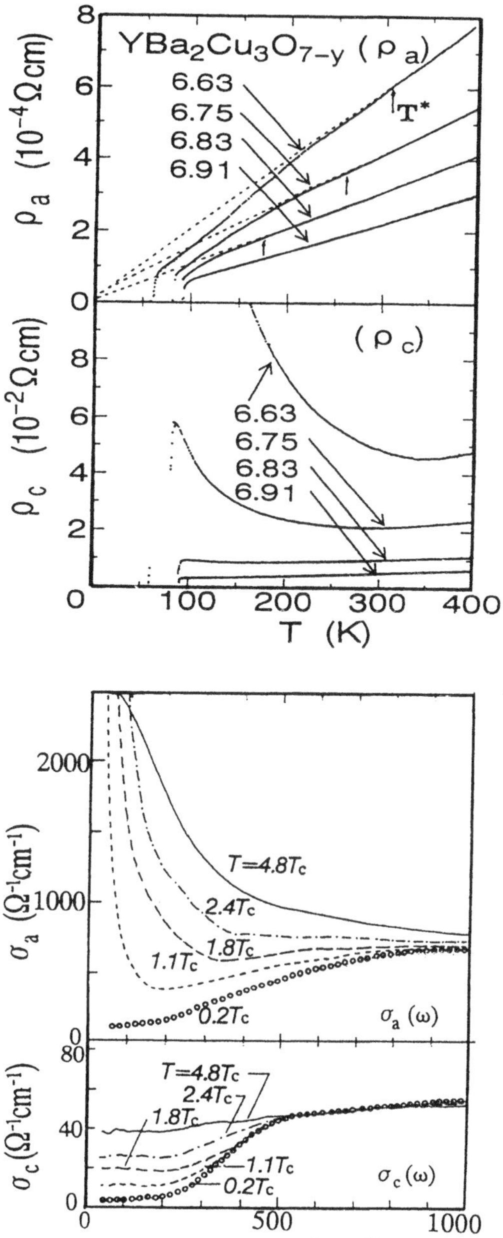

Figure 2. Temperature dependences of the in-plane and c-axis resistivity for Y123 with various dopings. In-plane and c-axis optical conductivity spectra of an underdoped cuprate at various temperatures are shown in the lower panels.

$\alpha \sim 2.5$) below T^* [8]. This change is universal for any underdoped cuprate and for any doping. As displayed in Fig. 3, all the ρ_{ab} vs T characteristics for underdoped bilayer $YBa_2Cu_3O_{7-y}$ (Y123) and trilayer $HgBa_2Ca_2Cu_3O_{8+y}$ [16] fall into a single curve by scaling T/T^* and $\rho_{ab}(T)/\rho_{ab}(T^*)$. A single-layer system La214 also shows a good scaling when the residual resistivity component is subtracted from $\rho_{ab}(T)$ [17].

The change in the T dependence of ρ_{ab} at T^* and the scaling shown in Fig. 3 suggest that the effect of the spin gap might be on the carrier scattering for the in-plane charge dynamics. Actually, the development of a pseudogap in $\sigma_c(\omega)$ is correlated with a rapid narrowing of the Drude width in $\sigma_c(\omega)$ [10] (Fig. 2). A naive interpretation of this would be that the opening of the spin gap below T^* leads to a reduction in the scattering rate ($1/\tau$) between charge carriers and spin fluctuations. It is anomalous that the same physical effect acts on the c-axis and in-plane charge transport in the opposite direction, which points to totally different charge dynamics between two directions.

A consequence of the reduction in $1/\tau$ is that the Drude component in $\sigma_c(\omega)$ is isolated from the so-called mid-IR band. Its implication is a reduction of the Drude (or coherent) spectral weight which is linked with the low carrier density as will also be illustrated by the Zn doping experiment.

3. Zn DOPING EFFECTS

Zn atoms substituted for Cu sites in the CuO_2 planes greatly reduce the SC critical temperature and act as a potential scatterer for the in-plane charge transport. It is shown that a superconductor-insulator (SI) transition takes place in the underdoped regime, when the residual sheet resistance induced by the Zn doping reaches a value near the universal two-dimensional resistance $h/4e^2$ [18]. Reflecting the totally different charge dynamics in the two directions, the Zn doping has distinct effect on the charge dynamics, from which one can get deeper insight into the electronic state of high-T_c cuprates.

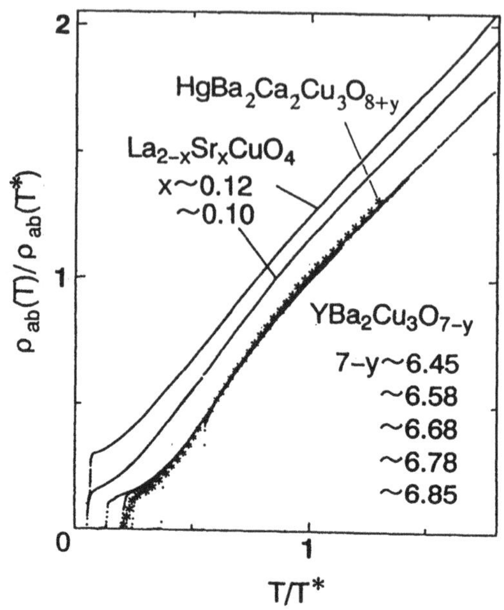

Figure 3. The scaling of the T dependence of the in-plane resistivity for various underdoped cuprates.

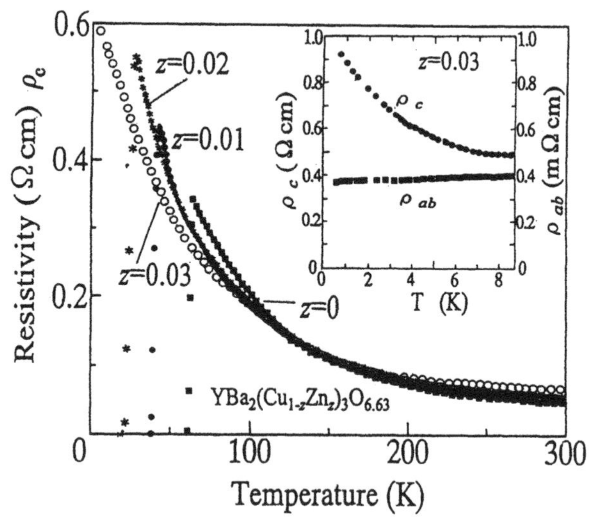

Figure 4. Temperature dependence of ρ_c for Zn-doped $YBa_2Cu_3O_{6.63}$. The inset shows the low-temperature part of ρ_c and ρ_{ab} for $z = 0.03$.

3.1. c-axis dynamics

The T dependence of ρ_c is shown in Fig.4 for a typical underdoped cuprate $YBa_2(Cu_{1-z}Zn_z)_3O_{6.63}$ with various Zn contents up to $z = 0.03$. $z = 0.03$ is close to the SI transition with T_c lower than 1K. As shown also in the inset, ρ_c continues to increase in the normal state when T_c is suppressed by Zn doping. For $z = 0.03$ the semiconducting ρ_c persists down to T=0.5K, the lowest temperature of the present experiment, and tends to diverge toward $T = 0$ whereas ρ_{ab} remains finite (metallic). Such contrasting behavior has also been observed for $Bi_2Sr_2CuO_{6+y}$ under sufficiently magnetic fields for T_c suppression [19,20].

One may claim that, as the Zn doping has a radical effect on T_c, it would have seriously altered the c-axis dynamics and the divergent ρ_c might be a consequence of this serious change. Certaintly, Zn or other entities that introduce disorder into the CuO_2 planes affect the low-energy in-plane dynamics [21]. However, from the study of the $\sigma_c(\omega)$-spectrum it turns out that Zn does not change the basic pseudogap feature in the low-energy region.

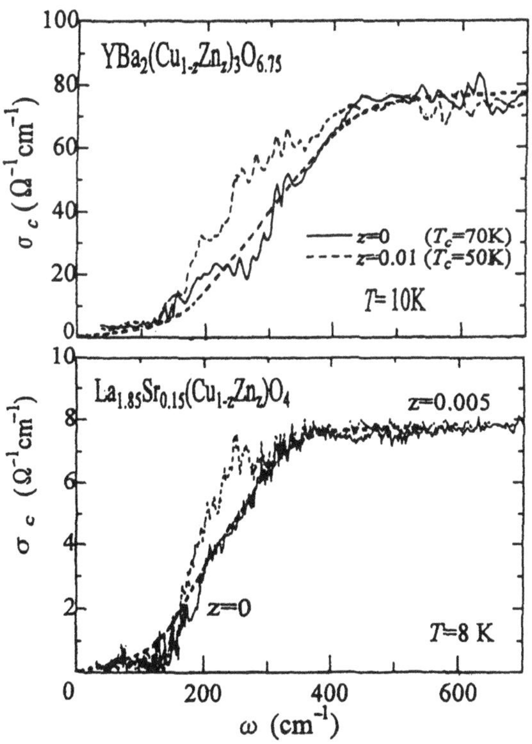

Figure 5. $\sigma_c(\omega)$ spectra with phonons removed for Zn-free and Zn-doped Y123 and La214 crystals.

The Zn doping effect on $\sigma_c(\omega)$ is shown in Fig. 5 for underdoped Y123 and La214. The onset temperature T^* for pseudogap is independent of Zn content [22], and the conductivity in the low-energy region (below $\sim$ 20meV) continues to be uniformly suppressed with decreasing temperature in the normal state (and even in the SC state). Zn induced changes in $\sigma_c(\omega)$ are visible only in the high-energy gap region. This result gives a credit to the result that the divergent ρ_c is an intrinsic property, and the contrasting behavior from the in-plane dynamics points toward the charge confinement [23].

3.2. In-plane dynamics

A primary effect of Zn doping on the in-plane dynamics is to produce a residual dc resistivity as an elastic potential scatterer. It is found that the residual resistivity ρ_0 is extremely large particularly in the underdoped regime, amounting to $\sim 100\mu\Omega$ cm at 1% Zn [18]. This value is the largest among so-far observed residual resistivity in any type of metals.

In the case of Zn impurity in the CuO_2 planes, the s-wave ($l = 0$) scattering is expected to dominate ρ_0, since Zn (Zn^{2+}) has closed shell of the d-orbitals and the relevant conduction band of the CuO_2 planes is non-degenerate. Then, the two-dimensional residual resistivity is given as $\rho_0 = 4(h/e^2)(n_i/n)\sin^2\delta_0$, where n_i is the impurity concentration, n the carrier density, and δ_0 is the s-wave phase shift [24]. The extremely large ρ_0 observed for underdoped compounds can only be fitted by putting $n = x$ (x ; the doped hole density), the lowest carrier density expected from theories, and $\delta_0 = \pi/2$ corresponding to the unitariry-limit scattering. It should be noted that, as long as $\delta_0(=\pi/2)$ is unchanged with increasing dopant concentration, ρ_0 is dependent only on the carrier density n. The experimental fact that ρ_0 for overdoped La214 ($x = 0.30$) is fitted with $\delta_0 = \pi/2$ and $n = 1 - x$ which is expected in a Fermi liquid state gives evidence that δ_0 remains to be $\pi/2$ all over the doping range

[18]. Therefore, one can estimate the evolution of the carrier density n with doping from the doping dependence of ρ_0 without ambiguity from the unknown x dependence of the effective mass m^* — the conductivity or Drude spectral weight is proportional to n/m^*.

The experimental results for La214 and Y123 show that ρ_0 rapidly decreases as the doping proceeds to the optical and overdoped regions, which is interpreted as a crossover from $n = x$ in the underdoped regime to $n = 1 - x$ in the overdoped regime [18]. In Fig. 6 is also plotted the Drude spectral weight N_D which is estimated by isolating the Drude component from the mid-IR band in $\sigma_c(\omega)$. N_D evolves with x just in harmony with the evolution of n in the underdoped regime. Note that N_D amounts only 20% or less of the total spectral weight (N_{eff}), Drude + mid-IR, which is transferred from the high-energy charge- transfer excitations upon doping and shows a distinct x dependence. From these results it is concluded that the spin gap in the underdoped regime radically suppresses the Drude component (or coherent part) of the in-plane optical conductivity which is linked to the low carrier density $n = x$ involved in the in-plane charge transport.

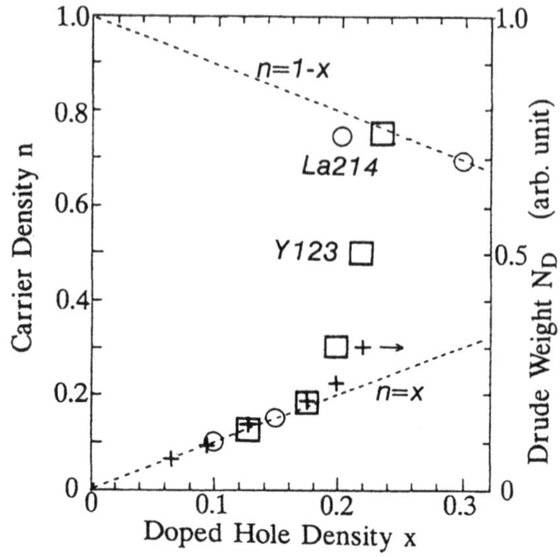

Figure 6. Variation of the carrier density, estimated from the Zn-induced residual resistivity, with doping. Also plotted the Drude spectral weight separated out from the mid-IR band in $\sigma_{ab}(\omega)$.

4. ADDITIONAL ISSUES

4.1. Effect on the Hall coefficient

The deviation from the T-linearity in $\rho_{ab}(T)$ correlates with the T dependence of the Hall coefficient (R_H) [8]. While $R_H \sim 1/T$ for the optimally doped compound, a deviation from the $1/T$ dependence is seen below T^* for the underdoped compounds and R_H shows a peak as T is reduced. The T dependent R_H in the high-T_c cuprates is not fully understood even at present. Anderson proposed a novel explanation by introducing another scattering time τ_H [25], assumed to be relevant to the Hall effect and directly connected to the Hall angle by $\cot\theta_H = 1/\omega_c\tau_H$ (ω_c : cyclotron frequency). The experimental result for the optimally doped Y123 indicates that $\tau_H^{-1} \sim T^2$ [24].

It is found that the T dependence of τ_H is basically unchanged in the underdoped regime and no significant feature is seen in $\tau_H(T)$ at T^* (Fig.7). This implies that $\tau_H \sim T^{-2}$ irrespective of the

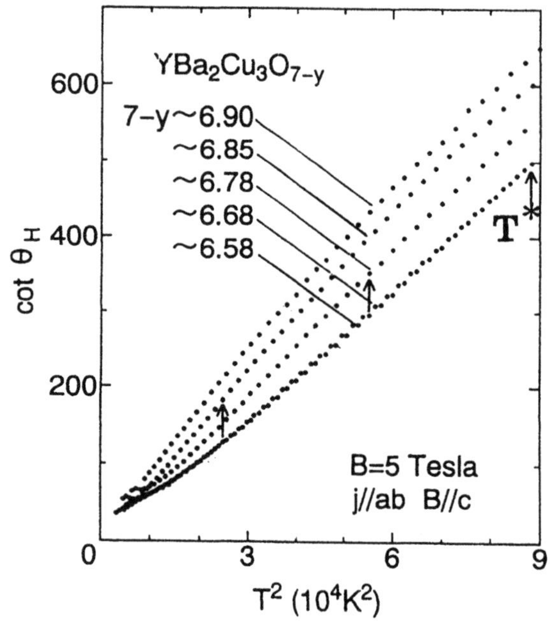

Figure 7. Hall angle plotted as a function of T^2 for Y123 with various dopings.

presence or otherwise of the spin gap and that the characteristic change of R_H at T^* is induced by that in the T dependence of τ, since $R_H \sim \tau_H/\tau$. However, the implication of the 'transverse' scattering time τ_H is still an open question.

4.2. Charge transport in $YBa_2Cu_4O_8$

$YBa_2Cu_4O_8$ (Y124) is an intrinsically underdoped compound. In fact, the in-plane resistivity (ρ_a) changes its T dependence at T^* in a similar manner to that observed for other cuprates [26], and $\sigma_c(\omega)$ shows a feature of pseudogap [27]. Recently, it has turned out that the T dependence of ρ_c is quite exceptional [28,29]. In contrast to other systems the semiconducting behavior of ρ_c is not seen, but ρ_c steeply decreases below $\sim T^*$. In this respect, the pseudogap feature observed in $\sigma_c(\omega)$ does not correlate with $\rho_c(T)$. Moreover, Zn doping into Y124 appears to destroy the pseudogap in $\sigma_c(\omega)$ [30]. These exceptional behaviors may be understood by dominant contribution from the Cu-O chains in Y124. The structure of the double Cu-O chains is almost perfect with negligible oxygen deficiency, and they are highly conductive along the chain direction. It appears that the chains might be also conductive in the c-direction, so that the chain contribution to ρ_c would have dominated, since the plane contribution should have been suppressed owing to the spin gap.

5. SUMMARY

One of the most anomalous properties of high-T_c cuprates is the presence of a gap in the spin excitations with no gap in the charge excitations. In the charge excitation spectrum the corresponding feature appears as a pseudogap only in the c-axis direction, and the pseudogap is responsible for the divergent c-axis dc resistivity toward $T = 0$.

On the other hand, the in-plane spectrum is dominated by a Drude term. The spin gap effect is seen by a rapid narrowing of the Drude peak or a rapid enhancement of the Drude peak height below a characteristic temperature T^*, in correspondence with the faster decrease in the in-plane resistivity. As a consequence, the Drude weight is isolated from the dominant mid-IR band, and the reduced Drude weight is linked to the low carrier density involved in the in-plane charge transport.

These results are generic to the underdoped cuprates and demonstrate that the fundamental features of high-T_c cuprates, two-dimensionality and low carrier density, result from the spin gap.

ACKNOWLEDGEMENTS

I thank N. Nagaosa, T. Ito, K. Takenaka, Y. Fukuzumi and K. Tamasaku for discussions. The work was supported by Grants for Priority Area and for COE Research from Mombusho, by a NEDO Grant for International Joint Research, and by Nissan Science Foundation.

REFERENCES

1. H. Yasuoka, T. Imai and T. Shimizu, in Strong Correlation and Superconductivity, H. Fukuyama, S. Maekawa and A. P. Malozemoff (eds.) Springer, 1989, p.254.
2. J. Rossat-Mignod *et al.*, Physica C **185-189** (1991) 86.
3. H. Yasuoka, this conference.
4. M. -H. Julien *et al.*, Phys. Rev. Lett. **76** (1996) 4238.
5. A. G. Loeser, D. S. Dessau and Z. -X. Shen, Physica C**263** (1996) 208.
6. C. C. Homes *et al.*, Phys. Rev. Lett. **71** (1993) 1645.
7. K. Tamasaku, Y. Nakamura and S. Uchida, Phys. Rev. Lett. **69** (1992) 1455.
8. T. Ito, K. Takenaka and S. Uchida, Phys. Rev. Lett. **70** (1993) 3995.
9. K. Takenaka *et al.*, Phys. Rev. B**50** (1994) 6534.
10. S. Uchida *et al.*, J. Low Temp. Phys.**105** (1996) 723.
11. H. Ding *et al.*, Nature **382** (1996) 51.
12. K. Tamasaku *et al.*, Phys. Rev. Lett. **72** (1994) 3088.
13. A. G. Rojo and K. Levin, Phys. Rev. B**48** (1993) 16861.
14. N. Nagaosa, Phys. Rev. B**52** (1995) 10561.
15. N. P. Ong, Science **273** (1996) 32.
16. A. Carrington *et al.*, Physica C**234** (1994) 1.

17. Y. Nakamura and S. Uchida, Phys. Rev. B**47** (1993) 8369.
18. Y. Fukuzumi *et al.*, Phys. Rev. Lett. **76** (1996) 684.
19. Y. Ando *et al.*, Phys. Rev. Lett. **77** (1996) 2065.
20. G. S. Boebinger *et al.*, Phys. Rev. Lett. **77** (1996) 5417, for the underdoped La214 both ρ_{ab} and ρ_c are divergent when SC is suppressed by magnetic fields. This exceptional behavior may be understood as fluctuation effect of the stripe order discovered by J. M. Tranquada *et al.*, Nature **375** (1995) 561.
21. D. N. Basov *et al.*, Phys. Rev. B**49** (1994) 12165.
22. K. Mizuhashi *et al.*, Phys. Rev. B**52** R3884 (1995).
23. P. W. Anderson, Phys. Rev. Lett. **67** (1991) 3844.
24. T. R. Chien, Z. Z. Wang and N. P. Ong, Phys. Rev. Lett. **67** (1991) 2088.
25. P. W. Anderson, Phys. Rev. Lett. **67** (1991) 2092.
26. B. Bucher *et al.*, Phys. Rev. Lett. **70** (1993) 2012.
27. D. N. Basov *et al.*, Phys. Rev. B**50** (1994) 3511.
28. B. Bucher *et al.*, a report at the Meeting of the American Physical Society, San Jose, 1995.
29. J. -S. Zhou *et al.*, Phys. Rev. Lett. **77** (1996) 4253.
30. D. N. Basov *et al.*, Phys. Rev. Lett. **77** (1996) 4090.

ELSEVIER

Physica C 282-287 (1997) 19-26

PHYSICA C

Vortex Matter

G. Blatter [a]

[a]Theoretische Physik, ETH-Hönggerberg, CH-8093 Zürich, Switzerland.

We present a tutorial review of the recent developments in the phenomenology of high temperature superconductors, covering aspects such as single vortex physics, the phase diagram of vortex systems, and the influence of quenched disorder.

The phenomenology of type II superconductors goes back to Abrikosov [1] and centers around the well known mean-field H-T phase diagram comprising the Meissner-Ochsenfeld phase at low fields $H < H_{c_1}$, the mixed or Shubnikov phase in the intermediate field regime $H_{c_1} < H < H_{c_2}$, and the normal metallic phase at $H > H_{c_2}$, see Fig. 1. Within the Shubnikov phase the magnetic field H enters the superconductor in the form of flux lines or vortices — these topological excitations arrange themselves into the (triangular) Abrikosov lattice and are central in determining the physical properties of the Shubnikov phase. The sharp jump in the specific heat typical for the transition into the superconducting state and observed in many experiments on conventional superconductors leads to the conclusion that fluctuations are small in this problem. Quantitatively, this observation is confirmed through the smallness of the Ginzburg number, $Gi \sim 10^{-8}$ typically. The situation has changed dramatically with the appearance of the copper-oxide based high temperature superconductors in 1986 [2]. With a transition temperature T_c enhanced by one order of magnitude and an anisotropy $\varepsilon^2 = m/M \sim 10^{-2} - 10^{-4}$ due to the layered structure, the Ginzburg number $Gi \propto T_c^4/\varepsilon^2$ is enhanced by a factor 10^6 at least and fluctuations become relevant. These strong fluctuations imply not only the existence of an interesting and experimentally accessible critical regime close to the transition line H_{c_2} but rather entail a major reorganization of the H-T phase diagram with new phases and phase transitions appearing in various regions, see Fig. 2. Rather than viewing the vortex lattice as a system of rigid rods as appropriate within the mean field theory applicable to conventional superconductors, the vortices then should be considered as a system of flexible interacting line objects generating a nontrivial statistical mechanics and thus providing the basis for a new state of matter, usually dubbed "vortex matter". In this tutorial lecture we shall attempt to develop a view on the properties of this new type of matter (for a review, see [3]).

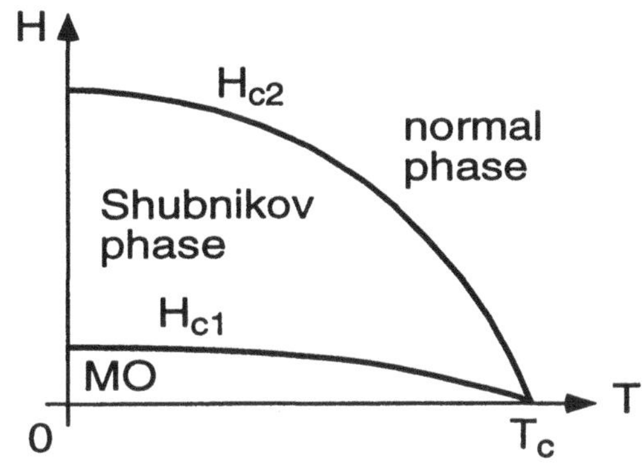

Figure 1. Mean field phase diagram of a conventional type II superconductor with small Ginzburg number $Gi = [T_c/H_c^2(0)\varepsilon\xi^3(0)]^2/2$.

A good starting point for our discussion is the basic constituent of the vortex matter, the individual flux line. Given the Ginzburg Landau functional $F(\Psi, \mathbf{A}) = \int d^3r\{\alpha(T)|\Psi|^2 + \beta/2\,|\Psi|^4 + (\hbar^2/2m)|[\nabla + (2\pi i/\Phi_\circ)\mathbf{A}]\Psi|^2 + (\nabla \wedge \mathbf{A})^2/8\pi\}$, the two basic length scales $\xi = (\hbar^2/2m|\alpha|)^{1/2}$ and $\lambda = (mc^2/16\pi e^2|\Psi_\circ|^2)^{1/2}$ follow from variations with respect to the fields Ψ^* and $\mathbf{A}$. The two parameters ξ and λ character-

PII S0921-4534(97)00197-4

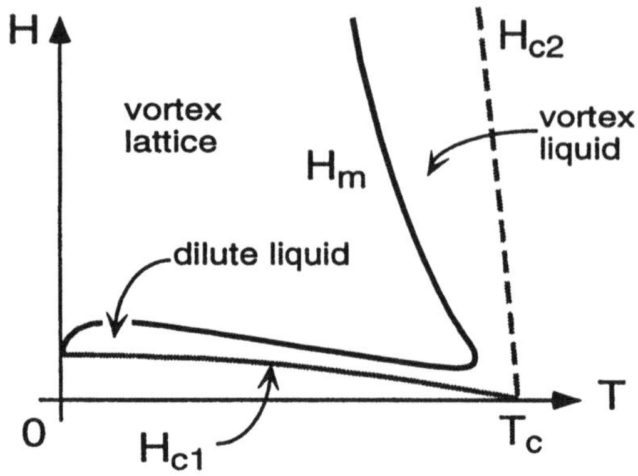

Figure 2. Phase diagram of an anisotropic high-T_c superconductor with large Ginzburg number. The vortex phase is melted over a large portion of the phase diagram, with a dense and a dilute vortex liquid phase appearing at large and low field values.

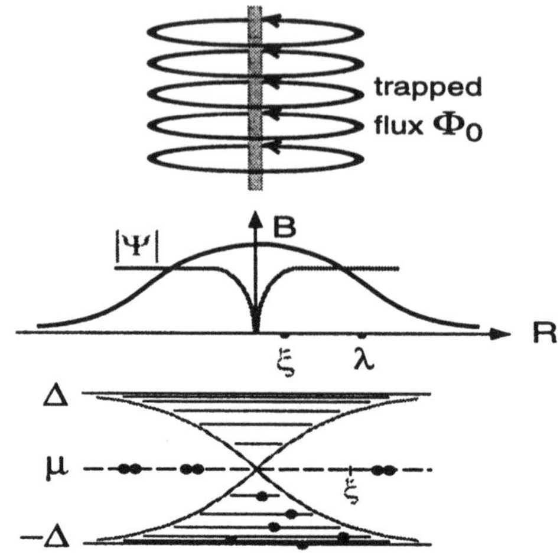

Figure 3. From top to bottom: current flow trapping a flux quantum $\Phi_\circ$ and vortex core, structure of the quantum liquid and the magnetic induction around a vortex, microscopic structure of the core region.

ize the superfluid ($\rightarrow \Psi$) and the electromagnetic ($\rightarrow \mathbf{A}$) response and determine the dimensions of the vortex core and the circular current flow of the vortex, $j = -(2e\hbar/m)|\Psi|^2[\nabla\varphi + (2\pi i/\Phi_\circ)\mathbf{A}]$, see Fig. 3. The latter traps the quantized magnetic flux $\Phi_\circ = hc/2e$. Within the phenomenology of type II superconductors the two structural elements (vortex core and vortex current) play the following roles:

The circular *current flow* on scale λ produces the vortex line energy $\epsilon_l = \varepsilon_\circ \ln(\lambda/\xi)$, the line elasticity $\varepsilon_l \propto \varepsilon_\circ$, and carries the interaction potential $V(R) = 2\varepsilon_\circ K_0(R/\lambda)$, where $\varepsilon_\circ = (\Phi_\circ/4\pi\lambda)^2$ is the basic energy scale in the problem. Whereas in an isotropic superconductor the elasticity straightforwardly is given through the increase in the length of the vortex (implying $\varepsilon_l \approx \varepsilon_\circ \ln[\ldots]$), in anisotropic superconductors, such as the copper oxides, the anisotropy parameter $\varepsilon^2 = m/M \ll 1$ strongly reduces the line elasticity to the value $\varepsilon_l^J \approx \varepsilon^2\varepsilon_\circ \ln[\ldots]$, where we have indicated the presence of logarithmic corrections implying a weak k_z dispersion. Within a layered (Lawrence-Doniach) model the presence of the anisotropy parameter ε in the vortex elasticity ε_l^J is traced back to the smallness of the Josephson currents screening the phase mismatch in subsequent layers when the vortex line is deformed. In addition, a strongly dispersive electromagnetic contribution to the line tension becomes relevant when the vortex line is distorted on large scales ($k_z < 1/\varepsilon\lambda$), $\varepsilon_l = \varepsilon_l^J + \varepsilon_l^{em}$, with $\varepsilon_l^{em} \approx (\varepsilon_\circ/2\lambda^2 k_z^2)\ln(1 + \lambda^2 k_z^2)$. In a (strongly) layered material such as $Bi_2Sr_2Ca_1Cu_2O_8$ (BiSCCO), the line elasticity is dominated by this strongly dispersive electromagnetic contribution, over a large regime in k_z space.

The *vortex core* plays a major role in the dynamical behavior of the individual flux line, besides contributing a correction $\varepsilon_\circ/2$ to the vortex line energy and a term $\varepsilon^2\varepsilon_\circ/2$ to the elasticity. In a conventional superconductor the quasiparticles trapped within the vortex core region (see Fig. 3 and [4]) can be modelled as a normal metallic region implying a dissipative dynamics $\eta_l \partial_t \mathbf{u} = -\nabla V$, where $\mathbf{u}$ denotes the vortex coordinate, $\eta_l \approx \Phi_\circ^2/2\pi\xi^2 c^2 \rho_n$, and $V(\mathbf{u})$ is a driving potential, see Fig. 4. This dissipative response of the vortices limits the technological applicability of type II superconductors and entails the quest for the development of "hard" superconductors,

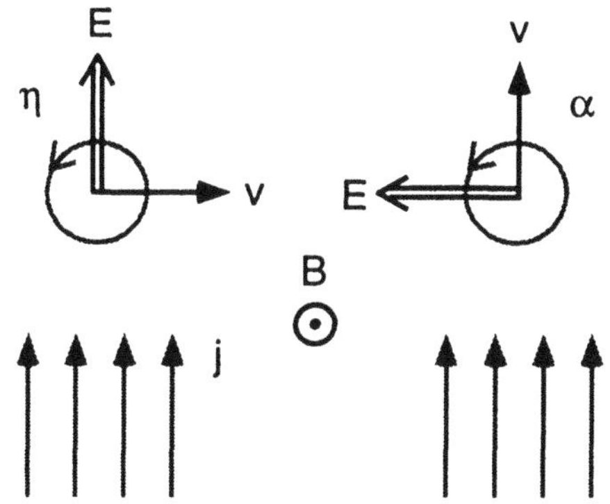

Figure 4. Vortex dynamics in conventional and superclean superconductors: The electric field **E** is aligned with the applied current density **j** leading to dissipation (left hand side). In a superclean material the vortex swims with the current, **v** || **j** (right hand side).

where vortices are pinned by material defects. In a (super)clean superconductor, where the quasi-particle levels are not smeared by finite lifetime effects, the vortex dynamics changes its character to a hamiltonian one. The equation of motion is that of an electron in a magnetic field, $\alpha_l \partial_t \mathbf{u} \wedge \mathbf{n} = -\nabla V$, $\alpha = \pi\hbar n$, and the vortex "swims" with the superflow, see Fig. 4. Rather than a dissipative longitudinal electric field $\mathbf{E} \parallel \mathbf{j}$, a Hall type transverse field $\mathbf{E} \perp \mathbf{j}$ is generated in the latter case. An interesting debate has emerged in the last few years with the experimental observation that vortices in various high-T_c materials appeared to show an upstream flow component leading to a sign change in the Hall effect [5]. Theoretically this phenomenon can be understood as arising from two competing contributions to the Hall effect, one from quasi-particles and a second term arising from the vortex dynamics. Breaking particle-hole symmetry, e. g., via band structure effects, then can explain this puzzling sign change in a natural way [6].

With growing evidence that superconductivity in the copper oxides is characterized by d-wave symmetry, recent interest concentrates on the structure of individual vortex lines. On the Ginzburg Landau level the basic functional is modified by the appearance of additional terms coupling various symmetry channels. E. g., the vortex-induced current in the d-wave channel in general drives an s-wave order parameter in the vicinity of the vortex center, leading to distortions of the order parameter in the core region. On the microscopic level the interest concentrates on the quasi-particle structure within the core region of a d-wave vortex, the main issue being the modifications of the quasi-particle spectrum [7] due to the zeros of the order parameter along the planar diagonals. On the phenomenological level such refinements play a minor role regarding the elasticity and the interaction between the flux lines, however, they may be relevant with respect to the dynamical behavior of the vortices.

Turning to a system of vortex lines the immediate task arises to construct the analog of the p-T phase diagram for conventional matter, which is the *H-T phase diagram* of the vortex matter (the magnetic field variable H plays the role of the pressure p). An important difference between ordinary and vortex matter is observed in the generic shape of the interaction potential — whereas atoms/ions exhibit an attractive component in $V(\mathbf{r})$, the vortex interaction $V(\mathbf{R})$ is purely repulsive (a notable exception is mentioned below). As a consequence, the density of vortex matter is determined by the external magnetic field H (the "pressure"), and, contrary to ordinary matter, there is no generic density for the vortex-solid or -liquid phases. As usual, at low temperatures the interaction between the vortex lines wins over the entropy and a (triangular) lattice is formed. With increasing temperature, fluctuations increase and a first-order melting transition into a vortex liquid phase takes place. Whereas in conventional material this transition takes place only very close to the upper critical field line H_{c_2}, in the high-T_c materials the enhanced importance of thermal fluctuations suppresses the melting line down to low field values, see Fig. 2. The most simple argument giving the shape of the melting line is a dimensional one: Consider the London free energy of the vortex system $F[\{\mathbf{r}_\mu\}] \approx (\varepsilon_\circ/2T) \sum_{\mu,\nu} \int d\mathbf{r}_\mu \cdot d\mathbf{r}_\nu \exp[-|\mathbf{r}_\mu - \mathbf{r}_\nu|/\lambda]/|\mathbf{r}_\mu - \mathbf{r}_\nu|$ (corresponding to the (screened) interaction between current segments in magetostatics). Using the lattice constant $a_\circ \approx (\Phi_\circ/B)^{1/2}$ we go over to dimensionless variables $\mathbf{x}_\mu = \mathbf{r}_\mu/a_\circ$ and drop the screening factor at high fields where $a_\circ \ll \lambda$. As a result we obtain a free energy of the form $F[\{\mathbf{x}_\mu\}] \approx (\varepsilon_\circ a_\circ/2T)f[\{\mathbf{x}_\mu\}]$. The shape of the melting line then is given by equating the only dimensionless parameter to a constant, $[\varepsilon_\circ a_\circ/2T]_{\rm melting} =$ const. A Lindemann argument shows that this constant has to be identified with $1/c_L^2$, where $c_L \approx 0.1 - 0.3$ is the Lindemann number (the lattice melts when thermal fluctuations can drive elastic distortions with an amplitude $\langle u^2\rangle = c_L^2 a_\circ^2$, $T = \varepsilon_\circ\langle u^2\rangle/L$, with the length $L = a_\circ$ following from the competition between shear and tilt energies in the vortex lattice). Generalizing the result via scaling theory [8] to anisotropic superconductors and using the temperature and field dependence of the magnetic penetration depth $\lambda^2 = \lambda_0^2/[1 - T/T_c - B/B_{c_2}(0)]$ one arrives at a melting line given by the relation $B_m \sim (c_L^2 T_c/T)^2/Gi\,[1 - T/T_c - B/B_{c_2}(0)]^2 B_{c_2}(0)$, with $Gi \sim [T_c/\varepsilon\varepsilon_\circ(0)\xi(0)]^2$. A notable feature of this melting line is its negative slope $dB_m/dT < 0$, see Fig. 2, implying that the vortex fluid at high magnetic fields is denser than the vortex solid: the vortex lattice melts like ice. That the vortex lattice indeed melts via a first-order transition into a denser vortex liquid phase has been demonstrated in a series of beautiful experiments by Zeldov et al. [9], Welp et al. [10], and Schilling et al. [11]. Note that the vortex crystal in layered superconductors such as BiSCCO melts at extremely low field values $B < \Phi_\circ/\lambda^2$. The above analysis for the shape of the melting line then has to be refined by accounting for the electromagnetic interaction in the vortex system [12]. At very low fields $B \ll \Phi_\circ/\lambda^2$ the shear modulus in the vortex lattice decreases exponentially as the circular current flow of the individual vortex vanishes $\propto K_0(R/\lambda) \propto \exp(-R/\lambda)$ and the melting line exhibits a reentrant behavior resulting in a dilute liquid phase, see Fig. 2. In layered and strongly anisotropic superconductors thermal fluctuations generate an attractvie van der Waals-type interaction between the vortex lines, leading to a very dilute vortex solid/liquid phase with a generic density corresponding to an induction $B \approx \Phi_\circ/(20\lambda)^2 \sim 1$ G [13].

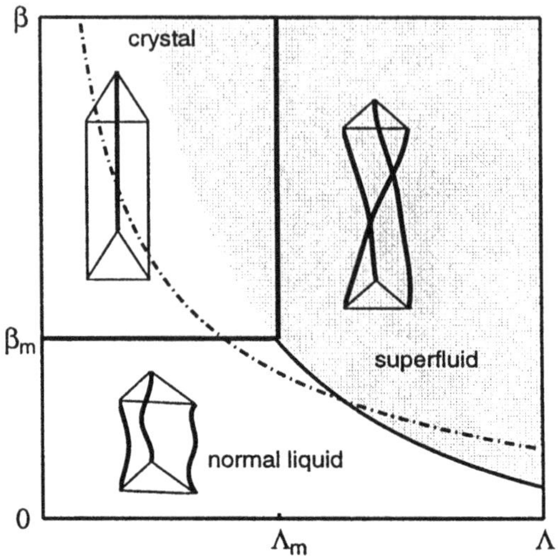

Figure 5. Phase diagram for a system of charged 2D bosons. At low temperatures $\beta \to 0$ and small densities ρ, i. e., small de Boer parameter $\Lambda = (\hbar^2\rho/mV)^{1/2}$, the bosons arrange in a crystal. The latter melts into a classical liquid with increasing temperature and into a quantum liquid with increasing density. Within the shaded area quantum fluctuations dominate over thermal fluctuations. The dash-dotted line marks a constant field trajectory in the vortex problem for a finite length sample.

The next question we wish to address is that of the *nature of the vortex liquid phase* resulting upon melting. An interesting concept has been introduced by Nelson [14], who has mapped the classical statistical mechanics of vortex lines to the quantum statistical mechanics of two dimensional bosons. In this mapping the elastic energy for a vortex line with elasticity ε_l is equivalent to the imaginary time action of a boson with mass m, $\mathcal{F}/T = (1/T)\int dz(\varepsilon_l/2)(\partial\mathbf{u}/\partial z)^2 \to (1/\hbar)\int d\tau(m/2)(\partial\mathbf{R}/\partial\tau)^2 = \mathcal{S}/\hbar$. The interaction potential maps trivially (the limit $\lambda \to \infty$ corresponds to charged bosons) and the chemical potential for the bosons is related to the applied magnetic field, $\mu = (\Phi_\circ/4\pi)(H - H_{c_1})$. Defining the two dimensionless parameters $\beta = V/T_B$ (inverse bosonic temperature) and $\Lambda^2 = \hbar^2\rho/mV$ (Λ is the de Boer parameter) we can construct the β-Λ phase diagram for the charged boson system, see Fig. 5: At low temperatures and small densities ρ (small $\hbar$) the potential energy wins and the bosons arrange in a "Wigner" crystal. Increasing the temperature (decreasing β) thermal fluctuations melt the crystal resulting in a classical normal fluid. On the other hand, increasing density (or $\hbar$) the relative importance of classical and quantum fluctuations is interchanged when $\beta\Lambda^2 \sim 1$, the latter relation defining the *entanglement line* beyond which the system is dominated by quantum fluctuations. At low temperatures (large β) quantum fluctuations transform the crystal via a melting transition into a superfluid liquid, whereas at high temperatures the normal fluid turns into a superfluid as the entanglement line is crossed (note that the melting transition is determined through the competition between the interaction V and the kinetic energy as quantified by the de Boer parameter Λ and does not match up with the entanglement line — rather the entanglement is inhibited within the crystal and freed only upon melting). Going back to the vortex system we observe the following analogy between the various phases: The Wigner crystal at finite bosonic temperature T_B corresponds to the Abrikosov lattice in a sample of finite length $L = \beta T/\varepsilon_o$. Decreasing the sample length we arrive at a disentangled vortex liquid phase, the analog of the normal Bose fluid. Increasing the vortex density $\rho = 1/a_o^2$ or the vortex temperature T, the Abrikosov lattice melts into an entangled vortex liquid phase corresponding to the bosonic superfluid. From this mapping one concludes that the vortex liquid phase in a thick sample is macroscopically entangled, i. e., the vortex fluid phase exhibits macroscopic transverse loops. The latter produce a dissipative response when an external current is applied parallel to the magnetic field, showing that the entangled vortex liquid is a normal metallic phase (note that a disentangled vortex liquid would exhibit a superfluid response). It has been suggested [15,3] that in a charged boson system the normal fluid phase may extend down to low temperatures, implying that the vortex crystal melts in a sequence

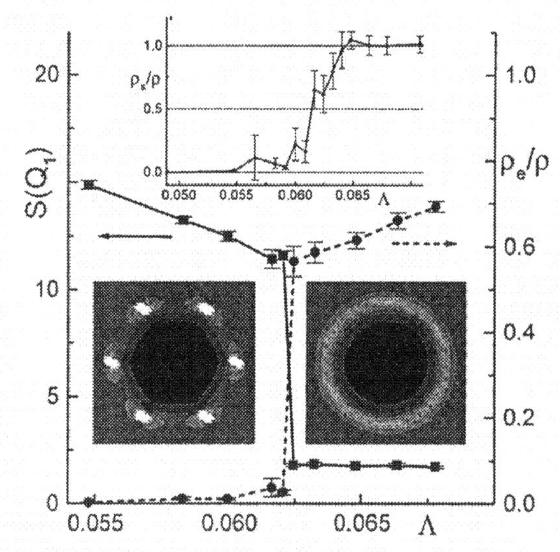

Figure 6. Structure factor $S(Q)$ and relative entanglement ρ_e/ρ of a system of 2D charged bosons at the $\beta \to \infty$ quantum melting transition (classical melting of the vortex system in a long sample). In this simulation of 64 particles one sharp transition is observed where the liquid entangles immediately upon melting. The inset shows the superfluid density ρ_s/ρ calculated for a sample of 36 particles.

Abrikosov lattice → disentangled liquid → entangled liquid, thus breaking translational order and longitudinal gauge in two subsequent transitions. However, in recent path-integral quantum Monte Carlo simulations of charged 2D Bosons [16] only one sharp transition from a crystal to a superfluid phase could be found, leading to the conclusion that the vortex lattice directly melts into an entangled liquid, see Fig. 6, where the structure factor $S(Q)$ and the entanglement density ρ_e/ρ are shown.

The last topic we wish to address is *quenched disorder.* This topic is of great technological relevance, as vortex pinning by material defects prevents vortices from moving under the action of the Lorentz force $\mathbf{F}_L = \Phi_o\mathbf{j} \wedge \mathbf{n}/c$ generated when an external current density $\mathbf{j}$ is applied to the superconductor. The various kinds of disorder studied in the recent past can be classified in two groups, see Fig. 7: *i)* Uncorrelated disorder due to point

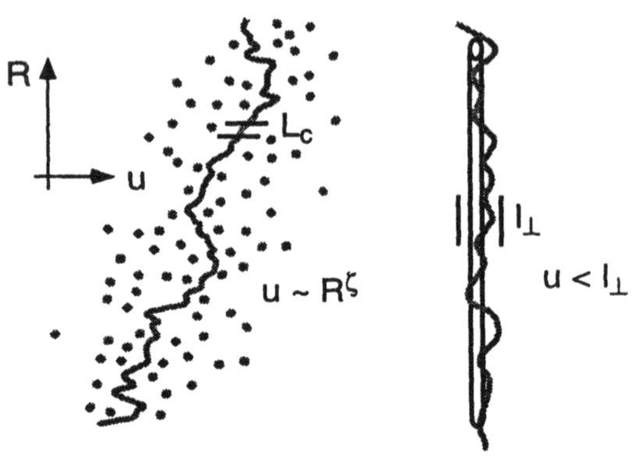

Figure 7. Uncorrelated point-like disorder leading to flux line wandering (left) and correlated disorder in the form of a columnar defect leading to localization of vortices (right).

defects, e. g., oxygen vacancies, leads to line wandering, with the flux lines being displaced by optimal defect configurations resulting in a power law growth of the mean displacement, $u \sim R^\zeta$. The determination of the wandering exponent ζ is a highly nontrivial task which has been solved exactly only for the directed-polymer/domain-wall problem in 2D [17] (the "1 + 1" problem corresponding to a string moving in a plane). *ii)* Correlated disorder, as provided by the twin boundaries in $Y_1Ba_2Cu_3O_{7-\delta}$ (YBCO) or generated artificially by irradiation of the sample with swift heavy ions producing columnar tracks, lead to the localization of the vortices on the defects and thus provides the most efficient way to immobilize the flux lines. Both types of disorder have been studied in detail and we give a brief account of the most important results.

The effect of *weak uncorrelated disorder* can be studied using the elastic free energy $F[\mathbf{u_k}] = (1/2)\int d^3k/(2\pi)^3\{c_{66}|\mathbf{K}_\perp\cdot\mathbf{u_k}|^2 + c_{44}(\mathbf{k})|k_z\mathbf{u_k}|^2 + c_{11}(\mathbf{k})|\mathbf{K}\cdot\mathbf{u_k}|^2\}$ as a starting point. Quenched disorder is introduced through a random disorder potential $U_{\rm pin}(\mathbf{r})$ with vanishing mean and short range correlations, $\langle U_{\rm pin}(\mathbf{r})U_{\rm pin}(\mathbf{r}')\rangle = \gamma\delta(\mathbf{r}-\mathbf{r}')$. This potential acts on the vortices and produces a pinning potential $E_{\rm pin}[\mathbf{r},\mathbf{u}(\mathbf{r})]$ which in turn exhibits short range correlations $\propto \gamma\xi^4/a_\circ^2$. The random potential acting on the flux lines destroys the long range translational order of the lattice [18] and transforms the crystal into a glass. The glassyness of the new phase can be quantified by calculating the displacement correlater $\langle u^2(R)\rangle_d \equiv \langle[\mathbf{u}(\mathbf{R})-\mathbf{u}(\mathbf{0})]^2\rangle_d$. At small distances this correlator grows algebraically with distance, $\langle u^2(R)\rangle_d^{1/2} \sim \xi(R/R_c)^\zeta$ (perturbative and elastic manifold regime [19]) but then turns to a logarithmic behavior at large distances, $\langle u^2(R)\rangle_d^{1/2} \sim a_\circ \ln(R/R_a)$ [20,21]. The resulting glass phase still shows quasi-long-range-order and (algebraic) Bragg peaks are found in the structure factor $S(\mathbf{Q})$. The name "Bragg glass" phase has been coined and the glass is characterized as "weak". It can be shown that for weak enough disorder the Bragg glass phase is stable against creation of dislocations, which guarantees consistency with the starting point given by the elastic free energy. A very recent idea is, that this Bragg glass phase can be "melted" not only by thermal fluctuations but also by quenched disorder [21]. To fix ideas, consider an individual vortex line trapped in the cage of its neighbors, see Fig. 8. When the disorder is strong enough to drive the flux line out of the cage, a transition to a more strongly disordered glass phase is postulated [22]. This idea can be quantified with the help of a Lindemann type argument and produces a phase diagram as shown in Fig. 9. Following this idea, uncorrelated quenched disorder seems to assist thermal fluctuations and one expects that the melting line shifts downward. For weak enough disorder the first-order nature of the flux lattice melting transition is preserved. Experiments show, that at higher fields/lower temperatures the first-order line either terminates in a critical point, with an additional second order transition (vortex glass transition) possibly emerging from a critical endpoint [23].

Strong correlated disorder can be studied most efficiently by exploiting the mapping between vortices and 2D Bosons [24] (see the above discussion on the nature of the vortex liquid phase). The correlated disorder then maps to a static random potential acting on the bosons ("dirty bosons") and tends to localize them onto the defects (note

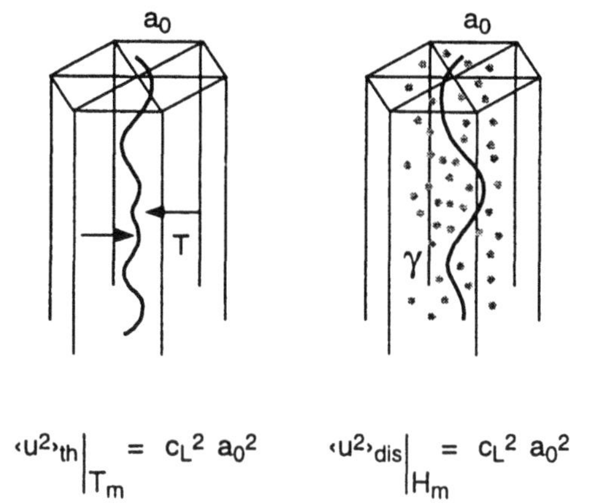

Figure 8. Cage model allowing to estimate the position and shape of the vortex lattice melting transition: The vortex can be driven to leave the cage either by thermal fluctuations (left) or by quenched disorder (right).

that a hard core repulsion prevents the bosons from collapsing onto the most favorable pin). The strong disorder completely destroys the original lattice order and the new resulting phase has been named the "Bose glass" phase. The thermal fluctuations of the vortex lines map to quantum fluctuations which tend to delocalize the bosons as their individual wavefunctions start to overlap at large values of $\hbar$, i. e., high temperatures in the vortex language. This delocalization transition (Bose glass transition) then plays the role of the melting transition in the defect-free sample. Compared to the first-order melting transition in pristine samples the (second-order) Bose glass transition line is shifted upward to higher temperatures and fields and thus enlarges the regime where the material can be used in technological applications. This feature is particularly useful in strongly layered material such as BiSCCO, where the original melting line is suppressed down to field values of the order of a few 100 G. In anisotropic material such as YBCO the upward shift of the transition can be calculated perturbatively, $T_{BG}(B) = gT_{\rm m}(B)+(1-g)T_{c_2}(B)$, with g quantifying the strength of the disorder

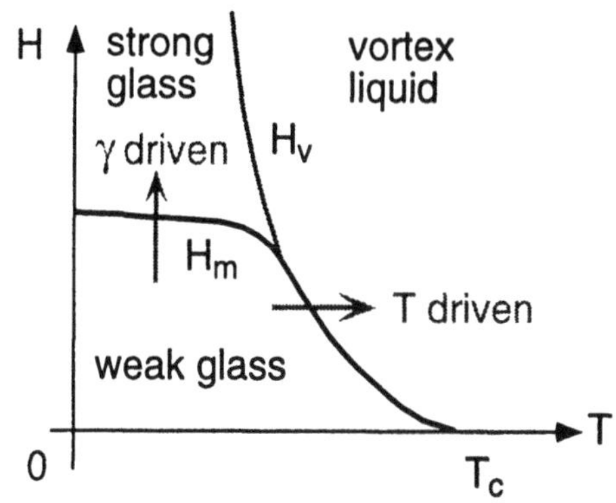

Figure 9. Possible phase diagram indicating the positions of the thermal and disorder driven melting transitions in the vortex lattice.

[25].

The presence of a disorder potential pinning the vortices has important consequences for the technological usefulness of type II superconductors: Pinning immobilizes the vortex lines and prevents them from moving under the action of the driving Lorentz force produced by an applied external current. A second important issue in the transport phenomenology of the strongly anisotropic copper oxides is vortex creep: Thermal fluctuations allow the flux lines to jump over the pinning barriers into new favorable states. The glassy nature of the Bragg/Vortex- and Bose-glass phases strongly modifies the vortex creep with respect to the usual classical picture: A point-particle moving in a tilted random potential will always exhibit linear (ohmic) response at small driving forces; with v denoting the velocity one obtains $v = v_\circ\{\exp[-U_\circ(1 - F/F_\circ)/T] - \exp[-U_\circ(1 + F/F_\circ)/T]\} = 2v_\circ \exp[-U_\circ/T]\sinh(U_\circ F/F_\circ T) \propto F/F_\circ \to 0$. This is just the thermal activated flux flow response (TAFF) of the vortex liquid phase [26]. For an elastic manifold the barriers for creep grow as the driving force is decreased and one obtains a highly non-linear response of the form $v \propto \exp[-(U_\circ/T)(F_\circ/F)^\mu]$, with μ the glass exponent with typical values around unity. For the hard type II superconductors this results in a highly non-linear current voltage characteristic of the glass phase, $E \propto \exp[-(U_\circ/T)(j_\circ/j)^\mu]$, see Fig. 10, and one concludes that the low temperature glass phases indeed define a new superconducting phase different from the vortex liquid [27].

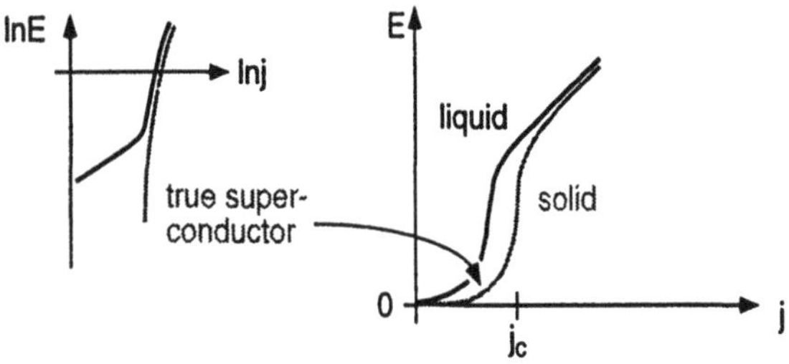

Figure 10. Current-voltage characteristics distinguishing the high temperature vortex liquid with its ohmic TAFF regime (normal response) from the low temperature superconducting vortex-/Bose-glass state with a non-linear glassy behavior.

REFERENCES

1. A. A. Abrikosov, Zh. Eksp. Teor. Fiz. **32**, 1442 (1957) [Sov. Phys. JETP **5**, 1174 (1957).
2. J. G. Bednorz and K. A. Müller, Z. Phys. B **64**, 189 (1986).
3. G. Blatter, M. Feigel'man, V. Geshkenbein, A. Larkin, and V. Vinokur, Rev. Mod. Phys. **66**, 1125 (1994).
4. C. Caroli, P. G. De Gennes, and J. Matricon, Physics Letters **9**, 307 (1964).
5. J. Harris, Y. Yan, O. Tsui, Y. Matsuda, and N. Ong, Phys. Rev. Lett. **73**, 1711 (1994).
6. A. van Otterlo, M. Feigel'man, V. Geshkenbein, and G. Blatter, Phys. Rev. Lett. **75**, 3736 (1995).
7. I. Maggio-Aprile, Ch. Renner, A. Erb, E. Walker, and Ø. Fischer, Phys. Rev. Lett. **75**, 2754 (1995).
8. G. Blatter, V. Geshkenbein, and A. Larkin, Phys. Rev. Lett. **68**, 875 (1992).
9. E. Zeldov, D. Majer, M. Konczykowski, V. Geshkenbein, V. Vinokur, and H. Shtrikman, Nature **375**, 373 (1995).
10. U. Welp, J. Fendrich, W. Kwok, G. Crabtree, and B. Veal, Phys. Rev. Lett. **76**, 4809 (1996).
11. A. Schilling, R. Fisher, N. Phillips, U. Welp, D. Dasgupta, W. Kwok, and G. Crabtree, Nature **382**, 791 (1996).
12. G. Blatter, V. Geshkenbein, A. Larkin, and H. Nordborg, Phys. Rev. B **54**, 72 (1996).
13. G. Blatter and V. Geshkenbein, Phys. Rev. Lett. **77**, 4958 (1996).
14. D. R. Nelson, Phys. Rev. Lett. **60**, 1973 (1988).
15. M. Feigel'man, V. Geshkenbein, L. Ioffe, and A. Larkin, Phys. Rev. B **48**, 16641 (1993).
16. H. Nordborg and G. Blatter, preprint, ETH-TH/96-52, cond-mat/9612025.
17. D. Huse, C. Henley, and D. Fisher, Phys. Rev. Lett. **55**, 2924 (1985).
18. A. I. Larkin, Zh. Eksp. Theor. Fiz. **58**, 1466 (1970) [Sov. Phys. JETP **31**, 784 (1970)].
19. M. Feigel'man, V. Geshkenbein, A. Larkin, and V. Vinokur, Phys. Rev. Lett. **63**, 2303 (1989).
20. T. Nattermann, Phys. Rev. Lett. **64**, 2454 (1990).
21. T. Giamarchi and P. Le Doussal, Phys. Rev. B **52**, 1242 (1995).
22. D. Ertaş and D. Nelson, Physica C **272**, 79 (1996).
23. H. Safar, P. Gammel, D. Huse, D. Bishop, W. Lee, and D. Ginsberg, Phys. Rev. Lett. **70**, 3800 (1993).
24. D. Nelson and V. Vinokur, Phys. Rev. Lett. **68**, 2398 (1992).
25. L. Krusin-Elbaum, L. Civale, G. Blatter, A. Marwick, F. Holtzberg, and C. Feild, Phys. Rev. Lett. **75**, 187 (1995).
26. P. Kes, J. Aarts, J. van den Berg, C. van der Beek, and J. Mydosh, Supercond. Sci. Technol., **1**, 242 (1989).
27. M. P. A. Fisher, Phys. Rev. Lett. **62**, 1415 (1989).

ELSEVIER

Physica C 282–287 (1997) 27–34

PHYSICA C

Introduction to the Structure and Chemistry of Superconducting Materials

Robert J. Cava

Department of Chemistry and Materials Institute, Princeton University, Princeton NJ 08540

The copper oxide superconductors are the latest and in some ways the most exotic of the series of new superconducting materials discovered during many decades of scientific research in the materials aspects of superconductivity. This tutorial lecture, summarized in part in the following, was designed to place the cuprate superconductors, for the non-expert, within the context of the superconducting materials which came before and bring all up to date on the latest developments in cuprates, where even after a decade of intensive research, new materials are being discovered which challenge our understanding.

1. INTERMETALLIC SUPERCONDUCTORS

It was no doubt the contrast between the chemistry of high temperature superconductors and the best superconducting materials which came before which was one of the stunning aspects of their discovery. These earlier materials were primarily intermetallic compounds, which in themselves of course followed a path since the discovery of superconductivity in Hg in 1911 toward increasing complexity, increasing T_c and improved superconducting properties. In one sense, the best of these from a practical point of view are based on the superconductivity in elemental Nb. As a chemist looking back on the way they developed between the 1940s and the 1970s, they would have been the result of relatively simple additions of elements to Nb to see what happened to the properties, resulting in the stabilization of the β tungsten structure for certain combinations of Nb with metalloid elements in what is the famous and very important A15 family of very high T_c and high critical field intermetallic superconductors of formula Nb_3X.

The crystal structure of this family of compounds is shown in figure 1.: It is based on the simple body-centered packing of the metalloid element (X) with Nb-Nb pairs on all the faces of the cube. Superconductors are found for Nb, Ta and V based compounds with T_cs generally between 15 and 20 K, the highest reported for the compound Nb_3Ge which can apparently only be prepared with a T_c of 23.2K as a sputtered film. This fact illustrates one of the important relations between the chemical stability and transition temperatures of superconducting materials - as the forces which give rise to superconductivity increase and the T_cs rise, the same forces tend to make the crystal structures unstable, a reasonable general rule which describes the materials which are superconducting due to electron phonon coupling (as the electron-phonon coupling increases, then structural distortions such as charge density wave behavior become more stable) Although electron-phonon coupling is not responsible for superconductivity in the cuprates, the same arguments have been made by many in their case as well - the Hg

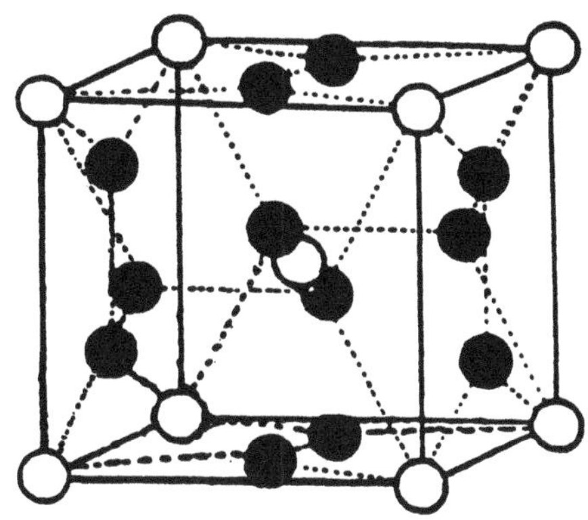

Fig 1. The A15 Crystal Structure, Nb_3X. Nb closed circles, X open circles.

PII S0921-4534(97)00200-1

cuprates for example, seem so much less thermodynamically stable than do low T_c materials such as La_2CuO_4.

The concept of charge counting is critical to the understanding of the behavior of the copper oxide superconductors - the doping away from the Cu2+ state, to an average copper valence of 2.2+ is one of the important empirical rules chemists need to design new superconducting materials and one of the important characteristics of the materials which must be successfully explained by any successful microscopic theory for the superconducting mechanism. The idea of charge counting, however, is a very old one in superconductivity, harking back to the earliest days of trying to understand the behavior of simple binary solid solutions of metals. The A15 structure superconductors are a particularly dramatic illustration of that principle in intermetallic superconductors, shown here in figure 2. The T_cs peak dramatically near 4.5 and 6.5 electrons per metal. In intermetallics this kind of behavior reflects the peaks in the electronic densities of states at those values of band filling - probably not happening for the same reason in the cuprates, but still reflecting an important characteristic of all superconducting materials regardless of the underlying superconducting mechanism.

Of course, with more and more research on intermetallic superconductors over the years, a dramatic variation in chemistry, and structure

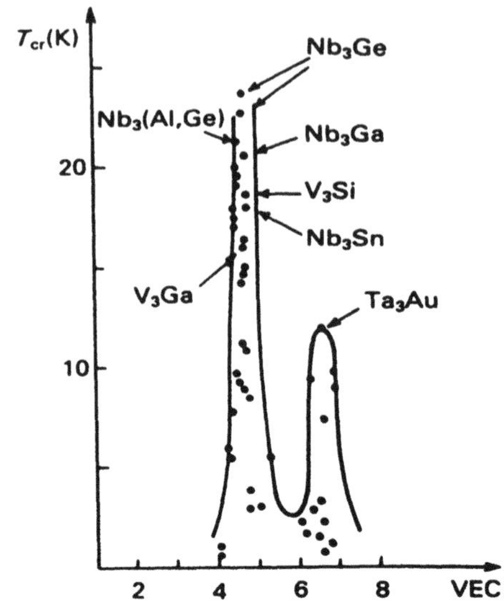

Fig. 2 The variation in T_c with electron count in A15 superconductors.

type was found, and materials of widely varying chemical and structural complexity and superconducting properties are known. Some examples which illustrate this point are shown in table 1, selected from the hundreds of intermetallic and complex compounds now known to be superconducting.

Table 1: Some interesting examples of Intermetallic or compound superconductors

Compound	Symmetry	Structure Type	Tc
Nb_3Ge	cubic	A15	23.2
BiNi	hexagonal	$B\text{-}8_1$	4.2
Au_5Ba	hexagonal	D-2	0.7
NbN	cubic	B-1	17.3
PdTe	trigonal	C-6	1.5
$PbMo_6S_8$	rhombohedral	Mo_3Se_4	15.2
$ScRu_4B_4$	tetragonal	$LuRu_4B_4$	7.2
$LaRh_3Si_5$	orthorhombic	$U_2Co_3Si_5$	4.4

2. "OLD" OXIDE SUPERCONDUCTORS

With the important exception of $BaPb_{.75}Bi_{.25}O_3$ (BPBO), described further in a subsequent section, all oxide superconductors discovered before 1986 appear to behave like conventional electron-phonon coupled superconductors, and have T_cs which do not stand out with respect to their intermetallic competitors of the same historical period. Their densities of states are often lower, as expected, but so are their T_cs. Inspection of their crystal structures and electronic properties suggests that the critical factor which was missing in these early oxide superconductors, with the exception of BPBO, is the extensive metal-oxygen covalency in the electronic band at the Fermi Energy found extensively in the cuprates.

The crystal structure of the highest T_c oxide superconductor which does not include Bi or Cu, $LiTi_2O_4$, is shown in figure 3. This (Spinel) structure is based on the cubic closest packing of oxygens in which one half of the available octahedrally coordinated interstitial sites are occupied by Ti and one eighth of the available tetrahedral sites are occupied by Li. The TiO_6 octahedra share edges to form infinite one-dimensional chains. The chains, however, intersect in three directions, making the structure fully three-dimensional. The Ti-Ti separation across the shared edges is short enough so that there is direct metal-metal bonding and the conduction band is of primarily Ti 3d character.

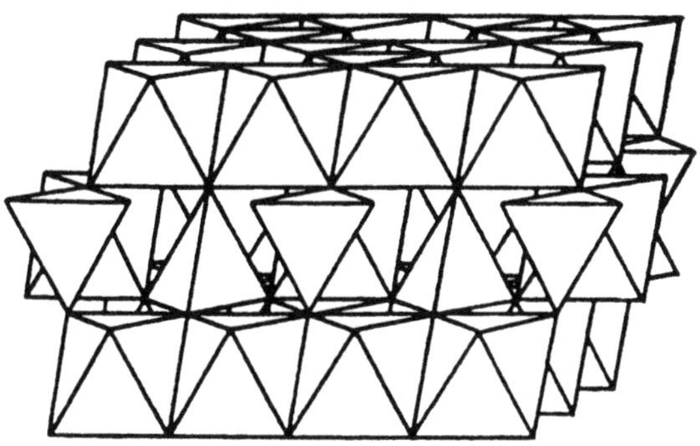

Fig 3: The spinel crystal structure of $LiTi_2O_4$. Shown are LiO_4 tetrahedra and TiO_6 octahedra.

Two other old oxide superconductors, NbO and TiO, with T_cs of 1K or below, also show clear, extensive metal-metal bonding. These can be considered as defective rock-salt structures, but are more easily seen as a network of MO_4 squares sharing corners, as shown in figure three. The metal-metal separation between the faces of some of the squares is very small and so a conduction band of metal d character is formed.

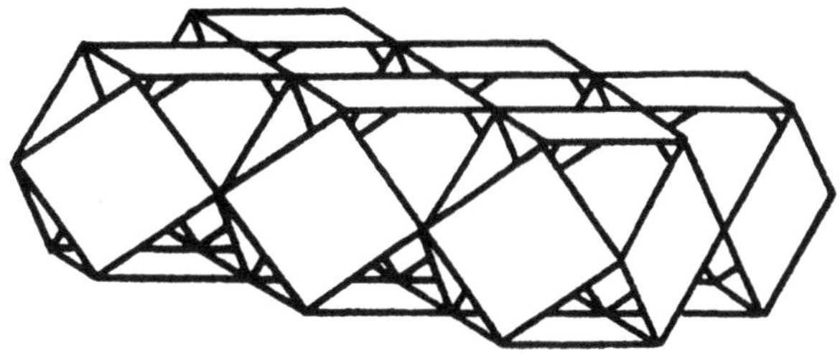

Fig 4: The crystal structure of NbO. The NbO_4 square coordination polyhedra are represented as simple squares: oxygens at the corners and Nb in the centers of the squares not shown for clarity.

Perhaps the most extensively studied of the "old" oxide superconductors are those of the type A_xWO_3 which have the hexagonal tungsten bronze structure, shown in figure 5. The structure occurs most commonly for the large alkali A ions K, Rb, and Cs. The crystal structure consists of planes of WO_6 octahedra sharing corners in the three-fold pattern shown in the figure: this pattern of octahedra is repeated in all layers, stacked one directly on top of the next. The triangular pattern brings the metals closer to one another than in the perovskite structure, and the greatly extended orbitals of tungsten make these compounds have orbital character at the Fermi level dominated by the metal orbitals. Many materials are known to be superconducting in this family, with T_cs ranging from approximately 2K to 7K. A variable alkali ion content in the tunnel is possible, which allows the adjustment of the electron count to optimize T_c. In spite of the wide range of work done on these materials there remains in the literature considerable disagreement about the composition dependence of their physical properties, which might be worth reconsideration.

Fig 5: The perovskite and hexagonal tungsten bronze structures. Shown is one elementary layer, made of MO_6 octahedra sharing corners, with the large ions in the cavities.

The most unusual of the oxide superconductors known before 1986 is $BaPb_{.75}Bi_{.25}O_3$, with a T_c of 12K. This material has the simple perovskite crystal structure (fig. 5), with a small tilting of the $(Pb/Bi)O_6$ octahedra, and Pb and Bi disordered among the octahedra. $BaPbO_3$ is a metal because covalent Pb-O bonding gives rise to a broad Pb 6s-O2p band which overlaps in energy with another more narrow band dominated by oxygen character. $BaBiO_3$, on the other hand, is an insulator with the perovskite structure. The insulating state forms due to the electronically driven charge disproportionation of Bi^{4+} to Bi^{3+} and Bi^{5+} (speaking for simplicity's sake from a "formal valence" point of view - the real situation is more complicated, involving holes on the oxygen p orbitals as well.) and localization of charge. The original experiments seeking superconductivity in the $BaPbO_3$ - $BaBiO_3$ solid solution were motivated by the prevailing views art the time that superconductivity might be induced in such s system exactly where one passed from the conducting to insulating state, which turned out to be correct in this case, as it is in the cuprates. When Bi or Sb is added to $BaPbO_3$ their electrons are donated to the broad Pb-O band, and superconductivity is induced at partial substitutions of 25%, with T_cs of 12 and 3.5K respectively, near the composition induced metal-insulator transition. The scattering from the Bi and Sb cores is not sufficient to cause localization in the three-dimensional materials. Starting from $BaBiO_3$ itself, metallic conduction and superconductivity can be induced by doping with K on the Ba site, thus leaving the electronically active sites alone, resulting in a T_c of 30K. Ba has one more electron than K so the K substitution drives the electron count in the same direction as does Bi substitution when starting from $BaPbO_3$. The optimal T_cs are found at considerably different electron concentrations in the two cases, however, as shown in figure 6. The overriding opinion in the community at the present time is that these materials are conventional electron-phonon coupled superconductors, due to a body of evidence which includes significant oxygen isotope effect. I for one am not so convinced, remembering how easy it was in the beginning to dismiss the Tcs of cuprates as being "explainable" by conventional BCS when their T_cs were on the order of 30K. Layered materials such as Ba_2PbO_4, $Ba_3Pb_2O_7$ and $Ba_4Pb_3O_{10}$ are known in this family, but their electrical properties on doping seem to be dominated by the effects of scattering of the Bi on the Pb sites, as they all become more insulating on doping.

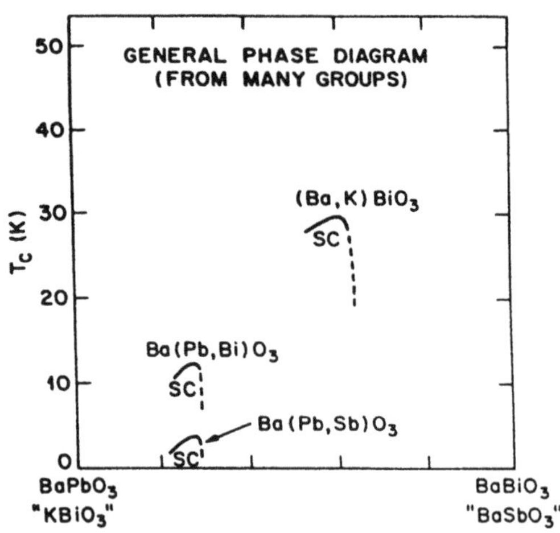

Fig 6: A composite Tc vs. Electron count/composition diagram for the $BaPbO_3$-$BaBiO_3$ based family of perovskite superconductors.

3. COPPER OXIDE SUPERCONDUCTORS

The cuprate superconductors entered the public spotlight in November of 1986, with the historic report of superconductivity in La-Ba-Cu-O at 28K. Although strictly from a temperature point of view this was not astoundingly higher than the highest T_c known for intermetallic superconductors, the fact that it occurred in a ceramic material was truly remarkable. The discovery is a tribute to the creativity of George Bednorz and Alex Muller - they looked for superconductivity where no-one else expected to find it. With the discovery of superconductivity in Y-Ba-Cu-O at 90K, the "Holy Grail of Solid State Physics", in the laboratories of M.K. Wu and C.W. Paul Chu soon thereafter, the world of condensed matter physics was never the same.

Many reviews have been written about the structures and crystal chemistry of the layered copper oxide based compounds which give rise to high temperature superconductivity, and so I will not repeat that here. The structures, briefly, consist of layers of CuO_4 sharing corners in the familiar checkerboard geometry, alternating with what are now commonly called "charge reservoir layers" which the chemist uses to control the amount of hole-doping on the CuO_2 planes. The complexity of these materials has increased with time - as chemists adjust the sizes and chemical bonding in the charge reservoir and intermediary layers to synthesize compounds by design - to obtain members of structural families which they expect to exist but which had not been observed in simpler chemical combinations.

It would have been useful indeed in the early days of the field to have set up a "commission" to set some minimum standard of data quality and reproducibility for reporting new superconductors. An almost countless number of "false alarms" have been reported in the past decade, some truly spectacular. Koichi Kitazawa from the University of Tokyo coined these reports "USOs", for Unidentified Superconducting Objects, in a clever cross-cultural double entendre likening them to UFOs (Unidentified Flying Objects, which certainly are their equivalent in many ways) and to "lies" in the Japanese translation of USO. These have caused great excitement on occasion, but more often distress. It is important, however, to keep in mind what a report of superconductivity at 130K in a ceramic material two decades ago might have looked like to rational people if it came out of the blue sky with no precedent. That having been said, it is true that **all** the reports of superconductivity in new materials which were later confirmed to be true did conform to some minimum standard of reproducibility and data quality. I have tried to keep up with which of the reports have turned out to be true and which haven't. There have been two common problems: 1. Experimental error- due, generally, to inexperienced investigators unfamiliar with measurement methods or what is required to show that a material is superconducting. This has become more rare as the field matures. And 2.: "New" superconductors are claimed in chemical systems already known to have superconductors containing some subset of the components. This is common even now, and can be difficult for even experienced researchers to avoid. The previously known superconductor is present in small proportions, sometimes in lower T_c form due to impurities added by the experimentalist trying to make a new compound. In a particularly nasty variation on this, sometimes extra components not intentionally added are present - such as Al from crucibles or CO_2 from exposure to air some time during the processing. I wish I had a dollar for every false report of superconductivity in a Nb containing oxide where the authors had unintentionally made NbN in small proportions.

Table 2 represents what I believe to be the structurally and chemically unique superconducting copper oxides discovered since 1986. If you know of any which I have missed, and which have confirmed superconductivity, please let me know (with references). Some are grouped together and counted as a single superconductor: for instance the 123 structure forms for Ba-Y-Cu but not for Sr-Y-Cu, where an additional element is required on the "chain" site for stabilization: about once every 6 months a "new" superconductor is reported with a different stabilizing ion present, I count them all together as one new superconductor.

Finally, I show in figure 7 the rate of new copper oxide superconductor discovery vs. time in the past decade. What might not be generally recognized by the non-expert is the fact that the discovery rate is dual peaked. The first peak represents the dramatic appearance in 1988 of the new Bismuth and Thallium based families of superconductors. The second peak, somewhat unexpected after so much work had already been done, reflects the appearance in many laboratories around the world of high pressure synthetic apparatus - and the resulting discovery of many new materials. Included in the secondary peak are the mercury-based and carbonate based intermediary layer materials, both of which benefited greatly from the high pressure synthetic techniques. This technique was critical in the discoveries of all the new superconductors found in 1996.

Table 2: Copper Oxide Superconductors Found Since 1986

$Ba_2YCu_3O_7$
$Ba_2YCu_4O_8$
$Ba_4Y_2Cu_7O_{15}$

"$Bi_2Sr_2CuO_6$"
$Bi_2Sr_2CaCu_2O_8$
$Bi_2Sr_2Ca_2Cu_3O_{10}$

$Tl_2(Ba,Ln)_2CuO_6$
$Tl_2Ba_2CaCu_2O_8$
$Tl_2Ba_2Ca_2Cu_3O_{10}$
$Tl(Ba,Ln)_2CuO_5$
$TlBa_2CaCu_2O_7$
$TlBa_2Ca_2Cu_3O_9$
+ (Tl,Pb)Sr equivalents

$Sr_{1-x}CuO_2$
$(Sr,Nd)CuO_2$

$(X,Cu)(Eu,Ce)_2(Eu,Sr)_2Cu_2O_8$ X=Pb,Ga

$Hg(Ba,La)_2CuO_{4+\delta}$
$HgBa_2CaCu_2O_{6+\delta}$
$HgBa_2Ca_2Cu_3O_{8+\delta}$
$Hg_2Ba_2CaCu_2O_{7+\delta}$
$(La,Ba)_2CuO_4$
$La_2CuO_{4+\delta}$
$(Nd,Ce)_2CuO_4$

$Nd_2CuO_{4-d}F_\delta$
$Sr_2CuO_3F_x$
$(Nd,Ce,Sr)_2CuO_4$
$(La,Sr)_2CaCu_2O_6$
$(Eu,Ce)_2(Eu,Sr)_2Cu_3O_9$

$Pb_2(Sr,Ln)_2Cu_2O_6$
$PbSr_2(Ln,Ca)Cu_3O_8$

$(Cu,X)Sr_2(Y,Ca)Cu_2O_7$
X= Pb, CO_2, SO_3, BO_3,
Bi, Mo

$GaSr_2(Y,Ca)Cu_2O_7$
$NbSr_2(Nd,Ce)_2Cu_2O_{10}$

$(Cu,CO_2)(Ba,Sr)CuO_{3+\delta}$
$(Cu,CO_2)(Ba,Sr)CaCu_2O_{5+\delta}$
$(Cu,CO_2)(Ba,Sr)Ca_2Cu_3O_{7+\delta}$

$TlO_x(CO_3)_y(Sr,Ca)_{n+1}Cu_nO_z$

$Sr_{0.4}Ca_{13.6}Cu_{24}O_{41}$

$(Sr,K)_2CuO_2Cl_2$
$(Sr,K)_3Cu_2O_4Cl_2$

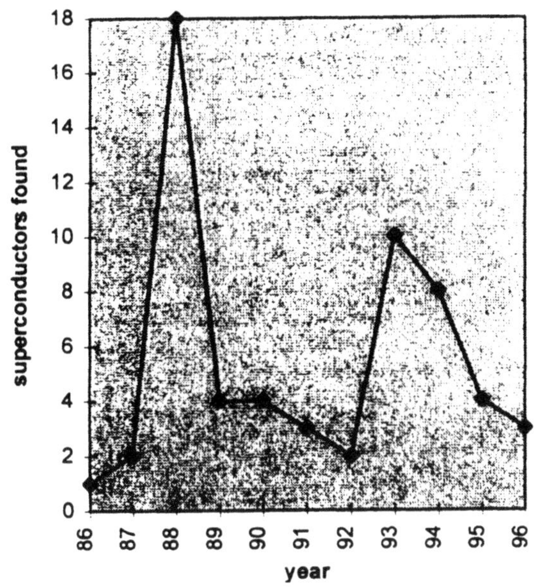

Fig 7: The Rate of discovery of new Copper Oxide Superconductors

4. CONCLUSIONS

The most important of the materials discovered in 1996, if all is confirmed to be true as originally reported, is the $Ca_{13.6}Sr_{0.4}Cu_{24}O_{41}$ superconductor. This material does not have the CuO_2 plane in the same geometry as do all other cuprate superconductors, but rather a Cu_2O_3 plane with a "2-leg ladder" geometry. Discussion of this complex material is not possible in the remaining space for this article, and may be premature at this point, but it is certainly highly possible that it has opened the door to a new kind of cuprate superconductor which may prove to be the subject of many studies in the future and may have an important influence on the theoretical models for cuprate superconductivity. It is a sign, especially, that new materials still wait to be discovered, and, as in the whole history of this field, will help to guide the intellectual and technological directions which solid state physics will follow in the decades to come.

REFERENCES

There are many excellent general references on superconducting materials. Some of them used to prepare this review are:

1. "Superconducting Materials", J. Etourneau, in *Solid State Chemistry: Compounds*, A.K. Cheetham and Peter Day, eds., Clarendon Press, Oxford, 1992, pp. 60-111

2. *Topics in Current Physics - Superconductivity in Ternary Compounds*, (in two volumes), O. Fischer and M.B. Maple, eds., Springer-Verlag, Berlin, 1982

3. *Chemistry of Superconductor Materials*, T.A. Vanderah, ed., Noyes Publications, New Jersey, 1992

4. *Concise Encyclopedia of Magnetic and Superconducting Materials*, Jan Evetts, ed., Pergamon Press, Oxford, 1992

Invited Papers

ELSEVIER

Physica C 282–287 (1997) 37–40

PHYSICA C

Novel Two-Dimensional Perovskites

J. Georg Bednorz

IBM Research Division, Zurich Research Laboratory, 8803 Rüschlikon, Switzerland

Using the example of rare earth titanium oxides, a new route for tailoring superlattices in bulk oxides will be discussed. Starting with well-known ABO_3 compounds, series of artificial structures can be created in a systematic way by oxygen intercalation, i.e. by an ordered incorporation of excess oxygen to form ABO_{3+y} compositions. This intercalation process slices the three-dimensional ABO_3 structure into subunits of various thickness, each representing a defined oxygen composition. These basic units can be combined in multiple ways to form new superlattices whose physical character varies between semiconductor and ferroelectric insulator. These variations can be achieved without changing the chemical character of the compound, in contrast to the usual procedure of extrinsic doping, which creates disorder.

With the discovery of high-T_c superconductivity [1], perovskites and related layered oxides have experienced an unexpected revival in physics, chemistry, and materials science. The numerous two-dimensional (2D) compounds under investigation in the past were created by applying either the well-known Ruddlesden–Popper scheme [2] or modifications thereof. The latter result in the formation of structures in which single or multiple transition-metal (TM) oxygen octahedra layers—in contrast to the former—may consist only of fragments (ordering of oxygen defects) and are separated by one or more block layers [3]. The common feature, however, is that the transformation from a 3D to a 2D structure is achieved by separating the TM polyhedra within a (100) perovskite plane. Only a few pure oxide compounds indicated that an ABO_3 structure can be layered by cutting it along (110) planes (Fig. 1) and inserting an additional oxygen layer such as in rare earth (RE) titanates and alkaline earth (AE) niobates or tantalates of the general composition $A_2B_2O_7$ [4–6]. $La_2Ti_2O_7$ [7], $Ca_2Nb_2O_7$ [8], and $Sr_2Nb_2O_7$ [9], to mention but a few, are isostructural and exhibit ferroelectricity with transition temperatures >1300 K. The drastic change in electronic properties occurring when transforming the 3D perovskite base compound into a layered structure is demonstrated by the example of a $LaTiO_x$ phase diagram (Fig. 2), which has been studied in detail for the range from x = 3.0 to 3.5 [10]. The single crystals used throughout our studies are obtained by applying the floating zone melting process in an argon atmosphere to ceramics in which the respective oxygen stoichiometry

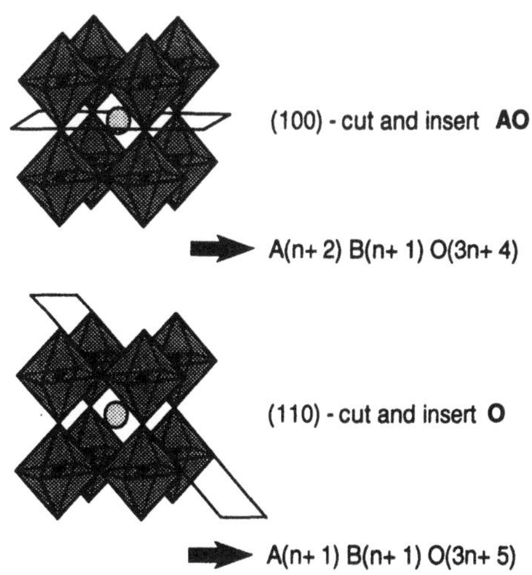

Fig. 1. Process to create layered structures from a 3D perovskite: Homologous series $A_{(n+2)}$-$B_{(n+1)}O_{(3n+4)}$ with AO = RE or AE oxide and changing the anion and cation ratio according to the Ruddlesden–Popper scheme (top). New homologous series $A_{(n+1)}B_{(n+1)}O_{(3n+5)}$ whose structure and electronic properties change only with oxygen stoichiometry at a constant cation ratio (bottom).

PII S0921-4534(97)00201-3

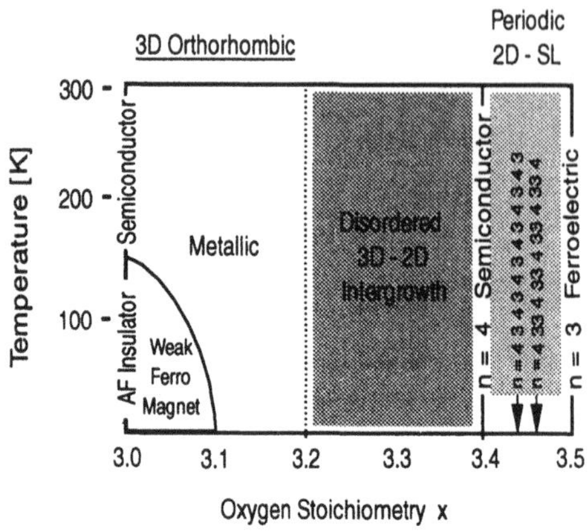

Fig. 2. Phase diagram of $LaTiO_x$ for oxygen stoichiometry x between 3.0 and 3.5 indicating structural changes from 3D orthorhombic to 2D periodic superlattices (SL) and electronic changes from antiferromagnetic (AF) insulator to metal, semiconductor, and ferroelectric insulator.

had been adjusted by employing appropriate ratios of TM oxides in various states of oxidation.

$La_2Ti_2O_7$ represents the fully oxidized end member (Ti^{4+}:$3d^0$ configuration) in the $LaTiO_x$ phase diagram at $x = 3.5$. Its structure is obtained from the 3D perovskite by making cuts parallel to a (110) plane every four Ti-oxygen octahedra layers and refilling the free positions of the resulting defect octahedra with excess oxygen ($y = 0.5$). Every other perovskite slab (consisting of 4 octahedra layers) is shifted by one-half a unit in [001] direction. Although distortions of the octahedra lead to a low symmetry unit cell, i.e. either they are monoclinic or the canting within adjacent perovskite slabs in the niobates causes the unit cell to double, it will be sufficient in the remaining discussion to treat the layers as idealized structural subunits.

Stoichiometric $LaTiO_{3.0}$ (Ti:$3d^1$) has an orthorhombic distorted, perovskite structure ($GdFeO_3$) and exhibits semiconductor-like transport properties below 300 K. At 150 K the compound undergoes a transition to an antiferromagnet, where spin canting superimposes weak ferromagnetism [11]. Approaching T_c, the system exhibits a crossover to an insulator. The phase diagram clearly displays the system's high sensitivity to excess oxygen, with the transition temperature rapidly decreasing to zero in a narrow range of x, i.e. $3.0 < x < 3.1$. Within this range the temperature dependence of the resistivity also changes gradually to metal-like with increasing excess oxygen; minima at T_c suggest a metal-insulator transition. The 3D structure, though increasingly disordered, remains stable up to $x = 3.2$, as revealed by transmission electron microscopy (TEM) and X-ray powder diffraction. When this threshold is exceeded, measurements of the electronic properties will be misleading because the microstructure of samples up to $x = 3.4$ indicates the occurrence of phase separation. As a result, a disordered intergrowth between the 3D and a new layered structure in the form of irregular lamellae is observed. Only the adjustment of an oxygen stoichiometry of 3.4 allows the growth of the new phase in bulk crystalline form. Like the case of $La_2Ti_2O_7$, its structure consists of perovskite slabs but its thickness has increased by one octahedra layer. Therefore, according to the terminology of the homologous series $A_{(n+1)}B_{(n+1)}O_{(3n+5)}$, the former compound would represent the $n = 3$ ($La_4Ti_4O_{14}$) and the latter the $n = 4$ ($La_5Ti_5O_{17}$) member [12].

A major change, however, occurs in the electronic nature of the $n = 4$ compound with the formation of a mixed valent state (Ti^{3+}/Ti^{4+}) leading to semiconducting transport properties with thermally activated behavior [13]. The activation energy at low temperature obtained from resistivity measurements is of the order of only a few millielectron-volts. Although the anisotropy of the structure is reflected by the resistivity ratio $\rho_\perp : \rho_\parallel$ ($\approx 10^2$) at room temperature (RT), its decrease with decreasing temperatures (≈ 10 at $T < 12$ K) can explain why no significant difference is observed in the respective activation energies.

The general trends described by the $LaTiO_x$ phase diagram, with only some differences in the magnetic behavior of the 3D phases [14,15], are also found for the Nd, Pr, and Ce titanates [16]. In the case of the AE niobates, a continuous variation of oxygen, starting from the 3D host, has

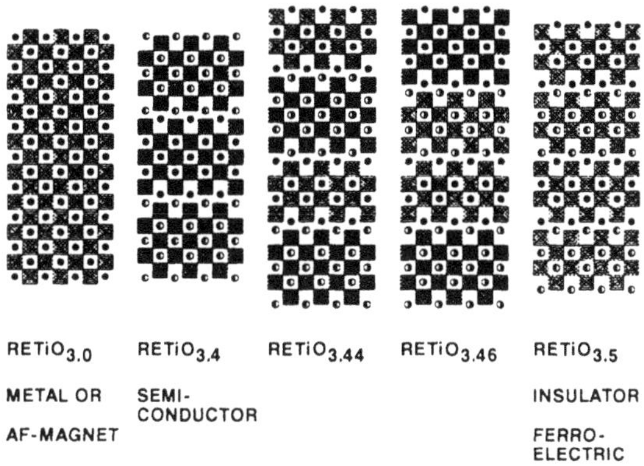

Fig. 3. Schematic representation (idealized) of structures obtained by slicing 3D perovskites along a (110) plane and subsequent oxygen intercalation. Projection along the [001] perovskite axis. Filled and half-filled circles represent La positions at different heights (0 and 1/2), which also applies to the level of the surrounding octahedra of the respective slabs.

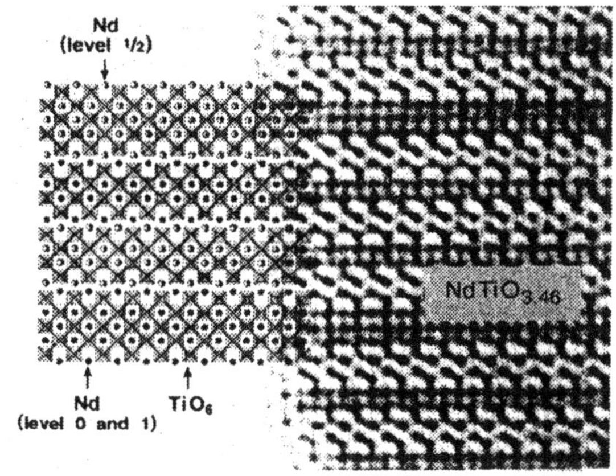

Fig. 4. TEM image of a periodic superlattice in the $NdTiO_x$ system composed of $n = 3$ and $n = 4$ subunits leading to a sequencing of $n =$ 433433433433 etc. Overlay of idealized structure is a guide to the eye.

not yet been possible, owing to the presence of defect structures with Ca and Sr deficiencies, but the concept of oxygen intercalation is indeed applicable [13,17]. The range of x between 3.4 and 3.5 is of particular interest in all these compounds because, in addition to the members of the homologous series ($n = 3$ and $n = 4$), a combination of their respective subunits (Fig. 3) allows new superlattices to be created [18]. These subunits represent the basic building blocks of an insulator ($n = 3$) and a conductor ($n = 4$), which are combined in a bulk preparation process in a ratio of 1:1 or 2:1 (the latter of which is shown in Fig. 4) without any interfaces. The consequence of the progressive decoupling experienced by the "semiconductor" layers as the combination ratio is increased is reflected by an enhanced anisotropy over the entire temperature range between RT and 4 K. Hence, in perovskite-type oxides, it is possible to create a pure 2D confinement only a few atomic layers thick. This may eventually allow the various properties of the 3D perovskites to be studied in an ideal two-dimensionality when the method of oxygen intercalation can be applied to a larger variety of 3D compounds.

In conclusion, it is shown that, by slicing the 3D perovskite along (110) and performing subsequent oxygen intercalation, the crystal structure and the electronic structure can be varied without changing the cation ratio. In addition to creating a new homologous series, new semiconductor superlattices with strong 2D confinement can be obtained for a series of RE titanates and AE niobates.

REFERENCES

1. J. G. Bednorz and K. A. Müller, Z. Phys. B **64**, 189 (1986).
2. S. N. Ruddlesden and P. Popper, Acta Cryst. **11**, 54 (1958).
3. Y. Tokura and T. Arima, Jpn. J. Appl. Phys. **29**, 2388 (1990).
4. M. Gasperin, Acta. Cryst. B **31**, 2129 (1975); K. Scheunemann and H.-K. Müller-Buschbaum, J. Inorg. Nucl. Chem. **37**, 1897 (1975).
5. K. Scheunemann and H.-K. Müller-Buschbaum, J. Inorg. Nucl. Chem. **36**, 1965 (1974).
6. N. Ishizawa, F. Marumo, T. Kawamura and M. Kimura, Acta Cryst. B **31**, 1912 (1975).
7. S. Nanamatsu, M. Kimura, K. Doi, S. Matsushita and N. Yamada, Ferroelectrics **8**, 511

(1974).
8. R.D. Rosner and E.H. Turner, Appl. Opt. **7**, 171 (1968).
9. S. Nanamatsu, M. Mimura, K. Doi and M. Takahashi, J. Phys. Soc. Japan **30**, 300 (1971).
10. F. Lichtenberg, D. Widmer, J.G. Bednorz, T. Williams and A. Reller, Z. Phys. B **82**, 211 (1991).
11. D.A. MacLean and J.E. Greedan, Inorganic Chem. **20**, 1025 (1981).
12. T. Williams, H. Schmalle, A. Reller, F. Lichtenberg, D. Widmer and J.G. Bednorz, J. Solid State Chem. **93**, 534 (1991).
13. F. Lichtenberg, T. Williams, A. Reller, D. Widmer and J.G. Bednorz, Z. Phys. B **84**, 369 (1991).
14. D.A. MacLean, K. Seto and J.E. Greedan, J. Solid State Chem. **40**, 241 (1981).
15. D.A. MacLean and J.E. Greedan, Inorg. Chem. **20**, 1025 (1981).
16. J.G. Bednorz, K. Wachtmann, R. Broom and D. Ariosa, in Proc. NATO–ASI on Materials Aspects of High-T_c Superconductivity: 10 Years after the Discovery, ed. by E. Liarokapis and E. Kaldis (Kluwer, Dordrecht, 1997) in press.
17. H.W. Schmalle, T. Williams, A. Reller, F. Lichtenberg, D. Widmer and J.G. Bednorz, Acta Cryst. C **51**, 1243 (1995).
18. T. Williams, F. Lichtenberg, D. Widmer, J.G. Bednorz and A. Reller, J. Solid State Chem. **103**, 375 (1993).

ELSEVIER

Physica C 282–287 (1997) 41–44

PHYSICA C

Strategy of the Synthesis of New Layered Cuprates and Oxycarbonates : from the Ceramics to the Thin Films

B. Raveau, M. Hervieu, C. Michel

Laboratoire CRISMAT, ISMRA et Université de Caen, UMR 6508 associée au CNRS, Bd du Maréchal Juin - 14050 Caen Cedex - France

The present paper describes the recent results obtained for the synthesis of new layered cuprates. It emphasizes two directions of research that appear most promising : the introduction of oxyanions (CO_3, BO_3, CrO_4, VO_4 etc...) between copper layers and the application of shearing mechanisms to already known layered cuprates.

Ten years after the discovery of superconductivity at high temperature in copper oxides, almost hundred original superconductors have been discovered, so that it appears most difficult to generate new materials. There is no doubt that the next generation of copper oxides, will involve metastable materials and consequently will require specific methods for the synthesis, or soft chemistry. We would like to emphasize herein, two routes that appear as most promising, taking into consideration the recent results obtained in this field.

The first direction of research is based on the fact that the infinite layer structure $Ca_{0.85}Sr_{0.15}CuO_2$ [1] should be considered as the parent structure of most of the other copper superconductors. The latter consists indeed of $[CuO_2]_\infty$ layers that are absolutely necessary for the appearance of superconductivity, so that all the other cuprates can be derived from this structure by introducing additional $[SrO]_\infty$ or $[BaO]_\infty$, and $[TlO]_\infty$ or [BiO], or $[Hg]_\infty$ layers between these layers (see for a review ref. 2). A similar mechanism

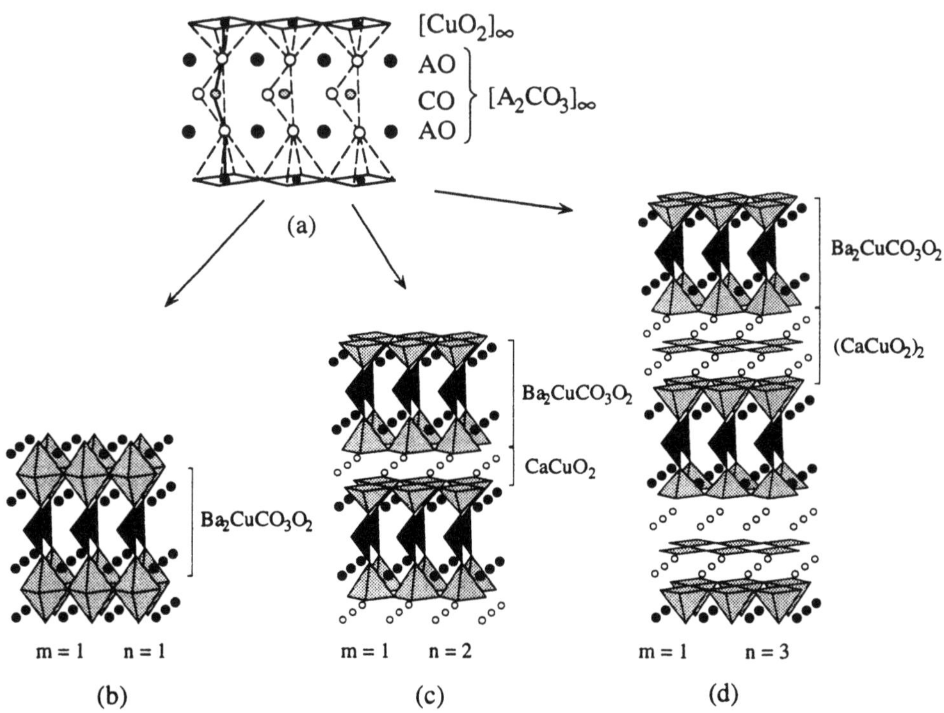

Figure 1. Substitution of $[A_2CO_3]_\infty$ layers for $[Ca]_\infty$ layers in the IL structure (a) leading to the thin film superconductors $(A_2CO_3)_mCa_{n-1}(CuO_2)_n$; as show for "m = n = 1", $Ba_2CuO_2CO_3$ (b), "m = 1, n = 2", $Ba_2CaCu_2O_4CO_3$ (c) and m = 1, n = 3", $Ba_2Ca_2Cu_3O_6CO_3$ (d).

PII S0921-4534(97)00203-7

can be envisaged by considering the possibility to introduce layers of oxyanions between the $[CuO_2]_\infty$ layers. The introduction of rows of carbonate groups between two $[CuO_2]_\infty$ layer obeys to this mechanisms. In that case two apices of each CO_3 group form the apex of CuO_5 pyramids and by the way the $[AO]_\infty$ layers ; consequently, these groups ensure the cohesion between the successive $[CuO_2]_\infty$ layers (Fig. 1a). Thus the possibility of existence of a series of layered oxycarbonates with the formulation $(A_2CO_3)_mCa_{n-m}(CuO_2)_n$ has to be considered. The recent deposition of superconducting thin films, by laser ablation has allowed such oxides to be synthesized as various intergrowths for A = Ba [3]. These compounds, for which $T_{c(onset)}$ = 110 K and $T_{c(R=0)}$ = 75 K appear as very promising superconductors. The m = n = 1 member ($Ba_2CuO_2CO_3$) does not superconduct : it consists of octahedral copper layers interconnected by rows of CO_3 groups (Fig. 1b) ; such a structure has been prepared as a bulk material for $Sr_2CuO_2CO_3$ [5]. The "m = 1 n = 2" member, $Ba_2CaCu_2O_4CO_3$ (Fig. 1c) and the "m = 1 n = 3" member $Ba_2Ca_2Cu_3O_6CO_3$ (Fig. 1d) are considered as being at the origin of superconductivity at high temperature in these compounds. Note that such an ideal formulation should involve Cu(II) only ; superconductivity is in fact easily explained by a slight substitution of CuO_4 groups for CO_3 groups leading to the actual formulation $(A_2CO_3)_{m-\delta}(A_2CuO_3)_\delta Ca_{n-1}(CuO_2)_n$ ($\delta < 0.10$).

Such an exploration should be extended to the boron compounds owing to the geometry of the BO_3 groups that are very similar to the CO_3 groups. But most important is to consider the possible introduction of tetrahedral anionic species between the $[CuO_2]_\infty$ layers. The synthesis of the superconducting oxychromate $Tl_3(CrO_4)Sr_8Cu_4O_{16}$ [6-8], that exhibits a T_c of 40 K is of great interest since it opens the route of the synthesis of a new series of oxyanion materials. This compound derives indeed from the classical "1201" structure by an ordered substitution of CrO_4 groups for thallium oxygen polyhedra, chromium being hexavalent. This original structure (Fig. 2) is closely related to that of the oxycarbonates. Like for the latter, one observes that two successive $[CuO_2]_\infty$ layers are waving in phase opposition, so that this confirms that the planar character of such layers

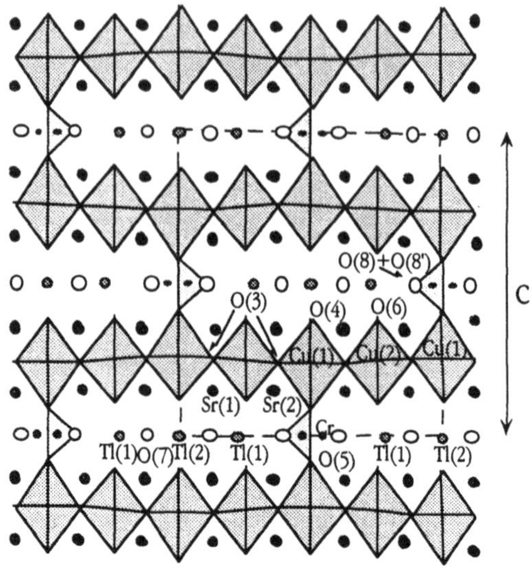

Figure 2. Projection of the structure of $Tl_3CrO_4Sr_8Cu_4O_{16}$ onto the (100) plane. The unit cell is imaged with dashed lines. Between the two possible orientations of the CrO_4 groups, one has been arbitrarily selected.

is not necessary for the appearance of superconductivity. In contrast, the $[(Tl,Cr)O]_\infty$ layers are rigourously planar, and the thallium and oxygen atoms form triple ribbons of edge-sharing TlO_4 tetrahedra and TlO_6 octahedra running along $\vec{a}$. The CrO_4 tetrahedra form also rows parallel to $\vec{a}$, but are disconnected from each other and from the other polyhedra in the layer. Clearly, the great flexibility of the two kinds of layers, suggests that other members, corresponding to different ordered substitutions of CrO_4 tetrahedra for other metallic species should also be possible : the recent substitution of chromium for univalent copper in the cuprate $Pb_2Sr_2(Ca,Y)Cu_3O_8$ [9] supports strongly this viewpoint.

The introduction of oxyanionic species should not be limited to CO_3, BO_3 or CrO_4 groups, but can be extended to other MO_n species. This is particularly the case of the mercury based cuprates where the stabilisation of such superconductors by different metallic species may correspond to the introduction of oxyanions. This is illustrated for the "1201" superconductor $Hg_{1-y}(VO_4)_y(Ba,Sr)_2CuO_{4-2y+\delta}$ [10], whose single crystal X-ray diffraction study, coupled with electron microscopy has shown for the first time the presence of VO_4 tetrahedra replacing partly the mercury atoms.

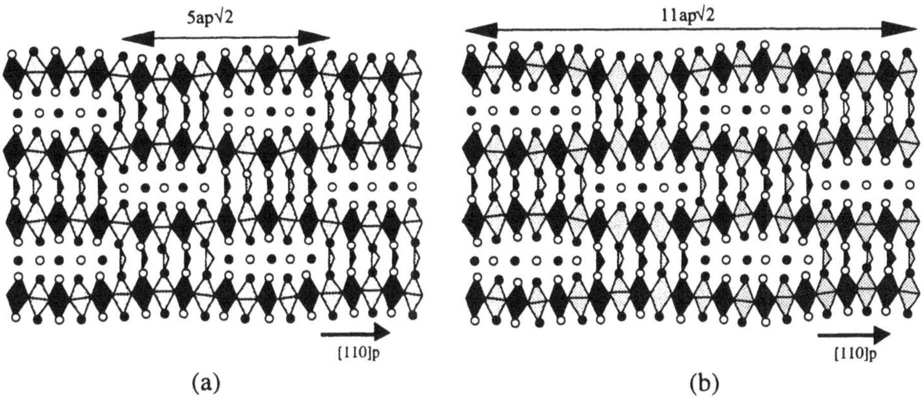

Figure 3. Idealized drawing of the (110)-collapsed structures : (a) $(Tl,M)Sr_4Cu_2O_7CO_3$ and (b) $HgBa_2Sr_2Cu_2(CO_3)O_{6+z}$.

The second direction of research concerns the great ability to apply shearing mechanism perpendicularly to the $[CuO_2]_\infty$ layers. In this way numerous collapsed oxycarbonates, involving thallium or mercury have been previously synthesized (see for review ref. 11-12). These materials are all superconductors, with Tc ranging from 50 K to 70 K. The shearing phenomena take place either along (100) or along (110), so that two kinds of superconductors named (100) or (110)-collapsed cuprates are obtained. The idealized structures of two (110)-collapsed cuprates are shown as examples on Figure 3 for $(Tl,M)Sr_4Cu_2CO_3O_7$ and $HgBa_2Sr_2Cu_2CO_3O_{6+z}$. Note that the shearing phenomenon appears every five octahedra along $[110]_p$ for $(Tl,M)Sr_4Cu_2CO_3O_7$ (Fig. 3a), whereas it takes place every five and every six octahedra alternately for $HgBa_2Sr_2Cu_2CO_3O_{6+z}$ (Fig. 3b). It results in a waving of the $[CuO_2]_\infty$ layers, confirming the very high flexibility of the structure. These results suggests that by modifying the experimental conditions, various members should be synthesized in the future, that would correspond to shearing phenomena every n octahedra $[100]_p$ or along $[110]_p$.

The application of the shearing mechanisms is not limited to the oxycarbonates. It can indeed be applied to the layered cuprates. This is the case of the collapsed "2212" cuprate $Bi_{16}Sr_{28}Cu_{17}O_{69+\delta}$ [13] and of the "1201" shear like phase $Tl_5Ba_3Sr_5Cu_3O_{19}$ [14]. The structure of the first one (Fig. 4) derives from the "2212" bismuth cuprate by a double translation in the $\{011\}_{2212}$ shear plane. It consists of "2212" ribbons that are nine

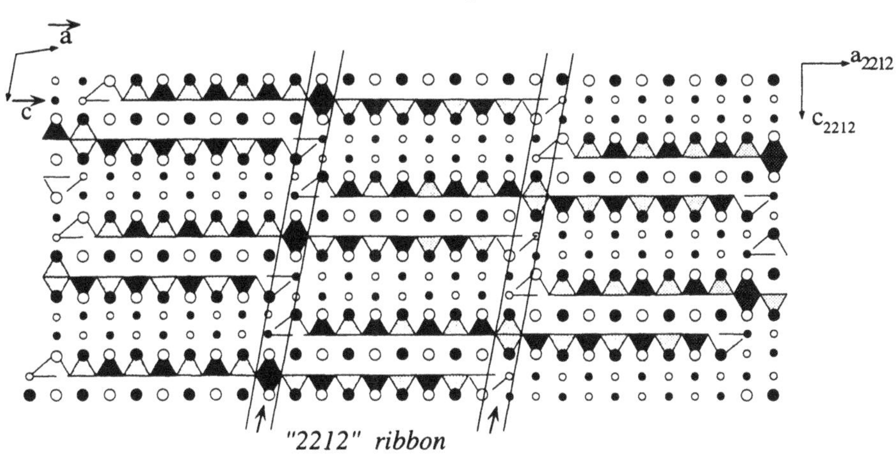

Figure 4. Idealized drawing of the double collapsed structure of $Bi_{16}Sr_{28}Cu_{17}O_{69+\delta}$.

polyhedra wide (8 CuO_5 pyramids + 1 CuO_6 octahedron), so that it can be considered as the member n = 9 of the series. Note, however that the junction between two "2212" slices is formed of CuO_6 octahedra and CuO_5 pyramids and exhibits a different arrangement from that of the "2212" structure. The ideal formula of this compound is in fact $Bi_{18}Sr_{26}Cu_{17}O_{70}$ in agreement with the nominal composition, but the actual structure is more complex than described here, involving a modulation similar to that observed for $Bi_2Sr_2CaCu_2O_{8+\delta}$. The structure of $Tl_5Ba_3Sr_5Cu_3O_{19}$ (Fig. 5) derives from the "1201" structure by applying a shearing mechanism perpendicularly to the copper layers every three octahedra, and introducing excess cations (Tl, Ba, Sr) at the junction between the so formed ribbons. In fact, this structure consists of "1201" ribbons parallel to (010) which are three CuO_6 octahedra wide ; the latter are interconnected through $Tl_2A_2O_4$ ribbons (A = Sr,Ba) whose cationic configuration is that of the rock salt structure.

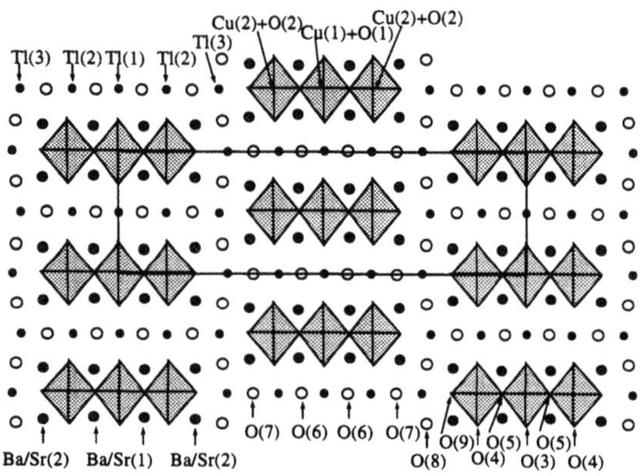

Figure 5: Projection of the structure of $Tl_5Ba_3Sr_5Cu_3O_{19}$ onto the (100) plane.

It appears clearly that this structure can be considered as the m = 3 member of a new series with the generic formulation $(TlA_2CuO_5)_m.Tl_2O_4$. Consequently, it should be possible to generate all the other members of the series. It is worth pointing out that all these shear-like cuprates, are not superconductors, as well for bismuth as for thallium. The absence of superconductivity is easily explained by the breaking of the $[CuO_2]_\infty$ layers. Although they do not superconduct, these "collapsed" cuprates are of great interest for superconductivity. Their structure is indeed closely related to that of the high T_c cuprates, so that it should be possible to introduce such structures in a coherent way as extended defects in the matrix of superconducting cuprates, in order to increase the pinning.

REFERENCES

1. T. Siegrist, S.M. Zahurak, D.W. Murphy and R.S. Roth, Nature 3354, 231 (1988).
2. B. Raveau. Nato-Asi "Materials aspects of high Tc Superconductivity 10 years after the discovery" European Cultural Center Delphi - Greece - 19-31 August 1996.
3. J.L. Allen, B. Mercey, W. Prellier, J.F. Hamet, M. Hervieu and B. Raveau, Physica C 214, 58 (1995).
5. D.V. Formichev, A.L. Kharlanov, E.V. Antipov and L.M. Koub, Superconductivity 3, 126 (1990).
4. W. Prellier, A. Tebano, J.L. Allen, J.F. Hamet, B. Mercey, M. Hervieu and B. Raveau, Chem. of. Mater. 9, 119 (1997) ; ibid 9, 319 (1997).
6. Z.E. Zheng, Y.F. Li, Y.Q. Tang, Z.Y. Cheng and D.O. Pederson, Solid State Comm. 83, 205 (1992).
7. C. Martin, F. Letouzé, A. Maignan, R. Seshadri, C. Michel, M. Hervieu and B. Raveau, Chem. of Mater 8, 865 (1996).
8. C. Michel, F. Letouzé, C. Martin, M. Hervieu and B. Raveau, Physica C 262, 159 (1996).
9. R. Seshadri, A. Maignan, M. Hervieu, C. Martin and B. Raveau, J. Solid State Chem. accepted.
10. K. Knized, S. Malo, C. Michel, M. Hervieu, A..Maignan and B. Raveau, Physica C accepted.
11. B. Raveau, C. Michel, and M. Hervieu, C.R. Acad. Sc. Paris 322, IIb, 609 (1996).
12. B. Raveau, C. Michel, B. Mercey, J.F. Hamet and M. Hervieu, J. Allogr. and Compounds 229, 134 (1995) ; J. Mater. Chem. 5, 803 (1995).
13. M. Hervieu, M.T. Caldès, D. Pelloquin, C. Michel, S. Cabrera and B. Raveau, J. Solid State Chem. 119, 169 (1995).
14. F. Letouzé, C. Martin, M. Hervieu, C. Michel, A. Maignan and B. Raveau, J. Solid State Chem. accepted

ELSEVIER

Physica C 282–287 (1997) 45–48

PHYSICA C

Superconducting properties of homologous layered cuprates synthesized under high pressure

K. Tanabe[a], T. Tatsuki, A. Tokiwa-Yamamoto, T. Tamura, X. -J. Wu, and S. Adachi

[a]Superconductivity Research Laboratory, International Superconductivity Technology Center, 1-10-13 Shinonome, Koto-ku, Tokyo 135, Japan

We review high-pressure synthesis of homologous layered cuprates and their superconducting properties, in particular, focusing on recently discovered Pb-12(n-1)n and (Hg_{1-x},Tl_x)-22(n-1)n homologous series. The synthetic way to control the number of CuO_2 plane and the way to control the doping level in each member of the series are described. The correlation of structural features of various homologous series with their superconducting properties such as T_c and the irreversibility field is discussed.

1. INTRODUCTION

The discovery of Bi- and Tl-based cuprate superconductors in 1988 [1,2] revealed that there is a series of layered cuprates which have the same charge reservoir block and only differ in the number of CuO_2 planes contained in the formula unit. The existence of such a homologous series in layered cuprates was further supported by Adachi *et al.* in 1993 who identified $Sr_2Ca_{n-1}Cu_nO_y$ (n = 2 - 4) in samples synthesized under high pressure [3]. A particularly interesting point is that the La-214 compound is considered as one of the members of this series with a single CuO_2 plane and its Sr site mostly substituted by La.

The general structure of hole-doped layered cuprates is schematically shown in Fig. 1. They consist of an alternate stack of charge reservoir (CR) block and so-called infinite-layer (IL) block in which n CuO_2 planes and (n-1) alkaline-earth (typically Ca) atomic planes are alternately stacked. There are two A-O (A is Ba, Sr, La *etc.*) planes with the rock-salt structure on both sides of the CR block. Thus all the layered cuprates can be represented by the general formula, M-m2(n-1)n, where M and m are the cation element and the number of M-O planes contained in the CR block, respectively.

Since the discovery of 02(n-1)n series in Sr-Ca-Cu-O system mentioned above, many homologous series with different M or m in the CR block have been synthesized so far by a high-pressure technique. These include Cu-12(n-1)n series with small anisotropy [4–6], Pb-12(n-1)n series which links Tl- and Bi-based series [7,8], M-12(n-1)n series with relatively small cations (M= B, C, Al, and Ga) [9–11], 02(n-1)n series with apical Cl or F [12,13], and (Hg_{1-x},Tl_x)-22(n-1)n series with an Hg-based bilayer [14–16].

In this paper, we first address some issues of high-pressure synthesis such as the way to control the number of CuO_2 planes n and the hole doping state, mainly focusing on the Pb-12(n-1)n and (Hg_{1-x},Tl_x)-22(n-1)n series recently discovered at SRL. Then the superconducting properties such as T_c and the irreversibility field for various homologous series are discussed in conjunction with their structural features.

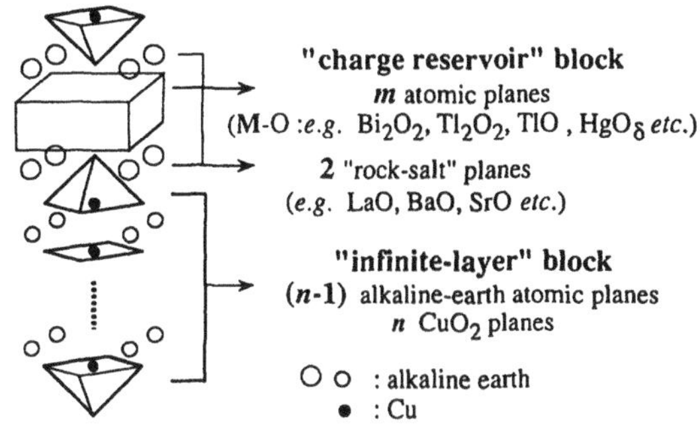

Figure 1. Schematic structure of layered cuprate superconductors.

0921-4534/97/$17.00
PII S0921-4534(97)00204-9

Table 1
T_c's of homologous layered cuprates synthesized under high pressure.

	T_c (K)					
M-m(AE)	n=1	n=2	n=3	n=4	n=5	References
Hg-1(Ba)[a]	96	127	135.4	125	110	19, 20
(Hg,Tl)-2(Ba)	(40)	90	120	117	94	14-17
Tl-1(Ba)[a]	40	109	120	122	117	21
Tl-2(Ba)[a]	90	110	130	120		2, 22
Pb-1(Sr)	39	70	122	107		7, 8
Bi-2(Sr)[a]	26	90	115	106		1
"Zero" (Sr)	(80?)	79	109	86		3
"Zero" (Sr)-Chloride	26	80	(46)	(ns)		12
Cu-1, (Cu,C)-1(Ba)	50	96	120	117	100	4-6, 18
(C,B)-1(Sr)	50	105	115			9
Ga-1(Sr)	(ns)	(70?)	(70)	107		10
Al-1(Sr)	(ns)	(ns)	(78)	110	83	11

The values in the parentheses may not be optimized yet.
[a] Most members of this series are synthesized under ambient pressure.

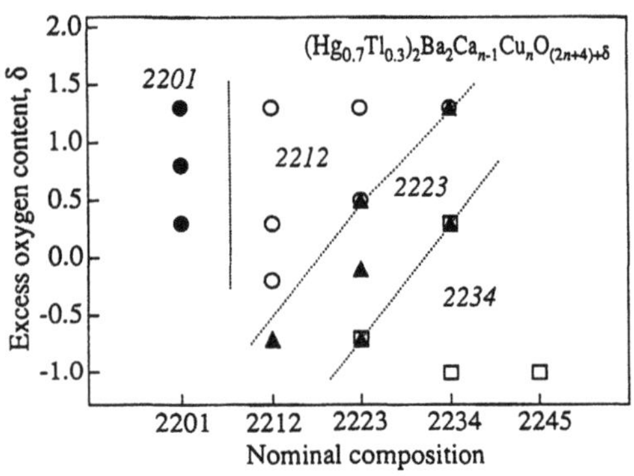

Figure 2. Relationship between observed phases and the nominal composition of starting materials, $(Hg_{0.7}Tl_{0.3})_2Ba_2Ca_{n-1}Cu_nO_{(2n+4)+\delta}$.

2. HIGH-PRESSURE SYNTHESIS

Most of the homologous layered cuprates listed in Table 1 are synthesized under typical conditions of 900 -1200 °C and 5 - 6 GPa in highly oxidizing atmosphere using $KClO_4$ or some peroxides as an oxidizing agent. One of the most important issues on the synthesis is the stabilization of CR blocks. Since the range of hole content allowed in each CuO_2 plane is not so large (0 - 0.3), the range of total minus charge in the CR block is also limited. For instance, the small valences of Cu and Hg lead to a substantial amount of oxygen vacancy in the CR block and thus structural instability. Recent success in stabilizing the Cu-O monolayer and Hg-O bilayer blocks by doping tetravalent C and trivalent Tl [6,14] is probably explained in this context.

Once the CR block of a certain homologous series is stabilized, it is necessary to control the nominal oxygen content of starting materials to obtain the phase with the desired number of CuO_2 planes. The nominal oxygen content of starting materials is often controlled by annealing precursors in oxidizing or reducing atmosphere. We have recently found an easier method to control the oxygen content which is varied over a wide range by changing the ratio of Cu metal and CuO in a powder mixture of simple oxides [16]. Figure 2 shows the relationship between obtained phases and the nominal oxygen as well as cation composition in the synthesis of $(Hg_{1-x}Tl_x)$- 22(n-1)n homologous series [16]. It is clearly seen that a larger oxygen content leads to formation of the phase with a smaller n for fixed cation compositions. This tendency seems reasonable since higher oxygen coordination of copper is probably favored under higher oxygen partial pressure.

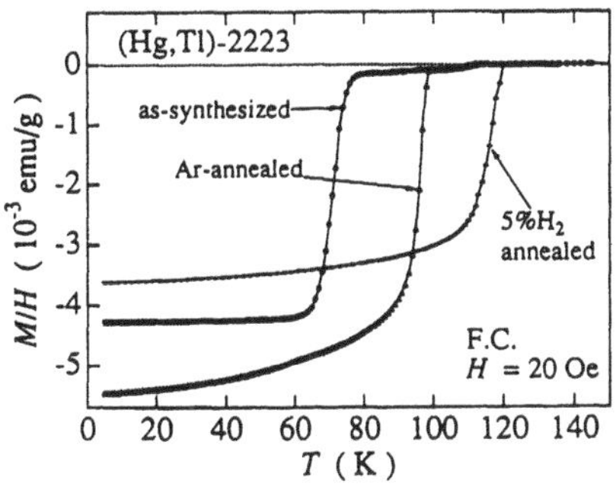

Figure 3. Temperature dependence of field-cooled dc magnetic susceptibility for as-synthesized and reduced ($Hg_{0.5}Tl_{0.5}$)-2223 phase.

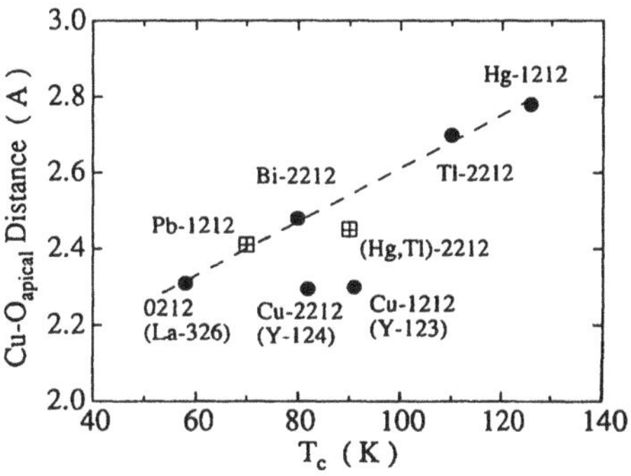

Figure 4. Correlation between T_c and the Cu-O_{apical} distance for n = 2 members of various homologous series.

3. SUPERCONDUCTING PROPERTIES

3.1. OPTIMIZATION OF HOLE DOPING

Since the above homologous layered cuprates are synthesized in highly oxidizing atmosphere, as-synthesized phases are overdoped in many cases. Typical example is the Cu-12(n-1)n series in which a significant increase in the T_c of the 1223 phase from 70 K to 120 K by reducing treatments has recently been reported [18]. Figure 3 shows the variation in superconducting transition for the ($Hg_{1-x}Tl_x$)-2223 phase subjected to reducing treatments [17]. The Ar annealing raises the T_c to 90 K and a more remarkable increase to 120 K is observed after annealing in 5% H_2/Ar. The rather sharp transition with a large volume fraction confirms that the high T_c actually originates from the 2223 phase. A substantial increase in T_c to 122 K by Ar annealing has also been observed in the Pb-1223 phase [8].

For the $n = 1$ and 2 members of each homologous series which tend to be more overdoped, the doping state can be extensively varied by substituting rare-earth elements for the A-site in the A-O plane and the Ca site in the IL block, respectively. On the other hand, the doping state of the members with $n \geq 4$ can be hardly modified even by annealing in highly reducing or oxidizing atmosphere. This is probably because the IL block is too rigid to tolerate the change in the lattice dimension accompanied with the change in the oxygen content in the CR block.

3.2. COMPARISON OF OPTIMIZED T_c

T_c values for various homologous series are summarized in Table 1. The maximum T_c in most of the series appears at $n = 3$. There are some exceptions such as Tl-12($n-1$)n and Al-12($n-1$)n series in which the $n = 4$ members have the maximum T_c's, though a possibility of further optimizing T_c's for $n = 3$ cannot be excluded. It has been reported that the T_c's of the $n = 2$ members in the series synthesized under ambient pressure (AP series) well correlate with the Cu-O_{apical} bond length and the degree of buckling of the CuO_2 plane [23]. In Figure 4, T_c is plotted versus the Cu-O_{apical} distance for some series synthesized under high pressure (HP series) as well as some AP series. The Cu-O_{apical} distances for the HP series are substantially shorter than those of Hg-12($n-1$)n and Tl-22($n-1$)n series, which seems consistent with their lower T_c of 70 - 90 K.

T_c's for the $n = 3$ members in many HP series are nearly the same. Although they are lower than those for Hg-12($n-1$)n and Tl-22($n-1$)n series, the difference is much smaller than the case for $n = 2$. This can be explained by the fact that the $n = 3$ members contain an additional flat CuO_2 plane. It is noted that each member of the (Hg,Tl)-22($n-1$)n series exhibits a similar

T_c to that of the corresponding member of the Cu-12$(n-1)n$ series. Both the series have relatively short Cu-O_{apical} distances, indicating that structural features rather than the difference in the constituent elements determine T_c.

3.3. IRREVERSIBILITY FIELD

Figure 5 shows the irreversibility field (B_{irr}) as a function of reduced temperature for various homologous superconductors. They can be roughly classified into three groups. Y-123 and the higher member of the same homologous series, Cu-1223 or (Cu,C)-1223 [25], have extremely high B_{irr}. The second group with an intermediate B_{irr} includes Pb-1223 [26], 0212-chloride as well as Tl-1223 and La-214 (0201). (Hg,Tl)-2223, Tl-2223, and Bi-2223 with two M-O atomic planes in the CR block exhibit the lowest B_{irr}. It was previously reported that the smaller anisotropy or the smaller thickness of the CR block leads to a higher B_{irr} [27]. The slightly larger B_{irr} of (Hg,Tl)-2223 than that of Bi-2223 is probably explained by the same reason. However, B_{irr} also depend substantially on the doping state or the conductivity of the CR block as well as the existence of strong pinning centers in materials. Thus further investigation is required to find a possibility of new materials with higher B_{irr} and J_c values.

ACKNOWLEDGMENTS

The authors would like to thank C. -Q. Jin, R. Puzniak, as well as H. Yamauchi for useful discussions. This work was supported by NEDO for the R & D of Industrial Science and Technology Frontier Program.

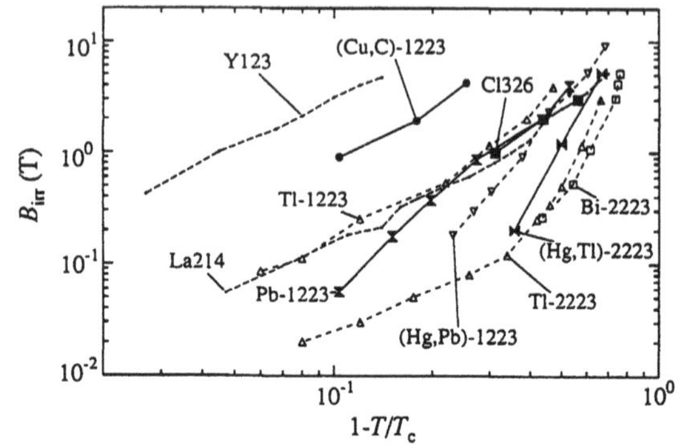

Figure 5. Irreversibility field as a function of reduced temperature for various homologous layered cuprates.

REFERENCES

1. H. Maeda *et al.*, Jpn. J. Appl. Phys. 27 (1988) L209.
2. Z. Z. Sheng and A. M. Hermann, Nature 332 (1988) 55.
3. S. Adachi *et al.*, Physica C 212 (1993) 164.
4. H. Ihara *et al.*, Jpn. J. Appl. Phys. 33 (1994) L503.
5. C. -Q. Jin *et al.*, Physica C 223 (1994) 238.
6. T. Kawashima *et al.*, Physica C 224 (1994) 69.
7. H. Yamauchi *et al.*, Jpn. J. Appl. Phys. 34 (1995) L349, and references therein.
8. T. Tamura *et al.*, Physica C (1997) in press.
9. M. Uehara *et al.*, Physica C 229 (1994) 310.
10. E. Takayama-Muromachi and M Isobe *et al.*, Jpn. J. Appl. Phys. 33 (1994) L1399.
11. Isobe *et al.*, Physica C 234 (1994) 120.
12. C. -Q. Jin *et al.*, Nature 375 (1995) 301.
13. T. Kawashima *et al.*, Physica C 257 (1996) 313.
14. A. Tokiwa-Yamamoto *et al.*, Physica C 259 (1996) 36; presented at this conference.
15. T. Tatsuki *et al.*, Jpn. J. Appl. Phys. 35 (1996) L205.
16. T. Tatsuki *et al.*, Advances in Superconductivity IX, Springer, Tokyo,1997, in press.
17. T. Tatsuki *et al.*, Physica C (1997) in press.
18. C. Chaillout *et al.*, Physica C 266 (1996) 215.
19. A. Schlling *et al.*, Nature 363 (1993) 56.
20. K. Isawa *et al.*, Physica C 222 (1994) 33.
21. H. Ihara *et al.*, Nature 334 (1988) 510.
22. T. Kaneko *et al.*, Physica C 178 (1991) 377.
23. J. D. Jorgensen *et al.*, Lecture Notes in Physics, Springer, Berlin, 1996.
24. T. Ohta *et al.*, presented at this conference.
25. H. Kumakura *et al.*, Physica C 226 (1994) 222, and references therein.
26. T. Tamura *et al.*, Advances in Superconductivity IX, Springer, Tokyo, 1997, in press.
27. K. Kishio *et al.*, Physica C 235-240 (1994) 2775.

N·H
ELSEVIER

Physica C 282-287 (1997) 49-52

PHYSICA C

The high-pressure synthesis and superconductivity of $Pr_{0.5}Ca_{0.5}Ba_2Cu_3O_z$

Y. S. Yao[a], Y. F. Xiong[a,b], D. Jin[b], J. W. Li[a], F. Wu[a] , J. L. Luo[b], Z. X. Zhao[a]

[a]National Laboratory for Superconductivity, Institute of Physics, and Center for Condensed Matter Physics, Chinese Academy of Sciences, Beijing 100080, China

[b]Cryogenic Laboratory, Chinese Academy of Sciences, Beijing 100080, China

The bulk superconductors of Ca-doped Pr-123 phase, $Pr_{0.5}Ca_{0.5}Ba_2Cu_3O_z$ were prepared at 1000°C under high pressure of 5GPa. The superconducting properties of the sample depend on the synthesis conditions. The sample synthesized under high oxygen pressure was a nearly single orthorhombic 123 phase, and shows a superconducting transition at 52K with a Meissner signal of 12%. However, the sample prepared under a reduced environment is a high purity tetragonal 123 phase and has a much higher critical temperature (T_c=97K) than that in orthorhombic phase. The results demonstrate that suppression of superconductivity by Pr in bulk Pr-123 phase can be compensated by appropriate hole doping with Ca.

1. INTRODUCTION

The problem for absence of superconductivity in $PrBa_2Cu_3O_z$ has been extensively studied but is the least understood [1-3]. To explain the insulating character of this compound, some mechanisms involving hole localization/filling or pair breaking have been proposed [2,3]. Norton et al. [4] reported that the $Pr_{1-x}Ca_xBa_2Cu_3O_z$ thin films ($0.4<x<0.5$) grown by pulsed-laser deposition exhibited a superconductivity with onset temperature as high as 47 K. Their results seem to support that hole localization is responsible for the suppression of superconductivity by Pr in 123 phase. However, no superconductivity has been found in any bulk $Pr_{1-x}Ca_xBa_2Cu_3O_z$ samples to date [5]. The single phase bulk samples with $x<0.3$ prepared under ambient pressure were all semiconducting, while the samples with $x>0.3$ were multiphase and metastable [6]. Recently, Blackstead et al. [7] claimed that they observed inhomogeneous superconductivity in $PrBa_2Cu_3O_7$ at ≈92K, and assigned the nonsuperconductivity of $PrBa_2Cu_3O_7$ to pair breaking by Ba-site magnetic Pr. But their model can not give a reasonable interpretation of Norton et al.'s results. In this paper, we report the bulk superconductivity with the transition temperature T_c=97K in $Pr_{0.5}Ca_{0.5}Ba_2Cu_3O_z$ sample synthesized under high pressure. Our results provide a convincing evidence that hole localization is the primary cause of the suppression of superconductivity by Pr in $RBa_2Cu_3O_z$.

2. EXPERIMENTAL

Precursor of $Pr_{0.5}Ca_{0.5}Ba_2Cu_3O_z$ used for the high-pressure synthesis was prepared by the conventional solid-state reaction method. The stoichiometric quantities of high-purity Pr_6O_{11}, $CaCO_3$, $BaCO_3$ and CuO powders were mixed and thoroughly ground. The mixture was fired in air at temperature of 900°C for 80 hours with several intermediate grindings. High-pressure synthesis was performed using a belt-type apparatus. The precursor was reground and pressed into pellets. Then the pellets were wrapped in gold foil and sintered under a pressure of 5 GPa at 1000°C for 0.5h. In order to make an appropriate oxygen atmosphere pressure, about 5 wt% of $KClO_4$ was added into some precursor pellets.

X-ray powder diffraction was carried out on a

0921-4534/97/$17.00
PII S0921-4534(97)00206-2

18 KW self-rotational anode x-ray diffractometer (MXP 18A-HF type) with Cu Kα radiation at room temperature. Lattice parameters were determined from least-squares fits of the diffraction lines. The microstructure of the sample was characterized by transmission electron microscopy (H-9000NA type). Electrical resistance was measured by the standard four-probe method. The DC magnetic susceptibility measurement was carried out by a superconducting quantum interference device (SQUID) magnetometer under field-cooling (FC) mode.

3. RESULTS AND DISCUSSION

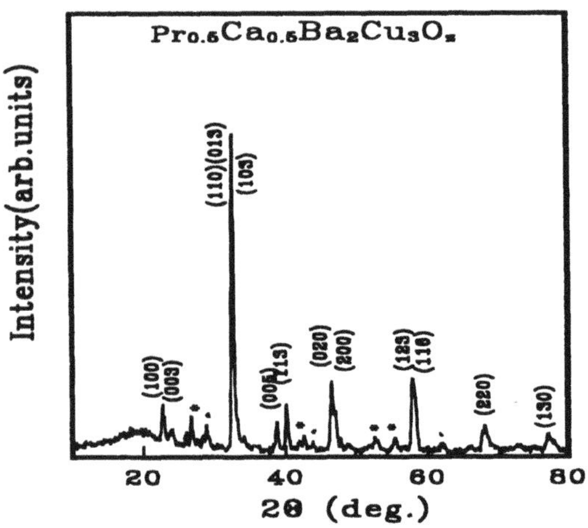

Figure 1. Rome-temperature powder x-ray-diffraction patterns of high-pressure sample of $Pr_{0.5}Ca_{0.5}Ba_2Cu_3O_z$. Several weak peaks are due to impurities (*-BaO_2 ; •-$BaCuO_2$).

X-ray diffraction reveals that the precursor sample was a complicated multiphase of tetragonal 123-phase, $BaCuO_2$ and other phases as reported by other authors[6]. Fig. 1 shows typical x-ray diffraction pattern for the bulk sample of $Pr_{0.5}Ca_{0.5}Ba_2Cu_3O_z$ treated under high pressure with $KClO_4$. It exhibits nearly single phase of an orthorhombic 123 structure with lattice constants of *a*=0.3877 nm , *b*=0.3905 nm and *c*=1.1610 nm. Small amounts of BaO_2 and $BaCuO_2$ impure phases were found, but no Cu-1212 or infinite layer structure, which is superconducting, was detected. This result strongly demonstrates that bulk sample $Pr_{0.5}Ca_{0.5}Ba_2Cu_3O_z$ of single 123 phase can be synthesized under high pressure and the high pressure synthesis can make the bulk compound of $Pr_{0.5}Ca_{0.5}Ba_2Cu_3O_z$ as stable as the thin films reported in other literature[4].

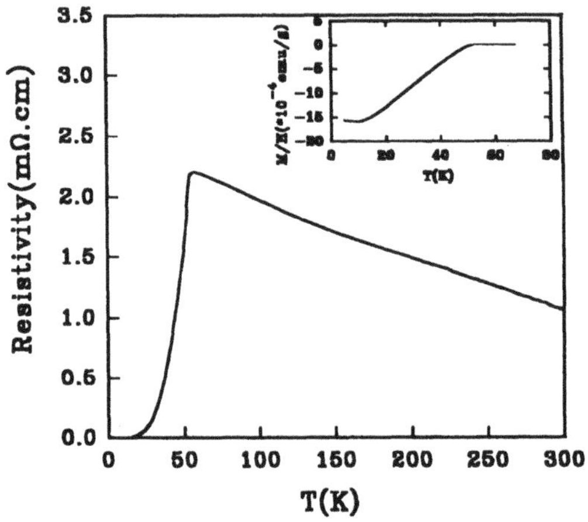

Figure 2. Resistivity as a function of temperature for the high-pressure sample of $Pr_{0.5}Ca_{0.5}Ba_2Cu_3O_z$. The inset shows the magnetic susceptibility-temperature curve of it.

The temperature dependence of electrical resistivity for the orthorhombic $Pr_{0.5}Ca_{0.5}Ba_2Cu_3O_z$ sample is shown in Fig.2. A superconducting transition is clearly found for a obvious dropping at Tc(onset) =52 K and reaching zero at about 15 K. The superconducting transition of the high-pressure bulk sample of $Pr_{0.5}Ca_{0.5}Ba_2Cu_3O_z$ was also confirmed by the DC magnetic susceptibility measurement, as shown in the inset of Fig.2. A distinctive drop of magnetic susceptibility for high-pressure sample appears at 52 K, which coincides well with the Tc onset value observed in ρ-T curve. The critical temperature is about 5K higher than that in thin films with the same composition. The superconductivity volume fraction estimated at 10 K is ~ 12%. This indicates that the bulk high-pressure sample is a bulk superconductor.

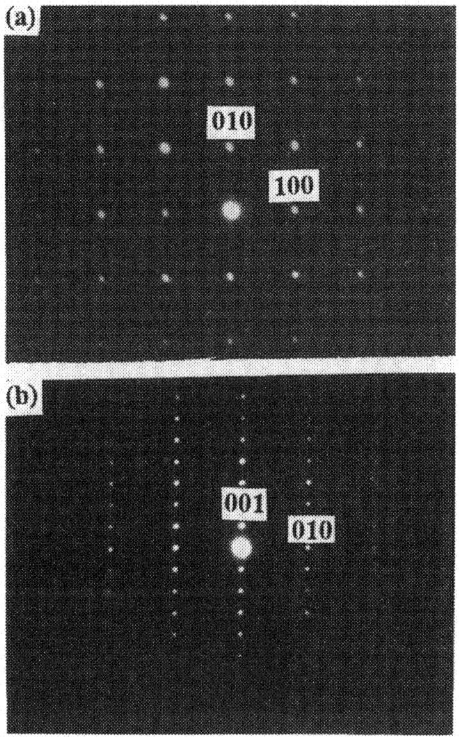

Figure 3. Electron-diffraction patterns for the high pressure synthesized tetragonal $Pr_{0.5}Ca_{0.5}Ba_2Cu_3O_z$ phase. (a) taken along the [001] direction and [b] taken along the [100]direction.

The observations above indicate that the high pressure synthesis method has an ability to properly incorporate large amounts of Ca, which exceeded the solubility limit under ambient pressure, into the bulk compound. The results reported in the literature[8,9] about Ca^{2+} substitution for R^{3+} in $RBa_2Cu_3O_7$ reveal that such a substitution will generate excess hole and Tc is suppressed by the overdoping effect. The Ca substitution essentially results in the additional holes into the CuO_2 planes, therefore, the presence of superconductivity in $PrBa_2Cu_3O_{7-\delta}$ by Ca substitution on Pr site strongly supports the view that Tc suppresses by the localization of holes, due to the hybridization of Pr 4f and O 2p orbitals, can be compensated by an appropriate Ca substitution. So, our experiments provide evidence that hole localization, instead of pair breaking must be the primary cause of the absence of superconductivity in PBCO.

It must be pointed out that the superconductivity of high-pressure $Pr_xCa_{1-x}Ba_2Cu_3O_z$ compound is sensitive not only to the Ca concentration, but also to the synthesized conditions, such as the temperature and oxidizing atmosphere. Both x-ray diffraction and electron diffraction reveal that the $Pr_{0.5}Ca_{0.5}Ba_2Cu_3O_z$ sample synthesized under high pressure but without adding any $KClO_4$, providing a strong reduced environment, displays a high purity tetragonal 123 structure with a=0.3884 nm, c=1.1646 nm. Fig. 3 shows the selected-area-diffraction patterns of the tetragonal phase. It is known that RBCO undergoes an orthorhombic to tetragonal transition when it was heavily deoxygenated, but oxygen removal occurs only in the interbarium CuO chains. Comparing the lattice parameters, it can be seen that the c-axis parameter expands in tetragonal $Pr_{0.5}Ca_{0.5}Ba_2Cu_3O_z$ phase. This c-parameter expansion was also found in $YBa_2Cu_3O_{7-y}$ system, which was explained with an enlargement of the planar copper to apical oxygen [Cu(2)-O(1)] distance[10].

Figure 4 shows resistivity as function of temperature for high pressure synthesized tetragonal $Pr_{0.5}Ca_{0.5}Ba_2Cu_3O_z$ sample. A superconducting transition was clearly observed at T_{conset} =97K. Corresponding with the drop of resistivity curve, the dc magnetic susceptibility becomes negative showing bulk superconductivity. The critical temperature not only much higher than that in high pressure synthesized orthorhombic $Pr_{0.5}Ca_{0.5}Ba_2Cu_3O_z$ sample, but also a little higher than that in other superconducting 123 family members. Neumeier *et al.*[3], based upon their result of the $(Y_{1-x-y}Ca_y)Pr_xBa_2Cu_3O_{7-\delta}$ system, suggested the fully oxygenated 123 system is overdoped and the optimal critical temperature should be 97K. This prediction is quite consistent with our result. The reason why tetragonal phase has much higher T_c than orthorhombic phase may be due to the c-parameter expansion which reduces the hybridization of Pr 4f and O 2p orbitals.

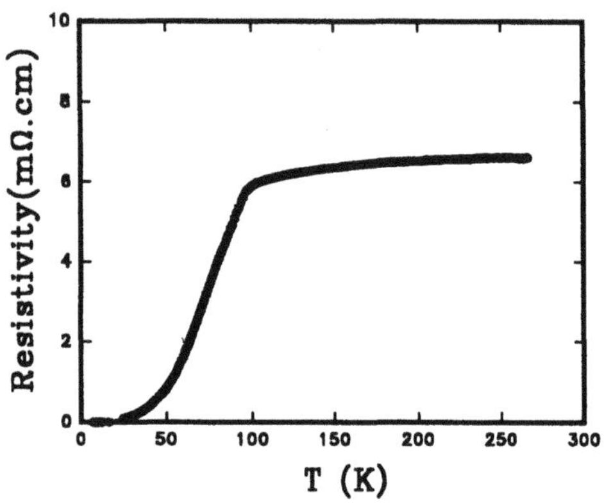

Figure 4. The temperature dependence of resistivity for the $Pr_{0.5}Ca_{0.5}Ba_2Cu_3O_z$ sample synthesized under high pressure without adding oxidizer.

3. CONCLUSIONS

Using high-temperature and high-pressure technique, the nearly single 123 orthorhombic structure superconductor with T_c=52K and 123 tetragonal structure superconductor with T_c=97K for $Pr_{0.5}Ca_{0.5}Ba_2Cu_3O_z$ have been synthesized for the first time. The results provide evidence that hole localization, instead of pair breaking must be the primary cause of the absence of superconductivity in $PrBa_2Cu_3O_{7-\delta}$.

REFERENCES

1. D.H.Lownders, D.P.Norton, and J.D.Budai, Phys. Rev. Lett. 65, 1160(1990).
2. J.Fink, N.Nucker, H.Romberg, M.Alexander, M.B.Maple, and J.J.Neumeier, Phys. Rev. B 42, 4823(1990).
3. J.J.Neumeier, T.Bjϕrnholm, M.B.Maple, and J.K.Schuller, Phys. Rev. Lett. 63, 2516(1989).
4. D.P.Norton, D.H.Lownders, B.C.Sales, J.D.Budai, B.C.Chakoumakos, and H.R.Kerchner, Phys. Rev. Lett. 66, 2516(1989).
5. H.D.Yang, M.W.Lin, C.H.Luo, H.L.Tsay, and T.F.Young, Physica C 203, 320(1992).
6. D.P.Norton, D.H.Lownders, B.C.Sales, J.D.Budai, E.C.Jones, and B.C.Chakoumakos, Phys. Rev. B 49, 4182(1994).
7. H.A.Blackstead, J.D.Dow, D.B.Chrisey, J.S.Horwitz, M.A.Black, P.J.McGinn, A.E.Klunzinger, and D.B.Pulling, Phys. Rev. B 54, 6122(1996).
8. J.C.Chen, Y.H. Xu, M.K.Wu, and W.Y.Guan, Phys. Rev. B 53, 5839(1996).
9. B.Fisher, J.Genossar, G.C.Kuper, L.Patlagan, G.M.Reisner, and A.Knizhnik, Phys. Rev. B 47, 6054(1993).
10. R.J.Cava, B.Batlogy, K.M. Rabe, E.A.Rietman, P.K.Gallagber, and L.W.rupp, Physica C 156, 323(1988).

ELSEVIER

Physica C 282–287 (1997) 53–56

PHYSICA C

Higher T_c's by chemical substitutions in layered cuprates containing Hg in the reservoir blocks

M. Marezio and F. Licci

MASPEC-CNR, via Chiavari 18/A, 43100 Parma, Italy

In order to explain why Tc of all Hg-12(n-1)n compounds increases with pressure and decreases with chemical substitutions, the variation of the Hg-1223 structure as a function of external pressure is compared to that induced in the same structure by total substitution of Ba with Sr and partial substitution of Hg with Re. The introduction of Sr in the structure seems to induce all the structural changes needed to simulate the structure stable under pressure except the thickness of the superconducting blocks, which decreases with pressure and increases by Sr substitution. The Re substitution does not seem to play an important role except for the stability of the structure and seems to contribute to produce overdoped samples.

1. INTRODUCTION

One of the strategies which may lead to the optimization of T_c, for a given superconducting system, is the stabilization at ambient pressure of the phases responsible for the higher T_c's obtained by the application of external pressure. For example, resistance measurements under pressure on the Hg-12(n-1)n series show that T_c increases as a function of pressure for every compound studied, provided that it is optimally doped or underdoped. For example, at 35 GPa Hg-1223 exhibits the record highest T_c (164 K (onset)) ever recorded [1]. These in-situ high pressure studies, carried out together with in-situ structural studies, may indicate the strategy to obtain 'higher T_c' materials by appropriate chemical substitutions. The stabilization of the structure responsible for these higher T_c at ambient pressure might prove to be very rewarding. However, these substitutions have proved to cause more problems than anticipated as all of them have induced so far decreases of T_c. By comparing the structure of Hg-1223 at 8.5 GPa and that of a substituted Hg-1223 at ambient pressure, the origin of these problems is discussed.

2. THE STRUCTURE OF Hg-1223 AS A FUNCTION OF PRESSURE

The tetragonal structures of Hg-12(n-1)n compounds contain the layer sequence: $(HgO_\delta)(BaO)\{(CuO_2)(Ca)\}_{n-1}(CuO_2)(BaO)$. They have been refined by neutron powder diffraction data at ambient pressure and as a function of pressure up to 8 GPa. They all possess similar structural features. These are: nearly flat CuO_2 layers, long apical Cu-O distances, short Hg-O distances forming the dumbbell, and very long in-plane Hg-O distances. Another important feature is the partial oxygen-depleted HgO layer. This partial occupancy is actually the doping necessary to obtain the appropriate Cu valence. The optimal value of δ for each member, which increases with n, can be reached for n=1-3. For n>3 the maximum value of δ that can be inserted in the structure seems to be smaller than the optimal value to appropriately dope the CuO_2 layers. This might prevent the optimization of Tc of these compounds.

PII S0921-4534(97)00207-4

Table 1 gives the most important distances for Hg-1223 at ambient pressure and as a function of pressure [2]. The value of Tc at the pressure at which the structural refinement was carried out, was read from the Tc(P) curve for Hg-1223 [3]. The details of the structural determination as function of pressure are reported in reference [2]. From the values shown in Table 1 it can be seen that the decrease in both a and c lattice parameters is essentially linear, $da/dP \approx -0.011$Å/GPa, $dc/dP \approx -0.090$Å/GPa, with some deviation from linearity at very high pressures. Note that the decrease of the c parameter is 8 times that of the a parameter. At ambient pressure the Cu-O apical bond distance is very large (2.72 Å) and decreases to 2.44 Å at 8.5 GPa, which corresponds to a change of 10%, while the Hg-O apical bond distance reduces by only 2%. Both these two distance lie along the highly compressible c axis. The decrease in the Cu-O distance corresponds to 38% of the total decrease of the c axis. The Cu-O in-plane bond lengths for the square and pyramidal Cu cations as well as the in-plane Hg-O distance decrease by about 2%. This small decrease of the in-plane bond lengths is the source of the small contraction of the a lattice parameter. The shortening of the Cu-O in plane bond lengths should be regarded as the result of additional doping induced by pressure.

From this behavior it can be surmised that one could simulate the effect of pressure on the structure of Hg-1223 by substituting Ba by Sr. This substitution, being Sr smaller than Ba, would induce a shortening of the lattice parameters. At the same time, the Sr cation should move closer to the oxygen located in the middle of the Hg mesh with the consequent movement of the apical oxygen toward the pyramidal Cu cation, movement needed to readjust the electrostatic equilibrium.

An attempt to substitute Ba with Sr in Hg-1201 was made by Subramanian & Whangbo [4] who prepared samples of the series $Hg(Ba_{2-x}Sr_x)CuO_{4+\delta}$ under a pressure of 0.1 GPa. These authors have been able to obtain single-phase samples up to x = 0.7; however, a decrease of T_c is obtained for increasing x while both lattice parameters decrease. Similar results have been obtained by Loureiro et al. [5] with Hg-1223. Since these measurements were not accompanied by structural refinements, it is not possible to single out the structural changes causing the T_c decrease.

3. THE STRUCTURE OF Re- AND Sr-SUBSTITUTED Hg-1223

Several other groups have used a double substitution in order to stabilize that of Ba with Sr. The second cation is either a 3d or 4d element, Ga, Pb, Bi, Tl, or Re. With the exception of Pb and Bi all the other elements are usually in their highest valence state. Every substituted sample of this type has exhibited a decrease of T_c. As in the case of the single substituted samples, no structural refinement has been carried out for these samples except for the (Re, Sr)-substituted Hg-1212 and Hg-1223 [6] and (Cr, Sr)-substituted Hg-1212 [7].

For the $(Hg_{0.75}Re_{0.25})Sr_2Ca_2Cu_3O_{8+\delta}$ samples Chmaissem et al. found that the as-prepared sample is orthorhombic with the Re cations ordered on the Hg layers. One of the two possible 1:3 orderings in the Hg layer is orthorhombic with a basal unit cell equal to (2a x 4a), whereas the other is tetragonal with a (2a x 2a) unit cell. Electron diffraction photographs showed reflections corresponding to the (2a x 4a x 2c) unit cell (a and c are those of the unsubstituted Hg-1223). All oxygen sites of the (Hg, Re) layer are occupied, but they are displaced from the ideal (1/2,1/2,0) position along the diagonal in order to form the appropriate octahedral coordination around the Re cations. This sample was found to become superconducting at 107 K. By vacuum-annealing it, an appreciable increase of T_c (from 107 to 120 K) was obtained. Table 2 gives selected interatomic distances found for the two phases by neutron powder diffraction. The values for the unsubstituted samples are given for comparison.

Table 1
Lattice parameters, interatomic distances (Å) and T_C (K) of Hg-1223 as a function of pressure

	Applied Pressure (GPa)				
	0.0	0.5	3.5	6.0	8.5
a	3.8516(0)	3.8481(3)	3.8089(3)	3.7822(5)	3.7608(6)
c	15.7644(4)	15.740(3)	15.400(4)	15.180(6)	15.018(7)
Tc	135	138	142	144	146
Hg-O (dumbell)	1.98(1)	1.98(1)	1.95(2)	1.93(2)	1.97(2)
Hg-O (in-plane)	2.72(1)	2.72(2)	2.69(2)	2.67(3)	2.66(3)
Cu1-O (in-plane)	1.926(1)	1.924(1)	1.905(1)	1.891(1)	1.880(1)
Cu2-O (in-plane)	1.926(1)	1.926(1)	1.906(2)	1.892(2)	1.883(2)
Cu2-O (apical)	2.727(7)	2.70(2)	2.61(2)	2.57(3)	2.44(3)
Cu2-Cu2*	6.35(2)	6.38(3)	6.27(3)	6.18(3)	6.20(3)
(BaO)-(BaO)**	4.66	4.59	4.51	4.45	4.44

* Distance between two Cu cations of the same superconducting block, it gives its thickness.
** This represents the average distance between two BaO layers of the same reservoir block.
Reprinted from references [2, 3]

Table 2
Lattice parameters and interatomic distances in as-prepared and vacuum annealed $(Hg_{0.75}Re_{0.25})Sr_2Ca_2Cu_3O_{8+\delta}$ samples.

	Hg-1223	as prepared	vacuum annealed
a	3.8478(0)	3.8224 (av)	3.8265(1)
c	15.7782(3)	15.1709(6)	15.2126(5)
Cu2-O apical	2.741(6)	2.43 (av.)	2.406(6)
Cu2-O in plane	1.924(0)	1.914 (av)	1.914(0)
Hg-O dumbell	1.972(3)	1.99(1)	1.931(5)
Cu2-Cu2*	6.35(1)	6.54(1)	6.54(1)
(SrO)-(SrO)*	4.71 (av.)	4.53 (av.)	4.51 (av.)

*These distances are defined in Table 1.
Reprinted from reference [6].

4. DISCUSSION

It can be seen from the distances reported in Table 1 that the effect of pressure on the structure of Hg-1223 is the shortening of all cation-anion bond lengths. However, the compressibility of the unit cell is highly anisotropic. For example, the ratio of the compressibility along c and that along a has an average value greater than 2. Moreover, bonds along the same axis exhibit very different compressibility. For example, the compressibility of the apical Cu-O bond is more than 5 times that of the apical Hg-O bond. Both of the bonds lie along the c axis. One can state that the large increase of T_c observed in all Hg-12(n-1)n compounds is accompanied by the large decrease of the apical Cu-O bond length, whereas the apical Hg-O bond is only slightly compressible. Thus, in order to increase T_c one has to envisage a structure in which the c axis is shorter than that of unsubstituted Hg-1223 and as shown by Subramanian et al. [4] this is obtained by substituting Ba with Sr.

Gao et al. [1], from the Tc(P)-Tc(P=0) versus P plot, showed that the behavior of T_c for Hg-1201, Hg-1212, and Hg-1223 is nearly identical. The effect of pressure on the structures is also similar [2, 8]. For example, Subramanian et al. [4] found for $Hg(Ba_{1.3}Sr_{0.7})CuO_{4+\delta}$ a = 3.847 and c = 9.324 Å while the corresponding lattice parameters of the unsubstituted compound are a = 3.885 and c = 9.521 Å. The Sr substitution gives the appropriate variation of the lattice parameters, but the corresponding decrease of T_c strongly indicates that at least another structural feature must have changed inducing the observed T_c decrease.

A possible explanation for this behavior can be deduced from the data reported in Table 2 where some structural parameters for $(Hg_{0.75}Re_{0.25})Sr_2Ca_2Cu_3O_{8+\delta}$ are given. If we take into consideration the results of the vacuum annealed sample, we notice that, at least qualitatively, the a and c parameters and the two apical bond lengths decrease, so that the effect of the chemical substitution is equivalent to that of the external pressure. The thickness of the reservoir block, (BaO)(HgOd)(BaO), also decreases with both the substitution and the application of pressure. The thickness of this block is the distance between two (BaO) layers and since Ba and O have different z parameters, the average of the two z's is considered. On the other hand, the Cu-Cu distance, which measures the thickness of the superconducting block, increases with the substitution whereas it decreases for increasing pressure. It should be this structural feature, that is the increase of the thickness of the superconducting layers, which is responsible for the decrease of T_c in the double substituted samples and by analogy in the single substituted samples.

5. CONCLUSION

From this analysis it is clear that a double substitution is needed for the simulation of the structure responsible for the high T_c's obtained by the application of external pressure. One substituant should be Sr. This cation induces all the structural changes needed, except one, that is the increase in thickness of the superconducting block. The second substituant should be chosen in such a way as to prevent this increase. Small cations such as Re, Mo, and Cr, likely in the 6+ state, seem to induce the wrong effect, and besides they tend to produce samples in the overdoped state as shown by the small values for the in-plane Cu-O distances (see Table 2).

REFERENCES

1. Gao et al., Phys. Rev. B 50 (1994) 4260.
2. Armstrong et al., Phys. Rev. B 52 (1995) 15551.
3. Nunez-Regueiro et al., Science 262 (1993) 97.
4. Subramanian et al., JSSC 109 (1994) 410.
5. Loureiro et al., unpublished results.
6. Chmaissem et al., Phys. Rev. B 53 (1996) 14647
7. Chmaissem et al., Phys. Rev. B 52 (1995) 15636
8. Hunter et al., Physica C 221 (1994) 1.

ELSEVIER

Physica C 282–287 (1997) 57–60

PHYSICA C

Search for Novel High Temperature Superconducting Materials: A New 126 K System, $Ba_2Ca_{n-1+x}Cu_{n+y}O_z$

C. W. Chu[*†], Y. Cao[*†], Z. L. Du[*], L. Gao[*†], K. Ross[*], I. Rusakova[*], Y. Y. Sun[*], N. L. Wu[†‡] and Y. Y. Xue[*†]

[*]Texas Center for Superconductivity and [†]Department of Physics, University of Houston, Houston, TX 77204-5932

An approach in the search for high temperature superconductors with a T_c above 120 K without volatile toxic elements is proposed and its rationale is given. The compound system $Ba_2Ca_{n-1+x}Cu_{n+y}O_z$ = $(Ca_x,Cu_y)Ba_2Ca_{n-1}Cu_nO_z$ with a T_c up to 126 K has thus been successfully synthesized under high pressure and characterized. Another three superconducting phases at 90, 105, and 116 K, respectively, were also detected. The compound system offers a possibility for high temperature superconducting (HTSg) thin-film devices operable above 120 K, reachable in space by a passive cooler, when its stability is improved; and an opportunity to investigate the dimensionality effect on high temperature superconductivity.

1. RATIONALE

The search for novel superconducting materials had been an important part of superconductivity research for decades prior to the advent of high temperature superconductivity (HTSy). Novel superconductivity compounds found had had an enormous impact on our understanding of superconductivity and the development of superconducting devices. During the search, many novel physical phenomena were also detected and new theories to account for them proposed. The discovery of HTSy a decade ago has not changed the important role of the search for high temperature superconducting (HTSg) materials in the development of HTSy science and technology.

In the last decade since the discovery of HTSy, more than 100 compounds have been found with a transition temperature (T_c) above 23 K, the record for the conventional intermetallic low temperature superconductors. All high temperature superconductors (HTSrs) with a T_c above 77 K belong to the cuprate family, which displays a layered structure [1]. Several salient features of HTSrs emerge following the study of these compounds. For instance, HTSy appears to evolve from an insulating state after introducing carriers to the CuO_2-layers of the cuprates. Therefore, a great majority of the theoretical models [2] advanced to date is based on the two-dimensional CuO_2-layer system, with special emphasis on the 1/2-spin Cu-ions for some. A question arises subsequently as to the necessity of a two-dimensional carrier system with copper and/or oxygen for HTSy to occur. The search for new HTSg compounds will extend the material base, hopefully to beyond the cuprates, for the development of a comprehensive microscopic theory of HTSy.

The large anisotropy associated with the layered structure of cuprate HTSrs poses great challenges to the fabrication of HTSg devices for both small- and large-current applications [3]. Additional challenges appear when one attempts to take advantage of the superior superconductivity properties of those HTSrs with a T_c above 120 K. This is because of the volatile toxic elements that exist in these compounds (*i.e.* Tl in Tl-2223 and Hg in Hg-1212 and -1223), making it impossible to prepare multilayer thin

[‡]On leave from National Taiwan University

This work is supported in part by NSF DMR 95-00625, EPRI RP-8066-04, the Texas Center for Superconductivity at the University of Houston, the T. L. L. Temple Foundation and the John J. and Rebecca Moores Endowment.

PII S0921-4534(97)00209-8

films with the required smooth surface morphology and difficult to melt-texture conductors by needed high temperature processing. Presently, YBCO and BSCCO are the only viable HTSrs for thin-film devices and conductors, respectively. The success of discovering HTSg compounds with a high T_c, less anisotropy, less brittleness, and/or without volatile toxic elements will make the development of many HTSr applications currently being developed easier.

Therefore, we have decided to search for novel HTSg cuprate compounds with a T_c above 120 K but without the presence of any volatile toxic elements. Here we report our successful synthesis and characterization of a novel compound system, $Ba_2Ca_{n-1+x}Cu_{n+y}O_z$ (BCCO), with a T_c up to 126 K for n = 3, x ~ 0.9±0.1, and y ~ 0.7±0.1, the highest T_c of a superconductor without volatile toxic elements. It is also the first time that an element of the IIA-family in the periodic table has become a major part of the charge reservoir of a HTSr. The compound system offers a possibility for thin-film devices operated at above 120 K, reachable by a passive cooler in space, once the stability of BCCO is improved. BCCO may also provide an opportunity to investigate the dimensionality effect on HTSy by adjusting the thickness of the charge reservoir through simple intercalation.

2. APPROACHES

It has been shown [1] that all cuprate HTSrs display a layered structure with a common stacking sequence of (charge-reservoir block)(active block). The charge-reservoir block consists of two BaO- or SrO-layers separated by other cation-oxide layers and the active block contains CuO_2-layers separated by cation-layers. The layered structure permits the introduction of charge carriers to the CuO_2-layers in the active block without introducing detrimental defects to them, *via* anion and/or cation doping of the charge-reservoir block. Compounds without a charge-reservoir, such as the so-called infinite-layered $(SrCa)CuO_2$, can never be superconducting. It has also been shown that T_c increases with the number (n) of CuO_2-layers in the active block and peaks at n = 3 or 4. The T_c-drop with n > 3 or 4 has been attributed to the possible inhomogeneous distribution from one CuO_2-layer to another in the active block [4] and/or the inability to achieve optimal doping for a maximum T_c [5]. We have therefore chosen to search for high T_c compounds without volatile toxic elements by modifying the charge-reservoir block.

Our examination of the existing T_c-data reveals that compounds with a simpler charge-reservoir block, compounds whose charge-reservoir block contains Ba, and compounds without rare earth elements usually have a higher T_c. For instance, $YBa_2Cu_3O_7$ displays a T_c ~ 93 K, in contrast to $YBa_2Cu_4O_8$, which has a more complex charge-reservoir block and shows a T_c ~ 80 K, and to $YSr_2Cu_3O_7$, which has SrO double layers in the charge-reservoir block and shows a T_c < 60 K. A simpler charge reservoir may be the easier way to retain the integrity of and improve the coupling between the CuO_2-layers in the active block. The higher T_c for compounds with double BaO-layers than with double SrO-layers may be associated with the possible higher polarizability of Ba than Sr or the possible mixing of Sr with Ca in the active block. It is interesting to note that most Ba-based perovskite ferroelectrics have a higher Curie point than the Sr-based ones. Given the structural similarity between HTSg cuprates and perovskite ferroelectrics, the above correlation of T_c with the Curie temperature may not be accidental. Therefore, we decided to synthesize HTSrs with a simple charge-reservoir block consisting of double BaO-layers, *i.e.* HTSrs in the Ba-Ca-Cu-O system.

3. RESULTS AND DISCUSSIONS

The obvious candidate would be $Ba_2Ca_2Cu_3O_{8-\delta}$, which includes a simple $(BaO)_2$ charge-reservoir and an active block of three CuO_2-layers separated by two Ca-layers. Unfortunately, such a compound is too unstable to exist even under high pressure and the valence of Cu is too difficult to adjust for maximum T_c. By inserting some Ca-ions between the double BaO-layers to reduce the Coulomb repulsion and to adjust the Cu-valence, we succeeded in synthesizing under high pressure $Ba_2Ca_{n-1+x}Cu_{n+y}O_z$ (BCCO) with T_c up to 126 K, when n = 3, x ~ 0.9±0.1, and y ~ 0.7±0.1. Details of compound synthesis will be published elsewhere [6]. Briefly, compacted pellets of binary and ternary oxides with a nominal composition $Ba_2Ca_{2+x}Cu_{3+y}O_z$ were calcinated in air or flowing O_2 between 650 and 900 °C for 40 to 48 hrs, with two intermittent pulvarizations and compactions. The pellets were subsequently powdered, mixed with AgO as oxidizer, and heated in a multi-anvil Walker module at 5-6 GPa and 800-1000 °C for 0.5 to 24 hrs. Great caution has to be taken to avoid unintended C-contamination from air or heating furnaces used for high pressure synthesis since such contamination was found to destroy the 126 K phase. The X-ray diffraction (XRD) pattern of the compound so obtained is shown in Fig. 1a. It gives a body-centered tetragonal structure with an I4/mmm symmetry and has lattice parameters of a = 3.85 Å and c = 28.2 Å. The wave dispersive spectrometer and energy dispersive spectrometer for X-ray analysis measurements show

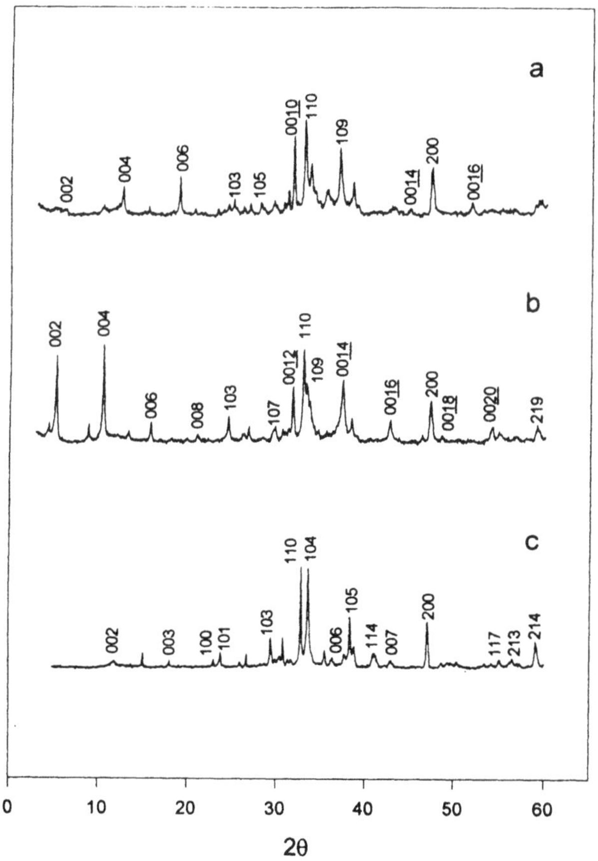

Figure 1. XRD patterns of $Ba_2Ca_{2+x}Cu_{3+y}O_z$ before (a) and after (b) exposure to humid air, and of (c) $(Cu,C)Ba_2Ca_2Cu_3O_x$.

a cation ratio of the compound to be Ba:Ca:Cu = 2:2.5±0.1:3.5±0.1. We therefore propose a schematic atomic arrangement displayed in Fig. 2a for $Ba_2Ca_{2.5\pm0.1}Cu_{3.5\pm0.1}O_8$, representing a stacking

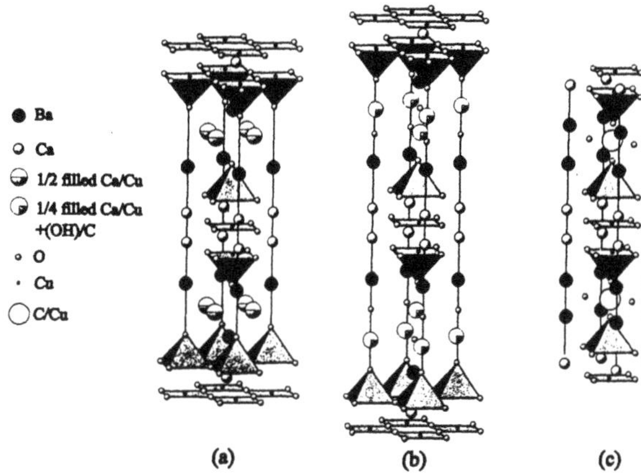

Figure 2. The proposed atomic arrangements for $Ba_2Ca_{2+x}Cu_{3+y}O_z$, before (a) and after (b) exposure to humid air, and for (c) $(Cu,C)Ba_2Ca_2Cu_3O_x$.

sequence of [(BaO)(Ca,Cu)(BaO)]-[(CuO_2)(Ca)(CuO_2)(Cu)(CuO_2)], with the (Ca,Cu)-layer only partially filled to form the charge-reservoir block originally intended. Detailed structural analysis of the compound will be discussed by Y. Y. Xue *et al.* in this Conference. The superconducting transition at 126 K is clearly evident in the dc magnetic and resistive measurements shown in Fig. 3. The BCCO-compound with a T_c = 126 K appears to be unstable in humid air and transforms to a phase with a T_c ~ 90 K as also shown in Fig. 3. The XRD results of this transformed phase in Fig. 1b show an expanded c = 33.8 Å due to the intercalation of (CO_3) and/or (OH) radicals in the charge-reservoir block. The proposed atomic arrangement is given in Fig. 2b.

It should be noted that the Ba-Ca-Cu-O system has been studied previously by several groups [7-12] and found to form $(Cu_{1-y}C_y)Ba_2Ca_{n-1}Cu_nO_x$ for n = 3, 4, or 5 under high pressure, only after partial C-substitution for Cu, with a T_c only up to 120 K [12], although a trace of superconductivity above 120 K was reported [8-12] and ascribed to a possible modified yet unknown [10], or an optimized $(Cu,C)Ba_2Ca_2Cu_3O_x$ [12] which has a primitive tetragonal structure with a P4/mmm symmetry

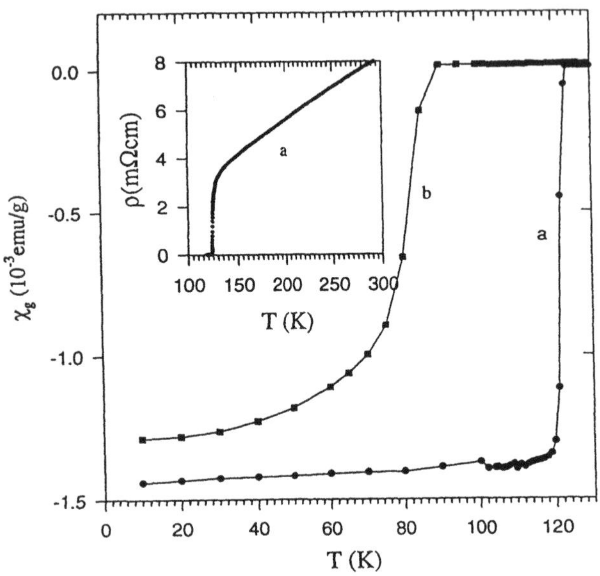

Figure 3. The magnetic susceptibility and resistivity of $Ba_2Ca_{2+x}Cu_{3+y}O_z$ before (a) and after (b) exposure to humid air.

(Figs. 1c and 2c), very different from the 126 K phase discovered here. Furthermore, prolonged annealing under different equilibrium conditions has shown [13] that the 126 K-phase and $(Cu,C)Ba_2Ca_2Cu_3O_x$ display different T_c's. All of these are indicative of the basic difference between the two phases.

4. CONCLUSIONS

In summary, we have proposed that layered cuprates have a higher T_c when they possess a simpler charge-reservoir consisting of double BaO-layers instead of double SrO-layers (perhaps due to the more polarizable nature of Ba). A novel HTSg compound system $Ba_2Ca_{n-1+x}Cu_{n+y}O_z$ has therefore been discovered and characterized with a T_c up to 126 K for n = 3, the highest for HTSrs without any toxic elements. This offers a possible HTSg compound for use in thin-film devices operable at a temperature higher than that for YBCO thin-film devices, and an opportunity for the study of the dimensionality effect on HTSy by intercalating the charge-reservoir of the compounds. Given the flexibility in adjusting the voids for cations between the double BaO-layers in the charge-reservoir block, many new HTSg compounds are expected of this system. By varying the synthesis conditions, new phases with T_c at 90, 105, and 116 K have been obtained. Their structures are being determined.

REFERENCES

1. See for example, C. W. Chu, Proceedings of the 10th Anniversary HTS Workshop on Physics, Materials and Applications, ed. by B. Batlogg, C. W. Chu, W. K. Chu, D. Gubser and K. A. Müller (World Scientific, Singapore, 1996) p. 17.
2. See for example, J. Ruvalds, Supercond. Sci. Technol. 9, 905 (1996).
3. See for example, C. W. Chu, Encyclopedia of Applied Physics, ed. by R. G. Lerner and G. L. Trigg (VCH Publishers, Inc., 1997) to appear.
4. M. Di Stasio, K. A. Müller and L. Pietronero, Phys. Rev. Lett. 64, 2827 (1990).
5. C. W. Chu, History of Original Ideas and Basic Discoveries in Particle Physics, ed. by H. B. Neuman and T. Ypsilanti (Plenum, New York, 1996) p. 793; and Q. M. Lin, Ph.D. Thesis, 1996.
6. Z. L. Du *et al.*, to be published.
7. H. Ihara *et al.*, Jpn. J. Appl. Phys. 33, L503 (1994); X. J. Wu *et al.*, Physica C 223, 243 (1994); C. Q. Jin *et al.*, Physica C 223, 238 (1994); T. Kawashima *et al.*, Physica C 224, 69 (1994); and *ibid.* 227, 95 (1994).
8. M. A. Alario-Franco *et al.*, Physica C 235-240, 975 (1994).
9. M. Núñez-Regueiro *et al.*, Physica C 235-240, 2093 (1994).
10. L. Gao *et al.*, Mod. Phys. Lett. B 9, 1397 (1995).
11. M. Jaime *et al.*, Sol. State Comm. 97, 131 (1996).
12. C. Chaillout *et al.*, Physica C 26, 215 (1996).
13. C. W. Chu *et al.*, Philosophical Magazine 75, 15 (1997)

ELSEVIER

Physica C 282–287 (1997) 61–64

PHYSICA C

Structural Features, Oxygen and Fluorine Doping in Cu-based Superconductors

E.V. Antipov[a], S.N. Putilin[a], R.V. Shpanchenko[a], V.A. Alyoshin[a], M.G. Rozova[a], A.M. Abakumov[a], D.A. Mikhailova[a], A.M. Balagurov[b], O. Lebedev[c], and G. Van Tendeloo[c]

[a]Department. of Chemistry, Moscow State University, Moscow 119899 Russia

[b]Laboratory of Neutron Physics, JINR, Dubna 141980 Russia

[c]EMAT, University of Antwerp (RUCA), Groenenborgerlaan 171, B-2020 Antwerp, Belgium

The variation of structures and superconducting properties by changing extra oxygen or fluorine atoms concentration in Hg-based Cu mixed oxides and $YBa_2Cu_3O_{6+\delta}$ was studied. The data obtained by NPD study of Hg-1201 can be considered as an evidence of the conventional oxygen doping mechanism with 2δ holes per (CuO_2) layer. The extra oxygen atom was found to be located in the middle of the Hg mesh only. Different formal charges of oxygen and fluorine inserted into reduced 123 structure results in its distinct variations. The fluorine incorporation into strongly reduced $YBa_2Cu_3O_{6+\delta}$ causes a significant structural rearrangement and the formation of a new compound with a composition close to $YBa_2Cu_3O_6F_2$ (tetragonal a=3.87Å and c≈13 Å), which structure was deduced from the combined results of X-ray diffraction, electron diffraction and high resolution electron microscopy. Fluorination treatment by XeF_2 of nonsuperconducting 123 samples causes an appearance of bulk superconductivity with T_c up to 94K.

1. INTRODUCTION

The variation of an anion concentration in the crystal structures of Hg-based layered Cu mixed oxides and 123 phases allows to change a hole concentration and T_c of these superconductors. However, different location and formal charges of oxygen and fluorine can cause distinct modifications of the crystal structures of these two groups of superconductors. Oxygen insertion into Hg-based superconductors does not change essentially their crystal structures, because these extra anions are placed in the middle of the Hg mesh, and form relatively weak bonds with surrounding cations (Fig.1a).

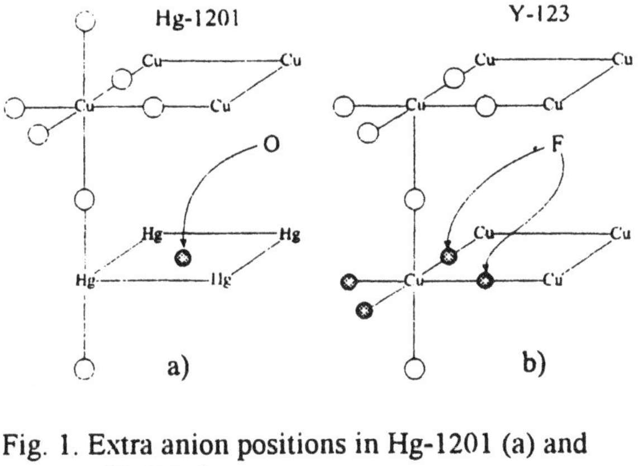

Fig. 1. Extra anion positions in Hg-1201 (a) and Y-123 (b)

Extra oxygen atoms in $YBa_2Cu_3O_7$ structure are located in the middle of the edge forming a square coordination for the Cu1 cations. Ordering of these oxygen causes orthorhombic distortion of the unit cell. Based on simple considerations of the F^- and O^{2-} similarities, one can expect that a fluorine substituted $YBa_2Cu_3O_6F_2$ phase (Fig.1b) should have tetragonal symmetry because of the replacement of a one-dimensional Cu-O chain by two-dimensional Cu-F layers. Moreover, in this fluorinated structure a structural rearrangement should occur due to the instability of the octahedral coordination of the Cu cation with practically equivalent $Cu\text{-}O_{apical}$ and $Cu\text{-}F_{in\text{-}plane}$ bonds.

The goal of the present work was to investigate the relationship between the change of the extra anions concentration, structure modifications and superconducting properties of these two groups of compounds.

PII S0921-4534(97)00210-4

2. NPD STUDY OF Hg-1201

The three-temperature method of vapor composition control allowed to adjust Hg-1201 synthesis parameters. This synthetic technique was used to prepare monophasic Hg-1201 samples at temperature T(3) [1]. $P(O_2)$ was adjusted by the mixture of metal oxides Co_3O_4/CoO at temperature T(1). P(Hg) was regulated by the temperature T(2) > T(3) from the three-phase heterogeneous equilibrium:

$$HgBa_2CuO_{4+\delta} \Leftrightarrow Ba_2CuO_{3+x} + Hg(g) + + (1+\delta-x)/2\ O_2$$

Three samples of $HgBa_2CuO_{4+\delta}$ (A, B, and C) with T_c = 71 K, 83 K, and 98 K, respectively, were studied by NPD [2]. The cation composition of one of the monophase sample with cell parameters close to the sample C was determined by ICP-AES technique. This gave a good agreement with a stoichiometric cation composition: $Hg_{0.99(1)}Ba_{2.00(1)}Cu_{0.97(1)}O_{4+\delta}$. The formal oxidation state of copper was determined by iodometric titration. Based on the iodometric data, and assuming $V_{Hg}=V_{Ba}$= +2 and V_O = −2, the formal charges of copper atoms were found to be: V_{Cu} = +2.02 (A), +2.25 (B), +2.16 (C) with statistical errors of 0.02.

Neutron diffraction experiments were performed with the high resolution Fourier diffractometer (HRFD) at the IBR-2 pulsed reactor in Dubna. The refined temperature factors of Hg and O, which form a 'dumb-bell' configuration of $Hg-O_2$ are quite large. The two parameters, n(Hg) and B(Hg), correlate strongly, that does not lead to a conclusion about their real values. The refinements carried out at fixed values of B(Hg) for sample C showed that the R_w-factor exhibited a gently sloping minimum at B(Hg) = 1.0 - 1.8 Å^2 and, accordingly, at n(Hg) = 0.9 - 1.0. A possible explanation for a low occupancy factor or a large thermal factor of Hg can be mercury substitution by copper, or a carbonate. The authors of Refs. [3,4] have proved the substitution of Hg by Cu by finding the additional extra oxygen atoms located near the middle of the edge of the Hg mesh with appropriate square coordination for the Cu cations. We tried to find this additional oxygen by calculating the experimental scattering density distribution and drawing maps for the various cross sections of the unit cell. However, only the O3 atom located in the middle of the mesh was revealed. Consequently, the noticeable substitution of Hg by Cu was not present in our samples, and we can now conclude that the large value of B(Hg) and B(O2) are the consequence of large oscillations of the HgO_2 fragment as a whole. The most probable reason for this phenomenon is the different static displacements of the HgO_2 dumbbell from 4-fold axis due to an interaction with an inserted oxygen.

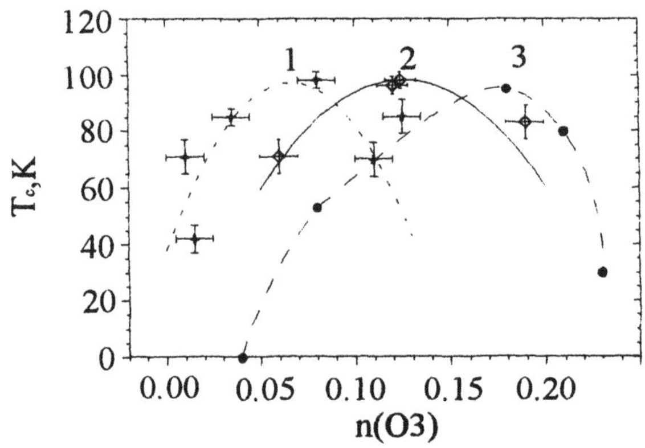

Fig.2 The behavior of T_c vs. δ from NPD (2), together with data from Ref. [5] (3) and our data of iodometric titration (1).

Special attention was made to the extra oxygen O3 occupancy refinement. The refined n(O3) parameters were found to be significantly smaller than those determined by Huang et al. [5] for the $HgBa_2CuO_{4+\delta}$ superconductors with close T_c's (Fig.2). The $T_c(\delta)$ dependence for our samples where the δ content was determined by iodometric titration, assuming $V_{Ba}=V_{Hg}$= +2, V_O= −2, is shown also in Fig.2. The parabolic-like dependencies are shifted from each other along the δ-axis with δ≈0.05. The maximum discrepancy in the δ-values for the samples with close T_c's exists between the iodometric data and the neutron diffraction data from [5]. From our analysis, including the analysis of correlation between occupancy and thermal factor of the O3 atom, we believe that the systematic error for oxygen content in our samples does not exceed 0.02. Therefore we can conclude that the optimal δ values for the Hg-1201 superconductor are in the range δ=0.08 – 0.12. The data obtained in the present work support the traditional mechanism of hole creation in the conducting band, with a concentration close to 2δ.

3. THE NEW OXYFLUORIDE: $YBa_2Cu_3O_6F_2$

The preparation of Cu-based oxyfluorides is embarrassed by the high stability of simple fluorides or oxyfluorides. The use of soft chemistry techniques is a powerful tool to overcome this problem and prepare new materials. We have showed already that XeF_2 may be successfully used as a fluorinating reagent [6] at temperatures below 300°C to prepare $Sr_2CuO_2F_{2+\delta}$, which had been discovered by Al-Mamouri et al. [7]. The main goal of this work was to perform a fluorination of reduced 123 phase using XeF_2 in order to induce the superconducting properties.

Initially $YBa_2Cu_3O_{6.95}$ and $YBa_2Cu_3O_{6.11}$ were used for fluorination. All operations were carried out in a glove box in dried N_2 atmosphere which excluded the presence of O_2. Syntheses were carried out in Ni-crucibles placed in a hermetically closed steel or Ni-container. Time and temperature of annealing were varied from 3 h till 70 h and from 100°C to 350°C.

The fluorination of $YBa_2Cu_3O_{6.95}$ did not lead to any significant changes on the X-ray diffraction pattern and in the T_c value (93K). However, the fluorination of $YBa_2Cu_3O_{6.11}$ revealed a strong dependence of the properties on the synthetic conditions. Significant changes were detected on the X-ray powder pattern of samples with a synthesis temperature above 200°C. The increase of temperature and time of fluorination led to a dramatic decrease of the intensity of reflections with *l*≠*0*. No splitting of the *hk0* reflections was observed. These X-ray patterns suggest a well established in-plane order, along with a strong disorder along the *c*-direction or the presence of numerous defects in the microstructure of the samples.

Since X-ray diffraction only allows to deduce information on the cell dimensions in the *a-b* plane, an electron diffraction was used to determine the *c*-parameter. The combination of electron diffraction (ED) and high resolution microscopy (HREM) allows to determine the approximate structure, while EDX unambiguously confirms the presence of F even on a nanometer scale. [001]* ED patterns confirm the tetragonal symmetry with a=3.87 Å. The [100]* zone (Fig.3a), however, reveals a strong disorder along the c^*-direction. Careful analysis of such ED patterns revealed the existence of at least two phases with different *c*-parameters. Real space investigation indeed confirms the presence of two closely interlinked phases with the *c*-parameters ≈11.7 Å (F1) and ≈13 Å (F2) (Fig.3b). Both phases do contain fluorine, but the F-content is definitely smaller in the former one than in the areas with an

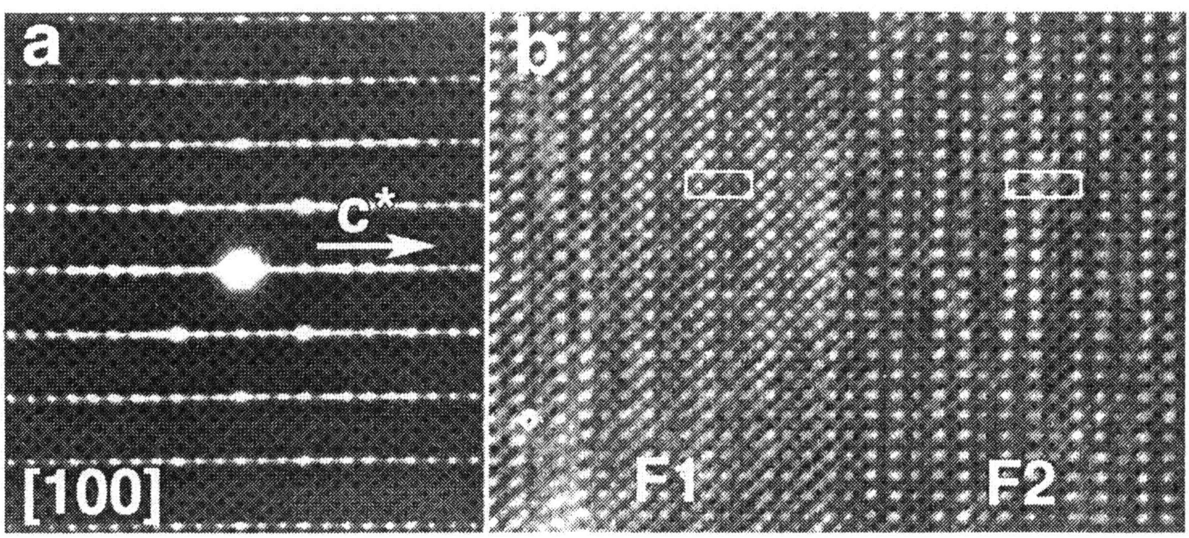

Fig. 3. [100]* ED pattern (a) and corresponding HREM image (b).

enlarged c-parameter. The latter phase either exists as noticeable bands included within the main structure or occurs as isolated defects with a limited extension. We assume that the composition of this phase is close to $YBa_2Cu_3O_6F_2$, with fully occupied anion positions in the Cu1 layer by F-atoms. HREM image simulations confirm this idea and demonstrate that fluorine has been successfully introduced into the 123 structure with the formation of a new compound.

The insertion of fluorine into a 123 structure was also evidenced from the T_c dependence upon the annealing temperature. The initial tetragonal phase is non-superconducting (V_{Cu} = +1.74), but an annealing at 250°C during 1 day already reveals a superconducting transition. The highest value of T_c= 94K was found after 50 h at 350°C (Fig.4, V_{Cu} = +2.14). The fluorinated samples exhibited a large volume fraction of superconducting phase, and this confirms the bulk nature of observed superconductivity. It should be noted that the

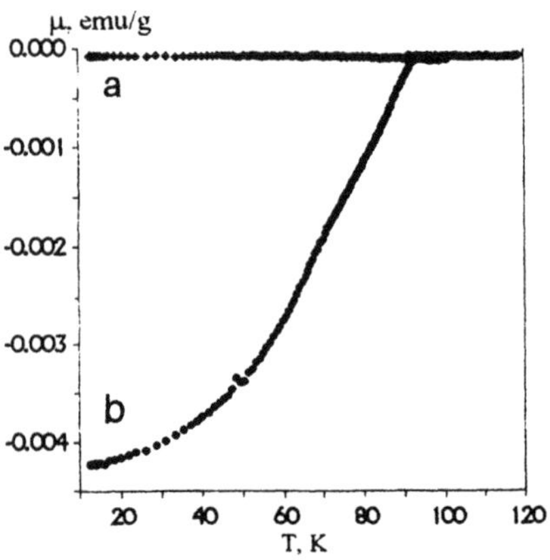

Fig. 4. χ (T) dependencies for initial (a) and fluorinated Y-123 (b)

orthorhombic 123 phase was not observed neither on X-ray patterns nor on ED and HREM images. Therefore, we can conclude that F insertion into reduced 123 structure induces superconductivity, and the revealed phase $YBa_2Cu_3O_6F_2$ might be responsible for this phenomenon.

The insertion of fluorine into the 123 structure may occur at two different positions. The most probable is an occupation of the positions in the Cu1-layer. The presence of two F atoms in this plane results in the formation of an octahedral coordination for Cu1 atoms. Such an arrangement should lead to a distortion of the octahedron due to a Jahn-Teller effect and a substantial increase of the Cu1-Cu2 separation. An increase of the Cu1-$O_{apic.}$ distance (from 1.85 Å in reduced 123 to 2.3-2.5 Å in the fully fluorinated phase) will lead to a c-parameter for the new phase of 12.8-13.2 Å. Obviously, such fluorination will result in an oxidation of the Cu atoms and keep the superconducting CuO_2 layers unaltered. All previous studies of fluorinated 123 phases reported only a decrease of the c-parameter with increasing fluorine concentration in the lattice, which suppose only a partial incorporation of F anions. A highly fluorinated 123 phase with a composition close to $YBa_2Cu_3O_6F_2$ and a significant elongation of the c-parameter of the unit cell has never been observed. To a limited extent, we cannot exclude, however, oxygen-fluorine anion intermixing between different positions.

We can conclude that the fluorination of the non-superconducting Cu mixed oxides can be successfully used to induce superconductivity, and for the synthesis of new materials as well.

Acknowledgment

This work was supported by NATO (HTECH.LG 960325), RFBR-INTAS (00639 I-96) and the Russian Council on Superconductivity.

REFERENCES

1. V.A. Alyoshin, D.A. Mikhailova, E.V. Antipov, Physica C 271 (1996) 197.
2. V.L. Aksenov, A.M. Balagurov, V.V. Sikolenko, V.G. Simkin, V.A. Alyoshin, E.V. Antipov, A.A. Gippius, D.A. Mikhailova, S.N. Putilin, F. Bouree, Phys. Rev. B 55 (1997) 3966.
3. J.L. Wagner, P.G. Radaelli, D.G. Hinks, J.D. Jorgensen, J.F. Mitchell, B. Dabrowski, G.S. Knapp, M.A. Beno, Physica C 210 (1993) 447.
4. A. Asab, A.R. Armstrong, I. Gameson, P.P. Edwards, Physica C 255 (1995) 180.
5. Q. Huang, J.W. Lynn, Q. Xiong, C.W. Chu, Phys. Rev. B 52 (1995) 462.
6. E.I. Ardashnikova, S.V. Lubarsky, D.I. Denisenko, R.V. Shpanchenko, E.V. Antipov, G. Van Tendeloo, Physica C 253 (1995) 59.
7. Al-Mamouri, P.P. Edwards, C. Greaves, M. Slaski, Nature 369 (1994) 382.

ELSEVIER

Physica C 282–287 (1997) 65–68

PHYSICA C

High Pressure Synthesis and Structure of the Superconducting Mercury Cuprates $(Hg_{1-x}M_x)Ba_2Ca_{n-1}Cu_nO_{2+2n+\partial}$ with M= C, S.

P.Bordet[1], S.Loureiro[1], S.LeFloch[1], J.J.Capponi[1], C.Chaillout[1], J.Chenavas[1], J.L.Tholence[2], E.V.Antipov[3], P.G.Radaelli[4], M.Marezio[1].

[1] Laboratoire de Cristallographie, BP 166, 38042 Grenoble Cedex 09, France

[2] LEPES, CNRS, BP 166, 38042 Grenoble Cedex 09, France

[3] Inorganic Chemistry Department, Moscow State University, 119899, Moscow , Russian Federation

[4] ILL, BP 156, 38042 Grenoble Cedex 09, France.

In mercury superconducting cuprates synthesized at high pressure, partial carbon substitution on the mercury site occurs when the precursors are not absolutely carbon free. Hg-12(n-1)n samples made from carbonate-containing precursors show lower T_cs and smaller c parameter. A model of the structural arrangement of the CO_3^{-2} groups is proposed from neutron diffraction data and crystal-chemistry considerations. The third oxygen of the oxyanion is situated in the (Hg,C)O_∂ layer and hinders the incorporation of extra-oxygen atoms.

The $(Hg_{1-x}C_x)Ba_2CuO_{4+\partial}$ phase diagram shows three different phases, namely $HgBa_2CuO_{4+\partial}$, $Hg_{0.5}C_{0.5}Ba_2CuO_{4+\partial}$ and $CBa_2CuO_{4+\partial}$. The latter two phases are not superconducting due to the insufficient hole concentration on the superconducting layer.

Sulphur can also partially replace mercury in $Hg_{1-x}S_xBa_2CuO_{4+\partial}$ and forms a solid solution up to x=0.15. Neutron diffraction shows that the sulphur atoms arrange as SO_4^{-2} oxyanions. The larger S-O distances induce steric limitations to the presence of other sulphate anions in the same mesh.

1. Introduction

In the structures of the mercury cuprates of general formula $HgBa_2Ca_{n-1}Cu_nO_{2n+2+\partial}$, oxygen-deficient perovskite blocks alternate with oxygen-deficient NaCl blocks. Tc first increases with n up to n=3 where it reaches the record value of 135 K, and then progressively decreases. Bulk samples of compounds with n≤6 have been synthetised, and their structures have been determined, mainly by using neutron powder diffraction. The specific structural features of these compounds can be summarized as follows : i) The Cu-O apical distances are very large (≈2.8 Å) compared to other high Tc cuprates. ii) The Hg cations adopt a dumbbell-type oxygen coordination along the c axis with two apical oxygens. iii) The oxygen site O_∂ in the Hg layer is located at the center of the mesh and is only partially occupied. This occupancy ∂ controls the doping of the superconducting layers and the maximum occupancy depends on n. The in-plane Hg-O distance is very long (≈ 2.75 Å), and the oxygen anion is practically only bonded to two Ba cations along the c axis.

Due to the linear coordination of Hg cations, these compounds show a great ability for the partial replacement of Hg cations by various chemical species, such as other cations or oxyanions. Numerous studies have been devoted to the effects of such substitutions on the structural and physical properties of mercury cuprates. However, the bad consequence of this ability, is the difficulty to prepare substitution-free samples which are needed for precise investigations of the physical and structural properties. It also complicates the comparison between results obtained on samples prepared with different synthesis techniques. The possibility of Hg substitution was first suspected due to the low occupancy of high thermal parameter values obtained for Hg in diffraction studies. For example, Wagner et al. [1] first reported the subtitution of Cu for Hg in Hg-1201 prepared by the sealed tube technique. Alexandre et al. [2], working with samples prepared by the high pressure-high temperature (HP-HT) technique, indicated that such substitution either did not take place, or would only exist in a very narrow range. Based on x-ray diffraction and iodometric titration results, they concluded that replacement of Hg by carbon was likely to occur in such samples. Kopnine et al. [3] also reported on the detrimental effect on Tc of the substitution of C for Hg in HP-HT samples.

0921-4534/97/$17.00
PII S0921-4534(97)00211-6

In order to understand the effect of this substitution on the structural and physical properties, we have studied the $(Hg_{1-x},C_x)Ba_2CuO_y$ $(0\leq x\leq 1)$ solid solution on samples prepared by the HP-HT technique.

A similar study was carried out for the replacement of Hg by S, which allows to compare the effects of different oxyanions.

2. The $(Hg_{1-x},C_x)Ba_2CuO_y$ solid solution

We have decided to synthesize by the HP-HT technique, samples of the $[Hg_{1-x}C_x]Ba_2CuO_y$ system with large stoichiometric range $0.00\leq x\leq 1.00$. A precursor of nominal composition Ba_2CuO_y was prepared by mixing high purity oxides in an agate mortar. The mixtures were then fired at 830°C for 10h under oxygen flow. This precursor was subsequently mixed in an argon filled dry box with different amounts of HgO, CuO and $BaCO_3$ in order to prepare the samples with general formula $Hg_{1-x}C_xBa_2CuO_y$ and the eight different stoichiometries $0.00\leq x\leq 0.50$, $\Delta x=0.10$, x=0.75 and x=1.00. The HP-HT conditions for the preparation of the sample with x=0.00 were 18kbar/920°C/0.6h. These conditions were kept the same for the first attempts to prepare the carbon-substituted phases but the results were unsatisfactory. Temperature, pressure and time were then varied in order to improve the synthesis yield.

The sample with x=0.00 contained a non-superconducting Hg-1201 phase with lattice parameters **a**=3.8676(2)Å, **c**=9.470(1)Å. The absence of superconductivity and relatively short lattice parameters indicated that the phase was overdoped. The samples with nominal compositions x=0.10, x=0.20 and x=0.30 which were synthesized under the same conditions as those used for the sample with x=0.00 contained the Hg-1201 phase with lattice parameters similar to those found for the x=0.00 phase, together with ≈5% of a new substituted $[Hg_{1-x}C_x]$-1201 phase, denoted as $HgBa_4Cu_2CO_3O_y$. The samples also contained impurities such as unreacted $BaCO_3$ and CuO. The percentage of these impurities increased with increasing nominal x. Initially, the samples with nominal composition x=0.50 were also synthesized under 18kbar/920°C/0.6h just as the samples with x ranging from 0 to 0.40. These attempts resulted in samples containing $HgBa_4Cu_2CO_3O_y$ as the predominant phase together with a minor Hg-1201 phase and the impurities quoted above. Samples with x=0.75 and x=1.0 were prepared under the same synthesis conditions. The x=0.75 sample contained $HgBa_4Cu_2CO_3O_y$ as major component, but a phase whose powder pattern could be indexed on the lattice of $CO_3Ba_2CuO_y$ [4] was also present. The Hg-1201 phase was absent. For x=1.00, this $CO_3Ba_2CuO_y$ phase [C-1201] appeared to be the major component of the sample, but the amount of impurities was approximately 50%. Since the sample with x=0.50 proved the existence of a new 1201 phase with approximate 50% replacement of Hg by C, the synthesis conditions for this composition were optimized. The best results were obtained when the time was increased to 2h while the pressure and temperature were kept at the same value. For x=1, in order to obtain a sample containing at least 90% of the $CO_3Ba_2CuO_y$ phase the pressure had to be raised to 40kbar. In fig. 1 a column diagram shows the approximate percentages of the various phases present in the sample as a function of x.

The lattice parameters of the phase obtained for the x=0.50 composition sample were found to be a=3.9271(3)Å, c=8.676(2)Å. The replacement of Hg by C induced a significant increase of the a-parameter (1.5%) and a major decrease of the c-parameter (8.4%). A sample with x=0.5 was studied by HRTEM. Superlattice peaks corresponding to an incommensurate modulation of wave vector ≈ (b*/0.33 + c*/2) can clearly be observed in electron diffraction images taken along the (100) zone axis. The corresponding electron micrograph indicates that the modulation is due to an ordering between

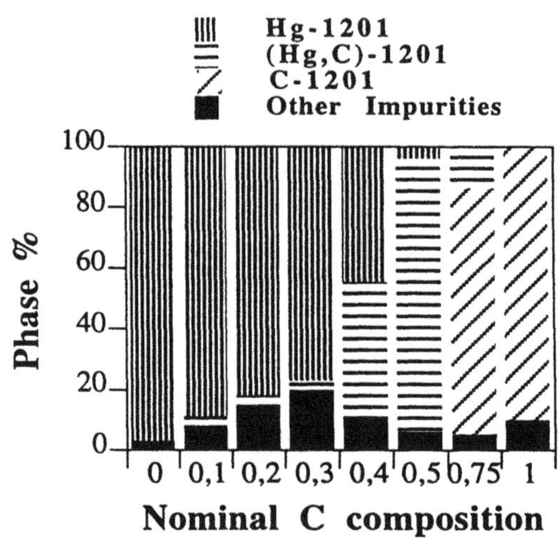

Figure 1 : Phase type % vs nominal carbonate content in the $[Hg_{1-x}C_x]Ba_2CuO_y$ system.

Hg and C cations. Similar results have also been reported by Uehara et al. [5].

The $CO_3Ba_2CuO_y$ phase obtained in the sample with nominal composition x=1 exhibited lattice parameters a=4.0044(3)Å, c=7.909(1)Å. The a and c parameters continued to increase and decrease respectively.

The a.c. susceptibility of the sample with x=0.5 indicated superconductivity at ≈63K. However, the small superconducting volume fraction indicates that this transition has to be attributed to the Hg-1201 impurity phase, rather than to $HgBa_4Cu_2CO_3O_y$, which represented ≈90% of the sample volume. For the x=0.75 and x=1 samples, no diamagnetic signal was detected down to 4K. This result confirms that the $HgBa_4Cu_2CO_3O_y$ and $CO_3Ba_2CuO_y$ phases are non-superconducting.

From these results, it clearly appears that the $(Hg,C)Ba_2CuO_y$ system is constituted of three main phases, namely: $HgBa_2CuO_{4+\delta}$, $Hg_{1-x}C_xBa_2CuO_y$ (x≈0.50) and $CO_3Ba_2CuO_y$, which appear in different proportions according to the percentage of substitution and the applied synthesis pressure. Very probably, each of these phases can accomodate a range of variation of the Hg/C substitution ratio around its average composition. The replacement of Hg by C induces major structural variations that correspond to a drastic expansion and contraction of the a- and c-parameters, respectively. The modulation observed for the x≈0.5 phase indicates a tendency toward ordering and separation of the Hg cations and carbonate groups. These structural variations lead to a suppression of superconductivity in the $HgBa_4Cu_2CO_3O_y$ phase, which may be caused by strong distortions of the conducting (CuO_2) layers or unsufficient carrier concentration.

3. The $(Hg_{1-x},S_x)Ba_2CuO_y$ solid solution

We have prepared samples by HP-HT with nominal stoichiometries $[Hg_{1-x}(SO_4)_x]Ba_2CuO_{4+y}$, having 0.00≤x≤0.30 [Δx=0.05], x=0.50, 0.75 and 1.00. The synthesis was carried out at 18kbar and 920°C.

The samples with nominal composition 0.00≤x≤0.15 were monophasic. Starting at x=0.20, small quantities of $BaSO_4$ and CuO appeared and their content increased with the percentage of sulphate substitution. The evolution of the a- and c-parameters vs x is shown in figure 2. For x≤0.2, a increases and c decreases, then both values stabilize, and impurity phases appear. This can be interpreted

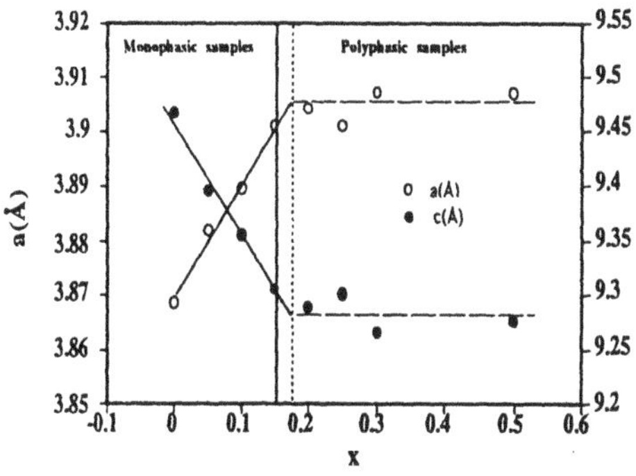

Figure 2 : Cell parameters vs. x for as-prepared $Hg_{1-x}S_xBa_2CuO_{4+2x+\delta}$ samples with 0.00≤x≤0.50

as a solid solution limit lying at x≈0.2, as indicated by the vertical line in figure 2.

Samples with x=0.00 and x=0.05 were found to be non-superconducting by a.c. susceptibility measurements. This result is consistent with highly overdoped non-superconducting Hg-1201 phases. Superconductivity appeared when the substitution reached x=0.10. The Tc(onset) is at 41K but the transition is quite broad. The sample with x=0.15 exhibited already a sharp transition with an onset at 62K. The Tc(onset) increases slightly to 65K for x=0.20, but for x=0.25 it decreased to 63K. The sample with nominal composition x=0.30 exhibited a sharp transition with an Tc(onset) at 73K. This phase is then the most reduced of the obtained phases. The sample with x=0.50 had a much smaller superconducting volume and a Tc(onset) at 71K. The increase of Tc vs x is very probably due to the decrease of the average valence of Cu in the precursor. In order to introduce sulfate anions, $BaSO_4$ and CuO are added to the Ba_2CuO_y phase for which the Cu valence is higher than 2+, as revealed by the non superconducting nature of the x=0 and 0.05 samples. The average Cu valence of the precursor is thus decreased for higher sulfate contents.

Time-of-flight neutron powder diffraction data were obtained on the sample with x=0.15 using the POLARIS diffractometer at the ISIS facility. Structural parameters were refined by the Rietveld method, using the GSAS program. Bragg peaks from the vanadium can and a small amount of $BaSO_4$ were fitted as additional phases. Initial cycles of refinement with the Hg-1201 model were followed by a different Fourier synthesis, which revealed the

Table 1 :
Positional parameters for $(Hg_{1-x}S_x)Ba_2CuO_y$ with x=0.15. P4/mmm, a=b=3.8986(2), c=9.2466(6)Å. R_{wp}=1.26%; $R(F^2)$=10.2%; χ^2=2.54

Atom	x	y	z	B (Å^2)
Hg	0	0	0	0.89(7)
Ba	0	0	0.2927(2)	0.71(5)
Cu	0	0	0.5	0.44(5)
O1	0	0	0.5	0.53(5)
O2	0	0	0.2153(4)	1.0(1)
O3	0.5	0.5	0	1.0
S	0	0.139(4)	0	1.0
O4	0.330(3)	0.372(5)	0	1.0
O5	0	-0.095(5)	0.138(1)	1.0

Table 2 :
Interatomic distances from the results of table 1.
† : constrained to be equal.

Bond Pair	Bond Length (Å)
Hg-O2	2.001(4)
Hg-O3	2.75672(13)
Ba-O1	2.7413(16)
Ba-O2	2.8491(11)
Ba-O3	2.7209(22)
Cu-O1	1.94930(9)
Cu-O2	2.647(4)
S-O4	1.574(16)†
S-O5	1.574(16)†

presence of additional peaks around the Hg cation, along the c axis and in the Hg layer. The relative positions of these peaks were consistent with a tetrahedral coordination for sulfur cations substituting for Hg. The refinement was then carried on by describing the SO_4 anions in the rigid body constraint model, for which only the S-O distance and the position and occupancy of the sulfate ion are refined, while its tetrahedral geometry is constrained. The results of this refinement are given in Table 1. Table 2 shows the interatomic distances.

The refined occupancy values are consistent with a sulfate content of 0.18(1) mols/f.u. A substantial amount of additional oxygen (0.15) was found in the "interstitial" position O3 indicating that the sample is overdoped. The sulfur is displaced by 0.5Å from the [0,0,0] position, and is at the center of an O_4 tetrahedron, with S-O distances of 1.57(2)Å. This distance is longer by a few standard deviations than typical S-O distances in other sulfates. The "displaced apical oxygen" O5 and the "displaced interstitial oxygen" O4, which form the SO_4 tetrahedron, are both shifted by about 0.8Å with respect to the "ideal" sites O2 and O3.

On the basis of the structural determination, the correlation between sulfate content and T_c for the as-prepared samples can be discussed in greater detail. Since the substitution is isovalent, the increase of Tc with x (i.e. the decrease of the Cu valence, since the x=0 compound is overdoped) has to be due to the decrease of the interstitial oxygen content O3 at the (1/2 1/2 0) position. Neutron diffraction data indicate that two of the oxygen atoms belonging to the sulfate anions are located in the Hg plane. Consequently, as sulfate groups are introduced in the structure, less sites are available for the "doping" oxygen anions. A similar argument is probably also valid to explain the solid solution limit. Due to the presence of oxygen anions from the sulfate groups, it is logical to assume than no more than one Hg cation per mesh can be replaced by sulfur cations, which puts the solid solution limit at ≈0.25, close to the value experimentally observed.

Conclusion

The $(Hg,C)Ba_2CuO_y$ and $(Hg,S)Ba_2CuO_y$ solid solutions have been investigated. For the former, the substitution range extends from x=0 to 1. The tendency of ordering of carbonate groups leads the formation of the $HgBa_4Cu_2CO_3O_y$ phase for x≈0.5. Samples with intermediate compositions are mixtures of the x=0 and x=0.5 phases for x<0.5 and of the x=0.5 and x=1 phases for 0.5<x. The detrimental effect on Tc is due both to the heterovalent substitution of C for Hg, and to the decreasing amount of doping interstitial oxygen.

In the case of sulfate substitution, a solid solution limit is observed at x≈0.2. Two of the oxygen anions forming the SO_4 tetrahedron are located in the (Hg,S) layer. This leads to a repulsive effect opposed to the presence of neighboring sulfate anions and interstitial oxygens. It is probably the reason for the relatively low solid solution limit, and for the decrease of the CuO_2 plane doping level.

References

[1] J.L. Wagner et al., Physica C210 (1993) 447
[2] E.T. Alexandre et al., Physica C245 (1995) 207
[3] E.M. Kopnine et al., Physica C243 (1994) 222
[4] A.R. Armstrong, H.S. Obhi, P.P. Edwards, J. Solid State Chem. 106 (1993) 120
[5] M. Uehara et al., Physica C222 (1994), 25

ELSEVIER

Physica C 282–287 (1997) 69–72

PHYSICA C

Quaternary borocarbides - exotic superconducting and magnetic materials

R. Nagarajan

Tata Institute of Fundamental Research, Bombay 400 005, INDIA

Discovery of superconductivity (SC) in quaternary borocarbides, R-T-B-C (R = Sc, Y, rare earth, U, Th; T = d- element), was one of the highlights of M^2S-HTSC-IV. Several exciting features in the properties of this family of materials, such as, high T_C for intermetallics, coexistence of SC and magnetism with high T_C and T_N, have generated wide interest. Some of our recent work along with other works from literature will be reviewed here.

Discovery of SC in Y-Ni-B-C system [1,2], and finding of quaternary family of superconducting borocarbide materials, with high T_C (> 10K) for intermetallics [3,4] were part of the highlights of the preceding conference of this series, M^2S-HTSC-IV [5]. A large body of important work from many laboratories [6], since then, has established SC in quaternary borocarbides, as a subject in its own right [7]. These materials have been synthesised with some of the combinations of R = Sc, Y, rare earth, U, Th and T = d- elements. They form in the homologous series $(RC)_m(NiB)_n$ [8]. Largest number (~30) of compounds have been found in the series RT_2B_2C [6]. With 3d elements, the structure is stable only for Ni and limit of structural stability seems to reach for R = Sc, as $ScNi_2B_2C$ (T_C ~ 15K) is a metastable phase [9]. The only known superconductos in the other series are: LuNiBC (T_C ~2.9K) [10] and the *boronitride*, $La_3Ni_2B_2N_3$ (T_C ~12.5K) [11]. Though, initially $La_3Ni_2B_2N_3$ was found to be unstable [11], recent studies have shown that the instability arises due to the presence of impurities and that the pure compound is quite stable [12]. We review here, some of the recent highlights in quaternary borocarbides both from our work and from literature.

Even though SC in the family appears to be BCS type, some interesting features have been observed in YNi_2B_2C. Though, the structure of RNi_2B_2C (R = Sc, Y, Lu, Tm, Er, Ho, Dy) is tetragonal with large structural anisotropy (c/a ~3), SC properties are found to be essentially isotropic. Theoretical investigations have suggested that there is a peak in the electronic density of states at the Fermi energy [13], and that it essentially has Ni d-character. However, XPS results do not show such a peak [14]. ARPES studies may resolve the conflict.

We have observed rf SQUID effect in a polycrystalline sample of YNi_2B_2C, arising from the grain boundary weak links [15]. This is surprising, considering the relatively large coherence length in this material (~100Å) (compared to ~15Å in *ab* plane in cuprates). Though this may point to a d-wave symmetry of the order parameter, it is necessary to examine the phase purity of grain boundaries before a conclusion is reached.

Inelastic neutron scattering on a large single crystal of YNi_2B_2C has revealed a phonon mode arising below T_C [16]. The intensity of this peak is found to depend on externally applied magnetic field and vanishes at H_{C2} [16], which establishes that this mode is related to SC. This is a direct demonstration of electron phonon coupling in a superconductor. The energy of the peak (~ 4 meV) is close to the SC gap estimated from BCS theory, but temperature dependence of the peak position is found to be weaker than that expected for BCS behaviour [16].

Our ^{11}B NMR studies show that temperature dependence of spin-lattice relaxation rate does not follow the well known Korringa

PII S0921-4534(97)00213-X

relation but could be interpreted in terms of presence of antiferromagnetic fluctuations [17]. Others have also reported a similar behaviour in their studies [18], but interpret their results on the basis of band structure effects. However, it must be noted that they have attributed part of the signal as due to impurity [18]. In this context, it should be noted that in our μSR studies of $TmNi_2B_2C$ [19,20], long lived magnetic fluctuations have been observed, well above the magnetic ordering temperature of the material.

$YbNi_2B_2C$ has turned out to be a unique member of the family due to absence of SC. Our synthesis and study of $YbNi_2B_2C$ show that though according to lattice parameter and magnetic susceptibility results, valence of Yb is close to 3+ state, it does not superconduct down to 2K [21]. This is a very surprising result, as, considering the systematic of T_C amongst RNi_2B_2C (R = Lu, Tm, Er, Ho, Dy), one would expect $YbNi_2B_2C$ to exhibit SC with T_C ~12K. We find that it evolves into a heavy fermion system [22], which implies that hybridisation of Yb- 4f-electrons with conduction electrons suppresses SC in $YbNi_2B_2C$. In fact, we find an anomalous depression of T_C even by dilute Yb in YNi_2B_2C. T_C in $Y_{0.9}Yb_{0.1}Ni_2B_2C$ is depressed by ~12K [23] - the largest depression of T_C by Yb substitution. YbNiBC, which we succeeded in synthesising [24], is also an interesting material. It does not superconduct down to 2.5K (we recall that LuNiBC has a T_C of ~2.9K). But the material orders magnetically at ~4K which is rather high for a Yb system as such, and also on the basis of that expected from de Gennes scaling of T_N (GdNiBC:T_M ~33K [24]).

Coexistence of SC and magnetism with high coexistence temperatures and all combinations of T_C and T_N (T_C (~10.5K) > T_N (~7K) in $ErNi_2B_2C$, T_C (~8K) ~ T_N (~8K) in $HoNi_2B_2C$, and T_C (~6.5K) < T_N (~ 11K) in $DyNi_2B_2C$), is one of the exotic aspects of quaternary borocarbides. Such large T_Ns suggest magnetic interaction being mediated through conduction electrons via RKKY mechanism, thus indicating a much more intimate interplay of SC and magnetism than ever encountered before. Observation of change of magnetic relaxation rate at T_C in ^{166}Er Mössbauer results [25] is also suggestive of this fact. Observation of square flux lattice and its rotation with respect to external field on the onset of magnetic order [26] is a direct evidence of interplay of SC and magnetism observed in a magnetic superconductor.

Our ^{11}B NMR results in YNi_2B_2C had suggested that that SC may be sensitive to stoichiometry of B and C or its site occupation, [17]. Sensitivity of SC to stoichiometry is more in magnetic superconductors of quaternary borocarbides and the sensitivity seem to increase as T_C approaches T_N. In the case of $HoNi_2B_2C$ one observes a *double* re-entrant behaviour, not seen in any other system. This re-entrance and its stoichiometry dependence in $HoNi_2B_2C$ has been experimentally demonstrated [27].

When T_C < T_N, stoichiometry effects can be severe - even non magnetic impurities can depress T_C [28], due to breaking up of antiferromagnetic spin pairs. Such effects have been observed in $DyNi_2B_2C$ (T_C ~6.5K, < T_N ~11K). 10% substitution of Lu in $DyNi_2B_2C$ depresses T_C by ~4K [29], compared to a depression of 0.3K on 10% Lu substitution in YNi_2B_2C [23]. Probably, it is due to such a reason that, initially, SC was not clearly indicated in this material, but was found in later investigations [30]. This also prompts reinvestigations for search for SC in magnetically ordered materials where SC was expected but was not found.

Considering that only two other such materials are known and their T_C is much less than 4.2K, $DyNi_2B_2C$ becomes an important and model system to study subtle effects of T_C < T_N [31]. For example, one could look for Meissner expulsion at atomic scale. Our μSR experiments [32] on $DyNi_2B_2C$ in this regard seem to give a hint of presence of such an effect. Onset of magnetic order could be seen in the loss of muon polarisation. The precession frequency could not be measured, as the field sensed by muon was too large. Our preliminary findings show a small decrease in the initial asymmetry of the muon polarisation, just in the limit of the accuracy of our measurements, at the SC transition(T_C=6.5K)

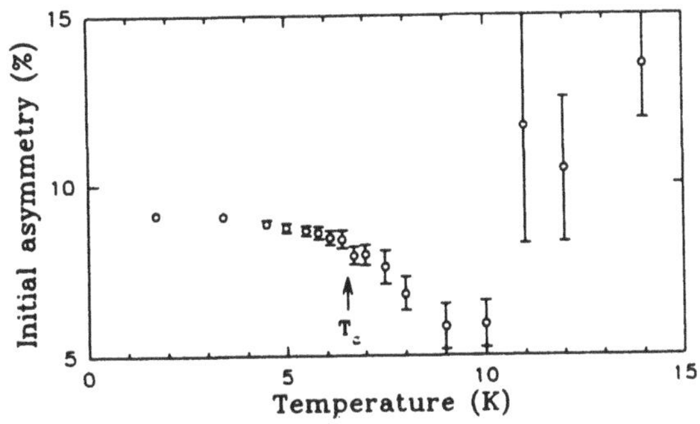

Fig. 1. Initial asymmetry of muon polarisation as a function of temperature in $DyNi_2B_2C$ [32].

(Fig. 1). A larger effect may not be seen because of large penetration depth and isotropic nature of SC in $DyNi_2B_2C$. Another interesting feature in our result is the unexpected temperature dependence of muon polarisation in the paramagnetic state of $DyNi_2B_2C$ (not shown in the figure). Implications of these findings are being examined.

While our ^{161}Dy Mössbauer results [33] indicate that Dy-C planes are insulating which imply that conduction electrons in the system arise from Ni-B planes, very recent finding that intercalation of Li in YNi_2B_2C [34] suppresses T_C and changes valence of carbon (inferred from XPS studies [34]) suggests a role for Y-C plane in SC of borocarbides, which is rather interesting.

Finally, we comment upon the potential for high T_C in quaternary borocarbides. Partial substitutions in $(RC)(MB)_2$ (M = Ni, Pd, Pt) series have not resulted in increase of T_C However, partial substitution of Cu for Ni in non superconducting (down to ~2K) YNiBC has produced SC with T_C as high as 8.9K [35]. This gives hope for increased T_Cs with appropriate substitution.

The phase in Y-Pd-B-C system which has record T_C (23K) for intermetallics [3], is still not synthesised in single phase form and, therefore, its structure is not clear. Early works [36,37] attributed the phase to 1221-type tetragonal structure. Considering that the SC phase is a metastable phase, one would expect the fraction of the SC phase to

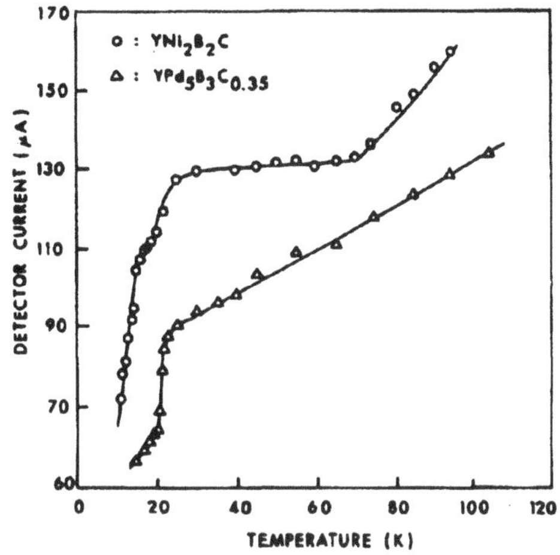

Fig. 2. The temperature dependence of Microwave absorption for YNi_2B_2C and $YPd_5B_3C_{0.35}$ (from Ref. 40).

enhance in rapidly quenched samples which in turn might yield a better information on the structure. Recent results on melt spun Y-Pd-B-C sample suggest that the SC phase might have a composition of 1221 but with a cubic structure (a~4.15Å) [38]. In our samples of $YPd_2B_{1+x}C$, we do find a cubic phase (a~4.14Å), but with weak SC [39].

SC in Ni-based materials with T_C ~ 13K [1,2] was already a surprise. Our microwave absorption studies of YNi_2B_2C had indicated a minority SC phase with T_C ~23K in Y-Ni-B-C system (Fig. 2) [40]. Recently, SC with T_C ~24K and zero resistance and diamagnetism has been observed in a sample of YNi_2B_2C prepared by high pressure route [41]. This is the most exciting result as it sets a new record for T_C in intermetallics and keeps alive the hopes of finding even higher T_Cs in intermetallics. Establishing the composition and structure of 24K SC phase in Y-Ni-B-C and so also those of SC Pd-quaternaries, including the phase with T_C of 10K [42], may point to fresh directions for materials which may have even higher T_Cs.

Acknowledgements: The author acknowledges the other members of the borocarbide work at TIFR, L.C. Gupta, S.K. Dhar, Z. Hossain and Chandan Mazumdar.

REFERENCES

1. Chandan Mazumdar et al., Solid. State Commun., 87 (1993) 413.
2. R. Nagarajan et al., Phys. Rev. Lett., 72 (1994) 274.
3. R.J. Cava et al., Nature, 367 (1994) 146.
4. R.J. Cava et al., Nature, 367 (1994) 252.
5. K. Kitazawa, Physica C, 235-240 (1994) xxiii.
6. See for example, contributions on quaternary borocarbides in the Proc. of Int. Confs: M^2S-HTSC-IV, Physica C 235-240 (1994); SCES'94, Physica B 206-207 (1995); SCES'95, Physica B 223-224 (1996); MOS'96, J. Low Temp. Phys. (1997) (to appear); LT21, Czech. J. Phys. 46 Suppl. S1-S6 (1996), SCES'96, Physica B (to appear).
7. PACS of AIP, U.S.A., for 1996, has a new classification: 74.72.Ny - Quaternary and multinary borocarbides.
8. T. Siegrist et al., Nature, 367 (1994) 254.
9. H.C. Ku et al., Phys. Rev., B50 (1994) 351.
10. L. Gao et al., Phys. Rev., B50 (1994) 9445.
11. R.J. Cava et al., Nature, 372 (1994) 245.
12. H. Michor et al., Phys. Rev. B54 (1996) 9408.
13. W.E. Pickett and D.J. Singh, Phys. Rev. Lett., 72 (1994) 3702.
14. A. Fujimori et al., J. Phys. Chem. Solids, 56 (1995) 1875.
15. Neeraj Khare et al., Appl. Phys. Lett., 69 (1996) 1483.
16. H. Kawano et al., Phys. Rev. Lett., 77 (1996) 4628.
17. T. Kohara et al., Phys. Rev., B51 (1995) 3885.
18. B.J. Suh et al., Phys. Rev., B53 (1996) R6022.
19. D.W. Cooke et al., Phys. Rev., B52 (1995) R3864.
20. S.J. Blundell et al., Physica B, 223 & 224 (1996) 69.
21. Z. Hossain et al., Physica B, 223 & 224 (1996) 99.
22. S.K. Dhar et al., Solid State Commun., 98 (1996) 985.
23. Z. Hossain et al., Physica B (in press).
24. Z. Hossain et al., SSP DAE Symp., Bombay, India, 39C (1996) 116.
25. P. Bonville et al., Z. Phys., B101 (1996) 511.
26. U. Yaron et al., Nature 382 (1996) 236.
27. H. Schmidt et al., Physica C 246 (1995) 177.
28. M.J. Nass et al., Phys. Rev. B 25 (1982) 4541.
29. B.K. Cho et al., Phys. Rev. Lett., 77 (1996) 163.
30. Z. Hossain et al., IEEE Trans. Magn.; 31 (1995) 4133; Z. Hossain et al., Physica B, 223 & 224 (1996) 99 and references therein.
31. L.C. Gupta, Physica B, 223 & 224 (1996) 56.
32. μSR experiments performed in colloba-tion with L.C. Gupta, R. Nagarajan, Z. Hossain (T.I.F.R., Bombay, India); S.F.J. Cox and S.P. Cottrell (ISIS, Oxfordshire, U.K.); S.J. Blundell, F.L. Pratt, Th. Jestadt (Clarendon Laboratory, Oxford, U.K.); J.L. Smith, D.W. Cooke (LANL, Los Alamos, U.S.A.).
33. J.P. Sanchez et al., Phys. Rev., B54 (1996) 9421.
34. W. Zhaoxiang et al., Abstract No. SA-PS-252, M^2S-HTSC-V, Beijing, China, 1997.
35. A.K. Gangopadhyay and J.S. Schilling, Phys. Rev., B54 (1996) 1010.
36. H.W. Zandbergen et al., Physica C, 226 (1994) 365.
37. S. Ikeda et al., Jpn. J. Appl. Phys., 33 (1994) 3896.
38. V. Ström et al., J. Mater. Res., 11 (1996) 572.
39. C. Godart et al., 12th Int. Symp. boron, borides and related compounds, Baden, Austria, 25-30 Aug. 1996.
40. R.M. Kadam et al., Physica C, 232 (1994) 359.
41. H. Szillat et al., High T_c Update,June 1, 1996; Poster No. PM143 MOS'96, Karlsruhe, 1996.
42. Z. Hossain et al., Solid State Commun., 92 (1994) 341.

ELSEVIER

Physica C 282–287 (1997) 73–76

PHYSICA C

Superconductivity in Ru-based Double Perovskite - The Possible Existence of a New Superconducting Pairing State*

D.Y. Chen[a], F. Z. Chien[a], D.C. Ling[b], J.L.Tseng[b], S.R. Sheen[b], M.J. Wang[b] and M.K. Wu[b]

[a]Department of Physics, Tamkang University, Tamsui, Taiwan

[b]Materials Science Center and Department of Physics, National Tsing Hua University, Hsinchu, Taiwan

We have studied the double perovskite structure $A_2Y(Ru_{1-x}Cu_x)O_6$ (A = Ba or Sr) system. The parent compounds are antiferromagnetic insulators. Partial substitution of the Ru-ion by Cu was found to induce superconductivity in these compounds. We also observed in these materials the coexistence of superconductivity and anomalous magnetic orders. Resistive measurements in high magnetic fields show that the field dependence of the transition temperatures is significantly different from those observed in high T_c cuprates. Our results suggest that the pairing state of these new materials may be of odd parity such as p-wave.

1. INTRODUCTION

After the discovery of high T_c [1] $YBa_2Cu_3O_7$ (YBCO), a clean superconducting $YSr_2Cu_3O_y$ (YSCO) with 123-structure was not achieved until Chien et al. [2] used Mo to partially replace Cu in $YSr_2Cu_{3-x}Mo_xO_y$. The presence of Mo atom was found to decrease the reaction temperature and is essential for the formation of pure high-T_c phase. Similar phase stabilization was also made by the introduction of Fe, Al, Co [3] and several other transition metals. Our recent detailed systematic study on YSCO by doping different transition metal ions [4] led to the formulation of an empirical rule which provides us important information that are relevant to the mechanism responsible for the structural stability of the superconducting YSCO system. More importantly, this empirical rule provided us a possible direction for the search of new superconducting oxides.

In an attempt to test the validity of the empirical rule, we looked into the possibility of fine tuning the materials such that the T_c ~ 80K superconductivity observed by Wu et al. [5] in mixed phase YSrCuOy can be reproduced. A calculation showed that the substitution of Ru-ion by Cu in $YSr_2(Cu_{3-x}Ru_x)O_y$ satisfies the requirement for the stabilization of the 123 phase. Indeed, our study on $YSr_2(Cu_{3-x}Ru_x)O_y$ system [4] did show that it is superconducting with an onset temperature ~ 80K (T_c-onset). However, detailed crystal structure analysis suggest that the sample contains, in addition to the 123 phase, another perovskite structure system with stoichiometry Sr_2YRuO_6. The double perovskite compound A_2YRuO_6 (2116) [6], where A stands for divalent alkaline earth, is a well known antiferromagnetic insulating system. In this paper we report the observation of superconductivity in A_2YRuO_6, where A = Ba or Sr, when Ru ion is partially substituted by Cu ion. A preliminary results on the Sr-based $YSr_2(Cu_{1-x}Ru_x)O_y$ compound has been published earlier [7].

2. EXPERIMENTALS

All powder samples used in this study were prepared through solid state reaction route. Appropriate amounts of ACO_3, A = Ba or Sr, RuO_2 (or Ru metal), Y_2O_3 and CuO are uniformly mixed, calcined at 1000°C. The reacted powder were then sintered at 1460°C for Ba-based 2116 (1380°C for Sr-based 2116) under atmosphere of Ar(30%)/O_2(70%) mixture. We have also successfully grown Ba-based single crystals by high temperature solution growth technique with PbO-PbF_2 as solvent. The as grown crystals were not superconducting, possibly due to insufficient Cu-doping. Superconductivity appears

0921-4534/97/$17.00
PII S0921-4534(97)00214-1

after the crystals annealed at about 1400°C for 6 hours. All samples were characterized by SEM, high resolution x-ray refinement for and dc magnetization for superconducting transitions.

3. RESULTS AND DISCUSSIONS

Figure 1 shows the high resolution x-ray diffraction patterns of $A_2Y(Ru_{0.9}Cu_{0.1})O_6$ (A = Sr or Ba) compounds [9]. It clearly shows single phase 2116 structure for both systems. However, they exhibit different sub-lattice crystal symmetry. The $Sr_2YRu_{1-x}Cu_xO_6$ (Sr-2116) is best fit with a random sub-lattice, while $Ba_2YRu_{1-x}Cu_xO_6$ (Ba-2116) with rock-salt symmetry. Figure 2 displays the temperature dependence of the resistance for $Sr_2YRu_{0.9}Cu_{0.1}O_6$ and $Ba_2YRu_{0.9}Cu_{0.1}O_6$. Their resistivity are considerably higher than those of the high T_c cuprates. We often observe a small drop in the resistance at ~85 K, for Sr-2116 samples, preceding the sharp onset transition at ~60 K and zero resistance at ~ 25K). On the other hand, the Ba-2116 samples show a broad resistive transition with the onset temperature as high as 100K and zero resistance at ~ 30K.

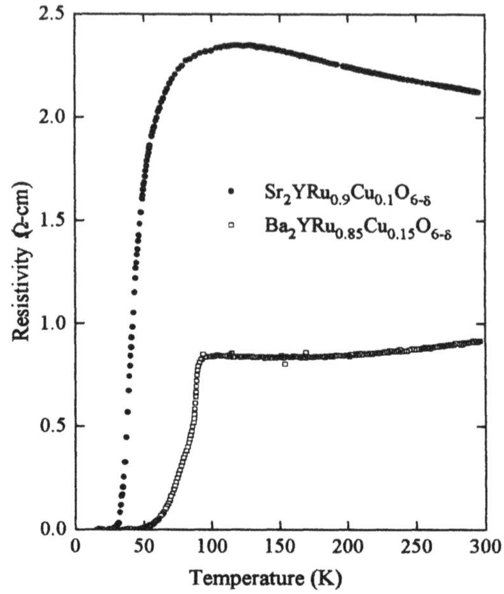

Figure 2. Temperature dependence of resistivity for $A_2Y(Ru_{0.9}Cu_{0.1})O_6$ (A=Ba or Sr).

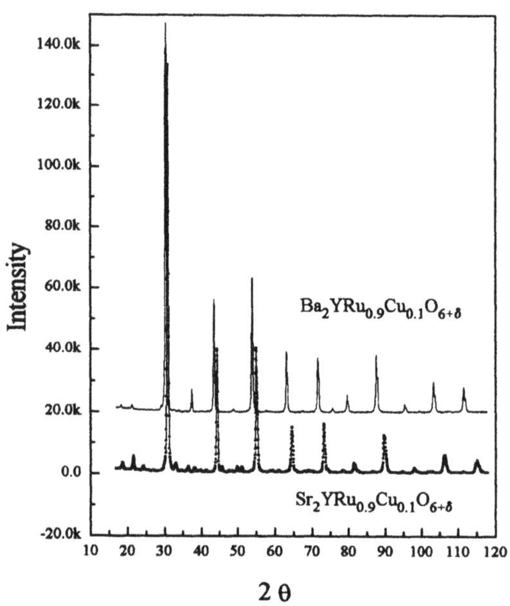

Figure 1. X-ray diffraction patterns of double perovskites $A_2Y(Ru_{0.9}Cu_{0.1})O_6$ (A=Ba or Sr).

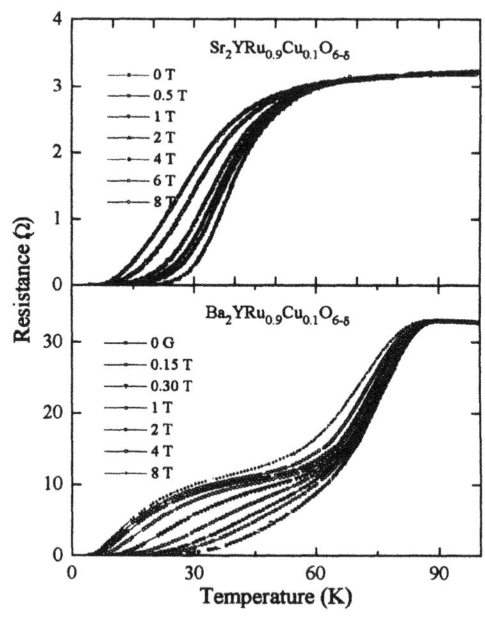

Figure 3. Resistivity of $A_2Y(Ru_{0.9}Cu_{0.1})O_6$ (A=Ba or Sr) in magnetic fields.

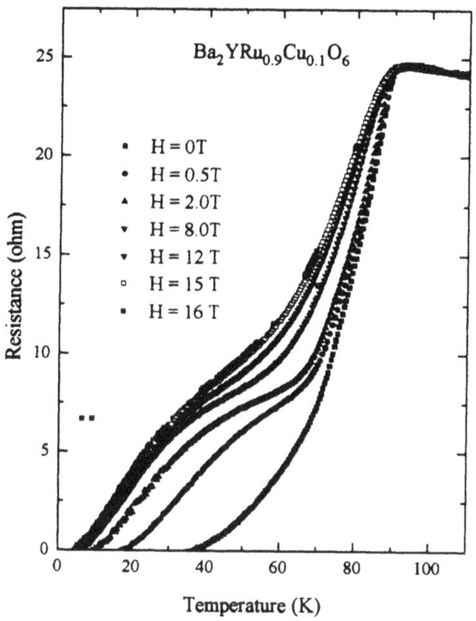

Figure 4. Resistivity of $Ba_2Y(Ru_{0.9}Cu_{0.1})O_6$ in magnetic field up to 16 Tesla.

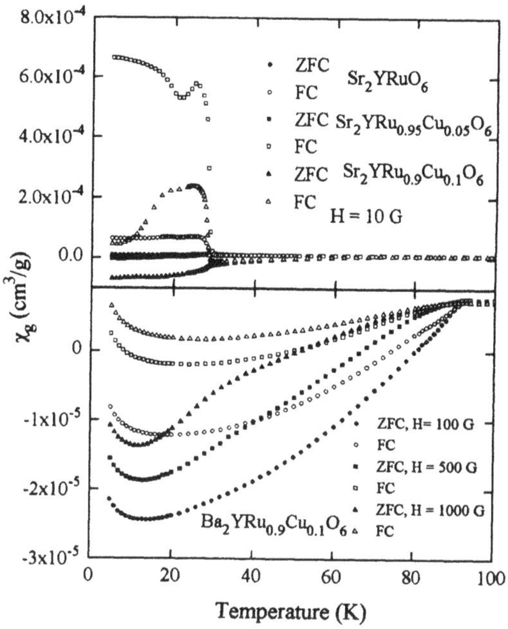

Figure 5. Temperature dependence of magnetic susceptibility for $A_2Y(Ru_{0.9}Cu_{0.1})O_6$ (A=Ba or Sr).

Figure 3 displays resistive transitions for both Ba-2116 and Sr-2116 samples measured up to 8 Tesla. We observe a broad resistive transition with a sharp onset temperature at ~60K, which decreases with field, for Sr-2116 sample (3(a)). The zero-resistance temperature first shows similar decrease to that of the onset transition but stays almost unchanged for fields higher than 6 Tesla. For Ba-2116 sample, a broad transition with an onset temperature as high as 90K is observed. At a field larger than ~500 Oe, it develops a shoulder, which appears at ~55K, in the resistive transition. The field dependence of this second transition is quite different from that of the onset transition. Figure 4 shows the resistive transition of a Ba-2116 measured up to 16 Tesla. The resistive transition remains essentially unchanged as the field is higher than 15 Tesla. These results clearly demonstrate that the field dependence of the zero resistance temperature for Ba-2116 and Sr-2116 are very much different from those observed in the high T_c cuprates, suggesting the vortex dynamic of Cu-doped 2116 may be intrinsically different from that of the known superconductors.

The ZFC (zero field cool) and FC (field cool) magnetic susceptibility curves for Sr-2116 samples of different cu-doping are shown in Figure 5(a). The susceptibility of an undoped 2116 sample is also included for reference. The ZFC curves for Sr-2116 samples show, below ~30 K, a relatively weak diamagnetic response. On the other hand, the FC curves exhibit a sharp rise in susceptibility and a kink at 26 K that is similar to the one, due to antiferromagnetic order, observed in the parent compound. Upon further cooling the susceptibility rises again that resembles to a ferromagnetic transition. A broad superconducting transition is observed in Cu-doped Ba-2116, as shown in figure 5(b). There appears in magnetic susceptibility a second transition with onset at ~50K in higher magnetic field which is consistent with the resistive shoulder observed in high fields.

The above results strongly suggest that antiferromagnetism persists in Cu-doping samples. The temperature dependence of the lower critical field $B_{c1}(T)$ for both samples, particularly for Sr-2116 sample, fit better with an exponential functional form in contrast to the linear or power laws for conventional or high T_C superconductors. This observation further leads us to believe the simultaneous presence of magnetic order in these double perovskite superconductors. Because of the intrinsic coexistence of the ferromagnetism and superconductivity, we suspect that the paring symmetry in these double perovskites may be of odd parity such as p-wave. The abnormal flux dynamics observed in high fields can be qualitatively understood if the superconductivity is p-wave.

Table 1. Bond Valence Sum of Ru-Y and total BVS of samples in different synthesis conditions

$Sr_2YRu_{1-x}Cu_xO_6$	x=0.01	x=0.02	x=0.03	x=0.05	x=0.1
BVS of Ru-Y	4.009	4.035	3.748	3.820	4.014
Total BVS	11.99	11.95	12.01	11.93	12.09
Atmosphere			1atm. O_2		
Superconductivity	No	No	No	No	Yes
$Sr_2YRu_{1-x}Cu_xO_6$	x=0.04	x=0.05	x=0.07	x=0.08	x=0.1
BVS of Ru-Y	4.061	4.215	4.894	4.751	4.889
Total BVS	12.17	12.40	14.16	13.49	13.80
Atmosphere			70% O_2 +30% Ar		
Superconductivity	Yes	Yes	Yes	Yes	Yes

The antiferromagnetic order of the double perovskite A_2YRuO_6 is known to originate from the superexchange among Ru ions that are in 5+ valence state with a localized moment of total spin J = 3/2 [6]. A possible explanation to our observed anomalous magnetic and superconducting properties in Cu-doped samples is that partial doping of the Ru ion by Cu provides mobile holes to the lattice to form a formal Ru^{6+} of spin 1. This conjecture is supported by our detailed structure studies. As shown in the table, the bond valence sum derived from various bond lengths reveals that the average Ru valence in samples exhibit superconductivity is larger than 5. The mobile holes give rise to "double exchange"[8] that results in the simultaneous existence of both ferromagnetic and antiferromagnetic orderings [9]. Our low temperature specific heat measurements [10] also support such a complex magnetic structure. It is noted that de Gennes has proposed a superconducting mechanism [11] based on the double exchange and a canted spin state. The Cooper pairs must be p-wave if this picture is correct. It is noted that a recent model calculation has suggested the possible existence of p-wave superconducting state in La-Ca-Mn-O [12] perovskites. Whether the observed superconductivity in $A_2Y(Ru_{1-x}Cu_x)O_6$ is the manifest of a double exchange mechanism or a single spin mechanism remains to be investigated.

REFERENCES

*The authors acknowledge fruitful discussions with Profs. F.C. Zhang, C.C. Chi, H.C. Ren and T.K. Lee. This work is supported by the ROC National Science Council grants No. NSC85-2212-M-007-005PH.

1) M.K. Wu, J.R. Ashburn, C. J. Torng, P. H. Hor, R. L. Meng, L. Gao, Z. J. Huang, Y. Q. Wang and C. W. Chu, Phys. Rev. Lett. 58, 908 (1987).
2) M. K. Wu, C. Y. Huang and F. Z. Chien, Superconductivity and Applications, ed. by Y. H Kao and H. S. Kwok, Plenum Publishing Co. 1990.
3) S. A. Sunshine, L. F. Schneemeyer, T. Siegrist, D. C. Douglass, J. V. Waszczak, R. J. Cava, E. M. Gyorgy and D. W. Murphy, Chemistry of Materials, 1, 331, (1989).
4) D.H. Chen, Ph.D. Thesis, Natioanl Tsing Hua University, Taiwan 1995; D.Y. Chen, Ph.D. Thesis, Tamkang University, Taiwan 1996.
5) M. K. Wu, J. R. Ashburn, C. A. Higgins, B. H. Loo, D. H. Burns, A. Ibrahim, T. D. Rolin, F. Z. Chien and C. Y. Huang, Phys. Rev. B37, 9765 (1988).
6) For a review, see Anderson, M. T., Greenwood, K. B., Taylor, G. A. and Poppelmeier: K. R., Prog. Solid State Chem. **22**, 197 (1993).
7) M.K. Wu, D.Y. Chen, F.Z. Chien, S.R. Sheen, D.C. Ling, C.Y. Tai, G.L. Tseng, D.H. Chen and F.C. Zhang, Z. Physik B102, 37 (1996).
8) Zener, C.: Phys. Rev. **82**, 403 (1951); Anderson, P.W. and Hasegawa, H.: Phys. Rev. **100**, 675 (1955).
9) de Gennes, P.G.: Phys. Rev., **118**, 141 (1960).
10) M.K. Wu et al., to be published.
11) de Gennes, P. G.: C. R. Acad. Sci. Paris, **305**, 345(1987).
12) W. E. Pickett, Phys. Rev. Lett., **77**, 3185 (1996).

ELSEVIER

Physica C 282-287 (1997) 77-80

PHYSICA C

HIGH GAS PRESSURE CRYSTAL GROWTH AND PROPERTIES OF $HgBa_2Ca_{n-1}Cu_nO_{2n+2+\delta}$, $Sr_{0.73}CuO_2$ AND $Ca_XA_{1-X}CuO_2$

J.Karpinski, H.Schwer, E.Kopnin, R.Molinski, G.I.Meijer, K.Conder
Laboratorium für Festkörperphysik ETH 8093-Zürich

Abstract: An overview is presented of crystal growth and physical investigations of: $HgBa_2Ca_{n-1}Cu_nO_{2n+2+\delta}$, $Ca_{1-x}A_xCuO_2$ and $Sr_{1-x}CuO_2$ compounds synthesized at high pressure. The Hg-12(n-1)n (n=1-5) single crystals have a T_c onset up to 130K. Thin films of Hg-12(n-1)n have been grown from a melt. At a high oxygen pressure a new phase $Sr_{0.73}CuO_2$ has been synthesized. It contains one-dimensional edge-sharing CuO_2 infinite chains instead of corner-sharing infinite layers, typical for cuprate superconductors.

1. Introduction

Pressure is a very important physical parameter and can change considerably various properties of materials. In general two kinds of pressure effects should be taken into account:
1. Hydrostatic pressure itself, applied by inert medium like argon, or a solid medium.
2. Thermodynamic equilibrium between a condensed phase and a gas, when the gas is an active component of the system.

Hydrostatic pressure itself can change properties of materials during crystallization or annealing by: a) structural transitions at high pressure, b) reduced evaporation of volatile components. An example of a) is the synthesis of infinite layer $Ca_xA_{1-x}CuO$ and of b) $HgBa_2Ca_{n-1}Cu_nO_{2n+2+\delta}$ compounds [1]. Dense Ar gas under pressure of 10-15 kbar reduces evaporation of Hg and HgO from the melt. This allowed us to grow single crystals and layers of the HgBaCaCuO compounds. Thermodynamic equilibrium between a condensed phase and a gas under pressure should be taken into acount when the gas is one of components of the synthesized compound. The application of a high oxygen pressure lead us to the synthesis of single crystals and polycrystalline samples of superconductors: $YBa_2Cu_4O_8$ (124), $Y_2Ba_4Cu_7O_{15-x}$(247) and infinite chain compound $Sr_{0.73}CuO_2$.

2. Solid- and gas-medium pressure techniques.

A very important aspect of the high pressure experiments is a control of partial pressures of volatile components. Many high-pressure solid-medium experiments published in literature show that this is difficult or impossible. The growth of phase pure single crystals is the best solution for investigating the true intrinsic properties of new compounds. This is generally not possible in solid-medium high pressure systems used to synthesize superconductors, because of the lack of a gas atmosphere and of well defined partial pressures of the volatile components. The gas-phase high-pressure technique seems to be a good solution of these problems. An gas atmosphere with a controlled partial oxygen pressure leaves free space for the single-crystal growth. A temperature gradient in a multizone furnace is relatively easy to control. A maximum sample volume is several cm^3.

3. Thermodynamics of Hg-12(n-1)n

One of the main reasons of the difficulties with the synthesis of Hg-12(n-1)n is the low thermal stability of the mercury compounds. Due to difficulties with the control of the Hg, HgO and O_2 partial pressures it is very difficult to obtain stoichiometric samples. Therefore, a high pressure technique with a solid-medium, usually pyrophylite have been frequently applied. However, due to the above mentioned difficulties this technique has many limitations. Therefore, we have applied a gas-phase high-pressure technique. Although many papers appeared on the synthesis of the Hg superconductors, there is a little information in literature concerning thermodynamics of this system. Recently, two groups in Houston [2] and Moscow [3] investigated partial pressures above Hg-12(n-1)n compounds. V.A.Alyoshin et al. [3] measured equilibrium partial pressures of Hg and O_2 above Hg1201 for the reaction:

PII S0921-4534(97)00216-5

(1) $HgBa_2CuO_{4+\delta} \rightleftharpoons Ba_2CuO_{3+x} + Hg(g) + (1+\delta-x)/2O_2$

This is the lower boundary of $HgBa_2CuO_{4+\delta}$ existance. Upper boundary is determined by the formation of HgO, $BaHgO_2$ and $BaCuO_2$ at higher P_{Hg} and Po_2. Fig.1 shows partial pressures of Hg, HgO and O_2 in the closed crucible containing Hg-1201 sample. Curve (a) is the mercury equilibrium partial pressures above Hg-1201 compound as a function of temperature. Here $Po_2={}^1/_2P_{Hg}$ and $\delta=x=0$ were asumed. Curve (b) shows P_{Hg} under assumption of initial Po_2=2 bar in the crucible. These calculations consider P_{Hg} and Po_2 in equilibrium with solid 1201 and Ba_2CuO_3 phases. The situation changes when the temperature increases up to the melting temperature. These compounds melt incongruently. As HgO is the volatile chemical component of the system we have calculated partial pressures caused by decomposition of HgO (Fig.1 (c),(d),(e)). Depending on the free volume of the crucible and the amount of the material, these pressures can reach more than hundred bar at the synthesis temperature. By a variation of the ratio between the free volume and the mass of the sample one can control partial pressures of the components. Calculations for 3.2 cm^3 free volume in the crucible and 0.87g HgO correspond to conditions of our experiment in which Hg-12(n-1)n crystals have been grown. At the temperature of synthesis (1200-1340K) a solid HgO and a liquid Hg disapeared from the sample and remained only as vapour. This causes bending of the curves above 1000 K. In the cool zone of the crucible HgO can condensate. Fig.2 shows the three-dimensional phase diagram with a surface corresponding to the three phase equilibrium of the reaction (1), calculated from data of [3]. The hached area shows conditions of Hg-1201 crystals and layers growth in our experiments. The arrow corresponds to the variation of partial presures ($Po_2={}^1/_2P_{Hg}$) during heating of the sample.

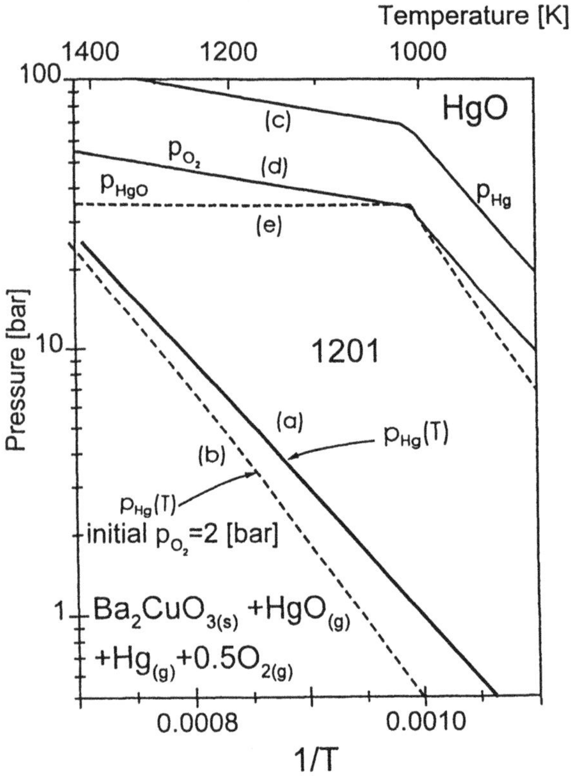

Fig.1 Partial pressures above Hg-1201 compound as a function of temperature and oxygen pressure: (a) heating of a closed ampoule with Hg-1201 sample, (b) with initial Po_2=2 bar. (c),(d),(e): partial pressures created in a closed crucible of a free volume 3.2 cm^3 by a decomposition of: 0.87g HgO.

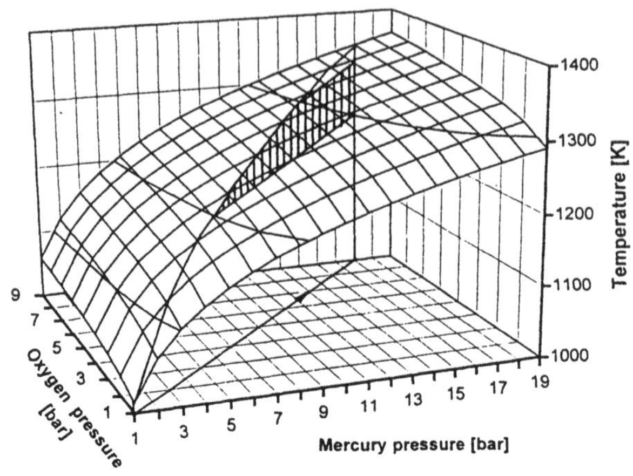

Fig.2. Three-dimensional phase diagram P_{Hg}, Po_2, T with a surface corresponding to the three phase equilibrium (1) calculated from data of [3].

4. Crystal growth of Hg-12(n-1)n.

Hg-12(n-1)n compounds melt peritectically, but at ambient pressure they decompose before melting and the volatile components evaporate. Our goal was to prevent decomposition by an encapsulation of the sample with a high hydrostatic inert gas pressure until the peritectic temperature is reached. All experiments have been performed at Ar pressure of 10-15 kbar. Due to the high density of Ar gas at this pressure ($\varrho > 1 g/cm^3$) the evaporation of Hg is strongly supressed. Additionally, application of a flux with a lower melting point allows growth of single crystals below the peritectic decomposition temperature. As flux we used PbO or $BaCuO_2$-CuO mixture. The crystallization temperature was $930° < T < 1070°C$. Two crucible materials have been applied: alumina and yttria. Crystal growth in the

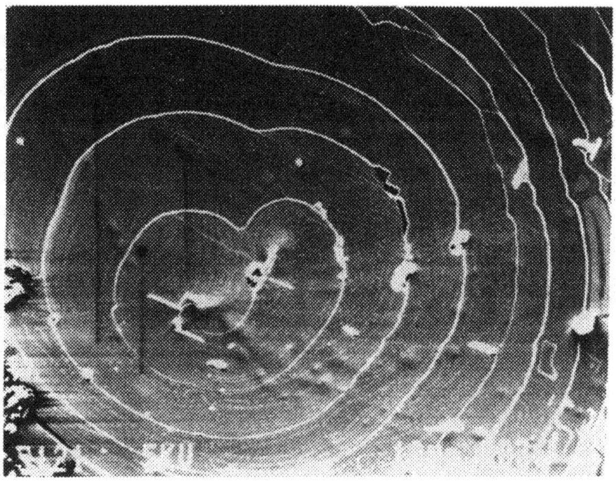

Fig.4. Hg-1223 single crystal with a double-spiral growth-steps.

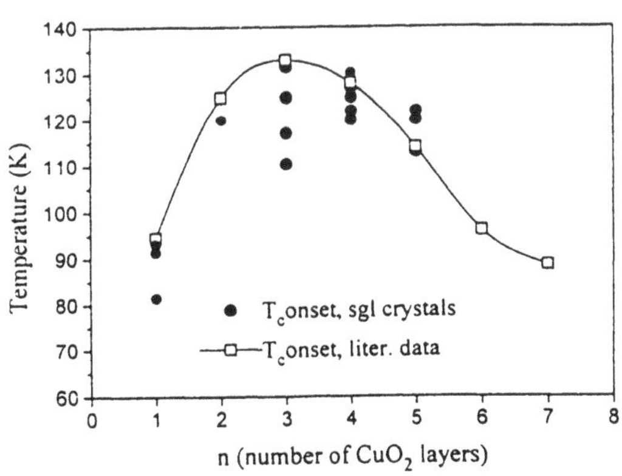

Fig.5 . T_c values of single crystals with various n in comparison with literature data [5] for ceramics.

alumina crucible leads to a considerable doping with Al and mixed (HgAl)BaCaCuO have been obtained. In yttria crucibles, we have crystallized Hg-12(n-1)n for n=1 to 5. The detailes of the crystal growth procedure have been published previously [1,4].

There is a relatively good agreement of T_c values for our single crystals and those reported in the literature [5] for ceramic samples (Fig.5).

5. $Hg_{1-x}Pb_xBa_2CuO_{4+\delta}$ single crystals.

$Hg_{1-x}Pb_xBa_2CuO_{4+\delta}$ crystals have been grown from PbO flux or from a stoichiometric melt. Undoped crystals have an almost perfect structure with full occupation (97%) of the Hg position and a maximum T_c=94K. PbO flux leads to Pb substitution for Hg up to 10%. Single crystal X-ray investigations [6] show, that oxygen and mercury atoms are shifted off their ideal positions and oxygen atoms are distributed in a ring-like configuration around the ($^1/_2$,$^1/_2$,0) site. This is due to the formation of short Pb-O bonds (~2.1 A). This has consequences for the physical properties: Pb doping decreases T_c to 80 K (Fig.6). It changes also the pinning properties of the crystals. Torque measurements indicate a strong fish-tail effect for Pb-doped crystal. The anisotropy γ measurements on undoped crystals show a value of 25 for the 94K crystals, which increases upon reduction in argon to γ=60 for 80K crystals [7].

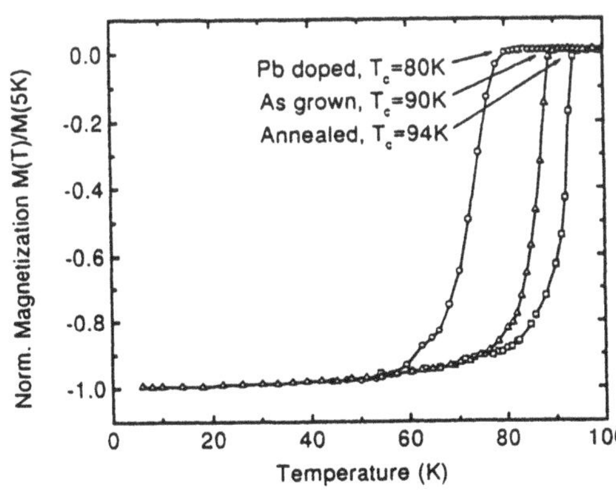

Fig.6. DC magnetization of two undoped and one Pb-doped Hg-1201 single crystals.

6. Growth of Hg1201 layers on $BaZrO_3$ substrates

By application of a similar technique it is possible to grow layers of Hg-12(n-1) superconductors from the melt. Whereas this process is very similar to the crystal growth described above, we used the fact that during partial melting of the HgPb-12(n-1)n pellets the liquid can flow over $BaZrO_3$ substrates placed in the crucible. By slow cooling the liquid solidifies and a layer 1-2 μm of the pure Hg-1201 phase forms on the surface of the substrate (Fig.7). The layer is (001) oriented. As we have used polycrystalline $BaZrO_3$ substrates the layers consist of domains with no orientation in the ab plane. The size of these domains is up to 100 μm^2. Up to now $BaZrO_3$ material is the only compatible substrate for the deposition of such layers [8]. The in-plane resistivity shows an onset temperature T_c=94K. The shape of R(T) is linear, like in optimally doped superconductors.

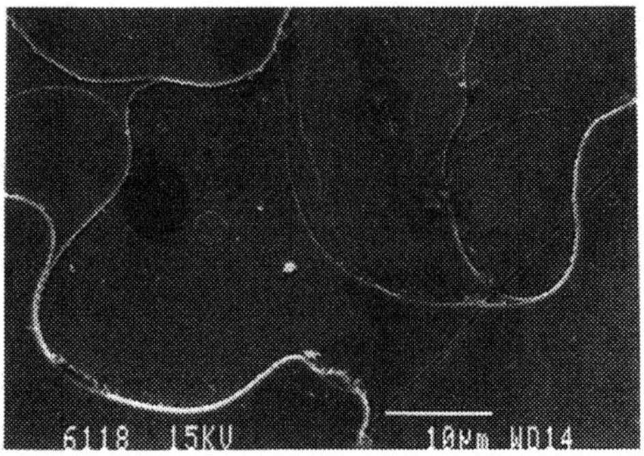

Fig.7. Morphology of Hg-1201 layer grown on $BaZrO_3$ polycrystalline substrate.

7. Infinite-chain and -layer compounds.

Since the discovery of the HTS much attention has been paid to the low-dimensional electron systems and especially to the two-dimensional CuO_2 plane which is the common feature to all HTS. By application of a high pressure during synthesis it is possible to modify CuO_2 planes. We have synthesized at Po_2=2000 bar a new compound in the Sr-Cu-O system, namely $Sr_{0.73}CuO_2$, containing only CuO_2 chains. This compound is closely related to the $Sr_{14}Cu_{24}O_{41}$ (14,24) phase but made only of edge sharing CuO_2 chains separated by pure Sr planes [9]. We have investigated the structure of $Sr_{0.73}CuO_2$ (Fig.8) single crystals and the physical properties of the polycrystalline material. It appears to be an insulator with an antiferromagnetic exchange coupling and a sharp transition in the magnetic susceptibility at 10K [9]. At higher magnetic fields the amplitude of the transition becomes smaller and broader. The origin of the transition is now investigated.

The "infinite-layer" (IL) structure is the simplest one

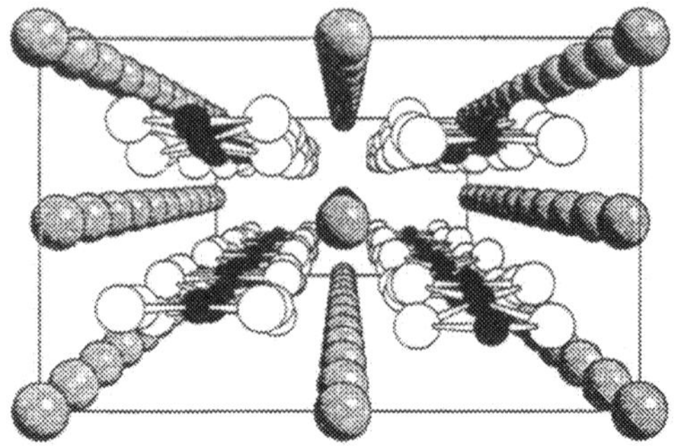

Fig.8. Perspective view of the $Sr_{0.73}CuO_2$ structure.

which contains (CuO_2) layers. We have grown single crystals of $CaCuO_2$, $Ca_{1-x}La_xCuO_2$(x_{max} = 0.016) and $Ca_{0.47}Sr_{0.53}CuO_2$ with IL structure at 1120-1160°C and 14-16 kbar Ar pressure. The substitution of La^{3+} for Ca^{2+} has been performed in order to induce n-type superconductivity. The non-doped $CaCuO_2$ crystals showed a diamagnetic signal with Tc = 78 K probably due to some intergrowth defects which could not be detected by X-ray structure analysis or EDAX data. The resistivity measurements revealed the semiconducting type of conductivity. La-doping suppresed this effect. It can be explained in terms of Cu-O in-plane distances. These values are 1.928-1.93Å and close to those of hole-doped superconductors. Cu-O distances for electron-doped superconductors are significantly larger (about 1.96-1.98 Å). Probably, in $Ca_{1-x}La_xCuO_2$ it is not possible to create the appropriate concentration of antibonding electrons to induce superconductivity. Single crystal $Ca_{0.47}Sr_{0.53}CuO_2$ was not diamagnetic. On contrary, $Sr_{0.7}Ca_{0.3}CuO_2$ (a=3.9024Å, c=3.3507Å), $(Sr_{0.8}Ca_{0.2})_{0.9}La_{0.1}CuO_2$ (a = 3.9215Å, c = 3.3650Å) and $(Sr_{0.8}Ca_{0.2})_{0.9}Nd_{0.1}CuO_2$ ceramic samples revealed diamagnetic onsets at 38,37 and 33 K, respectively.

References.

1. J.Karpinski et al. Physica C 234 (1994) 10.
2. Y.Y.Xue et al. submitted to Physica C.
3. V.A.Alyoshin et al. Physica C 271 (1996) 197.
4. J.Karpinski et al., in "Lecture in Physics" Vol.475, Springer Verlag 96, Ed.J.Klamut et al.83.
5. B.A.Scott et al., Physica C 230 (1994) 239.
6. H.Schwer et al. Physica C, accepted.
7. J.Hofer et al., to be published.
8. A.Erb et al., Physica C 245 (1995) 245.
9. J.Karpinski et al. Physica C, accepted.

ELSEVIER

Physica C 282–287 (1997) 81–84

PHYSICA C

Single Crystal Growth of RE-123 Superconductor and Its Superconductive Properties

Y. Shiohara[a], M. Nakamura[a], X. Yao[a], M. Kambara[b] and T. Umeda[b]

[a] Superconductivity Research Laboratory, ISTEC, 1-10-13 Shinonome, Koto-ku, Tokyo 135, Japan
[b] Department of Metallurgy, Graduate School of Engineering, The University of Tokyo, 7-3-1 Bunkyo-ku, Tokyo 113, Japan

Single crystals of $Nd_{1+x}Ba_{2-x}Cu_3O_{7\pm\delta}$ (Nd123) have been successfully grown by the top-seeded solution-growth (TSSG) method. BaO-CuO solvent was used in a Nd_2O_3 crucible. Furthermore, compositions of 123 compounds could be controlled using two different methods which were the oxygen partial pressure control of the atmosphere during growth and controlling liquid composition. Using the BaO-CuO solvent with 3/5 of Ba/Cu ratio, the almost stoichimetric Nd123 single crystals were grown under low oxygen partial pressure atmosphere ($P(O_2)$=0.01 atm). In the air atmosphere, the Ba enriched liquid with Ba/Cu ratio of 0.78 was used to obtain the almost stoichimetric Nd123 single crystal.

Moreover, to investigate whether for Nd123 the anomalous peak effect in J_c-H curve is an intrinsic property or not, different series of heat treatment were performed using almost stoichiometric Nd123 single crystals grown by the above-mentioned TSSG method.

1. INTRODUCTION

$REBa_2Cu_3O_{7-\delta}$ [RE123 (RE: La, Nd, Sm)] superconductors have been known to show relatively low T_c because RE ions easily substitute into Ba sites[1-3]. Accordingly, it is difficult to synthesize RE123 superconductors having high T_c (>90K). Especially, it has not been successful to fabricate melt textured RE123 bulk superconductors with high T_c and a sharp superconductive transition in dc magnetization measurements.

Recently, Nd123 and Sm123 samples prepared by the oxygen controlled melt growth (OCMG) method were reported to have increased T_c (96K) values and higher critical current density (J_c) especially under high magnetic fields region than those of the melt processed Y123, due to existence of the anomalous peak effect in J_c-H curve[4-6]. However, the mechanism of this anomalous peak effect in J_c-H curve has not been clarified yet. Therefore high quality, large Nd123 single crystal is necessary for further understanding of fundamental physical properties and especially an anomalous peak effect in the J_c-H curve.

In this paper, first, single crystal growth of Nd123 in different oxygen partial pressure atmospheres ($P(O_2)$=0.01, 0.21, and 1.00 atm) was attempted by the top-seeded solution-growth (TSSG) method using a BaO-CuO solvent with 3/5 of Ba/Cu ratio in a Nd_2O_3 crucible. Second, in the air atmosphere, crystal growth of Nd123 was performed using different Ba/Cu ratios (0.41~1.0 of Ba/Cu ratio) of the liquid by the same TSSG method. Finally, the Nd123 anomalous peak effect in J_c-H curve was investigated by means of different series of heat treatments using almost 123 stoichiometric Nd123 single crystals.

2. CRYSTAL GROWTH

Single crystal growth of Nd123 in different oxygen partial pressure atmospheres ($P(O_2)$=0.01, 0.21, and 1.00 atm) was attempted by the top-seeded solution-growth (TSSG) method using a BaO-CuO solvent with 3/5 of Ba/Cu ratio in a Nd_2O_3 crucible[7]. Nd_2O_3 was selected as a crucible material to prevent contamina-

PII S0921-4534(97)00217-7

tion from the crucible. The Nd solute is supplied by dissolution of the Nd_2O_3 crucible. Crystals were grown at pulling rates of 0~0.25 mm/h under the growth temperature of 1020°C in $P(O_2)$=0.01 atm, 1075°C in $P(O_2)$=0.21 atm and 1095 °C in $P(O_2)$=1.0 atm, respectively. The crystal rotation speed was 120 rpm. The seed crystals were Nd123 thin films deposited on (100) MgO single crystals[8].

The average compositions of the Nd123 single crystals were analyzed by inductively coupled plasma spectrometry (ICP) and were confirmed to be Nd:Ba:Cu=1.02:1.97:3.00 in $P(O_2)$=0.01 atm, Nd:Ba:Cu=1.07:1.95:3.00 in $P(O_2)$=0.21 atm and Nd:Ba:Cu=1.10:1.90:3.00 in $P(O_2)$=1.0 atm, respectively. Nd substitution into Ba sites in the grown crystal increases with increasing oxygen partial pressure atmosphere during crystal growth. Figure 1 shows the temperature dependence of dc magnetization for Nd123 single crystals, which were grown in different oxygen partial pressures and annealed at 340 °C for 200 h in O_2. The Nd123 single crystal grown in $P(O_2)$=0.01 atm shows a steep superconductive transition with ΔT less than 0.5 K at around 96.5 K. Nd substitution into Ba sites in crystal degrades superconducting properties with increasing oxygen partial pressure during crystal growth. This result indicates that low oxygen partial pressure atmosphere during crystal growth is found to be effective for minimizing the substitution of Nd ion into Ba sites. Figure 2[9] displays the Nd-Ba-Cu-O quasi-ternary phase diagram around the Nd123 and liquid coexisting temperature region of 1073 °C in air. Figure 2 shows that the solid solubility range for Nd/Ba substitution is sensitive to liquid composition. With an increase in the Ba/Cu ratio of the liquid the Nd substitution content reduces. In the air atmosphere, a stoichiometric Nd123 crystal is expected to be obtained by selecting Ba enriched liquid with Ba/Cu ratio of more than 3/5.

Crystal growth of Nd123 in the air atmosphere was performed using the different Ba/Cu ratios of the liquid by the same TSSG method[10]. The different Ba/Cu ratios of the BaO-CuO solvent were used for adjusting the Ba/Cu ratio of the liquid from 0.41~1.00. Crystals were grown at pulling rates of 0.10~0.25 mm/h under the growth temperature of 1058~1070 °C. The

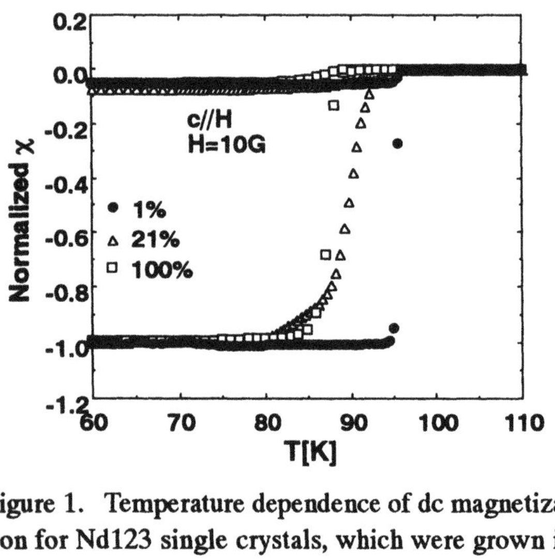

Figure 1. Temperature dependence of dc magnetization for Nd123 single crystals, which were grown in different oxygen partial pressures and annealed at 340 °C for 200 h in O_2.

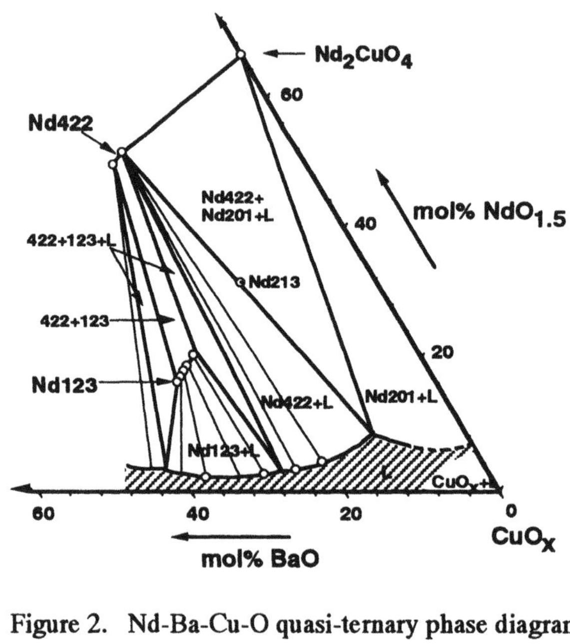

Figure 2. Nd-Ba-Cu-O quasi-ternary phase diagram around the Nd123 and liquid coexisting temperature region of 1073 °C in air.

rotation speed was in the range of 70~120 rpm. When the ratio of Ba/Cu ranged 0.55 to 0.85 in the liquid, Nd123 single crystals were produced. Figure 3 shows the temperature dependence of dc magnetization for Nd123 single crystals, which were grown from the liquid with different ratios of Ba/Cu and annealed at 340 °C for 200 h in O_2. With increase of Ba/Cu ratio of the liquid the T_c increases and the transition width ΔT becomes narrow, indicating that the ratio of Ba to Cu in the liquid plays a significant role in controlling the T_c of Nd123 superconductor. The crystal grown from the liquid with the Ba/Cu ratio of 0.78 resulted in formation of 123 stoichiometric compound.

3. SUPERCONDUCTIVE PROPERTIES

To investigate whether for Nd123 the anomalous peak effect in J_c-H curve is an intrinsic property or not, different series of heat treatment were performed using almost 123 stoichiometric Nd123 single crystals grown in P(O_2)=0.01atm by the above-mentioned TSSG method[11]. It is noteworthy that the crystal growth temperature was constant, and as-grown crystals were removed from the furnace immediately after crystal growth was finished. The typical time required to remove a crystal was approximately 10 min, corresponding to the cooling rate of about 100 °C/min. Different series of heat treatments in a pure oxygen gas flow were performed using the Nd123 single crystals produced separately.

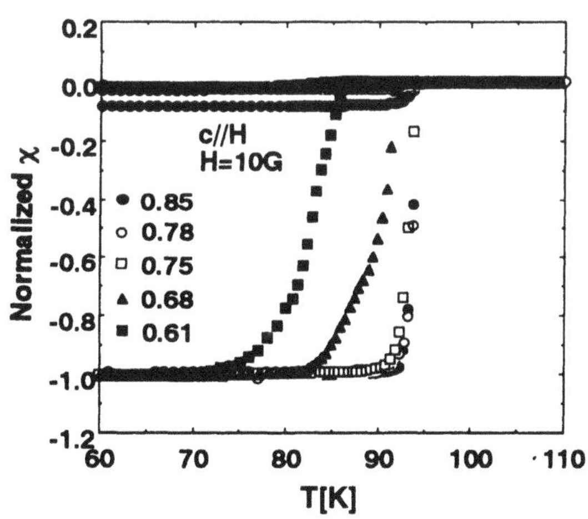

Figure 3. Temperature dependence of dc magnetization for Nd123 single crystals, which were grown from the liquid with different ratios of Ba/Cu and annealed at 340 °C for 200 h in O_2.

Figure 4 shows the J_c-H curve with an $H//a$-axis and an $H//c$-axis at 77 K for the Nd123 crystal heat treated at 340 °C for 200 h. No peak effect with either the $H//a$-axis or the $H//c$-axis in the J_c-H curve was

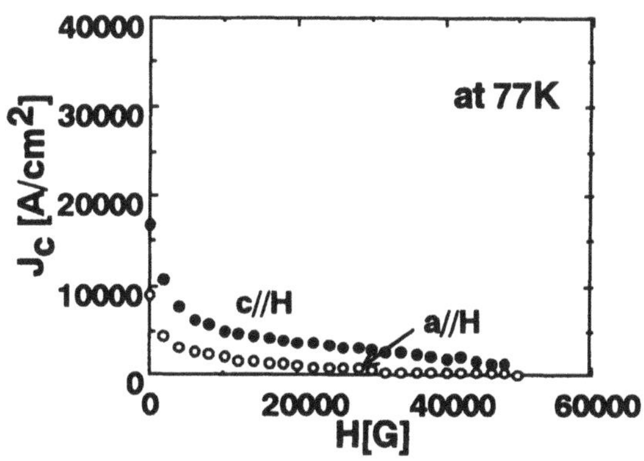

Figure 4. J_c-H curve with an $H//a$-axis and an $H//c$-axis at 77 K for the Nd123 crystal heat treated at 340 °C for 200 h

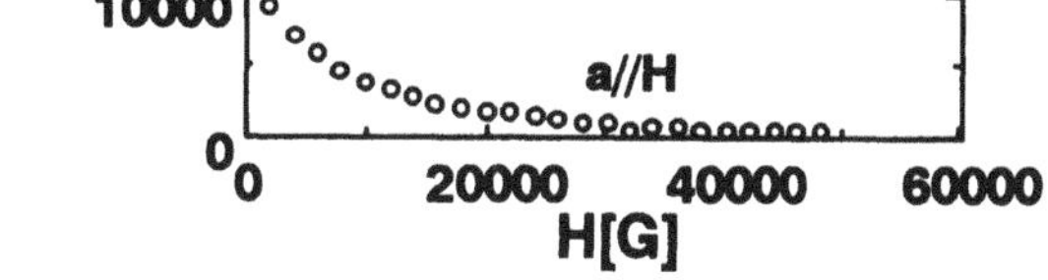

Figure 5. J_c-H curve with the $H//a$-axis and the $H//c$-axis at 77K for the Nd123 crystal subjected to the two-step heat treatment (500 °C for 100h and 340 °C for 200 h.)

observed and J_c for this Nd123 single crystal was as small as those for Y123 single crystals.

In contrast, figure 5 shows the J_c-H curve with the $H//a$-axis and the $H//c$-axis at 77K for the Nd123 crystal subjected to the two-step heat treatment (500 °C for 100 h and 340 °C for 200 h). The J_c-H curve with only the $H//c$-axis shows the peak effect, although with the $H//a$-axis shows no peak effect. These results indicate that for Nd123 the peak effect in the J_c-H curve is considered to be not an intrinsic property but process dependent. Further annealing (900 °C for 100 h, quenched, 340 °C for 200 h) of the Nd123 crystal with the peak effect resulted in disappearance of its peak effect. This result indicates that this solid reaction is found to be a reversible solid state reaction. The modulated structure (tweed image) was observed by TEM in the Nd123 crystals shown the peak effect[11]. The wavy compositional fluctuations of Nd/Ba within a range of several tens nanometers scale was observed by TEM with EDX compositional analyses in the Nd123 crystals shown the peak effect[12]. These results suggest that existence of phase separation, i.e. spinodal decomposition of the unstable solid solution, causes compositional fluctuation of Nd/Ba. Compositional fluctuation of Nd/Ba might be thought to be responsible for the peak effect in J_c-H curve.

4. CONCLUSION

Single crystals of Nd123 have been successfully grown by the TSSG method. BaO-CuO solvent was used in a Nd_2O_3 crucible. Furthermore, a 123 stoichiometric compound of Nd123 single crystal could be grown by using two different methods which were the oxygen partial pressure control of the atmosphere during growth and controlling liquid composition. Using the BaO-CuO solvent with 3/5 of Ba/Cu ratio, the stoichiometric Nd123 single crystal were grown under low oxygen partial pressure atmosphere ($P(O_2)$=0.01 atm). In the air atmosphere, the Ba enriched liquid with Ba/Cu ratio of 0.78 was used to obtain the stoichiometric Nd123 single crystal.

The origin of the anomalous peak effect in J_c-H curve for Nd123 is thought to be compositional fluctuation of Nd/Ba due to the spinodal decomposition of the unstable solid solution.

ACKNOWLEDGMENTS

This work was supported by NEDO for the R&D of Industrial Science and Technology Frontier Program and the Proposal-Based Advanced Industrial Technology R&D Program.

REFERENCES

1. A. Maeda, T. Yabe, K. Uchinokura and S.Tanaka, Jpn. J. Appl. Phys. 26 (1987) L1368.
2. S. I. Yoo and R. W. McCallum, Physica C 210 (1993) 147.
3. T. Wada, N. Suzuki, T. Maeda, A. Maeda, S. Uchida, K. Uchinokura and S. Tanaka, Appl. Phys. Lett. 52 (1988) 1989.
4. M. Murakami, S. I. Yoo, T. Higuchi, N. Sakai, J. Weltz, N. Koshizuka and S. Tanaka, Jpn. J. Appl. Phys. 33 (1994) L715.
5. S. I. Yoo, N. Sakai, H. Takaichi, T. Higuchi and M. Murakami, Appl. Phys. Lett. 65 (1994) 633.
6. S. I. Yoo, M. Murakami, N. Sakai, T. Higuchi and S. Tanaka, Jpn. J. Appl. Phys. 33 (1994) L1000.
7. M. Nakamura, H. Kutami and Y. Shiohara, Physic C 260 (1996) 297.
8. S. Yuhya, K. Kikuchi and Y. Shiohara, J. Mater. Res., 7 (1992) 2673.
9. M. Kambara, M. Tagami, X. Yao, E. A. Goodilin, Y. Shiohara and T. Umeda, submitted to J. Am. Ceram. Soc.
10. X. Yao, M. Kambara, M. Nakamura T. Umeda and Y. Shiohara, submitted to Appl. Phys. Lett.
11. M. Nakamura, Y. Yamada, T. Hirayam, Y. Ikuhara, Y. Shiohara and T. Tanaka, Physic C 259 (1996) 295.
12. T. Hirayama, M. Nakamura, Y. Yamada, K. Ikuhara and Y. Shiohara: J. Mater. Res., in print.

ELSEVIER

Physica C 282–287 (1997) 85–88

PHYSICA C

Low-frequency spin fluctuations in the superconducting $La_{2-x}Sr_xCuO_4$

K.Yamada[a], C.H. Lee[a], Y.Endoh[a], G. Shirane[b], R.J. Birgeneau[c], and M.A. Kastner[c]

[a]Department of Physics Tohoku University, Aramaki Aoba, Sendai 980-77, Japan
[b]Department of Physics, Brookhaven National Laboratory, Upton, NY 11973, USA
[c]Department of Physics, Massachusetts Institute of Technology, Cambridge, Massachusetts 02139,USA

Recent progress in low-energy neutron scattering study on the spatially modulated or so-called incommensurate spin fluctuations is reviewed. Well-defined incommensurate spin fluctuations are observed beyond x≈0.05 and up to the highest Sr-concentration x=0.25 in the present study. The incommensurability δ(x) is saturated at around 1/8 in the overdoped region and linearly scaled with the upper limit of $T_c(x)$ between the underdoped and optimally doped region. A characteristic energy E^* below which the dynamical magnetic susceptibility dramatically diminishes in the superconducting state can be determined to be around 7 meV in the energy spectrum of the spin fluctuations for x=0.15 and x=0.18 crystals. The disorder effect induces a low-frequency component of the incommensurate spin fluctuations below E^*.

1. INTRODUCTION

Spatially modulated spin correlations or so-called incommensurate spin fluctuations in hole-doped La_2CuO_4 provide us fruitful information on the hole and spin state on the CuO_2 plane and the role of magnetic interactions for the high-T_c superconductivity. However, neither the origin nor the frequency dependence of such spin fluctuations is fully understood yet. For example, there still exist several competing models to describe the incommensurate spin correlations such as nesting of Fermi surface[1-3], stripe ordering of doped holes[4] and percolation of cluster spin glass[5] and so on. Furthermore, the spin excitation gap reflecting the superconducting one has been observed only for optimally doped crystal. And the reason for the missing of energy gap or pseudo-gap in the underdoped or pair-broken sample is not understood. Although neutron inelastic scattering is one of the unique technique to study both spatial correlation and dynamics of spin fluctuations, no systematic studies on the doping dependence have ever been performed possibly due to the difficulty in preparing large single crystals. In this report, we present recent progress in our neutron scattering studies on Sr-doped La_2CuO_4 over a wide doping region between the underdoped and overdoped boundary of superconducting phase. Based on the obtained results we discuss the effect of intrinsic/extrinsic disorder in this system on the low-energy spin fluctuations.

2. EXPERIMENTAL

We have grown sizable single crystals of Sr-doped $La_{2-x}Sr_xCuO_4$ with x=0.04, 0.06, 0.08, 0.10, 0.12, 0.15, 0.18 and 0.25 by using a travelling-solvent-floating-zone

PII S0921-4534(97)00218-9

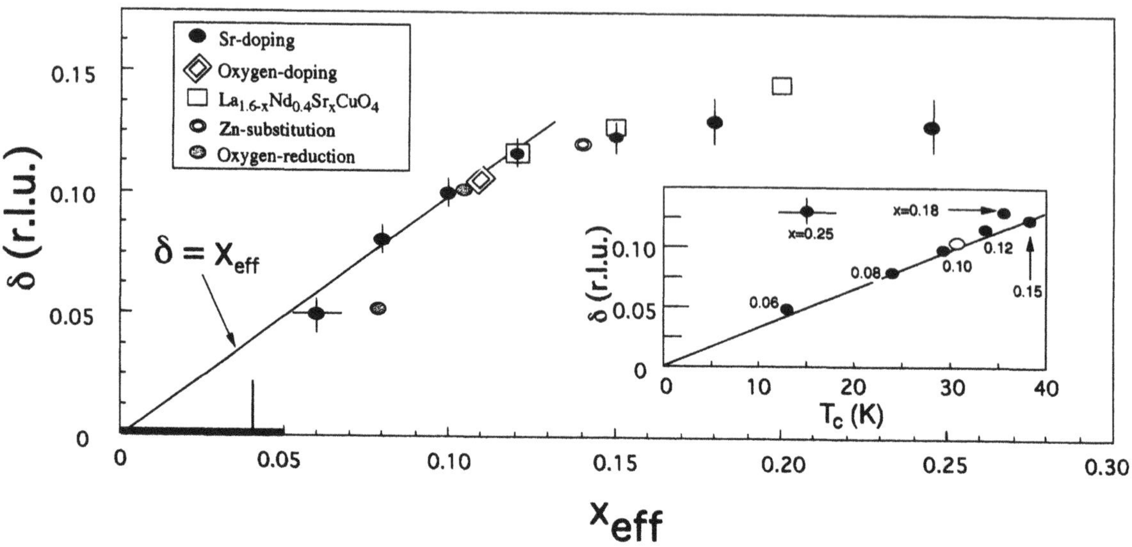

Figure 1. Incommensurability δ of the spatial spin modulation in $La_{2-x}Sr_xCuO_4$, oxygen-doped superconducting La_2CuO_{4+y}[12)], deoxygenated or Zn-doped nonsuperconducting $La_{2-x}Sr_xCuO_4$[13)] and $(La,Nd)_{2-x}Sr_xCuO_4$[14)] as a function of x_{eff}, which denotes Sr concentration x for $(La,Nd)_{2-x}Sr_xCuO_4$ and x-2y for La_2CuO_{4-y}. The inset shows δ(x) as a function of T_c(x) which is determined by a SQUID magnetometer. The attached numbers denote the Sr concentration for $La_{2-x}Sr_xCuO_4$.

method.[6)] All the annealed crystals except x=0.04 exhibit bulk superconductivities. For x=0.18 and x=0.25 T_c is lower by 1 and about 20 degrees, respectively compared to the maximum T_c of 37.5±0.5 K at x=0.15. One of the optimally doped single crystal was annealed under Ar gas-flow to prepare the pair-broken non-superconducting sample. The effective doping rate x_{eff} is defined as x_{eff}=x-2y where y is defined oxygen non-stoichiometry in reduced sample.

Inelastic neutron scattering experiments were performed on the triple-axis spectrometers of TOPAN and HER installed at JRR-3M in JAERI.

3. RESULTS

3.1. q-spectrum

Four magnetic peaks at (1/2(1±δ),1/2,0) and (1/2,1/2(1±δ),0) in the tetragonal (h,k,0) reciprocal plane are observed in all samples for 0.06≤x≤0.25. In Fig.1, we summarize the doping dependence of the incommensurability δ of the spin fluctuations at low energies (below 3.5 meV) and temperatures as a function of Sr concentration x. The δ becomes noticeable beyond x≈0.05, increases linearly with x up to x≈0.12 and finally is saturated in the overdoped region. As shown in Fig.1, a universal curve for δ(x) exists in this 2-1-4 cuprate system where δ(x) depends only on the doping rate and is not drastically altered by pair-breaking nor localization of doped holes. On the other hand, the peak-width of the q-spectrum depends on the disorder induced by oxygen-reduction. It is also noted that the peak-width *increases* with decreasing Sr concentration.

As shown in the inset of Fig.1, we found a linear relation between δ(x) and T_c(x) up to the optimal doping. However, in the overdoped region, $T_c(x)$ is much lower than that expected from the linear $\delta(x)$-$T_c(x)$ relation.

3.2. Energy spectrum

As shown in Fig.2, a characteristic energy E* of about 7 meV for the energy spectrum of the spin fluctuations is clearly determined for x=0.15 and x=0.18 below which the dynamical magnetic susceptibility $\chi''(q,\omega)$ dramatically diminishes in the superconducting state. Previously, the energy-gap E_{gap} was defined by the energy below which $\chi''(q,\omega)$ completely disappears at T=0K.[7] With the same definition, E_{gap} for x=0.18 is found to be around 4.5 meV which is *larger* than that for x=0.15.

4. DISCUSSIONS

Doping dependence of δ is presented for the wide doping region between the underdoped and overdoped superconducting boundary. However, for the quantitative interpretation of this doping dependence one should consider the relation between the number as well as the character of the inplane carriers and dopant concentration. Here, we just remark an interesting similarity in the doping dependence between δ and the so-called N_{eff} on the CuO_2 planes evaluated from the energy-integrated spectral weight of optical reflectivity on $La_{2-x}Sr_xCuO_4$.[8] Since the doping dependence of the N_{eff} may reflect a crossover behavior of carrier number n from "n=x" to "n=1-x" relation, the non-linear and saturating x-dependence of δ may suggest the gradual change in the electronic state of carrier for x>0.12.

The finite δ observed in the non-superconducting samples seems to be in conflict with the linear $\delta(x)$-$T_c(x)$ relation. We interpret, however, that $\delta(x)$ correlates with upper limit of $T_c(x)$, $T_c^{max}(x)$ although $T_c(x)$ is generally lower than $T_c^{max}(x)$ due to extrinsic pair-breaking such as Zn substitution or oxygen reduction. In other words, the above relation indicates an identical doping dependence between δ and $T_c^{max}(x)$. It is very important to find out the mechanism which controls the incommensurability, degree of spatial spin modulation. The deviation from the linear $\delta(x)$-$T_c(x)$ relation in the overdoped region suggests the existence of inherent or additional pair-breaking mechanism in this region. On the other hand, the linear $\delta(x)$-$T_c^{max}(x)$ relation in the underdoped region

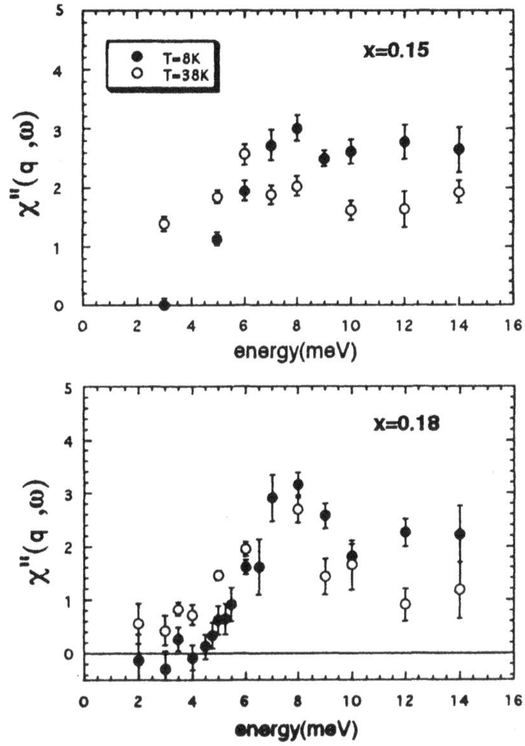

Figure 2. The energy spectra of dynamical susceptibility at the incommensurate peak position for x=0.15 and 0.18 below(closed circle) and around (open circle) T_c.

indicates that $T_c^{max}(x)$ is controlled only by the number of inplane doped holes and the low-frequency spin fluctuations below E* may not induce additional pair-breaking effect in this region.

The larger value of E_{gap} for x=0.18 compared to x=0.15 is very difficult to be interpreted because T_c is lower for the former. If E_{gap} corresponds to the superconducting gap, there should exist pair-breaking mechanism in the overdoped region decoupled with the size of superconducting gap. On the other hand, it is necessary to consider the possibility of E* as a superconducting gap as is already pointed out by Mason et al.[9] from a similar measurement on the gapless sample(E_{gap}=0). Assuming the $d(x^2-y^2)$-type anisotropy for E*, we evaluated the maximum value of the energy gap 2Δ to be about 35 meV ($\approx 11T_c$ at x=0.15) which is comparable to the superconducting- or pseudo-gap proposed from optical conductivity measurement.[10]

It should be noted that E* may exist above T_c as indicated in Fig.2. Furthermore, the recent neutron scattering measurement on the underdoped(x=0.1)[11] and highly overdoped (x=0.25)[13] revealed a signature of E* around 7 meV. We speculate therefore, that the size of E* weakly depends on doping as well as temperature and correlates with characteristic pairing interaction on CuO_2 planes. For the underdoped, oxygen-reduced or Zn substituted sample E* is not well defined due to the enhancement of the low-frequency spin fluctuations below E*. The detailed information on the doping as well as the temperature dependence of E* is highly required.

The authors wish to thank H. Fukuyama, and S. Uchida for valuable discussions and to S. Hosoya, M. Onodera, K. Nemoto, J. Wada, S. Wakimoto, K. Kurahashi, H. Kimura and S. Ueki for their valuable contributions for neutron scattering measurements as well as single crystal growth. The present work was supported by Grant-in-Aid for Scientific Research on Priority Areas, "Anomalous Metallic State near the Mott Transition" (07237101) Ministry of Education, Science and Culture, Grant of Shinko Chousei of Science and Technology Agency, and US-Japan collaboration program on neutron scattering.

REFERENCES

1) N. Bulut, D. Hone, D.J. Scalapino et al., Phys. Rev. Lett., 64, 2723 (1990).
2) Q. Si, Y. Za, K. Levin, J.P. lu and J.H. Kim, Phys. Rev. B 47, 9055 (1993).
3) T. Tanamoto, H. Kohno and H. Fukuyama, J. Phys. Soc. Jpn. 63, 2739 (1994)
4) S.A. Kivelson and V.J. Emery, Proc. of "Strongly Correlated Electronic Materials: The Los Alamos Symposium 1993," 619; V.J. Emery and S.A. Kivelson, Physica 235- 240, 189 (1994).
5) A. Aharony, R.J. Birgeneau, A. Coniglio et al., Phys. Rev. Lett., 60, 1330 (1988); R.B. Gooding (private communication).
6) S. Hosoya, C.H. Lee, S. Wakimoto, et al., Physica C 235-240, 547 (1994).
7) K. Yamada, S. Wakimoto, G. Shirane, C.H. Lee, M.A. Kastner, S. Hosoya, M. Greven et al., Phys. Rev. Lett., 75, 1526 (1995).
8) S. Uchida, T. Ido, H. Takagi, T. Arima, et al., Phys. Rev. B 43, 7942 (1991).
9) T. E. Mason, G. Aeppli, et al., Physica B 199&200,284(1994);T.E. Mason, A.Schroder, et al., Phys. Rev. Lett. 77, 1604 (1996).
10) K.Tamasaku, T. Ito, H. Takagi and S. Uchida, Phys. Rev. Lett. 72(1994) 3088.
11) Y. Kim and M. Greven (unpublished data).
12) R. J. Birgeneau, R. J. Christianson, et al., Proc. ICPTM-96 (Osaka 1996) .
13) K. Yamada (unpublished data).
14) J. M. Tranquada, J. D. Axe, et al., Phys. Rev. Lett. 78, 338 (1996).

ELSEVIER

Physica C 282-287 (1997) 89-92

PHYSICA C

IMPROVEMENTS IN CRYSTAL GROWTH AND CRYSTAL HOMOGENEITY AND ITS IMPACT ON PHYSICS

Andreas Erb, Eric Walker, Jean-Yves Genoud and René Flükiger

Département de Physique de la Matière Condensée, Université de Genève, 24, quai Ernest Ansermet, 1211 Genève 4, Switzerland

With the development of the novel crucible material $BaZrO_3$ a new class of 123 - single crystals has become available, the purity of which reaches now semiconductor grade (> 99.995 %). Due to the high purity the experiments are not longer obscured by residual impurities and the influence of different oxygen content and distributions could be studied in detail. This has given way to several new findings and to better basic understanding of the physics of high Tc superconductors. Examples for the impact of this new crystal generation on physics are the reversible suppression of the so called fishtail effect, which is accompanied with the absence of the Schottky anomaly in low temperature specific heat and the calorimetric observation of the vortex lattice melting . On such samples it is possible to observe the vortex lattice by STM without any surface conditioning and to study the influence of pinning centers such as twin boundaries and oxygen vacancies on vortices.

1. METALLIC IMPURITIES

Ten years of history of high T_c superconductivity is also a story of ten years of sample preparation and crystal growth. After some early successes, the use of ZrO_2 crucibles has been established as state of the art, even though the single crystals obtained in such experiments did not reach a purity better than 99.9 %. In the Table the reported impurity values for different crucible materials are given for comparison. After that crystal growth and along with it sample quality of high T_c superconductor single crystals did not develop further. The reason of this stagnation was that the main problem has been left unsolved for all the years - the finding of an inert crucible material or the development of a container free crystal growth method. Since all crucible materials known up to now, namely Al_2O_3, SnO_2, MgO, Au, Pt, ThO_2, Y_2O_3, or also the ZrO_2 crucibles, which have to be stabilised with CaO or Y_2O_3, react with the $BaCuO_2$ - CuO mixture, commonly used as solvent for the crystal growth experiments, this corrosion induces in all cases impurities in the grown crystals. It should be noted , that also the use of e.g. Y_2O_3 does not solve this problem, even in the case of crystal growth of the $YBa_2Cu_3O_{7-\delta}$ - compound, because the commercially available crucibles of Y-oxide, as well as many other ceramics contain additional elements (typically Al, Ti, Si, Mn, Zn, Sn) necessary as sintering aids.

Crucible	**Impurities in Crystals**
Al_2O_3 [e.g. 1]	up to 5 at. % Al **Purity ≤ 95 at. %**
Gold [2]	1.5 at. % Au **Purity ≤ 98.5 at. %**
MgO [3]	1 at. % Mg **Purity ≤ 99 at. %**
ZrO_2 (Y - stab.) [4]	Mg, Zn, Al, Fe, Mn, Ti all between 0.07 and 0.14 at. % **99.5 at. % ≤ Purity ≤ 99.9 at. %**
$BaZrO_3$ [5]	La, Sr, Zr between 0.0005 and 0.0014 at. % **Purity ≥ 99.995 at. %**

0921-4534/97/$17.00
PII S0921-4534(97)00219-0

With the development of the $BaZrO_3$ crucibles in our laboratory [1] a new class of single crystals has become available. Crystals grown in $BaZrO_3$ crucibles exhibit a superior purity even with respect to those grown in ZrO_2 crucibles by more than an order of magnitude. It must be mentioned that the purity of the starting materials (5N) used in the experiments of Ref. [4] does not reflect in the purity of the crystals, which is deteriorated to less than (3N) due to crucible corrosion. The problem of impurities is even more severe if RE-123 crystals others than $YBa_2Cu_3O_{7-\delta}$ are grown, since then the corrosion of the Y - stabilised ZrO_2 leads to crystals with the nominal composition $RE_{1-x}Y_xBa_2Cu_3O_{7-\delta}$, where the value for x is typically in the range of 0.1 to 0.2.

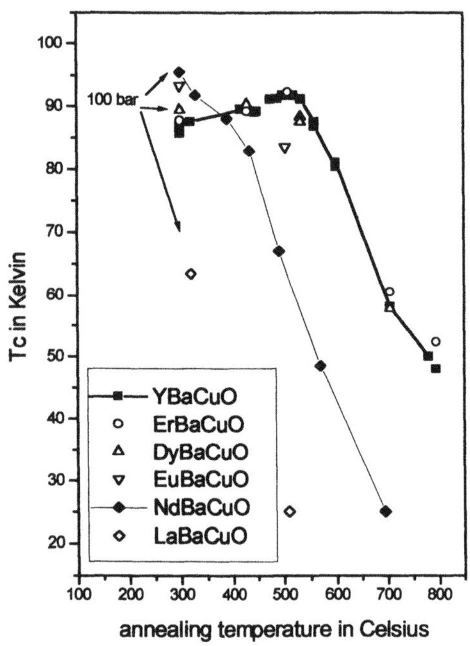

Fig. 1 T_c - values for different RE - 123 single crystals in dependence of the annealing temperature for 1 bar oxygen atmosphere (except the 300 °C treatment)

With the $BaZrO_3$ crucibles also this problem can be solved and crystals can be grown without this undesired doping. The differences for the T_c vs. annealing temperature for the different rare earth elements are remarkable. We found that the light rare earth element 123-crystals have to be oxygenated at much lower temperatures to obtain the highest transition temperature (Fig 1). Residual impurities, however can have strong influence on the microstructure and hence the physical properties of these compounds as it has been shown for instance by Storey et al. [6] for the case of gold doped crystals, where a clustering effect of the gold atoms has been observed by TEM studies. Since the coherence length for the 123 compounds is just in the range of an unit cell, such non-superconducting regions within a crystal produce ideal pinning centres, leading to an increase in the critical current density as determined from magnetisation measurements. Thus, measurements on imperfect samples may only represent the situation for a particular specimen and do not show the intrinsic property of the 123 compound.

2. OXYGEN DISTRIBUTION

The standard oxidation annealing to obtain high transition temperatures well above 90 K in $YBa_2Cu_3O_{7-\delta}$ single crystals consists of a treatment under 1 bar of oxygen at around 510 °C for 100 - 200 hours, followed by quenching to room temperature. Such a treatments result in oxygen deficient samples expressed by δ values around 0.08. Taking into account that the oxygen sites in the copper oxide planes do not contribute to the oxygen deficiency an oxygen deficiency expressed by δ= 0.08 means simply that 8 % of the crystal consist of $YBa_2Cu_3O_{6.0}$ unit cells. Allowing half filled cells of $YBa_2Cu_3O_{6.5}$ this value doubles to 16 %. A crystal with 16 % of oxygen deficient cells however should not be referred to be nearly fully oxygenated but is in fact a heavily doped sample. If these oxygen deficient regions are large enough to form effective pinning centres they could be responsible for the enhanced critical current density with increasing magnetic field, often referred to as fishtail effect [7].
According to Lindemer et. al [8], the same oxygen content can be obtained as well by annealing at higher temperatures in high pressure oxygen atmospheres. Depending on the annealing temperature however a different microstructural

arrangement of the oxygen vacancies can be expected as sketched in Fig. 2. At high temperatures as high as 650 °C the oxygen deficient regions are more likely to be randomly distributed, while at lower temperatures a clustering effect of oxygen deficient unit cells may occur due to the smaller entropy term in the free energy [9, 10 and references therein].

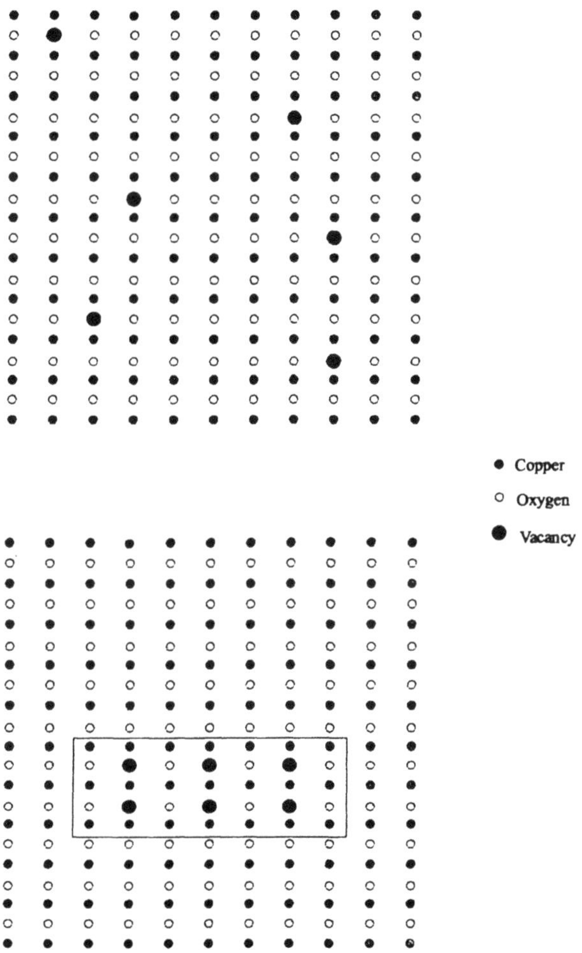

Fig. 2 Different arrangements of the oxygen vacancies in the basal plane of $YBa_2Cu_3O_{6.94}$; the upper one corresponding to higher temperatures than the lower one

We therefore performed high pressure oxygen treatments at 650 - 750 °C in 100 bar of O_2 for periods of 12 - 40 h inside a water-cooled autoclave with an internal oxygen pressure of up to 110 bar, followed by quenching to room temperature, so that the high temperature oxygen state can be maintained.

The crystals used for the oxygenation experiments had a size in the range of about 1.5 x 1.5 x 0.1 mm^3 and showed the usual twinning within the a-b plane. The magnetisation measurements were performed in a SQUID Magnetometer (Quantum Design) with the field parallel to the c-axis of the crystal. The result of an oxygenation treatment at 650 °C for 12.5 h at 100 bar can be seen in Fig. 1, where M(H) is shown for a temperature of 70 K.

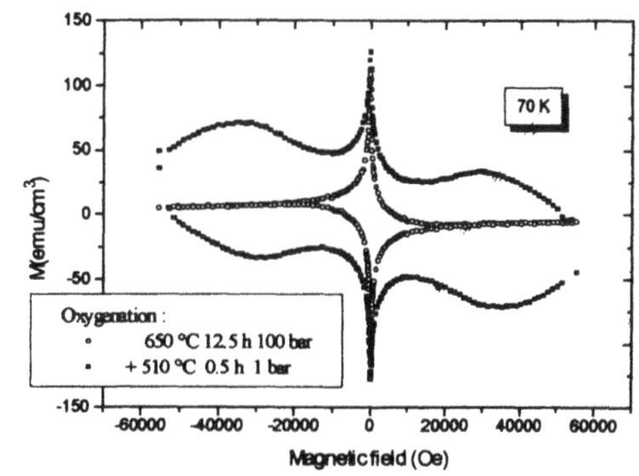

Fig 3. Magnetisation of $YBa_2Cu_3O_{6.92}$ after different annealing regimes

No fishtail behaviour is present after this oxygenation treatment. Re-annealing the same crystal for only 30 minutes at a temperature similar to the standard oxygenation (510 °C, under 1 bar oxygen), leads to the reappearance of the fishtail anomaly (also shown in Fig 3). Note, that according to Ref. 8 the overall oxygen content of the sample has the same value of 6.92 for the two treatments and that T_c is unchanged. Also the short re-annealing time speaks against a change in the overall oxygen content. The treatments have no influence on the state of twinning.

Thus, we conclude that only a locally altered distribution of the oxygen vacancies like the formation of clusters of oxygen deficient regions, for instance of $YBa_2Cu_3O_{6.5}$ unit cells in the sample, must be responsible for this effect. For such a short range reorganisation of the oxygen deficient regions it is only necessary that the oxygen diffuses over a few lattice parameters, which can be done even during short (0.5 h in Fig.3) heat treatments.

Another test for this explanation consisted of a low temperature annealing at high pressures, in our case at 300 °C under 100 bar oxygen pressure. We observe no fishtail behaviour since in this case the equilibrium oxygen content for such a low temperature annealing treatment is nearly 7.0 [8], thus the amount of oxygen vacancies is very small and it is not possible for those oxygen vacancies to form clusters of $YBa_2Cu_3O_{6.5}$ unit cells, which can act as effective pinning centres on increasing magnetic field.

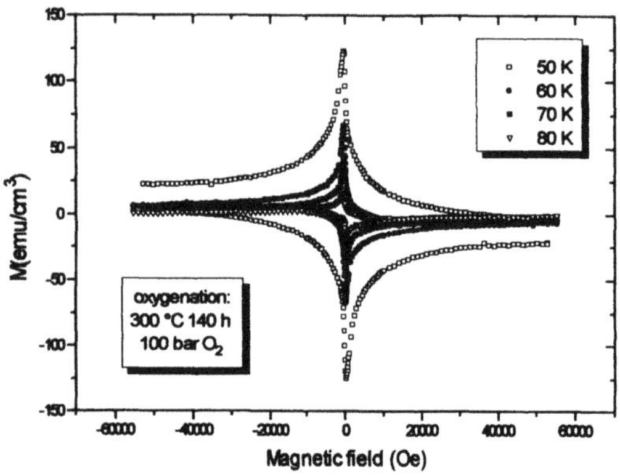

Fig 4. Magnetisation of a $YBa_2Cu_3O_7$ single crystal

Further support for this explanation of the fishtail anomaly as being a due to clusters of oxygen deficient regions arises from the absence of this anomaly in $BaZrO_3$ grown single crystals of the naturally untwinned and stoichiometric high pressure compound $YBa_2Cu_4O_8$ [11].
The same heat treatment which suppresses the fishtail effect in the $BaZrO_3$ grown crystals fails to suppress this anomaly in crystals grown from ZrO_2.The failure of the suppression of the fishtail anomaly in the latter case is attributed to the presence of residual metallic impurities (see Tab. 1), which can prevent the free mobility of the oxygen atoms as it is the case for aluminium.

3. CONCLUSION

The reversible suppression of the fishtail effect turned out to be a key experiment. Those crystals exhibiting weak pinning were found to be ideal samples for specific heat studies. Low temperature specific heat experiments revealed that for fully oxygenated $YBa_2Cu_3O_{7.0}$ crystals the suppression of the fishtail effect is accompanied with the vanishing of the Schottky anomaly [12], which therefore can also be attributed to the existence of free Cu valances due to oxygen vacancies. Another direct correlation with the fishtail anomaly is made with the direct observation of a first order-like melting transition of the vortex lattice in calorimetric measurements [13,14] in twinned crystals without fishtail anomaly. Unlike crystals from other container materials the twinned state did not hinder this direct observation. The explanation for this difference is that the small number of residual pinning centers, others than twins, which recently have been proven to be the only still effective ones [15], allows this observation. Since the absolute number of pinning centers needs to be below a certain threshold value, crystals with a larger number of other pinning centers have to be detwinned to obtain the melting transition.

REFERENCES

[1] A. Erb, E. Walker, R. Flükiger, Physica C 245 (1995) 245
[2] H. Ikuta and D.M. Ginsberg, J. of Supercond. 9 (1996) 259 and ref. therein
[3] C. Krauns, Ph. D. thesis, University Göttingen, Germany (1996)
[4] H. Casalta, P. Schleger, P. Harris, B. Lebech, N. H. Andersen, R. Liang, P. Dosanjh, W. N. Hardy, Physica C, 258 (1996) 321
[5] A. Erb, E. Walker, R. Flükiger, Physica C 258 (1996) 9
[6] B. G. Storey et al., Physica C 246 (1995) 46
[7] M. Däumling, J. M. Seuntjens, D. C. Larbalestier, Nature 346, 332 (1990)
[8] T. B. Lindemer et al., J. Am. Ceram. Soc., 72, 1775 (1989)
[9] J. L. Vargas and D. C. Larbalestier, Appl. Phys. Lett. 60, 1141 (1992)
[10] A. Erb, J.-Y. Genoud et al. J. Low. Temp. Phys. 105, 1023 (1996)
[11] J.-Y. Genoud et al., this conference
[12] J.-Y. Genoud et al., Proc. of the LT21 conference in Prague 1996, to be published in Czech. J. Phys.., Supplement
[13] M. Roulin, A. Junod, A. Erb, E. Walker, J. Low. Temp. Phys. 105, 1099 (1996)
[14] A. Junod et al., Physica C 275, 384 (1997) and this conference
[15] I. Maggio-Aprile et al., submitted to Nature

ELSEVIER

Physica C 282–287 (1997) 93–96

PHYSICA C

Microscopic x-ray characterization of inhomogeneities in YBCO crystals with sharp superconducting transitions

S. B. Qadri,[a] E. F. Skelton,[a] M. S. Osofsky,[a] V. M. Browning,[a] J. Z. Hu,[b] P. R. Broussard,[a] M. E. Reeves,[c] and W. Prusseit[d]

[a]Condensed Matter and Radiation Sciences Division and Materials Science and Technology Division, U.S. Naval Research Laboratory, 4555 Overlook Avenue, S.W., Washington, DC 20375-5320

[b]Geophysical Laboratory and Center for High Pressure Research, Carnegie Institution of Washington, 5241 Broad Branch Road, Washington, DC 20015-1305

[c]Department of Physics, The George Washington University, Wahsington, DC 20052

[d]Physik Department E10, TU Munchen, James-Franck Str. 1, 85747 Garching, Germany.

High resolution x-ray diffraction measurements of YBCO single crystals with optimal transport properties reveal gross structural inhomogeneities, such as domains of different oxygen content and high dislocation densities. Thin epitaxial YBCO films show similar structural inhomogeneities.

1. INTRODUCTION

The most widely studied of the high temperature superconductors (HTS) is $YBa_2Cu_3O_{7-\delta}$ (YBCO). The research responsible for this article has been motivated by hopes of understanding the physics of HTS, as well as potential applications. Significant progress has been achieved in fabricating epitaxial films with superior superconducting properties by a variety of deposition techniques, such as rf sputtering, molecular beam epitaxy, and pulsed laser deposition. Similarly, single crystals of YBCO have been synthesized with "optimal" superconducting properties. To date, many unresolved questions concerning the interpretation of many experimental results remain, in addition to an inability to exploit this material for active electronic device applications.

It is widely accepted that the coherence lengths (the decay distance of the superconducting order parameter) of the HTS are short, on the order of tens of Angstroms or less. Therefore, sample inhomogeneities (such as cation inhomogeneities, oxygen vacancies, dislocations and small angle grain boundaries) on length scales comparable to a unit cell will affect the measured superconductive properties of the system. Nevertheless, it has been generally assumed that, based on electrical transport measurements, sample properties are homogeneous and optimal. Structural studies, which are rarely reported, provide an independent test of this assumption. We have shown that structural inhomogeneities may exist in YBCO single crystals which would be considered homogeneous with optimal transport properties.[1,2] Similar structural inhomogeneities are shown to exist in YBCO epitaxial films.

The objective of this investigation was to perform the aforementioned x-ray measurements on YBCO samples that were considered "good," based on their transport properties. These are the most extensive structural and electrical transport results on the same crystals reported to date.

2. EXPERIMENTAL DETAILS

Three different types of x-ray diffraction measurements were used to structurally characterize the HTS samples: high resolution diffractometer scans, topographs, and energy dispersive measurements. The first two are carried out using Cu $K\alpha_1$ x-rays. They probe the entirety of the films and the first micron or two of the surfaces of the single crystals. Very high energy (> 60 keV) x-rays are used for the third method. These are generated on a superconducting wiggler beamline (X17C) at the National Synchrotron Light Source, Brookhaven National Laboratory. Because of the extreme brightness of this radiation, a very narrow (< 20 μm) pencil of radiation is all that is needed, thus leading to a very high spatial resolution. And, because of the greater penetrating power of these photons, the entire crystal volume can be sampled.

0921-4534/97/$17.00
PII S0921-4534(97)00223-2

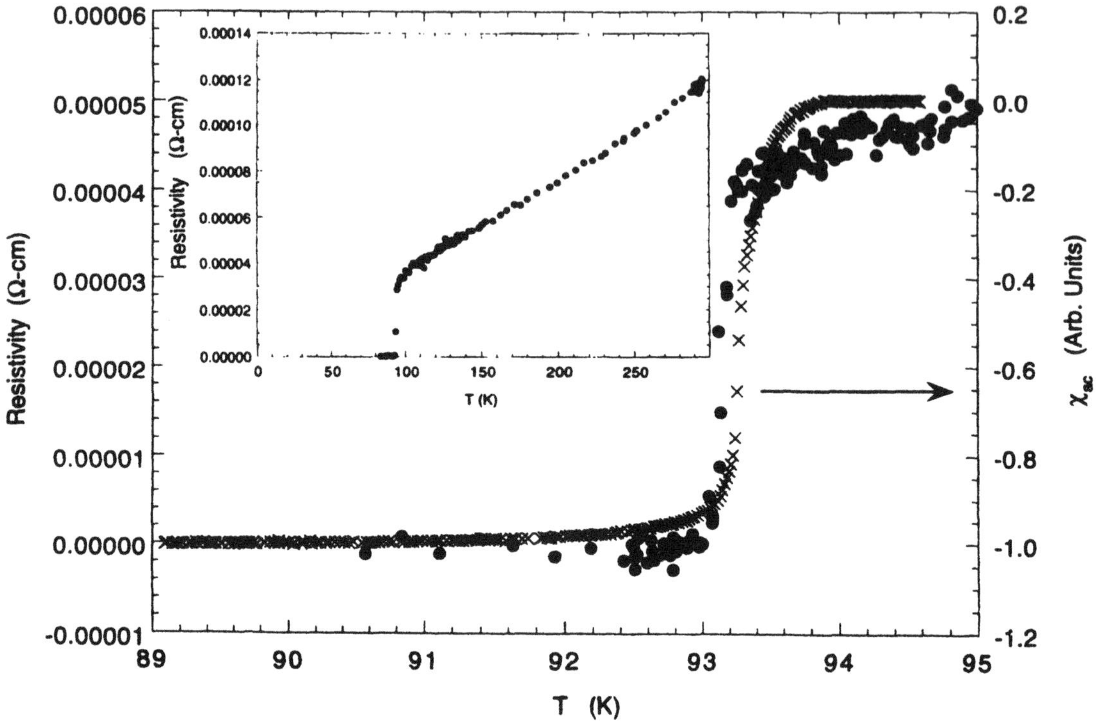

Fig. 1. Superconductivity transition as measured by ac susceptibility (x) and resistivity (●).

Extensive research on bulk ceramic samples by others has led to a quantifiable relation between the x-ray measurements and the oxygen content, *i.e.*, the term (7-δ). From the bulk data, we have identified an empirical relation linking the length of the *c*-axis of the unit cell with (7-δ) for YBCO crystals. For YBCO epitaxial films, the oxygen stoichiometry is estimated from the ratio of the integrated intensities of the I(005)/I(006) diffraction peaks.[3] For ease of interpretation, it is assumed that only oxygen defects control the value of the *c*-parameter. However, it has been shown by others that cation disorder also can alter the x-ray diffraction patterns in a similar manner. This result only changes the details of how the defects affect the measured properties.

3. RESULTS AND DISCUSSIONS

T_C's were measured by ac susceptibility and ac resistivity. The susceptibility measurements were made at 10 kHz using a pancake coil configuration with an excitation field amplitude of 0.1 G. The ac resistivity data were obtained with 135 μA current at 486 hz using a four probe configuration. The susceptibility results were comparable for the crystals studies. The transitions for one crystal are plotted in Fig. 1. The crystals studied had low room temperature resistivities and high, sharp T_C's, e. g., $\rho = 1.3 \times 10^{-4}$ Ω-cm at 300 K, $T_c > 93$ K and $\Delta T_c < 300$ mK. These features alone are often cited as evidence for a crystal's quality. The crystal's magnetotransport properties then were measured, resulting in a set of data which were identical to those attributed in the literature to "high quality" samples. Nevertheless, the crystals showed significant compositional variations within few micrometers of the surface and in the interior.

We performed high resolution x-ray studies on two YBCO crystals to determine the variation of the *c*-parameter, and hence oxygen content, and the distribution of dislocations and the defect density. The YBCO crystals were grown following a procedure reported elsewhere,[4] followed by annealing at 500° C in an oxygen atmosphere for 21 days.

In order to investigate the surface quality of the crystals, two types of x-ray measurements were performed: 2θ-scans and rocking curves. Both were on the (0,0,6)-reflection. The 2θ-scans reveal variances in the unit cell parameter; the rocking curves give information about the crystal quality. Multiple peaks were observed for both scans on these crystals. These are shown for one crystal in Figures 2 and 3, respectively.

Multiple peaks in the rocking curves may be due to different oxygen concentrations and small angle grain boundaries between individual domains or grains. The full width at half maximum (FWHM) values range between 864 and 132 arcseconds. One can assume the broadening is due to presence of random arrays of dislocations, the density of which can be estimated from standard expressions.[5,6] From these, a lower limit to the dislocation density is estimated to be between 2.5 x 10^8 to 1.1 x 10^{10} lines/cm^2.

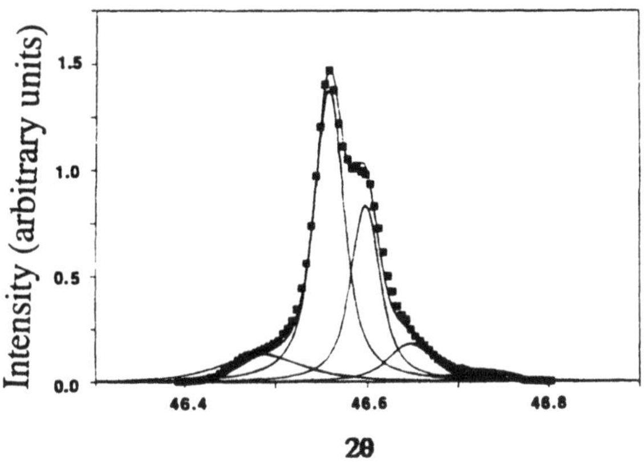

Fig. 2. 2θ-scan of the (0,0,6)-reflection

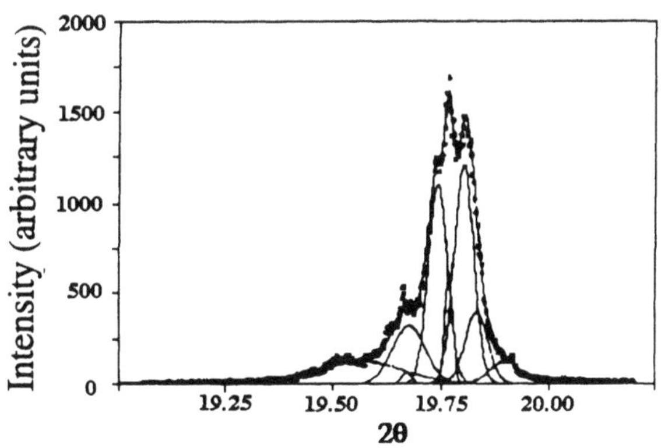

Fig. 3. Rocking curve of the (0,0,6)-reflection

A double-crystal topograph also was obtained using the (0,0,6)-reflection. This is shown in Fig. 4. The domain structure of the crystal is clearly revealed in the topograph. The topograph was recorded at a 2θ setting of 46.558°, which corresponds to a value of (7-δ) of 6.82±0.04. The dark areas in Fig. 4 correspond to the portions of the surface with (7-δ) = 6.82. Thus, by taking topographs in conjunction with 2θ-scans, we can image the region of the sample with a specific oxygen content.

In order to probe the interior of these crystals, the very high energy x-rays available at the National Synchrotron Light Source (NSLS), at Brookhaven National Laboratory were used. The energy dispersive diffraction facility on Beamline X17C at NSLS, which was used for this work, has been described elsewhere.[7] A variation of the *c*-axis as measured throughout the bulk of the crystal on a 20 μm spatial scale is clearly seen. [Fig. 5] This is interpreted to be primarily due to a variance in oxygen.

Fig. 4. Topograph of the (0,0,6)-reflection

Structural inhomogeneities in the thin single-crystal films of YBCO were characterized using high-resolution x-ray diffraction. Results of these measurements show a distribution in the *c*-axis lattice parameter and variations in the integrated intensities of (00ℓ) diffraction peaks. Both effects can be interpreted as due to a variance in the oxygen content. Films prepared under identical growth conditions, when subjected to different post annealing treatments showed varying structural

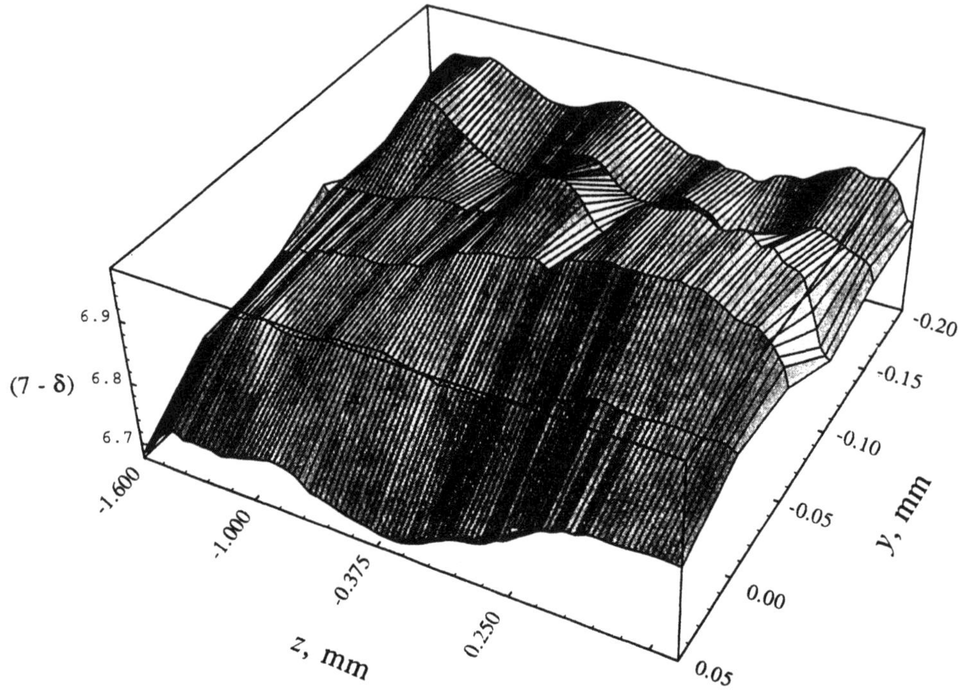

Fig. 5. Three dimensional plot of (7-δ) versus (*y*,*z*)-position in the crystal

and superconducting properties. A film with 7-δ close to 7.0 had a transition temperature (T_c) of 91 K, a width (ΔT_c) of 0.6 K, and a critical current density (J_c) of 6.5 x 10^6 A/cm^2, whereas a film in which the oxygen content varied between 6.86 - 6.96 had a T_c of 87 K, a ΔT_c of 0.5 K, and a J_c of 8.17 x 10^6 A/cm^2. For each of the samples measured, the full width at half maximum (FWHM) of the (006) diffraction peak ranged between 340 and 2100 arcseconds and was dependent on the method of preparation and the substrate. The J_c's of these films were not sensitive to the quality of the film, as determined from the FWHM of the (006) peak, and the oxygen content, as determined from the (005)-to-(006) integrated intensity ratio. High resolution x-ray topographs showed uniform microstructure of the films, indicating that oxygen distribution is finely dispersed in the film.

CONCLUSIONS

Both single crystals and epitaxial films which would be classified as "optimal" based on their transport properties, exhibit structural inhomogeneities when analyzed by sophisticated x-ray diffraction techniques. The presence of these features points out that electrical transport measurements are inadequate for determining sample quality and brings into question whether results of experiments performed on such samples can be attributed to intrinsic behavior. Futhermore, these x-ray techniques can be used to identify helpful and harmful structural features in epitaxial films that will be important in characterizing samples to be used in superconducting electronics applications.

REFERENCES

1. S. B. Qadri, M. S. Osofsky, V. M. Browning, E. F. Skelton and T. A. Vanderah, Appl. Phys. Lett. **68**, 2729 (1996).
2. V. M. Browning, E. F. Skelton, M. S. Osofsky, S. B. Qadri, J. Z. Hu, L. W. Finger, and P. Caubet, to be pub. in Phys. Rev. (1997).
3. J. Ye and K. Nakamura, Phys. Rev.-B **48**, 7554 (1993).
4. T. A. Vanderah, C. K. Lowe-Ma, D. E. Bliss, M. W. Decker, M. S. Osofsky, E. F. Skelton, and M. M. Miller, J. Cryst. Growth **118**, 385 (1992).
5. P. Gay, P. B. Jirsch, and A. Kelly, Acta Metall. **1**, 315 (1953).
6. S. B. Qadri and J. H. Dinan, Appl. Phys. Lett. **47**,1066(1985).
7. E. F. Skelton, S. B. Qadri, M. S. Osofsky, A. R. Drews, P. R. Broussard, J. Z. Hu, L. W. Finger, T. A. Vanderah, D. Kaiser, J. L. Peng, S. A. Anlage, R. L. Greene, and J. Giapintzakis, "Direct Observation of Microscopic Inhomogeneities in High Tc Superconductors," in SPIE Proc. Vol. 2516, (1995), pp. 160-169.

ELSEVIER

Physica C 282–287 (1997) 97–100

PHYSICA C

Defects that Control the Properties of Tl- and Hg-based Superconductors

J. D. Jorgensen[a], O. Chmaissem[b], J. L. Wagner[b,c], W. R. Jensen[c], B. Dabrowski[a,d], D. G. Hinks[a], and J. F. Mitchell[a]

[a]Materials Science Division, Argonne National Laboratory, Argonne, IL 60439

[b]Science and Technology Center for Superconductivity, Argonne National Laboratory, Argonne, IL 60439

[c]Department of Physics, University of North Dakota, P. O. Box 7129, Grand Forks, ND 58202

[d]Physics Department, Northern Illinois University, DeKalb, IL 60115.

Defects that affect T_c in $Tl_2Ba_2CuO_{6+\delta}$ and $HgBa_2CuO_{4+\delta}$ have been characterized by neutron powder diffraction. In $Tl_2Ba_2CuO_{6+\delta}$, the dominant defect is interstitial oxygen between the two Tl-O planes, but Cu substitution on the Tl site also affects properties and there is evidence for a second oxygen defect for compositions in the vicinity of maximum T_c. In $HgBa_2CuO_{4+\delta}$, there are two competing oxygen defects in the Hg layer. The relative concentrations of these defects switches upon passing from the underdoped region, through the maximum T_c, to the overdoped region. This remarkable behavior could result from a change in the topology of the Fermi surface upon passing through the van Hove singularity.

1. INTRODUCTION

The $Tl_2Ba_2CuO_{6+\delta}$ (Tl-2201) [1] and $HgBa_2CuO_{4+\delta}$ (Hg-1201) [2] superconducting compounds are of particular importance for the study of the physics of HTS because these materials exhibit the highest T_c's observed for any material with one CuO_2 layer, they have simple crystal structures, and the doping can be varied over a wide range. For Tl-2201, the maximum T_c is ~94K and the compound can be overdoped into the normal-metal regime. For Hg-1201, the maximum T_c is ~98K and the compound can be both underdoped and overdoped.

To enable the reproducible synthesis of these materials and to interpret the physical measurements, one must understand the defects that control the properties. Multiple defects form in both compounds. In this paper, we describe these defects and their effects on T_c. We also comment on what this complex defect chemistry may teach us about the underlying electronic structure — specifically, that the competition between slightly different oxygen defects can be strongly affected by crossing the van Hove singularity.

2. $Tl_2Ba_2CuO_{6+\delta}$

Two defects have previously been described in the literature for Tl-2201 [3]. T_c is affected most strongly by an interstitial oxygen defect O4 located between the two Tl-O layers. T_c is maximum (~94K) for the minimum amount of defect oxygen and decreases as more oxygen is added, eventually reaching the normal metallic state.

The second defect is Cu substitution at the Tl site. This defect was discussed in some of the earliest papers about Tl-2201, but the experimental evidence was not strong. More recently, Aranda et al. [4] used anomalous x-ray diffraction (at the Cu edge) to prove the existence of this defect. By comparing two samples, they showed that the unit cell tends to have a smaller orthorhombic strain (or appear tetragonal within experimental resolution) for higher concentrations of this defect. However, they did not correlate superconducting properties to the concentration of this defect.

To investigate the defect chemistry more fully, we prepared samples of Tl-2201 by a conventional solid-state-reaction technique for various Tl/Cu starting compositions from 2.3/1.0 to 2.0/1.1 [5]. Portions of each of these samples were then annealed in oxygen at 400° C and various oxygen partial pressures to vary the oxygen contents. This synthesis technique yielded samples with a wide range of structural and superconducting properties. For each sample, the amount of oxygen on the O4 interstitial site was determined by Rietveld refinement using neutron powder diffraction data.

PII S0921-4534(97)00225-6

The T_C's of these samples vs. the oxygen content on the O4 site are shown in Fig. 1. From the wide scatter in the data, it is immediately obvious that O4 oxygen content is not the only factor controlling T_C. The T_C's and also the structural data show that these differences result from the different Tl/Cu starting compositions. We assume, but cannot prove directly from the neutron diffraction data, that these samples differ in the amount of Cu on the Tl site. The variation of orthorhombic strains vs. Tl/Cu ratio for our samples (see Fig. 2) agrees with the results of Aranda et al. [4], supporting our assumption. Samples with more Cu on the Tl site have smaller orthorhombic strains. For these samples, the maximum T_C that can be achieved by varying the oxygen content is also slightly (5-10° K) lower.

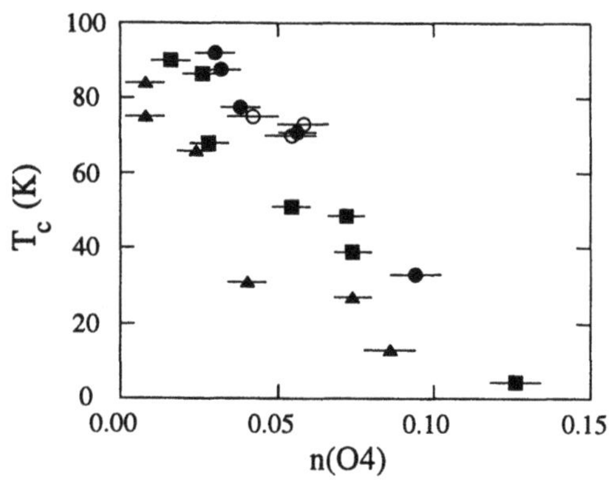

Fig. 1. T_C vs. oxygen content at the O4 site for Tl-2201. The different symbols are for samples with different Tl/Cu starting compositions.

At oxygen contents near those that give the maximum T_C, we see evidence for a second oxygen defect. In the vicinity of maximum T_C, with continuously decreasing oxygen content, the c-axis lattice parameter increases, passes through a maximum, and then decreases again (see Fig. 3). Such behavior is not consistent with a monotonically changing concentration of a single defect. By comparing the oxygen content on the O4 site (determined by neutron diffraction) with the total oxygen uptake determined by TGA, we conclude that, for these low oxygen concentrations, there is apparently an oxygen defect that is not at the O4 site. We have been unable to locate this second oxygen defect by neutron diffraction (which is not surprising, due to its small concentration). However, the observation of a similar phenomenon in Hg-1201 at compositions in the vicinity of maximum T_C (to be discussed in the next section) suggests that our conclusions are correct and may provide some clues about what different kind of oxygen defect is being formed.

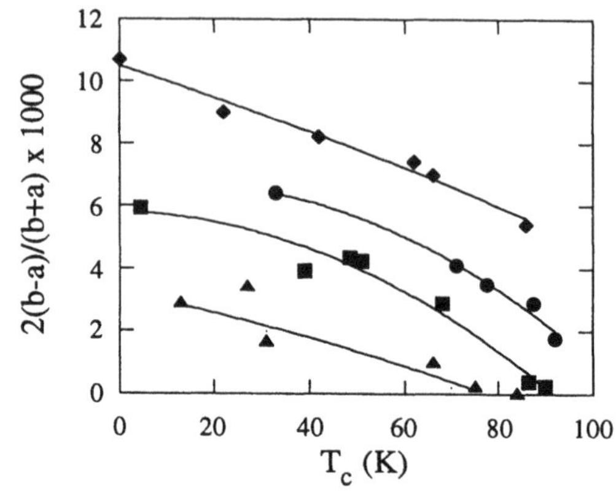

Fig. 2. Orthorhombic strain vs. T_C for Tl-2201 samples with different Tl/Cu starting compositions and oxygen contents: closed circles and closed squares, Tl/Cu=2.0/1.0, closed triangles 2.0/1.1; diamonds are the data of Ref. 3.

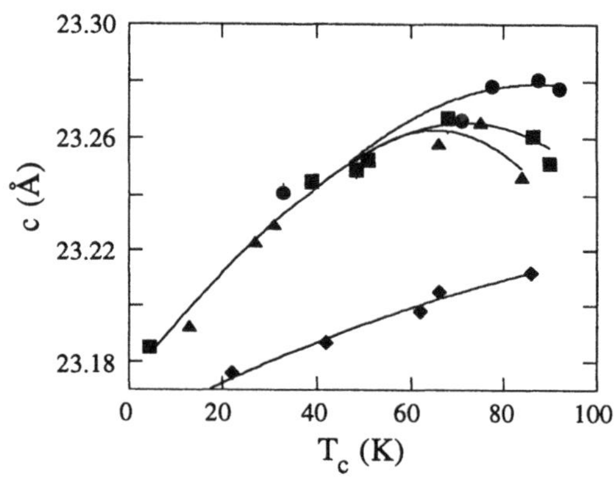

Fig. 3. c-axis length vs. T_C for Tl-2201 samples (same format as Fig. 2).

3. $HgBa_2CuO_{4+\delta}$

Less is known about the defect chemistry of Hg-1201 than for Tl-2201. There is general agreement that the defect that most influences T_C is an interstitial oxygen defect in the center of the Hg plane. There are conflicting reports that Cu may substitute at the Hg site with additional oxygen bonded to this Cu [6-8]. In this work, we do not investigate this Hg-site substitutional defect. We focus on the oxygen defect chemistry as the oxygen content is varied over a wide range extending from the underdoped to the overdoped region. We see a remarkable change in the oxygen defect chemistry upon passing through the maximum T_C.

To avoid confusion from possible differences in metal-site defect chemistry, the measurements reported here were performed by oxygen annealing small samples from a single larger sample. Repeated annealings of the same small samples confirmed that the results are reproducible and do not depend significantly on sample history. The oxygen anneals were performed at 300-400° C in oxygen pressures from 10^{-7} to 150 atm. Structural parameters were obtained by Rietveld refinement using neutron powder diffraction data obtained on the SEPD at IPNS [9].

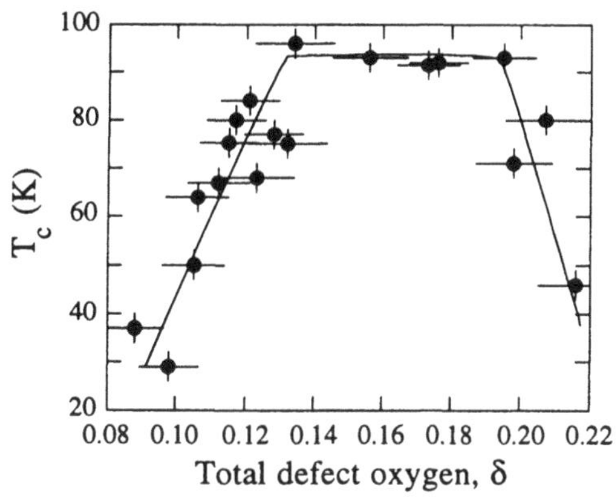

Fig. 4. T_C vs. total defect oxygen content, δ, for Hg-1201 samples annealed in various oxygen environments.

Two slightly different oxygen defects were identified in the Rietveld refinements. One of these is the previously reported interstitial oxygen site in the Hg plane (site O3) [2,6-8]. The second site (O4) is also in the Hg plane, but is at an (x,x,0) site that is displaced towards one of the Hg atoms, and the Hg atom is displaced slightly in the same direction, to form a Hg-O4 bond 2 Å in length.

The T_C vs. total defect oxygen content, δ (combined occupancies of the O3 and O4 sites determined from neutron powder diffraction), is shown in Fig. 4. Data from the present samples are in good agreement with those from the literature, when total defect oxygen contents are used. A surprising feature of the data is the plateau at maximum T_C, where T_C is nominally constant (>90K) for total oxygen contents from 0.13 to 0.19.

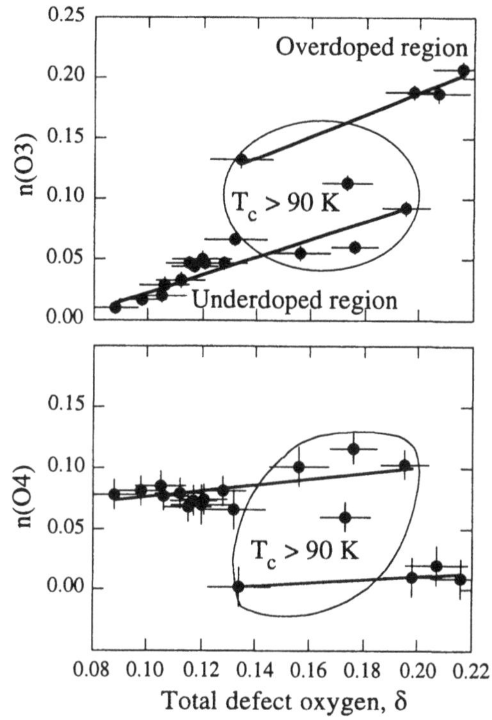

Fig. 5. O3 and O4 oxygen defect concentrations vs. total oxygen defect concentration [δ=n(O3)+n(O4)] for Hg-1201.

This plateau results from an unusual competition between the two oxygen defects, illustrated in Fig. 5. For underdoped samples (δ<0.13), the dominant defect is O4; for overdoped samples (δ>0.19), the dominant defect is O3. This unusual behavior in the defect chemistry also manifests itself in other structural parameters such as cell parameters and positions of the apical oxygen (O2) and Ba atoms. For example, the unit cell volume vs. δ is shown in Fig. 6. The monotonic behavior in the underdoped and overdoped regions is separated by a region (corresponding to T_C>90K) where the cell volume is nominally constant while the total oxygen content is changing.

From the bond lengths, the charge on Cu can be estimated by the bond valence sum method. The result of such a calculation is shown in Fig. 7. The charge on Cu increases with increasing oxygen content in both the underdoped and overdoped regions, but remains nominally constant for the plateau region (0.13≤δ≤0.19) where T_C is constant. Thus, the

T_C is controlled by the number of carriers and the number of carriers remains constant upon passing through the range of oxygen contents where the dominant defect switches from O4 to O3.

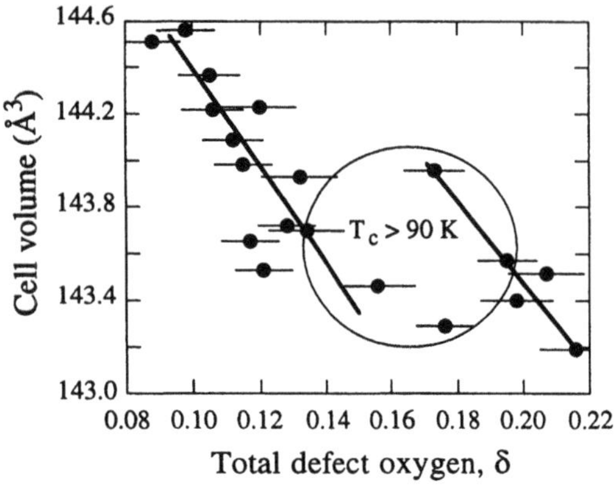

Fig. 6. Unit cell volume vs. total oxygen defect content, δ, for Hg-1201.

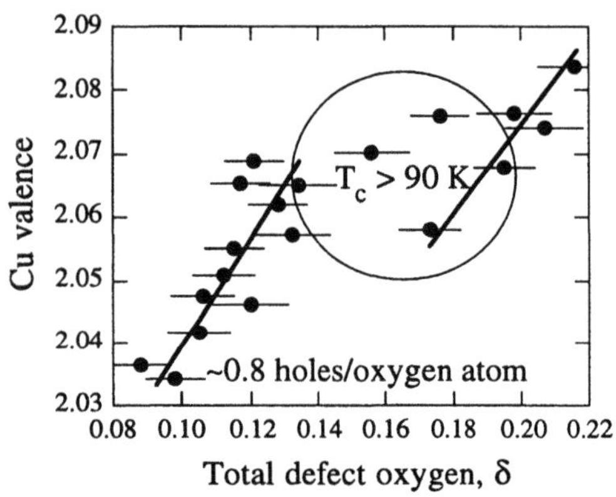

Fig. 7. Bond valence sum for Cu in Hg-1201 vs. total oxygen defect content, δ.

4. CONCLUSIONS

These data suggest that the formation energies for O3 and O4 oxygen defects in Hg-1201 differ by only a small amount and that this delicate energy balance is modified upon passing through the region of maximum T_C. A similar behavior (the appearance of a second oxygen defect in the region of maximum T_C) is suggested by our data for the Tl-2201 system, but the data are much less clear in that case because the underdoped region cannot be accessed. Such behavior could result from crossing the van Hove singularity at maximum T_C [10]. The resulting topological transition in the Fermi surface could affect the defect chemistry by altering bands involving Hg (or Tl). This conjecture remains to be validated; but, if it is correct, the present data may be the first evidence for crossing the van Hove singularity at maximum T_C by varying the chemical composition. Such transitions have not been predicted to result in first-order changes in structural parameters. But, the remarkable behavior observed here could result from the first-order response of the delicately balanced defect chemistry to the changes in electronic structure.

ACKNOWLEDGEMENTS

We would like to thank Simine Short for her help with neutron data collection. This work was supported by the US Dept. of Energy, Basic Energy Sciences - Materials Sciences, contract No. W-31-109-ENG-38, and the NSF Office of Science and Technology Centers, grant No. DMR 91-20000.

REFERENCES

1. Z.Z. Sheng, and A.M. Hermann, Nature, 332 (1988) 55.
2. S.N. Putilin, E.V. Antipov, O. Chmaissem, and M. Marezio, Nature, 362 (1993) 226.
3. Y. Shimakawa, Physica C, 204 (1993) 247.
4. M.A.G. Aranda, D.C. Sinclair, J.P. Attfield, and A.P. Mackenzie, Phys. Rev. B, 51 (1995) 12747.
5. J.L. Wagner, O. Chmaissem, J.D. Jorgensen, D.G. Hinks, P.G. Radaelli, B.A. Hunter, and W.R. Jensen, Physica C, in press.
6. J.L. Wagner, P.G. Radaelli, D.G. Hinks, J.D. Jorgensen, J.F. Mitchell, B. Dabrowski, G.S. Knapp, and M. Beno, Physica C, 210 (1993) 447.
7. A. Asab, A.R. Armstrong, I. Gameson, and P.P. Edwards, Physica C, 255 (1995) 180.
8. O. Chmaissem, Q. Huang, A. Santoro, S.N. Putilin, and M. Marezio, Physica C, 212 (1993) 259.
9. J.D. Jorgensen, J. Faber, Jr., J.M. Carpenter, R.K. Crawford, J.R. Haumann, R.L. Hitterman, R. Kleb, G.E. Ostrowski, F.J. Rotella, and T.G. Worlton, J. Appl. Crystallogr., 22 (1989) 321.
10. D.L. Novikov, M.I. Katsnelson, Jaejun Yu, A.V. Postnikov, A.J. Freeman, Phys. Rev. B, 54 (1996) 1313, & the references therein.

ELSEVIER

Physica C 282–287 (1997) 101–104

PHYSICA C

Electron crystallographic study of superconducting oxides and related compounds

F. H. Li[a], H. F. Fan[a] and Z. X. Zhao[b]

[a]Institute of Physics & Center for Condensed Matter Physics, Chinese Academy of Sciences, Beijing 100080, P. R. China

[b]National Laboratory for Superconductivity, Center for Condensed Matter Physics, Chinese Academy of Sciences, Beijing 100080, P. R. China

A review on crystal structure determination of high temperature superconductors using electron diffraction or the combination of electron diffraction and high resolution electron microscopy is given. The two-stage image processing technique and the multi-dimensional direct method are employed. Four examples including the determination of commensurate and incommensurate modulated structures are reported.

1. INTRODUCTION

High temperature superconductors often exist together with some other compounds in multiphase samples and the crystal size is usually very small. It is difficult to obtain single crystals large enough for X-ray diffraction analysis, while electron diffraction patterns and high resolution electron microscope images of good quality can easily be taken from such small crystals. The crystal structure determination can be carried out by electron diffraction [1,2] or by high resolution electron microscopy (HREM) [3] independently, or by combining them together [4].

Principally, in all cases it is essential to have a very thin sample so that the kinematical diffraction theory and the weak-phase-object approximation are valid [1,3]. Practically, the high resolution electron microscope images of crystals much thicker than a weak-phase-object can reveal the projected crystal structure with correct atomic positions, although the peak area might not linear to the atomic weight. The pseudo-weak-phase-object approximation theory [5] indicates that the image is similar to the structure when the crystal thickness is below a critical value which depends on the constituent atoms and the electron wavelength and is usually about tens Angstrom. The resolution of the structure determined by the HREM is limited by the resolution of the electron microscope and hence is not high enough to reveal all atoms.

The two-stage image processing technique based on the combination of HREM and electron diffraction was developed to obtain the structure with a resolution close to the diffraction resolution limit. The first stage of the technique is image deconvolution carried out on the basis of the maximum entropy [6] or the direct method [7] In this stage the original image is transformed into the low resolution structure image in which heavy atoms are usually seen. The second stage is the phase extension by using the direct method. The resolution of the final image is close to the diffraction resolution limit so that almost all atoms can be seen [8]. Because the sample area for taking a diffraction pattern is much larger then that for the image, the average crystal thickness for the former is larger than that for the latter. Hence, the electron diffraction data is inevitablly suffered the modulation of dynamical scattering effect. It was shown that the dynamical scattering effect can be partly removed by an empirical method [9]. In the case of incommensurate modulated structures the dynamical scattering effect is weaker then usual [10].

The incommensurate modulated structure can be described as the three-dimensional section of a multi-dimensional periodic structure [11]. The multi-dimensional direct method was developed

0921-4534/97/$17.00
PII S0921-4534(97)00227-X

to determined the incommensurate modulated structure by electron diffraction [12].

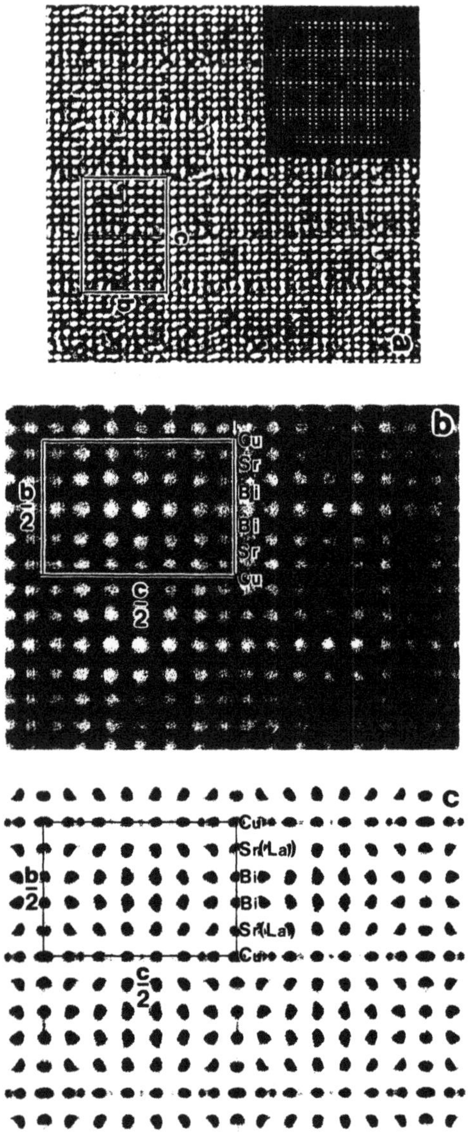

Figure 1. (a) [100] images of $Bi_4(Sr_{0.75}La_{0.25})_8Cu_5O_y$, (b) image deconvoluted from (a) and (c) final image obtained by phase extension.

In the present paper, the structure determination of four crystals possessing commensurate or incommensurate modulated structures by using the two stage image processing technique and/or multi-dimensional direct method are reported.

2. $Bi_4(Sr_{0.75}La_{0.25})_8Cu_5Oy$

The crystal structure of $Bi_4(Sr_{0.75}La_{0.25})_8Cu_5O_y$ belongs to the orthorhombic system with the lattice parameters $a = 5.39, b = 23.5$ and $c = 34.0$ Å, which are close to those of $Bi_4Sr_8Cu_5O_y$. The crystal structure was determined by using the two-stage image processing technique [13]. Both image deconvolution and phase extension were based on the direct method.

Two high resolution electron microscope images with the contrast almost reversed each other (one of them is shown in Fig. 1(a)) were deconvoluted. The same structure image (Fig 1(b)) was obtained. In this image all metallic atoms appear black and the arrangement of black dots is the same as that of metallic atoms in the crystal structure of $Bi_4Sr_8Cu_5O_y$. Fig. 1(c) is the final image obtained by phase extension after three times correction of diffraction intensities. By comparing Fig. 1(c) with the structure of $Bi_4Sr_8Cu_5O_y$, it can be concluded that a partial replacement of Sr by La leads to imperfect Cu-O layers parallel to the (**a**,**c**) plane, while those parallel to the (**a**,**b**) plane remain to be perfect.

3. $Bi_2Sr2CaCu2Oy$ (Bi-2212)

The crystal of Bi-2212 has the one-dimensional incommensurate modulated structure. The average structure belongs to the orthorhombic system and the lattice parameters are $a = 5.39, b = 5.40$ and $c = 30.6$ Å. The modulation wave vector is $\mathbf{q} = 0.21\mathbf{b}^* + \mathbf{c}^*$. The incommensurate modulated structure was determined by the two-stage image processing technique and the high-dimensional direct method [14]. Fig.2(a) shows the [100] electron diffraction pattern and 2(b) the image digitized from Ref. [15]. Fig. 2(c) is the deconvoluted average image revealing the projected average structure. The Fourier transform of Fig. 2(c) yields 17 independent main reflections. Their phases were employed as the initial phases to carry out the phase extension and phases of 21 additional main reflections were obtained. 114 phases of satellite reflections were then derived based on the phases of main reflections. Fig. 2(d) shows the final image in which all the atoms, excepting those overlapped due to the projection along the **a** axis, are well separated. Modulation waves of individual atoms are revealed intuitively.

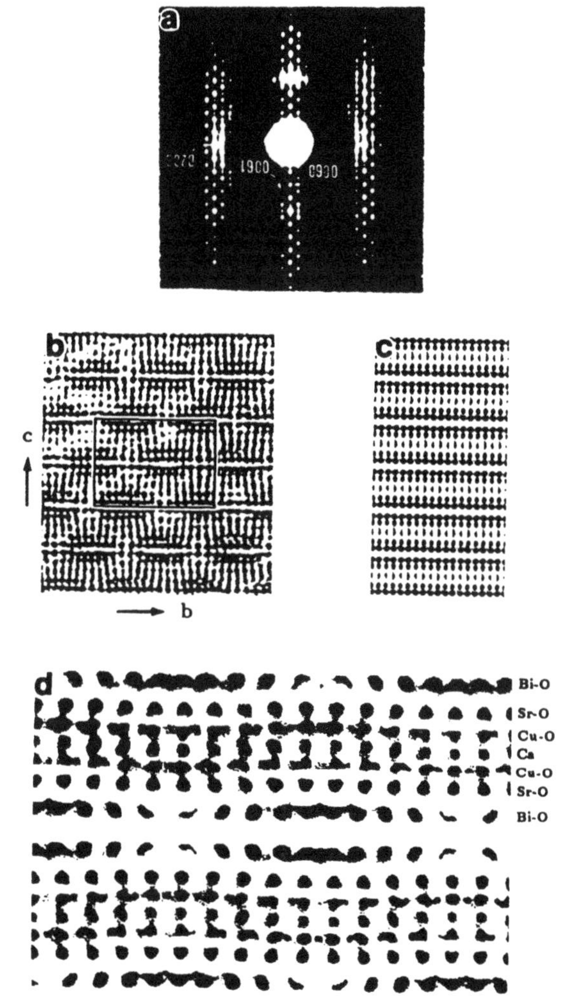

Fig.2 (a) [100] electron diffraction pattern of bi-2212, (b) digitized image, (c) average structure image and (d) final image obtained by phase extension. The absolute area of (d) is the same as the framed region in (b).

4. Bi-2212 ABOVE AND BELOW T_c

The [100] electron diffraction data of Bi-2212 were collected at 200K and 40K, and the corresponding diffraction patterns are shown in the left and right of Fig. 3, respectively. The moduration in superconducting state becomes weaker. The one-dimensional incommensurate modulated structure was determined by the multi-dimensional direct method [16]. The initial main reflections were calculated based on the average structure reported in Ref. [17]. From the two obtained projected potential maps (Fig. 4) it can be seen that the modulation features above and below T_c are different from each other. The positions of Ca with a high occupancy corresponds to those of of Bi also with the high occupancy in the case above T_c but corresponds to positions of Bi with the low occupancy in the case below Tc. However, the high occupancy of Ca always leads to a displacement of two adjacent Cu-O and Su-O layers towards the Bi-O layers in both cases above and below T_c.

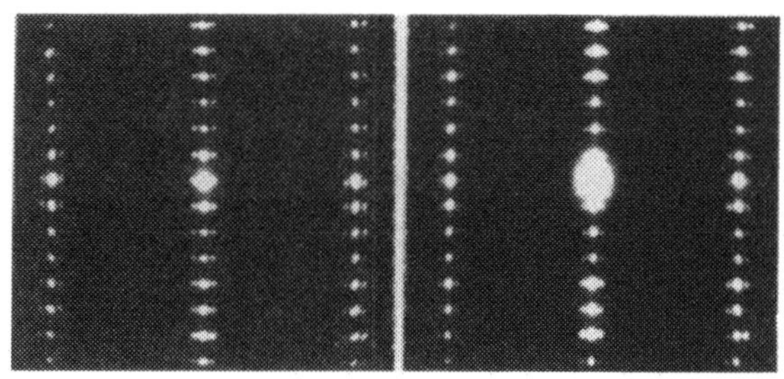

Fig. 3 [100] electron diffraction patterns of Bi-2212 taken at 200 K (left) and 40 K (right)

5. Pb DOPED Bi-2223

The crystal structure of Pb doped Bi-2223 is one-dimensional incommensurate modulated. Its average structure belongs to the orthorhombic system and the lattice parameters are $a = 5.49, b = 5.41$ and $c = 37.1$ Å. The modulation wave vector is $\mathbf{q} = 0.117\mathbf{b}^*$. The structure was determined based merely on the diffraction data by using the direct method [18]. The average structure model reported in Ref. [19] was used to calculate the phases of main reflections. Fig. 5 shows the obtained contour map projected along the **a** axis. Oxygen bridges can be found among three oxygen layers.

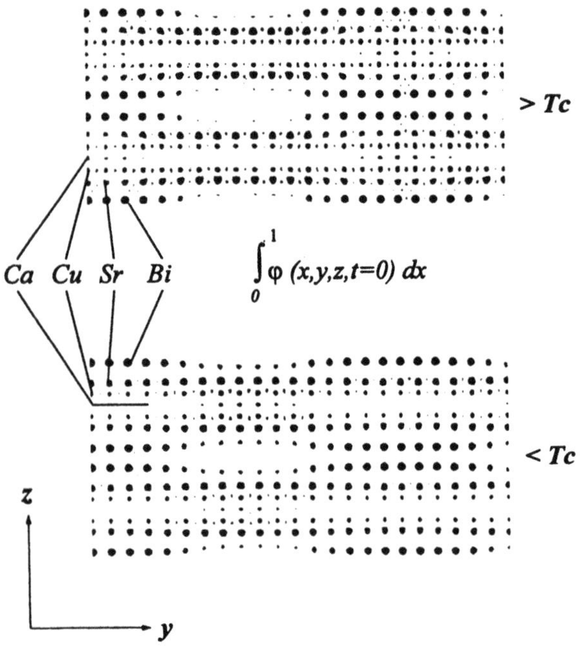

Fig. 4 Final [100] projected potential maps of Bi-2212 corresponding to 200 K (top) and 40 K (bottom).

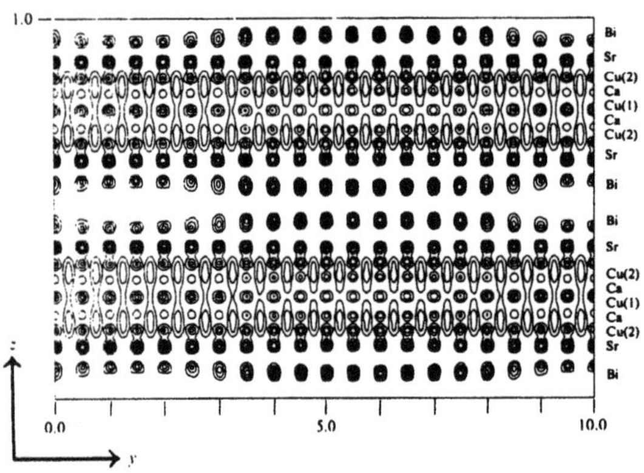

Fig. 5 The countour potential map of Bi-2223 projected along the **a** axis.

REFERENCES

1. B. K. Vainshtein, Structure Analysis by Electron Diffraction, Academy Science, USSR, 1964.
2. D. L. Dorset, Structure Electron Crystallography, Plenum press, 1995.
3. P. R. Buseck, J. M. Cowley and L. Ering, High Resolution Electron Microscopy, Oxford University Press, 1988.
4. F. H. Li, Proc. 13th Intern. Congr. on Electron Microscopy, Paris (1994) Vol. 1, 481.
5. F. H. Li and D. Tang Acta Cryst. A41 (1985) 376.
6. J. J. Hu and F. H. Li, Ultramicroscopy, 35 (1991) 339.
7. F. S. Han, H. F. Fan and F. H. Li, Acta Cryst. A42(1986) 353.
8. H. F. Fan, S, B. Xiang, F. H. Li, Q. Pan, N. Uyeda and Y. Fujiyoshi, Ultramicroscopy, 36 (1991) 361.
9. D. X. Huang, W. Liu, Y. X. Gu, H. F. Fan and F. H. Li, Acta Cryst. A (1996)
10. Y. Li et al ...., unpublished.
11. P. M. de Wolf, Acta Cryst. A30 (1974) 777.
12. S. B. Xiang, X. J. Wu, F. H. Li and H. F. Fan, Acta Cryst., A46 (1990) 929.
13. B. Lu, F. H. Li, Z. H. Wan, H. F. Fan and Z. Q. Mao, 1997 submitted to Ultramicroscopy.
14. Z. Q. Fu, D. X. Huang, F. H. Li, J. Q. Li, Z. X. Zhao, T. Z. Cheng and H. F. Fan, Ultramicroscopy 54 (1994) 229.
15. Y. Matsui and S. Horiuchi, Jpn. J. Appl. Phys. 27 (1988) L2306.
16. Z. H. Wan, J. Q. Li, H. F. Fan, Z. X. Zhao, unpublished.
17. Y. Gan, P. Lee and D. P. Coppens, Phys. Rev. B, 42 (1990) 387.
18. Y. D. Mo, T. Z. Cheng, H. F. Fan, J. Q. Li, B. D. Sha, C. D. Zheng, F. H. Li and Z. X. Zhao, Phys. Scr. T 42 (1992) 18.
19. A. Sequeira, J. V. Yakhmi, R. M. Iyer, H. Rajagopal and P. V. Sastry, Physica C, 167 (1990) 291.

ELSEVIER

Physica C 282–287 (1997) 105–106

PHYSICA C

Scanning tunneling microscopy and spectroscopy study of substitution- and defect-doped films of the infinite layer compound*

Ron Feenstra[a], Takuya Matsumoto[b] and Tomoji Kawai[b]

[a]Solid State Division, Oak Ridge National Laboratory, P.O. Box 2008, Oak Ridge, Tennessee 37831-6057, USA

[b]Institute of Scientific and Industrial Research, Osaka University, 8-1 Mihogaoka, Ibaraki, Osaka 567, Japan

Scanning tunneling microscopy and spectroscopy (STM-STS) has been used to study doping induced changes in the local electronic structure of CuO_2 planes at the surface of thin epitaxial films of the infinite layer compound $Sr_{1-x}A_xCuO_2$, chemically modified via substitutions A (Ca, La) on the Sr site. It is observed that doping produces states within the insulating parent band gap, reflecting contributions from dopants (La) as well as composition dependent defects (Sr vacancies).

1. INTRODUCTION

In STM-STS [1], a tunneling current is established between a specimen and the tip upon the application of a bias potential V which, for a given tip position, depends on the density of states at energy $(eV - E_F)$ and (exponentially) on the spacing relative to the tip (E_F is the Fermi energy of the probed electronic structure, e: electron charge). STM-STS for copper-oxide superconductors previously has been used to study a range of topics, including surface morphologies, spectroscopy of the superconducting gap [2], ordering effects in the chain-layer of $YBa_2Cu_3O_{7-\delta}$ [3] etc., while on-edge experiments (probing the *a-c* plane) have confirmed the electronically active role of the CuO_2 planes [2]. To study the *normal-state* properties of CuO_2 planes, however, contributions from unrelated oxide layers at the vacuum interface should be essentially eliminated, a prerequisite difficult to meet for complex HTS compounds.

The infinite layer compound $Sr_{1-x}A_xCuO_2$, on the other hand, has a simple defect perovskite structure [4] in which monolayers of alkaline (rare) earth atoms are the only layers separating the CuO_2 planes [5,6]. As reported in ref. [7], STM-STS on the (a,a) surfaces of thin epitaxial films of this compound (suitably grown by laser-MBE and preserved without exposure to air) probes the outermost CuO_2 planes apparently without interference from intermediary layers, making it possible to study the intrinsic electronic properties. Here, variations in bond chemical parameters were induced via substitutions A (Ca, La) on the Sr site, producing doping effects on the CuO_2 planes but leaving the basic atomic structure intact. From related atomic layer epitaxy film growth experiments, it is inferred that the probed surfaces consist of CuO_2 planes covered with a layer of (Sr,A) adatoms. The adatoms presumably are bonded to O squares of the CuO_2 plane, providing a full coordination without blocking Cu from the vacuum interface [7]. For the present study, $Sr_{1-x}A_xCuO_2$ films with reproducible compositions were grown by laser-MBE using composite targets. A SrO monolayer was deposited on the (100) $SrTiO_3$ substrates to induce the desired layering sequence. Details regarding the film preparation and multi-chamber ultra high vacuum STM–laser-MBE system have been described in ref. [7].

*Research sponsored by the Yamada Science Foundation, Osaka, Japan, and the Oak Ridge National Laboratory, managed by Lockheed Martin Energy Research Corp. for the U.S.D.O.E. under contract no. DE-AC05-96OR22464.

0921-4534/97/$17.00 Elsevier Science B.V
PII S0921-4534(97)00228-1

2. RESULTS: Doping induced in-gap states

STS data averaged over various surface sites for a range of film compositions are presented in Fig. 1. $V > 0$ corresponds to vacant electron states and $V < 0$ to occupied states. A clear gap is observed for energies close to E_F ($V = 0$), which is identified as the correlation gap [8]. The gap is wide (~ 1.8 eV) for the insulating composition $Ca_{0.7}Sr_{0.3}CuO_2$, but it is narrower for $SrCuO_2$. For the electron-doped superconductor series $Sr_{1-x}La_xCuO_2$ there is a trend whereby states of the unoccupied conduction band (CB) gradually approach E_F and the top of the valence band (VB) shifts to lower energies. Using the VB and CB gap edges of the insulating parent (curve 1A) as a reference for undoped CuO_2 planes, these composition dependent gaps may be interpreted as *doping induced in-gap states.* Then, one finds that there exists a close correlation between the occurrence and nature of these states and predictions from a bond valence sum analysis [9] regarding doping effects in the CuO_2 planes and composition dependent structural instabilities.

For example, the in-gap states for $SrCuO_2$ presumably are due to Sr vacancies, induced by a Sr–O vs. Cu–O bond length mismatch, giving rise to holes on O that overcompensate for electrons on Cu. Electron injection upon La substitution expands the Cu–O bonds [6] which diminishes the intrinsic stress responsible for the Sr defect formation. Thus it is inferred that for $Sr_{0.9}La_{0.1}CuO_2$, the only active dopants are the La solutes, giving rise to in-gap states below the original CB. On the other hand, the wide gap for $Sr_{0.95}CuO_2$ is predicted to result from a compensation of O holes and Cu electrons, apparently producing undoped planes. Local variations in electronic properties typically were less than those induced by the chemical substitutions. An exception is depicted by curves 3*a*,*b* in Fig. 1 observed for some films after vacuum annealing. It is surmised that spectrum 3*b* (containing a localized peak at $V \simeq 1.5$ V) corresponds to a local O deficiency of the outermost CuO_2 plane.

In conclusion, systematic variations have been observed in the width of the insulating gap for CuO_2 planes at the surface of $Sr_{1-x}A_xCuO_2$ epitaxial films. The energies relative to E_F at which in-gap states are formed depend on the polarity of the doped charge (holes or electrons).

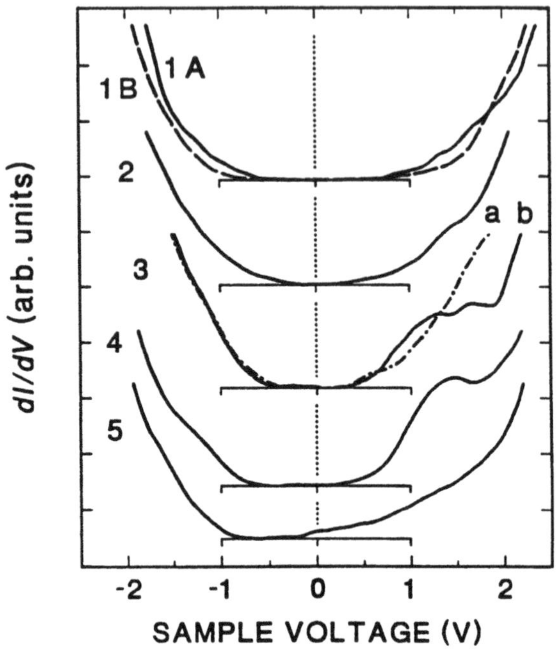

Figure 1. Derivative of spatially averaged current-voltage characteristics obtained by room temperature STM-STS on the growth surface of thin epitaxial films of the infinite layer compound: (**1A**) $Ca_{0.7}Sr_{0.3}CuO_2$; (**1B**) $Sr_{0.95}CuO_2$; (**2**) $SrCuO_2$ oxidized; (**3**) $SrCuO_2$, reduced; (**4**) $Sr_{0.95}La_{0.05}CuO_2$; (**5**) $Sr_{0.9}La_{0.1}CuO_2$.

REFERENCES

1. R. M. Feenstra and J. A. Stroscio, *J. Vac. Sci. Technol. B* **5** (1987) 923.
2. T. Hasegawa et al., *J. Phys. Chem. Solids* **53** (1992) 1643.
3. H. L. Edwards et al., *Phys. Rev. Lett.* **75** (1995) 1387.
4. T. Siegrist et al., *Nature* **334** (1988) 231.
5. M. Takano et al., *Physica C* **159** (1989) 375.
6. M. G. Smith et al., *Nature* **351** (1991) 549.
7. K. Koguchi et al., *Science* **267** (1995) 71; *Physica C* **258** (1996) 47.
8. S. Uchida et al. *Phys. Rev. B* **43** (1991) 7942.
9. I. Brown, *J. Solid State Chem.* **82** (1989) 122.

ELSEVIER

Physica C 282–287 (1997) 107–110

PHYSICA C

YBCO Film Deposition on Very Large Areas Up to 20×20 cm²

H. Kinder, P. Berberich, W. Prusseit, S. Rieder-Zecha, R. Semerad, and B. Utz

Phys. Dept. E10, Technische Universität München, D-85747 Garching, Germany

In the last decade we have developed thermal reactive co-evaporation as a technique to produce high quality YBCO and other oxide films of very large size up to 9 inches in diameter. This was achieved by intermittent deposition and reaction with oxygen using a heater which rotates the substrate in and out of an oxygen pocket. Even larger substrates, e. g. coated conductors, cannot be rotated. Therefore we have recently developed a new setup where the substrate is held fixed, and the oxygen pocket is set in linear reciprocation. This technique allows simultaneous deposition on a square of 20×20 cm². Moreover, we have developed an instant refill mechanism for the thermal boats, and stable rate control by atomic absorption spectroscopy (AAS), in order to obtain a continuous process suitable for small scale mass production.

1. INTRODUCTION

A high production rate is required for the industrial fabrication of YBCO thin films to make optimum use of investments and manpower. Since the growth rate is limited by a loss of quality, the only way to bring the production rate up is by depositing on a large area simultaneously. Thermal reactive co-evaporation of the metals [1] is particularly well suited for this purpose because it is inherently homogeneous due to the ballistic propagation of the constituent metals. A difficulty arose from the fact that the metals have to react with oxygen of sufficiently high pressure, and this leads to strong scattering of the metal atoms.

We have overcome this difficulty by intermittent deposition and oxidation, using an oxygen pocket in front of the film which is in constant motion. [2] In our first setup, we have accomplished the motion by rotating the substrate and keeping the oxygen pocket in a fixed position. This allows deposition on a circular area of 23 cm diameter.[3]

However, some applications require even larger substrates which cannot be rotated. Therefore we have developed a new setup where the substrate is held fixed, and the oxygen pocket is set in linear reciprocation. This technique allows simultaneous deposition on a square, presently 20×20 cm².

To achieve a continuous production process, we have also developed an instant refill mechanism for the thermal boats using vibrator conveyors, and a new method of rate control using compensated atomic absorption spectroscopy (AAS).[4]

2. SETUP WITH ROTATING SUBSTRATES

A sketch of the deposition setup with sample rotation is shown in Fig. 1. The metals Y, Ba, and Cu are evaporated from resistively heated boats which are mounted in a vacuum chamber with a base pressure below 2×10^{-6} mbar. The evaporation rate of each metal is controlled by an individual quartz crystal monitor with a collimator tube in front of it to confine its sight to only one boat. The rates have to be controlled to approximately 1 % accuracy to avoid outgrowth and to get high quality films reproducibly.

For the formation of YBCO films an oxygen partial pressure of some 10^{-3} mbar is needed. At the same time the chamber pressure must be around 10^{-5} mbar at least to ensure that the metal vapor can reach the substrate nearly ballistically from larger distance. To meet both requirements the substrates are placed on a rotating disk in this setup, and alternated between a zone of the metal deposition on one side and an oxygen pocket on the other. At a rotation speed of 5 cycles per second, about one monolayer of metal atoms condenses on each turn

PII S0921-4534(97)00230-X

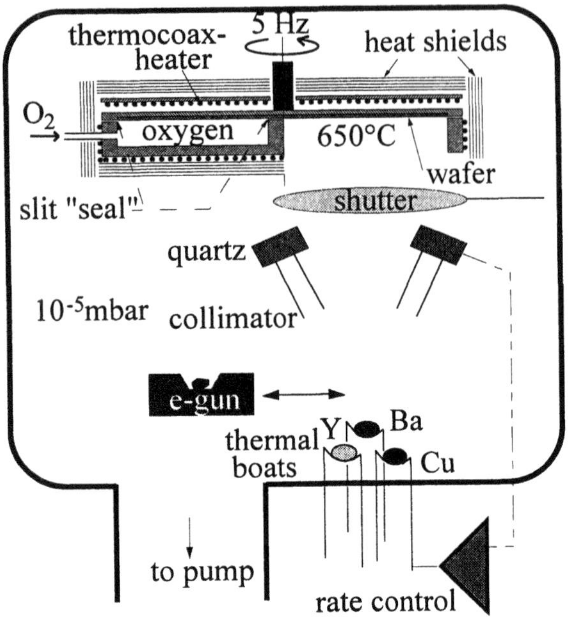

Figure 1. Deposition scheme for thermal co evaporation combined with an oxygen pocket heater and rotating sample holder.

and consecutively oxidizes in the pocket to form YBCO. The oxygen is effectively sealed from the vapor zone by keeping the clearance between the pocket and the rotating disk very narrow, typically 0.3 mm. An oxygen pressure drop of 1:200 is easily achieved. The substrate is heated radiatively from the rear, the sides and front (part covered by the pocket) by thermocoax heater wire to a temperature of 650 °C. YBCO films are deposited typically with rates of 20 to 30 nm/min. The film thickness is generally in the range of 300 nm to 2 µm.

To adjust the ratio of the evaporation rates we analyzed the film composition with heavy ion RBS. Once the stoichiometry has been fixed, the process is stable on a months time scale. Small variations in composition allow to optimize the films, e.g. in respect to high T_C or very smooth surfaces. These small deviations can be easily monitored by changes in the surface morphology as observed in high resolution scanning electron microscopy (SEM) images.

A fourth boat next to the other ones is used to cap YBCO films with in-situ gold overlayers to achieve good contact resistances below 10^{-8} Ωcm^2 which are desired for contacts in power- and RF-applications. In addition oxide buffer layers like MgO, CeO_2 or YSZ can be deposited in-situ by an e-beam gun which can be moved above the boats, see Fig. 1. This system allows us to deposit YBCO films on a variety of substrates including MgO, YSZ, $LaAlO_3$, CeO_2 buffered sapphire, YSZ buffered Si and even MgO buffered GaAs.

The heater is designed for 9" diameter sample trays. This area can be flexibly filled either with one single 8" wafer or higher numbers of smaller samples. Insets for square shaped substrates accompanied by small substrates to monitor the deposition success can be used as well.

The evaporation zone as seen from below is a sector whose size is only about half of the full 9" circle. This fact, and the constant rotation of the samples effectively homogenize the evaporation process. So the films prepared with this setup are usually excellently homogeneous and reproducible.

3. RESULTS OF ROTATING SETUP

Sapphire substrates are available with diameters of 8". To demonstrate the ability of our system we deposited some YBCO films on such large substrates. The films were characterized by a spacially resolved J_c- Measurement which was done by an inductive technique as described earlier. Although the substrates were only optical grade, i. e. not specified for epitaxial growth, and the data were taken one year after the fabrication of the films, the result was still acceptable, see Fig. 2.

Due to the radiation heating, films on the back side of the substrates are not contaminated. This fact allows us to produce double sided films by simply flipping the substrates over ex-situ after a first deposition step and fabricating another film on the other side of the wafer without degrading the first one. Such films are usually applied for RF filters in cellular and satellite communications. The J_c-maps of a double sided 2" $LaAlO_3$ wafer shows Fig. 3.

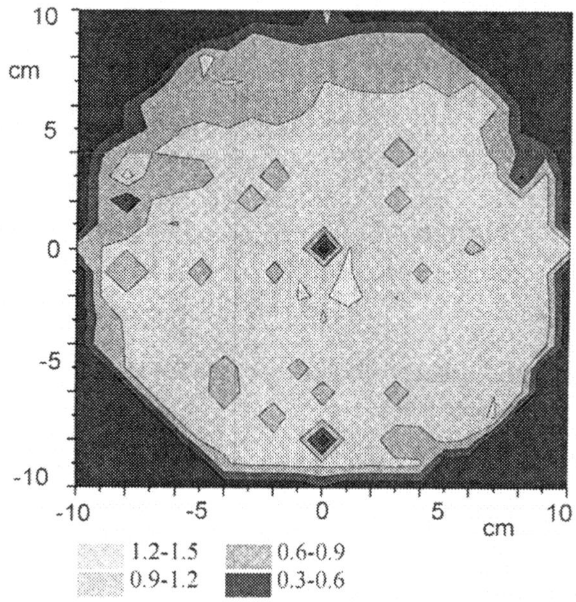

Figure 2. J_C(77K)-map of 8" film on sapphire.

4. SETUP WITH RECIPROCATING POCKET

Not all substrates can be rotated. Examples are large ceramic slabs as substrates of thin film resistive fault current limiters, and coated conductors for power transmission which must be wound up to bulky coils. So we have reversed the principle and hold the substrate fixed while the oxygen pocket is moved in a new setup.

Obviously, a first thought would be using a rotary motion of the pocket. From the mechanical point of view this is simpler than a linear motion.

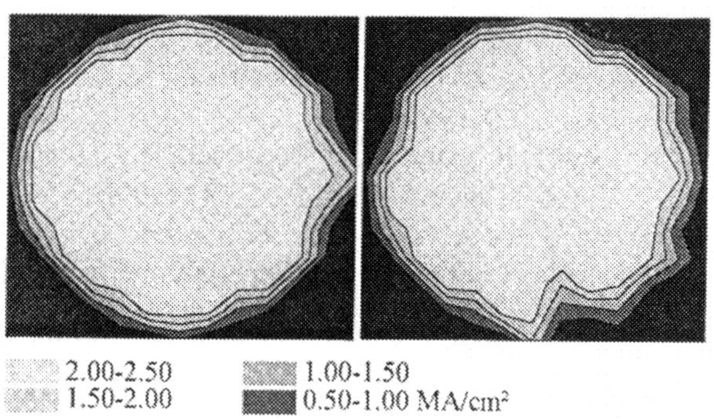

Figure 3. J_C(77K)-map of double sided 2" films on $LaAlO_3$.

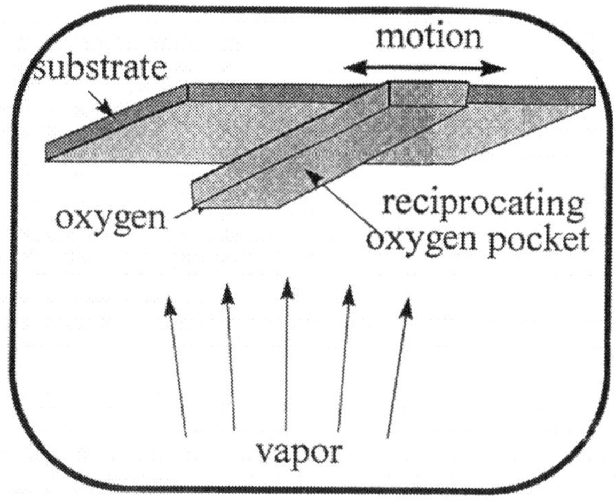

Figure 4. New design with fixed substrate allows deposition on a square of 20cm × 20cm.

However, one has to keep in mind that the moving pocket is partly shading the substrate behind it, and the shading time should be the same on all parts of the substrate in order to achieve a homogeneous thickness of the film. In this respect, the linear motion of a rectangular object is the simplest solution, although the slide motion in vacuum is a little more tricky. Fig. 4 shows a schematic picture of this new setup.

The clearance between the substrates and the oxygen pocket is small enough to maintain an oxygen pressure of 10^{-2} mbar inside the pocket at a chamber pressure of 4×10^{-5}mbar. The chamber is pumped at 2200 l/s, and the speed of the pocket is 2

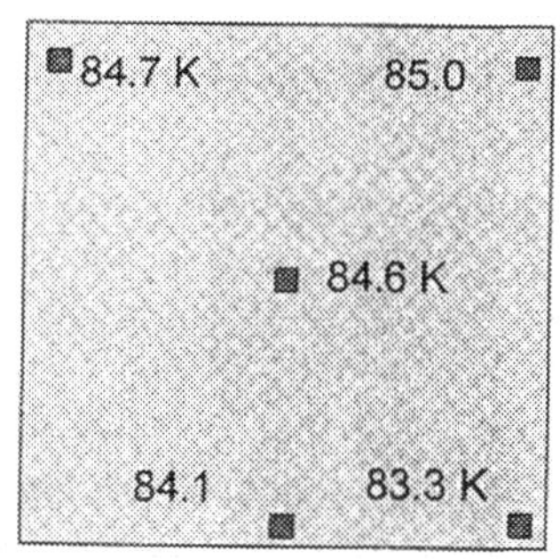

Figure 5. First results obtained with small MgO test samples distributed on 20×20 cm² square

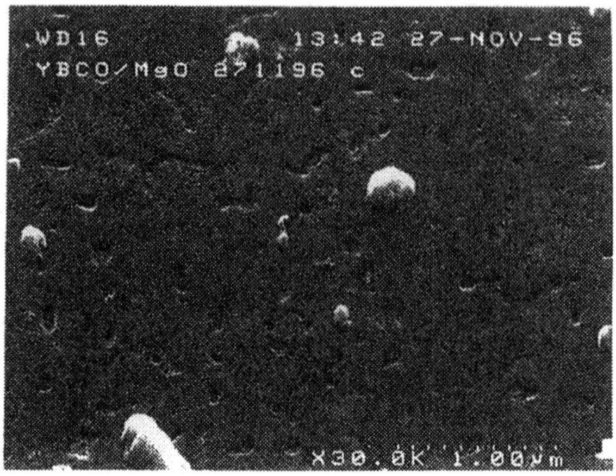

Figure 6. SEM picture of middle film of Fig. 5.

to 3 cps.

To test the homogeneity, we have distributed small (1×1cm²) MgO samples over the 20×20cm² area. The critical temperatures are indicated at the sample positions in Fig. 5. The homogeneity is already satisfactory, while the T_C is still somewhat low. RBS measurements have shown that we did not hit the stoichiometry with sufficient accuracy yet. The surface morphology is already acceptable, as shown in Fig. 6.

Technical films should not only be large, but also as thick as possible. The amount of material that can be deposited in one run is limited in our case by the capacity of the thermal boats, so that our maximum YBCO film thickness is around 600 nm. One way to make thicker films is by interrupting the evaporation, flooding the chamber while leaving the heater on, and refilling the boats manually. This way we have reached YBCO thicknesses up to 1.3 μm without degradation. However this technique is not very elegant. So we built an in-situ refill device.

A refill device should move new material into each boat at a predictable rate to allow dosing. We found that vibrator conveyors are best suited for this purpose. We use a helical type which can be conveniently filled, see Fig. 7. From the outlet, the metal grains fall onto a slide which ends above the evaporation source. The barium conveyor can be flooded by dry nitrogen to prevent oxidation.

A last drawback of the coevaporation method

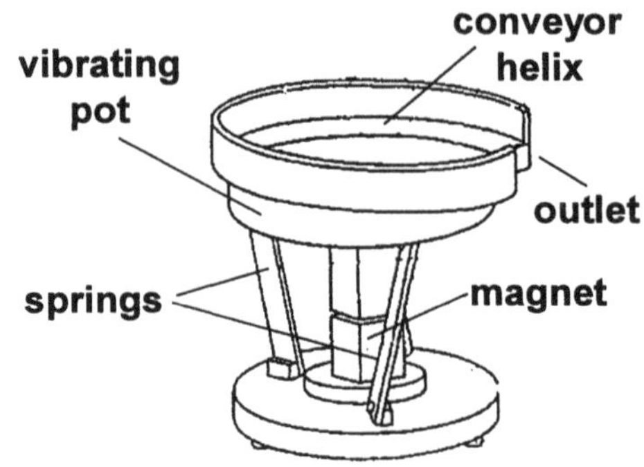

Figure 7. Helical vibrator conveyor for instant refill of thermal sources.

which prevented a continuous process was the use of quartz crystal monitors for rate control. The quartz crystals saturate after a few runs and have to be replaced. Therefore, we have developed atomic absorption spectroscopy (AAS) for rate control. AAS has two problems, one is the saturation of the lines at high evaporation rates and for wide atomic beams, and the second is a signal drift with time. We find that the first problem can be easily solved by calibration and computer control. The second problem is serious only for yttrium and barium, and we have resolved it by a compensation technique using two different absorption lines. This way we have achieved an accuracy of a few percent which allows sufficient control of the film properties. We have already used this AAS rate control to prepare the films shown in Figs. 5 and 6.

REFERENCES

1. P. Berberich, J. Tate, W. Dietsche, and H. Kinder, Appl. Phys. Lett. 53 (1988) 925.
2. P. Berberich, B. Utz, W. Prusseit, and H. Kinder, Physica C 219 (1994) 497.
3. H. Kinder, R. Semerad, P. Berberich, B. Utz, and W. Prusseit, SPIE proceedings 2697 (1996) 154.
4. B. Utz, S. Rieder-Zecha, and H. Kinder Proceedings of the Applied Superconductivity Conference 1996 Pittsburgh, in print.

ELSEVIER

Physica C 282–287 (1997) 111–114

PHYSICA C

Epitaxial ferroelectric/superconductor heterostructures

Yu. A. Boikov[a,b], Z.G. Ivanov[a], E. Olsson[a], and T. Claeson[a]

[a] Physics and Eng. Physics, Chalmers Univ. Techn. and Göteborg University, S-41296 Göteborg, Sweden

[b] Ioffe Physico-Technical Institute RAS, 194021 St. Petersburg, Russia

The growth, structure and dielectric parameters of epitaxial multilayers of superconducting and polarizable dielectric films grown by pulsed laser deposition were investigated: $SrTiO_3/YBa_2Cu_3O_{7-\delta}$, $KTaO_3/YBa_2Cu_3O_{7-\delta}$, $PbZr_yTi_{1-y}O_3/YBa_2Cu_3O_{7-\delta}$ and $SrBi_2Nb_2O_9/YBa_2Cu_3O_{7-\delta}$.

1. Introduction

The combination of low loss superconducting films with thin layers of polarizable dielectrics looks attractive for applications in tunable microwave components and fatigue free non volatile memories.

To manufacture high Q tunable filters, phase shifters and delay lines, low loss superconducting films have to be combined with paraelectric materials having strong dependences of dielectric permittivity ε on an electric field. It is the change in ε that causes the variation of the electromagnetic wave phase velocity in the superconducting transmission line. The tunability of the microwave components based on superconductor/ferroelectric combinations is determined by the dielectric non linearity of the ferroelectric layer while the surface resistance of the superconducting films and the dielectric losses of the ferroelectric layer cause large microwave losses.

The strongest dependence of ε on an external electric field is typically observed around the Curie point of a ferroelectric material. Cubic perovskites, like $SrTiO_3$ (STO) and $KTaO_3$ (KTO), are presently the main candidates for applications in microwave components. STO and KTO are incipient ferro-electrics and their remnant polarization P_r is zero down to 4.2K. $Ba_xSr_{1-x}TiO_3$ and $KNb_xTa_{1-x}O_3$, however, are ferroelectrics and their Curie points may be adjusted around 77K if $x \approx 0.05$-0.08.

Stoichiometric deviations at grain boundaries and the ferroelectric/electrode interfaces may strongly influence the effective values of ε and P_r in the ferroelectric layer. The similarities in the crystalline structures of the perovskite type ferroelectrics with those of oxide high-T_c superconductors imply that a thin epitaxial ferroelectric layer may be inserted between two epitaxial $YBa_2Cu_3O_{7-\delta}$ (YBCO) films with a minimum of distortion.

Sapphire seems to be one of the best candidates as a substrate for superconductor/ferroelectric based microwave components due to its excellent properties at high frequencies.

2. Experiment

STO/YBCO, KTO/YBCO, $PbZr_yTi_{1-y}O_3$/YBCO (PZT/YBCO) and $SBi_2Nb_2O_9$/YBCO (SBNO/YBCO) multilayers were grown in-situ by laser ablation (KrF, l =248 nm, t =30 ns) on the surface of $(1\underline{1}02)Al_2O_3$ buffered by CeO_2 (CeO). Features of CeO growth on the R-plane of sapphire have been discussed previously [1]. The oxygen pressure p_o during the depositions was 0.2 mbar.

The YBCO and STO layers were grown at substrate temperature T_s= 750°C. To decrease the chemical interaction between YBCO and the ferroelectric layer, T_s was lowered to 660-700°C during KTO and SBNO depositions. To counteract a Pb (PbO) reevaporation, PZT films were grown at 530°C.

x-ray ($\theta/2\theta$, ϕ−scan), transmission electron

* The study was supported by the Swedish Materials Consortium on Superconductivity and project 95-02-04186a of the Russian Foundation of Basic Research.

PII S0921-4534(97)00231-1

microscopy and atomic force microscopy were used to control structure, phase composition and surface morphology of the layers in the heterostructures.

To investigate the influence of electrode-ferroelectric interfaces on the effective ε of STO and KTO layers, parallel plane capacitors with YBCO electrodes or with a combination of YBCO as bottom and Ag as top electrodes were formed.

To measure e and P_r of PZT and SBNO layers, the ferroelectric layers were sandwiched between bottom YBCO and top Ag electrodes. Ag films were formed by thermal evaporation with T_s at room temperature. The top electrodes, 300x300 μm^2, were patterned by ion milling.

3. Experimental results and discussion

3.1. STO/YBCO heterostructures

The YBCO/STO/YBCO and STO/YBCO heterostructures were grown epitaxially at the surface of $(1\bar{1}02)Al_2O_3$ buffered by (001)CeO; (100)[010]STO//(001)[010]YBCO. There were no outgrowths and spirals detected at the free surface of the YBCO films, but separate a-axis particles ([001]YBCO//(001)CeO) with pronounced azimuth orientation. The YBCO/STO interface was sharp, with typical steps about 12Å, and free from second phase inclusions as follows from the results of a transmission electron microscopy study, see.Fig.1.

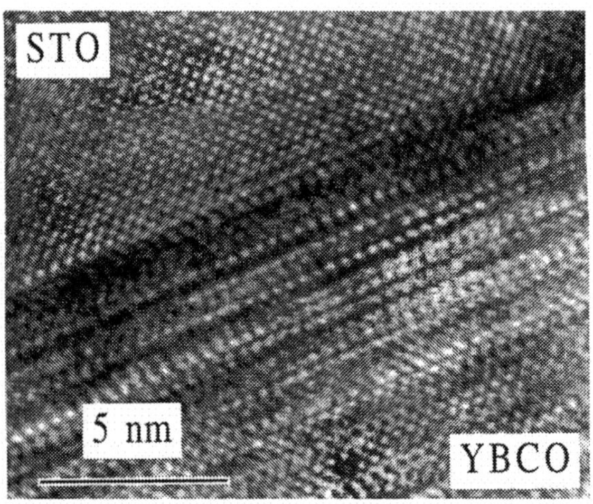

Figure 1. A transmission electron macroscopy image of the cross section of an YBCO/STO epitaxial heterostructure.

The lattice parameter of the intermediate STO layer (a=3.905 Å), sandwiched between the YBCO films, agreed well with the value for bulk STO single crystals. The c-axis parameter of the YBCO films grown on CeO/Al_2O_3 was smaller (11.66 - 11.67 Å) than values usually obtained for epitaxial films grown on substrates with a smaller difference in the thermal expansion coefficient.

The superconducting transition temperatures T_c of YBCO films on CeO on sapphire as well as top YBCO films in YBCO/STO/YBCO were 88-90 K and critical current density $j_c > 10^6$ A/cm^2 at 77 K.

The values and temperature dependences of ε for STO in YBCO/STO/YBCO trilayers differed considerably from those of Ag/STO/YBCO as shown in Fig. 2.

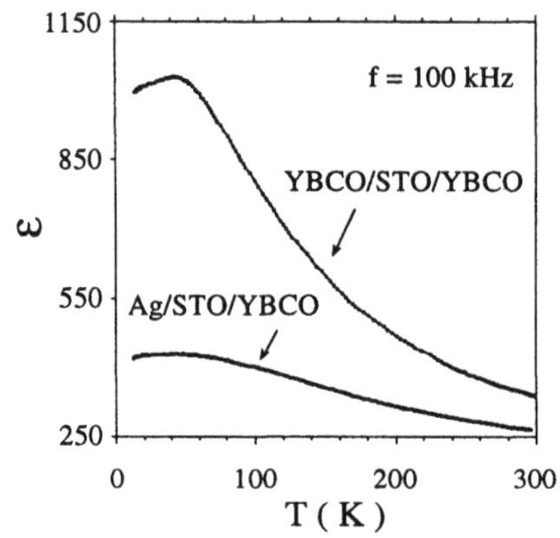

Figure 2. Temperature dependence of ε for STO intermediate layers in YBCO/STO/YBCO and Ag/STO/YBCO capacitance structures

ε was about 30% higher at room temperature in YBCO/STO/YBCO capacitors than in corresponding Ag/STO/YBCO and more than 100% higher at low temperature. The effective dielectric parameter depends strongly on interface properties.

The dielectric losses of STO were higher in trilayers where both the electrodes were YBCO than in a capacitor with an Ag top electrode (tanδ= 0.01-0.04, f=100 kHz, T=77K). tanδ for STO films does not increase with frequency as for bulk single crystals, but is frequency independent up to the microwave region [5]. The higher tanδ for STO with YBCO electrodes may be explained by an interdiffusion between components of the superconducting and ferroelectric layers at the enhanced substrate temperatures during deposition.

The conductance of an STO film between YBCO electrodes is typically much higher than the one of a bulk single crystal. High densities of structural defects and impurity atoms may be responsible for the degradation of tanδ in the ferroelectrics films.

There was a much stronger influence on ε by polarization, induced by external bias voltage V_b, in the case of YBCO electrodes than for Ag ones. The strongest dielectric non-linearity was observed at temperatures around the phase transition point (65 K for STO [2]), Fig. 3.

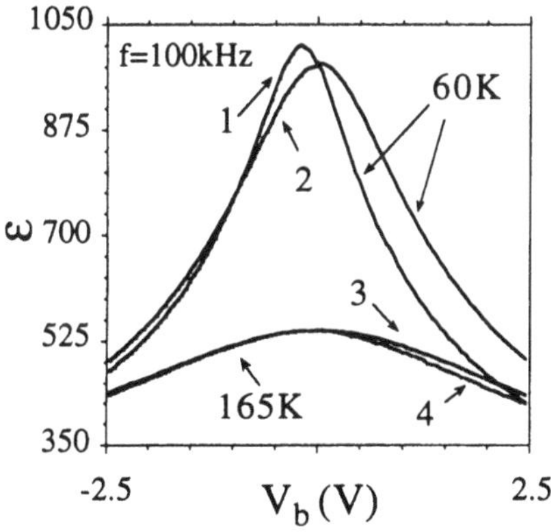

Figure 3. Measured (1,4) and fitted (2,3) ε-V_b dependencies (using rel.(1)) for the STO layer in YBCO/STO/YBCO. T= 60K and 165K

The electric field E dependence of ε for a paraelectric phase with cubic perovskite structure may be presented as [3]

$$\varepsilon(E,T) = \varepsilon(0,T)/[\ 1+ (E/E_o)^2\ ]^{1/3} \quad (1)$$

where E_0 is an electric field independent parameter. The ε-V_b dependencies calculated in accordance with relation (1) for 60K and 165K are shown in Fig.3. E_0 was determined from experimental data for ε at V_b= -1V ($E=V_b/d$, where d is the film thickness). As seen from the figure, the theoretical data are well fitted to the experimental ones.

There was no sign of hysteresis in the ε-V_b dependencies within the temperature range of 15-300 K. The unsymmetry relatively zero bias of the ε-V_b curve was much less for YBCO than for Ag top electrodes. The small shift in the maximum of ε towards negative V_b values for the YBCO/STO/ YBCO capacitance structure, see Fig.3, may be attributed to a difference in the structure at the interface between STO and superconducting top and bottom electrodes. The c-axis parameter in the bottom YBCO film was considerably longer than the one in the top superconducting film, as follows from x-ray data.

3.2. YBCO/KTO/YBCO heterostructures

In contrast to STO, KTO interacts with YBCO at processing temperatures of about 700°C. A thin layer with distorted structure is formed at the YBCO/KTO interface [4]. A parallel plane capacitor based on an YBCO/KTO/YBCO trilayer can be considered as consisting of two capacitors in series. The first one represents the contribution of the interface regions and the second one the bulk volume of the ferroelectric layer. The effective ε is essentially determined by the dielectric properties of the materials at the interfaces. The temperature dependence of ε for a KTO layer sandwiched between two YBCO electrodes was similar to that for a Ag/KTO/YBCO trilayer [4], but the maximum was shifted about 20K to a higher temperature.

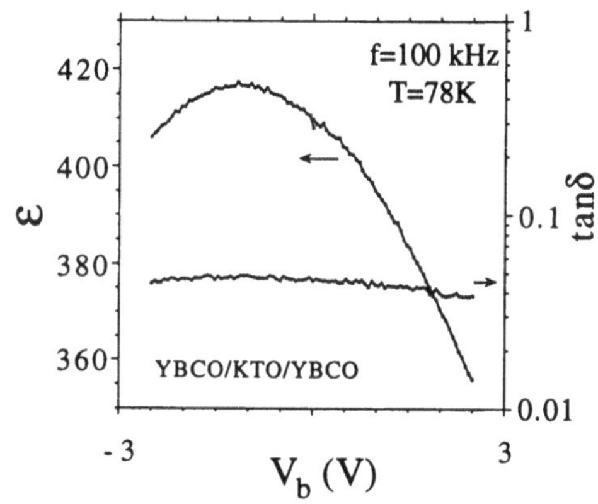

Figure 4. Bias voltage dependence of ε and tanδ for the KTO layer in an YBCO/KTO/YBCO capacitor.

The bias voltage dependence of ε was roughly the same for YBCO/KTO/YBCO and Ag/KTO/YBCO capacitors. ε changes about 20% for V_b=2.5V (d=400 nm) at T=78 K.

The tanδ for KTO intermediate layers in the Ag/KTO/YBCO was in the range 0.007-0.01 at 77K. It was 3-5 times higher in the case of all YBCO electrodes.

3.3. PZT/YBCO epitaxial heterostructures

Conducting oxide electrodes may be used to enhance the endurance of polarization of PZT films. (100)PZT (y≥0.54) films were grown epitaxially at the surface of (001)YBCO at T_s=530°C. The structure of PZT films with y < 0.54 was tetragonal. PZT films with y=0.4 were polycrystalline with (001) and (100) preferential orientations of the grains. ε for a PZT film was proportional to temperature in the range 20-300K. The maximal value of ε (T=300K) was observed at y=0.54, Fig.5.

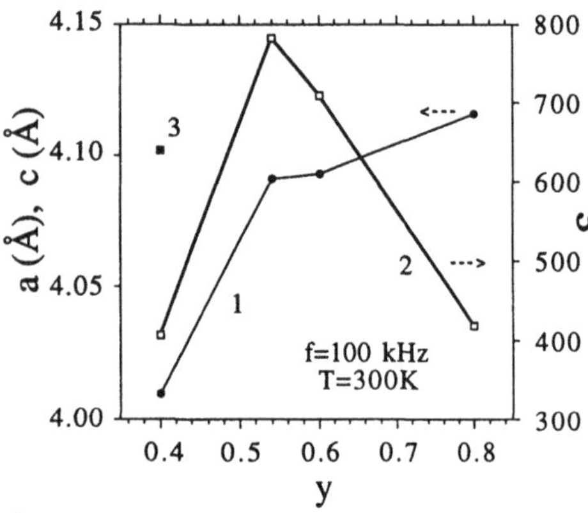

Figure 5. Lattice parameter a (1) and ε (2) for PZT layers as functions of y; the c-axis value of the tetragonal unit cell for PZT, y=0.4, is given by (3).

The P_r for PZT (y=0.4-0.8) was in the range 8-10 $\mu C/cm^2$ and the coercive field (25-35) kV/cm. An interaction between Zr and Ba at the YBCO/PZT interface may be a reason of the decrease in ε and P_r relatively bulk PZT.

3.4. SBNO/YBCO heterostructures

Films of layered Bi-based ferroelectrics may be alternatives to PZT ones in non-volatile memories. Epitaxial (001)SBNO films were grown on (001)YBCO at T_s=660°C. A layer by layer growth of (001)SBNO films was observed in the case of weak chemical interaction with the substrate. A pronounced interaction between the YBCO bottom electrode and Nb from SBNO was detected by x-rays when processing temperatures had been higher than 720°C. ε=120 and tanδ=0.04 were measured along the c-axis for SBNO films at 300K. The ε for SBNO films increases with temperature in the range 20-700K. A pronounced maximum in ε was observed at temperatures around the Curie point (T=680 K), see Fig.6.

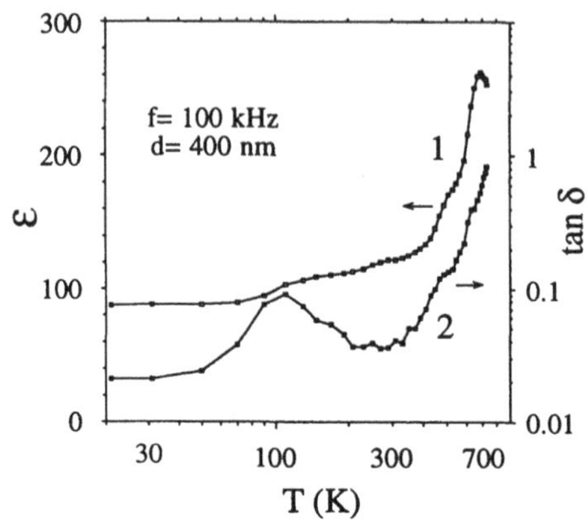

Figure 6. The temperature dependences of ε (1) and tanδ (2) for a SBNO film, measured along the c-axis.

The P_r value along the c-axis for SBNO epitaxial films was less than 0.5$\mu C/cm^2$ at 300K [6].

4. Summary

Thin layers of STO or KTO sandwiched between two YBCO films may possess high dielectric non linearities at temperatures around 77K. To decrease losses (tanδ) in the heterostructure, the density of structural defects and impurity atoms in the ferroelectric layer has to be decreased substantially.

Chemical interactions of Zr (Ti) from PZT and Nb from SBNO with Ba from YBCO may essentially influence the dielectric parameters of the ferroelectric layers.

References

1. Yu.A. Boikov, and T. Claeson, J.Appl.Phys. (1997) , in press
2. R.C. Neville, B. Hoeneisen, and C.A. Mead, J.Appl.Phys.**43**, (1972), 2124
3. A. Outzourhit, and J.U. Trefny, J.Mat.Res.**10**, (1995), 1418
4. Yu.A. Boikov, Z.G. Ivanov, A.L. Vasiliev, I. Pronin, E. Olsson, and T. Claeson, Appl.Phys.Lett.**67**, (1995), 2708
5. D. Galt, J.C. Price, J.A. Beal, and T.E. Harvey, IEEE Trans.Appl.Supercond.**5**, (1995), 2575
6. Yu.A. Boikov, Z.G. Ivanov, D. Erts, and T. Claeson, Phys.Solid State.**39**, (1997), in press

ELSEVIER

Physica C 282–287 (1997) 115–118

PHYSICA C

Studies of high temperature superconducting thin films by transmission electron microscopy

L. P. Guo, Y. Yang, Y. Q. Zhou, Y. Wu, Y. L. Qin, O. Jin and L. Li.

State Key Laboratory for Superconductivity, Institute of Physics, Chinese Academy of Sciences, Beijing, 100080, China.

High temperature superconducting thin films have been studied by transmission electron microscopy. Observations showed many superconducting and non-superconducting particles formed on the $YBa_2Cu_3O_7$ (YBCO) thin films synthesized by pulsed laser deposition, which increased the surface microwave resistance of the film. However, if the deposition conditions are selected properly, epitaxial single crystalline YBCO films can be obtained. The effect of resputtering deviates the deposited film from stoichiometric composition for the DC magnetron sputtering. An off-axis sputtering configuration can compensate this shortcoming. The low T_c (~98K) $Tl_2Ba_2CaCu_2O_8$ films showed many defects, but if the post annealing procedure is performed carefully, good quality Tl-2212 thin films can be obtained.

1. INTRODUCTION

Since the rapid development of high T_C superconducting thin film devices, the quality of the thin films become more and more important. The films not only require high T_C and J_C but also low surface microwave resistance R_S. Importance should be emphasized on the microstructural studies of the high T_C thin films, which is the link between fabrication process and the superconducting properties of the films. Transmission electron microscopy (TEM) is a powerful technique to study the microstructure of the high T_C superconducting thin films. In our laboratory, a H9000NA analytical electron microscope was stalled in 1989, this microscope is equipped with an energy dispersive X-ray spectroscopy (EDAX) and a double tilting specimen holder which can rotate 45° in two directions, these analytical apparatus are useful for the structural determination of the thin films and the chemical composition and structure of the particles formed in the film. So it is useful for the study of the high T_C superconducting thin films.

In this report we present the results of TEM observation on (1) Single layer $YBa_2Cu_3O_7$ (YBCO) thin films prepared by pulsed laser deposition (PLD) and DC magnetron sputtering, (2) $Tl_2Ba_2CaCu_2O_8$ (Tl-2212) thin films prepared by DC magnetron sputtering followed by post annealing[1].

2. EXPERIMENTAL

To study the microstructure of the thin films by TEM, the most important procedure is to prepare the TEM samples without any artifact. After achieving this operation successfully, nearly half of the work is been done. Both plane-view (PTEM) and cross-sectional (XTEM) samples were prepared. The samples were hand ground to ~30μm in thickness and then ion thinned till perforation occur. Usually thin films transparent to the electron beam could be found attached around the hole. However due to the fragileness of the film and the hardness of the substrate, an expertise is needed to achieve this operation without much failure.

3. RESULTS AND DISCUSSION

3.1 Single layer YBCO thin films on (100) $SrTiO_3$ substrate

Up to date, PLD is still the best method for preparing high T_C YBCO thin films. It is versatile, easy to obtain YBCO thin films of stoichiometric composition, and the deposition rate is very fast. But the drawback is that many large (~ 0.5μm) particles

0921-4534/97/$17.00
PII S0921-4534(97)00232-3

may form on the surface of the thin films if the preparation process is not at its optimum condition, the film surface becomes very rough, and is detrimental to R_S. Two types of particles were found, one is the superconducting particles, such as the a-axis YBCO grains and deviated from c-axis YBCO grains. The other is non-superconducting particles, they usually are Cu-rich particles, such as CuO, Ba_2CuO_3, $BaCuO_2$, $CuYO_2$, and Y_2BaCuO_3 compounds. Figure 1 shows a CuO particle nucleated on the surface of the YBCO film. From its selected area diffraction pattern (SAED), no crystallographic relation has been found between the film and the particle. The split double spots in the SAED indicated that twin structure exits in this CuO. Many other growth modes of the CuO particles were found, but all the CuO particles are single crystal in structure. Locquet et al[2] have shown that the origin of large Cu-rich particles is due to formation of CuO_2 crystallites, which nucleate extremely easily on top of $DyBa_2Cu_3O_7$ and act as very effective sinks for the rest of the deposited copper and form large crystals which may revert to monoclinic CuO upon cooling. Figure 2 shows a Ba_2CuO_3 particle nucleated on top of the YBCO film surface[3], which also has no crystallographic relation with the film.

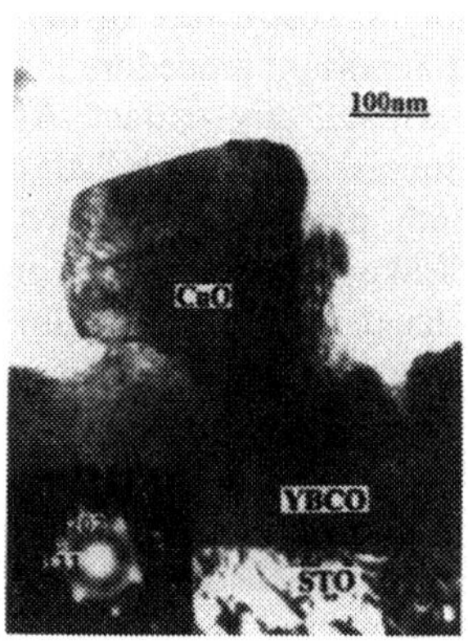

Figure 1. A twin structure CuO particle nucleated on the surface of the YBCO film matrix, the inlet is its SAED.

The disadvantage of the magnetron sputtering is the resputtering effect which tend to deviate the composition of the film from stoichiometric composition, however, an off axis configuration can compensate this shortcoming. Fewer particles are

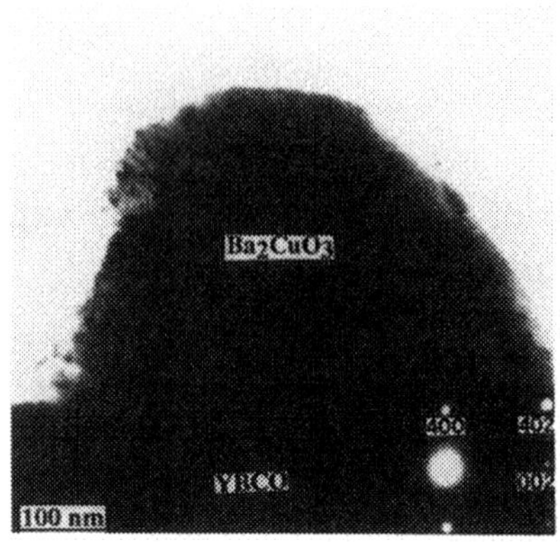

Figure 2. A large Ba_2CuO_3 particle nucleated on the surface of the YBCO film. The inlet is its SAED.

found on the film surface, most of them are a-axis YBCO grains and small sized (~10nm) CuO/Y_2O_3 particles embedded in the film matrix. Figure 3 shows a XTEM of YBCO thin film. The fine a-axis YBCO grains are embedded in the c-oriented film matrix, however, the a-axis grain grows much faster than the c-axis YBCO film, so frequently it formed as a particle protrude out of the film. Figure 4 shows this result. Figure 5 is a small Y_2O_3 particle embedded in the film matrix, the small particles could increase the critical current density perpendicular to the applied magnetic field[4].

3.2 Tl-2212 thin films on (100) $LaAlO_3$ substrate

The precursor films were prepared first by DC magnetron sputtering, then followed by post annealing to make them superconducting. The samples we have chosen for TEM examination have T_{C0} of 98K~105K. For the sample with T_{C0}=98K, the surface looked porous, a general scan of TEM at low magnification showed many defects such as grain boundaries, domain boundaries, stacking faults, cracks and even voids, impurity particles. Figure 6 shows a XTEM of the film/substrate interface, an amorphous transition layer is broken off by a big void at the center. At the right side of the void the layer is about 40 nm in thickness, at the left side of the void is only about 4 nm in thickness. These transition layers can only be detected intermittently at the interface. Figure 7 shows the morphology of A, B, C, D impurities. Micro-diffraction technique is used to determine the structure of the impurities. After indexing, A, C, D are determined as $BaCO_3$, B is a c-axis-mis-oriented Tl-2212 grain. From Figures

6 and 7, we deduce that the $BaCO_3$ impurities may be present in the Tl-target material for post annealing. During the heat treatment of the precursor film the impurities adhered to

Figure 3. HREM of a XTEM YBCO sample showing a-axis YBCO grains embedded in the c-oriented film matrix.

the film and incidentally become impurity particles in the Tl-2212 film. The amorphous transition layer at the interface could be due to the post annealing temperature was not high enough to convert the amorphous precursor film to a crystalline film, and the big voids could be formed by the dropout of the

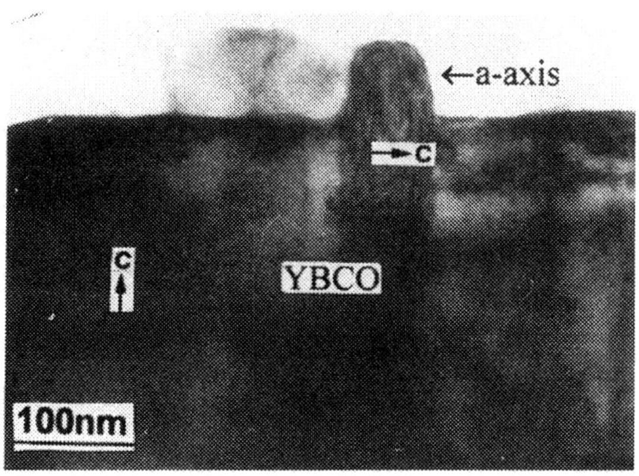

Figure 4. Micrograph showing the a-axis YBCO grain protrude above the c-axis YBCO film as a particle.

$BaCO_3$ during ion thinning of the TEM specimen. From the above microstructural results, the quality of the film is poor, and is not an epitaxial film and is

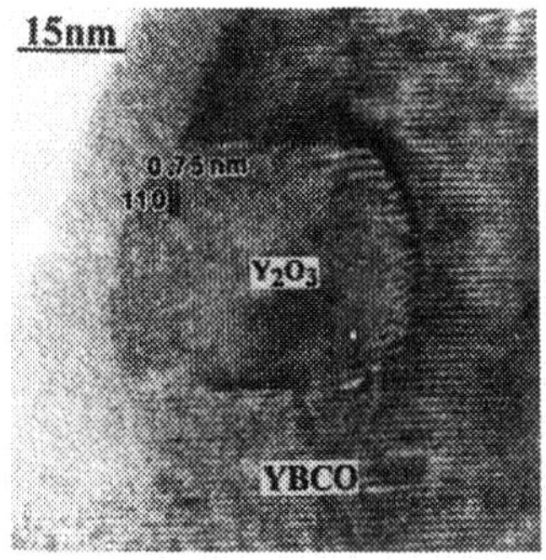

Figure 5. HREM of the Y_2O_3 particle embedded in the YBCO film matrix.

not suitable for step edge Josephson junctions. The Tl-2212 film with T_{C0}=105K showed very shiny surface. TEM examination of the XTEM specimens showed that the film is relatively uniform with only some stacking faults. Lattice fringes of spacing ~1.5 nm can be seen everywhere, this is the (002) plane of Tl-2212, which means that the film is purely c-axis oriented. Figure 8 shows the typical lattice image of the film. This kind of Tl-2212 film is suitable for making step edge Josephson junctions and other superconducting devices. But unfortunately, the Tl-2212 films we have examined frequently contained amorphous layer and impurities. We suggest that the low temperature post annealing procedure should be performed more carefully.

4. CONCLUSION

The microstructural studies of TEM showed that PLD synthesized YBCO films have more non-superconducting particles than that of the DC magnetron sputtered films. But if the deposition condition is at its optimum condition even epitaxial single crystalline films can be obtained. In order to obtain good quality Tl-2212 films, the post annealing of the precursor films should be performed carefully.

REFERENCES

1. P. Jocquet, Y. Jaccard, C. Gerber and E. Machler, Appl. Phys. Lett., 63 (1993) 1426.

2. Yan, L. Fang, M. S. Si, H. L. Cao, Q. X. Song, J. Yan, X. D. Zhou, J. M. Hao, Supercond. Sci. Technol. 7 (1994) 681.
3. L. P. Guo, Y. J. Tian, J. Z. Liu, L. Li, Z. X. Zhao, S. F. Xu, H. B. Lu, Y. L. Zhou, Z. H. Chen, G. Z. Yang, Physica C, 241 (1995) 30.
4. Y. J. Tian, L. P. Guo, L. Li, Y. Q. Zhou, Y. Yang, Z. X. Zhao, S. F. Xu, H. B. Lu, Y. L. Zhou, Z. H. Chen, D. F. Cui and G. Z. Yang, Appl. Phys. Lett., 65 (1994) 234.

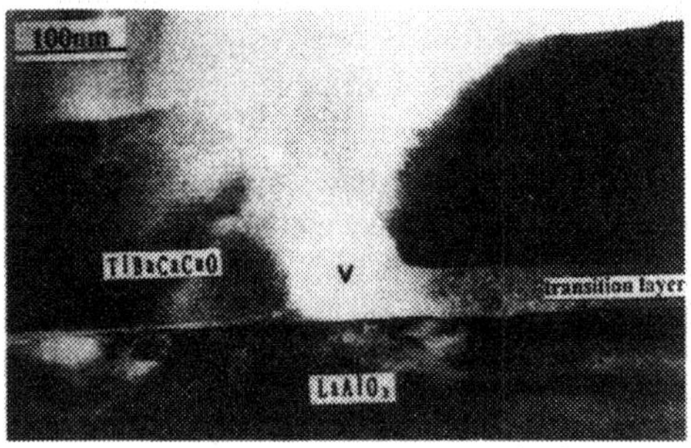

Figure 6. The XTEM of Tl-2212 showing an amorphous transition layer at the film/substrate interface. The big void breaks the layer into two parts.

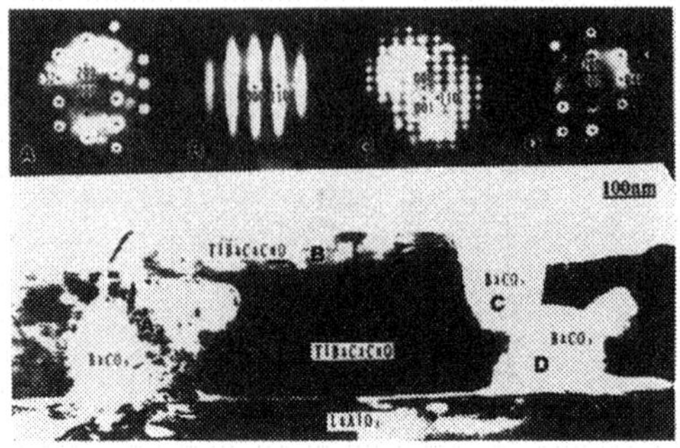

Figure 7. The micro-diffraction patterns of A, B, C, D, and their morphology. A, C, D are determined as $BaCO_3$.

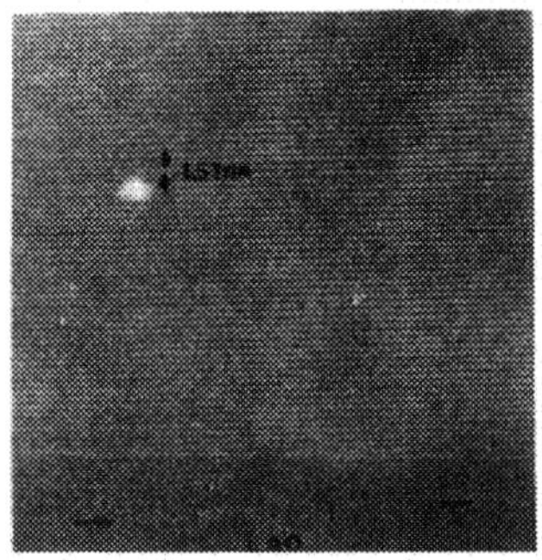

Figure 8. The lattice image of a T_{C0}=105K Tl-2212 thin film.

ELSEVIER

Physica C 282–287 (1997) 119–122

PHYSICA C

Pseudo Spin Gap in High-T_c's Manifested by the NMR Experiments

Hiroshi Yaśuoka

Institute for Solid State Physics, University of Tokyo, Roppongi, Minato-ku, Tokyo 106, JAPAN

From the temperature dependence of the ^{63}Cu nuclear spin-lattice relaxation rate, $1/T_1T$, in the planar Cu(2) sites, it is well established by now that highly enhanced and strongly temperature dependent relaxation process due to antiferromagnetic Cu spin fluctuations exhibits the opening of a pseudo gap in the low-lying magnetic excitations with an energy comparable to the superconducting gap, particularly for the so-called low doping region. It is also found that, irrespective of the single-layer or the bi-layrer system, the temperature at which the spin-gap opens, T_{sg}, determined as the peak of $1/T_1T$ vs. T, has a linear decrease with increasing of the doping concentration. A general spin-gap phase diagram is proposed.

1. Introduction

Perhaps the most prominent feature in low energy spin excitations seen in the normal state of high-T_c's is that strongly temperature dependent antiferromagnetic spin fluctuations manifested in the nuclear spin-lattice relaxation rate, $1/T_1T$, of the planar Cu(2) sites stop increasing or even decrease with decreasing temperature at a certain temperature above T_c. We have associated this behavior with the opening of a pseudo gap in the antiferromagnetic spin excitation spectrum, the so-called "pseudo spin-gap" (hereafter we simply abbreviate it as the spin-gap), which should be distinguished from the superconducting gap [1]. It has also been argued that this spin-gap temperature depends on the doping level with a characteristic way. In contrast to this T_1 behavior, the temperature variations of the indirectly coupled nuclear spin-spin relaxation rate, $1/T_{2g}$, and the anisotropic Knight shifts K_a have a quite independent signature; the former shows continuous increase toward T_c and the latter in low doping regime decreases from temperatures much higher than those where the spin-gap is seen in $1/T_1T$. In this paper, we extend our original arguments by showing the results for the $1/T_1T$, in the single layer materials, $HgBa_2CuO_{4+\delta}$ (Hg1201) and $La_{2-x}Sr_xCuO_4$ (La214) systems. We conclude here that the spin-gap behavior is a unique characteristic for the individual CuO_2 plane, and present a general "spin-gap phase diagram" in high-T_c's.

2. Pseudo spin gap and its doping dependence

In contrast to the Korringa process seen in non-superconducting but metallic Cu oxides, the normal state relaxation process for all of the lightly or optimally doped high-T_c's is governed by the Curie-Weiss (CW) type temperature dependent antiferromagnetic spin fluctuations. In order to discuss this more quantitatively, we have already selected the case of $YBa_2Cu_4O_8$, since we could obtained very reliable T_1 data at high temperatures [2]. The basic features are summarized below.

As is discussed in ref.2, the temperature dependence of $1/T_1T$ for the ^{63}Cu in the Cu(2) sites of $YBa_2Cu_4O_8$ is well fitted to the Curie-Weiss form above 250K with,

PII S0921-4534(97)00234-7

$$\frac{1}{T_1T} = 0.28 + \frac{1454}{(T+29)} \ (\text{sec}.K)^{-1}, \qquad (1)$$

where the first and the second terms in the right hand side are due to the Korringa (mostly due to the $q = 0$ component) and the antiferromagnetically enhanced relaxation processes, respectively. However, further decreasing temperature, $1/T_1T(T)$ shows a maximum at around $T^* \cong 160$K and decreases rapidly toward the value at T_c of 81K where the superconducting gap starts to open. We have analyzed this rapid reduction of $1/T_1T$ in the temperature range between T_c and T^* by fitting the data to an excitation to the antifrromagnetically enhanced states. We found a good fit of the data to the following activation type dependence,

$$\left(\frac{1}{T_1T}\right)_{obs} = \left(\frac{1}{T_1T}\right)_{CW} \times A\exp\left(\frac{-\Delta}{kT}\right) \quad (T_c < T < T^*), \qquad (2)$$

with the activation energy $\Delta/k = 145$ K. The same kind of analysis has been made for other low doping systems, and the activation energy of about 13~18 meV has been extracted. It should be stressed here that this gap is not directly related to the superconducting gap, since, most of the cases, a noticeable decrease of $1/T_1$ due to the superconducting gap opening dose exist at T_c that is well below T^*. Here, we define T^* as the spin-gap temperature, T_{sg}.

The doping dependence of $1/T_1T(T)$ has also been studied in several systems. Utilizing data for $Y(Ba_{1-x}La_x)Cu_3O_y$ [3], $LaBa_2Cu_3O_y$ [4] and $(Y_{1-x}Pr_x)Ba_2Cu_3O_7$ [5] systems, the spin-gap temperatures are plotted against the estimated hole number per CuO_2 sheet in Fig.1. It is evident from these plots that the spin-gap temperature decreases linearly with increasing hole concentration and reaches to T_c at the optimal doping, while T_c increases with increasing the hole concentration. We believe the general phase diagram shown in Fig.1 presents a strong constraint to making a theoretical model in the low doping regime. The only theory compatible with the present experimental phase diagram is seems to be based on the spin-charge separation picture developed by Fukuyama's group[6] or Lee's group[7]. The former associates the spin-gap behavior with the singlet formation of spinons called as the singlet RVB state within the individual CuO_2 plane, while the latter emphasize the importance of bi-layer coupling.

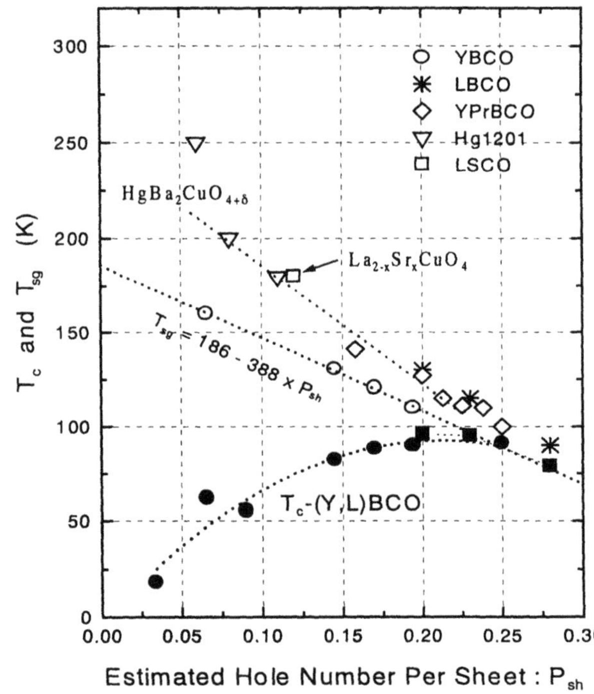

Fig. 1. *A spin-gap phase diagram. The spin-gap temperatures, T_{sg}, are defined as the temperature where $1/T_1T(T)$ has a maximum. The doping dependence of T_c for bi-layer systems are also shown by black circles.*

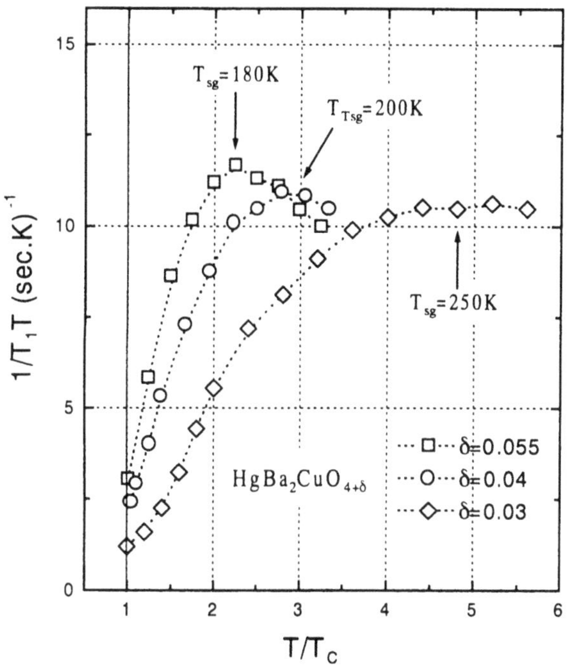

Fig. 2. *Cu NMR relaxation rates, $1/T_1T$, for CuO_2 plane of Hg1201 system are shown as a function of the reduced temperature T/T_c. Tc for δ=0.055, 0.04 and 0.03 are 80K, 72K and 50K, respectively.*

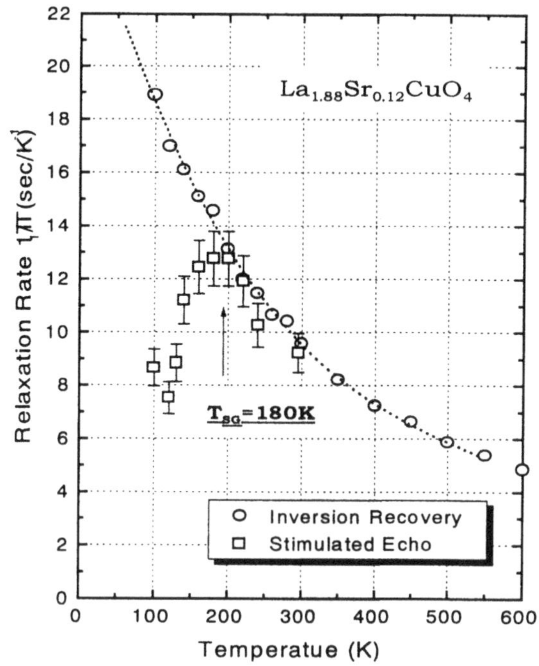

Fig. 3. *Temperature dependence of the Cu NMR relaxation rate $1/T_1T$ for CuO_2 plane of $La_{1.88}Sr_{0.12}CuO_4$. A clear spin-gap behavior is seen in the data obtained by the stimulated echo method.*

To solve this single-layer vs. bi-layer problem, we decided to examine the T_1 data for two representative single layer materials: Hg1201 and La214 systems.

For the Hg1201 case, we have observed very clear spin-gap behavior as is shown in Fig. 2 [8]. The doping dependence is also shown in Fig. 1.

As to the La214 case, we have already demonstrated that there exists definite temperature range where $1/T_1T$ data start to deviate from the CW increase at high temperatures. This temperature range has been associated with the onset of the spin-gap in this system [9]. The reason why the spin–gap is not clearly seen in La214 is probably due to some microscopic randomness in the electronic states of the CuO_2 planes. This random distribution of doped holes manifests itself a broad and peculiar NQR spectrum [10], and makes difficult to obtain the intrinsic relaxation process by the conventional inversion recovery method. To avoid this difficulty, we have utilized the stimulated echo method where the spin diffusion process across the inhomogeneously broadened NQR spectra can be eliminated experimentally. Although the detail will be published elsewhere [11], a clear spin-gap behavior in $1/T_1T(T)$ at 180K has been observed in $La_{1.88}Sr_{0.12}CuO_4$, as is shown in Fig. 3.

Therefore, together with the recent report of the spin-gap in the three-layer system $HgBa_2Ca_2Cu_3O_{8+\delta}$ [12], we strongly believe that the accumulating NMR results are concluding the importance of formation of the spin-gap without necessity of bi-layer coupling.

Finally, it should be stressed that the spin-gap behavior discussed here is one of the most important characteristics to describe the so-called unconventional metallic state near the metal-insulator transition boundary, and it is likely to be related with the mechanism of the high-T_c.

3. Acknowledgment

We would like to acknowledge stimulating discussion with Prof. H. Fukuyama, Dr. A. Goto, Dr. M. Itoh and Dr. T. Machi. We are also very much obliged to Professor Y. Ueda and Dr. K. Tanabe for their excellent material synthesis. This work has been partially supported by a Grant-in-Aid for Scientific Research from the Ministry of Education, Science and Culture.

References

1. H. Yasuoka, T. Imai and T. Shimizu: in Springer Series in Solid State Sciences, **89**, "*Strong Correlation and Superconductivity*", ed. by H. Fukuyama, S. Maekawa and A.P. Malozemoff (Springer-Verlag, 1989) p.254
2. H. Yasuoka, S. Kambe, Y. Itoh and T. Machi, Physica B **199** (1994) 278
3. M. Matsumura, Y. Sakamoto, T. Fushihara, Y. Itoh and H. Yamagata, J. Phys. Soc. Jpn. **64** (1995) 721
4. A. Goto, H. Yasuoka, K. Otzschi and Y. Ueda, J. Phys. Soc. Jpn. **64** (1995) 367
5. A.P. Reys, D.E. MacLaughlin, M. Takigawa, P.C. Hammel, R.H. Heffner, J.D. Thompson and J. E. Crow, Phys. Rev. B **43** (1991) 2989
6. Y. Suzumura, Y. Hasegawa and H. Fukuyama, J. Phys. Soc. Jpn. **57** (1988) 2768
7. M. U. Ubbens, P. A. Lee, Phys. Rev. **B50** (1994) 438
8. Y. Ttoh, T.Machi, A. Fukuoka, K. Tanabe and H. Yasuoka, J. Phys. Soc. Jpn. **65** (1996) 3751
9. H. Yasuoka, Springer Series in Solid State Sciences, **119**, "*Spectroscopy of Mott Insulators and Correlated Metals*" ed. by A. Fujimori and Y. Tokura (Springer-Verlag, 1995) p.213
10. H. Yasuoka, Hyperfine Interactions, **20** (1996) 1
11. S. Fujiyama, H. Yasuoka and Y. Ueda, in preparation
12 M. H. Julien, P. Carreta, M. Horvatic, C. Berthier, Y. Berthier, P. Segransan,A. Carrington and D. Colson, Phys. Rev. Lett., **76** (1996) 4238

ELSEVIER

Physica C 282–287 (1997) 123

PHYSICA C

Extended Abstract

Recent photoemission data from underdoped Bi2212 single crystals

Zhi-Xun Shen

Department of Applied Physics and Stanford Synchrotron Radiation Laboratory, Stanford University, Stanford, CA 94305, USA

We summarize our recent angle-resolved photoemission data from underdoped Bi2212 single crystals and thin films [1-5]. Photoemission data from underdoped Bi2212 has also been reported by Ding et al.[6,7].

The data from the underdoped samples reveal an interesting evolution of the normal state electronic structure as a function of doping. In the slightly overdoped regime, Bi2212 has a large Fermi surface centered the (π,π) point of the Brillouin zone. As the sample being underdoped, one still sees a Fermi surface crossing along the (0, 0) to (π,π) line. On the other hand, the Fermi surface crossing along the (π, 0) to (π,π) line disappears, indicating the opening of an energy gap in this region [1, 2].

There are two aspects to the energy gap near (π, 0). If one characterizes the gap by the leading edge of the spectra that represents the lowest energy excitation, the gap has very similar magnitude and momentum dependence as that of the superconducting gap measured by ARPES. Furthermore, the gap size measured at low temperature in the underdoped regime does not decrease with Tc reduction and the gap becomes un-measurable only in a temperature range much higher than Tc [3]. These results can not be reconciled with the conventional mean-field BCS-Eliashberg theory. On the other hand, if one characterizes the gap by the centroid of the broad feature in the (π, 0) spectra from underdoped Bi2212, the energy scale of the "gap" is much higher than the leading edge gap. This "gap" appears to evolve smoothly from the behavior seen in the insulator $Sr_2CuO_2Cl_2$[8] where the energy difference between (π, 0) and (π/2, π/2) is about 2.5J to the behavior seen in overdoped Bi2212 where the energy at (π, 0) is basically the same as that of the Fermi surface crossing point near (π/2, π/2).

The lineshape of the (π, 0) spectrum changes rapidly from a broad feature in the underdoped case to a much sharper feature in the overdoped case. This change is correlated with the change of the size of the superconducting gap which is a measure of the pairing strength[4]. Analysis of the photoemission lineshape and the size of the superconducting gap provides strong evidence for the pairing interaction being peaked near Q=(π, π) [5].

REFERENCES:

1. D.S. Marshall, D.S. Dessau, A.G. Loeser, C.H. Park, A.Y. Matsuura, J. Eckstein, I. Bozovic, P. Fournier, A. Kapitulnik, W.E. Spicer and Z.-X. Shen; Phys. Rev. Lett. 76, 4841 (1996).
2. A.G. Loeser, Z.-X. Shen, D.S. Dessau, D.S. Marshall, C.H. Park, P. Fournier and A. Kapitulnik; Science, Vol. 273, 325 (1996).
3. J.M. Harris, A.G. Loeser, D.S. Marshall, M.C. Schabel, Z.-X. Shen, J.N. Eckstein, I. Bozovic; Phys. Rev. B54, R15665 (1996).
4. P.J. White, Z.-X. Shen, C. Kim, J.M. Harris, A.G. Loeser, P. Fournier and A. Kapitulnik; Phys. Rev. B54, R15669 (1996).
5. Z.-X. Shen and J.R. Schrieffer; Phys. Rev. Lett., 78, 1771 (1997).
6. H. Ding, T. Yokoya, J.C. Campuzano, T. Takahashi, M. Randeria, M.R. Norman, T. Mochiku, K. Kadowaki, and J. Giapintzakis; Nature 382, 51 (1996).
7. H. Ding, M.R. Norman, T. Yokoya, T. Takeuchi, R. Randeria, J.C. Campuzano, T. Takahashi, T. Mochiku and K. Kadowaki; Phys. Rev. Lett. 78, 2628 (1997).
8. B.O. Wells, Z.-X. Shen, A. Matsuura, D.M. King, M.A. Kastner, M. Greven, and R.J. Birgeneau; Phys. Rev. Lett. 74, 964 (1995).

PII S0921-4534(97)00235-9

PHYSICA C

ELSEVIER

Physica C 282–287 (1997) 124–127

The Spin Gap and Extended t-J Model

Hidetoshi Fukuyama and Hiroshi Kohno

[a]Department of Physics, University of Tokyo, 7-3-1 Hongo, Bunkyo-ku, Tokyo 113, Japan

Key experimental facts of high T_c cuprates have been summarized and theoretical understanding of these systems viewed as doped Mott insulators is given based on the slave-boson mean-field theory of the extended t-J model. A special emphasis is placed on the spin-gap phenomena. Effects of fluctuations around this mean field theory treated as a gauge field are also briefly discussed.

1. Experimental facts on Spin Gap

Among various experiments so far carried out for high-T_c cuprates, the disclosures of the existence of the spin gap are most noteworthy. This has originally been noted by Yasuoka *et al.* [1] in the NMR rate in the underdoped region of YBCO(123), as the maxima of the rate, $(T_1T)^{-1}$, at a temperature T_{SG} well above the superconducting critical temperature, T_c. This T_{SG} will be called the spin-gap temperature in the following. In early stage, the existence of T_{SG} has been clearly identified only in the underdoped YBCO(123) [1,2] and YBCO(248) [3], which have bilayer structures, and then there were some discussions that the spin-gap phenomena are particular only to the bilayers. However, later experiments on three-layer Hg(1223) [4] and a single-layer Hg(1201) [5] also demonstrated the existence of T_{SG}. Moreover, very recent experiment by Yasuoka [6] has disclosed that T_{SG} actually existed even in LSCO(214). All these experiments have demonstrated that the spin gap is intrinsic to all high-T_c cuprates with CuO_2 planes. The dependences of T_{SG} on the doping rate, δ, is seen to have very characteristic features as indicated in Fig.1; T_{SG} appears to merge into the optimal T_c.

In this paper we propose a possible theoretical explanation to this remarkable experimental facts on the basis of the slave-bosen mean field theory to the extended t-J model [7,8]. As it turns out, this theory also offers an understanding of not only the spin gap seen in NMR rate, $(T_1T)^{-1}$, but also the onset of anomalous frequency shift

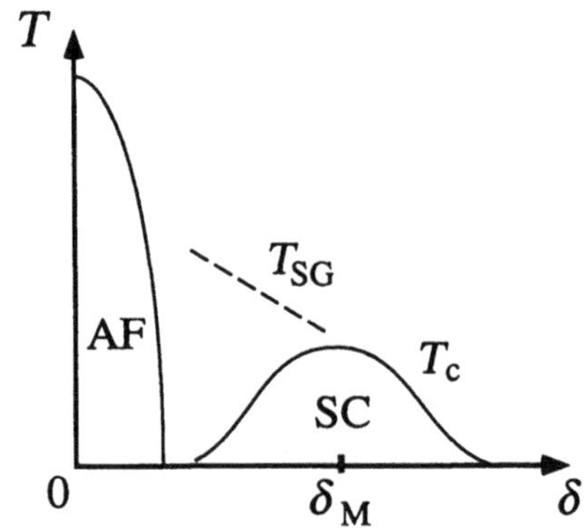

Figure 1. Phase diagram of high-T_c cuprates deduced from experiments. T is the temperature and δ is the hole doping rate.

of some particular phonon modes and moreover the observation of the pseudogap above T_c in the angle-resolved photoemission spectroscopy. Though less directly, the spin gap appears to affect transport and optical properties as well [9–11].

2. Extended t-J Model and Slave-Boson Mean Field Theory

The low-lying excitations of high-T_c cuprates, whose parent ($\delta = 0$) compounds are charge-transfer type insulators, are described by the effective single-band model [12] and explicitly by the t-J model [13–18]

$$H = \sum_{i,j,\sigma} t_{ij} a^{\dagger}_{i\sigma} a_{j\sigma} + J \sum_{<i,j>} \vec{S}_i \cdot \vec{S}_j \qquad (1)$$

0921-4534/97/$17.00
PII S0921-4534(97)00237-2

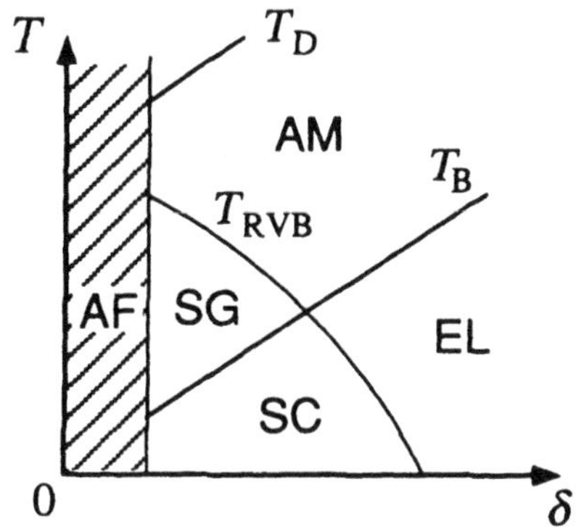

Figure 2. Schematic phase diagram of the (extended) t-J model. T is the temperature and δ is the hole doping rate.

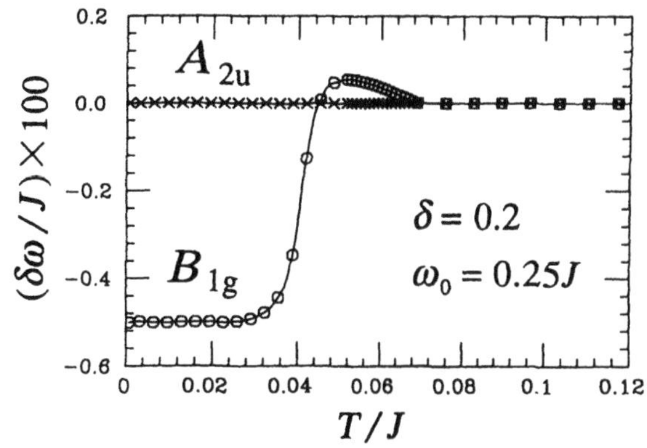

Figure 3. Temperature dependence of the phonon frequency shift $\delta\omega$ of the in-plane oxygen mode with bare frequency $\omega_0 = 0.25$.

where the double occupancy of each lattice site is excluded, and t_{ij} and J represent the transfer integrals of the Zhang-Rice singlet and the super-exchange interaction, respectively. Here we note that t_{ij} is very sensitive to the local structural properties which leads to different values of t_{ij} for different compounds [15]. Hence the characteristic features of each compound having different Fermi surfaces are represented by the different spatial extent of t_{ij}, which is termed as extended t-J model [8].

The mean-field theory to the t-J model in the slave-boson framework, which is developed on the original idea of RVB [12,19] has predicted [7,20] a phase diagram as shown in Fig.2, where AF means antiferromagnetic phase realized at low doping.

In Fig.2, T_D, T_{RVB} and T_B are the characteristic temperatures of the onset of the coherent motion of spinons and holons, the formation of spinon singlet pairing and the bose condensation of holons, respectively. By these characteristic temperatures, the phase diagram is decomposed into various states. For $T > T_B$ the spin and charge are separated, while for $T < T_B$ spin and charge are combined to form ordinary electron liquid (EL). The superconducting critical temperature, T_c, in this theory is the lower one of T_{RVB} and T_B. In the anomalous metallic (AM) state described as a uniform RVB (u-RVB) state, spinons and holons move independently and coherently. In the spin gap (SG) state, where spin and charge are also separated, the spinons form short-range singlet described as a singlet RVB (s-RVB). This singlet has the $d_{x^2-y^2}$ symmetry [20] which is the same as in the superconducting state. It has been shown that the fluctuations around the present mean-field solution treated as gauge field make T_{RVB} and T_B above T_c crossover temperatures [21].

3. Spin Excitation

From Figs.1 and 2, it will be quite natural to associate experimental T_{SG} with theoretical T_{RVB}. Actually the results of calculations demonstrate a clear peak of $(T_1T)^{-1}$ at T_{RVB} [8].

Moreover, the detailed calculations of spin excitation spectra for LSCO offered explanation of the experimental results [22]; (a) the existence of gap at the incommensurate wavevector at low temperatures (below T_c), (b) the doping dependence of the degree of incommensurability. For the case of YBCO, however, there exists a clear discrepancy between our theoretical prediction and the experimental results, especially as regards the '41meV peak' [23–26], to which there is an interesting proposal [27]. (See also Ref.[28].)

4. The Spin-Gap Phenomena

The identification of T_{RVB} with T_{SG} is further supported by the experimental results of the onset of the anomalous frequency shift of in-plane oxygen modes vibrating along the c-axis seen in YBCO(248) and underdoped YBCO(123). In the optimally doped YBCO(123) the neutron scattering [29] on B_{1g} ($340cm^{-1}$) mode has revealed the sudden drop of frequency at T_c. This fact has been explained theoretically by taking account of the existence of the buckling [30]. The results are shown in Fig.3. Here B_{1g} mode has a sharp change at $T = T_{RVB}$ (whereas A mode does not), which should be identified with T_c at the optimal doping. If our identification of T_{RVB} with T_{SG} is correct, a similar onset of anomalous frequency shift of B_{1g} mode should be expected at temperature well above T_c in samples located in the underdoped region. This is actually observed by Raman scattering in YBCO(248) at around 150K [31], which is T_{SG} seen in NMR rate. Similar behaviours of B_{2u} mode was seen in underdoped YBCO(123) by neutron scattering [32].

Another important experimental support for $T_{RVB} = T_{SG}$ is due to the angle-resolved photoemission spectroscopy (ARPES). Loeser *et al.* first observed [33] the onset of $d_{x^2-y^2}$ like gap formation in underdoped BSCCO even above T_c, which fact has been subsequently confirmed by Ding *et al.* [34].

5. Summary and Discussion

Various experiments in underdoped region of high-T_c cuprates point to the existence of the characteristic temperature above T_c, especially the spin-gap temperature, T_{SG}, characterized by the maximum of NMR rate, $(T_1T)^{-1}$. It has been pointed out that the phase diagram derived by the slave-boson mean field theory for the extended t-J model has qualitative correspondence at least to a part of the experimental results, and that, especially, the identification of T_{SG} with the formation of the short range singlet RVB gives a coherent explanation among spin excitations, anomalous frequency shift of some particular phonon modes, and the pseudo-gap formation above T_c seen in ARPES.

The present report is mainly on spin excitations based on the mean field theory. There are, of course, fluctuations around mean-field solutions, which can be treated as gauge fields [35,36,21]. The studies on the roles played by the gauge field fluctuations are under way [37–39] . It has been pointed out in ref.[39] that large gauge-field fluctuation can transform the statistics of quasiparticles and then holons will not be pure bosons. In this case gauge-field fluctuations will be greatly reduced due to the Chern-Simons term and at the same time the charge transport properties will be modified from the slave-boson description. Even in this case, spin excitations will be little affected if spinon excitations are close to fermionic ones. This is due to the existence of large cut-off energy, the Fermi energy.

In the present view, the driving force of the high T_c in cuprates is the strong superexchange interaction, J, which in general will lead to magnetic orderings. However, in cuprates, such magnetic orderings are hindered by strong quantum fluctuations due to single band, two-dimension and $S = 1/2$, and spin-liquid state is stabilized. If the charge is introduced into such spin-liquid state, which has been described as RVB, the superconductivity will be a natural consequence. All of these basic features are in the original indication by Anderson [12].

REFERENCES

1. H. Yasuoka, T. Imai and T. Shimizu, in *Strong Correlation and Superconductivity*, (Springer Verlag, 1989) ed. H. Fukuyama, S. Maekawa and A.P. Malozemoff, p.254.
2. H. Yasuoka, to appear in Hyperfine Interaction (1996).
3. K. Asayama, Y. Kitaoka, G.-q. Zheng and K. Ishida, Prog. in Nuclear Magnetic Resonance Spectroscopy **28** 221 (1996).
4. M.H. Julien, P. Carretta, M. Horvatic, C. Berthier, Y. Berthier, P. Ségransan, A. Carrigton and D. Colson, Phys. Rev. Lett. **76** 4238 (1996).
5. Y. Itoh, T. Machi, A. Fukuoka, K. Tanabe and H. Yasuoka, J. Phys. Soc. Jpn. **65** 3751 (1996).
6. H. Yasuoka, private communication.
7. Y. Suzumura, Y. Hasegawa and H. Fukuyama, J. Phys. Soc. Jpn. **57** 401, 2768

(1988).
8. T. Tanamoto, H. Kohno and H. Fukuyama, J. Phys. Soc. Jpn. **62** 717 (1993); **63** 2739 (1994).
9. S. Uchida, K. Takenaka and K. Tamasaku, in *Proceedings of the 10th Anniversary HTS Workshop*, (World Scientific, 1996) p.300.
10. B. Bucher, P. Steiner, J. Karpinski, E. Kaldis and P. Wachter, Phys. Rev. Lett. **70** 2012 (1993).
11. C.C. Homes, T. Timusk, R. Liang, A. Bonn and W.N. Hardy, Phys. Rev. Lett. **71** 1645 (1993).
12. P.W. Anderson, Science **235** 1196 (1987).
13. F.C. Zhang and T.M. Rice, Phys. Rev. B **37** 3759 (1988).
14. L.H. Tjeng, H. Eskes and G.A. Sawatzky, in *Strong Correlation and Superconductivity* (Springer Verlag, 1989) ed. H. Fukuyama, S. Maekawa and A.P. Malozemoff, p.33.
15. H. Matsukawa and H. Fukuyama, J. Phys. Soc. Jpn. **58** 2845; 3687 (1989); **59** 1723 (1990).
16. T. Tohyama and S. Maekawa, J. Phys. Soc. Jpn. **59** 1760 (1990)); S. Maekawa, Y. Ohta and T. Tohyama, in *Physics of High-Temperature Superconductors* (Spring Verlag, 1992) ed. S. Maekawa and M. Sato, p.29.
17. M.S. Hybertsen, E.B. Stechel, M. Schluter and D.R. Jennison, Phys. Rev. B **41** 11068 (1990).
18. G. Dopf, A. Muramatsu and W. Hanke, Physica **185-189C** 1495 (1991); J. Wagner, W. Hanke and D.J. Scalapino, *ibid* 1617.
19. G. Baskaran, Z. Zou and P.W. Anderson, Solid State Commun. **63** 973 (1987).
20. G. Kotliar and J. Liu, Phys. Rev. B **38** 5142 (1988).
21. N. Nagaosa and P.A. Lee, Phys. Rev. Lett. **60** 2450 (1990); P.A. Lee and N. Nagaosa, Phys. Rev. B **46** 5621 (1992).
22. K. Yamada, S. Wakimoto, G. Shirane, C.H. Lee, M.A. Kastner, S. Hosoya, M. Greven, Y. Endoh and R.J. Birgeneau, Phys. Rev. Lett. **75** 1626 (1995).
23. J. Rossat-Mignot, L.P. Regnault, C. Vettier, P. Bourges, P. Burlet, J. Bossy, J.Y. Henry and G. Lapertot, Physica **185-189C** 86 (1991).
24. H.A. Mook, M. Yethiraj, G. Aeppli, T.E. Mason, T. Armstrong, Phys. Rev. Lett. **70** 3490 (1993); P. Dai, M. Yethiraj, H.A. Mook, T.B. Lindemer and F. Doğan, Phys. Rev. Lett. **77** 5425 (1996).
25. H.F. Fong, B. Keimer, P.W. Anderson, D. Reznik, F. Doğan and I.A. Aksay, Phys. Rev. Lett. **75** 316 (1995); H.F. Fong, B. Keimer, D.L. Milius and I.A. Aksay, Phys. Rev. Lett. **78** 713 (1997).
26. P. Bourges, L.P Regnault, Y. Sidis and C. Vettier, Phys. Rev. B **53** 876 (1996).
27. E. Demler and S.C. Zhang, Phys. Rev. Lett. **75** 4126 (1995).
28. H. Kohno, B. Normand and H. Fukuyama, in the present volume.
29. N. Pyka, W. Reichardt, L. Pintschovius, G. Engel, J. Rossat-Mignod and J.Y. Henry, Phys. Rev. Lett. **70** 1457 (1993).
30. B. Normand, H. Kohno and H. Fukuyama, Phys. Rev. B **53** 856 (1996).
31. M. Käll, A.P. Litvinchuk, P. Berastegui, L.G. Johansson and L. Börjesson, Physica **225C** 317 (1994).
32. H. Harashina, K. Kodama, S. Shamoto, M. Sato, K. Kakurai and M. Nishi, Physica **263C** 257 (1996).
33. A.G. Loeser, D.S. Dessau and Z.X. Shen, Physica **263C** 208 (1996); Science **273** 325 (1996).
34. H. Ding, T. Yokoya, J.C. Campuzano, T. Takahashi, M. Randeria, M.R. Norman, T. Mochiku, K. Kadowaki and J. Glapintzakis, Nature **382** 51 (1996).
35. G. Baskaran and P.W. Anderson, Phys. Rev. B **37** 580 (1988); G. Baskaran, Physica Scripta **T27** 53 (1988).
36. L.B. Ioffe and A.I. Larkin, Phys. Rev. B **39** 8988 (1989).
37. X.G. Wen and P.A. Lee, Phys. Rev. Lett. **76** 503 (1996).
38. P.A. Lee, N. Nagaosa, T.K. Ng and X.G. Wen, unpublished.
39. F. Basco, H. Kohno, H. Fukuyama and G. Baskaran, J. Phys. Soc. Jpn. **65** 687 (1996); submitted to Phys. Rev. B.

ELSEVIER

Physica C 282–287 (1997) 128–131

PHYSICA C

Probing the Symmetry of the Pairing State of Unconventional Superconductors by SQUID Interferometer Measurements*

D. J. Van Harlingen

Department of Physics and Materials Research Laboratory, 104 S. Goodwin Avenue, Urbana, IL 61801

Phase-sensitive SQUID interferometer experiments have emerged as the most powerful technique for determining the symmetry of the pairing state of unconventional superconductors. This approach uses Josephson tunneling to probe specific directions in k-space and long-range phase coherence to measure the phase anisotropy of the order parameter, allowing a direct and definitive determination of the symmetry. Developed for testing the symmetry of the cuprates, corner SQUID and junction experiments indicate a sign change in the order parameter in orthogonal in-plane directions in YBCO, strong evidence for $d_{x^2-y^2}$ pairing. Now, this technique is being applied to other classes of superconductors suspected to be unconventional.

INTRODUCTION

One of the primary success stories of high temperature superconductivity has been the determination that the cuprates exhibit unconventional pairing with a $d_{x^2-y^2}$ symmetry. At the first three M²S meetings, there was little indication that order parameter symmetry would be an important issue. The possibility of d-wave symmetry arising from magnetic interactions had been raised [1], but had initially been thought to be too weak to account for the high transition temperatures of the cuprates. By M²S-IV in Grenoble in 1994, however, the situation had changed dramatically. Detailed calculations suggested that magnetic spin fluctuation pairing in a $d_{x^2-y^2}$ channel could account for many of the normal and superconducting state properties of the cuprates [2], motivating intense experimental activity directed toward testing the symmetry. Key measurements of NMR relaxation rates [3], the low temperature penetration depth [4], and angle-resolved photoemission spectroscopy [5] gave strong evidence for substantial anisotropy in the order parameter magnitude, consistent with $d_{x^2-y^2}$ symmetry. More than any other experiments, however, it was the development of the SQUID interferometer technique [6,7], with its unique ability to probe the *phase* anisotropy of the order parameter, that unambiguously established $d_{x^2-y^2}$ as the dominant pairing symmetry in the cuprates. Thus, as we meet in Beijing for M²S-V, there is now general agreement on the symmetry, and the focus now turns to issues of the microscopic mechanism and implications of the unconventional symmetry on the properties and applications of the high temperature superconductors.

In this paper, I will focus on illuminating some of the physical concepts underlying the phase-sensitive interferometer experiments, summarize the evidence that demonstrates the $d_{x^2-y^2}$ symmetry, and discuss briefly the application of this technique to other materials that are candidates for unconventional superconductivity.

1. PAIRING SYMMETRY CANDIDATES

The most direct approach to distinguishing different pairing symmetries is to measure the anisotropy of the phase of the order parameter. The value of this is illustrated in Figure 1, which shows the magnitude and phase of the order parameter vs. k-space angle for candidate states. Whereas the magnitude anisotropy is rather similar for all but the isotropic s-wave state, distinguishable only by

* In collaboration with D. M. Ginsberg, A. J. Leggett, D. A. Wollman, and B. D. Yanoff. Research supported by the Science and Technology Center for Superconductivity under grant NSF-DMR-91-20000.

PII S0921-4534(97)00238-4

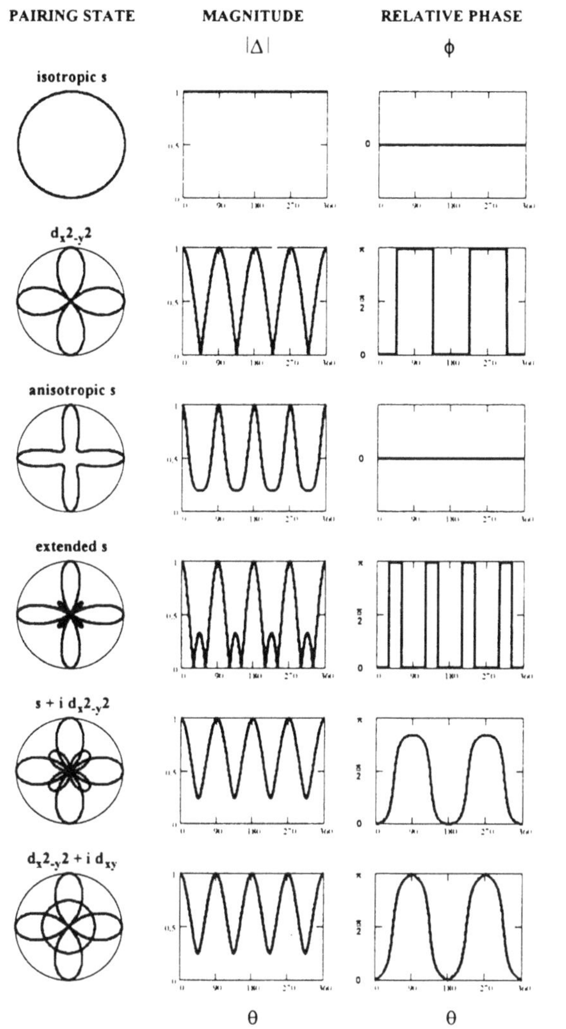

Figure 1. Magnitude and phase anisotropy of the order parameter for some candidate pairing symmetries.

the behavior near the nodes, the phase anisotropy is distinctly different. The objective of the SQUID interferometer technique is to measure the relative phase in orthogonal directions. This single measurement is sufficient to distinguish s-wave from d-wave, and can also identify a complex mixture which breaks time-reversal symmetry.

2. The Corner SQUID Experiment

The design of the corner SQUID experiment is shown in Figure 2(a). The circuit is a bimetallic dc SQUID, in which Josephson junctions on the a and b faces of the cuprate crystal are connected by a loop of a conventional superconductor. The junctions probe the order parameter in orthogonal directions, and the response of the SQUID to applied magnetic fields is sensitive to a phase shift inside the crystal arising from the symmetry. In this regard, the corner SQUID is simply an interferometer in which the internal order parameter phase drop shifts the diffraction pattern, as in a two-slit optical interference experiment. Figure 2(b) shows the shift of the critical current modulation for the d-wave state compared to a conventional s-wave SQUID, demonstrating the sensitivity to the order parameter phase anisotropy. Figure 2(c) shows measured results for 21 corner SQUIDs and 11 edge SQUIDs (control samples with both junctions on the same face). The phase shift of order π in all corner devices is strong evidence for pairing in a d-wave channel; the spread is due to vortices trapped near the SQUID loop that couple a flux up to $0.2\Phi_0$ into the loop.

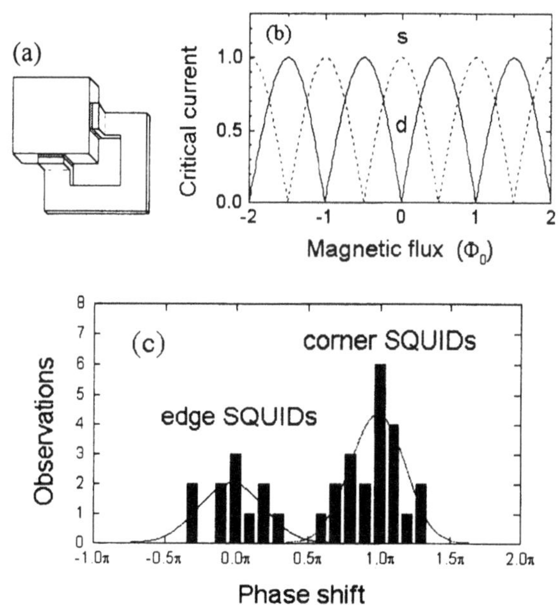

Figure 2. Corner SQUID experiment: (a) plan, (b) critical current modulation vs. symmetry, (c) results.

Several key principles are important in this experiment, many of which have not been fully understood or appreciated:

(a) *Domain structure.* The corner experiment measures the global phase shift between orthogonal k-space directions but between spatially-separated regions of the crystal adjacent to the junctions in real space. As a result, the range of the order parameter domains and, in particular, how the order parameter behaves across grain and twin boundaries are important. The only way to determine this is to measure the phase response in both twinned and untwinned single crystals. Our measurements in crystals of these types [6] have established that the order parameter has gyroscopic symmetry, that is, the order parameter maintains the same sign orientation across grain and twin boundaries, forming a single domain throughout the crystal. This result is crucial for the SQUID interferometer technique to make a definitive determination of the microscopic symmetry, especially in heavily-twinned thin film samples in which subsequent versions of the interferometer experiment have been carried out.

(b) *Directionality.* The measurement of the phase anisotropy depends on probing the order parameter in specific k-space directions. This relies on the directionality of the Josephson tunneling orthogonal to the junction barrier. Although this is known to occur in insulator-barrier junctions for which the charge transport is solely by tunneling, it also applies to any planar junction (such as the SNS junctions used in our experiments) since Josephson tunneling involves phase-coherent electron-hole pair conduction. The grain boundary junctions used in thin film interferometers certainly do not strictly maintain directionality because of the wandering of the boundary, but their ability to resolve the orthogonal phase shift indicates that directionality is maintained on average.

(c) *Origin of phase shift.* The phase shift observed in the corner SQUID experiment and its variations occurs *inside* the crystal between orthogonal directions. A common misconception is that the phase shift can be described in terms of π-junctions, Josephson junctions in which the minimum coupling energy is at junction phase π, corresponding to a negative critical current. This concept was introduced [8] to describe Josephson junctions in which the tunneling occurs via magnetic spin-flip scattering, adding a π-phase shift inside the junction. To date, such junctions have not been realized, and have nothing to do with the phase shifts and the spontaneous circulating currents observed in interferometer experiments. The junctions in these experiments are all ordinary Josephson junctions, with their minimum energy at zero phase. In particular, it is clear in the SQUID experiments that π-junctions cannot account for the observed phenomena since there are two identical junctions.

3. VARIATIONS

Following our initial corner SQUID tests, several subsequent experiments based on the same principle were carried out, contributing further evidence for the d-wave symmetry.

3.1 Single corner junction

Perhaps the most qualitatively clear indication of the d-wave symmetry comes from measurements of the critical current modulation pattern of a single Josephson junction that straddles the corner of a cuprate crystal [9], as in Figure 3(a). This is the single junction equivalent of the corner SQUID experiment, essentially a SQUID without the loop. For s-wave symmetry, such a junction exhibits the ordinary Fraunhofer diffraction pattern familiar from single slit optical interference. However, in YBCO-Pb junctions, we observe the dramatically modified pattern in Figure 3(b), characteristic of a phase shift of π between the orthogonal segments. The most telling feature is the precipitous drop in the critical current at zero field, demonstrating the cancellation of the critical currents through the two sections. Recent observation of complex critical current modulation patterns of 45° grain boundaries can be understood in terms of multiple corner junctions formed by the wandering boundary [10].

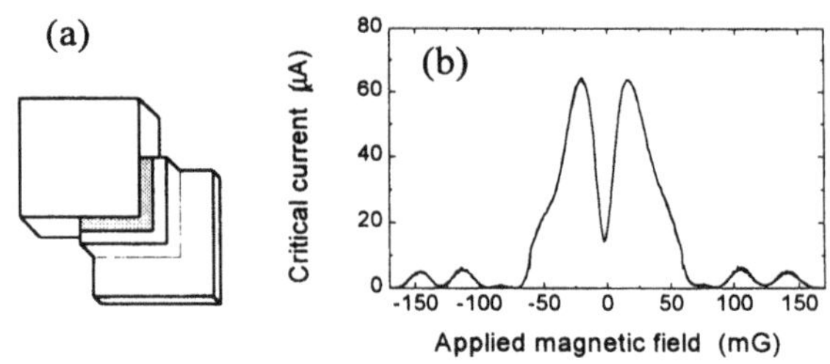

Figure 3. Design (a) and measured diffraction pattern (b) for a single YBCO-Pb corner Josephson junction.

3.2 Spontaneous circulating currents

An alternate approach to measuring the modified critical current modulation of the SQUID (or single junction) in a d-wave superconductor is to detect the spontaneous (zero applied field) circulating current required to maintain phase coherence around the SQUID loop. This has been achieved using SQUID magnetometry in the corner SQUID geometry by the Maryland group [11] and first in a tricrystal ring geometry by the IBM group [12]. The ring sample consists of a cuprate thin film grown on a tricrystal substrate, forming three superconducting regions separated by grain boundary Josephson junctions. This system is exactly equivalent to the corner SQUID, except that the magnetic field response depends on whether one, two, or all three of the cuprate sections exhibit a π-phase shift (i.e., straddle a node). In both configurations, the spontaneous circulating currents only flow if the loop inductance is sufficiently large to make it energetically-favorable to generate a current (as opposed to flipping the phase on one junction to its higher-energy zero-current state). The generated magnetic flux approaches $\frac{1}{2}\Phi_0$ for large loop inductance.

3.3 c-axis supercurrents

A somewhat different set of experiments involves Josephson tunneling into the c-axis face of YBCO samples. The observed finite Josephson supercurrents [13] are inconsistent with a state of pure d-wave symmetry since such currents are expected to average to zero, implying the existence of an s-wave component in the order parameter. Recent measurements of single c-axis junctions straddling a twin boundary exhibit the same critical current diffraction patterns as we observed in corner junctions (Figure 3), indicating that this s-wave component is largely the result of the orthorhombic structure of YBCO [14].

4. FUTURE DIRECTIONS

The results of the corner SQUID, corner junction, spontaneous current, grain boundary diffraction, and c-axis tunneling experiments now all point conclusively to a state with at least predominately $d_{x^2-y^2}$ symmetry. With the unconventional nature of the pairing symmetry in the cuprates established, attention is now being focused in three directions:

(a) *Determining the microscopic mechanism.* Surprisingly and somewhat disappointingly, knowing the symmetry has not significantly accelerated the search for the microscopic mechanism. Although the $d_{x^2-y^2}$ symmetry was first predicted from magnetic pairing mechanisms, it is now recognized that many interactions can result in pairing in a d-wave channel.

(b) *Implications of unconventional symmetry.* The nodes and sign change associated with the d-wave pairing affect many thermodynamic, transport, optical, and interface properties of the cuprates. Work is only beginning to explore the exciting science and potential technological impact resulting from the d-wave symmetry.

(c) *Application to other superconductors.* The interferometer experiments we pioneered can also be used to test the symmetry of other candidates for unconventional superconductivity. We are currently carrying out experiments designed to determine the symmetry of the heavy fermion superconductor UPt_3, suspected to be p-wave, and the organic superconductor κ-$(ET)_2Cu[N(CN)_2]Br$, which may be also be d-wave.

REFERENCES

1. N. E. Bickers, D. J. Scalapino, and S. R. White, Phys. Rev. Lett. **62**, 961 (1989).
2. P. A. Monthoux, A. Balatsky, and D. Pines, Phys. Rev. B **46**, 14803 (1992).
3. J. Martindale *et al.*, Phys. Rev. B **47**, 9155 (1993).
4. W. N. Hardy *et al.*, Phys. Rev. Lett. **70**, 3999 (1993).
5. T. P. Devereaux *et al.*, Phys. Rev. Lett. **72**, 396 (1994).
6. D. A. Wollman *et al.*, Phys. Rev. Lett. **71**, 2134 (1993).
7. D. J. Van Harlingen, Rev. Mod. Phys. **67**, 515 (1995).
8. L. N. Bulaevski, V. V. Kuzii, and A. A. Sobyanin, JETP Lett. **25**, 290 (1977).
9. D. A. Wollman *et al.*, Phys. Rev. Lett. **74**, 797 (1995).
10. J. Mannhart *et al.*, Z. Phys. B **101**, 175 (1996).
11. C. C Tsuei *et al.*, Phys. Rev. Lett. **73**, 593 (1994).
12. A. Mathai *et al.*, Phys. Rev. Lett. **74**, 4523 (1995).
13. A. G. Sun *et al.*, Phys. Rev. Lett. **72**, 2267 (1994).
14. J. Clarke *et al.*, preprint.

ELSEVIER

Physica C 282–287 (1997) 132–135

PHYSICA C

Symmetry of the Order Parameter: Implications for the Transport Properties of Grain Boundaries

J. Mannhart[a], H. Hilgenkamp[a] and Ch. Gerber[b]

[a]Electronic Correlations and Magnetism, University of Augsburg, Universitätsstraße 1, D-86135 Augsburg, Germany.

[b]IBM Research Division, Zurich Research Laboratory, Säumerstr. 4, CH-8803 Rüschlikon, Switzerland.

Implications of a $d_{x^2-y^2}$ symmetry component of the order parameter for the transport properties of grain boundaries in high-T_c superconductors are discussed. Among other effects, the $d_{x^2-y^2}$ symmetry causes intrinsic weak links, inhomogeneous current distributions in the junctions and reduced I_cR_n products.

Various experiments have revealed that the order parameter of many high-T_c cuprate superconductors has a predominant $d_{x^2-y^2}$ symmetry component, see e.g. [1,2]. For this symmetry, the amplitude of the gap-function is maximal along the a and b directions of the crystal unit-cell. In addition, the phase of the macroscopic wavefunction that describes the superconducting state differs by π between these directions. Besides the $d_{x^2-y^2}$ geometry, the order parameter symmetry is likely to include other symmetry components, e.g. due to the orthorhombicity of many high-T_c cuprates, such as $YBa_2Cu_3O_{7-x}$.

The unconventional order parameter symmetry has significant implications for the characteristics of high-T_c Josephson junctions [3-9]. By considering the angle-dependent magnitude of the gap in the direction perpendicular to the barrier plane, Sigrist and Rice [3] have derived an expression for the dependence of the Josephson current on the orientations of two $d_{x^2-y^2}$ superconducting electrodes, separated by an insulating barrier layer. Subsequently, it was shown by Barash *et al.* [4] that, in addition to these density-of-states effects, also the $d_{x^2-y^2}$ symmetry can give rise to a depression of the order parameter in the superconducting electrodes abutting an insulating or normal-conducting barrier layer. The magnitude of this depression depends on the orientations of the superconductors with respect to the boundary plane and on the transparency of the barrier.

In addition, we have proposed that a boundary region with a depressed order parameter is created intrinsically at the interface between two superconductors with anisotropic order parameters, that form a contact at an angle θ [10]. The order parameter depression results from pair-breaking effects caused by the different crystallographic orientations of the superconductors and the finite size of the coherence length ξ. Typically this depression will occur over distances of the order of ξ from the boundary, with its magnitude depending among other parameters on θ, temperature, and on the materials involved. Considering [001]-tilt boundaries in $d_{x^2-y^2}$ superconductors, the depression is expected to be strongest for a misorientation angle of 45°, for which the maxima of the order parameter of one superconductor coincide with the nodes in the gap function of the other.

For the mechanism of misorientation-induced order parameter depression to be effective, no insulating or normal-conducting barrier layer has to be present *a priori*, but as a consequence of a complete depression of the order parameter at the boundary a normal-conducting barrier layer may arise intrinsically.

The suppression of the superconductivity according to the mechanism described above may help to understand why grain boundaries in high-T_c superconductors are weak links or Josephson junctions. Clean grain boundaries in isotropic superconductors, such as Nb, do not behave as weak links.

It has been found, in particular for the most

0921-4534/97/$17.00
PII S0921-4534(97)00240-2

commonly studied [001]-tilt grain boundaries, that the critical current density J_c decreases exponentially with θ increasing from 0 to 45° [11,12]. Various reasons have been proposed for this behavior [13], which mainly rely on the occurrence of a non-superconducting barrier layer originating from dislocations and stress at the boundary [14] and/or non-stoichiomety, in particular oxygen deficiency [15].

These mechanisms are important factors controlling the transport properties of the grain boundaries. Considering them in detail, it seems surprising why the drop of J_c with increasing misorientation has been found, with no exception, in all samples and for all high-T_c compounds for which such measurements have been made. It is also startling why low-Σ grain boundaries are not characterized by exceptionally high J_c-values. Therefore we suggest that in addition to the above-mentioned microstructural causes for the weak link properties of grain boundaries in high-T_c superconductors, other mechanisms play an important role in controlling their transport properties. The proposed misorientation-induced order parameter depression presents a possible candidate for such a mechanism, which has to be studied quantitatively.

The normal-state properties of the grain boundaries give rise to another interesting question. The normal-state resistivity of $YBa_2Cu_3O_{7-x}$ -based grain boundary junctions varies typically between 10^{-9} Ωcm² and 10^{-7} Ωcm², which is much higher than the resistivity expected for a boundary-layer just consisting of $YBa_2Cu_3O_{7-x}$ with a completely depressed order parameter. For example, for a non-superconducting $YBa_2Cu_3O_{7-x}$ layer with a resistivity of 100 μΩcm extending over a distance of 2 nm on both sides of the boundary, a grain boundary resistivity of 4×10^{-11} Ωcm² would be expected naïvely.

At present it is a matter of debate which mechanisms cause the high resistivity values. Non-stoichiometric layers at the grain boundaries and disorder leading to quasiparticle scattering will provoke large grain boundary resistivities. In addition, the mere misorientation of the grains will be a contributing factor. An example herefore is enhanced quasiparticle-scattering originating from the misorientation of the Fermi-surfaces of the two superconducting crystals abutting the boundary.

As these mechanisms differ to some extend from the mechanisms controlling the supercurrent transport, it seems understandable that the critical current I_c and normal-state resistance R_n are only partly correlated.

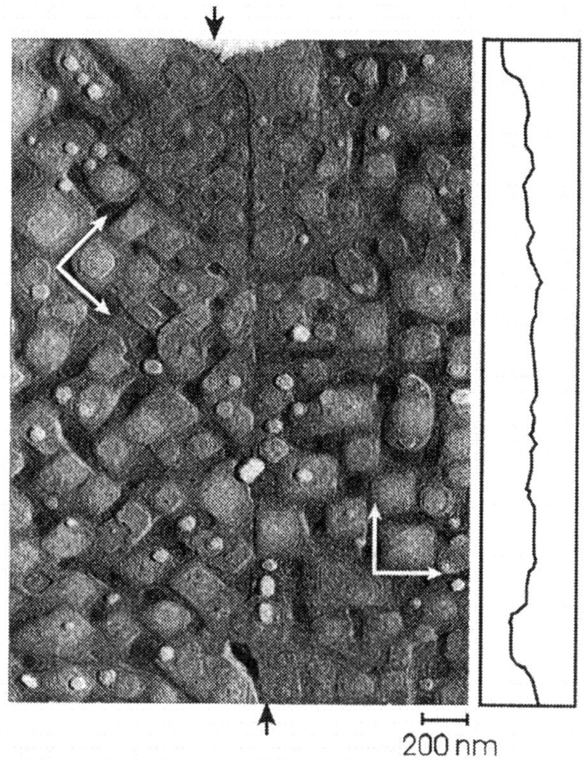

Figure 1. AFM image of the surface of a 150 nm thick $YBa_2Cu_3O_{7-x}$ film with an asymmetric 45° [001]-tilt boundary. The orientation of the grains is clearly visible by the orientation of the facets of the growth islands. The location of the grain boundary is indicated by the arrows. The meandering path of the grain boundary has been replotted at the right of the figure. From [18].

Thin-film grain boundaries are not straight, but generally consist of facets of various orientations and with typical dimensions < 100 nm. In combination with the $d_{x^2-y^2}$ symmetry component of the order parameter this leads to an intrinsically inhomogeneous distribution of the Josephson current in the junctions [16,17], including regions for which an additional π phase shift occurs. The inhomogeneity is most prominent for asymmetric 45° [001] tilt boundaries ($\theta_1 = 0$, $\theta_2 = 45°$) for which an AFM image is depicted in fig. 1.

Using the expression proposed by Sigrist and Rice for the Josephson critical current density between two $d_{x^2-y^2}$ superconductors [3], the spatial distribution of the critical current density, J_c, was calculated for the boundary depicted in fig. 1. The result is shown in fig. 2. It follows clearly from this figure that the current distribution is very inhomogeneous, with about half of the junction having a critical current density with opposite sign.

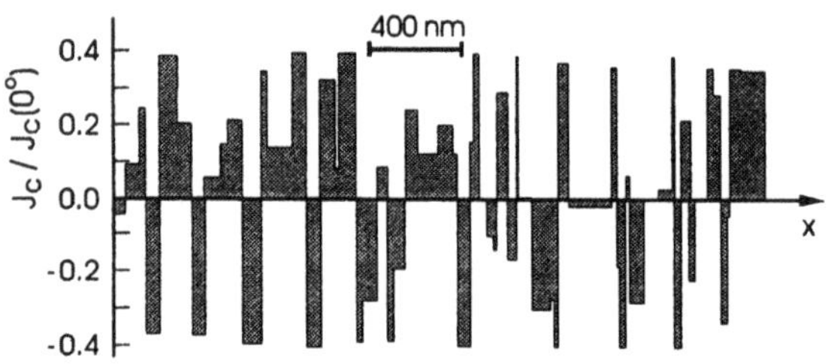

Figure 2. Ratio of the grain boundary critical current density J_c and the zero-angle critical current density $J_c(0°)$ calculated for the asymmetric 45° [001] tilt grain boundary of fig. 1, according to the expression given in [3].

This inhomogeneity results in an anomalous dependence of the critical current I_c on an applied magnetic field H_a [16,17,19]. A striking characteristic is the possible absence of a global maximum of I_c at $H_a = 0$ (see Fig. 3). The highly symmetric characteristic gives evidence that the anomaly is not caused by magnetic flux quanta trapped in, or close to, the boundary.

Figure 4 shows the $I_c(H_a)$ dependence calculated from the J_c-distribution of Fig. 2, with $J_c(0°)=10^7$ A/cm^2. The characteristic features observed in the experiments, such as the distorted periodicity and the absence of a global maximum of I_c at $H_a = 0$, are also found in the calculated $I_c(H_a)$ dependences. It is noted that the difference in the magnetic field scale results from the fact that the boundary of fig. 1 is about an order of magnitude smaller than the one used for the experiments of fig. 3 and because flux focusing has not been taken into consideration.

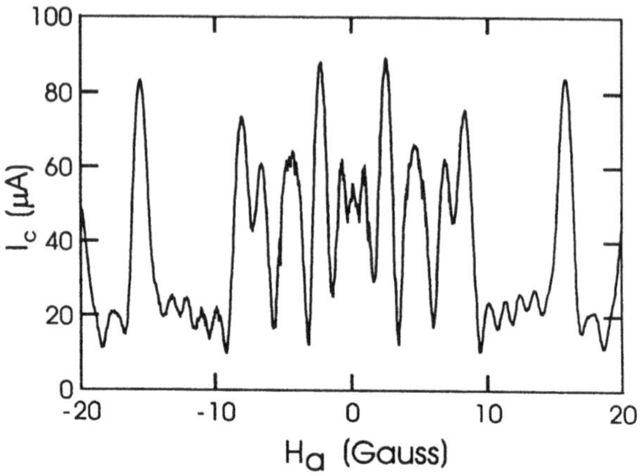

Figure 3. Critical current (at 4.2 K) as a function of applied magnetic field in a 10-µm wide and 180-nm thick $YBa_2Cu_3O_{7-x}$ line straddling an asymmetric 45° [001] tilt grain boundary.

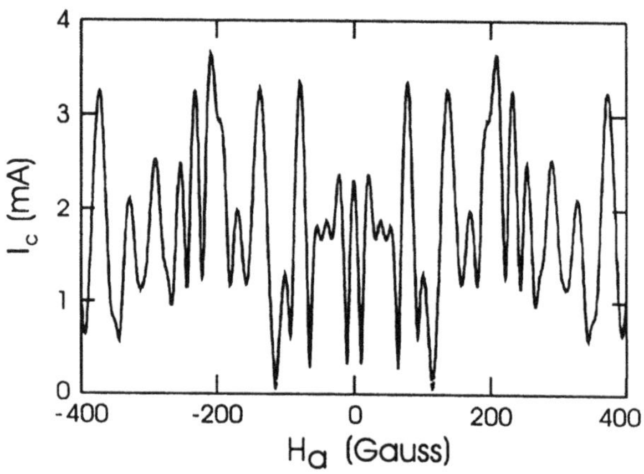

Figure 4: Magnetic field dependence of the critical current calculated for the grain boundary of fig. 1, using the spatial distribution of J_c shown in fig. 2.

Transport properties measured in practical grain boundary experiments reflect the averaged behavior of these inhomogeneous junctions and are therefore no direct measure of intrinsic grain boundary properties. The reduction of the junction critical current due to the occurrence of regions with negative Josephson current gives rise to reduced I_cR_n products. Furthermore, the local Josephson penetration depth λ_J can be substantially smaller than the value derived from the junction critical current density according to $\lambda_J \propto J_c^{1/2}$.

The appearance of regions in the junction with π-phase shifts give rise to spontaneous generation of magnetic flux in the boundary, as has been shown by scanning SQUID microscopy [18]. The magnetic flux is a potential source of noise for grain boundaries and for their applications, for example in SQUID magnetometers. Furthermore the spontaneous magnetization can modify the current-phase relation of the junctions [20] and enhance critical current fluctuations. In addition, the flux will cause a paramagnetic component to the grain boundary susceptibility.

The authors acknowledge with pleasure discussions with many collegues, in particular J.R. Kirtley.

REFERENCES

1. D.J. Van Harlingen, Rev. Mod. Phys. 67 (1995) 515.
2. C.C. Tsuei, J.R. Kirtley, C.C. Chi, L.S. Yu-Jahnes, A. Gupta, T. Shaw, J.Z. Sun and M.B. Ketchen, Phys. Rev. Lett. 73 (1994) 593.
3. M. Sigrist and T.M. Rice, Rev. Mod. Phys. 67 (1995) 503.
4. Yu.S. Barash, A.V. Galaktionov and A.D. Zaikin, Phys. Rev. B 52 (1995) 665.
5. G. Deutscher and R. Maynard, Europhys. Lett. 30 (1995) 361.
6. S. Kashiwya, Y. Tanaka, M. Koyanagi, H. Takashima and K. Kajikura, Phys. Rev. B 51 (1995) 1350.
7. C. Bruder, A. van Otterlo and G.T. Zimanyi, Phys. Rev. B 51 (1995) 12904.
8. Y. Tanaka and S. Kashiwaya, Phys. Rev. B 53 (1996) 9371.
9. Y. Tanaka and S. Kashiwaya, Phys. Rev. B 53 (1996) R11957.
10. H. Hilgenkamp and J. Mannhart, Appl. Phys. A, in press.
11. D. Dimos, P. Chaudhari and J. Mannhart, Phys. Rev. B 41, 4038 (1990).
12. Z.G. Ivanov, P.Å. Nilsson, D. Winkler, J.A. Alarco, T. Claeson, E.A. Stepantsov and A.Ya. Tzalenchuk, Appl. Phys. Lett. 59, 3030 (1991).
13. P. Chaudhari, D.Dimos and J. Mannhart, in *Superconductivity*, J.G. Bednorz and K.A. Müller eds. (Springer, Heidelberg, 1990).
14. J.A. Alarco and E. Olsson, Phys. Rev. B 52 (1995) 13625.
15. D.M. Kroeger, A. Choudhury, J. Brynestad, R.K. Williams, R.A. Padgett and W.A. Coghlan, J. Appl. Phys. 64 (1998) 331.
16. C.A. Copetti, F. Rüders, B. Oelze, Ch. Buchal, B. Kabius and J.W. Seo, Physica C 253 (1995) 63.
17. H. Hilgenkamp, J. Mannhart and B. Mayer, Phys. Rev. B 53 (1996) 14586.
18. J. Mannhart, H. Hilgenkamp, B. Mayer, Ch. Gerber, J.R. Kirtley, K.A. Moler and M. Sigrist, Phys. Rev. Lett. 77 (1996) 2782.
19. J. Mannhart, B. Mayer and H. Hilgenkamp, Z. Phys. B 101 (1996) 175.
20. E. Il'Ichev, V. Zakosarenko, R.P.J. IJsselsteijn, V. Schultze and H.-G. Meyer, unpublished.

ELSEVIER

Physica C 282–287 (1997) 136–139

PHYSICA C

Pairing symmetry in single-layer tetragonal $Tl_2Ba_2CuO_{6+\delta}$ from in-plane torque anisotropy

C. Rossel, [a] M. Willemin,[a,b] J. Hofer,[a,b] H. Keller, [b] Z.F. Ren [c] and J.H. Wang [c]

[a] IBM Research Division, Zurich Research Laboratory, 8803 Rüschlikon, Switzerland

[b] Physik-Institut, Universität Zürich, Winterthurerstr. 190, 8057 Zürich, Switzerland

[c] Superconductive Materials Laboratory, State University of New York, Buffalo, NY 14260

The in-plane torque anisotropy of the high-T_c cuprate $Tl_2Ba_2CuO_{6+\delta}$ gives evidence of a mixed $s+d$ pairing state allowed by the possible breaking of its tetragonal symmetry at low temperature. The validity of our result is supported by the absence of any in-plane anisotropy in the classical superconductors NbN and the layered 2H-$NbSe_2$ measured under the same conditions.

Many experiments for testing the pairing symmetry in high-T_c superconductors were performed, yielding increasing evidence in favor of $d_{x^2-y^2}$ symmetry [1,2], but the question whether it is a pure d-wave or a combination of s- and d-wave is still debated. The possible coexistence of s- and d-wave has recently been proposed by K.A. Müller [3]. Most of the phase-sensitive measurements using Josephson junctions were done on $YBa_2Cu_3O_{7-\delta}$. The orthorhombic structure of this compound, with the presence of CuO chains, of two CuO_2 planes per unit cell and the formation of twin boundaries, is certainly a disadvantage when testing the symmetry of the order parameter. A better suited material is $Tl_2Ba_2CuO_{6+\delta}$ (Tl2201), which has a tetragonal structure, free of chains, and with only one CuO_2 layer sandwiched between the thick thallium oxide layers. Moreover, no evidence for a superstructure along its a-axes has been reported. This material was recently used by Tsuei et al. [4] when they repeated their tricrystal experiment on spontaneous half-integer flux quanta originally done on YBCO rings.

Several studies using the a-b plane anisotropy of the magnetization [5] or of the upper critical field $H_{c2}(\theta)$ [6,7] provided evidence of the $d_{x^2-y^2}$ pairing symmetry. Angular-dependent torque magnetometry is a very sensitive technique applied recently by Ishida et al. [8] to rectangular untwinned YBCO crystals. In their a-b plane torque anisotropy curves $\tau(\theta)$, they observed a fourfold rotational symmetry attributed to the $d_{x^2-y^2}$ state in addition to a twofold one attributed to orthorhombicity and geometrical effects. Here θ is the angle that the applied field B makes with the a-axis. Won and Maki [9] worked out the Ginzburg–Landau (G–L) equations for a d-wave orthorhombic cuprate, deriving expressions for the free energy F, for $H_{c2}(\theta)$ and for the torque given by $\tau(\theta) = -\partial F/\partial\theta \sim \partial H_{c2}/\partial\theta$. A model for mixed $d+s$ wave superconductivity was also proposed by Maki and Beal–Monod [10,11] which is characterized by a gap function of the form $\Delta(\mathbf{k}) = \Delta\,[\cos(2\varphi) + r]$, where r measures the amount of the s-wave contribution and would vanish in the absence of orthorhombic distortion. Calculating $H_{c2}(\theta)$ for B parallel to the a-b plane, they derived the torque angular dependence [10]

$$\tau(\theta) = -BH_{c2}(\pi/4,t)(8\pi\kappa^2\beta_A)^{-1} \times \ln(\eta H_{c2}(\theta,t)/B)X\sin(2\theta) \times [1+X\cos(2\theta)]^{-3/2}, \quad (1)$$

where η is a constant of order unity, κ and β_A are the G–L and the Abrikosov parameters, respectively. Here X is a measure of the degree of anisotropy and mixing between the s- and d-wave components. The in-plane anisotropy of H_{c2} is given by $H^b_{c2}/H^a_{c2} = \sqrt{(1+X)/(1-X)}$.

0921-4534/97/$17.00 Elsevier Science B.V.
PII S0921-4534(97)00241-4

In tetragonal Tl2201 one would in principle expect only a fourfold symmetry component, owing to the four-leaf-clover shape of the order parameter. To test this point and eliminate any artifacts due to structure and sample geometry, we have performed in-plane torque measurements on high-quality c-axis Tl2201 films grown on $SrTiO_3$ (001) by rf magnetron sputtering [12] and shaped into disks (2 mm diameter). The structural tetragonal symmetry of the films at room temperature was carefully checked [12]. Two films, 200 and 500 nm thick, with respective transition temperatures $T_c = 82.5$ and 69 K were measured using a capacitive torquemeter based on thin (20 μm) Si cantilevers (see inset of Fig. 1) with sensitivity $\Delta\tau \simeq 10^{-12}$ Nm. Particular attention was paid to avoid any out-of-plane precession of the rotating horizontal field B. A specific procedure was developed to maintain the field misalignment within 0.07° and to cancel all non-planar contributions, such as large hystereses due to vortex pinning along the c-axis [13]. The resulting torque, measured by increasing ($\tau^+(\theta)$) or decreasing ($\tau^-(\theta)$) the angle θ, is found to be fully reversible and nonzero. A typical reversible torque curve $\tau_{rev}(\theta) = [\tau^+(\theta) + \tau^-(\theta)]/2$ measured at 37.7 K with $B = 0.3$ T on the 500-nm-thick film is shown in Fig. 1. In principle such an

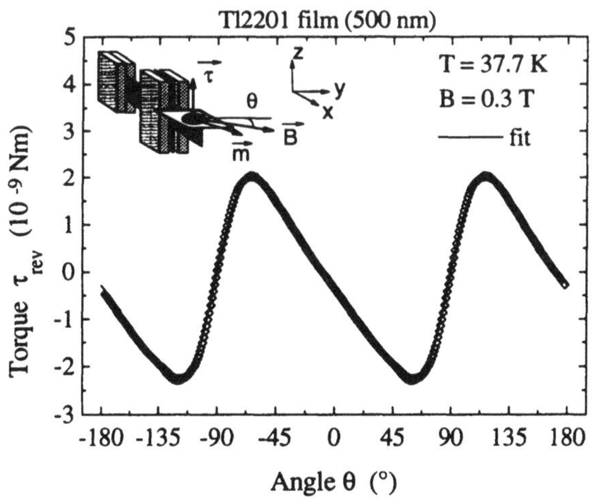

Figure 1. In-plane reversible torque $\tau_{rev}(\theta)$ of a 500-nm-thick Tl2201 film, shaped into a disk, upon rotation of the external field B around the c-axis. Solid line: fit to the data with Eq. (1).

angular dependence can be fitted with an expression like $\tau(\theta) = \tau_0 + \tau_2 \sin 2(\theta - \theta_2) + \tau_4 \sin 4(\theta - \theta_4)$, where τ_2, τ_4 are the amplitudes and θ_2, θ_4 the phases of the two- and fourfold components, respectively [8,9]. The general trend of the ratio τ_4/τ_2 is a monotonous increase with temperature. It reflects the progressive dominance of the fourfold term as T_c is approached.

We found that the above expression for $\tau(\theta)$ is less suited to fit our data than Eq. (1) derived for a d+s wave superconductor [10]. An excellent fit with Eq. (1) is shown in Fig. 1, which yields the value $X = 0.427$. Measurements made on the two Tl2201 films at different temperatures give the monotonic increase of $X(t)$ shown in Fig. 2(a), where $t = T/T_c$. The similar behavior for the two films having different thicknesses and T_c's demonstrates the reproducibility of the experiment. The upper critical field anisotropy ratio H^b_{c2}/H^a_{c2} plotted in Fig. 2(b) is found to increase by 55% in the measured temperature range $0.25 \leq T/T_c \leq 0.65$.

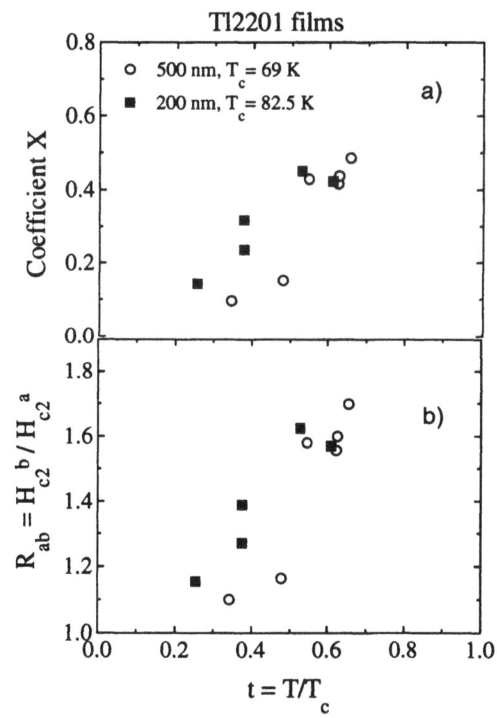

Figure 2. Temperature dependence (a) of the coefficient X (see Eq. (1)) and (b) of the upper critical field anisotropy ratio H^b_{c2}/H^a_{c2}.

In order to check the validity of the data on Tl2201 and to rule out the artifactual origin of the twofold component, we have measured two classical superconductors under the same conditions: one is the face-centered cubic NbN material grown as a 200-nm thin film on Si(100) with $T_c = 15.7$ K and shaped into disks. The other is a single crystal of layered hexagonal 2H-$NbSe_2$ with $T_c = 7.2$ K. In both cases no anisotropy is found in $\tau(\theta)$, in agreement with the expected s-wave pairing for these compounds. As an example, the data for the NbN disk are shown in Fig. 3. Without the alignment correction of the applied rotating field, one observes a large hysteresis between $\tau^+(\theta)$ and $\tau^-(\theta)$ due to the remaining pinning contribution along the z-direction (Fig. 3a). This misalignment effect has also been observed in all Tl2201 film measurements. The resulting curves after proper correction are almost flat, having only little structure left (Fig. 3b), in such a way that the derived $\tau_{rev}(\theta)$ is approximately zero (Fig. 3c). This holds true in the entire temperature range investigated.

Our results on Tl2201 thus point towards the existence of a mixing of the s- and d-wave pairing states (s+d), which is surprising in view of the tetragonality of its structure. In addition to the expected fourfold symmetry, an additional term modifies the angular dependence of $\tau_{rev}(\theta)$. As we have proven through the measurements on 2H-$NbSe_2$ and NbN that all non-in-plane contributions and geometrical effects can be fully cancelled, this term must arise from an intrinsic admixture of the s- and d-wave pair states. One explanation for the mixing in Tl2201 could be the occurrence of a low-temperature structural transition. It is reasonable to expect that the presence of an orthorhombic distortion above T_c already influences the normal-state properties [10,11]. Another possibility could be the breaking of the local symmetry through thermal fluctuations, disorder or strains present in the material. The field anisotropy $R_{ab} = H^b_{c2}/H^a_{c2}$ that we derived from our data increases between 1.1 and 1.7 in the measured temperature range. The relation between field anisotropy and effective mass anisotropy given by $\gamma_{ab} = \sqrt{m^*_a/m^*_b} = \lambda_a/\lambda_b$ is not straightforward. Assuming an isotropic Fermi

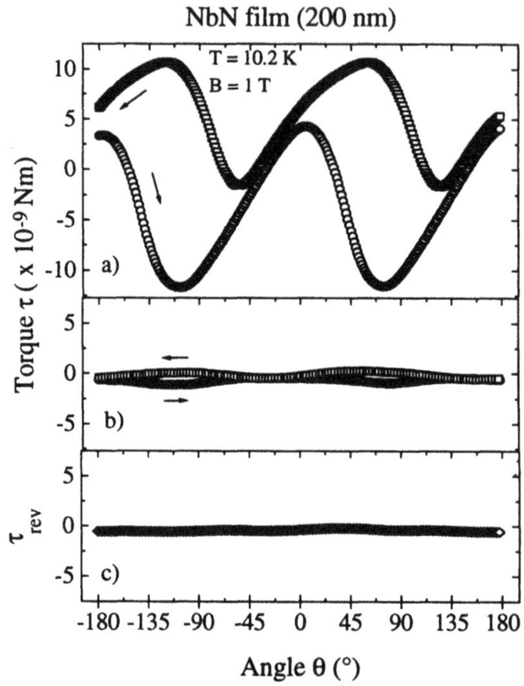

Figure 3. Torque angular dependence $\tau(\theta)$ of a NbN film (200 nm) patterned into a disk, upon rotation of the field $B = 1$ T around the vertical z-axis at 10.2 K. (a) Before proper in-plane alignment of B, (b) after alignment corrections, and (c) reversible torque $\tau_{rev}(\theta)$.

velocity in the a-b plane and an anisotropic gap Δ, one can approximately write $\Delta_b/\Delta_a \simeq \sqrt{R_{ab}}$, which thus takes the experimental values 1.05 to 1.30. These values seem reasonable and comparatively smaller than $\gamma_{ab} = 1.55$ reported for YBCO [14]. In Tl2201 the a- and b-axes are in principle not distinguishable. The monotonic increase of $R_{ab}(T)$ is not yet clear and cannot be simply explained in terms of ratio between the d- and s-wave components. A detailed X-ray study of the presumed orthorhombic distortion of the tetragonal lattice at low temperature is still needed. The a-c anisotropy was also measured in these films. By analyzing the reversible torque obtained close to T_c with a 3D London anisotropic model [15] we determined the effective mass ratio to be $\gamma_{ac} \simeq 40$ [13].

In conclusion, we found a clear in-plane anisotropy of the torque in Tl2201 films, which supports the idea of a real $s+d$-wave superconducting state. The most plausible candidate seems to be an admixture of a $d_{x^2-y^2}$ component with a s^+ one [1]. Our result is confirmed by the comparison we made to two classical superconductors, but the detailed interpretation remains dependent on the model used. The exact origin of the orthorhombic distortion in Tl2201, if really needed for a mixed pairing state, is still to be clarified.

Acknowledgments

We gratefully acknowledge stimulating discussions with K. Maki, T. M. Rice, R. Griessen, K. A. Müller, D. Ariosa, and T. Schneider.

REFERENCES

1. For a review, see J. Annett, N. Goldenfeld and A.J. Leggett, Physical Properties of High Temperature Superconductors, Vol. 5, D.M. Ginsberg (ed.) (World Scientific, Singapore, 1996)
2. D.J. Scalapino, Physics Reports **250** (1995) 329
3. K.A. Müller, Nature **377** (1995) 133
4. C.C. Tsuei et al., Science **271** (1996) 329
5. J. Buan et al., Phys. Rev. Lett. **72** (1994) 2632
6. Y. Koike et al., Phys. Rev. B **54** (1996) R776
7. T. Hanaguri et al., Physica B **165** (1995) 1449
8. T. Ishida et al., Physica C **263** (1996) 260 and Czech. J. Phys. **46** (1996) 1217
9. H. Won and K. Maki, Europhys. Lett. **34** (1996) 453
10. K. Maki and M.T. Beal–Monod, to appear in Phys. Rev. B
11. M.T. Beal–Monod and K. Maki, Phys. Rev. B **53** (1996) 5775 and Europhys. Lett. **33** (1995) 309
12. Z.F. Ren, J.H. Wang and D.J. Miller, Appl. Phys. Lett. **69** (1996) 1798
13. M. Willemin et al., in preparation
14. K. Zhang et al., Phys. Rev. Lett. **73** (1994) 2484
15. V. G. Kogan, Phys. Rev. B **38** (1988) 7049

ELSEVIER

Physica C 282–287 (1997) 140–144

PHYSICA C

Andreev reflections from $La_{2-x}Sr_xCuO_4$ single crystals

Guy Deutscher*, N.Achsaf*,D.Goldschmidt*, A.Revcolevschi+and A.Vietkine+
*School of Physics and Astronomy, Tel Aviv University, 69978 Ramat Aviv, Israel
+Laboratoire de Chimie des Solides, Universite Paris-Sud, 91405 Orsay, France

We have measured the conductance of Point Contacts between LSCO single crystals and Gold tips. The initial increase of the conductance as a function of applied bias rules out the presence of nodes in the gap. The data is best fitted by an extended s-wave symmetry gap, with a ratio of about three between its extrema values.

1 Introduction

Because of its simple structure, and the possibility to dope it at any wished level, $La_{2-x}Sr_xCu_4O$ (LSCO) is of special importance for our understanding of the fundamental properties of the High Tc cuprates. This is particularly true in reference to the symmetry of the order parameter: the fact that in this compound there is only one CuO plane per unit cell, and no chains, precludes all sorts of complications that may obscure the basic symmetry, itself linked, it is generally assumed, to the high Tc mechanism. However, experiments conducted in recent years to determine the gap symmetry, that have largely concluded to a d-wave symmetry, have been conducted mostly on other compounds. This is particularly true of interference experiments: none has been performed, to the best of our knowledge, on LSCO.

In this contribution, we wish to expand on a previous work (1), where we reported that Andreev reflections data obtained on a LSCO single crystal, could not be fitted to a d-wave symmetry gap. In the present contribution we show that experiments, using Sharvin contacts, have more generally the capability to establish the presence or absence of nodes in the gap, as well as their position. Our conclusion is, in fact, that there are no nodes in the gap of LSCO, not just that the data cannot be fitted to a d-wave symmetry. This conclusion holds for optimally doped, as well as overdoped and underdoped samples.

It was first pointed out by Hu (2), that in the case of a d-wave symmetry gap, surface bound states are expected at the Fermi energy, when the surface is perpendicular to the [110] direction. Tanaka (3) extended this result, and showed that for all surface orientations - except for surfaces perpendicular to the [100] orientation - such surface bound states exist, and that they lead to an enhancement of the zero bias conductance. Furthermore, expanding on the classical work of Blonder, Tinkham and Klapwijk (4), he calculated the conductance for arbitrary interface barriers. A central result of his calculations is that, in the case of a d-wave symmetry, the conductance is in general, a *de-*

PII S0921-4534(97)00242-6

creasing function of the applied bias. This result is in contrast with that of BTK. They predict that in the case of an s-wave symmetry gap, for any finite interface barrier, the conductance is first an *increasing* function of the applied bias, before going through a maximum and then returning to its normal state value.Thus, the initial bias dependence of the conductance allows to distinguish between a d-wave and a s-wave gap. We shall show here that the significance of the bias dependence is in fact more general, and that it reflects the presence, or the absence, of nodes in the gap, rather than its symmetry.

2 EXPERIMENTAL

We have studied the properties of contacts between a gold point and LSCO single crystals, whose surface is perpendicular to the [110] direction. We have studied particularly contacts that are in the Sharvin limit, i.e. whose conductance in the normal state is limited by the number of independent transport quantum channels, rather than by an interface tunneling barrier. The usual reason for using Sharvin contacts is to avoid heating effects in the contact region. A more specific one in our case is that in the presence of a tunneling barrier, the transmission coefficient has a strong angular dependence, decreasing exponentially for angles off the perpendicular orientation. This angular dependence was not taken into account by Tanaka (3). Instead of probing the gap all around the Fermi surface, the experiment would then sense the gap only over a small fraction of this surface. This would prevent probing bound states that occur precisely because of interference effects between gap values that change sign around the Fermi surface, as in the case of a d-wave symmetry. An additional reason for our preference for barrier-free Sharvin contacts, is that they have a better chance of being clean of magnetic impurities, that may by themselves be at the origin of zero-bias anomalies.

The cleaning procedure for obtaining low resistance contacts has been described elsewhere (1). The qualifying condition for the contact to be in the Sharvin limit is that its size should be smaller than the mean free path in the electrodes. At low temperatures (4.2K), where all measurements were taken, this mean free path has been estimated to be as large as 0.5 micron (5). The Au tip itself is of micron-size, but the actual contact size is usually much smaller. We cannot measure it directly,but we can estimate it by noting that, when the ratio of the conductances at small and large bias is close to the ideal Andreev value of 2, the barrier is very weak, in which case the conductance of the contact is close to the universal conductance ($\frac{e^2}{\hbar}$), times the number of channels, which is on the order of $(k_F d)^2$, where k_F is the Fermi momentum in the electrode that has the smaller one, and d is the contact size. Since the cuprates have a rather large Fermi surface, we estimate the number of channels to be on the order of $(\pi d/a)^2$, where a is the lattice paremeter. We conclude that contacts that are close to the ideal Andreev behavior, and whose resistance values are more than 10Ω, have sizes less than about 100 A, and thus qualify as Sharvin contacts. It is safe to assume that the principles of Tanaka's calculation (3) apply to them, with the entire gap structure around the Fermi surface being probed.

3 RESULTS

All conductance characteristics show an initial rise, a maximum at a voltage of 3 to 5 meV, with a slow return to the normal state -high voltage value, completed at 10 to 15 meV. The ratio of the zero bias to the high bias conductance is in the range 1.8 to 1.9 for many contacts.

Although the general shape of the characteristics is in accordance with the predictions of BTK, two of their features do not allow a quantitative fit to this model: the conductance peak is too wide, and the return to the normal state conductance is too slow. If, for instance, one determines the gap value from the peak position, and the barrier parameter from the ratio of the zero bias to high bias conductances, the fit is rather poor (Fig.1). Thus, the data cannot be fitted quantitatively to an isotropic gap.

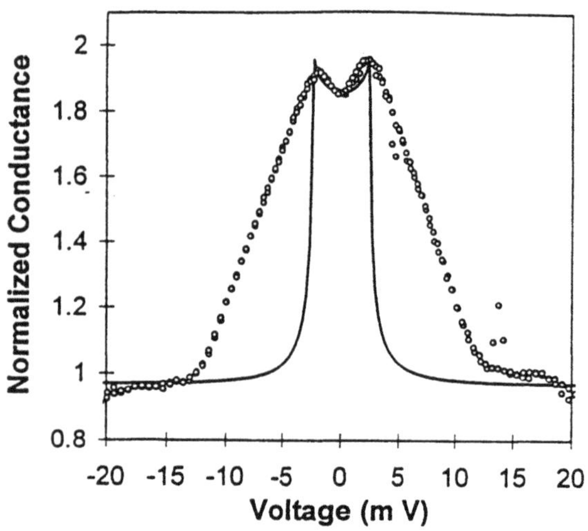

Fig.1.s-wave BTK symmetry fit

If one now tries to fit the data to a d-wave symmetry gap, one immediatly encounters the problem that the low bias conductance behavior is *qualitatively wrong:* for the [110]orientation investigated, the conductance should initially decrease, while it increases. This conclusion remains unchanged if the orientation is not exactly [110], which may very well be the case since in fact we do not know the exact orientation of the actual contact area, which as explained above is significantly smaller than the size of the Au tip. The only case where there can be an initial rise of the conductance with a d-wave gap, is if the orientation is *exactly* [100], which is quite unlikely in our case since the crystal surface was cut perpendicular to the [110] axis. And even if this were the case, an initial conductance increase is only possible if the low bias to high bias conductance ratio is smaller than 1.5, which is in contradiction with the data shown. And even disregarding this disagreement, the conductance peak should occur near the voltage value where the conductance returns to its normal state value, not at a much smaller value as seen in the data. The best possible fit is obtained by assuming a nearly perfect interface (Fig.2). The high voltage fit is then reasonable, but of course not the small bias one. Thus, there is no way that the data shown, which is representative of a large number of measurements, can be fitted to a d-wave symmetry gap.

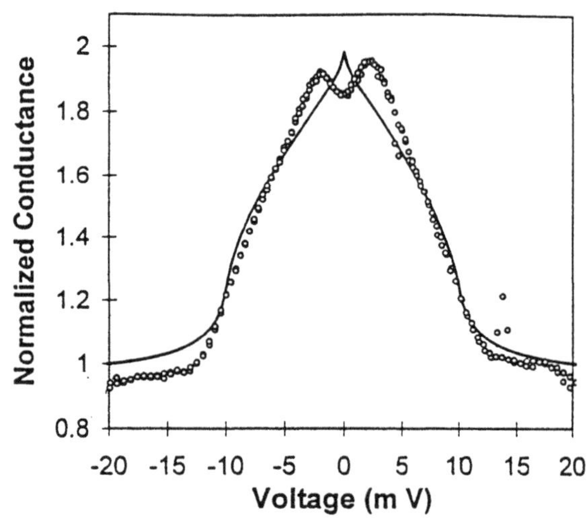

Fig.2 d-wave symmetry fit.

4 Discussion

One can even exclude, more generally, the presence of nodes in the gap. For instance, we have used Tanaka's formalism to calculate

s-wave symmetry gap, with nodes. Again, for most orientations and arbitrary values of the interface barrier parameter, an initial decrease of the conductance is obtained (Fig.3). This is not surprising, since the surface bound states arise directly from the interference between gaps of opposite signs around the Fermi surface, and not specifically from the d-wave symmetry. In fact, our data excludes for LSCO all gap structures with nodes, including for instance a (s+d) gap with a dominant d part.

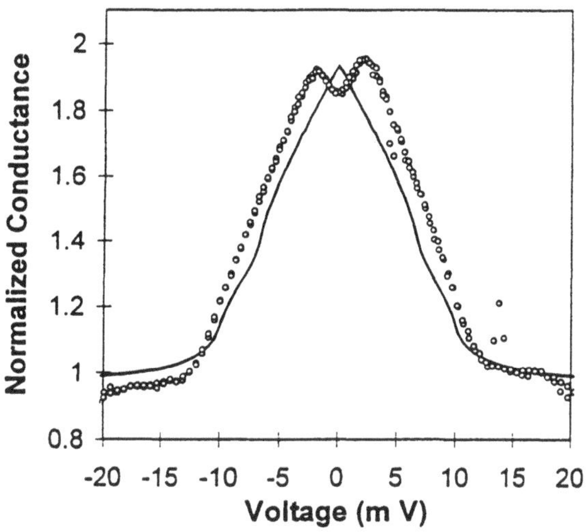

Fig.3. Fit to an extended s-wave gap with nodes $\Delta = \Delta_0 + \Delta_1 \cos 4\theta, \Delta_0 = 0.2\Delta_1$

We have tried to fit our data with two gap forms: an extended s-wave form $\Delta(\theta) = \Delta_0 + \Delta_1 \cos 4\theta$,with $\Delta_0 > \Delta_1$; and a (s+d) gap, $\Delta(\theta) = \Delta_0 + \Delta_1 \cos 2\theta$, again with the same condition. An excellent fit can be obtained for both forms, with however an important difference. The good fit with the extended s form is obtained assuming that the surface of the sample is perpendicular to the [110] direction, as is the case experimentally; while the fit to the (s+d) form can only be obtained if one assumes that the surface is cut at a $\pi/8$ angle, rather than at $\pi/4$ as is the case. This shows how the fit is sensitive to the detailed form of the gap. We cannot exclude the (s+d) form, although the extended s form is in better agreement with the actual cut of the sample.

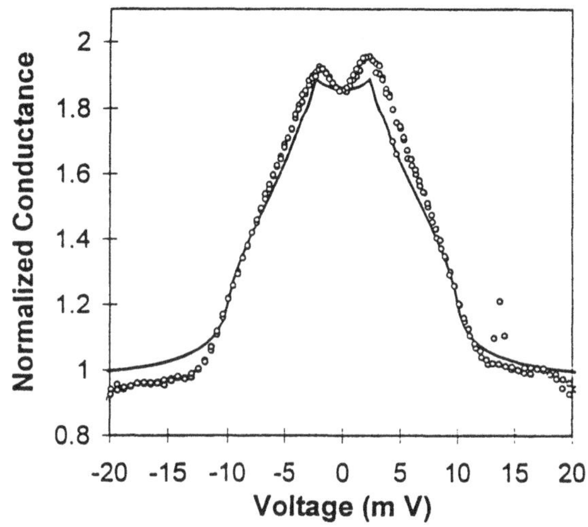

Fig.4. Fit to an (s+d) gap, with a dominant s part

We have performed some measurements on a number of samples of different compositions, from underdoped to overdoped. A more detailed account of these measurements will be given elsewhere, but we have not noted any drastic change in the shape of the characteristics as a function of composition. In particular, the conductance peak is seen at all compositions, at roughly the same voltage.

At optimum doping, we find an anisotropy ratio between the maximum and minimum gap values of about 3. This is compatible with the weak screening interpretation of the anisotropy (6), if the screening parameter $(q_0 a)$ is in the range 0.1 to 0.2 (7), where q_0 is the inverse screening length. Such a small value is quite interesting, as it introduces a fundamental difference between the cuprate and usual metals. A possible interpretation for it is the quasi-two dimensional character of the cuprates, although at the moment we lack first principle calculations to which this value of $(q_0 a)$ can be compared.

5 CONCLUSIONS

The absence of nodes in LSCO poses evidently a serious problem for our general comprehension of the cuprates. One would be strongly tempted to state that the gap symmetry should be universal amongst them. It is true that there is so far no evidence for a d-wave symmetry in the electron-doped compounds, but at least one would expect a universal symmetry amongst the hole-doped ones. We can see only two possibilities to reconcile our result on LaSrCuO with those obtained on other hole doped curprates, if we wish to conserve universality. The first is that the general symmetry is extended s, without or with nodes. The difficulty with this hypothesis is that the geometries used in the interference experiments that show the existence of nodes in the other cuprates, fit a d-symmetry, rather than an extended s-symmetry (8,9). The other possibility is that of a (s+d) combination. Our own experiment does not rule it out, although an extended s-symmetry is more in keeping with our geometry. If this were to be the right answer, the lower Tc cuprates would have a dominant s component (no nodes), and the higher Tc ones a dominant d component.More work is needed to clarify the situation, involving for instance Sharvin contact interference experiments on higher Tc cuprates, and/or "conventional" multi-junction interference experiments on LSCO crystals.

This work was supported in part by the Israel National Science Foundation, the Heinrich Hertz - Minerva Center for High Temperature Superconductivity, and the Oren Family Chair for Experimental Solid State Physics.

REFERENCES

1)N.Achsaf et al., Journal of Low Temperature Physics 105 (1996)329

2)C.R.Hu,Phys.Rev.Lett.72(1994)1526

3)Y.Tanaka and S.Kashiwaya, Phys.Rev.Lett.74(1995)3451

4)G.E.Blonder, M.Tinkham and T.M.Klapwijk, Phys.Rev.B25(1982)4515

5)J.M.Harris et al.,Journal of Superconductivity,105(1996)877

6)See J.Bok and J.Bouvier in these Proceedings.

7)J.Bouvier, G.Deutscher and J.Bok, to be published

8)D.Van Harlingen, Rev.Mod.Phys.67(1995)515

9)C.C.Tsuei et al., Science 271(1996)288

ELSEVIER

Physica C 282–287 (1997) 145–148

PHYSICA C

Anisotropic Penetration Depth Measurements of High-T_c Superconductors

C. Panagopoulos[a], J.R. Cooper[a], T. Xiang[a], G.B. Peacock[b], I. Gameson[b], P.P. Edwards[b], W. Schmidbauer[c] and J.W. Hodby[c]

[a]IRC in Superconductivity, University of Cambridge, Madingley Road, Cambridge CB3 OHE, UK

[b]School of Chemistry, University of Birmingham, Edgbaston, Birmingham, B15 2TT, UK

[c]Clarendon Laboratory, University of Oxford, Parks Road, Oxford, OX1 3PU, UK

A systematic study of the anisotropic magnetic penetration depth λ for several high-T_c superconductors is presented. It is shown that the existence of the linear term in the low temperature in-plane penetration depth is independent of the number of CuO_2 planes per unit cell, the carrier concentration and the anisotropy. The anisotropy, however, is found to have strong effects in the out-of-plane component. The effects of Zn substitution on λ and the importance of the CuO_2 planes for $d_{x^2-y^2}$ pairing are discussed.

1. INTRODUCTION

The high-T_c superconductors (HTSC) containing CuO_2 planes are layered materials with anisotropic physical properties. Evidence is now growing that the energy gap in HTSC has "*d*-wave" symmetry with nodes in the order parameter for certain directions in ***k*** space [1]. Detailed knowledge of the behaviour of the superfluid density and the symmetry of the energy gap are important for understanding various properties of the superconducting state and the pairing mechanism in these materials. One of the best probes of the superfluid density and the energy gap morphology at the Fermi surface is the penetration depth, λ.

Here we present our recent experimental results for the T dependence and the absolute values of the in-plane, λ_{ab}, and out-of-plane, λ_c, penetration depths for high quality grain-aligned $HgBa_2CuO_{4+\delta}$, $HgBa_2Ca_2Cu_3O_{8+\delta}$, $YBa_2Cu_3O_{7-\delta}$ (δ = 0.0, 0.30 and 0.43) and $YBa_2(Cu_{1-x}Zn_x)_3O_7$ (x = 0.00, 0.02, 0.03 and 0.05) which were all obtained using the low field ac-susceptibility technique. The isotropic perovskite oxide superconductor $Ba_{0.6}K_{0.4}BiO_3$ was also studied for comparison.

2. EXPERIMENTAL

Details of sample preparation and characterisation can be found in Refs [2-6]. Special care was taken to obtain grains with clean surfaces after grinding [7]. Low field ac susceptibility measurements were performed in the Meissner state over a wide temperature range, 1.2 K $\leq T \leq$ 150 K, using both commercial equipment and a home built susceptometer. The data were analysed using London's model [8], and the variation of λ with temperature was obtained for the applied field parallel and perpendicular to the *ab* plane [4, 5].

3. RESULTS AND DISCUSSION

3.1 $HgBa_2CuO_{4+\delta}$ (Hg-1201)

The T dependencies of λ_{ab} and λ_c for Hg-1201, which has one CuO_2 plane per unit cell, are shown in Figs. 1(a) and 1(b). The linear term in $\lambda_{ab}(T)$, characteristic of *d*-wave superconductivity, has a slope of 6.5 Å/K. In contrast the data for $\lambda_c(T)$ are much flatter at low T. The $\lambda_c(T)$ data fit the expression suggested by Xiang and Wheatley for tetragonal HTSC [9]: $\lambda_c(T) \sim \lambda_c(0)\,[1+\alpha(T/T_c)^5]$, with $\alpha \sim 3$. The parameter α depends strongly on the ratio of the superconducting transition temperature to the zero temperature energy gap, $\alpha =$

0921-4534/97/$17.00
PII S0921-4534(97)00244-X

$225(T_c/\Delta_0)^5$. The experimental value of α yields $(\Delta_0/T_c) \sim 2.37$ which is very close to the weak coupling BCS value, $(\Delta_0/T_c) = 2.14$, for a *d*-wave superconductor. The weaker T dependence of λ_c compared to λ_{ab} is a consequence of the unusual *c*-axis electronic structure of high-T_c oxides. In clean tetragonal samples, the *c*-axis hopping of holes mainly occurs through large radius Cu 4*s* orbitals [10]. In this case, the *c*-axis hopping constant is a function of the in-plane momentum $k_{//} = (k_x, k_y)$ and vanishes at $k_x = k_y$. This unusual $k_{//}$ dependence of the *c*-axis hopping constant leads to the T^5 dependence in λ_c [9].

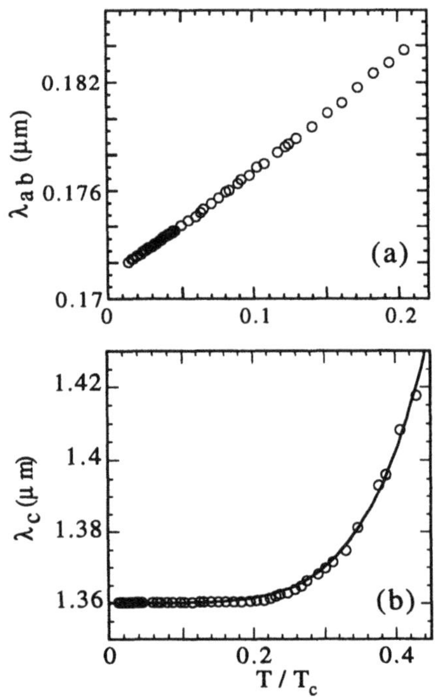

Figure 1. λ_{ab} (a) and λ_c (b) versus T/T_c (T_c = 93 K) for Hg-1201. λ_{ab} (0) ≅ 0.171 μm and λ_c (0) ≅ 1.36 μm. The solid line in (b) shows the T^5 behaviour predicted theoretically [9].

3.2. $HgBa_2Ca_2Cu_3O_{8+\delta}$ (Hg-1223)

Figures 2(a) and (b) show the T dependencies of λ_{ab} and λ_c for Hg-1223, which has three CuO_2 planes per unit cell. The linear term in $\lambda_{ab}(T)$ has a slope of 4.2 Å/K. $\lambda_c(T)$ is much flatter than $\lambda_{ab}(T)$ and depends quadratically on T. This T^2 behaviour could be due to the high anisotropy of Hg-1223. It has been shown theoretically [9] that for weakly coupled planes, disorder can give rise to a T^2 term in $\lambda_c(T)$ without significantly affecting the T^1 term in $\lambda_{ab}(T)$. However, a careful analysis is required to distinguish this behaviour from Josephson coupling effects [4].

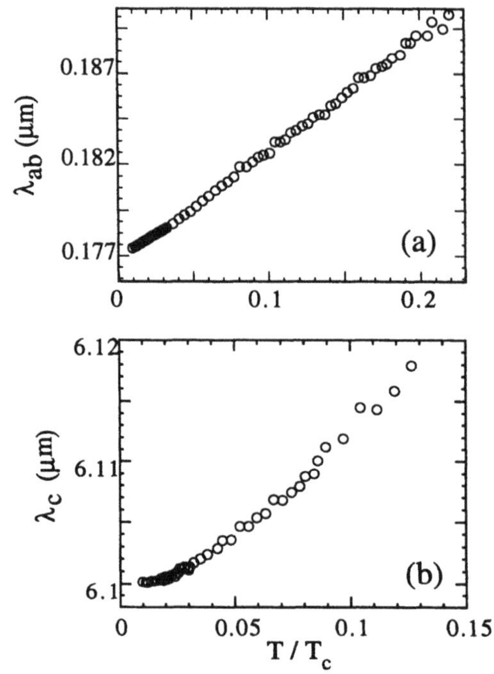

Figure 2. λ_{ab} (a) and λ_c (b) versus T/T_c (T_c = 134.5 K) for Hg-1223. λ_{ab} (0) ≅ 0.177 μm and λ_c (0) ≅ 6.1 μm.

3.3. $YBa_2Cu_3O_{7-\delta}$ (δ = 0.0, 0.3 and 0.43)

The effects of oxygen concentration on the anisotropic magnetic penetration depth of $YBa_2Cu_3O_{7-\delta}$ with $\delta = 0.0$ ($YBCO_7$, T_c = 92.5 K), 0.3 ($YBCO_{6.7}$, T_c = 66 K), and 0.43 ($YBCO_{6.57}$, T_c = 55 K), were also studied [Figs. 3(a) and (b)]. $\lambda_{ab}(0)$, $\lambda_c(0)$ and the anisotropy ratio $\gamma = [\lambda_c(0) / \lambda_{ab}(0)]$ were found to increase with δ.

$\lambda_{ab}(0) \cong 0.14$, 0.21, 0.29 μm and $\lambda_c(0) \cong 1.26$, 4.53, 7.17 μm for δ = 0.0, 0.3, and 0.43, respectively. This is in contrast to previous work [11] in which the surfaces of the particles were probably not as clean and the degree of grain alignment was lower.

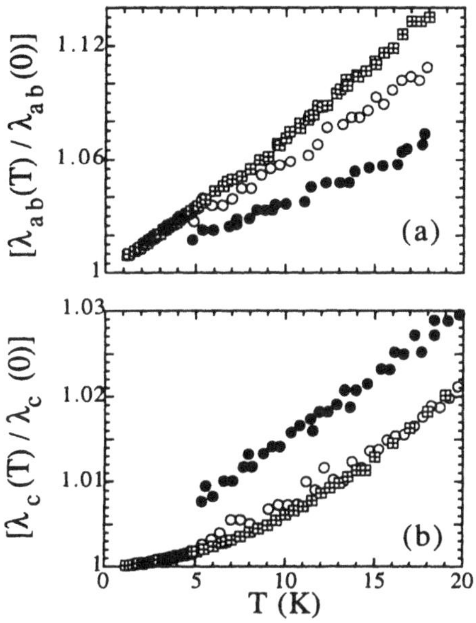

Figure 3. $[\lambda_{ab}(T) / \lambda_{ab}(0)]$ (a) and $[\lambda_c(T) / \lambda_c(0)]$ (b) versus T for $YBCO_7$(closed circles), $YBCO_{6.7}$ (open circles) and $YBCO_{6.57}$ (crossed squares). The T_c, $\lambda_{ab}(0)$ and $\lambda_c(0)$ values are given in the text.

For the above values of δ, both $\lambda_{ab}(T)$ and $\lambda_c(T)$ follow the behaviour expected for a superconductor with nodes in the energy gap. The linear term in $\lambda_{ab}(T)$ increases with oxygen deficiency, namely, from 4.5 to 12 and 20 Å/K, for δ = 0.0, 0.3 and 0.43, respectively. For the less anisotropic $YBCO_7$ ($\gamma \sim 6.5$, using $\lambda_{ab}(0) = 0.187$ μm, i.e., the value from the planes only [5]), we observe a linear T dependence in λ_c at low temperatures but the relative change is about a factor of two smaller than in $[\lambda_{ab}(T) / \lambda_{ab}(0)]$. This probably arises from the effects of hopping between planar and chain bands [12]. By removing oxygen from $YBCO_7$ one can reduce the effects of the chains. $\lambda_c(T)$ of $YBCO_{6.7}$ ($\gamma \sim 22$) and $YBCO_{6.57}$ ($\gamma \sim 25$) vary as T^2, like Hg-1223 which has similar anisotropy [9,12].

3.4. $YBa_2(Cu_{1-x}Zn_x)_3O_7$ (x = 0.00, 0.02, 0.03 and 0.05)

Zinc substitution is of much interest in the study of the physical properties of the cuprates because it replaces the planar Cu atoms and rapidly suppresses T_c. Our $\lambda_{ab}(T)$ and $\lambda_c(T)$ results for $YBa_2(Cu_{1-x}Zn_x)_3O_7$ with x = 0.00, 0.02, 0.03 and 0.05 are shown in Figs. 4(a) and (b).

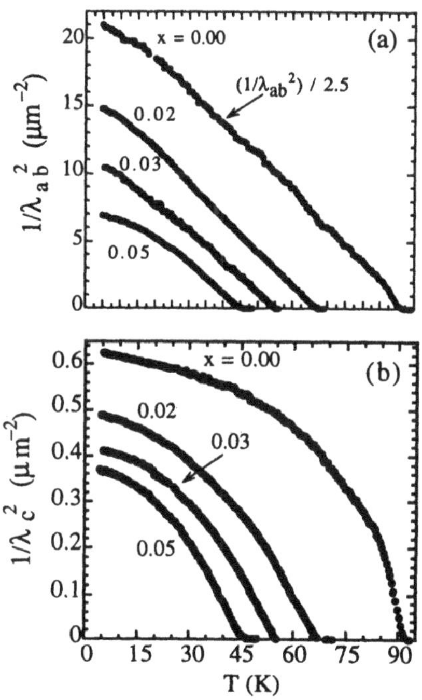

Figure 4. $(1 / \lambda_{ab})^2$ (a) and $(1 / \lambda_c)^2$ (b) versus T for $YBa_2(Cu_{1-x}Zn_x)_3O_7$ with x = 0.00, 0.02, 0.03 and 0.05. The $(1 / \lambda_{ab})^2$ data for x = 0.00 in (a) have been divided by 2.5 for clarity. The $\lambda_{ab}(0)$ and $\lambda_c(0)$ values are given in the text.

The $\lambda_{ab}(0)$ and $\lambda_c(0)$ values were found to increase strongly with increased Zn concentration whereas γ decreased slightly from 9 to 4.4.

Namely, $\lambda_{ab}(0) \cong 0.14, 0.26, 0.30, 0.37\ \mu m$ and $\lambda_c(0) \cong 1.26, 1.42, 1.55, 1.64\ \mu m$ for $x = 0.00$, 0.02, 0.03 and 0.05, respectively. A fast increase of $\lambda_{ab}(0)$ with Zn is consistent with an increase in the residual density of states [5,13]. The slower increase in $\lambda_c(0)$ may be due to an enhanced c-axis coupling caused by Zn doping. The low T dependencies of $\lambda_{ab}(T)$ and $\lambda_c(T)$ change from T to T^2 with the addition of Zn due to impurity scattering. For $x > 0$ both $[\lambda_{ab}(0) / \lambda_{ab}(T)]^2$ and $[\lambda_c(0) / \lambda_c(T)]^2$ gradually approach each other and for $x = 0.05$ they have almost the same behaviour. This suggests that Zn doping progressively smears out the anisotropy of the $d_{x^2-y^2}$-wave gap.

3.5. $Ba_{0.6}K_{0.4}BiO_3$ (BKBO)

It is interesting to make a comparison between the above HTSC cuprates and the isotropic oxide superconductor BKBO. $\lambda(0)$ of BKBO was found to be 0.35 μm. The temperature dependence of the normalised superfluid density of BKBO resembles that expected from weak-coupling BCS theory for s-wave superconductors [14], except near T_c where the deviation probably arises from thermodynamic fluctuations (Fig. 5) [4]. This difference between the high T_c cuprates and BKBO underlines the importance of the CuO_2 planes for $d_{x^2-y^2}$-wave superconductivity.

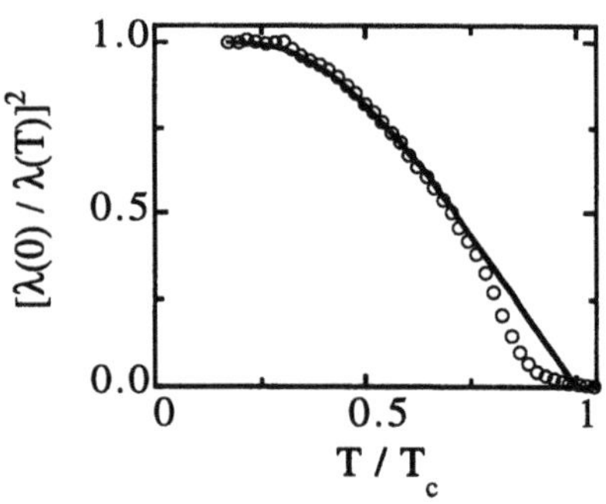

Figure 5. $[\lambda(0) / \lambda(T)]^2$ versus T/T_c ($T_c = 25$ K) for BKBO. Data (open circles); s-wave weak coupling BCS fit (solid line). $\lambda(0) \cong 0.35\ \mu m$

4. SUMMARY

In summary, we have found that the presence of the linear term in $\lambda_{ab}(T)$ is independent of the number of CuO_2 planes per unit cell, the carrier concentration and the anisotropy. It was shown experimentally that for HTSC with tetragonal symmetry and moderate anisotropy the Cu 4s orbital is important for the c-axis hopping of holes. Namely a T^5 dependence of $\lambda_c(T)$ was observed for Hg-1201 and a T^2 law for more anisotropic materials (Hg-1223, $YBCO_{6.7}$, $YBCO_{6.57}$). $YBCO_7$ shows a flatter but linear T dependence in $\lambda_c(T)$ at low T. Zinc substitution in $YBCO_7$ affects the anisotropy and the residual density of states. Finally, results for BKBO emphasise the importance of the CuO_2 planes for $d_{x^2-y^2}$ pairing symmetry.

REFERENCES

[1] For example, J.F. Annet, N. Goldenfeld, and A. Leggett, in *Physical Properties of High Temperature Superconductors* V edited by D.M. Ginsberg (World Scientific, Singapore, 1996), cond-mat / 9601060.
[2] G.B. Peacock, *et al.* (in preparation); G.B. Peacock *et al.*, Adv. Mater. (in press).
[3] W. Schmidbauer *et al.*, Physica C **235-240**, 759 (1994).
[4] C. Panagopoulos *et al.*, Phys. Rev. B **53**, R2222 (1996).
[5] C. Panagopoulos *et al.*, Phys. Rev. B **54**, R12721 (1996).
[6] J. Chrosch *et al.*, Physica C **265**, 233 (1996).
[7] C. Panagopoulos *et al.*, Physica C **269**, 157 (1996).
[8] D. Shoenberg, *Superconductivity* (Cambridge University Press, Cambridge, 1954), p.164.
[9] T. Xiang and J.M. Wheatley, Phys. Rev. Lett. **77**, 4632 (1996).
[10] O.K. Andersen *et al.*, J. Phys. Chem. Solids, **56** 1573 (1995).
[11] N. Athanassopoulou *et al.*, Physica C **259**, 326 (1996).
[12] T. Xiang and J.M. Wheatley, Phys. Rev. Lett. **76**, 134 (1996).
[13] J.W. Loram *et al.*, Physica C **171**, 243 (1990).
[14] B. Mühlschlegel, Z. Phys. **155**, 313 (1959).

ELSEVIER

Physica C 282–287 (1997) 149–152

PHYSICA C

Impurity effects on a 2-leg spin ladder compound $SrCu_2O_3$

M. Takano[a], M. Azuma[a], Y. Fujishiro[a], M. Nohara[b], H. Takagi[b], M. Fujiwara[b], H. Yasuoka[b], S. Ohsugi[c], Y. Kitaoka[c] and R. S. Eccleston[d]

[a]Institute for Chemical Research, Kyoto University, Uji, Kyoto-fu 611, Japan

[b]Institute for Solid State Physics, University of Tokyo, Roppongi Minato-ku, Tokyo 106, Japan

[c]Department of Material Physics, Faculty of Engineering Science, Osaka University, Toyonaka, Osaka 560, Japan

[d]ISIS Facility, Rutherford Appleton Laboratory, Chilton, Didcot, Oxfordshire OX10 OQX, United Kngdom

Measurements of magnetic susceptibility, specific heat and inelastic neutron scattering were performed on a Zn-substituted spin-1/2 Heisenberg 2-leg ladder compound $Sr(Cu_{1-x}Zn_x)_2O_3$ ($x \leq 0.08$) to investigate nonmagnetic impurity effects on the quantum spin system with a large spin gap of about ~ 400 K. A clear anomaly attributed to the onset of antiferromagnetic ordering was observed both in the magnetic susceptibility and specific heat data even at only 1 % of Zn. The Néel temperature showed a systematic Zn-concentration dependence with a broad maximum at around 4 % substitution (T_{Nmax} ~ 8K). A finite amount of T-linear term with a coefficient of γ ~ 3.5 mJ/K^2 was observed in the specific heat above T_N for $x = 0.02$ and 0.04, suggesting the existence of a finite density of states at $E = 0$ in the spin excitation spectrum as supported by the inelastic neutron scattering measurement.

It is well-known that the spin excitation spectrum of a one-dimensional (1D) quantum antiferromagnet reflects its structural feature and the spin degree of freedom. A uniform Heisenberg half-integer spin chain has a gapless spin excitation spectrum, while alternation in the intra-chain interaction leads to the opening of a spin gap as typically seen in a spin-Peierls material $CuGeO_3$ [1]. On the other hand, a uniform integer-spin chain has a "Haldane gap" as experimentally confirmed for $Ni(C_2H_8N_2)_2NO_2(ClO_4)$ (NENP) typically[2]. Recent theoretical [3] and experimental works have added another type of gapped quasi-1D antiferromagnets, that is spin-1/2 Heisenberg 2-leg ladders found in $SrCu_2O_3$ [3-8]. In $SrCu_2O_3$, the ladders made of antiferromagnetic (AF) Cu-O-Cu linear bonds are connected with each other spatially so that they form 2D Cu_2O_3 sheets [4], but the ladders are separated from each other magnetically because of the inter-ladder 90° Cu-O-Cu bond causing spin frustration at the interface. The existence of a wide spin gap of 420 K was found through measurements of magnetic susceptibility [5], $1/T_1$ of Cu NMR [6] and inelastic neutron scattering [7]. Contrarily, $Sr_2Cu_3O_5$ comprising 3-leg ladders was found to be gapless [5] and a magnetic long-range ordering was found to set in at 52 K by a μSR study [8].

An even-leg ladder is believed to have a short-range RVB ground state [9] with a spin correlation length of only a few lattice spacings. If so, an impurity ion would create a free spin-1/2 localized nearby, without affecting the gapped nature of the host, while an impurity-substituted odd-leg ladder would provide the free spin with a counter spin to form a new singlet pair because of its longer spin correlation length. We have studied how $SrCu_2O_3$ and $Sr_2Cu_3O_5$ are influenced by different kinds of impurity ions. To be reported

PII S0921-4534(97)00245-1

increases with Zn content up to $x = 0.04$ and then decreases as shown in Fig. 2 (b). Another point to be remarked is that a tiny but finite difference was observed between the ZFC and FC data below the temperature of anomaly. Specific heat C_p was therefore measured to further clarify the origin of the anomaly. The C_p/T data are plotted against T and T^2 (inset) for $x = 0$, 0.02 and 0.04 in Fig. 3. A λ-type anomaly was seen for every Zn-substituted sample around the temperature of magnetic anomaly, indicating that it is an AF transition, not a spin glass freezing. The transition temperatures estimated from the specific heat data are also plotted in Fig. 2 (b), which are in good agreement with the temperatures of the magnetic anomaly. The intrinsic nature of the magnetic ordering has been further confirmed by a microscopic measurement, NQR, very recently [11].

For $x = 0$, since the temperature range studied was much lower than the gap width, it is reasonable to assume that we measured only the lattice contribution (C_l) in the specific heat. Indeed, as shown in the inset of the figure, C_p/T decreased almost linearly toward zero with decreasing T^2 as expected from its gapped magnetic excitation spectrum. A small upturn at the lowest temperatures was assigned to a

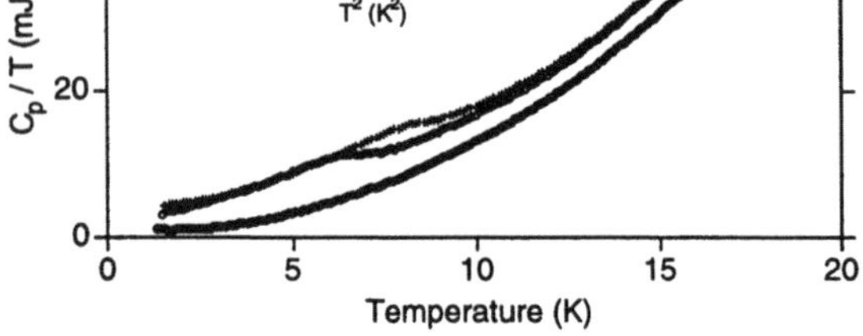

Figure 3. Total specific heat divided by temperature C_p/T vs$T$ and T^2 (inset) of $Sr(Cu_{1-x}Zn_x)_2O_3$ ($x = 0$, 0.02 and 0.04)

small amount of an unidentified impurity. The lattice term was fitted as $C_l = \beta_1 T^3 + \beta_2 T^5$ with $\beta_1 = 1.3 \times 10^{-1}$ mJ/K^4 Cu+Zn mole and $\beta_2 = 2.1 \times 10^{-5}$ mJ/K^6 Cu+Zn mole. The value of β_1 corresponds to a Debye temperature $\Theta_D \sim 400$ K. Another interesting point seen in the specific heat data is that, in the Zn doped samples, an almost parallel shift by 3.5 mJ/K^2 Cu+Zn mole was observed above the temperatures of the λ-like anomaly. This implies the appearance of a finite amount of T-linear term, indicating a finite density of states at $E = 0$ in the spin excitation spectrum. The coefficient, γ, is almost constant at 3.5 mJ/K^2 Cu+Zn mole for $x = 0.02$ and 0.04. This x-independent value of γ appears to suggest that the T-linear term originates from the spin ladder as a whole rather than from the spin degree of freedom localized near the Zn ions. It may be interesting to note that the presence of such a T-linear term is characteristic of isolated 1D Heisenberg AF system, where $\gamma = 2Nk_B{}^2/3J$ for a homogeneous isolated spin-1/2 chain. Applying this equation to the present case assuming N is equal to the number of Cu ions, rather than to that of Zn ions, a J/k_B value of about 1600K was obtained. Though the validity of this equation is not certified for the present ladder, the value of J appears reasonable in order of magnitude by comparing with those in other 1D Cu-O chain compounds, supporting the picture that the γT term arises from the whole ladder.

An inelastic neutron scattering with a TOF technique was performed to investigate the spin excitation spectrum of the impurity-introduced ladder directly [7]. Figure 4 (a) shows the magnitude of the spin gap for $x = 0$, 0.003, 0.006, 0.01, 0.02 and 0.04. The size of the gap is almost independent of Zn concentration. This might seem to be against the assertion that the Zn-substitution makes the ladder gapless. On the other hand, the integrated intensity of the magnetic scattering exhibits a different behavior. As shown in Fig. 5 (b), it decreases monotonically and the gap can hardly be seen at $x = 0.04$. Therefore, on account of the total-sum rule, we suggest that the in-gap density of states at $E = 0$ is finite and grows with increasing x. Note that this leads to the T-linear

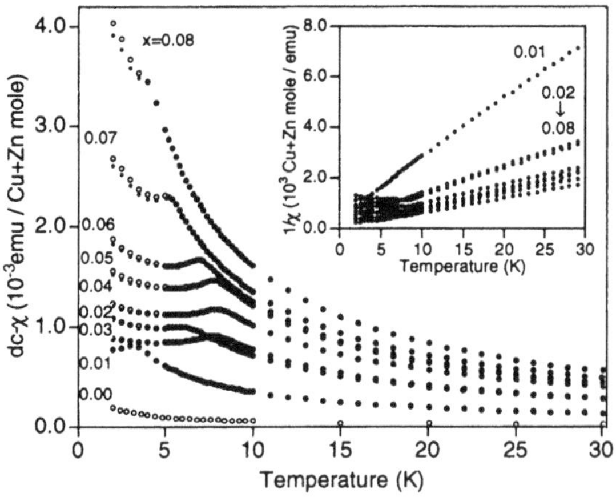

Figure 1. Temperature dependence of the dc magnetic susceptibility of $Sr(Cu_{1-x}Zn_x)_2O_3$ measured in an external field of 100 Oe on heating (closed circles) and cooling (open circles). Inset shows temperature dependence of the inverse susceptibilitiey for x = 0.01 to 0.08 from the top to the bottom.

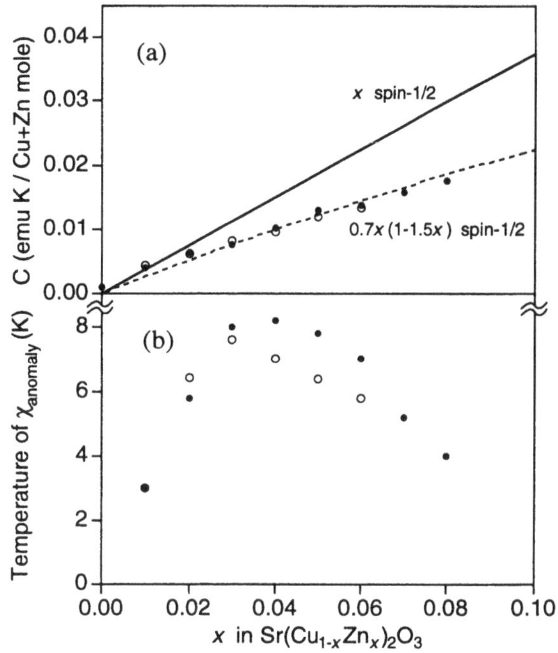

Figure 2. (a) Estimated Curie constant of $Sr(Cu_{1-x}Zn_x)_2O_3$ plotted against Zn content. The solid line stands for the Curie constant expected from an assumption that every Zn ion produces a spin- 1/2. See the text about the dotted line. (b) Néel temperatures estimated from susceptibility (closed circles) and specific heat (open circles) data.

here is a surprising result that the singlet spin liquid state with a large spin gap in $SrCu_2O_3$ is switched to an antiferromagnetically ordered state even by 1 % of Zn (x = 0.01)[10].

Plotted in Fig. 1 are the magnetic susceptibility for x = 0 to 0.08 measured in a external field of 100 Oe scaled by 1 mole of Cu+Zn. The increase of susceptibility at low temperatures is indicative of the existence of localized moments, the amount increasing with Zn content. As suggested from the linearity of the inverse-χ vs. T plot shown in the inset of Fig. 1, the data between 15 and 30 K followed the Curie-Weiss law with a temperature-independent component included : $\chi(T) = C / (T - \theta) + \chi_0$. In all the samples examined, χ_0 was found to be as small as ~ 10^{-5}emu /Cu+Zn mole. The Curie constant, C, thus estimated is plotted against x in Fig. 2 (a). As mentioned above, one Zn ion was naively expected to give rise to a spin-1/2 in a 2-leg ladder. In the lightly doped region $x < 0.03$, the experimentally obtained Curie constant is roughly consistent with the naive picture represented by the solid line in Fig. 2 (a), while that for $x \geq 0.03$ is considerably smaller than expected. Consideration of the formation of Zn pairs occupying two adjacent sites which do not break singlet pairs makes the agreement slightly better, but the experimental value is still about 0.7 times as small as the expected value shown with the dotted line in the same figure. Here, each site has three intra-ladder nearest neighbors, and therefore the probability of the Zn-pair formation is approximately equal to $1.5x$. The Weiss temperature, θ, showed a small negative value (~ -2 K) for all the samples suggesting that the induced spins interact with each other very weakly.

At lower temperatures below 10 K, however, we found an unexpected cusp-like anomaly suggesting the onset of antiferromagnetic ordering. The anomaly can be seen very clearly even at x = 0.01. The cusp temperature

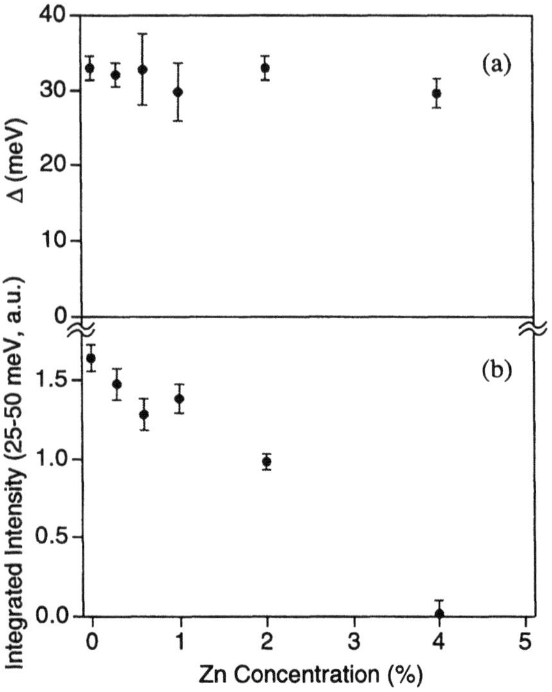

Figure 4. (a) Magnitude of the spin gap of $Sr(Cu_{1-x}Zn_x)_2O_3$ (x = 0, 0.003, 0.006,0.01, 0.02 and 0.04) estimated from the inelastic neutron scattering data. (b) Integrated intensity of magnetic scattering observedbetween 25 and 50 meV.

term in specific heat.

Based upon these results, we propose the following picture of impurity effects. Spins localized nearby the impurity ions might remain isolated in the gapped matrix at the low concentration limit, while within the composition region of $x < 0.02$, there appear in-gap states, without affecting the size of the original spin gap, for the whole ladder so that interactions between the localized moments can be mediated. The T_N first increases with increasing x, while it next descends because the impurity ions work to cut the correlation.

In summary, our magnetic susceptibility and specific heat study on $Sr(Cu_{1-x}Zn_x)_2O_3$ established the presence of long range antiferromagnetic ordering induced by a very small amount of nonmagnetic impurity. Magnetic specific heat of Zn-substituted samples is linear to temperature above T_N, indicating the existence of a finite density of states at $E = 0$ in the spin excitation spectrum, which is in consistency with the inelastic neutron scattering data.

This work was partly supported by a Grant-in Aid for Scientific Research on Priority Areas, "Anomalous metallic state near the Mott transition", of Ministry of Education, Science and Culture, Japan and CREST (Core Research for Evolutional Science and Technology) of Japan Science and Technology Corporation (JST).

REFERENCES

1. M. Hase, I. Terasaki and K. Uchinokura, Phys. Rev. Lett. **70**, 3651 (1993).
2. J. P. Renard, M. Verdaguer, L. P. Regnault, W. A. C. Erkelens, J. Rpssat-Mignod and W. Stirling, Europhys. Lett. **3**, 945 (1987).
3. E. Dagotto and T. M. Rice, Science **271**, 618 (1996).
4. Z. Hiroi, M. Azuma, M. Takano and Y. Bando, J. Solid State Chem. **95**, 230 (1991).
5. M. Azuma, Z. Hiroi, M. Takano, K. Ishida and Y. Kitaoka, Phys. Rev. Lett. **73**, 3468 (1994).
6. K. Ishida, Y. Kitaoka, K. Asayama, M. Azuma, Z. Hiroi and M. Takano, J. Phys. Soc. Jpn. **63**, 3222 (1994); K. Ishida, Y. Kitaoka, Y. Tokushige, S. Matsumoto, K. Asayama, M. Azuma, Z. Hiroi and M. Takano,Phys. Rev. B **53** 2827 (1996)
7. M. Azuma, R. S. Eccleston, M. Takano preprint
8. K. Kojima, A. Keren, G. M. Luke, B. Nachumi, W. D. Wu, Y. J. Uemura, M. Azuma and M. Takano, Phys. Rev. Lett. **74**, 2812 (1995).
9. S. R. White, R. M. Noack and D. J. Scalapino, Phys. Rev. Lett. 73, 886 (1994).
10. M. Azuma, Y. Fujishiro, M. Takano, M. Nohara and H. Takagi, Phys. Rev. B **55** (1997) in press
11. S. Ohsugi, private communication.

ELSEVIER

Physica C 282–287 (1997) 153–156

PHYSICA C

Superconductivity in the ladder compound $Sr_{2.5}Ca_{11.5}Cu_{24}O_{41}$ (single crystal)

T. Nagata[a], M. Uehara[a], J. Goto[a], N. Komiya[a], J. Akimitsu[a], N. Motoyama[b], H. Eisaki[b], S. Uchida[b], H. Takahashi[c], T. Nakanishi[d], and N. Môri[d]

[a]Department of Physics, Aoyama-Gakuin University, Chitosedai, Setagaya-ku, Tokyo 157, Japan

[b]Department of Superconductivity, The University of Tokyo, Bunkyo-ku, Tokyo 113, Japan

[c]Department of Physics, College of Humanities and Sciences, Nihon University, Sakurajosui, Setagaya-ku, Tokyo 156, Japan

[d]Institute for Solid State Physics, The University of Tokyo, Minato-ku, Tokyo 106, Japan

We report the superconductivity in the single crystal of $Sr_{2.5}Ca_{11.5}Cu_{24}O_{41}$ at $T_{c(\mathrm{onset})} = 6.5$ K under an optimal pressure of about 4.5 GPa. Normal resistivity along ladder direction ρ_c, gradually changes from T-linear to T^2-dependence with increasing an applied pressure. The normal resistivity ratio ρ_a/ρ_c in the ladder plane indicates the crossover from 1D to 2D with increasing an applied pressure.

1 INTRODUCTION

Recently, much attention has been focused for the spin ladder compounds, being composed of the two leg Cu-O chains. Of particular interest are the theoretical predictions of a spin gap in the ladder compounds with an even number of legs and of a possible *d*-wave superconductivity in the doped metallic state[1] . Recently, superconductivity has been achieved under high pressure in the $Sr_{0.4}Ca_{13.6}Cu_{24}O_{41}$[2] , which consists of two-leg Cu_2O_3 ladders and edge-sharing CuO_2 chains. They are stacked alternately along the *b*-axis. Its maximum transition temperature is about 12 K in a very narrow pressure range between 3.0 GPa and 4.5 GPa and superconductivity disappears above 6 GPa. Very recently, we succeeded in growing the single crystal of $Sr_{2.5}Ca_{11.5}Cu_{24}O_{41}$, showing the superconductivity at 6.5 K under an optimal pressure of 4.5 GPa. We briefly summarize here the results of electrical resistivity and magnetic susceptibility measured on the single crystal of $Sr_{2.5}Ca_{11.5}Cu_{24}O_{41}$.

2 EXPERIMENTAL

A single crystal of nominal composition $Sr_{2.5}Ca_{11.5}Cu_{24}O_{41}$ was grown by the traveling solvent floating zone (TSFZ) method under oxygen gas at 10 atm. The single crystals were characterized by X-ray diffraction and were found to be a single domain. A cubic-anvil type apparatus was used for the high-pressure measurements installed at I.S.S.P.. Electrical resistivity measurements were carried out using the standard four-probe method. The dimensions of the samples for ρ_c(resistivity along ladder (*c*-axis) direction) and ρ_a(resistivity perpendicular to the ladder (*a*-axis) direction in the ladder plane) measurements were about $0.10\times0.48\times1.02\mathrm{mm}^3$ and $0.35\times0.38\times1.18\mathrm{mm}^3$, respectively. Magnetic susceptibility data were obtained from AC susceptibility carried out using primary and secondary coils wound around the single crystal. The data were collected as output signals of a lock-in amplifier.

PII S0921-4534(97)00247-5

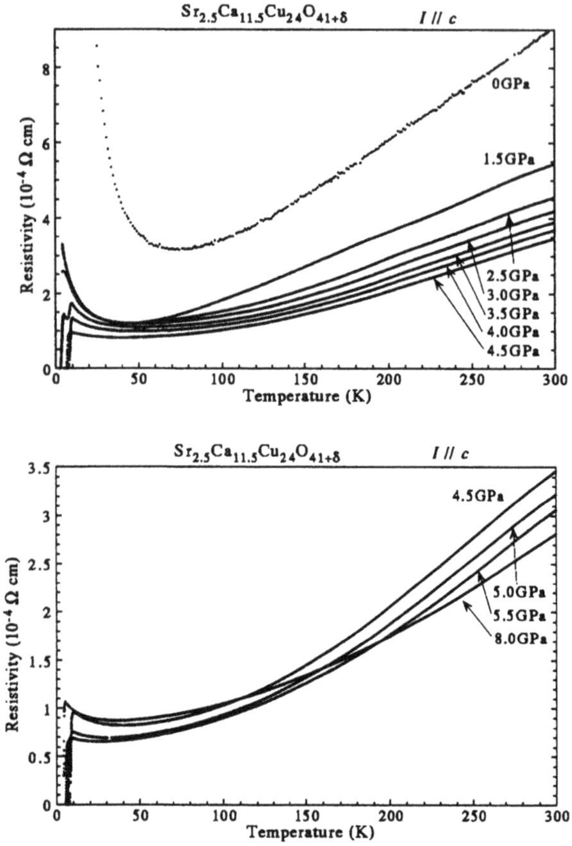

Figure 1: Temperature dependence of resistivities along c-direction under various pressures in $Sr_{2.5}Ca_{11.5}Cu_{24}O_{41+\delta}$

3 RESULTS AND DISCUSSION

Figure 1 shows the temperature dependence of ρ_c under pressure between 0 GPa and 8 GPa. The value of ρ_c at 0 GPa shows a metallic ($d\rho_c/dT > 0$) behavior characterized by T-linear dependence above 80 K and changes to semiconducting characterized by T^{-2}dependence below 80 K. With increasing the applied pressure, the $d\rho_c/dT > 0$ region is expanded and the T-linear dependence varies to T^α ($\alpha > 1$) as would be expected from Fermi liquid theory. The superconductivity was observed below 10 K under 3.5 GPa $\sim$ 8 GPa. The gradual change from T-linear to $T^\alpha (\alpha \approx 2)$ dependence could be clearly seen in Figure 2.

Figure 2 shows the typical example of the T-dependence of ρ_c under 1.5 GPa, 4.5 GPa and 8 GPa. We should emphasize here that the T-linear resistivity region was decreased with increasing an applied pressure, and T^2 resistivity was clearly seen at 8 GPa over a wide temperature range below 300 K. Such a contrasting T dependence corresponds to the carrier density change from the underdoped to overdoped region in the high-T_c superconductor. This seems to suggest that the increasing applied pressure corresponds to the increase of the carrier density in the ladder planes.

Figure 3 shows the temperature dependence of ρ_a under pressure between 0 GPa and 8 GPa. The value of ρ_a at room temperature decreases until under the pressure of 5 GPa and increases again under the pressure over 5 GPa. The ρ_a shows the semiconducting behavior between 0 GPa and 1.5 GPa below 300 K, whereas the ρ_a at 3 GPa become metallic above 110 K, slightly upturn below 110 K, and finally shows metallic behavior under 5 GPa at all temperature. The temperature dependence of ρ_a under low pressure gives evidence that the electronic state for this system is highly one-dimensional. However, the two-dimensional nature increases as inter-ladder coupling increases for increasing pressure.

Figure 4 shows the temperature dependence of the ratio of resistivity ρ_a/ρ_c for the various pressures. ρ_a/ρ_c at 0 GPa is increased to $\sim$ 85 at about 65 K and is decreased to 20 below 65 K. The ρ_a/ρ_c maximum value is increased from 85 at 0 GPa to 100 at 1.5 GPa and it decreases with increasing an applied pressure. The temperature dependence of anisotropy ρ_a/ρ_c becomes almost flat and ρ_a/ρ_c at 10 K shows the lowest value at 4.5 GPa, in which the superconducting volume fraction shows maximum. These results suggest that the carriers are confined in the ladder under the pressure of 0 GPa and 1.5 GPa and the pressure increase induces the 2D metallic behavior, resulting in the crossover from 1D to 2D under pressure.

Figure 5 shows the temperature dependence of the output signals of the lock-in amplifier (AC susceptibility) under the various pressures. We

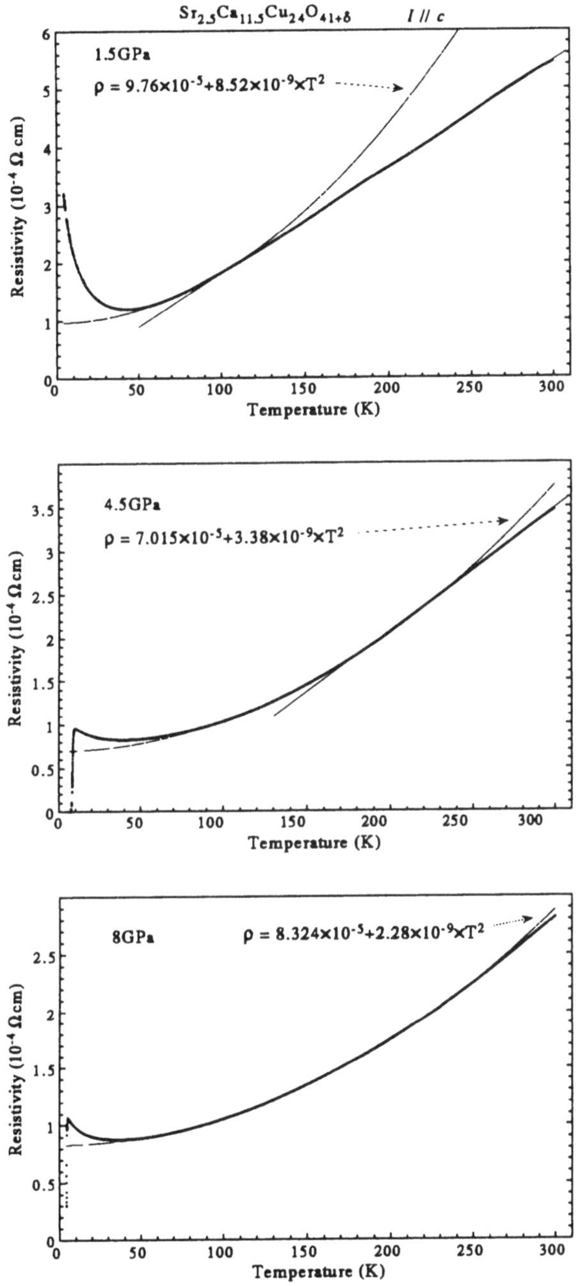

Figure 2: Temperature dependence of resistivity along c-direction under 1.5 GPa, 4.5 GPa and 8 GPa in $Sr_{2.5}Ca_{11.5}Cu_{24}O_{41+\delta}$. Fitting lines show the T-linear and T^2-dependence, respectively.

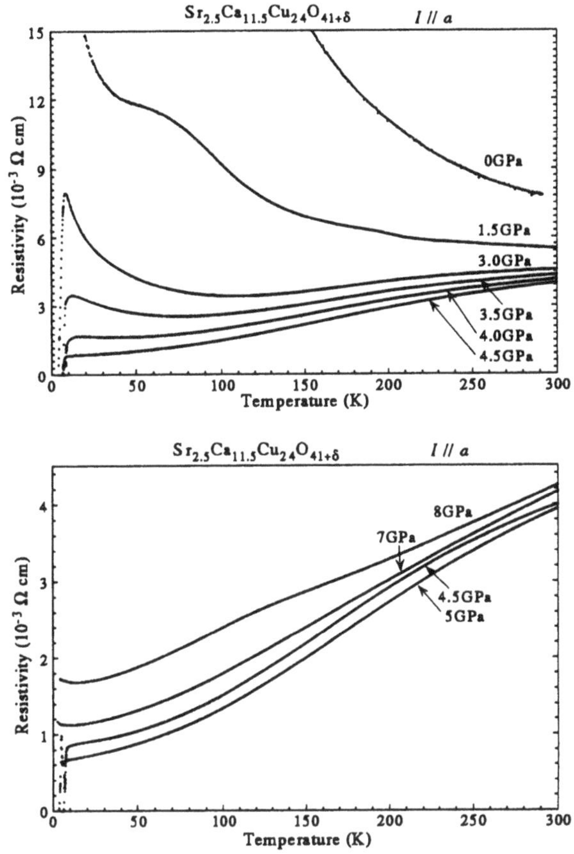

Figure 3: Temperature dependence of resistivity along a-direction under various pressures in $Sr_{2.5}Ca_{11.5}Cu_{24}O_{41+\delta}$

could observe the small diamagnetic signal at 3.5 GPa and then signal became larger with increasing an applied pressure, and the largest diamagnetic signal was observed at 4.5 GPa with $T_{c(\mathrm{onset})}$ ~6 K. The superconducting volume fraction is about 10 % at 2.3 K under 4.5 GPa by measuring the superconductivity of Pb under same condition. The systematic changes of T_c with applying pressure correspond to that observed in the resistivity measurements. In the case of $Sr_{0.4}Ca_{13.6}Cu_{24}O_{41+\delta}$, the largest diamagnetic signal was observed at 3.5 GPa with $T_{c(\mathrm{onset})} =$ 9 K. Therefore, it can be concluded that the optimal pressure is increased and maximum $T_{c(\mathrm{onset})}$

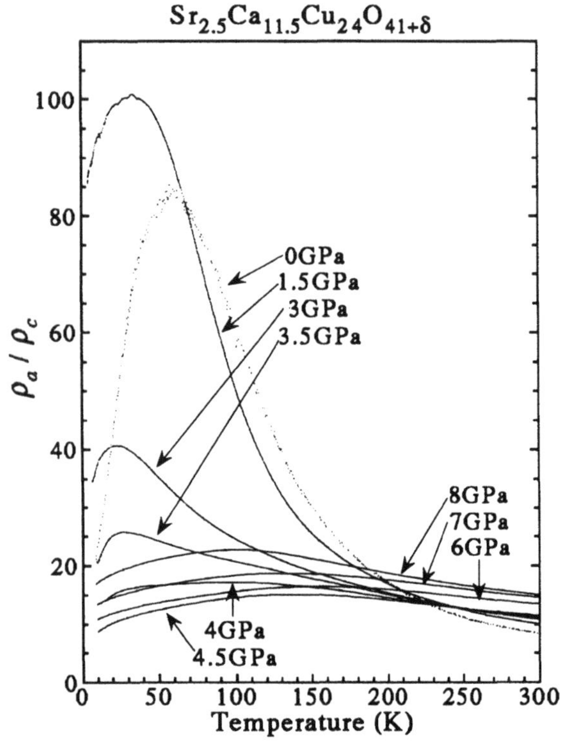

Figure 4: ρ_a/ρ_c vs. temperature under various pressure.

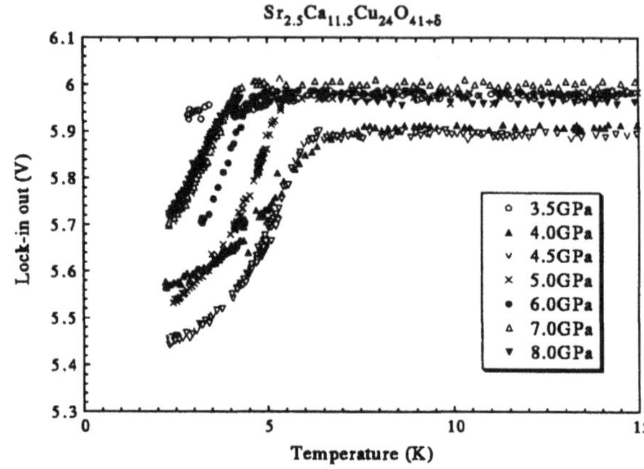

Figure 5: Magnetization vs. temperature under various pressure.

is decreased by reducing an amount of the Ca substitution.

In summary, we have observed the bulk superconductivity in $Sr_{2.5}Ca_{11.5}Cu_{24}O_{41}$ single crystal by applying a pressure over 3.5 GPa and the optimal pressure is about 4.5 GPa with $T_{c(\mathrm{onset})} =$ 6 K. The ρ_c shows the gradual change from T-linear to T^2-dependence with increasing an applied pressure. The ratio of the normal state resistivity, ρ_a/ρ_c indicates the crossover from 1D to 2D with increasing an applied pressure and the anisotropy become lowest at the optimal pressure of 4.5 GPa, in which the superconductivity appears.

REFERENCES

[1] E. Dagotto and T. M. Rice, Science **271** (1996) 618. References are herein.

[2] M. Uehara, T. Nagata, J. Akimitsu, H. Takahashi, N. Môri and K. Kinoshita, J. Phys. Soc. Jpn. **65** (1996) 2764.

ELSEVIER

Physica C 282–287 (1997) 157–161

PHYSICA C

Numerical Studies of the 2-Leg Hubbard Ladder*

D.J. Scalapino

Department of Physics, University of California,
Santa Barbara, CA 93106–9530, U.S.A.

Materials containing Cu–O ladders exhibit a number of properties which are similar to the high T_c cuprates. Here we review recent numerical results for the 2-leg Hubbard model and discuss what they imply about the pairing mechanism.

The recently discovered Cu–O ladder materials provide an important testing ground for theories of strongly correlated electron systems.[1,2] The initial suggestion,[3] that an insulating 2-leg ladder could have a spin-gap and upon doping exhibit pairing correlations, has been supported by a variety of analytic[4,5] and numerical[6] results. In addition, a spin gap has been observed[1] in the insulating 2-leg ladder material $SrCu_2O_3$, and superconductivity under pressure has been reported in $(Sr,Ca)_{14}Cu_{24}O_{41}$, which at atmospheric pressure has a structure containing alternate layers of 2-leg and 1-leg ladders separated by a (Sr,Ca) layer.[2] Thus we believe that a better understanding of the pairing mechanism in the 2-leg ladder problem can provide insight into the high T_c cuprates. Here we review results from recent numerical density matrix renormalization group (DMRG)[7] and Monte Carlo[8] calculations for the 2-leg Hubbard ladder.

The Hamiltonian we have studied for the 2-leg Hubbard ladder is

$$H = -t \sum_{i,\lambda,\sigma} \left(c^\dagger_{i\lambda\sigma} c_{i+1\lambda\sigma} + \text{h.c.} \right) - t_\perp \sum_{i,\sigma} \left(c^\dagger_{i1\sigma} c_{i2\sigma} + \text{h.c.} \right) + U \sum_{i\lambda} n_{i\lambda\uparrow} n_{i\lambda\downarrow}. \quad (1)$$

Here t and $t_\perp$ are the intra- and inter-leg hopping matrix elements respectively, and U is the on-site Coulomb interaction. The operator $c^\dagger_{i\lambda\sigma}$ creates an electron with spin σ on the ith rung of the λth leg with $i = 1, \ldots, L$ and $\lambda = 1$ and 2.

The magnetic structure factor of the Hubbard ladder,

$$S(\mathbf{q}) = \sum_{\boldsymbol{\ell}} e^{i\mathbf{q}\cdot\boldsymbol{\ell}} \left\langle m^z_{\mathrm{i}+\boldsymbol{\ell}} m^z_{\mathrm{i}} \right\rangle, \quad (2)$$

with $m^z_{\mathrm{i}} = (n_{\mathrm{i}\uparrow} - n_{\mathrm{i}\downarrow})$ is plotted for the DMRG ground state of a 2×32 ladder with $U/t = 8$ in Fig. 1. The structure factor has a Lorentzian

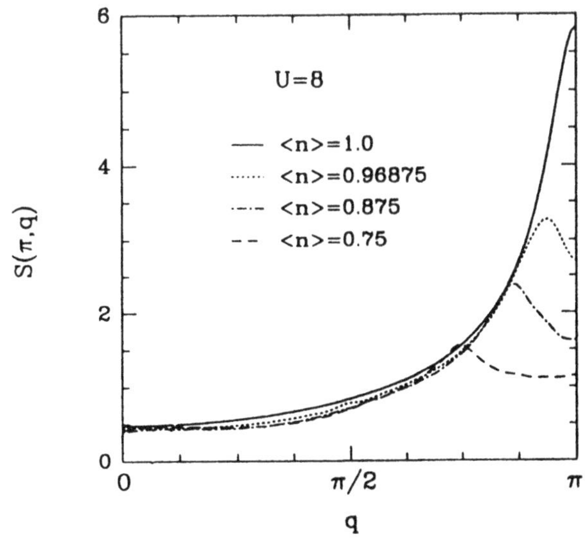

Figure 1. The magnetic structure factor $S(q_x, \pi)$ versus q_x for various fillings computed for a 2×32 ladder with $U/t = 8$ and $t_\perp/t = 1$.

*Pieces of this work were carried out with N. Bulut, T. Dahm, R. Noack, and S.R. White. This research was supported in part by the National Science Foundation under Grant No. DMR95–27304 and by the Department of Energy under Grant. No. DE–FG03–85ER45197. A number of the numerical computations were carried out at the San Diego Supercomputer Center, and we are grateful to them for their continuing support.

PII S0921-4534(97)00249-9

line shape reflecting the exponential decay of the spin-spin correlations due to a spin gap. It clearly shows that the doped ladder has strong short-range incommensurate antiferromagnetic correlations. At half-filling, $\langle n \rangle = 1$, the system has a spin gap $\Delta_s \simeq 0.18\,t$ which for sufficiently large U/t approaches the Heisenberg result $\Delta_s = 0.51\,J$. When the system is doped away from half-filling, a spin gap remains and reflects the energy to break a singlet hole pair.

When holes are doped into the half-filled system, power law pair-field correlations are found to develop. The operator

$$\Delta_i^\dagger = c_{i,1\uparrow}^\dagger c_{i,2\downarrow}^\dagger - c_{i,1\downarrow}^\dagger c_{i,2\uparrow}^\dagger \tag{3}$$

creates a singlet pair on the ith rung between legs 1 and 2, and the ground state equal time *rung–rung* pair-field correlations function

$$D(\ell) = \left\langle \Delta_{i+\ell}^\dagger \Delta_i \right\rangle , \tag{4}$$

calculated using a numerical density matrix renormalization group (DMRG) technique is shown in Fig. 2. The DMRG technique uses open end boundaries, and $D(\ell)$ falls off when ℓ becomes comparable to the ladder length. However, as seen in Fig. 2, the results for the 2×32 ladder are reliable to distances $\ell \lesssim 20$ over which this correlation falls as a power law $\ell^{-\theta}$ with $\theta \simeq 1$. A similar calculation of the singlet pair-field *rung–leg* correlation function shows a similar decay but with a negative sign, indicating that we are dealing with a $d_{x^2-y^2}$-like pair field. This can also be seen from the amplitude,

$${}_{N-2}\langle \psi_0 | \left(c_{i\uparrow} c_{j\downarrow} - c_{i\downarrow} c_{j\uparrow} \right) | \psi_0 \rangle_N , \tag{5}$$

for creating a singlet hole pair between the ith and j^{th} site shown in Fig. 3. This illustrates the $d_{x^2-y^2}$-like internal structure of the Cooper pair wave function.

Another view of the pairs which form in the doped 2-leg ladder is shown in Fig. 4, where we plot the rung charge density

$$\langle n_i \rangle = \sum_{\lambda_s = 1,2} \langle n_{i,\lambda_s} \rangle . \tag{6}$$

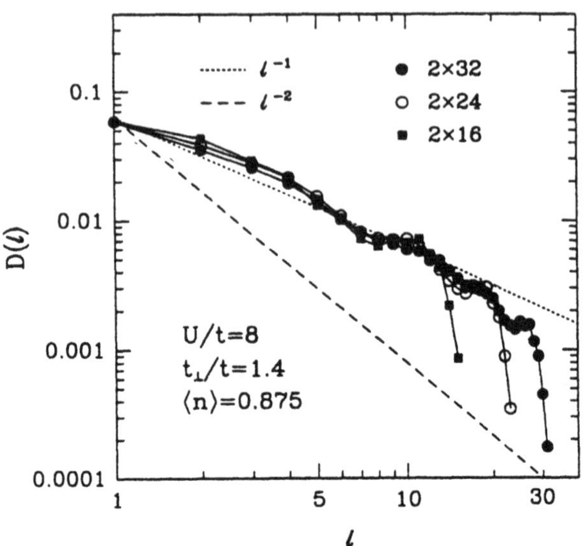

Figure 2. The rung-rung pair field correlation $D(\ell)$ versus ℓ for $U/t = 8$, $t_\perp/t = 1.4$, $\langle n \rangle = 0.875$ on $2 \times L$ ladders with $L = 16$, 24, and 32. The dashed line varies as ℓ^{-2} and the dotted line as ℓ^{-1}.

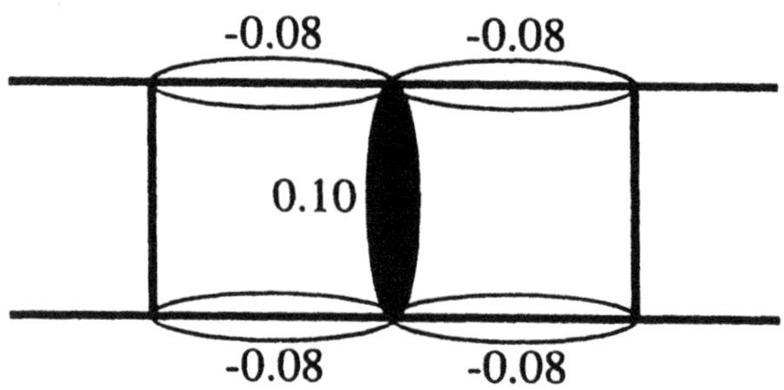

Figure 3. Schematic of the pair wave function showing the values of ${}_{N-2}\langle \psi_0 | \left(c_{i\uparrow} c_{j\downarrow} - c_{i\downarrow} c_{j\uparrow} \right) | \psi_0 \rangle_N$ for creating a singlet pair between various near-neighbor sites. The internal pair wave function extends over four to five rungs.

As pairs of holes are doped into the 2-leg half-filled system, one clearly sees that they behave as pairs, giving rise to a CDW structure pinned by the open end boundary conditions of the ladder.

To examine in more detail the dependence of the pairing on the parameters of the model, we adopt as a measure of the strength of the pairing the average

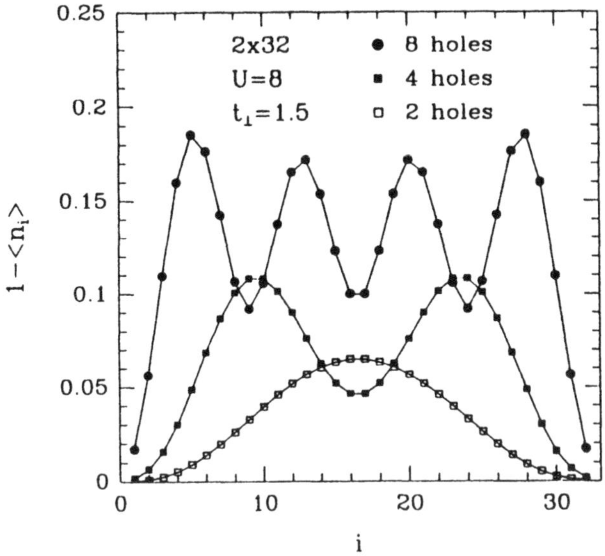

Figure 4. The local hole density $1-\langle n_i\rangle$ as a function of position along the chain i for 2, 4, and 8 holes ($\langle n\rangle = 0.96875, 0.9375, 0.875$) on a 2×32 chain for $U=8$ and $t_\perp = 1.5$.

$$\overline{D} = \frac{1}{5}\sum_{\ell=8}^{12} D(\ell). \quad (7)$$

The dependence of $\overline{D}$ on U/t and $t_\perp/t$ for a filling of $\langle n\rangle = 0.9375$ is shown in Fig. 5. The strength of the pairing, as measured by $\overline{D}$, increases until U is of order the bandwidth $U/t \simeq 4$. As U becomes larger, the antiferromagnetic exchange $J = 4t^2/U$ decreases and the pairing response is reduced.

The dependence of the pairing strength on $t_\perp/t$ arises from the behavior of the single-particle spectral weight distribution

$$A(p,\omega) = -\frac{1}{\pi}\,\mathrm{Im}\ G(p,\omega). \quad (8)$$

Results for $A(p,\omega)$ obtained from a maximum entropy analytic continuation of Monte Carlo data for a 2×32 ladder with periodic boundary conditions are shown in Fig. 6. Here $U/t=4$, $\langle n\rangle = 0.875$, and the temperature $T=0.25\,t$. For $t_\perp/t = 1.6$, the spectral weight of the bonding band $(p_x, 0)$ peaks at the Fermi surface for large values of p_x,

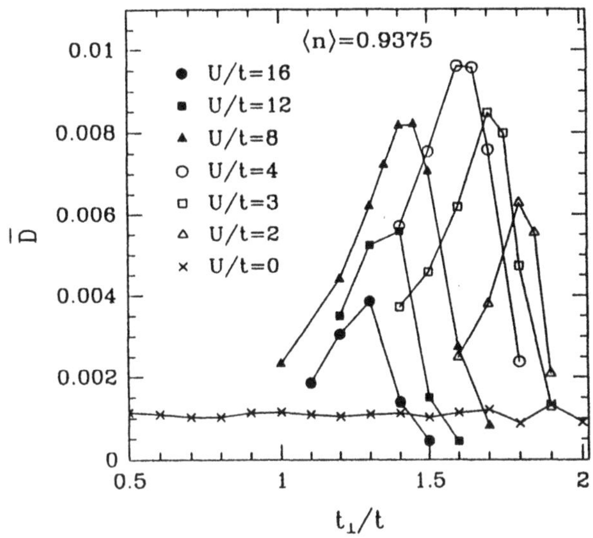

Figure 5. $\overline{D}$ versus $t_\perp/t$ for $\langle n\rangle = 0.9375$ and various values of U/t.

while it peaks for the antibonding band (p_x,π) for p_x near zero. The flatness of the quasi-particle dispersion is a many-body effect arising from the proximity to the metal-insulator transition and is responsible for the peak in $\overline{D}$ shown in Fig. 5 for $t_\perp/t \simeq 1.6$ and $\langle n\rangle$ near half-filling. As seen in Fig. 6, this occurs as the system begins to develop what will become the two-peak quasi-particle spin density wave-like dispersion characteristic of the insulating state.[9,10] Note that in the hole doped state the chemical potential lies near the top of the lower energy peak in Fig. 6.

Using Monte Carlo techniques, we have also calculated the irreducible pairing interaction in the singlet channel $V(p'-p)$. In Fig. 7(a) we plot V versus $q_x = p'_x - p_x$ for $q_y = p'_y - p_y = \pi$. Here we have set the energy transfer to zero. The three curves correspond to temperatures $T = 1.0\,t, 0.5\,t$, and $0.25\,t$. The q_x momentum dependence of the magnetic susceptibility

$$\chi(q) = \frac{1}{N}\int_0^\beta d\tau \sum_{\boldsymbol{\ell}} e^{i\mathbf{q}\cdot\boldsymbol{\ell}} \left\langle m^-_{i+\ell}(\tau) m^+_i(0)\right\rangle, \quad (9)$$

with $q_y = \pi$, is shown for comparison in Fig. 7(b). From this it is clear that the pairing interaction in the ladder is associated with the development of short-range antiferromagnetic correlations.

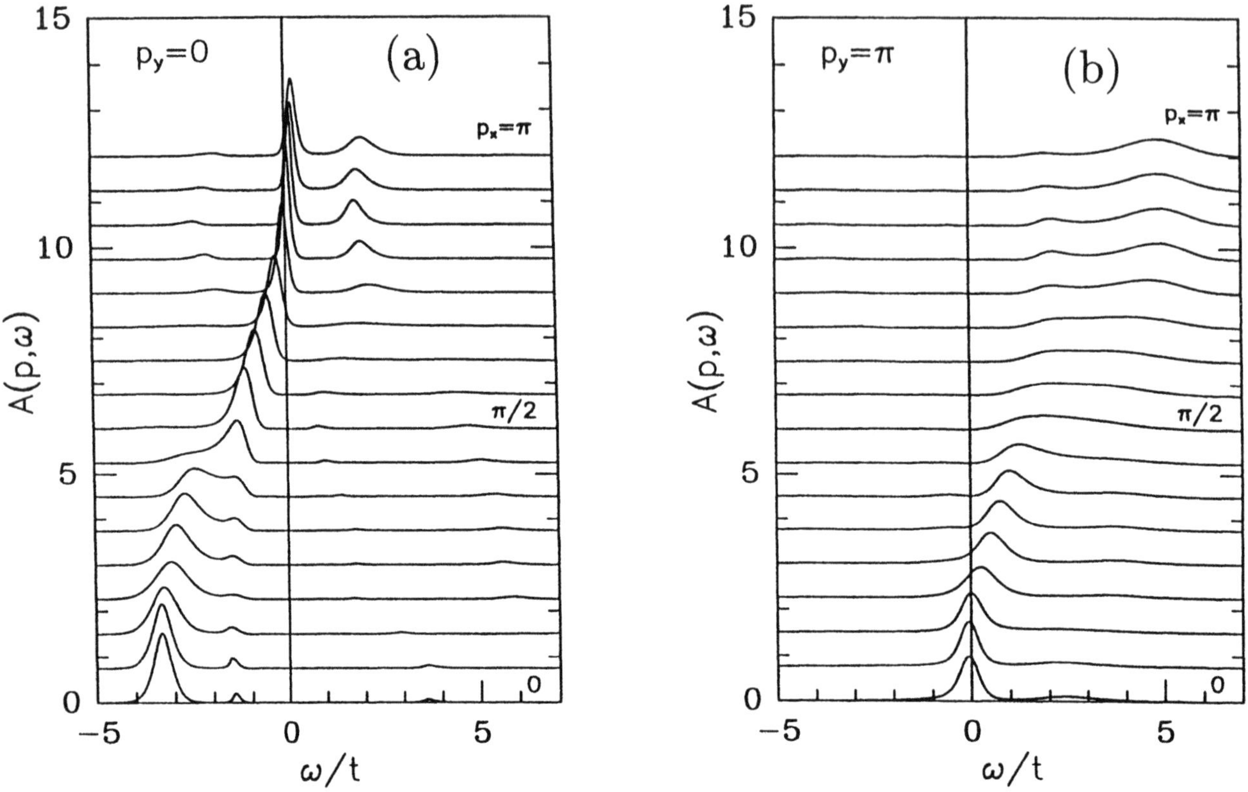

Figure 6. The single-particle spectral weight $A(p,\omega)$ for $U/t = 4$, $\langle n \rangle = 0.875$ at $T = 0.25\,t$: (a) for the bonding band $(p_x, 0)$, (b) for the antibonding band (p_x, π). Here p_x increases from 0 to π as one moves up the figure with successive $A(p,\omega)$ spectra displaced by 0.75.

From these results we conclude[11] that the pairing in the 2-leg Hubbard ladder arises from short-range antiferromagnetic correlations and is enhanced near the metal-insulator transition by the flat, high density of spectral weight which develops, due to these interactions, in regions near $(\pi, 0)$ and $(0, \pi)$. Furthermore, we have found that the pairing correlations, as measured by $\overline{D}$, are strongest for intermediate U/t values becoming weaker for both small and large ratios of U/t.

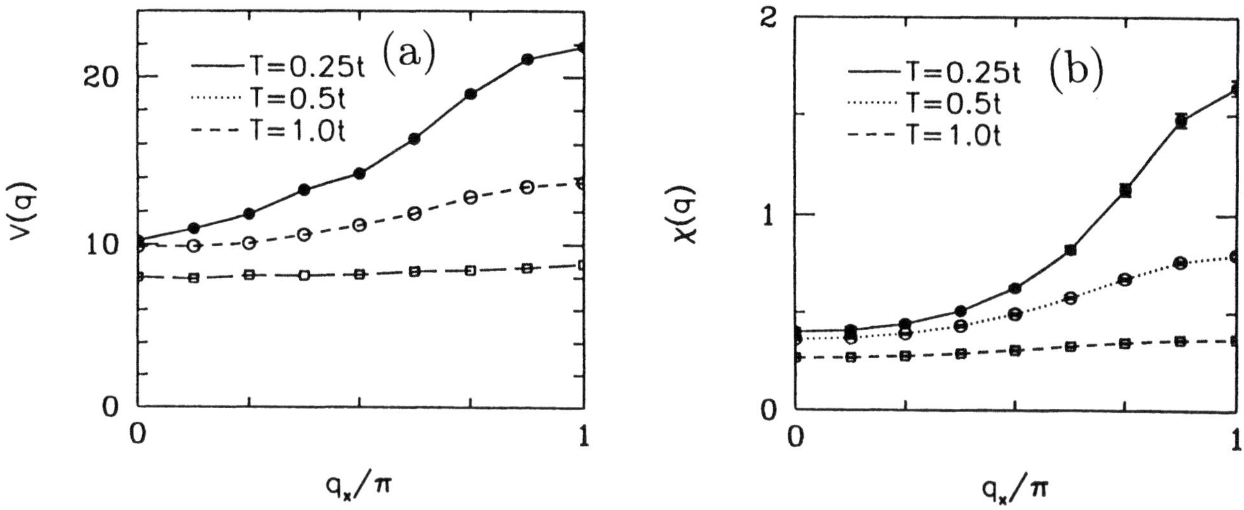

Figure 7. The momentum dependence of (a) the irreducible pairing interaction $V(q_x, \pi)$ and (b) the magnetic susceptibility $\chi(q_x, \pi)$ versus q_x for $U = 4t$, $t_\perp/t = 1.5$, and $\langle n \rangle = 0.875$.

References

1. Z. Hiroi and M. Takano, *Nature* **377**, 41 (1995).

2. M. Uehara *et al.*, *J. Phys. Soc. Jpn.* **65**, 2764 (1996).

3. E. Dagotto, J. Riera, and D.J. Scalapino, *Phys. Rev. B* **45**, 5744 (1992).

4. T.M. Rice, S. Gopalan, and M. Sigrist, *Europhys. Lett.* **23**, 445 (1993); H. Tsunetsuga, M. Troyer, and T.M. Rice, *Phys. Rev. B* **49**, 16078 (1994); S. Gopalan, T.M. Rice, and M. Sigrist, *Phys. Rev. B* **49**, 8901 (1994).

5. L. Balents and M.P.A. Fisher, *Phys. Rev. B* **53**, 12133 (1996).

6. R.M. Noack, S.R. White, and D.J. Scalapino, *Phys. Rev. Lett.* **73**, 882 (1994); *Europhys. Lett.* **30**, 163 (1995).

7. R.M. Noack, N. Bulut, D.J. Scalapino, and M.G. Zacher, preprint Cond-Mat/9612165.

8. N. Bulut, T. Dahm, and D.J. Scalapino, preprint Cond-Mat/9612166V2.

9. A.P. Kampf and J.R. Schrieffer, *Phys. Rev. B* **42**, 7967 (1990).

10. R. Preuss, W. Hanke, and W. von der Linden, *Phys. Rev. Lett.* **75**, 1344 (1995).

11. D.J. Scalapino, *Turkish J. Phys* **20**, 560 (1996).

ELSEVIER

Physica C 282–287 (1997) 162–165

PHYSICA C

Transport properties of MBE grown Cuprate Spin Ladders

M. Laguës[a], J.Y.Laval[b], C.F.Beuran[a], C. Deville Cavellin[a], B.Eustache[c], J.B.Moussy[c], C.Partiot[a] and X.Z.Xu[a]

[a] *Laboratoire Surfaces et Supraconducteurs - Université Paris 12 - CNRS UPR5,*
[b] *Laboratoire de Physique des Solides - CNRS UPR5, Ecole Supérieure de Physique et de Chimie Industrielles - 10 rue Vauquelin, 75005, Paris, France*
[c] *Wintici SA - 17 rue Jean Moulin, 94300 Vincennes, France*

Thin films belonging to the family of Spin Ladders $(Sr,Ca)_{h-1}Cu_{h+1}O_{2h}$ are grown by MBE. The deposition is performed under atomic oxygen using real time control by the RHEED intensity. The structural parameters measured by four-circle X-ray diffraction agree with the values observed for cuprate Spin Ladders in bulk compounds from various laboratories. HRTEM images shows the double Cu rows within the CuO planes which are characteristic of ladder structures. The room temperature resistivity ranges from 1 mΩ.cm for $Ca_2Cu_4O_6$ to 1 Ω.cm for $Sr_2Cu_4O_6$. Doping with BiO planes was tested. A superconducting transition at 40K was observed in a BiSrCaCuO compound with a composition compatible with two-leg ladders.

1. INTRODUCTION

Cuprate Spin Ladders were first synthesized under high pressure by Hiroi and Takano.[1] They describe electrical and magnetic properties of the two first members $Sr_2Cu_4O_6$ and $Sr_4Cu_6O_{10}$, of the $Sr_{h-1}Cu_{h+1}O_{2h}$ homologous series. Figure 1 compares their unit cell with the structure of the Infinite Layer Compounds[2] (ILC). These compounds are referred as *Spin Ladders* because the CuO_4 tetragonal units are distributed along stripes looking like ladders. The quasi one-dimensionnal magnetic properties of these compounds[3,4] are interesting. It was suggested by Rice[5] that they could lead to superconductivity with high critical temperatures. It was reported[6] that the ladder compound $(La,Sr)CuO_{2.5}$ shows no superconductivity. A related phase, $Sr_{14}Cu_{24}O_{41}$, is suggested to be the intergrowth of the Spin Ladder unit cell $Sr_2Cu_4O_6$ (figure 1) with a $Sr_7Cu_{10}O_{20}$ block containing CuO_2 chains[7,8]. One may also notice the relation between this phase and the superconducting behavior observed but not well described in some Infinite Layer Compounds : the $Sr_{14}Cu_{24}O_{41}$ phase shows up often as a secondary phase when superconductivity is observed in ILC, both in the case of n-type[9] and p-type doping[10,11].
A recent report describes the superconductivity under high pressure for a $(Sr,Ca)_{14}Cu_{24}O_{41}$ compound[12] with a Tc of 12K. All these ladder compounds are bulk materials prepared under very high pressure.

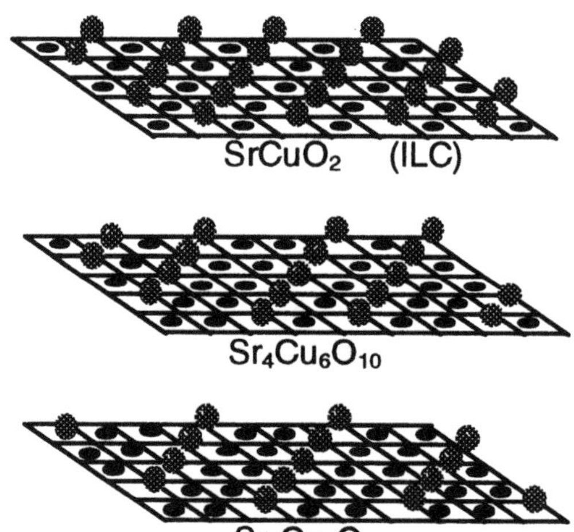

Fig. 1 Structure of the unit cell for the first two members (h=3 and h=5) of the Spin Ladder family $Sr_{h-1}Cu_{h+1}O_{2h}$ compared to the ILC structure.

The compounds presented here are thin films prepared by atomic engineering. Depending on the composition we observe low resistivities and metallic behavior -but no superconducting properties as yet- on thin films belonging to the family $(Sr,Ca)_{h-1}Cu_{h+1}O_{2h}$. Their room temperature resistivity ranges from 1 mΩ.cm for $Ca_2Cu_4O_6$ to

PII S0921-4534(97)00251-7

1 Ω.cm for $Sr_2Cu_4O_6$. A BiSrCaCuO compound with a composition [4,4,3,10] compatible with a two-leg Spin Ladder shows superconducting properties.

2. EXPERIMENTAL

Films with a typical thickness of 30 nm were grown by the Sequentially Imposed Layer Epitaxy (SILE). Using such SILE deposition with our conditions, we find that the growth of compounds belonging to the system (Sr,Ca,Cu,O) is more stable for Spin Ladder structures than for infinite layer compounds[13-15]. The deposition chamber is a Riber Eva32 equipped with an atomic oxygen plasma source. The substrate were MgO or $SrTiO_3$ (100). During the deposition the local pressure near the sample is kept to 10^{-4} Torr of atomic oxygen. The substrate temperature is in the range 500 to 600°C for SrCaCuO compounds, and 700 to 800°C for BiSrCaCuO. Bismuth, strontium and calcium are deposited from Knudsen cells. Copper is evaporated by an electron gun. The surface structure of the film is monitored in real time using the diffraction pattern of a 35 keV electron beam. The oscillations of the Reflection High Energy Electron Diffraction (RHEED) intensity are used to stop the deposition of each atomic layer when an extremum is reached. As the sticking coefficients of the different species change during the deposition of typically the first 5 nm,[16] the deposition sequence is adjusted in real time for these first few atomic layers. A deposition sequence controlled by computer is then used to complete the film thickness up to around 30nm. Rutherford Backscattering allows *absolute* measurements of the average amount of each element (except for the oxygen) deposited during one layering sequence. The oxygen content used in the following compositions, i.e. $Ca_2Cu_4O_6$, is a formal value without quantitative meaning.

3. STRUCTURE

Figure 2 shows a typical High Resolution Trnasmission Electron Microscopy (HRTEM) picture observed in $Sr_2Cu_4O_6$ films grown on $SrTiO_3$. The double Cu row observed here in the Cu_2O_3 planes is characteristic of the two-leg cuprate ladders. A detailed study of the structure

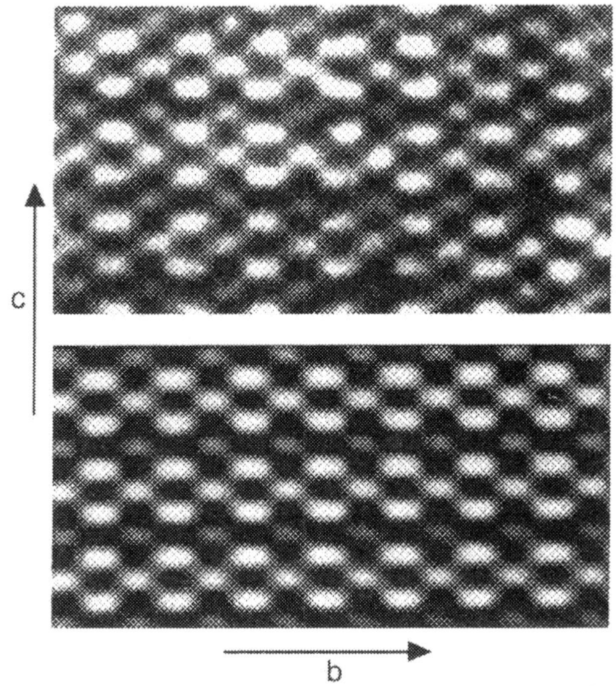

Fig 2 HRTEM picture of $Sr_2Cu_4O_6$ showing the double copper rows. Upper: raw image, lower : after Wiener filtering.

shows a c-axis cell parameter corresponding to four Cu_2O_3 planes [17]. This corresponds to the 13.43 Å measured by X-ray diffraction for the *(00l)* lattice parameter. In this structure, the copper-copper distance is modulated from one Cu_2O_3 plane to the following between 3.25 3.45Å . The measurement of the in-plane structure on a four-circle diffractometer[18] shows the expected lattice parameters for a two leg ladder, a = 3.95 Å and b=11.57 Å . The same structural study is under progress for $Ca_2Cu_4O_6$ films on MgO, in which a = 3.90 Å and b=11.40 Å, and c = 12.40 Å.

Copper rich unconventional BiSrCaCuO compounds were sequentially deposed including BiO reservoir block layers. One can use a general ladder scheme to describe their composition : it is assumed that the copper layers exhibit a $Ca_{1-1/h}Cu_{1+1/h}O_2$ composition where 1/h equals zero for conventional compounds. Compared to the conventional $Bi_2Sr_2Ca_{n-1}Cu_nO_x$ we expect to observe a general composition :

$$[Bi_2Sr_2Ca_{n-1}]_{1-1/h}[Cu_n]_{1+1/h}O_y \qquad [1]$$

We assume here that the reservoir block-layer builds itself by epitaxy onto the copper-oxide ladders planes, with a reduced surface density as do the Ca layers. The sample for which electric and magnetic properties are described in the following exhibits an ***absolute*** composition for each cell of $Bi_{1.3}Sr_{1.4}Ca_{0.9}Cu_{3.3}O_y$ in units of 0.675 atoms/cm². (In these units the cuprate plane composition is simply CuO_2). Such a copper rich composition without presence of copper oxide phases in the X-ray pattern, corresponds precisely to the composition of a two-leg ladder structure (h=3) with an intergrowth having equal parts of n=2 and n=3. The composition of such a sample according to [1] should $Bi_{1.3}Sr_{1.3}CaCu_{3.3}O_y$, $Bi_4Sr_4Ca_3Cu_{10}O_y$ for three cells . HRTEM observations are presently in progress in order to verify the presence of ladders in these compounds.

4. TRANSPORT and MAGNETIC PROPERTIES

Figure 3 shows the temperature dependence of the resistivity for $Sr_2Cu_4O_6$ grown on $SrTiO_3$ and

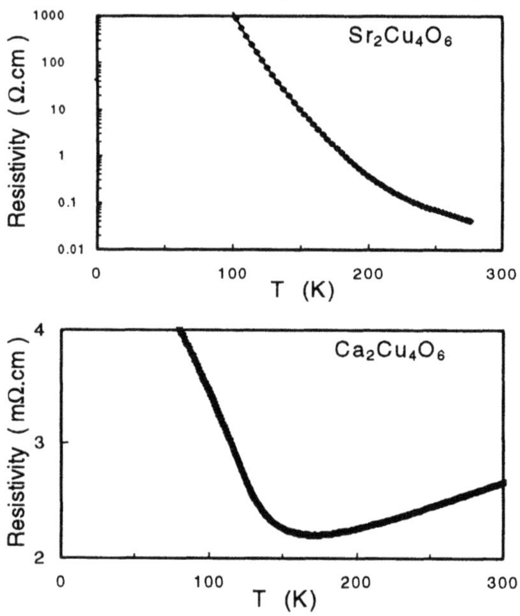

Fig 3 Resistivity of $Sr_2Cu_4O_6$ and $Ca_2Cu_4O_6$ films grown on MgO

$Ca_2Cu_4O_6$ grown on MgO. The Sr two-leg ladder sample shows an insulating behavior with an activation energy of about 160 meV for T>130 K. For T>150K a metallic behavior is observed for Ca ladders without any special doping process. Figure 4 presents the temperature dependence of the resisitivity for the unconventional Bi [4,4,3,10] sample. Figure 5 presents the susceptibility of this last sample for which a zero resistance is observed at 40K. These properties were already presented elsewhere.[19] The diamagnetic onset corresponds well with the resistive transition, as shown in figure 5.

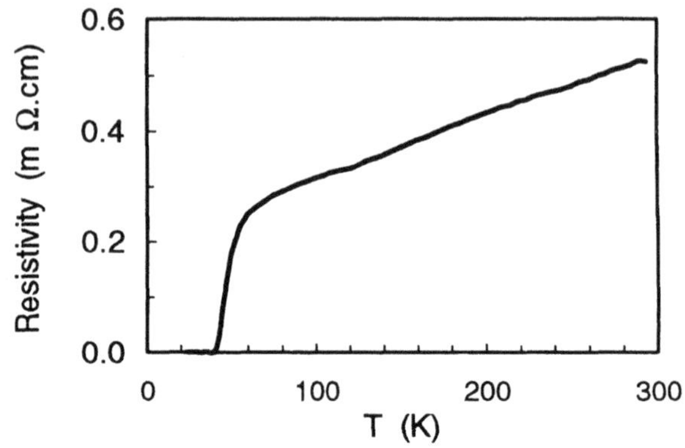

Fig 4 Resistivity of the Bi [4,4,3,10] sample

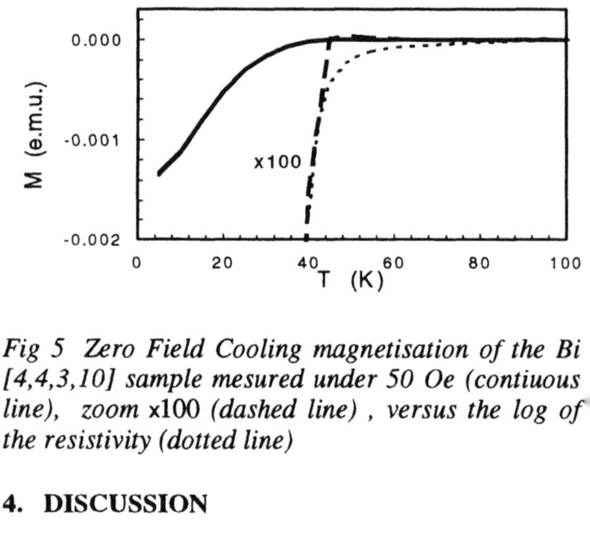

Fig 5 Zero Field Cooling magnetisation of the Bi [4,4,3,10] sample mesured under 50 Oe (contiuous line), zoom x100 *(dashed line) , versus the log of the resistivity (dotted line)*

4. DISCUSSION

A detailed HRTEM structural study was performed for $Sr_2Cu_4O_6$ grown on $SrTiO_3$. The in plane parameters a and b agree with those of bulk compounds[1] but the c-axis parameter of our films corresponds to four Cu_2O_3 planes. This compound is thus different from the bulk compound which shows c=3.40 Å. HRTEM structural study is currently under progress for $Ca_2Cu_4O_6$ and the unconventional Bi compounds. Polarized X-ray

Absorption Fluorescence studies were also performed and measurements of the Cu L3-edge lead to an average Cu valence of 2.15 for the unconventional Bi compounds and 2.26 for $Sr_2Cu_4O_6$. This high oxidation level, corresponding to a composition $Sr_2Cu_4O_{6.5}$, could explain the difference of the c-axis structure between this compound and the bulk $Sr_2Cu_4O_6$ Spin Ladder[20] : the film shows c = 11.40Å leading to an average distance of 3.35Å between the Cu_2O_3 planes, while in the bulk compound this distance is 3.43Å.

5. CONCLUSION

Thin films belonging to the family of Spin Ladders $(Sr,Ca)_{h-1}Cu_{h+1}O_{2h}$ are grown by MBE. In plane lattice parameters measured by X-ray diffraction and HRTEM are in agreement with the values observed for cuprate Spin Ladders in bulk compounds. The room temperature resistivity ranges from 1 mΩ.cm for $Ca_2Cu_4O_6$ -showing a metallic behavior above 150K- to 1 Ω.cm for $Sr_2Cu_4O_6$. A superconducting transition at 40K was observed in an unconventional BiSrCaCuO compound whose composition could correspond to a two-leg spin ladder. The presence of ladders in this sample has to be verified, for instance by HRTEM.

Acknowledgments
SQUID measurements were performed by N.Bontemps at ENS (Paris). Cu-edge XAS in collaboration with F.Studer, and four-circle diffraction in collaboration with J.P.Lauriat were performed at LURE (Orsay). RBS measurements were performed in Université Paris 7 with the support of the GRD86 of CNRS. This work was financially supported by Soffinova, Anvar, Conseil Régional d'Ile de France and CRITT Chimie Ile de France .

6. REFERENCES

1. .Z.Hiroi, and M.Takano, *Physica C* **235-240** (1994) 29
2. T.Tatsuki, S.Adachi, M.Itoh, T.Tamura, X.J.Wu, C.Q.Jin, H.Yamauchi, N.Koshizuka and S.Tanaka, *Jpn. J. Appl. Phys.* **34** (1995) L615
3. M.Azuma, Z.Hiroi, M.Takano, K.Ishida and Y.Kitaoka, *Phys.Rev.Lett.* **73** (1994) 3463
4. T.M.Rice, *Europhys. Lett.* **23** (1993) 445
5. Z.Hiroi and M.Takano, Nature 377 (1995) 41
6. M.Laguës, C.F.Beuran, C.Coussot, C.Deville Cavellin, B.Eustache, C.Hatterer, P.Laffez, V.Mairet, X.M. Xie and X.Z.Xu *Coherence in superconductors* p70 Ed. G.Deutscher & A. Revcoleschi (World Scientific 1996)
7. N.Sugii , M.Ichikawa, K.Kubo, M.Ichikawa, K.Yamamoto, H.Yamauchi and S.Tanaka, *Physica C* **196** (1992) 129
8. X.Zhou , J.W.Li, F.Wu, J.Q.Li, B.Yin, S.L.Jia, Y.S.Yao and Z.X.Zhao, *Physica C.* **223** (1994) 30
9. C.Prouteau , P.Strobel, J.J.Capponi, C.Chaillout and J.L.Tholence, Physica C 228 (1994) 63
10. Mc Carron, M.A Subramanian, J.C.Calabrese and R.L.Harlow, *Mat. Res. Bull.* **23** (1988) 1355
11. H. Shaked, Y.Shimakawa, B.A. Hunter, R.L.Hitterman, J.D. Jorgensen, P.D.Han and D.A. Payne, *Phys. Rev. B* **51** (1995) 11784
12. M.Uehara, T.Nagata, J.Akimitsu, H.Takahashi, N.Mori, K.*Kinoshita* *J.Phys.Soc Jpn* **65** (1996) 2764
13. V. Mairet PhD thesis Univ. Paris VI P. & M.Curie (21 december 1995)
14. C.Hatterer PhD thesis Univ. Paris VI P. & M.Curie (11 july 1996)
15. B.Eustache, C.F.Beuran, C.Deville Cavelin, Ch. Hatterer, V.Mairet, C.Partiot, X.Z.Xu, P.Germain, M.Laguës *J. of Alloys and compounds* (1997) in press
16. X.Z.Xu, M.Viret, H.Tebbji and M.Laguës *Applied Superconductivity* **1** (1993) 755
17. J.Y.Laval & al. to be published
18. C. Deville Cavellin, B.Eustache, C.J.Hatterer, J.-P. Lauriat, X.Z. Xu, V. Mairet, F. Beuran and M. Laguës (this meeting)
19. M. Laguës, N.Bontemps, C.Partiot, B.Eustache, J.B.Moussy, C.F.Beuran, X.Z.Xu and C. Deville Cavellin MRS meeting (Boston , USA, 1-5 december 1996)
20. Z.Hiroi, M.Azuma,M.Takano and Y.Bando *J.Sol.Stat.Chem.* **95** (1991) 230

ELSEVIER

Physica C 282–287 (1997) 166–169

PHYSICA C

Charge stripes and spin correlations in copper-oxide superconductors

J. M. Tranquada[a*]

[a]Physics Department, Brookhaven National Laboratory, Upton, NY 11973, USA

Recent neutron diffraction studies have yielded evidence that, in a particular cuprate family, holes doped into the CuO_2 planes segregate into stripes that separate antiferromagnetic domains. Here it is shown that such a picture provides a quantitatively consistent interpretation of the spin fluctuations measured by neutron diffraction in $La_{1.85}Sr_{0.15}CuO_4$ and $YBa_2Cu_3O_{6+x}$.

1. INTRODUCTION

To obtain superconductivity in a layered copper-oxide compound, it is necessary to introduce charge carriers into the antiferromagnetic CuO_2 planes. Recent neutron diffraction studies of the system $La_{1.6-x}Nd_{0.4}Sr_xCuO_4$ [1,2] provide evidence that the dopant-induced holes choose to segregate into periodically-spaced stripes which separate antiferromagnetic domains, in a manner similar to that found in hole-doped La_2NiO_4 [3]. The charge and spin stripe modulations are identified by the appearance of scattering at incommensurate positions. In the Nd-doped system, elastic scattering is observed, corresponding to static stripes. In pure $La_{2-x}Sr_xCuO_4$, the magnetic scattering that is observed is purely inelastic [4]. Where samples with and without Nd, but with the same Sr concentration, have been measured, the incommensurate (IC) splittings of the magnetic signal are found to be essentially identical [2,4,5]. It has been proposed that the spin correlations in the two systems are fundamentally the same, thus implying similar charge correlations. The static nature of the stripes in the Nd-doped system is attributed [1] to pinning of the otherwise dynamic correlations by a special distortion of the lattice [6]. That distortion is driven by purely ionic interactions and is stabilized by the smaller ionic radius of the substituted Nd.

To strengthen the case for the charge-stripe interpretation, it is necessary go beyond a simple comparison of incommensurabilities. In this paper I will show that the neutron scattering measurements of spin fluctuations in superconducting $La_{1.85}Sr_{0.15}CuO_4$ and $YBa_2Cu_3O_{6.6}$ are quantitatively consistent with what one should expect based on the stripe picture. The general expectations are described in section 2. Specific analyses for the two different superconductors are given in sections 3 and 4, respectively. Section 5 contains some further discussion.

2. MODEL

First recall the situation for the 2D Heisenberg model [7]. At $T = 0$, where long-range order exists, the spin-wave dispersion is linear at small $\mathbf{q}$ ($= \mathbf{Q} - \mathbf{Q}_{AF}$), and can be expressed as $\omega = c_0 q$. At finite temperatures, thermal fluctuations result in a finite spin-spin correlation length ξ. Spin waves are overdamped for $q < 1/\xi$, but spin-wave dispersion should still be detectable for $q > 1/\xi$ (as long as kT is less than the superexchange energy J). The characteristic energy Γ_0 separating the over- and underdamped regimes is then given by

$$\Gamma_0 \approx \hbar c_0/\xi. \tag{1}$$

As for the dynamical susceptibility, $\chi''(\mathbf{Q},\omega)$, Γ_0 acts somewhat like a gap energy [7]. If χ'' is integrated over $\mathbf{Q}$ to give $\tilde{\chi}''(\omega)$, then there is a peak at $\hbar\omega \approx \Gamma_0$, with $\tilde{\chi}''(\omega)$ going to zero as ω goes to zero.

Next consider an ordered stripe phase. The charge stripes need not affect the superexchange

*Work at Brookhaven is supported by Contract No. DE-AC02-76CH00016, Division of Materials Sciences, U.S. Department of Energy.

0921-4534/97/$17.00
PII S0921-4534(97)00252-9

energy within the antiferromagnetic domains, but they certainly will weaken the magnetic coupling between domains. Castro Neto and Hone [8] have considered the effect of the anisotropic coupling on the spin wave velocity. For a large anisotropy, the velocity c for propagation parallel to the stripes is equal to $c_0/\sqrt{2}$; it is more difficult to estimate an effective velocity perpendicular to the stripes because the stripe modulation should result in several optical branches as well as the acoustic mode. An experimental test of this situation is given by recent measurements of spin waves in stripe-ordered $La_2NiO_{4.133}$ [9]. That study found $c = 0.6c_0$ for propagation parallel to the stripes. Dispersion was less well defined but comparable in the perpendicular direction. (Note that the spin waves disperse out of the IC points in reciprocal space that characterize the magnetic order.) Thus, it seems reasonable to take $c = c_0/\sqrt{2}$ as an estimate of the effective spin-wave velocity in an ordered stripe phase.

Finally, consider a stripe phase with a finite spin-spin correlation length induced by fluctuations of the charge stripes. The characteristic energy separating over- and underdamped modes should now be

$$\Gamma \approx \hbar c_0/\sqrt{2}\xi. \tag{2}$$

Excitations at $\hbar\omega > \Gamma$ should look spin-wave-like. On the other hand, a scan at a fixed excitation energy smaller than Γ through the IC points characterizing the stripe correlations should show a broad peak centered at each IC point, with the width varying inversely as ξ. To obtain an analytic formula describing the q-dependence of the scattering, consider a model in which the spins have a uniform correlation within each domain, but where the correlations between domains fall off exponentially with distance. If the charge stripes are separated by n lattice spacings, then the intensity along $\mathbf{Q} = (h, \frac{1}{2}, 0)$ should be proportional to

$$I(h) \sim |F|^2 \frac{1-p^2}{1+p^2-2p\cos(2\pi nh)}, \tag{3}$$

where $p = (-1)^{n+1}\exp(-na/\xi)$, and F is the structure factor for a single domain. (The factor of -1 included in p takes care of the antiphase relationship between neighboring domains.) For the calculations below, it will be assumed that the spin correlations also fall off exponentially parallel to the stripes, with the same correlation length ξ.

3. $La_{1.85}Sr_{0.15}CuO_4$

Recently, Hayden *et al.* [10] reported inelastic neutron scattering measurements of $\chi''(\mathbf{Q},\omega)$ in $La_{1.86}Sr_{0.14}CuO_4$, covering a very large energy range. They found that χ'' at $\hbar\omega > 100$ meV could be modelled reasonably well using spin wave formulas, with only a slight reduction in J. At lower energies, they showed that $\tilde{\chi}''(\omega)$ has a peak at $\Gamma = (22 \pm 5)$ meV. If this latter result is combined with the value $\hbar c_0 = (850 \pm 30)$ meV-Å [11], then by inverting Eq. (2) one obtains the estimate $\xi = (27 \pm 6)$ Å $= (7.0 \pm 1.5)a$.

Figure 1 shows an inelastic scan, at a fixed energy transfer of 3 meV, through the IC peaks found in a single crystal of $La_{1.85}Sr_{0.15}CuO_4$ [12]. The peak positions are consistent with a charge stripe period of $n = 4$. The solid line through the data points is a calculation using Eq. (3) and the estimate $\xi = 7a$—only the amplitude and the linear background have been adjusted to fit the data. (The calculated intensity, taking into account signal from all 4 IC peaks, was also convolved with the experimental resolution function.) The calculation clearly gives an excellent description of the data.

4. $YBa_2Cu_3O_{6.6}$

In contrast to the 214 system, measurements of χ'' in $YBa_2Cu_3O_{6.6}$ do not show well resolved IC peaks [13]. As shown in Fig. 2, the scattered intensity is peaked at $\mathbf{Q}_{AF}$, but with a peak shape that has a rather flat top and steep sides. Despite the lack of clear evidence for incommensurability, it is possible nevertheless to describe the data with the same stripe model discussed above. To give the stripe picture an honest test, ξ will again be estimated using Eq. (2). The value of Γ is approximately (30±5) meV [13], and $\hbar c_0 = (670 \pm 30)$ meV-Å [14,15], which together yield $\xi = (16 \pm 3)$ Å $= (4.1 \pm 0.8)a$. To estimate

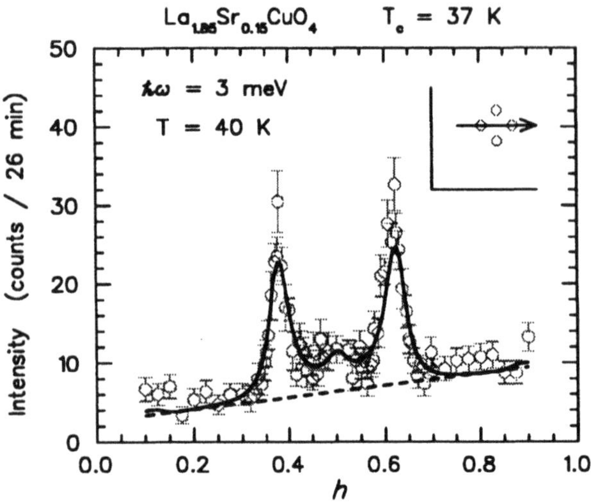

Figure 1. Inelastic scan at $\hbar\omega = 3$ meV through the incommensurate points (see inset) measured at $T = 40$ K on a crystal of $La_{1.85}Sr_{0.15}CuO_4$ with $T_c = 37$ K. Solid line is the calculation described in the text; dashed line is the fitted background. Scan is along $\mathbf{Q} = (h, \frac{1}{2}, 0)$, within the $(hk0)$ zone, as indicated schematically in the inset.

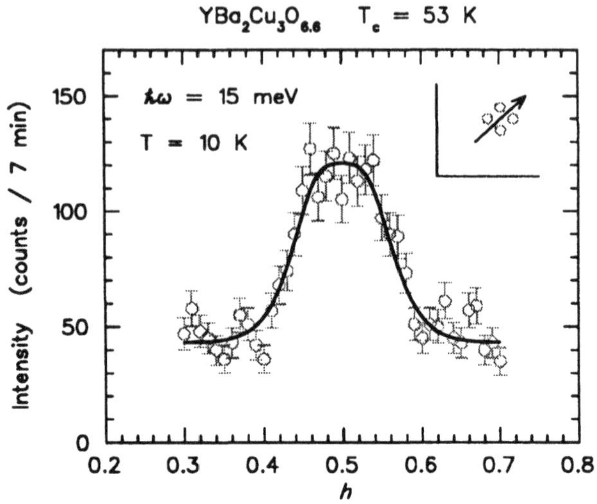

Figure 2. Inelastic scan at $\hbar\omega = 15$ meV along the zone diagonal (see inset) measured at $T = 10$ K on a crystal of $YBa_2Cu_3O_{6.6}$ with $T_c = 53$ K. The solid line is the calculation described in the text. The scan was measured along $\mathbf{Q} = (h, h, -1.8)$ within the (hhl) zone. The inset shows the scan direction within the $(h, k, -1.8)$ zone.

the stripe spacing, assuming that the hole density within the charge stripes is the same as in the 214 system [1], it is first necessary evaluate the hole concentration within a plane. Inverting the formula [16]

$$T_c/T_{c,max} = 1 - 82.6(p - 0.16)^2, \qquad (4)$$

yields $p \approx 0.088$ for the $T_c = 53$-K sample, which corresponds closely to $n = 6$. The solid line through the data in Fig. 2 represents a calculation using the same stripe model as above, with the parameter values $n = 6$ and $\xi = 4.1a$. The calculated function was convolved with the spectrometer resolution function, and only the amplitude and background were adjusted to fit the data.

For the calculated curve, the filling in of weight between the IC peaks occurs because ξ is less than the stripe spacing na. In this case the weak correlation between neighboring antiphase domains results in little interference. At $\mathbf{Q}_{AF}$ the scattering from an individual domain does not get cancelled by the weak contributions from neighboring domains. In contrast, the situation for $La_{1.85}Sr_{0.15}CuO_4$ is $\xi > na$. That is the necessary for condition for obtaining well resolved IC peaks. Although the correlations between antiphase domains are perhaps less well defined in 123, the smaller value of ξ suggests greater stripe fluctuations, a condition that appears to correlate with an increased T_c [2].

5. DISCUSSION

It has been shown that the $\mathbf{Q}$ dependence of scattered intensity from low-energy spin fluctuations in $La_{1.85}Sr_{0.15}CuO_4$ and $YBa_2Cu_3O_{6.6}$ can be consistently interpreted on the basis of dynamic antiphase antiferromagnetic domains. The correlation lengths ξ used to model the scattering measurements were calculated from experimental values of Γ and $\hbar c_0$. It was pointed out that the IC peaks become unresolvable when $\xi < na$, where na is the charge-stripe spacing.

There are, of course, other neutron scattering results that are consistent with the stripe picture. A comparison of $\tilde{\chi}''(\omega)$ in superconducting $YBa_2Cu_3O_{6.6}$ and antiferromagnetic $YBa_2Cu_3O_{6.15}$ shows that, despite a difference in ω dependence, the overall weight within the range 5 to 50 meV is comparable [14]. Similar results were found in a comparison of $La_{1.85}Sr_{0.15}CuO_4$ and La_2CuO_4 [10], although the spectral weight for the superconductor is somewhat reduced when the comparison is extended up to 300 meV. With a strong segregation of holes of density n_h, the density of Cu moments contributing to the low-energy (< 300 meV) spin fluctuations is roughly $1 - 2n_h$, so one may expect to see relatively little change in overall spectral weight compared to the undoped antiferromagnet.

Measurements on Zn-doped $La_{1.86}Sr_{0.14}CuO_4$ have shown that the positions and Q-widths of the IC peaks are essentially unaffected by the presence of the Zn [17]. Similarly, doping Zn into $YBa_2Cu_3O_{6.6}$ does not change the **Q** dependence of the scattering, but it does shift spectral weight down to low-energy, so that $\tilde{\chi}''(\omega)$ looks more like that of an antiferromagnet [18]. This suggests that doping Zn into the CuO_2 planes destroys superconductivity by pinning the stripes.

Recently there has been considerable excitement over observations of resonant enhancements of the magnetic scattering below T_c [19,20]. If the enhanced scattering is assumed to come from itinerant electrons, then it is certainly quite remarkable. On the other hand, if the magnetic scattering all comes from antiferromagnetic domains between charge stripes, the change in the scattering might in some way be associated with a reduction in damping of the spin fluctuations due to the onset of superconductivity. In any case, it will be interesting to see how models of these effects evolve.

REFERENCES

1. J. M. Tranquada *et al.*, Nature **375**, 561 (1995); J. M. Tranquada *et al.*, Phys. Rev. B **54**, 7489 (1996).
2. J. M. Tranquada *et al.*, Phys. Rev. Lett. **73**, 338 (1997).
3. C. H. Chen, S.-W. Cheong, and A. S. Cooper, Phys. Rev. Lett. **71**, 2461 (1993); J. M. Tranquada, D. J. Buttrey, V. Sachan, and J. E. Lorenzo, *ibid.* **73**, 1003 (1994).
4. S.-W. Cheong *et al.*, Phys. Rev. Lett. **67**, 1791 (1991).
5. K. Yamada *et al.* (preprint).
6. M. K. Crawford *et al.*, Phys. Rev. B **44**, 7749 (1991); J. D. Axe *et al.*, Phys. Rev. Lett. **62**, 2751 (1989).
7. A. Auerbach and D. P. Arovas, Phys. Rev. Lett. **61**, 617 (1988); S. Tyč, B. I. Halperin, and S. Chakravarty, *ibid.* **62**, 835 (1989); G. J. Mata and G. B. Arnold, Phys. Rev. B **39**, 9768 (1989).
8. A. H. Castro Neto and D. Hone, Phys. Rev. Lett. **76**, 2165 (1996).
9. J. M. Tranquada, P. Wochner, and D. J. Buttrey, cond-mat/9612007.
10. S. M. Hayden *et al.*, Phys. Rev. Lett. **76**, 1344 (1996).
11. G. Aeppli *et al.*, Phys. Rev. Lett. **62**, 2052 (1989).
12. B. J. Sternlieb and G. Shirane (unpublished); K. Yamada *et al.*, Phys. Rev. Lett. **75**, 1626 (1995).
13. J. M. Tranquada *et al.*, Phys. Rev. B **46**, 5561 (1992).
14. S. Shamoto *et al.*, Phys. Rev. B **48**, 13817 (1993).
15. S. M. Hayden *et al.*, Phys. Rev. B **54**, R6905 (1996).
16. M. R. Presland *et al.*, Physica C **176**, 95 (1991); S. D. Obertelli, J. R. Cooper, and J. L. Tallon, Phys. Rev. B **46**, 14928 (1992).
17. M. Matsuda *et al.*, J. Phys. Soc. Jpn. **62**, 443 (1993).
18. K. Kakurai *et al.*, Phys. Rev. B **48**, 3485 (1993).
19. T. E. Mason *et al.*, Phys. Rev. Lett. **77**, 1604 (1996).
20. H. F. Fong, B. Keimer, D. L. Milius, and I. A. Aksay, Phys. Rev. Lett. **78**, 713 (1997); P. Dai *et al.*, *ibid.* **77**, 5425 (1996).

ELSEVIER

Physica C 282–287 (1997) 170–173

PHYSICA C

Phase separation, Staging and Spin Fluctuations in La_2CuO_{4+y}

Y.Endoh[a], R.J.Birgeneau[b], M.A.Kastner[b], Y.S.Lee[b], G.Shirane[c], S.Wakimoto[a], B.O.Wells[b] and K.Yamada[a]

a Department of Physics, Tohoku University, Aoba-ku, Sendai 980-77, Japan
b Department of Physics and Center for Material Science and Engineering, Massachusetts Institute of Technology, Cambridge, MA 02139, USA
c Department of Physics, Brookhaven National Laboratory, Upton NY 11973, USA

Oxygen doped La_2CuO_{4+y} single crystals have been successively synthesized and studied their structural, magnetic and superconducting characters. Phase separation due to the existence of the miscibility gap as well as the oxygen staging were detected by systematic neutron diffraction studies. Spin correlations and fluctuations were also studied in details. Spin fluctuations in the oxygen doped La_2CuO_{4+y} can be compared with those of the Sr doped crystals and is understood in terms of the general 'hole' doping.

1. INTRODUCTION

The doped La_2CuO_4 is regarded as an important class of materials due to the fact that each CuO_2 layer is well separated by intervening block layers so that additional complications like an interlayer coupling can be neglected [1]. Furthermore sizable single crystals with high quality can be supplied to various experiments, in particular neutron magnetic scattering experiment [2]. Spin fluctuations or dynamical aspect of spin correlations is now recognized to act an essential role for the unusual high temperature superconductivity.

Recent neutron scattering studies of the $La_{2-x}Sr_xCuO_4$ showed a clear crossover in spin correlations at around x = 0.05. In a lower region of x (x ≤ 0.04), spin correlations shows a 2 dimensional (2D) antiferromagnetic characteristics of which the correlation length ξ at 0 K is well determined by the average distance of the doped hole separation, $a/\sqrt{x}$. In x ≥ 0.05 where superconductivity appears, on the other hand, the spin correlations becomes incommensurate without a static long range order. Spin fluctuations with the tremendous inelasticity is characteristics, which was observed by recent high energy neutron spectroscopy [3]. Spin dynamics cannot be well characterized by ξ. Then it becomes a current issue how to interpret this unique dynamical spin structure [4].

Careful studies by using high quality single crystals revealed that intrinsic inhomogeneity effect by Sr doping is enhanced in a critical concentration region near the insulator-metal transition. The spin glass phase and the charge localization appear, which makes detailed investigations very difficult [5]. In this respect,

PII S0921-4534(97)00254-2

the study by using oxygen doped La_2CuO_{4+y} is shed light on , because the random effect might not be the same as the Sr doping [6]. Namely, the annealing effect becomes evident in the oxygen doped La_2CuO_{4+y}, while there seems to be no annealing effect in Sr doped crystals, although the superconductivity is seriously affected by the oxygenation or reduction in $La_{2-x}Sr_xCuO_4$. Therefore we have systematically investigated the effect of the oxygen doping on the crystal structure, spin fluctuations and superconductivity by preparing rich oxygen doped single crystals of La_2CuO_{4+y} [6]. Then experimental results of this system are compared with the Sr doping studies in order to extract a common key role of hole doping on the superconductivity at least to find the effect of the intrinsic inhomogeneity.

2. PHASE SEPARATION and OXYGEN STAGING

Unlike oxygen doping into La_2NiO_{4+y} single crystals, the oxygen doping into La_2CuO_{4+y} single crystals is not easy requiring high temperature annealing under very high oxygen gas pressure or electro-chemical techniques. Recently oxygen-rich La_2CuO_{4+y} single crystals were successfully prepared by using the latter technique, which was described in details by B.O.Wells et al [6]. Careful experiments revealed that the phase separation is associated with the oxygen staging, which was first proposed by J.M.Tranquada from the systematic studies of La_2NiO_{4+y} single crystals [7]. The phase separation and oxygen staging are similar in both systems, but there are quantitative differences. For instance, the time scale for the ordering of oxygen into the staging phase is different; it takes several hours in La_2NiO_{4+y} systems, but much faster in La_2CuO_{4+y}. The oxygen disordered phase could be realized by a rapid cooling of the annealed, stage 2, $La_2NiO_{4.11}$ single crystal [8]. The fact of the wider range of the staging in the latter system reflects the difference of the ordering time of doped excess oxygen. Then a relatively large miscibility gap between an oxygen-poor antiferromagnetic phase and oxygen-rich superconducting phase of stage 6.6 occurs between y=0.01 and y=0.055 in La_2CuO_{4+y}, where the antiferromagnetic long range order might disappear. All of the samples studied are multi-phase single crystals.

In this respect, Neel temperatures of flux grown La_2CuO_4 single crystals are lower than that of pure La_2CuO_4 single crystal and also a single phase with lower Neel temperatures is available by oxidation of the flux grown La_2CuO_{4+y} [9]. This suggests that a small amount of impurities molten from Pt crucible or the change of the grown condition, possibly a lower melting temperature at the grown status makes a large effect on oxygen ordering. Then we applied this interesting phenomenon to oxygen doping into a $La_2Bi_xCuO_{4+y}$ system [10], where x is around 0.05 - 0.1 and Bi can be trivalent. The experimental results are shown in the following section.

Structural model constructed from neutron diffraction experiments revealed that doped oxygen atoms occupying the center of La tetrahedron (tetrahedral site) form a 2D sheet, which regularly stacks along c direction. The stacking sequence is determined by the staging like an intercalation in the graphite. Note that neutron diffraction detects the super cell structure separated by a series of antiphase domain boundaries between reversed tilt directions of the CuO_6 octahedra, not the doped oxygen staging itself. Temperature dependence of the superlattice reflection also revealed that the phase transition due to the oxygen staging is of the first order and the transition temperature is weakly dependent on the staging number, which is ranged 220K - 250 K.

3. SPIN CORRELATIONS

The hole doping effect on 2D antiferromagnetic correlations is so dramatic that the long range order is easily suppressed by an only 1% doping, although the 2D antiferromagnetic network is not destroyed. Hence thermal evolution of ξ is approximated by the following heuristic relation.

$$\xi_x(T) = \xi_x(T=0) + \xi_{x=0}(T) :$$
$$x = \text{hole or Sr concentration}$$

However as mentioned above, this relation could not be tested for the oxygen doping sample due to the existence of a large miscibility gap, but this relation was confirmed to hold in the samples of the flux grown single crystals [9]. Then S.Wakimoto et al. also showed that spin correlations in single crystals of $La_{1.95}Bi_{0.05}CuO_{4+y}$, where single phase samples were successfully synthesized with continuous changing of y, holds this this relation. These systems don't have a miscibility gap at around x=0.02-0.07, in which region the insulator-metal transition occurs. This search will be continued to answer the question whether a quantum critical point exists.

2D antiferromagnetic spin correlations in the lightly doped single crystals changes the signature, of which the sharp peaks appear in the 2D reciprocal lattice of $\pi/a(1\pm\delta,1)$ and $\pi/a(1, 1\pm\delta)$ with varying δ as a function of hole doping, at a certain concentration of Sr doping (x=0.05) We define the incommensurate spin fluctuations, since it essentially possesses inelastic nature. Then the interpretation of the incommensurate spin fluctuations has become an important issue [4]. K.Yamada et al. [11], showed that the incommensurability or the wave vector δ starts from the critical value of x = 0.05 of $La_{2-x}Sr_xCuO_4$. Furthermore they observed another heuristic relation that δ is linearly related to Tc in x < 0.15, where the highest Tc (= 38 K) realizes in this system. We have plotted δ of the stage 6.3 crystal with Tc = 31 K and y = 0.055. Then δ=0.105 falls perfectly on the straight line defined by the $La_{2-x}Sr_xCuO_4$ crystals [11]. The doping effect on δ can be mapped in a unified curve of δ vs. x as is shown by K.Yamada in this conference [12]. According to this mapping, the saturation behavior becomes evident when x exceeds the optimal value around x=0.15. It is important to note here that the effective charge number or hole concentration also saturates in this x range [13]. Therefore, all the data of δ fall on the straight line with respect to the effective charge number. On the other hand , the incommensurability δ in $La_{2-x}Sr_xNiO_{4+y}$ system gives a straight line with respect to the dopant charge concentration, although this system is all non-metallic in the doping region studied and also the incommensurate structure is different. The different feature of the incommensurability in both systems can not be explained by the charge-stripe model in a simple way.

Furthermore, when we compared $\chi''(q,\omega)$ with the previously observed data of $La_{2-x}Sr_x$-CuO_4, thermal evolution of $\chi''(q=Q_\delta,\omega)$, where Q_δ is $\pi/a(1\pm\delta,1)$ or $\pi/a(1, 1\pm\delta)$, is identical to those of $La_{1.85}Sr_{0.15}CuO_4$ with almost equivalent Tc = 33 K crystal. The data are characterized that $\chi''(q=Q_\delta,\omega)$ at lower energies show a salient peak at around Tc, below which intensities drop substantially but remain finite at the lowest temperature. Since the optimal or over doping crystal show a clear energy gap of 3.5 meV in the superconducting state, the remaining magnetic scattering at lower energies might give an important hint for another issue whether it is intrinsic for under doped samples with poor screening or is so-called a random effect due to the crystal inhomogeneity, If the random effect is significantly different in two systems, the fact of the identical character of $\chi''(q=Q_\delta,\omega)$ in the two different systems might suggest intrinsic. In this respect, we have been recently succeeded to prepare an under doped superconducting $La_{1.95}Bi_{0.05}CuO_{4+y}$ single crystal (y=0.035), which presents Tc=15K. Since the issue of the incommensurate spin fluctuations still be unsolved, further challenge will be required by using better quality and rich-oxygen doped La_2CuO_{4+y} single crystals.

Work at Tohoku university was carried out under the Research Grant of Monbusho-Japan of Priority Area " Anomalous Metallic State near the Mott Transition" and of Promotion of Scientific Research Grant of the Science Technology Agency in Japan. Work at MIT was supported by the MRSEC Program of the US-NSF under the award number DMR 94-00334. Research at BNL was carried out under the contract No.DE-AC02-76CH00016, Division of Material Sciences, US-DOE. This work was also supported under the Neutron Scattering Research Project of US-Japan Collaboration program sponsored by both Monbusho-Japan and US-DOE.

REFERENCES

1 G.Shirane, R.J.Birgeneau, Y.Endoh and M.A.Kastner, Physica **B 197** 158 (1994)

2 S.Hosoya et al., Physica **C235 -240**, 547 (1994)

3 S.M.Hayden et al., Phys.Rev.Lett. **76**, 1344 (1996), Y.Endoh et al., unpublished

4 J.M.Tranquada : this micro-symposium, T.Tanamoto, H.Kohno and H.Fukuyama, J.Phys.Soc.Jpn. **63**, 2739 (1994), V.J.Emery and S.A.Kievelson, Physica **C 235 -240**, 189 (1994) A.Aharony et al., Phys.Rev.Lett., **60**, 1330 (1988)

5 F.C.Chou et al., Phys.Rev.Lett. **75**, 2204 (1995)

6 B.O.Wells et al., Z.Phys. **B 100**, 535 (1996)

7 J.M.Tranquada, D.J.Buttrey, V.Sachan and J.D.Axe, Phys.Rev.Lett., **73**, 1003 (1994)

8 K.Nakajima et al., J.phys.Soc.Jpn. in press

9 B.Keimer et al., Phys.Rev.**B 45**, 7430 (1992)

10 S.Wakimoto et al., J.Phys.Soc.Jpn. **65**, 581 (1996)

11 K.Yamada unpublished

12 K.Yamada et al., this conference

13 S.Uchida et al., Phys.Rev.**B 43**, 7942 (1991)

ELSEVIER

Physica C 282–287 (1997) 174–177

PHYSICA C

Pairing and Phase Coherence in High Temperature Superconductors

V. J. Emery, [a] * S. A. Kivelson, and O. Zachar [b] †

[a]Dept. of Physics, Brookhaven National Laboratory, Upton, NY 11973

[b]Dept. of Physics, University of California at Los Angeles, Los Angeles, CA 90095

Mobile holes in an antiferromagnetic insulator form a slowly fluctuating array of quasi one-dimensional metallic stripes, which induce a spin gap or pseudogap in the intervening Mott-insulating regions. The mobile holes on an individual stripe acquire a spin gap via pair hopping between the stripe and its environment; *i.e.* via a magnetic analog of the usual superconducting proximity effect. This process is the analog of pairing in conventional superconductors. At non-vanishing stripe densities, Josephson coupling between stripes produces a dimensional crossover to a state with long-range superconducting phase coherence. In contrast to conventional superconductors, the superconducting state is characterised by a high density of (spin) pairs, but the phase stiffness, which is determined by the density and mobility of holes on the stripes, is very low.

1. Introduction

Superconductivity in metals requires pairing and long-range phase coherence [1]. In clean homogeneous conventional superconductors, pairing involves a relatively small fraction of the conduction electrons [2], but the superfluid density (which determines the phase stiffness) involves all of them. Here we argue that, in the high temperature superconductors [4], it is just the other way round: most of the holes are involved in pairing, but the superfluid density is proportional to the density of doped holes.

The poor phase stiffness of the high temperature superconductors is well known, and it implies that the transition temperature T_c is much less than the pairing temperature, $\sim \Delta_0/2$, where Δ_0 is the energy gap at zero temperature. This separation of the energy scales has been demonstrated on phenomenological grounds [3]. In underdoped and optimally doped materials, T_c is determined by the onset of phase coherence, which occurs at a temperature of about $\hbar^2 n_s(0)/4m^*$. Here $n_s(0)$, the two-dimensional superfluid density at zero temperature, determines the phase stiffness, and it is low because the high temperature superconductors are doped Mott insulators.

The need for a different approach to pairing is clear in view of the difficulty of achieving a high transition temperature via the conventional mechanisms. The problem is that a high pairing scale requires a strong attractive interaction which may favor other instabilities or, alternatively, produce a large mass renormalization which depresses the phase coherence temperature. Moreover, as the pairing energy increases, retardation becomes less effective, so it is all the more difficult to overcome the Coulomb repulsion. This problem is especially acute for the high temperature superconductors, which are doped Mott insulators with relatively poor screening. Angle resolved photoemission spectroscopy [5] (ARPES) suggests that the energy gap has the form $\cos k_x - \cos k_y$. This implies that, in real space, the gap function (and hence the pairing force) has a range of one lattice spacing, where the bare Coulomb interaction is very large.

In short, a theory of high temperature superconductivity must show how to obtain a large temperature scale for local superconductivity, without detriment to global phase coherence, despite poor screening of the Coulomb interaction. There are several phenomenological constraints on such a theory. First of all the order param-

*Supported by the Division of Materials Science, U. S. Department of Energy under contract No. DE-AC02-76CH00016.

†Supported in part by the National Science Foundation grant number DMR93-12606.

0921-4534/97/$17.00
PII S0921-4534(97)00255-4

eter has charge $2e$ [6], *i.e.* there is some kind of pairing. However ARPES experiments, which show that the chemical potential is near the center of the bare hole band, rule out real-space pairing, which anyway is implausible for a d-wave superconductor with a strong and poorly-screened Coulomb repulsion between electrons. More generally, the conventional view of superconductivity as a Fermi surface instability resulting from an attractive interaction between quasiparticles is inapplicable, since, according to analyses of resistivity [7] and ARPES data [8] there are no well-defined quasiparticles or Fermi surface in the normal state of high temperature superconductors.

2. Spin-gap Proximity effect

We have proposed [9] that the high temperature superconductors satisfy these constraints in a unique manner. The mechanism of pairing is a form of internal magnetic proximity effect in which a spin gap is generated in *Mott-insulating* antiferromagnetic (AF) regions through spatial confinement by charge stripes, and communicated to the stripes by pair hopping. Many of the problems listed above are avoided because the order parameter relies on spin-pairing and does not require the existence of *bound* pairs of holes. The successive steps in the argument are as follows:

An AF insulator tends to expel holes [10]. For neutral holes this leads to phase separation into hole-rich and hole-free regions [11], whereas, for charged holes, it gives rise to *local* charge inhomogeneity, typically in the form of topological doping [12], in which metallic stripes are separated by insulating antiphase AF regions [13]. There is much experimental evidence of ordered or fluctuating structures of this kind [14,15].

Topological doping [12] is a general feature of doped Mott insulators, and it amounts to a strong (anti)correlation of spin and charge. However, within a metallic stripe or the intervening undoped regions there is a separation of spin and charge [16], as in the one-dimensional electron gas (1DEG). In the 1DEG, the boson representation of the operator $\psi_{1,\uparrow}\psi_{2,\downarrow} - \psi_{1,\downarrow}\psi_{2,\uparrow}$ (which annihilates pairs of fermions at opposite Fermi points, labelled by 1,2) may be written in the form $\exp(-i\theta_c)\cos\phi_s$, where θ_c and ϕ_s are associated with the charge and spin degrees of freedom respectively. This is an *operator relation*, and θ_c, the superconducting phase field, is a property of the charge modes. On the other hand, pairing (or, equivalently a well-defined amplitude) corresponds to a finite expectation value of $\cos\phi_s$, which requires a spin gap [17].

A large spin gap (or pseudogap) arises naturally in a spatially-confined AF region, such as the medium between stripes. This behavior is well documented for clusters [18], frustrated spin chains [19], and spin ladders [20]. In general, if there are N spins in a unit cell, all spin excitations are gapped if N is even, but one excitation is ungapped if N is odd, *i.e.* there is a pseudogap. Such a spin gap does not conflict with the Coulomb interaction since the energetic cost of having localized holes in Cu $3d$ orbitals has been paid in the formation of the material.

The spin degrees of freedom of the 1DEG acquire a spin gap by pair hopping between the stripe and the AF environment [9]. Because of the *local* separation of spin and charge, the spin-gap fixed point is stable even in the presence of strong Coulomb interactions, and there is no mass renormalization to depress the onset of phase coherence. Thus the phase stiffness of the mobile holes on the stripes is large enough to give a high superconducting transition temperature.

Both superconducting and charge density wave correlations develop on a given stripe. They compete at longer length scales, although they may coexist in certain regions of the phase diagram.

2.1. Symmetry of the order parameter:

If stripe order breaks the four-fold rotational symmetry of the crystal, the superconducting order will have strongly mixed extended-s and $d_{x^2-y^2}$ symmetry. This will happen in a stripe-ordered phase, such as in $La_{1.6-x}Nd_{0.4}Sr_xCuO_4$, or in a possible "stripe nematic" phase, in which the stripe positional order is destroyed by quantum or thermal melting or quenched disorder, but the stripe orientational order is preserved. Such phases also would display large induced asymmetries in the electronic response in the ab-plane.

In tetragonal materials, the order parameter

must have a pure symmetry, but the way in which it emerges from the short-distance physics is very different from more conventional routes. If doping is not too high, $d_{x^2-y^2}$ order should give the long distance behavior because the extended-s order parameter ($\cos k_x + \cos k_y$) is small on the Fermi surface of the noninteracting system. Nevertheless, if the stripe fluctuations exhibit substantial orientational order at intermediate length scales, the interplay between the two types of superconducting order may be more subtle than in conventional, homogeneous materials.

2.2. Phase diagram

A feature of the stripe model is that, in lightly doped materials, the temperature scale, T_{pair}, at which pairing occurs on a single stripe is parametrically larger than the superconducting transition temperature, T_c, which is governed by the Josephson coupling between stripes. Moreover, both T_{pair} and T_c must be less than the temperature scale, T_{AF}, at which the *local* AF correlations are developed. The crossovers observed experimentally in underdoped high temperature superconductors have tentatively been identified [22,9] with these two phenomena.

3. Evidence for Spin Pairing

The idea that there is pairing in a large range of temperatures above T_c is an immediate consequence of the fact that T_c is significantly suppressed by phase fluctuations; *i.e.* that T_c is close to the value obtained from the phase stiffness derived from the London penetration depth at zero temperature. Moreover the idea that "pairing" means a high density of singlets between stripes, rather than *bound* Cooper pairs provides a very natural explanation of the spin-gap behavior that has been widely observed in planar copper NMR measurements in underdoped cuprates [23]. The interpretation of the spin gap as a superconducting gap has recently received considerable support from ARPES experiments [24] which find that the magnitude and wave vector dependence of the pseudogap above T_c is similar to that of the gap seen well below T_c in both underdoped and optimally doped materials. The temperature above which this gap structure disappears correlates well with the pairing scale deduced from NMR. The in-plane optical response, which shows a narrowing of the "Drude-like" peak, rather than a pseudogap structure [25], also is consistent with the idea of spin pairing.

3.1. Spectroscopic evidence

The most direct evidence for the proposed pairing mechanism would be the observation of an isolated spin-1, charge zero excitation, with an energy of order the superconducting gap, which could be identified as the low-energy excitation of a small undoped region [26]. Neutron scattering experiments on optimally-doped $YBa_2Cu_3O_{7-\delta}$ [27] have indeed found a spin triplet excitation, with wave vector $(\pi/a, \pi/a)$ and energy 40meV which first appears in the neighborhood of T_c. At lower doping [28], the energy of the mode decreases, and it first appears above T_c, although its intensity is enhanced below T_c. Recently, ARPES data in $Bi_2Sr_2CaCu_2O_{8+\delta}$ have been interpreted as evidence for such a mode [29].

It is natural to interpret these experiments in terms of excitations of the pairs that give rise to high temperature superconductivity [27]. However it is important to note that the observed cross section and Cu-Cu bilayer modulation of the triplet mode are very close to those of AF spin waves in the undoped antiferromagnet. (See *e.g.* Fig. 4 of ref. [27].) This strongly suggests that the mode is associated with undoped regions of the material and not with mobile holes. The point is that the wave function of the mobile holes has a strong admixture of O(2p) orbitals and therefore a different magnetic form factor than in the undoped material, in which Cu(3d) orbitals predominate. Furthermore, in a conventional BCS picture, the intensity of the peak would be quite small because it is proportional to $N(0)\Delta_0$.

In our model [9], there is a composite order (involving the stripe and the AF regions), which does not break gauge invariance, but associates a superconducting phase factor $\exp(-i\theta_c)$ with singlets in the AF regions. Thus, the triplet mode should appear below T_{AF}, and its intensity should begin to increase below T_{pair} (where significant phase fluctuations begin), then grow substantially near T_c (where long-range phase order is established). In optimally-doped $YBa_2Cu_3O_{7-\delta}$, T_{AF}, T_{pair} and T_c are very close together [22,9], so it is reasonable that the mode is first observed close to T_c. The position and width of the peak reflect the wave function of the singlets; it occurs at $(\pi/a, \pi/a)$ because the singlets involve spins on opposite sublattices, and it it corresponds to a singlet size of a few lattice spacings. A more detailed description of this behavior will be given in a future publication.

REFERENCES

1. J. Bardeen, L. N. Cooper, and J. R. Schrieffer, Phys. Rev. **106**, 162 (1957).
2. The density of electrons involved in particle-hole mixing is $N(0)\Delta_0$, where $2N(0)$ is the density of states at the Fermi level.
3. V. J. Emery and S. A. Kivelson, Nature **374**, 434-437 (1995).
4. J. G. Bednorz and K. A. Müller, Z. Phys. B **64**, 189 (1986).
5. Z.-X. Shen *et al.*, Phys. Rev. Lett. **70**, 1553 (1993): H. Ding *et al.* Phys. Rev. B **54**, R9678 (1996).
6. C. E. Gough *et al.*, *Nature* **326**, 855 (1987).
7. V. J. Emery and S. A. Kivelson, Phys. Rev. Lett. **74**, 3253 (1995).
8. J. C. Campuzano *et al*, Phys. Rev. B **53**, R14737 (1996).
9. V. J. Emery, S. A. Kivelson, and O. Zachar, Phys. Rev. B, submitted.
10. J. R. Schrieffer, S. C. Zhang, and X. G. Wen, Phys. Rev. Lett. **60**, 944 (1988).
11. V. J. Emery, S. A. Kivelson, and H.-Q. Lin, Phys. Rev. Lett. **64**, 475 (1990); C. S. Hellberg and E. Manousakis, preprint (cond-mat/9611195).
12. S. A. Kivelson and V. J. Emery, Synthetic Metals, **80**, 151-158 (1996).
13. U. Löw, V. J. Emery, K. Fabricius, and S. A. Kivelson, Phys. Rev. Lett. **72**, 1918 (1994); L. Chayes, V. J. Emery, S. A. Kivelson, Z. Nussinov, and J. Tarjus, Physica **A225**, 129 (1996).
14. Experimental evidence for stripes is given in J. M. Tranquada's paper at this conference.
15. For other perspective on the origin and implications of stripes, see the papers of C. Castellani and J. Zaanen at this conference.
16. A. Luther and V. J. Emery, Phys. Rev. Lett. **33**, 589 (1974).
17. V. J. Emery, Phys. Rev. B **14**, 2989 (1976); G. T. Zimanyi, S. A. Kivelson, and A. Luther, Phys. Rev. Lett. **60**, 2089 (1988); S. A. Kivelson and G. T. Zimanyi, Molec. Cryst. Liq. Cryst. **160**, 457-481 (1988).
18. S. Tang and H. Q. Lin, Phys. Rev. B **38**, 6863 (1988).
19. C. K. Majumdar and D. K. Ghosh, Math. Phys. **10**, 1388 (1969).
20. For a review see E. Dagotto and T. M. Rice, Science, **271**, 618 (1996).
21. D. A. Wollman *et al.*, Phys. Rev. Lett. **71**, 2134 (1993); A. G. Sun *et al.*, Phys. Rev. B **50**, 3266 (1994); D. A. Brawner *et al.*, Phys. Rev. B **50**, 6530 (1994); A. Mathai *et al.*, Phys. Rev. Lett. **74**, 4523 (1995).
22. B. Batlogg and V. J. Emery, *Nature* **382**, 20 (1996).
23. W. W. Warren *et al*, Phys. Rev. Lett. **62**, 1193 (1989); H. Yasuoka, T. Imai, and T. Shimizu, in *Strong Correlation and Superconductivity*, edited by H. Fukuyama, S. Maekawa, and A. P. Malozemoff, (Springer-Verlag, Berlin, 1989) p. 254.
24. A. G. Loeser *et al.*, *Science* **273**, 325 (1996); H. Ding *et al.*, *Nature* **382**, 51 (1996).
25. J. Orenstein *et al.*, Phys. Rev. B **42**, 6342 (1990); Z. Schlesinger *et al.*, Phys. Rev. Lett. **65**, 801 (1990); L. D. Rotter *et al.*, Phys. Rev. Lett. **67**, 2741 (1991); D. Basov *et al.*, Phys. Rev. B **52**, R13141 (1995).
26. V. J. Emery and S. A. Kivelson, Physica **C 209**, 597 (1993).
27. H. F. Fong *et al.*, Phys. Rev. B **54**, 6708 (1996), and references therein.
28. J. M. Tranquada *et al.*, Phys. Rev. B **46**, 556 (1992); P. Dai *et al.*, Phys. Rev. Lett. **77**, 5425 (1996); H. F. Fong *et al.*, Phys. Rev. Lett. **78**, 713 (1997);
29. Z. X. Shen and J. R. Schrieffer, unpublished; M. R. Norman *et al.*, preprint (cond-mat/9702144).

ELSEVIER

Physica C 282–287 (1997) 178–181

PHYSICA C

Dynamical stripe correlations and the spin fluctuations in cuprate superconductors

J. Zaanen and W. van Saarloos[a]

[a]Institute Lorentz for Theoretical Physics, Leiden University,
P.O. Box 9506, NL-2300 Leiden, The Netherlands

It is conjectured that the anomalous spin dynamics observed in the normal state of cuprate superconductors might find its origin in a nearly ordered spin system which is kept in motion by thermally meandering charged domain walls. 'Temperature sets the scale' finds a natural explanation, while a crossover to a low temperature quantum domain wall fluid is implied.

1. Introduction

Much excitement is generated by the observation by Tranquada *et al.* of the striped phase in the cuprates[1], which appears to be in a tight competition with the superconducting state as long as the doping level is not too high[2]. Taken together with the observation of the persistent gap in the normal state[3] and the finding that a large magnetic field stabilizes an insulating state[4], this points to a bosonic physics. The electrons pair up above T_c, forming an interacting boson system and as in, e.g., ^4He the crystalline ('charge-ordered') and superconducting state appear as the natural candidates for the ground state.

The real novelty of the striped phase is that it is at the same time a Néel spin condensate. At least, it is a bound state of charge and spin[5], with the specialty that the charge sector couples into the spin sector in the form of *disorder operators*: the charged stripes are anti-phase boundaries in the Néel state. In fact, all evidence for persisting dynamical stripe correlations in the superconductors rests entirely on the spin sector[2].

Initially we considered the possibility of stripe-like correlations, motivated by the anomalous behavior of the spin-fluctuations in the cuprates[6]. Both inelastic neutron scattering[8,9] and NMR[7] reveal that these fluctuations show a characteristic time- (τ) and length-scale (ξ) which are both precisely inversely proportional to temperature. The lack of an intrinsic scale is reminescent of a (quantum) critical regime. This is the underlying idea behind the 'nearly antiferromagnetic Fermi liquid' idea of Pines and coworkers[10]: the system is on its way to a zero temperature ordered spin state, and superconductivity intervenes at the last moment.

There is a conceptual difficulty with this scenario. It is expected that temperature and doping play the role of control parameters: how can it be that the critical regime extends over a temperature- and doping range of $\sim$ 1000 K and $\sim$ 10 %, respectively? At least classical critical regimes do not behave in this way. Assuming that the system is on its way to the striped phase, the relative insensitivity to doping dependence is trivially explained. In contrast to normal antiferromagnets, the striped phase is an incommensurate solid which can exist over a large doping range. The non-trivial part is to explain the temperature dependences of the characteristic dynamical scales.

The problem of thermally fluctuating domain walls in two dimensions ('incommensurate fluids') has been studied in great detail[11]. In such systems, the phenomenon of *entropic repulsions* is common. For increasing temperature, the thermal meandering motions of individual domain walls increases, as well as the number of collisions between domain walls. Collisions cost entropy and the net effect is that the stiffness of the system as a whole increases. Due to these entropic repulsion effects, temperature becomes an impor-

0921-4534/97/$17.00
PII S0921-4534(97)00186-X

tant parameter in the dynamics. It is emphasized that the basic effect is unrelated to criticality and it can therefore appear over a large temperature range. As we showed in an earlier paper[6], the behaviors of the characteristic time- and length scales of the spin fluctuations match in a natural way with the expectations for a thermally fluctuating incommensurate fluid in two dimensions.

Although the essence of our earlier work is unchanged, the relationship between the domain wall dynamics and the observed spin dynamics was treated too casually in this earlier paper[6]. After summarizing the basics, we will make this connection more precise. This is used, together with a refined experimental characterization of the normal state magnetic fluctuations[9], to deduce a more complete picture of the domain wall fluid.

2. Incommensurate fluids in two dimensions.

The standard theory of incommensurate fluids[11] starts with the assumption that single domain walls (DW) behave like Gaussian strings. It is assumed that its motions can be completely parametrized in terms of a transversal sound mode $\omega_q = c_W q$, where q is the wave number along the string, while $c_W = \sqrt{\Sigma/\rho}$ (Σ is the string tension and ρ the mass density). In addition, it is convenient to first only consider hard core string-string interactions. At finite temperatures, DW will be subject to meandering motions. The mean square fluctuation of a single string in 2D in the transverse z direction between two points separated by a distance l diverges as $\langle (\Delta z_l)^2 \rangle \simeq \frac{k_B T}{\Sigma} l$. When the walls have an average distance d, they will on average collide when $\langle (\Delta z_{l_c})^2 \rangle \simeq d^2$. The typical distance between collision points follows immediately,

$$l_c \simeq \frac{\Sigma d^2}{k_B T} \tag{1}$$

This is the most important length scale in this problem. It sets (a) the crossover length where the system changes from single wall dynamics to that of the domain wall fluid, (b) it determines the collision density and thereby the strength of the entropic repulsions, and (c) it corresponds to the shortest spin-spin disordering length in the special case of the striped fluid. For a further elaboration of the statistical physics we refer to the literature[11]. The essence is that the domain wall system can be described as a 2D elastic medium with an elastic constant which itself is proportional to temperature: $F \sim T \int d\vec{r} (\Delta \vec{n})^2$. This is like a XY spin system with $J \sim T$, and it follows that it is characterized by a *zero temperature Kosterlitz-Thouless transition*: at every finite temperature, free dislocations proliferate and the system is a fluid.

In order to adress the temporal aspects of the dynamics we suggested that the single wall dynamics is *coherent* on length scales $\leq l_c$: the single string sound oscillations are assumed to be underdamped. If this is the case, the relative motion of two points a distance l apart becomes important after a time l/c_W, and likewise the mean average fluctuations of two points a distance l_c apart deviates significantly from the value d after a time of order l_c/c_W. Using Eq. (1), it follows that the collision frequency is of order[6],

$$\hbar\Gamma = \mu k_B T, \quad \mu = \frac{\pi\hbar}{\rho c_W d^2} \tag{2}$$

Hence, we find both a characteristic length- and time scale which vary as $1/T$, with the obviously important meaning that they both refer to the crossover from single wall to many wall behavior. It remains to be demonstrated that these characteristic scales govern the spin dynamics as well.

3. Charged domain walls and spin fluctuations.

The unique feature of the stripes is that it corresponds with an incommensurate solid/fluid problem in the charge/longitudinal sector, coupled into a spin problem: when the stripes are ordered one is still left with a Heisenberg antiferromagnet. Besides the modes of the previous section, one has to deal with the transversal fluctuations associated with the rotational invariance of the spin system. At least on the classical level, *the domain wall- and the tranversal spin fluctuations appear to decouple in the infrared.* Although this is theoretically not well understood, this appears to be implied by the fact that the charge- and spin ordering transitions are decoupled[1,12]. This can be stated more precisely as follows: as long as the charged domain walls are intact, the spin system is unfrustrated. Calling the exchange interaction inside the spin domains J and that between spins on opposite sides of the wall J', the (classical) spin system appears identical for every possible configuration of domain walls when $J = J'$. Apparently, $J - J'$ is an irrelevant operator. Hence, although microscopically $J \neq J'$ (and frustration), the spin- and domain wall dynamics decouple in the approach to the long wave length limit, although the numerical factors (like the spin wave velocity) have to be adjusted.

Does this mean that the domain wall dynamics would be completely invisible in a measurement of the spin dynamical form factor? In fact, in this way one measures a *convolution* of (transversal) spin- and domain wall dynamics. One can define two distinct equal time correlators ($\vec{M}_i = (-1)^i \vec{S}_i$ is staggered spin),

$$S(R) = \langle \vec{M}_{i+R} \cdot \vec{M}_i \rangle \sim e^{-R/\xi} \quad (3)$$

$$S_T(R) = \langle \vec{M}_{i+R} \left[\Pi^{i}_{l=i+R} e^{i\pi \hat{n}_l} \right] \vec{M}_i \rangle \sim e^{-R/\xi_T} \quad (4)$$

where the last proportionalities hold when the spin system is disordered. In Eq. (4), $\hat{n}$ measures the charge associated with a single domain wall unit cell (e.g., in the nickelates $\hat{n}$ corresponds with the hole number operator, while in the half-filled cuprate walls $\hat{n} = 1$ corresponds with half a hole). Hence, the charge sector appears in a *disorder operator* form in the spin correlator Eq. (4): regardless of the location of the domain walls, the change of sign of the spin order parameter upon crossing the walls is undone by the term between square brackets. It is this correlator which measures the transversal spin correlations. On the other hand, in the experiment one measures the correlator Eq.(3), which is also sensitive for the disorder induced by the domain wall fluid. This general wisdom should also apply to unequal time correlations.

With the above observations, it becomes possible to deduce some characteristics of the $\vec{q}$ and ω dependence of the spin dynamical structure factor. Let us consider the case that Eq. (4) would show near-long-range order, while the charge sector is an incommensurate fluid ($\xi_T >> \xi$). To address the temporal aspects of the dynamics it is convenient to consider the imaginary part of the *local* dynamical spin susceptibility $\chi''(\omega) = \int d\vec{q} \chi''(\vec{q}, \omega)$: every time a domain wall passes a particular site, the spin at this site is flipped and $\chi''(\omega)$ is dominated by this dissipation source. It is easy to see that the domain wall scale Γ, Eq. (2), should appear as a characteristic scale in χ'': (a) when $\omega >> \Gamma$ the domain wall system appears as static and the frequency dependence should be similar to that of the pure spin system, (b) when $\omega \to 0$ the spin-flip rate is non-zero as long as the domain walls are in a fluid state (even when $\xi_T \to \infty$) and on hydrodynamic grounds $\chi''(\omega) \sim \omega$. In ref.[6] we present a more detailed hydrodynamic argument showing that in fact $\chi''(\omega, T) \sim \omega/T$. The crossover regime between the high- and low frequency regimes should be governed by the single scale parameter Γ, and $\chi''(\omega, T) \sim F(\omega/\Gamma) = F(\omega/(\mu k_B T))$, and this is the scaling observed in neutron scattering [8]. Hence, we suggest that this crossover regime is actually probed in the inelastic neutron scattering experiments. As an interesting ramification, the prefactor μ in Eq. (2) is also the dimensionless quantum of action of the incommensurate fluid: the characteristic energy scale for quantum fluctuations $\hbar\Gamma_Q = (\hbar c_W/a)\exp(-1/\mu)$[6]. If, and only if, μ is of order unity a crossover is expected at a reasonable temperature from a high temperature classical fluid to a low temperature quantum regime (like in ^4He) and this appears as a consistency requirement in the present context. Experimentally, $\mu \simeq 1$ [8].

A little more can be said regarding the spatio-temporal appearance of the magnetic fluctuations. Assuming the near decoupling of the transversal- and domain wall dynamics, the spin wave (transversal mode) spectrum of the static striped phase appears as a sensible zeroth order: the goldstone modes attach to the incommensurate wave vectors and the branches should come together at the (π, π) point, while gaps $\sim J - J'$ are found at the new Brillioun zone boundaries at smaller momenta[13]. When the domain walls are disordered, one expects that the spin waves start to propagate when their wave length becomes of order of, or less than the disordering length of the domain wall fluid. Longer wavelength spin waves will be overdamped. Hence, as a function of frequency a crossover has to occur from an overdamped response at low frequencies to an underdamped spin wave response at higher frequencies. The frequency where this occurs is $\omega_c \sim c\pi/l_{m.f.}$, where the mean-free path of the spin waves is expected to be proportional to the domain wall disordering length: $l_{m.f.} \simeq \alpha\xi_c$, where α is a spin wave-domain wall scattering cross section ($0 < \alpha < 1$). At frequencies much less than ω_c, the width in momentum space should be set by $\pi/l_{m.f.}$. As long as the spin-wave cones are unresolved, the width of the peaks in $S(\vec{q},\omega)$ should behave as a simple geometric average of the inverse domain wall disordering length ($1/(\alpha\xi_c)$) and the apparent ω dependent width coming from the unresolved spin-wave cone ($\kappa_\omega = \omega/c$). The domain wall disordering length is nothing else than the collision length (Eq.1) and we expect for the frequency- and temperature dependence of the width of the incommensurate peaks,

$$\kappa(\omega,T) = \sqrt{\frac{(k_B T)^2}{(\alpha \Sigma d^2)^2} + \frac{\omega^2}{c^2}} \tag{5}$$

where c is the spin-wave velocity.

Except for a residual width, this is the behavior found in the experiment[9]. Interestingly, Aeppli *et al* find that temperature and frequency appear with common prefactors. Together with $\mu \simeq 1$ this implies that $\alpha c_W \simeq c$: the product of the spin wave-domain wall cross section and the wall transversal sound velocity is of order of the spin wave velocity ($c \simeq 0.2$ eV Å). This observation might hint at the microscopic mechanism of stripe formation.

Further experimentation is needed to see if the above physics is the correct one. Central to our analysis is the assertion that the collective dynamics in the normal state is in a *classical* regime. General arguments indicate that the difference between Eqs. (3,4) acquires a different meaning in the quantum system, and it is tempting to speculate that the coherent spin mode observed by Mason et al [14] reflect the onset of quantum coherence in the domain wall fluid upon entering the superconducting regime.

Aknowledgements. We acknowledge stimulating discussions with G. Aeppli, A. V. Balatski, D. Pines, M. Horbach, S. Kivelson, D. H. Lee and J. Tranquada. JZ acknowledges financial support by the Dutch Academy of Sciences (KNAW).

REFERENCES

1. J. M. Tranquada *et al*, Nature **375**, 561 (1995).
2. J. M. Tranquada *et al*, Phys. Rev. Lett. **78**, 338 (1997).
3. B. G. Levi, Physics Today **49 (6)**, 17 (1996).
4. G. S. Boebinger *et al*, *Ground state of superconducting* $La_{2-x}Sr_xCuO_4$ *in 61-Tesla magnetic fields*, Bell Labs. preprint (1996).
5. J. Zaanen and O. Gunnarsson, Phys. Rev. B **39**, 236 (1989).
6. J. Zaanen, M. L. Horbach and W. van Saarloos, Phys. Rev. B **53**, 8671 (1996).
7. T. Imai *et al*, J. Phys. Soc. Jpn. **59**, 3846 (1990).
8. T. E. Mason *et al*, Phys. Rev. Lett. **68**, 1414 (1992).
9. G. Aeppli *et al*, *Magnetic fluctuations in the normal state of the high* T_c *superconductor* $La_{1.86}Sr_{0.14}CuO_4$, NEC preprint (1996).
10. P. Monthoux, A. V. Balatski and D. Pines, Phys. Rev. Lett. **67**, 3448 (1991).
11. see S. N. Coppersmith *et al*, Phys. Rev. B **25**, 349 (1982) and refs. therein.
12. J. M. Tranquada *et al*, Phys. Rev. Lett. **73**, 1003 (1994); J. M. Tranquada *et al*, *Charge-driven stripe order of holes and spins in* $La_{2-x}Sr_xNiO_4$ *with* $x = 0.225$, Phys. Rev. B (in press).
13. J. M. Tranquada *et al*, *spin dynamics in an ordered stripe phase*, cond-mat/9612007.
14. T. E. Mason *et al*, Phys. Rev. lett. **77**, 1604 (1996).

ELSEVIER

Physica C 282–287 (1997) 182–185

PHYSICA C

Local structure study of the underdoped-overdoped transition in $YBa_2Cu_3O_x$ (x=6.806-6.984) by yttrium EXAFS

J. Röhler * [a], P.W. Loeffen[b], K. Conder[c] and E. Kaldis [c]

[a]Universität zu Köln, II. Physikalisches Institut, Zülpicherstr. 77, D-50937 Köln, Germany

[b]European Synchrotron Radiation Facility, B.P. 220, F-38043 Grenoble, France

[c]Laboratorium für Festkörperphysik ETH, CH-8093 Zürich, Switzerland

We have measured the x-ray absorption-fine-structure (EXAFS) at the Y-K edge of $YBa_2Cu_3O_x$ for x=6.806, 6.826, 6.886, 6.947, 6.968, 6.984 at T=20-300 K. Optimum doping is found to be a notable point in the x-T phase diagram, also concerning the c-axis related displacements of the planar O2,3 and Cu2 atoms. The O2,3–Cu2 spacing along c is found to be largest near x_{opt}. In the underdoped regime doping increases the O2,3–Cu2 spacing by displacing the Cu2 layer from the O2,3 layer. In the overdoped regime doping reduces the O2,3–Cu2 spacing by displacing the O2,3 layer towards the Cu2 layer. Possible implications of these c-axis related displacements for the electronic structure of doped CuO_2 layers are dicussed.

1. INTRODUCTION

The CuO_2 planes in $YBa_2Cu_3O_x$ are not flat but dimpled. In contrast to the buckled CuO_2 planes in the 214 compounds the planar oxygens in $YBa_2Cu_3O_x$ are all displaced in the same direction. The O2,3 layers are located closest to the Y layer, and the Cu2 layers closest to the Ba layer (*cf.* Fig.1). Therefore the geometry of the CuO_2 planes in $YBa_2Cu_3O_x$ may be seen as stacked O2,3 and Cu2 layers separated in c direction. The O2,3–Cu2 spacing is $\simeq$ 0.275 Å in superconducting and optimum doped $YBa_2Cu_3O_{6.92}$ [1], but only $\simeq$ 0.200 Å in nonsuperconducting $YBa_2Cu_3O_6$ [2]. Neutron diffraction showed the O2,3–Cu2 spacing to increase in the underdoped regime of $YBa_2Cu_3O_x$ for $x \rightarrow x_{opt}$ [1], and to decrease in overdoped $Y_{1-y}Ca_yBa_2Cu_3O_7$, $0 \leq y \leq 0.2$ [4]. A step-like discontinuity in the x-dependency of the O2,3–Cu2 spacing occurs near $x_{opt}=6.92 \pm 0.01$ in the overdoped regime of Ca-free $YBa_2Cu_3O_x$ $x_{opt} \leq x \leq 7$ [1]. Here the vibrational dynamics of the planar oxygens O2,3 changes dramatically: the in-phase A_{1g} Raman mode) softens by $\simeq$ 40% for $x_{opt} \geq x \rightarrow 7$ [7]. The joint variation of the O2,3–Cu2 spac-

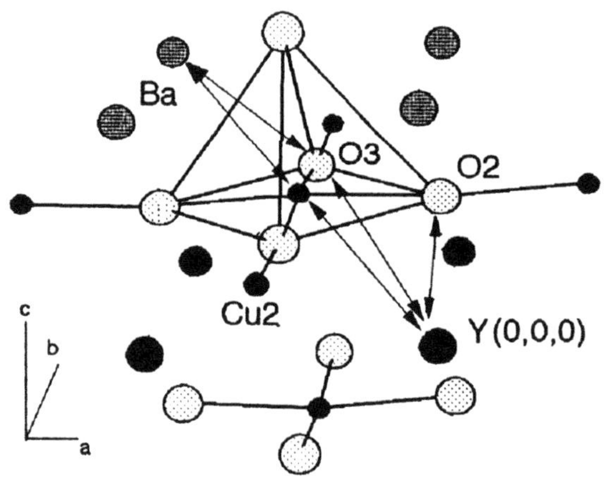

Fig. 1. Atomic cluster projected from the unit cell of $YBa_2Cu_3O_x$. Shown are two dimpled CuO_2 plaquettes separated by the Y layer. One Ba layer is indicated on the top including the apical oxygen of the CuO_5 pyramid. Arrowed lines indicate schematically some of the important Y EXAFS scattering configurations: Y–O2,3 ("O2,3"), Y–Cu2 ("Cu2"), and Y–Cu2–Ba ("6.2"), Y–O2,3–Ba ("5"), discussed in the text.

*Corresponding author. Email:abb12@rs1.rrz.Uni-Koeln.de

PII S0921-4534(97)00188-3

ing with the superconducting transition temperature points to an important relationship of the c-axis related displacements of the planar O2,3 and Cu2 atoms with the electronic structure at E_F. A relationship of the superconductivity in $YBa_2Cu_3O_x$ with c-axis related displacements in the CuO_2 plane may be also deduced from the significant increase of the O2,3–Cu2 spacing found by systematic comparison of superconducting $LnBa_2Cu_3O_7$ (Ln=Yb–La) to nonsuperconducting $LnBa_2Cu_3O_6$. Most interestingly the O2,3–Cu2 spacing in fully oxygenated but nonsuperconducting $PrBa_2Cu_3O_7$ turns out to be close that in oxygen depleted nonsuperconducting $LnBa_2Cu_3O_6$ [2].

The photoexcited yttrium atom turns out to be an ideal observer of the local atomic structure of the CuO_2 planes. Here we report on the x-dependency of the O2,3–Cu2 spacing in $YBa_2Cu_3O_x$ across the underdoped–(oxygen) overdoped transition (x=6.806-6.984) as measured by Y EXAFS. We discuss briefly possible implications of our results for the electronic structure at E_F.

2. EXPERIMENTAL DETAILS

The Y EXAFS probe the positions and vibrational dynamics of a atomic cluster centered at the Y site (Fig.1) with $d \simeq 16$ Å, which is close to the coherence length, ξ_{ab}, including fewer than about five superconducting pairs.

Fig.2 *(top)* exhibits typical EXAFS Fourier-transform spectra (k=2-16 Å^{-1}, Gaussian broadened by 0.2 Å^{-1}) of all samples under investigation at T=90 K. "O2,3" and "Cu2" denote the nearest neighbour Y–O2,3 and next nearest neighbour Y–Cu2 single scattering (SS) peaks. The peak "3.5" arises from a mixture of many single and multiple scattering (MS) configurations which we discuss elsewhere [3]. To measure the x-dependent variation of the O2,3–Cu2 spacing we take particular advantage of the strongly enhanced scattering amplitudes from the nearly collinear and well resolved Y–O2,3–Ba ("5") and Y–Cu2–Ba ("6.2") three-body configurations (Fig.2, *bottom*). The positions of the intervening O2,3 and Cu2 atoms deviate from collinearity only by about 14° and 5°, respectively. Therefore even tiny variations of the forward scattering geometry have very strong effects on the effective amplitudes [5], which we may determine directly from the peak heights. Calculations using the FEFF6 code [6] yield the height of peak

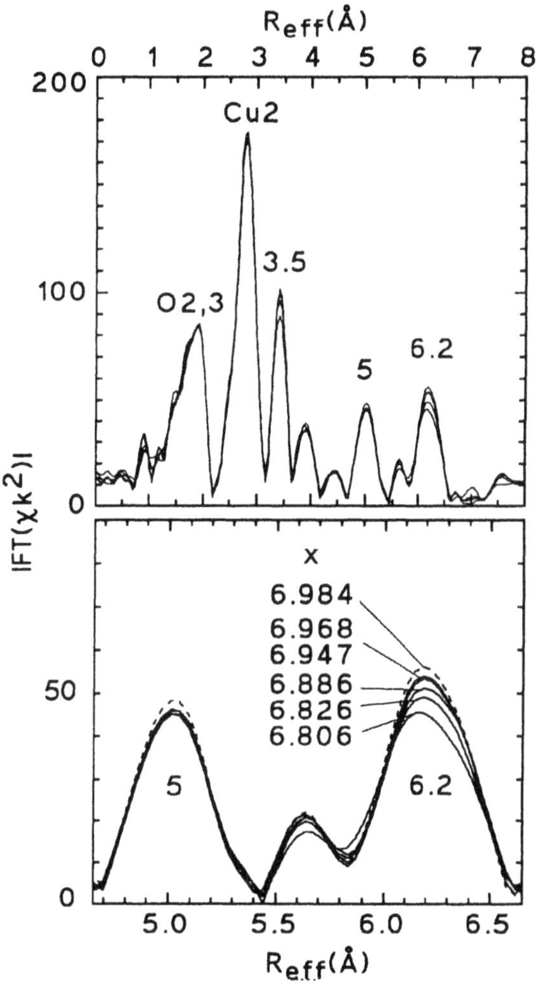

Fig. 2. *Top:* Yttrium K EXAFS in $YBa_2Cu_3O_x$ as a function of oxygen concentration at T=90 K: co-plotted moduli of the Fouriertransform spectra, | FT(χk^2) | using k=2–16 Å^{-1}. *Bottom:* Zoomed from *Top.* Note the reduced heights of peak "6.2" (Y–Cu2–Ba) in the underdoped regime, x=6.806-6.886, and the unaltered heights of peak "5" (Y–O2,3–Ba).

"6.2" to vary almost linearily as a function of the Cu2 displacement along c by $1/\delta R$=294 $Å^{-1}$ (forward scattering angle $\Theta \simeq 7.5° \pm 2.5$, no Gaussian broadening, and for the given choice of the FT parameters). Since both, the "5" and "6.2" three body scattering configurations, include the same Ba layer, the x- and T-dependency of the O2,3–Cu2 spacing may be read from the ratio |FT(6.2)/FT(5)|. The x-ray absorption spectra (T=20-300 K) were recorded at the European Synchrotron Radiation Facility (ESRF) using the double crystal spectrometer BL18(BM29). Details of the spectroscopic measurements and of the data analysis are given in Ref. [3].

The polycrystalline preparations ($BaCO_3$ precursor, CAR method) had the oxygen contents: x=6.806, 6.826, 6.886 ("underdoped"), x=6.947 ("optimum doped"), x=6.968, 6.984 ("overdoped"). All samples were from the same batches characterized earlier by magnetization and x-ray diffraction. Near x_{opt} no inversion in the x-dependency of the c-axis could be resolved [1].

3. RESULTS

It is clearly visible in Fig.2 *(top)* that doping affects the height of peak "6.2" (Y–Cu2–Ba) but almost not the height of peak "5" (Y–O2,3–Ba). (Peak "3.5" is discussed elsewhere [3]). Fig.2 *(bottom)* zooms the peaks "5" and "6.2". Doping increases the height of peak "6.2" because the Cu2 atoms are displaced along c towards the Ba layer resulting in a decrease of the Y–Cu2–Ba forward scattering angle. The height of peak "5" is almost independent on x, except for the overdoped sample x=6.984 (dashed line). We conclude that doping does not affect the average position of the O2,3 layer along c, except in the overdoped regime for x=6.984. Fig.3 *(bottom)* plots the heights of the peaks "5" and "6.2" *vs.* x. The increase of "6.2" for x=6.806→ 6.984 corresponds to a relative displacement of Cu2 by δR=0.051 Å. The ratio of the peak heights in Fig. 3 *(top)* exhibits a maximum near x=6.947, indicating a maximum of the O2,3–Cu2 spacing. It is important to note that the decrease of the O2,3–Cu2 spacing in the overdoped regime is caused by displacements of the average O2,3 layer towards the Cu2 layer.

Temperature dependencies are reported elsewhere [3]. In the underdoped regime we find a change of slope near the so-called spin gap temperature $T^* \simeq 160$ K.

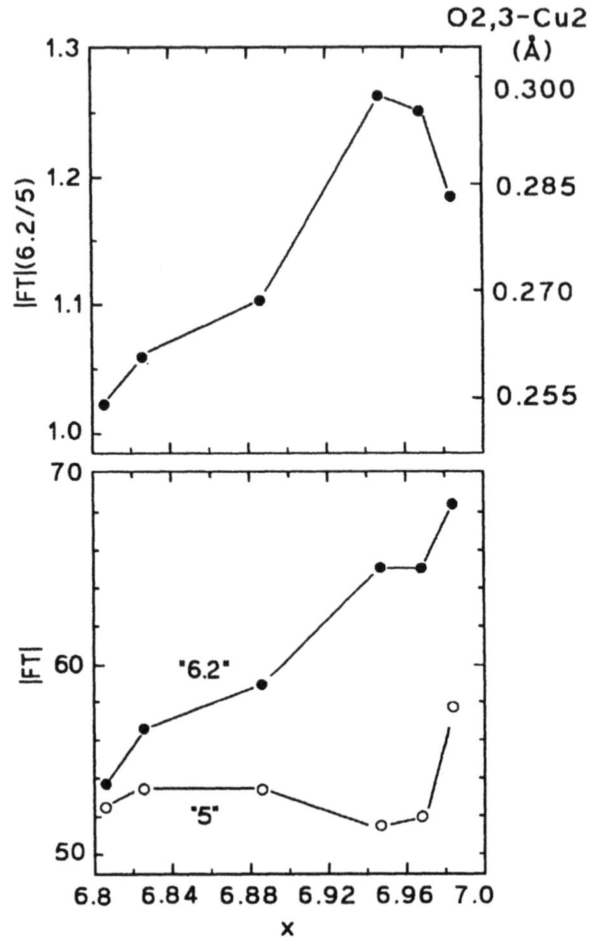

Fig. 3. The underdoped-overdoped transition in $YBa_2Cu_3O_x$ at T=25±5 K as determined from the Y–Cu2–Ba ("6.2") and Y–O2,3–Ba ("5") MS scattering. Lines are guides to the eyes. *Top:* Ratio of the peak heights shown below. The scale of the O2,3–Cu2 spacing is adjusted to the neutron data at x=6.806 [1]. *Bottom:* Heights of the peaks "6.2" (closed circles) and "5" (open circles) as a function of x.

4. DISCUSSION

Bonding between nominal Cu^{2+} and O^{2-} involves the Cu $3d$ and O $2p$ orbitals and has both ionic and covalent character. The holes introduced into the CuO_2 planes are located in the Cu–O bonds, and the electronic states near E_F have both, $3d$ and $2p$ character, with a dominant weight of the latter. If the Cu2 atoms are displaced along c out of the CuO_2 plane, coupling between the σ-type Cu $3d_{x^2-y^2}$ and the π-type O2 $2p_z$ and O3 $2p_z$ orbitals becomes allowed, and the electronic states near E_F may be significantly affected. The O2,3–Cu2 spacing may also activate significant overlap of the π-type Cu $3d_{xz}$, Cu $3d_{yz}$, and Cu $4s$ orbitals with the π-type oxygen orbitals. LDA band structure calculations using a "chemical Hamiltonian" with 8 orbitals yielded detailed insights in the possible effects of plane dimpling on the energy bands of stoichiometric $YBa_2Cu_3O_7$ [8]. From the 8 orbitals per CuO_2 plane (four σ-type orbitals: Cu $3d_{x^2-y^2}$, O2 $2p_x$, O3 $2p_y$, Cu $4s$; four π-type orbitals: Cu $3d_{xz}$, Cu $3d_{yz}$, O2 $2p_z$, O3 $2p_z$) the hybridized Cu $4s$ orbital is found to be crucial for the anomalous bandstructure of $YBa_2Cu_3O_7$ (extended saddle-points around E_F in the Brillouin zone). It is interesting to note that the Cu $4s$ band has some Cu $3d_z^2$ character and mixes with the conduction band.

In a local description of the electronic structure the mixing between Cu $4s$ and Cu $3d_z^2$ may be described as a ionic instability of the covalently bound copper ions: $3d^9 4s^1 L \leftrightarrow 3d^{10} 4s^0 L$ [10]. This configurational instability may occur, because the $d^9 s^1 - d^{10} s^0$ energy separation is low (2.7 eV in the free ion [9]) and both, the Cu $4s$ and Cu $3d_z^2$, combine with the neigbouring O2,3 $2p_z$ orbitals (indicated by L). Dimpling of the CuO_2 planes in $YBa_2Cu_3O_x$ activates Cu $3d_z^2 - 4s$ mixing, and thereby tends to increase the $3d_z^2$ character of the conduction band. Very recently nonsuperconducting $Y_{0.77}Ca_{0.23}Ba_2Cu_3O_{6.0}$ was found from XAS to exhibit no O $2p$ out-of-plane hole states, whereas superconducting $YBa_2Cu_3O_{6.5}$ does [11]. It is tempting to speculate that the symmetry of the doped charge carriers might distinguish the underdoped from the overdoped regime: doping of underdoped cuprates mixes states with $3d_z^2$ character in the conduction band, whereas addition of holes beyond n_{opt} suppresses the admixed $3d_z^2$ character of the charge carriers.

Acknowledgements Beamtime at the ESRF was under the proposals HC362 and HS45. We are grateful to S. Müllender and S. Link for efficient help in an early stage of this work.

REFERENCES

1. K. Conder, D. Zech, Ch. Krüger, E. Kaldis, H. Keller, A.W. Hewat and E. Jilek. In *Phase Separation in Cuprate Superconductors*, E. Sigmund and K.A. Müller, eds., p. 210, Springer, Berlin, 1994, and E. Kaldis in Ref. 2.
2. M. Guillaume, P. Allenspach, W. Henggeler, J. Mesot, B. Roessli, U. Staub, P. Fischer and A. Furrer. J. Phys.: Cond. Matter **6**, 7963 (1994).
3. J. Röhler, P.W. Loeffen, S. Müllender, K. Conder and E Kaldis. In *Workshop on High-T_c Superconductivity 1996: 10 Years after the Discovery*, E. Kaldis, E. Liarokapis and K.A. Müller, eds., NATO ASI Series, Kluwer, Dordrecht, 1997. In the press.
4. G. Böttger, I. Mangelschots, E. Kaldis, P. Fischer, Ch. Krüger and F. Fauth J. Phys.; Cond. Matter **8**, 8889 (1996).
5. J. Röhler and R. Crüsemann. In *Anharmonic Properties of high-T_c Cuprates*, D. Mihailovic, G. Ruani, E. Kaldis and K.A. Müller, eds., p. 86, World Scientific, Singapore, 1995.
6. S.I. Zabinsky, J.J. Rehr, A. Ankudinov, R.C. Albers and M.J. Eller. Phys. Rev. B **52**, 2995 (1995).
7. E. Liarokapis, N. Poulakis, D. Palles, K. Conder, E. Kaldis and K.A. Müller. Physica C **219-220**, 139 (1996).
8. O.K. Andersen, O. Jepsen, A.I. Liechtenstein and I.I. Mazin. Phys. Rev. B **49**, 4145 (1994).
9. L.E. Orgel. J. Chem. Soc. **48**, 4186 (1958), and **49**, 3815 (1959).
10. J. Röhler. J. Superconductivity **9**, 457 (1996).
11. N. Nücker *et al.*. To be published.

ELSEVIER

Physica C 282–287 (1997) 186–189

PHYSICA C

Evidence for Two-Component Superconductivity in the Femtosecond Optical and Transient Photoconducting Response of $YBa_2Cu_3O_{7-\delta}$.

D.Mihailovic[1], C.J.Stevens[2], B.Podobnik[1], J.Demsar[1], M.Zavrtanik[1], D.Smith[2] and J.F.Ryan[2]

[1] J. Stefan Institute, 1001 Ljubljana, Slovenia

[2] Clarendon Laboratory, University of Oxford, Oxford OX1 3PU, U.K.

Femtosecond time-resolved spectroscopy and transient photoconductivity measurements have been used to investigate the time-dynamics, frequency dependence and symmetry of electronic excitations associated with the opening of the superconducting gap in $YBa_2Cu_3O_{7-\delta}$. Two clearly identifyable relaxation components are observed operative on nanosecond and picosecond timescales respectively, implying the co-existence of localized states and band-like carriers both in the superconducting and normal state. From the temperature dependence of the two components it is clear that both relaxation phenomena are directly connected with the formation of the superconducting condensate. The spectral response for both excitations closely follows the spectrum observed in thermal modulation spectroscopy and implies that the electron-boson coupling function includes a charge-transfer component at optical excitation energies in the region of 1.5-2 eV.

Although the comparison of the measured normal - to - superconducting state optical conductivity ratio σ_s/σ_n in the infrared region of the optical spectrum when compared with theory has been one of the most decisive factors influencing the acceptance of BCS theory for "normal" superconductors, in high-temperature superconductors the optical response has been found to be much more complicated, not least because of the high anisotropy and complex electronic structure of these materials. Thus the temperature dependence of these spectra did not reveal a simple gap picture, but rather showed large temperature-dependent features which were highly anisotropic and in addition to the expected changes associated with the onset of a superconducting transition, significant changes in the optical spectra were observed also above T_c [1] and at energies well above the gap frequency[2]. Because localized state spectra - this applies to optical spectrosopies as well as ARPES for example - are indistinguishable from those originating from itinerant electron bands, the identification of particular spectral features is very difficult on the basis of its spectrum alone. This problem has inevitably resulted in controversy and whereas the low-frequency component has by now almost universally accepted as the electrodynamic (ED) response of in-plane excitations, the so-called mid-infrared feature in the optical conductivity $\sigma(\omega)$ remained controversial. Of all the possible interpretations, the polaronic one has been the most common[3][4], with the implication that localized states states exist in both the normal and superconducting state.

Recently we have found that measurement of the carrier time-relaxation dynamics at different energies and on different timescales using optical spectroscopies can give qualitatively new information regarding the issue of localized versus itinerant states. In recent ultrafast Raman carrier relaxation measurements we have unambiguously found a signature of localised states in the normal state of optimally doped superconducting YBCO and LaSCO samples[5]. To pin down the origin of the excitations more effectively, we have since performed time-resolved measurements of optical reflectivity, transmission and photocurrent on different $YBa_2Cu_3O_{7-\delta}$ samples as a function of temperature both in the normal and superconducting state - the former also as a function of energy in the mid-infrared region - revealing a significant amount of qualitatively new information regarding the dynamics of the super-carriers in these materials. The time-resolved femtosecond measurements on optimally doped YBCO

PII S0921-4534(97)00190-1

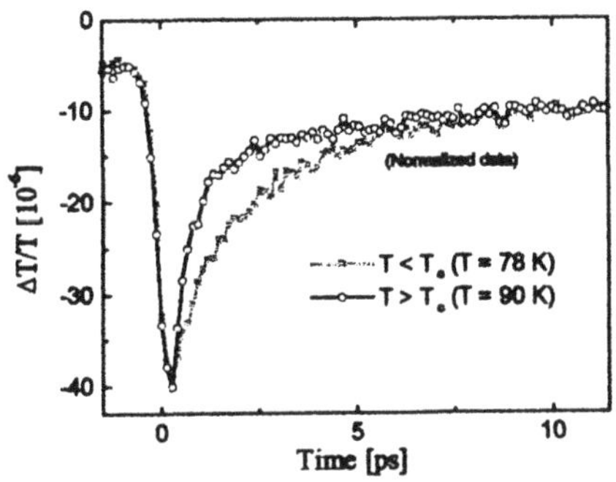

Figure 1. The transmission through an underdoped YBaCuO sample ($T_c = 85$K) as a function of time after photoexcitation by a 1.5 eV laser pulse below and above T_c. (The peak amplitude has been normalized.)

were recently described[6]. The data show a ubiquitous presence of two distinct carrier relaxation processes which occur on very different timescales. The faster relaxation component occurs on picosecond timescales typical of band-carrier relaxation, whereas the slow component displays non-exponential relaxation dynamics on timescales extending from nanoseconds to hundreds of microseconds and beyond[7], characteristic of localized or polaronic states. In Fig. 1 we show the relaxation of the transmission through an YBCO sample above and below T_c. Apart from the almost 10 fold longer lifetime of the fast band-like carrier relaxation in the superconducting state, the amplitude of this signal increases dramatically below T_c and in the optimally doped sample is described quite well by a BCS curve for $T < T_c$ as seen in the bottom panel of Figure 2. Similar measurements on underdoped and entirely undoped samples (see Fig. 2) show a dramatic difference in temperature dependence, and the amplitude can no longer be described by anything resembling a BCS curve. In fact the signal amplitude shows a much slower drop with increasing temperature and eventually drops to zero at near $T = 150$ K. We note that in this sample ($T_c = 85$ K) this temperature $T = 150$ K corresponds closely to the pseudogap temperature T^* found by other measurement techniques. Measurements on different samples show that this T^* is very sensitive to the O concentration. For this reason a large number of samples have been measured with different O concentrations and on different substrates. Although it is not possible to discuss these results in detail in this paper, the measurements confirm qualitatively the behaviour outlined above.

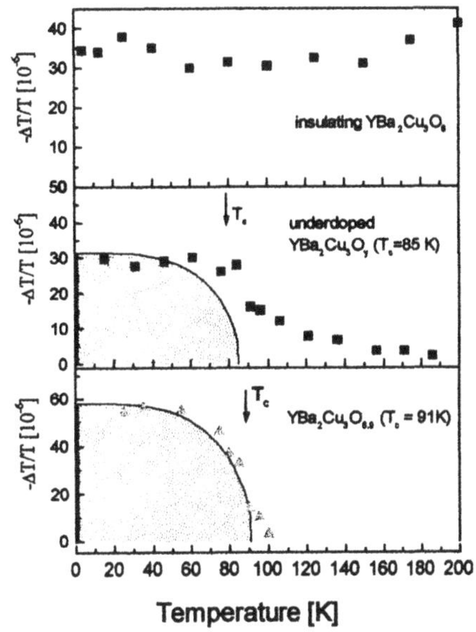

Figure 2. The temperature dependence of the amplitude of the fast carrier relaxation process for undoped (top), underdoped (middle) and an optimally doped sample (bottom).

The temperature dependence of the *slow* signal is qualitatively different than of the fast signal, and in the optimally doped sample the temperature dependence below T_c can be described in terms of an single thermally - activated temperature dependence *below* T_c and a rapid drop in amplitude *above* T_c to a small constant, i.e. temperature-independent value. Consideration of possible processes which can give rise to such behavior has lead us to the conclusion that the long-lifetime states which we are probing in these experiments are thermally activated quasiparticle states, not least because the activation energy in the optimally doped samples $E_a = 2\Delta = 4.0 \pm 0.5 k_B T_c$. For the underdoped sample (middle panel in Figure 3) the peak in the amplitude at T_c is augmented by an additional temperature-independent signal. In the undoped $YBa_2Cu_3O_6$ sample the temperature dependence of the slow signal is essentially monotonic, as expected (see top panel of Figure 3). In this case also, just as in the case of the fast signal discussed above, the measurements on a variety of samples give essentially the same results.

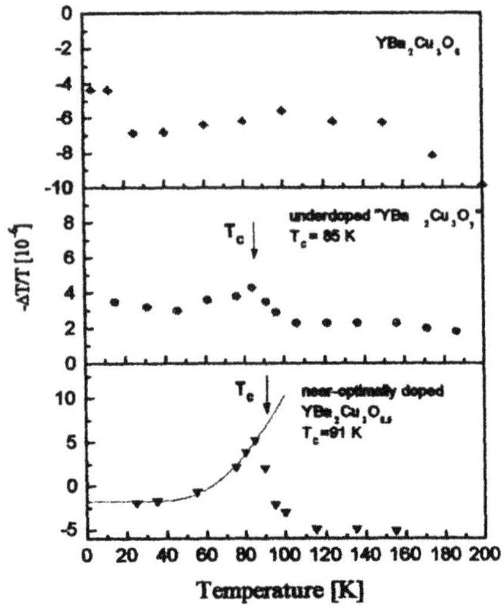

Figure 3. The temperature dependence of the amplitude of the *slow* carrier relaxation process for undoped (top), underdoped (middle) and near-optimally doped samples (bottom). The amplitude of the signal follows thermally activated behaviour below T_c with $E_a = 4 \pm 0.5kT_c$.

The optical data on quasiparticle relaxation above are complemented by measurements of the photoconducting transient signal recorded as the voltage across a YBCO microbridge after excitation by a short laser pulse. In this case the laser pulse drives the superconductor normal which results in the appearance of a voltage transient across the superconductor. The time-relaxation dynamics are essentially determined by the quasiparticle relaxation dynamics. The most surprising feature of the observed photoresponse signal shown in Fig. 4 is that the signal voltage shows first a fast transient and subsequently drops to a negative value and then relaxes on an unusually long timescale. Detailed analysis and careful modelling shows that this behaviour cannot be described in terms of any kind of bolometric signal. As shown in Figure 4, both the calculated bolometric signal arising from the kinetic inductance and the direct bolometric signal due to change in resistivity of the sample with temperature show qualitatively different time-dependences than the observed signal and have been consequently excluded as possible mechanisms for explaining the observed photoconducting transient response. The long lifetime kinetics are thus apparently independent evidence of long quasiparticle lifetimes already observed in the optical relaxation measurements. We should also mention here that although the dynamics of resistive vortex flow on these timescales is still not understood in detail, we estimate that this contribution to the signal is too small and of the wrong sign to explain the "inductive" signal observed in the photoconductivity.

The general conclusion from these experiments is that there exist two different types of quasiparticle relaxation processes in these materials. The existence of two components in the relaxation do not necessarily mean that we are probing two independent carrier bands, but rather, the data give a clear indication that the supercarriers have a dual character, judging especially from the fact that both the fast and the slow signal have a similar spectral response [6]. This inference of the existence of a dual or composite quasi particle is entirely consistent with the hypothesis discussed

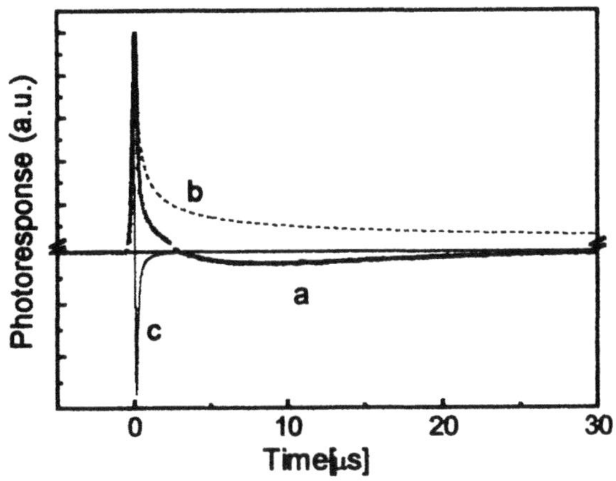

Figure 4. a) The photoconducting response following a laser pulse incident on a microstrip of YBCO. b) is the calculated bolometric response and c) is the calculated kinetic inductance. Clearly the experiment (a) cannot be reconciled with either (b) or (c).

in the introductory paragraph regarding the polaronic and itinerant carrier components in the optical conductivity $\sigma(\omega)$. Moreover the two different relaxation experiments (optical and transport) compare rather well regarding quasiparticle lifetimes, in spite of the fact that the details of the relaxation processes in the two cases are necessarily different.

An important point regarding the spectral dependence of the optical signal is that it matches rather closely the spectral shape of the thermal differential reflectance (TDR) peak at 1.6 eV measured on the same material[2]. In both the time-resolved and TDR measurements the feature at 1.6 eV appears only in the superconducting phase (for optimally doped samples) and is thus naturally assigned to the electrodynamic response of the pairs or the condensate. The corollary is that the non-zero signal in the underdoped samples at temperature above T_c is naturally interpreted as a sign of pairs - possibly bipolarons - in the underdoped materials at intermediate temperatures as suggested by some theories[9].

We wish to acknowledge J.E.Evetts and V.I.Dediu for kindly providing YBCO samples for these experiments.

REFERENCES

1. A.V.Puchkov, P.Fournier, T.Timusk, N.N.Kolesnikov, Phys. Rev. Lett., **77**, 1853 (1996), *ibid*, Phys. Rev. Lett., **77**, 1853 (1996).
2. M.J. Holcomb, et al, Phys. Rev. Lett. **73**, 2360 (1994), M.J. Holcomb, C.J. Perry, J.P. Collman and W.A. Little, Phys. Rev.B **53**, 6734 (1996)
3. D.Tanner and T.Timusk, in "*Physical Properties of High-T_c Superconductors I and III*" Ed. D.M.Ginsberg (World Scientific, 1989,1992)
4. The two-component approach to the ED response is discussed in more detail in the paper by D.Mihailovic and K.A.Müller, Proc. NATO ASI "*Materials aspects of High-T_c Superconductivity: 10 Years after the Discovery*", Delfi, Greece 1996 (to be published by Kluwer, 1997)
5. D.Mihailovic, T.Mertelj, I.Poberaj, J.Demsar and C.Chen, *Journal of Supercon.*, **8**, 531 (1995), T.Mertelj, J. Demsar, B. Podobnik, I. Poberaj, D. Mihailovic, Phys. Rev. B **55** (9) (1997).
6. C.J.Stevens, D.C.Smith, B.Podobnik, C.Chen, J.F.Ryan, D.Mihailovic, G.Wagner and J.E. Evetts, *Phys. Rev. Lett*, **78** (8) (1997)
7. T.N.Thomas, C.J.Stevens, S.Choudhary, J.F.Ryan, D.Mihailovic, L.Foro, G.Wagner and J.E.Evetts, *Phys. Rev.* B **53**, 12436 (1996)
8. J.Demsar, M.Zavrtanik, B.Podobnik, V.I.Dediu, D.Mihailovic, *to be published*
9. See for example J.Ranninger in "Anharmonic Properties of High-T_c Cuprates", Eds. D.Mihailovic, G.Ruani, E.Kaldis, K.A.Muller (World Scientific, 1995), N.F.Mott *Adv. Phys.*, 39, 55 (1990) and V.J.Emery and S.A.Kivelson, *Nature* **374**, 434 (1995)

ELSEVIER

Physica C 282–287 (1997) 190–193

PHYSICA C

EPR evidence for Jahn-Teller polarons in $La_{1-x}Ca_xMnO_{3+y}$

A. Shengelaya, Guo-meng Zhao, H. Keller and K. A. Müller

Physik-Institut der Universität Zürich, CH-8057 Zürich, Switzerland

We present the results of the electron paramagnetic resonance (EPR) study in the mixed valence perovskite $La_{1-x}Ca_xMnO_{3+y}$ for ^{16}O and ^{18}O isotope substituted compounds. The characteristic differences observed in EPR intensity and linewidth for the two isotope samples can be explained by a model in which a bottlenecked spin relaxation takes place from the exchange-coupled constituent Mn^{4+} ions via the Mn^{3+} Jahn-Teller ions to the lattice. For x=0.2 the ferromagnetic exchange energy J exhibits a $^{16}O/^{18}O$ oxygen isotope effect of ~-10%. The observed isotope effects provide the *microscopic* evidence for the formation of Jahn-Teller polarons in doped manganese perovskites.

1. Introduction

Recently, the ferromagnetic systems $La_{1-x}Me_xMnO_{3+y}$ (where Me=Ca,Sr,Ba) have become the focus of scientific and technological interest because of the colossal magnetoresistance (CMR) effects found in these materials [1]. Doped manganese perovskites are mixed-valent systems containing Mn^{3+} and Mn^{4+} ions. The magnetic and electronic properties in these compounds have traditionally been examined with the double exchange (DE) model, which considers the transfer of an electron between neighboring Mn^{3+} and Mn^{4+} ions through the Mn-O-Mn path [2]. However, recent theoretical considerations indicated that DE alone does not explain the CMR, and that polaronic effects due to a very strong electron-phonon coupling should be included [3,4]. The strong electron-phonon coupling is expected because the electronic ground state of the Mn^{3+} ions is degenerate and this degeneracy is removed by a spontaneous distortion of the surrounding lattice, known as the Jahn-Teller (JT) effect [5]. A recent demonstration of a giant oxygen isotope shift of > 20K on the ferromagnetic transition temperature T_c by Zhao et al. [6] provided direct experimental evidence of the strong coupling of the charge carriers to JT lattice distortions and of JT polaron formation [7] in $La_{1-x}Ca_xMnO_{3+y}$.

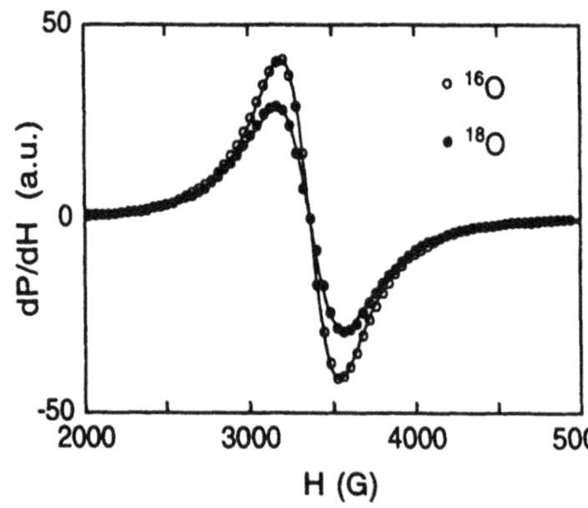

Figure 1. EPR signal of ^{16}O and ^{18}O samples of $La_{0.8}Ca_{0.2}MnO_{3+y}$ measured at T=300 K under identical experimental conditions. The fits with Lorentzian line shape are indicated by solid lines.

2. Experiments

We measured the temperature dependence ($T_c < T < 3T_c$) of the Electron Paramegnetic Resonance (EPR) for ceramic powder samples of $La_{1-x}Ca_xMnO_{3+y}$ with x=0.1, 0.2 which were substituted by different oxygen isotopes (^{16}O and ^{18}O). The samples used here are the same as those studied by Zhao et al. [6]. A strong symmetric EPR signal with a lineshape very close to Lorentzian and g-value of 2.0 was observed

0921-4534/97/$17.00
PII S0921-4534(97)00192-5

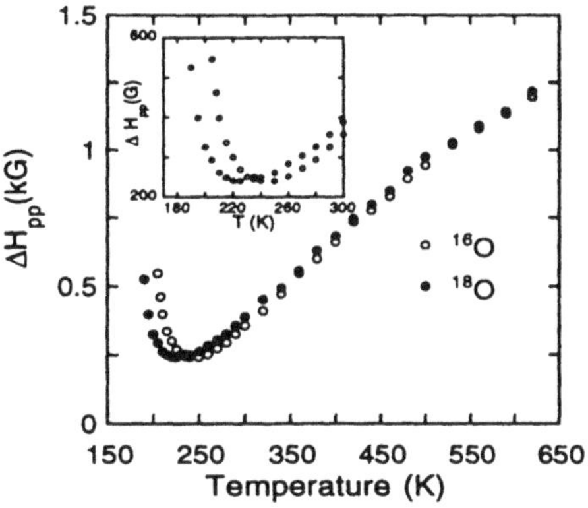

Figure 2. Temperature dependence of the peak-to-peak EPR linewidth ΔH_{pp} for ^{16}O and ^{18}O samples of $La_{0.8}Ca_{0.2}MnO_{3+y}$. The inset shows the low temperature region on an enlarged scale.

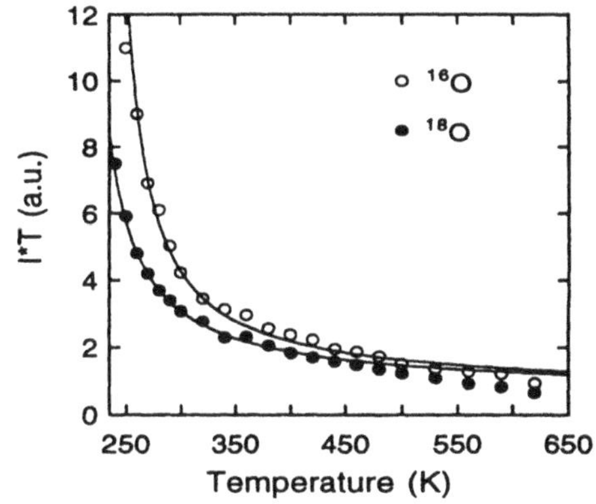

Figure 3. Temperature dependence of the integral intensity of EPR signal times temperature ($I \times T$) for ^{16}O and ^{18}O samples of $La_{0.8}Ca_{0.2}MnO_{3+y}$. The solid lines represent the best fit to Eq.(1) in the temperature range 250 K≤T≤500 K.

over the whole range of temperature investigated. Two typical EPR signals are shown in Fig. 1.

The temperature dependence of the linewidth for the x=0.2 sample with different oxygen isotopes is shown in Fig. 2 . With decreasing temperature the linewidth decreases, passes through a minimum at a temperature T_{min} and increases on further cooling to T_c. It is interesting that T_{min} in the ^{18}O sample is shifted to lower temperatures in comparison with the ^{16}O sample and that there are significant differences in linewidths below T_{min}. The integral intensity I of the EPR signal decreases with temperature much faster than would be expected according to the Curie law. To show this, we plot in Fig. 3 the product $I \times T$ versus temperature. From Fig. 3 one can see that the intensity of the EPR signal in the ^{16}O sample is higher than in the ^{18}O sample.

3. Discussion

In order to understand the striking differences of resonance linewidths and intensities in different oxygen isotope samples, it is necessary to clarify first the origin of the EPR signal in manganese perovskites. We propose that the EPR signal observed in $La_{1-x}Ca_xMnO_{3+y}$ is primarily due to Mn^{4+} ($3d^3$ with S=3/2) ions. In an octahedral anion crystal electric field this ion has a ground state, corresponding to an orbital singlet A_2. Consequently the spin-lattice relaxation is weak and this makes EPR of Mn^{4+} easy to observe even at high temperatures [8]. The Mn^{3+} ($3d^4$ with S=2) is unlikely to have an observable EPR signal as it exhibits a large zero-field splitting and strong spin-lattice relaxation (the ground state of Mn^{3+} ion is the orbital doublet) [9].

However, it is clear that the observed signal cannot be attributed to isolated Mn^{4+} ions. To construct a model of paramagnetic centers responsible for the EPR signal, it is important to point out that doped manganese perovskites are mixed valence compounds with Mn^{4+} and Mn^{3+} ions and strong ferromagnetic DE interaction between them. Thus, we should consider the EPR response of the system to contain three distinct components: Mn^{4+} ions: s, Mn^{3+} ions: σ, and the lattice: L. Fig. 4 shows a standard schematic picture for such a system, with arrows indicating possible relaxation paths between components. The theory to describe such a system was developed in connection with the EPR of localized

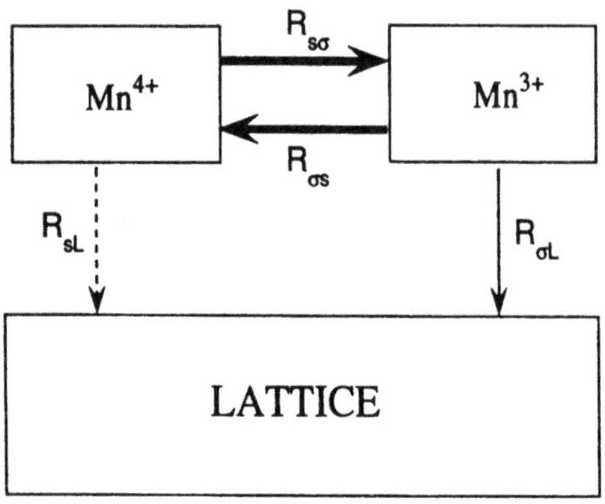

Figure 4. Block diagram showing the energy flow paths for the Mn^{4+} and Mn^{3+} spin subsystems and the lattice. The relaxation rate R_{ab} represents relaxation from subsystem a to subsystem b. The thickness of the arrows is a measure of the magnitude of the particular relaxation rate R_{ab}.

magnetic moments in metals [10]. Assuming that the relaxation rate $R_{\sigma L}$ of the Mn^{3+} spins to the lattice is much smaller than the exchange-induced cross relaxation rate $R_{\sigma s}$ (and back $R_{s\sigma}$) between Mn^{3+} and Mn^{4+}, and if the direct relaxation of Mn^{4+} ions to the lattice R_{sL} is negligible ($R_{\sigma L}, R_{sL} \ll R_{\sigma s}, R_{s\sigma}$), then a so-called "bottleneck" effect will take place in the transfer of energy between the spin subsystems [10]. In this limit magnetic energy which is transfered from the Mn^{4+} to the Mn^{3+} spin system is quite likely to be returned back rather than passed on to the lattice. Consequently, the relaxation of the system is dominated by the bottleneck due to the slow Mn^{3+}-lattice relaxation process.

Let us consider first the temperature dependence of the linewidth presented in Fig. 2. Below T_{min} the broadening is associated with a critical "slowing down" of the spin fluctuations in ferromagnets when T_c is approached from above [12]. The mechanism of line broadening above T_{min} is different. It is caused by the spin-lattice relaxation. However, as was mentioned above, the spin-lattice relaxation for the Mn^{4+} is expected to be weak. It seems that the most relevant mechanism is one involving the exchange coupling of the Mn^{4+} to Mn^{3+} ions that quickly relax into the lattice.

It is also possible to explain why the EPR signal in $La_{1-x}Ca_xMnO_{3+y}$ is much stronger than what is expected from individual Mn^{4+} ions, as well as its sharp decrease with temperature above T_c. The intensity of the EPR signal is proportional to the static magnetic susceptibility of the ions responsible for the signal ($I \propto \chi_s$) [9]. In the bottleneck regime χ_s is renormalized due to the exchange coupling [10]:

$$\chi_s = \chi_s^0 \frac{1 + \lambda\chi_\sigma^0}{1 - \lambda^2\chi_\sigma^0\chi_s^0}. \quad (1)$$

Here χ_s^0 and χ_σ^0 are the bare (without exchange) susceptibilities of the Mn^{4+} and Mn^{3+} ions, respectively. The factor λ is a dimensionless exchange coupling constant between them:

$$\lambda = \frac{zJ}{N g_s g_\sigma \mu_B^2}, \quad (2)$$

where z is the number of nearest neighbors, N is the total number of Mn spins per cm^3, g_s and g_σ are the g-factors of Mn^{4+} and Mn^{3+} ions, respectively. Equations (1) and (2) show that in the present case of ferromagnetic coupling ($J > 0$), the intensity of the EPR signal is enhanced in comparison to that expected from individual Mn^{4+} ions. To show that these equations also correctly describe the temperature dependence of our EPR signal intensity, we analyzed the experimental results presented in Fig. 3 in terms of this model. We assumed that the bare susceptibilities of Mn^{4+} and Mn^{3+} ions follow the Curie law. The result of this fit using only a single free parameter, namely, the exchange integral J, is shown in Fig. 3. The agreement with the experimental data is fairly good below 500K. The fit yields $J = 78(1)$K for ^{16}O and $J = 71(1)$K for ^{18}O. As can be seen from Fig.3, above 500K the intensity of the EPR signal drops faster than is predicted by Eq. (1). This deviation can be associated with a gradual transition from the bottleneck to the isothermal regime where Eq. (1) is no longer valid.

Before we consider the EPR response of samples with different oxygen isotopes (^{16}O and ^{18}O) in terms of our model for spin dynamics in

$La_{1-x}Ca_xMnO_{3+y}$, let us summarize its main features. We assume that the spin system of the Mn^{4+} and Mn^{3+} ions displays a collective motion of their total magnetic moments (bottleneck regime), which is a consequence of a strong ferromagnetic coupling between them. The relaxation of these ions to the lattice is ineffective to destroy this collective mode, at least below 500K. In terms of this model it is easy to explain the striking changes of the EPR linewidth and the intensity with oxygen isotope substitution if we assume that the exchange integral J is larger in ^{16}O samples than in ^{18}O samples. Indeed, according to Eq. (1), for ferromagnetic exchange ($J > 0$) a larger J corresponds to a higher EPR signal intensity (see Fig. 3). It is worthy to note that a fit to the data in Fig. 3 using Eq.(1) gives a value of J for the ^{16}O sample which is about 10% larger than for the ^{18}O sample. This difference in J should be compared with the 10% difference of T_c observed in these samples ($T_c \approx$ 207K and 186K for the ^{16}O and ^{18}O samples, respectively) [6]. The shift of T_{min} to lower temperatures in the ^{18}O sample (see Fig. 2) can also be easily understood. This sample has a lower T_c due to the smaller J, thus critical broadening starts at lower temperatures in comparison with the ^{16}O sample. Here we should note that qualitatively the same results were obtained for the x=0.1 sample. The EPR linewidth and intensity also depend on the oxygen isotope mass, but the changes observed were slightly smaller then in the x=0.2 sample, in accordance with the smaller oxygen isotope effect found for x=0.1 [6].

The analysis of our results leads to the conclusion that the ferromagnetic exchange integral between Mn^{4+} and Mn^{3+} ions depends on the oxygen mass and $J(^{16}O)>J(^{18}O)$. This important fact can be understood in terms of JT polaron formation in $La_{1-x}Ca_xMnO_{3+y}$, which was considered by Zhao et al. [6] to explain the giant oxygen isotope effect on T_c observed in this compound. In the JT polaron model the exchange integral J is proportional to the effective bandwidth W_{eff} of JT polarons, which in turn depends on the isotope mass [6]:

$$W_{eff} \propto W exp(-\gamma E_{JT}/\hbar\omega), \qquad (3)$$

where W is the bare conduction bandwidth, E_{JT} is the JT stabilization energy, ω is the characteristic frequency of the optical phonons depending on the isotope mass M ($\omega \propto M^{-1/2}$). The dimensionless parameter γ is a function of E_{JT}/W with $0 \leq \gamma \leq 1$. According to Eq.(3), W_{eff}, and in turn J, decreases with enhanced oxygen isotope mass, in agreement with the present experimental observations.

4. Acknowledgments

This work was supported by the Swiss National Science Foundation.

REFERENCES

1. K.Chahara, T.Ohno, M.Kassai, and Y.Kozono, Appl. Phys. Lett. **63**, 1990 (1993).
2. C.Zener, Phys.Rev. **82**, 403 (1951); P.G.DeGennes, Phys.Rev. **118**, 141 (1960).
3. A.J.Millis, P.B.Littlewood, and B.I.Shraiman, Phys. Rev.Lett. **75**, 5144 (1995).
4. H.Röder, Jun Zang, and A.R.Bishop, Phys.Rev.Lett. **76**, 1356 (1996).
5. H.A.Jahn and E.Teller, Proc.R.Soc.Lond.A **161**, 220 (1937).
6. Guo-meng Zhao, K.Conder, H.Keller, and K.A.Müller, Nature (London) **381**, 676 (1996).
7. K.-H.Höck, H.Nickisch, and H.Thomas, Helv.phys.Acta **50**, 237 (1983).
8. K.A.Müller, Phys.Rev.Lett. **2**, 341 (1959).
9. A.Abragam and B.Bleaney, Electron Paramagnetic Resonance of Transition Ions (Clarendon Press, Oxford, 1970).
10. S.E.Barnes, Adv.Phys. **30**, 801 (1981).
11. J.E.Gulley and V.Jaccarino, Phys.Rev.B **6**, 58 (1972).
12. K.Kawasaki, Progr.Theor.Phys. **39**, 285 (1967).

ELSEVIER

Physica C 282–287 (1997) 194–197

PHYSICA C

Bose-Einstein to BCS Crossover Picture for High-T_c Cuprates

Y. J. Uemura[a]

[a]Dept. of Physics, Columbia University, New York, NY 10027, U. S. A.

Combining (1) the universal correlations between T_c and n_s/m^* (superconducting carrier density / effective mass) and (2) the pseudo-gap behavior in the underdoped region, we obtain a picture to describe superconductivity in cuprate systems in evolution from Bose-Einstein to BCS condensation. Overdoped and Zn-substituted cuprate systems show signatures of reduced superfluid density in a microscopic phase separation. Scaling of T_c to the superfluid volume density n_s in all these cases indicate importance of Bose condensation.

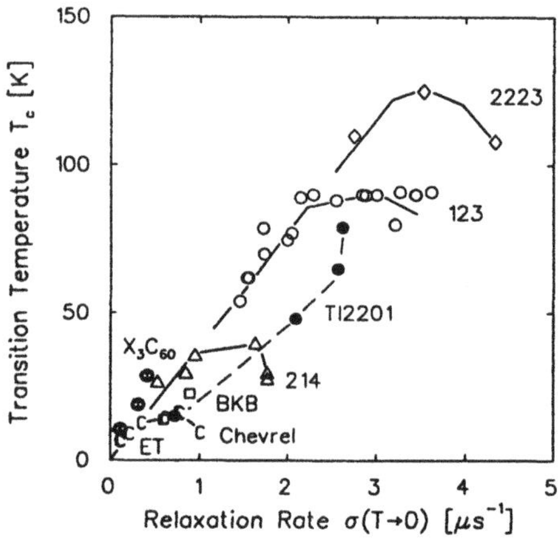

Figure 1. Muon spin relaxation rate $\sigma(T \to 0)$ in various superconductors plotted versus T_c [1-3].

The magnetic field penetration depth λ reflects the spectral density of superfluid as $1/\lambda^2 \propto n_s/m^*$ (the superconducting carrier density / effective mass). The absolue values of λ can be determined accurately by Muon Spin Relaxation (μSR) measurements where the relaxation rate $\sigma \propto 1/\lambda^2 \propto n_s/m^*$ if $H_{c1} \ll H_{ext} \ll H_{c2}$. Figure 1 shows our μSR results in various cuprate and other superconductors in a plot of T_c versus $\sigma(T \to 0)$ [1-3]. This figure demonstrates that: (A) T_c is proportional to n_s/m^* (m^* predominantly reflects the in-plane value m^*_{ab} in cuprates) in the underdoped region with the slope common to many different cuprates (123, 214, 2223 systems etc.) – universal correlations [1]; (B) T_c shows saturation in the "optimum doping" region [1]; (C) n_s/m^* *decreases* with increasing carrier doping in the overdoped region (Tl2201) [3,4]; and (D) not only the cuprates but also organic BEDT, doped C_{60} and some other superconductors have ratios between T_c and n_s/m^* [2] comparable to those of the cuprates.

Implications of Fig. 1 become clearer if we convert the horizontal axis into an "effective" Fermi energy ϵ_F [2], with ϵ_F in 2-d systems obtained as $\epsilon_F \propto n_{s2d}/m^* \propto \sigma \times c_{int}$ where n_{s2d} stands for the areal 2-d carrier density on each CuO_2 plane and c_{int} the average interplanar distance between the CuO_2 planes. For 3-d systems, σ can be combined with the Sommerfelt constant $\gamma \propto n^{1/3}m^*$ to obtain $\epsilon_F \propto \sigma^{3/4}\gamma^{-1/4}$. Figure 2 shows the resulting plot, together with the broken line which corresponds to the Bose-condensation temperture T_B for 3-dimensional non-interacting bosons of boson density $n_s/2$ and mass $2m^*$.

This figure demonstrates that: (E) cuprates, organic BEDT, and some other "exotic" suprconductors have kT_c/ϵ_F as high as 0.01-0.1, much higher than those of conventional BCS superconductors; (F) the linear relationship $T_c \propto n_s/m^*$ may originate from that of Bose-Condensation, as the points lie parallel to the broken line; (G) T_c of cuprates is about 4-5 times smaller than T_B. The feature (G) can be attributed partly to the overlap of bosons, as several pairs exist per ξ^2_{ab} (ξ_{ab} is the in-plane coherence length) on the conducting planes in the cuprates. Even for liquid ^{4}He, finite size of bosons and their interactions make the lambda transition at $T_\lambda = 2.2$ K significantly reduced from $T_B \sim 3.2$ K. The 2-dimensional character of the cuprates would further reduce T_c, as we discuss later. Figure 2 provides a phenomeno-

PII S0921-4534(97)00194-9

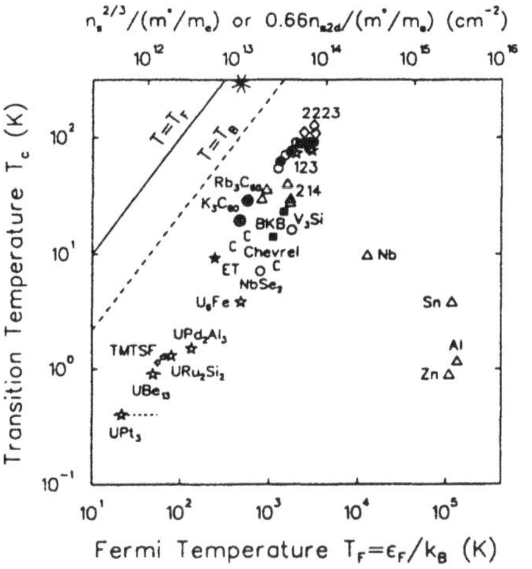

Figure 2. A plot of T_c versus the effective Fermi energy ϵ_F calculated from σ [2]. T_B denotes BE condensation temperature for non-interacting bosons with density $n_s/2$ and mass $2m^*$. The * symbol shows T_θ^{max} of ref. [14] for K_3C_{60}.

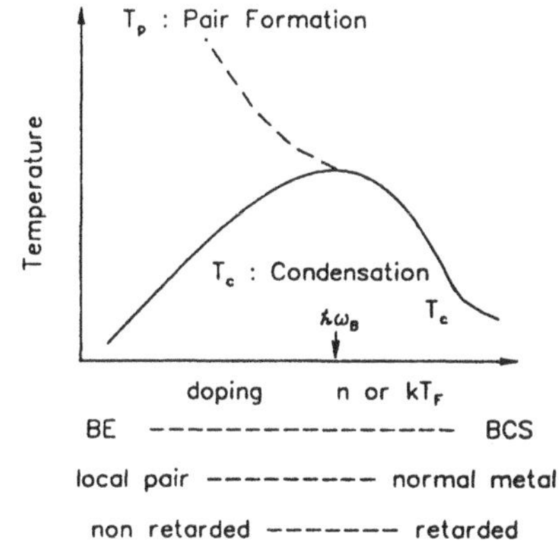

Figure 3. Phase diagram describing BE-BCS crossover with increasing carrier concentration n. This phase diagram can be mapped to that of the cuprates by assuming that the pseudo gap temperature T^* corresponds to the formation of normal-state pairs. [5,6]

logical method to classify superconductors in evolution from Bose-Einstein (BE) condensation (in real space with non-retarded strong interaction: close to the broken line) and BCS condensation (in momentum space with retarded weak interaction: at large ϵ_F and moderate T_c).

As an interpretation of these results, we presented a picture for the high-T_c cuprates in terms of a crossover from BE to BCS condensation in 1994 [5,6]. This picture stems from the following observations and concepts [5,6]: (A) and (F) mentioned above; (H) the loss of mangtic response below the pseudo-gap temperature T^* observed by NMR and neutrons [7] may infer a formation of singlet pairs (i.e. pre-formed bosons); (I) the insulating behavior observed in c-axis conductivity below T^* [8] may be due to reduction of hopping/tunneling probability between CuO_2 planes for a 2e particle compared to that of a single fermion. Then the situation in the cuprates can be mapped to a more general phase diagram in Fig. 3 [5,6] expected for BE-BCS crossover with increasing carrier doping.

When the energy scale of attractive interaction is higher than T_B in the low density region, the singlet pairs will be first formed at T^* and then they undergo BE condensation at a much lower temperature. With increasing carrier doping, T_c will increase as $T_c \propto T_B$ while T^* would decrease. T_c and T^* would merge in the "crossover region", which corresponds to the "optimum T_c" region of the cuprates. The higher density side is characterized by the simultaneous pair-formation and condensation at T_c, i.e., the most fundamental feature of BCS condensation. We further conjectured that (J) the crossover will occur where ϵ_F becomes comparable to the energy scale $\hbar\omega_B$ of the exchange bosons which mediate the pairing interaction [5,6], thus characterizing the underdoped cuprates with retarded coupling. Assuming (J), we can estimate $\hbar\omega_B/k_B$ to be comparable to $\epsilon_F/k_B \sim 2,000$ K of the "optimally doped" cuprates. We recently pointed out [9] that: (K) this energy scale matches well with that of the antiferromagnetic exchange interaction J, providing a support for pictures that spin fluctuations indeed mediate the pairing interaction; and (L) the mid infrared reflection (MIR) in optical conductivity of the cuprates, appearing with energy scales comparable to $\hbar\omega_B$, could originate from the pairing interaction, in [6] contrast to the Drude part representing the translational motion of carriers (with an energy scale ϵ_F).

This picture is re-inforced by recent observation of photo-emission studies [10] that the superconducting gap in the "optimal" T_c region evolves smoothly to the pseudo gap in the underdoped region with the same geometrical symmetry. General concepts of negative-U Hubbard model and BE-BCS crossover have been considered by number of authors [11,12]. In 1990, Doniach and Inui pointed out that the pseudo gap behavior may be interpreted in terms of phase fluctuations of the superconducting order parmeter [13]. In 1995, Emery and Kivelson [14] discussed that the universal correlations between T_c and n_s/m^* (A) is expected when T_c is determined by the phase fluctuations. Although they argued this as "re-interpretation" of (A), the present author believes that their phase-fluctuation picture is essentially identical to our BE-BCS crossover picture.

Emery and Kivelson calculated the kinetic energy for phase fluctuations, and estimated the maximum condensation temperature $T_\theta^{max} \propto n_s/m^* \times y$, where y is a parameter representing distance. For 2-d systems they substituted the interplanar distance c_{int}, which lead to $T_\theta^{max} \propto T_B$. Their choice of the coherence length ξ as y for 3-d systems led to unrealistically high value of $T_\theta^{max} \gg T_B$, as shown in Fig. 2 for the example of K_3C_{60}. (Note that for $\xi \to \infty$, $T_\theta^{max} \to \infty$.) The ultimate upperlimit for condensation temperature should be given by T_B, since any interaction, dimensionality, and other factors would reduce actual T_c from T_B. These considerations indicate that y=ξ for 3-d systems in ref. [14], which led to an apparent difference between 2-d and 3-d systems in their Table 1, is a bad choice. Instead, if one substitutes the interparticle distance for y for 3-d systems, Table 1 of ref. [14] reduces to Fig. 3 of ref. [2].

The role of the 2-dimensional nature of cuprates is an interesting issue. Figure 1 does not give much information on this, as the horizontal axis represents the 3-d carrier density n_s as well as the 2-d density $n_{s2d} = n_s \times c_{int}$ for the 123, 214 and 2223 systems, all having the interlayer distance $c_{int} \sim 6$Å. Recently, we have performed μSR measurements in Hg1201 [15] systems, with $c_{int} = 9.5$ Å. Our previous data of nearly-optimally doped Tl2201 system [3] represents the case with $c_{int} = 11.5$ Å. All these results and those of underdoped 214, 123 and 2223 systems lie approximately on the same straight line when we plot T_c versus $\sigma \propto n_s/m^*$. However, when plotted versus $\sigma \times c_{int} \propto n_{s2d}/m^*$ in Fig. 4, we find that: (M) systems with smaller c_{int} have higher T_c for a given n_{s2d}/m^*, following a relation $T_c \propto 1/c_{int}$. This result is consistent with the reduction of T_c for increasing thickness of $PrBa_2Cu_3O_7$ (PBCO) layer in thin films of $YBa_2Cu_3O_7$ (YBCO) having a single unit-cell stacked in the c-axis direction with non-superconducting PBCO [16] (see Fig. 4).

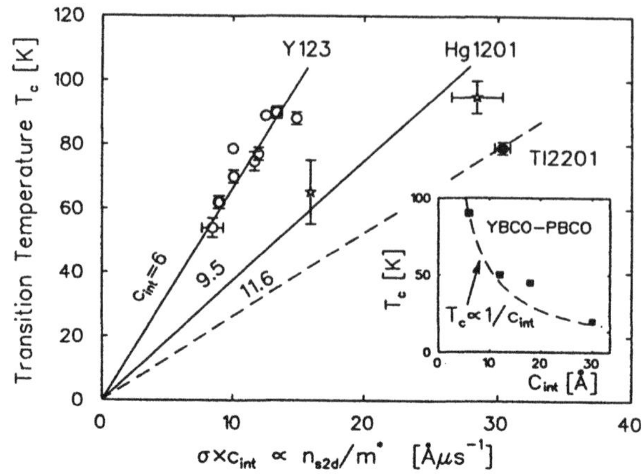

Figure 4. Plot of T_c versus $\sigma \times c_{int} \propto n_{s2d}/m^*$ for the underdoped region of Y123 [1], Hg1201 [15] and neally-optimally doped Tl2201 [3] systems. Inset: T_c vs. c_{int} in YBCO-PBCO [16].

This feature $T_c \propto n_{s2d}/c_{int}$ ($=n_s$) indicates: (N) a Kosterlitz-Thouless transition in its simplest form, where n_{s2d}/m^* is the sole factor for T_c, in not sufficient for predicting T_c; and (O) BE condensation in quasi 2-d systems gives a better explanation since T_c dependes logarithmically as $T_c \propto n_{s2d}/m^* \times [1/\ln(1/I_c)]$ on the interlayer tunnelling probability I_c [17] while I_c should depend exponentially on c_{int} as $I_c \propto \exp(-c_{int})$, leading to $T_c \propto 1/c_{int} \times n_{s2d}/m^*$.

The decrease of n_s/m^* with increasing doping in the overdoped Tl2201 systems (feature (C) in Fig. 1) can *not* be expected in the BCS model with retarded interaction, where the nomal state carrier density n should be equal to n_s. In ref. [3], we proposed (P) a spontaneous microscopic phase separation between superconducting and normal regions ($n_s < n$) as a possible explanation. Recently, we found a similar situation in Zn-substituted 214 and 123 systems [18], where the reduction of n_s/m^* with increasing Zn doping suggests that: (Q) carriers within the region $\pi\xi_{ab}^2$ on the CuO_2 planes around each Zn impurity may be excluded from the superfluid – like "swiss cheese". Specific heat in overdoped Tl2201 and Zn-doped 123/214 systems both exhibit increasing T-linear term at $T \to 0$ with increasing carrier/Zn doping [19], further supporting our models (P) and (Q) with phase separation.

The short coherence length ξ in cuprates will help reduce the energy cost for creating phase boundaries, facilitating phase separation. Indeed the swiss cheese model for Zn-doped systems is reminiscent of the behavior of superfluid ^{4}He in porous media [20]. For overdoped cuprates, we propose the following scenario: (R) if an abrupt decrease of effective attractive interaction occurs in the cuprates as the doping exceeds the "optimal" concentration, then energy gain for maintaining superconductivity might win over energy cost for phase separation: the "cheese" part composed of regions with local n_s near the "optimum concentration", while its volume fraction decreasing as further doping.

In underdoped (M), overdoped (P) and Zn-doped (Q) cuprates, T_c is proportional to the (spatially averaged) 3-dimensional volume density n_s of superfluid. This demonstrates fundamental importance of BE condensation.

This study is supported by NSF (DMR-95-10453, 10454) and NEDO (Japan). The author is indebted to Oleg Tchernyshyov for discussions leading to the interpretations (N) and (O).

REFERENCES

1 Y.J. Uemura *et al.*, Phys. Rev. Lett. **62** (1989) 2317.
2 Y.J. Uemura *et al.*, Phys. Rev. Lett. **66** (1991) 2665.
3 Y.J. Uemura *et al.*, Nature **364** (1993) 605.
4 Ch. Niedermayer *et al.*, Phys. Rev. Lett. **71** (1993) 1764.
5 Y.J. Uemura, in *Proceedings of the Workshop in Polarons and Bipolarons in High-T_c Superconductors and Related Materials, Cambridge, 1994* ed. by E. Salje *et al.* (Cambridge Univ. Press, 1995) p.p. 453.
6 Y.J. Uemura, in *Proceedings of the CCAST Symposium on High-T_c Superconductivity and the C_{60} Family, Beijing, 1994* ed. by S. Feng and H.C. Ren (Gordon and Breach, New York, 1995) pp. 113.
7 H. Yasuoka, T. Imai and M. Shimizu, in *Spin Correlations and Superconductivity*, ed. by H. Fukuyama (Springer Verlag, 1989), pp. 254; J. Rossat-Mignot *et al*, Physica **C185-189** (1991) 86.
8 T. Ito *et al.*, Nature **350** (1991) 569
9 Y.J. Uemura, in *Proceedings of the 10th Anniversary HTS Workshop, Houston, 1996*, ed. by B. Batlogg *et al.*, (World Scientific, 1996) pp. 68.
10 A.G. Loeser *et al.*, Science **273** (1996) 325; H. Ding *et al.*, Nature **382** (1996) 51.
11 R. Micnâs, J. Ranninger and S. Rabaskiewicz, Rev. Mod. Phys. **62** (1990) 113.
12 M. Randeria, in *Bose-Einstein Condensation*, ed. by A. Griffin, D.W. Snoke and S. Stringari (Cambridge Univ. Press, 1995) pp. 355.
13 S. Doniac and M. Inui, Phys. Rev. **B41** (1990) 6688.
14 V.J. Emery and S. Kivelson, Nature **374** (1995) 434.
15 B. Nachumi *et al.*, Hyperfine Interact. (1997), in press.
16 O. Fischer *et al.*, Physica **B169** (1991) 116.
17 R. Friedberg, T.D. Lee and H.C. Ren, Physica **A152** (1991) 423.
18 B. Nachumi *et al.*, Phys. Rev. Lett. **77** (1996) 5421.
19 J.W. Loram *et al.*, Physica **C235-240** (1994) 134.
20 J.D. Reppi, J. Low Temp. Phys. **87** (1992) 205.

ELSEVIER

Physica C 282–287 (1997) 198–201

PHYSICA C

EXPERIMENTAL STUDIES OF THE ISOTOPE EFFECT IN HIGH T_c SUPERCONDUCTORS

J.P. Franck

Department of Physics, University of Alberta
Edmonton, AB, Canada T6G 2J1

A brief review of isotope effect studies in high T_c cuprates is given. It is shown that in all systems a close connection exists between the isotope effect and the pressure dependence of T_c. Both charge transfer and non-charge transfer mechanisms contribute to these effects.

1. INTRODUCTION

A brief survey of results of studies of the isotope effect in various high T_c superconductor systems is given. Illustrative examples of the effect in $La_{2-x}Sr_xCuO_4$, $YBa_2(Cu_{1-x}Zn_x)_3O_7$, and oxygen-deficient $YBa_2Cu_3O_{7-\delta}$ are presented. A comparison with pressure effect studies on these systems is given, and close similarities between the two effects are found.

2. DEFINITIONS

The isotope effect in superconductors is conventionally expressed through the partial isotope exponents

$$\alpha_i = -\Delta\ln T_c/\Delta\ln m_i \qquad (1)$$

where m_i refers to the various atomic species of the superconducting compound, and T_c is considered a function of these masses, $T_c=T_c(m_i)$. In high T_c superconductors, most investigations refer to the oxygen isotope effect. Replacement of ^{16}O by ^{18}O up to 90 to 95% can be achieved. Copper isotope effects can also be studied, this requires the careful parallel production of comparison samples with either ^{63}Cu or ^{65}Cu. In this case close to complete (99.7%) change in the isotope is possible.
The isotope effect was studied in a large number of high T_c superconductors [1], in the following we will briefly discuss three different systems.

3. THE SYSTEM $La_{2-x}Sr_xCuO_4$

In this system, both the oxygen and copper isotope effects were studied, [2,3] It was found that near optimal doping ($x=0.15$, $T_c=38K$), a sizable isotope effect exists, with both α_O and α_{Cu} near 0.15. For doping at $x=0.125$ (1/8 doping), the isotope effect becomes anomalously large (α_O and α_{Cu} near 0.9). This local maximum in α_O and α_{Cu} at 1/8 doping is related to the special electronic properties of this system, which also leads to a local depression in T_c. A similar effect has not been seen in other cuprate superconductor systems. Apart from this effect, α_O increases smoothly both below and above optimal doping. The data on the oxygen and copper isotope effect in this system are shown in Fig. 1

4. THE SYSTEM $YBa_2(Cu_{1-x}Zn_x)_3O_7$

In optimally doped YBCO, a small positive oxygen isotope exponent of $\alpha_O=0.02$ to 0.05 is found. No copper isotope shift is discernible, with $\alpha_{Cu}<0.01$. Zn rapidly depresses the transition temperature at a rate of $dT_c/dx_{Zn}=-10K/(\%Zn)$. The isotope effect was studied over the whole range of transition temperatures, down to $T_c\cong 3.0K$ (in zero field), and $T_c = 1.5K$ (in 18T) [4, 5]. It is found that the observed isotope shift in T_c is small for all levels of Zn doping ($\Delta T_c \approx 0.3K$), it decreases to $\Delta T_c \approx 0.10K$ in a 9% Zn sample in a field of 18T ($T_c \approx 1.5K$). In Fig. 2 we show resistive transitions

0921-4534/97/$17.00
PII S0921-4534(97)00196-2

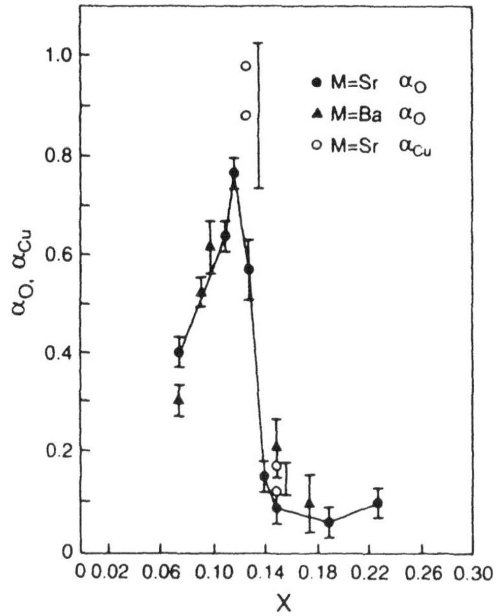

Figure 1. The oxygen and copper isotope effect in $La_{2-x}Sr_xCuO_4$, as function of doping x, from [2], and [3]. The oxygen isotope effect in $La_{2-x}Ba_xCuO_4$ is also shown, from [2].

for this sample pair. The calculated isotope exponent at such low transition temperatures is quite sensitive to the choice of T_c. If we use the resistive onset, then we find values of α_O near 0.5 for the lowest T_c samples in this system. α_O drops rapidly to much lower values for higher T_c samples, Fig. 3. It is interesting to note, that the istope effect persists to extremely low temperatures, and high fields, and apparently approaches 0.5 in this case.

5. THE OXYGEN-DEFICIENT SYSTEM $YBa_2Cu_3O_{7-\delta}$

In this system the copper isotope effect was studied from fully oxygenated YBCO ($\delta \approx 0.06$) to $\delta \approx 0.60$, [6]. The transition temperature shows the well-known plateau at 60K. A negative α_{Cu} (i.e. $T_c(^{65}Cu) > T_c(^{63}Cu)$) is found at all levels of oxygen deficiency. α_{Cu} is smallest in the 60K plateau, but not zero. Dependent on the production of oxygen deficiency, values between -0.14 and -0.34 are found in the plateau. The oxygen isotope effect in this system was studied by Zech et al.[7].

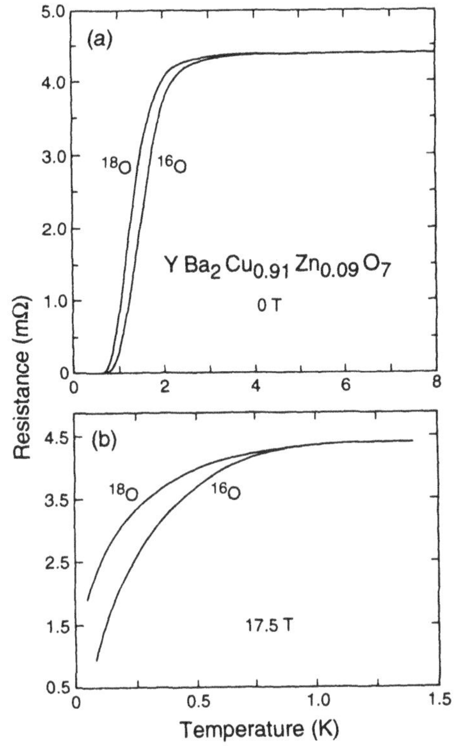

Figure 2. The resistive transition for a pair of ^{16}O and ^{18}O substituted $YBa_2(Cu_{0.91}Zn_{0.09})_3O_7$.
a) in zero field, $\Delta T_c \approx 0.14K$,
b) in 17.5T, $\Delta T_c \approx 0.10K$.

6. OVERVIEW AND COMPARISON WITH PRESSURE EFFECT RESULTS

Comparison of the isotope effect in various high T_c superconductor systems shows certain regularities. It is always observed that α_O, or α_{Cu}, are at a minimum at the optimum doping level. When doping away from this condition α_O, and α_{Cu} increase. In Fig. 3 we show this for the oxygen isotope effect in various doped YBCO systems. It is clear that no universal dependence of α_O on the transition temperature T_c is found. Fe doping of YBCO gives the steepest observed increase in α_O with doping [8], and Zn doping gives the lowest. In Fig. 3 we compare this with observed pressure effects in this system, and find the same tendency [10, 11].

In the La214 system we can make a similar comparison. Hydrostatic measurements of dT_c/dP

ELSEVIER

Physica C 282–287 (1997) 202–205

PHYSICA C

Oxygen isotope effects in the manganites and cuprates: Evidence for polaronic charge carriers

Guo-meng Zhao, M. B. Hunt, K. Conder*, H. Keller and K. A. Müller

Physik-Institut der Universität Zürich, CH-8057 Zürich, Switzerland

*Laboratorium für Festkörperphysik ETH Zürich, CH-8093, Switzerland

We present various oxygen-isotope effects in the manganites from magnetization and thermal-expansion measurements. We found that the ferromagnetic transition temperature T_C, the magnitude of the jump in the thermal-expansion coefficient $\Delta\beta(T_C)$ at T_C all depend strongly on the oxygen mass. Upon replacing ^{16}O with ^{18}O, T_C can be lowered by as much as 20 K, and $\Delta\beta(T_C)$ can be raised by about 20%. The observed novel isotope effects clearly demonstrate that charge carriers in these materials are strongly coupled to the local Jahn-Teller distortions, so that polarons are formed.

We also report the experimental evidence for polaronic supercarriers in the cuprate superconductors $La_{2-x}Sr_xCuO_4$. From magnetization and thermal-expansion measurements, we can quantitatively determine the oxygen-isotope effects on the in-plane penetration depth $\lambda_{ab}(0)$ and on the carrier density n. We find a negligible isotope effect on n, but a large effect on $\lambda_{ab}(0)$. Our results show that polaronic charge carriers exist and condense into Cooper pairs in the cuprate superconductors.

1. Oxygen isotope effects in the manganites

Recently, very large (colossal) magnetoresistance has been observed in the manganese-based perovskites $Ln_{1-x}A_xMnO_3$ (where Ln is a trivalent rare earth ion and A is a divalent alkali earth ion) [1,2]. Because of their unusual magnetic properties and potential applications, these materials have recently attracted intensive research interest. The physics in manganites has primarily been described by the double-exchange model [4,3]. However, Millis, Littlewood and Shraiman [5] have pointed out that double-exchange alone cannot fully explain the data of $La_{1-x}Sr_xMnO_3$. They proposed that lattice-polaronic effects due to strong electron-phonon coupling (arising from a strong Jahn-Teller effect) should be involved.

In most materials, magnetic phenomena at room temperature and below are essentially unaffected by lattice-vibrations because the electronic and lattice subsystems are decoupled according to the Born-Oppenheimer adiabatic approximation. The atoms can usually be considered as infinitely heavy and static in theoretical descriptions of the magnetic phenomena. However, this approximation would break down in compounds where there is a strong Jahn-Teller (JT) effect. Höck *et al.* [6] studied JT ions in a conductor within a linear chain model. They showed that small JT polarons can be formed when the JT stabilization energy is comparable with the bare conduction bandwidth. Polarons are not 'bare' charge carriers, but are carriers dressed by local-lattice distortions. In other words, the electronic and lattice subsystems are no longer decoupled, so one would expect that lattice vibrations should affect the electronic quantities, and thus there should be effects of varying isotope mass on some magnetic properties such as the ferromagnetic transition temperature T_C.

To be more specific, the polaronic nature of the conduction carriers can be demonstrated by the isotope effect on the effective bandwidth W_{eff} of polarons, which in turn depends on the isotope mass M [7]:

$$W_{eff} \propto W \exp(-\gamma E_b/\hbar\omega), \quad (1)$$

where E_b is the binding energy of polarons, and is independent of the isotope mass M, γ is a

0921-4534/97/$17.00
PII S0921-4534(97)00198-6

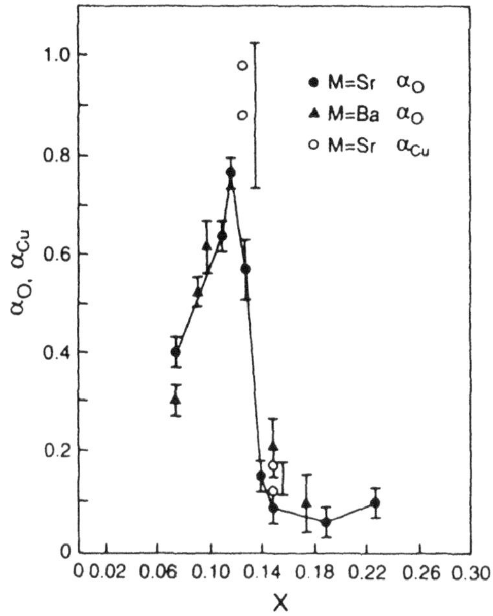

Figure 1. The oxygen and copper isotope effect in $La_{2-x}Sr_xCuO_4$, as function of doping x, from [2], and [3]. The oxygen isotope effect in $La_{2-x}Ba_xCuO_4$ is also shown, from [2].

for this sample pair. The calculated isotope exponent at such low transition temperatures is quite sensitive to the choice of T_c. If we use the resistive onset, then we find values of α_O near 0.5 for the lowest T_c samples in this system. α_O drops rapidly to much lower values for higher T_c samples, Fig. 3. It is interesting to note, that the istope effect persists to extremely low temperatures, and high fields, and apparently approaches 0.5 in this case.

5. THE OXYGEN-DEFICIENT SYSTEM $YBa_2Cu_3O_{7-\delta}$

In this system the copper isotope effect was studied from fully oxygenated YBCO ($\delta \approx 0.06$) to $\delta \approx 0.60$, [6]. The transition temperature shows the well-known plateau at 60K. A negative α_{Cu} (i.e. $T_c(^{65}Cu) > T_c(^{63}Cu)$) is found at all levels of oxygen deficiency. α_{Cu} is smallest in the 60K plateau, but not zero. Dependent on the production of oxygen deficiency, values between -0.14 and -0.34 are found in the plateau. The oxygen isotope effect in this system was studied by Zech et al.[7].

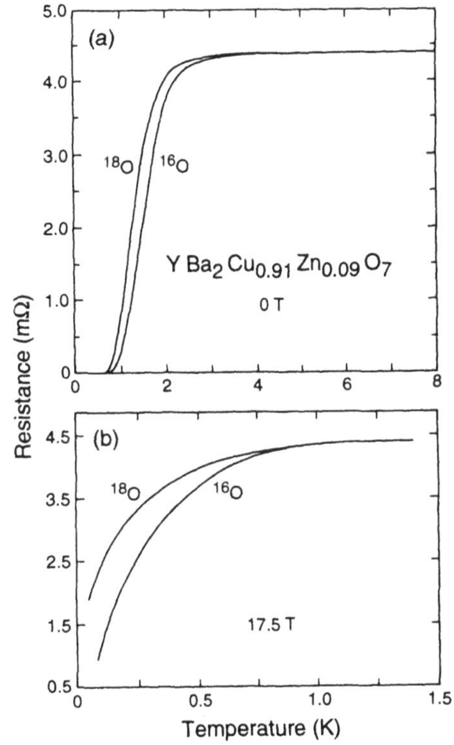

Figure 2. The resistive transition for a pair of ^{16}O and ^{18}O substituted $YBa_2(Cu_{0.91}Zn_{0.09})_3O_7$.
a) in zero field, $\Delta T_c \approx 0.14K$,
b) in 17.5T, $\Delta T_c \approx 0.10K$.

6. OVERVIEW AND COMPARISON WITH PRESSURE EFFECT RESULTS

Comparison of the isotope effect in various high T_c superconductor systems shows certain regularities. It is always observed that α_O, or α_{Cu}, are at a minimum at the optimum doping level. When doping away from this condition α_O, and α_{Cu} increase. In Fig. 3 we show this for the oxygen isotope effect in various doped YBCO systems. It is clear that no universal dependence of α_O on the transition temperature T_c is found. Fe doping of YBCO gives the steepest observed increase in α_O with doping [8], and Zn doping gives the lowest. In Fig. 3 we compare this with observed pressure effects in this system, and find the same tendency [10, 11].

In the La214 system we can make a similar comparison. Hydrostatic measurements of dT_c/dP

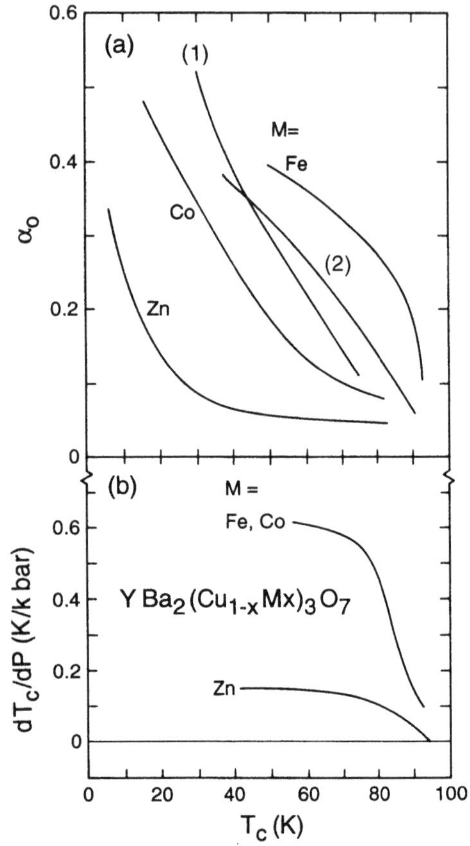

Figure 3. (a) oxygen isotope exponent as function of T for various doped YBCO systems.
(1) $(Y_{1-x}Pr_x)Ba_2Cu_3O_7$;
(2) $YBa_{2-x}Ca_xCu_3O_7$ [9];
Fe doped YBCO from [8]; all others see [1, 4, 5].
b) Pressure effect dT_c/dP, from [8, 9].

by Tanahashi et al. [12] show a pronounced minimum near optimal doping. A similar effect is seen in uniaxial pressure effect measurements [13].

A very detailed comparison can be made in the case of oxygen-deficient YBCO, since pressure effects have been published both for hydrostatic and for uniaxial pressure effects. Hydrostatic measurements show a very small dT_c/dP near optimal doping, a pronounced maximum near $O_{6.7}$ and a smaller value in the 60K plateau [14, 15]. Uniaxial pressure effect studies show, however, that even for optimal doping, there is a large positive dT_c/dP for compression along the b axis and a large negative value for compression doping the a axis. c-axis compression at optimal doping has almost no effect on T_c [16]. These effects nearly cancel at optimal doping. For oxygen-deficient YBCO, Kraut et al [17] find very different behaviour for a, b, and c-axis compression. c-axis compression can be seen to transfer carriers into the (a,b) plane, and, therefore, change T_c.a, and b axis compression is strongly dependent on oxygen concentration. This is shown in Fig. 4. These effects cannot be understood on the basis of charge transfer, since they persist, and are large, at optimal T_c, where T_c is independent of carrier concentration. The total hydrostatic pressure effect is also shown in Fig. 4, it can be seen to track the copper isotope exponent closely. In particular, one sees that the large value of dT_c/dP at $O_{6.7}$ is due to the large combined values of a and b axis compression. It is clear, therefore, that both pressure and isotope effect in this system are similar, and are to a large extent due to effects that are not related to charge transfer.

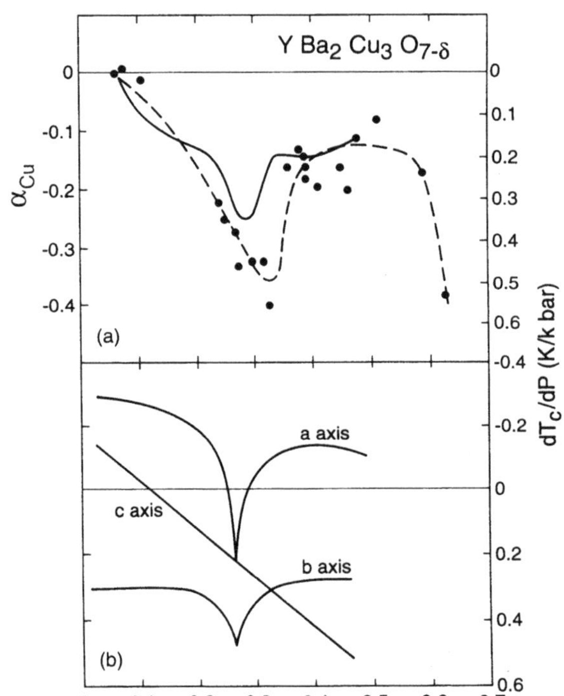

Figure 4. (a) The copper isotope exponent of $YBa_2Cu_3O_{7-\delta}$ as function of δ (dotted line and points, [6]), compared with the hydrostatic pressure dependence (solid line, [17].
b)Uniaxial pressure coefficients from [17].

A close connection between isotope effect and pressure effect has been previously noted. It is generally discussed in terms of the theory of Schneider and Keller [18]. In this theory α is given by $\alpha = \alpha_\omega + \alpha_n$, where α_ω is a true (phonon related) isotope effect and α_n is related to the parabolic dependence of T_c on carrier concentration:

$$\alpha_n = -\left(\frac{d\ln T_c}{d\ln n}\right)\left(\frac{d\ln n}{d\ln m}\right) \tag{2}$$

Values of dlnn/dlnnm near -0.125 have been obtained by fitting to observed data. Zech et al.[7] found α_n=-0.25 by fitting their data on the oxygen isotope effect in oxygen-deficient YBCO. Since T_c depends on n/m*, m*= apparent carrier mass, it is also possible that m* is mass-dependent. Zhao and Morris [19] give $d\ln m^*/d\ln m_o = 0.61$ in YBCO. Our analysis of the results on the copper isotope effect in oxygen-deficient YBCO shows that the results cannot be fully understood on this basis. We find that in the 60K plateau very large values of $d\ln n/d\ln m_{Cu}$ are required to fit the data, in the range +0.5 to 1.2. This is, of course, due to the fact that a very sizable α_{Cu} is observed where T_c is independent of doping level, i.e., independent of n. Effects other than a mass dependent change transfer must, therefore, be present. It appears that the large uniaxial pressure effect for compression in the a and b axis are related to this. Compression in these directions changes the interaction in the "hot spots" on the Fermi surface, where large saddle point Van Hove singularities are observed. Isotopic substitution will also change this interaction, and a combined analysis of isotope and pressure effects may identify this.

CONCLUSION

We find that the dependence of the isotope effect on T_c is not universal, but is system dependent. Similar system dependencies are found in the pressure effect. Contributions to α and dT_c/dP, from mass, or pressure dependent charge transfer exist, but non-charge transfer contributions are large and can dominate in certain systems. The non-charge transfer contributions are related to the (a,b) axis symmetry, and therefore, related to the d-wave picture. Interpretation of the isotope effect should relate effects common to mass and pressure change.

ACKNOWLEDGMENT

This research was supported by grants from the Natural Sciences and Engineering Research Council of Canada (NSERC).

REFERENCES

1. J.P. Franck, in "Physical Properties of High Temperature Superconductors IV" ed. D.M. Ginsberg, World Scientific (1994), p.189.
2. M.K. Crawford et al., Phys. Rev. B41 (1990) 182; Science (1991) 1390.
3. J.P. Franck et al., Phys. Rev. Lett. 71 (1993) 283.
4. J.P. Franck et al., in "Lattice Effects in High T_c Superconductors", et Y. Bar Yam et al., World Scientific (1992), p.148.
5. G. Soerensen and S. Gygax, Phys. Rev. B51 (1995) 11848.
6. J.P. Franck and D.D. Lawrie, J. of Superconductivity 8 (1995) 591; Physica Scripta T66 (1996) 220.
7. D. Zech, Physica B 219-220 (1996) 136.
8. H.J. Bornemann, et al., Physica C185-189 (1991) 1359.
9. H.J. Bornemann and D.E. Morris, Phys. Rev. B44 (1991) 5322.
10. M. Kurisu et al, Physica C152 (1988) 339.
11. C. Meingast et al., Physica C173 (1991) 309.
12. N. Tanahashi et al., Jap. J. Appl. Phys. 28 (1989) L762.
13. F. Gugenberger et al., Phys. Rev. B49 (1994) 13137.
14. W.H. Fietz et al., Physica C 235-240 (1994) 1785.
15. B. Bucher et al., Europhys. Lett. 34 (1996) 391.
16. U. Welp et al., Phys. Rev. Lett. 69 (1992) 2130.
17. O. Kraut et al., Physica C 205 (1993) 139.
18. T. Schneider, and H. Keller, Phys. Rev. Lett. 69 (1992) 3374.
19. G. Zhao and D.E. Morris, Phys. Rev. B51 (1995) 16487.

ELSEVIER

Physica C 282–287 (1997) 202–205

PHYSICA C

Oxygen isotope effects in the manganites and cuprates: Evidence for polaronic charge carriers

Guo-meng Zhao, M. B. Hunt, K. Conder*, H. Keller and K. A. Müller

Physik-Institut der Universität Zürich, CH-8057 Zürich, Switzerland

*Laboratorium für Festkörperphysik ETH Zürich, CH-8093, Switzerland

We present various oxygen-isotope effects in the manganites from magnetization and thermal-expansion measurements. We found that the ferromagnetic transition temperature T_C, the magnitude of the jump in the thermal-expansion coefficient $\Delta\beta(T_C)$ at T_C all depend strongly on the oxygen mass. Upon replacing ^{16}O with ^{18}O, T_C can be lowered by as much as 20 K, and $\Delta\beta(T_C)$ can be raised by about 20%. The observed novel isotope effects clearly demonstrate that charge carriers in these materials are strongly coupled to the local Jahn-Teller distortions, so that polarons are formed.

We also report the experimental evidence for polaronic supercarriers in the cuprate superconductors $La_{2-x}Sr_xCuO_4$. From magnetization and thermal-expansion measurements, we can quantitatively determine the oxygen-isotope effects on the in-plane penetration depth $\lambda_{ab}(0)$ and on the carrier density n. We find a negligible isotope effect on n, but a large effect on $\lambda_{ab}(0)$. Our results show that polaronic charge carriers exist and condense into Cooper pairs in the cuprate superconductors.

1. Oxygen isotope effects in the manganites

Recently, very large (colossal) magnetoresistance has been observed in the manganese-based perovskites $Ln_{1-x}A_xMnO_3$ (where Ln is a trivalent rare earth ion and A is a divalent alkali earth ion) [1,2]. Because of their unusual magnetic properties and potential applications, these materials have recently attracted intensive research interest. The physics in manganites has primarily been described by the double-exchange model [4,3]. However, Millis, Littlewood and Shraiman [5] have pointed out that double-exchange alone cannot fully explain the data of $La_{1-x}Sr_xMnO_3$. They proposed that lattice-polaronic effects due to strong electron-phonon coupling (arising from a strong Jahn-Teller effect) should be involved.

In most materials, magnetic phenomena at room temperature and below are essentially unaffected by lattice-vibrations because the electronic and lattice subsystems are decoupled according to the Born-Oppenheimer adiabatic approximation. The atoms can usually be considered as infinitely heavy and static in theoretical descriptions of the magnetic phenomena. However, this approximation would break down in compounds where there is a strong Jahn-Teller (JT) effect. Höck *et al.* [6] studied JT ions in a conductor within a linear chain model. They showed that small JT polarons can be formed when the JT stabilization energy is comparable with the bare conduction bandwidth. Polarons are not 'bare' charge carriers, but are carriers dressed by local-lattice distortions. In other words, the electronic and lattice subsystems are no longer decoupled, so one would expect that lattice vibrations should affect the electronic quantities, and thus there should be effects of varying isotope mass on some magnetic properties such as the ferromagnetic transition temperature T_C.

To be more specific, the polaronic nature of the conduction carriers can be demonstrated by the isotope effect on the effective bandwidth W_{eff} of polarons, which in turn depends on the isotope mass M [7]:

$$W_{eff} \propto W \exp(-\gamma E_b/\hbar\omega), \qquad (1)$$

where E_b is the binding energy of polarons, and is independent of the isotope mass M, γ is a

0921-4534/97/$17.00
PII S0921-4534(97)00198-6

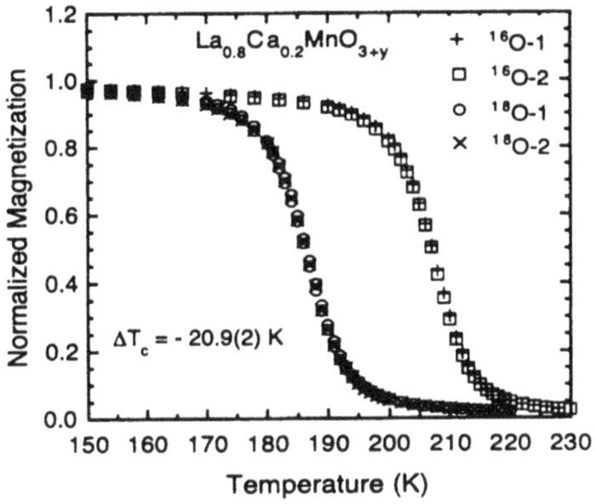

Figure 1. Oxygen isotope effect on the Curie temperature of $La_{0.8}Ca_{0.2}MnO_{3+y}$.

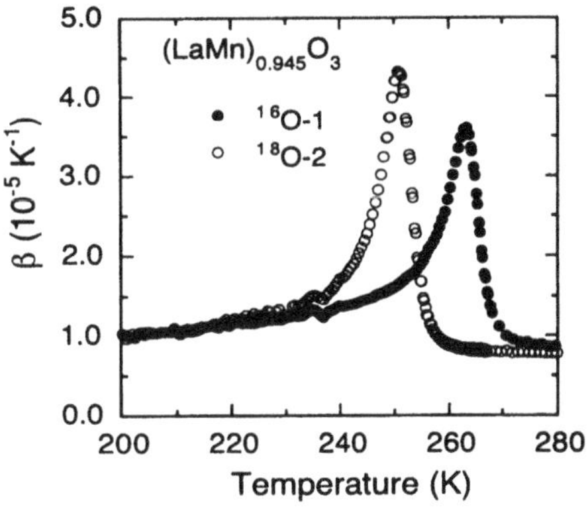

Figure 2. The linear thermal-expansion coefficient β(T) for the ^{16}O and ^{18}O samples of $(LaMn)_{0.945}O_3$.

dimensionless constant ($0 \leq \gamma \leq 1$) depending on E_b/W, ω is the characteristic frequency of the optical phonons ($\omega \propto 1/\sqrt{M}$). In the ferromagnetic manganites, it is easy to study the isotope effect on W_{eff}. This is due to the fact that in the strong-coupling limit, where the Hund's rule coupling J_H is very large compared to W_{eff} ($J_H \gg W_{eff}$), the Curie temperature $T_C \propto W_{eff}$ (Ref. [3,5]).

1.1. Giant oxygen-isotope shift of T_C in $La_{1-x}Ca_xMnO_{3+y}$

In Fig. 1 we show the magnetization (normalized to the magnetization well below T_C) for pairs of ^{16}O and ^{18}O samples of $La_{0.8}Ca_{0.2}MnO_{3+y}$ (i.e., 2 for each isotope). The detailed procedures for sample preparation and isotope exchange were described in Ref. [8]. For the ferromagnet $La_{0.8}Ca_{0.2}MnO_{3+y}$ with a strong JT effect, the ^{18}O samples have lower T_C's than the ^{16}O samples by ~21 K. On the other hand, no oxygen isotope effect on T_C could be detected in the ferromagnet $SrRuO_3$ with a negligible JT effect [8].

Since there is no oxygen-isotope effect on T_C in the ferromagnet $SrRuO_3$ with a negligible JT effect, the giant isotope effect observed in the Ca-doped ferromagnets is very likely related to the strong JT effect in this system. For a compound with a strong JT effect, the electron-phonon interaction is usually large, leading to the formation of JT polarons [6].

1.2. Oxygen-isotope effect on thermal expansion in the manganites

In Fig. 2 we show the linear thermal-expansion coefficient β(T) for the ^{16}O and ^{18}O samples of $(LaMn)_{0.945}O_3$. The thermal-expansion was measured using a capacitance dilatometer with a length resolution of ~0.1Å (Ref. [9]). The thermal-expansion coefficient has a sharp asymmetric peak at T_C, corresponding to a large thermal-expansion jump $\Delta\beta(T_C)$. It is surprising that the jump $\Delta\beta(T_C)$ depends strongly on the oxygen-mass; $\Delta\beta(T_C)$ for the ^{18}O sample is larger than that for the ^{16}O sample by 24(3)%. A similar isotope effect was also observed in $La_{0.67}Ca_{0.33}MnO_3$ (Ref.[9]). We also showed [9] that the T_C can be changed by varying either the Mn^{4+} concentration or the oxygen mass, while $\Delta\beta(T_C)$ can be varied only by changing the oxygen mass. This implies that the observed large oxygen-isotope effects on both T_C and the thermal-expansion jump are intrinsic.

The conventional theory of ferromagnetism cannot explain the oxygen-isotope effects observed here. In order to find an explanation, we may relate $\Delta\beta(T_C)$ to the pressure effect ($d\ln T_C/dP$). If the ferromagnetic transition is of second order, then

$$d\ln T_C/dP = 3\Delta\beta(T_C)/\Delta C_P(T_C), \qquad (2)$$

where $\Delta C_P(T_C)$ is the jump of the specific heat. From equation 2 we see that there must be a corresponding dependence of $d\ln T_C/dP$ on the oxygen mass M provided that $\Delta C_P(T_C)$ is independent of M. Indeed we showed that [9] $\Delta C_P(T_C)$ depends strongly on M.

2. Evidence for polaronic supercarriers in $La_{2-x}Sr_xCuO_4$

Now it is known that high-T_c superconductivity involves the pairing of charge carriers, forming spin-singlet Cooper pairs, but the precise nature of these carriers remains unclear. The cuprates are known to exhibit a strong Jahn-Teller effect which would lead to the formation of Jahn-Teller polarons [6]. Still, direct evidence that Jahn-Teller polarons exist in the superconducting state of the cuprates has been lacking, although some indirect evidence comes from their recent discovery [8] in the structurally similar but nonsuperconducting manganite $La_{1-x}Ca_xMnO_3$. Moreover, there has recently been increasing evidence that polaronic charge carriers are present in the normal-state of the layered cuprates [10,11]. However, it is not clear whether these normal-state polaronic carriers should condense into Cooper pairs. To show that this occurs, one has to demonstrate that the effective supercarrier mass along the CuO_2 planes (m^*_{ab}) depends strongly on ionic mass M in these layered cuprate superconductors.

Since the magnetic penetration depth $\lambda(0)$ is proportional to $\sqrt{m^*/n_s}$, then

$$\Delta m^*/m^* = 2\Delta\lambda(0)/\lambda(0) + \Delta n_s/n_s, \qquad (3)$$

where Δ means any small change of a quantity upon isotope substitution. Thus the isotope dependence of m^* can be determined if one can independently measure the isotope dependences of $\lambda(0)$ and of n_s.

The isotope dependence of $\lambda(0)$ can be determined from that of the Meissner fraction $f(0)$ which, for decoupled and fine-grained samples, depends on the penetration depth $\lambda(0)$ and on the average grain radius R, as seen from the Shoenberg formula for spherical grains [12]:

$$f(T) = \frac{3}{2}[1-3(\frac{\lambda(T)}{R})\coth(\frac{R}{\lambda(T)})+3(\frac{\lambda(T)}{R})^2], \qquad (4)$$

where $\lambda(T) = \sqrt[3]{[(\lambda_{ab}(T)]^2\lambda_c(T)}$ for layered compounds. From equation 4, one can see that a change in $\lambda(0)$ will lead to a change in $f(0)$, so the isotope dependence of $\lambda(0)$ can be determined from the isotope dependence of $f(0)$.

The isotope dependence of n_s should be equal to the isotope dependence of the normal carrier density n in clean superconductors. Since the structural phase transition temperature T_s (from the tetragonal to orthorhombic phase) in the $La_{2-x}Sr_xCuO_4$ system is very sensitive to n, a very small difference in n will lead to a huge difference in T_s. Thus the isotope dependence of n can be measured precisely if one can accurately determine the isotope shift of T_s. It has also been found that there is an anomaly at T_s in the thermal-expansion coefficient of this system [13]. Therefore the isotope shift of T_s can be precisely determined by a thermal-expansion measurement.

Fig. 3 shows the temperature dependence of the Meissner effect for the ^{16}O and ^{18}O samples of $La_{1-x}Sr_xCuO_4$ with $x = 0.105$. The results for other compositions were shown in Ref. [14,15]. The T_c of the ^{18}O sample is lower than that of the ^{16}O sample by ~1.5 K, and the low-temperature Meissner fraction of the ^{18}O sample is reduced by 6(1)% relative to the ^{16}O sample. This is in agreement with the result reported in Ref. [15], indicating the excellent reproducibility of these isotope experiments.

In Fig. 4 we show the linear thermal-expansion coefficient β(T) for the another set of ^{16}O and ^{18}O samples of $La_{1.895}Sr_{0.105}CuO_4$, which shows the same oxygen isotope effects as the one shown in Fig. 3. One can see that there is an anomaly in β(T) at a temperature of ~265 K (see arrow), which corresponds to the tetragonal-orthorhombic transition temperature T_s [13]. It is clear that the two curves nearly overlap, implying that the two isotope samples have the same T_s

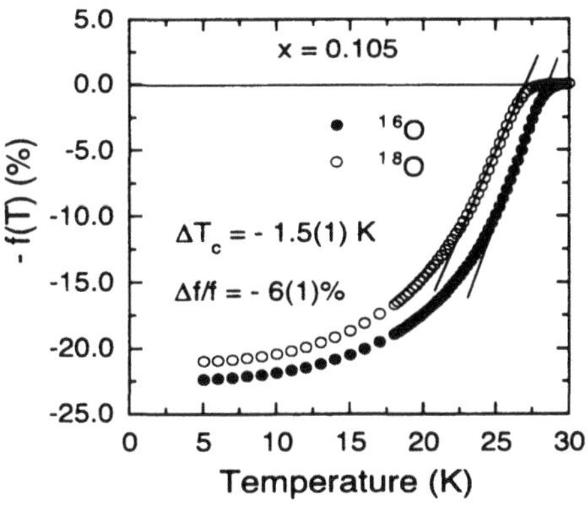

Figure 3. Temperature dependence of the Meissner effect for ^{16}O and ^{18}O samples of $La_{2-x}Sr_xCuO_4$ with $x = 0.105$.

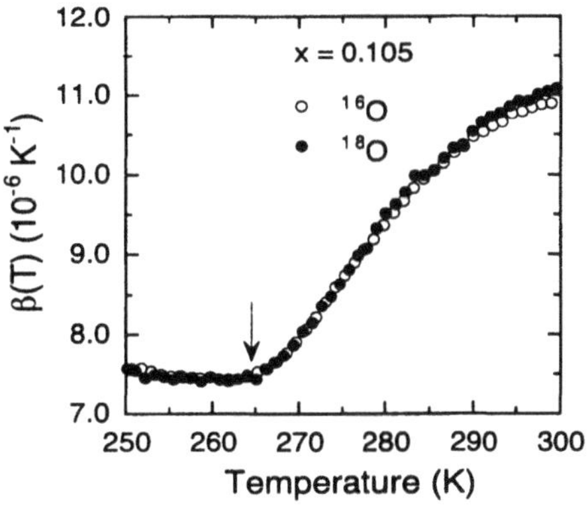

Figure 4. The linear thermal-expansion coefficient β(T) for the ^{16}O and ^{18}O samples of $La_{1.895}Sr_{0.105}CuO_4$.

within the uncertainty of our measurements ($\leq$0.3 K). In addition, we also showed [14] that $dT_s/dn \approx dT_s/dx$ = B = 2900 K. Using $\Delta T_s \leq$0.3 K and n = 0.105, we obtain $|\Delta n/n| = |(1/nB)\Delta T_s| \leq$ 0.1%. Thus, there is a negligible oxygen-isotope effect on n. Moreover, we have shown that the ratio of the normal-state susceptibilities (including a small Curie term contributed from some localized carriers) for the two isotope samples is temperature-independent down to a temperature near T_c. This implies that the two isotope samples must have the same mobile carrier concentration near T_c even if charge localization occurs. Since $n_s = n$ for clean superconductors, this result also implies that there is a negligible oxygen-isotope effect on the supercarrier density n_s.

The observed large oxygen-isotope effect on $f(0)$ and negligible effect on n_s imply that the effective supercarrier mass m^*_{ab} depends strongly on the oxygen mass. This suggests that polaronic charge carriers exist and condense into Cooper pairs in the cuprate superconductors. The quantitative data analysis can be seen in Ref. [14].

REFERENCES

1. S. Jin *et al.*, Science **264**, 413 (1994)
2. R. von Helmolt*et al.*, Phys. Rev. Lett. **71**, 2331 (1993)
3. P. W. Anderson, and H. Hasegawa, Phys. Rev. **100**, 675 (1955)
4. C. Zener, Phys. Rev. **82**, 403 (1951)
5. A. J. Millis *et al.*, Phys. Rev. Lett. **74**, 5144 (1995)
6. K.-H, Höck, H. Nickisch, and H. Thomas, Helv. Phys. Acta. **56**, 237 (1983)
7. A. S. Alexandrov, and N. F. Mott, Int. J. Mod. Phys. **8**, 2075-2109 (1994)
8. Guo-meng Zhao, K. Conder, H. Keller, and K. A. Müller, Nature **381**, 676 (1996)
9. Guo-meng Zhao, M. B. Hunt, and H. Keller, Phys. Rev. Lett. **78**, 955 (1997).
10. Proc. Int. Workshop on "Anharmonic Properties of High-T_c Cuprates", edited by D. Mihailowic, G. Ruani, E. Kaldis, and K. A. Müller (World Scientific, Singapore, 1994)
11. X. X. Bi, P. C. Eklund, Phys. Rev. Lett. **70**, 2625 (1993)
12. D. Shoenberg, Proc. R. Soc. London, Ser. **A175**, 49 (1940)
13. Y. Okajima, S. Hashimoto, and K. Yamaya, Physica C **235-240**, 1317 (1994)
14. Guo-meng Zhao, M. B. Hunt, H. Keller, and K. A. Müller, Nature **385**, 236 (1997)
15. Guo-meng Zhao *et al.*, Phys. Rev. **B52**, 6840 (1995)

ELSEVIER

Physica C 282–287 (1997) 206–209

PHYSICA C

Electronic States of the Superconductor Sr_2RuO_4

Y. Maeno

Department of Physics, Kyoto University, Kyoto 606-01

We discuss the characteristics of the electronic states of the copper-free layered perovskite superconductor Sr_2RuO_4, which exhibits a sharp transition with T_c up to 1.35 K. The normal state behavior is characterized as an essentially two-dimensional Fermi liquid, consistent with the Fermi surface determined by the de Haas-van Alphen oscillations. The specific heat jump across T_c is consistent with the gap with line nodes, and the unexpectedly large residual T-linear term is suggestive of exotic pairing.

1. INTRODUCTION

It has been more than two years since superconductivity of Sr_2RuO_4 was first reported [1]. With the recent improvement of the sample quality, T_c = 1.35 K has reproducibly been obtained. To the best of our knowledge, this compound is still the only Cu-free layered perovskite superconductor. Extensive studies have extracted remarkable characteristics of its electronic states, which we will discuss in this paper. At the same time this paper is intended to serve as a guide to more detailed descriptions in the literature. For this purpose we have included a large number of references at the end, although they are by no means a complete listing of the available literature on this subject. Other treatments of the overview are found in refs. 2-5.

2. FERMI SURFACE CONTROVERSY

The crystal structure of Sr_2RuO_4 is the tetragonal K_2NiF_4-type [6-10]. According to the band-structure calculations [11-15], the electronic states near the Fermi level are derived mainly from four 4*d* electrons of Ru in the t_{2g} orbitals (the low-spin configuration). The hybridization with the oxygen 2*p* electrons is substantial and results in antibonding $dp\pi^*$ states.

The quantum oscillations (both the de Haas-van Alphen and Shubnikov-de Haas oscillations) by Mackenzie *et al*. [16] showed that the Fermi surface consists of three pieces of two-dimensional (2D) sheets, α, β and γ. The observed cyclotron masses are in excellent agreement with the observed electronic coefficient of the specific heat, γ_N =39 ± 2 mJ/K^{2}mol [1, 17, 18], and all enhanced from the respective expectations of the band structure calculations by a factor of three to four. We identify the surface α to be the small hole pocket, and β and γ to be the electron pockets. This assignment is also consistent with the negative sign of the Hall coefficient at low temperatures [19, 20].

Significantly different Fermi surface has been observed by angle-resolved photoemission spectroscopy (ARPES) by two independent research groups [21-23]. These ARPES results are in agreement with each other and indicate the presence of an extended van Hove singularity (a flat band), much like the ones commonly observed in high-T_c cuprates. The reported Fermi surface consists of *one electron pocket* and *two hole pockets*, in sharp contrast to the results of quantum oscillations. Such Fermi surface would give the de Haas-van Alphen frequency of only 47 % of the actual observation for γ. Since ARPES has been a powerful and unparalleled method of providing useful information on the electronic states near the Fermi level for high-T_c cuprates, it is quite important to resolve this controversy on this layered perovskite material. ARPES is a surface sensitive technique and in our opinion the consistency with the bulk properties such as the de Haas-van Alphen effects should be confirmed for each compound under investigation.

3. TWO-DIMENSIONAL FERMI-LIQUID BEHAVIOR

In contrast with the metallic in-plane resistivity ρ_{ab} [24], the out-of-plane resistivity ρ_c takes a maximum at about 130 K and changes to non-metallic

0921-4534/97/$17.00
PII S0921-4534(97)00199-8

temperature dependence ($d\rho_c/dT<0$) at higher temperatures [25]. Therefore 3D-2D crossover of metallic conduction occurs. Below about 25 K both ρ_{ab} and ρ_c exhibit the T-squared dependence quite precisely: $\rho = \rho_0 + AT^n$ with n=2.0 [1, 17]. This dependence implies the dominance of the Umklapp process of the electron-electron scattering at low temperatures. The origins of the coherent interlayer transport at low temperatures and of 3D-2D metallic crossover have semi-quantitatively been explained in terms of the Fermi surface parameters [16, 26]. The coherent out-of-plane charge transport at low temperatures has also been confirmed by optical conductivity [27].

The Kadowaki-Woods ratio $R_{KW} = A/\gamma_N^{\,2}$ is phenomenologically known to be universal among strongly correlated systems with the value of a_0 = 1.0 x 10^{-5} μΩcm/(mJ/Kmol)^2. For Sr_2RuO_4 the ratio with the in-plane resistivity is 0.3-0.5a_0, consistent with the universality. The ratio with the out-of-plane resistivity, however, is 300-500a_0. This severe deviation from the universality reflects the formation of the essentially two-dimensional Fermi liquid [17].

For other electronic properties, the Seebeck coefficient has been reported by Yoshino *et al,* [28] and the magnetoresistance by Hussey *et al.* [29]. The electronic Raman scattering has revealed [30] the broad excitation spectrum similar to the one known for high-T_c cuprates.

4. BAND-WIDTH CONTROL VS CARRIER DOPING

The results of annealing on T_c under various oxygen partial pressures ranging from 10^{-3} to 10^2 bar [31] indicated that, in contrast with $La_2CuO_{4+\delta}$, the stoichiometric Sr_2RuO_4 without carrier doping is metallic as well as superconductive.

The integrated ultraviolet photoemission spectra (UPS) by Inoue *et al.* [32, 33] indicated the states reminiscent of the lower Hubbard band. Other groups have drawn similar conclusions [34, 35]. The on-site Coulomb energy of U = 2.4 eV is estimated from the spectra and in conjunction with the one-particle band width of W = 1.4 eV [11], it gives a large ratio of U/W = 1.7. Therefore, despite the highly metallic conductivity, Sr_2RuO_4 is already near the Mott transition.

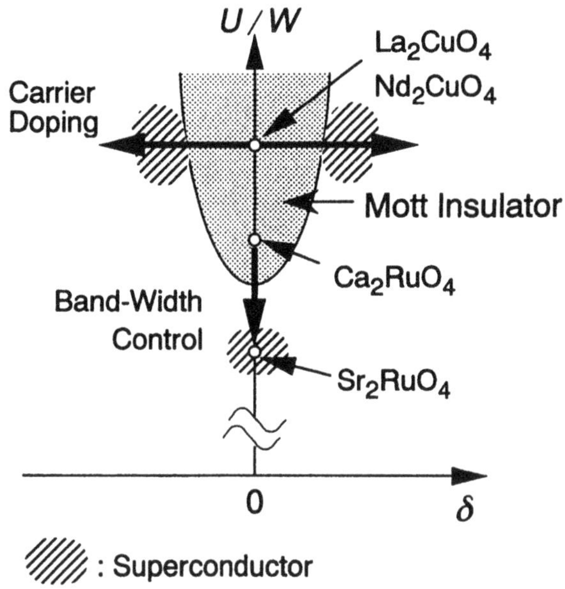

Figure 1. Schematic phase diagram in the plane of the effective correlation U/W and the deviation from the integer filling δ. The horizontal arrows indicate the "carrier-doping" route, whereas the vertical arrow the "band-width control" route to metallize the Mott insulators.

We have recently synthesized a new Mott insulator, Ca_2RuO_4 [36]. There are two structural variants, but both order magnetically below 110-150 K. Presence of the lattice distortions in Ca_2RuO_4 leads to the reduction in the overlap of the electronic wave functions and thus in the band width W, resulting in the increase in U/W. This qualitatively explains the difference in metal-insulator characteristics between Sr_2RuO_4 and Ca_2RuO_4 (Fig. 1). Therefore, while the metallic phase of the high-T_c cuprate is induced by "carrier doping" into the parent insulator, the superconductive Sr_2RuO_4 is obtained by "band-width control" of the Mott insulator.

5. UNCONVENTIONAL SUPERCONDUCTIVITY

The determination of the coherence lengths [37] shows that the superconductivity of Sr_2RuO_4 is more anisotropic than that of $La_{2-x}Sr_xCuO_4$. It is nevertheless of 3D because the interlayer coherence length ξ_c= 4.0 nm is substantially longer than the interlayer spacing d = 0.64 nm. The anisotropy in the flux pinning and the peak effects have also been reported by Yoshida *et al.* [38].

The specific heat [39] shows a clear jump at T_c. By requiring the entropy balance, we estimate $\Delta C_P/\gamma_J T_c = 0.96$ for a crystal with a sharp transition centered at T_c = 1.13 K. Here $\gamma_J \equiv \gamma_N - \gamma_0$ is the component that exhibits the jump and γ_0 is the residual value for the electronic specific-heat coefficient. This value is to be compared with 0.95 expected for the gap with line nodes in a system with cylindrical Fermi surface.

On the basis of the entropy balance, the specific heat divided by temperature, C_P/T, contains a large contribution from γ_0. Although γ_0 normalized by γ_N for crystals with different T_c tends to decrease systematically with increasing T_c, it tends to retain a large value of about 50 %. It is suggestive of the presence of low lying excitations in the unconventional superconducting state, either induced by the impurities or of the intrinsic origin.

When resonance impurity scattering occurs as considered in unconventional superconductors, we expect the presence of large γ_0 without suppressing T_c completely. In that case, however, we expect a correspondingly large value for the quantity $\Delta C_P/\gamma_J T_c$ [40]. Since the observed specific-heat jump takes the nearly intrinsic (impurity-free) value, the simultaneous presence of the large γ_0 is not attributable to the impurity-induced effects alone.

Rice and Sigrist [41] have made an enlightening suggestion for the spin triplet correlation for the superconductivity of Sr_2RuO_4. Their argument is based on the strong Hund's coupling which tends to align the spins of the two holes of Ru^{4+}, and also on the fact that the related perovskite ($SrRuO_3$) exhibits the ferromagnetic metallic state. If the triplet pairing state is realized in Sr_2RuO_4, it is tempting to associate the large residual γ_0 with the intrinsic origin expected for the nonunitary spin-triplet state which gives $\gamma_0 = \gamma_N/2$ [42, 43].

Recently, Ishida *et al.* [44] have studied Ru-NMR and NQR, and also revealed the evidence for unconventional superconducting state in Sr_2RuO_4 [45].

The magnetic interaction in the *insulating* Ca_2RuO_4 is essentially antiferromagnetic because of the superexchange interaction. To gain more insights in the spin correlations in *metallic* Sr_2RuO_4, studies of the itinerant-electron magnetism of another layered perovskite $Sr_{3-x}Ca_xRu_2O_7$ [46-51], and the simple perovskite $Sr_{1-x}Ca_xRuO_3$ [52] may prove helpful.

6. CONCLUSIONS

Although there are pieces of strong evidence for unconventional pairing state in Sr_2RuO_4, they have not provided the decisive information concerning the specific pairing symmetry yet. Various experimental techniques, especially the ones developed in the study of heavy Fermion superconductors and high-T_c cuprates, should be applied to Sr_2RuO_4.

Studies of the layered ruthenates have shown that even in 4*d* transition-metal oxides with more extended wave functions compared with 3*d* systems, the effects of strong electron correlations can still play a crucial role in determining their unusual electronic properties.

ACKNOWLEDGEMENTS

The author is grateful to his students, S.Ikeda, S.Nishizaki, S.Nakatsuji, and Y.Mori for their contributions. The results presented here are obtained though enjoyable collaborations with many researchers as indicated in the references. He is especially indebted to T.Fujita, K.Yoshida, and A.P.Mackenzie for many valuable discussions and suggestions. He is also grateful to T.Ishiguro for the support in setting up the laboratory at Kyoto University. This work has been supported by Grant-in-Aid for Scientific Research from the Ministry of Education, Science, Sports and Culture, and by Core Research for Evolutionary Science and Technology (CREST) of Japan Science and Technology Corporation.

REFERENCES

1. Y.Maeno, H.Hashimoto, K.Yoshida, S.Nishizaki, T.Fujita, J.G.Bednorz and F.Lichtenberg, Nature **372** (1994) 532.
2. Y.Maeno, in "Advances in Superconductivity **VIII**" (eds. H.Hayakawa and Y.Enomoto, Springer-Verlag, Tokyo, 1996) pp.15-20.
3. Y.Maeno, S.Nishizaki, K.Yoshida, S.Ikeda, and T.Fujita, J. Low Temp. Phys. **105** (1996) 1577.
4. Y.Maeno and K.Yoshida, Czechoslovak J. Phys. **46** (1996) Suppl. S6, 3097.
5. Y.Maeno and S.Ikeda, to be published in the proc. "Material Aspects of High T_c Superconductivity," NATO-ASI, 19-31 August, 1996, Delphi, Greece (eds. E.Kaldis and E.Liarokapis, Kluwer, Dordrecht, 1997).
6. Q.Huang, J.L.Soubeyroux, O.Chmaissem, I.Natali Sora, A.Santoro, R.J.Cava, J.J.Krajewski, and W.F.Peck, Jr., J. Solid State Chem. **112** (1994) 355.

7. J.S.Gardner, G.Balakarishnan, D.McK.Paul, and C.Harworth, Physica C **265** (1996) 251.
8 M.Braden, A.H.Moudden, S.Nishizaki, Y.Maeno, and T.Fujita, Physica C **273** (1997) 248.
9. Y.Inoue, M.Hara, Y.Koyama, Y.Maeno, and T.Fujita, in "Advances in Superconductivity VIII" (eds. H.Hayakawa and Y.Enomoto, Springer-Verlag, Tokyo, 1996) pp. 243-246.
10. M.Udagawa, T.Minami, N.Ogita, Y.Maeno, F.Nakamura, T.Fujita, J.G.Bednorz, and F.Lichtenberg, Physica B **219&220** (1996) 222.
11. T.Oguchi, Phys. Rev. B **51** (1995) 1385.
12. D.J.Singh, Phys. Rev. B **52** (1995) 1358.
13. G.L.McMullan, M.P.Ray, and R.J.Needs, Physica B **223&224** (1996) 529.
14. I.Hase and Y.Nishihara, J. Phys. Soc. Jpn. **65** (1996) 3957.
15. T.M.Mishonov, I.N.Genchev, R.K.Koleva, and E.P.Penev, J. Low Temp. Phys. **105** (1996) 1611.
16. A.P.Mackenzie, S.R.Julian, A.J.Diver, G.J.McMullan, M.P.Ray, G.G.Lonzarich, Y.Maeno, S.Nishizaki, and T.Fujita, Phys. Rev. Lett. **76** (1996) 3786. The values for cyclotron masses quoted in this paper need slight corrections. In unit of the bare electron mass, they should be 3.4, 7.5, and 14.6, respectively, for the Fermi surface α, β, and γ.
17. Y.Maeno, K.Yoshida, H.Hashimoto, S.Nishizaki, S.Ikeda, M.Nohara, T.Fujita, A.P.Mackenzie, N.E.Hussey, J.G.Bednorz, and F.Lichtenberg, to appear in J. Phys. Soc. Jpn. **66**, No.5 (1997).
18. A.P.Mackenzie, S.R.Julian, A.J.Diver, G.G. Lonzarich, N.E.Hussey, Y.Maeno, S.Nishizaki, and T.Fujita, Physica C **263** (1996) 510.
19. A.P.Mackenzie, N.E.Hussey, A.J.Diver, S.R.Julian, Y.Maeno, S.Nishizaki, and T.Fujita, Phys. Rev. B **54** (1996) 7425.
20. N.Shirakawa, K.Murata, Y.Nishihara, S.Nishizaki, Y.Maeno, T.Fujita, J.G.Bednorz, F.Lichtenberg, and N.Hamada, J. Phys. Soc. Jpn. **64** (1995) 1072.
21. T.Yokoya, A.Chainani, T.Takahashi, H.Katayama-Yoshida, M.Kasai, and Y.Tokura, Phys. Rev. Lett. **76** (1996) 3009; Physica C **263** (1996) 505.
22. D.H.Lou, M.Schmidt, T.R.Cummins, S.Schuppler, F.Lichtenberg, and J.G.Bednorz, Phys. Rev. Lett. **76** (1996) 4845; J. Low Temp. Phys. **105** (1996) 1587.
23. T.Yokoya, A.Chainani, T.Takahashi, H.Ding, J.C.Campuzano, H.Katayama-Yoshida, M.Kasai, and Y.Tokura, Phys. Rev. B **54** (1996) 13311.
24. The temperature dependence of ρ_{ab} reported in ref. 1 contained contribution from ρ_c due to a technical error. See refs. 2 and 19 for the intrinsic behavior.
25. F.Lichtenberg, A.Catana, J.Mannhart, and D.G.Schlom, Appl. Phys. Lett. **60** (1992) 1138.
26. K.Yoshida, Y.Maeno, S.Nishizaki, S.Ikeda, and T.Fujita, J. Low Temp. Phys. **105** (1996) 1593.
27. T.Katsufuji, M.Kasai, and Y.Tokura, Phys. Rev. Lett. **76** (1996) 126.
28. H.Yoshino, K.Murata, N.Shirakawa, Y.Nishihara, Y.Maeno, and T.Fujita, J. Phys. Soc. Jpn. **65** (1996) 1548.
29. N.E.Hussey, A.P.Mackenzie, J.R.Cooper, Y.Maeno, S.Nishizaki, and T.Fujita, preprint (1996).
30. A.Yamanaka, N.Asayama, M.Sasada, K.Inoue, M.Udagawa, S.Nishizaki, Y.Maeno, and T.Fujita, Physica C **263** (1996) 516.
31. S.Nishizaki, Y.Maeno, and T.Fujita, J. Phys. Soc. Jpn. **65** (1996) 1876.
32. I.H.Inoue, Y.Aiura, Y.Nishihara, Y.Haruyama, S.Nishizaki, Y.Maeno, T.Fujita, J.G.Bednorz, and F.Lichtenberg, Physica B **223&224** (1996) 516; J. Electron Spectroscopy and Related Phenom. **78** (1996) 175.
33. I.H.Inoue, H.Makino, Y.Aiura, I.Hase, Y.Maeno, Y.Haruyama, S.Nishizaki, T.Fujita, and Y.Nishihara, Czechoslovak J. Phys. **46** (1996) Suppl. S5, 2699.
34. T.Yokoya, A.Chainani, T.Takahashi, H.Katayama-Yoshida, M.Kasai, Y.Tokura, N.Shanthi, and D.D.Sarma, Phys. Rev. B **53** (1996) 8151.
35. M.Schmidt, T.R.Cummins, M.Bürk, D.H.Lu, N.Nücker, S.Schuppler, and F.Lichtenberg, Phys. Rev. B **53** (1996) R14761.
36. S.Nakatsuji, S.Ikeda, and Y.Maeno, preprint (1997).
37. K.Yoshida, Y.Maeno, S.Nishizaki, and T.Fujita, Physica C **263** (1996) 519.
38. K.Yoshida, Y.Maeno, S.Nishizaki, and T.Fujita, J. Phys. Soc. Jpn. **65** (1996) 2220.
39. S.Nishizaki, Y.Maeno, S.Farner, S.Ikeda, and T.Fujita, Physica C, this volume (1997).
40. P.J.Hirschfeld, P.Wölfle, and D.Einzel, Phys. Rev. B **37** (1988) 83.
41. T.M.Rice and M.Sigrist, J Phys.: Condens. Matter **7** (1995) L643.
42. M.Sigrist and M.E.Zhitomirsky, J. Phys. Soc. Jpn. **65** (1996) 3452.
43. K.Machida, M.Ozaki, and T.Ohmi, J. Phys. Soc. Jpn. **65** (1996) 3720.
44. K.Ishida, Y.Kitaoka, K.Asayama, S.Ikeda, Y.Maeno and T.Fujita, Czechoslovak J. Phys. **46** (1996) Suppl. S2, 1093.
45. K.Ishida, Y.Kitaoka, K.Asayama, S.Ikeda, Y.Maeno and T.Fujita, preprint (1996).
46. R.J.Cava, H.W.Zandbergen, J.J.Krajewski, W.F.Peck,Jr., B.Batlogg, S.Carter, R.M.Fleming, O.Zhou, and L.W.Lupp,Jr., J. Solid State Chem. **116** (1995) 141.
47. S.Ikeda, Y.Maeno, M.Nohara, and T.Fujita, Physica C **263** (1996) 558.
48. S.Ikeda, Y.Maeno, H.Muranishi, and T.Fujita, J. Low Temp. Phys. **105** (1996) 1599.
49. Y.Inoue, M.Hara, Y.Koyama, S.Ikeda, Y.Maeno, and T.Fujita, to appear in "Advances in Superconductivity IX" (Springer-Verlag, Tokyo, 1997).
50. G.Cao, S.McCall, and J.E.Crow, Phys. Rev. B **55** (1997) R672.
51. G.Cao, S.McCall, J.E.Crow, and R.P.Guertin, Phys. Rev. Lett. **78** (1997) 1751.
52. For example: T.Kiyama, K.Yoshimura, K.Kosuge, Y.Ikeda, and Y.Bando, Phys. Rev. B **54** (1996) R756.

ELSEVIER

Physica C 282–287 (1997) 210–213

PHYSICA C

Spin fluctuation and unconventional pairing in Sr_2RuO_4

Y. Kitaoka, K. Ishida, K. Asayama[a], S. Ikeda, S. Nishizaki, Y. Maeno[b], K.Yoshida and T.Fujita[c]

[a]Dept. of Material Physics, Faculty of Engineering Science, Osaka Univ, Toyonaka, Osaka 560, Japan

[b]Dept. of Physics, Faculty of Science, Kyoto University, Kyoto 606-01, Japan

[c]Dept. of Physics, Faculty of Science, Hiroshima University, Higashi-Hiroshima 739, Japan

Sr_2RuO_4 with T_c =0.7 K with the same layered perovskite structure as La_2CuO_4 is the exchange-enhanced Pauli paramagnet with the two dimensional (2D) electronic structure. The superconductivity of has been found to be unconventional from the nuclear relaxation behavior exhibiting a sharp decrease with no coherence peak just below T_c and the T_1T=constant behavior well below T_c. In comparison with the high-T_c $Bi_2Sr_2CaCu_2O_{8+y}$ (Bi2212) and the 3D analogue of exchange-enhanced Pauli paramagnet $CeCo_2$, the similarity and dissimilarity are presented on the normal and superconducting states.

Recently, Maeno *et al* discovered superconductivity with an onset temperature around 1 K in a ruthenious oxide Sr_2RuO_4, which has the same layered perovskite structure as $La_{2-x}(Sr, Ba)_x CuO_4$ [1]. In spite of their structural similarity, many differences in the electronic and magnetic properties between both have been addressed from the de Haas-van Alphen [2], specific heat, magnetic susceptibility and transport measurements [3]. These experimental results revealed the two dimensional (2D) character of the electronic structure and the substantial mass enhancement, $m^*/m_0 \sim 4$ pointing to a signature of strong electron correlations[2], if the cyclotron masses are compared to those of the band structure calculation [4]. This mass renormalization of quasiparticles due to the electron correlation was suggested to result in an enhancement of the linear coefficient of the specific heat, γ=39 mJ/K^{2}mol as compared with 10 mJ/K^{2}mol estimated from the band structure calculation [3,4]. Furthermore, recent Ru NMR study by Ishida et al has clarified that the spin susceptibility deduced from the Knight shift is enhanced by $\sim$ 5 times larger than the value estimated from the band calculation [5]. Then the Wilson ratio was extracted to be R_w=1.28, which is smaller than R_w=3.3(4) for the liquid ^{3}He at ambient pressure (at 27 bar). From the latter fact, the spin fluctuation in this compound is suggested to be a unique in a sense that any specific structure is not distinct in the q dependence of the susceptibility associated with the 2D electronic structure, even though the electron correlation effect enhances the susceptibility significantly. In contrast to the experimental progress on the normal state property, the Ru relaxation experiment has been the first to provide a clue for the superconducting nature in Sr_2RuO_4 to be in an unconventional regime. As shown in Fig. 1, the nuclear spin lattice relaxation rate, $1/T_1$ of ^{101}Ru measured in zero field by the NQR technique exhibited no coherence peak just below T_c accompanied by the T_1T=constant behavior [5].

In this paper, in order to reinforce the novel relaxation behavior in Sr_2RuO_4, we compare with the relaxation behaviors in the 3D exchange- enhanced Pauli paramagnet $CeCo_2$ and the high-T_c Bi2212 with the 2D superconducting characteristics. The C15 Laves phase compound, $CeCo_2$ is a superconductor with $T_c \sim$1.5K. Recent transport and specific measurements on the high-quality single crystal $CeCo_2$, revealed the influence of the electron correlations likewise Sr_2RuO_4 [6]. Remarkably, the measured susceptibility indicated the T dependence upon cooling affected by possible ferromagnetic spin fluctuations. Therefore, $CeCo_2$ was suggested to be a superconductor with the exchange-enhanced Pauli paramagnet. From the specific heat data below T_c exhibiting the jump of the electronic specific heat C_e at T_c (ΔC_e/

PII S0921-4534(97)00202-5

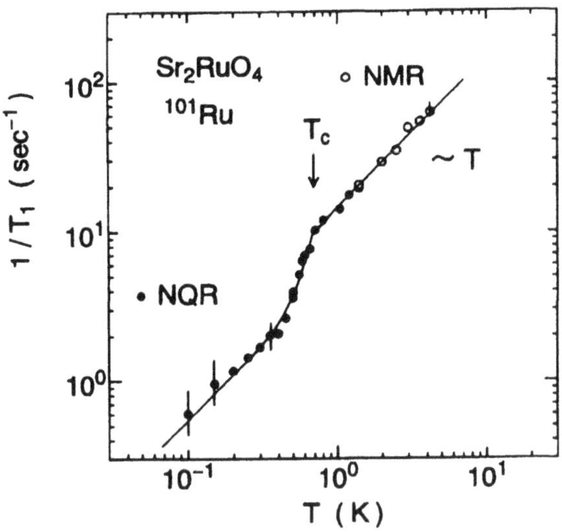

Figure 1. T-dependence of $1/T_1$ of ^{101}Ru in Sr_2RuO_4 [5]. Arrow indicates T_c. The solid curve below T_c is the calculation by using the anisotropic pairing model with line node ($\Delta(\theta) = \Delta_0 cos(\theta)$) with $2\Delta_0/k_BT_c$=7 and N_{res}/N_0=0.62 in two dimensional cylindrical fermi surface.

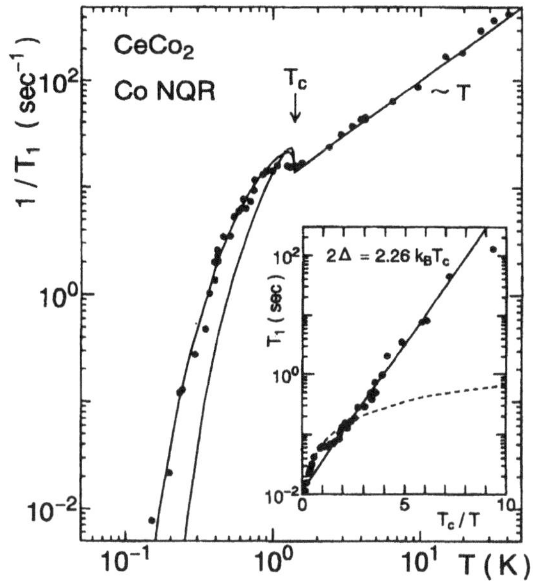

Figure 2. T-dependence of $1/T_1$ of ^{59}Co in $CeCo_2$ [8]. Solid curves below T_c are the calculations based on the anisotropic s-wave model and the BSC model with 2Δ=2.26 and 3.5 k_BT_c, respectively (see in text). The inset shows the Arrhenius plot T_1 vs. T_c/T.

$\gamma T_c \sim 0.85$) with a smaller value than that of the BCS value of 1.43, and approximately following to a T^2 at low temperatures, it was proposed that the anisotropic superconducting gap disappears along line [7]. This was, however, not consistent with the ^{59}Co relaxation result measured in zero field by the Co-NQR technique. Figure 2 shows T-dependence of $^{59}(1/T_1)$ in the T-range between 0.15 K and 50 K [8]. In the normal state, $1/T_1$ obeys a T_1T=constant relation down to T_c. In the superconducting state, the Hebel-Slichter peak in $1/T_1$ was observed just below T_c followed by the exponential decrease characteristic for the ordinary s-wave superconductor [8]. From the Arrhenius plot shown in the inset of Fig.2, the value of the superconducting gap, 2Δ was evaluated to be $2.26k_BT_c$ much smaller than that predicted by the BCS theory, $2\Delta = 3.5k_BT_c$. Solid curves calculated by the anisotropic s-wave model are drawn in Fig.2, in which the anisotropic superconducting gap model where $\Delta(\theta) = <\Delta> (1 + r\cos(2\theta))$ was assumed with $2<\Delta>/k_BT_c$ =2.26 and r = 0.2 by incorporating the anisotropy of the phonon spectrum and/or the band structure. Although the depression of the coherence peak of $1/T_1$ and the smaller energy gap than the BCS value may be brought about by the electron correlation effects in $CeCo_2$, the s-wave pairing nature was well established from the relaxation behavior. On the other hand, the novel relaxation behavior in Sr_2RuO_4 is consistent with an unconventional pairing nature. As displayed in the Fig.3 where the $1/T_1$ normalized by the value at T_c is plotted against T/T_c in a linear scale, the differences between two compounds are substantial in the T dependences just below T_c and well below T_c. In spite of many differences in the normal state properties, it should be noted that the T dependence of $1/T_1$ in Sr_2RuO_4 and high-T_c cuprates share similar features of the steep decrease without the coherence peak just below T_c and the $(T_1T)_s$=constant relation well below T_c. As an example, the T-dependences of $^{63}(1/T_1)$ of ^{63}Cu in Bi2212 with T_c=86 K is shown in Fig.4 [9]. The

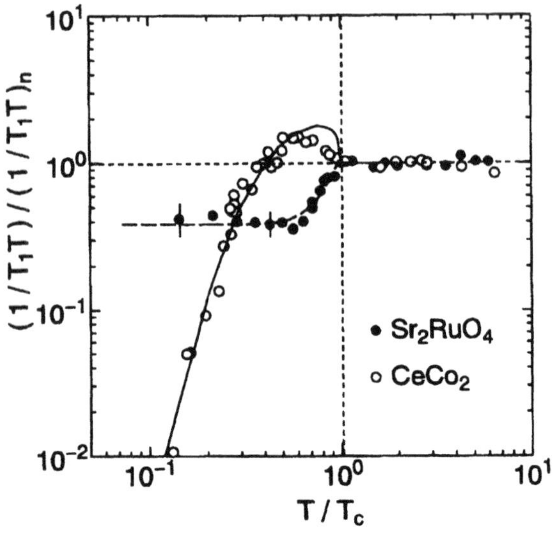

Figure 3. Comparison of relaxation behavior of Sr_2RuO_4 with that of $CeCo_2$ in a linear scale where $(1/T_1T)_s/(1/T_1T)_n$ is plotted against T/T_c.

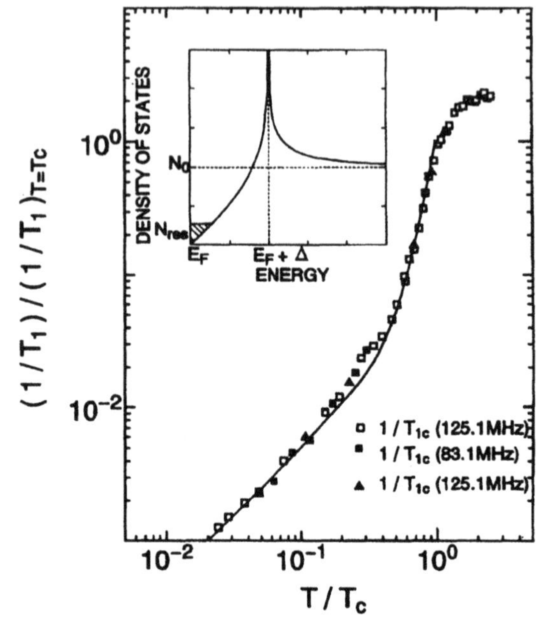

Figure 4. T dependence of $^{63}(1/T_1)$ in Bi2212 with T_c=86 K [9]. Inset is the quasiparticle DOS for the gapless $d_{x^2-y^2}$ pairing state with residual DOS, N_{res}. Solid line is the calculation with $2\Delta = 8k_BT_c$ and N_{res}/N_0=0.24.

absence of the coherence peak just below T_c and the $(T_1)_s$= constant law are commonly observed for both, giving evidences for the unconventional superconductivity. The fact that T_1 was uniquely determined without any distribution even at low temperatures does not mean that the $(T_1T)_s$= constant relation is not due to the presence of some non-superconducting part, but the presence of the residual density of states (RDOS) at the Fermi level in the superconducting state, providing an evidence for the gapless superconductivity [5,9]. A fraction of the RDOS in Sr_2RuO_4 deduced from the $(T_1T)_s$=constant relation well below T_c was in agreement with the value of the residual electronic specific heat, $\gamma_0/\gamma_N = N_{res}/N_0$ $\sim$0.67 estimated tentatively from the entropy balance [3]. For a p- and d-wave pairing, the DOS at the gap edge is suppressed due to the anisotropic gap and the anomalous DOS associated with the coherence effect that appears in the s-wave pairing is absent. These lead to the suppression of the coherence peak of $1/T_1$ just below T_c as reported in the heavy Fermion and high-T_c cuprate superconductors so far. Therefore, the absence of the coherence peak of $1/T_1$ in Sr_2RuO_4 was reasonably interpreted in terms of either anisotropic p- or d-wave model.

Furthermore, it should be noted that the low-lying excitation originating from nodes of the energy gap is affected significantly by crystal imperfections and/or distortions, producing the RDOS at the Fermi level. When the impurity scattering is treated in terms of the unitarity scattering for the anisotropic superconductor with nodes in the energy gap, the decrease of T_c is associated with the increase of the RDOS. The transition temperature T_{c0} in case that the pair-breaking effect is absent can be estimated from the relation between T_c and RDOS, N_{res}/N_0[10]. By this, T_{c0} is estimated to be $\sim$1.5 K for pure Sr_2RuO_4, from a value of $N_{res}/N_0 \sim 0.62$ and T_c=0.7 K. However, the recent specific heat measurement has revealed that a large fraction of the residual value, $\gamma_0/\gamma_N \sim 0.57$ is still found even in a sample with relatively high T_c of $\sim$ 1.2 K [3], suggesting no appreciable correlation between T_c and

N_{res}/N_0. Furthermore, the quality of the present Sr_2RuO_4 seems to be in a clean limit because the Schubnikov-de Haas oscillation is observed[2]. Accordingly, the $(T_1T)_s$=constant relation pointing to the presence of RDOS seems to be not ascribed to the impurity scattering, but is considered to be relevant to the intrinsic superconducting characteristics in Sr_2RuO_4. As a matter of fact, by assuming that the energy gap vanishes along line at the Fermi level and has the RDOS, a fairly reasonable fit to $1/T_1$ in a whole T range is possible as drawn by the solid curve in Fig.4 based on the polar state where $\Delta(\theta) = \Delta_0 \cos(\theta)$ with $2\Delta_0/k_BT_c = 7$ and the RDOS, N_{res}/N_0=0.62 for the cylindrical Fermi surface.

Recently, Sigrist and Zhitomirsky [11] and Machida et al [12] have proposed that nonunitary p-wave states within possible superconducting states with the odd parity, have gapless quasiparticle excitations, which would interpret the large RDOS in the superconducting state. In this picture, it is expected that the RDOS at the Fermi level would be intrinsic with $N_{res}/N_0 = 1/2$. However, since the RDOS arises from gapless excitations of one state in equal spin pairing as in the superfluid A_1 phase of ^{3}He, the spin flip process of a nuclear spin is not allowed by the gap formation of the non-unitary state. Namely, a $(T_1T)_s$=constant relation would be expected only in case when both states with up- and down-equal spin pairing are mixed with each other by the spin-orbit interaction as pointed out theoretically [11,12]. On the other hand, the band calculation showed that the orbital current induced by the spin-orbit interaction is negligible, suggesting that the spin state is well defined in Sr_2RuO_4 [4], which hence contradicts with above scenario of the non-unitary pairing state. At the present stage, it is not decisive yet to identify the symmetry of the order parameter in Sr_2RuO_4. A further systematic Ru NQR/NMR investigations on single crystals with higher T_c will enable us to gain a more detailed insight into the pairing state in Sr_2RuO_4, which are now in progress.

In conclusion, the normal state properties are different in many respects from the cuprate oxides in which the AF spin fluctuations play important roles. A large enhancement of the uniform susceptibility is substantial in Sr_2RuO_4. Unexpectedly, a possible pairing symmetry belongs to an unconventional class, well evidenced from such novel relaxation behaviors as the sharp decrease without the coherence peak of $(1/T_1)$ just below T_c and the $(T_1T)_s$=constant relation well below T_c. Since these behaviors cannot be explained by the s-wave model, the anisotropic p- or d-wave pairing is considered to be realized in Sr_2RuO_4.

REFERENCES

1. Y.Maeno, et al, Nature **372**, 532 (1994).
2. A.P.Mackenzie et al, Phys. Rev. Lett. **76**, 3786 (1996) .
3. Y. Maeno et al, *Proc. Int. Conf. on MOS' 96.*
4. T.Oguchi. Phys. Rev. **51**, 1385 (1995).
5. K. Ishida et al, submitted to Phys. Rev. B.
6. H. Sugawara et al, J. Phys. Soc. Jpn. 64 (1995) 4726.
7. Y. Aoki et al, to be published in Physica B.
8. K. Ishida et al, to be published in Physica B.
9. K. Ishida et al, J. Phys. Soc. Jpn. 63 (1994) 1104.
10. K. Miyake, private communication
    T.Hotta, private communication.
    Y. Kitaoka, K. Ishida and K. Asayama, J. Phys. Soc. Jpn. **63**, 2052 (1994).
11. T.M.Rice and W.Sigrist, J. Phys. Condens. Matter. **7**, 643 (1995); M. Sigrist and M. E. Zhitomirsky, J. Phys. Soc. Jpn. **65**, 3452 (1996).
12. K.Machida, M.Ozaki and T.Ohmi, J. Phys, Soc. Jpn 65, 3720 (1996).

ELSEVIER

Physica C 282–287 (1997) 214–217

PHYSICA C

Sr_2RuO_4 - A Realization of Spin Triplet Pairing?

M.Sigrist, D. Agterberg, T. M. Rice [a] and M.E. Zhitomirsky [b c].

[a]Theoretische Physik, ETH-Hönggerberg, CH-8093 Zürich, Switzerland.

[b]Department of Physics, University of Toronto, Toronto, Ontario, Canada M5S 1A7.

[c] Landau Institute for Theoretical Physics, Moscow, Russia

The recently discovered superconductor Sr_2RuO_4 is an essentially two-dimensional Fermi liquid with strong correlation effects. The similarity with ^{3}He and the presence of strong ferromagnetic spin fluctuations suggests that the superconducting state could be odd-parity, i.e. spin triplet pairing. The possible superconducting states are classified and experiments are discussed which could lead to a clear symmetry identification. The presence of a large residual density of states in the superconducting phase may indicate that the superconducting state is nonunitary. However, it is shown that an alternative scenario involving superconductivity of different strength on different sheets of the Fermi surface could also lead to similar behavior.

Nearly three years ago Maeno and coworkers found superconductivity in the compound Sr_2RuO_4 with $T_c = 1K$ [1]. Like La_2CuO_4, which is the parent compound of the high-temperature superconductors $La_{2-x}Sr_xCuO_4$, this compound forms a layered perovskite structure. While La_2CuO_4 is a Mott-insulator with antiferromagnetic long-range order, Sr_2RuO_4 displays metallic behavior without doping and is a Fermi liquid at low temperatures. The layered structure yields a rather low conductivity along the c-axis so that the system behaves essentially like a two-dimensional (2D) system. The metallic behavior originates from the nature of the 4d-2p-orbital hybridization of Ru and O ions, which generates three bands crossing the Fermi level. These bands are due to $4d_{xy}$, $4d_{zx}$ and $4d_{yz}$-orbitals of the Ru^{4+} ions. LDA calculations show three nearly cylindrical Fermi surfaces [2,3] which are consistent with the sheets observed by de Haas-van Alphen oscillations [4]. The $4d_{xy}$-orbital yields an electron-like Fermi surface, while the other two orbitals lead via hybridization to an electron- and a hole-like Fermi surface (Fig.1).

Strong correlation effects give rise to the enhancement of the linear specific heat coefficient and the Pauli spin susceptibility compared with the LDA values [1–3]. A comparison of the cyclotron masses measured by quantum oscillation experiments with those found by the LDA calculations yields $m^*/m_0 \sim 4$ [4]. The Pauli spin susceptibility is also enhanced by a factor of about 8 (Wilson ratio $\pi^2 k_B^2 \chi / 3\mu_B^2 \gamma = 1.8$) [1]. These corrections are nearly the same as those found in ^{3}He. Therefore we may argue that Sr_2RuO_4 represents an electronic two-dimensional analogue of ^{3}He. The similarity with ^{3}He suggests that the superconducting state could be due to odd-parity (spin triplet) pairing [5]. There are various indications that indeed spin triplet pairing might be favored in this compound. Hund's rule coupling tends to form a local $S = 1$ configuration among the holes in the (d_{xz}, d_{yz})-orbitals. Furthermore, the related three-dimensional compound $SrRuO_3$ undergoes a transition to a ferromagnetic state at $T_c \approx 160K$ with a saturation moment of $0.85\mu_B$. Therefore, strong ferromagnetic spin fluctuations favoring pairing with parallel spins are expected to be important in Sr_2RuO_4. Although there is no direct experimental verification, recent experimental data could indicate odd-parity pairing. One piece of evidence is the rather large residual linear term in the temperature dependence of the specific heat in the superconducting state [7]. Additional evidence is found in NQR measurements which show no Hebel-Slichter peak and Korringa behavior at low temperatures [8].

We continue here assuming that odd-parity

PII S0921-4534(97)00205-0

pairing is realized and analyze the properties of possible superconducting states. The gap function for odd-parity pairing is described by a 2×2 matrix $\Delta_{ss'}(\mathbf{k})$ in spin space which may be represented by a 3D vector $\mathbf{d}(\mathbf{k})$, i.e. $\hat{\Delta}(\mathbf{k}) = i\sum_\mu d_\mu(\mathbf{k})\hat{\sigma}_\mu\hat{\sigma}_y$ with $\hat{\sigma}_\mu$ being the Pauli matrices. If $\hat{\Delta}$ is unitary, ($\hat{\Delta}^\dagger\hat{\Delta}$ is proportional to the 2×2 unit matrix $\hat{\sigma}_0$), the quasi-particle spectrum is given by

$$E_{\mathbf{k}} = \sqrt{\varepsilon_{\mathbf{k}}^2 + |\mathbf{d}(\mathbf{k})|^2} \tag{1}$$

where $\varepsilon_{\mathbf{k}}$ denotes the electron band energy measured relative to the Fermi level as a function of the momentum $\mathbf{k}$ ($\mathbf{k}$ restricted to the Fermi surface). The nonunitary state has the property $\hat{\Delta}^\dagger\hat{\Delta} = |\mathbf{d}|^2\hat{\sigma}_0 + i(\mathbf{d}^*\times\mathbf{d})_\mu\hat{\sigma}_\mu$ which yields a quasi-particle spectrum with two branches,

$$E_{\mathbf{k},\pm} = \sqrt{\varepsilon_{\mathbf{k}}^2 + |\mathbf{d}(\mathbf{k})|^2 \pm |\mathbf{d}^*(\mathbf{k})\times\mathbf{d}(\mathbf{k})|}, \tag{2}$$

and breaks time-reversal symmetry. The gap of the second branch $E_{\mathbf{k},-}$ is smaller than that in $E_{\mathbf{k},+}$ and could even vanish over the whole Fermi surface.

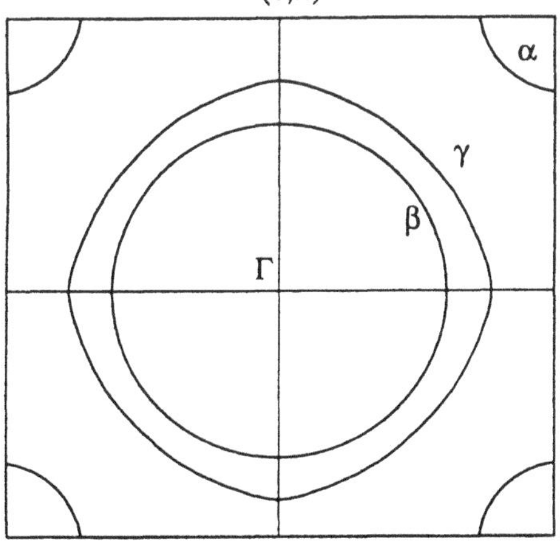

Figure 1. A schematic representation of the three Fermi surfaces of Sr_2RuO_4 for the $4d_{xy}$-orbital (γ) and the hybridized $(4d_{zx}, 4d_{yz})$-orbitals (α,β).

For the following classification of possible odd-parity states we assume a single cylindrical Fermi surface. Then there are six pairing basis functions: $\mathbf{d}(\mathbf{k}) = \{\hat{x}k_x, \hat{y}k_x, \hat{z}k_x, \hat{x}k_y, \hat{y}k_y, \hat{z}k_y\}$ where $\hat{a}$ denotes the unit vector along a-direction. We consider the case of large spin-orbit coupling, but we neglect crystal field effects for the moment. We find that pairing states can be listed in the following pairs of degenerate binding energy, i.e. transition temperature T_c [10],

$$\begin{aligned}
&\left.\begin{aligned}\mathbf{d}_1(\mathbf{k}) &= \hat{x}k_x + \hat{y}k_y\\ \mathbf{d}_2(\mathbf{k}) &= \hat{x}k_y - \hat{y}k_x\end{aligned}\right\} \quad T_c = T_{12}\\
&\left.\begin{aligned}\mathbf{d}_3(\mathbf{k}) &= \hat{x}k_x - \hat{y}k_y\\ \mathbf{d}_4(\mathbf{k}) &= \hat{x}k_y + \hat{y}k_x\end{aligned}\right\} \quad T_c = T_{34}\\
&\left.\begin{aligned}\mathbf{d}_5(\mathbf{k}) &= \hat{z}(k_x + ik_y)\\ \mathbf{d}_6(\mathbf{k}) &= \hat{z}(k_x - ik_y)\end{aligned}\right\} \quad T_c = T_{56}.
\end{aligned} \tag{3}$$

We restrict ourselves to one of these pairs, the one with the highest T_c, e.g. $(\mathbf{d}_1, \mathbf{d}_2)$, and consider the complex two component order parameter (η_1, η_2) defined by $\mathbf{d}(\mathbf{k}) = \eta_1\mathbf{d}_1(\mathbf{k}) + \eta_2\mathbf{d}_2(\mathbf{k})$ while the others are assumed to be irrelevant (much lower T_c). For this two-dimensional order parameter we can write a generalized Ginzburg-Landau (GL) theory to analyze the possible superconducting phases ($\vec{\eta} = (\eta_1, \eta_2)$),

$$\begin{aligned}\mathcal{F}_{12} = \;& a(T - T_{12})\vec{\eta}^*\cdot\vec{\eta} + \beta_1(\vec{\eta}^*\cdot\vec{\eta})^2\\ &+\beta_2|\eta_1|^2|\eta_2|^2 + \beta_3(\eta_1^{*2}\eta_2^2 + \eta_1^2\eta_2^{*2})\end{aligned} \tag{4}$$

where the coefficients a and β_i are real material dependent constants. Depending on the relation among the β's three different phases are stabilized for temperature lower than T_c (see Table 1). Note, that all three phases are two-fold degenerate. The phase A is nonunitary (time-reversal symmetry breaking), while the B and C phases are unitary. In the weak coupling limit the coefficients β_i are such that the phases B and C are degenerate. All three phases are so-called equal spin pairing (ESP) states. In the nonunitary state the quasiparticle spectrum has one branch with a isotropic gap and another with completely gapless excitations, which correspond to spin up and spin down pairing. The B phase has line nodes in the gap and the C phase has an isotropic gap.

If both spin-orbit coupling and crystal field effects (tetragonal point group symmetry) are

Table 1
Possible superconducting phases in Eq.(6).

	$\mathbf{d}(\mathbf{k})$	β_2', β_3'
A	$(\hat{x} \pm i\hat{y})(k_x \mp ik_y)$	$\beta_2' < 2\lvert\beta_3'\rvert, \beta_3' > 0$
B	$(\hat{x} \pm \hat{y})(k_x \mp k_y)$	$\beta_2' < 2\lvert\beta_3'\rvert, \beta_3' < 0$
C	$\hat{x}k_x + \hat{y}k_y, \hat{x}k_y - \hat{y}k_x$	$\beta_2' > 2\lvert\beta_3'\rvert$

strong, then the degeneracy of the pairs $(\mathbf{d}_1, \mathbf{d}_2)$ and $(\mathbf{d}_3, \mathbf{d}_4)$ in (5) is lifted and we can obtain states belonging to five irreducible representations of D_{4h} (see Tab. 1). The states of Γ_{1-4}^- belong to the class of equal spin pairing (ESP) states, i.e.: $|\Delta_{\uparrow\uparrow}| = |\Delta_{\downarrow\downarrow}| > 0$ and $\Delta_{\uparrow\downarrow} = \Delta_{\downarrow\uparrow} = 0$. The Γ_1^--state is the two-dimensional analogue of the B-phase of ^{3}He. The Γ_5^--state is two-fold degenerate and corresponds to the A-phase of ^{3}He which breaks the time-reversal symmetry. All states are unitary and have a constant (nodeless) quasiparticle gap.

Now let us briefly list some properties which could lead to the identification of superconducting state realized in Sr_2RuO_4. One of the key features of an odd-parity (spin triplet) pairing state (in particular ESP) is the unusual temperature dependence of the spin susceptibility. The Pauli spin susceptibility does not follow the standard Yosida-law for certain directions of the applied field [5,6]. The verification of this property by NMR Knight shift measurements is very difficult due to the low value of the critical magnetic fields. However, μSR measurements appear to be more promising. An alternative probe is the observation of collective modes analogous to ^{3}He as described in Ref.5 and 6. Furthermore, phase sensitive test experiments could give decisive information like in the case of high-temperature superconductors. [9,5] To our knowledge none of these experiments has been performed yet.

There are, however, two experiments mentioned above which could be indicative for the symmetry of the superconducting state. The large residual zero-energy density of states (DOS) in the superconducting phase, which was observed consistently in the specific heat [7] and the Korringa-behavior in NQR [8]. It was shown that the ratio of the DOS in the normal and the superconducting state towards zero temperature

Table 2
List of possible triplet pairing states including effects of spin-orbit coupling and crystal field. J and J_z denote the total angular momentum of the pairing state and its z-component, respectively.

Γ	(J, J_z)	$\mathbf{d}(\mathbf{k})$
Γ_1^-	$(0, 0)$	$\hat{\mathbf{x}}k_x + \hat{\mathbf{y}}k_y$
Γ_2^-	$(1, 0)$	$\hat{\mathbf{x}}k_y - \hat{\mathbf{y}}k_x$
Γ_3^-	$(2, \pm 2)$	$\hat{\mathbf{x}}k_x - \hat{\mathbf{y}}k_y$
Γ_4^-	$(2, \pm 2)$	$\hat{\mathbf{x}}k_y + \hat{\mathbf{y}}k_x$
Γ_5^-	$(1, \pm 1)$	$\hat{\mathbf{z}}(k_x \pm ik_y)$

approaches about half for purified samples. This behavior rules, of course, states out with a complete gap in the quasiparticle spectrum. Examining the list of the superconducting states we find that the nonunitary A phase in Tab.I has the property that half of the quasiparticles have a completely gapless spectrum [10,11]. This state shows similarity with the A_1-phase of superfluid ^{3}He.

There are two kinds of these nonunitary phases resulting from a combination of the basis states $(\mathbf{d}_1, \mathbf{d}_2)$ and $(\mathbf{d}_3, \mathbf{d}_4)$, respectively. Unfortunately, the nonunitary states are difficult to stabilize by microscopic mechanisms. In the weak-coupling limit states with an isotropic gap are favored. The standard spin fluctuation feedback mechanisms are also insufficient to stabilize the nonunitary state [10]. Sugiyama and Ohmi suggested that a nonunitary state could be realized for strong particle-hole asymmetry and strong ferromagnetic spin fluctuations [12]. There is a good chance that these conditions are satisfied in Sr_2RuO_4. In particular, the latter is visible in the large Stoner enhancement [8]. The two-dimensionality as well as spin-orbit effects might play an important role to stabilize the nonunitary state here. To the best of our knowledge this problem has not been analyzed so far.

The large residual density of states could also be due to orbitally dependent superconductivity [13]. The $4d_{xy}$-orbital is symmetrically clearly distinct from the other two (hybridizing) Ru-4d-orbitals. This is the case because the invariance of the Hamiltonian H under a reflection σ_z

about the Ru-O plane implies the matrix element $\langle xy\ \mathbf{k}|H|\alpha z\ \mathbf{k}'\rangle$ ($\alpha = x, y$) is zero for $\mathbf{k}$ and $\mathbf{k}'$ in the basal plane. Note as well that the interband pair hopping matrix element between the d_{xy} band and the other two bands is determined by $\langle xy\ \mathbf{k}, xy\ \ -\mathbf{k}|v_{eff}|\alpha z\ \ -\mathbf{k}', \alpha z\ \mathbf{k}'\rangle$. The z dependence of the d_{xy} orbitals limits the integrations along the z direction in this matrix element to a distance on the order of $d_{Ru-Ru}/7$ where d_{Ru-Ru} is the distance between two neighboring Ru ions. Consequently, the contribution to non-isotropic pairing interactions is expected to be significantly smaller for the interband $d_{xy} \leftrightarrow \{d_{xz}, d_{yz}\}$ matrix elements than the intraband pairing matrix elements. This is again due to the different symmetry properties of these bands under σ_z. These considerations lead to the conclusion that the gap on one or more of the Fermi surface sheets can be significantly smaller than that on the remaining sheets [13]. Consequently, well into the superconducting state there will exist essentially gapless quasiparticle excitations until the temperature is on the order of the smaller gap. These excitations will appear as a residual density of states. Furthermore, the density of states of the γ surface is approximately half the total density of states which agrees well with the experimentally observed residual density of states in the superconducting state.

In summary Sr_2RuO_4 is a good candidate for unconventional superconductivity. Strongly enhanced ferromagnetic fluctuations and other similarities with the quantum liquid ^{3}He suggest that Cooper pairing in the odd-parity channel is likely. At present the information from experimental side is not sufficient to make any unambiguous conclusion about the symmetry of the superconducting order parameter. A number of test experiments is necessary to verify our proposal. We believe that the observation of a large residual DOS in the superconducting indicates some exotic type of superconducting state. The fact that Sr_2RuO_4 has a comparatively simple Fermi surface and behaves like a strong correlated Fermi liquid may allow for an easier theoretical and experimental analysis of its properties than most of the heavy fermion systems.

We would like to thank to Y. Maeno, Y. Kitaoka, K. Machida and T. Ohmi for helpful discussions. We are grateful for financial support from the Swiss Nationalfonds (in particular M.S. for the PROFIL fellowship). D.F.A. also acknowledges the financial support of the Natural Sciences and Engineering Research Council of Canada.

REFERENCES

1. Y. Maeno, H. Hashimoto, K. Yoshida, S. Nishizaki, T. Fujita, J.G. Bednorz and F. Lichtenberg, Nature **372** (1994) 532.
2. T. Oguchi, Phys. Rev. **B51** (1995) 1385 (1995).
3. D.J. Singh, Phys. Rev. **B52** (1995) 1358.
4. A.P. Mackenzie, S.R. Julian, A.J. Diver, G.G. Lonzarich, Y. Maeno, S. Nishizaki and T. Fujita, Phys. Rev. Lett. **76** (1996) 3786.
5. T.M. Rice and M. Sigrist: J. Phys., Condens. Matter **7** (1995) L643.
6. M. Sigrist and T.M. Rice, Proceedings of the Inauguration Conference of the APCTP, Seoul (1996) preprint.
7. Y. Maeno, S. Nishizaki, K. Yoshida, S. Ikeda and T. Fujita, J. Low Temp. Phys. **105**, 1577 (1996).
8. K. Ishida, Y. Kitaoka, K. Asayama, S. Ikeda, K. Yoshida, S. Nishizaki, Y. Maeno and T. Fujita, preprint.
9. V.B. Geshkenbein, A.I. Larkin and A. Barone, Phys. Rev. B**36**, 235 (1987).
10. M. Sigrist and M.E. Zhitomirsky, J. Phys. Soc. Jpn. **65**, 3452 (1996).
11. K. Machida, M. Ozaki and T. Ohmi, J. Phys. Soc. Jpn. **65**, 3720 (1996).
12. T. Sugiyama and T. Ohmi, J. Phys. Soc. Jpn. **64** (1995) 2746.
13. D.F Agterberg, T.M. Rice, and M. Sigrist, in preparation.

ELSEVIER

Physica C 282–287 (1997) 218–221

PHYSICA C

Angle-resolved photoemission study of Sr_2RuO_4; An extended van-Hove singularity in non-cuprate superconductor

T. Takahashi[a], T. Yokoya[a], A. Chainani[a], H. Ding[b,c], J.C. Campuzano[b,c], M. Kasai[d], and Y. Tokura[d,e]

[a]*Department of Physics, Tohoku University, Sendai 980-77, Japan*
[b]*Material Science Division, Argonne National Laboratory, 9700 South Cass Avenue, Argonne, Illinois 60439*
[c]*Department of Physics, University of Illinois at Chicago, 845 West Taylor Street, Chicago, Illinois 60607*
[d]*Joint Research Center for Atom Technology (JRCAT), Tsukuba, Ibaraki 305, Japan*
[e]*Department of Applied Physics, The University of Tokyo, Tokyo 113, Japan*

We have performed high-resolution angle-resolved photoemission spectroscopy of a non-cuprate layered superconductor Sr_2RuO_4 to study the electronic structure near the Fermi level (E_F). The result shows an extended van Hove singularity (vHs) along the Ru-O bonding direction just like the cuprate, regardless of the character of the electronic states at E_F. This suggests that the extended vHs is a general electronic structure of a two dimensional correlated *d*-electron metal. We also have found that the observed Fermi surface topology (one electron-like Fermi surface sheet centered at Γ point and two hole-like ones centered at X point) is qualitatively different from the calculated Fermi surface topology (two electron-like Fermi sheets centered at Γ point and one hole-like one centered at X point) in the consequence of the existence of the vHs.

1. INTRODUCTION

The extended van-Hove singularity (vHs) derived from energy bands of the CuO_2 planes has been regarded as one of the characteristic electronic structure of high temperature superconductors (HTSC's), especially after the direct observation with high-resolution angle-resolved photoemission spectroscopy (HR-ARPES)[1-2]. Several models based on the vHs have been proposed to explain the high superconducting transition temperature and/or the T-linear dependence of the normal-state resistivity [3]. Sr_2RuO_4 is a non-cuprate layered Perovskite oxide superconductor (T_c ~ 1 K), having the same crystal structure as La_2CuO_4 with RuO_2 layers instead of CuO_2 in cuprate HTSC's [4]. The resistivity exhibits T-square (T^2) dependence below 25 K. This material would give us an unique opportunity to study the roll of CuO_2 plane.

In this paper, we report a high-resolution angle-resolved photoemission study of Sr_2RuO_4. We found an extended vHs near the Fermi level (E_F) in Sr_2RuO_4, which is quite similar to the vHs in cuprates, We discuss the correlation between the vHs and the superconducting and normal-state properties. We also mapped out the Fermi surface of Sr_2RuO_4.

2. EXPERIMENT

Single crystals were prepared by the floating zone method. X-ray diffraction was used to characterize the structure of the samples. Magnetic susceptibility measurements confirmed that superconductivity occurs below 0.9 K, as reported earlier [4]. HR-ARPES were carried out at the Synchrotron Radiation Center, Wisconsin, using a 4-meter normal incidence monochromator. The total energy and angular resolution were set at 20 meV (15 meV) and ±1°, respectively. The measurements were performed on several clean surfaces obtained by *in-situ* cleaving at 13 K with 21.2 eV photon energy. All spectra labeled in terms of the value of (k_x, k_y), where k_x and k_y are the wave vectors parallel to the Ru-O bonding direction (in units of $Å^{-1}$). A Pt foil in electrical contact with the sample was used as a Fermi level reference.

PII S0921-4534(97)00208-6

3. RESULTS AND DISCUSSION

Figures 1(a) and (b) show HR-ARPES spectra measured along ΓZ (Ru-O bonding direction) and XX in the Brillouin zone, respectively. Broken lines in the spectra are guide to the eyes. Measurement points in the Brillouin zone are shown with open and filled circles in Fig. 1(c), together with the calculated Fermi surface (dashed lines). In Fig. 1(a), we find that a band crosses E_F at k_y=0.58 and the same band enters from unoccupied states at k_y=1.04. But we do not observe further crossings between the two points. Instead, we find that a flat band is situated very close to E_F over a wide portion of the Brillouin zone centered around the mid point

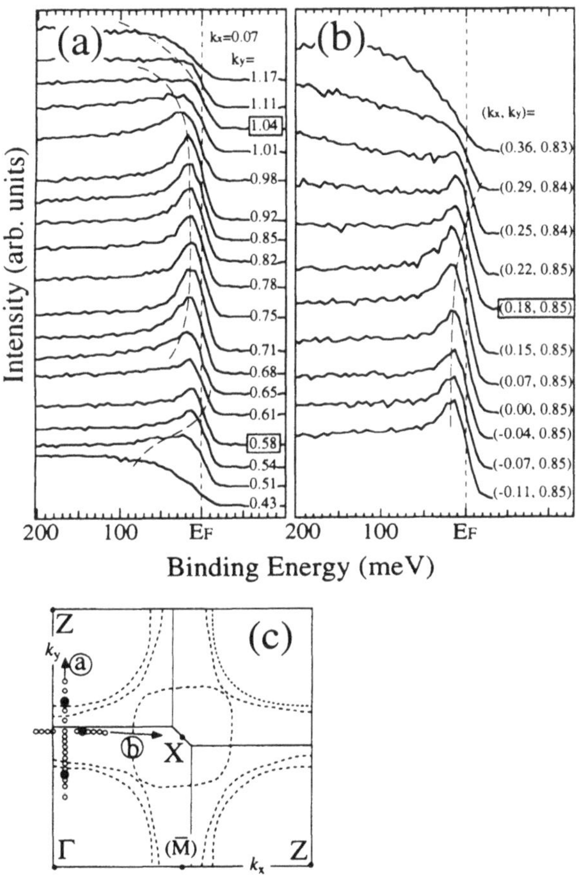

Fig. 1. HR-ARPES spectra near E_F along two directions (a) along ΓZ, and (b) along XX in the Brillouin zone. Measurement points in the Brillouin zone are shown with open and filled circles in Fig. 1(c).

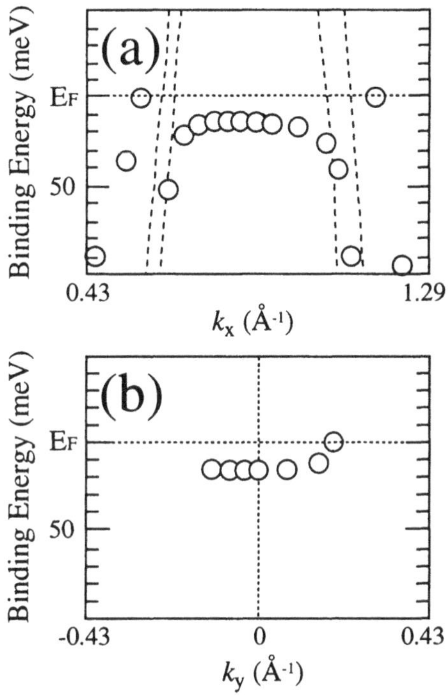

Fig. 2. Experimental band structure near E_F of Sr_2RuO_4 (open circles) determined by the present ARPES measurement. The band structure calculation [5] is shown by broken lines for comparison.

between the Γ and Z points (M point). To further characterize this flat band, we performed another set of HR-ARPES measurement along XX [Fig. 1(b)]. We find that a sharp peak just below E_F at [(k_x, k_y)=(0.00,0.85)] moves gradually toward E_F with increasing k_x and appears to enter into unoccupied states at [(k_x, k_y)=(0.18,0.85)].

The band dispersions obtained from the spectra in Figs. 1(a) and (b) are shown in Figs. 2(a) and (b), respectively. Dashed lines in Fig. 2(a) are the band calculation for comparison [5]. While two bands cross E_F in the calculation, only one band crosses it in the experiment as shown in Fig. 2(a). Another band near E_F shows a flat maximum over a substantial range of k values along ΓZ and also shows a flat minimum perpendicular to it, forming an extended saddle point. A similar extended vHs has been reported in several cuprate HTSC's [1-2]. For example, a HR-ARPES study of $YBa_2Cu_4O_8$ observed an extended vHs at 19 meV binding

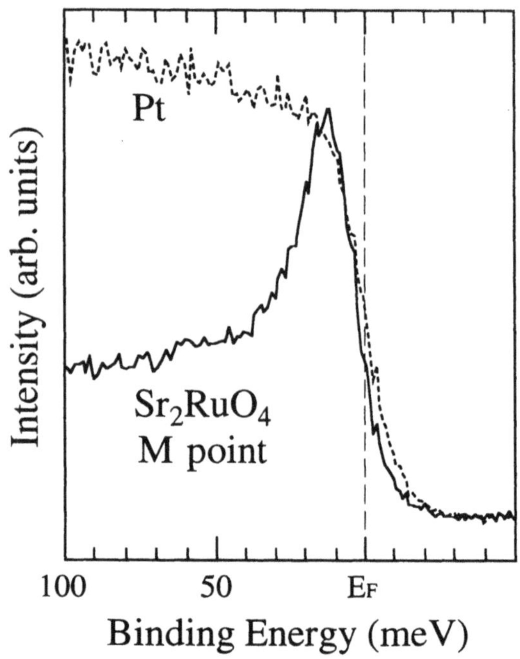

Fig. 3. Higher-resolution ARPES spectrum of Sr_2RuO_4 at the M point in the Brillouin zone measured at 13 K using 15 meV energy resolution, compared with that of a Pt foil measured under the same condition.

energy along the Cu-O bonding direction [2].

To accurately check the position of the extended vHs, we performed a higher-resolution measurement (15 meV total energy resolution) at the M point. As shown in Fig. 3, the peak is found at about 12 meV, and since its leading edge is sharper than the Fermi function as seen in the spectrum of Pt, it is clearly below E_F. The observation of the extended vHs with a smaller binding energy in a low T_c material raises a question as to the validity of the simple van-Hove scenario for high temperature superconductivity.

The van-Hove scenario also explains some anomalous normal-state properties such as the T-linear resistivity in terms of the critical energy of the vHs [3]. For example, King et al. have pointed out that $Nd_{2-x}Ce_xCuO_4$ shows T^2-dependent resistivity with its vHs located about 350 meV below E_F, while $Bi_2(Sr_{0.97}Pr_{0.03})_2CuO_6$ and Bi2212 show T-linear resistivity with vHs' being just below E_F [1]. However, in Sr_2RuO_4 this correlation does not hold. Therefore the simple explanation of anomalous normal-state properties in the van-Hove scenario has to be reexamined.

As a consequence of the presence of the extended vHs which is not predicted from the band calculation [5], the observed Fermi surface topology would be different from calculated one. In Fig. 4, we plot E_F-crossing points in the extended Brillouin zone determined by the HR-ARPES measurements (gray open, and filled circles). Dashed lines show the Fermi surface predicted by the band calculation [5]. The observed Fermi surface consists of one electron-like Fermi surface sheet centered at the Γ point and two hole-like ones centered at the X point. A remarkable difference between the experiment and the calculation is that one of the calculated electron-like sheets appears to transform into a hole-like sheet centered at the X point.

Such a drastic change in the Fermi surface topology could be lead from a change of the electron count, as discussed by Singh [6]. In Table 1, we listed the electron count of each and total Fermi surface sheets from our measurement (ARPES-1) and compare it with those from the dHvA result by

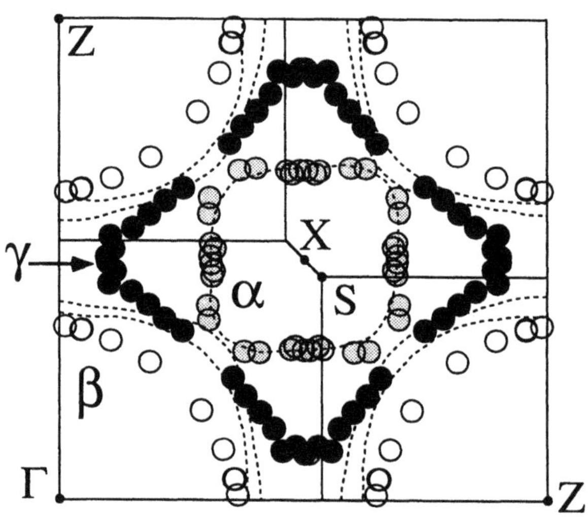

Fig. 4. Experimental Fermi surface of Sr_2RuO_4 obtained by the present ARPES measurement (α:gray, β:open and γ:filled circles). It consists of one electron-like Fermi surface sheet centered at the Γ point (β) and two hole-like ones centered at the X point (α and γ). The Fermi surface predicted by the band calculation [5] (broken lines) is also shown for comparison.

Table 1

Measured volumes of the electron count observed by three different measurements for Sr_2RuO_4:(a) present study, (b) A. P. Mackenzie et al.(dHvA) [7], and (c) D. H. Lu et al. (ARPES-2) [8]. The character of Fermi surface sheets obtained by the measurements is indicated in parenthesis (electron:e, hole:h).

	ARPES-1	dHvA	ARPES-2
α	1.75 ± 0.05 (h)	1.784 (h)	1.8 (h)
β	0.78 ± 0.08 (e)	0.914 (e)	0.7 (e)
γ	1.37 ± 0.09 (h)	1.334 (e)	1.5 (h)
Total	3.90 ± 0.22	4.032	4.0

Mackenzie et al.[7] and a recent ARPES result by Lu et al.[8] (ARPES-2). We found the total electron count from our measurement is 3.90±0.22, which is consistent with that expected for a Ru^{4+} ion and satisfies the Luttinger theorem. The electron count of each Fermi surface sheet shows a good agreement between the present ARPES study and the dHvA measurements, while the character (electron- or hole-like) of one Fermi surface sheet (γ Fermi sheet) is different between the two experiments. It is also noted that two independent ARPES studies show a good agreement with each other even for the character. The ARPES results seem consistent with the Hall coefficient measurement [9], which reported the sign change around 10 K from the negative to the positive value, indicative of the subtle balance of electron and hole counts as found by ARPES. The present result shows that the hole count (0.88±0.14) is close to the electron count (0.78±0.08). It is most likely that the electron correlations modify the band dispersion near E_F, and as a result give a different Fermi surface topology with an extended vHs just below E_F.

4. CONCLUSION

We have observed an extended vHs very close to E_F along the Ru-O bonding direction in analogy to the cuprate superconductors regardless of the difference of the character of band near E_F. This result suggests that an extended vHs may be a general feature of two-dimensional correlated metals. On the other hand, the large difference of T_c between $S_{r2}RuO_4$ and the cuprates, as well as qualitatively different transport properties, requests a reexamination of the van-Hove scenario and its relation to the normal and superconducting properties. As a consequence of the vHs, the experimental Fermi surface is different from that of the band calculation, although the total electron count is the same.

ACKNOWLEDGMENTS - This work was supported by grants from the CREST and the Ministry of Education, Science and Culture of Japan, and the U. S. Dept. of Energy, Basic Energy Sciences, under contract W-31-109-ENG-38.

REFERENCES

[1] D.M. King et al., Phys. Rev. Lett. 70 (1993) 3159; D.M. King et al., Phys. Rev. Lett. 73 (1994) 3298.
[2] K. Gofron et al., Phys. Rev. Lett. 73 (1994) 3302.
[3] D.M. Newns et al., Phyisica (Amsterdam) 186B (1993) 801.
[4] Y. Maeno et al., Nature 372 (1995) 532.
[5] T. Oguchi, Phys. Rev. B51 (1995) 1385.
[6] D.J. Singh, Phys. Rev B52 (1995) 1358.
[7] A.P. Mackenzie et al., Phys. Rev. Lett. 76 (1996) 3786.
[8] D.H. Lu et al., Phys. Rev. Lett. 76 (1996) 1845.
[9] N. Shirakawa et al., J. Phys. Soc. Jpn. 64 (1996) 1072.

ELSEVIER

Physica C 282–287 (1997) 222–225

PHYSICA C

Raman Scattering by Electronic Excitations in High T_c Superconductors

M. Cardona, T. Strohm, and Xingjiang Zhou[a]

[a]Max-Planck-Institut für Festkörperforschung, Heisenbergstraße 1, 70569 Stuttgart, Germany, EU

Electronic excitations have been observed in the normal and the superconducting states of most high T_c superconductors by means of Raman spectroscopy. The absolute scattering efficiencies measured in the various polarization configurations suggest that the scattering vertex is related to the effective masses at the Fermi surface. The interpretation of the "normal" scattering requires the introduction of a phenomenological scattering time which also accounts for the observed "superconducting scattering" well above the gap region. In the gap region, the scattering in polarized configurations contains information about the dependence of the gap on $\boldsymbol{k}$-vector. Particularly informative in this respect is the asymptotic behavior at low frequencies. The peaks observed near gap frequencies, however, are difficult to interpret under the assumption of gaps with a simple dependence on $\boldsymbol{k}$ (such as $k_x^2 - k_y^2$). The state of the art and outstanding problems are reviewed.

Since the work of Lyons *et al.* [1], who demonstrated that gap-like structure can be seen in the Raman spectra of high T_c superconductors, scattering by electronic excitations in HT_cSC has received considerable attention. The scattering in the normal state, which should vanish at low frequencies for a Fermi liquid, is found to be finite and is usually interpreted on the basis of mass fluctuations plus scattering around the Fermi surface [2]. The measured absolute scattering efficiencies correspond roughly to the values calculated from the mass fluctuations of LMTO band structures which are given by (to a factor of $\simeq 1$) [3,4]

$$B \simeq \frac{r_0^2}{\pi} N(E_F) \left\langle \left| \hat{e}_L \cdot (\boldsymbol{\mu}^{-1} - \langle \boldsymbol{\mu}^{-1} \rangle) \cdot \hat{e}_S \right|^2 \right\rangle, \quad (1)$$

being typically a factor of three higher than calculated. This discrepancy could, in principle, be attributed to a weak resonance enhancement of the mass vertex [5]. The normal state scattering is, at least in optimally doped materials, surprisingly independent of frequency ω and temperature T. This led Varma [6] to postulate the marginal Fermi liquid (MFL) model, in which the broadening of electronic states is proportional to either ω or T (whichever is larger). More recently, evidence has been given showing that in the underdoped case the excitations very close to the Fermi surface become Fermi-liquid-like [7,8].

Raman scattering in the superconducting state shows a peak in the 2Δ region ($\Delta \simeq$ superconducting gap) even if the lifetime of the unpaired electrons is infinite. The qualitative reason is that finite excitation energies $E_{\boldsymbol{k}}$ appear even for $\boldsymbol{k} = 0$ (we recall that $E_{\boldsymbol{k}} = \left(\varepsilon_{\boldsymbol{k}}^2 + \Delta^2\right)^{1/2}$ which does not vanish for $\varepsilon_{\boldsymbol{k}=0} = 0$). The efficiency, however, does vanish for $\omega = 2E_{\boldsymbol{k}}$ tending to either zero or infinity: the latter case should reproduce the results of the normal state and thus can be made finite by assuming a finite electron lifetime, i.e., patching up the theory with phenomenological considerations. According to the theory of Raman scattering in the superconducting state, developed by Klein and Dierker [9], Eq. (1) for the scattering efficiency must be replaced by

$$\frac{\partial^2 S}{\partial\omega\partial\Omega} = \frac{r_0^2}{\pi} Im \left\{ \langle \mu^{-2} F \rangle_F - \frac{\langle \mu^{-1} F \rangle_F^2}{\langle F \rangle_F} \right\} \quad (2)$$

where F is the so-called Tsuneto function. $Im\ F$ represents the density of pair-breaking excitations:

$$Im\ F(\omega; \boldsymbol{k}) = \pi N(E_F) \frac{4\,|\Delta(\boldsymbol{k})|^2}{\omega\sqrt{\omega^2 - 4\,|\Delta(\boldsymbol{k})|^2}} \quad (3)$$

for $\omega > 2\Delta$ and zero for $\omega \leq 2\Delta$. The real part of (3) can be obtained by Kramers-Kronig (KK) transformation. Some errors which have crept into evaluations of (3) found in the literature [10,11] seem to have resulted from errors in the KK evaluation. In (2) $\langle \mu^{-2} F \rangle_F$ represents a Fermi surface average while μ^{-1} is the inverse

PII S0921-4534(97)00212-8

effective mass tensor contracted with the incident and scattered polarization vectors $\hat{e}_L$ and $\hat{e}_S$, respectively (see Eq. (1)). The second term of (2) represents screening due to Coulomb interaction. It is zero unless the polarization configuration contains fully symmetric terms (A_{1g} symmetry), e.g., for $\hat{e}_L \parallel \hat{e}_S$, and completely cancels the first term if μ^{-1} has only one, fully isotropic value. Hence scattering of A_{1g} symmetry should be larger if the material has FS's with several sheets or, more generally, if there are more than one CuO_2 planes (chains may also increase the A_{1g} scattering). A_{1g} screened scattering thus arises from *mass fluctuations*.

After the original euphoria concerning the ability of Raman scattering to detect the symmetry of $\Delta(\boldsymbol{k})$ [12], some circumspection set in [13,14]. This was partly due to the observation that the Tsuneto function (3) depends only on $|\Delta(\boldsymbol{k})|^2$ and, contrary to Josephson tunneling experiments, is not able to detect the phase of $\Delta(\boldsymbol{k})$ (it always has an A_{1g} component). This problem is, however, common to many experiments which are often adhuced to prove the $d_{k_x^2-k_y^2}$ symmetry of the gap (e.g. photoemission, penetration depth vs. T, etc.). Additional objections are discussed below.

The Raman spectra observed in the SC-phase exhibit features common to most high T_c superconductors and, except for details, are agreed upon by most experts. This is specially true of the B_{1g} ($\hat{e}_L \perp \hat{e}_S \parallel [110]$) and B_{2g} ($\hat{e}_L \perp \hat{e}_S \parallel [100]$) components ($D_{4h}$ point group notation is used unless otherwise specified for cases of orthorhombic symmetry). The B_{2g} spectra are weak and display a broad peak with a maximum well below that of the much stronger and sharper B_{1g}. This result is reproduced by most numerical evaluations of (2) with a $d_{k_x^2-k_y^2}$-like gap and therefore is compatible with gaps of that symmetry. It is, however, also compatible with $\Delta \sim \left|k_x^2 - k_y^2\right|$, i.e., A_{1g} symmetry. The only objection which can be made against this possibility is that *nature abhors nonanalytic functions*. The problem becomes more serious for the A_{1g} spectra which show a sharp peak near that of B_{2g}, much lower and stronger than that of B_{1g}. This is not found in any of the calculations performed with μ's derived from *ab initio* band structures for gaps of $d_{k_{x^2}-k_{y^2}}$ symmetry [14–16]. In order to try to

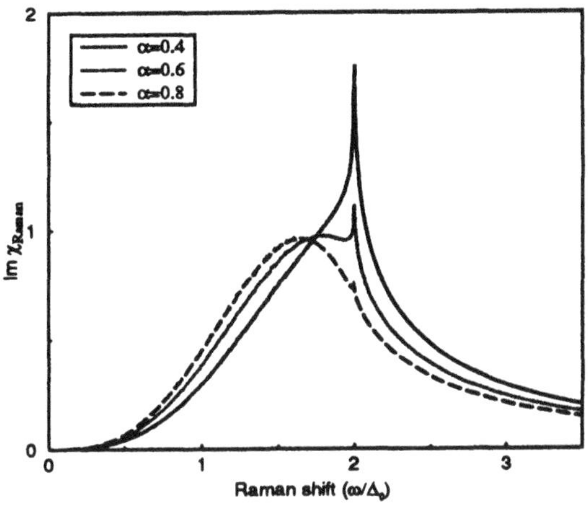

Figure 1. Raman spectra calculated for the B_{1g} scattering configuration assuming $\Delta = \Delta_0 \cos 2\varphi$ and a mass vertex $\mu^{-1} = \sqrt{2/1+\alpha^2}(\cos 2\varphi - \alpha \cos 6\varphi)$. Note the downshift of the maximum with increasing α and the ω^3 dependence for $\omega \rightarrow 0$.

explain this result, parametrized expansions of μ in terms of angular harmonics, such as:

$$\mu^{-1}(\boldsymbol{k}) \sim \cos 4\varphi - \alpha \cos 8\varphi \qquad (4)$$

have been used [11]. The position of the peak in the A_{1g} spectrum can then be reproduced for $\alpha \simeq 0.1$ but depends very critically on the value of α. This is easy to understand since $\alpha > 0$ in (4) decreases the value of μ^{-1} for $\boldsymbol{k}$ along the axes (where $\Delta(\boldsymbol{k}) \sim (k_x^2 - k_y^2)$ has maxima) and pushes its weight to regions of the FS where Δ is smaller. The need to introduce rather unstable parameters of dubious physical meaning impares the predictive power of the theory. It has been pointed out that only the A_{1g} component is sensitive to higher terms in the expansion of μ [18]. Figure 1 suggests that this is not true. Using the B_{1g} mass vertex:

$$\mu^{-1}(\boldsymbol{k}) \sim \cos 2\varphi - \alpha \cos 6\varphi \qquad (5)$$

we find that the peak position shifts down with increasing α. We thus conclude that at this point we do not have any robust explanation for the relative behavior of A_{1g} and B_{1g}. The simple, but not very plausible explanation is that resonance effects enhance $\mu^{-1}(A_{1g})$ along certain directions of $\boldsymbol{k}$-space. The problem is, however, found for nearly all HT$_c$SC measured and it is hard to believe that they all will have the same resonance behavior. A way out of this dilemma may appear if the hopping between planes turns out to be incoherent, thus eliminating $3D$ band structure effects. The resonances could then be the same for all materials since the relevant, decoupled two-dimensional blocks (CuO_2 planes) are the same. However, relative changes of A_{1g} and B_{1g} when changing the laser frequency have not been reported.

Resonance measurements [5], as well as calculations of scattering efficiencies in both normal and SC state [3,4,16], suggest that the Raman vertices in both cases are indeed the same, i.e., the inverse masses μ^{-1} and their fluctuations. As already mentioned, strong mass fluctuations of A_{1g} symmetry appear naturally from the existence of several CuO_2 planes. The question then arises as to whether materials with several CuO_2 planes per unit cell should have stronger A_{1g} scattering than those with only one. Recent measurements by Kang *et al.* for $Tl_2Ba_2CuO_{6+\delta}$ support this statement [19] although contradictory evidence exists [20,21].

Another possible source of strong mass fluctuations is the presence of van Hove singularities near the Fermi surfaces which make the bands strongly nonparabolic. In this case, the integration over the FS involved in (2) may not be accurate because of the rapid changes in μ^{-1} which are obtained when moving away from the FS. Averaging over the whole BZ can be performed to overcome this problem, using the $3D$ version of (2). This has been accomplished for the $2D$ [22] and $3D$ cases [16]. In $2D$ the van Hove singularities result in additional sharp peaks in the near-gap structure which, as expected, is smoothed out considerably in the $3D$ case. Such structure has not yet been reported for measured spectra although some evidence of it may have been observed in $HgBa_2Ca_2Cu_3O_{8+\delta}$ [23]. BZ integration leads to calculated spectra quantitatively very similar to those obtained when FS integration is performed (except for the appearance of van Hove singularities in the case of BZ-integration).

Another aspect of these spectra which contains information about the $\boldsymbol{k}$ dependence of $\Delta(\boldsymbol{k})$ is the asymptotic behavior for $\omega \rightarrow 0$ [10]. Let us consider (2) in this limit and for the B_{1g} polarizations. The corresponding average $\langle \mu^{-1} F \rangle$ contains two quantities which vanish along the $[k, \pm k, k_z]$ directions under the assumption of a $(k_x^2 - k_y^2)$-like gap. A simple calculation leads, in this case, to an asymptotic behavior $\propto \omega^3$. The same thing holds for any other gap function with nodes along $[k, \pm k, k_z]$. For B_{2g} and A_{1g} polarizations, however, the nodes of μ^{-1} and those of Δ do not coincide, a fact which leads to an asymptotic behavior $\propto \omega$. Figure 2 shows that the ω^3-like behavior holds rather accurately for the B_{1g}-like scattering of $HgBa_2Ca_2Cu_3O_{8+\delta}$ crystals [23], a material which has the tetragonal D_{4h} point group. The scattering of this material exhibits asymptotic behavior linear in ω for the B_{2g} and A_{1g} polarizations, in agreement with the discussion above.

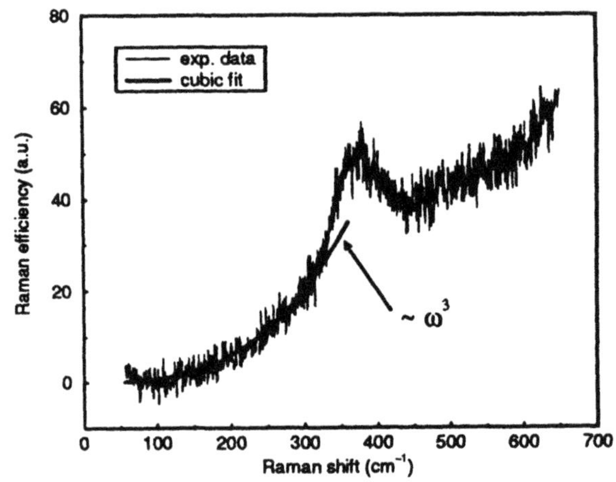

Figure 2. Measured B_{1g} spectrum of $HgBa_2CaCu_2O_{6-\delta}$ in the superconducting state showing the asymptotic dependence $\propto \omega^3$ found for small ω (solid line) [23].

Of particular interest are the orthorhombic materials $YBa_2Cu_3O_{7-\delta}$ and $Bi_2Sr_2CaCu_2O_{10-\delta}$. Both have D_{2h} point group symmetry although the unit cell of the Y-compounds is P-orthorhombic while that of Bi-compound is C-(centered)orthorhombic. This implies that the axes of the former are along the Cu-O bonds while those of the latter are rotated by 45°. The B_{1g} configuration of the tetragonal CuO_2 planes thus becomes A_g (fully symmetric) in the D_{2h} case of the Y-compound but remains asymmetric (B_{1g}) for the Bi-compound. Hence, in the latter, the nodes of a $(k_x^2 - k_y^2)$-like gap and those of $\mu^{-1}(B_{1g})$ still coincide and the asymptotic behavior remains ω^3-like for the B_{1g} configuration ($\hat{e}_L \perp \hat{e}_S \parallel [110]$). In the case of $YBa_2Cu_3O_{7-\delta}$, however, the $\hat{e}_L \perp \hat{e}_S \parallel [110]$ configuration is fully symmetric since the diagonals of the CuO_2-squares do no longer correspond to reflection planes. This means that the nodes of $(k_x^2 - k_y^2)$-like gaps and those of the corresponding μ^{-1} (for $\hat{e}_L \perp \hat{e}_S \parallel [110]$) do not coincide and the ω^3 asymptotic behavior acquires now a small component linear in ω. This is illustrated in Fig. 3 for both, the experimental and the calculated spectra of $YBa_2Cu_3O_{7-\delta}$. Similar results have been found for $YBa_2Cu_4O_8$ [16].

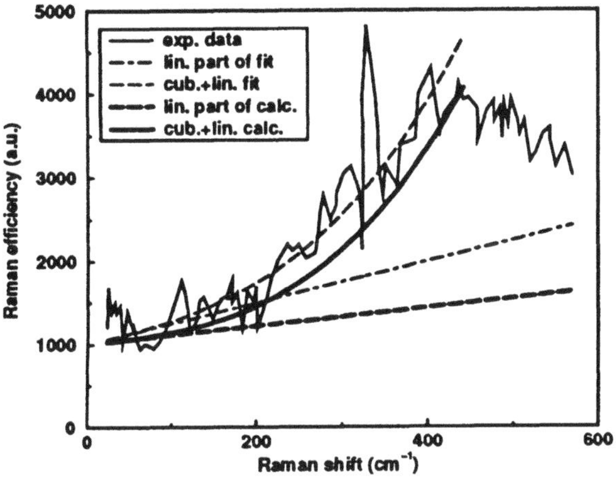

Figure 3. Measured "B_{1g}"-like spectrum of $YBa_2Cu_3O_{7-\delta}$ ($\hat{e}_L \perp \hat{e}_S \parallel [110]$ configuration) showing the asymptotic behavior at low frequency $\propto A\omega + \omega^3$ (with $A = 9 \cdot 10^4$ cm^{-2}). Thick lines correspond to the calculation. [16,17].

In conclusion, the B_{1g} and B_{2g} spectra measured in the SC phase suggest a gap function with nodes in the $[k, \pm k, k_z]$ planes, compatible with $(k_x^2 - k_y^2)$-symmetry but not excluding other possibilities. The position *and* relative intensities of the A_{1g}-spectra are difficult to account for theoretically. The asymptotic behaviors found for the various spectra in tetragonal materials (D_{4h}) for $\omega \to 0$ agree with theoretical predictions for similar types of gaps. In the case of orthorhombic $YBa_2Cu_3O_{7-\delta}$, a linear term in ω has to be added to the ω^3 dependence found in the D_{4h} case. Note that in the discussion above we have assumed rather complex dependences of μ^{-1} on $\boldsymbol{k}$ while the corresponding dependence of Δ on $\boldsymbol{k}$ was assumed to be rather simple. Several authors have tried to explain the discrepancies encountered here by taking more complex dependences of Δ on $\boldsymbol{k}$, such as gaps of different symmetries on the different FS sheets [14,15].

REFERENCES

1. K.B. Lyons *et al.*, Phys. Rev. B **36**, 5592 (1987).
2. A. Zawadowski and M. Cardona, Phys. Rev. B **42**, 10732 (1990).
3. M.C. Krantz *et al.*, Phys. Rev. B **51**, 5949 (1995).
4. M. Cardona *et al.*, in *Spectroscopic Studies of Superconductors*, edited by I. Bozovic and D. van der Marel, Proc. SPIE 2696, 182 (1996).
5. Moonsoo Kang *et al.*, Phys. Rev. Lett. **77**, 4434 (1996).
6. C.M. Varma, Phys. Rev. Lett. **75**, 898 (1995).
7. C.M. Varma, Phys. Rev. Lett. **63**, 1996 (1989).
8. T. Zhou *et al.*, Solid State Commun. **99**, 669 (1996).
9. M.V. Klein and S.B. Dierker, Phys. Rev. B **29**, 4976 (1984).
10. T.P. Devereaux and D. Einzel, Phys. Rev. B **51**, 16336 (1995).
11. T.P. Devereaux and D. Einzel, Phys. Rev. B **54**, 15547 (1996).
12. T.P. Devereaux *et al.*, Phys. Rev. Lett. **72**, 396 (1994).
13. M. Krantz and M. Cardona, Phys. Rev. Lett. **72**, 3290 (1995).
14. M. Krantz and M. Cardona, J. Low Temp. Physics **99**, 205 (1995).
15. M. Krantz, Phys. Rev. B **54**, 1334 (1996).
16. T. Strohm and M. Cardona, Phys. Rev. B, in press.
17. R. Hackl *et al.*, Phys. Rev. B **38**, 7133 (1988).
18. T.P. Devereaux and A.P. Kampf, Int. J. of Modern Physics B, in press.
19. Moonsoo Kang *et al.*, to be published.
20. L.V. Gasparov *et al.*, Phys. Rev. B **55**, 1223 (1997).
21. X.K. Chen *et al.*, Phys. Rev. Lett. **73**, 3290 (1994).
22. D. Branch and J.P. Carbotte, Phys. Rev. B **52**, 603 (1995).
23. X.J. Zhou *et al.*, reported at this conference.

ELSEVIER

Physica C 282-287 (1997) 226

PHYSICA C

Extended Abstract

NMR Studies of the original magnetic properties of the cuprates: influence of impurities and defects

H. Alloul[a], J. Bobroff [a], A. Mahajan[a], P. Mendels[a], Y. Yoshinari[a], G. Collin[c], and J. F. Marucco[b]

[a]LPS, UA2 CNRS ; [b]LCNS , Universite Paris Sud 91405 Orsay, France; [c] LLB, CE Saclay, CEA-CNRS, 91191, Gif sur Yvette, France.

NMR and neutron scattering experiments have revealed that the cuprates display original magnetic properties, both in their insulating and metallic phases. In underdoped metallic systems, a pseudo gap in the magnetic q=0 excitations is observed from local susceptibility measurements, as well as from specific heat and transport properties. A pseudo spin-gap in the AF excitations at the AF wave vector is detected both by NMR and inelastic neutron scattering. These magnetic anomalies are shown to occur in underdoped bilayer as well as in single layer $HgBa_2CuO_{4+\delta}$ cuprates, and cannot be ascribed to a magnetic coupling within the bilayer[1].

The modifications of magnetic properties induced by substitutions or defects in the planes, which do not modify appreciably the charge transfer, have been studied. The observations done are quite original and do again stress the importance of electronic correlations in the metallic state.

The case of Zn which drastically reduces Tc has been examined in some detail. NMR reveals that the substitution of this $3d^{10}$ non magnetic atom on a Cu site induces a Curie like contribution to the local susceptibility on the near neighbor coppers [2,3]. The effective induced moment decreases with hole doping, and becomes rather weak for optimal doping [4].

The spin polarization induced by impurities in the CuO_2 band has been monitored through detailed experimental studies of the ^{17}O NMR. The linewidth of the ^{17}O located in the CuO_2 planes increases at low temperature. This broadening is associated with the long range modification of the spin polarization induced by the impurity moment. Even for the relatively simple case of Ni substitutions for Cu, large qualitative differences between the underdoped and slightly overdoped YBCO are evidenced [5]. This allows us to suggest that the induced spin polarization is associated with the T-dependence of the antiferromagnetic staggered spin susceptibility $\chi'(Q_{AF})$ intrinsic to the CuO_2 planes. We show that the ^{17}O linewidth probes the amplitude of $\chi'(Q_{AF})$ rather than the AF correlation length.

The deduced variation of $\chi'(Q_{AF})$ is found somewhat more T dependent than that deduced from measurements of the transverse spin relaxation T_2 of ^{63}Cu. Similar data for T_2 taken on ^{17}O in these systems allow us to demonstrate that the measurements of $^{63}T_2$ have to be corrected for a dynamic contribution associated with Cu nuclear spin fluctuations [6]. We conclude that all these data might be reconciled if the AF correlation length is assumed T independent, which would agree with the observations on $\chi''(q)$ done by neutron scattering experiments.

REFERENCES:

1. J. Bobroff, H.. Alloul, V. Viallet, J. F. Marucco and D. Colson , Phys. Rev. Lett., May 1997.
2. A. V. Mahajan, H. Alloul, G. Collin and J. F. Marucco, Phys. Rev. Lett. 72, 3100 (1994).
3. K. Ishida, Y. Kitaoka, K. Yamazoe, K. Asayama and Y. Yamada, Phys. Rev. Lett., 76, 531(1996); H. Alloul, J. Bobroff and P. Mendels, ibidem, 78, 2494 (1997).
4. P. Mendels et al, Phys. Rev. B49, 10035(1994).
5. J. Bobroff, H. Alloul, Y. Yoshinari, A. Keren, P. Mendels, N. Blanchard, G. Collin, J.F. Marucco, submitted. See also J. Bobroff et al, this conference.
6. A. Keren, H. Alloul, P. Mendels, Y. Yoshinari, Phys. Rev. Lett., May 1997.

PII S0921-4534(97)00215-3

ELSEVIER

Physica C 282–287 (1997) 227–230

PHYSICA C

Cross-over temperatures and spin-gap in High T_c Cuprate Superconductors An NMR approach

C. Berthier, M.-H. Julien, O. Bakharev, M. Horvatić and P. Ségransan[a]

[a]Grenoble High Magnetic Field Laboratory, Max-Plank Institut für Festkörper forschung and Centre National de la Recherche Scientifique, BP 166, 38042 Grenoble Cedex 9, France
Laboratoire de Spectrométrie Physique, Université Joseph Fourier Grenoble I, BP 87, 38402 Saint-Martin d'Hères Cedex, France

We present NMR experiments in the normal phase ($T > T_c$) in underdoped $HgBa_2Ca_2Cu_3O_{8+\delta}$ single crystals ($T_c = 115$ K) and $YBa_2Cu_3O_{6.5}$ ($T_c = 46$ K in zero field). Through these examples, and on a more general basis, we discuss the signature of the pseudo-gap as observed in NMR experiments and how it compares with other techniques, like transport, specific heat, inelastic neutron scattering and photoemission.

1. INTRODUCTION

The striking difference between the normal state properties of underdoped (UD) and overdoped high T_c cuprate superconductors was first revealed by the analysis of NMR data [1–3]. The two major features of the UD regime are:

i) a decrease of the copper nuclear spin-lattice relaxation rate (T_1^{-1}) divided by the temperature (T) below some temperature T^*, well above T_c. This behavior, quite early identified as the opening of a spin-gap [4], is now recognized as the signature of a loss in the spectral weight of low energy spin excitations with wavevectors close to antiferromagnetic (AF) one Q_{AF}, as confirmed by inelastic neutron scattering experiments in UD $YBa_2Cu_3O_{6+x}$ compounds [5,6]. However, a clear NMR evidence for the opening of this gap or pseudogap requires in addition a continuous increase (or a flattening) of the gaussian part of the spin-spin relaxation rate (T_{2G}) as T decreases down to T_c [3].

ii) a continuous decrease with T of the static spin susceptibility $\chi(q = 0)$ observed through the magnetic hyperfine shift of CuO_2 planar copper and oxygen (or yttrium) nuclei below some temperature T_0, always higher than T^* [3,7]. Whether the cross-over temperature T_0 corresponds to the opening of a spin-gap [8–10], or have a different origin [11–13], is still an open question. Later on, the opening of a spin gap in the UD regime has become widely recognized, and various cross-over temperatures have been deduced from transport [9] or specific heat measurements [14]. More recently, photoemission experiments in UD and overdoped BISCO compounds [15–17] have shown that the gap opening above T_c in the UD regime has the same symmetry, i.e., $d_{x^2-y^2}$, and the same magnitude as the superconducting gap. Its temperature onset, and its magnitude at low T are almost constant or slightly increasing when reducing both the hole concentration and the value of T_c [17], in striking disagreement with BCS theory, and strongly suggesting the existence of preformed pairs. The purpose of this paper is to reexamine the more plausible interpretation of the cross-over temperatures T^* and T_0 determined from experiments, as well as to compare the NMR and inelastic neutron scattering data. We shall consider experimental data in the UD three layers $HgBa_2Ca_2Cu_3O_{8+\delta}$ and two layers $YBa_2Cu_3O_{6+x}$ compounds, but we shall extend the discussion to UD cuprates in general.

2. UNIFORM versus STAGGERED SUSCEPTIBILITY

The temperature T_0 at which the spin susceptibility starts to decreases, or passes through a maximum, is usually much higher than T^*, the temperature corresponding to the maximum of

PII S0921-4534(97)00220-7

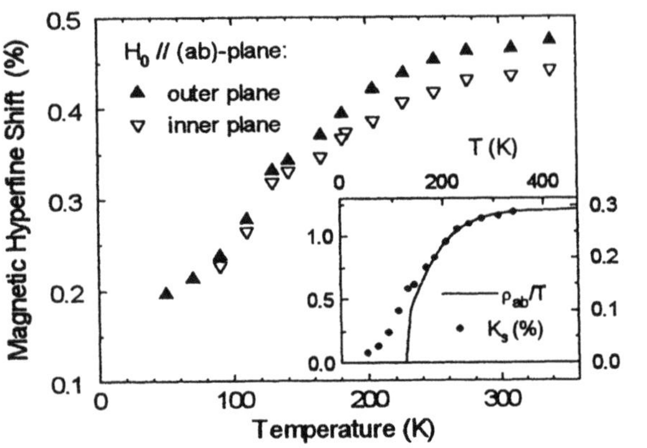

Figure 1. ^{63}Cu magnetic hyperfine shift in $HgBa_2Ca_2Cu_3O_{8+\delta}$. Inset: Spin part of the shift (dots, right scale in %), and in-plane resistivity devided by T (solid line, left scale in $\mu\Omega$ cm K^{-1}.

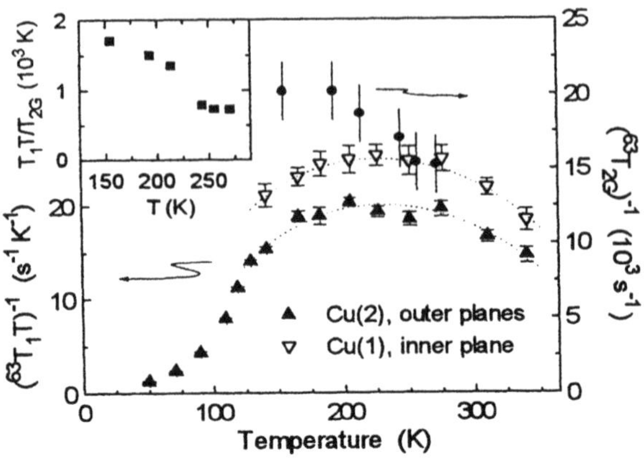

Figure 2. Copper $(T_1T)^{-1}$ and $(T_{2G})^{-1}$ in $HgBa_2Ca_2Cu_3O_{8+\delta}$. The ratio T_1T/T_{2G} is shown in the inset.

the quantity $(T_1T)^{-1}$ or to a sharp increase of the ratio $(T_1T)/T_{2G}$ [3]. A good example can be found in the UD compound $HgBa_2Ca_2Cu_3O_{8+\delta}$ studied by Julien *et al.* [18]. Fig. 1 shows the T dependence of the ^{63}Cu magnetic hyperfine shifts (^{63}K) for inner and outer CuO_2 plane which both start to decrease around 350 K. It is worth to note that the conductivity starts to deviate from linear T behavior at the same temperature T_0, and that the resistivity divided by the T $\rho(T)/T$ follows the same T dependence as ^{63}K. On the other hand, the temperature T^* is seen in Fig. 2 to be approximately equal to 250 K, that is much lower than T_0, although being the highest value observed among the high T_c cuprates. Perhaps the most striking difference between T_0 and T^* can be found in the $La_{2-x}Sr_xO_4$ where it has been thought by long that there was no decrease of $(T_1T)^{-1}$ before T_c [19]. This has been the starting point of a long standing debate to decide whether the opening of a gap above T_c could only occur in presence of a transverse AF exchange coupling between adjacent CuO_2 planes [20]. However, $(T_1T)^{-1}$ and T_{2G} data in an UD tri-layer mercury compound $HgBa_2Ca_2Cu_3O_{8+\delta}$ first revealed the spin-pairing responsible of this pseudogap should necessarily occur *within* a layer [18]. More recently, the T dependence of $(T_1T)^{-1}$ [21] and that of the oxygen magnetic hyperfine shift [22] in UD single layer $HgBa_2CuO_{4+\delta}$ have been shown to be strikingly similar to that observed in UD $YBa_2Cu_3O_{6+x}$ compounds, thus definitely settling the controversy. Furthermore, recent data in $La_{2-x}Sr_xO_4$ compounds clearly indicate the existence of a maximum of $(T_1T)^{-1}$ around 50 K[23]. This leads to think that the magnitude of T^* observed in the phase diagram of a high T_c cuprate as a function of hole doping is intimately related to the optimal value of T_c, as already suggested by Julien *et al.* [18].

We shall now consider the dependence of T^* on the hole concentration n_h within the UD regime of $YBa_2Cu_3O_{6+x}$ compounds. This latter is rather weak, as shown in Fig. 3, presenting $^{63}(T_1T)^{-1}$ measurements for various single crystals and oriented powders with T_c values varying from 50 K to 91 K. Note that the flattening of $(T_1T)^{-1}$ in the optimally doped compound do no reflect the opening of a pseudo gap in the AF excitations, since the ratio $(T_1T)/T_{2G}$ stay constant[24]. This weak variation of T^* has to be compared with the recent photoemission results in UD BISCO compounds[17], which clearly indicate that the magnitude of the gap observed in the normal state is weakly dependent on the value of n_h (or T_c) in the UD regime. On the other

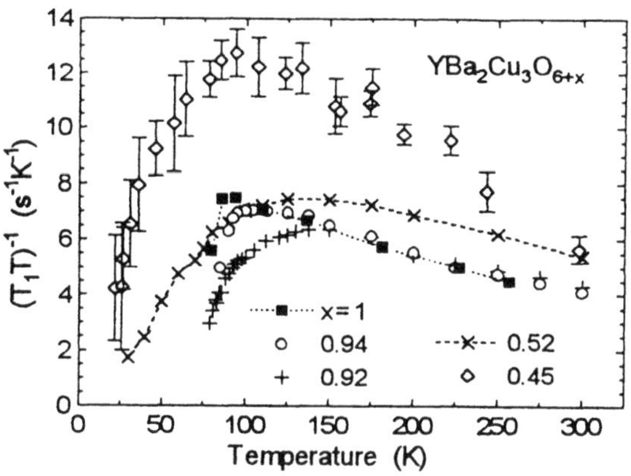

Figure 3. Copper $(T_1T)^{-1}$ in $YBa_2Cu_3O_{6+x}$ as a function of doping.

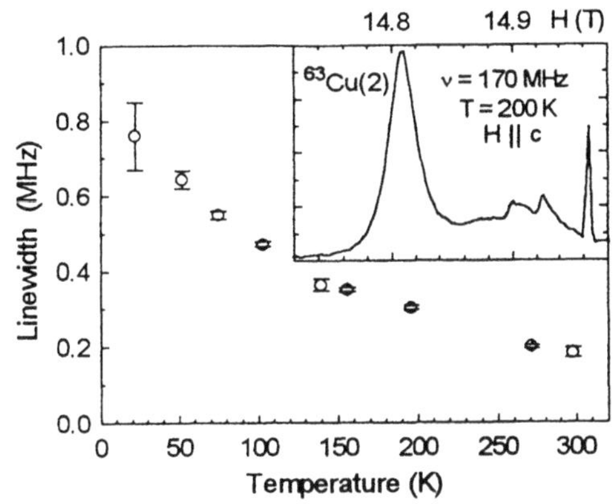

Figure 4. Full width at half maximum of ^{63}Cu NMR line (shown in the inset) in $YBa_2Cu_3O_{6.45}$.

hand, the strong dependence of T_0 on n_h, including the case of LASCO [9], seems to indicate that this cross-over temperature is not directly related to the gap observed by photoemission. It is more likely than T_0corresponds to the onset of sizable AF correlations [11-13,25], growing at the expenses of the uniform susceptibility. This growing of AF correlations in the UD regime can be evidenced through the T depedence of T_{2G} [26,27,3], from which one can estimate the magnitude and the T dependence of the real part of the staggered susceptibility $\chi(Q_{AF})$ divided by the correlation length of AF fluctuations ξ. In a system like $YBa_2Cu_3O_{6+x}$, both magnitude and T dependence of this quantity are strongly increasing when reducing n_h. Close to the metal-insulator transition, T_{2G} becomes so short than it can no longer be measured. However, the second moment of the copper NMR line obtained by sweeping the magnetic field at fixed frequency, becomes an alternative way to access to the T dependence of $\chi(Q_{AF})/\xi$, as shown in Fig. 4. This increase of $\chi(Q_{AF})$, accompanied by a decrease of $\chi(q=0)$ as n_h and (or) T decrease is a strong and robust evidence of the existence and the importance of AF correlations in the whole UD regime, and is quite reminiscent of the behavior of $\chi(q=0)$ and $\chi(q=2k_F)$ in quasi-1D fermion systems.

3. NMR and INELASTIC NEUTRON SCATTERING

Inelastic neutron scattering (INS) experiments [5,6,28] in UD $YBa_2Cu_3O_{6+x}$ have shown that the imaginary part of the dynamic spin susceptibility $\chi''(q,\omega)$ is peaked around the AF wave vector $Q_{AF}=(\pi/a,\pi/a)$ and its width is T independent and increasing with n_h. The situation in $La_{2-x}Sr_xCuO_4$ looks quite different[29,30] where $\chi''(q,\omega)$ consists of four peaks centered at incommensurate value $\pi/a(1\pm\delta,1)$ and $\pi/a(1,1\pm\delta)$. In contrast with $YBa_2Cu_3O_{6+x}$, the sharpness of the peaks was found strongly T dependent. At the same time, the quantity δ has been shown to be proportional to n_h [31]. In both systems, using neutron data on $\chi''(q,\omega)$ has led to a quantitative agreement as far as the planar copper nuclear spin-lattice relaxation is concerned, but to a severe failure in the case of the planar oxygen, for which the contribution of the AF fluctuations is found experimentally much less than the computed one [32,3,33]. The opening of a pseudogap around T^*, as observed by NMR, has currently been associated to a transfer from low to higher energy in the spectral weight of the AF excitations, corresponding to a decrease of the initial slope of $\chi''(Q_{AF},\omega)$ measured by inelastic neutron scattering [3,5,24,34]. However, we would like to stress here that for $YBa_2Cu_3O_{6+x}$

compounds with T_c lower than 60 K, this initial slope has been found to continuously increase with decreasing T [5,35], although the NMR signature of the opening of a pseudogap is present [36,37]. Another strange discrepancy which deserves to be underlined is the weak dependence of the magnitude of $^{63}(T_1T)^{-1}$ (Fig. 3) with n_h in the whole UD regime, which contrasts with the intensity at low energy (10 meV and below) of $\chi''(Q_{AF},\omega)$ measured by inelastic neutron scattering[5,28]. So a lot of progress is still to be made in the microscopic description of spin fluctuations in the cuprates.

4. CONCLUSION

We have analyzed the difference between the two cross-over temperatures observed by NMR in the underpoped regime of the cuprates. We propose that T^*, corresponding roughly to the maximum of the copper nuclear-spin relaxation divided by T, should be the relevant quantity to compare with gap observed by photoemission experiments.

REFERENCES

1. T. M. Rice, in "Physics and Chemistry of Oxide Superconductors", eds. Y. Iye and H. Yasuoka, (Springer Verlag, Berlin 1991) p. 313.
2. C. Berthier *et al.*, Appl. Magn. Reson. **3** (1992) 449.
3. For a review on NMR in the normal state of High T_c superconductors, see C. Berthier, *et al*, J. Phys. (France) **6** (1996) 2025.
4. H. Yasuoka, T. Imai and T. Shimizu., in "Strong Correlations and Superconductivity", eds. H. Fukuyama *et al.*,(Springer Verlag, Berlin, 1989) p. 254.
5. J. Rossat-Mignod *et al.*, Physica C **185-89** (1991) 86.
6. J. Tranquada *et al.*, Phys. Rev. B **46** (1992) 5561.
7. H. Alloul, T. Ohno and P. Mendels, Phys. Rev. Lett. **63** (1989) 1700.
8. F. Mila and T.M. Rice, Phys. Rev. B **40** (1989) 11382.
9. B. Battlog *et al.*, Physica C **235-40** (1994) 130.
10. G.V.M. Williams *et al.*, preprint.
11. T. Tanamoto, H. Kohno and H. Fukuyama, J. Phys. Soc. Jpn. **62** (1993) 2591.
12. A. V. Chubukov and S. Sachdev, Phys. Rev. Lett. **71** (1993) 169; A. Sokol and D. Pines, Phys. Rev. Lett. **71** (1993) 2813.
13. V.J. Emery, S.A. Kivelson and O. Zachar, preprint cond-mat/9610094
14. J. Loram *et al.*, Physica C **235-40** (1994) 134.
15. D.S. Marshall *et al.*, Phys. Rev. Lett. **76** (1996) 4841.
16. H. Ding *et al.*, Nature **382** (1996) 51.
17. Harris, Phys. Rev. B **54**, (1996) R15665.
18. M. H. Julien *et al.*, Phys. Rev. Lett. **76** (1996) 4238; B. L. Altshuler *et al.*, Phys. Rev. B **53** (1996) 1870 and references therein.
19. S. Ohsugi, Y. Kitaoka, K. Ishida and K. Asayama, J. Phys. Soc. Jpn. **60** (1991) 2351.
20. A.J. Millis and H. Monien, Phys. Rev. Lett. **70** (1993) 1870.
21. Y. Itoh *et al.*, preprint.
22. J. Bobroff *et al.*, preprint.
23. Y. Itoh *et al.*, preprint.
24. C. Berthier *et al.*, Physica C **235-40** (1994) 67.
25. D.C. Johnston, Phys. Rev. Lett. 62 (1989) 957.
26. C.H. Pennington *et al.*, Phys. Rev. B **39** (1989) 274.
27. Y. Itoh *et al.*, J. Phys. Soc. Jpn. **61** (1992) 1287.
28. L.P. Regnault *et al.*, Physica C **235-40** (1994) 59.
29. T.E. Mason, G. Aeppli and H.A. Mook, Phys. Rev. Lett. **68** (1992) 1414.
30. G. Aeppli *et al.*, preprint.
31. K. Yamada *et al.*, preprint.
32. M. Horvatić *et al.*, Phys. Rev. B **48** (1993) 13848.
33. R.E. Walstedt, B. S. Shastry, S.-W Cheong, Phys. Rev. Lett. **72** (1994) 3610.
34. C. Berthier *et al.*, Phys. Scr. T **49** (1993) 131.
35. L.P. Regnault, private communication.
36. M.-H. Julien *et al.*, to be published.
37. A. Goto, H. Yasuoka and Y. Ueda, J. Phys. Soc. Jpn **65** (1996) 3043

ELSEVIER

Physica C 282–287 (1997) 231

PHYSICA C

Magnetic coherence in the superconducting and normal states of $La_{2-x}Sr_xCuO_4$

G. Aeppli[a], T.E. Mason[b], H.A. Mook[c], A. Schroeder[d], S.M. Hayden[e]

[a]NEC, Princeton NJ 08540,USA; [b]Univ. of Toronto, Toronto M5S 1A7,Canada; [c]Oak Ridge National Lab.,TN 37831,USA; [d]Riso National Lab., Roskilde 4000, DK; [e]University of Bristol, Bristol BS8 1TL,UK.

We describe magnetic neutron scattering experiments which probe changing magnetic coherence effects in the superconducting and normal states of the single-layer high-Tc material $La_{2-x}Sr_xCuO_4$. In the normal phase, we have measured the temperature dependence of the incommensurate peaks[1] which dominate the magnetic response. While the peak positions do not change, their amplitudes and widths undergo a dramatic temperature-dependent evolution[2]. In particular, the widths, which are inversely proportional to a magnetic coherence length increase in proportion to the temperature and energy transfer added in quadrature, while the amplitudes vary as $1/T^2$. We have thus, for the first time, established the nearly singular behavior of magnetic fluctuations in the normal state of a superconducting cuprate, a result which puts the spin sector more on a par with the charge sector, which has long been known to display the nearly singular properties gathered under the margin Fermi liquid rubric.

In addition to allowing the magnetic fluctuations to be probed in the normal state, magnetic neutron scattering can be used as a spectroscopy of the superconducting gap function. In the latter role, the technique, while suffering from relatively low count rates, has the tremendous advantage of a simple and exactly known cross-section and the ability to access essentially arbitrary momentum and energy transfers. In accord with the BCS theory, within which opposite spin electrons are paired to obtain a collective singlet state, the magnetic response measured by neutron scattering at low frequencies is much reduced on entry of the superconducting state[3]. The momentum dependence of this suppression can be used to establish the momentum dependence of the superconducting gap function[4]. So far, in spite of much effort, neutron scattering has failed to reveal non-trivial Q-dependence at low frequencies, a failure which is likely due to the unclean nature of the large crystals employed so far in neutron experiments. Very recently, however, we have discovered a non-trivial change in the momentum dependence of the magnetic scattering at high frequencies in our rather dirty sample of $La_{1.86}Sr_{0.14}CuO_4$. Specifically, we find that superconductivity causes an increase in the magnetic response above a threshold of 7meV[5]. While sum rule considerations make an increase necessary at some high frequencies and certain momenta because of the need to compensate for the loss of fluctuations at low frequencies, the sharp momentum dependence of the excess scattering found in the superconducting phase comes as surprise. Indeed, the excess is resolution-limited, implying the existence of a new coherence length indistinguishable from infinity which characterizes the superconductivity in our single-layer high-Tc material. Beyond describing the new work on $La_{1.86}Sr_{0.14}CuO_4$, we review earlier results on marginal Fermi liquid phenomenology near the metal-insulator transition[6], our measurements of the complete magnetic fluctuation spectrum for $La_{2-x}Sr_xCuO_4$[7], as well as our work on the both the 41meV 'resonance'[8] and bilayer coupling in $YBa_2Cu_3O_x$[9].

REFERENCES

1. S-W. Cheong et al., Phys.Rev.Lett.67, 1791(1991).
2. G. Aeppli et al., Preprint(1994).
3. T.E. Mason et al., Phys.Rev.Lett.68,1414(1992).
4. T.E. Mason et al., Phys.Rev.Lett.71,919(1993).
5. T.E. Mason et al., Phys.Rev.Lett.77,1604(1996).
6. S.M. Hayden et al., Phys.Rev.Lett.66,821(1991).
7. S.M. Hayden et al., Phys.Rev.Lett.67,3622(1996).
8. H.A. Mook et al., Phys.Rev.Lett.70,3490(1993).
9. S.M. Hayden et al., Phys.Rev.B54,R6905(1996).

PII S0921-4534(97)00221-9

ELSEVIER

Physica C 282–287 (1997) 232–235

PHYSICA C

Doping Dependence of the Magnetic Resonance Peak in $YBa_2Cu_3O_{6+x}$

B. Keimer[a*], H.F. Fong[a], S.H. Lee[b], D.L. Milius[c], I.A. Aksay[c]

[a]Dept. of Physics, Princeton University, Princeton, NJ 08544, USA

[b]Reactor Division, NIST, Gaithersburg, MD 20899, USA

[c]Dept. of Chemical Engineering, Princeton University, Princeton, NJ 08544, USA

We report inelastic neutron scattering experiments on the doping dependence of the energy and spectral weight of the sharp magnetic resonance peak in $YBa_2Cu_3O_{6+x}$. These measurements also shed light on the relationship between the magnetic excitations in the normal and superconducting states.

Magnetic excitations in high temperature superconductors have been intensively studied experimentally and theoretically for a number of years as they provide a direct and incisive probe of correlation effects in the cuprates. These efforts have been redoubled after the discovery of a sharp magnetic collective mode in $YBa_2Cu_3O_7$ by inelastic neutron scattering [1–3]. This mode is strongly coupled to superconductivity in this material; in fact, it is only present in the superconducting state and disappears in the normal state [1]. Two different mechanisms, with various modifications, have been proposed to explain this observation. First, it may be a consequence of the pileup of electronic states above the superconducting energy gap which compensates for the loss of states below the gap. Both a *d*-wave BCS gap function with strong Coulomb correlations [4–6] and the (non-BCS) gap function resulting from the interlayer pair tunneling model of superconductivity [7] can account for the sharpness of the mode in both wavevector $\mathbf{q}$ and energy $\hbar\omega$. Second, superconductivity may provide a matrix element (through particle-hole mixing) that couples a preexisting collective mode to the external probe, magnetic neutron scattering [8].

Further experimental information is clearly necessary in order to distinguish between these fundamentally different interpretations. Since the doping dependence of the superconducting energy gap has recently been determined independently by angle-resolved photoemission [9], the doping dependence of the collective mode may provide additional insights into the mechanism reponsible for coupling of the spin excitations to superconductivity. Further, the normal-state spin susceptibility in $YBa_2Cu_3O_7$ is too small to be reliably determined by present neutron scattering techniques. In underdoped $YBa_2Cu_3O_{6+x}$, however, a sizeable normal-state susceptibility has been observed in previous experiments, so that the relationship between the sharp resonance resonance mode in the superconducting state and the normal-state excitation spectrum can be probed.

The experiments were conducted on the H4M, H7 and H8 thermal triple-axis spectrometers at the High Flux Beam Reactor at Brookhaven and at the SPINS cold-neutron triple axis spectrometer at NIST. At Brookhaven we used neutrons with 30.5 meV final energy and adjusted the energy resolution to $\sim$ 7 meV by collimating the beam. At NIST we used neutrons with 3.5 meV final energy, with typically $\sim$ 1 meV energy resolution. In order to convert the observed intensity to the dynamical susceptibility $\chi''(\mathbf{q},\omega)$ in absolute units, the magnetic cross section was calibrated to transverse acoustic phonons around the (004) nuclear Bragg reflection at low energies and to an oxygen vibration of energy 42.5 meV at high energies.

Our primary interest was the influence of superconductivity on the spin excitation spectrum. In Figs. 1 and 2 we have therefore plotted the

*Supported by NSF-DMR94-00362.

PII S0921-4534(97)00222-0

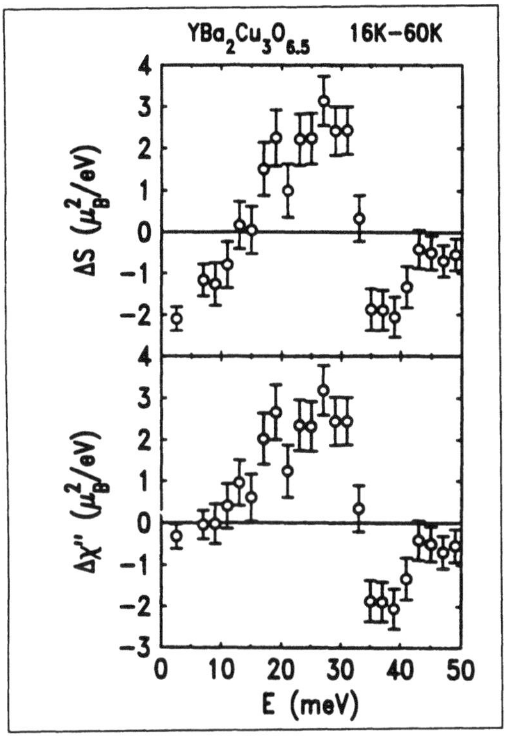

Figure 1. Difference of $S(\mathbf{q},\omega)$ (upper panel) and $\chi''(\mathbf{q},\omega)$ (lower panel) at $\mathbf{q} = (\pi,\pi)$ above and below T_c = 52K for $YBa_2Cu_3O_{6.5}$.

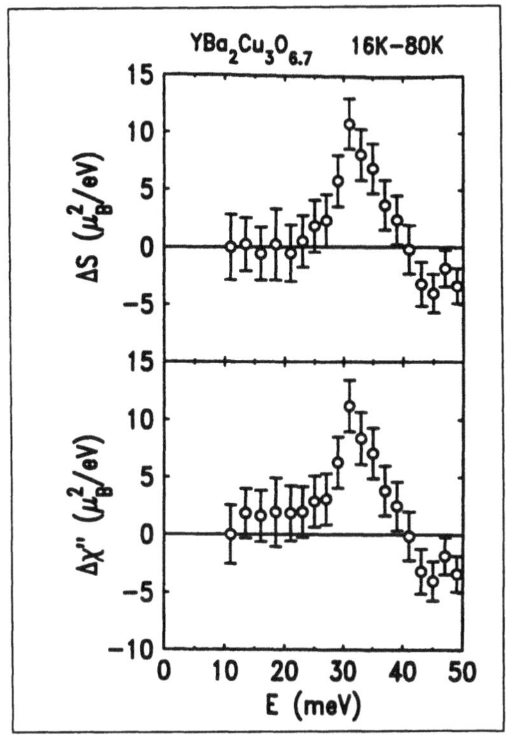

Figure 2. Difference of $S(\mathbf{q},\omega)$ (upper panel) and $\chi''(\mathbf{q},\omega)$ (lower panel) at $\mathbf{q} = (\pi,\pi)$ above and below T_c = 67K for $YBa_2Cu_3O_{6.7}$.

difference of the intensity measured at low temperatures below T_c and the normal-state intensity above T_c for two $YBa_2Cu_3O_{6+x}$ crystals of different oxygen concentrations, x ~ 0.5 (T_c=52K) and x ~ 0.7 (T_c=67K). The data were taken at the in-plane wavevector $\mathbf{q} = (\pi,\pi)$, and at the out-of-plane wavevector corresponding to the maximum of the sinusoidal magnetic structure factor. (For further discussions of the structure factor, see, *e.g.*, Ref. [1].) A subset of these data has been reported in Ref. [10], in arbitrary units. The data of Fig. 2 are also consistent with those of Ref. [11]. Here we have plotted the data in two different ways, both as the scattering function $S(\mathbf{q},\omega)$ which is directly proportional to the magnetic cross section, and as the imaginary part of the dynamical susceptibility which is related to $S(\mathbf{q},\omega)$ through the fluctuation-dissipation theorem: $\chi''(\mathbf{q},\omega) = [1 - exp(-\hbar\omega/k_B T)]S(\mathbf{q},\omega)$. (The unit conventions are the same as in Ref. [12] and differ by $3\pi\mu_B^2/2$ from those of Ref. [1].)

Clearly, the dynamical susceptibility below T_c exceeds that above T_c over a certain energy range, centered around 25 meV for $YBa_2Cu_3O_{6.5}$ and 33 meV for $YBa_2Cu_3O_{6.7}$. Fig. 3 shows that for both doping levels the enhancement of the susceptibility is strongly correlated to the onset of superconductivity. For $YBa_2Cu_3O_{6.7}$ the energy range over which this enhancement is observed is limited only by the experimental resolution. The intrinsic linewidth of the enhanced part of the dynamical susceptibility, $\chi''_+(\mathbf{q},\omega)$, is therefore indistinguishable from zero, as it is for $YBa_2Cu_3O_7$ where $\chi''_+(\mathbf{q},\omega)$ is centered

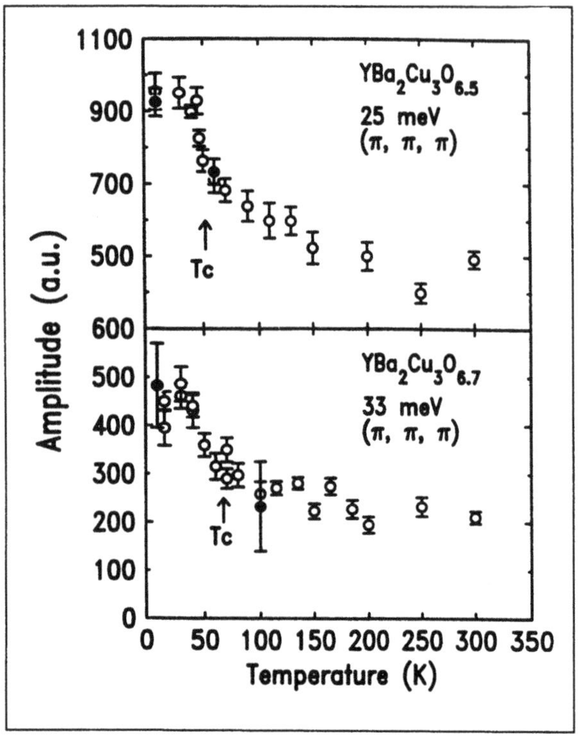

Figure 3. Temperature dependence of $\chi''(\mathbf{q},\omega)$ at $\hbar\omega = 25$ meV for $YBa_2Cu_3O_{6.5}$ (upper panel) and at $\hbar\omega = 33$ meV for $YBa_2Cu_3O_{6.7}$ (lower panel). The closed (open) symbols represent data taken with a polarized (unpolarized) beam.

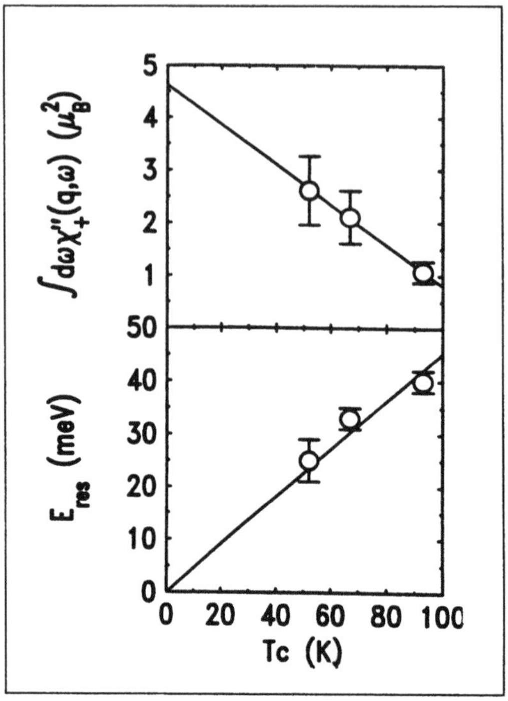

Figure 4. T_c-dependence of the enhanced part of the dynamical susceptibility, $\int d\omega\chi''_+(\mathbf{q},\omega)$, and the resonance energy E_{res}.

around 40 meV. By contrast, $\chi''_+(\mathbf{q},\omega)$ is somewhat broadened in $YBa_2Cu_3O_{6.5}$.

In Fig. 4 we have summarized our observations. The integrated spectral weight of the enhanced part of the susceptibility, $\int d\omega\chi''_+(\mathbf{q},\omega)$, decreases as a function of increasing T_c, whereas E_{res}, the energy around which $\chi''_+(\mathbf{q},\omega)$ is peaked, increases. While Fig. 3 shows that the ratio of the resonance spectral weight to the spectral weight of the normal-state spin excitations increases strongly with increasing doping, it is interesting to note that the resonance spectral weight actually decreases on an absolute scale with increasing carrier concentration. The functional dependence of E_{res} on T_c (or doping, which depends monotonically on T_c) is obviously not well defined by the three data points, but the qualitative trend contrasts sharply with the weak doping dependence of the energy gap directly determined by photoemission spectroscopy [9]. This discrepancy was predicted in the model of Zhang and collaborators [8] where the resonance energy is not tied to the gap but is directly related to the doping level. It does not rule out the gap model, however. Millis and Monien [5] have shown that the resonance energy can be lower than the gap energy due to final state interactions of the quasiparticle-quasihole pair. These interactions are expected to increase as the doping level is reduced. A more clearcut picture may emerge when the present data are compared in detail to quantitative predictions of the resonance energy and integrated spectral weight [5,8].

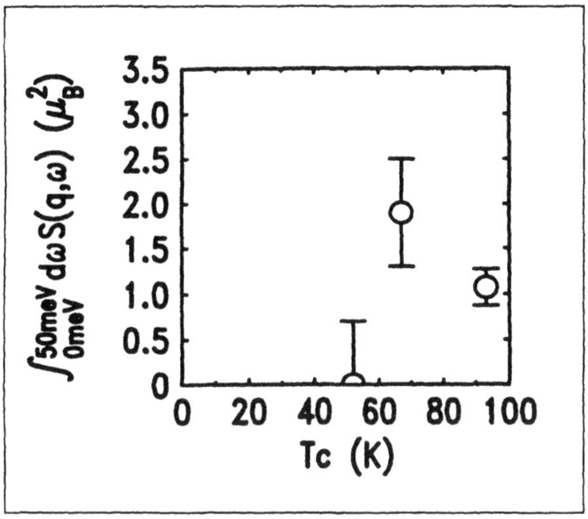

Figure 5. Energy integral of $S(\mathbf{q},\omega)$ at $\mathbf{q} = (\pi,\pi)$ and $0 \leq \hbar\omega \leq 50$ meV, for different doping levels.

Finally, we turn to the relation between the normal-state spin excitations and the resonance in the superconducting state. We are motivated by the total moment sum rule which requires that the the integral $\int d\mathbf{q}d\omega S(\mathbf{q},\omega)$ is weakly temperature dependent (and temperature independent for local-spin models). The difference of $S(\mathbf{q},\omega)$ in the normal and superonducting states is plotted in the upper panels of Figs. 1 and 2 for $\mathbf{q} = (\pi,\pi)$. For $YBa_2Cu_3O_{6.5}$ the positive and negative areas are equal to within our resolution, that is, the total moment sum rule is exhausted for $\mathbf{q} = (\pi,\pi)$ and in the energy range investigated in out experiment ($0 \leq \hbar\omega \leq 50$ meV). By contrast, in both $YBa_2Cu_3O_{6.7}$ and $YBa_2Cu_3O_7$ the positive part of the difference clearly exceeds the negative part in this domain of $(\mathbf{q},\omega)$. This situation is summarized in Fig. 5. For the total moment sum rule to be satisfied, the intensity of the resonance mode must be drawn from a much wider range of $(\mathbf{q},\omega)$. Further investigations, especially at higher energies, are clearly warranted.

REFERENCES

1. H.F. Fong, B. Keimer, P.W. Anderson, D. Reznik, F. Dogan and I.A. Aksay, Phys. Rev. Lett. 75 (1995) 316; B. Keimer, H.F. Fong, D. Reznik, F. Dogan and I.A. Aksay, J. Phys. Chem. Solids 56 (1995), 1927; H.F. Fong, B. Keimer, D. Reznik, D. L. Milius and I.A. Aksay, Phys. Rev. B 54 (1996) 6708.
2. P. Bourges, L.P. Regnault, Y. Sidis and C. Vettier, Phys. Rev. B 53 (1996) 876.
3. For earlier neutron scattering work on the 40 meV mode, see J. Rossat-Mignod *et al.*, Physica C 185-189 (1991), 86; H.A. Mook *et al.*, Phys. Rev. Lett. 70 (1993) 3490.
4. D.Z. Liu, Y. Zha and K. Levin, Phys. Rev. Lett. 75 (1995) 4130; I.I. Mazin and V.M. Yakovenko, *ibid.* 75 (1995) 4134; F. Onufrieva, Physica C 251 (1995) 348; Y. Zha, V. Barzykin and D. Pines, Phys. Rev. B 54 (1996) 7561; N. Bulut and D.J. Scalapino, *ibid.* 53 (1996) 5149; G. Blumberg, B.P. Stojkovic and M.V. Klein, *ibid.* 52 (1995) 15741.
5. A.J. Millis and H. Monien, Phys. Rev. B 54 (1996) 16172.
6. For earlier related theoretical work, see K. Maki and H. Won, Phys. Rev. Lett. 72 (1994) 1758; P. Monthoux and D.J. Scalapino, *ibid.* 72 (1994) 1874; H. Fukuyama, H. Kohno and T. Tanamoto, J. Low Temp. Phys. 95 (1994) 309; F. Onufrieva and J. Rossat-Mignod, Phys. Rev. B. 52 (1995) 7572.
7. L. Yin, S. Chakravarty and P.W. Anderson, Report No. cond-mat/9606139; P.W. Anderson, J. Phys: Cond. Matter 8 (1996) 10083.
8. E. Demler and S.C. Zhang, Phys. Rev. Lett. 75 (1995) 4126; S.C. Zhang, Science 275 (1997) 1089.
9. J.M. Harris *et al.*, Phys. Rev. B 54 (1996) R15665.
10. H.F. Fong, B. Keimer, D.L. Milius and I.A. Aksay, Phys. Rev. Lett. 78 (1997) 713.
11. P. Dai, M. Yethiraj, H.A. Mook, T.B. Lindemer and F. Dogan, Phys. Rev. Lett 77 (1996) 5425.
12. S.M. Hayden *et al.*, Phys. Rev. Lett. 76 (1996) 1344.

PHYSICA C

ELSEVIER

Physica C 282–287 (1997) 236–239

Phase Separation, Pseudogap and Impurity Scattering in the HTS Cuprates

J.L. Tallon[1], G.V.M. Williams[1], N.E. Flower[1] and C. Bernhard[2]

[1]New Zealand Institute for Industrial Research, P.O. Box 31310, Lower Hutt, New Zealand.
[2]Max Planck Institüt für Festkörperforschung, D-70506, Stuttgart, Germany.

NMR and phase studies on a wide range of cuprates as a function of hole doping reveal many new systematic features. The 60K plateau in 123-like materials is consistent with generic electronic phase separation rather than oxygen vacancy ordering. The normal-state pseudogap has a d-wave-like character and the same energy scale as superconductivity. Contrary to ARPES data, the pseudogap energy is T-independent and does not vanish at T=T*. The suppression of T_c by Zn substitution is consistent with unitarity scattering with a d-wave order parameter in the presence of a pseudogap.

1. INTRODUCTION

Recently, attention has focussed on the underdoped region ($p<p_{opt}$) of HTS cuprates where correlations of some sort result in a gap in the normal-state density of states (DOS) widely referred to as the *pseudogap*. Because of its universal occurrence amongst HTS the pseudogap is evidently of central importance to an understanding of HTS, however little is yet known about its systematics or origin. It is generally accepted that the pseudogap opens below a temperature T* as suggested by recent ARPES studies [1] resulting in an enhanced thermopower, diminished resistivity, a temperature-dependent Hall coefficient and a depression in the electronic entropy and spin susceptibility [2]. Here, (i) we suggest that the underdoped cuprates exhibit a common phase separation for p<0.11 which underlies the 60K plateau in Y-123; (ii) the cuprates exhibit a universal scaled phase behaviour in T_c, the pseudogap and spin-glass freezing; (iii) we show from NMR Knight shifts that the pseudogap is not temperature dependent, opening at T*, but rather T* is just the energy scale for the gap; and (iv) using a simple pseudogap model we quantitatively explain the systematics of the depression in T_c by impurity scattering.

2. THE "60K PLATEAU" IN Y-123

We have investigated $T_c(\delta)$ for $YBa_{2-z}La_zCu_3O_{7-\delta}$ and $Y_{1-y}Ca_yBa_2Cu_3O_{7-\delta}$ with z=0.2 and 0.1 and y=0, 0.1 and 0.2. The series follows a progressive increase in hole concentration, p, for a fixed value of δ and we find each exhibits a "60K plateau" centred at δ = 0.11, 0.22, 0.45, 0.62 and 0.73, respectively. The plateau cannot therefore be associated primarily with oxygen ordering at δ=0.5 although in the pure compound this clearly has an additional effect. Using measurements of the room-temperature thermoelectric power we have determined p for each La- and Ca-substituted compound [3] across a range of δ-values and $T_c(p)$ for each is plotted in Fig. 1. For the sake of clarity each curve has been displaced up or down by 20K from the adjacent curve. The plateaux in $T_c(\delta)$, which are so widely separated in δ, are not only preserved in $T_c(p)$ but come together at p≈0.11 showing that the plateau is a p-dependent effect not a δ-dependent effect. It may be related to incipient stripe formation near p=0.125 as found in $La_{2-x}Ba_xCuO_4$. We note that at p=0.11 many properties exhibit dramatic changes. The superconducting gap rapidly fills in and

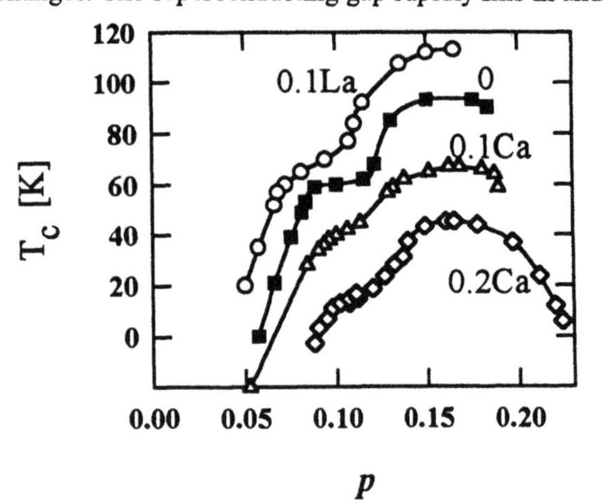

Fig. 1. T_c versus p for La- and Ca-substituted Y-123. Curves are each displaced apart by 20K for clarity.

PII S0921-4534(97)00224-4

both the spin susceptibility, χ_s, determined from NMR studies and S/T (where S is the electronic entropy) suddenly become quite temperature independent. At this point NMR line shapes abruptly narrow. Starting also at p=0.11 there occurs a spin-glass phase where antiferromagnetism and superconductivity coexist (see Fig. 2) [4]. The constancy of T_c, the flatness of χ_s and the coexisting magnetism and superconductivity all point to a picture of dynamic phase separation which appears to be common to La-214 and Y-123. The appearance of a miscibility gap in $La_2CuO_{4+\delta}$ for $0 \le \delta \le 0.055$ [5] seems to confirm this and we note that p=0.11 at the upper limit. On the other hand, for p>0.11 the detailed changes in thermodynamic properties for these systems point to a homogeneous electronic system there [4].

3. COMMON SCALED PHASE BEHAVIOUR

In order to compare the phase diagram of Y-123 with La-214 it is necessary to determine the hole concentration in the former, a process which is complicated by the uncertain distribution of carriers between the chains and the planes. Bond valence sums calculated from atomic-coordinate refinements from neutron diffraction and incorporating both copper and oxygen contributions have been used successfully to determine p for the CuO_2 planes [3]. In Fig. 2 we compare the phase diagrams of (a) $La_{2-x}Sr_xCuO_4$ with p=x and (b) $Y_{1-y}Ca_yBa_2Cu_3O_{7-\delta}$ with p=x/2 for samples with δ=1 and $p=V\_=2+V_{Cu}-V_{O2}-V_{O3}$ for samples with $\delta<1$. The spin-glass freezing temperatures T_g are determined from μSR studies [4]. The phase behaviour for T_c and T_g appears identical in each case and just scales with $T_{c,max}$ as does the pseudogap energy as shown in §4. The 60K plateau is also evident in the neutron data thus confirming the results in §2 obtained from thermopower estimates of p.

4. THE PSEUDOGAP DOES NOT VANISH AT T*

Recent ARPES measurements [1] appear to confirm the common view that in the underdoped cuprates the pseudogap opens below a temperature $T^* > T_c$. This picture is given substance by the model of Emery and Kivelsen [6] which proposes that the pseudogap correlations are short-range pairing correlations which set in below a mean-field transition temperature T* but do not develop into long-range phase coherence until a lower temperature T_c due to phase fluctuations. Our NMR studies show that this cannot be the case. It is easily shown that if χ_s is the spin susceptibility then above T_c the temperature dependence of $\chi_s \times T$ will be

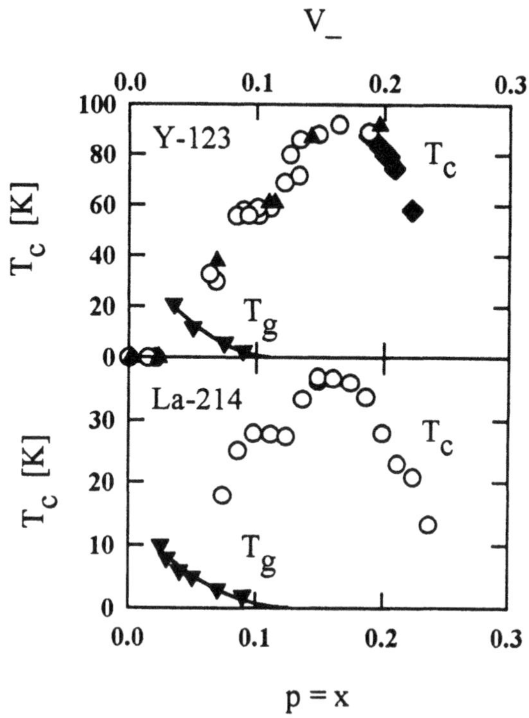

Fig. 2. T_c and T_g versus p or V_ for Y-123 and La-214.

linear through the origin. With the opening of a d-wave-like pseudogap with gap energy E_g the product χ_sT is depressed downwards due to the reduced DOS. If E_g is temperature independent χ_sT is displaced downwards in parallel fashion while if the pseudogap opens at T* then χ_sT returns to the parent linear curve, meeting it at T*. The latter situation is not observed in Knight shift measurements for any of a broad range of HTS cuprates we have investigated. With progressive underdoping these curves form a series of downward-moving parallel lines representing the increasing pseudogap energy. Similar behaviour has been demonstrated for the entropy, S, in Y-123 [7]. If indeed E_g is temperature independent (at least over the range 0 to 300K) then χ_s should scale to a single universal curve for all p and T. The susceptibility is given by

$$\chi_s = \mu_B^2 \int g(E)\,[-df/dE]\,dE \qquad (1)$$

where g(E) is the 2-D DOS and f(E/kT) is the Fermi function. Taking g(E) for a d-wave pseudogap with a temperature-independent E_g one finds that $\chi_s(T) = \chi_o F(kT/E_g)$ and thus χ_s has a universal dependence upon kT/E_g. Fig. 3 shows that this is indeed the case. The figure shows Knight shift data for a broad range of HTS cuprates (Y-123, Y-124, Y-247, (Y,Ca)-123,

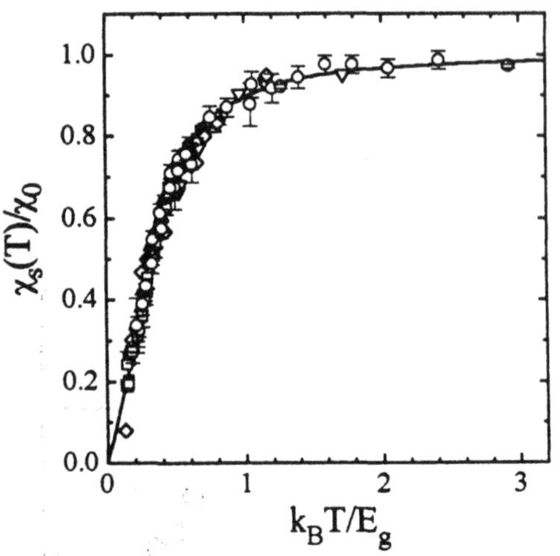

Fig. 3. Scaling of the spin susceptibility versus kT/E_g.

La-2126, Tl-2212, Hg-1201 and Hg-1223) over a wide range of hole concentrations. The scaling behaviour is exceptionally good. Notably also, the linearity of χ_s at low temperature is a clear signature of the d-wave-like DOS for the normal-state pseudogap thus confirming earlier ARPES [1] and NMR [8] data. E_g values, scaled by $T_{c,max}$, fitted for these different cuprates at different doping levels are plotted in Fig. 4. Also plotted is the superconducting energy gap Δ_o determined from Raman [9] and heat capacity studies [10]. It can be seen that the $E_g/kT_{c,max}$ values lie on a single curve showing that the energy scale for the pseudogap is the same as that for superconductivity. In spite of this, and their same symmetry, E_g and Δ_o have quite different

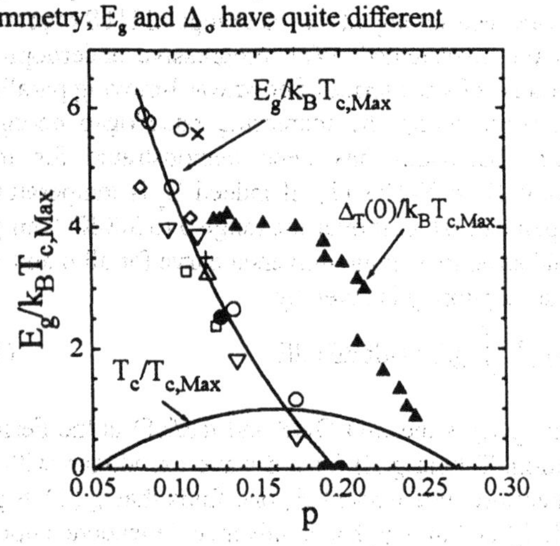

Fig. 4. E_g, T_c and Δ_o scaled by $T_{c,max}$ plotted versus hole concentration for a variety of different HTS cuprates.

doping dependences thus showing that the pseudogap correlations cannot be the same pairing correlations that underlie superconductivity. We note that a charge-density wave competing with superconductivity may have the same energy scale and exhibits a phase diagram very similar to that shown in Fig. 4 [11].

5. PSEUDOGAP AND IMPURITY SCATTERING

The rapid suppression of superconductivity with substitution of Zn for Cu has been the subject of continuing speculation since the discovery of HTS cuprates. In spite of this the p-dependent systematics across the superconducting phase curve do not appear to have been fully investigated. We have synthesized two series which span the phase curve: (a) $La_{2-x}Sr_xCuO_4$ with 0, 1, 2 and 3% Zn substitution and (b) $Y_{0.8}Ca_{0.2}Ba_2Cu_3O_{7-\delta}$ with 0, 2, 4 and 6% Zn substitution. Fig. 5 shows the p-dependence of T_c for each. In the former case p=x while, in the latter, p is determined from thermoelectric power measurements. The roughly parabolic phase curves are seen to collapse around the pseudogap line $E_g(p)$ shown by the solid line in Fig. 5 (b) and the dashed line in 5 (a) which, in the absence of direct data for La-214, is simply estimated by scaling with $T_{c,max}$ as indicated in §4. Both data sets indicate that the assymetric collapse of the phase curves is associated with the presence of the pseudogap on the underdoped

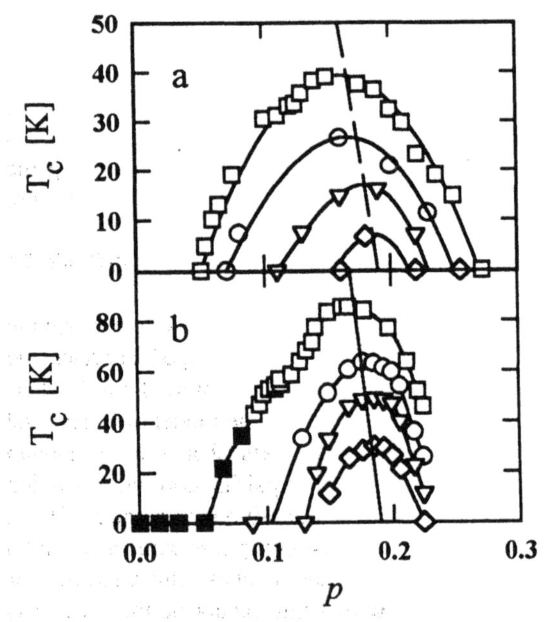

Fig. 5. T_c versus p for Zn doped La-214 and (Y,Ca)-123.

side. In Fig. 6 we plot our experimental data for the p-dependence of $[dT_c/dy]_{y=0}$ for $Y_{0.8}Ca_{0.2}Ba_2Cu_{3-3z}Zn_{3z}O_{7-\delta}$ where y=1.5z is the fraction of Zn on the CuO_2 planes. In the overdoped region dT_c/dy remains constant but, on reducing hole concentration starting at p=0.19, begins to rise sharply due to the appearance of the pseudogap there. We have modelled this behaviour using the Sun and Maki theory [12] for d-wave impurity scattering in the unitarity limit. Here the scattering rate $\Gamma = n_i/\pi g(E_F)$ where $g(E_F)$ is the DOS per spin at the Fermi level and n_i (=2y/abc) is the volume density of impurity scatterers.

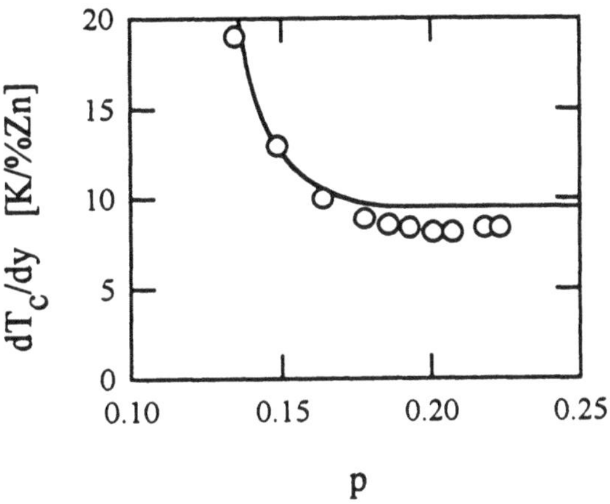

Fig. 6. The rate of reduction of T_c with Zn content plotted as a function of p for (Y,Ca)-123. The solid curve is as calculated in the text.

For weak coupling the critical scattering rate Γ_c=0.88T_{c0}=0.4122Δ_{00} but many spectroscopies [9,10] indicate that HTS are in the strong-coupling regime with $2\Delta_o/k_BT_c \approx 8$. We therefore take $\Gamma_c = 0.4122\Delta_{00}$ = 1.649T_{c0}. The DOS may be determined from the temperature coefficient of the heat capacity, $\gamma = C_V/T$:

$$\gamma = (2/3)\pi^2 k_B^2 g(E_F)\,. \quad (2)$$

Thus, for $Y_{0.8}Ca_{0.2}Ba_2Cu_3O_{7-\delta}$ we find Γ = 34.8 (y/γ) [K] where γ is in units of J/mol.K^2 and dT_c/dy = 15.2/γ. From heat capacity studies on these samples [10] in the overdoped region where the pseudogap is absent γ = (13-δ) × 1.9 mJ/mol.K^2. Thus dT_c/dy = 9.53 K/%, in quite good agreement with the observed value 8.35 K/%. In the underdoped region we account for the pseudogap by noting that if Δ_o is interpreted as the magnitude of the order parameter, Δ_o' then this is reduced by the presence of the pseudogap. If the quasiparticle excitations are modelled by $E^2 = \epsilon_k^2 + \Delta^2 = \epsilon_k^2 + \Delta'^2 + E_g^2$ then $\Delta'^2 = \Delta^2 - E_g^2$. Thus dT_c/dy = 15.2/γ ≈ 15.2/[1.9(13-δ)$\sqrt{1-(E_g/\Delta_o)^2}$]. We note from §4 that E_g/Δ_o may be a universal function of p for all HTS. Heat capacity measurements on these samples show Δ_o=320K across the underdoped region. Taking this and the p-dependence of E_g shown in Figs. 4 or 5(b) we obtain the solid curve for dT_c/dy plotted in Fig. 6. This gives excellent agreement with the observed data.

This result resolves the long-standing debate over the interpretation of impurity doping on the CuO_2 planes. The p-dependent systematics become at once quite simple although an understanding of the origins and nature of the pseudogap is still lacking. One clear implication is that there is a well-defined hole concentration at which the pseudogap opens, namely p≈0.19. This negates models of the pseudogap in which a supposed temperature T*, at which the pseudogap opens, merges with the T_c(p) phase curve on the overdoped side. Finally, we note the striking correspondence between the pseudogap line E_g(p), shown in Figs. 4 and 5(b), and the crossover line between insulating and metallic conduction observed by Boebinger *et al* [13] by suppressing superconductivity in ultra-high pulsed magnetic fields. This crossover appears to be consistent with thermal activation over a d-wave pseudogap.

One of the authors (JLT) acknowledges Royal Society of New Zealand funding under a James Cook Fellowship.

REFERENCES

1. J.M. Harris *et al.*, Phys. Rev. B **54**, R15665 (1996).
2. J.L. Tallon *et al.*, Phys. Rev. Lett. **75**, 4114 (1996).
3. J.L. Tallon *et al.*, Phys. Rev. B **51**, 12911 (1995).
4. Ch. Niedermayer *et al.*, (to be published); J.L.Tallon *et al.*, Sup. Sci.Tech. (in press).
5. P.G. Radaelli *et al.*, Phys. Rev. B **49**, 6239 (1994).
6. V.J. Emery and S.A. Kivelson, Nature **374**, 434 (1995).
7. J.W. Loram *et al.*, Phys. Rev. Lett. **71**, 1740 (1993).
8. G.V.M. Williams *et al.*, Phys. Rev. Lett. **78**, 721 (1997).
9. M. Kang *et al.*, Phys. Rev. Lett. **77**, 4434 (1996).
10. J.W. Loram *et al.*, Proc. M²S-HTSC-V, Beijing, March 1997.
11. C.A. Balsiero and L.M. Falicov, Phys. Rev. B **20**, 4457 (1979).
12. Y. Sun and K. Maki, Phys. Rev. B**50**, 6059 (1995).
13. G.S. Boebinger *et al.*, Phys. Rev. Lett. **77**, 5417 (1996).

ELSEVIER

Physica C 282–287 (1997) 240–243

PHYSICA C

Low-temperature normal-state resistivity of high-T_c cuprates

Yoichi Ando[a*], G.S. Boebinger[a], A. Passner[a], N.L. Wang[b], C. Geibel[b], F. Steglich[b], T. Kimura[c], M. Okuya[c], J. Shimoyama[c], K. Kishio[c], K. Tamasaku[d], N. Ichikawa[d], and S. Uchida[d]

[a]Bell Laboratories, Lucent Technologies, Murray Hill, New Jersey 07974, USA

[b]Institut für Festkörperphysik, Technische Hochschule Darmstadt, D-64289 Darmstadt, Germany

[c]Department of Applied Chemistry, University of Tokyo, Hongo, Tokyo 113, Japan

[d]Superconductivity Research Course, University of Tokyo, Yayoi, Tokyo 113, Japan

The results of low-temperature normal-state resistivity measurements of $La_{2-x}Sr_xCuO_4$ (LSCO) and $Bi_2Sr_{2-x}La_xCuO_y$ (Bi-2201) using 61-T pulsed magnetic fields are summarized. Measurements of LSCO with various Sr doping reveal (1) an insulator-to-metal crossover in the normal state that takes place near optimum doping and (2) unusual $\log(1/T)$ divergence of both ρ_{ab} and ρ_c of insulating underdoped samples. In particularly clean Bi-2201 crystals, metallic in-plane resistivity and insulating c-axis resistivity are found to coexist down to the lowest experimental temperature. Recent measurements of the low-temperature normal-state Hall effect on Bi-2201 find an essentially temperature-independent Hall coefficient below 20 K, which suggests that the unusual insulating behavior does not come from conventional disorder-enhanced electron interactions.

1. Introduction

The in-plane resistivity ρ_{ab} of high-T_c cuprates near optimum doping shows a metallic behavior where ρ_{ab} changes linearly with temperature, while the out-of-plane resistivity ρ_c often shows "semiconducting" behavior. Whether or not this contrasting behavior of normal-state ρ_{ab} and ρ_c extends below T_c has been an unanswered question, mostly due to the extremely high H_{c2} of the cuprates. By suppressing superconductivity with the use of a pulsed magnetic field of 61 T, we have successfully measured the normal-state ρ_{ab} and ρ_c of $La_{2-x}Sr_xCuO_4$ (LSCO) single crystals and La-doped $Bi_2Sr_2CuO_y$ (Bi-2201) single crystals down to 0.66 K. The Bi-2201 data suggests a two-dimensional non-Fermi-liquid ground state of the cuprate [1], while the insulating divergence observed in LSCO [2] might arise from an unusual quasiparticle localization mechanism and/or suppression of the two-dimensional density-of-states near the Fermi energy [3,4]. To investigate the origin of this insulating behavior, we have set out to measure the normal-state Hall effect down to low temperatures using the 61-T magnetic field pulses, since the Hall effect is known to be useful in separating weak localization from electron interaction effects in disordered metals [5]. In this paper, we summarize the result of our measurements of the low-temperature normal-state resistivity of both LSCO and Bi-2201 and present our recent result of low-temperature Hall measurements of Bi-2201.

*Present address: Central Research Institute of Electric Power Industry, Komae, Tokyo 201, Japan.

2. Experimental

The LSCO single crystals are grown by the floating-zone method with variety of x values. Here we present, as a summary, the data for Sr concentrations x=0.08 (underdoped), 0.15 (slightly underdoped), and 0.22 (overdoped). The high quality crystals with x=0.08 and 0.22 were grown by the Kishio group [6] and samples with x=0.15 were grown by the Uchida group [7].

The Bi-2201 single crystals are grown by the flux method [8] at Darmstadt. To obtain nearly optimally doped crystals, La is doped onto the Sr site, giving the composition of $Bi_2Sr_{2-x}La_xCuO_y$.

0921-4534/97/$17.00
PII S0921-4534(97)00226-8

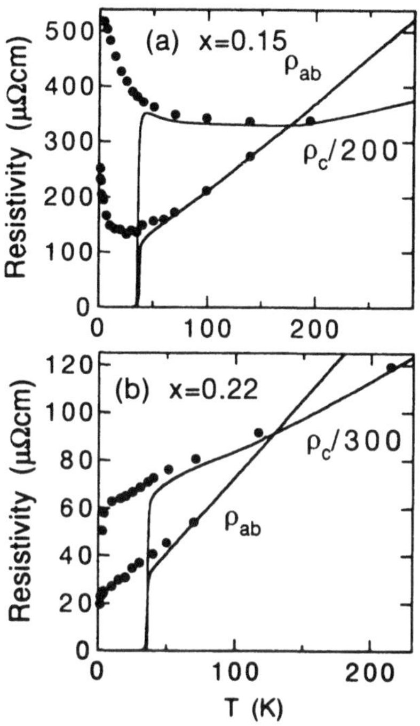

Figure 1. Normal-state ρ_{ab} and ρ_c (scaled as indicated) of LSCO crystals with (a) x=0.08 and (b) x=0.22 to represent the behavior of underdoped and overdoped samples, respectively. The lines (circles) are 0 T (60 T) data

Here we concentrate on the samples of nominal x=0.05 (actual concentration is about 0.3), in which T_c (midpoint) is around 13 K, which corresponds to slightly overdoped samples.

The techniques to measure the anisotropic resistivity in pulsed magnetic fields have been described in Ref. [1,2]. Magnetic fields are applied parallel to the c axis. For the Hall measurements, longitudinal and transverse voltages are measured simultaneously. Two magnetic-field sweeps with opposite polarities are taken for each temperature. We interpret the asymmetrical component of the transverse voltage from the two sweeps as the Hall voltage. In the absence of superconductivity ($T > T_c$) the Hall voltage is linear in magnetic field, giving confidence in the Hall data.

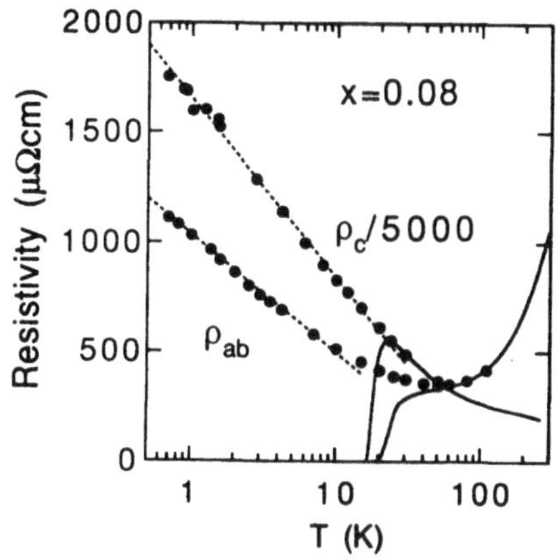

Figure 2. ρ_{ab} and ρ_c (scaled as indicated) vs $\log T$ for x=0.08 LSCO crystal. The lines (circles) are 0 T (60 T) data and the dotted lines emphasize the $\log(1/T)$ dependence.

3. Results and Discussion

Figures 1(a) and 1(b) present ρ_{ab} and ρ_c of LSCO crystals with x=0.15 and 0.22, respectively. Figure 1(a) represents the behavior of underdoped samples, in which both ρ_{ab} and ρ_c show insulating behavior at low temperatures when the superconductivity is suppressed with the 61-T magnetic field. For x=0.15 [Fig. 1(a)], ρ_{ab} in zero field shows the linear temperature dependence above T_c and yet shows an insulating behavior below 20 K. On the other hand, in overdoped samples where both ρ_{ab} and ρ_c are metallic above T_c, the low-temperature normal-state ρ_{ab} and ρ_c stay metallic down to low temperatures [Fig. 1(b)]. Both the normal-state in-plane and c-axis resistivities cross from insulating to metallic behavior near optimum doping ($x\simeq0.16$) as the carrier concentration is increased [9].

One of our findings in LSCO is that the insulating normal-state resistivity of underdoped samples changes as $\log(1/T)$ at low temperatures [2]. This $\log(1/T)$ dependence is found in *both* ρ_{ab} and ρ_c, which is represented in Fig. 2 for x=0.08. All superconducting underdoped samples show the logarithmic temperature dependence in both ρ_{ab} and ρ_c [4]. Note that non-superconducting

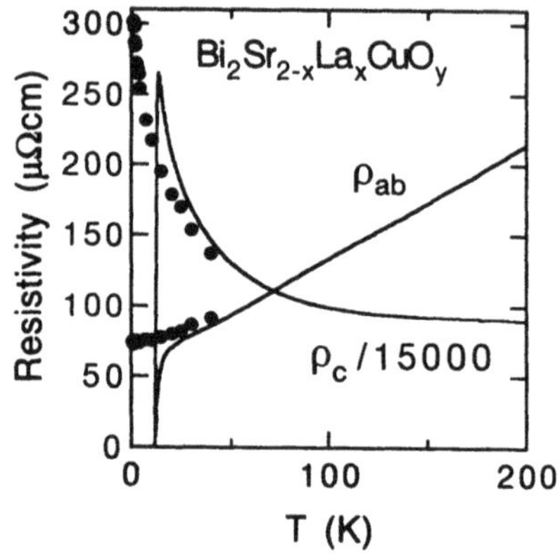

Figure 3. Normal-state ρ_{ab} and ρ_c (scaled as indicated) of a clean Bi-2201 crystal. The lines (circles) are 0 T (50 T) data. The contrasting behavior of the normal-state ρ_{ab} and ρ_c persists to the lowest experimental temperature in this sample.

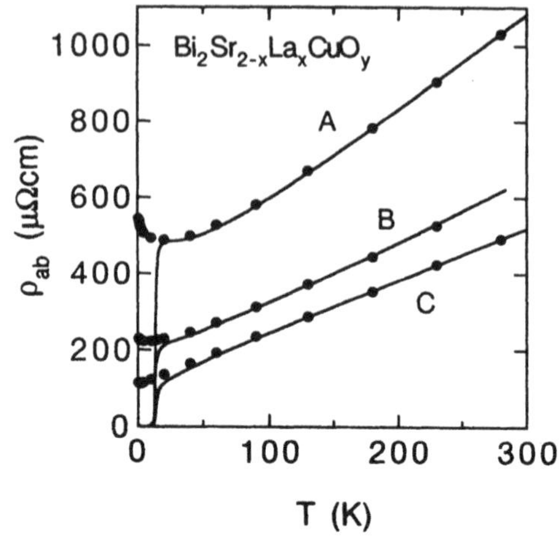

Figure 4. Normal-state ρ_{ab} of three Bi-2201 crystals with the same La concentration (nominally 0.05). The lines (circles) are 0 T (57 T) data.

samples ($x \leq 0.05$) [3] and weakly superconducting samples ($T_c < 1$ K) [10] show rapidly diverging resistivity which is consistent with variable-range hopping (VRH) conduction. The $\log(1/T)$ dependence is observed only in strongly superconducting samples in the underdoped regime.

In slightly overdoped Bi-2201, qualitatively different behavior is observed. Figure 3 shows ρ_{ab} and ρ_c of the least resistive sample (out of more than ten samples measured). In this sample, ρ_{ab} stays metallic down to the lowest experimental temperature, 0.66 K, while ρ_c continues to diverge. This observation strongly suggests that the metallic in-plane conduction and the insulating out-of-plane conduction can indeed coexist in the zero-temperature limit in high-T_c cuprates, although we can not exclude the possibility that ρ_{ab} will cross over to insulating behavior below the experimental temperature range. The Bi-2201 crystals tend to have naturally introduced disorders, which have been discussed to be the oxygen vacancies in the CuO_2 planes [11], and ρ_{ab} varies significantly even among crystals grown in the same crucible. Figure 4 shows ρ_{ab} of three different crystals, all have the same nominal composition (x=0.05). The ρ_{ab} of sample C, which is nearly as clean as the crystal of Fig. 3, stays metallic down to the lowest temperature. In apparently more disordered samples A and B, ρ_{ab} at low temperatures exibits insulating behavior, of which the temperature dependence is consistent with $\log(1/T)$. This result suggests that the logarithmically divergent ρ_{ab} arises, in part, from the presence of disorder in the sample [4].

The normal-state Hall effect was measured for samples A, B and C at the same time as the resistivity data of Fig. 4. Figure 5 shows the Hall coefficient R_H at 57 T of the three samples. At high temperatures, all the samples show a broad peak at around 100 K, consistent with previous works [11]. What is remarkable is that R_H is essentially independent of temperature below roughly 20 K within the experimental resolution for all three samples, despite the qualitative difference in the behavior of ρ_{ab}.

Although it is not clear how to properly interpret the low-temperature Hall coefficent in Bi-2201 [12], for disordered 2D metals, Hall coefficients in the presence of weak localization effects and electron interaction effects have been calculated [5]. Weak localization gives a temperature-independent R_H, while disorder-enhanced elec-

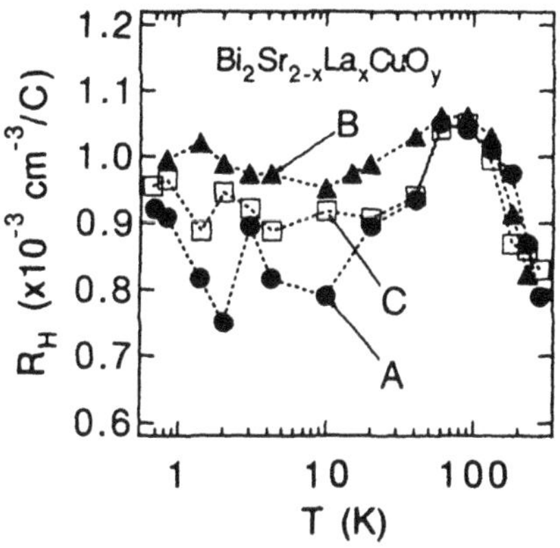

Figure 5. T dependence of the Hall coefficient at 57 T from the three Bi-2201 crystals of Fig. 4.

tron interactions suppress the density of states at the Fermi level, giving an R_H that diverges as $\log(1/T)$. A logarithmically diverging ρ_{ab} accompanied by a temperature-independent R_H is not consistent with the conventional electron interaction model.

4. Conclusion

In LSCO, both ρ_{ab} and ρ_c show a crossover from insulating to metallic behavior near optimum doping $x \simeq 0.16$ [9] and the insulating ρ_{ab} and ρ_c of underdoped samples show the characteristic $\log(1/T)$ divergence [2,4]. In Bi-2201, the contrasting behavior of metallic ρ_{ab} and insulating ρ_c can coexist down to the lowest experimental temperature, 0.66 K, when the sample resistivity is small enough [1]. In increasingly disordered crystals, the low-temperature ρ_{ab} show increasingly strong insulating behavior. Despite the qualitative difference in ρ_{ab} with changing disorder, the normal-state Hall coefficient R_H is essentially temperature-independent at low temperatures. Such temperature independence of R_H suggests that the unusual insulating behavior does not come from conventional disorder-enhanced electron interactions.

REFERENCES

1. Y. Ando, G.S. Boebinger, A. Passner, N. L. Wang, C. Geibel, and F. Steglich *Phys. Rev. Lett.* **77**, 2065 (1996).
2. Y. Ando, G.S. Boebinger, A. Passner, T. Kimura, and K. Kishio, *Phys. Rev. Lett.* **75**, 4662 (1995).
3. Y. Ando, G.S. Boebinger, A. Passner, R.J. Cava, T. Kimura, J. Shimoyama, and K. Kishio, in *Proceedings of 10th Anniversary HTS Workshop on Physics, Materials and Applications*, ed. by B. Batlogg, C.W. Chu, W.K. Chu, D.U. Gubser and K.A. Mueller (Singapore, World Scientific, 1996).
4. Y. Ando, G.S. Boebinger, A. Passner, K. Tamasaku, N. Ichikawa, S. Uchida, M. Okuya, T. Kimura, J. Shimoyama, and K. Kishio, *J. Low Temp. Phys.* **105**, 867 (1996).
5. B. L. Altshuler and A. G. Aronov, in *Electron-Electron Interactions in Disordered Systems*, ed. M. Pollak and A. L. Efros (North-Holland, 1985), pp.1-153.
6. T. Kimura, K. Kishio, T. Kobayashi, Y. Nakayama, N. Motohira, K. Kitazawa, and K. Yamafuji, *Physica C* **192**, 247 (1992).
7. Y. Nakamura and S. Uchida, *Phys. Rev. B* **47**, 8369 (1993).
8. N.L. Wang, B. Buschinger, C. Geible, and F. Steglich *Phys. Rev. B* **54**, 7449 (1996).
9. G.S. Boebinger, Y. Ando, A. Passner, K. Tamasaku, N. Ichikawa, S. Uchida, M. Okuya, T. Kimura, J. Shimoyama, and K. Kishio, *Phys. Rev. Lett.* **77**, 5417 (1996).
10. K. Karpinska, A. Malinowski, M.Z. Cieplak, G. Guha, S. Gershman, G. Kotliar, T. Skoskiewicz, W. Plesiewicz, M. Berkowski, and P. Lindenfeld, *Phys. Rev. Lett.* **77**, 3033 (1996).
11. A.P. Mackenzie, S.D. Hughes, J.R. Cooper, A. Carrington, C. Chen, and B.M. Wanklyn, *Phys. Rev. B* **45**, 527 (1992).
12. For example, in the Luttinger liquid model, it has been suggested that R_H is constant at temperatures sufficiently low that spin-charge separation does not occur: P.W. Anderson, T.V. Ramakrishnan, S. Strong, and D.G. Clarke, *Phys. Rev. Lett.* **77**, 4241 (1996).

ELSEVIER

Physica C 282–287 (1997) 244–247

PHYSICA C

Unusual Field Dependence of Thermal Conductivity in $La_{2-x}Sr_xCuO_4$: Suppression of Quasi-Particle Coherence.

N. P. Ong, * K. Krishana [a] and T. Kimura [b] †

[a]Joseph Henry Laboratories of Physics, Princeton University, Princeton, New Jersey 08544, U.S.A.

[b]Department of Applied Chemistry, University of Tokyo, 7-3-1 Hongo, Bunkyo-ku, Tokyo 113, Japan

The anomaly in the thermal conductivity κ of $La_{2-x}Sr_xCuO_4$ is very sensitive to field B. Below 14 Tesla, it is quite well described by $Y(x) = \psi(\frac{1}{2}+\frac{1}{x}) + \ln(x)$ where $\psi(x)$ is the digamma function. The experiment suggests that the field suppression is a quantum decoherence effect involving quasi-particles diffusing in the disordered CuO_2 plane. Understanding how B affects qp diffusion may provide a sensitive probe of coherence effects using the heat current.

The normal state of the cuprate superconductors is characterized by an intense inelastic scattering rate of the charge carriers. Below T_c, evidence from transport [1–3] and photoemission [4] show that the intense rate is strongly suppressed. This striking anomaly distinguishes the cuprates from low-T_c superconductors. While the transport properties above T_c are well studied, the behavior of quasi-particles (qp) below T_c is difficult to investigate because of the strong screening effects of the Meissner current. Among the few tools available, the thermal conductivity may be the most useful because it is a probe of the bulk.

In most of the cuprates a prominent anomaly is observed in the in-plane thermal conductivity κ below T_c. This peak is well investigated in optimum $YBa_2Cu_3O_7$ (YBCO),[2,3] which represents the clean limit in bilayer cuprates. We report a detailed investigation on the field dependence of κ in the single-layer cuprate $La_{2-x}Sr_xCuO_4$ (LSCO). In LSCO, the strong disorder severely limits the quasi-particle mean-free-path (mfp). Zero-field measurements by Nakamura et al. [5] on crystals over a wide range of x show that the disorder introduced by Sr strongly suppresses κ. As in YBCO, κ in LSCO exhibits a pronounced anomaly. We show that this anomaly is very sensitive to field, decreasing as $\log B$ in moderately weak fields. Two crystals were measured in fields up to 14 Tesla. Sample I is close to optimum (x = 0.17, T_c = 38 K) while Sample II is underdoped (x = 0.08, T_c = 24 K). At our resolution (3 parts in 10^4), thermocouples cannot be used directly because their field sensitivity (a few percent at 1 T and 10 K) swamps the signal. To avoid this large error, we adopted a bridge balancing technique in which the thermocouple (chromel-constantan) serves only as a *null* detector.

In the inset of Fig. 1 we display the profiles of κ versus temperature in the two crystals. Above T_c, the monotonic decrease of κ with decreasing T is consistent with dominant phonon current in a highly disordered lattice. Superconductivity in Sample I is accompanied by a pronounced zero-field anomaly in κ. Application of a 14-Tesla field **H** || **c** suppresses the anomaly, leaving a background (symbols) that is a smooth extrapolation of the normal state curve. A similar suppression is observed in Sample II. In agreement with Nakamura [5], the zero-field κ in II is half as large as in I, consistent with a higher degree of disorder in the underdoped regime.

In view of the large disorder, the anomaly in κ is unexpectedly sensitive to field. In Sample I, 0.5 T is enough to decrease κ by 5 % at 15 K. The cusp-like suppresion in weak fields is most prominent at low temperatures, but is observed up to the 35 K. The sharpness of the cusp implies

*Supported by U.S. Office of Naval Research and the National Science Foundation.

†Present address: JRCAT, National Inst. Adv. Interdisc. Res., 1-1-4 Higashi, Tsukuba, Ibaraki 305, Japan

PII S0921-4534(97)00229-3

a field-scale less than 0.1 T at 30 K. In Sample II similar curves are obtained, except that the field scales are 10 times larger.

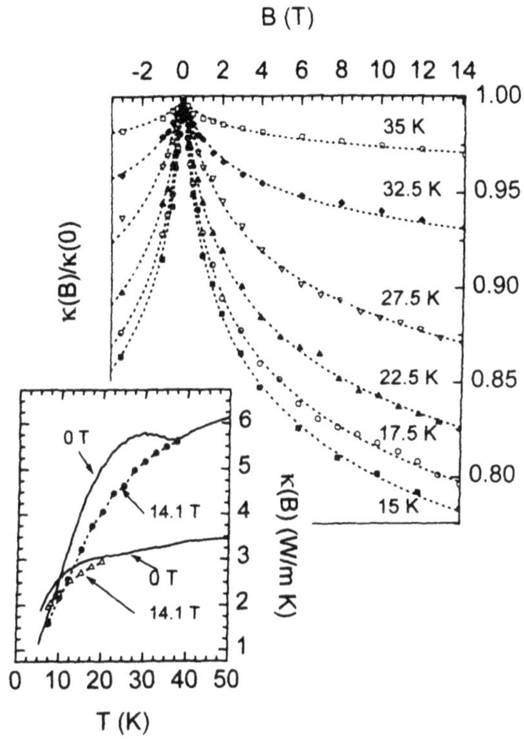

Figure 1 (Main panel) The field dependence normalized $\kappa(B)$ in LSCO (Sample I, $x = 0.17$) from 15 to 35 K ($\mathbf{H} \parallel \mathbf{c}$). The dashed lines are fits to Eq. 2. The inset shows the κ vs. T at 0 and 14.1 T for Samples I (upper curves) and II (lower).

The conventional explanation of the decrease in κ with field is that vortices strongly scatter phonons and/or electronic excitations (quasiparticles) which carry the heat current. Our experiment suggests that this picture is inapplicable in the cuprates. Rather, the log B dependence points to a quantum decoherence effect below T_c.

Evidence against the scattering of phonons by vortices (as the source of the field dependence) is obtained by comparing the fractional decrease of κ *per vortex* at 10 K where $\kappa(0)$ is the same in the two crystals (2.7 W/mK) [Fig. 2]. Since the heat capacities are nearly equal, the phonon mean free paths ℓ_{ph} are similar. In the phonon scattering scenario, the initial fractional decrease $\Delta\kappa/\kappa(0)B$ is given by $\ell_{ph}\sigma/\Phi_0$, where σ is the vortex cross-section for phonon scattering and Φ_0 the flux quantum.

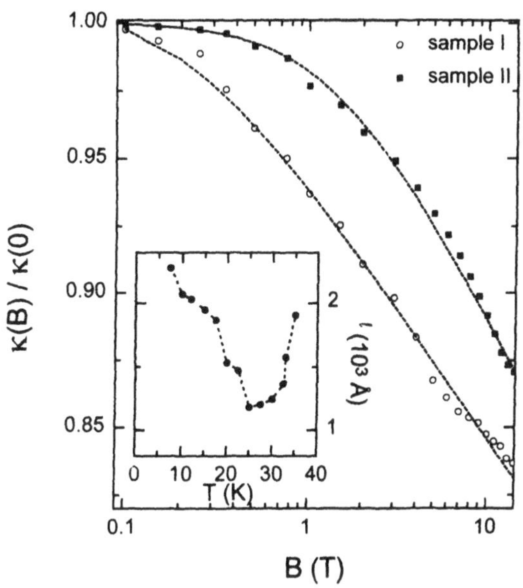

Figure 2 Comparison of κ vs. B in Samples I and II at 10 K. The fractional decrease is much larger in the optimum sample. The inset shows the unphysical behavior of the *mfp* vs. T extracted by fitting to the independent vortex scattering formula Eq. 1.

Because the vortex core size is larger in the underdoped crystal (II), we expect a larger fractional decrease. Instead, we observe quite the opposite. The measured $\Delta\kappa/\kappa(0)B$ is considerably larger in the optimum sample (I). Thus, we conclude that the scattering of phonons by vortices is insignificant. The pronounced field dependence in Fig. 1 is predominantly electronic in origin.

Next, we consider the scattering of qp by vortices. In the frequently used semi-classical model of independent scattering, κ is given by

$$\kappa(B) = \kappa(0)/[1 + |B|/B_S] + \kappa_0, \quad (1)$$

with the field scale $B_S \equiv \Phi_0/\sigma_{tr}\ell$, where σ_{tr} is the transport cross-section, and ℓ the *mfp* of quasiparticles in zero field. κ_0 is a field-independent background (phonons). Equation 1 is capable of fitting the data at low to moderate fields (deviations above 7 T are significant).

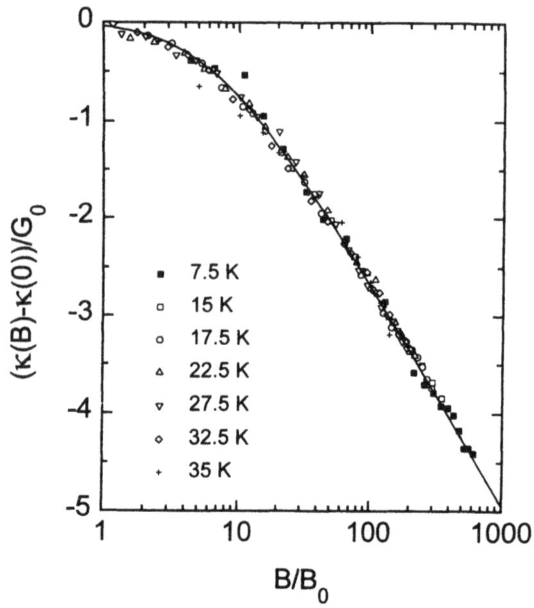

Figure 3 Rescaled plot of $\kappa(0) - \kappa(B)$ versus the reduced field $\frac{B}{B_0}$. All data between 7.5 K and 35 in Sample I may be rescaled to fit Eq. 2 (solid line).

However, the behavior of the derived ℓ is unphysical (Fig. 2, inset). With increasing T, it falls to a broad minimum near 25 K, and then increases rapidly to 2,000 Å, as $T \to T_c$. The discontinuous jump in ℓ across T_c is especially problematic, since the normal-state value is < 50Å). The unusual profile for ℓ should be reflected in the T dependence of $\kappa(0)$ which is proportional to ℓ (the *qp* density remains quite high just below T_c). The observed $\kappa(0)$ would appear to rule this out.

This unphysical behavior motivated us to replot our data in a variety of ways. We have found that κ actually decreases linearly as $-\log B$ in high fields (Fig. 3). Below a cross-over field (which is strongly T dependent) κ deviates from $-\log B$ and approaches its zero field value. The field dependence is reminiscent of the electrical conductivity in the 2D weak localization problem [6] expressed by the function $Y(x) \equiv \psi(\frac{1}{2} + \frac{1}{x}) + \ln(x)$, where $\psi(\frac{1}{2} + \frac{1}{x})$ is the digamma function. Close fits (up to 14 Tesla) are obtained with the equation

$$\kappa(B,T) - \kappa(0,T) = -G_0(T)Y(B/B_0), \qquad (2)$$

with the adjustable parameters G_0 and B_0. An alternate way to display the fit to Eq. 2 is to collapse all the data between 7.5 and 35 K in Sample I. In Fig. 3, we plot the change in conductivity, scaled by G_0, against B/B_0. The plot shows that Eq. 2 provides an excellent fit to the data over nearly 3 decades of reduced field.

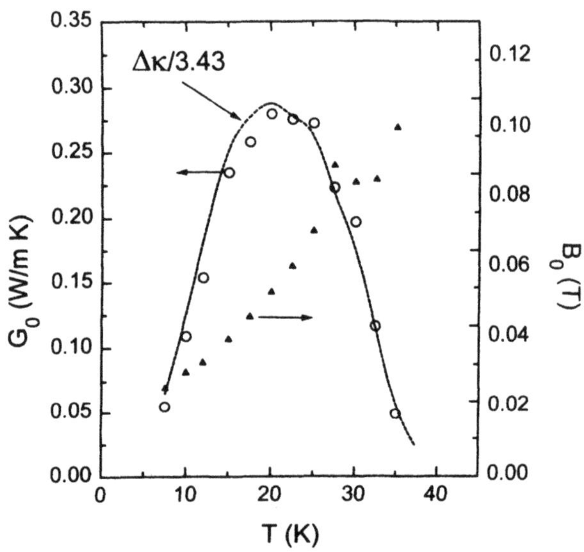

Figure 4 The temperature dependence of fit parameters G_0 (open circles) and B_0 (solid triangles) in Sample I. The smooth curve is the change in conductivity $\Delta\kappa$ at 14.1 Tesla reduced by the scale factor 3.43.

The temperature dependence of G_0 and B_0 are shown in Fig. 4. The prefactor G_0 (open symbols) rises to a broad maximum near 22 K, and appears to mimic the T dependence of the thermal anomaly in zero field. We find that G_0 is nearly proportional to $\Delta\kappa \equiv \kappa(0) - \kappa(14T)$ (the difference of the two curves in each sample in the inset of Fig. 1). Reducing $\Delta\kappa$ by a factor of 3.4, we find that the two quantities match each other to within the measurement uncertainty (open circles and dashed line in Fig. 4). The maximum value of $\Delta\kappa$ is close to the Wiedemann-Franz value calculated from the normal state resistivity. In the low T limit, both $\Delta\kappa$ and G_0 decrease to zero because of the entropy factor. In the other limit, $T \rightarrow T_c$, the decrease of G_0 reflects the vanishing of the superconducting gap parameter.

In contrast with the unphysical T dependence of B_S derived from Eq. 1, the new field scale B_0 is more physically sensible. Starting at 7.5 K, B_0 increases monotonically from 0.02 T to 0.10 T at 35 K. In view of the large disorder in the samples, the smallness of the field scale is striking.

The function $Y(x)$ is natural in problems involving electron diffusion in disordered 2D systems, because it describes how a weak field destroys interference effects. In weak localization [6] coherent interference between two closed paths that are time-reversed partners is destroyed when the external field inserts half a flux quantum within each path. The dephasing field B_0 is given by $\hbar/4De\tau$, where D is the diffusion constant, τ the time between dephasing collisions, and e the electronic charge. The field produces an increase in conductivity $\Delta\sigma$ proportional to $Y(B/B_0)$.

Interplay between electron diffusion and superconducting fluctuations in 2D systems also leads to a field dependence of the conductivity (above T_c) described by $\Delta\sigma \sim -Y(\frac{B}{B_0})$, with $B_0 = \hbar/4De\tau$ as well [7,8].

Thus, the observation of $Y(x)$ in κ vs. B in LSCO may not be surprising. The experiment suggests the following scenario. Superconductivity produces a large enhancement of the electronic diffusion constant D in zero field. The enhancement is readily suppressed by $\mathbf{B} \parallel \mathbf{c}$. The weak field scale precludes the conventional semiclassical picture of qp scattering by individual vortices (Eq. 1). Rather, the experiment points to a decoherence effect in which B destroys quantum interference effects that appear below T_c. The weak field scale B_0 corresponds to length scales (between dephasing collisions) $L = \sqrt{(D\tau)}$ that vary from 400 Å at 35 K to 1,050 Å at 7 K. Since these are diffusion lengths between inelastic collisions, there is no contradiction with the short mfp above T_c. We note further that the field acts to restore the semiclassical current by destroying quantum interference effects. Thus, it leads to a decrease in κ in the present experiment (whereas it produces an enhancement in σ in the weak localization problem).

The enhancement in the diffusion constant D appears closely related to the dramatic sharpening of the qp spectral weight observed in photoemission, a process dubbed 'recovery of coherence'. Our new result is that this coherence is easily suppressed by field. The form of the field suppression implies that the enhancement originates from a coherent interference effect that is rapidly decohered by the external flux. Unravelling this interference should provide a powerful way to probe the excitations of the condensate. Lastly, we point out that the field effect is probably unrelated to the destruction of the superconducting order parameter. In that case, the field scale should be much larger, and *increase* with decreasing T, in conflict with Fig. 4.

We have benefitted from many discussions with P. W. Anderson, P. A. Lee and T.V. Ramakrishnan.

REFERENCES

1. D. A. Bonn et al., Phys. Rev. Lett. **68**, 2390 (1992); D. A. Bonn et al. Phys. Rev. B **47**, 11314 (1992).
2. R.C. Yu et al., Phys. Rev. Lett. **69**, 1431 (1992).
3. K. Krishana, J. M. Harris and N. P. Ong, Phys. Rev. Lett. **75**, 3529 (1995).
4. For a review see Z.X. Shen and D. S. Dessau, Phys. Rep. **253**, 1 (1995).
5. Y. Nakamura, S. Uchida, T. Kimura, N. Motohira, K. Kishio, K. Kitazawa, T. Arima ad Y. Tokura, Physica C **185-189**, 1409 (1991).
6. For a review see Patrick A. Lee and T. V. Ramakrishnan, Rev. Mod. Phys. **57**, 287 (1985).
7. A.I. Larkin, Pis'ma Zh. Eksp. Teor. Fiz. 31, 239 (1980) [JETP Lett. **31**, 219 (1980)].
8. J.M.B. Lopes dos Santos and E. Abrahams, Phys. Rev. B **31**, 172 (1985).

ELSEVIER

Physica C 282–287 (1997) 248–251

PHYSICA C

The role of density of states fluctuations in the normal state anomalies of high T_c superconductors

A. A. Varlamov[a]

[a]Forum: Laboratory for Condensed Matter Theory of INFM, Pisa - Florence, Dipartimento di Fisica, Università di Firenze, L.E.Fermi,2 Firenze 50125;
Landau Network - Centro "A.Volta", University of Pavia, Pavia, Italy
Department for Theoretical Physics, Moscow Institute for Steel and Alloys, Moscow, Russia

The role of the DOS fluctuations in the normal properties anomalies of HTS is analyzed. It is shown how, taking into account this effect, many puzzling and long debated properties of HTS materials (e.g. the steep increase of the electrical resistivity along the c axis just above T_c, the anomalous magnetoresistance, the effects of the magnetic field on the resistive transition along the c-axis, the c-axis far infrared absorption spectrum, NMR spectra around the critical temperature etc.) can be understood, leading to a simple consistent description in terms of fluctuation theory.

1. INTRODUCTION

During the last decade a lot of efforts to explain the unusual normal state properties of high temperature superconductors (HTS) in the framework of unconventional theories based on strongly interacting electrons, preformed Cooper pairs, polaron mechanism of superconductivity etc. have been undertaken. A different approach to this problem is possible developing the perturbation theory for interacting electrons in the normal phase of strongly anisotropic superconductors without specifying the origin of this interaction. The Cooper channel of interelectron interaction is equivalent to the account of superconducting fluctuations. We show that such peculiarities of HTS as strong anisotropy, low carriers concentration, high values of T_c lead not only to the increase of the magnitude of the fluctuations but also frequently can change the hierarchy of the different fluctuation contributions and even the sign of the effect observed.

The first and well known result consists in the fact that now one has the "preformed Cooper pairs" automatically, *ab initio* from calculations: taking into account the thermal fluctuations leads to the appearance of a non-zero density of fluctuating Cooper pairs (with finite lifetime $\tau_{GL} = \frac{\pi\hbar}{8k_BT_c\epsilon}$, $\epsilon = (T-T_c)/T_c$) within layers without the establishment of the long range order in the system. Very important that in $2D$ systems the fluctuating Cooper pairs density decreases with temperature very slowly ($\sim \ln \frac{T_c}{T-T_c}$) and the effect of Cooper channel interelectron interaction manifests itself in the wide range of temperatures above T_c. The formation of these pairs of normal electrons leads to the decrease of the density of one-electron states (DOS renormalization) at the Fermi level and this effect turns out to be the key one in our discussion. Because of its weak singularity the DOS contribution was considered as negligible for the traditional superconductors normal properties. Nevertheless even there it can be dominant when by some reasons the more singular Aslamazov-Larkin and Maki-Thompson contributions are weak (as it happens with thermoelectric power). The domination of the DOS contribution may be pronounced especially in c-axis transport properties of HTS materials where due to their strong anisotropy and pair-breaking the latter two contributions turn out to be suppressed considerably.

2. C-AXIS RESISTANCE PEAK

The first and illustrative example of the DOS contribution manifestation is the fluctuation conductivity along the c-axis in the anisotropic lay-

PII S0921-4534(97)00233-5

ered superconductor. The tunneling character of the electron motion leads to the strong suppression of the paraconductivity. Because of the necessity of two particle tunneling it turns out to be proportional to the square of the interlayer hopping probability W. The same concerns the anomalous Maki-Thompson process which is additionally suppressed by strong phase-breaking expected in HTS (in principle MT process, being the precursor phenomena for the Josephson current, can manifest itself in the first order of the barrier transparency, but this case can be realized in the absence of the pair-breaking only). The DOS contribution, related with the decrease of the one-electron concentration because of the involvement of the electrons in the intra-layer fluctuation pairings, is weakly singular but turns out to be proportional to the first power of W only [1]:

$$\sigma_{fl}^{\perp} \sim \alpha W^2 \frac{1}{\epsilon^2} - \beta W \ln \frac{1}{\epsilon} \tag{1}$$

where α and β are the coefficients which can be calculated in the framework of the exact microscopic theory [1]. As the result in strongly anisotropic materials the c-axis resistance can increase in the wide enough range of the temperatures and only in the immediate vicinity of T_c the concurrence of the growing positive AL contribution leads to the formation of the maximum and further decrease of resistance. The theoretical results fit very good the shape of the experimental curves obtained for BSCCO epitaxial films and single crystals with different oxygen contents [2].

The magnetic field applied along c-axis does the effect even more pronounced leading to the growth of the maximum and its shifting to lower temperatures. The reason of such behavior is the additional suppression of the AL contribution by the magnetic field while the DOS contribution turns out to be much less sensitive to it [3].

3. MAGNETORESISTANCE ABOVE T_c

While the ab-plane magnetoresistance (MR) above T_c is well known to have a positive sign due to the suppression of the AL paraconductivity, the c-axis MR of BSCCO single crystals has been found to be negative and growing anomalously with the decrease of temperature in the range of $95 - 140K$ [4]. This effect was attributed [5] to the manifestation of the DOS fluctuations in MR. Really, the suppression of the DOS term, in contrast to the AL one, leads to the decrease of the resistivity, so in the region of temperatures where the DOS contribution dominates a negative c-axis MR is conceivable. The fitting of the experimental results mentioned above with formula of [3] gave the excellent results both in its magnetic field ($\sim H^2$) and temperature dependencies. Moreover, the analysis of the temperature dependence of the total fluctuation correction to MR at fixed magnetic field led to the non-monotonous, sign alternating function [5]. The physical origin of this change of sign is the same as for the appearance of the peak discussed above: relatively far from T_c the positive AL MR is suppressed by its proportionality to the square of the interlayer transparency and the negative DOS contribution dominates, while close enough to T_c the very singular temperature dependence of the negative AL contribution ($\sim \epsilon^{-4}$) makes it prevail on the less singular DOS one ($\sim \epsilon^{-2}$), in spite of the linear dependence on the transparency of the latter.

Shortly after the problem of the transverse magnetoresistivity was examined [6] basing on the idea of its fluctuation origin. In all cases a good agreement between the fluctuation theory and the measured data has been found, both for BSCCO and YBCO samples. In particular, in [6] the transverse magnetoresistance was measured down to temperatures close to T_c. Its temperature dependence was found to follow very closely the behavior predicted in [5] and the existence of a sign reversal temperature was experimentally confirmed.

4. OPTICAL CONDUCTIVITY

The most puzzling experiments widely discussed recently are related with the observation of the pseudogap-like behavior in different properties of HTS materials above the critical temperature. One of them is the observation of the pseudogap opening in the c-axis far infrared conductivity [7] which we discuss here basing on the same idea of the importance of Cooper channel interelectron interaction [10]. Let us start from the analysis of each fluctuation contribution separately and then we will discuss their interplay in $\mathrm{Re}\,[\sigma_{\perp}(\omega)]$.

The AL contribution describes the fluctuation condensate response to the electromagnetic field applied. The component of the current associated with it can be treated as the precursor phenomenon of the screening currents in the superconducting phase. Above T_c the virtual Cooper pairs binding energy is of the order of $T - T_c$, so it is not surprising that at higher frequencies the AL contribution decreases fast with the further increase of ω. Actually $\omega_{AL} \sim T - T_c$ is the only relevant scale for σ^{AL}: its frequency dependence doesn't contain T, τ_{ϕ} and τ. The independence from the latter manifests mathematically the fact that elastic impurities do not represent obstacles for the motion of Cooper pairs.

Another effect related with the formation of the fluctuation Cooper pairs, but on the self-intersecting trajectories (like the weak localization correction), is described by the MT *anomalous* contribution. Being the contribution related with the Cooper pairs electric charge transfer it doesn't depend on the elastic scattering time but it turns out to be extremely sensitive to the phase-breaking mechanisms. So two characteristic scales turn out to be relevant in the frequency dependence of MT contribution: $T - T_c$ and τ_{ϕ}^{-1}. In the case of HTS, where τ_{ϕ}^{-1} has to be estimated as at least $0.1T_c$, for temperatures up to $5 \div 10\,K$ above T_c the MT contribution is overdamped, but in 2D regime it is still dependent on temperature.

The density of states fluctuation renormalization gives quite different contribution to $\mathrm{Re}\,[\sigma(\omega)]$ with respect to the others. At low frequencies ($\omega \ll \tau^{-1}$) the lack of electron states on the Fermi level leads to an opposite effect in comparison with AL and MT contributions: $\mathrm{Re}\,[\sigma^{DOS}(\omega)]$ turns out to be negative and this means the increase of the surface impedance, or, in other words, the decrease of reflectance. Nevertheless, the electromagnetic field applied affects the electron distribution and at high frequencies $\omega \sim \tau^{-1}$ the DOS contribution changes its sign. It is interesting that DOS contribution, as one-electron effect, depends on the impurity scattering similarly to the normal Drude conductivity. The decrease of $\mathrm{Re}\,[\sigma^{DOS}(\omega)]$ starts already at frequencies $\omega \sim \min\{T, \tau^{-1}\}$ which for HTS are much higher than $T - T_c$ and τ_{φ}^{-1}.

The scenario of $\mathrm{Re}[\sigma_{\perp}^{tot}]$ ω-dependence may be presented as following. The positive AL and MT effects, in their ω-dependence, are well pronounced at low frequencies on the background of the DOS contribution which remains, in this region, a negative constant. Then at $\omega \sim T - T_c$ the former decays and the $\mathrm{Re}\sigma_{\perp}$ remains negative up to $\omega \sim \min\{T, \tau^{-1}\}$. The DOS correction changes its sign at $\omega \sim \tau^{-1}$ and the following high frequency behavior is governed by the Drude law. So one can see that the characteristic pseudogap-like behavior in the frequency dependence of the optical conductivity takes place in the range $\omega \in [T - T_c, \tau^{-1}]$. The depth of the window increases logarithmically with ϵ when T tends to T_c.

5. NQR-NMR RELAXATION RATE

One more example of the possible manifestation of the DOS is related to the NMR-NQR relaxation rate. The decrease of the electron density of states on the Fermi level evidently leads to the decrease of the relaxation rate with respect to Korringa law [8]. The opposite sign effect could be expected from paraconductivity-like contribution, but the necessity of the spin flip in the process of the nucleus-electron interaction excludes the AL term from consideration (after the unique act of such interaction one of the fluctuation Cooper pair electrons should change its spin orientation, which destroyed the pair). Nevertheless the positive competitor for the DOS contribution appears as one sophisticated version of the anomalous MT process with the spin flip. Electron can move along the self-intersecting trajectory with the nucleus in the vicinity of the intersection point. After the passing along the trajectory once, electron can interact with the nucleus, change the spin orientation and the momentum on the opposite ones and then to pass the same trajectory again in the opposite direction, interacting with itself in "the past" (what is possible due to the retarded character of superconducting attraction). This, purely quantum, process exist in the s-wave scenario of pairing. It leads to the increase of the relaxation but, typically for MT processes, turns out to be very sensitive for the phase-breaking and is considerably suppressed in HTS. The concurrence of two mentioned terms was observed on experiment on YBCO samples. Moreover, the positive MT contribution was suppressed by the additional magnetic field applied [9]. The last fact could be treated as the inderect evidence of the presence of the s-wave component in the order parameter.

6. THERMOELECTRIC POWER

Another example of the domination of the DOS contribution on the other fluctuation terms, even in the case of the isotropic electron spectrum, is the electron part of the thermoelectric power at the edge of superconducting transition which can be demonstrated from the following qualitative consideration (we will refer to the most interesting 2D case) [11].

The thermoelectric coefficient may be estimated through the electrical conductivity σ as $\beta \sim (\epsilon^*/eT)f_{as}\sigma$, where ϵ^* is the characteristic energy involved in thermoelectric transport and f_{as} is the electron-hole asymmetry factor, which is defined as the ratio of the difference between numbers of electrons and holes to the total number of particles. Conductivity can be estimated as $\sigma \sim e^2\mathcal{N}\tau^*/m$, where $\mathcal{N}$, τ^* and m are the density, lifetime and mass of charge (and heat) carriers, respectively. In the case of AL contribution the heat carriers are non-equilibrium Cooper pairs with energy $\epsilon^* \sim T - T_c$ and density $\mathcal{N} \sim p_F^d \frac{T}{E_F} \ln \frac{T_c}{T-T_c}$ and characteristic lifetime is given by the Ginzburg-Landau one. Thus in 2D case $\Delta\beta^{AL} \sim (T-T_c)/(eT_c)f_{as}\Delta\sigma^{AL} \sim ef_{as} \ln \frac{T_c}{T-T_c}$. One can easily get that the fluctuation correction due to AL process is less singular (logarithmic in 2D case) with respect to the corresponding correction to conductivity and does not depend on impurity scattering.

The analogous consideration of the single-particle DOS contribution ($\epsilon^* \sim T$, $\tau^* \sim \tau$) evidently results in the estimate of the same singularity as AL one: $\beta^{DOS} \sim ef_{as}T_c\tau \ln \frac{T_c}{T-T_c}$. The detailed comparison of the coefficients shows that the DOS contribution exceeds AL by the large parameter both in dirty and clean cases.

The fluctuation contribution provides a positive correction $\Delta\beta$, resulting in the decrease of the absolute value of Seebeck coefficient S at the edge of superconducting transition ($\Delta\beta/\beta_0 < 0$). To the result of the same sign the account of fluctuation correction to conductivity leads. So one can conclude that the thermodynamical fluctuations always reduce the overall S as temperature approaches T_c. So the very sharp maximum in the Seebeck coefficient experimentally observed in few papers seems to be unrelated to fluctuation effects within our simple model.

7. ACKNOWLEDGEMENT

The support from Cariplo Foundation for Scientific Research is gratefully acknowledged.

REFERENCES

1. L. B. Ioffe , A. I. Larkin, A. A. Varlamov, L. Yu, Phys. Rev. B47 (1993) 8936.
2. G. Balestrino, E. Milani, and A. A. Varlamov Phys. Rev. B54 (1996) 3628.
3. V. Dorin, R. Klemm, A. Varlamov, D. Livanov, A. Buzdin, Phys. Rev. B48 (1993) 12951.
4. Y. F. Yan, P. Matl, J. M. Harris, N. P. Ong, Phys. Rev, B52 (1995) R751
5. G. Balestrino, E. Milani, and A. A. Varlamov, Soviet JETP Letters 61 (1995) 833.
6. J. Axnas, W. Holm, Yu. Eltsev, O. Rapp, Phys. Rev. Lett. 77 (1996) 2280.
7. D. N. Basov, T. Timusk, B. Dabrovski, H. A, Mook, Phys. Rev. B52 (1995) 13141; Phys. Rev. B50 (1994) 3511.
8. M. Randeria and A. A. Varlamov, Phys. Rev. B50 (1994) 10401.
9. P. Carretta D. Livanov, A.Rigamonti, A. Varlamov, Phys.Rev. B54 (1996) R9682.
10. F. Federici, A. A. Varlamov, JETP Letters 64 (1996) 497.
11. A. Varlamov, D. Livanov, F. Federici, JETP Letters 65, (1997) N 2.

ELSEVIER

Physica C 282–287 (1997) 252–255

PHYSICA C

Nonlinear electrodynamics in cuprate superconductors

J. Orenstein*, J. Bokor*, E. Budiarto*, J. Corson*, R. Mallozzi*, I. Bozovic#, and J. N. Eckstein#

*University of California at Berkeley and Lawrence Berkeley National Laboratory

#E. L. Ginzton Research Center, Varian Associates, Inc.

We present two types of experiments which probe the high-frequency conductivity of BSCCO films in the presence of large superfluid velocities. In the first the superflow is the static velocity field associated with the vortex lattice. In the second the velocity is dynamic, induced by high-intensity single-cycle electromagnetic pulses. We show that intrinsic pair-breaking in a *d*-wave superconductor can provide a unified explanation for both effects.

1. Nonlinear electrodynamics

The high-frequency response of the cuprate superconductors, from microwave to infrared frequencies, has been subject to intense study. Despite great progress, two important aspects of the problem remain unresolved. One concerns the effect of static magnetic fields ($H \gg H_{c1}$) on the electrodynamic response functions, $\sigma(\omega)$ or $\rho(\omega)$. The second aspect is the change in response as the amplitude of the high-frequency field is increased. In this paper we present recent experimental results which shed light on both of these problems.

Traditionally, these two aspects of superconducting electrodynamics have been viewed as unrelated problems. The response in large static fields is generally ascribed to the motion of vortices[1]. The deviation from linear response as the high-frequency field amplitude increases is often assigned to the nonlinear response of internal weak links[2]. It is important to recognize that *d*-wave superconductivity introduces intrinsic nonlinear mechanisms which can provide a unifying explanation for both effects.

In a clean *d*-wave superconductor at low temperature the superfluid density is predicted to decrease with increasing superfluid velocity[3],

$$\rho_s(v_s) \cong \rho_s(0)\left(1 - v_s/v_c\right) \qquad (1)$$

where v_c is the depairing velocity Δ_0/p_F. This leads to a nonlinear Meissner effect. Although *s*-wave superconductors are intrinsically nonlinear as well, their nonlinearity vanishes as $T \to 0$. In the following we discuss the role of Eq. 1 in determining $\sigma(\omega)$ in static magnetic fields.

In the absence of a field the conductivity of the superfluid condensate is given by $\sigma_s(\omega) = \rho_s\left[\pi\delta(\omega) + i/\omega\right]$. Two mechanisms lead to changes in $\sigma_s(\omega)$ in a static field: vortex motion and condensate depletion. To understand the vortex dynamics contribution, recall that ρ_s is the phase stiffness of the superconductor. The vibration of pinned vortices decreases ρ_s by an amount proportional to H/κ, where κ is the pinning force constant. The reduced spectral weight in the δ-function part of $\sigma(\omega)$ appears in a Drude-like peak of width κ/η, where η is the vortex viscosity.

There is a second mechanism for the reduction of ρ_s in the vortex state. The quasiparticles trapped in the core of each vortex deplete the superfluid. For an *s*-wave superconductor this transfer of spectral weight is small, of order H/H_{c2}. Whenever κ is less than its maximal value given by the condensation energy, the collective mechanism described in the previous paragraph will dominate.

The presence of nodes in the gap function and low-lying quasiparticle excitations in *d*-wave superconductors change this picture drastically. While the vortex motion mechanism remains essentially the same, the depletion of superfluid density due to quasiparticle creation is far larger. Using arguments first presented by Volovik[4], Parks[5] *et al.* calculated the fractional reduction in ρ_s in the clean, isotropic limit by averaging Eq. 1 over the velocity field of the vortex lattice. The result is $(H/H_{c2})^{1/2}$ rather than H/H_{c2} as in an *s*-wave superconductor.

0921-4534/97/$17.00
PII S0921-4534(97)00236-0

Notice that the fractional reduction per vortex *diverges* as $1/H^{1/2}$ in the low-H limit. At T=0, this divergence is cutoff only by the finite vortex size.

2. Terahertz conductivity in a field

Terahertz time domain spectroscopy provides an accurate means of measuring the complex conductivity in the vortex state over a broad frequency range. In the measurements on single crystal BSCCO films reported here we determined the response from 100-250 GHz, the low-frequency range of our technique. Measurements extending to 1.5 THz will be reported elsewhere.

Figure 1 shows the change in $\sigma(\omega)$ with magnetic field for a broad range of temperatures below T_c. The top panel shows the decrease in the imaginary part σ_2, while the bottom panel shows the increase in σ_1, both on a linear scale. Figure 2 shows again the variation of σ_2 with field, this time on a double logarithmic scale and for several frequencies at a representative temperature of 50K. The most striking aspect of these results is the sublinear dependence of σ_2 on H over a broad range of field, frequency, and temperature.

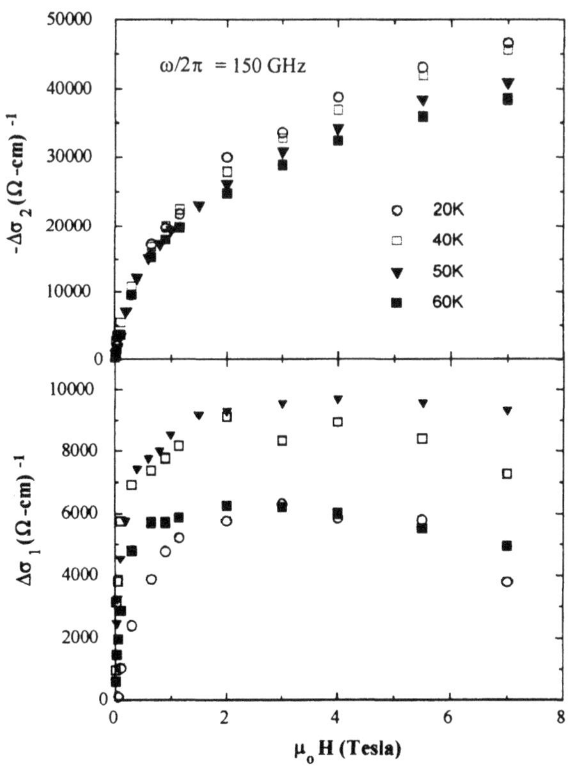

Figure 1. Conductivity change versus field.

The change in σ_2 is proportional to the decrease in ρ_s with field. The change in $\Delta\rho_s$ with H is remarkably close to the clean *d*-wave prediction of $H^{1/2}$. Attempting to explain this result in terms of vortex motion leads to nonphysical assumptions for the relevant parameters. For example, since the reduction in ρ_s in vortex dynamical models is proportional to H/κ, κ must increase with magnetic field, as approximately $H^{1/2}$. However theory predicts and experiments confirm that pinning decreases with increasing field.[1,6,7,8].

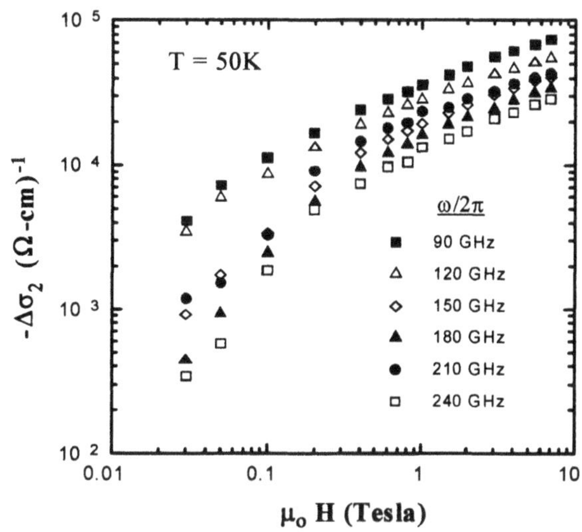

Figure 2. $\Delta\mathrm{Im}(\sigma)$ versus field, various ω.

The *d*-wave electrodynamics, on the other hand, not only provides a natural explanation for the power law dependence of σ_2 on field, but accounts for the magnitude of the change quantitatively. Recently, we have shown that the measured $\sigma_2(\omega,H)$ can be analyzed to yield $\Delta x_n(r)$, the local normal fluid fraction created by the superflow around a vortex, as a function of distance from the vortex center[9]. The results of that analysis are shown in Figure 3, which is a plot of Δx_n, the normal fluid fraction, as a function of $1/r$. We choose $1/r$ for the horizontal axis because it is proportional to the local shift in quasiparticle creation energy. The energy shift is plotted on the top horizontal scale.

In this analysis Δx_n is extracted using only zero-field parameters available from other measurements: v_F, from photoemission [10], and the total spectral weight ω_p^2 from linear spectroscopy.

The dashed line which passes through the data points is the clean *d*-wave prediction, which is simply a line whose slope is $(\partial\Delta/\partial\theta)^{-1}$ evaluated near the gap node. Here $\Delta(\theta)$ is the gap as a function of angular deviation from the nodal direction. We have used a value of 54meV for $\partial\Delta/\partial\theta$ suggested by recent photoemission data[11]. The coincidence between the data at energies above 3meV and this line indicates a "no-adjustable parameter" agreement with the clean *d*-wave model for vortex-state electrodynamics.

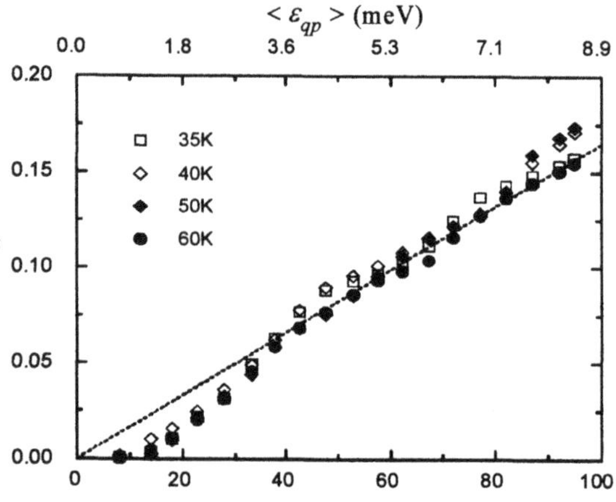

Figure 3. Normal fraction created around a vortex.

The deviation from the line for energies below 3meV corresponds to a deviation from the clean *d*-wave theory. Referring to the raw data shown in Fig. 2, the deviation corresponds to the change in slope of log σ_2 vs. log H at fields below 0.2 Tesla. There are several possible sources of this effect. The most obvious is a disorder-induced change in the density of states below about 3 meV. Another possibility is that as the energy scale approaches the incident photon energy (~1 meV), the measurement no longer simply probes the normal fluid fraction.

3. Terahertz nonlinear transmission

Recently we have begun a new type of measurement to further explore the role of the nonlinear Meissner effect in electrodynamic properties. In this measurement we again explore the change in electrodynamic response due to large superfluid velocities. However, now the superfluid velocities are dynamic, induced by intense terahertz fields rather than static magnetic fields.

In these experiments we measure the transmission coefficient of intense, single-cycle electromagnetic pulses, with duration of order 1 picosecond, through the same thin film samples as used in the high-field measurements. The single-cycle terahertz radiation is generated by photoexcitation of a voltage-biased GaAs wafer using an amplified Ti:sapphire laser. The THz pulse is focused onto the thin-film, and the transmitted pulse is refocused into a bolometric detector which measures the integrated energy per pulse.

In Figure 4, we plot the energy per pulse transmitted through a BSCCO film as a function of the incident energy for several temperatures. The slope of this line in the limit of low incident pulse energy defines the linear transmission coefficient T_L. Each set of data in Fig. 4 has been scaled so that the slopes coincide at low energy. Above T_c the data points fall on a straight line, indicating nearly linear electrodynamics in the normal state. Below T_c the data diverge from the line with increasing incident energy, *i.e.* T increases with increasing incident field. In this scaled plot the deviation from the straight line is proportional to $\Delta T/T_L$, where $\Delta T \equiv T - T_L$. As Fig. 4 shows clearly, $\Delta T/T_L$ increases with decreasing temperature below T_c. Notice that the size of the increase in T is extremely large. If interpreted in terms of superfluid depletion, each pulse reduces the superfluid density by roughly 1/3.

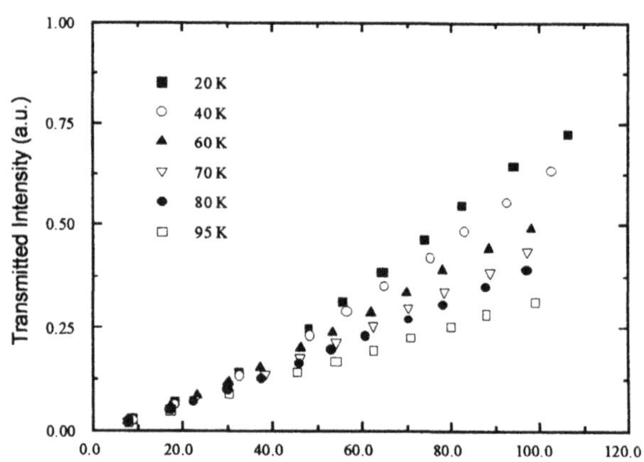

Figure 4. Transmitted versus incident pulse energy.

Figure 5 shows $\Delta T/T_L$ as a function of current density at low temperature for two orientations of the current with respect to the *a-b* axes of the twinned film. Within experimental uncertainty of about 10%, $\Delta T/T_L$ increases quadratically with current and is independent of orientation. The curvature of the parabolic fits to these data define a characteristic current scale $J_c \approx 10^8$ Amp-cm^{-2}.

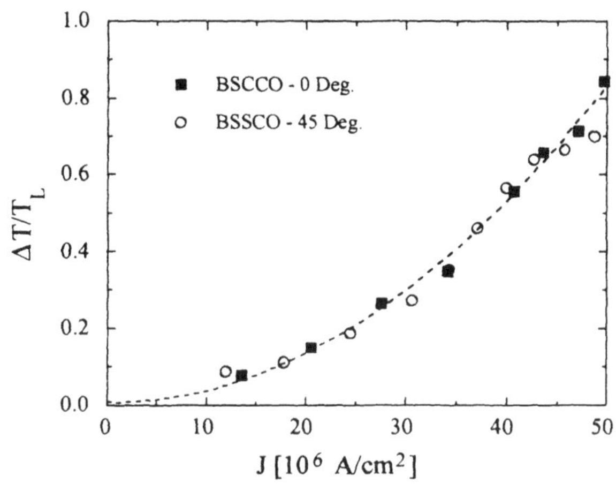

Figure 5. Fractional change in transmission versus current density in film.

We have considered several models to explain the nonlinear transmission. We can eliminate sample heating because of the lack of correlation between the temperature dependence of $\Delta T/T_L$ and the derivative of T_L with respect to temperature. Furthermore, we can estimate the terahertz induced temperature rise using the experimentally determined σ_1 and literature values of the specific heat[12]. The increase in temperature corresponds to 3K for T=10K and 0.1K for T=30K, far too small to account for the nonlinear response.

The nonlinearity in cuprate films seen in microwave experiments near 10 GHz is often attributed to internal weak links[2]. However the characteristic current we observe is much closer to the intrinsic depairing current density J_{pb} than critical current densities associated with weak links, 3×10^6 Amp-cm^{-2}[2]. J_{pb} is defined by equating the depairing energy $p_F v_s$ with Δ_0. It can be written in terms of well-known parameters as follows, $J_{pb}=\Delta_0/\mu_0(\lambda_L)^2 e v_F$. With Δ_0=25meV, λ_L=300nm, and $v_F=10^7$cm-s^{-1} we find a depairing current of 1.4×10^8 Amp-cm^{-2}, in excellent agreement with the experimentally determined J_c.

Because of this agreement, we conclude that the terahertz frequency nonlinear response is determined by intrinsic pairbreaking due to the large induced superfluid velocities. The results differ however with the expectations of the YS theory, which predicts $\Delta T/T_L$ linear in J/J_{pb} and a $2^{1/2}$ anisotropy for the two orientations shown in Fig. 5. The discrepancy could arise from the assumption that the YS quasistatic theory applies to the high-frequency response. In the quasistatic theory, both temperature and disorder provide crossover energy scales below which the nonlinearity is quadratic in superfluid velocity. It is reasonable to expect that the photon energy might also provide a crossover. As for the lack of anisotropy in the response, YS assumed a circular Fermi surface with uniform and radial Fermi velocities. For the real Fermi surface, which is highly anisotropic, the predicted orientation dependence may only be observable for energies very close to ε_F and vanish rapidly at nonzero temperature, disorder, or frequency.

4. Conclusion

We have presented two types of experiments which probe the high-frequency conductivity of BSCCO films in the presence of large superfluid velocities. Although in one the flow is static and the other dynamic, intrinsic pairbreaking in a *d*-wave superconductor can provide a unified explanation for both effects.

We acknowledge Jim Sauls for valuable discussions.

References

1. For a review see M. Golosovsky *et al.*, Supercond. Sci. Technol. **9**, 1 (1996).
2. P.P Nguyen, *et al.*, Phys. Rev. B **48**, 6400 (1993).
3. S.K. Yip, and J.A. Sauls, Phys. Rev. Lett. **69**, 2264 (1992).
4. G.E. Volovik, JETP Lett. **58**, 469 (1993).
5. B. Parks *et al.*, Phys. Rev. Lett. **74**, 3265 (1995).
6. T.T.M Palstra, B. Batlogg, *et. al.*, Appl. Phys. Lett. **54**, 763 (1989)
7. J.Z. Sun *et al.*, Appl. Phys. Lett. **54**, 663 (1989)
8. J.T. Kucera *et al.*, Phys. Rev. **B** 46, 11004 (1992).
9. R. Mallozzi *et al.*, to be published.
10. D.S. Dessau, Phys. Rev. Lett. **71**, 2781 (1993).
11. H. Ding *et al.*, Phys. Rev. **B** 54, R9678 (1996)
12. A. Junod, in Physical Properties of High Temperature Superconductors, vol II, ed. by D. Ginsberg (World Scientific, 1990) p. 13

ELSEVIER

Physica C 282–287 (1997) 256–259

PHYSICA C

Evidence for multi-component superconducting order parameter from microwave measurements of $YBa_2Cu_3O_{7-\delta}$ single crystals *

S. Sridhar, H. Srikanth, Z. Zhai, B.A. Willemsen, T. Jacobs [a] and A. Erb, E. Walker and R. Flukiger[b]

[a]Physics Department, Northeastern University, 360 Huntington Avenue, Boston, MA 02115

[b]DPMC, Universite de Geneve, CH-1211 Geneva, Switzerland

The microwave complex conductivity in the superconducting state ($\sigma_s = \sigma_1 - i\sigma_2$) of high quality $YBa_2Cu_3O_{7-\delta}$ single crystals measured at $10GHz$ using a high-Q Nb cavity reveal two new features - (i) A new peak in σ_1 at $\sim 80K$ and (ii) enhanced pair conductivity σ_2 below $\sim 65K$. The results indicate two pairing temperature scales and point to a multi-component superconducting order parameter in $YBCO$.

Electrodynamic measurements of the complex conductivity ($\sigma_s = \sigma_1 - i\sigma_2$) yield important information regarding the quasiparticle density of states and the nature of the superconducting condensate [1,2]. Microwave experiments have played an important role in understanding fundamental issues in both low (LTS) and high temperature superconductivity (HTS).

In HTS, improvement of materials quality over the years has gradually led to many experimental results which reflect the intrinsic properties untainted by defects. The recent growth of $YBa_2Cu_3O_{7-\delta}$ single crystals in $BaZrO_3(BZO)$ crucibles has ushered in a new generation of high quality samples [3]. This growth method avoids the critical problem of crucible corrosion and leads to single crystals with extremely clean surfaces and purity exceeding 99.995%.

We have carried out precision measurements of the surface impedance ($Z_s = R_s + iX_s$) of these new high quality $YBa_2Cu_3O_{7-\delta}$ single crystals grown in BZO crucibles. The experiments were done in a superconducting Nb cavity resonator operating at $10GHz$ [4]. Three single crystals from different batches grown in BZO crucible and one crystal of comparable dimensions grown in YSZ crucible were measured to study the systematics in the different samples. The crystals grown in BZO had $T_c = 93.4K$ and the one grown in YSZ had $T_c = 92.4K$.

In Figs.1 and 2, the typical temperature dependence of $\sigma_2(T)$ and $\sigma_1(T)$ are plotted for $YBCO/BZO$ and $YBCO/YSZ$ samples. Two striking features emerge clearly from a comparison of the data:

(i) $\sigma_2(T)$ shows two distinct regions of variation for $T < 65K$ and $65K < T < T_c$ respectively in the $YBCO/BZO$ samples. Since the estimated $\lambda(0)$ is lower in these crystals, this feature implies enhanced pair conductivity indicating an onset of additional pairing below $65K$.

(ii) The normal conductivity $\sigma_1(T)$ shows a new peak at around $80K(\sim 0.9T_c)$ in $YBCO/BZO$ crystals which is absent in $YBCO/YSZ$.

Very near T_c, a third sharp peak is present in $\sigma_1(T)$ of all four samples and its sharpness attests to the quality of the crystals. The sharpness of the superconducting transition is evident in the R_s data which indicated a transition width of $0.2K$ at $10GHz$ [5], and is confirmed by the sharpness of the σ_1 peak. A comparison of the inelastic scattering rates extracted from the normal state conductivity of all the samples shows that the second peak at $80K$ is prominent in samples with lower inelastic scattering in the normal state.

The data indicate that in addition to the pairing at $93K$, additional pairing develops below around $65K$. This strongly suggests the presence of multiple pairing interactions in $YBa_2Cu_3O_{6.95}$. A peak in σ_1 around $30K$ is seen in all YBCO crystals and to account for its location, an argument based on a precipitous drop in quasiparticle scat-

*Work supported by NSF-DMR-9623720

PII S0921-4534(97)00239-6

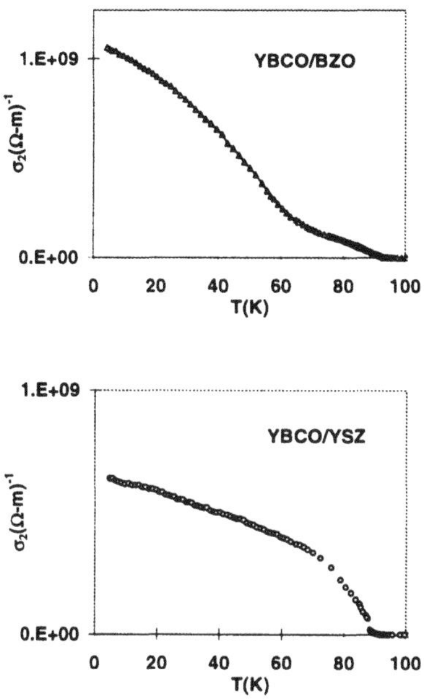

Figure 1. Pair conductivity $\sigma_2(T)$ for the YBCO/BZO sample and YBCO/YSZ sample. Note the enhanced pairing turning on below $65K$ for YBCO/BZO shown in the top panel.

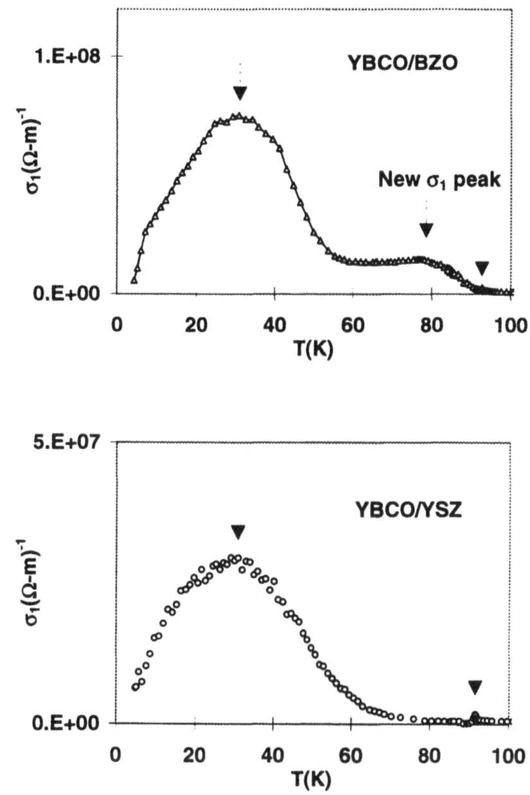

Figure 2. Normal conductivity $\sigma_1(T)$. The arrows denote the location of the conductivity peaks. The new peak observed is marked in the YBCO/BZO data.

tering rate below T_c has been proposed [6]. In light of our present results, it now becomes clear that the $30K$ conductivity peak in all samples is really associated with pairing at $65K$ and not with pairing at $93K$, since we are now able to observe and associate the new $80K$ conductivity peak with pairing at $93K$. This two component picture provides a satisfying explanation of the $30K$ peak without invoking a rapid temperature dependence of the scattering rate of a single component.

It is rather difficult to reconcile the data with a single order parameter of any symmetry. One needs two, possibly coupled, superconducting components to account for this data. Simple models which incorporate decoupled superconducting components with separate pairing temperature scales around $65K$ and $93K$ and in which the conductivities of the two components are treated additively describe the observed data for σ_2 and σ_1 [5]. In order to reproduce the low T behavior which is linear, at least one of the components has to have nodes in the gap. We have found good agreement with the σ_2 data for a model using one s-wave and one d-wave component .

However the two components are likely to be coupled. A Ginzburg-Landau calculation can be carried out (*eg.* [7]) using order parameters Ψ_A

and Ψ_B with pairing temperatures T_{cA} and T_{cB} ($> T_{cA}$). For weak coupling between the components, the calculation yields a single superconducting transition near T_{cB}. As the temperature of the superconductor is lowered, the superconducting state would have symmetry corresponding to Ψ_B just below T_{cB}, with mixed symmetry $A+B$ appearing below T_{cA} which could be dominated by the symmetry of Ψ_A. This will be reflected in measurements of the penetration depth as well as many other properties, which will show a crossover at T_{cA} well below T_{cB}, similar to the σ_2 data in Fig. 1.

Earlier a two-component order parameter was suggested in $Bi:2212$ on the basis of photoemission data [8]. However penetration depth data on $Bi:2212$ crystals [2] do not resemble the present data on $YBCO/BZO$ and instead look similar to the $YBCO/YSZ$ results. It should be pointed out here that materials quality played a crucial role in the case of $Bi:2212$ too. Although prior measurements indicated a T^2 dependence of the low temperature penetration depth [9,10], data on purer crystals grown using the Traveling Solvent Floating Zone method (TSFZ) yielded a linear-T dependence similar to the case of $YBCO$ [2,11].

The present results are consistent with several other experimental observations in $YBCO$ and also other cuprates. Earlier measurements on $YBCO$ thin films yielded penetration depth data which are similar to the present results and were there interpreted in terms of a 2-gap model [12]. We note that STM tunneling measurements on similar $BaZrO_3$ grown $YBCO$ crystals show multiple structure at around 5 mV and 20–25 mV [13], close to the pairing energy scales in the present work. If indeed this correspondence is relevant, then our results would suggest that these features should have different temperature dependences which should be explored.

The possibility of multiple pairing energy scales in YBCO is not surprising in view of the crystal structure. The Fermi surface of $YBCO$ is believed to consist of multiple sheets associated with two plane bands and a chain band. This raises the possibility of superconductivity with multiple pairing energies. Recent thermal conductivity studies on $YBCO$ have also revealed enhanced pairing below $55K$ which has been interpreted as evidence for chain superconductivity [14].

We now remark on the sharp peak in σ_1 very near T_c which is present in all samples [15] and has been commonly attributed to inhomogeneties [16,17] or fluctuations [18,19]. The multicomponent scenario raises a new possibility for this sharp peak near T_c as being due to a third superconducting component (C). Such a sharp peak can be obtained from Mattis-Bardeen calculations assuming an s-wave order parameter when a very large gap value $\Delta_C(0)/kT_c \sim 10-15$ is used. This speculation receives support from the fact that structure at very high energies of $100mV$ is also observed in STM tunneling characteristics [13] which may match with the high energy scale of the sharp peak near T_c.

While phase sensitive experiments [20,21] suggest a dominant d-wave superconducting state, an $s+d$ state is likely in $YBCO$ because of its orthorhombic crystal structure. This would be consistent with c-axis tunneling experiments [22]. However all of these experiments appear to be carried out at low temperatures near $4.2K$. Our present results would suggest that the underlying symmetry could change with temperature. Hence careful experiments should be performed above and below $60K$ in high quality $YBCO$ samples.

Several theories [23–26,7,27,28] have addressed the issue of mixed order parameter symmetry or multi-component superconductivity in cuprates and scenarios based on many of these ideas would be consistent with our striking microwave results.

In summary, these data represent direct observation of the presence of *multiple* components in the superconducting state of $YBa_2Cu_3O_{6.95}$ [5].

REFERENCES

1. W. N. Hardy et al., Phys. Rev. Lett. **70**
2. T. Jacobs et al., Phys. Rev. Lett. **75** (1995) 4516
3. A. Erb, E.Walker and R. Flukiger, Physica C **258** (1996) 9
4. S. Sridhar, D.-H. Wu and W. L. Kennedy, Phys. Rev. Lett. **63** (1989) 1873
5. H. Srikanth et al., cond-mat preprint 9610032 (1996)

6. D. A. Bonn, P. Donsajh, R. Liang and W. N. Hardy, Phys. Rev. Lett. **68** (1992) 2390
7. J. Betouras and R. Joynt, Europhys. Lett. **31** (1995) 119
8. J. Ma, C. Quitmann and M. Onellion, Science **267** (1995) 862
9. Z. Ma et al., Phys. Rev. Lett. **71** (1993) 781
10. S. Oxx et al., Physica C **235-240** (1994) 889
11. S. F. Lee et al., Phys. Rev. Lett. **77** (1996) 735
12. N. Klein et al., Phys. Rev. Lett. **71** (1993) 3355
13. I. Maggio-Aprile et al., Phys. Rev. Lett. **75** (1995) 2754
14. R. Gagnon, S. Pu, B. Ellman and L. Taillefer, Phys. Rev. Lett. **78** (1997) 1976
15. K. Holczer et al., Phys. Rev. Lett. **67** (1991) 152
16. H. K. Olsson and R. H. Koch, Phys. Rev. Lett. **68** (1992) 2406
17. A. A. Golubov et al., J. Phys. I France **6** (1996) 2275
18. M. L. Horbach, W. van Sarloos and D. A. Huse, Phys. Rev. Lett. **67** (1991) 3464
19. S. M. Anlage, J. Mao and J. L. Peng, Phys. Rev. **B 53** (1996) 2792
20. J. R. Kirtley, C. C. Tsuei and L. S. Yu-Jahnes, Nature **373** (1995) 225
21. D. J. Van Harlingen, Rev. Mod. Phys. **67** (1995) 515
22. A. G. Sun, D. A. Gajewski, M. B. Maple and R. C. Dynes, Phys. Rev. Lett. **72** (1994) 2267
23. P. N. Spathis, M. P. Soerensen and N. Lazarides, Phys. Rev. B **45** (1992) 7360
24. C. O'Donovan and J. P. Carbotte, Phys. Rev. B **52** (1995) 16208
25. I. I. Mazin et al. Phys. Rev. Lett., **75** (1995) 2574
26. R. Combescot and X. Leynoras, Phys. Rev. Lett. **75** (1995) 3732
27. X. Leyronas and R. Combescot, cond-mat preprint 9612145 (1996)
28. M. Beal-Monod and K. Maki, preprint (1997)

ELSEVIER

Physica C 282–287 (1997) 260–263

PHYSICA C

The charge-density-wave quantum-critical-point scenario

C. Castellani, C. Di Castro, and M. Grilli[a*]

[a]Istituto di Fisica della Materia e Dipartimento di Fisica, Università di Roma "La Sapienza", Piazzale A. Moro 2, 00185 Roma, Italy

The presence of a charge-driven quantum critical point near optimal doping in the superconducting cuprates is shown to account for the non-Fermi liquid normal state properties, *d*-wave superconductivity, strong doping dependence of the superconducting critical temperature

1. INTRODUCTION

The understanding of the anomalous normal-state properties of the superconducting cuprates requires singular scattering processes at low energy. Singular scattering can arise from gauge field fluctuations[1]. An alternative point of view is related to the existence of a quantum critical point (QCP), where critical fluctuations can mediate singular interactions between the quasiparticles, providing at the same time a strong pairing mechanism. The proximity to the critical point at zero temperature is characterized by the absence of any energy scale besides the temperature itself. Various proposals have been put forward on the possible nature of the QCP, ranging from an antiferromagnetic (AF) mechanism[2], to charge-transfer[3], or to incommensurate charge-density-wave (ICDW) [4,5].

The peculiarities of optimal doping make this point a natural candidate where to place a QCP relevant for the superconducting and the non-Fermi liquid (non-FL) properties of the cuprates. As far as the nature of this critical point is concerned, it is natural that charge degrees of freedom should play the major role, since the disordered region of this QCP coincides with the highly metallic overdoped regime. The existence of an ICDW-QCP is not alternative to the existence of an AF-QCP and the two QCPs control the behavior of the system at different doping. The ICDW-QCP sets up the maximum superconducting critical temperature and can constitute the substrate to substain AF fluctuations far away from the AF ordered phase, by allowing for hole-rich and hole-poor "stripes".

After phase separation (PS) was shown to be present in the phase diagram of the t-J model [6], we pointed out that PS commonly occurs in strongly correlated models with short-range interactions and stressed its relation with superconductivity[7,8]. Emery and Kivelson [9] suggested that, although long-range Coulomb (LRC) forces spoil PS as a static thermodynamic phenomenon, the frustrated tendency towards PS may lead to the formation of a phase with hole-rich and hole-poor stripes. This latter result is on the same line of the finding of Refs. [8,4], where an ICDW phase was shown to arise in models where PS is spoiled by LRC forces. Our finding is then that in all these models there exists a QCP associated to the formation of ICDW.

Several experimental findings provide support for the existence of a QCP at (or near) optimal doping. This is seen in recent transport[10] experiments in $La_{2-x}Sr_xCuO_4$ (LSCO) with high magnetic fields, which allow to access the normal phase hidden by superconductivity. Indications in the same sense are provided by neutron scattering[11] and by the qualitative changes of behavior at optimal doping detected by optical spectroscopy, NMR, susceptibility, neutron scattering, photoemission, specific heat, thermoelectric power, Hall coefficient, resistivity.

Many indications exist that the above QCP is related to charge ordering. A direct observation of charge-driven ordering was possible by neutron

*Part of this work was carried out with the financial support of the Istituto Nazionale di Fisica della Materia, Progetto di Ricerca Avanzata 1996.

PII S0921-4534(97)00243-8

scattering[12], in $La_{1.48}Nd_{0.4}Sr_{0.12}CuO_4$ where the low temperature tetragonal lattice structure is believed to pin the CDW (see also the case of $La_{1.88}Ba_{0.12}CuO_4$). Increasing the Sr content at fixed Nd concentration, the pinning effect is reduced leading to metallic and superconducting behavior. I this latter case, the existence of dynamical ICDW fluctuations is suggested by the presence of dynamical incommensurate spin scattering. In this regard, also $La_{2-x}Sr_xCuO_4$ is expected to display dynamical charge fluctuations with doping-dependent spatial modulation as indeed observed in the magnetic scattering[13]. ICDW have been proposed from extended X-ray absorption fine structure (EXAFS) experiments both in optimally doped LSCO[14](a) and $Bi_2Sr_2CaCu_2O_{8+x}$ (Bi-2212)[14](b). Superstructures have also been detected in Bi-2212 from X-ray diffraction[14](c).

2. THE ICDW-QCP AND THE PHASE DIAGRAM

According to the scenario outlined in Section I, the ICDW-QCP is the crucial ingredient characterizing the physics of the cuprates all over their metallic (under-, optimally, and over-doped) regime. The ICDW-QCP is characterized by singular interactions, which in the RPA-like treatment assumed the form[4]

$$\Gamma(\boldsymbol{q},\omega) \approx \tilde{U} - \frac{1}{4}\sum_{\alpha} \frac{V}{\kappa^2 + \omega^{\alpha}_{\boldsymbol{q}} - i\gamma\omega} \qquad (1)$$

where the sum is over the four equivalent vectors of the CDW instability $\boldsymbol{q}^{\alpha} = (\pm q_c, 0), (0, \pm q_c)$ and $\omega^{\alpha}_{\boldsymbol{q}} = 2(2 - \cos(q_x - q^{\alpha}_x) - \cos(q_y - q^{\alpha}_y))$ (all q's are in units of inverse lattice spacing). Approaching the ICDW-QCP the mass term κ^2 vanishes as $(\delta - \delta_c)$. From our analysis of specific models (the Hubbard-Holstein in Ref.[4]) we found that the instabilities are at or close to the (1,0) or (0,1) directions with $|\boldsymbol{q}_c| \approx 1$. However, the scattering is quite strong, although non-singular, in all directions for $|\boldsymbol{q}| \approx |\boldsymbol{q}_c|$. The functional dependence in Eq.(1) is not strictly related to the specific origine of the QCP, while it could depend on the approximate (nearly-mean-field) treatment. For the sake of definiteness and simplicity, we will assume the form (1) (therefore the dynamical critical index $z = 2$) and a simple Gaussian behavior of the QCP.

2.1. The ICDW-QCP in the absence of pairing

From the theory of QCP's[15,16], one can schematically draw the phase diagram, which would be valid *in the absence of any superconducting pair formation* (thin lines in Fig. 1) The

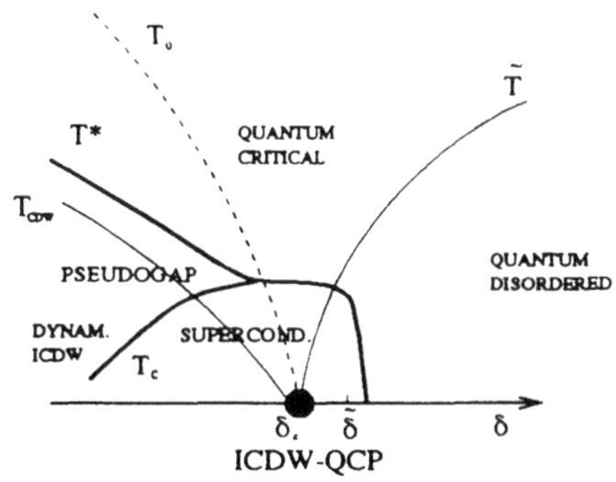

Figure 1. Schematic structure of the phase diagram around the ICDW-QCP in the presence of superconducting pairing.

overdoped region on the right of the QCP corresponds to the quantum disordered regime, where $\kappa^2 = \xi^{-2} \sim a(\delta - \delta_c)$. Increasing the temperature above the crossover temperature $\tilde{T}$ we enter in the so-called classical gaussian regime where κ^2 becomes a function of T, $\kappa^2 \sim bT$.

The region on the left would generically correspond to the ordered ICDW phase occurring below a critical temperature $T_{CDW}(\delta)$ starting from the QCP at $\delta_c(T = 0)$. The true critical line is naturally depressed with respect to its mean-field expression T_0, which contains the quantum dynamical reduction controlled by the proximity to the QCP. The region between the two curves is dominated by strong precursor effects.

The quantum disordered region on the right corresponds to the overdoped region of the cuprates with FL behavior. The classical gaussian region around optimal doping has no energy scales, but the temperature. Here the best non-FL behavior is obtained. In particular, with the scattering of the form given in Eq. (1), a linear-in-T resistivity is expected in $d = 2$ [4].

The scenario presented so far should find a physical correspondence whenever the superconducting pair formation is forbidden. Indeed the transport experiments in LSCO under strong magnetic field[10] display a metal-insulator crossover, ending at T=0 at a QCP at optimal doping (see Fig. 3 in Ref.[10]). The experimental line separating the planar metal from the insulator would correspond in our picture to the "true" T_{CDW} critical temperature as a function of doping. We find quite remarkable that for filling close to the "magic" value 1/8, the experimental temperature for the insulator is substantially higher (see Fig.1(a) of Ref.[10]) than for nearby filling values. This indicates that commensuration effects at this particular doping pin the thermal ICDW fluctuations leading to a high T_{CDW}, strengthening the indications of a charge-ordering phenomenon.

As far as T_0 is concerned, this temperature marks the onset of the ICDW precursors and is characterized by a loss of spectral weight at low energies giving rise to a uniform decrease of the density of states near the Fermi energy. This shows up as the well known decrease of the uniform magnetic susceptibility below a characteristic temperature, which vanishes by approaching the optimal doping[18]. In underdoped $YBa_2Cu_3O_{6+x}$ (YBCO) compounds, this last temperature has also been put in correspondence[17] with the temperature below which the planar resistivity ρ_{ab} deviates from its linear behavior, while the interplane resistivity ρ_c acquires a non metallic behavior[20]. This finding is then compatible with the further identification of our T_0 line with a second line present in the phase diagram of LSCO in high magnetic field [10], separating the metallic from non-metallic ρ_c. Consistently with our scenario this crossover line also ends on the QCP at optimal doping.

2.2. The ICDW-QCP in the presence of pairing

The above scenario is drastically modified, once superconducting pairing is considered[5]. In particular it is found that the singular interaction of Eq.(1) is also present in the particle-particle channel, thus providing a strong pairing mechanism by approaching the critical point. Pairing is of d-wave symmetry and, being mediated by an interaction rapidly varying with $\kappa^2 \simeq Max[a(\delta - \delta_c), bT]$, is strongly depends on doping. The maximal critical temperature is obtained in correspondence to the quantum-disordered/quantum-critical crossover ($\delta_{opt} \approx \tilde{\delta}$) [5].

This analysis of pairing is confined to the over- and optimally doped regions of the phase diagram, where the fluctuations are not strong enough to destroy the homogeneous character of the system. In the underdoped region the ICDW fluctuations become critical in the proximity of the line $T_{CDW}(\delta)$ where the ICDW transition would occur in the absence of superconducting pairing. Approaching T_{CDW} the attractive fluctuations would lead to formation of (local) pairs along a curve $T^*(\delta)$. As a consequence of strongly paired quasiparticles, below $T^*(\delta)$ pseudogap effects would show up as seen in many experiments (NMR, ARPES, optical conductivity, specific heat, ...). However, despite the strong pairing, the true superconducting critical temperature is lower than T^* and it decreases inside the underdoped pseudogap region. This occurrence is schematically depicted in Fig. 1 by the bifurcation of the heavy line. It is worth noting that the low-temperature gap in the underdoped cuprates weakly depends on doping, while T^* increases fast by decreasing δ[19,21]. We believe that this peculiar behavior requires a remarkable temperature dependence of the pairing potential as implied by our ICDW scenario in the underdoped region. To get an insight on the physics of the pseudogap phase, we introduced in κ^2 in Eq. (1) a modified temperature dependence given by the distance from the critical line T_{CDW} and added a contribution due to the presence of a local superconducting gap $\kappa^2 \equiv Max\,[|\Delta_{Max}(T)|, c(T - T_{CDW})]$. We aim, in this way, to introduce in the pairing potential the stabilizing effect of local superconducting order on the ICDW instability. Δ_{Max} is the maximum value of $\Delta(\boldsymbol{k})$ in k-space. The decreasing of the fluctuations above the ICDW critical line is considered via $c(T - T_{CDW})$. For simplicity, we assumed the BCS equation to be valid to estimate the (local) d-wave gap function. The result[22] for Δ_{Max}, is reported in Fig. 2 and bears a striking resemblance with the analogous quantity recently measured with ARPES in underdoped Bi-2212 samples[21]. Of course the above BCS treat-

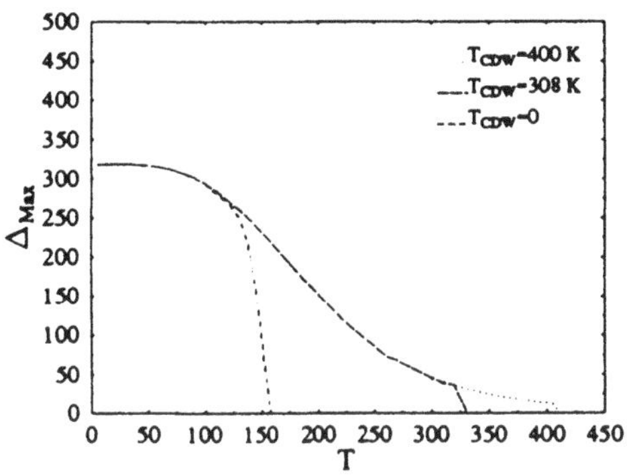

Figure 2. Δ_{Max} vs T for different values of T_{CDW} In the pairing potential $V = 0.30$ eV are used.

ment only deals with the amplitude of the gap and says nothing on the way a true superconducting phase coherence is established at $T_c < T^*$. For this we have to invoke phase fluctuations as in the usual (large) negative-U Hubbard model. Here the strongly peaked q-dependence of the pairing potential, which only couples few points on the Fermi surface, and precursors of (1D) stripe formation are expected to give rise to strong phase fluctuations.

We also emphasize that the occurrence of local pairing prevents the actual establishing of the ICDW long-range order, so that T_{CDW} merely encloses the area where pairing and strong local-dynamical ICDW order interplay.

In conclusion the marked non-FL character of the optimally doped quantum disordered region, the d-wave symmetry of the superconducting order and the peculiar doping dependence of the superconducting critical temperature in the over- and optimally doped regions are natural consequences of the ICDW-QCP scenario. It is also remarkable that the appearance of a temperature dependence for the pairing potential implied by this scenario accounts well for the peculiar(so far unexplained) temperature behavior of the local gap and the non-trivial relation between T^* and $\Delta_{Max}(T = 0)$.

REFERENCES

1. N. Nagaosa and P. A. Lee, Phys. Rev. Lett. **64**, 2450 (1990)
2. P. Montoux and D. Pines, Phys. Rev. B **B50**, 16015 (1994) and references therein.
3. C. M. Varma, cond-mat 9607105 and references therein.
4. C. Castellani, C. Di Castro, M. Grilli, Phys. Rev. Lett. **75**, 4650 (1995).
5. A. Perali, C. Castellani, C. Di Castro, and M. Grilli, Phys. Rev. B **54**, 16216 (1996).
6. V. J. Emery, *et al.* Phys. Rev. Lett. **64**, 475 (1990); M. Marder, *et al.* Phys. Rev. B **41**, 6920 (1990).
7. C. Di Castro and M. Grilli, Phys. Scr. T **45**, 81 (1992) and references therin.
8. R. Raimondi, *et al.*, Phys. Rev. B **47**, 3331 (1993).
9. V. Emery and S. Kivelson, Physica C **209**, 597 (1993); V. Emery, *et al.* cond-mat 9610094.
10. G. S. Boebinger,*et al.*, Phys. Rev. Lett. **77**, 5417 (1996).
11. T. Mason, *et al.*, Phys.Rev.Lett. **77**, 1604 (1996).
12. J.M. Tranquada, *et al.* Nature **375**, 561 (1995); Phys. Rev. B **56**, 7689 (1996).
13. K. Yamada, *et al.* preprint 1996.
14. (a) A. Bianconi, *et al.*, Phys. Rev. Lett. **76**, 3412 (1996); (b) Phys. Rev. B **54**, 12018 (1996); (c) *ibid.*, 4310 (1996).
15. S. Sachdev and J. Ye. Phys. Rev. Lett. **69**, 2411 (1992).
16. A. J. Millis, Phys. Rev. B **48**, 7183 (1993).
17. T. Ito, *et al.*, Phys.Rev.Lett. **70**, 3995 (1993); B. Batlogg, *et al.*, Physica C **235**, 130 (1994).
18. D. Johnston, Phys.Rev.Lett. **62**, 957 (1989).
19. J. Loram, *et al.* Physica C **235**, 134 (1994).
20. K. Takenaka, *et al.*, Phys. Rev. B **50**, 6534 (1994).
21. J. M. Harris, *et al.*, cond-mat 9611010; H. Ding, *et al.*, Nature **382**, 51 (1996).
22. C. Castellani, C. Di Castro, M. Grilli, and A. Perali, unpublished.

ELSEVIER

Physica C 282–287 (1997) 264

PHYSICA C

Extended Abstract

Marginal Fermi-liquid state and pairing instability in a general model of copper-oxides

C. M. Varma

Bell Laboratories, Lucent Technologies
700 Mountain Avenue, Murray Hill, NJ 07974, USA

A model of copper-oxygen bonding and anti-bonding bands with the most general two-body interactions allowable by symmetry is systematically solved. The model has a continuous transition as a function of hole-density χ and temperature T to a phase in which a current circulates in each unit cell. This terminates at a quantum critical point at $\chi = \chi_c$, T = 0. Near this point the logarithm of the frequency of the current fluctuations scales with their momentum. This appears necessary to derive the microscopic basis for the marginal Fermi liquid phenomenology and to understand the observed long wavelength transport anomalies near $\chi = \chi_c$. They symmetry of the current fluctuations is such that the associated magnetic field fluctuations are absent at oxygen sites. Several experiments are proposed to test the theory. Details of this work are given in C. M. Varma, Phys. Rev. (to be published), available on http://www.lanl.gov as cond-mat/9607105. The application of the theory to NMR is given in C. M. Varma, Phys. Rev. Lett., **77**, 3431 (1966). Effects of impurity scattering are given in Phys. Rev. Lett. (submitted/available on http://www.lanl.gov as cond-mat/9703122.

0921-4534/97/$17.00
PII S0921-4534(97)00246-3

ELSEVIER

Physica C 282–287 (1997) 265–268

PHYSICA C

The $SO(5)$ theory of high T_c superconductivity

Shou-Cheng Zhang[a]

[a]Department of Physics, Stanford University, Stanford, CA 94305

This paper gives a simple introduction to the $SO(5)$ theory of high T_c superconductivity. Current status and relation to experiments are summarized.

One of the most strikingly universal properties of high T_c superconductivity is the close proximity between the superconducting (SC) and the antiferromagnetic (AF) phases. The AF exchange coupling J is responsible for the AF phase at half-filling. The same couping could lead for formation of spin singlets, a prerequesite for superconductivity. On the other hand, while the origins of these two phenomena appear to be related, it is hard to imagine a greater difference in the physical properties between a insulator and a superconductor. Is it possible that behind the apparent difference, these two phases enjoy a deeper and fundamental unity?

Let us take the example of electricity and magnetism in vacuum, two of the best understood phenomena in physics. Before the 18th century, it is widely believed that these two phenomena are fundamentally distinct. However, the works of Faraday and Maxwell showed that they are in fact deeply related. In the theory of relativity, the electric field $\vec{E}$ and the magnetic field $\vec{B}$ are unified into a electromagnetic field tensor $F_{\mu\nu}$. The magnetic field $\vec{B}$ is nothing but the electric field $\vec{E}$ viewed from a rotated coordinate system in the four dimensional Minkowski space. Is it possible, that the two basic phases of the cuperates, AF and SC phases, are in fact related to each other by a simple rotation in some higher dimensional space?

A recently proposed theory of high T_c superconductivity unifies AF and the d−wave SC phases and treat them on equal footing[1]. The AF phase is described by a three dimensional order parameter N_α, the staggered magnetization. Therefore, it has spin 1, charge 0 and total momentum (π, π). On the other hand, a spin singlet d-wave SC phase is described by a complex order parameter Δ with two real components, which has spin 0, charge ± 2 and total momentum 0. The idea of the $SO(5)$ theory is to group these five components together into a object called superspin, $n_a = (Re\Delta, N_x, N_y, N_z, Im\Delta)$ and ask if there exists well-defined rotation operators which can transform AF into SC and vice versa. Such operators must have spin 1, charge ± 2 and total momentum (π, π) in order to patch up the difference between the AF and SC order parameters. These quantum numbers determine the form of the operator uniquely up to a form factor. One of them[2] is given by $\pi^\dagger = \sum_{\mathbf{k}}(\cos k_x - \cos k_y) c^\dagger_{\mathbf{k}+\pi,\uparrow} c^\dagger_{-\mathbf{k},\uparrow}$ Since this operator has spin 1, one can obviously define three of them $\pi^\dagger_\alpha$, where α is a spin index. This operator also has charge +2, one can therefore define its hermitian conjugate π_α which has charge −2. Remarkably, these operators rotate AF order parameter into the d-wave order parameter

$$[\pi^\dagger_\alpha, N_\beta] = i\delta_{\alpha\beta}\Delta^\dagger \tag{1}$$

and vice versa. Together with the total spin operators S_α and the total charge operator Q, the six π's can be identified with the 10 components of a antisymmetric tensor $L_{ab} = -L_{ba}$ in five dimensions, and they form the generators of the five dimensional rotation group, $SO(5)$. The $SO(5)$ group contains the familiar $SO(3) \times U(1)$ spin and charge symmetry as a subgroup.

We have thus apparently accomplished the task of unifying AF with SC: they are grouped into a five dimensional object and SC is nothing but AF viewed in some rotated coordinates and vice

PII S0921-4534(97)00248-7

versa! This construction looks a bit similar to the unification of $\vec{E}$ and $\vec{B}$ by the Lorentz group. But so far this is only a mathematical fantasy, we haven't asked if Mother Nature approves the $SO(5)$ symmetry or not. In the high T_c problem, Mother Nature is very complicated, but we can check the $SO(5)$ symmetry within some simple models. One can easily check the $SO(5)$ symmetry by evaluating the commutator between the Hamiltonian with the π operators. Analytical[2] and numerical[3] works show that that the π operators are approximate eigen-operators of the Hubbard Hamiltonian, in the sense that

$$[\mathcal{H}, \pi_\alpha^\dagger] = \omega_0 \pi_\alpha^\dagger \quad (2)$$

where eigen-frequency ω_0 is of the order of J, and proportional to the number of holes. This relation reminds us of the commutation relation between the transverse spin components in a magnetic field and the Zeeman Hamiltonian. Therefore, the $SO(5)$ symmetry is broken explicitly by the chemical potential, but the pattern of explicit symmetry breaking is simple and familiar, and therefore easy to handle. In particular, the Casimir operator $\sum_{a<b} L_{ab}^2$ (a generalization of the total spin operator $\vec{S}^2$) of the $SO(5)$ group commutes with the Hamiltonian if (2) is valid. The approximate relation (2) is highly nontrivial. One could ask if a similar relation would exist for a modified π operator which rotates AF into s-wave SC order parameters. The answer is negative[2,3]. Therefore, there is only a approximate symmetry between AF and d-wave SC near half filling.

The ideas of the $SO(5)$ symmetry can now be used to construct a poor man's model of a high T_c superconductor. The reason that the high T_c problem appear to be hopelessly complicated is because fermion problems are hard to deal with. But the recent experimental discovery of the pseudogap[4] gives a ray of hope. At a temperature T_{MF} higher than the true T_c, the fermions already appear to be gapped. Better yet, these two temperature scales go in opposite direction as one approaches the AF phase from the SC side. This means that for low temperature properties in the underdoped regime, one can forget about the fermions and concentrate on the collective degrees of freedom, and map the whole problem into a "effective magnetic problem" involving the $SO(5)$ superspin.

We can define T_{MF} to be the temperature below which the superspin acquires a finite magnitude, but its orientation is still not fixed. Below this temperature, the fermionic degrees of freedom are quenched. One can always rescale the parameters so that the constraint $n_a^2 = 1$ is satisfied. We can define superspin vectors locally and end up with a quantum rotor model with some moment of inertia and gradient coupling. At half filling, where the chemical potential is defined to vanish, we assume that the $SO(5)$ symmetry is broken in such a way that AF phase is favored. This is similar to a magnetic probelm with "easy directions". The most important question is to ask what happens when the chemical potential deviates from zero.

The chemical potential μ couples to the charge, one of the symmetry generators in the $SO(5)$ theory. It does not couple to either AF or SC order parameters directly. Therefore, one's naive expectation is that the superspin direction is unaffected. However, a close inspection shows that the rotation generators are not independent of the order parameters. The simplest example of this kind of relation is the constraint between the total spin S_α and the Neel vector N_α of a antiferromagnet: $S_\alpha N_\alpha = 0$. Neel first derived this result by expressing S_α and N_α in terms of the sum and the difference of the sublattice magnetization. However, this orthogonality relation has a simple geometric interpretation. The $\vec{N}$ vector is confined to lie on a sphere. The $\vec{S}$ operator is a generator of the rotation of the $\vec{N}$ vector, therefore it has to lie in a plane tangent to the sphere. The Neel orthogonality relation has a extremely important physical consequence. If we apply a uniform magnetic field to a antiferromagnet, it induces a total spin along the field direction. The constraint relation immediately tells us that the Neel vector has to be perpendicular to the field direction. This phenomenon is called a spin-flop transition induced by a uniform magnetic field.

How does this picture generalize to higher component spins? In higher dimensions, the rotation generators can not be described as vectors, but can always be described by antisymmetric tensors L_{ab}. The index ab specify a plane tangent to a sphere traced out by the n_c vector, and the generalized orthogonality relation reads

$$L_{ab}n_c + L_{bc}n_a + L_{ca}n_b = 0 \quad (3)$$

This equation can be proved by expressing the angular momentum operator in the usual fashion, $L_{ab} = n_a p_b - n_b p_a$, where p_a is the momentum conjugate of n_a, and substituting it back into (3). This simple equation expresses a geometric property of a hyper-sphere, but interpreted in the $SO(5)$ theory, it provides a powerful constraint between three most important physical quantities in the high T_c problem: doping (which is the convention of [1] is identified with L_{15}), AF (n_2, n_3, n_4) and SC (n_1, n_5). Since only the charge operator L_{15} couples to a external field, namely the chemical potential, the expectation values of all other generators vanish. The constraint equation (3) immediately tells us that at finite doping, the superspin has vanishing AF components and a finite SC component. Therefore, while the Neel constraint equation tells us about the physics of the spin flop, the $SO(5)$ constraint equation (3) tells us about the physics of the superspin flop, *i.e.* the transition from a AF phase to the SC phase induced by the chemical potential. In the $SO(5)$ theory, the mysterious transition from AF to SC upon doping is explained by a simple geometric property of a sphere is five dimensions!

Once the analogy with the spin flop problem and the superspin flop problem is realized, one can simply borrow the spin-flop phase diagram to construct the phase diagram of the high T_c superconductors in the temperature and chemical potential plane, see figure 1. When the chemical potential μ is less than some critical value, say μ_c, the superspin prefers to lie in the AF direction. The chemical potential plays the role similar to the uniform magnetic field in the spin flop problem. Beyond the critical μ_c, the superspin flops from the AF direction to the SC plane. When temperature is raised at a constant chemical potential, the AF state undergoes a second order transition at T_N, while the SC state undergoes a second order transition at T_c. These two second order transition lines meet at a bicritical point T_{bc}. At this point, the $SO(5)$ symmetry becomes exact due to critical fluctuations. Since this point has the most thermal and quantum fluctuations in the entire phase diagram, both T_c and T_N are depressed near μ_c. The $SO(5)$ theory predicts that both second order lines merge into the first order lines tangentially, with a behavior close to a square root singularity. Because of the materials difficulty in the underdoped regime, it is experimentally unclear if T_c and T_N actually meet at a single point or are detached from each other. More experimental work in this direction is clearly desired to test this crucial prediction of the $SO(5)$ theory.

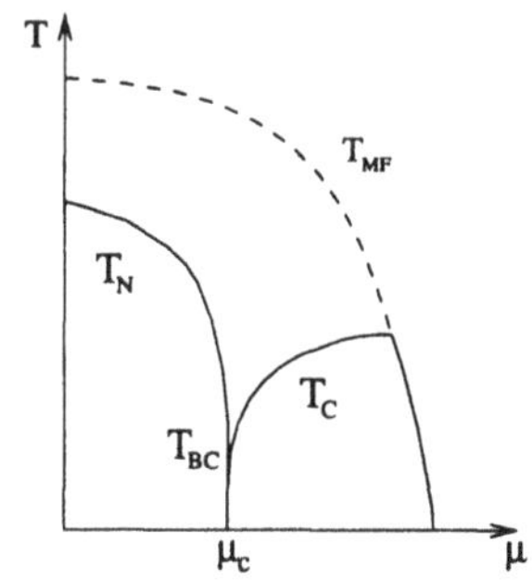

It is important to note that we plotted the phase diagram as a function of μ, not doping x. Because the density jumps discontinuously across a first order transition line, the plot as a function of doping x would contain a coexistence region. Physics in the region maybe very interesting in itself. The long range Coulomb interaction could lead to a stripe order of alternating AF and SC phases.

So far the most direct evidence of the $SO(5)$ symmetry come from the resonant neutron scattering peaks in the $YBCO$ superconductors below T_c[6]. These resonances have spin 1, momentum (π, π), and resolution limited peaks at $41meV$, $33meV$ and $25meV$ for materials with $T_c = 92K$, $T_c = 67K$ and $T_c = 52K$ respectively. The resonance energy scales with T_c, but is not simply related to the size of the SC gap, since recent photoemission experiments show that the

SC gap increases with decreasing doping[5].

These resonance peaks have a natural explanation within the $SO(5)$ theory. In the previous discussions, we argued that beyond a critical chemical potential μ_c, the superspin vector lies within the SC plane (n_1, n_5). However, this notion is a classical one, since the Heisenberg uncertainty relation does not allow the angle of the superspin to be sharply defined. In fact, there is zero point motion of the superspin into the AF directions (n_2, n_3, n_4). What are the appropriate coordinates describing this zero point motion? Let us recall the fundamental $SO(5)$ commutation relation (1), which tells us that the $\pi_\alpha^\dagger$ operator rotates AF into SC. In the SC state, we can approximate the right-hand-side of (1) by a c-number expectation value, and this equation can now be interpreted as the Heisenberg commutation relation between the canonical momentum p and coordinate q of a harmonic oscillator. The eigenfrequency of this oscillator can be simply read off from equation (2). This oscillator can be naturally identified with the resonant neutron peak observed in the $YBCO$ superconductors. It has momentum (π, π), spin 1 and a resonance energy which scales with the hole doping. Since the harmonic oscillator interpretation crucially depends on the SC order parameter having a finite expectation value, one would expect the mode to disappear above T_c, which is again consistent with the experiments in the $T_c = 92K$ and $T_c = 67K$ superconductors. The $T_c = 52K$ material shows a broad peak even above T_c. The situation with this system is a bit unclear, it could be related to the intrinsic disorder present in the $T_c = 52K$ material, but we do not have a simple theory for it at this moment.

A number of other experimental consequences of the $SO(5)$ are currently being worked out. The $SO(5)$ theory predicts that a SC vortex in the underdoped regime has a AF core. Such a configuration is called a "meron" in field theory literature. In this configuration, the superspin lies within the SC plane far away from the vortex core, but it rotates from the SC plane into the AF sphere as the vortex core is approached in the radial direction. A vortex lattice with AF core could be detected as satellite peaks in the elastic neutron scattering experiment. It can also be detected by muon spin resonance inside the vortex core. Since the AF vector lies in a plane perpendicular to the applied field, this could give a distinct signature in the μSR experiment. The $SO(5)$ theory also predicts a charge doublet excitation in the AF state, which is the cousin of the spin triplet excitation in the SC state. In a AF state, the superspin vector lies within the AF sphere, however, its zero point motion leads to a fluctuation into the SC plane. Similar to the spin triplet resonance, the charge doublet resonance should only appear below the Neel temperature T_N. Although the phase diagram of the $SO(5)$ theory is qualitatively similar to many other alternative theories, it distinctively predicts a direct first order phase transition from AF to SC and a bi-critical point where both T_N and T_c merge. It is hard to access the AF/SC transition region experimentally because of sample inhomogeneities. This is certainly a major challenge which can hopefully be resolved experimentally in the near future.

This work is heavily based in the insights gained from the previous theoretical works in the field. However, due to space limitations, readers are refered to reference [1] for detailed discussion of the relationships. This work is supported in part by the NSF under grant numbers DMR-9400372 and DMR-9522915.

REFERENCES

1. S.C. Zhang, *Science*, 275, xxx, (1997).
2. E. Demler and S.C. Zhang, *Phys. Rev. Lett.*, 75, 4126, (1995).
3. S. Meixner, W. Hanke, E. Demler and S.C. Zhang, *preprint*, cond-mat/9701217.
4. N.P. Ong, *Science*, 273, 321, (1995).
5. J. M. Harris *et al*, *preprint*.
6. J. Rossat-Mignod et al. *Physica C*, 185:86, 1991; H. Mook et al. *Phys. Rev. Lett.*, 70:3490, 1994; H.F. Fong et al. *Phys. Rev. Lett.*, 75:316, 1995; H.F. Fong et al. *Phys. Rev. Lett.*, 78:713, 1997; P. Dai et al. *Phys. Rev. Lett.*, 77,5425, 1996.

ELSEVIER

Physica C 282–287 (1997) 269–272

PHYSICA C

BIPOLARON KINETICS OF HIGH-Tc CUPRATES

A.S. Alexandrov

Department of Physics, Loughborough University, Loughborough LE11 3TU, UK,
and
IRC in Superconductivity, University of Cambridge, Cambridge CB3 OHE , UK.

Based on the generic Hamiltonian including the electron-phonon interaction and the direct Coulomb repulsion the ground state of cuprates is shown to be a charged 2e Bose liquid of small bipolarons if the electron-phonon coupling constant λ is about or larger than 0.5. Hole bipolaron band structure is derived for perovskite lattices. A microscopic theory of the normal state in-plane and c-axis transport of copper oxides is developed. The fundamental relationship between the anisotropy of resistivity and the spin susceptibility is derived. The temperature and doping dependence of the in-plane and out-of-plane resistivity as well as the spin susceptibility are found to be in remarkable agreement with the experimental data in underdoped, optimally and overdoped $La_{2-x}Sr_xCuO_4$ for the entire temperature regime from Tc up to 600 K. The normal state gap is explained and its doping and temperature dependence is clarified. Unusual logarithmic temperature dependence of resistivity at very low temperatures is quantitatively described as well. A new phase diagram of doped cuprates is proposed.

1. INTRODUCTION

Extensive experimental studies of high-T_c cuprates and doped fullerenes over the last decade showed that these materials belong to a family of doped semiconductors with a high carrier density, and a low mobility about the Mott-Ioffe-Regel limit [1]. Like in many transition metal oxides carriers form small polarons whose characteristic spectral function has been observed in infrared conductivity of the insulating and superconducting compounds [2,3]. Rather heavy small polarons readily form small bipolarons. In ionic solids the essential part of the Coulomb repulsion is screened by optical phonons, the deformation potential and molecular vibrations can therefore easily bind two small polarons at a short distance. There are hallmarks of small bipolarons such as a superfluid phase transition similar to that of He4 [4], a spin gap in the magnetic susceptibility, (if a singlet is the ground state [1]), a divergent upper critical field [5], and quite a few other bosonic features observed in cuprates [1]. In particular, the in-plane normal state transport as well as the normal state pseudogap observed with NMR [6] and infrared reflectivity [7] were well described by the preformed bosons [8,9,10]. Also the 'T_c-λ_H' plot [11] was explained by us [12] including the overdoped region.
The latest photoemission experiments [13] showed a pseudogap above T_c which we believe is the bipolaron binding energy. All these data invite an interpretation that there is normal state pairing of carriers in cuprates.

Here I propose the coherent theory of the normal state transport including an explanation of the logarithmic low-temperature resistivity, magnetic susceptibility and the normal state gap, and suggest a new phase diagram of high-T_c cuprates.

2. BIPOLARON CONFIGURATION AND TUNNELING IN CUPRATES.

The small polarons are paired above T_c in strongly correlated Mott-Hubbard insulators with the electron-phonon coupling constant above an intermediate value about 0.5 [1].

The ground and low-energy states are well described by mixture of the singlet pairs (small bipolarons) and thermally excited polarons. The proposed kinetics of cuprates is derived from the generic Hamiltonian, which includes the bare hopping integral, the electron-phonon interaction and the direct Coulomb repulsion.
It can be diagonalised exactly in the limit $\lambda = E_p/E_F = \infty$ [1], where E_p is the polaron level shift and E_F is the Fermi energy . The ground state bipolaron

PII S0921-4534(97)00250-5

configuration is found by the use of the lattice minimisation technique [14] fully taking into account the direct Coulomb repulsion. Then applying $1/\lambda$ perturbation theory the bipolaron effective mass tensor is readily derived [12].

The intersite oxygen-oxygen pairs appear to be the ground state in perovskite structures. Therefore the in-plane bipolaron tunneling is essentially one-particle with the in-plane effective mass m_{ab}^{**} is of the order of the small polaron mass m_{ab}^{*} about $10m_e$ each. On the other hand, the c-axis tunneling is only possible via a simultaneous two-particle hopping. Therefore, the singlet c -axis mass is strongly enhanced, $m_c^{**} >> m_{ab}^{**}, m_{ab}^{*}$. In that way we explain a large transport anisotropy ($>10^3$) of copper oxides at low temperatures when polarons are frozen out. The crucial point of our theory [15] is that polarons dominate in the c-axis transport at intermediate and high temperatures because they are much lighter in the c-direction than bipolarons. At the same time the polaron contribution to the in-plane transport is small at any temperature due to their low density compared with the bipolaron one. As a result we have a mixture of the nondegenerate quasi-two dimensional spinless bosons and the thermally excited fermions, which are capable of propagating between the planes. Only polarons contribute to the spin susceptibility which leads to a fundamental relation between the anisotropy and the magnetic susceptibility.

3.C-AXIS RESISTIVITY, MAGNETIC SUSCEPTIBILITY AND NORMAL STATE GAP.

Quite generally the in-plane bipolaron, σ_{ab}, c-axis polaron, σ_c conductivities and the uniform spin susceptibility χ_s are expressed as

$$\sigma = -\int dE\sigma(E)\partial f / \partial E$$
$$\chi_s = -2\mu_B{}^2 \int dEN(E)\partial f / \partial E \qquad (1)$$

where $f = [\exp\{(E-\mu)/T\} - 1]^{-1}$ for the in-plane conductivity (ab) and $f = f_p = [\exp\{(E + \Delta/2 - \mu/2)/T\} + 1]^{-1}$ for the c-axis transport and magnetic susceptibility. The bipolaron binding energy Δ is expected to be of the order of a few hundred K. Then polarons are not degenerate at any temperatures. Above T_c bipolarons are not degenerate as well. If the scattering mechanism is the same for polarons and bipolarons the ratio of the differential bipolaron $\sigma_{ab}(E)$ and polaron $\sigma_c(E)$ conductivities is independent of the energy and doping

$$\sigma_{ab}(E) / \sigma_c(E) = const.$$

We expect a highly anisotropic polaron energy spectrum [12] with a quasi one -dimensional polaron density of states as observed with the high resolution ARPES

$$N(E) \approx const / \sqrt{E}$$

Then the resistivity anisotropy ratio is expressed via the spin susceptibility as

$$\rho_c(T,x) / \rho_{ab}(T,x) = Cx / T^{1/2}\chi_s(T,x) \qquad (2)$$

with x is the number of holes, C is constant, and the chemical potential $\mu \sim T \ln(x)$, which corresponds to the carrier density increasing linearly with temperature in agreement with the Hall ratio.

The formula (2) provides a parameter-free fit to the experimental c-axis resistivity of $La_{2-x}Sr_xCuO_4$ [16] with the experimental magnetic susceptibility [17], $\chi = \chi_s(T,x) + 0.4 \times 10^{-7}$ emu/g, as shown in Fig.1. The linear temperature dependence of the ab resistivity has been explained within the same approach [8]. Thus the c axis resistivity is now understood as well.

The normal state gap $\Delta(T,x)$ can be determined from Eq.(1) by the use of the experimental values of $\chi_s(T,x)$ as

$$\Delta(T,x) = 2T \ln\{B_\infty / T^{1/2}\chi_s(T,x)\} \qquad (3),$$

where B_∞ is doping and temperature independent. The result, Fig.2, is about the same as the normal state-gap temperature T^* observed with different kinetic and thermodynamic measurements [18]. The doping and temperature dependence of the normal state gap are the result of screening of the electron-phonon interaction by nondegenerate carriers.

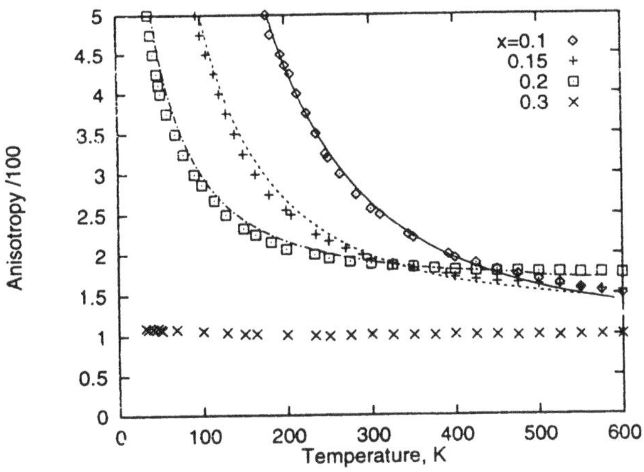

Fig. 1. Theoretical anisotropy (lines) compared with the experiment [16] for $La_{2-x}Sr_xCuO_4$.

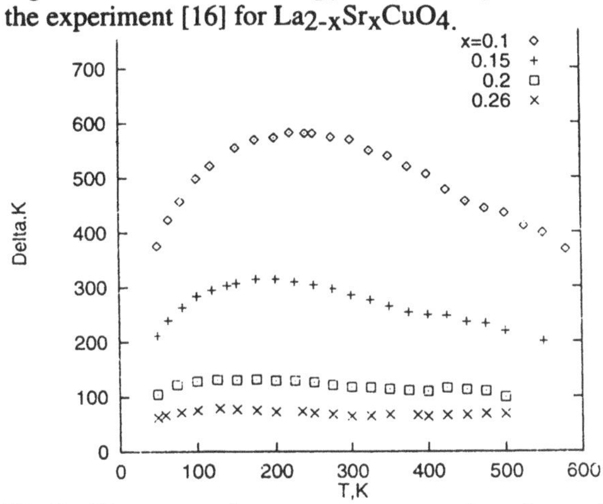

Fig.2. The normal state gap as a function of temperature and doping.

4. LOGARITHMIC RESISTIVITY AND THE PHASE DIAGRAM.

At very low temperatures the transport relaxation time of bipolarons is determined by the elastic boson-impurity scattering and single polarons are frozen out. The normal state of the bipolaronic superconductor is reminiscent of a nondegenerate semiconductor. The characteristic kinetic energy of carriers appears to be of the order of the temperature rather than of the Fermi energy of usual metals. The most effective scattering at low temperatures is then caused by the attractive shallow potential wells which for slow particles is described by the familiar Wigner resonance cross-section

$$\tau^{-1} = 2\pi n_{im} \frac{\sqrt{2E}}{(m^{**})^{3/2}(E+|\varepsilon|)} \tag{4}$$

where $n_{im}(U,a)$ is the number of impurities and ε is the virtual bound state energy, $\varepsilon = -\pi^2 (U - U_{\min})^2 / 16 U_{\min}$ with $U_{\min} = \pi^2 / 8 m^{**} a^2$. By averaging Eq.(4) with respect to the size (a) and the depth (U) of random potential wells one readily obtains the logarithmic transport relaxation time

$$\tau^{-1} \propto \ln(T_0 / E) \tag{5}$$

where $T_0 = \pi^4 / 128 m^{**} a_0^2$ (a_0 is a lattice parameter). Combining the elastic Eq.(5) and inelastic ($\sim T^2$) boson-boson scattering [8] and taking into account the temperature dependent density of carriers due to the Anderson localisation one arrives with

$$\rho_{ab}(T) \propto \frac{\ln(T_0/T) + (T/T_B)^2}{1 + T/T_L} \tag{6}$$

which describes well the experimental results of ref.[19] in the entire temperature range, Fig.3.

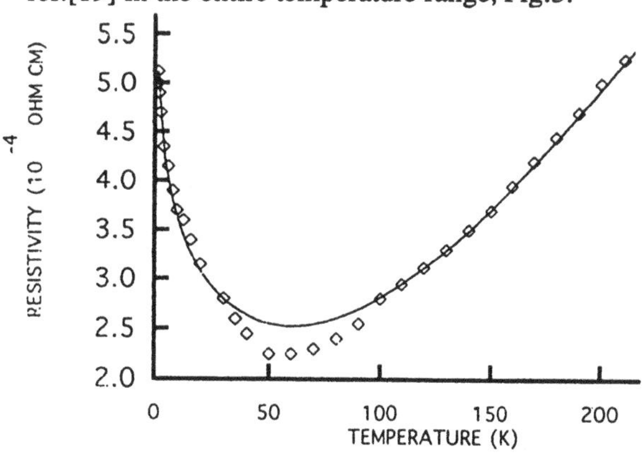

Fig. 3. In-plane resistivity of $La_{1.87}Sr_{0.13}CuO_4$ [19] described by Eq.(6) (line) with T_0=1900K, T_L=234K, and T_B=62K.

Our results together with the observation [18] of the semiconducting doping dependence of the dc resistivity ($\rho \sim 1/x$) suggest that doped cuprates are doped insulators irrespective to the level of doping where the carriers are small bipolarons. We believe that the phase diagram, Fig.4, does not include any metallic state 'normal' or 'strange'.

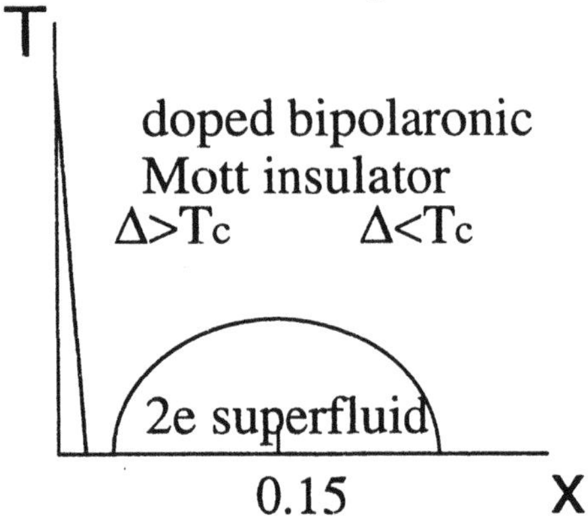

Fig.4. Suggested phase diagram of high-T_c cuprates.

If this is so, the superconducting state of cuprates is a charged (2e) Bose-Einstein condensate of bipolarons. The order parameter is then a real space condensate wave function with the d-wave symmetry resulting from the bipolaron band dispersion [12] rather than from any particular coupling mechanism.

The pair-distribution analysis of neutron scattering [20] suggest that the normal state `spin-gap' is consistent with the formation of a bipolaronic local singlet due to the electron-phonon interaction. The observation of the isotope effect on the carrier mass [21] provides an evidence for the polaronic carriers and the violation of the Migdal theorem in cuprates. We believe that the normal state `pseudogap' observed with magnetic, kinetic, thermodynamic and ARPES measurements has the bipolaronic origin, and a so-called ` pairing without phase coherence' is none other than the bipolaron formation.

REFERENCES

A.S. Alexandrov and N.F. Mott, Rep .Prog. Phys. **57**, 1197 (1994); `High Temperature Superconductors and Other Superfluids', Taylor and Francis, London (1994).

2. P.Calvani et al, in `Anharmonic Properties of high-T_c Cuprates', eds D. Mihailovic, G. Ruani, E. Kaldis and K.A. Muller, World Scientific, 139 (1995).
3. D. Mihailovic et al, Phys. Rev. B**42**, 7989 (1990).
4. A.S. Alexandrov and J. Ranninger, Solid St. Commun. **81**, 403 (1992).
5. A.S. Alexandrov, Phys. Rev. B**48**, 10571 (1993); A.S. Alexandrov, V.N. Zavaritsky, W.Y. Liang, and P.L. Nevsky, Phys. Rev. Lett. **76**, 983 (1996); A.S.Alexandrov,W.Y. Liang and V.N. Zavaritsky, Phys. Rev. Lett. **78**, 982 (1997).
6. M. Takigawa, Phys. Rev. B **42** 243 (1991); H. Zimmerman et al, Physica C**185-189**, 1145 (1991); Machi et al, Physica C **173**, 32 (1991).
7. Z. Schlesinger et al, Physica C**185-189**, 57 (1991).
8. A.S. Alexandrov, A.M. Bratkovsky, and N.F. Mott, Phys. Rev. Lett. **72**, 1734 (1994).
9. A.S. Alexandrov, J. Low Temp. Phys. **87**, 721 (1992).
10. A.S. Alexandrov and D.K. Ray, Phil. Mag. Lett. **63**, 295 (1991).
11. Y. J. Uemura, in'Polarons and Bipolarons in high-T_c Superconductors and Related Materials' eds E.K.H. Salje, A.S. Alexandrov and W.Y. Liang, Cambridge University Press, p.453 (1995).
12. A.S. Alexandrov, Phys. Rev. B**53**, 2863 (1996).
13. D.M.King et al, J. Phys. Chem. Solids **56**, 1865 (1996).
14. X. Zhang and C.R.A. Catlow, J. Mater. Chem.**1**, 233 (1991).
15. A.S. Alexandrov, V.V. Kabanov, and N.F. Mott, Phys. Rev. Lett. **77**, 4796 (1996).
16. Y. Nakamura and S. Uchida, Phys. Rev. B**47**, 8369 (1993).
17. T. Nakano et al, Phys. Rev. B**49**, 16000 (1994).
18. H.W. Hwang et al, Phys. Rev. Lett. **72**, 2636 (1994); B. Batlogg et al, Physica C{\bf 135-140}, 130 (1994).
19. Y. Ando et al, Phys. Rev. Lett. **75**, 4662 (1995); G.S. Boebinger et al, ibid **77**, 5417 (1996).
20. T.R. Sendyka et al, Phys.Rev. B**51**, 6747 (1996).
21. Guo-meng Zhao et al, Phys. Rev. B **54**, 14956 (1996); ibid, 14982.

ELSEVIER

Physica C 282–287 (1997) 273–278

PHYSICA C

Understanding High Temperature Superconductors: Progress and Prospects

David Pines

Center for Nonlinear Studies, LANL, and Physics Dept. and Science and Technology Center for Superconductivity, University of Illinois at Urbana-Champaign, 1110 West Green Street, Urbana, IL 61801-3080.

I review progress in measurements of the dynamic spin susceptibility in the normal state which yield a new phase diagram and discuss microscopic calculations which yield qualitative, and in many cases, quantitative agreement with the measured changes in the quasiparticle, transport, magnetotransport, and optical properties of the cuprate superconductors as one varies doping and temperature provided one describes the systems as nearly anti-ferromagnetic Fermi liquids in which the effective magnetic interaction between planar quasiparticles mirrors the dynamic spin susceptibility measured in NMR and INS experiments. Together with the demonstration that the NAFL pairing potential leads inexorably to a $d_{x^2-y^2}$,pairing state, this work provides a "proof of concept" for the NAFL description of high T_c materials. I review Eliashberg calculations of the mean-field behavior found in overdoped systems and discuss the extent to which the crossovers to pseudoscaling and pseudogap behavior found in the effective magnetic interaction and quasiparticle behavior in the optimally doped and underdoped systems may be derived microscopically. I conclude with a tentative scenario for the dependence of T_c on doping level and imperfections in different systems.

1. INTRODUCTION

Understanding high temperature superconductors means, first of all, answering the following questions: What is the nature and physical origin of their anomalous normal state behavior? How is the normal state best characterized? What is the mechanism for high T_c? What is the pairing state? Thanks to a large number of interrelated experimental and theoretical investigations carried out during the past three years, we have, as a community, made considerable progress toward obtaining the answers to these key questions and in what follows I present some highlights of that progress.

My answers may be simply stated. There is an intimate relationship between the anomalous evolution of planar quasiparticle spectra with temperature and doping and the anomalous spin and charge behavior measured in the normal state of optimally doped and underdoped systems: all these phenomena originate in a highly anisotropic magnetic interaction between the almost localized planar quasiparticles. The close approach to antiferromagnetism leads to three distinct kinds of normal state magnetic behavior: mean field, pseudo-scaling, and pseudogap; counterparts are found in the quasiparticle spectra, transport, magnetotransport, and optical behavior. The normal state may thus be characterized as a nearly antiferromagnetic Fermi liquid, with the dominant contribution to quasiparticle interaction coming from spin-fluctuation exchange, and so mirroring the momentum and frequency dependence of the dynamical spin susceptibility, $\chi(q,\omega)$, measured in NMR and INS experiments. Because $\chi(q,\omega)$ peaks sharply at wave vectors Q_i, in the vicinity of $Q \equiv (\pi,\pi)$, there are two distinct groups of quasiparticles: *hot quasiparticles*, located near the singular points on the Fermi surface which can be connected by Q_i, interact strongly and display highly anomalous (non-Landau Fermi liquid) behavior; *cold quasiparticles* are sufficiently far away from singluar points that their behavior is closer to that of a strongly coupled Landau Fermi liquid. Recent microscopic calculations based on the NAFL description of the doping and temperature dependence of the transport, magnetotransport, and optical properties of both overdoped and underdoped systems yield good qualitative, and in many cases, quantitative

PII S0921-4534(97)00253-0

agreement with experiment [1], while pseudogap behavior, and more generally the magnetic crossovers found in even optimally doped systems, is associated with the hot quasiparticles [2]. The highly anisotropic quasiparticle interaction found in NAFL's leads inexorably to a transition to a superconducting state with d_{x2-y2} pairing [3], and I shall present arguments which suggest that the doping dependence of T_c originates in a competition between the spin-fluctuation-induced pairing potential and the quasiparticle pseudogap. The enhancement of T_c on going from 1-2-3 systems to Tℓ- and Hg-based systems is explained, in the NAFL scenario, by differences in their spin fluctuation spectra [4], while the markedly different influence of the planar impurities, Zn and Ni, on the quasiparticle pseudogap in underdoped systems, and on T_c in optimally doped systems, can be quantitatively understood using the NAFL description [3][5], and provides strong support for the magnetic origin of high T_c, and of pseudogap behavior in the normal state.

2. SPIN-FLUCTUATION EXCITATIONS AND TEMPERATURE CROSSOVERS: A NEW PHASE DIAGRAM.

A quantitative fit to existing NMR and INS experiments on the 1-2-3 and 2-1-4 systems may be obtained with a phenomenological expression, $\chi_{NAFL}(q,\omega)$ for the dynamical spin susceptibility which takes into account the strong temperature dependent antiferromagnetic correlations between the itinerant Cu^{2+} spins [6]. χ_{NAFL} possesses in general four incommensurate peaks at wave vectors, Q_i, in the vicinity of (π,π) and for $(Q_i\text{-}q) \lesssim 1/\xi$ takes the form [7]

$$\chi_{NAFL}(q,\omega) = \frac{1}{4}\sum_i \frac{\chi_{Q_i}}{1+(Q_i - q)^2\xi^2 - i\omega/\omega_{SF}} \qquad (1)$$

elsewhere. $\chi(q,\omega)$ takes a more Fermi liquid-like form,

$$\chi_{FL}(q,\omega) = \frac{\chi_q}{1 - i\omega/\Gamma_q} \cong \frac{\chi_0(T)}{1 - i\omega/\Gamma_0} \qquad (2)$$

where $\chi_Q \equiv \alpha\xi^2$ is the static spin susceptibility, ξ the AF correlation length, $\omega_{SF} << (\Gamma_q,\Gamma_0)$ describes NAFL relaxational behavior, $\chi_0(T)$ is the uniform spin susceptibility, and $\Gamma_q \sim \Gamma_0$ is the effective bandwidth or Fermi energy. The ^{63}Cu spin-lattice relaxation time and spin echo decay experiments discussed in Refs. (2) and [7] show that for magnetically overdoped systems one finds mean-field behavior, $\omega_{SF} \sim \xi^{-2} \sim a + bT$, in the normal state. For optimally doped and underdoped systems, such behavior is found only above a characteristic temperature, T_{cr}, close to the temperature at which $\chi_0(T)$ exhibits a maximum in the normal state of these systems. Barzykin and I [7] argued that the crossover in system behavior at T_{cr} occurs when the AF correlations reach a critical strength, $\xi_{cr} \sim 2a$; thus $\xi(T_{cr}) \equiv 2a$ determines T_{cr}, a criterion which was confirmed by the subsequent INS experiments on 2-1-4 discussed in Ref. (8) and the $^{63}T_{2G}$ measurements

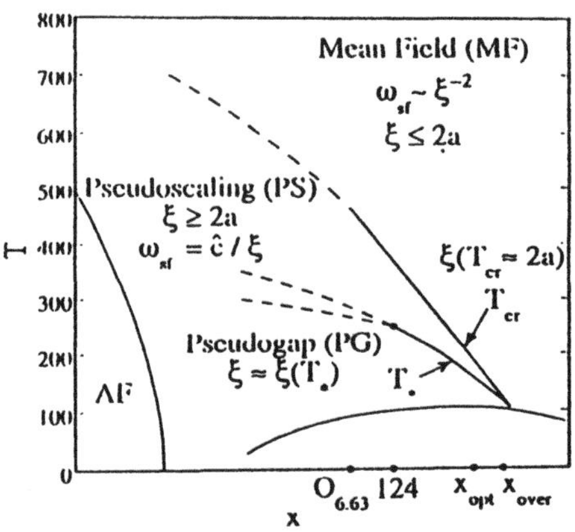

Fig. 1 Phase diagram, showing three distinct normal state phases for magnetically underdoped members of the 1-2-3 system. Extrapolations of the crossover temperatures, T_{cr} and T_*, to lower doping levels are denoted by dotted lines. Comparable phase diagrams have been established for the 2-1-4, Bi-based, Hg-based and Tℓ-based systems.

on $YBa_2Cu_4O_8$ reported in Ref. (9). As T decreases below T_{cr}, $(d\chi_0/dT) > 0$, and ω_{SF} exhibits pseudoscaling behavior, $\omega_{SF} = \hat{c}/\xi = a' + b'T$, until a second crossover temperature, T_* is reached. T_* marks the onset of pseudogap behavior, characterized by ξ~constant, and a rapid increase in $^{63}T_1T$ and ω_{SF} for $T_c \leq T \leq T_*$ [7]. The new phase diagram obtained from an analysis of NMR and INS experiments on the 1-2-3 and 2-1-4 systems is shown in Fig. 1. Note that T_{cr} increases rapidly as the doping level falls below that which is optimal for T_c while T_* increases more slowly. For optimally doped 1-2-3, one finds T_{cr} ~ 150K, T_* ~ 120K, while for $YBa_2Cu_4O_8$, Curro *et al.* [9] have shown that $T_{cr} \cong$ 500K, T_* ~ 200K. For $La_{1.85}Sr_{0.15}CuO_4$, one has T_{cr} ~ 300K [8], while Yasuoka [10] finds T_* ~ 200K. Note, too, the considerable variation from system to system in the strength of AF correlations at T_c; for $YBa_2Cu_3O_7$ one finds $\xi(T_c) = 2.2a$, while for $La_{1.86}Sr_{0.14}CuO_4$, one has $\xi(T_c) = 7.6a$ [8]. Equally striking is the increase in $\xi(T_c)$ within a given system as one goes to sub-optimal doping levels; as might be anticipated from Fig. 1, one finds $\xi(T_c) = 5.5a$ for $YBa_2Cu_4O_8$.

Note, too, that from a magnetic perspective, optimally doped systems are, in fact, underdoped; only when one goes to substantially higher doping concentrations (e.g., $La_{1.76}Sr_{0.24}CuO_4$) does one reach the magnetically overdoped regime, where χ_{NAFL} displays mean-field behavior, $(d\chi_0/dT) < 0$ down to T_c and $\xi(T_c) \leq 2$.

For magnetically overdoped systems, and in the mean-field regime for magnetically underdoped systems, microscopic strong coupling Eliashberg calculations of the kind carried out for $YBa_2Cu_3O_7$ [3], may be expected to provide a quantitative account of normal state properties. The non-linear feedback of changes in the quasiparticle spectrum on the effective quasiparticle interaction in the NAFL description,

$$V_{eff}(q,\omega) = g^2 V_{NAFL}(q,\omega) \qquad (3)$$

which determines their behavior brings about both the comparatively slow increase, $\xi(T) \sim (a+bT)^{-1/2}$, found above T_{cr}, and a moderate increase in $\chi_0(T)$ as T decreases, while the recent work of Chubukov and Monthoux [11] shows that vertex corrections are comparatively modest, are maximal for the hot quasiparticles, and act to enhance the pairing potential for superconductivity. However, as discussed in Ref. (2), and as the criterion, $\xi(T_{cr}) = 2$ implies, for magnetically underdoped systems below T_{cr} the spectra of the hot quasiparticles [which must be located near $(0,\pm\pi)$ and $(\pm\pi,0)$] begins to be modified by the strong AF correlations in such a way that one finds pseudo-scaling behavior for ω_{SF} between T_{cr} and T_*; the pseudogap behavior below T_*, represents the "end-point" of the impact of the strong AF correlations on quasiparticle properties. This gapping of the hot quasiparticles at T_* is seen directly in the rapid increase in ω_{SF} below T_*, in the BSCCO ARPES experiments [12], and in the specific heat and uniform susceptibility measurements [13] discussed at this conference.

The magnetic origin of the quasiparticle pseudogap has a number of observable consequences. First, only hot quasiparticles should become gapped. Second, both the gap magnitude and its onset temperature T_*, should be unrelated to T_c. Third, once in place, at temperatures slightly below T_*, the hot quasiparticle gap may be expected to be independent of T, *and should not be influenced by the superconducting transition.* Fourth, since the gapped hot quasiparticles do not participate in the superconducting transition, for magnetically underdoped systems, the superfluid density, ρ_s, will be distinctly less than the number of quasiparticles (~1-x), one estimates from the area of the large Fermi surface expected above T_{cr} (or T_*), and is plausibly ~x, the hole fraction.

3. HOT AND COLD QUASIPARTICLES: TRANSPORT, MAGNETOTRANSPORT, AND OPTICAL PROPERTIES

Microscopic calculations of the properties of hot and cold quasiparticles in a NAFL reveal dramatically different behavior, and show how these differences give rise to the measured anomalous dependence of transport, magneto-transport, and optical properties as one varies the hole concentration and temperature [1]. One finds quite generally that hot quasiparticle scattering rates are considerably larger than those for cold quasiparticles with

$$\frac{1}{\tau_{hot}} \sim T\xi \tag{4}$$

while the cold quasiparticle lifetime depends on its distance, Δk, from the nearest hot spot;

$$\frac{1}{\tau_{cold}} \sim \frac{T^2}{(\Delta k)^3 \xi^2 \omega_{SF}} \qquad T << \frac{\omega_{SF}(\Delta k)^2 \xi^2}{\pi} \tag{5a}$$

$$\frac{1}{\tau_{cold}} \sim \frac{T}{(\Delta k)} \qquad T >> \frac{\omega_{SF}(\Delta k)^2 \xi^2}{\pi} \tag{5b}$$

The main contribution to the longitudinal conductivity, σ_{xx}, comes from the cold regions of the FS; up to logarithmic corrections, one finds [1]

$$\rho_{xx} \sim \frac{T^2}{T_0 + T} \frac{1}{(\Delta k)_{max}} \tag{6}$$

where the crossover temperature, T_0, is

$$T_0 = (\Delta k_{max}) \xi^2 \frac{\omega_{SF}}{2\pi} \tag{7}$$

and $\Delta k_{max} \sim 1$ for a large FS. For magnetically underdoped systems, Fermi surface evolution leads to $\Delta k_{max} \lesssim 0.3$; hence for a wide choice of parameters $T_0 < T_*$, and one finds a linear in T resistivity down to T_*, while the rapid increase in T_0 below T_* leads to the crossover to T^2 behavior seen in $YBa_2Cu_3O_{6.63}$ below 160K and in $La_{1.9}Sr_{0.1}CuO_4$ below 150K. The cold quasiparticles likewise dominate the Hall conductivity, σ_{xy}, so that one finds, for $T \lesssim 2\pi T_0$,

$$\mathrm{ctn}\theta_H \sim \frac{T^2}{T_0} \tag{8}$$

and one recovers the famous T^2 dependence of ctn θ_H for optimally doped and overdoped systems. In Ref. (1) it is shown how the highly anisotropic hot and cold quasiparticle mean-free paths enable one to extract τ_{cold} and τ_{hot} from measurements of σ_{xx} and σ_{xy}; in so doing, one finds for $YBa_2Cu_3O_7$, $YBa_2Cu_4O_8$, $YBa_2Cu_3O_{6.63}$, as well as for $La_{1.9}Sr_{0.1}CuO_4$ and $La_{1.85}Sr_{0.15}CuO_4$, quantitative agreement with the theoretical expressions, Eqs. (4) and (5).

On extending these results to finite frequencies, ω, the terms involving both ω and τ cancel; Eqns. (5) then become

$$\frac{1}{\tau_k(\omega, T)} \sim \frac{1}{\Delta k}\left[\frac{\pi T^2}{T + T_0} + \frac{\omega^2}{\omega + \pi T_0}\right]. \tag{9}$$

At typical frequencies and temperatures one then recovers the marginal Fermi liquid result,

$$\frac{1}{\tau_k(\omega, T)} = A\omega + B\pi T. \tag{10}$$

One finds thereby agreement with optical experiments on materials as disparate as single layer Tℓ 2201 ($T_c = 23K$) and $YBa_2Cu_3O_7$ [1].

4. WHAT DETERMINES T_c?

As discussed in detail in Refs. (3), in a NAFL the peaking of the highly anisotropic quasiparticle interaction at wave vectors near (π,π) leads inexorably to a $d_{x^2-y^2}$ pairing state, while strong coupling (quasiparticle lifetime) corrections do not prevent one from obtaining a $T_c \sim 90K$ with parameters appropriate to $YBa_2Cu_3O_7$ and a coupling strength which yields a $\rho_{xx}(T)$ comparable to that found experimentally. This "proof of concept" for the NAFL description led Monthoux and me to predict a $d_{x^2-y^2}$ state for all cuprate superconductors, at a time when only NMR experiments supported this pairing assignment. As we heard at Grenoble, and in still greater detail here, our prediction has now been supported by essentially all experiments which are sensitive to the symmetry of the order parameter.

For the optimally doped and overdoped systems for which our Eliashberg calculations are applicable, we found that

$$T_c \sim \omega_{SF}\, \xi^2 \tag{11}$$

so that for comparable values of $\xi(T_c)$, T_c should scale with $\omega_{SF} \sim {}^{63}T_1T$. This scaling relation has been invoked to explain both the substantially higher transition temperatures found in the $T\ell$- and Hg-based systems, where one finds values of $\xi(T_c)$ comparable to those of $YBa_2Cu_3O_7$, but values of $\omega_{SF}(T_c)$ some 50% larger [4], and the variation of T_c with Ni doping measured in$YBa_2Cu_3O_7$ [5]. On examining the sensitivity of our results to $\xi(T_c)$, we found that for $\xi \lesssim 1$, T_c plummetted to zero, so that, as might be expected of a mechanism which depends on a near approach to AF behavior, overdoping, which weakens the AF correlations, brings out the reduction in T_c seen experimentally.

On the other hand, how can we account for the observed maximum in $T_c(x)$, since in the pseudoscaling regime, Eq. (11) leads one to expect monotonic behavior, with $T_c \sim \xi$? The answer of course lies in the pseudogap, which by removing hot quasiparticles from the superconducting arena, weakens substantially the overall NAFL pairing potential. On this scenario, as one enters the magnetically underdoped regime, initially the increasing strength of the AF correlations wins out over the pseudogap, but below the optimal doping concentration, the pseudogap wins out, and T_c falls off as x decreases. Obviously, further calculations are required to support this scenario, in which the dimensionless coupling constant, N(0)g, is assumed to vary weakly with hole concentration.

Support for the above scenario comes from experiments on the influence of the planar impurities, Ni, and Zn, on T_c. For $YBa_2Cu_3O_7$ Monthoux and I argued that since Zn does not possess a spin which can align with neighboring Cu^{2+} spins, it will change the magnetic pairing potential and so act as a "superunitary" scatterer, while Ni, which enters the plane in such a way as to look like a spin 1/2 to its neighboring Cu^{2+} spins, will be a subunitary scatterer. In this way we could explain the fact that Zn is some three times more effective than Ni in suppressing T_c for $YBa_2Cu_3O_7$, and argued that this effect serves as a "smoking gun" for the magnetic mechanism [3]. But how then to explain the fact that both Ni and Zn exert a somewhat larger influence on T_c in $YBa_2Cu_4O_8$? The answer lies in the fact that the pseudogap is well developed in this system so that the density of states at T_c is significantly reduced; hence the increased effectiveness of both Ni and Zn impurities in reducing T_c in underdoped systems. Because it changes the pairing potential as well, Zn continues to be a more effective pair-breaker than Ni.

5. SOME NEXT STEPS

Although we possess a good understanding of optimally doped and overdoped systems, much work remains to be done before we possess a comparable understanding of the normal state properties and superconducting transition temperature of the magnetically underdoped systems. Some next steps include the development of a microscopic description of hot spot, pseudoscaling, and pseudogap behavior, which makes evident the interplay between the influences of Fermi surface evolution and pseudogap formation on the measured quantities, ω_{SF} and χ_Q, and establishes a quantitative relationship between ρ_s, χ_0, and the doping level. Experiments on the same samples of BSCCO and other materials are needed to verify the correctness of the two-gap scenario proposed in Sec. 2 and to establish the connection between features measured in ARPES and INS experiments and crossovers measured in NMR experiments. With regard to microscopic theories, at present we know what doesn't work: neither one band Hubbard models with U or nearest neighbor exchange or the t-J model yield a quantitative account of the pseudoscaling and pseudogap behavior measured in underdoped materials, while the predictions of the non-linear sigma model plus holes are inconsistent with the doping dependence of pseudoscaling behavior. What might work is far less clear. A one or three band Hubbard model with a non-local restoring force which takes into account local SDW formation at $\xi \geq 2a$ is a promising candidate, but much work will be required before its viability can be assessed.

It gives me pleasure to acknowledge stimulating conversations on these and related topics with my collaborators, A. Balatsky, V. Barzykin, A. Chubukov, P. Monthoux, J. Schmalian, and B. Stojkovic and with my UIUC experimental colleagues, G. Blumberg, D. Ginsberg, L. Greene, M. Klein, C. P. Slichter, R. Stern, and D. Van Harlingen, and to thank CNLS and CMS at Los Alamos, and STCS NSF grant DMR91-20000 at UIUC for the support provided to this work.

REFERENCES

1. B. Stojkovic and D. Pines, Phys. Rev. B, in the press (April 1, 1997 issue); in preparation.
2. A. Chubukov, D. Pines and B. Stojkovic, J. Phys. Cond. Matt. **8** (1996) 1; D. Pines, Z. Phys. B, in the press (1997).
3. P. Monthoux and D. Pines, Phys. Rev. B **47** (1993) 6069; Phys. Rev. B **49** (1994) 4261; D. Pines, Physica C **235-236** (1994) 113; D. Pines and P. Monthoux, J. Phys. Chem. Solids **56** (1995) 1651.
4. G-q. Zheng *et al.*, J. Phys. Soc. Jpn. **64** (1995) 2524; K. Magishi *et al.*, J. Phys. Sco. Jpn **64** (1995) 4561; M. Julien *et al.*, Phys. Rev. Lett. **76** (1996) 4238.
5. Y. Tokunaga *et al.*, these proceedings.
6. A. Millis, H. Monien and D. Pines, Phys. Rev. B **42** (1990) 167.
7. V. Barzykin and D. Pines, Phys. Rev. B **52** (1995) 13585.
8. Y. Zha, V. Barzykin and D. Pines, Phys. Rev. B **54** (1996) 7567.
9. C. P. Slichter *et al.*, Phil. Mag. **74** (1996) 545; N. Curro et al., Phys. Rev. B, in the press (1997).
10. H. Yasuoka, these proceedings.
11. A. V. Chubukov and D. Morr, Physics Reports, in the press (1997); P. Monthoux, Phys. Rev. B, in the press (1997).
12. Z-X. Shen, these proceedings; H. Ding, *ibid.*
13. J. Loram, these proceedings; J. Tallon, *ibid.*

ELSEVIER

Physica C 282–287 (1997) 279–282

PHYSICA C

The Unusual Normal and Superconducting States of Underdoped Cuprates

Patrick A. Lee[a]

[a]Department of Physics, Massachusetts Institute of Technology, Cambridge, MA 02139

There is increasing experimental evidence that the superconducting energy gap Δ_0 in the underdoped cuprates is independent of doping concentration x while the superfluid density is linear in x. We show that under these conditions, thermal excitation of the quasiparticles is very effective in destroying the superconducting state, so that T_c is proportional to $x\Delta_0$ and part of the gap structure remains in the normal state. We then estimate H_{c2} and predict it to be proportional to x^2. We also discuss to what extent the assumptions that go into the quasiparticle description can be derived in the U(1) and SU(2) formulations of the t-J model.

It has become clear in the past several years that the underdoped cuprates deviate strongly from conventional materials, both in the normal and superconducting states. NMR,[1,2] neutron and c-axis optical conductivity[3] indicate the existence of a pseudogap in the normal state, and photoemission experiments[4,5] reveal that the pseudogap is of the same size and $\boldsymbol{k}$ dependence as the superconducting gap. Furthermore, as the doping concentrating x is reduced from optimal doping, the superconducting gap is constant or may be slightly increasing, while the transition temperature T_c is reduced. Thus a strong deviation from the BCS ratio between energy gap and T_c is to be expected. At the same time, London penetration depth and optical conductivity[6] show that the Drude weight in the normal state as well as the superfluid density in the superconducting state are proportional to x, and a linear relation between T_c and the superfluid density have been noted.[7]

In order to understand these data, a number of workers start with strongly correlated models such as the t-J model and interpret the pseudogap as the spin excitation gap in some RVB singlet state. In particular, this scenario is realized in a decomposition of the electron into a fermion which carries spin and a boson which represents a vacancy in order to enforce the constraint of no double occupation of the t-J model. At the mean field level, spin and charge are separated. In the underdoped region of the mean field phase diagram the fermions are paired in some intermediate temperature range, and become a d-wave superconductor only below the Bose condensation temperature of the bosons.[8,9] Very recently, a modification of this mean field theory was proposed, which incorporates an SU(2) symmetry known to be important at half filling.[10] It was argued that the new SU(2) formulation allows a smoother connection to half-filling and includes low lying fluctuations ignored in the original U(1) formulation. Features in the photoemission data are qualitatively explained by this approach, but so far discussions have been limited to the normal state. In this paper we report on two recent works based on the SU(2) formulation.[11,12]

In the first work[11] we calculate the spin susceptibility χ and the specific heat C_V in the normal state (the spin gap phase). In the SU(2) formulation, this is interpreted as the staggered flux phase. The fermion spectrum is the same as that of a d-wave state and consists of 4 Dirac points centered at $(\pm\pi/2, \pm\pi/2)$. The dispersion near the Dirac points can be approximated by

$$E_{\boldsymbol{k}} = (v_f^2 k_1^2 + v_2^2 k_2^2)^{1/2} \qquad (1)$$

where k_1 and k_2 is the momentum coordinate normal to parallel to the Fermi surface. In terms of a maximum gap of Δ_0 located at (π, π), $v_2 = \Delta_0 a/\sqrt{2}$. If the fermion tight binding band is parametrized by a nearest neighbor hopping in-

PII S0921-4534(97)00256-6

tegral t_f, we have $v_f = 2\sqrt{2}t_f a$. If the fermions are treated as free, we find that χ is linear in T and $C_V = \gamma T$, where γ is also linear in T. By introducing an energy cutoff T_{uv} to simulate the upper limit of the applicability of the Dirac spectrum, we find $\chi_0 = \frac{4T}{\pi v_f v_2} F(T_{uv}/2T)$ where $F(x) = \int_0^x y/\cosh^2 y dy$. If the energy gap Δ_0 does not change as we enter the superconducting state, this simple theory predicts that the same χ_0 describes the spin gap and the superconducting states, in clear disagreement with experiments. We find that in the staggered flux phase, gauge fluctuations are very important. The inclusion of gauge fluctuations in the presence of bosons which describe hole doping lead to a renormalization of the velocities in the Dirac spectrum. This in turn gives rise to a logarithmic in T correction to χ and to the specific heat coefficient which takes the form at low temperatures:

$$\chi = \frac{4}{\pi} \ln(2) T \left(1 + \frac{4}{\pi^2} \ln \frac{T_{uv}}{T} + ...\right) \tag{2}$$

$$C_V = \frac{9}{\pi}(1.202) 2T^2 \left(1 + \frac{4}{\pi^2} \ln \frac{T_{uv}}{T} + ...\right) \tag{3}$$

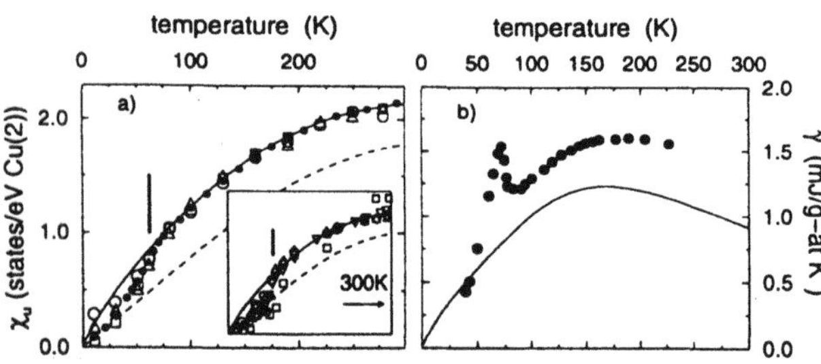

Figure 1. (a) Knight shift data of $YBa_2Cu_3O_{6.63}$ [ref. 1] and $YB_2Cu_4O_8$ [insert, from ref. 2]. Solid line is fit to our theory. Dashed line is χ_0 of free Dirac fermions. (b) $\gamma(T)$ of $YB_2Cu_3O_{6.67}$ from Liang *et al.*, *Physica* C **263**, 277 (1996). Solid line is theory using the same parameters in Fig. 1(a).

In Fig. 1(a) we fit the Knight shift data[1] of $YBCO_{6.6}$ to our theory. A shift of the zero by 0.25 states/eV-Cu(2) was made, which is within the error bars for the uncertainty of the orbital contribution. The fit has two parameters: $\sqrt{v_2 v_f} = 0.76Ja$ and $T_{uv} = 0.63J$ where $J = 1500$ K. Also shown is the $\chi_0(T)$ corresponding to free fermions. We expect that in the superconducting state the gauge fluctuations are quenched due to the Higgs mechanism. The fermions become well defined quasiparticles and the low temperture χ should be given by χ_0. Thus we expect a cross-over from $\chi(T)$ to $\chi_0(T)$ as we go below T_c. This is in qualitative agreement with the data.

In Fig. 1(b) we show the specific heat coefficient γ. The solid line is the theory using the same two parameters used in Fig. 1(a). The qualitative agreement is quite encouraging.

In the second work[12] we explore further the nature of the superconducting state, first in a phenomenological way based on a quasiparticle picture, and then we attempt to make connections to microscopic theory. The idea is that for $T < T_c$, the excitations are quasiparticles with dispersion given by Eq. (1). We now make the key assumption that in the presence of an electromagnetic field the quasiparticle spectrum is shifted according to

$$E(\boldsymbol{k}, \boldsymbol{A}) = E(\boldsymbol{k}) + \frac{e}{c} \boldsymbol{v_k} \cdot \boldsymbol{A} \tag{4}$$

where $\boldsymbol{v_k} = d\varepsilon_k/dk$ is the normal state velocity. Equation (4) is correct to first order in $\boldsymbol{A}$ in the BCS theory. It has the consequence that the current carried by the quasiparticle is $-cdE/dA = -ev_k$, i.e. it is the same as the current in the normal state. For high T_c superconductivity, we do not have the equivalence of BCS theory, so Eq. (4) must be regarded as an assumption, albeit a very reasonable one.

We next argue that the superfluid tensor defined by $j_\mu = \frac{e}{c}\frac{\rho^s_{\mu\nu}}{m} A_\nu$ can be written as $\frac{1}{m}\rho^s_{\mu\nu} = \frac{1}{m}\rho^s(T=0)\delta_{\mu\nu} - \frac{1}{m}\rho^n_{\mu\nu}$ where $\rho^s(T=0)/m = x/a^2 m$ is directly measured by the weight of the Drude peak in the normal state and by λ_L^{-2} where λ_L is the Landau penetration depth in the superconducting state. By taking $\lambda_L = 1600$Å for $YBCO_{6.95}$ and $x = 0.2$, we find $m = 2.1m_e$. It is convenient to fit this mass to the bottom of a tight binding bandwidth with hopping integral t_h, and we have $t_h = (2ma^2)^{-1} = 0.122$ eV which happens to be very close to J. On the other hand, $\frac{1}{m}\rho^n_{\mu\nu}$ is given by the quasiparticle response to the $\boldsymbol{A}$ field. This we can calculate by writing the free energy in terms of non-interacting quasiparticles, and differentiating twice with respect to $\boldsymbol{A}$.

$$\frac{1}{m}\rho^n_{\mu\nu} = -2\sum_{\boldsymbol{k}} \frac{dE}{dA_\mu}\frac{dE}{dA_\mu}\frac{\partial f}{\partial E} \quad . \tag{5}$$

Using Eq. (4), we replace dE/dA_μ by the normal state velocity $\boldsymbol{v}_\mu$ and $\rho_s(T)$ can be evaluated in a straightforward way

$$\frac{\rho^s}{m}(T) = \frac{x}{a^2 m} - \alpha T \tag{6}$$

where $\alpha = (2\ln(2)/\pi)v_f/v_2 = (8\ln(2)/\pi)t_f/\Delta_0$. We see that for small x, the quasiparticle excitation is an effective way of destroying the superconducting state by deriving ρ^s to zero. By extrapolating Eq. (6) to $\rho^s = 0$, we can estimate T_c as

$$kT_c \approx 1.13x\Delta_0(t_h/t_f) \quad . \tag{7}$$

If we assume that Δ_0 is independent of x for underdoped cuprates, we see that T_c is proportional to x (or more precisely to $\rho_s(T=0)/m$), thus providing an explanation of Uemura's plot.[7]

Another important implication is that superconductivity is destroyed when only a small fraction of the quasiparticles (with energy $\leq x\Delta_0$) are thermally excited. Thus the gap near $(0,\pi)$ must remain intact in the normal state, leaving a strip of thermal excitations which extend a distance proportional to x from the nodal points. This is qualitatively in agreement with the photoemission experiment. Of course our phenomenological picture does not provide a description of the normal state. It simply states that the normal state gap is an inescapable consequence of a finite Δ_0 and a vanishingly small superfluid density as $x \to 0$.

The fact that $d\rho_s/dT$ is independent of x and that both ρ_s and T_c are proportional to x means that a scaled plot of $\rho_s(T)/\rho_s(0)$ vs T/T_c should be independent of x for small T/T_c. In fact, such a scaled plot for $YBCO_{6.95}$ and $YBCO_{6.60}$ shows a remarkable universality over the entire temperature range.[13] We can use the data to extract the ratio v_f/v_2 using Eq. (6). Using the $YBCO_{6.95}$ data, we obtain a velocity anisotropy $v_f/v_2 = 6.8$.

We can use the value $\sqrt{v_2 v_f} = 0.76J$ quoted earlier from fitting the χ data to combine with $v_f/v_2 = 6.8$ to extract Δ_0 and v_f separately. We find $\Delta_0 = 50$ meV, a bit too high but not unreasonable and $v_f \approx 1.4 \times 10^7$ cm/sec. It will be interesting to obtain more precise measurements of the low T χ or γ to fix the quantity $v_2 v_f$. Combined with the $\rho_s(T)$ data, v_2 and v_f can be precisely determined.

Similar arguments allow us to estimate the vortex core size by considering the distance at which the supercurrent flow around the vortex induces sufficient quasiparticle excitations at the Fermi energy to drive the current to its critical value. We estimate the core size in the $\hat{x}$ direction to be

$$R_1 = \frac{1}{x}(v_f/\pi\Delta_0)\left(t_f/\sqrt{2}t_h\right) \tag{8}$$

Note that it is greater than the BCS coherence length $v_f/\pi\Delta_0$ by the factor x^{-1}. On the other hand, using the T_c estimate in Eq. (7), we can write

$$R_1 = v_f/(1.25\pi kT_c) \quad . \tag{9}$$

which is quite close to the BCS coherence length written in terms of T_c instead of Δ_0. The two ways of writing the coherence length are of course equivalent in BCS theory, but very different for underdoped cuprates. A main conclusion of this paper is that Eq. (9) is the proper expression for the coherence length.

When the core is approached from the (1,1) direction, similar considerations show that the core size is given by $\sqrt{2}R_1$. Thus the core takes on an approximately square shape. We estimate H_{c2} by assuming that the square vortex cores are closed-packed, so that

$$H_{c2} = (hc/2e)/4R_1^2 \tag{10}$$

Due to the crudeness of the extrapolation process, we expect both the T_c and H_{c2} expressions to be overestimates. An estimate may be made by using Eq. (9) for R_1 and using the experimental T_c and we obtain instead $R_1 \approx 38$ Å and $H_{c2} \approx 34$ T. While the absolute value of H_{c2} is quite uncertain, the prediction that H_{c2} is proportional to x^2 (or more accurately to $\rho_s^2(T=0)$), as long as Δ_0 is constant for underdoped cuprates, should be amenable to experimental test.

Next we comment on whether existing microscopic models can reproduce the assumptions of the quasiparticle description. In the U(1) formulation of the t-J model the normal state in the underdoped limit is described by d-wave pairing of fermions, so that there exists an energy gap Δ_0 which remains finite as $x \to 0$ in the normal state.[11,12] Superconductivity is driven by condensation of bosons and well defined quasiparticles are formed. The superconducting T_c occurs as an energy scale of $4\pi x t_h$ at the mean field level, and may be suppressed by gauge fluctuations.[14] In this theory the superfluid density is given by the Ioffe-Larkin rule,[15] $\rho^s(T)^{-1} = \rho_f^{s-1} + \rho_b^{s-1}$. Since the energy gap appears in the fermion spectrum, we expect $\rho_f^s(T) = (1-x) - T/\Delta_0$ while $\rho_b^s \approx x$ with a higher power in T corrrection. Then the U(1) theory predicts $\rho^s(T) = x - x^2 T/\Delta_0$. While the $T=0$ value is correctly given to be x, the temperature dependence is in strong disagreement with Eq. (6) and with experiment in that α is suppressed by x^2. The origin of this difficulty is that the fermion does not couple directly to $\boldsymbol{A}$, but to the U(1) gauge field $\boldsymbol{a}$ while the bosons couple to $\boldsymbol{A} + \boldsymbol{a}$. The external $\boldsymbol{A}$ produces a finite $\boldsymbol{a}$ but its magnitude is reduced by x. In the quasiparticle language, the shift of the spectrum in the presence of $\boldsymbol{A}$ is smaller than that given in Eq. (1) by x.

In the SU(2) theory the normal state is a state which fluctuates between d-wave pairing and the staggered flux phase of fermions and the fermion spectrum acquires a gap Δ_0 which is finite for small x.[10] Unlike the U(1) case, quantum fluctuations of the gauge fields are important even at $T=0$. The coupling of the quasiparticles $\boldsymbol{A}$ can be calculated including these quantum fluctuations and we find that the shift of the quasiparticle spectrum is of the form given by Eq. (1), except that the $\boldsymbol{v}_k \cdot \boldsymbol{A}$ is multiplied by a numerical constant of order unity. Thus the SU(2) theory incorporates the main ingredients underlying the present paper, i.e., a finite gap Δ_0, a superfluid density proportional to x, and a quasiparticle spectrum given by Eq. (1). Details of this microscopic theory will be given elsewhere.

The work I reported here was done in collaboration with D.H. Kim and X.G. Wen. I acknowledge support by NSF-MRSEC Grant No. DMR-94–00334.

REFERENCES

1. See M. Takigawa *et al.*, *Phys. Rev.* B. **43**, 247 (1991).
2. M. Bankay *et al.*, *Phys. Rev.* B **50**, 6416 (1994).
3. C.C. Homes *et al.*, *Phys. Rev. Lett.* **71**, 1645 (1993).
4. A.G. Loeser *et al.*, *Science* **273**, 325 (1996); J. Harris *et al.*, preprint.
5. H. Ding *et al.*, *Nature* **382**, 51 (1996).
6. J. Orenstein *et al.*, *Phys. Rev.* B **42**, 6342 (1990).
7. Y. Uemura *et al.*, *Phys. Rev. Lett.* **62**, 2317 (1989); *ibid.* **66**, 2665 (1991).
8. G. Kotliar and J. Liu, *Phys. Rev.* B **38**, 5142 (1988).
9. H. Fukuyama, *Prog. Theo. Phys.* Suppl. **108**, 287 (1992).
10. X.G. Wen and P.A. Lee, *Phys. Rev. Lett.* **76**, 503 (1996); P.A. Lee, N. Nagaosa, T.K. Ng, and X.G. Wen, preprint.
11. D.H. Kim, P.A. Lee, and X.G. Wen, Cond-mat/9702218.
12. P.A. Lee and X.G. Wen, Cond-mat/9702119.
13. D.A. Bonn *et al.*, *Czech. J. Phys.* **46** S6, 3195 (1996); *J. Phys. Chem. Solids* **56**, 1941 (1995).
14. P. Lee and N. Nagaosa, *Phys. Rev.* B **45**, 966 (1992).
15. L. Ioffe and A. Larkin, *Phys. Rev.* B **39**, 8988 (1989).

ELSEVIER

Physica C 282-287 (1997) 283-285

PHYSICA C

Effects of Impurities in Spin Gap State of Cuprates

N. Nagaosa[a*]

[a]Department of Applied Physics, University of Tokyo,
Bunkyo-ku, Tokyo 113, Japan

The effect of non-magnetic impurities is studied for the cuprates including high-Tc superconductors and two-leg ladder compounds. Two topics are discussed. One is the anomalous Kondo effect in metallic high-Tc cuprates, and the other is the impurity induced antiferromagnetism in the insulators.

1. Introduction

The anomalous normal state properties, especially in the underdoped region, have been the key issue to understand the mechanism of the superconductivity. Up to now, however, no final thoery has been approved, and the decisive experimetnal tests of the theories are highly desirable. In this paper, I propose that the response of the system to the nonmagnetic impurities is a promising candidate for that and the anomalous features already observed in high-Tc cuprates and two-leg ladder compounds are explained in terms of the gauge theory for strongly correlated systems. In Section 2, I discuss the single-impurity Kondo effect in high-Tc cuptrates where the host system is metallic. In Section 3, the effect of the impurities in the insulating spin gap system is studied. In this case the random Berry phase introduces the quantum decoherence and the AFLRO is induced.

2. Kondo Effect in High-Tc Cuprates

The nonmagnetic impurities, e.g., Zn, replacing Cu atoms in the conducting layers induce the following anomalous properties.

1. The formation of the magnetic moments due to Zn has been revealed by magnetic susceptibility [1,8], NMR [2-5], μSR [6] and EPR [7], whose number is roughly proportional to the Zn concentration and an almost full moment appears in the underdoped region , i.e., $0.8\mu_B$ in $La_{1.85}Sr_{0.15}Cu_{1-x}Zn_xO_4$ [8] and $0.86\mu_B$ in $YBa_2(Cu_{1-x}Zn_x)_3O_{6.64}$ [5] per Zn ion. The magnitude of the induced local moments is strongly dependent on the hole concentration x and becomes smaller or even vanishes as x increases [5,8].

2. The residual resistivity can be described in terms of the classical expression in terms of the Boltzmann transport theory [8], i.e.,

$$\rho_{\rm res} = 4(\hbar/e^2)(n_{\rm imp}/n)\sin^2\delta_0 \qquad (1)$$

which is independent of the effective mass of the carriers and is determined only by the phase shift δ_0, the impurity concentration $n_{\rm imp}$ and the carrier density n. It has been found that Zn acts as a strong scatterer at unitary limit, i.e., $\delta_0 = \pi/2$. The interesting thing is that the carrier number n is the hole concentration x in the underdoped region, and rather rapidly crosses over to the electron number $1-x$ in the overdoped region.

I now discuss that these anomalous features can be understood in terms of the guuge theory. In the slave boson mean field theory and the gauge field theory based on it, the electron is represented as the composite particle of a spinon and a holon [9,10]. Then the transport and magenetic properties are determined in different ways in terms of the spinons and holons. The magnetic properties are dominated by the spinons, and the

*This work is supported by Grant-in-Aid for COE research and Priority Areas Grants from Monbusho, Japan

0921-4534/97/$17.00
PII S0921-4534(97)00257-8

formation of the local spin moment has been attributed to the spin gap formation in the underdoped region [11]. Due to the linear density of states, the spinons can not screen the local moment induced by the nonmagnetic impurity, while the Kondo screening occurs in the optimally- or overdoped region without the spin gap. Let us turn to the residual resistivity [12]. The residual resistivity is determined by the phase shift for the spinons and holons. In the optimally- and overdoped region, the usual argument in terms of the Friedel sum rule and Kondo effect applied to the spinons, while the holon phase shift is 0 or π because there is no spin label for holons. Then the residual resistivity is predicted to be $\rho_{res} = 4(\hbar/e^2)(n_{imp}/(1-x)$ in agreement with experiments. In the underdoped region the situation is more anomalous because the carrier number n in eq.(1) is x, which contradicts with the Fermi liquid picture. Furthermore $\rho_{res.}$ in the underdoped region is also inconsistent with the small hole pocket picture in the Neel state.

Recently an SU(2) theory for the underdoped cuprates has been developed, which takes into account the low lying excitations related to the SU(2) symmetry, a symmetry known to be exact at half-filling [13]. This theory resolves several difficulties in U(1) formulation, and describes the smooth crossover form the small pockets to the large Fermi surface. A feature of this theory is that the holon is the two component SU(2) doublet, and the SU(2) constraint is given by

$$\sum_\sigma f^\dagger_{i\sigma} f_{i\sigma} + b^\dagger_{1i} b_{1i} - b^\dagger_{2i} b_{2i} = 1 \qquad (2)$$

Then the local moment formation means $\Delta \sum_\sigma f^\dagger_{i\sigma} f_{i\sigma} = 1$ when we consider the large sphere including the Zn atom. This condition togeher with the neutrality condition determines the holon phase shift uniquly as $\delta_{b_2} = -\delta_{b_1} = \pi/2$, which results in $\rho_{res} = 4(\hbar/e^2)n_{imp}/x$ in agreement with experiments [12]. Hence we believe that the SU(2) formulation is essential to explain the residual resistivity in the underdoped region.

3. Impurity-induce Antiferromagnetism

Now we turn to the insulating case, where the static holes and/or nonmagnetic impurities act as the vacancy for the spin system. In the nonlinear sigma model formulation, the cancelation of the Berry phases in the two-leg ladder and two-dimensional Heisenberg models is violated by the depleted spins, and the Berry phase term A_{Berry} becomes [14]

$$\begin{aligned} A_{\text{Berry}} &= iS \sum_{i \neq i_n} \omega(\{\mathbf{S}(x_i, \tau)\}) \\ &= iS \sum_i \omega(\{\mathbf{S}(x_i, \tau)\}) - iS \sum_n \omega(\{\mathbf{S}(x_{i_n}, \tau) \\ &= -iS \sum_n (-1)^{i_n} \omega(\{\mathbf{\Omega}(x_{i_n}, \tau)\}), \end{aligned} \qquad ($$

where $\mathbf{S}$ is the spin and the $\mathbf{\Omega}$ is the staggerd magnetization field in the continuum approximation. The index i_n specifies the site of a depleted spin. This modification leads to a substantial change in the quantum mechanical behavior of the system, because this Berry phase term gives the destructive interference between the configurations of the quantum fluctuations, i.e., quantum decoherence. This enhances the classical nature of the staggered magnetization, and hence the generalized susceptibility $\chi_\Omega(T)$. For the 1D two-leg ladder system with nonmagnetic impurities, the detailed analysis using the real space RG resutls [15], we conclude $\chi_\Omega(T) \sim T^{-1-2\alpha}$ with $\alpha \cong 0.22$ [14]. Considering that $\chi_\Omega(T)$ remains finite for the pure two-leg ladder system, this is the huge enhancement. Actually this divergence is even stronger than that in $S = 1/2$ AF Heisenberg chain.

Essentially similar phenomenon has been observed experimentally and studied thoeretically in the spin-Peierls system [16]. Also the quantum decoherence and the enhanced antiferromagnetism is expected in 2D. Experimentally increasing Zn doping eventually destroys the suoperconductivity and the system becomes an insulator. In the insulating phase, the AFLRO is observed [17]. We regard this as the manifestation for the role of the quantum Berry phase of the spin, i.e., the gauge field, in the strongly correlated electronic systems.

The results in Section 2 is the work with P.A.Lee, T.K.Ng, X.G.Wen, and those in Section 3 with A.Furusaki, M.Sigrist, and H.Fukuyama. I aknowledge these people for collaboration, and S.Uchida, N.P.Ong, H.Takagi, Y.Kitaoka for useful discussions.

REFERENCES

1. G.Xiao, M.Z.Cieplak, J.Q.Xiao, and C.L.Chien, Phys. Rev. **B35**,
2. H.Alloul et al., Phys. Rev. Lett. **67**, 3140 (1991).
3. G.Zheng et al., J. Phys. Soc. Jpn. **62**, 2591 (1993).
4. K.Ishida et al., J. Phys. Soc. Jpn. **62**, 2803 (1993).
5. A.V.Mahajan, H.Alloul, G.Collins, and J.F.Marucco, Phys. Rev. Lett. **72**, 3100 (1994).
6. P.Mendels et al., Phys. Rev. **B49**, 10035 (1994).
7. A.M.Finkelstein et al., Physica **C168**, 370 (1990).
8. K. Mizuhashi et al., Phys. Rev. **B52**, R3884 (1995); Y. Fukuzumi et al., Phys. Rev. Lett. **76**, 684 (1996); T. R. Chien, Z. Z. Wang, and N. P. Ong, Phys. Rev. Lett. **67**, 2088 (1991).
9. G. Baskaran, Z. Zou, and P. W. Anderson, Solid State Commun. **63**, 1233 (1991).
10. N. Nagaosa and P. A. Lee, Phys. Rev. Lett. **64**, 2450 (1990); P. A. Lee and N. Nagaosa, Phys. Rev. **B46**, 5621 (1992).
11. N. Nagaosa and T. K. Ng, Phys. Rev. **B51**, 15588 (1995).
12. N. Nagaosa and P. A. Lee, preprint (cond-mat/9702103).
13. X. G. Wen and P. A. Lee, Phys. Rev. Lett. **76**, 505; P. A. Lee, N. Nagaosa, T. K. Ng, X. G. Wen, cond-mat/9701168.
14. N. Nagaosa, A. Furusaki, M. Sigrist, and H. Fukuyama: J. Phys. Soc. Jpn. **65** (1996) 3724.
15. E. Westerberg, A. Furusaki, M. Sigrist and P. A. Lee: Phys. Rev. Lett. **75** (1995) 4302.
16. H. Fukuyama, T. Tanimoto and M. Saito: J. Phys. Soc. Jpn. **65** (1996) 1183 and references therein.
17. S. Uchida, private communications.

ELSEVIER

Physica C 282–287 (1997) 286–289

PHYSICA C

Electronic Excitations in Strongly Correlated Systems; 1D vs. 2D

Sadamichi Maekawa[a b] and Takami Tohyama[c],

[a]Department of Applied Physics, Nagoya University, Nagoya 464-01, Japan

[b]Institute for Materials Research, Tohoku University, Sendai 980-77, Japan

[c]Department of Physics, Faculty of Education, Mie University, Tsu 514,Japan

The single-particle, spin and charge excitation spectra in one- (1D), two-dimensional (2D) and two-leg-ladder t-J models are presented in the exact diagonalization technique. In 1D, the spin-charge separation is obtained. The single-particle spectrum at half-filling is in accord with the recent angle-resolved photoemission spectrum in the 1D cuprate $SrCuO_2$ and is in strong contrast with that in the 2D one $Sr_2CuO_2Cl_2$. In 2D, the behaviors of these excitation spectra are strongly dependent on electron concentration. Far away from half-filling, the single-particle spectrum includes both spin and charge excitations as expected in a Fermi liquid. On the other hand, near half-filling, the spectra are rather similar to those in 1D, indicating that the spin and charge degrees of freedom obey the different dynamics. The spectra in a two-leg ladder are similar to those in 2D.

Electronic excitations in low-dimensional strongly correlated systems have attracted considerable attention since the discovery of high temperature superconductors. In the one-dimensional (1D) systems, it is known that the spin-charge separation occurs and the spin and charge degrees of freedom of electrons obey the different dynamics. The elementary excitations of spin and charge are called spinons and holons, respectively. The nature of spin and charge degrees of freedom in the two-dimensional (2D) systems is one of the central issues in the study of high temperature superconductors. Recent advances in angle-resolved photoemission spectroscopy (ARPES) offer an opportunity of testing the theory of the strongly correlated systems. Here, we present the single-particle, spin and charge excitation spectra in 1D and 2D t-J models in the exact diagonalization technique. The spectra in a two-leg t-J ladder are also shown in order to examine effects of dimensionality.

The Hamiltonian is given by

$$H_{tJ} = -t \sum_{\langle i,j\rangle\sigma} (\tilde{c}^{\dagger}_{i\sigma}\tilde{c}_{j\sigma} + \mathrm{H.c.}) + J \sum_{\langle i,j\rangle} \mathbf{S}_i \cdot \mathbf{S}_j \ , \quad (1)$$

where $\tilde{c}_{i\sigma} = c_{i\sigma}(1 - n_{i-\sigma})$ is the annihilation operator of an electron with spin σ at site i with the constraint of no double occupancy, $n_{i\sigma} = c^{\dagger}_{i\sigma}c_{i\sigma}$, and $\mathbf{S}_i$ is the spin operator at site i with $S = 1/2$. The summation $\langle i,j\rangle$ runs over nearest neighbor pairs. The single-particle spectral function, $A(\mathbf{k},\omega)$, is expressed as

$$A(\mathbf{k},\omega) = A_-(\mathbf{k},\omega) + A_+(\mathbf{k},\omega) \ , \quad (2)$$

$$A_{\pm}(\mathbf{k},\omega) = \sum_{\nu\sigma} \left|\langle \nu \left| a_{\mathbf{k}\sigma} \right| 0\rangle\right|^2 \delta(\omega \mp E_\nu \pm E_0) \ , (3)$$

where $a_{\mathbf{k}\sigma} = c^{\dagger}_{\mathbf{k}\sigma}$ and $c_{\mathbf{k}\sigma}$ for A_+ and A_-, respectively. $|\nu\rangle$ is the ν-th eigenvector with eigenvalue E_ν. $|0\rangle$ denotes the ground state with E_0.

Fig. 1(a) shows the spectrum obtained by the standard Lanczos procedure for a 1D 22-site t-J model in the insulating state. Here, the parameters t and J were taken to be $t = 0.6$ eV and $J = 0.2$ eV in order to compare the spectrum with the recent ARPES data in the 1D cuprate, $SrCuO_2$ [1]. We also calculate the dynamical spin and charge correlation functions, $S(q,\omega)$ and $N(q,\omega)$, expressed as

$$S(q,\omega) = \sum_{\nu} \left|\langle\nu| S^z_q |0\rangle\right|^2 \delta\left(\omega - E_\nu + E_0\right) \ , \quad (4)$$

$$N(q,\omega) = \sum_{\nu} \left|\langle\nu| N_q |0\rangle\right|^2 \delta\left(\omega - E_\nu + E_0\right), \quad (5)$$

with $S^z_q = L^{-1/2}\sum_i e^{iqR_i} S^z_i$ where S^z_i is the z component of the spin operator and L is the total

PII S0921-4534(97)00260-8

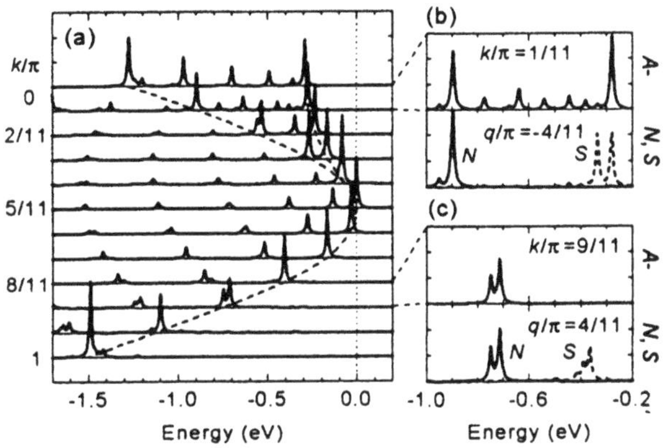

Figure 1. (a) Single-particle spectral function $A_-(k,\omega)$ for a 1D 22-site t-J model with $J = 0.2$ eV and $t = 0.6$ eV at half-filling. The momentum k is defined as $n\pi/11$, n being an integer. The energy ω is measured from the highest energy peak at $k = 5\pi/11$. The δ functions are convolved with a Lorentzian broadening of $0.01t$. The broken lines guide the eye. (b) $A(k,\omega)$ with $k = \pi/11$, and $S(q,\omega)$ (dashed line) and $N(q,\omega)$ (solid line) with $q = k - 5\pi/11$. The energy ω of S and N is converted into $-\omega$. (c) The same as (b), with $k = 9\pi/11$.

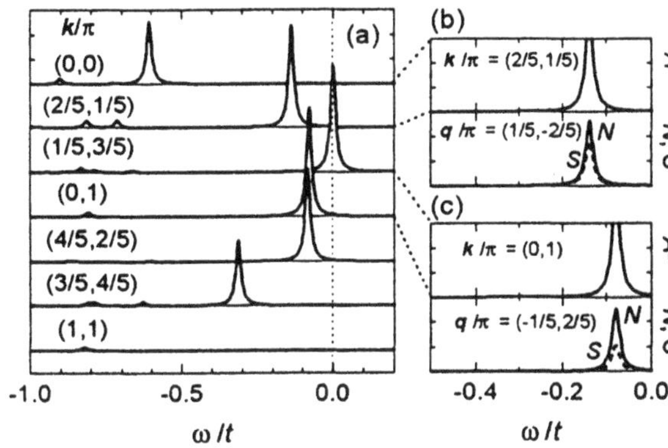

Figure 2. (a) Single-particle spectral function $A_-(\mathbf{k},\omega)$ for a 2D 20-site t-J model ($J/t = 0.4$) at half-filling. The energy ω is measured from the highest energy peak at $\mathbf{k} = (\pi/5, 3\pi/5)$. The δ functions are convolved with a Lorentzian broadening of $0.01t$. (b) $A(\mathbf{k},\omega)$ with $\mathbf{k} = (2\pi/5, \pi/5)$, and $S(\mathbf{q},\omega)$ (dashed line) and $N(\mathbf{q},\omega)$ (solid line) with $\mathbf{q} = \mathbf{k} - (\pi/5, 3\pi/5)$. The energy ω of S and N is converted into $-\omega$. (c) The same as (b), with $\mathbf{k} = (0, \pi)$.

number of sites, and $N_q = L^{-1/2}\sum_i e^{iqR_i}(n_i - n)$ where $n_i = n_{i\uparrow} + n_{i\downarrow}$ and n is the electron number per site. Here, the ground state $|0\rangle$ in Eqs. (4) and (5) is the eigenstate for the single-particle excitation spectrum with lowest energy [2]. Therefore, in this case, $S(q,\omega)$ [3] and $N(q,\omega)$ are calculated in the system with 21 electrons, the ground state of which has $k = 5\pi/11$. In Fig. 1(b), $A_-(k,\omega)$ with $k = \pi/11$ is compared with $S(q,\omega)$ and $N(q,\omega)$ where $q = k - 5\pi/11$ is the momentum transfer. We find that the high energy edge of $A_-(k,\omega)$ is related to the spin excitation, while the low energy one is due to the charge excitation. In Fig. 1(c), $A_-(k,\omega)$ with $k = 9\pi/11$ is replotted together with $S(q,\omega)$ and $N(q,\omega)$. The significant weight of $A_-(k,\omega)$ is given by the charge excitation, in contrast with that for $k \leq 3\pi/11$. We note that although there exists spin excitation around -0.3eV, the intensity of $A_-(k,\omega)$ is very weak. We find that the spectrum in Fig. 1(a) agrees quantitatively with the ARPES data in $SrCuO_2$ [1].

In Fig. 2(a), $A_-(\mathbf{k},\omega)$ for a 2D 20-site t-J model with $J/t = 0.4$ in the insulating state is presented. In the following, energy is measured in units of t unless noted. $S(\mathbf{q},\omega)$ and $N(\mathbf{q},\omega)$ are also plotted for comparison in Figs. 2(b) and (c). As seen in the Figures, $A_-(\mathbf{k},\omega)$ includes both spin and charge excitations in contrast with Fig. 1 in the 1D system.

The ladder compounds provide an opportunity to examine effects of dimensionality on electronic excitations [5,6]. In Fig. 3(a), $A_-(\mathbf{k},\omega)$ for a 2×10 site t-J model of two-leg ladders with $J/t = 0.4$ for both legs and rungs in the insulating state is presented. $S(\mathbf{q},\omega)$ and $N(\mathbf{q},\omega)$ are also plotted for comparison in Figs. 3(b) and (c). We find that $A_-(\mathbf{k},\omega)$ in the ladder includes both spin and charge excitations. This means that the excitations in the ladder are qualitatively similar to those in 2D.

Let us move on to the metallic 2D systems. We first consider the case of low electron density

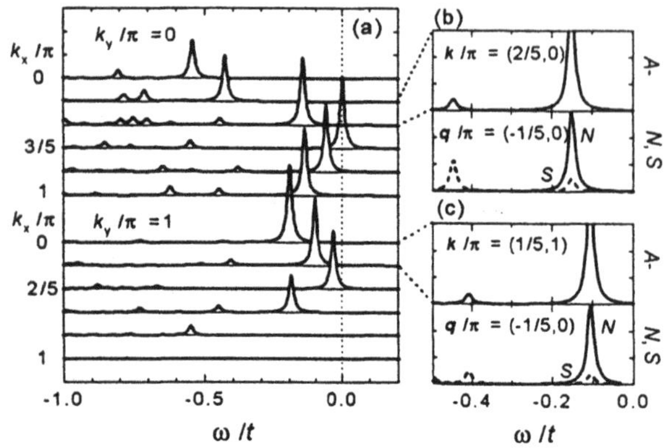

Figure 3. (a) Single-particle spectral function $A_{-}(\mathbf{k},\omega)$ for a 2×10 site t-J model of two-leg ladders at half-filling with $J/t = 0.4$ for both legs and rungs. The x and y axes are taken along leg and rung directions, respectively. The energy ω is measured from the highest energy peak at $\mathbf{k} = (3\pi/5, 0)$. The δ functions are convolved with a Lorentzian broadening of $0.01t$. (b) $A(\mathbf{k},\omega)$ with $\mathbf{k} = (2\pi/5, 0)$, and $S(\mathbf{q},\omega)$ (dashed line) and $N(\mathbf{q},\omega)$ (solid line) with $\mathbf{q} = \mathbf{k} - (3\pi/5, 0)$. The energy ω of S and N is converted into $-\omega$. (c) The same as (b), with $\mathbf{k} = (\pi/5, \pi)$ and $\mathbf{q} = \mathbf{k} - (2\pi/5, \pi)$.

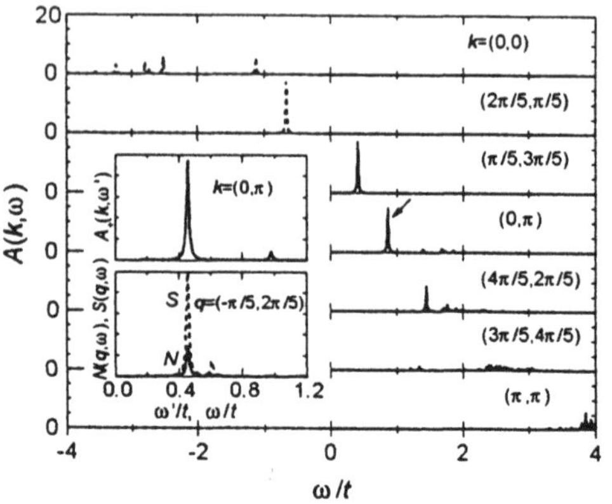

Figure 4. Single-particle spectral function $A(\mathbf{k},\omega)$ for the 2D 20-site t-J model ($J/t = 0.4$) at $n = 0.5$ (10 electrons). The dashed and solid lines correspond to A_{-} and A_{+}, respectively. The δ functions are convoluted with a Lorentzian broadening of $0.01t$. $\mathbf{k}_F^-$ and $\mathbf{k}_F^+$ are located at $(2\pi/5, \pi/5)$ and $(\pi/5, 3\pi/5)$, respectively. Inset shows the single-peak structure of A_{+} at $\mathbf{k} = (0, \pi)$ denoted by the arrow, ω' being measured from the 11-electron ground state, and $N(\mathbf{q},\omega)$ (solid line) and $S(\mathbf{q},\omega)$ (dashed line) at $\mathbf{q} = (-\pi/5, 2\pi/5)$ calculated for the 11-electron system.

(i.e., overdoped case). Figure 4 shows $A(\mathbf{k},\omega)$ for a 20-site ($\sqrt{20} \times \sqrt{20}$) cluster with 10 electrons ($n = 0.5$). The momentum to which an electron is removed (added) with minimum cost in energy is located at $(2\pi/5, \pi/5)$ ($(\pi/5, 3\pi/5)$). We denote this momentum as $\mathbf{k}_F^-$ ($\mathbf{k}_F^+$). We observe a single peak at $\mathbf{k}_F^{\pm}$. We also note that $A_{+}(\mathbf{k},\omega)$ at $\mathbf{k} = (0, \pi)$ close to $\mathbf{k}_F^+$ exhibits a large single peak at $\omega = 0.87t$. Both the corresponding $S(\mathbf{q},\omega)$ and $N(\mathbf{q},\omega)$ have dominant excitations at the same energy, as shown in the inset in Fig. 4. This situation is consistent with that for a Fermi liquid having well-defined quasi-particles with both spin and charge characters.

The high-density case close to half-filling (i.e., underdoped case) is in strong contrast with the low-density case. Figure 5 shows $A_{-}(\mathbf{k},\omega)$ for a 20-site cluster with 2 holes ($n = 0.9$). $\mathbf{k}_F^-$ for A_{-} is located at $(\pi/5, 3\pi/5)$. There is a two-peak structure, marked by the arrows, at $\mathbf{k} = (2\pi/5, \pi/5)$ close to $\mathbf{k}_F^-$. In order to characterize the two peaks, we again calculate $S(\mathbf{q},\omega)$ and $N(\mathbf{q},\omega)$ with $\mathbf{q} = (\pi/5, -2\pi/5)$. It is seen in Fig. 5(b) that $N(\mathbf{q},\omega)$ has much weight for the low-energy peak in A_{-} at $\omega = -0.16t$, whereas $S(\mathbf{q},\omega)$ has much weight for the high-energy peak at $\omega = -0.29t$. This situation is qualitatively similar to that for the 1D t-J model. Here we note that finite mixing of spin and charge degrees of freedom is seen in the two-peak structure in Fig. 5. Although finite-size scaling is impossible at present, we consider that this mixing indicates approximate decoupling of spin and charge degrees of freedom, unlike in the completely decoupled 1D case.

We briefly comment on the large single peak at $\mathbf{k} = (0, \pi)$ and $\omega = -0.1t$ in Fig. 5. For non-interacting electrons, there is no spectral weight

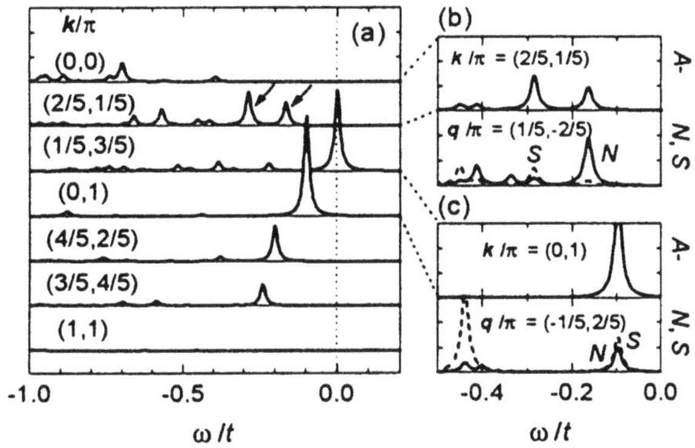

Figure 5. (a) Single-particle spectral function $A_-(\mathbf{k},\omega)$ for a 2D 20-site t-J model ($J/t = 0.4$) with 2 holes. The energy ω is measured from the highest energy peak at $\mathbf{k} = (\pi/5, 3\pi/5)$. The δ functions are convoluted with a Lorentzian broadening of $0.01t$. The arrows denote two-peak structure. (b) $A(\mathbf{k},\omega)$ with $\mathbf{k} = (2\pi/5, \pi/5)$, and $S(\mathbf{q},\omega)$ (dashed line) and $N(\mathbf{q},\omega)$ (solid line) with $\mathbf{q} = \mathbf{k} - (\pi/5, 3\pi/5)$. The energy ω of S and N is converted into $-\omega$. (c) The same as (b), with $\mathbf{k} = (0,\pi)$.

at $\mathbf{k} = (0,\pi)$ in A_-. Thus, the peak is not related to the Luttinger sum rule, but may be due to strong antiferromagnetic correlation existing in the system [4]. We also find that the peak has both spin and charge characters (Fig. 5(c)). The nature of the peak is, therefore, different from that of the two-peak structure discussed above.

In summary, we have investigated the single-particle excitation spectra in the 1D and 2D t-J models, and compared them with the spin and charge ones. In 1D, the spin-charge separation is clearly seen. The theoretical results agree quantitatively with the ARPES data in $SrCuO_2$. In 2D, the single-particle excitation spectrum depends strongly on electron concentration. Near the Mott transition, the spectrum has an anomalous two-peak structure which indicates the different dynamics of spin and charge degrees of freedom. We also found that the electronic excitations in the ladder compounds are similar to those in 2D.

In our finite-size-cluster calculation, the resolutions of momentum and energy are estimated to be $\Delta k \approx \pi/L$ and $\Delta E \approx t/L$, respectively. Taking $L \approx 20$ and $t \approx 0.5$ eV, we have $\Delta k \approx 0.05\pi$ and $\Delta E \approx 0.025$ eV. Since those in the ARPES are considered to be $\Delta k \approx 0.01\pi$ and $\Delta E \approx 0.07$ eV [1], the two-peak structure may be observed as a broad one in ARPES, if the spin-charge separation occurs in 2D. Although the behavior in the limit of low energy is still controversial [2,7,8], the exact diagonalization study of finite-size clusters suggests that there is a crossover from Fermi liquid-like behavior at low electron density to anomalous behavior near the Mott transition in the energy range which is seen in the finite-size clusters.

This work was supported by Priority-Areas Grants from the Ministry of Education, Science and Culture of Japan and NEDO. The computation in this work was done using the facilities of the Supercomputer Center, Institute for Solid State Physics, University of Tokyo.

REFERENCES

1. C. Kim, A. Y. Matsuura, Z.-X. Shen, N. Motoyama, H. Eisaki, S. Uchida, T. Tohyama and S. Maekawa: Phys. Rev. Lett. **77** (1996) 4054.
2. T. Tohyama and S. Maekawa: J. Phys. Soc. Jpn. **65** (1996) 1902.
3. We neglect spin excitation from the space with total spin of 3/2 to which A does not couple.
4. S. Haas, A. Moreo and E. Dagotto: Phys. Rev. Lett. **74** (1996) 4281.
5. E. Dagotto and T. M. Rice: Science **271** (1996) 618.
6. M. Troyer, H. Tsunetsugu and T. M. Rice: Phys. Rev. B **53** (1996) 251.
7. P. W. Anderson: *A Carrer in Theoretical Physics*, World Scientific Series in 20th Century Physics, Vol.7 (1994).
8. H. Fukuyama and M. Ogata: J. Phys. Soc. Jpn. **63** (1994) 3923; *Spectroscopy of Mott Insulators and Correlated Metals*, ed. A. Fujimori and Y. Tokura, Springer Series in Solid-State Sciences, Vol. 119 (1995) 34.

ELSEVIER

Physica C 282–287 (1997) 290–293

PHYSICA C

Phenomenological aspects of phase fluctuation in high Tc superconductors: Cooper pair droplets

B.K. Chakraverty[a] and T.V. Ramakrishnan[b]

[a]Laboratoire d'Etudes des Proprietes Electroniques des Solides, CNRS, BP 166 -Grenoble Cedex 9, France, associe a l'Universite Joseph Fourier, Grenoble

[b]Department of Physics, Indian Institute of Science, Bangalore 560012, India

We point out how fluctuation of the phase of the superconducting order parameter can play a key role in our understanding of high T_c superconductors. A simple universal criterion is given which illustrates why all oxide superconductors in contrast to classical superconductors ought to behave as a lattice of cooper pairs .T_c is to be thought of as the temperature of phase coherence or the temperature above which the lattice of Cooperpair 'melts' into a phase of Cooper-pair droplets that starts forming at $T \approx T^*$.This is the pseudo-gap region.Quantum fluctuation of the phase predicts a superconductor to insulator phase transition for all underdoped materials.

There has been increasing evidence over the past years that in may respects high T_c materials do not seem to behave, like classical superconductors obeying a mean field B.C.S. behavior - these include appearance of a characteristic temperature T^* [1] way above T_c at which many of the physical properties are anomalous. There is also the characteristic hallmark of these high T_c materials in that they are either super-conducting or insulating in its ground state in the underdoped regime. Irrespective of the underlying microscopic mechanism of the high T_c superconductivity, many of these phenomena can be well understood on the basis of the fluctuation of the phase of the order parameter. A preliminary account [2] was given previously.

A simple universal criterion emerges as to whether phase fluctuation will play any role at all or not, if we compare two length scales ; the distance d between the Cooper-pairs and ξ_0, the Pippard coherence length, which measures the "size" of a Cooper-pair.

$$\frac{1}{d^3} = N(\varepsilon_F)\Delta \qquad (1)$$

where $N(\varepsilon_F)$ is the density of states at the Fermi level, and Δ, the half-gap, so that the right hand side gives us the number of Cooper-pairs per unit volume.

We show in table I, the inter Cooper-pair distance d calculated for a variety of superconductors, along with known ξ_0. We note immediately that all oxyde superconductors including the high T_c superconductors have $\frac{\xi_0}{d} \sim 1$; the table clearly shows an universal trend, going from $\frac{\xi_0}{d} >> 1$ for the classical superconductors to a ratio << 1, in the dilute Bose-gas limit. We may consider $\frac{\xi_0}{d} \leq 1$ as defining regime of applicability of real-space xy model as a lattice of cooper-pairs, centered on a site "i" of extension ξ_0. This description allows us to model the system, on a coarse-grained Ginzburg-Landau model (CGGL) which we write as

$$H = \sum_i -A\left|\psi_i\right|^2 + B\left[\psi_i\right]^4 + C\sum_{ij}\left|\psi_i - \psi_j\right|^2 \qquad (2)$$

where A, B and C are phenomenological coefficients. The complex order-parameter ψ_i given by :

$$\psi_i = |\psi_i| \exp i\,\theta_i \qquad (3)$$

where $|\psi_i|$ is the amplitude, with a local phase θ_i. CGGL in this form is completely equivalent to the superconducting grain models . Substituting (3) into (2), we obtain :

$$H = \sum_i H_0(|\psi_i|) + \sum_{ij} H_0(\theta_{ij}) \qquad (4)$$

0921-4534/97/$17.00
PII S0921-4534(97)00263-3

TABLE 1

SUPERCONDUCTOR	ESTIMATED $\xi_o(\hat{A})$	ESTIMATED $d(\hat{A})$	$\frac{\xi_o}{d}$	$\lambda \hat{A}$	$T_c^{xy}(^oK)$	$T_c^{exp}(^oK)$	$T_c^{BCS}(^oK)$
He^4	4	4	~1	-	1	2.17	-
$LaSr_{0.15}CuO_4$	12*	15	~1	3800	46	33	215
YBaCuO	8*	10	~1	1600	160	92	574
$Ba_{0.6}K_{0.4}BiO_3$	35	28	~1	3000	160	30	86
$Li_{0.9}Mo_6O_{17}$	48 *	100	~2	10000	8	2	28
K_3C_{60}	22	26	~1	5000	44	18	98
Intermetallic NbC	23	10	2.3	-	-	11	-
V_3Si	60	30	2	-	-	17	-
Nb_3Sn	60	23	2.6	-	-	18	-
Nb	115	25	4.6	-	-	9	-
Chevrel Phase $Cu_2Mo_6S_8$	500	100	5	-	-	10	-
He^3	300	20	15	-	75	2.6×10^{-3}	3×10^{-3}
Heavy Fermions UBe_{13}	1000	30	33	-	10000	0.93	-
Classical Superconductors Al	16000	100	160	157	8×10^6	1.15	1.35
Pb	830	28	30	370	4×10^6	7	14

* Space averaged
-Not calculated

where i j are near neighbours.

$$H_O(|\psi_i|) = (-A + 2zC)|\psi_i|^2 + B|\psi_i|^4 \quad (5)$$

z is the near-neighbor coordination and

$$\sum_{ij} H_O(\theta_{ij}) = -J\sum_{ij} \cos(\theta_i - \theta_j) \quad (6)$$

. Equations (4), (5) and (6) admit two mean-field solutions [4].

(a) A completely phase disordered phase at $T \approx T^*$ with $<\cos(\theta_i - \theta_j)> = 0$, as long as the coefficient of the quadratic term $-A + 2ZC < 0$; this is not a true phase transition but signifies appearance of a local pairing amplitude at T^* without any near-neighbor phase correlation.

(b) A true phase coherent state defining superconductivity at $T = T_C \approx zJ$, where phase locking between all sites occurs, $<\cos(\theta_i-\theta_j)>=1$. At all tempera-tures $T>T_C$, local pairing amplitude corre-lation $<\psi_i>$ remains non zero, up to $T \sim T^*$. T_C is the 'melting' of the Cooper pair lattice.

These results are also obtained from the static phase Hamiltonian (6), by applying to it a vector-potential A_{ij}.We get

$$H_J = -J\sum_{ij}\cos(\theta_i - \theta_j - A_{ij}) \quad (7)$$

Double derivative of the free energy with respect to the vector potential gives us the phase stiffness as

$$\rho_s = J\sum_{ij}\left\langle \cos\left(\theta_i - \theta_j\right)\right\rangle - \frac{1}{kT}\sum_{ij}\left\langle\left[J\sin\left(\theta_i - \theta_j\right)\right]^2\right\rangle \quad (7a)$$

where ρ_s is the phase stiffness.

The second term, which is a Josephson current-current correlation function, can be thought of as anharmonic contribution to the free-energy. Expression (7a) is shown sche-matically in figure 1 and exhibits complete cancellation of the two terms at $T = T_C$, to give vanishing superconductive stiffness at T_c and a non-zero $|\psi|^2$ beyond T_c upto $T = T^*$ where a local pairing amplitude nucleates in a region ~ ξ_0, at $T < T^*$. This is pseudogap region.

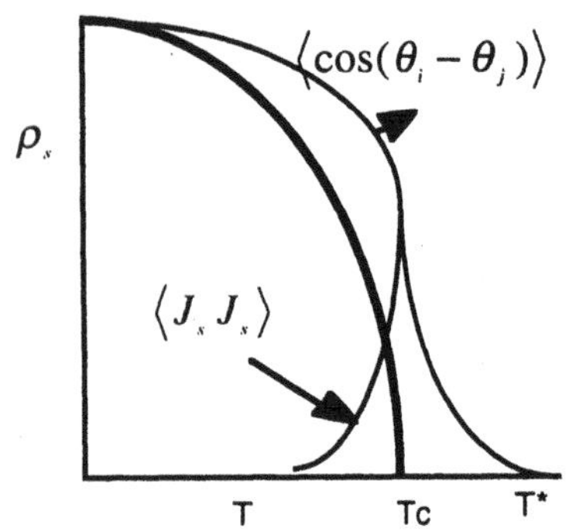

A long-wave length version of expression (6), allows us to write Tc as phase-stiffness of the 3d-XY model, in terms of known measurable parameters. We have

$$k_B T_c^{XY} = \frac{\phi_0^2}{15\pi^3} \frac{\xi_0 \eta}{\lambda^2} \quad (8)$$

where ϕ_0 is the flux quantum, λ the London penetration depth and η a lattice anisotropy parameters . This expression (8) can be used to calculate 3d T_c^{XY} and gives numbers quite close to the experimental Tc's of the high Tc materials [3](See TABLE 1).

We may note, that the B.C.S. Tc's in contrast, has $T_c^{B.C.S.}$ given by the empirical rule

$$T_c^{B.C.S} = 0.27 \frac{\hbar^2}{2m} \frac{n^{\frac{2}{3}}}{\xi_0} \quad (9)$$

and is inversely proportional to ξ_0. For the same parameters used to evaluate T_c^{xy} this gives unusually high values for the high T_c materials(TABLE1).

Starting from a microscopic Hamiltonian, we can derive [4], the following dynamic phase only Hamiltonian that includes quantum fluctuation of the phase, in the long wave-length limit.

$$H_{eff}(\theta) = \chi_\rho^o (\frac{\partial\theta}{\partial\tau})^2 + \rho_s^o (\frac{\partial\theta}{\partial x})^2 \quad (10)$$

Here the two coefficients, the charge susceptibility χ_ρ^o and the superfluid stiffness ρ_s^o are related respectively to the density-density correlation function $\langle\rho\rho\rangle$ and to the superfluid velocity-velocity correlation function $\langle v_s v_s \rangle$.

We can extract from the expression (10), the effect of the dynamic fluctuation term, on the phase-stiffness in several limits that we have studied [5].

If χ_ρ^o is assumed to be local, we can use the discretised version of equation (8), which becomes the quantum rotor problem [6] given by

$$H = \frac{U}{2} \sum \hat{n}_i^2 - J \sum_{ij} e^{-i\theta_i} e^{i\theta_j} \quad (16)$$

where $U \approx \frac{1}{\chi_\rho^o}$ represents a local Coulomb energy.

Since θ_i is the conjugate variable to the charge fluctuation operator $\hat{n}_i$ whichdescribes the deviation of the number of pairs from the average value which can be taken to be zero or a charge neutral state. The term in U suppresses charge fluctuation between sites i and j and as U becomes greater than J increased dynamic fluctuation of phase will lead to loss of phase coherence and charge localised insulating state.The mean field stability condition then gives, at low temperature,

$$\frac{U}{2} = 2zJ - T; at\ T_c = 0, U = U_c = 4zJ \quad (17)$$

At high temperature, ßU<1 and all the rotors can be considered to be independent so that, now the stability condition gives

$1 \cong (\frac{zJ}{T})(1 - \frac{U}{6T})$; *For* $U = 0$ We have the classical result $T_c = z.J$ (18)

The preceding mean field result tells us that at a critical value of local Coulomb interaction $U=U_c=4zJ$, the quantum fluctuation of phase has completely destroyed the superconductive phase coherence and $T_c=0$.This is a quantum critical point(QCP).Beyon $U>U_c$ a coulomb gap appears to add or to subtract a Cooper pair from a given domain.The system has been driven to an insulating state.

To understand the QCP let us define a quantum coupling constant g such that

$$g = \sqrt{\frac{U}{J}} \qquad g_c = 2\sqrt{z}$$

We also note that $\hbar\omega_J = \sqrt{JU}$ where ω_J is the Josephson plasma frequency.

Hence $g = \frac{h\omega_J}{J}$. where $g \approx \delta_0 - \delta$ where δ_o is the concentration for optimum T_C. (19)

This definition of g is very useful. This is the same definition used in a related magnetic phase transition of the nonlinear Sigma model by Chakravarty, Halperin and Nelson [7]. It can be shown that with this definition of the coupling constant g the dynamic hamiltonian in d-dimension of equation (16) divided by T is exactly equivalent at T=0 to a classical hamiltonian in d+1dimension given by

$$H = \frac{1}{g}\sum_{ij}\cos(\theta_i - \theta_j) \qquad (20)$$

This eqivalence is valid when there is isotropy of space and time as embodied in the hamiltonian of (16). The quantum coupling constant g is a fake temperature and is a measure of quantum phase fluctuation. As g increases to the critical value g_C, superconducting phase coherence gets eventually lost and we go to an insulating state. Zero energy charge fluc-tuation is no longer possible beyond $g=g_C$ and a Coulomb gap appears in the system. This charge gap corresponding to a local plasmon modewill also determine the quantum cross-over temperature T_Q below which the system is insulating with T_Q scaling as $T_Q \sim (g - g_C)^{\alpha}$ (21)
where α is a numerical coefficient ~ 0.67.
T_Q is a classical to quantum cross-over temperature, and at all $T > T_Q$ or $T > T_C$, the system ought to be in the quantum critical region. We note the contrasting behaviour of resitivity below T* and below T_Q. ***Below T* resitivity decreases while below T_Q resitivity increases in the ab plane.*** The generic phase diagram, for the underdoped regime is shown in figure 2.

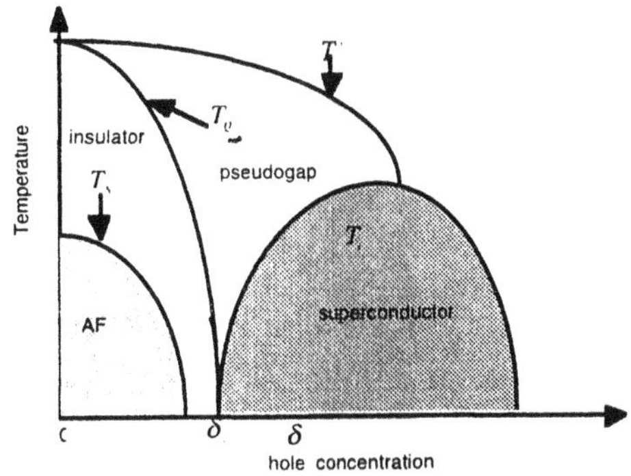

We have shown in this letter that there is an universal criterion which tells us that all oxide superconductors including the high T_C materials ought to behave as a lattice of Cooper-pairs and be governed by phase fluctuation of the superconducting order parameters irrespective of the underlying microscopic mechanism of superconductivity. The overdoped region is expected to be governed by amplitude rather than phase fluctuation. The possibility of simultaneous presence of extended s and d-wave symmetry [8] and the relative charge fluctuation between them may have profound consequence on the dynamic phase fluctuation term.

This work was financially supported by the Indo-French project # 502 of IFCPAR.

References

[1] B. Batlogg et al.
Physica C235-240, 130 (1994).
H.Ding et al Nature 382,51(1996).
[2] B.K. Chakraverty, A. Taraphder and M. Avignon
Physica C235-240, 2323 (1994).
[3] T. Schneider and H. Keller
Int. J. Mod. Phys. B 8, 487 (1993).
V.J. Emery and S. Kivelson
Nature 374, 434 (1995).
[4] T.V. Ramakrishnan
Physica Scripta T27, 24 (1989).
[5] B.K.Chakraverty and T.V.Ramakrishnan, to be published.
[6] S. Doniach
Phys. Rev. B24, 5063 (1981).
[7] S. Chakravarty, B. Halperin and D. Nelson
Phys. Rev. B39, 2344(1989).
[8] K. Alex Müller Nature 377, 133 (1995).

ELSEVIER

Physica C 282–287 (1997) 294–298

PHYSICA C

High Tc superconductors : the van Hove scenario, a review

J. Bok, J. Bouvier

Laboratoire de Physique du Solide, CNRS-ESPCI, 10, rue Vauquelin - 75231 Paris cedex 05 , France

We shall review 1/ the calculation of T_c in the framework of the v H scenario, 2/ the influence of the Coulon repulsion which decreases T_c strongly in narrow band cuprates, 3/ the calculation of the gap anisotropy using weakly screened electron-phonon interaction and the effect of doping, 5/ its application to tunneling spectroscopy a 6/ its application to the specific heat of HTCS.

1. CALCULATION OF Tc, THE LABBE BOK FORMULA

This formula was obtained using the following assumptions :

1- the Fermi level lies at the van Hove singularity

2- The B.C.S. approximations :

- The electron-phonon interaction is isotropic and so is the superconducting gap Δ.
- The attractive interaction V_p between electrons is non zero only in an interval of energy $\pm\hbar\omega_0$ around the Fermi level where it is constant. When this attraction is mediated by emission and absorption of phonons, ω_0 is a typical phonon frequency.

In that case, the critical temperature is given by

$$k_B T_c = 1.13 D \exp\left[-\left(\frac{1}{\lambda} + \ln^2\left(\frac{\hbar\omega_0}{D}\right) - 1.3\right)^{1/2}\right] \quad (1)$$

where $\lambda = \frac{1}{2} n_1 V_p$

A simplified version of formula (1), when $\hbar\omega_0$ is not too small compared to D is

$k_B T_c = 1.13 D \exp(-1/\sqrt{\lambda})$

The two main effects enhancing T_c are

1- the prefactor in formula (1) is an electronic energy much larger than a typical phonon energy $\hbar\omega_0$.

2- λ is replaced by $\sqrt{\lambda}$ in formula (1) so that in the weak coupling limit when $\lambda<1$, the critical temperature is increased. In fact it gives too high values of T_c, we shall see later that this is due to the fact that we have neglected Coulomb repulsion between electrons. Taking this repulsion into account we shall obtain values for T_c which are very close to the observed one.

As it is however, this approach already explains many of the properties of the high T_c cuprates near optimum doping.

- The variation of T_c with doping

The highest T_c is obtained when the Fermi level is exactly at the v.H.s. For lower or higher doping the critical temperature decreases. That is what is observed experimentally[1].

- The isotope effect

Labbé and Bok [2] showed using formula (1), that the isotope effect is strongly reduced for high T_c cuprates. Tsuei *et al* [3] have calculated the variation of the isotope effect with doping and shown that it explains the experimental observations

- Marginal Fermi liquid behaviour

In a classical Fermi liquid, the lifetime broadening $1/\tau$ of an excited quasiparticle goes as ε^2. The marginal Fermi liquid situation is the case where $1/\tau$ goes as ε. Theoretically marginal behaviour has been established in two situations (a) the half filled nearest-neighbour coupled Hubbard model on a square lattice and (b) the Fermi level lies at a v.H. singularity [3]. Experimental evidence of marginal Fermi liquid behaviour has been seen in angle resolved photoemission [4], infrared data [5] and temperature dependence of electrical resistivity [6]. Marginal Fermi liquid theory, in the frame work of v.H.s. predicts a resistivity linear with temperature T. This was observed by Kubo et al [6]. They also observe that the dependence of resistivity goes from T for high T_c material to T^2 as the system is doped away from the T_c maximum, which is consistent with our picture ; in lower T_c material the Fermi level is pushed away from the singularity

PII S0921-4534(97)00264-5

2. INFLUENCE OF THE COULOMB REPULSION

As soon as 1962 Anderson and Morel [7] have shown that the electron-electron repulsion plays a central role in superconductivity. Assuming a constant repulsive potential $V_{(k,k')} = V_c$ from 0 to E_F they find that T_c is given by :

$$T_c \cong T_o \exp\left[\frac{-1}{\lambda - \mu^*}\right]$$

With $\mu = N_o V_c$ and (2)

$$\mu^* = \frac{\mu}{1 + \mu \ln E_F / \omega_o}$$

Cohen and Anderson [8] assumed that for stability reasons μ is always greater than λ. Ginzburg [9] gave arguments that in some special circumstances μ can be smaller than λ. Nevertheless if we take $\mu \geq \lambda$, superconductivity only exists because μ^* is of the order of $\mu/3$ to $\mu/5$ for a Fermi energy of the order of 100 $\hbar\omega_0$. It is useless to reduce the width of the band W (E_F = W/2 for a half-filled band) because λ and μ vary simultaneously and μ^* becomes greater if E_F is reduced, thus giving a lower T_c. Superconductivity can even disappear in a very narrow band if $\lambda - \mu^*$ becomes negative.

We have shown [10] that nevertheless high T_c can be achieved in a metal containing almost free electrons (Fermi liquid) in a broad band, with a peak in the D.O.S. near the middle of the band.

Taking a D.O.S., which is a constant n_0 between energies - W/2 and W/2, (the zero of energy is at the Fermi level) and is $n(\varepsilon) = n_1 \ln\left|\frac{D}{\varepsilon}\right| + n_0$ between -D and +D we find for T_c, the following formula :

$$k_B T_c = \frac{D}{2} \exp\left[0.819 + \frac{n_0}{n_1} - \sqrt{F}\right]$$

where

$$F = \left(\frac{n_0}{n_1} + 0.819\right)^2 + \left(\ln\frac{\hbar\omega_0}{D}\right)^2 - 2 - \frac{2}{n_1}\left(n_0 \ln\frac{2.28\hbar\omega_0}{D} - \frac{1}{V_p - V_c^*}\right) \qquad (3)$$

$$V_c^* = \frac{V_c}{1 + V_c\left[\frac{n_1}{2}\left(\ln\frac{D}{\hbar\omega_0}\right)^2 + n_0 \ln\frac{W}{2\hbar\omega_0}\right]}$$

We can have a few limiting cases for this formula :
$n_1 = 0$: no singularity. We find the Anderson-Morel formula.
$V_c = 0$ and $n_0 = 0$: this gives the Labbé-Bok formula.
There are many effects enhancing T_c
$\lambda - \mu^*$ is reduced by the square root, down to $\sqrt{\lambda_1 - \mu_1^*}$ when n_1 is large enough.
As $\lambda - \mu^* < 1$ the critical temperature is strongly increased because this factor appears in an exponential.

The prefactor before the exponential is D, the singularity width instead of $\hbar\omega_0$. We expect $D > \hbar\omega_0$. For instance D may be of the order of 0.5 eV and $\hbar\omega_0$ about a few 10 meV ($D/\hbar\omega_0$ of the order of 5 to 10).

We have made some numerical calculations using formula (3) to illustrate the effect of Coulomb repulsion. We used two values of D : D = 0.9 eV corresponding to t = 0.25 eV and a much more smaller value D = 0.3 eV. This calculations are illustrated by figure (1).

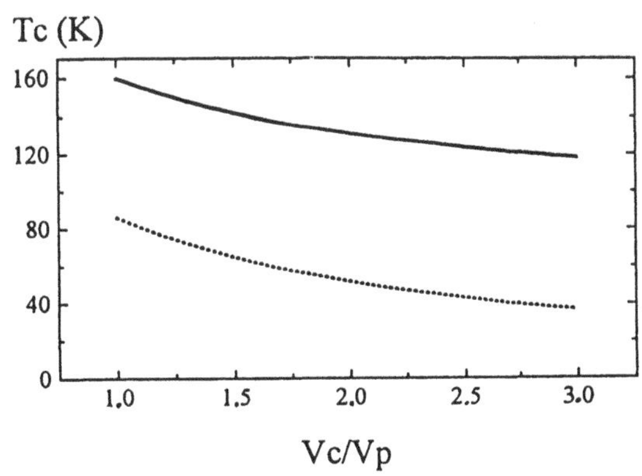

Fig.1 : Effect of Coulomb repulsion on T_c. The following numerical values have been used :
(solid line) D = 0.9 eV, n_0 = 0.3 states/eV/Cu atom, W = 2 eV, n_1 = 0.2
(dotted line) D = 0.3 eV, n_0 = 0.3 states/eV/Cu atom, W = 3 eV, n_1 = 0.16

These calculations show that the Coulomb repulsion does not kill superconductivity in the framework of the L.B. model. The general rule for high T_c in this model is to have a peak in the density of states near the middle of a broad band to renormalize the effective repulsion μ.

3. GAP ANISOTROPY

Bouvier and Bok [11] have shown that using a weakly screening electron-phonon interaction, and the band structure of the CuO_2 planes four saddle points: an anisotropic superconducting gap is found. For the superconducting gap, Δ_k, the B.C.S. integral equation (at T = 0K)

$$\Delta_{\vec{k}} = -\frac{1}{2}\sum_{\vec{k}'} \frac{V_{\vec{k}\vec{k}'}\Delta_{\vec{k}'}}{\sqrt{\varepsilon_{\vec{k}'}^2 + \Delta_{\vec{k}'}^2}} \qquad (4)$$

where $V_{kk'}$ is the electron-electron attraction mediated by phonons and ε_k the band structure given by

$$\varepsilon_k = -2t(\cos k_x a + \cos k_y a)$$

where t is a transfer integral between nearest neighbours. This corresponds to the case where the Fermi level E_F lies exactly at the singularity E_s ($E_F = E_s$). We take for $V_{kk'}$ a weakly screened attractive interaction :

$$V_{\vec{k}\vec{k}'} = \frac{-V}{q^2 + q_0^2}$$

$V > 0$ when $|\varepsilon_{kk'}| < \hbar\omega_0$, ω_0 being a typical phonon frequency. To solve eq (4) we use a Fourier expansion of $\Delta(\phi)$, where ϕ is the angle between $\vec{k}$ and the k_x axis :

$$\Delta(\phi) = \Delta_0 + \Delta_1 \cos 4\phi + \Delta_2 \cos(8\phi + \theta) + \ldots \qquad (5)$$

We have computed numerically the first two terms Δ_0 and Δ_1, by iteration. The results of Δ versus ϕ are presented in figure (2) for two sets of parameters. We insist that most of the parameters t, ω_0 are known experimentally. We have estimated the screening parameter q_0a, in a Thomas Fermi approximation, of the order of 0.2.

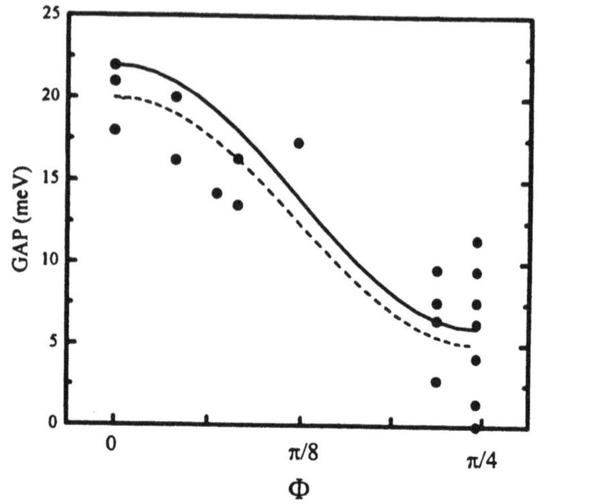

Fig. 2 : Angle-dependent calculated gap $\Delta(\phi)$ for two sets of parameters : Δ_0 = 14 meV, Δ_1 = 8 meV, for T_c= 88.5 k (solid line) and Δ_0 = 12,5 meV, Δ_1 = 7,5 meV, for T_c= 78.5 k (dashed line)

We also have computed the variation of Δ_k with temperature. On fig. (3) we have represented $\Delta_{Max} = \Delta_0 + \Delta_1$, $\Delta_{min} = \Delta_0 - \Delta_1$ and $\Delta_{AV} = \Delta_0$ as a function of temperature. We see that at zero temperature $2\Delta_{Max}/k_BT_c = 6$, $2\Delta_{min}/k_BT_c = 2.6$ and the B.C.S.value of the ration is obtained for Δ_{AV}. This may explain the various values $2\Delta/k_BT_c$ observed in experiments. Tunneling spectroscopy gives the maximum ratio and thermodynamic properties such as $\lambda(T)$ (penetration depth) gives the minimum gap.

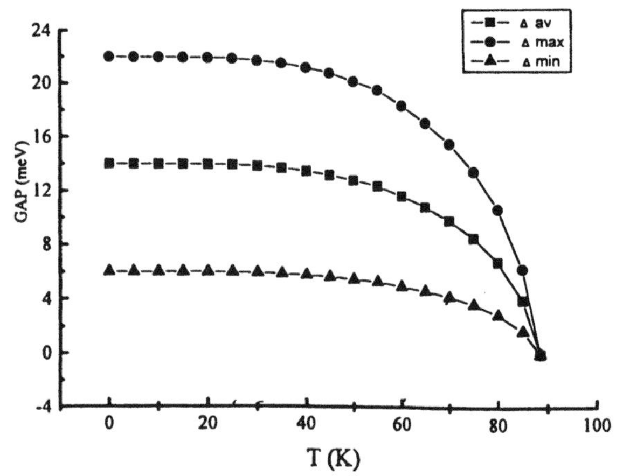

Fig. 3 : Temperature-dependent maximum (full circle), average (full square) and minimum (full triangle) gaps; for T = 0 K, for $\Delta_{av} = \Delta_0$ = 14 meV, Δ_1 = 8 meV , and for a Tc = 88.5 K.

4. TUNNELING SPECTROSCOPY

We have calculated the density of states (D.O.S) of quasiparticle excitations in the superconducting state of high T_c [12] cuprates using the model of anisotropic gap that we have recently developed [11]. We use a gap of the form given by equation (5) and the B.C.S. theory for the energies E_k of the excited states :

$$E_k^2 = \varepsilon_k^2 + \Delta_k^2$$

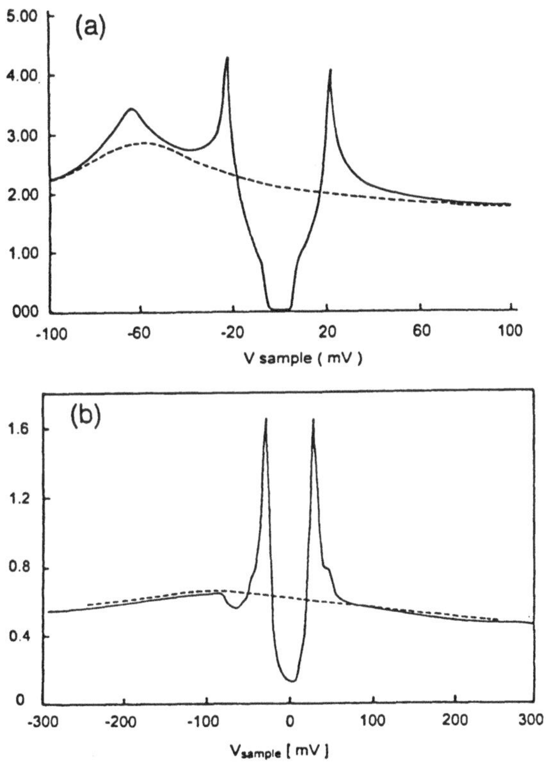

Fig. 4 : (a) Curves of the conductance calculated for a N-I-S junction. Solid line: in the superconducting state at T = 5 K with $\Delta_M = 22$ meV, $\Delta_m = 6$ meV, $\Gamma = 0.1$ meV, $t = 0.2$ eV and $\delta = -60$ meV, $\Gamma' = 5$ meV. Dashed line : in the normal state at T = 100 K with $\Delta_M = \Delta_m = 0$ meV, $\Gamma = 0.1$ meV, $t = 0.2$ eV and $\delta = -60$ meV, $\Gamma' = 5$ meV. (b) For comparison we show Fig. 7 of Ref. [13]. The maximum of the normal state conductance (or D.O.S.) at negative sample bias is well reproduced.

From these expressions of the D.O.S., we have computed the tunneling conductance dI/dV of tunnel junctions, both normal-insulator-superconductor (N.I.S.) and S-I-S junctions. We have also taken into account a possible finite lifetime of quasiparticle states, by introducing a phenomenological broadening Γ. The results are given in detail in ref [12].

Figure (4) represents the computed curve of a dI/dV versus V characteristic of a NIS junction and an experimental curve done by Renner and Fisher [13] on a BSSCO sample. We also give results corresponding to the case of a sample which is not at optimal doping, i.e. E_F-$E_S \neq 0$. In that case a dip is found is the D.O.S. Such a dip has been seen experimentally, both in photoemission [14]. and in the dI/dV characteristic [13].

5. SPECIFIC HEAT

We have also computed the specific heat of a superconductor in the framework of the van Hove scenario, i.e. using the D.OS. of excited states obtained before. Our calculations are detailed in another presentation in this conference (Bouvier, Bok). The main results are the following :

- the specific heat jump at T_c, $\Delta C/C$, is greater than the universal B.C.S. value 1.43, obtained with a constant density of state.

It varies from the value 3.5 when $E_F = E_S$ to the B.C.S. value for heavy doping $E_F - E_S > 70$ meV

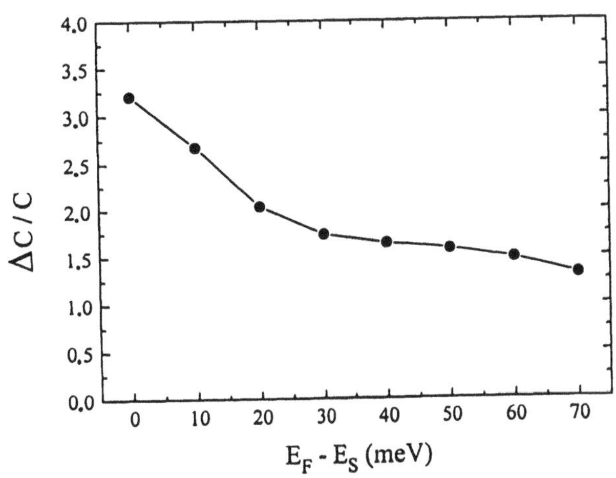

Fig. 5 : Variation of the jump in the specific heat at T = T_c, $\Delta C/C$, versus the doping D = E_F-E_s

CONCLUSION

In conclusion, we have shown that the van Hove scenario explains many properties of the high T_c cuprates.

- high T_c and anomalous isotope effect
- great variation of T_c, from 0 to 160 K, with band parameters
- the observed gap anisotropy for optimum, and overdoped samples (Bouvier, Bok, same conference).
- the observed tunneling conductance and its variation with doping
- the observed specific heat anomalies.

We want to stress that our approach neglects completely magnetic fluctuations, so that it does not apply to underdoped samples. It should apply to optimally and overdoped samples where magnetic fluctuations are very weak and even disappear.

REFERENCES

[1] Y. Koike et al, Physica C, **159**, 105 (1989)
[2] J. Labbé, J. Bok, Europhysics Letters, **3**, 1225 (1987)
[3] C.C. Tsuei, D.M. Newns, C.C. Chi et P.C. Pattnaik, Phys. Rev. Lett., **65**, 2724 (1990)
D.M. Newns, C.C. Tsuei, P.C. Pattnaik and C.L. Kane, Comments Cond. Mat. Physics, **15**, 273 (1992)
[4] G.G. Olson et al, Phys. Rev. B, **42**, 381 (1990)
[5] Z. Schlesinger et al, Phys. Rev. Lett., **67**, 1657 (1991)
[6] Y. Kubo et al, Phys. Rev. B, **43**, 7875 (1991)
[7] P. W. Anderson and P. Morel, Phys. Rev. **125**, 4, (1962)
[8] M.L. Cohen and P.W. Anderson in Superconductivity in d and f band Metals, edited by D.H. Douglass (A.I.P. New York) (1972)
[9] V. Ginzburg, Comtemporary Physics, **33**, 1, 15-23 (1992)
[10] L. Force et J. Bok, Solid State Comm., **85**, 975 (1993)
[11]J. Bouvier, J. Bok, Physica C, **249** (1995) 117
[12] J. Bok, J. Bouvier, Physica C, **274** (1997) 1
[13] Ch. Renner and O. Fisher, Phys. Rev. **B51** (1995) 9208
[14] Y. Huri et al, Phys. Rev. Lett. **67** (1991) 2573

ELSEVIER

Physica C 282–287 (1997) 299–302

PHYSICA C

Electron Correlation, Metal-Insulator Transition and Superconductivity in Quasi-2D Organic Systems, $(ET)_2X$

K. Kanoda

Institute for Molecular Science, Myodaiji, Okazaki 444, Japan

A family of quasi-two-dimensional organic systems, $(ET)_2X$, have been studied by NMR, susceptibility and specific heat measurements in the light of role of electron correlation. A systematic variation of antiferromagnetic spin fluctuations toward metal(superconductor)-to-insulator transition and the ordered spin structure in the insulating phase are uncovered. Based on the results, we propose a conceptual phase diagram for $(ET)_2X$, where the κ-type of family span a critical region of Mott transition. The superconductivity, which is enhanced near the Mott transition, is suggested to have line nodes in gap parameter by NMR relaxation and specific heat profiles.

1. INTRODUCTION

Organic systems, $(ET)_2X$, have layered structures with quasi-two-dimensional electronic bands like cuprates. (ET is an abbreviation of BEDT-TTF.) The holes provided by the monovalent X^- gives a quarter-filled band but in many cases the hole concentration is out of control, which provided fertile physics in the recent study of the transition metal oxides. Nevertheless, rich variation of the eletronic phases came out in $(ET)_2X$. The degree of freedom giving the variation is in the molecular arrangement. Figure 1 depicts the resistivity profiles for a family, κ-$(ET)_2X$, with the κ-type of ET arrangement (see the inset of Fig.4). Even a small structural modification within the same type of ET arrangement leads to a drastic change of electronic state.

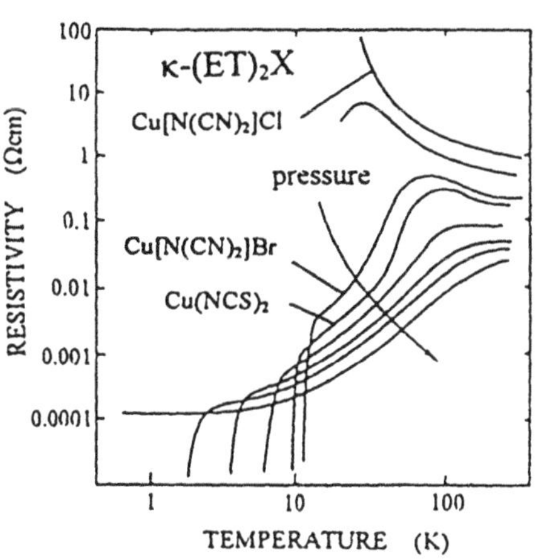

Figure 1. Resistivity profiles of κ-$(ET)_2X$

At ambient pressure, X=Cu[N(CN)$_2$]Cl salt is nonmetallic in a whole temperature range, while Cu(NCS)$_2$ and Cu[N(CN)$_2$]Br salts show a crossover from nonmetallic to metallic behaviors well below 100 K. The pressure interconnects these behaviors and further drives the system into the conventional metals with T_C decreased and vanishing eventually.

In this paper, our ^{13}C-NMR and specific heat studies toward understanding of the mechanisms of the M-I transition in κ-$(ET)_2X$ family with a parameter of X and superconductivity enhanced near the M-I transition are reviewed.

2. EXPERIMENTAL

For ^{13}C-NMR, the central carbon sites in BEDT-TTF molecules were enriched by ^{13}C isotope. The crystals of κ-$(ET)_2X$ [X=Cu(NCS)$_2$, Cu[N(CN)$_2$]Br and Cu[N(CN)$_2$]Cl] were grown by the conventional electrochemical oxidation. For X=Cu[N(CN)$_2$]Br, we also prepared a salt with deuterated BEDT-TTF molecule, which is referred to as d_8-Cu[N(CN)$_2$]Br while the nondeuterated salt is as h_8-Cu[N(CN)$_2$]Br hereafter. The ^{13}C NMR measurements were made for a single crystal, aligned crystals or a powder. The low-temperature specific heat was measured for a single crystal with the thermal relaxation method.

3. RESULTS AND DISCUSSION

3. 1. Antiferromagnetic spin fluctuations

As seen in Fig. 2, the spin susceptibility is temperature-insensitive at higher temperatures but

0921-4534/97/$17.00
PII S0921-4534(97)00266-9

exhibits rather steep decrease below 50 K for all of κ-$(ET)_2X$. The abrupt increases of the susceptibility around 30 K for $Cu[N(CN)_2]Cl$ salt and around 15 K for d_8-$Cu[N(CN)_2]Br$ salt were found to be due to canting of antiferromagnetically (AF) ordered spins by magnetization measurements [1]. Only for d_8-$Cu[N(CN)_2]Br$ salt, the ac susceptibility measurement indicated that in addition to the magnetic phase there exists a comparable fraction of superconducting phase even in a single crystal, suggesting that d_8-$Cu[N(CN)_2]Br$ salt is just on the border of the superconductor-magnetic insulator transition [2]. Accordingly, d_8-$Cu[N(CN)_2]Br$ salt exhibits separation of NMR spectra into two subsets below 35 K, as shown later. (The fraction of the magnetic phase in d_8-$Cu[N(CN)_2]Br$ salt, when cooled rapidly through 80 K, becomes nearly the full volume [2]. Hereafter, we mean the slowly cooled state with the two phases unless otherwise noted.)

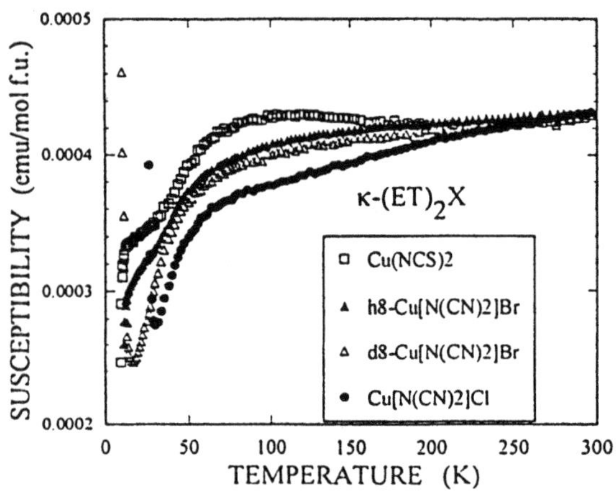

Figure 2. Spin susceptibility of κ-$(ET)_2X$

The powder-averaged values of ^{13}C nuclear spin-lattice relaxation rate divided by temperature, $(T_1T)^{-1}$, are plotted in Fig.2. As for d_8-$Cu[N(CN)_2]Br$ salt, T_1 was measured in respective positions of line below 35 K . The non-monotonous temperature dependence of $(T_1T)^{-1}$ means that the Korringa relation breaks down for all the salts. The divergent peaks at 27K for $Cu[N(CN)_2]Cl$ salt and at 15 K for d_8-$Cu[N(CN)_2]Br$ salt indicate magnetic ordering. The temperature dependence of $(T_1T)^{-1}$ above those is a manifestation of magnetic fluctuations. Above 60K, all the salts trace the same curve, meaning common nature of AF fluctuations [3].

For $Cu(NCS)_2$, h_8- and d_8-$Cu[N(CN)_2]Br$ salts, the peak temperatures of $(T_1T)^{-1}$, where depression of the AF fluctuations start, coincide with inflection points of resistivity crossover from nonmetallic to metallic behaviors; 60 K for $Cu(NCS)_2$ [4], 40-50 K for h_8-$Cu[N(CN)_2]Br$ [5] and 30-35 K for d_8-$Cu[N(CN)_2]Br$ [6]. Thus, the relaxation peak can be addressed as a crossover from a AF fluctuation regime with nature of localized spins to a Fermi liquid regime. It is emphasized that an order of magnitude enhancement of $(T_1T)^{-1}$ from the uncorrelated Korringa value just above T_C is an indication of highly correlated nature of the metallic phase.

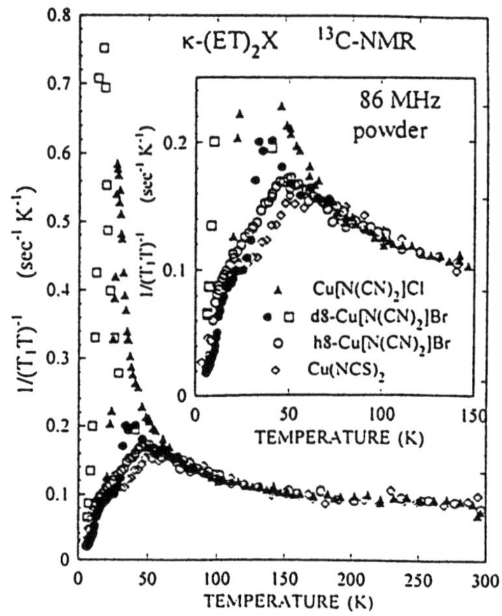

Figure 3. $(T_1T)^{-1}$ of κ-$(ET)_2X$

The magnetic phase in d_8-$Cu[N(CN)_2]Br$ salt is considered to touch the M-I boundary. The spin structure of this phase was deduced from analysis of the ^{13}C NMR spectra shown in Fig. 4, where a magnetic field of 8 T was applied to the crystallographic *a*-axis. It is known that this field geometry makes all ET molecules NMR equivalent and at the same time forms the magic angle with respect to the ^{13}C=^{13}C vector, giving ^{13}C-NMR spectra of two lines corresponding to the neighboring ^{13}C sites with different shifts but without dipolar splitting [7]. As expected, d_8-$Cu[N(CN)_2]Br$ shows two sharp NMR lines above 35 K. At lower temperatures, part of the spectra starts to shift and, below ~15 K, grows into a two-line spectrum well shifted only to the high frequency side in addition to the signal remaining near the origin of shift, as seen in Fig.4. The former corresponds to the magnetic phase, while the latter to the metallic (superconducting) one. The one-sided shift with the shape maintained allows only the simplest commensurate structure of the spins localized on ET dimer depicted in the inset. From the magnitude of the observed shift , the AF moment is estimated at about $0.3\mu_B$. The moment of Cu[N(CN)2]Cl salt is found to be greater than $0.4\mu_B$ [1].

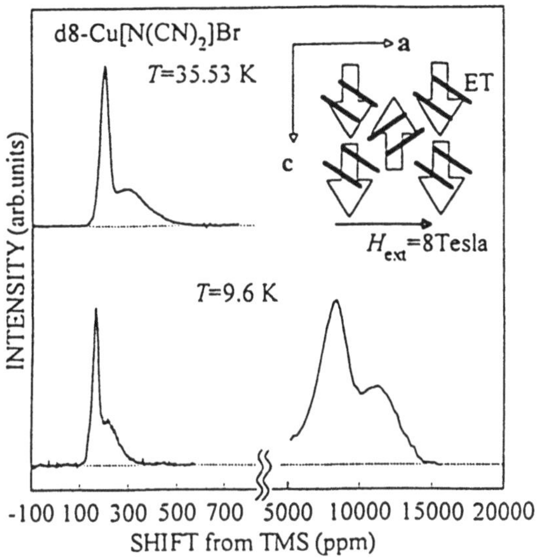

Figure 4. ^{13}C-NMR spectra of d_8-Cu[N(CN)$_2$]Br. Inset shows κ-type of ET arrangement (with the long axis of ET perpendicular to the sheet) and ordered structure of spins.

3.2. Phase diagram

Taking account of the whole results described above, we propose a conceptual phase diagram for this family as illustrated in Fig. 5 [8]. (A complicated M-I boundaries are reported in the pressure study of Cu[N(CN)$_2$]Cl salt [9].) The commensurate spin structure in the insulating side supports that the M-I transition shown by the shaded boundary is a Mott transition, not due to Fermi surface nesting. This is supported by the following considerations on band filling and U/W: While the band filling of this family is originally a quarter, the strong dimerization of ET molecules, as shown in the inset in Fig.4, tends to open a dimerization gap in the electronic band and result in half-filling of the holes in the upper subband. This corresponds to a manner to consider a dimer, ET_2, as a supermolecule with a hole. In this context, the on-site U to be compared with the subband width, W, is U_{ET2} rather than U_{ET}. When U_{ET} is much larger than the intradimer transfer energy, t_d, as in the present case ($U_{ET}\sim$ 1eV and $t_d\sim$0.25eV), U_{ET2} is given by $2t_d$ [8]. Then, U_{ET2} is estimated at 0.5eV, while W~0.4eV according to the band calculation [10]. Thus, U_{ET2}/W turns out to be of order of unity. The appearance of AF moment of $0.3\mu_B$ just on the M-I boundary implies that this Mott transition is of first order.

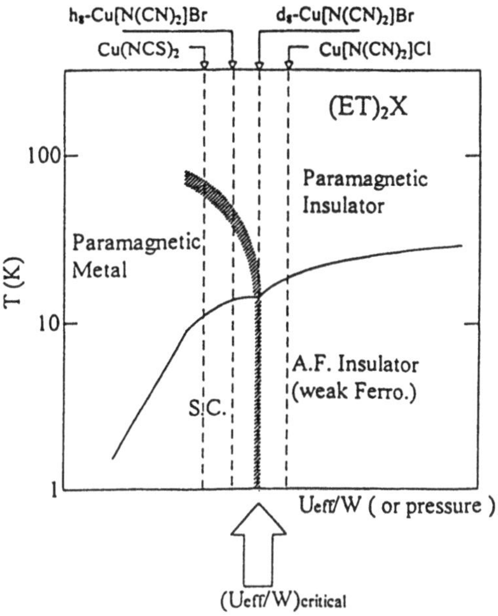

Figure 5. Conceptual phase diagram of $(ET)_2X$

3.3 Symmetry of electron pairing

The superconductivity in this family is most enhanced in the vicinity of the Mott transition. This fact intrigues interest in nature of electron pairing. We have approached this problem in two ways, ^{13}C-NMR and specific heat measurements.

The T_1^{-1} was measured for aligned crystals of h_8-Cu[N(CN)$_2$]Br with the layer parallel to an applied field to 2.3 T. The NMR relaxation rate in the vortex state contains the quasiparticle contribution and the field-fluctuation contribution due to vortex dynamics. The latter was selectively probed by ^{1}H-NMR and subtracted from the raw data [11]. This gives the former contribution, which is shown by open circles in Fig. 6. One consequence is absence of the coherence peak just below T_C. Another is that T_1^{-1} follows a cubic temperature dependence. These two characteristics indicate that the superconductivity in h_8-Cu[N(CN)$_2$]Br salt is an unconventional one with line nodes in gap parameter on the Fermi surface or highly anisotropic [7,11,12].

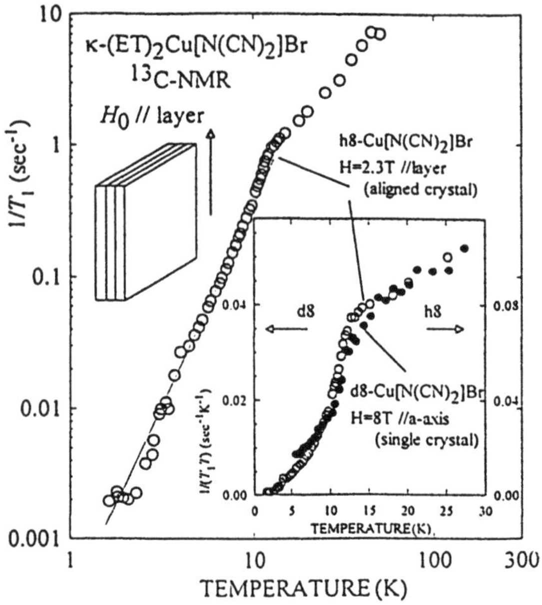

Figure 6. ^{13}C nuclear spin-lattice relaxation rate of Cu[N(CN)$_2$]Br salts in parallel fields. The solid line represents a T^3 dependence.

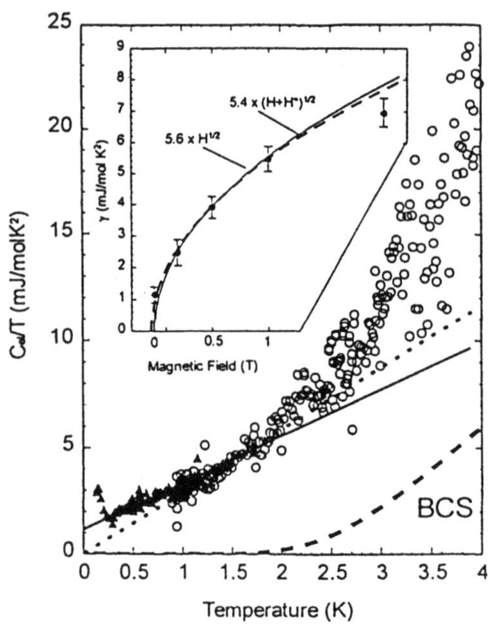

Figure 7. Electronic specific heat of h_8-Cu[N(CN)$_2$]Br. Inset shows field dependence of T-linear coefficient, γ.

The $(T_1T)^{-1}$ behavior of d_8-Cu[N(CN)$_2$]Br salt, shown in the inset of Fig. 6, is noteworthy. A gradual decrease of $(T_1T)^{-1}$ with temperature is observable from higher temperatures well above T_C with the kink at T_C less prominent. This behavior and the susceptibility decrease in Fig.2 are reminiscent of the cuprates in an underdoped region.

The low-temperature electronic specific heat, C_{el}, of h_8-Cu[N(CN)$_2$]Br salt is shown in Fig. 7. As the lattice part, we used the specific heat of rapidly cooled d_8-Cu[N(CN)$_2$]Br salt, which was found to be a nonmetal with vanishing γ in the nearly whole volume [2,13]. The C_{el} shows a quadratic temperature dependence different from the activation behavior expected in fully gapped superconductors [14]. This observation supports the gap profile of line nodes. Inset shows field dependence of γ. The observed square root variation is also consistent with the unconventional pairing .

4. CONCLUSION

The metal(superconductor)-to-insulator transition in κ-(ET)$_2$X is viewed as a Mott transition, where U/W is the parameter varied by modification of molecular arrangement under a fixed band filling. The superconductivity occurring neat the Mott transition is suggested to be unconventional.

AKNOWLEDGEMENTS

The author is grateful to Y. Nakazawa, K. Miyagawa and A. Kawamoto for their collaboration on this study.

REFERENCES

1. K. Miyagawa *et al*, Phys. Rev. Lett. **75** (1995) 1174.
2. A. Kawamoto *et al*, submitted for publication.
3. A. Kawamoto *et al.*, Phys. Rev. B **52** (1995) 15522; Phys. Rev. Lett. **74** (1995) 3455.
4. K. Murata *et al.*, Solid State Commun. **76** (1990)377.
5. L. I. Buravov *et al.*, J. Phys. (France) I **2** (1992) 1257.
6. H. Sato *et al.*, Physica C **185-189**, (1991) 2679 .
7. S. M. De Soto *et al*, Phys. Rev. B **52** (1995) 10364.
8. K. Kanoda, Hyperfine Interactions, in press.
9. H. Ito *et al.*, J. Phys. Soc. Jpn. **65** (1996) 2987.
10. Y.-N. Xu et *al.*, Phys. Rev. B **52** (1995) 12946.
11. K. Kanoda *et al.*, Phys. Rev. B **54** (1996) 76.
12. H. Mayaffre *et al.*, Phys. Rev. Lett. **75** (1995) 4122.
13. Y. Nakazawa *et al.*, Phys. Rev. B **53** (1996) R8875.
14. Y. Nakazawa *et al.*, Phys. Rev. B **55** (1997) in press.

ELSEVIER

Physica C 282-287 (1997) 303-306

PHYSICA C

The Normal and Superconducting States of $CeIn_3$ Near the Border of Antiferromagnetic Order

I. R. Walker, F. M. Grosche,
D. M. Freye and G. G. Lonzarich

Cavendish Laboratory, University of Cambridge,
Cambridge, UK CB3 0HE

Close to the border of antiferromagnetic order at high pressure the cubic stoichiometric compound $CeIn_3$ displays an unconventional normal-state resistivity followed by superconductivity at low temperatures. The superconducting transition temperature varies rapidly with pressure near the critical pressure where the Néel temperature collapses towards absolute zero.

1. Introduction

The nature of the normal states of magnetic metals above their Curie or Néel temperatures is a long standing problem with some possible relevance to the study of the normal state of short coherence length superconductors [1-2]. For the simplest case of the ferromagnetic *d*-metals with weak spin-orbit interactions, the behaviour in the limit of a vanishing Curie temperature (T_C) and in the absence of crystalline disorder, is expected to be that of an elementary marginal Fermi liquid [2-3]. At sufficiently low temperatures, and for crystals with inversion symmetry and long carrier mean free paths, this state may transform into an odd parity superconducting phase analogous to that of superfluid ^3He.

The normal state of antiferromagnetic metals with vanishing Néel temperature (T_N) can also display some properties at variance with that of a simple Fermi liquid, and under suitable circumstances this nearly antiferromagnetic state may be unstable to the formation of an anisotropic even parity form of superconductivity.

Several nearly *ferro*magnetic metals have been extensively investigated and many properties are indeed consistent with the above marginal Fermi Liquid model [3]. On the other hand, the description of almost antiferromagnetic metals, and of the possible superconducting instabilities for both the incipient ferromagnets and antiferromagnets, remains illusive. Progress in clarifying the intrinsic behaviour of such systems has been hampered partly by the scarcity of well characterised and pure metallic specimens in which the magnetic transition temperature can be depressed continuously towards absolute zero in a controlled manner, i.e., systems in which the nature of the normal states and any superconducting instabilities in the limit $T_C \rightarrow 0$ or $T_N \rightarrow 0$ may be expected to become most apparent.

We report a preliminary study of a relatively simple example of a pure antiferromagnetic metal, $CeIn_3$, in which T_N can be depressed towards absolute zero with pressures in the range available with conventional piston and cylinder cells [4,5]. $CeIn_3$ crystallises from a congruent melt into a simple cubic (Cu_3Au) structure and has been prepared in a relatively pure state with residual resistivities in the low temperature limit of below 1 μΩcm, i.e., nearly two orders of magnitude smaller than the maximum of the resistivity observed near 50 K. At ambient pressure, $CeIn_3$ orders in an antiferromagnetic state below 10.1 K with a low temperature moment of 0.4 μ_B per Ce atom that is substantially below the value expected from the effects of the static crystal field alone. The Ce moments, in keeping with the most elementary example of antiferromagnetism in the simple cubic lattice, are aligned antiferromagnetically in adjacent (111) ferromagnetic planes.

The following summarises the preliminary results of a study of the pressure and temperature dependence of the

PII S0921-4534(97)00267-0

resistivity of $CeIn_3$ carried out in an extension of an earlier programme focusing on cubic *d*-metals that differ from $CeIn_3$ notably in being essentially ferromagnetic and having a weak spin-orbit interaction.

2. Results

At ambient pressure the resistivity versus temperature exhibits a broad peak around 50 K and a cusp near 10 K that marks the onset of antiferromagnetic order (see Fig. 1 and inset). The Néel temperature T_N as inferred from the position of this cusp decreases rapidly with pressure and apppears to fall toward absolute zero just above 25 kbar where the cusp becomes ill defined (see Fig. 2). The broad peak at a higher temperature T_{sf}, on the other hand, appears to move only slowly with pressure. This behaviour is similar to that observed in the *d*-metals, but is in sharp contrast to that found in the *f*-metal antiferromagnet $CePd_2Si_2$ in which T_{sf} and T_N are comparable at ambient pressure and diverge rapidly with increasing pressure [6].

At ambient pressure the resistivity varies as T^3 over a wide range below T_N, as

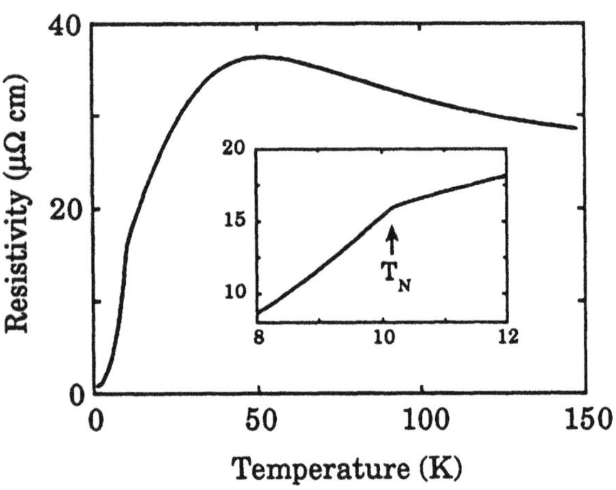

Fig. 1 Temperature dependence of the resistivity of $CeIn_3$ at ambient pressure. The inset is an expanded view in the vicinity of the antiferromagnetic transition near 10 K. Samples with residual resistivities of below 1 μΩcm were prepared by induction melting in water cooled copper boats in an UHV chamber with up to 8 atmospheres of very high purity Ar gas.

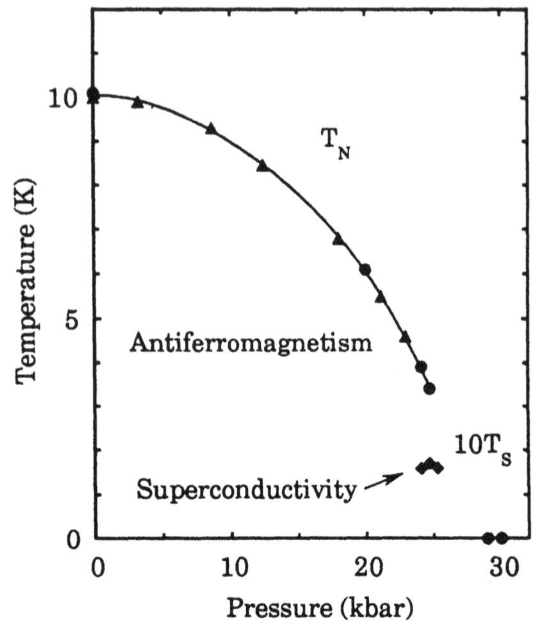

Fig. 2. The temperature versus pressure phase diagram inferred from resistivity data measured in piston and cylinder cells with pressures of up to 30 kbar. The data for T_N represented by filled triangles (all below 25 kbar) is taken from ref. [5], but with the pressure scaled by a factor of 0.9. Our pressure measurements (for data represented by filled circles and diamonds) are based on the procedure given in ref. [3]. The superconducting transition temperature T_S (filled diamonds) has been scaled by a factor of 10. The solid line represents a polynomial fit to the data for T_N.

expected from the effect of antiferromagnons. Also it varies as T^2 below approximately 10 K for pressures about 30 kbar, in agreement with the effect of scattering due to antiferromagnetic spin fluctuations in a Fermi liquid state well away from the critical pressure p_c where $T_N \to 0$. In the vicinity of p_c, however, the resistivity exponent increases with decreasing temperature and remains well below the Fermi liquid value of 2 over a wide range below 10 K (see Fig. 3). The rapid recovery of the Fermi liquid state above p_c is in sharp contrast to the behaviour observed in the *d*-metal ferromagnets [3] and in the *f*-metal antiferromagnets such as $CePd_2Si_2$ [6].

In the vicinity of the critical pressure and below 200 mK this nearly antiferromagnetic normal state appears to undergo a superconducting transition (see Figs 2 & 4).

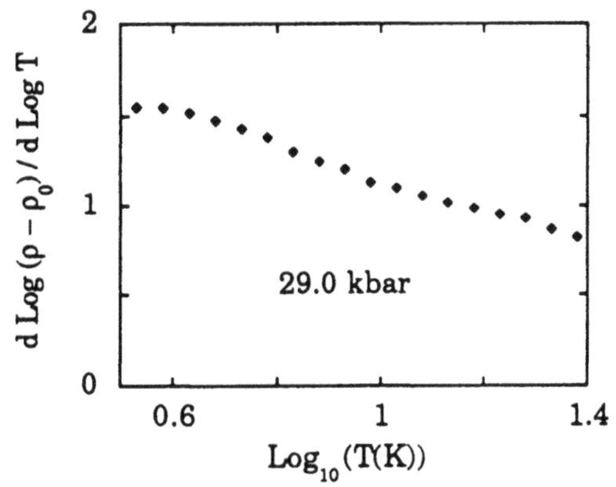

Fig. 3 The resistivity exponent in the normal state near the critical pressure for antiferromagnetic order. In contrast to the tetragonal antiferromagnet $CePd_2Si_2$ [6], the exponent increases monotonically with decreasing temperature and does not lock into a fixed value in our experimental temperature range.

In the absence of, e.g., upper critical-field data, the nature of this superconductivity is as yet unclear. However it is significant that the superconducting transition temperature T_S varies rapidly near the border of antiferromagnetism. A similar behaviour close to the critical pressure for antiferromagnetic order has also been found in the tetragonal compounds $CePd_2Si_2$ [6] and $CeRh_2Si_2$ [7]. These systems, however, have complex metallurgical and magnetic properties and different normal-state behaviours than that of $CeIn_3$ near p_c.

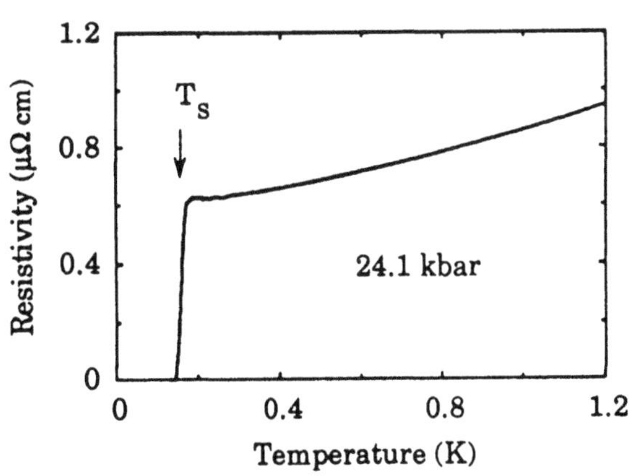

Fig. 4 The resistivity near the superconducting transition in the vicinity of the critical pressure for antiferromagnetic order.

3. Discussion

We limit our discussion below to the form of the temperature dependence of the resistivity of the normal state close to and above the critical pressure. A detailed analysis of the superconducting transition will be presented elsewhere. The experimental results have been compared with the predictions of a spin fluctuation model previously applied extensively to the d-transition metals. In this framework the resistivity exponent, defined as $\partial \log(\rho-\rho_o)/\partial \log T$, i.e., the temperature logarithmic derivative of the resistivity measured relative to the residual low temperature value, depends solely on the form of the spin fluctuation spectrum. In the normal state this is assumed to be described by a spin-fluctuation relaxation rate Γ_{Q+q} of the form

$$\Gamma_{\mathbf{Q+q}} = \gamma_{\mathbf{Q+q}} \chi^{-1}_{\mathbf{Q+q}} \tag{1}$$

where $\mathbf{Q+q}$ is the spin-fluctuation wavevector, $\mathbf{Q}$ is the ordering wavevector, $\gamma_{\mathbf{Q+q}}$ is approximated by a constant γ for an incipient antiferromagnet, and the last factor is the inverse generalised susceptibility

$$\chi^{-1}_{\mathbf{Q+q}} = A + c_{\mathbf{q}} \tag{2}$$

$$c_{\mathbf{q}} = \frac{2c}{a^2}(3 - \cos q_x a - \cos q_y a - \cos q_z a) \tag{3}$$

where c is a spin fluctuation stiffness parameter, a is the side of the simple cubic lattice, and A is the inverse staggered susceptibility. It follows from Eqs (2) and (3) that the inverse uniform susceptibility is given by

$$\chi^{-1} = A + c_{\mathbf{Q}} \tag{4}$$

where $c_{\mathbf{Q}} = 12\ c/a^2$ for $\mathbf{Q} = (\pi/a)$ [1, 1, 1]. Since $A(T)$ vanishes at T_N then from Eq. (4) c is determined by $\chi^{-1}(T_N)$ and a. Also, if c is independent of T, the temperature dependence of the staggered susceptibility is given by the change of $\chi^{-1}(T)$ from its value at T_N. In this case the temperature dependence of the uniform susceptibility fully determines $\Gamma_{\mathbf{Q+q}}$ apart from the constant γ which may be inferred from T_{sf}, *i.e.*, from the position of the shoulder of ρ versus T. Both parameters c and γ can also be determined independently from inelastic neutron scattering data. Preliminary calculations have been carried out given the assumption that c and γ are fixed and that, at least at not too low a temperature, $A(T)$ has a Curie-Weiss form α $(T + \theta)$ with an inverse Curie constant α obtained from the uniform susceptibility at ambient pressure. The pressure dependence of Γ_{Q+q} in this model is thus entirely due to θ which reduces to $(-T_N)$ below the critical pressure.

Our preliminary analyses (based on an isotropic approximation for c_q) suggest that the normal state of $CeIn_3$ near the critical pressure may be in reasonable accord with the prediction of this spin fluctuation model over a wide range in temperature. In particular the variation of the resistivity exponent shown in Fig. 2 is close to the form expected for the parameters obtained as described above if we assume θ to be of the order of a degree K. A full analysis will require a more realistic approximation in particular for $A(T)$ at low temperatures.

The rapid return to a Fermi liquid form $\rho \sim T^2$ above p_c is in sharp contrast to the behaviour in the ferromagnetic d-metals MnSi and $ZrZn_2$ and in the heavy fermion antiferromagnet $CePd_2Si_2$ [3,6]. The difference from the ferromagnets may be related to the difference in the value of the dynamical exponent z (defined by the dispersion relation $\Gamma_{Q+q} \sim q^z$ at small q which plays a key role in models of quantum critical phenomena [8]); i.e., to the appearance in the pure ferromagnetic metals of an additional factor of q in the relaxation spectrum (now measured relative to $\mathbf{Q} = \mathbf{0}$) due to Landau damping at long wavelengths. A quadratic temperature dependence of ρ can also arise in principle due to the short-circuiting of "hot spots" on the Fermi surface from states on the Fermi surface not strongly coupled by the antiferromagnetic spin fluctuations [9], or by a first order transition from an antiferromagnetic to a normal state.

The behaviour of heavy fermion systems such as $CePd_2Si_2$ remains enigmatic [10]. These differ from $CeIn_3$ in having lower values of the stiffness parameter c and more complex magnetic interactions in non-cubic lattices. Of particular importance may be competing ferromagnetic and antiferromagnetic couplings that can considerably complicate the nature of the spin fluctuation spectrum. $CeIn_3$ would seem to provide a link between the magnetic d-metals and these more complex antiferromagnets.

4. Acknowledgements

We wish to thank, in particular, S V Brown, F V Carter, J Flouquet, S R Julian, R K W Haselwimmer, S J S Lister, C. Pfleiderer, S S Saxena, and R Tamisari for their assistance in some parts of this work or for helpful discussions. This research has been supported by the EPSRC of the UK, the EC, and the Newton Trust in Cambridge.

5. References

1. See, e.g., Moriya T, Spin Fluctuations in Itinerant Electron Magnetism, Springer, Berlin, 1985

2. See, e.g., Lonzarich G G, The Magnetic Electron, Ch. 6 in The Electron, ed M Springford, Cambridge University Press, Cambridge, 1997, and references therein

3. Pfleiderer C, McMullan G J, Julian S R, and Lonzarich G G, Phys. Rev. B 55 (1997) 8330, and references cited therein

4. See, e.g., Lawrence J, J. Appl. Phys. 53 (1982) 2117, and references cited therein

5. Flouquet J, Haen P, Lejay P, Morin P, Jaccard D, Schweizer J, Vettier C, Fisher R A and Phillips N E, J. Magn. and Magn. Mat 90 & 91 (1990) 377, and references cited therein

6. Grosche F M, Julian S R, Mathur N D and Lonzarich G G, Physica B 223 & 224 (1996) 50

7. Movshovich R, Graf T, Mandrus D, Thompson JD, Smith JL and Fisk Z, Phys. Rev. B 50 (1996) 8241

8. See, e.g., Millis A J, Phys Rev B 48 (1993) 7183

9. Hlubina R and Rice T M, Phys. Rev. B 51 (1995) 9253

10. Julian S R, Pfleiderer C, Grosche F M, Mathur N D, McMullan G J, Diver A J, Walker I R and Lonzarich G G, J. Phys. Condens. Matter 8 (1996) 9675

ELSEVIER

Physica C 282–287 (1997) 307–310

PHYSICA C

Current issues in fulleride superconductivity

K. Prassides,[a] K. Tanigaki[b] and Y. Iwasa[c]

[a]School of Chemistry, Physics and Environmental Science, University of Sussex, Falmer, Brighton BN1 9QJ, United Kingdom

[b]NEC Fundamental Research Laboratories, 34 Miyukigaoka, Tsukuba 305, Japan

[c]Japan Advanced Institute of Science and Technology, Tatsunokuchi, Ishikawa 923-12, Japan

Alkali fullerides exhibit superconductivity with T_c's as high as 33 K at ambient pressure. The transition temperatures scale monotonically with unit cell size, an observation rationalized as arising from the modulation of the electronic density-of-states at the Fermi level, $N(\varepsilon_F)$ by the interfullerene spacing, a. Deviations from this relationship leading to suppressed T_c's are apparent at small interfullerene spacings for sodium fullerides, $Na_2(A,A')C_{60}$ and may be linked to the formation of quasi-one-dimensional bridged fulleride structures. Superconductivity is also suppressed at large interfullerene spacings for $(NH_3)K_3C_{60}$ whence magnetic origin effects play an important role and provide a connection with the established phenomenology in high-T_c and organic superconductors.

1. INTRODUCTION

Intercalation of solid C_{60} with electron donors, like the alkali metals, results in a wealth of intercalated fulleride salts with stoichiometries A_xC_{60} ($0 \le x \le 6$). Metallic compositions are encountered for stoichiometries A_3C_{60}. These salts become superconducting at critical temperatures as high as 33 K (at ambient pressure) for $RbCs_2C_{60}$, surpassed only by the high T_c cuprates. Superconductivity has been explained by both conventional electron-phonon and purely electronic models. In both models, pair binding is assumed to be dominated by intramolecular properties. Moreover, the observed monotonic scaling of the superconducting transition temperatures with the unit cell size [1] has been rationalised as arising from the modulation of the electronic density of states, $N(\varepsilon_F)$ by the interfullerene spacing. As the latter increases, the overlap between the molecules decreases; this leads to a reduced bandwidth, and, for a fixed band filling, to an increased density of states. Figure 1 collects together both ambient and high pressure data for various fulleride superconductors.

2. STRUCTURAL FAMILIES

$A_2A'C_{60}$ (A, A'= alkali metal) fullerides can be classified into three structural families, distinguished through the orientational state adopted by the C_{60}^{3-} ions in the crystal structure:

(i) K_3C_{60} and Rb_3C_{60} adopt a merohedrally disordered fcc ($Fm\bar{3}m$) structure. Geometrical considerations reveal that, while the size of the octahedral holes (~2.06 Å) is large enough to accommodate any alkali ion, the tetrahedral hole (~1.12 Å) is smaller than the ionic radii of K^+, Rb^+ and Cs^+. As a result, rotational motion of the fulleride ions is restricted and the A^+-C_{60}^{3-} orientational potential is dominated by repulsive interactions. T_c is affected in the same way by both physical and chemical pressure.

0921-4534/97/$17.00
PII S0921-4534(97)00271-2

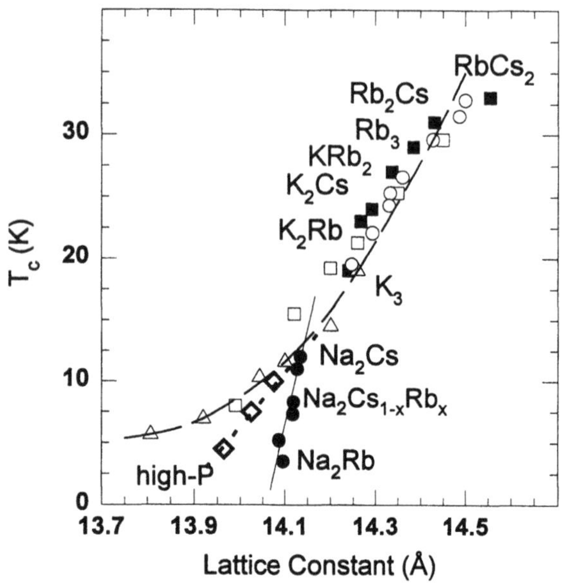

Figure 1. Evolution of the critical temperature, T_c of superconducting fullerides with the cubic lattice constant, a.

(ii) Na_2CsC_{60} and Na_2RbC_{60} adopt an orientationally ordered primitive cubic ($Pa\bar{3}$) structure [2]. Since the ionic radius of Na^+ (0.95 Å) is smaller than the size of the tetrahedral interstices, the C_{60}^{3-} ions are no longer confined to the two standard orientations found in K_3C_{60}. There is enough space for them to rotate in such a way as to optimise both the attractive Na^+-C_{60}^{3-} and the C_{60}-C_{60} interactions. The primitive cubic family displays a much steeper rate of change of T_c with interfullerene spacing, while at the same time, the effects of physical and chemical pressure on the superconducting properties are no longer identical (Figure 1).

(iii) Further deviations are encountered when the small highly-electropositive Li^+ ions are introduced in the tetrahedral holes. Li_2CsC_{60} and Li_2RbC_{60} are fcc ($Fm\bar{3}m$) and contain quasi-spherical C_{60}^{3-} ions but do not display superconductivity. This has been ascribed to enhanced Li^+-C interactions, resulting from the partial hybridisation of Li s and C p_z orbitals [3].

3. SODIUM FULLERIDES

The origin of the faster depression of T_c with decreasing interfullerene separation in $Na_2(A,A')C_{60}$ is of particular interest. Empirically it appears that the modified structure and intermolecular potential sensitively affect the electronic and conducting properties [2]. The changed orientational state of the ions in the $Pa\bar{3}$ structures may affect electron hopping between neighbouring ions and, as a consequence, may lead to a modified rate of change of $N(\varepsilon_F)$, and hence of T_c with interfullerene spacing (Figure 1). Indeed NMR measurements [4] have confirmed the suppressed values of $N(\varepsilon_F)$ for Na_2RbC_{60} and Na_2KC_{60}. However, both high pressure susceptibility measurements on Na_2CsC_{60} and the temperature dependence of its EPR spectra show little difference in the a-dependence of $N(\varepsilon_F)$ between Na_2CsC_{60} ($Pa\bar{3}$) and K_3C_{60} and Rb_3C_{60} ($Fm\bar{3}m$) [5], in agreement with the findings of tight-binding calculations [6].

However, recent X-ray and neutron diffraction, ^{13}C and ^{23}Na NMR, and susceptibility measurements [7] have

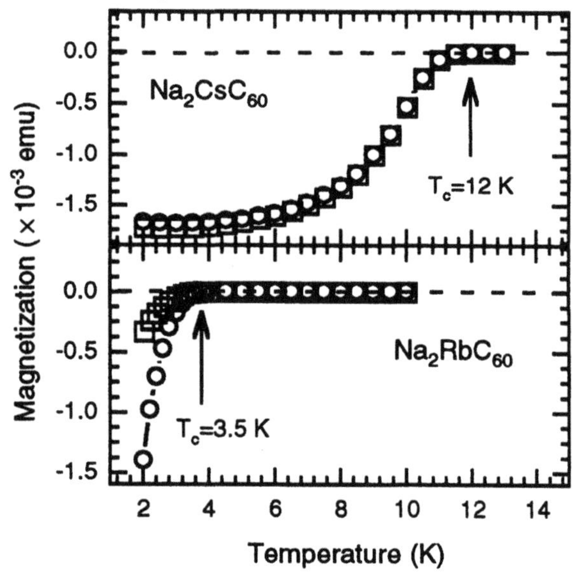

Figure 2. Magnetization of Na_2RbC_{60} and Na_2CsC_{60} at 50 Oe (ZFC). (O) rapid cooling, (□) slow cooling.

revealed a more complicated situation than hitherto appreciated. The properties of ternary, Na_2AC_{60} and quaternary, $Na_2(A,A')C_{60}$ fullerides are highly sensitive to the cooling protocol. While upon rapid cooling, the salts remain strictly cubic, slow cooling stabilises a monoclinic phase, comprising quasi-one-dimensional C_{60}^{3-} chains with short interball centre-to-centre distances of 9.38 Å and bridged by single C–C bonds. The influence of the formation of the polymer on the superconducting properties is profound. Figure 2 shows magnetisation measurements on Na_2RbC_{60} and Na_2CsC_{60} following different cooling procedures. The diamagnetic response in

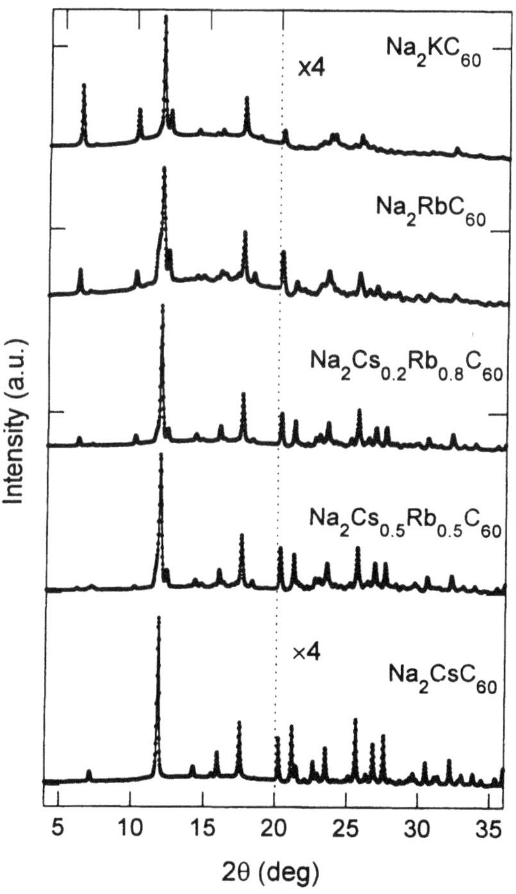

Figure 3. Synchrotron X-ray diffraction data of some sodium fullerides after slow cooling (λ= 0.873 Å, T= 200 K).

Na_2RbC_{60} is strongly suppressed on slow cooling, revealing that the polymer phase is not superconducting - the observed superconducting fraction arises from the untransformed cubic phase. The situation is different for Na_2CsC_{60} which is completely insensitive to thermal history. Figure 3 shows the synchrotron X-ray diffraction patterns of Na_2CsC_{60}, $Na_2Cs_{0.5}Rb_{0.5}C_{60}$, $Na_2Cs_{0.2}Rb_{0.8}C_{60}$, Na_2RbC_{60} and Na_2KC_{60}, obtained after slow cooling at 200 K. Na_2CsC_{60} remains strictly cubic ($Pa\bar{3}$), while the profiles of all other samples can be refined using a two-phase model of co-existing monoclinic and cubic phases. It thus appears that polymer formation occurs only below a critical interball separation. In addition, while superconductivity in sodium fullerides is confined to the isotropic 3D phases, there may be a connection between the suppressed values of T_c and polymer formation through the presence of intergrowths of non-superconducting polymer domains whose size grows as the lattice constant decreases. Such an interpretation is reminiscent of the detrimental effect on superconductivity of the formation of the LTT structure in the Ba-rich phases of the $La_{1.875}(Ba,Sr)_{0.125}CuO_4$ series [8].

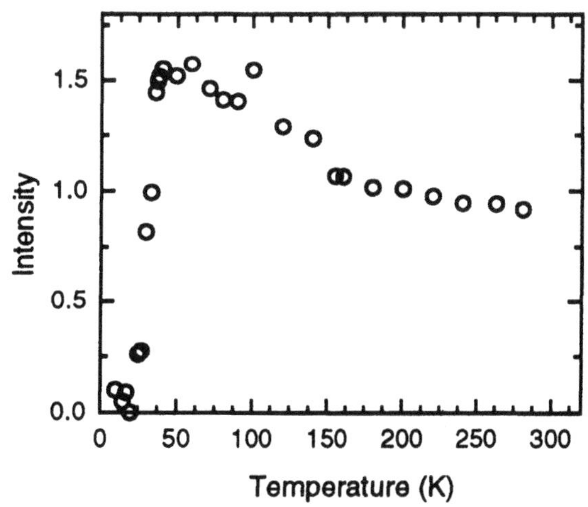

Figure 4. Temperature dependence of ESR intensity for $(NH_3)K_3C_{60}$ [10] (normalised to the K_3C_{60} value at room temperature).

4. $(NH_3)K_3C_{60}$

The synthesis of fullerides with large lattice constants - for instance, via intercalation of small molecules like NH_3 - has been actively pursued as a means of achieving higher T_c's. Indeed $(NH_3)K_3C_{60}$ is expected to show a T_c of ~30 K by extrapolation from the data in Figure 1. While no superconductivity has been observed at ambient conditions, application of pressure >1 GPa leads to T_c= 28 K [9]. As the structure of $(NH_3)K_3C_{60}$ is orthorhombic, the important question is raised whether the suppression of superconductivity is related to the structural distortion or is due to a Mott transition. Measurements of the electronic and magnetic properties by ESR and ^{13}C NMR have shown that at ambient pressure it is a narrow-band metal, exhibiting a transition to an insulating ground state at 40 K (Figure 4) [10]. While the evidence points to a vanishingly small spin susceptibility, χ_s for this state, surprisingly no broadening of the ESR and NMR lines, characteristic of an AF transition, was observed.

We have now performed a μ^+SR study of the magnetic properties of $(NH_3)K_3C_{60}$. μ^+SR spectroscopy has proven extremely powerful before in the field of small moment magnetism and in all instances when magnetic order is of a random, very short range, spatially inhomogeneous or incommensurate nature. In the present case, we observe in the absence of an applied external field a heavily damped oscillating (ν_μ= 0.51(2) MHz at 7.2 K) signal (Figure 5) which provides unambiguous proof of the existence of long range magnetic order below ~40 K. The strong relaxation found in zero field is the signature of spatial disorder and inhomogeneity effects. These results imply that the suppression of superconductivity at large interfullerene spacings is of magnetic origin and thus provide an important analogy with the well-established situation in high-T_c and organic superconductors.

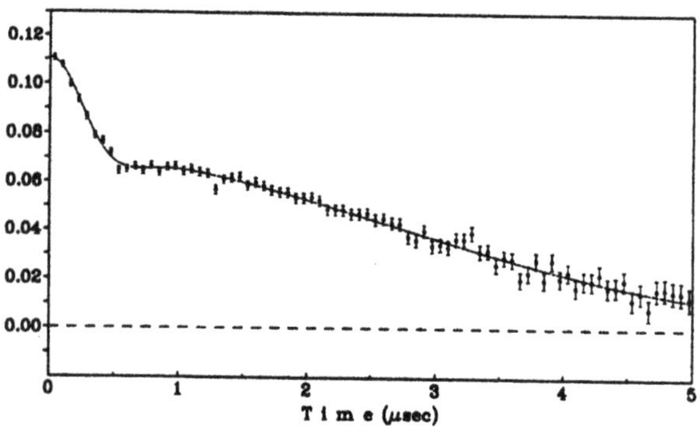

Figure 5. ZF-μ^+SR spectrum of $(NH_3)K_3C_{60}$ at 7.2 K.

Acknowledgement

Support by the EPSRC and British Council is gratefully acknowledged.

REFERENCES

1. R.M. Fleming et al., Nature, 352 (1991) 787.
2. K. Prassides et al., Science, 263 (1994) 950. T. Yildirim et al., Solid State Commun., 93 (1995) 269.
3. I. Hirosawa et al., Science, 264 (1994) 1294.
4. Y. Maniwa et al., Phys. Rev. B, 52 (1995) R7054.
5. K. Tanigaki and K. Prassides, J. Mater. Chem., 5 (1995) 1515.
6. O. Gunnarsson, Phys. Rev. B, 51 (1995) 3493.
7. K. Prassides et al., J. Am. Chem. Soc., 119 (1997) 834. L. Cristofolini et al., Chem. Commun. (1997) 375. G.M. Bendele et al., submitted.
8. A. Lappas et al., Physica C, 235-240 (1994) 1725.
9. M.J. Rosseinsky et al., Nature, 364 (1993) 425. O. Zhou et al., Phys. Rev. B, 52 (1995) 483.
10. Y. Iwasa et al., Phys. Rev. B, 53 (1996) R8836.

ELSEVIER

Physica C 282–287 (1997) 311–314

PHYSICA C

The Role of 2D Pancake Vortices in High-Temperature Superconductors

John R. Clem, Thomas Pe, and Maamar Benkraouda[a*]

[a]Ames Laboratory and Department of Physics and Astronomy, Iowa State University, Ames, IA 50011-3160 USA

To describe vortices in the high-temperature superconductors, a multilayer model consisting of alternating superconducting and nonsuperconducting layers is useful. Superconductivity within the layers is strong, while that between superconducting layers is weak. The Lawrence-Doniach model provides a helpful starting point, in treating the in-plane superconductivity via the Ginzburg-Landau theory and the out-of-plane superconductivity via the Josephson effect. With the help of this model, one may represent a vortex line threading through the layers as a stack of 2D pancake vortices connected by Josephson strings, whose axes are confined within the nonsuperconducting layers. The magnetic fields and currents generated by a single pancake vortex can be calculated analytically in the limit of extreme anisotropy or vanishing Josephson coupling. Here, we review the results of this calculation and discuss how pancake vortices magnetically interact with each other in the absence of Josephson coupling. In this limit, pancake vortices at zero temperature strongly prefer to align themselves perpendicular to the superconducting layers. We review recent results showing that a tilted stack of 2D pancake vortices in an infinite set of Josephson-decoupled superconducting layers is unstable when the angle of tilt is greater than 52°. We also discuss the behavior of pancake vortices in a finite set of N superconducting layers. The magnetic-field and current-density distributions generated throughout the multilayer structure by a single pancake vortex in an arbitrary layer can be calculated and used to examine the stability of tilted stacks of pancake vortices, where the tilt is maintained by application of transport currents in the top or bottom superconducting layers. At a sufficiently high transport current in the top layer, we find that the pancake vortex array in this layer shears off.

1. INTRODUCTION

The high-temperature superconductors are highly anisotropic, with strong superconductivity parallel to the CuO_2 layers and weak superconductivity perpendicular to the layers. To model the vortex structure in the most anisotropic high-temperature superconductors, it is useful to start with the Lawrence-Doniach model [1], in which one treats the in-plane superconductivity via the Ginzburg-Landau theory and the out-of-plane superconductivity via the Josephson effect. One then may consider vortices as consisting of *two-dimensional (2D) pancake vortices* [2], connected by Josephson strings. The 2D pancake vortices reside only in the superconducting layers, while the Josephson strings are confined to the nonsuperconducting layers. Electric fields parallel to the layers are associated with pancake-vortex motion and can be calculated using the Josephson relation, $E_{||} = V_{||}/L_{||} = h\nu_{||}/2eL_{||}$, where $\nu_{||}$ is the rate with which pancake vortices perpendicularly cross a line of length $L_{||}$ directed along the current flow parallel to the layers. Electric fields perpendicular to the layers are associated with Josephson-string motion and can be calculated using the Josephson relation, $E_{\perp} = V_{\perp}/L_{\perp} = h\nu_{\perp}/2eL_{\perp}$, where $\nu_{\perp}$ is the rate with which Josephson vortices perpendicularly cross a line of length $L_{\perp}$ directed along the current flow perpendicular to the layers.

The 2D pancake vortices are most easily visualized using a model in which there is no Josephson coupling between superconducting layers [2–4]. The magnetic-field and current-density distribution generated by an arbitrary vortex distribution then can be calculated by superposition. The magnetic-field and current-density distribution associated with just one pancake vortex can

*Ames Laboratory is operated for the U.S. Department of Energy by Iowa State University under Contract No. W-7405-Eng-82. This work was supported by the Director for Energy Research, Office of Basic Energy Sciences.

PII S0921-4534(97)00274-8

be readily computed by placing a vortex in only one superconducting layer at $z = 0$ (let us call this the central layer) and then accounting for the screening effect of all the other layers. For a vortex that carries magnetic flux up through the layers, the screening current in the central layer flows with a large magnitude in the counterclockwise direction when viewed from above. The screening currents in all the other layers flow with a much smaller magnitude in the clockwise direction. The highly concentrated current distribution, localized chiefly in the central layer, is reminiscent of a pancake, and this is the origin of the descriptive name [2].

A vortex line perpendicular to the layers thus can be represented as a straight stack of 2D pancake vortices, one on top of the next. For a stack centered on the z axis, the magnetic field, calculated by summing the individual pancake contributions, has only a z component [2],

$$b_z(\rho) = (\phi_0/2\pi\lambda_{||}^2)K_0(\rho/\lambda_{||}), \quad (1)$$

where $\phi_0 = hc/2e$ is the superconducting flux quantum, ρ is the radial coordinate, and $\lambda_{||}$ is the effective penetration depth for the decay of screening currents parallel to the layers (often called λ_{ab} in the high-temperature superconductors). Since the pancake vortices also must have cores of size approximately $\xi_{||}$ (or ξ_{ab}), in computing the energy per unit length of the vortex ϵ, one may cut off the kinetic energy integrals at $\rho = \xi_{||}$ to obtain

$$\epsilon(\theta = 0) = (\phi_0/4\pi\lambda_{||})^2 K_0(\xi_{||}/\lambda_{||}), \quad (2)$$

where the argument $\theta = 0$ indicates that the vortex line is not tilted with respect to the z axis.

For a straight stack of pancake vortices tilted at an angle θ relative to the z axis, the energy per unit length of the vortex line is, to logarithmic accuracy when $\lambda_{||} >> \xi_{||}$ [2],

$$\epsilon(\theta) = (\frac{\phi_0}{4\pi\lambda_{||}})^2 \ln[(\frac{\lambda_{||}}{\xi_{||}})\frac{(1+\cos\theta)}{2\cos\theta}]\cos\theta. \quad (3)$$

Equations (2) and (3) tell us that, in the absence of interlayer Josephson coupling, vortices energetically prefer to align themselves parallel to the z axis, even in the presence of an applied magnetic field $\mathbf{B}$ oriented at a finite angle θ relative to the z axis. The component of $\mathbf{B}$ perpendicular to the superconducting layers determines the intralayer spacing of 2D pancake vortices, which arrange themselves in a vortex lattice with the lattice in each layer perfectly aligned with the lattices in adjacent layers. The component of $\mathbf{B}$ parallel to the layers, on the other hand, penetrates nearly uniformly through the multilayer structure but has no other effect on the array of 2D pancake vortices.

In this paper we discuss some recent results concerning the stability of tilted stacks of magnetically coupled 2D pancake vortices in Josephson-decoupled multilayers. In Sec. 2 we point out that the form of Eq. (3) implies that a tilted straight stack in an infinite set of multilayers becomes unstable for angles larger than 52°. In Sec. 3 we briefly discuss some recent results on the behavior of tilted stacks of pancake vortices in a finite set of multilayers, and in Sec. 4 we present our conclusions.

2. INFINITE SAMPLE

The line tension $P(\theta)$ of a tilted stack can be computed from the relation [6,7]

$$P(\theta) = \epsilon(\theta) + d^2\epsilon(\theta)/d\theta^2, \quad (4)$$

where the second term on the right-hand side arises from the anisotropy of the system. From Eqs. (3) and (4) we obtain

$$P(\theta) = (\frac{\phi_0}{4\pi\lambda_{||}})^2 \frac{(\cos\theta - \sin^2\theta)}{\cos\theta(1+\cos\theta)}. \quad (5)$$

From this result we have shown [5] that the line tension and the tilt modulus are positive for small values of θ but change sign at $\theta = 52°$ and remain negative for angles $52° < \theta < 90°$.

This indicates that a tilted stack of magnetically coupled pancake vortices is unstable and that other arrangements of the pancake vortices are energetically preferred, even if the pancake vortices at the ends of the vortex line are fixed in position. We have shown [5], for example, that for sufficiently large values of θ a kinked structure, in which the pancake vortices are aligned perpendicular to the layers but kinked in the middle, is energetically favored over a uniformly tilted stack.

We next discuss vortex configurations in a sample of finite thickness.

3. SAMPLE OF FINITE THICKNESS

Consider a multilayer structure consisting of N superconducting layers centered on the planes $z_n = ns$, where $n = 0, 1, 2, ..., N-1$. We are ultimately interested in determining the configuration of a stack of N pancake vortices, one in each layer. To calculate the interaction of the pancake vortices, one first needs to derive a general expression for the current generated in layer n in response to a 2D pancake vortex in layer m, using a method [8] that extends our earlier approaches [2,9–11]. Consider, for example, a pancake vortex centered on the z axis at $z = ms$. In cylindrical coordinates (ρ, ϕ, z), the vector potential $\mathbf{a}$ has only a ϕ component, a_ϕ. Using the gauge in which the phase γ of the order parameter is $\gamma = -\phi$ in layer m and is constant in all the other layers, we may write [8]

$$a_\phi(\rho, z; z_m) = \int_0^\infty dq A(q; z_m) J_1(q\rho) Z(q, z; z_m), \quad (6)$$

where $J_1(q\rho)$ is a Bessel function of order 1, and $\rho = (x^2 + y^2)^{1/2}$. The function $Z(q, z; z_m)$ is determined by the requirement that Maxwell's equations must be satisfied everywhere and that the London equation must be satisfied within each of the superconducting layers.

We assume that the thickness d of each layer is much less than the bulk penetration depth λ_s within a layer. The sheet current density $\mathbf{K}$ (the integral of the current density $\mathbf{j}$ over the thickness d) also has only a ϕ component,

$$K_\phi(\rho, z_m; z_m) = -(c/2\pi\Lambda)[a_\phi(\rho, z_m; z_m) - \phi_0/2\pi\rho] \quad (7)$$

in the superconducting layer $n = m$ containing the pancake vortex, where $\Lambda = 2\lambda_s^2/d$, and

$$K_\phi(\rho, z_n; z_m) = -(c/2\pi\Lambda) a_\phi(\rho, z_n; z_m) \quad (8)$$

in all the other superconducting layers, $n \neq m$. It is convenient to take the limit as d and λ_s both vanish while $\Lambda = 2\lambda_s^2/d$ is held constant. Then $\Lambda = 2\lambda_{||}^2/s$, and K_ϕ in superconducting layer n is given by the discontinuity of the radial component b_ρ of the local magnetic field $\mathbf{b} = \nabla \times \mathbf{a}$ across the layer:

$$K_\phi(\rho, z_n; z_m) = (c/4\pi)[b_\rho(\rho, z_n^+; z_m) - b_\rho(\rho, z_n^-; z_m)]. \quad (9)$$

The function $A(q, z_m)$ is determined by applying Eqs. (7)-(9) and requiring continuity of a_ϕ and b_z at $z = ns$ for all n.

The Lorentz force the above vortex exerts on a second vortex at (ρ, z_n) in another layer $n \neq m$ has only a radial component [2],

$$F_\rho(\rho, z_n; z_m) = K_\phi(\rho, z_n; z_m)\phi_0/c. \quad (10)$$

It can be shown that this force is always negative, corresponding to a restoring force tending to align the two pancake vortices so that one is directly above the other. For example, when $\rho << \lambda_{||}$ and $0 < |n-m|s << \lambda_{||}$, we obtain, to good approximation [8],

$$F_\rho(\rho, z_n; z_m) = -\left(\frac{1}{\rho}\right)\left(\frac{\phi_0}{2\pi\Lambda}\right)^2[(\rho^2 + |z_n - z_m|^2 s^2)^{1/2} - |z_n - z_m|s]. \quad (11)$$

For the case that there is one pancake vortex in each of the N superconducting layers and no net current flows along any of the layers, the minimum-energy configuration (force-balance condition) occurs when the vortices form a straight line parallel to the z axis. A more interesting case is that for which the pancake vortex in the top layer ($n = N-1$) is displaced to the right (in the x direction) and the pancake vortex in the bottom layer ($n = 0$) is displaced by the same distance to the left. This could be accomplished by applying a sheet current K in the y direction in the top layer and a sheet current of the same magnitude but opposite direction in the bottom layer. We have solved the force-balance equations for this case and have found that the equilibrium configuration of pancake vortices is one in which the other $N-2$ vortices (in layers $n = 1, 2, ..., N-2$) remain close to the z axis and align themselves in a nearly straight stack approximately perpendicular to the layers, while the top and bottom pancakes have large displacements to the right and left. This configuration is very different from that of N balls connected together by nearest-neighbor springs. This difference arises because the lateral restoring force between magnetically coupled pancake vortices has a long range, set by the length scale $\lambda_{||}$, as seen in Eq. (11).

Another interesting result is that the magnitude of the restoring force on the top and bottom displaced vortices reaches a maximum when their displacement is of order $\lambda_{||}$. This means that there is a critical value K_d of the sheet current K, above which the top and bottom 2D pancake vortices are decoupled from the other pancakes in the stack. This is an effect similar to that studied theoretically [9,10] and experimentally [12,13] in Giaever dc transformers consisting of two superconducting films separated by an insulating layer thick enough to prevent Josephson coupling.

We have extended the above approach to calculate the behavior in the presence of a magnetic field B parallel to the z axis [8]. In this case there is a pancake vortex lattice in each of the N superconducting layers, and the minimum-energy configuration is that for which the vortex lattice in each layer is perfectly aligned with respect to the layers above and below it. We have solved the force-balance equations for the case of a sheet current K applied in the y direction in the top layer and a current of the same magnitude but opposite direction in the bottom layer. The behavior is similar to that described above, in that vortices in the top and bottom layers have very large displacements, while the vortices in the inner layers have small displacements and align themselves very nearly parallel to the z axis. As the displacements of the top and bottom vortex lattices grow, they shear away from the others at a field-dependent critical value K_d of the sheet current.

Recently, we also have generalized the above approach to calculate the dynamics of aligned arrays of pancake vortices when a sheet current K is applied only to the top layer. This has enabled us to calculate the dynamic critical decoupling current K_d and the time-averaged voltages V_{top} and V_{bot} along the top and bottom of the sample as the top vortex lattice in the top layer shears away from the others.

4. CONCLUSIONS

We have found that tilted stacks of 2D pancake vortices are unstable when subjected to large tilts. If the pancake vortices at the top and bottom of a finite stack are displaced in opposite directions, they slip off the stack, leaving the other pancakes aligned very nearly parallel to a line perpendicular to the layers.

5. ACKNOWLEDGMENTS

We thank V. G. Kogan for helpful discussions.

REFERENCES

1. W. E. Lawrence and S. Doniach, in Proc. of the Twelfth International Conference on Low Temperature Physics, E. Kanda (ed.), Academic Press of Japan, Kyoto, 1971, p. 361.
2. J. R. Clem, Phys. Rev. B 43 (1991) 7837.
3. S. N. Artemenko and A. N. Kruglov, Phys. Lett. A 143 (1990) 1971.
4. A. Buzdin and D. Feinberg, J. Phys. (Paris) 51 (1990) 1971.
5. M. Benkraouda and J. R. Clem, Phys. Rev. B 53 (1996) 438.
6. E. H. Brandt and U. Essmann, Phys. Status Solidi 144 (1987) 13.
7. A. Sudbo and E. H. Brandt, Phys. Rev. Lett. 66 (1991) 1781.
8. T. Pe, M. Benkraouda, and J. R. Clem, Phys. Rev. B (1997).
9. J. R. Clem, Phys. Rev. B 9 (1974) 898.
10. J. R. Clem, Phys. Rev. B 12 (1975) 1742.
11. J. R. Clem, Physica C 235-240 (1994) 2607.
12. J. W. Ekin, B. Serin, and J. R. Clem, Phys. Rev. B 9 (1974) 912.
13. J. W. Ekin and J. R. Clem, Phys. Rev. B 12 (1975) 1753.

ELSEVIER

Physica C 282–287 (1997) 315–318

PHYSICA C

Tunneling spectroscopy and STS observation of vortices on high temperature superconductors

Ø. Fischer, Ch. Renner, I. Maggio-Aprile, A. Erb, E. Walker, B. Revaz and J.-Y. Genoud.

Département de Physique de la Matière Condensée, Université de Genève
24, Quai E.-Ansermet, CH-1211 Genève 4, Switzerland

We report scanning tunneling spectroscopy investigations on the doping and temperature dependence of the tunneling spectra on $Bi_2Sr_2CaCu_2O_{8+\delta}$, and on the vortex lattice observation on $YBa_2Cu_3O_{7-\delta}$.

1. INTRODUCTION

In this paper we report on Scanning Tunneling Spectroscopy (STS) investigation of high temperature superconductors. We shall focus on two topics. The first concerns the doping and temperature dependence of the tunneling spectra in $Bi_2Sr_2CaCu_2O_{8+\delta}$ and the second deals with the observation of the vortices in $YBa_2Cu_3O_{7-\delta}$. A crucial feature of our investigation is that STS offers the possibility to carry out electron spectroscopy on a nanoscopic level and thus it allows to separate local effects from the average behaviour over the investigated surface. In our studies of $Bi_2Sr_2CaCu_2O_{8+\delta}$, this feature has allowed us to focalise on features in the spectra which are locally reproducible and systematically found over the whole investigated surface [1]. In the case of the observation of the vortex lattice we exploit the high spatial resolution of STS.

The high temperature superconductors all show a characteristic doping dependence of the critical temperature. An important question is whether the superconducting state is the same irrespective of the doping level or it changes its fundamental nature as doping is progressing. Another central question is related to the temperature dependence of the superconducting gap. Does it follow a BCS relation or does it have a completely different behaviour? If the latter is true, it would suggest that the superconducting state is indeed very different from the classical BCS state. We have started a detailed investigation and we report here some of the results.

Observation of the vortex lattice by STS in a high temperature superconductor has been a challenge since the success obtained on the conventional superconductors $NbSe_2$ and $Nb_{1-x}Ta_xSe_2$ [2]. This technique has the important advantage over all other direct vortex mapping techniques that it measures the quasi-particle density of states and thus basically reflects the superconducting order parameter, rather than the magnetic field, variations associated with the vortices. In particular, this technique can be used both at high and low magnetic field and it allows in addition to do spectroscopy inside and around the vortex cores. Recently, we succeeded to map the vortex lattice on an $YBa_2Cu_3O_{7-\delta}$ crystal [3].

2. EXPERIMENTAL

All the STS measurements presented below were performed in a low temperature set-up described elsewhere [4], in the

PII S0921-4534(97)00275-X

presence of a low He exchange gas pressure. Before cooling, $Bi_2Sr_2CaCu_2O_{8+\delta}$ crystals were cleaved in ultra high vacuum. No particular surface conditioning was done on $YBa_2Cu_3O_{7-\delta}$ crystals. Spectroscopic measurements were performed in a standard way, by sweeping the tunnel voltage over a specific range, maintaining a constant tip-sample distance. The tunneling resistance was kept as high as possible (ca. 1 GΩ) in order to ensure a sufficiently large tip-sample separation. STM tips were made of chemically etched iridium wire.

3. TUNNELING SPECTROSCOPY ON $Bi_2Sr_2CaCu_2O_{8+\delta}$

We have investigated the tunneling spectra of several $Bi_2Sr_2CaCu_2O_{8+\delta}$ samples with different oxygen doping δ. The central result that we are reporting here is that the overall shape of the spectra change only very little upon doping. In Fig. 1a we show an example of the spectrum dI/dV versus V, taken at 4.2K for a slightly overdoped sample with T_c = 74K. The main feature of this spectrum is the same as what is observed for optimum doping [1,5]. In particular we find that the background is essentially ohmic up to energies well above the gap. Well developed peaks appear at the gap energy. In spite of the high peaks there is considerable conductivity inside the gap the low energy part of the spectrum is consistent with a d-wave symmetry of the wave function if one assumes that impurity scattering can account for the finite conductivity at zero bias. As for optimum doping we also find an asymmetry of the conductivity just beyond the gap: There is a dip present at negative sample bias, situated at about twice the gap voltage. This dip is absent at positive sample bias. Our measurements of a strongly overdoped sample with T_c = 56K showed the same behaviour [5] and more recently we have found that an underdoped sample with a T_c = 83 K also display the same generic features [6]. The main difference between the various doping levels are to be found in the gap values. In the overdoped samples the $2\Delta/k_BT_c$ values decrease slightly from about 10-11 at optimum doping to 8-9 at the highest doping level [5]. Similar results have also been reported by Oda et al. [7]. In the underdoped sample the gap is essentially found to be the same as for optimum doping.

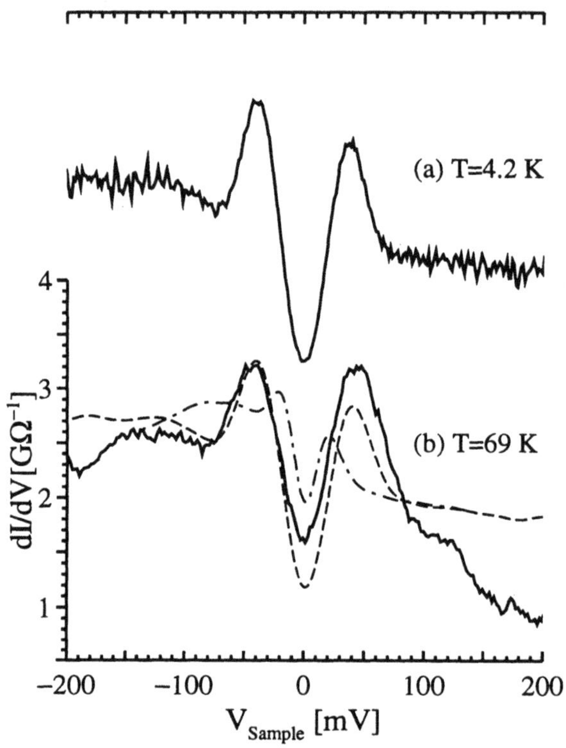

Figure 1. dI/dV(V) for an overdoped $Bi_2Sr_2CaCu_2O_{8+\delta}$ crystal. (a) T=4.2 K (b) T=69 K. Solid : experimental. Dashed : calculation assuming constant Δ. Dot-dashed : calculation assuming BCS temperature dependence.

We also note that there is a small but systematic evolution of the spectra at high voltages beyond the gap.

We now turn to the temperature dependence of the spectra and we focus our attention on the slightly overdoped sample for which we show the low temperature spectrum in Fig. 1a. The evolution of the spectra at higher temperatures will be reported in a forthcoming paper [6], here we shall concentrate on the spectra just below T_c.

The remarkable property of these spectra is they evolve only very little with temperature and in particular that the peaks basically stay at the same energy up to very close to T_c. The determination of the gap as a function of temperature from such spectra is not straight forward and it depends on the shape of the density of states. In order to extract an information about this temperature dependence we have assumed that the 4.2K spectra reflect the correct density of states and we have then calculated the spectra at higher temperature for the case that the gap stays constant and for the case that the gap follows a BCS temperature dependence. The result of this calculation for a temperature of T = 69K, (5K below T_c) is shown in Fig. 1b together with the experimental spectrum. As appears very clearly from this comparison, the gap stays to a very good precision equal to the low temperature gap. The main conclusion is that the gap is essentially temperature independent and does not seem to close at the critical temperature, but rather turn into a quasi-gap behaviour above T_c. Details about the behaviour above T_c will be presented in the forthcoming paper [6].

4. STS OBSERVATION OF VORTICES IN $YBa_2Cu_3O_{7-\delta}$

In absence of a magnetic field we observe on $YBa_2Cu_3O_{7-\delta}$ a multiple peak spectrum which is very different from the one observed on $Bi_2Sr_2CaCu_2O_{8+\delta}$ [3]. The main gap feature corresponds to a conduction peak close to ±20 meV with a second peak appearing in the 25 to 30 meV range. A weak structure is often observed at a few meV. This latter structure is not systematically seen. To observe vortices in the presence of a magnetic field we plot maps where the contrast is given by the ratio of the conductivity at 20 meV and at zero bias. If this is made at zero field we observe rather uniform behaviour as shown in Fig. 2a. If a magnetic field of 6 Tesla is added dark areas which we associate with the vortex cores appear in the maps (Fig. 2b).

If we follow the time evolution of these maps we observe that the vortex cores move

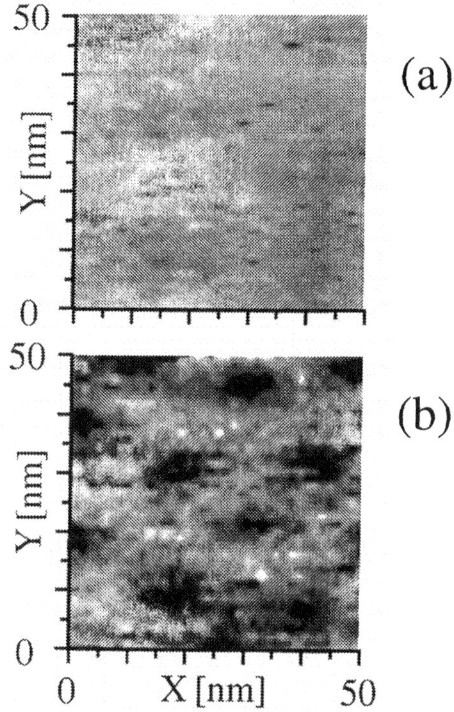

Figure 2. STS 50 x 50 nm^2 images acquired on a $YBa_2Cu_3O_{7-\delta}$ crystal at T=4.2 K. (a) B=0 T : no vortices present. (b) after a 6T field was applied. Dark areas represent the vortex cores.

around for about 1-2 days before they settle into a nearly regular lattice. In Fig. 3a we show a map taken 10 hours after the sample was cooled in a field of 6 Tesla. Fig. 3b show exactly the same area 36 hours later. Comparing the two we find that the lattice in Fig. 3a is still very irregular whereas the lattice shown in Fig. 3b has a nearly regular oblique structure. Further measurements showed no further evolution of the lattice as long as the temperature or field was not changed. Characteristic for the vortex lattice is its oblique shape and the elliptic shape of the vortex cores. The latter reflects the a-b plane anisotropy of the superconducting coherence length ξ. Averaging over a large number of vortices we find ξ_b/ξ_a = 1.5 ± 0.2. However, there are areas where we observe a

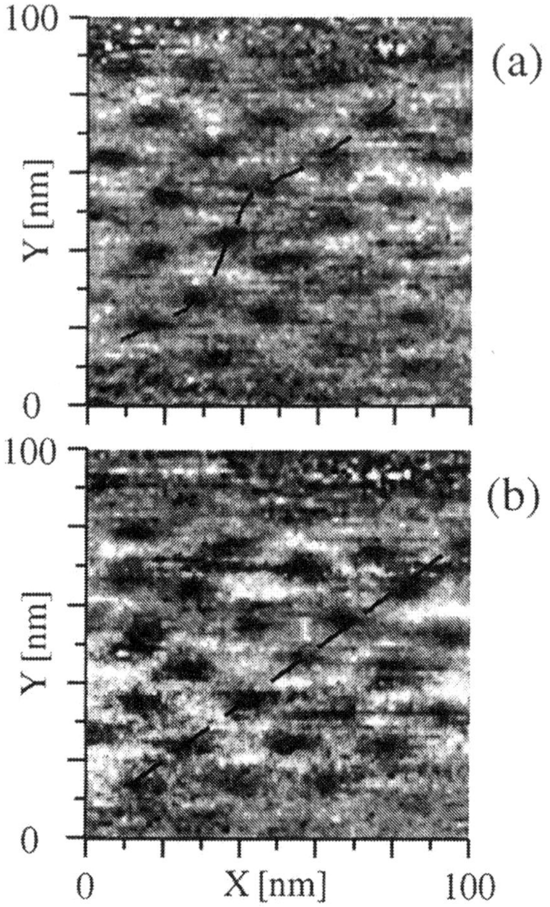

Figure 3. STS 100 x 100 nm^2 images of the vortex lattice acquired on a YBCO crystal at T=4.2 K, B=6 T. (a) 10 hours after field cooling (b) 36 hours later. Dashed lines are a guide to the eye to illustrate the degree of vortex lattice order.

much larger anisotropy with values as large as 2 [8]. The oblique symmetry of the lattice reflects the anisotropy of the superconducting penetration depth λ. Although we observe that the spatial variations of ξ correspond to a variation of the angle β of the oblique lattice we find that the value of β does not correspond the one expected if the isotropic case would be an hexagonal lattice. A good correspondence is found if one assumes that the lattice in the isotropic case would be square [8,9]. A remarkable result is the spectrum observed in the vortex core. The two main peaks disappears and two new peaks appear at ± 5.5 meV. These peaks have tentatively been associated with the first localized state in the vortex cores. This implies that the vortex cores are in an extreme quantum limit and that there is only one state localized in the gap. The precise interpretation of these spectra remains an open challenge.

The sample on which these measurements were carried out is twinned and the question about the influence of the twin boundaries on the shape of the observed vortex lattice has been brought up. The fact that we observe within one domain considerable changes of the anisotropy and thus of the vortex lattice shape, demonstrates that the effect of these twin boundaries on our observations is negligible. Recently we have been able to observe the vortices at and close to a twin boundary. These results which are reported in Ref. 10 confirm the above conclusion.

REFERENCES

1. Ch. Renner and Ø. Fischer, Phys. Rev. B **51**, 9208 (1995).
2. H.F. Hess et al., Phys. Rev. Lett. **62**, 214 (1989); Ch. Renner et al., Phys. Rev. Lett. **67**, 1650 (1991).
3. I. Maggio-Aprile, Ch. Renner, A. Erb, E. Walker and Ø. Fischer, Phys. Rev. Lett. **75**, 2754 (1995).
4. Ch. Renner et al., J. Vac. Sci. Technol. **A8**, 330 (1990) ;A.D. Kent et al., Ultramicroscopy **42-44**, 1631 (1992).
5. Ch. Renner, B. Revaz, J.-Y. Genoud, and Ø. Fischer, J. Low Temp. Phys. **105**, 1083 (1996).
6. Ch. Renner, B. Revaz, J.-Y. Genoud, K. Kadowaki, and Ø. Fischer, to be published.
7. M. Oda et al., these proceedings.
8. I. Maggio-Aprile, Ch. Renner, A. Erb, E.Walker and Ø.Fischer, J. Low Temp. Phys. **105**, 1129 (1996).
9. Ch. Renner, I. Maggio-Aprile, A. Erb, E.Walker and Ø.Fischer, SPIE Vol. 2696, p. 322 (1996).
10. I. Maggio-Aprile, Ch. Renner, A. Erb, E.Walker and Ø.Fischer, to be published.

ELSEVIER

Physica C 282–287 (1997) 319–322

PHYSICA C

Van der Waals Attraction of Vortices in Superconductors

G. Blatter[a] and V. B. Geshkenbein[b, a]

[a]Theoretische Physik, ETH-Hönggerberg, CH-8093 Zürich, Switzerland.
[b] L. D. Landau Institute for Theoretical Physics, 117940 Moscow, Russia

We show that in anisotropic and layered superconductors the fluctuations of vortex lines produces an attractive long-range vortex – vortex interaction of the van der Waals type. This attraction follows from the anisotropic screening properties of the material and has profound consequences for the low-field phase diagram of these materials.

Following general knowledge, vortices in type II superconductors repel one another on all scales [1]. The repulsion follows a logarithmic law on distances $r < \lambda$ and decays exponentially on scales $r > \lambda$, where λ denotes the magnetic penetration depth. This result remains true in the presence of thermal (or quantum) fluctuations of the vortex lines as long as the material is isotropic. In an anisotropic or layered material, however, fluctuations produce an attractive long-range interaction beyond the scale λ. This long-range attraction is of the van der Waals type and leads to a variety of interesting novel phenomena in dilute vortex systems. A first derivation of this new van der Waals force and its consequences for the low field phase diagram of layered superconductors can be found in Ref. [2]. The effects of disorder have recently been studied by Mukherji and Nattermann [3]. Here we give a discussion of this phenomenon exploiting the vortex $\leftrightarrow$ 2D boson analogy introduced by Nelson [4].

The occurrence of a van der Waals type attractive force between vortex lines is most simply illustrated in an extremely layered superconductor with no superconducting coupling between the layers: Consider two straight vortices 1 and 2 at a distance R. Displacing an individual pancake vortex in the stack 1 by u_1 is equivalent to placing a pancake–anti-pancake pair, i. e., a pancake dipole d_1, onto the vortex line. The pancake dipole d_1 induces a second dipole d_2 in vortex 2 within the same layer. With two pancakes interacting logarithmically within the same layer, the force from dipole d_1 acting on vortex 2 follows a $1/R^2$ law, hence $d_2 \propto d_1/R^2$. Finally, the interaction potential between two vortex dipoles follows a $1/R^2$ behavior as well and we obtain a long-range attractive potential $V_{\rm vdW} \propto 1/R^4$ between the vortex segments. The initial fluctuation d_1 can be induced thermally or, at low temperatures, quantum mechanically.

In an anisotropic material the same mechanism is at work with slight modifications. Crucial for the understanding of the effect are the different screening properties of the material parallel and perpendicular to the axis of anisotropy, the c-axis. Let $\varepsilon^2 = m/M \ll 1$ denote the anisotropy of the material, with m and M the effective masses parallel and perpendicular to the ab plane. The anisotropy parameter ε quantifies the Josephson coupling between the layers of the anisotropic material. Whereas currents flowing within the ab planes are screened on a scale λ, the out-of-plane currents decay only on the longer scale λ/ε. Placing two vortices parallel to the c-axis at a distance R, a fluctuation in vortex 1 again induces a "counter" fluctuation in vortex 2 with the transverse parts of the vortex lines attracting one another within the distance $R < \lambda/\varepsilon$. Within the range $\lambda < R < d/\varepsilon$, the Josephson coupling is irrelevant and the van der Waals interaction follows the behavior of the uncoupled system, $V_{\rm vdW} \propto 1/R^4$. At larger distances $d/\varepsilon < R < \lambda/\varepsilon$, the c-axis currents are relevant and the interaction drops faster, $V_{\rm vdW} \propto 1/R^5$. For strong coupling with $d/\varepsilon < \lambda$ only the latter regime survives.

It is quite instructive to compare the van der

0921-4534/97/$17.00
PII S0921-4534(97)00276-1

Waals attraction between vortices in anisotropic superconductors with the fluctuation induced attraction between neutral atoms. The latter decays on short distances with the famous $1/r^6$ power law, turning into a $1/r^7$ behavior at large distances when retardation effects come into play [5]. The changeover from a $1/R^4$ to a $1/R^5$ behavior in the vortex system is an analogous effect. This can be most simply understood when the vortex problem is mapped onto the imaginary time quantum problem of 2D charged bosons, see Refs. [4,6]. The vortex-vortex interaction then is expressed through a non-local interaction between the bosons described by an additional fictitious gauge field $\mathbf{a}$. In a uniaxially anisotropic superconductor the effective action reads (for notation, see [6])

$$\mathcal{S} = \int d\tau d^2R \left\{ i\varepsilon \left(\mathbf{j}_B - \frac{\nabla \wedge \mathbf{A}}{\Phi_0} \right) \cdot \mathbf{a} + \frac{\varepsilon^2 \mathbf{h}_\perp^2 + h_0^2}{8\pi\varepsilon_\circ} + \frac{(\nabla \wedge \mathbf{A})^2}{8\pi} \right\}, \quad (1)$$

where $\mathbf{j}_B$ is the boson current, $\mathbf{h} = \nabla \wedge \mathbf{a}$, $h_0 = h_z$ and $\varepsilon_\circ = (\Phi_0/4\pi\lambda)^2$. Integrating out the fictitious gauge field $\mathbf{a}$ in the absence of bosons we recover the anisotropic London free energy functional

$$\mathcal{F}[\mathbf{A}(\mathbf{r})] = \frac{1}{8\pi} \int d^3r \left[(\nabla \wedge \mathbf{A})^2 + \frac{\mathbf{A}_\perp^2}{\lambda^2} + \frac{\varepsilon^2 A_z^2}{\lambda^2} \right].$$

Alternatively, a useful relation between the superfluid properties of the imaginary time bosons and the response of the vortex structure to a tilt of the magnetic field follows from (1) when the integration over the bosons is performed. The polarizability Π^B of the boson system produces an additional term $\varepsilon^2 \Pi^B \mathbf{a}_\perp^2$ in the Lagrangean density for the gauge field [6]

$$L = \left(\frac{1}{8\pi\varepsilon_\circ} + \frac{\Pi^B \varepsilon^2}{2\mathbf{k}_\perp^2} \right) h_0^2 + \frac{i\varepsilon A_z}{\Phi_0} h_0 + \frac{\mathbf{B}^2}{8\pi} + \ldots,$$

where we have chosen the gauge $\nabla_\perp \cdot \mathbf{a}_\perp = 0$ and have dropped terms which do not depend on A_z. Integrating out h_0 we obtain the free energy density as a function of $\mathbf{B}$,

$$F = \frac{\mathbf{B}^2}{8\pi} + \frac{B_\perp^2}{8\pi} \frac{\varepsilon^2}{\lambda^2 k_\perp^2 + \varepsilon^2 \Pi^B \Phi_0^2 / 4\pi}, \quad (2)$$

where we have concentrated on the limit $k_z \to 0$. Variation of (2) with respect to the angle $\theta \approx B_\perp / B$ provides us with an expression for the long wavelength tilt modulus

$$c_{44}(\mathbf{k} = 0) = \frac{B^2}{4\pi} \left(1 + \frac{\varepsilon^2}{4\pi\lambda^2 n_s^B} \right), \quad (3)$$

where we have used $\Pi^B = n_s^B / m_0^B$ with n_s^B the bosonic superfluid density and $m_0^B = \varepsilon^2 \varepsilon_0$ the (bare) boson mass. In the entangled vortex liquid phase the bosonic superfluid density is given by the full density of vortices, $n_s^B = n_v = B/\Phi_0$, and we recover the usual expression for the long-wavelength tilt modulus as given by the anisotropic London theory, $c_{44} = B^2/4\pi + \varepsilon^2 B \Phi_0 / 16\pi^2 \lambda^2$. On the other hand, introducing correlated disorder in the form of twin boundaries or columnar tracks, the bosons tend to localize onto the defects and the superfluid density n_s^B will vanish on approaching the Bose glass transition from above. Accordingly, the tilt modulus (3) will diverge at the transition and we recover the locking phenomenon in the language of boson physics. We thus have related the superfluid density n_s^B of the 2D bosons to the directly measurable response of the vortex system to a transverse magnetic field in an anisotropic case.

To derive the van der Waals interaction between the vortices it is more appropriate to integrate out the real magnetic field in the action (1). Rescaling $\tau \to \varepsilon\tau$ and $a_0 \to a_0/\varepsilon$ [6] we obtain the effective Lagrangean density

$$L(\mathbf{j}_B, \mathbf{a}) = i\mathbf{j}_B \cdot \mathbf{a} + \frac{2\pi}{\Phi_0^2} \left[a_0^2 + \varepsilon^2 \mathbf{a}_\perp^2 + \lambda^2 (\nabla \wedge \mathbf{a})^2 \right].$$

Considering a_0 as a 'scalar potential' we find that the longitudinal interaction is screened on the length λ, whereas the transverse one is screened on the much larger length scale λ/ε. The quantum fluctuations of this transverse gauge field in the boson picture then map to thermal fluctuations in the vortex system and produce the van der Waals attraction between vortices on the scale λ/ε [2]

$$V_{\mathrm{vdW}} = \begin{cases} \dfrac{-\varepsilon_\circ}{\ln^2(\pi\lambda/d)} \dfrac{T}{d\varepsilon_\circ} \dfrac{\lambda^4}{R^4}, & \lambda < R < \dfrac{d}{\varepsilon}, \\ \dfrac{(-3\pi/32)\varepsilon_\circ}{\ln^2(\pi\lambda/\varepsilon R)} \dfrac{T}{\varepsilon\lambda\varepsilon_\circ} \dfrac{\lambda^5}{R^5}, & \lambda, \dfrac{d}{\varepsilon} < R < \dfrac{\lambda}{\varepsilon}. \end{cases}$$

This attractive interaction should then be added to the conventional repulsion $2\varepsilon_\circ K_0(a_\circ/\lambda)$ acting on the shorter scale $R < \lambda$. The two components add up to the effective vortex – vortex potential $V_{vv}(R)$ which exhibits a van der Waals minimum at the distance $a_{\rm vdW} \sim 20\lambda$.

The appearance of an attractive van der Waals type interaction between the vortex lines has interesting consequences for the low-field ($B \sim$ few Gauss) phase diagram of anisotropic/layered superconductors, see Fig. 1. A discussion in terms of the boson language of the physics involved is quite instructive: The attractive interaction between the bosons may lead to the absence of the gaseous phase. With a potential which is repulsive at short distances but turns attractive on large separations, the two-boson scattering amplitude can turn negative, leading to an instability of the dilute bose gas phase. The scattering amplitude f can be written as a sum of repulsive and attractive contributions, with the short-range repulsive part of the potential V_{vv} producing a term of order unity. The attractive part is treated within perturbation theory and we arrive at the following estimate for the total scattering amplitude f

$$f \sim 1 - \frac{m^B}{2\pi\hbar_B^2}\int_{<0} d^2R\, V_{vv}(R),$$

where $m^B = \varepsilon_\circ$ is the mass of the 'dressed' bosons which is enhanced due to the interaction with the gauge field and the integral is taken only over the negative part of the intervortex potential. The dilute bose gas turns unstable when the scattering amplitude changes sign, forming a generic liquid state with a density determined by the interaction between particles (the van der Waals minimum) rather than by the external pressure. In the original vortex problem this instability occurs at a temperature $T_g \sim (\varepsilon_\circ V_{\rm vdW}(a_{\rm vdW}))^{1/2} a_{\rm vdW}$. Note that even if the two particle scattering amplitude is repulsive (absence of two particle bound states), the dilute gas still may be unstable towards the formation of three- or multiple-particle bound states. This possibility may change T_g by a numerical factor of order unity.

The above description in terms of scattering of bosons maps to the classical treatment of the van

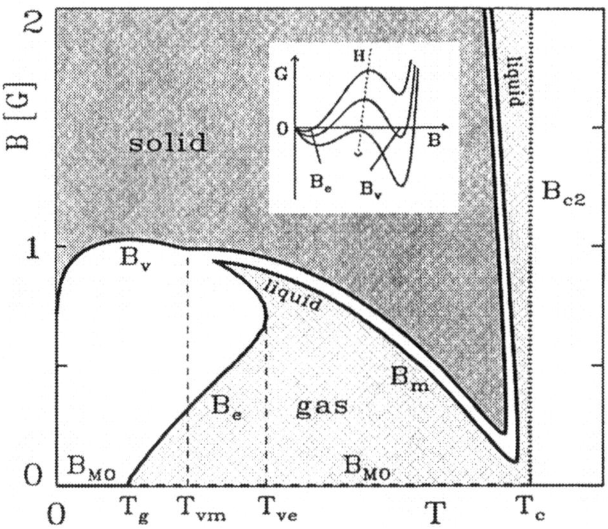

Figure 1. Low-field phase diagram for a layered superconductor. Parameters appropriate for $Bi_2Sr_2Ca_1Cu_2O_8$ have been used ($\lambda = 1800$ Å, $\varepsilon = 1/300$, $c_L = 0.3$). Continuous lines denote first-order transitions (note the associated jumps in B); the dashed line is a second order transition. At low temperatures $T < T_g$, T_{vm} the Meissner-Ochsenfeld (MO) phase (B_{MO}) becomes unstable towards a vortex solid ($B > B_v$) at $H_{c_1} < H^r_{c_1}$. At the temperature T_g a new vortex gas phase appears at low densities ($B < B_e$), rendering the MO-phase unstable at $H_{c_1} > H^r_{c_1}$. At a higher magnetic field $H_{cv} > H_{c_1}$ a transition to a solid phase takes place ($B > B_v$). The latter melts at T_{vm}, giving way to a sliver of vortex liquid in the high density region above B_v. At temperatures $T > T_{ve}$ larger than critical we cannot distinguish any more between the low density vortex gas and the high density vortex liquid regimes — we call the resulting phase a vortex gas. With increasing external field H this gas transforms into a vortex solid ($B > B_m$). With increasing anisotropy the critical temperatures T_g and T_{ve} move towards higher temperatures. The reentrant nose close to T_c is only a schematic drawing. The inset shows the free energy G versus induction B at fixed temperature $T > T_g$ for 3 values of the magnetic field H.

der Waals problem in terms of the virial expansion [7]. The contribution to the energy density of the vortex system due to the interaction is $\propto n_v^2 f$, where $n_v = B/\Phi_0$ is the vortex density. In the vortex language the positive (repulsive) part in f, due to close collisions of bosons, describes the so called entropic repulsion [4]. The attractive component of the interaction enters the energy in terms of an average of the potential over distances $R > \sigma$, where $V_{\rm vdW}(\sigma) = 0$. In [2] we compared the entropic repulsion term directly with the van der Waals interaction $V_{\rm vdW}$ evaluated at the mean interparticle separation, such that the repulsive and attractive components entered the free energy with a different dependence on vortex density. This approach is actually not correct [8], as both the effective attractive and the repusive part of the interaction exhibit the same dependence on the vortex density n_v, as demonstrated by the above scattering approach as well as by the usual virial expansion. The crucial point is that in the van der Waals problem the spacially averaged potential is the relevant quantity and not the potential at the averaged position.

To determine the phase diagram we analyze the Gibbs free energy density $G(H,T) = F(B,T) + B(H_{c_1}^\circ - H)/4\pi$, where $H_{c_1}^\circ = 4\pi\varepsilon_\circ \ln(\lambda/\xi)/\Phi_\circ$ is the unrenormalized lower critical field. The equilibrium is determined by the condition $\partial F/\partial B = (H - H_{c_1}^\circ)/4\pi$. Above T_g, the derivative $\partial F/\partial B$ is a nonmonotonic function of B and two solutions corresponding to the liquid and gas phase are found, leading to a positive slope in the phase line $B_e(T)$, see Fig. 1. In the end it turns out that the phase diagram obtained in [2] is qualitatively unchanged although the way it has been derived in [2] is subject to criticism [8].

It is interesting to note that the critical behavior near the liquid – gas critical point T_{ve} for the case of line objects has already been considered in [9]. Because of the line nature of the vortices it turns out that the critical behaviour at T_{ve} belongs to the universality class of the uniaxial ferroelectric, where only logarithmic corrections to the mean field result appear [10].

In conclusion, we have presented a discussion in terms of the boson language of the fluctuation-induced van der Waals-type long-range attractive vortex – vortex interaction. Accounting for this attractive component in the Gibbs free energy G, we have constructed the low-field phase diagram of strongly anisotropic and layered superconductors. We have found that the attraction between the vortices removes the dilute liquid phase at low temperatures and the vortex system enters the specimen in a solid state. For highly layered materials the residual dilute gas phase is removed alltogether as T_g, $T_{ve} \to T_c$, leading to a considerable simplification of the low-field phase diagram.

We wish to thank E. H. Brandt, H. Katzgraber, D. R. Nelson, and O. Wagner for illuminating discussions.

REFERENCES

1. M. Tinkham, *Introduction to Superconductivity* (Krieger, 1980).
2. G. Blatter and V. B. Geshkenbein, Phys. Rev. Lett., **77**, 4958, (1996).
3. S. Mukherji and T. Nattermann, unpublished.
4. D. R. Nelson, Phys. Rev. Lett. **60**, 1973 (1988).
5. L. D. Landau and E. M. Lifshitz, *Course in Theoretical Physics*, Vols. 4 and 9 (Pergamon Press, London/Paris).
6. G. Blatter *et al.*, Rev. Mod. Phys. **66**, 1125 (1994).
7. L. D. Landau and E. M. Lifschitz, *Course in Theoretical Physics*, Vol. 5 on *Statistical Physics* (Pergamon, London/Paris).
8. We wish to thank Orlando Wagner for pointing out an error in [2] originating from treating the entropic repulsion and the van der Waals interaction separately.
9. R. D. Kamien, and D. R. Nelson, J. Stat. Phys. **71**, 23 (1993).
10. A. I. Larkin and D. E. Khmel'nitskii, Zh. Eksp. Teor. Fiz. **56**, 2087 (1969) (Sov. Phys. JETP **29**, 1123 (1969)).

ELSEVIER

Physica C 282–287 (1997) 323–326

PHYSICA C

Vortex Matter Phase Transitions in $Bi_2Sr_2CaCu_2O_8$

R. A. Doyle[a,b], B. Khaykovich[a,c], M. Konczykowski[c], E. Zeldov[a], N. Morozov[a,c], D. Majer[a], P. H. Kes[d], and V. Vinokur[e]

[a]Department of Condensed Matter Physics, The Weizmann Institute of Science, 76100 Rehovot, Israel
[b]IRC in Superconductivity, University of Cambridge, Cambridge CB3 0HE, United Kingdom
[c]CNRS, URA 1380, Laboratoire des Solides Irradiés, École Polytechnique, 91128 Palaiseau, France
[d]Rijksuniversiteit Leiden, P. O. Box 9506, 2300 RA Leiden, The Netherlands
[e]Materials Science Division, Argonne National Laboratory, Argonne, Illinois 60439

Vortex-lattice phase transitions in $Bi_2Sr_2CaCu_2O_8$ crystals are studied using local magnetization measurements. The vortex matter is found to exhibit at least three distinct phases. A rather ordered quasilattice phase is present at low fields. At elevated temperatures the quasilattice melts (or sublimates) through a first-order phase transition, whereas at lower temperatures a transition to a highly disordered vortex solid occurs at the second magnetization peak. A very low density of columnar defects results in significant pinning of the quasilattice, demonstrating the finite shear modulus of this phase. Local ac measurements reveal that in the presence of columns the pinned quasilattice still melts through a first-order phase transition with an observable step in equilibrium magnetization at elevated temperatures.

1. INTRODUCTION

The high-temperature superconductors (HTSC) in the mixed state display a very complicated phase diagram. The nature of the different vortex phases and the transitions between them are still unclear. In the low temperature superconductors it was believed that only one thermodynamic phase, a solid Abrikosov lattice, exists in the mixed state. Since the discovery of HTSC it has became apparent that thermal fluctuations play a crucial role and may result in melting of the lattice, thus forming two phases, vortex solid and vortex liquid. However in highly anisotropic HTSC like $Bi_2Sr_2CaCu_2O_8$ (BSCCO) the vortex matter phase diagram is much more complicated and displays at least three distinct phases as shown in Fig. 1. We have incorporated a very low dose of columnar defects in order to elucidate the structure of the various phases. We find that low concentrations of columnar defects have an observable effect only in the low field quasilattice phase, resulting in significantly enhanced pinning in this state. The pinned quasilattice still displays a first-order melting transition at elevated temperatures.

2. EXPERIMENTAL

The experiments were carried out on several BSCCO crystals [1] with typical dimensions of $700 \times 300 \times 10\ \mu m^3$. The crystals were irradiated by 1 GeV Xe or 0.9 GeV Pb ions at GANIL (Caen, France), which produced uniform columnar defects parallel to the crystalline c-axis. We have investigated very low irradiation doses corresponding to matching fields B_ϕ in the range of 20 to 100 G ($B_\phi = n\phi_0$, where n is the density of the columnar tracks and ϕ_0 is the magnetic flux quantum). The local magnetization measurements were performed in applied field $H_a||z||c$-axis using arrays of $10 \times 10 \mu m^2$ GaAs/AlGaAs Hall sensors [2]. The sensors measure the perpendicular component of the local induction B_z at various locations across the sample. For ac studies a small ac field H_{ac} was superimposed onto the dc magnetic field H_{dc} and the resulting local ac induction B_{ac} across the crystal was measured using lock-in techniques.

3. PHASE TRANSITIONS

Figure 1 shows the phase diagram of the vortex matter in BSCCO as inferred from our local

0921-4534/97/$17.00
PII S0921-4534(97)00277-3

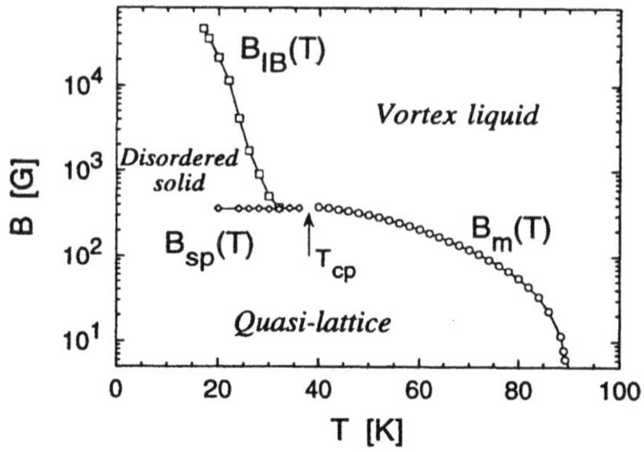

Figure 1: Vortex matter phase diagram in BSCCO displaying three distinct phases with three boundary lines. The quasilattice phase melts through a FOT at $B_m(T)$ and transforms into a disordered solid through a disorder-induced transition at B_{sp}. The disordered solid melts continuously along $B_{IB}(T)$.

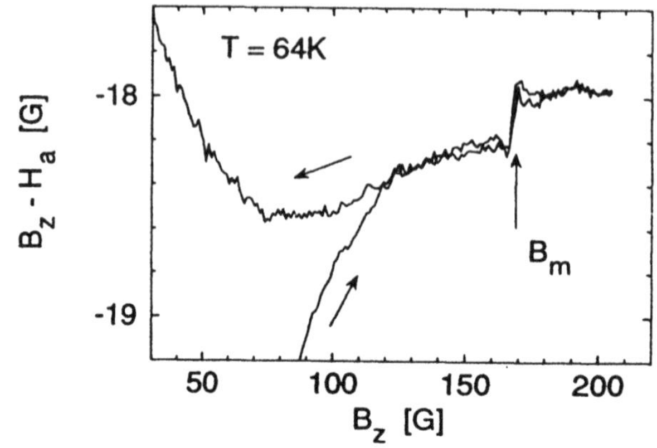

Figure 2: Local magnetization loop at elevated temperature showing the FOT manifested by the sharp step in equilibrium magnetization at $B_m(T)$.

magnetization measurements. There are three experimentally observed phase boundary lines. At elevated temperatures and low fields a first-order phase transition (FOT) is observed. Local magnetization measurements $B_z - H_a$ vs. B_z display a small positive sharp step in equilibrium magnetization at the transition as shown in Fig. 2. A similar step is observed by temperature scans [2]. The position of the FOT line $B_m(T)$ is shown in Fig. 1. The FOT line terminates abruptly at intermediate temperatures at a critical point T_{cp} as indicated by the arrow in Fig. 1.

At temperatures below the critical point an onset of a large second magnetization peak [3] is observed at $B_{sp}(T)$ as shown in Fig. 3. Analysis of the field profiles across the sample shows a transition from smooth dome-shaped profiles due to surface and geometrical barriers at low fields to Bean profiles due to the onset of significant critical currents at fields above B_{sp} [4]. Local magnetization measurements indicate that this transition is very sharp [5]. We therefore infer that the sharp change in the apparent pinning at the onset of the second peak is driven by an underlying thermodynamic phase transition.

It is interesting to note that the second peak line $B_{sp}(T)$ and the FOT line $B_m(T)$ meet at the critical point forming one continuous transition line. Similar investigations were carried out on oxygen annealed crystals with various anisotropy [5,6]. The position of $B_{sp}(T)$ and $B_m(T)$ is found to be strongly anisotropy dependent, however, in all cases the two lines always merge at the critical point [5,6]. This finding is a strong indication that the two lines represent one continuous vortex-lattice transition which is a first-order at high temperatures, but possibly becomes second-order at some sample-dependent T_{cp} as the temperature is decreased.

At fields above the B_{sp} line, significant bulk pinning is observed. Figure 4 shows the behavior of the apparent critical current in this region. It is important to note that in this region surface barriers also contribute significantly to the magnetic hysteresis. Therefore one cannot readily determine the critical current from the global magnetization measurements. In local measurements a proper analysis of the field gradient dB_z/dx can be made [4] which allows separation of the two contributions. Figure 4 thus represents the hysteretic behavior of the field gradient which reflects bulk pinning. The apparent critical current decreases rapidly with increasing temperature and

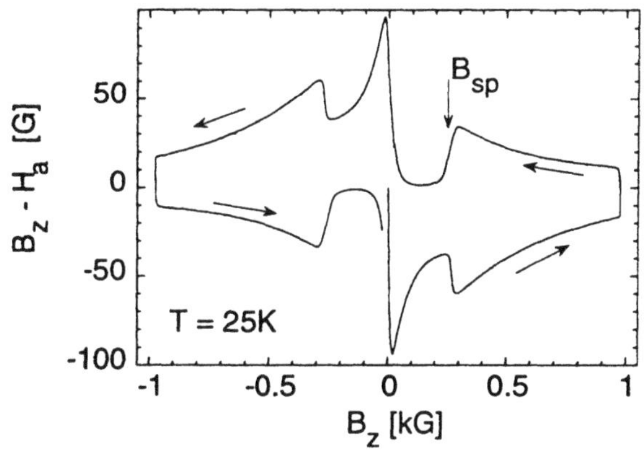

Figure 3: Local magnetization loop in the vicinity of the second magnetization peak B_{sp}.

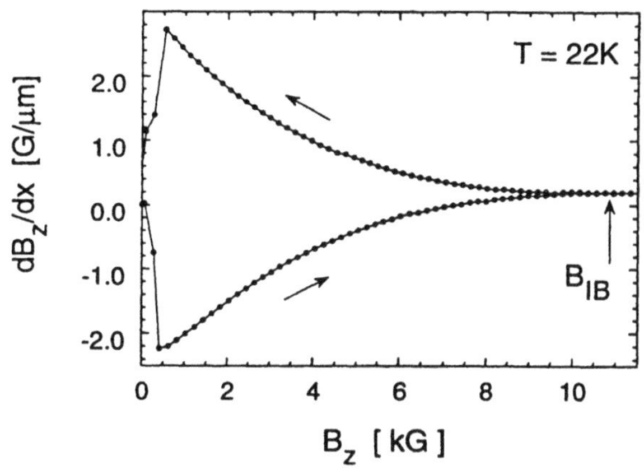

Figure 4: Field gradient dB_z/dx vs. the local induction B_z at low temperature. The locations of the corresponding sensors across the sample were chosen in order to reflect bulk pinning rather than surface barriers. B_{IB} thus indicates the position of the bulk irreversibility line.

field, and vanishes at $B_{IB}(T)$ [4]. We believe that this line in Fig. 1 approximately reflects the position of another phase boundary which extrapolates to the critical point T_{cp}.

4. COLUMNAR DEFECTS

In the as-grown BSCCO crystals the phase below $B_m(T)$ line in Fig. 1 is practically unpinned with effectively zero critical current, and therefore in principle could be either in a liquid or a solid phase. Hence there is a need for a direct test of the shear modulus in this phase. We have introduced a very low dose of columnar defects such that their density is much lower than the vortex density over a range of fields below $B_m(T)$ [7]. This case is similar to trying to tack a carpet with few nails. If the vortex matter is in a solid phase with a finite shear modulus the entire lattice will be pinned. On the other hand, in a liquid phase the unpinned vortices will flow freely around the few vortices trapped by the columns and the critical current will remain zero. Local dc magnetization measurements indicate that below $B_m(T)$ the shear modulus is indeed finite [7]. Here we present similar behavior measured by a local ac technique. Figure 5 shows the local ac response B'_{ac}/H_{ac} as a function of the applied dc field H_{dc} in the as-grown crystal along with a crystal irradiated to a dose corresponding to $B_\phi = 20$ G. In the as-grown sample the FOT manifests as a sharp paramagnetic peak which is the result of the thermodynamic step in the equilibrium magnetization at the transition [8]. Above B_m the sample exhibits no shielding, $B'_{ac}/H_{ac} = 1$, and the critical current is zero. The same is also true in the vicinity of the transition below B_m in as-grown samples, which means that in these samples the critical current is practically zero below the FOT. At lower fields, significant shielding is observed due to the presence of the geometrical barrier. This barrier also results in the unusual negative local response [9] as seen in Fig. 5 at $H_{dc} \simeq 20$ Oe. In the irradiated sample, however, the situation is dramatically different. Above B_m the sample is still unshielded, however full shielding is immediately established just below the transition, indicating a large critical current and the existence of a finite shear modulus in this phase which vanishes abruptly at the FOT. The same conclusions were recently reached on the basis of transport measurements [10,11]. It is interesting to note here that at elevated temperatures the FOT is still preserved in presence of low doses of columnar defects as demonstrated in Fig. 5 by the presence of the paramagnetic peak at the transition. The height of the peak is somewhat suppressed relative to the unirradiated sample. At lower temperatures this peak is not detected. The small difference in the position of $B_m(T)$ for the two samples in Fig. 5 is due to some difference in the crystals and mainly in their sizes which results in different values of the internal field B at a given H_{dc}.

5. VORTEX MATTER PHASE DIAGRAM

Based on the above results we conclude that the phase below the FOT is a solid lattice which according to the neutron diffraction and μSR data has a long range orientational order [12]. We therefore refer to this state as a quasilattice [13]. At elevated temperatures the quasilattice melts through a FOT (see Fig. 1). Recent flux transformer measurements [11] indicate that this FOT is a sublimation transition in which the solid quasilattice transforms directly into a gas of pancake vortices. At lower temperatures the quasilattice transforms into a highly disordered entangled vortex solid through a disorder-driven transition at the second magnetization peak $B_{sp}(T)$ [13,14]. This disordered solid eventually melts continuously near the depinning line $B_{IB}(T)$.

Various parts of this work were supported by the German-Israeli Foundation - GIF, by the US - Israel Binational Science Foundation, by Minerva Foundation, Munich/Germany, by contract CT1*CT93-0063 from the Commission of the European Union, by the Dutch Foundation for Fundamental Research on Matter (FOM), and by the U.S. Department of Energy, BES-Material Sciences, under Contract No. W-31-109-ENG-38.

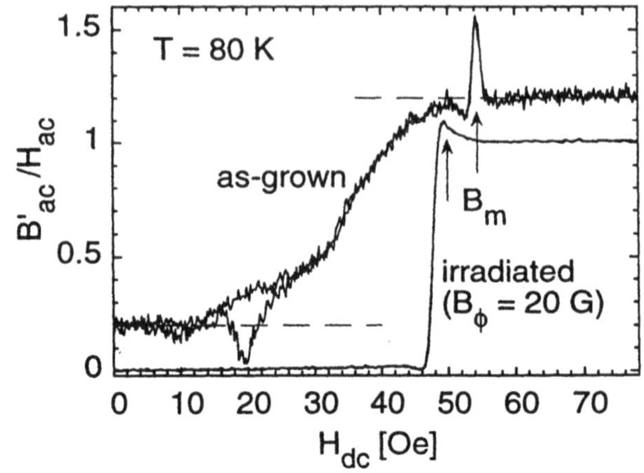

Figure 5: Normalized local ac response B'_{ac}/H_{ac} in the as-grown (shifted by 0.2 for clarity) and heavy-ion irradiated BSCCO crystals ($H_{ac} \approx 0.5$ G). The FOT is observed as a paramagnetic peak at $B_m(T)$ ($B'_{ac}/H_{ac} > 1$) in both samples. In irradiated crystal bulk pinning is present below the FOT and it vanishes sharply at the transition.

REFERENCES

1. N. Motohira *et al.*, J. Ceram. Soc. Jpn. Int. Ed. **97**, 994 (1989).
2. E. Zeldov *et al.*, Nature **375**, 373 (1995).
3. N. Chikumoto *et al.*, Phys. Rev. Lett. **69**, 1260 (1992); Physica C **185-189**, 2201 (1991); G. Yang *et al.*, Phys. Rev. B **48**, 4054 (1993); T. Tamegai *et al.*, Physica C **223**, 33 (1993); K. Kishio *et al.*, in "Proc. 7th Intnl. Workshop on Critical Currents in Superconductors" ed. H.W. Weber, World Sci. Pub., Singapore, p. 339 (1994).
4. E. Zeldov *et al.*, Europhys. Lett. **30**, 367 (1995); D. Majer *et al.*, Physica C 235-240, 2765 (1994).
5. B. Khaykovich *et al.*, Phys. Rev. Lett. **76**, 2555 (1996).
6. T. Hanaguri *et al.*, Physica C **256**, 111 (1996).
7. B. Khaykovich *et al.*, preprint.
8. N. Morozov *et al.*, Phys. Rev. B **54**, R3784 (1996); B. Schmidt *et al.*, Phys. Rev. B (in press).
9. N. Morozov *et al.*, Phys. Rev. Lett. **76**, 138 (1996).
10. H. Pastoriza and P.H. Kes, Phys. Rev. Lett. **75**, 3525 (1995); S. Watauchi *et al.*, Physica C **259**, 373 (1996); D. T. Fuchs *et al.*, Phys. Rev. B **54**, 796 (1996).
11. D. T. Fuchs *et al.*, Phys. Rev. B (in press).
12. R. Cubitt *et al.*, Nature **365**, 407 (1993); S. L. Lee *et al.*, Phys. Rev. Lett. **71**, 3862 (1993).
13. V. Vinokur *et al.*, preprint.
14. D. Ertas and D. R. Nelson, Physica C **272**, 79 (1996) and references therein.

Physica C 282–287 (1997) 327–330

CALORIMETRIC EVIDENCE FOR THE OCCURRENCE OF A FIRST-ORDER TRANSITION IN THE MIXED STATE OF UNTWINNED $YBa_2Cu_3O_{7-\delta}$

A. Schilling[1*§], R.A. Fisher[1§], N.E. Phillips[1§], U. Welp[2#], W.K. Kwok[2#], and G.W. Crabtree[2#]

[1]Lawrence Berkeley National Laboratory, Berkeley, CA 94720, U.S.A.
[2]Argonne National Laboratory, Argonne, IL 60439, U.S.A.

Using high-resolution differential thermal analysis (DTA) we have demonstrated directly, for the first time in a thermal experiment, that the vortex lattice of untwinned $YBa_2Cu_3O_{7-\delta}$ in the mixed state (for magnetic fields H parallel and perpendicular to the c axis of the crystal) undergoes a first-order phase transition. The phase-transition line $H_m(T)$ in the magnetic phase diagram coincides with the vortex-lattice melting line that had been deduced previously from resistivity and magnetization measurements. The value obtained for the latent heat, $L \approx 0.5\ k_BT$ per vortex per superconducting layer for $H//c$, is consistent with that inferred from magnetization data using the Clapeyron equation, thus proving thermodynamic consistency. The specific heat of the vortex fluid is larger than that of the vortex solid. This specific-heat difference, $\Delta C/T$, and the latent heat L, are independent of both the speed and the direction (up or down in T or H) of the traversal of the phase boundary $H_m(T)$, proving that the transition does not have the characteristics of a glass-like melting transition.

1. Introduction

Earlier experiments on twinned single crystals of $YBa_2Cu_3O_{7-\delta}$ [1] indicated slight but clearly resolvable step-like increases $\Delta C/T$ in the specific heat C/T within the vortex state, for external magnetic fields H parallel to the c axes of the crystals. Similar, but somewhat larger, changes in the specific heat of twinned single crystals were reported later [2]. Neither thermodynamic consistency nor thermodynamic equilibrium was demonstrated in either experiment, however. More recently we have shown in a thermal experiment that the phase transition from the vortex solid to the vortex-fluid phase in untwinned $YBa_2Cu_3O_{7-\delta}$ is a thermodynamic first-order transition associated with a latent heat $L \approx 0.5$ k_BT per vortex per superconducting layer for $H//c$, and have proven thermodynamic consistency using magnetization data on the same sample [3]. This estimate for L was later confirmed by the Geneva group [4,5]. However, in those measurements, made on twinned crystals, L was not detectable for $\mu_0H \leq$ 5 T. Here we report on more detailed specific-heat and magnetization investigations near the vortex-lattice melting line of untwinned $YBa_2Cu_3O_{7-\delta}$ that determine equilibrium values of both L and $\Delta C/T$.

The untwinned $YBa_2Cu_3O_{7-\delta}$ sample (3.3 mg, $T_c \approx 92$ K) had been used earlier to measure the latent heat of vortex-lattice melting for $H//c$ [3]. The magnetization $M(H,T)$ was measured with a SQUID magnetometer as described elsewhere [6]. The specific heat was investigated for $H//c$ and $H \perp c$ using a differential-thermal analysis (DTA) technique [7].

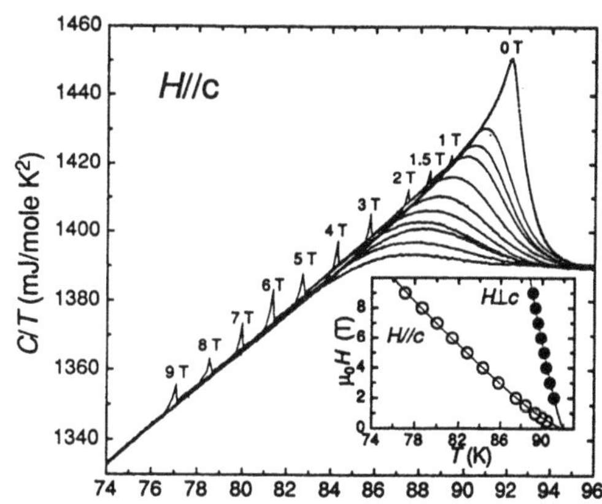

Fig. 1: C/T vs. T of untwinned $YBa_2Cu_3O_{7-\delta}$. The inset shows the melting lines for $H//c$ and $H \perp c$.

Supported by the Schweizerische Nationalfonds zur Förderung der wissenschaftlichen Forschung, NF 8220-042855 (*), and the US DOE Basic Energy Sciences-Materials Science, DE-AC03-76SF00098 (§) and W-31-109-ENG-38 (#), respectively.

PII S0921-4534(97)00278-5

2. Results

2.1. The latent heat of vortex-lattice melting

The specific heat of $YBa_2Cu_3O_{7-\delta}$ is displayed in Fig. 1 as C/T vs. T for $H//c$. To obtain these data, that are *virtually free of instrumental broadening*, we used data-evaluation techniques as described in Eq. 4 of Ref. 7. To summarize, a first estimate for the specific-heat difference ΔC_0 between the sample (temperature T_s) and the reference specimen (T_r) can be obtained using $\Delta C_0 = C_r(T_s-T_r)/(T_r-T_b)$, where C_r is the heat capacity of the reference sample and T_b is the temperature of the connected heat reservoir. The instrumental broadening in ΔC_0 can be very much reduced using $\Delta C \approx \Delta C_0 + T_\tau \mathrm{d}\Delta C_0/\mathrm{d}T$, which is valid for symmetrical heat links and $\Delta C_0 << C_r$ (a similar deconvolution formula can be derived also for less restrictive conditions). Here, $T_\tau = \tau \mathrm{d}T_b/\mathrm{d}t$, where τ is the time constant of the experiment [7] (see Figs. 2). At the melting temperature of the vortex lattice, a very sharp peak in ΔC develops, indicating a first-order transition. The entropy of melting, ΔS, can be obtained from constructions similar to those shown in Fig. 2b, and the corresponding latent heat is $L = T\Delta S$.

As an alternative to evaluate ΔS, we may fix T and vary H. We have shown earlier that there is a cooling (heating) effect on the sample due to the absorption (release) of latent heat, which is entirely symmetric and only changes sign when the magnetic field is ramped up or down. To correct the ΔT results for instrumental relaxation effects due to the presence of the finite heat links we can use the relation $\Delta T_{eff} = \Delta T + \int(\Delta T/H_\tau)\mathrm{d}H$. Here, $H_\tau = \tau \mathrm{d}H/\mathrm{d}t$ (see Figs. 3). ΔT_{eff} represents the *true* cooling effect that the sample would experience in the absence of the heat links. From the resulting shifts in temperature we again obtain $\Delta S = C\Delta T_{eff}/T$. From all these experiments, we can calculate the entropy of melting per vortex per superconducting layer, here taken as the thickness of one crystallographic unit cell in the c direction. We may compare these new results for $H//c$, which are *essentially free of potential instrumental broadening effects or of any assumptions about the transition width*, with our previously published values for ΔS [3], and with results from ΔM data taken on the same crystal and using the Clapeyron formula $\Delta S = -\Delta M \mathrm{d}H_m/\mathrm{d}T$ [3] (Fig. 4a). The coincidence of all these data clearly shows thermodynamic consistency. (Note that the ΔM data presented in Ref. 6 for a similar crystal represent intrinsic values. They had been carefully cross-checked with data obtained by a quasi-static dc measuring technique [8], i.e., without moving the sample or otherwise exposing it to unwanted ac fields).

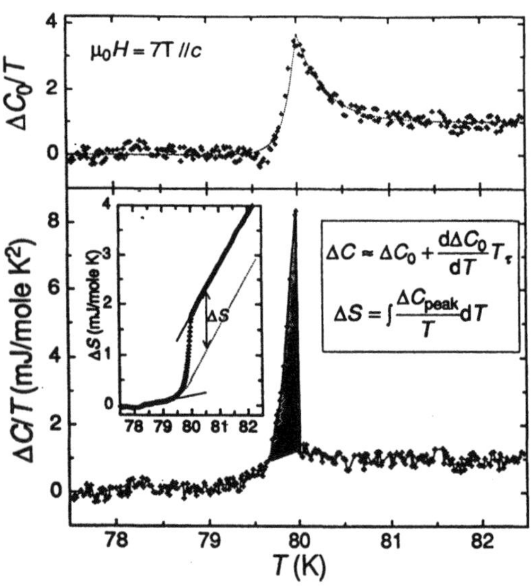

Figs. 2a-b: Representative ΔC_0 and ΔC data that we used to extract the entropy of melting, ΔS (see text).

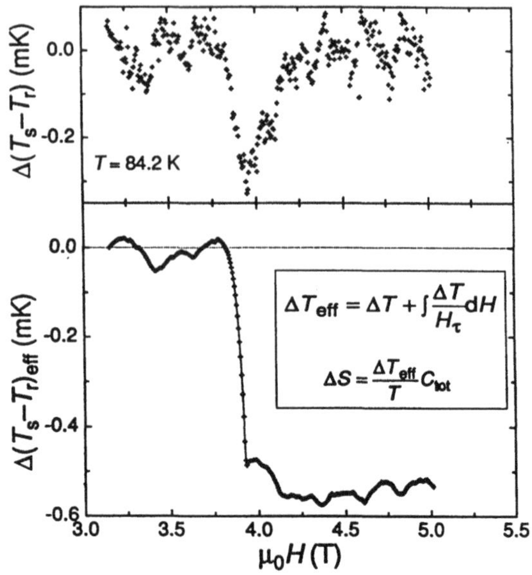

Figs. 3a-b: Cooling effect of the sample ($H//c$, increasing) due the absorption of the latent heat.

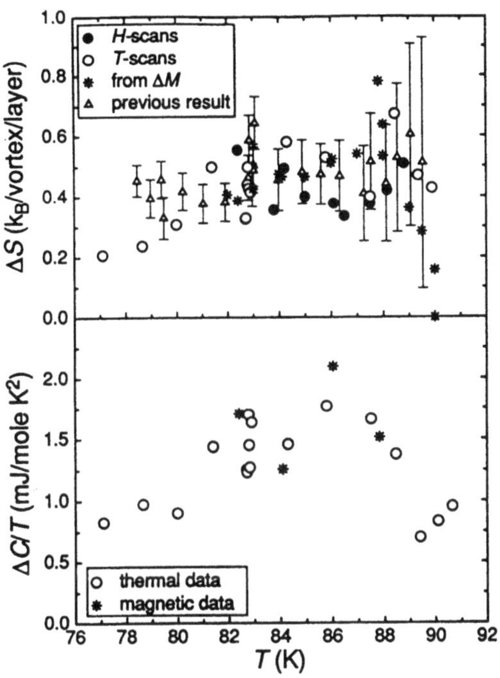

Fig. 4a: Entropies of vortex-lattice melting in $YBa_2Cu_3O_{7-\delta}$ from various experiments, for $H//c$.
Fig. 4b: Difference between C/T of vortex fluid and vortex-solid phases of $YBa_2Cu_3O_{7-\delta}$ for $H//c$.

Fig. 4a suggests that ΔS (typically ≈ 0.5 k_B per vortex per superconducting layer for $H//c$) slightly increases with increasing T. We may compare this value with entropies of melting for conventional solids, where $\Delta S \approx 1$ k_B/molecule (see Table 1). The fact that ΔS for vortex-lattice melting is of the order of 1 k_B per particle and almost constant over a wide range of H (i.e., of vortex densities), may suggest that the vortices in $YBa_2Cu_3O_{7-\delta}$ are essentially dissociated in the vortex-fluid phase.

In Figs. 5 we show $[C(H){-}C(0)]/T$ vs. T, for both $H//c$ and $H \perp c$. A small but clearly detectable latent heat is present also in the $H \perp c$ configuration for $\mu_0 H \geq 5$ T (with an angle $\Theta = 90 \pm 2°$ between H and c). The latent heat *per vortex* is approximately one order of magnitude smaller for $H \perp c$ than for $H//c$, while $H_m(T)$ is steeper by $\gamma \approx 7.8$, which we identify as the square root of the charge-carrier effective-mass anisotropy [10]. The corresponding phase-transition lines are displayed in the inset of Fig. 1.

Element	Melting point (°C)	Entropy of melting (k_B/molecule)
Al	658	1.37
Cd	321	1.03
Cu	1083	1.16
Hg	–39	1.22
H_2	–259	1.02
Ni	1435	0.97
N_2	–210	1.36
O_2	–219	0.98
Pb	327	0.95

Table 1: Entropies of melting for various solids.

2.2 Specific-heat difference between the vortex fluid and the vortex-solid state

From Fig. 5a it is obvious that the (first-order) liquefaction of the vortex lattice is accompanied by an *increase* in the specific heat C/T. This is not unexpected because the free-energy surfaces (as functions of T and H) of the solid and the fluid phases intersect in a line (the melting line), with finite differences in the first (entropy S, magnetization M) and the second (specific heat C/T, $\partial M/\partial T$) derivatives with respect to T and H, respectively. In conventional solids, a first-order melting transition is in most cases accompanied by an *increase* in C/T. We plotted the differences in C/T between the vortex fluid and the vortex-solid phase, as a function of temperature in Fig. 4b.

To check again thermodynamic consistency of the data we compare them with corresponding $M(T)$

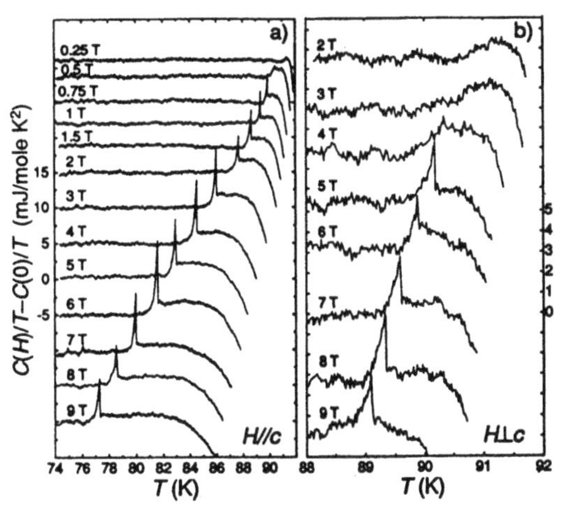

Figs. 5a-b: $[C(H){-}C(0)]/T$ vs. T of untwinned $YBa_2Cu_3O_{7-\delta}$, for both $H//c$ and $H \perp c$.

data measured on the same sample [9], using an appropriate thermodynamic relationship. The Ehrenfest formula $\Delta C/T = -\Delta\partial M/\partial T \cdot dH_m/dT$, where $\Delta\partial M/\partial T$ denotes the slope difference in $M(T)$ between the fluid and the solid phase at constant H, is valid only for pure second-order phase transitions and does not, therefore, apply here. For a first-order transition we must use instead [10]

$$\Delta\frac{C}{T}+\Delta\left(\frac{\partial M}{\partial T}\right)_H\frac{dH_m}{dT}+\frac{d\Delta M}{dT}\frac{dH_m}{dT}+\Delta M\frac{d^2H_m}{dT^2}=0,$$

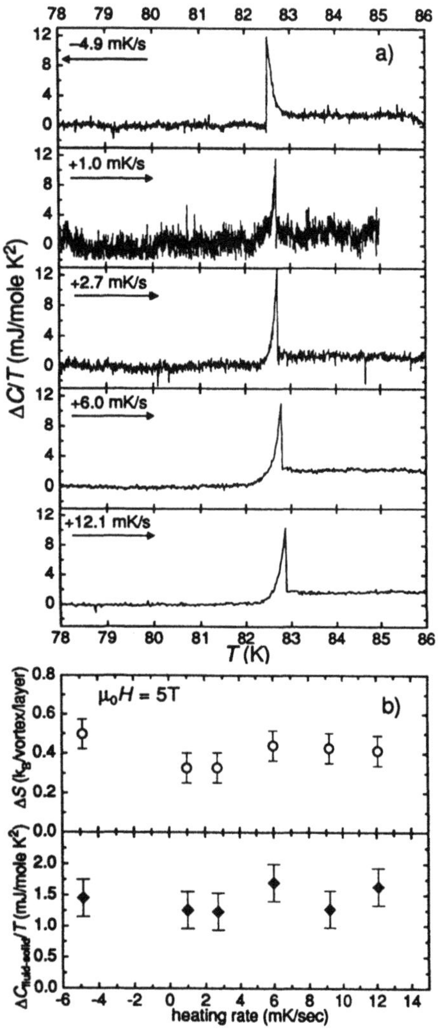

Figs. 6a-b: $[C(H)-C(0)]/T$ vs. T in μ_0H = 5 T //c, with the resulting ΔS and $\Delta C/T$ for various heating rates.

where $\Delta M(T)$ is the T-dependent discontinuity in magnetization at the melting line $H_m(T)$ as reported in Refs. 3 and 6. We obtain in this way an estimate for $\Delta C/T$ that is shown in Fig. 4b. The coincidence of the data suggests that also the $\Delta C/T$ steps reported here represent thermodynamic-equilibrium properties of the vortex matter of $YBa_2Cu_3O_{7-\delta}$.

2.3. Influence of the heating rate

It is known that first-order melting transitions can become glass-like in the presence of disorder. Therefore it is worthwhile studying whether or not the measured latent heat and the specific-heat changes at vortex-lattice melting are influenced by the thermal history of the sample and the speed of the measurement, as it is observed in glass transitions of conventional solids. We have already shown that the thermal features due to the latent heat are identical on heating and on cooling, or on ramping the magnetic field up or down, respectively [3]. We report here further systematic investigations varying the heating rate. Fig. 6a shows $[C(H)-C(0)]/T$ vs. T, taken in μ_0H = 5 T for various heating rates. It is obvious that the specific-heat feature associated with first-order melting does not change quantitatively when the heating rate is varied, although a slight hysteresis of unknown origin (of the order of 0.1 K) is observed. More importantly, both the evaluated ΔS and $\Delta C/T$ values do *not* depend on the rate of heating (Fig. 6b). The fact that the same result is obtained *on cooling* (or on ramping the magnetic field *down* [3]) rules out conventional glassy behaviour of the vortex matter in our crystal.

References

1. A. Schilling et al., Proceedings of the 10th Anniversary of HTSC Workshop, Houston, TX, (U.S.A.), March 12-16, 1996, pp. 349-352.
2. M. Roulin et al., Science **273** (1996) 1210.
3. A. Schilling et al., Nature **382** (1996) 791.
4. M. Roulin et al., J. Low Temp. Phys. **105** (1996) 1099.
5. A. Junod et al., in press in Physica C.
6. U. Welp et al., Phys. Rev. Lett. **76** (1996) 4809.
7. A. Schilling and O. Jeandupeux, Phys. Rev. B **52** (1995) 9714.
8. A. Schilling et al., Phys. Rev. B **46** (1992) 14253.
9. U. Welp et al., unpublished.
10. A. Schilling et al., submitted to Phys. Rev.

ELSEVIER

Physica C 282–287 (1997) 331–334

PHYSICA C

Properties of the vortex liquid in the classical and quantum regime

P.H. Kes, M.H. Theunissen and B. Becker

[a]Kamerlingh Onnes Laboratory, Leiden University, P.O. Box 9506, 2300 RA Leiden, The Netherlands.

The properties of vortex matter in thin films of amorphous NbGe is investigated in regimes that have not been studied before. By designing a device of alternating strong and weak pinning stripes we were able to measure the shear viscosity of the 2D flux line liquid above the 2D melting line. Good agreement is observed with the predictions of the theory of Nelson and Halperin. By studying the properties of an 18 nm thick film down to very low temperatures we entered the regime where quantum fluctuations influence the spectrum of vortex vibrations. It is seen that melting at $T = 0$ takes place well below the transition to the normal state at $B_{c2}(0)$. From resistivity measurements it follows that thermally activated creep in the liquid regime at relatively high temperatures crosses over to quantum creep at low temperatures. The crossover temperature as well as typical results for the energy barriers and Euclidean action are presented.

1. Introduction

Now the appearance of a 2D flux line liquid just below the upper critical field $B_{c2}(T)$ is well established down to 0.4 T_c [1], we want to focus our attention to (i) the temperature and field dependence of the shear viscosity $\eta(T, B)$ in the vortex liquid, (ii) the properties of the melting line at very low temperatures, i.e. below 0.2 $T_c \approx 500$ mK and (iii) the crossover from thermally activated creep to quantum creep in the vortex liquid regime.

2. Shear viscosity of the 2D vortex liquid

In a double layer of 500 nm amorphous NbGe ($T_c = 2.68$ K) and 50 nm NbN ($T_c = 11.4$) 300 parallel channels of width w were etched through the NbN top layer leaving weak pinning channels of 310 nm thick a-NbGe, see inset of Fig. 1. Several samples were prepared with different w ranging between 0.24 μm and 1.19 μm. Applying a sufficiently large current density j_s through this device in a perpendicular field B creates flow of vortex matter. Below the melting line flow starts when the plastic shear stress is surpassed, i.e. when $F_s = j_s B \approx 0.1 c_{66}/w$, where c_{66} is the shear modulus of the vortex lattice. Measured as a function of B commensurate–incommensurate transitions of the vortex solid with respect to the channel width causes oscillations in $F_s(B)$ from which w could be accurately determined [2]. In the vortex liquid phase, $c_{66} = 0$, the oscillations disappear and the flow velocity $\vec{v}$ is controlled by the Bardeen-Stephen friction γ_f and the shear viscosity η according to the hydrodynamic equation [3]:

$$-\gamma_f \vec{v} + \eta \nabla^2_{perp} \vec{v} + \vec{j} \times \vec{B} = 0 \qquad (1)$$

The velocity profile decays exponentially at the edges in a length scale $\delta = (\eta/\gamma)^{\frac{1}{2}}$ which together with w determines the flux flow resistivity ρ_f from

$$\rho_f = \rho_f^0 [1 - (2\delta/w) tanh(w/2\delta)] \qquad (2)$$

where $\rho_f^0 = B^2/\gamma_f$.

The crossover to the liquid phase is marked by a steep increase of the ac resistivity ρ_{ac}. It turnes out that both temperature and/or field at which this happens, depend on the channel width. We interpret this as a finite size effect of the melting transition which occurs when the correlation length $\xi_+(T, B)$, i.e. the average distance between free dislocation in the liquid near the melting line [4], equals $w + b_0 = w_{eff}$($b_0 = a_0\sqrt{3}/2$ and a_0 is the vortex spacing). This allows us to check the field-temperature dependence of ξ_+ which should behave like [4]

$$\xi_+ = \xi_{+0} exp[b'(T_m/(T_m(w) - T_m))^{\nu}] \qquad (3)$$

0921-4534/97/$17.00
PII S0921-4534(97)00285-2

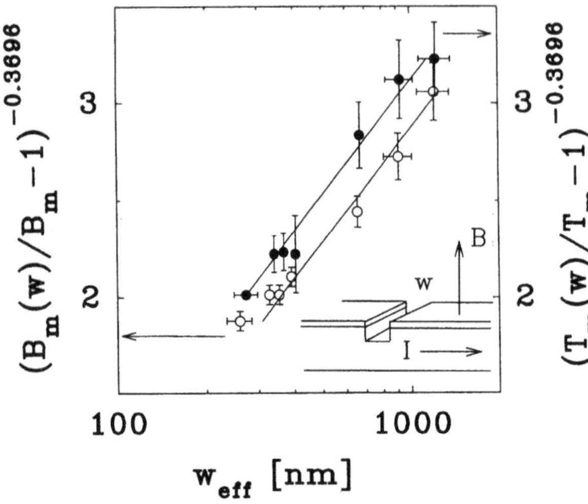

Figure 1. Finite size effect of the melting temperature $T_m(w_{eff})$ at $B = 1.00$ T (•) and the melting field $B_m(w_{eff})$ at $T = 1.70$ K (○). Identifying ξ_+ with the effective channel width w_{eff}, linear fits determine the nonuniversal constants b, b" and the typical length scale ξ_{+0} in Eq.(3). Inset shows a schematic view of one of the channels.

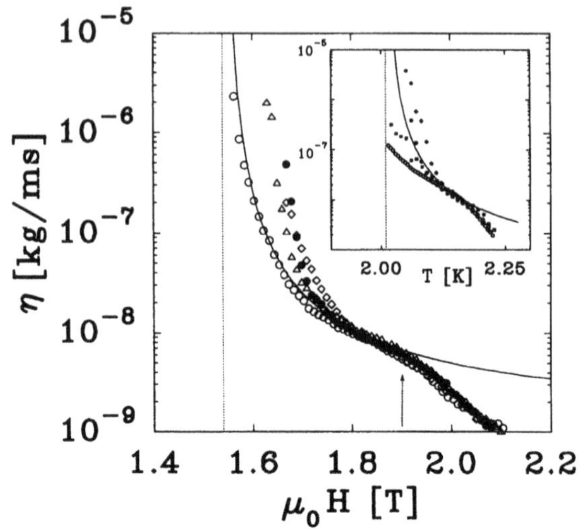

Figure 2. The shear viscosity $\eta(B)$ of the flux-line liquid at $T = 1.70$ K calculated from the resistivity data in Fig. 2 using Eq.(4) for $w =$ 1195 nm (□), 637 nm (○), 373 nm (△), 306 nm (•), and 245 nm (◇). The dashed line indicates the thermodynamic melting field B_m. The curve represents $\eta = \eta_0\xi_+^2/b_0^2$ with $\eta_0 = 2.5 \times 10^{-10}$ kg $m^{-1}s^{-1}$. The arrow is the possible location of the hexatic-to-(isotropic) liquid transition. The inset shows $\eta(T)$ results for the same samples at 1.00 T.

with $\nu = 0.36963$, b' a nonuniversal constant of order unity and $\xi_{+0} \approx b_0$ being the smallest length scale in the liquid. A similar expression holds in $B_m(w)$. T_m is the melting temperature for an infinite sample. This behavior is indeed observed as can be seen in Fig. 1. The lines represent Eq.(3) with $b' = 1.2$ and $\xi_{+0} = 26$ and 34 nm which should be compared to $b_0 = 33$ and 42 nm. By using Eq.(2) the shear viscosity was determined from the $\rho_{ac}(T, B)$ data in the liquid regime. For samples of different widths the results should coincide and follow the prediction $\eta = \eta_0(\xi_+/b_0)^2$ [4]. This is seen in Fig. 2 from which we deduce $\eta_0 \approx 2.5 \times 10^{-10}$ kg/ms yielding a diffusion constant $D_0 \approx 0.3$ mm^2/s and diffusion time $\tau_0 = D_0/a_0^2 \approx 2$ ns. The shoulder at 1.9 T could be a first indication of the transition of an hexatic to isotropic liquid. All these findings agree very well with the Nelson-Halperin theory [4]

3. Quantum corrections to the melting line

The phase diagram at very low temperatures was studied on a 18 nm thick a-NbGe film [5]. In Fig. 3a and 3b the results of ρ_{ac} measurements at zero dc bias current are displayed. The open symbols and dashed lines represent the experimental $B_{c2}(T)$ data and the dirty limit theory for $dB_{c2}/dT = -2.05T/K$ at T_c. The solid symbols denote the melting fields or temperatures determined from the crossover from non-Ohmic to Ohmic behavior. At high temperatures they coincide very well with the classical 2D prediction for dislocation mediated melting of a vortex lattice: $k_B T_m = A c_{66} a_0^2 d/4\pi$ [6] with a renormalization constant $A = 0.65$ (solid line). Below 250 mK the data clearly start to deviate from the classical behavior extrapolating to $B_m(0) = 3.5$ T well below $B_{c2}(0) = 4.2$ T. This deviation is ascribed to the effect of quantum fluctuations (or zero point motion) of the vortex lattice [7]. These quantum effects are predicted to become relevant when the ratio $R_\square/R_Q \approx 0.1$ with $R_\square = \rho_0/d$ and $R_Q = \hbar/e^2 \simeq 4.1k\Omega$ the quantum resistance, and d the thickness of the film. For our films this ratio is only 0.024. Nevertheless quantum effects do appear and the dash-dotted line in Fig. 3 follows from the Lindemann criterion for Gaussian fluctuations [7]. The Lindemann parameter that fits the data best is found to be $c_L \simeq 0.07$ which is not far from the more accepted value of 0.15. A fit with the latter value would require $R_\square/R_Q \simeq 0.1$ which is less likely. Similar results were reported for a 310 nm film of NbGe, where the fluctuations have a 3D character [7].

4. Quantum creep in the quantum liquid

The open squares in Fig. 3b represent the temperatures T^* where the resistivity plotted versus T^{-1} changes from an Arrhenius-like behavior to temperature independent creep, see Fig. 4 [5]. The solid line represents a simple interpolation between classical and quantum creep

$$\rho \simeq \tilde{\rho}_0[exp(-U_p/k_BT) + exp(-S_E/\hbar)] \quad (4)$$

Here $\tilde{\rho}_0$ is a resistivity of order of the normal state resistivity ρ_0, U_p is the energy barrier caused by the random pinning potential and S_E is the Euclidean action controlling tunneling through this barrier. At T^* both exponents are

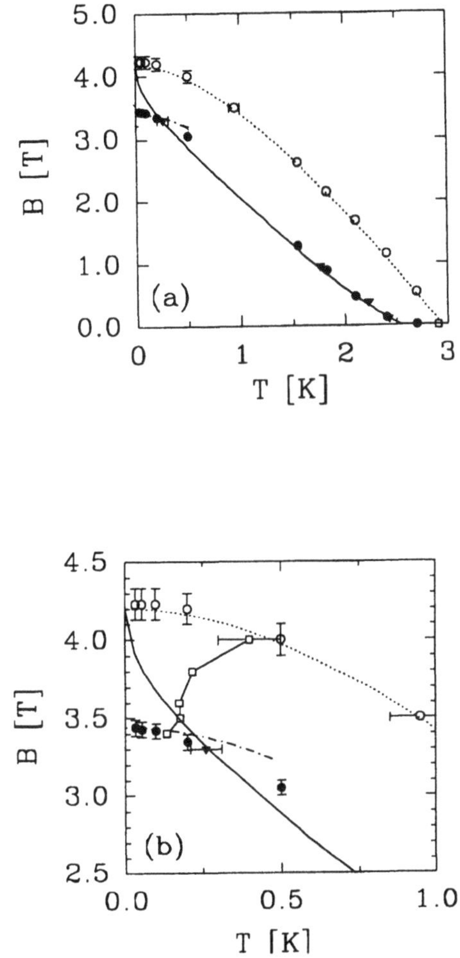

Figure 3. (a) and (b) Phase diagram of 18 nm thick a-NbGe film in perpendicular field. $B_{c2}(T)$ data and dirty limit theory are represented by open circles and dashed line, the classical 2D melting line is given by the solid line, the solid symbols denote experimental melting fields or temperatures. The open squares in (b) denote the crossover temperature between thermally activated creep and quantum creep.

equal. Note that these barriers are observed in

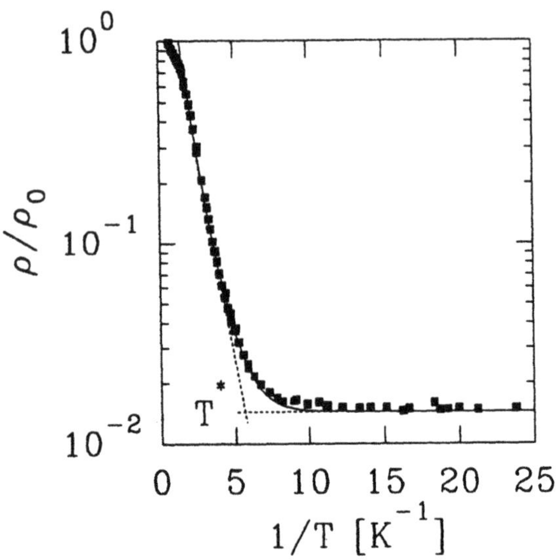

Figure 4. Arrhenius plot of normalized ρ_{ac} of 18 nm a-NbGe film in a perpendicular field of B = 3.5 T. The solid line represents Eq.(4) which describes the interpolation between thermally activated creep and quantum creep. The crossover temperature T^* is defined by the crossing point of the dashed lines.

the (quantum) liquid, presumably because of correlations in the liquid. From a detailed analysis [5] it follows that $U_p = U(0)[1-B/B_{c2}(T)]^2$ with $U(0) = \tilde{a}c_{66}a_0^2d$ and $\tilde{a} \approx 0.29$. Except from a bulk pinning scenario as the reason for the barriers, this result may also be explained by surface effects [8]. The Euclidean action can also be described by a $[1-B/B_{c2}(T)]^2$ dependence with a coefficient of about $140\,\hbar$.

Both the low temperature creep and melting phenomena show the presence of quantum effects in the vortex lattice.

This work has been supported by FOM.

REFERENCES

1. P. Berghuis, A. van der Slot and P.H. Kes, Phys. Rev. Lett **65,** 2583 (1990) and Phys. Rev. B **47,** 262 (1993).
2. M.H. Theunissen, E. van der Drift and P.H. Kes, Phys. Rev. Lett. **77,** 159 (1996). All details can be found in this paper.
3. M.C. Marchetti and D.R. Nelson, Phys. Rev. B **42,** 9938 (1990).
4. D.R. Nelson and B.I. Halperin, Phys. Rev. B **19**, 2457 (1979).
5. M.H. Theunissen, B. Becker and P.H. Kes, to be published.
6. B.A. Huberman and S. Doniach, Phys. Rev. Lett. **43**, 950 (1979).
7. G. Blatter et al., Phys. Rev. B **50**, 13013 (1994).
8. A.E. Koshelev, Physics C**223**, 276 (1994).

ELSEVIER

Physica C 282–287 (1997) 335–338

PHYSICA C

The Universal Scaling Law in Magnetic Phase Diagram of High Temperature Superconductors

K. Kitazawa, J. Shimoyama, H. Ikuta, T. Sasagawa and K. Kishio

Department of Superconductivity, University of Tokyo, Hongo, Bunkyo, Tokyo 113 Japan

An empirical scaling law $\gamma^2 B = F(T/T_C)$ is proposed to successfully describe the three phase boundary lines; flux lattice melting, decoupling, and irreversibility lines in the magnetic phase diagram of the high temperature superconductors. The characteristic field of the three boundary lines changes by orders of magnitude, depending upon the material system; YBCO, LSCO and Bi2212. However, it was found that all the three lines could be scaled by a single material parameter, the electromagnetic anisotropy factor γ^2 of each material. It is proposed that the three phase boundaries can be predicted for any HTSC system provided that its anisotropy factor is given.

1. INTRODUCTION

The anisotropic structure is thought to be one of the essential ingredients for the manifestation of the HTSC. The strong anisotropy, however, has turned out to be the major cause of the weak pinning behavior exhibited by the HTSC under a magnetic field. The anisotropy factor ranges over a wide variety; γ^2 = ca. 25 to 100 for YBCO (of T_C nearly 90K), 200 to 4000 for LSCO, and 3000 to 30000 for Bi2212 as typical examples, depending upon the carrier doping level. In association the characteristic magnetic fields also change over a wide range in the phase diagram.

So far there have been three boundaries experimentally recognized in the magnetic phase diagram, namely the flux lattice melting line (the 1st order phase transition line): B_{melt}-T, the magnetically induced 2D-3D crossover line where vortices are thought to undergo FLL inter-layer decoupling: $B_{2D\text{-}3D}$-T, and the irreversibility line where the magnetic hysteresis sets in: B_{irr}-T.

In the case of YBCO, it is not clear whether there is any decoupling line observable. For LSCO, only the irreversibility line is known. In the most anisotropic Bi2212, all the three boundaries have been extensively studied. Starting from the examination on Bi2212, we claim in this paper that all the three boundary lines can be scaled by a single material parameter, the anisotropy factor γ^2, irrespective of differences in material system or in composition.

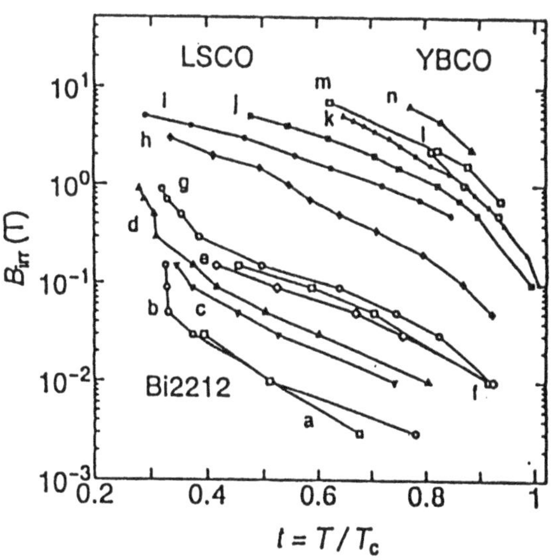

Fig.1 The irreversibility lines B_{irr} (H//c) for single crystals of Bi2212, LSCO and YBCO at various reduced temperatures.

PII S0921-4534(97)00258-X

2. THE UNIVERSAL SCALING LAW

The irreversibility field curves B_{irr}-T for Bi2212, LSCO and YBCO fall over a wide range as shown in Fig.1, depending upon the specific material system and the carrier density[1]. But if the results are plotted as $\log B_{irr}$-$\log\gamma^2$ for each reduced temperature as shown in Fig.2, it is noticed that a single straight line can be drawn at each T/T_C through the results of all the three compounds and of the various carrier densities in each compound[1]. Since the slope of the straight line can be set -1 as in the figure, we obtain an empirical scaling law approximated as[2]

$$B_{irr}[\mathrm{T}] \sim 3.34 \times 10^3 \gamma^{-2}(1-T/T_c)^{1.13} \quad (1).$$

Since this scaling law holds for the three materials of which the anisotropy factor ranges over more than 3 orders of magnitude, one might assume that it applies to all the HTSC compounds.

The FLL decoupling field $B_{2D\text{-}3D}$ of the Bi2212 system has been determined from the so-called "2nd peak" in the magnetization hysteresis curve and found to be nearly independent of temperature. Although the 2D-3D decoupling line at the low temperatures has not been reported in YBCO, very recently Junod et al. has observed the melting line up to 16T by specific heat measurements[3] and Nishizaki et al. observed reentrant peak effect in the magnetization hysteresis curve near and above 10T[4]. We tentatively assume that $B_{2D\text{-}3D}$, if it is present, should be slightly above 16 T for the fully oxygenated YBCO of the anisotropy factor $\gamma^2 \sim 28$. Fig.3 shows the relationship of thus estimated $B_{2D\text{-}3D}$ value with the anisotropy factor for Bi2212 and YBCO. Although the results are not very conclusive, one may still draw a straight line in the figure which has the slope -1. Namely,

$$B_{2D\text{-}3D}\,[\mathrm{T}] \sim 1.0 \times 10^{-16}\, \gamma^{-2} s^{-2} \quad (2)$$

where s is the inter-layer separation of the conductive planes; 1.54×10^{-9}m for Bi2212 and

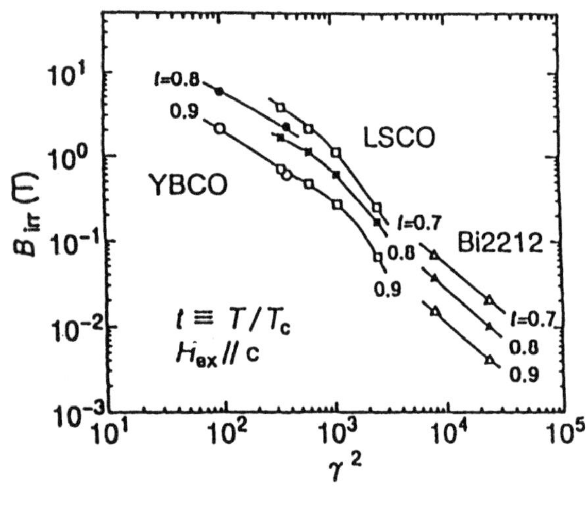

Fig.2 The irreversibility field B_{irr} as a function of the anisotropy factor ($\gamma^2 = \rho_c/\rho_{ab}$) at various reduced temperatures for single crystals of Bi2212, LSCO and YBCO.

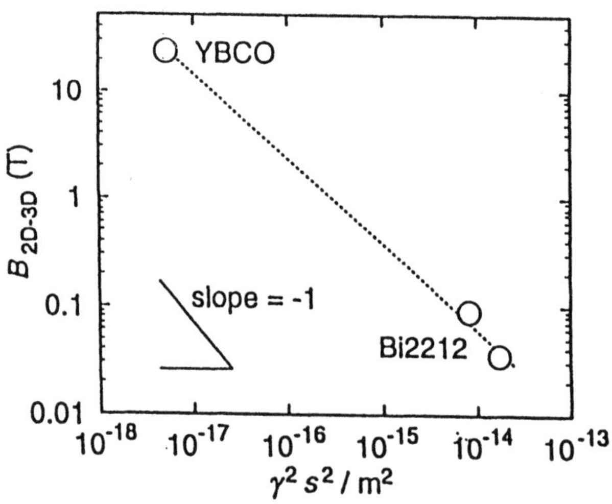

Fig.3 2D-3D decoupling field vs. product of anisotropy factor γ^2 and the square of inter-layer spacing s^2 for Bi2212 and YBCO.

0.43×10^{-9}m for YBCO. The theoretical model for the inter-layer decoupling criterion[5]; $B_{2D-3D} = \Phi_0 \gamma^{-2} s^{-2}$, predicts the coefficient in eq(2) to be 2.07×10^{-15}[T/m^2] and hence the agreement is reasonable.

Recently both the values of B_{2D-3D} and B_{melt} have been measured on the same Bi2212 samples with various carrier doping levels[6]. With the aid of eq.(2) one can obtain the relationship between $\log B_{melt}$ and $\log \gamma^2$ as shown in Fig.4. It is apparent that the γ^2-scaling law also holds for the results obtained in Bi2212 with different carrier doping levels. This encourages one to attempt extending the scaling law to YBCO with much smaller anisotropy factor. Fig.5 illustrates the relationship between $\gamma^2 B_{melt}$ and $(1-T/T_c)$ both for Bi2212 and YBCO. It is rather amazing that all the results nearly fall on a single line in spite of the great difference in the anisotropy facotor for the two different materials. Therefore, we obtain

$$B_{melt}[\mathrm{T}] \sim 2.7 \times 10^3 \gamma^{-2} (1-T/T_c)^{1.2} \quad (3).$$

Consequently, the γ^2-scaling law can also approximate the melting line as well.

3. DISCUSSION

It is rather amazing that all the characterisitc fields; B_{irr}, B_{2D-3D} and B_{melt} can be universally scaled by a single parameter, the anisotropy factor γ^2. Considering that YBCO and Bi2212 are the materials to exhibit the smallest and the largest anisotropy factors, respectively, we presume that the empirical scaling laws obtained as eqs.(1) to (3) are likely to hold for all the HTSC materials. Namely, once γ^2 is known, one may be able to roughly predict the phase diagram for the HTSC compound. Besides, γ^2 is related with the inter-layer separation s as[1,2]

$$\log \gamma^2 \sim 3.1 \times 10^9 s + 0.56 \quad (s\text{: in m}) \ (4)$$

for the optimally doped typical HTSC compounds.

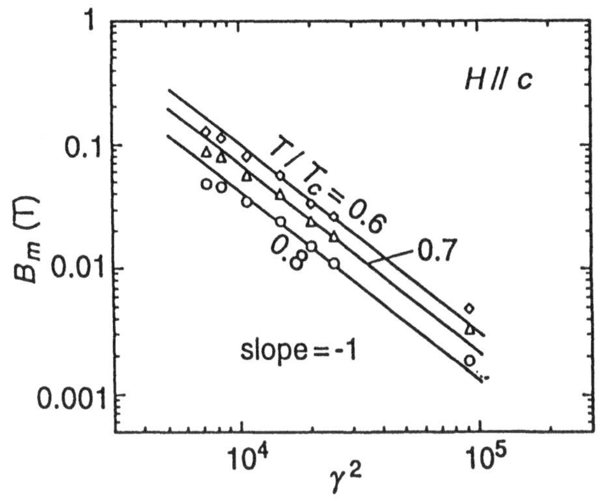

Fig.4 Flux lattice melting field B_{melt} in Bi2212 with various oxygen contents as a function of the anisotropy factor γ^2 at various reduced temperatures.

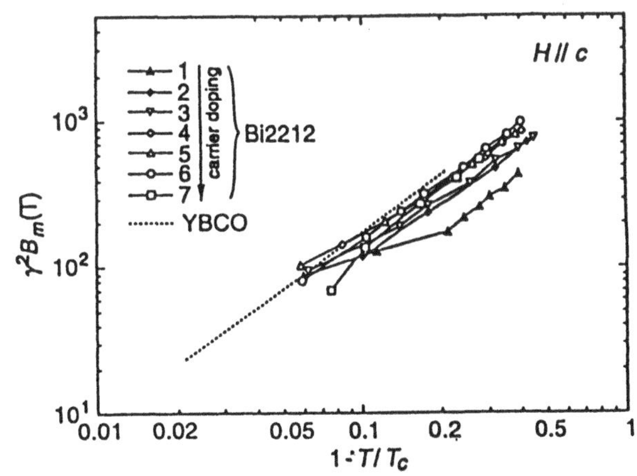

Fig.5 Product of anisotropy factor γ^2 and flux lattice melting field B_{melt} vs. reduced temperature for Bi2212 and YBCO. The results for YBCO are taken from ref.[3].

One may presume that the magnetic phase diagram of a given HTSC material can be estimated solely from the knowledge of the crystal structure at least for the optimally doped compositions.

However, the arguments for B_{irr} should not be accurate because the irreversibility line should be of extrinsic origin, depending upon the character of the pinning centers. Even so the fact that eq(1) applys to all the three material systems seems to suggest the similar nature of the pinning centers in the three pure single crystalline materials.

On the other hand, $B_{2D\text{-}3D}$ and B_{melt} are more of intrinsic nature. Hence the scaling laws proposed above imply that the two characteristic fields are determined essentially by the same mechanism. In this context, one may argue that if $B_{2D\text{-}3D}$ is due to decoupling of vortices across the layers, then B_{melt} should also be related with the decoupling process. That is, B_{melt} may rather be termed as the vortex sublimation field B_{subl} above which vortex lattice lines split up into decoupled pancake gases in addition to the dissappearance of the lattice due to thermal excitation. If one follows this argument, it is more readily understood why the $B_{melt}(B_{subl})$ curve always appears at the temperature where the $B_{2D\text{-}3D}$ curve disappears and that in a continuous manner[7]. Although there have not been extensive results available, we tend to believe that B_{subl} can better represent the first order phase boundary which has been widely termed as B_{melt}.

It should also be noted that one can control the anisotropy factor, significantly deviating from eq(4) by doping. The change in the carrier density significantly varies the anisotropy factor without changing the c-axis lattice parameter so much. Therefore, from the view points of the practical usage, the HTSC materials should be used in the slightly over-doped state to keep the pinning force up to as high a temperature as possible.

Furthermore, it has recently been found that a heavy doping on the "blocking layers" can significantly enhance the irreversibility field as in the case of heavy doping of Pb in Bi2212[8,9] and Re in Hg 1223 compounds[10,11], respectively. In these cases as well, we presume the anisotropy factor is significantly reduced due to the short-circuiting effect of the dopant ions which form octahedrally cordinated conducting paths in the blocking layers.

In conclusion, the three experimental (pseudo-)phase boundaries can be scaled universally in terms of the electromagnetic anisotropy factor γ^2.

REFERENCES

1.K.Kishio et al. **Physica C235-24**, 2775(1994)

2.K.Kitazawa, J.Shimoyama, H.Ikuta and K.Kishio; Proc.10th Anniversary HTS Workshop, Eds.B.Batlogg et al, Mar.12-16(1996) Houston (World Sci.Publ.) p.151

3.A.Junod, M.Roulin, J-Y. Genoud, B.Revaz, A.Erb and E.Walker, **Physica C275**, 245(1997)

4.T.Nishizaki, T.Naito and N.Kobayashi, in this proceeding.

5.V.M.Vinokur, P.H.Kes and A.E.Koshelev, **Physica C168**, 29(1990)

6.J. Shimoyama et al., unpublished results.

7.J.Shimoyama, K.Kishio, H.Ikuta, S.Watauchi, M.Okuya and K.Kitazawa, in this proceeding.

8.I.Chong, Z.Hiroi, M.Izumi, J.Shimoyama, Y.Makayama, K.Kishio, T.Terashima, Y.Bando, M.Takano, Science, in press.

9.J.Shimoyama, K.Kishio et al., Proc. Int. Workshop Crystal Chem. HTSC, (CC-HTSC '97) Mar.6-8(1997) Hefei, China

10.K.Kishio, J.Shimoyama, A.Yoshikawa et al., J.Low.Temp.Phys., **105**, 1359(1996)

11.J.Shimoyama, K.Kishio, A.Yoshikawa, K.Kitazawa, O.Chmaissem and J.D.Jorgensen, in this proceeding.

ELSEVIER

Physica C 282–287 (1997) 339–342

PHYSICA C

Real-time Observation of Vortex Interaction with Pinning Centers Using Electron Waves

Akira Tonomura

Advanced Research Laboratory, Hitachi, Ltd., Hatoyama, Saitama, Japan 350-03

Individual vortices in superconductors were directly observed as spots in a Lorentz electron micrograph using electron waves. The interaction of vortices with artificially produced pinning centers was dynamically investigated with this method.

1. EXPERIMENTAL METHOD

The experiments were carried out with a 300 kV field-emission transmission electron microscope we developed [1]. Vortices were detected as the phase shifts of the electron wave transmitted through the magnetic fields of vortices. These phase shifts are displayed through electron holography process [2] in the contour map of the transmitted wavefront, *i.e.*, the interference micrograph, where projected magnetic lines of force in h/e flux units are directly observed as contour fringes [3].

This method provides quantitative information about magnetic flux with high spatial resolution. However, when we are to observe the dynamics of vortices, Lorentz microscopy is more suitable. The phase shifts are manifested themselves as the intensity variations by defocusing the image [4]. The principle can also be explained classically as shown in Fig. 1: A superconducting thin film, set on a low-temperature stage, is tilted 45° to an incident beam of 300 kV parallel electrons, and an external magnetic field is applied in the horizontal direction. When an electron beam passes through the vortices, their magnetic fields deflect the electron beam. Therefore, when observed in the lower plane, a vortex can be viewed as a pair of black and

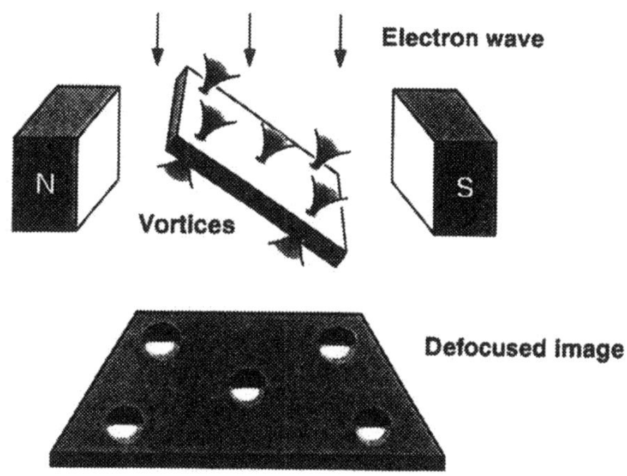

Fig. 1. Principle behind Lorentz microscopy of vortices.

PII S0921-4534(97)00259-1

white spots. In addition, defects in the thin film can be observed simultaneously though the images are blurred due to the defocusing. When vortices are made to move by changing the magnetic field, the dynamic interaction of vortices with defects can be directly observed.

2. OBSERVATION RESULTS

Lorentz microscopy was used to investigate the microscopic pinning effect of vortices in a Nb thin film which had square arrays of artificial defects.

The artificial defects were produced by Ga-ion irradiation [4]. Electron microscopy revealed that the focused ion beam produced a pit 400 Å in diameter surrounded by a region of entangled dislocations, both of which acted as pinning centers. The vortex dynamics when the applied magnetic field changed was observed for two cases where the spacings of the defect lattices were 3.3 μm and 0.83 μm. The two values of the spacings were selected for investigating two cases where the interaction between vortices trapped at defects was negligible and not negligible.

2.1 Sparse distribution of defects

In case of a 3.3 μm spacing, vortices did not form a single-crystalline lattice, but lattice domains. The reason for it is as follows: Since a vortex is trapped at each defect, vortices cannot form a closely packed triangular lattice.

An example of such a vortex configuration is shown in Fig. 2. Weak pinning centers existed in Nb films due to some irregular surface structures and their pinning effect was appreciable at T = 4.5 K. The effect became negligible above 7 K due to the

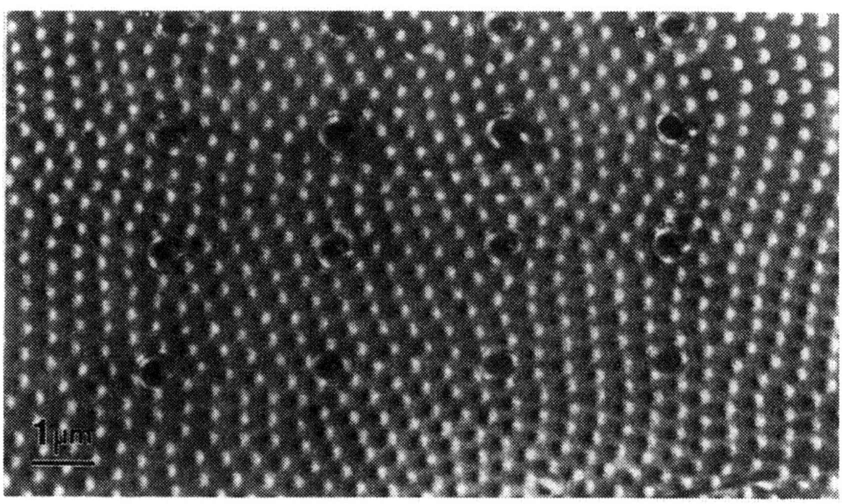

Fig. 2. Vortex configuration near defects (T = 7.5 K, B = 75 gauss).

thermal fluctuation of vortices, and consequently vortices tended to form a closely packed configuration in regions free of the defects. However, due to the existence of the artificial defects where vortices were trapped, the vortex lattices had domain boundaries and dislocations near the defects, as can be seen in the photograph.

When a force was applied to vortices by changing the applied magnetic field, the vortices resisted moving for a while until some vortices began to flow like "avalanches", resulting in "vortex rivers" along the boundaries. When vortices flowed, they became disordered. The flow was not continuous but intermittent. When they stopped flowing, however, the vortices formed a new configuration of lattices, and "vortex rivers" flowed along the new domain boundaries. This process was repeated until vortices reached the equilibrium state.

2.2 Dense distribution of defects

Situations were completely different, when the defect spacing was as short as 0.83 μm: At specific values of magnetic fields, vortices formed regular configurations determined by the square lattice of defects. In this case, vortices did not flow in rivers in this case, but hopped individually, or simultaneously in single lines.

Some Lorentz micrographs showing the configurations of vortices relative to defect positions (red) are shown in Fig. 3. At the matching magnetic field, H_1 (see Fig. 3(b)), all the defects are occupied by vortices without any vacancies or interstices thus forming a rigid square lattice. The peak effect of the critical current macroscopically observed can be explained microscopically: When vortices form a stable and regular lattice, even if a vortex is depinned from one pinning site it cannot find any vacant site to move to. As a result, a bigger force is required to move these vortices.

Regular lattices were formed not only at $H = H_1$, but also at $H = mH_1/n$ (n, m; integer) as in the case of $H = 4H_1$ (see Fig. 3(a)). In this vortex configuration, defect positions forming a square lattice are first occupied by vortices. Then two vortices aligned in the vertical direction are inserted at every interstitial site, and finally an additional vortex is inserted in the middle of two adjacent defects in the vertical direction. Figure 3(c) shows the case at $H = 1/4H_1$. Vortices occupy every fourth site in the horizontal direction thus forming a centered (4 × 2) rectangle lattice. The reason why the pinning force as a whole becomes stronger at the specific values of magnetic fields comes from the fact that vortices form rigid and regular lattices.

When "excess" or "deficient" vortices were produced at magnetic fields different from the specific ones, the vortices could hop by a weaker force just like "electrons" and "holes" in a semiconductor. While when a stronger force was applied to vortices forming a regular lattice, we observed quite different flow of vortices such as a simultaneous movement of vortices along a line.

3. CONCLUSION

We used a newly developed technique to dynamically observe the interaction of individual vortices with pinning centers. We observed various aspects of static configurations and dynamics of vortices near pinning centers. In particular, in case of sparsely distributed defects, vortices were found to flow like intermittent avalanches, forming vortex rivers along the domain boundaries of the vortex lattices. While in case of densely distributed defects, vortices formed rigid and regular lattices at specific values of magnetic fields thus producing peaks in the critical current.

REFERENCES

1. T. Kawasaki, T. Matsuda, J. Endo and A. Tonomura, Jpn. J. Appl. Phys. L508 (1980) 29.
2. A. Tonomura, *Electron Holography*, Springer, Heidelberg, 1994.
3. A. Tonomura, T. Matsuda, J. Endo, T. Arii and K. Mihama, Phys. Rev. Lett. 44 (1980) 1430.
4. K. Harada, T. Matsuda, J. E. Bonevich, M. Igarashi, S. Kondo, G. Pozzi, U. Kawabe and A. Tonomura, Nature 360 (5 November 1992) 51.
5. T. Matsuda, K. Harada, H. Kasai, O. Kamimura and A. Tonomura, Science 271 (8 March 1996) 1393.
6. K. Harada, O. Kamimura, H. Kasai, T. Matsuda, A. Tonomura and V. V. Moshchalkov, Science 274 (15 November 1996) 1167.

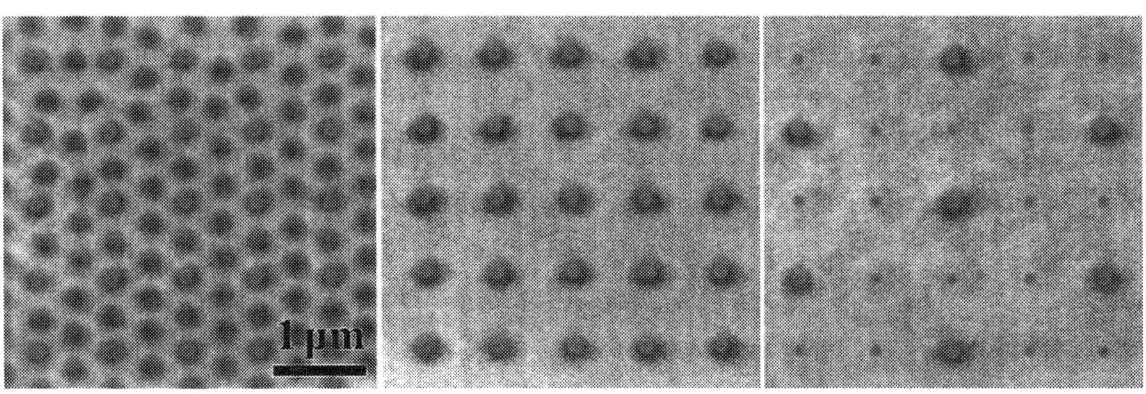

(a) (b) (c)

Fig. 3. Lorentz micrographs of vortices. (a) $H = 4H_1$, (b) $H = H_1$, (c) $H = 1/4H_1$.

PHYSICA C

ELSEVIER

Physica C 282–287 (1997) 343–346

Thick Superconductors in a Perpendicular Magnetic Field

Ernst Helmut Brandt[a]

[a]Max-Planck-Institut für Metallforschung,
D-70506 Stuttgart, Germany

The magnetic moment, ac susceptibility, current density $\mathbf{J}(\mathbf{r},t)$ and magnetic field $\mathbf{B}(\mathbf{r},t)$ in superconductors may be calculated for various geometries and material laws by time-integration of a one- or two-dimensional spatial integral over $\mathbf{J}(\mathbf{r},t)$. This is shown for disks or cylinders of arbitrary thickness in axial magnetic field.

1. INTRODUCTION

The determination of the electromagnetic properties of type-II superconducting materials from experiments requires reliable calculations of current and flux distributions in the superconductor and of the resulting magnetic moment or ac susceptibility. In most cases a continuum description is sufficient, obtained by averaging the microscopic currents and magnetic and electric fields over the Abrikosov vortices. Such continuum calculations may be classified according to which material laws are used and which geometry is considered.

2. MATERIAL LAWS

2.1. Reversible magnetization curve

The static linear and nonlinear magnetic response of a reversible material is characterized by a law $\mathbf{B}=\mathbf{B}(\mathbf{H})$ where $\mathbf{B}$ is the average induction (flux density) in a volume element which sees an applied field $\mathbf{H}$. For isotropic response one may write $\mathbf{B} = B(H)\mathbf{H}/H = \mu(H)\mathbf{H}$ with $\mu(H) = B(H)/H$ and $B(H) = \partial F/\partial B$, where $F(B)$ is the average free energy density. For reversible type-II superconductors one has $B(H) = 0$ at $H < H_{c1}$ (lower critical field) and $B(H) = \mu_0 H$ at $H > H_{c2}$ (upper critical field). For superconductors with large Ginzburg-Landau parameter κ one has $H_{c1} \ll H_{c2} \approx (2\kappa^2/\ln\kappa)H_{c1}$ and $B \approx \mu_0 H$ at $H \gg H_{c1}$.

Below we are mainly interested in the irreversible and time (t) dependent responses. For this one requires an additional material law, e.g., the current–voltage curve. In fact, in irreversible or dynamic problems in many cases (if everywhere in the superconductor $B > 2\mu_0 H_{c1}$) a good approximation is $B = \mu_0 H$, which is equivalent to assuming $H_{c1} = 0$. The magnetic moment $\mathbf{m}$ of the superconductor is then given by $\mathbf{m}(t) = \frac{1}{2}\int \mathbf{r}\times\mathbf{J}(\mathbf{r},t)\,d^3r$. From $\mathbf{m}(t)$ one may derive an ac susceptibility $\chi_{ac}(\omega, H_0)$ if $H_a(t) = H_0\sin\omega t$.

2.2. Bean model

A very useful and often realistic material law is the *static* Bean model [1] which assumes that the vortices do not move as long as $J = |\mathbf{J}|$ is less than a critical current density J_c. If $J > J_c$, the vortices rearrange such that $J \le J_c$ everywhere. In the simple Bean model J_c is assumed to be constant, but in general $J_c(B)$ depends on B. Of course, all material laws will depend also on the temperature T, but this does not affect isothermal calculations.

2.3. Nonlinear resistivity

A more general *dynamic* description uses the current–voltage characteristics $\mathbf{E} = \mathbf{E}(\mathbf{J})$ where $\mathbf{E} = \mathbf{B}\times\mathbf{v}$ is the local electric field, which is caused by the motion of vortices with velocity $\mathbf{v}$ (if annihilation of the vortices is disregarded). In the isotropic case and disregarding the Hall effect ones has $\mathbf{E} = E(J)\mathbf{J}/J$. A useful and realistic model is the power law $E(J) = E_c(J/J_c)^n$ with in general $J_c = J_c(B)$ and $n = n(B)$. Typically one has $n \gg 1$. In the limit $n \to \infty$ this power law yields the Bean model ($E = 0$ for $J < J_c$ and $E = \infty$ for $J > J_c$), while $n = 1$ means an Ohmic conductor with constant and real re-

0921-4534/97/$17.00
PII S0921-4534(97)00261-X

sistivity $\rho = E/J = E_c/J_c$. For $E \propto J^n$ a scaling law applies which states that the susceptibility $\chi_{ac}(\omega, H_0)$ in general depends only on a combination of frequency and amplitude, e.g., on $H_0/\omega^{1/\sigma} + \omega/H_0^\sigma$ with $\sigma = n - 1$. This scaling law nicely shows the transition from $\chi_{ac}(\omega)$ (Ohm, $\sigma = 0$) to $\chi_{ac}(H_0)$ (Bean, $\sigma \to \infty$).

2.4. Linear complex resistivity

A special case of $E(J)$ is $E_0 = \rho_{ac}(\omega) J_0$ where $\rho_{ac} = \rho' + i\rho''$ is a linear complex dispersive resistivity and $H_a(t) = H_0 \sin \omega t$, $E = Re\{E_0 e^{i\omega t}\}$, and $J = Re\{J_0 e^{i\omega t}\}$ are assumed, with $E_0(\mathbf{r})$ and $J_0(\mathbf{r})$ denoting complex amplitudes and Re the real part. So far we are not able to combine the nonlinear real $\rho(J) = E(J)/J$ with the linear complex $\rho_{ac}(\omega)$ into one general theory, but both resistivities may be used in the calculations below. The Meissner state with complete expulsion of magnetic flux is then realized, in spite of our assumption $H_{c1} = 0$, either at low H_a if $E(J) \propto J^n$ with $n \gg 1$, or, if $E/J = \rho_{ac}(\omega)$, at short times [for suddenly switched on $H_a(t)$] or high frequencies [for periodic $H_a(t)$].

3. GEOMETRIES

Various specimen shapes and orientations of the applied field H_a are used in experiments. The general three-dimensional (3D) problem in many cases may be approximated by a two- or even one-dimensional problem. If not specified else, we assume here that no current is fed in by contacts.

3.1. Parallel field, 1D and 2D

Infinitely long superconductors in parallel field H_a exhibit no demagnetizing effects. For infinite slabs or circular cylinders the problem is 1D, and for general cross-section (e.g. rectangular bars) it is 2D. The linear or nonlinear response is now obtained by solving a 1D or 2D linear or nonlinear diffusion equation. The resulting expressions are often used to evaluate experiments even when this longitudinal geometry does not apply.

3.2. Perpendicular field, 1D

Thin infinite strips and circular disks of width $2a$ and thickness $d = 2b \ll 2a$ in perpendicular $H_a \| y$ are described by 1D integral equations for the sheet current $\mathbf{J}_s = \int_{-b}^{b} \mathbf{J}(\mathbf{r}, t) dy$ which describe 1D nonlocal diffusion of J_s. Analytic solutions were obtained [3, 4] for the static Bean model extended to these 1D transverse geometries. The presence of a transport current in addition to the circulating screening current in a Bean strip was considered in [5]. These thin-limit results qualitatively differ from the longitudinal geometry. In particular, the sheet current does not vanish in the flux-free central zone and the perpendicular field $B_z(\mathbf{r})$ exhibits a logarithmic infinity at the specimen edge and a vertical slope at the penetrating flux front, while the longitudinal Bean model predicts constant slope of $B_z(\mathbf{r})$. The extension of the limit $d \to 0$ to finite thickness given below, can explain these features

The dynamic linear [6] and nonlinear [2] response, flux penetration [7] and flux creep [8] were computed by time integration of a 1D spatial integral for strips and disks. The linear susceptibility $\chi_{ac}(\omega)$ was calculated for general complex resistivity [9] $\rho_{ac}(\omega)$. These linear and nonlinear results recently were generalized to thin rings [2] and to the linear and nonlinear screening of a thin superconductor disk between two coils [10].

3.3. Perpendicular field, 2D

Thin flat superconductors of arbitrary shape in transverse (or arbitrarily oriented) H_a require the time integration of a 2D integral equation for the sheet current which describes nonlocal diffusion. Such first-principle computations were performed for thin squares and rectangles [11]. From the measured flux density $B_y(x, z)$ at the surface of a superconductor the sheet current may be obtained by inversion methods [12].

3.4. Bars and disks of arbitrary thickness

Infinite bars and circular disks of finite thickness in a perpendicular (or axial) field yield a $2D$ integral equation which still may be time-integrated on a Personal Computer. This method [13] applies to arbitrary cross-section (e.g. to spheres and elliptic bars) but here we consider only rectangular cross-section $2a \times 2b$ which includes, e.g., thin disks ($b \ll a$) and long cylinders ($b \gg a$). For brevity we consider here only cylindrical symmetry. The similar results for strips bars, and slabs are given in [13].

4. THICK DISKS IN AXIAL FIELD

The vector potential $\mathbf{A}(\mathbf{r},t) = \hat{\varphi}A(r,y,t)$ caused by the current density $\mathbf{J}(\mathbf{r},t) = \hat{\varphi}J(r,y,t)$ circulating in a disk of radius a and thickness $d = 2b$ and by the applied field $B_a(t) = \mu_0 H_a(t)$ (along y) is [13]

$$A(\mathbf{r}) = -\mu_0 \int_0^a dr' \int_0^b dy'\, Q_{\rm cyl}(\mathbf{r},\mathbf{r}')\, J(\mathbf{r}') - \frac{r}{2} B_a \quad (1)$$

with $\mathbf{r} = (r,y)$ and $\mathbf{r}' = (r',y')$. The kernel

$$Q_{\rm cyl}(\mathbf{r},\mathbf{r}') = f(r,r',y-y') + f(r,r',y+y'),$$
$$f(r,r',\eta) = \int_0^\pi \frac{(r'/2\pi)\cos\varphi\, d\varphi}{\sqrt{\eta^2 + r^2 + r'^2 + 2rr'\cos\varphi}} \quad (2)$$

was obtained by integrating the Green function of the 3D Laplace equation, $1/(4\pi|\mathbf{r}_3 - \mathbf{r}'_3|)$ with $\mathbf{r}_3 = (x,y,z)$, over the angle $\varphi = \arctan(z/x)$. If desired, $f(r,r',\eta)$ may be expressed in terms of elliptic integrals, but we found it more convenient to evaluate the φ integral numerically. With $\mathbf{B} = \nabla \times \mathbf{A}$ the induction law $\dot{\mathbf{B}} = -\nabla \times \mathbf{E}$ takes the form $\dot{A} = -E(J)$ since $\mathbf{E}(\mathbf{r},t) = \hat{\varphi}E(r,y,t)$. To fix ideas one may use $E(J) = E_c(J/J_c)^n$. The equation of motion for $J(r,y,t)$ is then

$$E[J(\mathbf{r},t)] = \mu_0 \int d^2r' Q_{\rm cyl}(\mathbf{r},\mathbf{r}')\dot{J}(\mathbf{r}',t) + \frac{r'}{2}\dot{B}_a. \quad (3)$$

This implicit equation may be inverted,

$$\dot{J}(\mathbf{r},t) = \mu_0^{-1} \int_0^a dr' \int_0^b dy'\, Q_{\rm cyl}^{-1}(\mathbf{r},\mathbf{r}') \Big(E - \frac{r'}{2}\dot{B}_a\Big) \quad (4)$$

with $E = E[J(\mathbf{r}',t)] = E_c J(r',y',t)^n / J_c^n$. The inverse kernel $Q_{\rm cyl}^{-1}(\mathbf{r},\mathbf{r}')$ may be computed by tabulating $Q_{\rm cyl}(\mathbf{r},\mathbf{r}')$ (2) on a discrete grid $\mathbf{r}_i$, $\mathbf{r}_j$ and inverting this matrix.

Figures 1 and 2 show the magnetic field lines during flux penetration into thick superconducting disks with aspect ratios $b/a = 1$ and $b/a = 0.25$, calculated by time integration of Eq. (4) with an exponent $n = 51$ (almost Bean limit). The depicted applied field values are $H_a/H_p = 0.1, 0.2, 0.4$, and 0.8, where

$$H_p = J_c b \ln\left[\frac{a}{b} + \Big(1 + \frac{a^2}{b^2}\Big)^{1/2}\right] \quad (5)$$

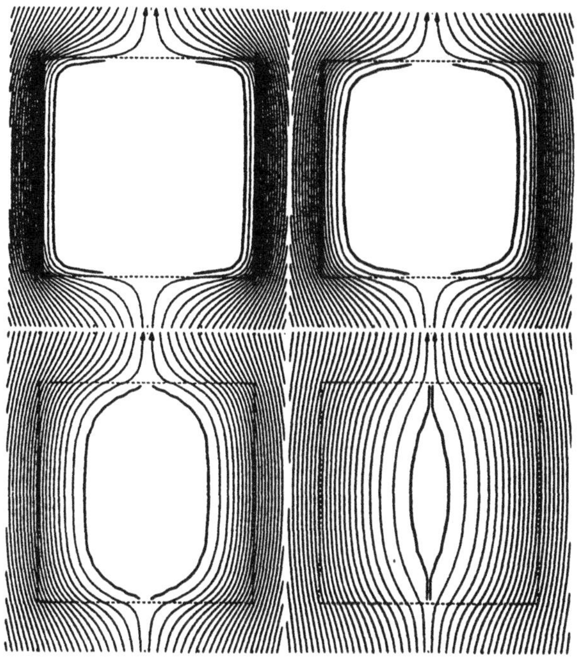

Figure 1. Field lines of the magnetic field penetrating into a thick Bean disk with $b/a = 1$ at applied fields $H_a/H_p = 0.1, 0.2, 0.4, 0.8$. The bold line marks the flux front, which separates the central core where $B = J = 0$ from the outer region with circulating current $J = J_c$.

is the field of full penetration [15] for thick disks and cylinders. These field lines look very similar to those of thick strips [13] with same b/a. Note that the lense-shaped field- and current-free central zone becomes isolated for H_a close to H_p.

Figure 3 shows the profiles of the perpendicular induction at the surface, $B_y(r,|y| = b)$, and at the middle plane, $B_y(r, y = 0)$, of a Bean disk with $b/a = 0.25$ in increasing applied field. For this thick disk the profiles at the surface still look similar to the profiles of a long Bean cylinder in parallel field (straight parallel lines), but the profiles on the central plane look more like in a thin disk in perpendicular field [3] (with steep front and with a cusp at the edge).

The resulting magnetization curves $m(H_a)$ fall between the limits for long cylinders [1,2] and thin disks [3,2]. When normalized to unity initial slope $m'(0) = 1$ and unity saturation value $m(H_a \geq$

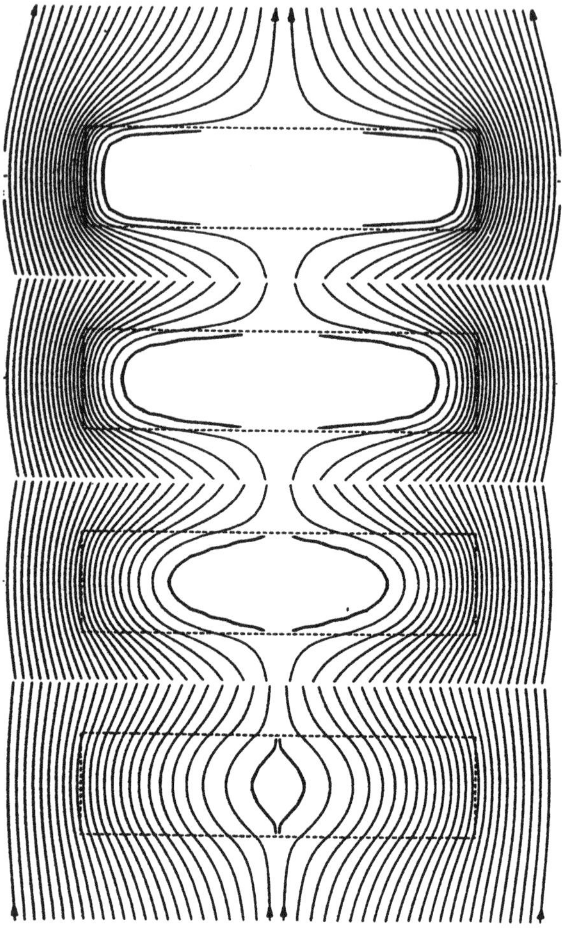

Figure 2. As Fig. 1 but for $b/a = 0.25$.

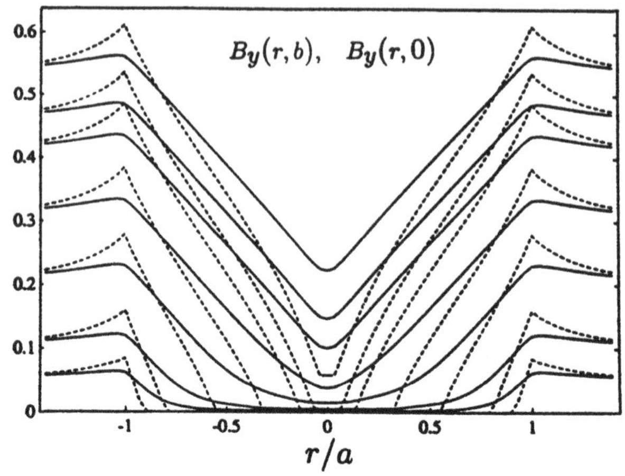

Figure 3. The profiles of the perpendicular magnetic field B_y at the surface $y = \pm b$ (solid lines) and in the central plane $y = 0$ (dashed lines) of a thick disk with $b/a = 0.25$ (like Fig. 2) during flux penetration. The depicted applied field values are (from bottom to top) $H_a/H_p = 0.1$, 0.2, 0.4, 0.6, 0.8, 0.9, 1.05. B is in units $\mu_0 J_c a$.

$H_p) = 1$ the magnetic moments $m(H_a)$ for all aspect ratios b/a differ only little.

REFERENCES

1. C.P. Bean, Rev. Mod. Phys. **36**, 31 (1964).
2. E.H. Brandt, Phys. Rev. B (submitted).
3. P.N. Mikheenko and Yu.E. Kuzovlev, Physica C **204**, 229 (1993).
4. E.H. Brandt, M. Indenbom, and A. Forkl, Europhys. Lett. **22**, 735 (1993).
5. E. Zeldov et al., Phys. Rev. B **49**, 9802 (1994); E.H. Brandt and M. Indenbom, Phys. Rev. B **48**, 12893 (1993).
6. E.H. Brandt, Phys. Rev. Lett. **71**, 2821 (1993); Phys. Rev. B **49**, 9024 (1994); **50**, 4034 (1994).
7. Th. Schuster, Phys. Rev. Lett. **73**, 1424 (1994); Phys. Rev. B **50**, 16684 (1994).
8. A. Gurevich and E.H. Brandt, Phys. Rev. Lett. **73**, 178 (1994); Phys. Rev. B (submitted); E.H. Brandt and A. Gurevich, Phys. Rev. Lett. **76**, 1723 (1996); E.H. Brandt, Phys. Rev. Lett. **76**, 4030 (1996).
9. J. Kötzler et al., Phys. Rev. B **50**, 3384 (1994); E.H. Brandt, Phys. Rev. B **50**, 13833 (1994).
10. J. Gilchrist and E.H. Brandt, Phys. Rev. B **54**, 3530 (1996).
11. E.H. Brandt, Phys. Rev. Lett. **74**, 3025 (1995); Phys. Rev. B **52**, 15442 (1995).
12. W. Xing et al., J. Appl. Phys. **76**, 4244 (1994); R. Wijngaarden et al., Phys. Rev. B **54**, 6742 (1996); E.H. Brandt, Phys. Rev. B **46**, 8628 (1992); A.E. Pashitski, A. Gurevich, et al. Science (Jan. 1997, in print); Ch. Joos et al., Phys. Rev. B (submitted).
13. E.H. Brandt, Phys. Rev. B **54**, 4246 (1996).
14. E.H. Brandt, Rep. Prog. Phys. **58**, 1465 (1995).
15. A. Forkl, Physica Scripta **T49**, 148 (1993).

ELSEVIER

Physica C 282–287 (1997) 347–350

PHYSICA C

Negative-μ vortex dynamics in high-T_c superconducting films

R. Griessen[a], A.F.Th. Hoekstra[a], H.H. Wen[a,b], G. Doornbos[a] and H.G. Schnack[a]

[a] Institute COMPAS and Faculty of Physics and Astronomy, Vrije Universiteit, De Boelelaan 1081, 1081 HV, Amsterdam, The Netherlands

[b] National Laboratory for Superconductivity, Institute of Physics, P.O.Box 603, Beijing 100080, P.R.China

Using a thermally activated vortex motion model with an activation energy $U(j)=(U_c/\mu)[(j_c/j)^{\mu}-1]$ in which U_c, μ and j_c are field and temperature dependent, we analyse both flux-creep data and current-voltage characteristics on the same footing. This analysis reveals the occurrence of negative-μ vortex dynamics in highly anisotropic superconductors such as $Tl_2Ba_2CaCu_2O_8$ or $Bi_2Sr_2CaCu_2O_8$ above a characteristic magnetic field. This is attributed to a dimensional cross-over and plastic creep.

1. INTRODUCTION

The small coherence length, the presence of weak and dense point pinning centres and the layered structure of high-T_c superconductors are the origin of the intricate nature of the mixed state phase diagram of these compounds[1]. They are also responsible for giant relaxation rates of the magnetic moment of the high-T_c superconductors. Numerous studies of flux creep have been carried out on bulk samples and single crystals. In general one finds that the data can reasonably well be explained within a model of thermally activated vortices collectively pinned by randomly distributed weak pinning centers. The activation energy U associated with this vortex motion is predicted by collective flux pinning (CFP) theories to diverge at small currents as $U(j) \propto (j_c/j)^{\mu}$ with μ and j_c depending on the dimensionality and the size of the vortex bundles under consideration (in 3D, $\mu = 1/7$, 3/2 and 7/9 for single vortices, small bundles and large bundles, respectively, while in 2D, $\mu = 0$, 7/4, 13/16 and 1/2 for single vortices, small, medium and large bundles, repectively). A similar divergence of U(j) at small j is also predicted by vortex-glass (VG) theories. However, the exponent μ is restricted to $0 \leq \mu \leq 1$. Below the glass temperature T_g, the system exhibits a glassy behaviour with $U(j) \propto (j_c/j)^{\mu}$ at large length scales (i.e. at relatively small currents) and a critical behaviour at small length scales (i.e. large currents) characterized by power-law current-voltage curves $E \propto j^{\alpha}$. Experimentally from current-voltage curve measurements near T_g one finds that $\alpha \approx 3$. In a log-log plot of E vs. j curves for various temperatures at a given magnetic field B, the $T=T_g$ line is therefore a straight line, the $T< T_g$ lines have a downward curvature and the curves at $T > T_g$ exhibit an upward curvature until an ohmic regime is reached at very small currents[2].

The purpose of this paper is twofold: i) to show that the slope and curvature of lnE vs lnj curves are closely related to flux creep data obtained from magnetic moment relaxation measurements and ii) that in strongly layered superconductors and films pinning is too strong at high magnetic fields to allow vortices to creep collectively. Instead, dislocation-mediated (plastic) creep is expected to be predominant. As far as we know, no direct evidence has been found for plastic creep except in a small part of the phase diagram (T > 81 K and B < 6 T) of a $YBa_2Cu_3O_7$ single crystal[3]. In a detailed investigation of the vortex dynamics of high quality $Tl_2Ba_2CaCu_2O_8$ thin films we found, however, that the B-T plane below the irreversibility line is separated into two regions by a $\mu = 0$ line with a remarkable shape. At low fields where $\mu > 0$, vortex motion occurs via elastic creep while plastic creep[4] with $\mu < 0$ is made possible at higher fields by a 3D to 2D transition around 0.7 T.

0921-4534/97/$17.00
PII S0921-4534(97)00262-1

2. RELATION BETWEEN FLUX CREEP AND CURRENT-VOLTAGE CHARACTERISTICS

The best method to determine the flux creep relaxation rate in films[5] is to measure the dynamic relaxation rate $Q \equiv d\ln j_s/d\ln(dB/dt)$ where j_s is the superconducting current flowing in a sample when the external field B is swept at a rate dB/dt. This can be done by sweeping the magnetic field up and down at a given rate around a certain field and measuring the width ΔM of the irreversible magnetic moment curve. For ring shaped samples, the current density j_s can be determined directly from ΔM by means of $j_s = \Delta M/(2\pi R^2 w d)$, with R the outer radius, w the width and d the thickness of the ring. Since $j_s \propto \Delta M$ and $E \propto dB/dt$ the measured ΔM vs. dB/dt curves can be converted into E vs. j_s curves (or I-V curves).

For a laser ablated $Y_{0.99}Pr_{0.01}Ba_2Cu_3O_{7-\delta}$ film we present in Fig. 1 typical data for the temperature dependence of the current density $j_s(T,B)$ measured at given fields B with dB/dt = 40 mT/s and of the corresponding relaxation rates Q. The Q vs. T curves exhibit a plateau over a large temperature interval at low fields as predicted by Malozemoff and Fisher[6] on the basis of a collective pinning model. At higher temperatures there is, however, a sharp increase which cannot be understood within a CFP model.

To proceed with the analysis and discussion of the measured data in Fig.1 we parametrize the activation energy $U(j_s,T,B)$ for thermally activated vortex motion in the form[7]

$$U(j_s,T,B) = \frac{U_c(T,B)}{\mu(T,B)}\left[\left(\frac{j_c(T,B)}{j_s(T,B)}\right)^{\mu(T,B)} - 1\right] \quad (1)$$

where both the exponent μ and the characteristic pinning energy U_c are explicitly allowed to vary with temperature and applied magnetic field B. This form is chosen because it contains the weak collective pinning models (μ = 1/7, 3/2, 7/9 etc.[1]), the Kim-Anderson model (μ = -1) and the Zeldov logarithmic model (i.e. $U(j) = U_c \ln(j_c/j)$ which corresponds to $\mu = 0$) as special cases. From the general creep relation[8]

$$U(j_s,T,B) = k_B T \ln\left(\frac{2 v_0 B}{R\, dB/dt}\right) \quad (2)$$

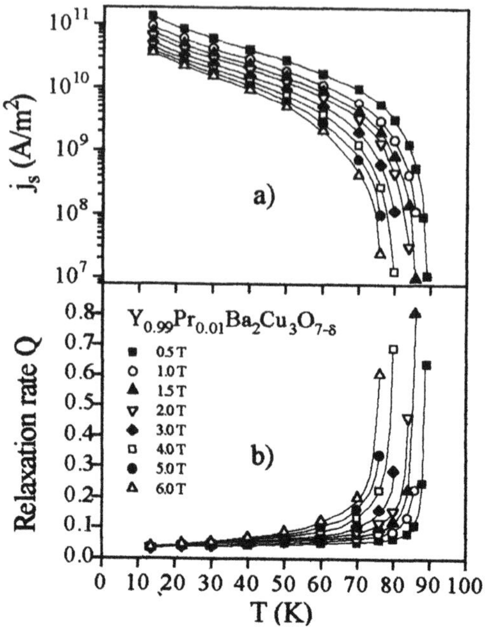

Figure 1. Temperature dependence of a) superconducting current density j_s and b) dynamic relaxation rate Q for various magnetic fields. The sample is a $Y_{0.99}Pr_{0.01}Ba_2Cu_3O_{7-\delta}$ film of 200 nm thickness laser ablated on $SrTiO_3$.

for a thin film ring of radius R, Eq.1 and the definition of the dynamic relaxation Q, one finds that

$$\frac{T}{Q(T,B)} = \frac{U_c(T,B)}{k_B} + \mu(T,B)\, C\, T \quad (3)$$

where $C = \ln[(2v_0B)/(R dB/dt)]$. The parameter C at a given field B can be determined from the low temperature slope of the $\ln j_s$ vs. T curves in Fig. 1a by means of $C = (T/Q)(-d\ln j_s/dT)$ after correction for quantum creep[8]. Here we find C = 20. In Fig.2a we show the temperature dependence of T/Q determined from the data in Fig. 1. At low fields T/Q varies linearly with T up to 60K as expected on the basis of Eq.3 with U_c and μ independent of temperature. As in all existing theories U_c is found to vary only weakly with temperature (e.g. $U_c \propto 1-(T/T_c)^4$ for δl-pinning in the single vortex regime[9]) we shall take $U_c(T) = U_c(0)$ for all data in Fig.2 and calculate the exponent μ by means of

$$\mu(T,B) = \frac{1}{C\, Q(T,B)} - \frac{U_c(0)}{C\, k_B\, T} \quad (4)$$

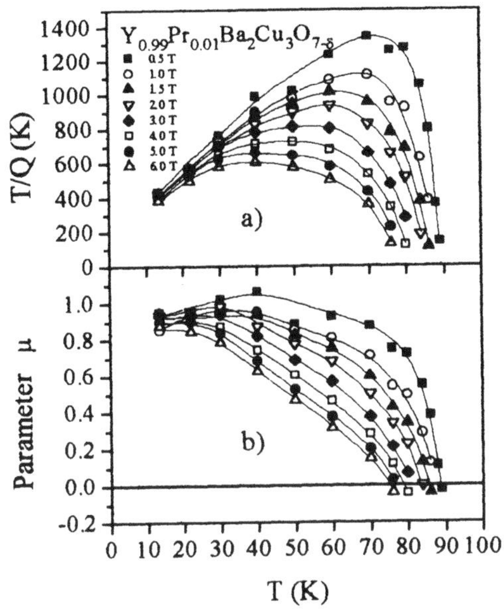

Figure 2. Temperature dependence of a) T/Q and b) the parameter μ obtained from the data in Fig. 1. The low temperature data (T < 10 K) have been omitted since they are influenced by quantum creep.

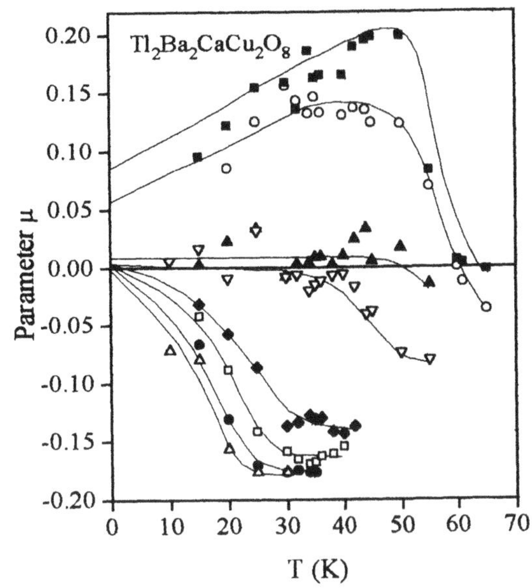

Figure 3. Temperature dependence of μ for a 0.7 T, $Tl_2Ba_2CaCu_2O_8$ film at magnetic fields 0.3 T, 0.4 T, 1.0 T, 2.0 T, 3.0 T, 4.0 T and 6.0 T (top to bottom).

The curves exhibit a plateau near $\mu = 0.85$ at low temperatures and at temperatures T* near 80 K, $\mu = 0$. Since $U_c(0) \approx 200$ K, $\mu(T^*) = 0$ in Eq.4 implies that $Q(T^*,B) \approx 0.4$. This leads to the interesting conclusion that the slope of the lnE vs. $\ln j_s$ curves should be 2.5 which is close to the value 2.95 found by Koch et al.[2] and Wöltgens et al.[10] by means of direct current-voltage curve measurements. This implies that our T* can in fact be associated with the glass temperature T_g. This is also consistent with the fact that the curvature of lnE vs. $\ln j_s$ curves following from Eq.1 is approximately given by $-\mu Q^2$. Thus $\mu = 0$ implies a linear lnE vs. $\ln j_s$ curve, or equivalently a power law $E \propto j^\alpha$ with $\alpha = kT_g/U_c(0)$. The identification of T* (defined by $\mu(T^*) = 0$) and T_g has been also tested with other data. Equation 4 provides us with a valuable tool to determine μ even at $T \ll T_g$ where the curvature of the lnE vs. $\ln j_s$ curves is very small since Q is typically of the order of a few percent. Our method is, therefore, complementary to direct current voltage measurements.

3. NEGATIVE μ VORTEX DYNAMICS

The data shown in Fig. 2b suggest that $\mu < 0$ at temperatures above T_g. This is consistent with the fact that the lnE vs. $\ln j_s$ curves[2] exhibit an upward curvature above T_g. It also suggests that at very high magnetic fields μ might become negative. In order to test the possibility of negative μ we investigated a $Tl_2Ba_2CaCu_2O_8$ thin film of 200 nm thickness deposited on a (100) $LaAlO_3$ substrate by means of DC magnetron sputtering. The μ-values determined from j_s and Q measurements at temperatures between 10 K and T_c in magnetic fields up to 7 T are shown in Fig. 3. At low fields (B < 0.4 T) the behaviour is similar to that of $Y_{0.99}Pr_{0.01}Ba_2Cu_3O_{7-\delta}$ shown in Fig. 2b. However, at $B \cong 0.7$ T, μ remains essentially zero over the whole temperature range. The $\mu > 0$ region corresponds to a 3D vortex glass while the $\mu < 0$ region corresponds to a 2D vortex glass with a vanishing glass temperature since at B> 0.7 T we find $\mu \approx 0$ at $T \approx 0$ K (see Fig. 3). From 0 to 50 K the 3D to 2D transition occurs at B = 0.7 T.

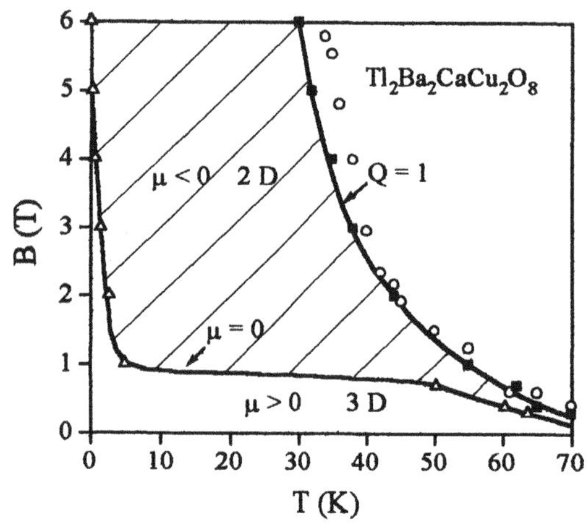

Figure 4. Phase diagram for a $Tl_2Ba_2CaCu_2O_8$ film determined on the basis of the μ-data in Fig. 3 and dynamic relaxation data. The Q = 1 -line (solid squares) defined in the text turns out to be similar to the irreversibility line for a current criterion $j_s = 10^7$ A/m^2 (open circles). Above the Q = 1 -line the E vs. j_s curves are ohmic.

At B = 0.4 T, $\mu = 0$ is reached around 58 K. At that temperature $Q \approx 0.38$ which implies that $E \propto j_s^{2.63}$. The exponent $\alpha = 2.63$ is again very similar to the typical value ($\alpha = 2.95$) found for the $YBa_2Cu_3O_7$ system. In the $\mu < 0$ region the activation energy $U(j) \propto 1-(j/j_c)^\beta$ with $\beta = |\mu|$. Such a dependence is analogous to that found for the thermally activated motion of dislocations in a crystal lattice[11], since the activation energy U_{dis} changes with the applied strain f (which is analogous to the externally applied current j_s in the vortex system) as $[1-(f/f_c)^a]^b$. The exponents a and b satisfy $0 < a \leq 1$ and $1 \leq b \leq 2$. The μ value found in our experiment (-0.2) at high fields near the irreversibility line implies a = 1/5 and b = 1. For dislocations in a solid this would correspond to the case of pinning sites with a homogeneous core and long-range interaction.

4. CONCLUSIONS

From the above discussion we conclude that i) the parametrisation of current and relaxation data within a thermally activated model with $U(j) = (U_c/\mu)[(j_c/j)^\mu - 1]$ allows us to treat flux-creep and current-voltage characteristic within the same framework. Relaxation data are especially valuable for $T < T_g$ in the high current limit.
ii) in 3D systems such as $YBa_2Cu_3O_7$ there is a small region above T_g where $\mu < 0$
iii) in 2D systems such as $Tl_2Ba_2CaCu_2O_8$ or $Bi_2Sr_2CaCu_2O_8$[12], $\mu < 0$ above a certain characteristic field and flux creep is dominated by the plastic motion of vortex dislocations.

ACKNOWLEDGMENTS

We are grateful to V.B. Geshkenbein, P.H. Kes and A. Koshelev for valuable discussions.

REFERENCES

[1] G. Blatter, M.V. Feigel'man, V.B. Geshkenbein, A.I. Larkin, V.M. Vinokur, Rev.Mod.Phys. 66, 1125 (1994), and references therein.
[2] R.H. Koch, V. Foglietti, W.J. Gallagher, G. Koren, A. Gupta, M.P.A. Fisher, Phys.Rev.Lett. 63, 1511 (1989).
[3] Y. Abulafia, A. Shaulov, Y. Wolfus, R. Prozorov, L. Burlachkov, Y. Yeshurun, D. Majer, E. Zeldov, H. Wuhl, V.B. Geshkenbein, V.M. Vinokur, Phys.Rev.Lett. 77, 1596 (1996).
[4] C.J. van der Beek, P.H. Kes, Phys.Rev.B 43, 13032 (1991).
[5] H.H.Wen, H.G.Schnack, R.Griessen, B.Dam, J.Rector, Physica C 241, 353 (1995).
[6] A.P. Malozemoff and M.P.A. Fisher, Phys.Rev.B 42 (1990) 6784.
[7] A.P. Malozemoff, Physica C 185-189, 264 (1991). In contrast to other forms such as $U(j) = U_c(j_c/j)^\mu$ or $U(j) = U_c\, sgn(\mu)\, [(j_c/j)^\mu - 1]$ used by others, the form in Eq.1 does not lead to unphysical results when $j \rightarrow j_c$ or $\mu \rightarrow 0$.
[8] H.G. Schnack, R. Griessen, J.G. Lensink, H.H. Wen, Phys.Rev.B 48, 13178 (1993).
[9] R. Griessen, Hai-Hu Wen, A.J.J. van Dalen, B. Dam, J. Rector, H.G. Schnack, Phys.Rev.Lett. 72, 1910 (1994).
[10] P.J.M. Wöltgens, C. Dekker, Phys.Rev.B 48, 16826 (1993).
[11] U.F. Kocks, A.S. Argon, M.F. Ashby, "Thermodynamics and kinetics of slip", Prog.Mater.Sci. 19 (1975), Chapts. IV, V.
[12] A.J.J. van Dalen, R. Griessen, M.R. Koblischka, Physica C 259 (1996) 157.

ELSEVIER

Physica C 282–287 (1997) 351–354

PHYSICA C

Non-solitary determination of vortex-glass transition temperatures in $YBa_2Cu_3O_{7-\delta}$ superconducting thin films

Hai-hu Wen[a], Xi-xian Yao[b], Rui-lan Wang[a], Hong-cheng Li[a], Shu-quan Guo[a], Zhong-xian Zhao[a]

[a] National Laboratory for Superconductivity, Institute of Physics and Center for Condensed Matter Physics, P.O.Box 603, Beijing 100080, P.R.China

[b] Physics Department, Nanjing University, Nanjing 210093, P.R.China

By accomplishing DC transport and magnetic sweep measurement on high quality $YBa_2Cu_3O_{7-\delta}$ superconducting thin films, we obtained E-j characteristic curves in a very large dynamic region (with E varied from 10 to 10^{-7} V/m). The data unambiguously show that, both positive curvature and negative curvature coexist in one double logarithmic curve log(E) versus log(j) at a fixed temperature and magnetic field. This indicates that the so-called vortex-glass transition temperature T_g as defined by the crossover temperature from negative curvature to positive curvature depends on the accessible electric field window. Further analyses show that our transport data can be approximately interpreted by the conventional flux-creep model. Finally we propose a model which shows that the flux motion can be changed from collective-creep or glassy behavior to the so-called thermally-assisted-flux-flow (TAFF) behavior by lowering the current.

1. INTRODUCTION

One interesting but heavily disputed argument in the recent research of high-T_c superconductors is the possible existence of the vortex-glass (VG) superconductivity in the mixed state [1,2]. According to Fisher [1] and Fisher et al.[2], in the mixed state, there may exist a continuing second order phase transition characterized by a specific temperature T_g. At T_g, the correlation length of the pairing field (l_{VG}) diverges, which leads to a true superconductivity (the VG solid state) and a zero linear resistivity $R = \left.\frac{V}{I}\right|_{I\to 0} \equiv 0$, in sharp contrast with the conventional flux creep model[3] which predicts a finite linear resistivity (except possibly at T = 0 K) due to thermal activation. The focus of this dispute lies on the dissipation manner predicted by different models in low temperature region ($T < T_g$). For example, at $T < T_g$, the VG theory envisions a power law dissipation relation $V \propto I^n$ (with n > 0) in the large current region and an exponential relation $\rho = \frac{E}{j} \propto \exp\left[-\left(\frac{j_T(T)}{j}\right)^{\mu}\right]$ in the small current limit, where $j_T(T)$ is a characteristic current density and $0 < \mu < 1$ for a 3D vortex-glass. Thus, in the frame of VG, the double logarithmic curve log(E) versus log(j) should always show a negative curvature at $T < T_g$. After the proposal of this possible new vortex state and vortex transition, many experimental investigations[4-9] have been made towards the verification of them. However, most of these measurements [4,6-9] were carried out in a DC transport way which probes E-j curves in a relatively narrow range of electric field (typically 10^{-4} V/m < E < 10^2 V/m). For example, Koch et al.[4] measured the I-V curves by means of DC transport and observed the negative curvature below a certain temperature. This result was regarded as the first experimental evidence for the existence of a second order vortex-glass to vortex-liquid phase transition. However, Coppersmith et al.[10] strongly argued that the conclusion of Koch et al. was drawn by artificially taking a limited window for the electric field. With

PII S0921-4534(97)00265-7

sufficiently low electric field, even at $T < T_g$, the upturn (positive curvature) of the log(E) vs. log(j) curve is anticipated by the flux-creep model in the small current limit, i.e., both the positive curvature and the negative curvature should coexist on one single isothermal double logarithmic curve log(E) vs. log(j). In addition, Coppersmith et al.[10] showed that, the scaling of all the E-j curves near T_g as done by Koch et al.[4] based on the VG ansatz can also be achieved by using the data generated by the conventional flux-creep model with typical values z = 13.5 and ν = 0.6. Thus, in order to clarify this issue, it is strongly needed to detect the E-j curves in a large dynamic region. In this paper, we present E-j curves measured by means of DC transport and magnetic sweep on two thin films with essentially the same qualities. In the magnetic sweep measurement, the electric field E went down to about 8×10^{-8} V/m, which is about 4 orders of magnitude lower than that in the conventional DC transport measurement.

2. EXPERIMENTAL

The samples were $YBa_2Cu_3O_{7-\delta}$ superconducting thin films cut from one original piece with the thickness around 0.1μm and sizes of 15×15 mm^2. Details about the fabrication of this thin film was described previously [11]. After the deposition, the film was cut into 10 smaller pieces with almost the same qualities (T_c = 89K, ΔT_c = 1K) as checked by measuring the temperature dependent magnetization under a field of 5 Oe. Two neighboring pieces with essentially the same qualities were chosen for our further transport and magnetic investigations. For the transport measurement, the film was patterned into a bridge with dimensions of 300 μm (length) × 30 μm (width). We use standard 4-probe method to measure the E-j curves.

The magnetic sweep measurement on a sample with dimensions of 3×3 mm^2 were accomplished on an Oxford Vibrating Sample Magnetometer (VSM 3001, 8T) with the resolution of about 10^{-6} emu. The field was continuously swept in a sweeping mode and the temperature stabilization is better than 0.1 K.

3. RESULTS

In Fig.1, presented are our central results: the E-j curves measured on the microbridge by transport method (top-left) and on the superconducting disc by the magnetic sweeping method (bottom-right). From the transport data, it is clear that, above a certain temperature (in this case 83K), a S-shaped curve appears, accompanied by a slope = dlog(I) / dlog(V) = 1 in the small current limit; below 83K, however, only the negative curvature appears. Indeed, according to the VG theory,

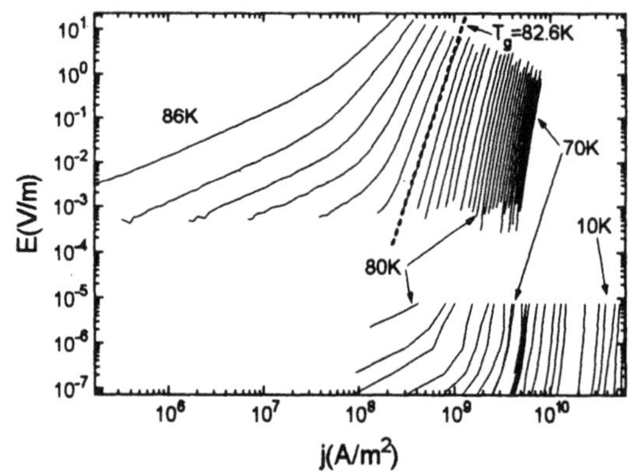

Figure 1. The E-j characteristic curves for the sputtered film determined from the transport measurement (top-left) and the magnetic measurement (bottom-right). For the transport data, E and j are simply determined via $E=V/l$ and $j=I/S$, respectively, with l the length and S the cross section area of the bridge at magnetic field B_e=0.8T and temperatures ranging from 70K to 86K (from right to left) with increments of 0.5K. For the data of magnetic sweep measurement, E is determined via $E=(R/2)(dB_e/dt)= 7.5\times10^{-4}$ (dB_e/dt), j is determined by multiplying ΔM (the width of the Magnetization hysteresis loops) by a constant (3.63×10^{11}) since we have assumed that $j(R) \propto \Delta M$. the magnetization hysteresis loops are measured around B_e=0.8T with 5 field sweep rates: 10, 3, 1, 0.3 0.1mT/s at temperatures (from right to left): 10, 15, 20, 25, 30, 40, 50, 53, 56, 60, 63, 64, 65, 66, 67, 68, 69.5, 70, 72, 73, 74, 76, 78, 79 and 80K, respectively. The positive curvatures can be easily observed at temperatures above 70K. The slope = 1 in the small current limit is quite peculiar since it looks very similar to the I-V curves obtained in the transport measurement at higher temperatures.

at $T < T_g$, only the negative curvature is observable, which means that, the upturn (or positive curvature) will never be observed with a whatever small current. To check this we performed the magnetic sweep measurement. During the magnetic field sweeping process, since the relaxation term L(dI/dt) is much smaller than the electromotive force [12], the electric field established at the perimeter of a circular disc is

$E(R) = \frac{R\times(dB_e/dt)}{2}$, with R the radius of the specimen, dB_e/dt the field sweeping rate, L the self-inductance. On the other hand, for a very thin superconducting disc, the total magnetic moment is mainly contributed by the current flowing in the region close to the perimeter, thus even there is a non-uniform distribution of current density as shown by van der Beek et al. [13], it is still a good estimation that $j(R) \propto \Delta M$, where ΔM is the width of the magnetization hysteresis loops (MHLs). Therefore, to measure the MHLs in different sweep rate, one essentially obtains the E(R)-j(R) curves.

4. DISCUSSION

From Fig.1, one can easily find that the temperature corresponding to the crossover from negative curvature to positive curvature of the curve log(E) vs. log(j) measured in the magnetic sweeping process is lower than that determined from the transport data. For example, the crossover temperature found in the transport measurement is $T_{cr} \approx 83$K, however, from the magnetic sweep, it is found that $T_{cr} \leq 73$ K. This indicates that the crossover temperature T_{cr} actually depends on the accessible electric field window. If we take the crossover temperature T_{cr} as the vortex-glass transition temperature T_g (the same definition as used by Koch et al.[4]), T_g cannot be solitarily determined. It is evident that even below T_g, the S-shaped curve appears again in lower temperature region with the peculiar slope = dlog(j)/dlog(E) = 1 in the small current limit, being qualitatively consistent with the conventional flux-creep model[3] and in sharp contrast with the VG model [1,2]. The reappearance of the positive curvature in lower temperature region and lower electric field region may manifest that the up-turn in the small current limit is unavoidable (perhaps for all temperatures). However, this may not postulate the non-existence of flux motion with collective-creep or glassy behavior in high current region.

To check whether our data can be explained in terms of flux-creep model, we compared our data with those generated by Coppersmith et al.[10] basing on flux-creep model. In their paper, many I-V curves were generated by using the flux-creep model. With these I-V curves, they tried to scale I to $|T-T_g|^{2\nu}$ and V to $I\times|T-T_g|^{\nu(z-1)}$, the same scaling method as used by Koch et al. [4]. In order to have a good scaling they found that z = 13.5 and ν = 0.6. In Fig.2, we present the same scaling on our transport data (31 isotherms), we found that z = 14 and ν = 0.7 can produce quite nice scaling on our data. The values found here z = 14, ν = 0.7 are very close to those found by Coppersmith et at.[10], which may indicate that our transport data can be approximately interpreted by the flux-creep model.

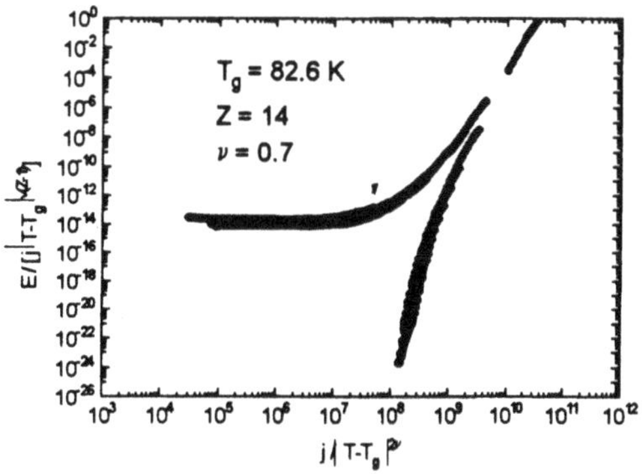

Figure 2. Scaling of 31 isothermal E-j curves measured in the DC transport experiment. The same scaling as used by Koch et al. [4] was used here. The critical values z = 14 and ν = 0.7 found here are very close to z = 13.5 and ν = 0.6 as found by Coppersmith et al. [10] in scaling the I-V curves generated by using the flux-creep model.

According to collective-creep theory[14] and VG theory[1,2], the increase of the effective activation energy U with lowering the current density j is mainly due to the growth of the optimal hopping length l_{opt}. For a sample with very large thickness (e.g., bulk sample) the optimal hopping length l_{opt} will hardly reach the thickness ($\geq 10000\ \xi$ with $\xi \approx 1$nm, the coherence length) and thus the activation energy will go to very high value when j is approaching 0. However, for a thin film sample, it has a quite small thickness ($\sim 100\ \xi$), thus when the optimal hopping length l_{opt} reaches the thickness, the effective barrier for flux hopping should have a cutoff instead of going to infinity. In this current region, the back hopping of flux lines is anticipated. Because of this back hopping, as described by Kes et al.[15], one can observe the ohmic behavior in the low current region. If this qualitative model is correct, for a thin film sample the positive curvature in the small current region will be unavoidable (perhaps for all temperatures). However, this does not postulate the non-existence of the glassy motion of vortices in high current region. What we want to say is that it seems tricky for one to draw the conclusion about the existence of a second order phase transition just from the E-j curves within limited window of electric field. By measuring the E-j curves with $YBa_2Cu_3O_{7-\delta}$ thin films of different thickness, Dekker et al. [16] found that the scaling for the second order phase transition will not work anymore for the thinner films ($\leq$ 50nm) since the correlation length along the c-axis is larger than the thickness of the film. Actually, in our point of view, all the scaling done for thin film samples suffer from a limited window of electric field. More detailed results and analysis will be published elsewhere.

Although we have shown that the second order " vortex glass – vortex liquid " phase transition cannot be illustrated just by simply doing the transport measurement in thin film samples, however, it by no means rules out the possible existence of such transitions. More elegant experimental data are strongly needed for clarifying this issue.

5. CONCLUSION

By measuring isothermal E-j characteristic curves in a very large dynamic region (with E varied from 10 to 10^{-7} V/m) on high quality $YBa_2Cu_3O_{7-\delta}$ superconducting thin films, we found that the positive curvature and negative curvature coexist on the double logarithmic curve log(E) vs log(j) at a fixed temperature and magnetic field and that the so-called vortex glass temperature depends on the accessible electric field. Our transport data can be approximately interpreted by the conventional flux-creep model. Finally we propose a model which shows that the flux motion can be glassy or collective-creep behavior in the high current region, while by lowering current density j, the so-called TAFF process will set in because of the limited thickness of the sample.

ACKNOWLEDGMENTS

We thank Prof. R.Griessen, Drs. G.Doornbos, K.Heeck, J.Rector, B.Dam at Free University at Amsterdam for valuable discussions.

REFERENCES

[1]. M.P.A. Fisher, Phys. Rev. Lett.62(1989) 1415.

[2]. D.S.Fisher, M.P.A.Fisher, and D.A. Huse, Phys. Rev. B43(1991)130.

[3]. P.W. Anderson, and Y.B. Kim, Rev. Mod. Phys. 36(1964)39.

[4]. R.H.Koch, V.Foglietti, W.J.Gallagher, G.Koren, A.Gupta, and M.P.A.Fisher, Phys. Rev. Lett.63(1989)1511. See also R.H.Koch, V.Foglietti, W.J.Gallagher, and M.P.A.Fisher, Phys. Rev. Lett.63(1989)2586.

[5]. P.L.Gammel, L.F.Schneemeyer, and D.J. Bishop, Phys. Rev. Lett.66(1991)953.

[6].T.K.Worthington, E.Olsson, C.S.Nichols, T.M.Shaw, and D.R.Clarke, Phys.Rev.B43 (1991)10538.

[7]. H.Safar, P.L.Gammel, D.J.Bishop, D.B.Mitzi, and A.Kapitulnik, Phys. Rev. Lett.68 (1992)2672.

[8]. C.Dekker, W.Eidelloth, R.H.Koch, Phys. Rev. Lett.68(1992)3373. C.Dekker, P.J.M.Woltgens, R.H.Koch, B.W.Hussey, and A.Gupta, Phys. Rev. Lett.69(1992)2717.

[9]. M.Friesen, J.Deak, L.F.Hou, and M.Mcelfrsh, Phys.Rev.B54(1996)3525.

[10].S.N.Coppersmith, M. Inui, and P.B.Littlewood, Phys. Rev. Lett.64(1990)2585. See also M.Inui, P.B.Littlewood, and S.N.Coppersmith, Phys. Rev. Lett. 63(1989) 2421.

[11].H.C.Li, H.R.Yi, R.L.Wang, Y. Chen, B.Yin, X.S. Rong, S.F.Cui, D.S.Dai, and L.Li, Appl.Phys.Lett.56(1990)2454.

[12].H.G.Schnack, R.Griessen, J.G.Lensink, C.J. van der Beek and P.H.Kes, Physca C197(1992)337.

[13].C.J. van der Beek, G.J. Nieuwenhuys, P.H. Kes, H.G. Schnack, and R.Griessen, Physica C197(1992)320.

[14].M.V.Feigel'man, V.B.Geshkenbein, A.I.Larkin, and V.M.Vinokur, Phys. Rev. Lett. 63(1989)2303.

[15].P.H.Kes, J.Aarts, J.van den Berg, C.J.van der Beek, and J.A.Mydosh, Supercon.Sci.Tech. 1(1989)242.

[16].C.Dekker, R.H.Koch, B.Oh, and A.Gupta, Physica C185-189(1991)1799.

ELSEVIER

Physica C 282–287 (1997) 355–358

PHYSICA C

The Superconducting Energy Gap and Vortex Lattice Structure in $LuNi_2B_2C$

Y. De Wilde[a], M. Iavarone[a] *, U. Welp[a], V. Metlushko[a], A. E. Koshelev[a], I. Aranson[a] †, G.W. Crabtree[a], P.L. Gammel[b], D.J. Bishop[b], and P.C. Canfield[c]

[a]Material Science Division-Science and Technology Center for Superconductivity, Argonne National Laboratory, Argonne, IL 60439, USA

[b]Bell Laboratories, Lucent Technologies, Murray Hill, NJ 07974, USA

[c]Ames Laboratory and Department of Physics and Astronomy, Iowa State University, Ames, IA 50011, USA

We present Scanning Tunneling Microscope (STM) observations of the superconducting energy gap and vortex lattice (VL) structure of the (001) face of a $LuNi_2B_2C$ single crystal. In zero field, the conductance versus voltage characteristics consistently show a peaked structure that can be fitted with a broadened BCS like density of states, giving a gap value of 2.2 meV and $2\Delta/kT_c = 3.2 \pm 0.1$, implying weak coupling superconductivity. In applied magnetic fields of 0.375 T and 1.5 T, we make the first STM observation of a square VL, with its unit cell rotated 45° with respect to the crystallographic unit cell. The VL structure is analyzed with Ginzburg-Landau theory, including an additional free energy term with four-fold symmetry arising from the tetragonal crystal structure. The square VL is found to be stable if the four-fold free energy term is large enough to cause the ratio of the in-plane upper critical fields $H_{c2}(45°)/H_{c2}(0)$ to be less than 0.97. Measurements of the in-plane upper critical field anisotropy by SQUID magnetometry show a ratio of 0.92 at 4.2 K, consistent with the Ginzburg-Landau analysis.

1. INTRODUCTION

The discovery of superconductivity in the tetragonal RNi_2B_2C (R=Lu, Y, Tm, Er, Ho, Dy) borocarbide family [1–3] stimulated many experiments probing the superconducting order parameter Δ in this new class of materials. In zero field, the gap has previously been measured by point contact tunneling on polycrystals [4] and single crystals [5], by break junctions [6], and by vacuum tunneling STM on thin films [7]. The gap values determined by the two former methods vary widely depending on the type of measurement and, in some cases, on the location in the sample. In contrast, vacuum tunneling STM measurements generally provide gap values which vary remarkably little with both the location on the sample and the value of the tunneling resistance which can be varied over a wide range in a controlled way [7,8]. Vacuum tunneling STM therefore increases substantially the precision of gap measurements. In a field, the spatial distribution of the vortex lines have been investigated on borocarbide samples by neutron scattering and magnetic decoration. Both techniques are sensitive to magnetic contrast on the scale of the London penetration depth λ and are therefore limited to low fields. In $ErNi_2B_2C$ single crystals, these experiments reveal a square vortex lattice (VL) in fields larger than 500 Oe which transforms into an hexagonal VL as the field is reduced [9,10]. In a similar way a square VL has also been observed in YNi_2B_2C and $LuNi_2B_2C$ single crystals [11,12] by neutron scattering. In contrast, STM is sensitive to the local density of states on the much smaller scale of the coherence length ξ and gives access to higher fields where the vortex spacing is smaller

*Permanent address : Dipartimento di Scienze Fisiche, Universitá di Napoli *Federico II*, Napoli, Italy

†Permanent address : Department of Physics, Bar Ilan University, Ramat Gan, Israel

0921-4534/97/$17.00 Elsevier Science B.V.
PII S0921-4534(97)00268-2

than λ. This technique has been successfully applied to vortices in $NbSe_2$[13] and YBCO[14].

In this paper we present STM measurements on the (001) surface of a $LuNi_2B_2C$ single crystal in zero and applied magnetic field at 4.2K. In zero field, we measure a gap value of 2.2 meV. In applied fields of 0.375 T and 1.5 T we make the first STM observations of the VL in this material, finding a square structure [8]. This square VL structure is correlated to an in-plane anisotropy of the upper critical field H_{c2} measured by SQUID magnetometry. The results are discussed within the framework of the Ginzburg-Landau theory.

2. RESULTS AND DISCUSSION

The experiments were performed on a $LuNi_2B_2C$ single crystal grown by a high temperature Ni_2B flux method using high purity elements [15]. The crystal is in the form of a platelet with the crystallographic c-axis perpendicular to the largest plate surface. A SQUID magnetometer was used to measure the critical temperature T_c and the upper critical field H_{c2} of the sample. We find $T_c = 15.8$ K and $H_{c2} = 5.7$ T at 4.5 K for $H \parallel c$. The STM used in the experiments is a home built system operating at 4.2 K in helium exchange gas combined with RHK electronics (RHK-Technology, STM 100). The tip is made from PtIr wire. In Fig. 1a we present a typical conductance vs. voltage normalized spectrum (dI/dV vs. V) obtained at zero field. Similar spectra are obtained with good reproducibility anywhere on the sample surface. To estimate the gap value, we fit the data with a broadened BCS-like density of states including a smearing parameter Γ [16,8]. The calculated curve presented in Fig. 1a corresponds to $\Delta = 2.2$ meV and $\Gamma = 0.4$ meV giving a BCS ratio $2\Delta/kT_c = 3.2 \pm 0.1$. This implies weak coupling superconductivity for $LuNi_2B_2C$.

For imaging the VL in $LuNi_2B_2C$ we fix the bias voltage at 3 mV and record the spatial variation of the conductance over the sample surface. At this bias, the conductance is expected to drop over a vortex core as the peak in the density of states disappears. Fig. 1b shows on a gray-scale

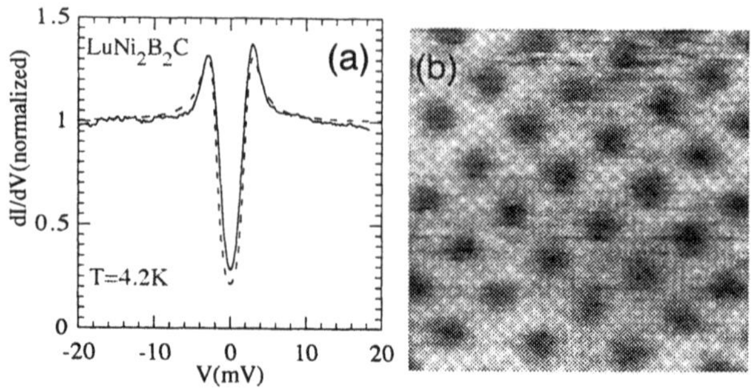

Figure 1. (a) Experimental conductance spectrum normalized to the quadratic normal state background (solid line) and calculated conductance spectrum (dashed line) using a smeared BCS density of states for which $\Delta = 2.2$ meV and $\Gamma = 0.4$ meV. (b) 2050Å × 2050Å conductance image (averaged data) showing the square vortex lattice obtained in a 1.5 T magnetic field at 4.2 K.

the recording of the spatial variation of the conductance for a field of 1.5 T after averaging over 6 well correlated regions on the sample. We observe a square VL with a lattice constant $a_0 = 370 \pm 10$ Å consistent with $a_0 = (\frac{\Phi_0}{B})^{\frac{1}{2}}$. The VL is oriented along the (110) direction of the crystal. STM observations in a field of 0.375 T also reveal a square lattice whose a_0 scales with $(\frac{\Phi_0}{B})^{\frac{1}{2}}$.

The occurrence of the square VL revealed by STM measurements can be analysed with Ginzburg-Landau theory including a fourfold term in the free energy which arises from the underlying tetragonal structure of the crystal. As described below, this analysis predicts a hexagonal structure at low field, with a second order phase transition to square structure at high field and low T. For microscopic approaches describing anisotropic effects in superconductors see Ref. [17]. The lowest order term in the Ginzburg-Landau expansion of the free energy which accounts for the square symmetry of the crystal, has the form $|D_x^2\Psi|^2 + |D_y^2\Psi|^2$, where $D_\alpha = \nabla_\alpha - \frac{2\pi i}{\Phi_0}A_\alpha$ is the gauge invariant derivative. With this term the energy functional $E[\Psi]$ can be written as

$$E[\Psi] = E_{GL}[\Psi] + E_{\Box}[\Psi] \quad (1)$$

where $E_{GL}[\Psi] = \int d^3r \left(-a_o\,|\tau|\,|\Psi|^2 + \frac{b_o}{2}|\Psi|^4 + +a_o\xi_o^2\,|\mathbf{D}\Psi|^2 + \frac{B^2}{8\pi}\right)$ is the conventional isotropic Ginzburg-Landau term with standard notations and
$E_{\Box}[\Psi] = \varepsilon_{\Box} a_o \xi_o^4 \int d^3r \left(\left|D_x^2\Psi\right|^2 + \left|D_y^2\Psi\right|^2\right)$
is the anisotropy term characterized by a strength coefficient $\varepsilon_{\Box}$. The equilibrium vortex configuration corresponding to the energy functionals in Eq. (1) is determined by minimizing the Abrikosov ratio $\beta = \overline{|\Psi|^4} / \left(\overline{|\Psi|^2}\right)^2$ Ref. [18,19]. A solution of the isotropic Ginzburg-Landau equation is first built for a rectangular cell with aspect ratio $R = L_x/L_y$ (see Fig. 2, inset b). We then calculate the correction to this solution caused by the square term in Eq. (1) and evaluate the corresponding correction to the Abrikosov ratio in first order with respect to $\varepsilon_{\Box}$. We obtain

$$\beta(R) = \beta_0(R) + \varepsilon_{\Box}\,|\tau|\,\frac{8b^2}{9b-1}\beta_{\Box}(R) \qquad (2)$$

with $b = H/H_{c2}$ and $\tau = (T - T_c)/T_c$ the reduced temperature. The dimensionless functions $\beta_0(R)$ and $\beta_{\Box}(R)$ are plotted in Fig. 2 (inset a). As first found in Ref. [19], $\beta_0(R)$ is minimum at $R = \sqrt{3}$ corresponding to the hexagonal lattice ($\beta_0 = 1.16$) and maximum at $R = 1$ corresponding to the square lattice ($\beta_0 = 1.18$). The function $\beta_{\Box}(R)$ is maximum at $R = 1$ and decreases monotonically, which implies that a minimum of $\beta(R)$ at $R = 1$ can only be found for negative values of $\varepsilon_{\Box}$. With $\varepsilon_{\Box} < 0$ the values of R which minimize (2) are always smaller than $\sqrt{3}$. This implies that any square term in (1) produces a distorted hexagonal VL [20] which transforms progressively (by decrease of the bond angle α) into a stable square VL tilted 45° with respect to the x-axis when crossing the phase transition line $b(\tau)$ defined by $\varepsilon_{\Box}\,|\tau|\,\frac{8b^2}{9b-1} = -0.09$. This transformation to a square VL occurs via a second order phase transition. Note that for $\varepsilon_{\Box} > 0$ it can be shown that a stable square VL can also be obtained but then oriented along the x-axis. An alternative description using nonlocal corrections to the London model is discussed in Ref. [21].

The energy functional (1) can also be used to calculate anisotropic properties of isolated vor-

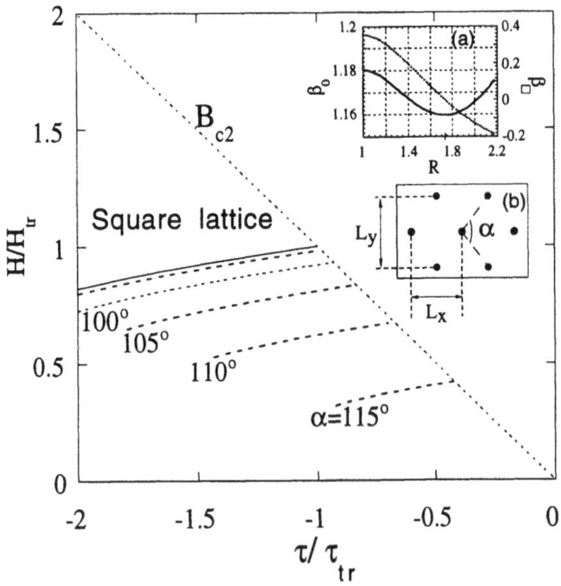

Figure 2. Phase diagram showing the transition from a distorted hexagonal to a square square VL. The reduced temperature τ is normalized to the square lattice transition temperature at H_{c2} and the field is normalized to the value of H_{c2} at this point. The dashed lines show contours of constant bond angle as defined in inset *b*. Inset *a* shows the two components of the Abrikosov ratio β.

tices. The square term disturbs the isotropic field and current distributions around the vortex core. These disturbances become stronger at smaller distances. For example, the disturbed field distribution of a single vortex can be written as
$B_v(\mathbf{r}) = \frac{\Phi_o}{2\pi\lambda^2}\left(K_o(r/\lambda) - \frac{5}{4}\varepsilon_{\Box}\xi_o^2\frac{\cos 4\phi}{r^2}\right)$
Fig. 3 represents a contour plot of the vortex field. At large distances a circular symmetry is recovered, similar to that found in an isotropic system. At smaller distances the field distribution adopts a progressively more pronounced fourfold symmetry. This shows that the square anisotropy becomes essential at high fields.

The square term in Eq. (1) also introduces an angular dependence of the in-plane upper critical field which can be expressed in the lowest order with respect to $\varepsilon_{\Box}$ by

$$H_{c2}^{\parallel}(\phi) = H_{c2}^{\parallel}(0)\left(1 + \frac{3\varepsilon_{\Box}\,|\tau|}{16}\,(1 - \cos 4\phi)\right) \qquad (3)$$

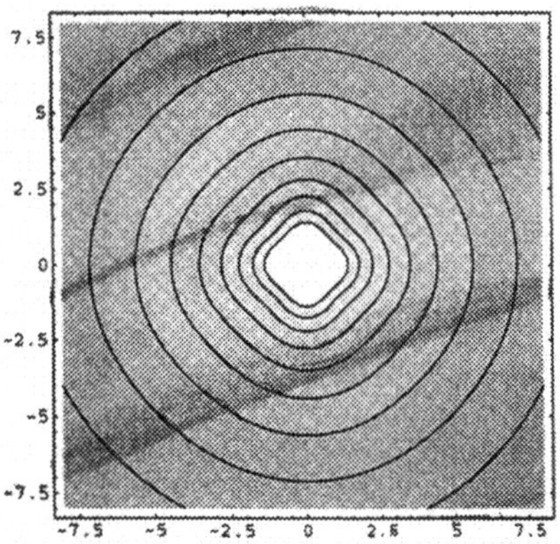

Figure 3. Contour plot of the field distribution of an isolated vortex calculated for $\varepsilon_{\square} = -0.5$. Distances are measured in units of ξ_o

According to this equation, at the transition line, we have $H_{c2}(45°)/H_{c2}(0) \approx 0.97$. In Fig. 4 we present $H_{c2}(T)$ of the sample as measured by SQUID magnetometry for an applied field parallel to the (001), (100), and (110) directions [22]. We find an in-plane anisotropy of H_{c2} which increases monotonously when lowering T. The transition point $H_{c2}(110)/H_{c2}(100) \approx 0.97$ occurs at $T = 14$ K implying $\varepsilon_{\square} \approx -0.8$. At $T = 4.5$ K we find an in-plane anisotropy of H_{c2} of the order of 0.92, consistent with the STM observation of a square VL.

This work was supported by the NSF Science and Technology Center for Superconductivity under contract # DMR 91-20000 (YDW, AEK, IA), by the US DOE Basic Energy Science-Material Science under contract # W-31-109-ENG-38 (MI, UW, VM, GWC) and contract # W-740-ENG-82 (PCC).

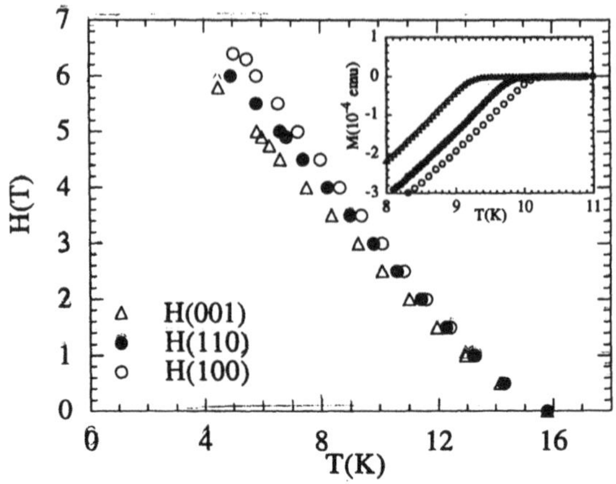

Figure 4. $H_{c2}(T)$ determined by SQUID magnetometry for a field applied along the (001)(△), (100)(○), and (110)(●) crystalline directions. The inset shows the temperature dependence of the magnetization at 3 T.

REFERENCES

1. R.J. Cava, *et al.*, Nature **367**, 252 (1994)
2. R. Nagarajan, *et al.*, Phys. Rev. Lett. **72**, 274 (1994)
3. T. Siegrist, *et al.*, Nature **367**, 254 (1994)
4. T. Hasegawa, *et al.*, Physica C **235-240**, 1859 (1994)
5. G.T. Jeong, *et al.*, Physica C **253**, 48 (1995)
6. T. Ekino, *et al.*, Phys. Rev. B **53**, 5640 (1996)
7. R. Vaglio, *et al.*, to be published.
8. Y. De Wilde, *et al.*, to be published.
9. U. Yaron, *et al.*, Nature **382**, 236 (1996)
10. P.L. Gammel, to be published.
11. M. Yethiraj, to be published.
12. P. L. Gammel, private communication.
13. H.F. Hess,*et al.*, Phys. Rev. Lett. **62**, 214 (1989); H.F. Hess, *et al.*, Phys. Rev.Lett. **69**, 2138 (1992); Ch. Renner, *et al.*, Phys. Rev. Lett. **67**, 1650 (1991); S. Behler, *et al.*, Phys. Rev. Lett. **72**, 1750 (1994).
14. I. Maggio-Aprile, *et al.*, Phys. Rev. Lett. **75**, 2754 (1995)
15. B.K. Cho, *et al.*, Phys. Rev. B **52**, R3844 (1995)
16. R.C. Dynes, *et al.*, Phys. Rev. Lett. **41**, 1509 (1978)
17. *Anisotropy Effects in Superconductors* ed. by H.Weber, Plenum, NY, 1977
18. A.A. Abrikosov, Soviet Phys. JETP **5**, 1174 (1957)
19. W.H. Kleiner, *et al.*, Phys. Rev. **133**, A1226 (1964)
20. D.K. Christen, *et al.*, Physica B **135**, 369 (1985)
21. V.G. Kogan, *et al.*, to be published.
22. V. Metlushko, *et al.*, to be published.

ELSEVIER

Physica C 282–287 (1997) 359–362

PHYSICA C

Thermal Fluctuations of Flux Lines and Nuclear Relaxation

F. Borsa [a,b], P. Carretta[a,c], M. Corti[a], A. Rigamonti[a] and B. J. Suh[b*]

[a]Dipartimento di Fisica "A. Volta", 27100 Pavia, Italy

[b]Ames Lab-USDOE and Department of Physics and Astronomy, ISU, Ames, Iowa, USA

[c]Grenoble High Magnetic Field Laboratory, CNRS-MPI, B.P. 166X, 30842 Grenoble CEDEX 9, France

The study of thermal motion of flux lines as evidenced in NMR measurements is reviewed in the light of our recent results on the nuclear spin-lattice and spin-spin relaxation rates in anisotropic HT_cSC $YBa_2Cu_4O_8$ and $HgBa_2CuO_{4+\delta}$

1. INTRODUCTION

NMR technique offers a tool of microscopic origin for the study of the spontaneous thermal fluctuations of the flux lines (FL) in type-II high-T_c superconductors (HT_cSC). The rigid lattice NMR spectrum, for $T \rightarrow 0$, reflects the local field distribution while the FL thermal fluctuations can cause narrowing of the NMR line and measurable contributions to the spin-lattice ($1/T_1$) and spin-spin ($1/T_2$) relaxation rates [1]. The narrowing of the NMR line can be observed if the thermal fluctuations reach an amplitude which is a sizeable fraction of the intervortex spacing and if the characteristic frequency becomes faster than the inhomogeneous linewidth itself (i.e. $\sim 10^3 - 10^5$ Hz). The detection of the thermal fluctuations and/or FL lattice melting from the temperature dependence of the linewidth is made difficult by several factors and one can extract significant information only when the amplitude and frequency of the fluctuations is sufficiently high and the temperature dependence of the penetration length $\lambda(T)$ known [2,3]. On the other hand, in cases in which the contribution to the nuclear relaxation related to the FL motion is not obscured by other relaxation mechanisms, then characteristic peaks or maxima are observed in the temperature dependence of $1/T_1$ and $1/T_2$. These peaks are a direct manifestation of the spontaneous fluctuations of the FL's. The characteristic frequency of the motion which can be detected ranges from very slow ($1 - 10^3$ Hz) in echo decay experiments [4] to $10^7 - 10^9$ Hz in T_1 measurements [5,6]. Here we report recent studies in $YBa_2Cu_4O_8$ and $HgBa_2CuO_{4+\delta}$ and we try to give a comprehensive description of the experimental observations in terms of longitudinal and transverse local field fluctuations due to different kind of motions of the FL's around the irreversibility temperature.

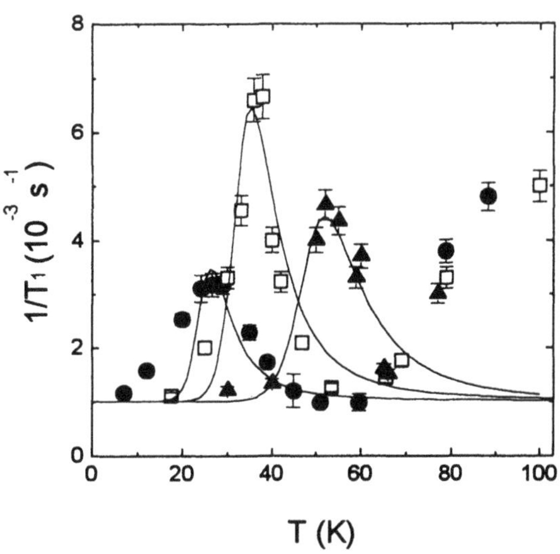

Figure 1. Temperature dependence of ^{89}Y spin-lattice relaxation rate in $YBa_2Cu_4O_8$ for $H = 5.9$ Tesla (triangles), 9.4 Tesla (squares) and $H = 14.9$ Tesla (circles), with $\vec{H} \parallel c$. The solid lines give the best fits according to Eq. 5, with $U = 576, 371$ and 264 K, respectively.

*present address Los Alamos National Lab., Los Alamos, USA

0921-4534/97/$17.00
PII S0921-4534(97)00269-4

2. FLUCTUATION OF LOCAL FIELDS AND NUCLEAR RELAXATION

The local field, in cylindrical coordinates, for a single 2D pancake vortex is given by [7,8]

$$h_\rho(z,\rho) = \frac{\Phi_o}{4\pi\lambda_{\|}^2}\frac{s}{\rho}\left[\frac{z}{|z|}e^{-|z|/\lambda_{\|}} - \frac{z}{r}e^{-r/\lambda_{\|}}\right] \quad (1)$$

$$h_z(z,\rho) = \frac{\Phi_o}{4\pi\lambda_{\|}^2}\frac{s}{r}e^{-r/\lambda_{\|}} \quad (2)$$

where $r = (\rho^2 + z^2)^{1/2}$ and s is the interplane separation. The fluctuations in the component of local magnetic field h_ρ, transverse to the applied dc field H, due to the time dependence of the FL position induce a contribution to the spin-lattice relaxation rate of the form

$$\frac{1}{T_1} = \frac{\gamma^2}{2}\int_{-\infty}^{\infty} < h_\rho(0)h_\rho(t) > e^{-i\omega_L t}dt \quad (3)$$

Regarding, instead, the spin-spin relaxation rate $1/T_2$, as measured in a two-pulse echo decay experiment, the effect of FL motion is more complex. If the characteristic frequency of the fluctuations of the FL position is higher than the strength of the perturbation itself (which is of the order of the NMR width for the rigid FL lattice) then the contribution to the irreversible decay of the transverse nuclear magnetization is given by

$$\frac{1}{T_2} = \frac{1}{2T_1} + \frac{\gamma^2}{2}\int_{-\infty}^{\infty} < h_z(0)h_z(t) > dt \simeq$$

$$\simeq \frac{1}{2T_1} + < |\delta\omega|^2 > \tau_z \quad (4)$$

When the motions of FL's become very slow (i.e. $\tau_z \sim 1 - 10^{-4}$ s) they can be studied by a different effect. Infact when the FL motion generates a change in the local field component $h_z(t)$ in the time interval τ_m which separates the two radiofrequency pulses generating the echo, one observes an additional contribution to the decay of the transverse nuclear magnetization, if the change in the local field is at least of the same order of magnitude as the other contributions to the decay of the echo [1,4]. In presence of more than one mechanism the decay of the transverse magnetization may have a functional form difficult to be characterized by a single parameter.

3. EXPERIMENTAL RESULTS

There are two anisotropic HT_cSC in which an unambiguous contribution to $1/T_1$ due to thermal FL motion has been observed below T_c on the top of the background relaxation due to charge carriers: $HgBa_2CuO_{4+\delta}$ (T_c = 96 K) [5] and $YBa_2Cu_4O_8$ (T_c = 81 K) [6]. ^{199}Hg spin-lattice relaxation rates in an aligned powder sample for the external magnetic field parallel to the c axis show an enhancement with maxima in $1/T_1$ vs. T which depend on the applied field H. For the results at $H = 5.9, 8.2$ and 9.4 Tesla we refer to Ref. 5. Similar but more pronounced peaks have been observed in the $1/T_1$ of ^{89}Y in $YBa_2Cu_4O_8$ (see Fig. 1).

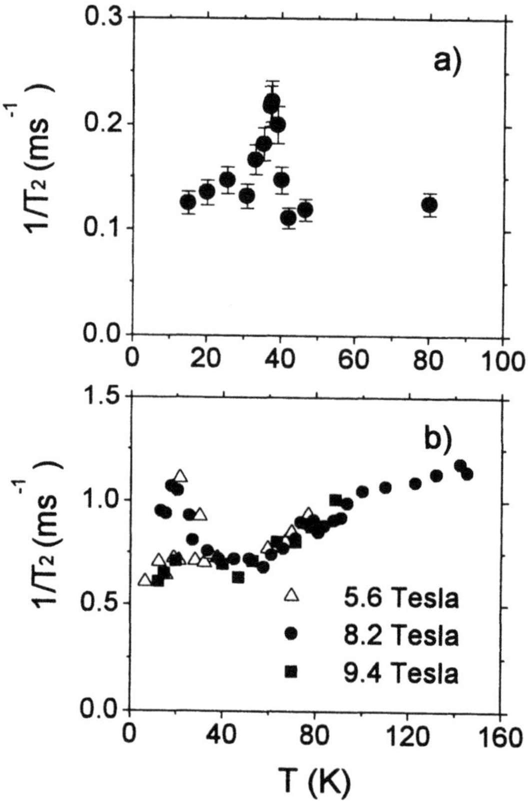

Figure 2. a) Temperature dependence of ^{89}Y $1/T_2$ in $YBa_2Cu_4O_8$ for H = 9.4 Tesla along the c axis. b) Temperature dependence of ^{199}Hg $1/T_2$ in $HgBa_2CuO_{4+\delta}$ for different field intensities. In both cases T_2 is given by the time at which the echo intensity decreases by $1/e$.

Data of spin-spin relaxation rate as a function of temperature are shown in Fig. 2a and Fig. 2b for $YBa_2Cu_4O_8$ and $HgBa_2CuO_{4+\delta}$, respectively. The functional form of the decay of the echo intensity is a combination of a Gaussian form and an exponential one with the exponential decay becoming prevalent at low temperature. For the sake of simplicity we report in Fig. 2 the inverse of the time at which the echo decays to $1/e$.

4. CORRELATION TIMES DESCRIBING THE FL SPONTANEOUS FLUCTUATIONS

For a rigid vortex line along the $c-axis$, corresponding to 3D correlated pancake vortices, the transverse field $h_\rho(t)$ in Eq. 3 is zero as a result of summation of contributions from pancakes in different planes, as given by Eq. 1. Thus, vibrational excitations of the FL characterized by wave vector $q_{||} = q_z = 0$ do not cause relaxation while a small contribution may arise from excitations at $q_{||} \neq 0$. On the other hand if the pancake vortices are weakly correlated the contributions from the planes above and below a given nucleus sum incoherently and 2D large amplitude diffusion-like motions are possible giving a large contribution to the relaxation rate. Thus the temperature region where the maxima in $1/T_1$ (see Fig. 1) are observed should be above the decoupling line describing the crossover from the solid 3D to the 2D pancake FL lattice. Furthermore the maximum of $1/T_1$ should occur close to the temperature where the characteristic motional frequency of the FL's is of the order of the Larmor frequency ω_L. A quantitative analysis of the data is conveniently done on the basis of a simple model describing the time dependence of $h_\rho(t)$ in Eq. 3 in terms of a pseudo-diffusive motion of the 2D pancake vortices characterized by a correlation time τ_e. The result obtained is [5,6]:

$$\frac{1}{T_1} = \frac{\gamma^2 s}{\xi_{||}^2 \lambda_{||}^2} k_B T ln(\frac{2\lambda_{||}^2}{\sqrt{3} l_e^2}) \tau_e ln(\frac{\tau_e^{-2} + \omega_L^2}{\omega_L^2}) \quad (5)$$

where l_e is the intervortex distance. We assume a temperature dependence for τ_e of the form $\tau_e = \nu_o^{-1} exp(U/T)$ where ν_o is the attempt frequency and U is a field dependent barrier where, being far from T_c we neglect possible temperature dependences of U. From the fits in Fig. 1 and from ref. 5 we find

$HgBa_2CuO_{4+\delta}$: $\nu_o \simeq 10^{11}$ s^{-1}; $U \simeq \frac{840}{H}$ (K/Tesla)

$YBa_2Cu_4O_8$: $\nu_o \simeq 10^{13}$ s^{-1}; $U \simeq \frac{3500}{H}$ (K/Tesla)

It is noted that the fit is done by first subtracting off from the measured $1/T_1$ the background contribution due to charge carriers. For $HgBa_2CuO_{4+\delta}$ the background contribution can be inferred from the $1/T_1$ data for $H \perp c$ where the contribution of FL's is negligible. For $YBa_2Cu_4O_8$ the background contribution is small and can be estimated from the extrapolation of the $1/T_1$ data above T_c and just below T_c.

As can be seen in Fig. 2a and 2b the effective spin-spin relaxation rate $1/T_2$ displays an anomalous enhancement at low temperatures. The temperature of the maximum appears to occur close to the irreversibility temperature T_{irr} while the maximum in $1/T_1$ occurs above T_{irr}. T_{irr} was estimated in the same samples by magnetization measurements [5,6].

Contrary to $1/T_1$, the spin-spin relaxation is affected by fluctuations of the longitudinal local field (Eq. 2). Thus the 3D motion of a vortex line for $q_z = 0$ can be quite effective in modulating $h_z(t)$ although not $h_\rho(t)$.

Therefore we postulate that the enhancement of $1/T_2$ is due to the 3D motion of the FL's below the irreversibility temperature. If the maximum in $1/T_2$ reflects the dynamical contribution related to spectral density at zero frequency of the longitudinal fluctuations of h_z one has from approximating Eq. 4, $1/T_2 \simeq \delta\omega^2 J(\omega \to 0) \simeq \delta\omega^2 \tau_z$. Thus by assuming that $\delta\omega$ is about 10 % of the rigid FL lattice linewidth one estimates $\tau_z(T = 20K) = 1.4 \times 10^{-6}$ s for $HgBa_2CuO_{4+\delta}$ and $\tau_z(T = 35K) = 5 \times 10^{-6}$ s for $YBa_2Cu_4O_8$. On the other hand, if the enhancement of $1/T_2$ where due to the dephasing of the transverse magnetization due to the change of position of the FL during the interval between the two radiofrequency pulses then τ_z should be of the order of T_2 namely $\tau_z \simeq 10^{-3}$ s at 20 K for $HgBa_2CuO_{4+\delta}$. To distinguish between the two cases one can look at the shape of the echo intensity decay. In the quasi-static case one should have a decay of the type $s(t) \propto exp[-\delta\omega^2 t^3/12\tau_z]$ [1] while for the dynamical case one expects an exponential decay. The experimental observation of a quasi exponential decay in the region of the peak in $1/T_2$ seems to favour the dynamical mechanism which implies a faster correlation time τ_z. A detailed analysis of the shape of the echo decay as a function of temperature and of the intensity of the radiofrequency field is necessary to reach a firm conclusion. This effort is under way.

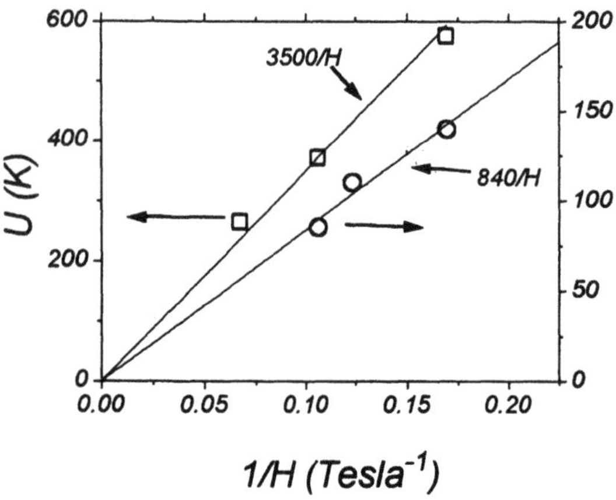

Figure 3. The amplitude of the activation energy is reported as a function of the inverse of the field intensity for $YBa_2Cu_4O_8$ (squares) and $HgBa_2CuO_{4+\delta}$ (circles).

5. CONCLUSIONS

We have shown that the analysis of the different NMR parameters $1/T_1$, $1/T_2$ and linewidth give access to different correlation times and different motional regimes for the FL's spontaneous fluctuations. Measurements of $1/T_1$ vs. T are most sensitive to thermal fluctuations of weakly correlated 2D pancake vortices described by a correlation time τ_e for the transverse local field fluctuations. We find an activated τ_e with an activation energy U inversely proportional to the applied magnetic field [9] as shown in Fig. 3. Measurements of $1/T_2$ are sensitive to longitudinal local field fluctuations described by τ_z and can therefore detect slow fluctuations even in the 3D correlated regime below the irreversibility temperature. We found a peak in $1/T_2$ vs T close to the irreversibility temperature (see Fig. 2a,b). It is tempting to ascribe the observed peak to the softening of the vibrational fluctuations of the 3D correlated pancakes close to the decoupling line leading to the 3D-2D crossover. Finally, linewidth measurements (described in Ref. 6) give access to a correlation time τ which results from a combination of motional effects. More measurements and theoretical analysis are necessary to be able to relate among themselves the different correlation times.

6. ACKNOWLEDGEMENTS

We thank F. Tabak for her help during NMR measurements. The research was carried out with the financial support of INFM (Istituto Nazionale di Fisica della Materia) and of INFN (Istituto Nazionale di Fisica Nucleare) (Project VORS). Partial support from NATO Grant No. CRG941204 and from CEE TMR program is gratefully acknowledged.

REFERENCES

1. For a review see F. Borsa, P. Carretta, F. Cintolesi, M. Corti, A. Rigamonti, B. J. Suh and D. R. Torgeson, Applied Magnetic Resonance 9, 149 (1995) and references therein
2. P. Carretta and M. Corti, Phys. Rev. Lett. 68, 1236 (1992); P. Carretta, Phys. Rev. B 45, 5760 (1992)
3. Y. Q. Song, W. P. Halperin, L. Tonge, T. J. Marks, M. Ledvij, V. G. Kogan and L. N. Bulaeskii, Phys. Rev. Lett. 70, 3127 (1993)
4. B. J. Suh, D. R. Torgeson and F. Borsa, Phys. Rev. Lett. 71, 3011 (1993)
5. B. J. Suh, F. Borsa, J. Sok, D. R. Torgeson, M. Corti, A. Rigamonti and Q. Xiaong, Phys. Rev. Lett. 76, 1928 (1996)
6. M. Corti, B. J. Suh, F. Tabak, A. Rigamonti, F. Borsa, M. Xu and B. Dabrowski, Phys. Rev. B 54, 9469 (1996)
7. J. R. Clem, Phys. Rev. B 43, 7837 (1991)
8. E. H. Brandt, Physica C 195, 1 (1992)
9. M. Tinkham, Phys. Rev. Lett. 61, 1658 (1988)

ELSEVIER

Physica C 282-287 (1997) 363-366

PHYSICA C

Bragg- and Moving-glasses: a theory of disordered vortex lattices

T. Giamarchi [a] * and P. Le Doussal[b] †

[a]Laboratoire de Physique des Solides, Université Paris-Sud, Bât 510, 91405 Orsay, France

[b]CNRS-Laboratoire de Physique Théorique de l'Ecole Normale Supérieure, 24 rue Lhomond, F-75231 Paris, France

We study periodic lattices, such as vortex lattices in type II superconductors in a random pinning potential. For the static case we review the prediction [1,2] that the phase diagram of such systems consists of a topologically ordered Bragg glass phase, with quasi long range translational order, at low fields. This Bragg glass phase undergoes a transition at higher fields into another glassy phase, with dislocations, or a liquid. This proposition is compatible with a large number of experimental results on BSCCO or Thalium compounds. Further experimental consequences of our results and relevance to other systems will be discussed.

When such vortex systems are driven by an external force, we show that, due to periodicity in the direction transverse to motion, the effects of static disorder persist even at large velocity[3]. In $d = 3$, at weak disorder, or large velocity the lattice forms a topologically ordered *glass* state, the "moving Bragg glass", an anisotropic version of the static Bragg glass. The lattice flows through well-defined, elastically coupled, static channels. We determine the roughness of the manifold of channels and the positional correlation functions. The channel structure also provides a natural starting point to study the influence of topological defects such as dislocations. In $d = 2$ or at strong disorder the channels can decouple along the direction of motion leading to a "smectic" like flow. We also show that such a structure exhibits an effective transverse critical pinning force due to barriers to transverse motion, and discuss the experimental consequences of this effect.

1. Introduction

To understand the effects of static substrate disorder on periodic media such as vortex lattices [4] is of paramount importance both for the technological applications of high-T_c materials and from a purely theoretical point of vue. Such a study also applies to many other physical problems such as charge density waves (CDW), Wigner crystals, colloids, magnetic bubbles.

For the vortex lattices it is generally agreed that disorder leads to a glass state with diverging barriers and pinning [5,6]. Although the precise nature of the glass state has been the subject of much debate, most theories agreed on the absence of translational order. General arguments also tended to prove that disorder would always favor the presence of dislocations [5]. Many points were not naturally fitting in the framework of these theories such as the existence of a first order transition [7,8], decoration experiments of showing remarkably large regions free of dislocations [9]. On the side of theory, scaling arguments [10] suggested, within a purely elastic decription, a slower, logarithmic, growth of deformations.

In a recent work we have obtained the first quantitative theory of the elastic vortex lattice [1] in presence of point disorder [11]. Contrarily to previous approaches, it provides a description valid at **all scales** and demonstrates that while disorder produces algebraic growth of displacements at short length scales, periodicity takes over at large scales and results in a decay of translational order *at most algebraic*. Moreover we showed [1] using energy arguments that dislocations are *not favorable* for weak disorder in $d = 3$. This implies the existence of a **thermodynamic** glass phase, as far as energy and very low current transport properties are concerned, retaining a **nearly perfect** (i.e. algebraic) translational order and thus Bragg peaks: the "Bragg glass".

*Laboratoire associated to CNRS; email: giam@lps.u-psud.fr

†email: ledou@physique.ens.fr

0921-4534/97/$17.00
PII S0921-4534(97)00270-0

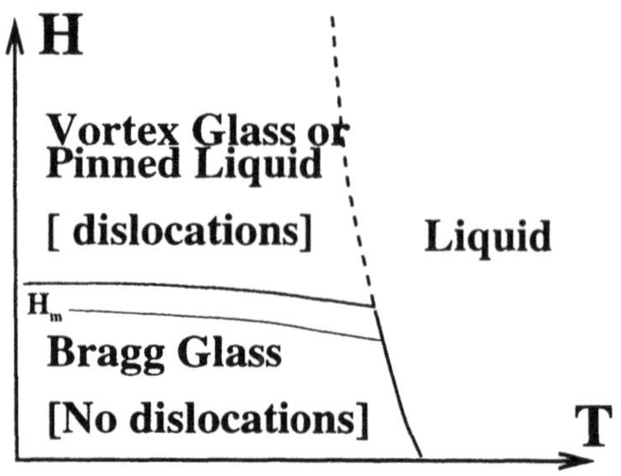

Figure 1. The stability region of the Bragg glass phase in the magnetic field H, temperature T plane is shown schematically. The thick line is expected to be first order, whereas the dotted line should be either second order or a crossover. Upon increasing disorder the field induced melting occurs for lower fields as indicated by the thin solid line.

Because it retains a "lattice" structure and Bragg peaks, this glass phase is radically different from the vortex glass picture. We proposed [1] that the phase seen experimentally at low fields was the Bragg glass, solving the apparent impossibility of a pinned solid. This accounted naturally for the first-order transition and the decoration experiments. Our prediction [1] received subsequent further support both from numerical [12] or analytical [13,14] calculations.

Upon raising the field, or equivalently increasing the disorder, the Bragg glass should disappear by spontaneous generation of dislocations, when the translational length R_a (the length for which relative displacements become of order of the lattice spacing a) is of the order of a. The critical point occuring on the melting line [15,8] was the end point of the transition line between the Bragg glass at low fields and a topologically disordered glassy phase (or a strongly pinned liquid) at higher field. The topology of the phase diagram proposed in [1] is as depicted in Fig. 1.

Several recent experiments can be interpreted to confirm the picture proposed in [1]. Neutron experiments can be naturally interpreted in term of the Bragg glass [16]. In BSCCO neutron peaks are observed at low fields and disappear upon raising the field [17]. The phase diagram of BSCCO [15] is also compatible with of our theory, the second magnetization peak line corresponding to the predicted field driven transition. A numerical estimate [2] of the melting field H_M for BiSCCO gives $H_M \sim 400G$ in good agreement with the observed experimental values [18]. This line is found to be relatively temperature independent at lower temperatures and to be shifted downwards upon increase of point disorder [18,19]. Similar types of phase diagrams are also observed in a variety of materials, including YBCO, organic superconductors and heavy fermion compounds. More experimental consequences and references can be found in [2].

Let us now turn to the dynamics of the driven system. It is crucial to determine how much of the glassy properties remain, and how translational and topological order behave [20]. Indeed several experiments suggest that a fast moving lattice is more ordered than a static one [21,16]. Arguments [22] based on large velocity expansion [23] concluded that at low T and above a certain velocity the moving lattice is a crystal at an effective temperature $T' = T + T_{sh}$, with bounded displacements and no glassy properties.

We showed [3] that in fact the perturbation theory [22] breaks down, even at large v. Some modes of the disorder are not affected by the motion and the periodic structure driven along x experiences a transverse **static** pinning force F_{stat}, perpendicular to the direction of motion.

$$F_\alpha^{\text{stat}}(r,u) = V(r)\rho_0 \sum_{K.v=0} iK_\alpha \exp(iK\cdot(r-u)) \quad (1)$$

where the random potential has correlations $\langle V(r)V(r')\rangle = \Delta(r-r')$ of range r_f. This force originates only from the periodicity along y and the uniform density modes along x [3,24]. It leads to a **moving glass** (MG).

We analyse here this result more quantitatively, using a functional RG to study the equation of motion. We only quote here the RG equation for the disorder term, for the periodic structure at

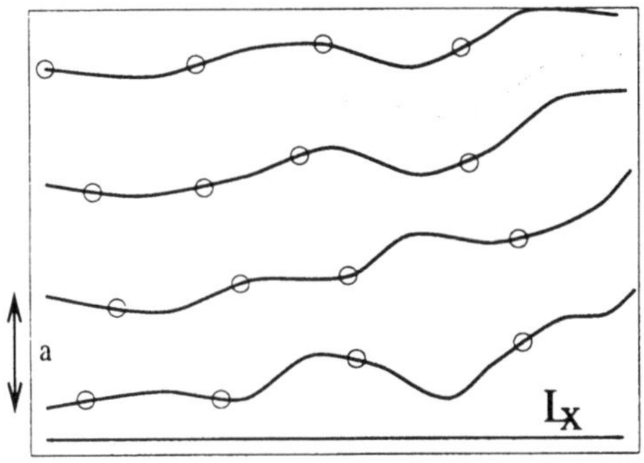

Figure 2. Motion in the moving glass occurs through static channels wandering at distance a over lengths $L_x \sim L_y^2$. If dislocations are present ($d = 2$ or strong disorder in $d = 3$) they should lead to a decoupling of channels, as indicated by the dotted line.

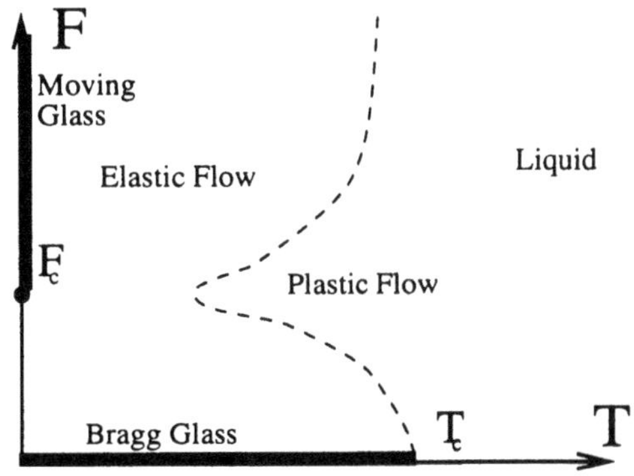

Figure 3. Phase diagram in force F, temperature T, for weak disorder and in $d = 3$. At zero external force the system is in the free Bragg glass state. At large velocities, in the moving Bragg glass one. This suggests that in this case the depinning transition could be purely elastic.

$T = 0$ (for the complete calculation see [25])

$$\frac{d\Delta(u)}{dl} = \Delta(u) + \Delta''(u)(\Delta(0) - \Delta(u)) \qquad (2)$$

where a factor $\frac{1}{4\pi v c}\epsilon$, with $\epsilon = 3 - d$ has been absorbed in $\Delta(u)$ (chosen to be of period 1). For $d > 3$ disorder renormalizes to zero and the moving system is a crystal. For $d < 3$ Δ flows to a new fixed point $\Delta^*(u) = \Delta(0)(l) + u^2/2 - u/2$, showing that the static disorder is still relevant in the moving structure (with the same conclusion in $d = 3$,[25]). The value of $\Delta(0)(l)$ grows unboundedly as $\Delta(0)(l) = \Delta(0)e^{\epsilon l}$ which indicates the existence of a random force along the y direction, generated under renormalization. A similar force is generated along x [25]. This does not spoil the above fixed point, since one can always separates $\Delta(0)$ and $\Delta(u) - \Delta(0)$. This has several consequences. In particular, for $d \leq 3$ the system remains a glass and motion occurs through elastic channels, as shown on figure 2. They are the easiest paths where particle follow each other in their motion. They form a manifold of elastically coupled, almost parallel lines or sheets (for vortex lines in $d = 3$) directed along x and characterized by some transverse wandering u_y. In the laboratory frame they are determined by the static disorder and do not fluctuate with time. Such channels were subsequently observed in numerical simulations [26] and in recent decoration in motion experiments [27]. In $d = 3$ diplacements only grow logarithmicaly, so the MG conserves quasi-long range translational order. Thus similarly to the statics, the MG in $d = 3$ at weak disorder or large velocity is expected to retain perfect topological order. This leads to the $F - T$ phase diagram of figure 3. In $d = 2$ however displacements grow algebraically and dislocations are likely to appear. The existence of channels [3] then *naturally* suggests a scenario by which dislocations affect the MG: when the periodicity along x is retained, e.g., presumably in $d = 3$ at weak disorder, the channels are coupled along x. Upon increasing disorder or decreasing velocity in $d = 3$, or in $d = 2$, decoupling between channels can occur, reminiscent of static decoupling in a layered geometry [14]. Dislocations are then inserted between the layers, naturally leading to a flowing smectic glassy state, recently observed in $d = 2$ numerical simulations [26]. Indeed, the transverse smectic order is likely to be more stable than topological order along x, because of particle conservation [25].

As an important consequence of the existence of the MG, barriers for transverse motion exist once the pattern of channels is established. Thus the response to an additional small transverse force F_y is very non linear with activated behaviour. At $T = 0$ and neglecting the dynamic part of the disorder a true transverse critical current J_y^c exists. This pinning force can directly be obtained from the RG equation (2). Since its fixed point has a *non analycity* at $u = 0$, (leading to $\Delta'(0^+) = 1/2$) there is a critical force, determined at the Larkin length L_y:

$$F_c = \int dq G(q) \Delta'(0^+) = \frac{\epsilon}{4\pi v c_y} \sim L_y^{-2} \quad (3)$$

(3) coincides with the more qualitative derivation consisting in balancing the pinning energy with the transverse Lorentz force acting on a Larkin domain [3]. Unlike in the $v = 0$ case, the critical force does not kill the random force in the FRG equation. The MG is dominated by the competition between the random force and the critical force.

These predictions can be tested in experiments on the vortex lattice, or other systems such as colloids, magnetic bubbles or CDW, or or numerical simulations. Additional physical consequences and references can be found in [3,25].

REFERENCES

1. T. Giamarchi and P. Le Doussal, Phys. Rev. Lett. **72**, 1530 (1994); Phys. Rev. B **52**, 1242 (1995).
2. T. Giamarchi and P. Le Doussal, Phys. Rev. B **55**, 6577 (1997).
3. T. Giamarchi and P. Le Doussal, Phys. Rev. Lett. **76**, 3408 (1996).
4. G. Blatter *et al.*, Rev. Mod. Phys. **66**, 1125 (1994).
5. M. P. A. Fisher, Phys. Rev. Lett. **62**, 1415 (1989); D. S. Fisher, M. P. A. Fisher, and D. A. Huse, Phys. Rev. B **43**, 130 (1990).
6. M. Feigelman, V. B. Geshkenbein, A. I. Larkin, and V. Vinokur, Phys. Rev. Lett. **63**, 2303 (1989).
7. M. Charalambous, J. Chaussy, and P. Lejay, Phys. Rev. B **45**, 5091 (1992).
8. H. Safar *et al.*, Phys. Rev. Lett. **70**, 3800 (1993).
9. D. G. Grier *et al.*, Phys. Rev. Lett. **66**, 2270 (1991).
10. T. Nattermann, Phys. Rev. Lett. **64**, 2454 (1990); J. Villain and J. F. Fernandez, Z Phys. B **54**, 139 (1984).
11. S. E. Korshunov, Phys. Rev. B **48**, 3969 (1993).
12. M. J. P. Gingras and D. A. Huse, Phys. Rev. B **53**, 15193 (1996); S. Ryu, A. Kapitulnik, and S. Doniach, Phys. Rev. Lett. **77**, 2300 (1996) and references therein.
13. J. Kierfeld, T. Nattermann, and T. Hwa, Phys. Rev. B **55**, 626 (1997).
14. D. Carpentier, P. Le Doussal, and T. Giamarchi, Europhys. Lett. **35**, 379 (1996).
15. E. Zeldov and al., Nature **375**, 373 (1995).
16. U. Yaron and al., Phys. Rev. Lett. **73**, 2748 (1994); T. Giamarchi and P. Le Doussal, Phys. Rev. Lett. **75**, 3372 (1995).
17. R. Cubbit and al., Nature **365**, 407 (1993).
18. B. Khaykovich and et al., Phys. Rev. Lett. **76**, 2555 (1996).
19. N. Chikumoto and al., Physica C **185-189** 2201 (1991); Phys. Rev. Lett. **69**, 1260 (1992).
20. M. J. Higgins and S. Bhattacharya, Physica C **257**, 232 (1996).
21. R. Thorel and al., J. Phys. (Paris) **34**, 447 (1973).
22. A. E. Koshelev and V. M. Vinokur, Phys. Rev. Lett. **73**, 3580 (1994).
23. A. Schmidt and W. Hauger, J. Low Temp. Phys **11**, 667 (1973); A. I. Larkin and Y. N. Ovchinnikov, Sov. Phys. JETP **38**, 854 (1974).
24. T. Giamarchi and P. Le Doussal, Phys. Rev. Lett. **78**, 752 (1997).
25. T. Giamarchi and P. Le Doussal, 1997, to be published.
26. K. Moon and al., Phys. Rev. Lett. **77**, 2378 (1997), S. Ryu et al. to be published.
27. M. Marchevsky and al., Phys. Rev. Lett. **78**, 531 (1997).

ELSEVIER

Physica C 282–287 (1997) 367–370

PHYSICA C

Vortex Dynamics within the BCS Theory

P. Ao[a*] and X.-M. Zhu[b]

[a] Department of Theoretical Physics, Umeå University, S-901 87, Umeå, SWEDEN

[b] Department of Experimental Physics, Umeå University, S-901 87, Umeå, SWEDEN

Based on the BCS theory we outline a conventional path integral derivation of the transverse force and the friction for a vortex in a superconductor. The derivation is valid in both clean and dirty limits at both zero and finite temperatures. The transverse force is found to be precisely as what has been obtained by Ao and Thouless using the Berry's phase method. The friction is essentially the same as the Bardeen and Stephen's result.

1. Introduction

In condensed matter there are two types of important excitations: the gapless and elementary ones, such as spin waves or phonons, which determine the local properties, and the topological ones, domain walls or dislocations, which determine the global properties. Vortices in superconductors belong to the second category. They have been under intensive studies since earlier sixties[1]. Still, there has no general agreement on the form of vortex dynamics equation.[1–8] We point out here that the main disagreement among many microscopic derivations lies in the use or not of the relaxation time approximation to account for the effects of impurities in the force-force correlation function. To provide a natural treatment of the both the transverse force and the friction and to avoid the use of this approximation, we present a path integral derivation within the BCS theory. The important results here are that the transverse force is the same as obtained by the Berry's phase method[2] and by the total force-force correlation function method[3], and is indeed insensitive to impurities, and that the friction coincides essentially with the Bardeen and Stephen result[4] but differs in some important details. Specifically, the nature of the friction is determined by the spectral function of the Hamiltonian. The relevance of the present theory to experiments has been discussed elsewhere.[9]

2. Path Integral Formulation

We begin with the standard BCS Lagrangian in the imaginary time representation for s-wave pairing. We will only consider neutral superconductors here, but the coupling to electromagnetic fields does no affect our results in the extreme type II case. The Lagrangian is

$$\begin{aligned} L_{BCS} = & \sum_\sigma \psi^\dagger_\sigma(x,\tau)\big(\hbar\partial_\tau - \mu_F \\ & - \tfrac{\hbar^2}{2m}\nabla^2 + V(x)\big)\,\psi_\sigma(x,\tau) \\ & - g\psi^\dagger_\uparrow(x,\tau)\psi^\dagger_\downarrow(x,\tau)\psi_\downarrow(x,\tau)\psi_\uparrow(x,\tau)\ , \end{aligned} \quad (1)$$

where ψ_σ describes electrons with spin $\sigma = (\uparrow, \downarrow)$, μ_F the chemical potential determined by the electron number density, $V(x)$ the impurity potential, and $x = (x, y, z)$. A vortex at x_v has been implicitly assumed. The partition function is

$$\begin{aligned} Z = & \int \mathcal{D}\{x_v, \psi^\dagger, \psi\}\times \\ & \exp\left\{-\tfrac{1}{\hbar}\int_0^{\hbar\beta} d\tau \int d^3x L_{BCS}\right\}\ , \end{aligned} \quad (2)$$

with $\beta = 1/k_BT$, and $d^3x = dxdydz$. Inserting the identity in the functional space,

$$\begin{aligned} 1 = & \int \mathcal{D}\{\Delta^*, \Delta\} \exp\Big\{ - \tfrac{g}{\hbar}\int_0^{\hbar\beta} d\tau \int d^3x \times \\ & \left|\psi_\downarrow\psi_\uparrow + \tfrac{1}{g}\Delta(x,\tau)\right|^2 \Big\}\ , \end{aligned}$$

into Eq.(2) we have

$$\begin{aligned} Z = & \int \mathcal{D}\{x_v, \psi^\dagger, \psi, \Delta^*, \Delta\}\times \\ & \exp\Big\{-\tfrac{1}{\hbar}\int_0^{\hbar\beta} d\tau \int d^3x \times \\ & \left(\psi^\dagger_\uparrow, \psi_\downarrow\right)(\hbar\partial\tau + \mathcal{H})\begin{pmatrix}\psi_\uparrow \\ \psi^\dagger_\downarrow\end{pmatrix} \\ & -\tfrac{1}{g}\int_0^{\hbar\beta} dt \int d^3x|\Delta|^2\Big\}\ . \end{aligned}$$

*Supported by Swedish NFR.

PII S0921-4534(97)00272-4

Here the Hamiltonian is defined as

$$\mathcal{H}(\Delta, \Delta^*) = \begin{pmatrix} H & \Delta \\ \Delta^* & -H^* \end{pmatrix}, \quad (3)$$

with $H = -(\hbar^2/2m)\nabla^2 - \mu_F + V(x)$. Integrating out the electron fields $\psi^\dagger_\sigma$ and ψ_σ first, then integrated out the the auxiliary(pair) fields Δ under the meanfield approximation, one obtains the partition function for the vortex

$$Z = \int \mathcal{D}\{x_v\} \exp\left\{-\frac{S_{eff}}{\hbar}\right\}, \quad (4)$$

with the effective action

$$\frac{S_{eff}}{\hbar} = -Tr \ln G^{-1} - \frac{1}{\hbar g}\int_0^{\hbar\beta} d\tau \int d^3x |\Delta|^2, \quad (5)$$

where Tr includes internal and space-time indices, and the Nambu-Gor'kov (NG) Green's function G defined by

$$(\hbar\partial_\tau + \mathcal{H})G(x,\tau;x',\tau') = \delta(\tau-\tau')\delta^3(x-x'), \quad (6)$$

together with the BCS gap equation, or the self-consistent equation,

$$\Delta(x,\tau) = -g < \psi_\downarrow(x,\tau)\psi_\uparrow(x,\tau) > . \quad (7)$$

A special attention should be paid to the equal time limit of the NG Green's function.[10]

We assume that the vortex is confined to move in a small regime around a point at x_0, which allows a small parameter expansion in terms of the difference between the vortex position x_v and x_0. We look for the long time behavior of vortex dynamics under this small parameter expansion. For the meanfield value of the order parameter, this expansion is

$$\begin{aligned}\Delta(x,\tau,x_v) = & \ (1 + \delta x_v(\tau)\cdot\nabla_0 \\ & + \tfrac{1}{2}(\delta x_v(\tau)\cdot\nabla_0)^2)\,\Delta_0(x,x_0) . \end{aligned} \quad (8)$$

Here $\delta x_v = x_v - x_0$. In Eq.(8) we have used the fact that when $x_v = x_0$ the vortex is static. The effective action for the vortex to the same order is, after dropping a constant term,

$$\begin{aligned}\frac{S_{eff}}{\hbar} = & -\tfrac{1}{2}Tr(G_0\Sigma')^2 + \frac{1}{\hbar g}\int_0^{\hbar\beta}\int d^3x \times \\ & \delta x_v\cdot\nabla_0\Delta_0^*\ \delta x_v\cdot\nabla_0\Delta_0 ,\end{aligned} \quad (9)$$

with

$$\Sigma' = \begin{pmatrix} 0 & \delta x_v\cdot\nabla_0\Delta_0 \\ \delta x_v\cdot\nabla_0\Delta_0^* & 0 \end{pmatrix}. \quad (10)$$

Here G_0 is the NG Green's function with $\Delta(\Delta^*)$ replaced by $\Delta_0(\Delta_0^*)$, and the gradient ∇_0 is with respect to x_0.

Now we construct the NG Green's function G_0 following the usual procedure. First, we consider the eigenfunctions of the Hamiltonian $\mathcal{H}_0 = \mathcal{H}(\Delta_0, \Delta_0^*)$. The stationary Schrödinger equation, the Bogoliubov-de Gennes equation, is

$$\mathcal{H}_0\Psi_k(x) = E_k\Psi_k(x) , \quad (11)$$

with

$$\Psi_k(x) = \begin{pmatrix} u_k(x) \\ v_k(x) \end{pmatrix}.$$

Since $\mathcal{H}_0$ is hermitian, all its eigenstates form a complete and orthonormal set, that is,

$$\int d^3x \Psi_k^\dagger(x)\Psi_{k'}(x) = \delta_{k,k'} ,$$

and

$$\sum_k \Psi_k(x)\Psi_k^\dagger(x') = \begin{pmatrix} 1 & 0 \\ 0 & 1 \end{pmatrix}\delta^3(x-x') ,$$

with $\Psi^\dagger(x) = (u^*(x), v^*(x))$. Furthermore, Eq.(11) has an interesting property:

$$\mathcal{H}_0\Psi(x) = E\ \Psi(x) \Rightarrow \mathcal{H}_0\overline{\Psi}(x) = -E\ \overline{\Psi}(x) ,$$

with

$$\overline{\Psi}(x) = \begin{pmatrix} v^*(x) \\ -u^*(x) \end{pmatrix}.$$

Finally, since $\delta x_v\cdot\nabla_0\mathcal{H}_0 = \Sigma'$, for $k \neq k'$ we have

$$\begin{aligned}&\int d^3x \Psi_k^\dagger(x)\Sigma'\Psi_{k'}(x) \\ &= (E_{k'} - E_k)\delta x_v\cdot\int d^3x\Psi_k^\dagger(x)\nabla_0\Psi_{k'}(x) ,\end{aligned}$$

which implies that the two ways of specifying the vortex coordinate, through the trapping potential or through the order parameter, are equivalent. Those relationships will be used below.

Given the eigenfunctions of Eq.(11), the NG Green's G_0 can be expressed as

$$G_0(x,\tau;x',\tau') = \sum_{n,k}\frac{-1}{\hbar\beta}\frac{e^{-i\omega_n(\tau-\tau')}}{i\hbar\omega_n - E_k}\Psi_k(x)\Psi_k^\dagger(x')$$

Here $\omega_n = n\pi/\hbar\beta$, with n odd integers. Assuming the rotational symmetry after the impurity average, a straightforward calculation leads to the following effective action

$$S_{eff} = \frac{1}{2}\int_0^{\hbar\beta} d\tau \Big[K\ |\delta x_v(\tau)|^2 + \int_0^{\hbar\beta} d\tau' F_{\|}(\tau-\tau')|\delta x_v(\tau) - \delta x_v(\tau')|^2 + \int_0^{\hbar\beta} d\tau' F_{\perp}(\tau-\tau')(\delta x_v(\tau)\times\delta x_v(\tau'))\cdot\hat{z}\Big] , \quad (12)$$

with the spring constant in the effective potential,

$$K = \frac{1}{g}\int d^3x |\nabla_0 \Delta_0^*(x, x_0)|^2 - \int_0^{\infty} d\omega \frac{J(\omega)}{\omega} , \quad (13)$$

the longitudinal correlation function,

$$F_{\|}(\tau) = \frac{1}{\pi}\int_0^{\infty} d\omega J(\omega)\frac{\cosh[\omega(\frac{\hbar\beta}{2}-|\tau|)]}{\sinh[\omega\frac{\hbar\beta}{2}]} , \quad (14)$$

and the transverse correlation function, in the long time limit, in terms of the virtual transitions,

$$F_{\perp}(\tau) = -\partial_{\tau-\tau'}\delta(\tau-\tau')\sum_{k,k'}\int d^3x \int d^3x' \ \hbar(f_k - f_{k'})\frac{1}{2}\hat{z}\cdot\Big(\Psi_k^{\dagger}(x')\nabla_0\Psi_{k'}(x')\times \nabla_0\Psi_{k'}^{\dagger}(x)\Psi_k(x)\Big) , \quad (15)$$

or in terms of the contribution from each state,

$$F_{\perp}(\tau) = -\partial_{\tau-\tau'}\delta(\tau-\tau')\sum_k \int d^3x\ \hat{z}\cdot \hbar(f_k\nabla_0 u_k^*(x)\times\nabla_0 u_k(x) -(1-f_k)\nabla_0 v_k^*(x)\times\nabla_0 v_k(x)) . \quad (16)$$

To reach Eqs.(12-16), following two identities have also been used:

$$\sum_n \frac{e^{-i\omega_n\delta}}{i\hbar\omega_n - E_k} = \begin{cases} \beta\, f_k\ , & \delta = 0^- \\ -\beta\,(1-f_k)\ , & \delta = 0^+ \end{cases} ,$$

and

$$\sum_{n-n'} \frac{-1}{\beta}\frac{e^{-i(\omega_n-\omega_{n'})\tau}}{i\hbar(\omega_n-\omega_{n'})-E} = \frac{1}{2}\frac{\cosh[\frac{E}{\hbar}(\frac{\hbar\beta}{2}-|\tau|)]}{\sinh[E\frac{\beta}{2}]} ,$$

with the Fermi distribution function $f_k = 1/(1+e^{\beta E_k})$, and the spectral function

$$J(\omega) = \frac{\pi}{4}\sum_{k,k'}\delta(\omega - |E_k - E_{k'}|)|f_k - f_{k'}|\times \frac{1}{\hbar}|\int d^3x \Psi_k^{\dagger}(x)\nabla_0\mathcal{H}_0\Psi_{k'}(x)|^2 . \quad (17)$$

In the following we discuss the implications of K, $F_{\|}$, and $F_{\perp}$ one by one, and show that Eq.(12) contains both the dissipative effect corresponding to the Bardeen-Stephen result[4] and the transverse force identical to the one by Berry's phase method[2] or the total force correlation function method[3].

3. Correlation Functions

3.1 Effective Spring Constant K

For the purpose of getting the friction and the transverse force, the precise value of K is irrelavent.

3.2 Longitudinal Correlation Function $F_{\|}$

This longitudinal correlation function contains all information on the vortex friction, revealed by the fact that Eq's.(12,14) are identical to the time-nonlocal action in the quantum dissipative dynamics[11]. The friction is determined by the low frequency behavior of the spectral function $J(\omega)$, Eq.(17). The temperature is naturally incorporated into the spectral function. This form of the action has been used in the study of the vortex tunneling.[12]

If we classify the eigenstates of the Bogoliubov-de Gennes equation, Eq.(11), according to core(localized) and extended states, the dissipation comes from all three parts: core-to-core, core-to-extended (or extended-to-core), and extended-to-extended contributions. Those contributions are added up, because they are all positive, which differs from the situation for the transverse correlation function.

For a clean and neutral superconductor, the core states are separated by finite energy gaps from each other. Their contributions to the dissipation will be exponentially small in the slow motion case. The dominant contribution in this case comes from the extended states, because their energy spectrum is continuous. This is also true for a thin film after the consideration of the electromagnetic field effects. Those contributions correspond to super-Ohmic cases.[13]

The presence of impurities mixes up all the clean limit eigenstates, and generates a continuous distribution of the energy, even for the core states, after the impurity average. A perturbative calculation shows that it is indeed the case, and the spectral function becomes Ohmic.[14] This suggests that in the clean limit the dissipation is super-Ohmic, and turns into Ohmic in the dirty limit. Hence in the dirty limit one should expect the Bardeen-Stephen result[4], because they have treated the core as in a normal state, corresponding to the uniform distribution of the energy eigenvalue.

3.3 Transverse Correlation Function $F_\perp$

In the long time limit the there are two equivalent forms for the correlation function: The virtual transition expression, Eq.(15), and the individual state contribution, Eq.(16). Their equivalence has already been discussed in Ref.[5] with the aid of conservation laws. Eq.(15) can be again casted into various equivalent forms because of the cancelations among those virtual transitions. For example, it can be expressed only in terms of the core-to-core transitions. In this case it is a statement of the spectral flow[5]. But one can also cast it into the forms of core-to-extended, or extended-to-extended transitions.

In Eq.(16), the counting of individual state contributions is expressed as an area integral of the momentum commutator. If we express it as the line integral of the momentum density far away from the core, the immediate conclusion, as drawn in Ref.[3], is that localized core states do not contribute, because the momentum density at the core is zero. The insensitive to the impurities is most transparent from Eq.(16): In the one-body density matrix both the electron number density and the phase $\theta(x - x_v)$ are all insensitive to impurities. Here the phase θ is defined through the order parameter $\Delta(x, t, x_v) \to |\Delta| e^{q\theta(x-x_v)}$, with $q = \pm 1$ describing the vorticity along the $\hat{z}$ direction and $\theta(x) = \arctan(y/x)$. Hence, as long as the localization effects due to impurities are negligible, such as in the cases of usual dirty superconductors, the transverse force will not be influenced by impurities. An analogy can be drawn to electrons in semiconductors and in metallic rings: The momentum is insensitive to band structures and disorder potentials.

4. Discussions

1. The major difference between the present approach and those in Ref's.[6,7] lies in the use or not of the relaxation time approximation in the calculation of the force-force correlation function. It is found in Ref's.[6,7] that this approximation is needed to makes the friction finite, which at the same time reduces the magnitude of the transverse force. Early studies of this approximation have shown that it is in general ambiguous and problematic in the force-force correlation function[15]. The source for the ambiguity is that it is usually unable to know *a priori* the transverse, the friction, and the fluctuating forces, although the total force is often well defined. This is exactly what happens in the present situation.

In the present approach without this approximation we are able to obtain the friction determined by the spectral function of the Hamiltonian, and demonstrate that the transverse force is insensitive to impurities.

2. The explicit counting of each state contribution, Eq.(16), confirms that the contribute to the transverse force comes from a region away from the core. One way to show it, as discussed above, is from the consideration of the momentum density, which is zero at the phase singular point, the vortex position. In an equivalent way, if the one-density density matrix, defined by $\{u_k, v_k^*\}$, is finite at the vortex position, as allowed by the Bogoliubov-de Gennes equation, the corresponding u_k or v_k^* will not have the phase factor θ.[16] Hence, there is no an additional contribution to the transverse force from the phase singular point as supposed in Ref.[8].

References

1. See, e.g., E.H. Brandt, Rep. Prog. Phys. **58**, 1465 (1995), for a recent review.
2. P. Ao and D.J. Thouless, Phys. Rev. Lett. **70**, 2158 (1993).
3. D.J. Thouless, P. Ao, and Q. Niu, Phys. Rev. Lett. **76**, 3758 (1996).
4. J. Bardeen and M.J. Stephen, Phys. Rev. **140**, A1197 (1965).
5. P. Ao, Phys. Lett. **A216**, 167 (1996).
6. N.B. Kopnin, Physica **B210**, 267 (1995).
7. E. Šimánek, Phys. Lett. **A194**, 323 (1994).
8. M.V. Feigel'man *et al.*, JETP Lett. **62**, 834 (1995).
9. P. Ao, J. Supercond. **8**, 503 (1995) and J. Low Temp. Phys. (in press); X.-M. Zhu *et al.*, Phys. Rev. Lett. **78**, 122 (1997).
10. J.R. Schrieffer, *Theory of Superconductivity*, revised, Addison-Wesley, New York, 1983.
11. A.O. Caldeira and A.J. Leggett, Ann. Phys.(NY) **149**, 374 (1983).
12. P. Ao and D.J. Thouless, Phys. Rev. Lett. **72**, 132 (1994).
13. Q. Niu *et al.*, Phys. Rev. Lett. **72**, 1706 (1994).
14. P. Ao and X.-M. Zhu, *to be published.*
15. R. Kubo *et al.*, *Statistical Physics II*, 2nd ed., Springer-Verlag, Berlin, 1985.
16. J. Bardeen, *et al.*, Phys. Rev. **187**, 556 (1969).

ELSEVIER

Physica C 282-287 (1997) 371-374

PHYSICA C

Flux pinning in melt processed RE-Ba-Cu-O

M. Murakami, N. Sakai, N. Chikumoto, H. Kojo, T. Higuchi, and S. I. Yoo

ISTEC, Superconductivity Research Laboratory, Div. VII
1-16-25 Shibaura, Minato-ku, Tokyo 105 Japan

RE-Ba-Cu-O (RE: rare earth element selected from Nd, Sm, Eu, Gd) superconductors exhibit high T_c and J_c values when they are melt processed in a reduced oxygen atmosphere (oxygen-controlled-melt-growth: OCMG process). Large J_c values in these systems can be attributed to finely dispersed RE rich 123 clusters, which can act as field-induced pinning centers, and thus leading to the second peak effect in *M-H* loops.
In this paper, we first propose the mechanism how RE rich clusters are formed during melt processing. Second we show that flux pinning depends on oxygen annealing conditions, because the optimum oxygenation conditions are different depending on the range of Nd-Ba substition. We also report that the addition of Nd422 is effective in raising J_C, although the mechanism is rather complicated owing to the fact that Nd422 also has a RE-Ba solid solution toward Ba rich direction.

1. INTRODUCTION

Unlike $YBa_2Cu_3O_7$, which is a stoichiometric point compound, RE-Ba-Cu-O (RE: La, Nd, Sm, Eu, Gd) supercondutors have a $RE_{1+x}Ba_{2-x}Cu_3O_y$ type solid solution [1]. When trivalent RE substitutes on bivalent Ba site, the carrier density of the compound decreases and thereby T_c is depressed. The presence of such RE-Ba substitution made us difficult to synthesize high T_c RE-Ba-Cu-O compounds by melt-processing in air, because RE rich 123 phase was always formed. Yoo et al. [2] found that such substitution could largely be suppressed by melt processing in a reduced oxygen atmosphere, which was later found to be attributed to the fact that the peritectic decomposition temperature of RE rich 123 phase is greatly depressed compared to that of the stoichiometric phase when oxygen partial pressure is reduced.

Furthermore, it was found that flux pinning of LRE-Ba-Cu-O superconductors is intrinsically superior to the Y-Ba-Cu-O even in the as-grown state without controlling microstructure [3]. Particularly in a high field and temperature region, J_c values of RE-Ba-Cu-O were higher than those of good quality melt-grown Y-Ba-Cu-O, which also resulted in enhanced irreversibility field. Microstructural observations revealed that RE rich clusters 30 - 50 nm in size are finely distributed in the LRE123 matrix, and these clusters act as field-induced pinning centers, leading to enhanced second peak effects in *M-H* loops of RE-Ba-Cu-O [4].

In this paper, we first propose the mechanism how RE rich clusters are formed during melt processing. Second we show that flux pinning depends on oxygen annealing conditions not only due to the presence of oxygen defects but also due to the fact that the optimum oxygenation conditions are different depending on the range of Nd-Ba substition. We also report that the addition of Nd422 is effective in raising J_c, although the mechanism is rather complicated owing to the fact that Nd422 also has a RE-Ba solid solution toward Ba rich composition.

2. EXPERIMENTAL

Single crystals of $NdBa_2Cu_3O_y$ were grown using a flux method under 1% PO_2, which were described elsewhere [5]. The bulk samples are prepared from mixtures of $NdBa_2Cu_3O_y$ + $Nd_4Ba_2Cu_2O_{10}$, and $SmBa_2Cu_3O_y$ + Sm_2BaCuO_5 powders to study the effect of the volume fraction of the second phase. The samples are then subjected to the OCMG process, the details of which are described in references [2, 3]. All the samples are oxgen-annealed in the temeprature range of 275 - 600°C in flowing oxygen for 20 - 300 hours.
T_c and J_c values were measured using a Quantum Design SQUID magnetometer. Microstructural characterizations were performed by optical and transmission electron microscopy.

PII S0921-4534(97)00273-6

3. RESULTS AND DISCUSSION

3.1. Flux pinning

Figure 1 shows J_c-B properties of Nd-Ba-Cu-O single crystal at various temperatures for fields parallel to the c axis. As shown in the figure, the enhanced second peak effects are observed in a high temperature region. The peak effects are also observed when the field is applied perpendicularly to the c axis. Similar peak effects are observable in bulk Nd and Sm samples.

We have confirmed that such peak effects are commonly observed in the RE-Ba-Cu-O systems with RE-Ba solid solution, and later we have shown that the peak effect is ascribed to field-induced pinning by RE rich clusters finely dispersed in the good RE123 matrix. However, the formation mechanism of such clusters is still unclear.

3. 2. Formation of RE rich clusters

We have first proposed that RE rich regions might be formed during solidification due to the presence of RE-Ba solid solution, in which RE rich regions are always produced like the solidification of two component alloys with solid solution. After solidification, Nd rich clusters will then be formed, because Nd ions occupying Ba site tend to form pairs to save extra oxygen, which sits on the anti-chain site as revealed by Neutron diffraction studies [6] (see Fig. 2).

Recent high temperature annealing experiments [7] suggested that such clusters tend to agglomerate and coarsen at temperatures close to the peritectic temperature, which was implied by the fact that the peak effect and J_c values were depressed without decaying T_c values after high temperature annealing. Such agglomeration has recently been confirmed with microstructural observations by transmission electron microscopy as shown in Fig. 3, where microclusters are no more present. Therefore, we may conclude that RE rich clusters are formed in the solid state owing to the tendency of Nd ions to form pairs, and the clusters have the tendency to grow and coarsen at high temperatures.

3. 3. Oxygen annealing

Oxygen annealing must be performed to achieve high T_c in the RE-Ba-Cu-O systems, although they

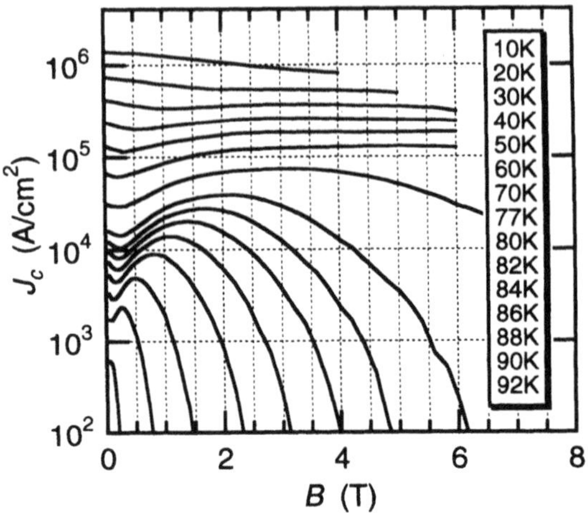

Fig. 1 J_c-B properties of Nd-Ba-Cu-O single crystal flux-grown in Ar-1%O_2 at various temperatures for $H//c$ axis.

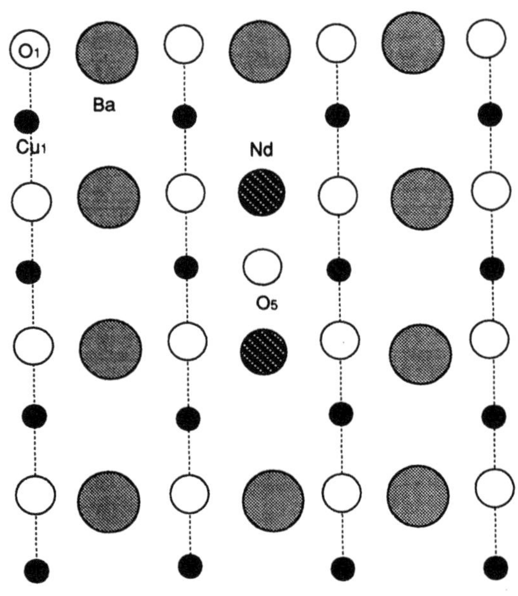

Fig. 2. Two-dimensional projection of the basal plane of Nd123. Nd substituting Ba accompanies an extra O in the O_5 site, so that Nd ions tend to form a pair to save extra oxygen.
(after Kramer et. al [6])

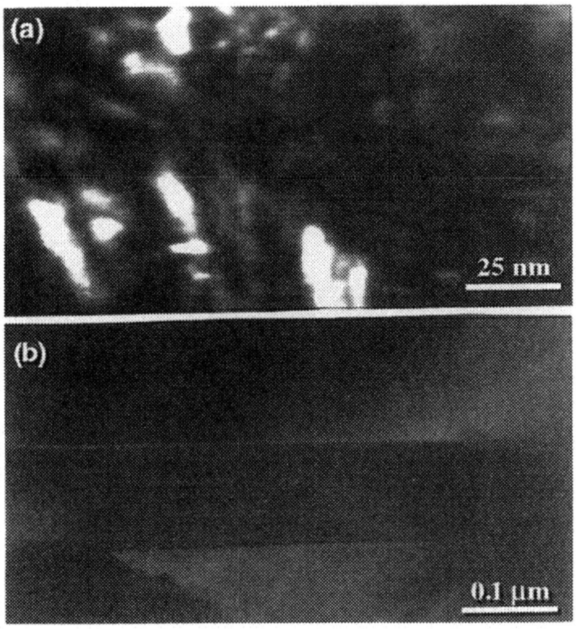

Fig. 3. Transmission electron microscopes for Nd-Ba-Cu-O (a) before and (b) after high temperature annealing. Note that the contrasted regions of Nd rich clusters disappears after annealing.

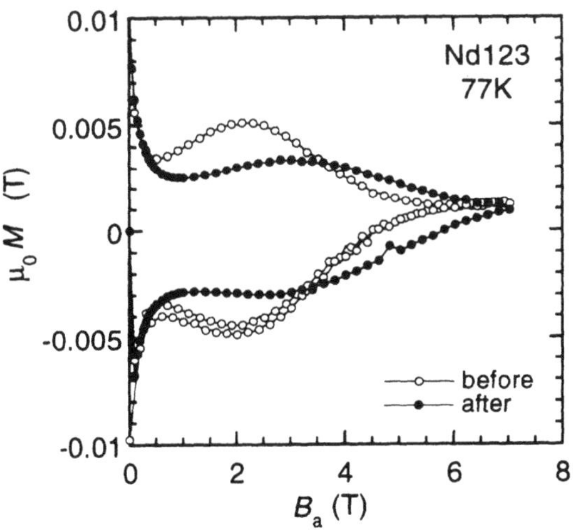

Fig. 4. Effect of high pressure oxygen annealing on the irreversibility field of Nd-Ba-Cu-O.

are melt-processed in low PO_2. We have found that the optimum annealing temperature is rather low and around 300°C for Nd and Sm systems. We have also found that despite low annealing temperature, oxygen can diffuse even into large bulk samples, implying that the oxygen diffusion rate is relatively fast in these systems. Like Y system, oxygen annealing conditions strongly affect flux pinning, although the second peak effect is not due to oxygen defects.

It is also important to note that the superconducting properties of Nd rich regions are also affected by oxygen annealing conditions. Full oxygenation of Nd rich regions are known to be diffcult due to the fact that a Nd ion which occupies Ba site always accompany an extra oxygen in the anti-chain site (O_5 site).

We have then performed oxygen annealing under high pressure [8]. It was found that both T_c and the irreversibility field are improved by high pressure oxygen annealing as shown in Fig. 4, suggesting that oxygen annealing was not sufficient under ambient pressure. It should also be noted that the peak field shifts to higher magnetic field, presumably due to the fact that T_c's of RE rich regions were also improved, which means that higher field is required to drive such regions normal conducting.

3. 4. The effect of second phase

In the Y system, it is common to distribute Y_2BaCuO_5 particles in the $YBa_2Cu_3O_7$ matrix, which contributes to the improvement of J_C-B properties and mechanical properties. It must be interesting to study the effect of $Nd_4Ba_2Cu_2O_{10}$ dispersion on J_c-B properties. Figure 5 shows field dependence of J_c for Nd-Ba-Cu-O with different Nd422 content. Although the size of Nd422 is relatively large compared with Y211, J_C values in a low field region can be improved. When the Nd422 content is low, the strong peak effect is observed, which suggests that field-induced pinning by Nd rich clusters and pinning by Nd422 can act simultaneously. However, when Nd422 content exceeds 20%, the peak effect disappears. Compositional analyses with transmission electron microscopy equipped with EDX revealed that the compostion of the matrix shifts toward Nd rich direction in such samples [9]. In the Nd system, Nd422 phase also has a Nd-Ba solid solution, and

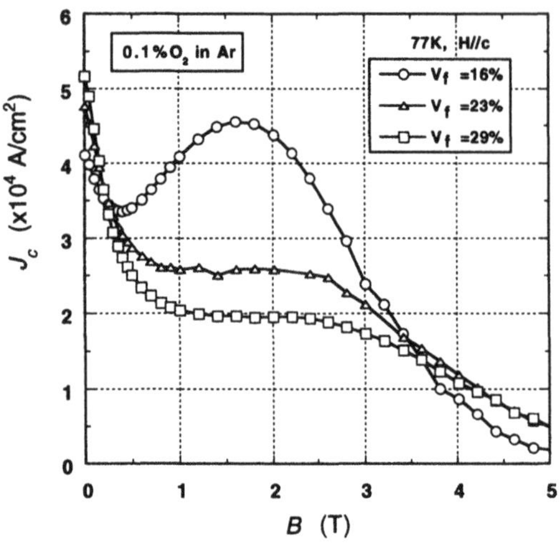

Fig. 5 Effect of Nd422 content on J_c-B properties (77K and $H//c$) of Nd123.

the tie line of Nd422-Nd123 moves toward Nd rich 123 region, because Ba mainly substitutes on Nd site in Nd422. Therefore, in order to optimize the pinning of Nd422, the control of Nd-Ba solid solution in the Nd422 phase is required.

4. CONCLUSIONS

In the RE-Ba-Cu-O systems (RE: Nd, Sm, Eu, Gd) melt-processed in a reduced oxygen atmosphere, RE rich clusters are finely distributed in the RE123 matrix. These clusters can act as field-induced pinning and contribute to strong pinning in a high H-T region. These clusters are formed due to the fact that RE ions occupying Ba site tend to form pairs. However, at high temperatures these clusters have the tendency to agglomerate and coarsen, which leads to reduced pinning and peak effect. Oxygen annealing affects flux pinning of the system partly because the T_c values of RE rich clusters are dependent on the oxygen content. Oxygen annealing under high pressure improved T_C and the irreversibility field and furthermore shifted the peak field to a high field region. The addition of Nd422 was effective in increasing J_c in a low field region and did not deteriorate the peak effect, although the addition >20% reduced the peak height, which was attributable to the presence of Nd-Ba solid solution in Nd422.

ACKNOWLEDGMENTS

The authors would like to thank J. Yoshioka, S. Ozawa and M. Otsuka of Shibaura Institute of Technology and U. Mizutani of Nagoya University for their discussions. This work was paratially supported by NEDO for the R & D of Industrial Science and Technology Frontier Program.

REFERENCES

1. S. I. Yoo and R. W. McCallum: Physica C, **210** (1993) 147.
2. S. I. Yoo, N. Sakai, H. Takaichi, T. Higuchi and M. Murakami: Appl. Phys. Lett., **65** (1994) 633.
3. M. Murakami, S. I., Yoo, T. Higuchi, N. Sakai, J. Weltz, N. Koshizuka and S. Tanaka: Jpn. J. Appl. Phys., **33** (1994) L715.
4. T. Egi, J. G. Wen, K. Kuroda, H. Unoki and N. Koshizuka: Appl. Phys. Lett., **67** (1995) 2406.
5. K. Sawada, S. I. Yoo, N. Sakai, T. Higuchi and M. Murakami: Advances in Superconductivity **VIII** (1996) 719.
6. M. J. Kramer, S. I. Yoo, R. W. McCallum, W. B Yelon, H. Xie, and P. Allenspach: Physica C, **219** (1994) 145.
7. J. Yoshioka, N. Chikumoto, M. Murakami and M. Otsuka: Advances in Superconductiviy **IX** (1997, Springer) ed. by Nakajima & Murakami to be published.
8. N. Chikumoto, S. Ozawa, M. Otsuka and M. Murakami:Advances in Superconductiviy **IX** (1997, Springer) ed. by Nakajima & Murakami to be published.
9. H. Kojo, S. I. Yoo, N. Sakai and M. Murakami: Advances in Superconductiviy **IX** (1997, Springer) ed. by Nakajima & Murakami to be published.

ELSEVIER

Physica C 282–287 (1997) 375–378

PHYSICA C

A PATH TO HIGH-FIELD PERFORMANCE OF CUPRATE SUPERCONDUCTORS ABOVE 100 K WITH FISSION INDUCED DEFECTS

L. KRUSIN-ELBAUM[a], D. LÓPEZ[a], J.R. THOMPSON[b], R. WHEELER[c], J. ULLMANN[d], C.C. TSUEI[a], C.W. CHU[e], and Q.M. LIN[e]

[a]IBM T.J. Watson Research Center, Yorktown Heights, New York 10598, USA

[b]Department of Physics, U. of Tennessee, Knoxville, Tennessee 37996, & Oak Ridge National Laboratory, Oak Ridge, Tennessee 37831, USA

[c]The Ohio State University, Columbus, Ohio 43210, USA

[d]Los Alamos National Laboratory, Los Alamos, New Mexico 87545, USA

[e]Texas Center for Superconductivity, Houston University, Houston, Texas 77204, USA

We demonstrate a method by which we expand the useful range of cuprate superconductors to above 100 K and enhance persistent currents by orders of magnitude in fields of several Tesla - namely *fission* of Hg nuclei within Hg-cuprates with 0.8 GeV protons. The fission process allows "doping" these cuprates with strongly pinning splayed columnar defects. The technique could be technologically relevant, since it is not limited by the short penetration range of heavier particles (or ions) and could permit modification of larger superconducting objects, such as magnets.

Large-scale applications of high-temperature superconductors require large critical current densities J_c in high magnetic fields. The chief factors limiting current conduction [1] are weak links at the grain boundaries in polycrystalline materials and high mobility of magnetic vortices in the bulk [1,2]. The ease of vortex motion is set by large thermal fluctuations (related to the high transition temperature T_c), intrinsically short superconducting coherence length ξ, large anisotropy [1], and possibly affected by a d-wave symmetry of the superconducting ground state [3].

A most efficient way to pin vortices is demonstrably with columnar defects [4]. A random array of nearly aligned tracks of amorphized material can be introduced by irradiation with energetic (~ 1 GeV) heavy ions (e.g. Au) - such columnar damage is known to shift the irreversibility line upward by tens of degrees and to expand the useful regime by several Tesla at high temperatures [4,5].

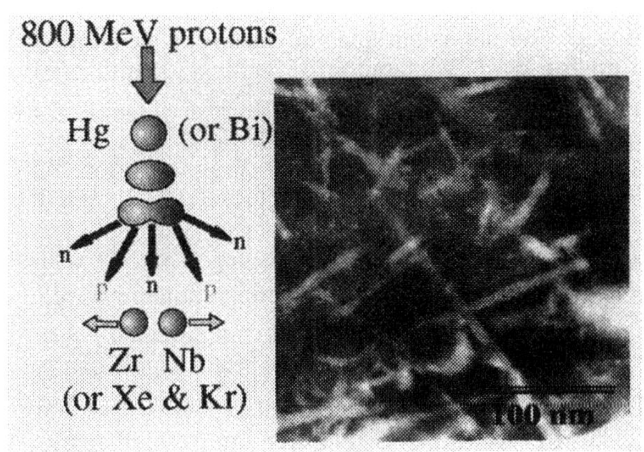

Fig. 1. A sketch of the fission process of Bi or Hg nuclei by 0.8 GeV protons. The fission fragments create randomly oriented (splayed) columnar defects, as illustrated in the TEM micrograph of a Hg-1212 bulk sample. Here the density of induced fission tracks is ~5.9×10^{10}/cm^2.

0921-4534/97/$17.00
PII S0921-4534(97)00279-7

This method, however, is not useful for large-scale applications, since typical heavy ions penetrate only ~ 20 – 30μm of the material [6]. Another way, which we have demonstrated recently for $Bi_2Sr_2CaCu_2O_8$ (Bi-1212), is by irradiation with *energetic* (0.8 GeV) *protons* [7]. Such protons can induce fission of $^{209}_{83}Bi$ nuclei, as sketched in Fig. 1. The fission cross-section of Bi nuclei is σ ~ 250 mbarns [8] (about 1/6 of the well known cross-section for $^{235}_{92}U$). Typical fission fragments in this case, $^{131}_{54}Xe$ or $^{84}_{36}Kr$ at ~ 100 MeV, are energetic enough to create extended, highly misaligned damage tracks [7,9].

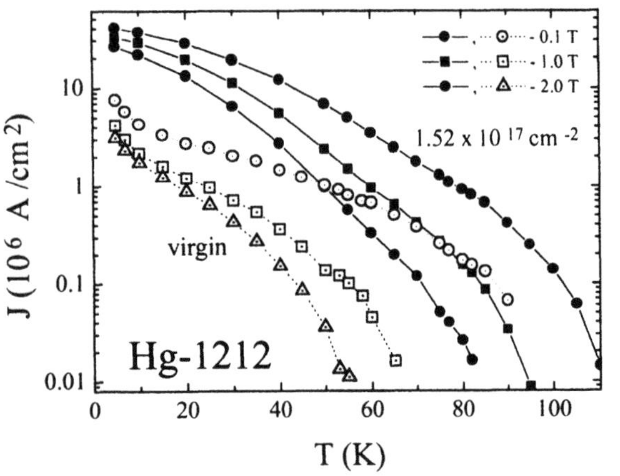

Fig. 2. The persistent current density **J** vs temperature for Hg-1212 sample as synthesized and after processing with $1.52\times10^{17}/cm^2$ proton fluence, for three values of applied magnetic field. **J** was obtained via Bean model from magnetization M measured with a SQUID magnetometer [12]. Bulk **J** is enhanced to well above $10^7 A/cm^2$. After irradiation **J** becomes large above 100 K, well *above* the irreversibility line of unirradiated sample.

Fission tracks in Bi-based cuprates significantly improve their current conduction in fields of a few Tesla [7], but the irreversibility line, which marks the useful region of non-zero J_c is limited by the T_c ~90 K and is only shifted to 77 K [7] for fields of 0.1 - 1 T. Clearly it would be desirable for a fission technique to work in mercury cuprates $HgBa_2Ca_{n-2}Cu_nO_{2n+2+\delta}$ with *n*=1, 2 or 3 adjacent CuO layers (Hg-1201, -1212, or -1223), the highest T_c (>130 K) materials to date [10,11].

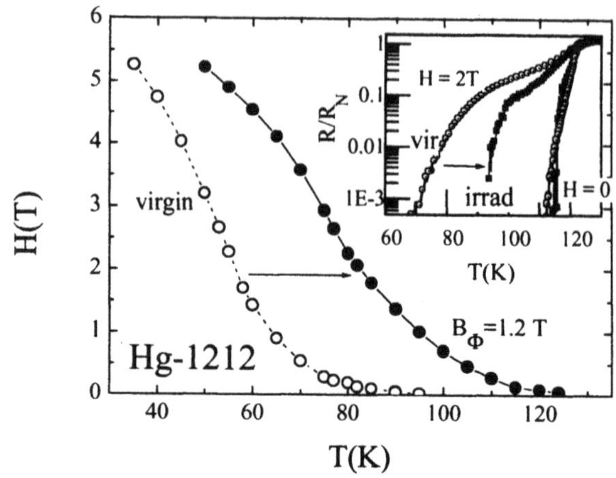

Fig. 3. Irreversibility lines of unirradiated and 0.8 GeV proton irradiated Hg-1212 (equivalent dose $B_\Phi \approx 1.2$ Tesla) obtained from the collapse of magnetic hysteresis. The arrow points to ~ 30 K shift at $\mu_0 H$ = 2 Tesla upon irradiation. Inset: the flux-flow resistance is significantly reduced by this process, as shown for $\mu_0 H$ = 2 T, while zero-field T_c is unchanged.

Briefly, the fission process is this. An incident 0.8 GeV proton penetrates a Hg nucleus. Absorption of a proton excites the nucleus, which first "boils off" several nucleons and then fissions into two fragments, each with a ~100 MeV energy [12]. Typical fission fragments here would be $^{90}_{40}Zr$ and $^{93}_{41}Nb$ at 95-100 MeV, sufficient to create columnar tracks [9,12].

The polycrystalline Hg-cuprates used in this study were prepared following procedures described in Ref. 13. The onset of a sharp resistive transition was measured at T_c ~ 95 K for Hg-1201, and ~ 120 K for Hg-1212. The virgin epitaxial films of Hg-1212 with T_c ~ 120-124 K were described in Ref. 11. The materials were exposed to 0.8 GeV proton fluences of $1.3-1.7\times10^{17}/cm^2$ at Los Alamos National Laboratory using the proton beam from LAMPF at the WNR Facility.

The defects resulting from such exposure are uniformly *splayed columnar tracks*, characteristic of a fission process [9]. The track diameter is found to be 8-9 nm [12]. Fig. 1 illustrates this in a crossectional TEM image of Hg-1212 sample exposed to a $1.52\times10^{17}/cm^2$ proton fluence. With the defect density obtained from TEM ~ $5.9\times10^{10}/cm^2$

and taking the calculated σ value for $^{200}_{80}Hg$ ~110 mbarns [8] and Hg density $\rho_{Hg} \approx 5.3\times10^{21}/cm^3$, we obtain the density of fission events at 8.9×10^{13} fiss./cm³. This gives us a mean track length of ~ 6.6μm for *two* fission fragments, consistent with the 8 - 9.5μm range for ~ 100 MeV Nb or Zr in Hg-1212 [12].

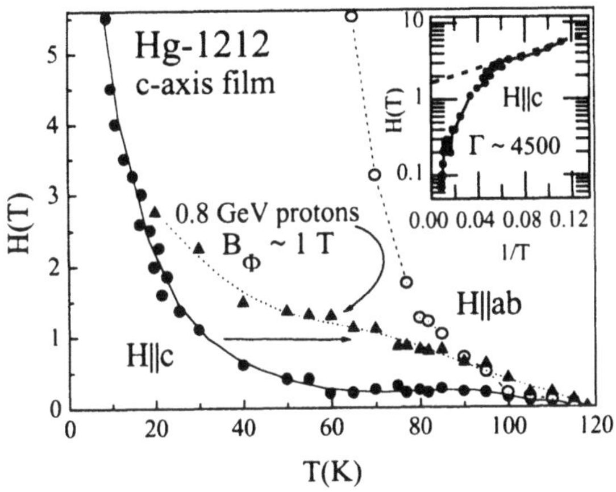

Fig. 4. Irreversibility lines for H||c and H||ab for a Hg-1212 c-axis oriented virgin film [11]. The line for H||c in the same film is shifted by the fission-induced tracks (B_Φ ~ 1T). Inset: the anisotropy deduced from the H||c line is Γ ~ 4500, in the range of Bi-1212 [7,11].

The temperature dependence of the persistent current density **J** in Hg-1212 before irradiation and with the fission density 8.9×10^{13} fiss./cm³ is shown in Fig. 2 for three different values of magnetic field [12]. In a virgin sample **J** drops rapidly at high fields, owing to a relatively low-lying irreversibility line, related to high anisotropy of this material [11]. At μ_0H = 2 T the drop exceeds three orders of magnitude at ~ $1/2T_c$ (~ 60 K), as is also seen in c-axis epitaxial Hg-1212 thin films [11]. The fission process halts the decrease in **J**(T) and the irreversible regime is expanded towards higher temperatures. At a 2 Tesla field and T ~60 K, the increase in **J** is over *two orders of magnitude* (see Fig. 2). At low fields **J** remains measurable up to 110 K .

The expansion of the irreversible regime for Hg-1212 is illustrated in Fig. 3. The irreversibility line was defined by the closing of the hysteresis loop M(H) at each temperature with the criterion of 5×10^{-6} emu, corresponding to ~200 A/cm² [12]. At 2 Tesla the shift is ~30 K. This moves $H_{irr}(T)$ from below liquid nitrogen temperature to more than 95 K. For fields μ_0H < 1 T, the irreversibility temperature T_{irr} shifts well above 100 K for the equivalent matching field [5] B_Φ of ~1.2 T. The shift of $H_{irr}(T)$ remains large up to 5.5 T. The suppressed vortex motion is also evident in the (linear) flux-flow regime - at high fields the linear resistance vanishes at much higher temperatures (see inset), corresponding to the shift of $H_{irr}(T)$. The zero-field resistive transition (T_c) is left essentially intact by the fission process for this particular fluence, which corresponds to B_Φ ~1.2 T.

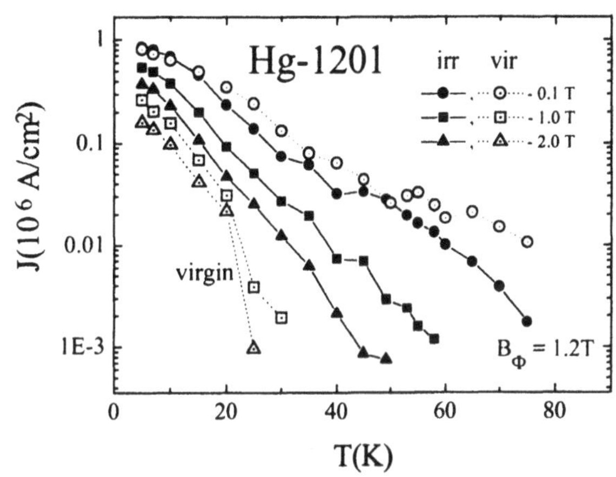

Fig. 5. The persistent current density **J** vs temperature for Hg-1201 sample before and after exposure to $1.52\times10^{17}/cm^2$ proton fluence. Bulk **J** is suppressed at low fields but enhanced at high fields.

A significant shift of $H_{irr}(T)$ is also observed in Hg-1212 c-axis epitaxial films, shown in Fig. 4. The films can support very high in-plane currents (H || c-axis) at low fields, but the current conduction declines very rapidly with field [11], suppressed by large anisotropy (a quasi-2D character) of the vortex structure (inset of Fig.4). The largest shift is found below μ_0H ~1 T, the value of the matching field B_Φ. Indeed, after irradiation, $H_{irr}(T)$ for the in-plane currents reaches the line measured for the intrinsic pinning configuration (H || ab).

The effect in Hg-1201 is illustrated in Fig. 5, which shows **J** *vs* T for three values of magnetic field before and after 0.8 GeV proton irradiation.

In this *less anisotropic* compound the uniformly splayed fission tracks somewhat *suppress* current conduction at the lowest fields (~0.1 T). (The regimes where splay is not beneficial were also reported in less anisotropic YBCO, where an optimal (10°) splay angle was established [14].) However, the situation is reversed at high fields. At $\mu_0H = 2$ T and T = 30 K, the persistent current is enhanced by about 2 orders of magnitude and $H_{irr}(T)$ (shown in Fig. 6) is shifted by as much as 35 K near B_Φ of 1.2 Tesla, and even more so at lower fields.

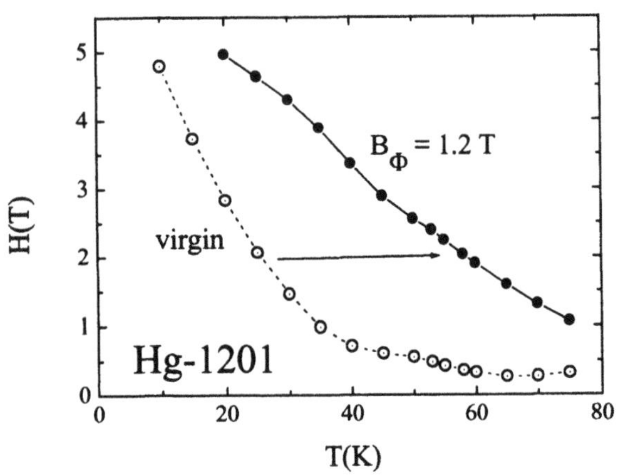

Fig. 6. Irreversibility lines of unirradiated and 0.8 GeV proton irradiated bulk Hg-1201 with the equivalent pin density $B_\Phi \approx 1.2$ Tesla. The shift of $H_{irr}(T)$ is over 35 K for fields less than B_Φ.

In summary, uniformly splayed fission tracks produce orders-of-magnitude improvement in current conduction of the two compounds described here; a single-CuO-layer Hg-1201 and a double-CuO-layer Hg-1212. The differences between the two Hg cuprates stem from the differences in their T_c's that limit the absolute range, differences in material parameters, and anisotropies [10 – 13]. So here we have the advantages of columnar tracks with a bonus of splay (dispersion) in the track directions, which further boosts J_c by forcing entanglement of vortices and limiting the escape of vortices from the tracks [15]. The long (~half-meter) penetration range of fast protons makes this method technologically appealing.

ACKNOWLEDGEMENTS: This work was supported in part by Electric Power Research Institute (EPRI Contract RP-8065-11), and by NSF GOALI Grant No. DMR-9510731. Work at ORNL was sponsored by the U.S. DOE under contract DE-AC05-96OR22464 with Lockheed Martin Energy Research Corp.

REFERENCES

1. Blatter, G. *et al.*, Rev. Mod. Phys. **66**, (1994) 1125, & references therein.
2. Larbalestier, D.C. *et al.*, in *Applied Superconductivity 1995* (Inst. of Physics Series No 148, Bristol & Philadelphia), **1**, p.29.
3. Mannhart, J. *et al.*, Phys. Rev. Lett. **77**, (1996) 2782.
4. Civale, L., *et al.*, Phys. Rev. Lett. **67**, (1991) 648; Konczykowski, M., *et al.*, Phys. Rev. B **44**, (1991) 7167.
5. Krusin-Elbaum, L., *et al.*, Phys. Rev. Lett. **72**, (1994) 1914.
6. J.F. Ziegler, J.B. Biersack, & U. Littlemark, *The Stopping Range of Ions in Solids*, (Pergamon Press, New York 1985), p.79.
7. Krusin-Elbaum, L., *et al.*, Appl. Phys. Lett. **64**, (1994) 3331.
8. Becchetti, F.D. *et al.*,, Phys. Rev. C **28**, (1983) 276.
9. Chadderton, L.T. & Torrens, McC.I., *Fission Damage in Crystals*, (Methuen & Co LTD, London 1969).
10. Putilin, S.N., Antipov, E.V., Chmaissem, O., & Marezio, M., Nature **362**, (1993) 226; Schilling, A., Cantoni, M., Guo, J.D., & Ott, H.R., *ibid.*, **363**, (1993) 56; Chu, C.W. *et al.*, *ibid.*, **365**, (1993) 323.
11. Krusin-Elbaum, L., Tsuei, C.C., & Gupta, A., Nature **373**, (1995) 679.
12. Krusin-Elbaum, L. *et al.*, preprint (1997).
13. Lin, Q.M. *et al.*, Physica C **254**, (1995) 207; Xiong, Q. *et al.*, J. Appl. Phys. **76**, (1994) 7127.
14. Krusin-Elbaum, L., *et al.*, Phys. Rev. Lett. **76**, (1996) 2563.
15. Hwa, T., LeDoussal, P., Nelson, D.R., & Vinokur, V.M., Phys. Rev. Lett. **71**, (1993) 3545.

ELSEVIER

Physica C 282–287 (1997) 379–382

PHYSICA C

Novel flux phases in superconductors with a periodic pinning array*

V.V. Moshchalkov, M. Baert, E. Rosseel, V.V. Metlushko, M. J. Van Bael, and Y. Bruynseraede[a]

[a] Laboratorium voor Vaste-Stoffysica en Magnetisme, Katholieke Universiteit Leuven, Celestijnenlaan 200D, B-3001 Leuven, Belgium.

Flux-line confinement by triangular and square regular arrays of microholes ("antidots") has been studied in superconducting films (Pb, WGe) and multilayers (Pb/Ge). For relatively large antidots sharp cusp-like magnetization anomalies appear at the matching fields Hm. These anomalies are caused by the formation of the multi-quanta vortex lattices at each subsequent Hm. The multi-quanta vortex lattices make possible a peacefull coexistence of the flux penetration at the antidots and the presence of a substantial superfluid density in the space between them. This leads to a very strong enhancement of the critical current density in films with an antidot lattice. For smaller antidots the vortices are forced to occupy the interstitial positions after the saturation of the pinning sites at antidots. This leads to the formation of the novel composite flux-line lattices consisting from the interpenetrating sublattices of weakly pinned interstitial single-quantum vortices and multi-quanta vortices strongly pinned by the antidots. When the interstitial flux-line lattice melts, it forms the interstitial flux liquid coexisting with the flux solid at antidots.

1. INTRODUCTION

In homogeneous superconductors the formation of the flux line (FL) lattices consisting of *multi-quanta* vortices is energetically unfavorable [1]. Contrary to this, *relatively large artificial pinning centers can stabilize multi-quanta vortices.* The maximum possible number of the FL trapped by the single insulating inclusion with the radius r is determined by the saturation number $n_S = r/2\xi(T)$, where $\xi(T)$ is the temperature dependent coherence length [2]. For regular arrays of artificial pinning centers the vortex-vortex interactions make the situation more complicated, but still multi-quanta FL lattices can exist, provided that the radius r is sufficiently large [3]. Following the pioneering work of Hebard et al. [4], we have shown recently [5–7] that the regular arrays of submicron holes (Fig. 1) ("antidots" (Fig. 1), this definition has been borrowed from the publications [8] on similar nanostructured semiconducting films) can be successfully used to stabilize novel flux phases consisting of multi-quanta and interstitial vortices in superconducting films with the antidot lattices. The formation of these novel phases leads to the appearance of the magnetization anomalies observed at matching fields $H = H_m$. Here the matching fields $H_m = m \cdot \Phi_0/S$ are determined by the unit cell area of the antidot lattice, S, and integer m. In what follows, we briefly characterize the main features of the novel flux phases, induced by the presence of regular pinning arrays.

2. MULTI-QUANTA VORTEX LATTICE

Superconducting films with relatively large antidots can successfully stabilize multi-quanta vortices. In this case puzzling sharp cusp-like magnetization anomalies appearing at H_m (Fig. 2) are somewhat similar to the well-known $M(H)$ cusp at lower critical field H_{c1}, but for the onset of multi-quanta vortices trapping by antidots : $2\Phi_0$ at $H = H_1$, $3\Phi_0$ at $H = H_2$, etc. Using the London limit expression for $M(H)$, we have shown [9] that exactly at the matching fields H_m, magnetization is given by the relation $M(H_m) \propto -m\Phi_0/2\Lambda^2(T)$, where $\Lambda(T)$ is the temperature dependent penetration depth in thin films. Between the matching fields $H_m < H < H_{m+1}$, $M(H)$ follows the logarithmic behavior $M \propto -\ln(H - H_m)$ typical for the system of strongly interacting FL. In other words, the

*Supported by the Belgian IUAP, Flemish FWO, GOA Programs and the Research Council of the K.U.Leuven.

PII S0921-4534(97)00280-3

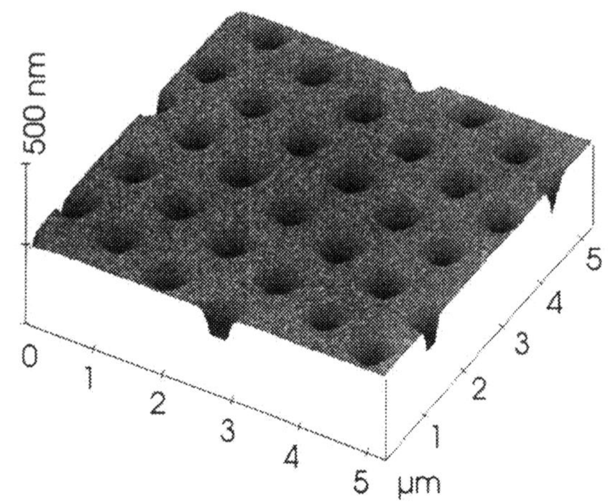

Figure 1. An AFM picture of a single Pb(500Å) film with a square lattice of submicron antidots (period $d = 1\mu m$, radius $r = 0.2\mu m$).

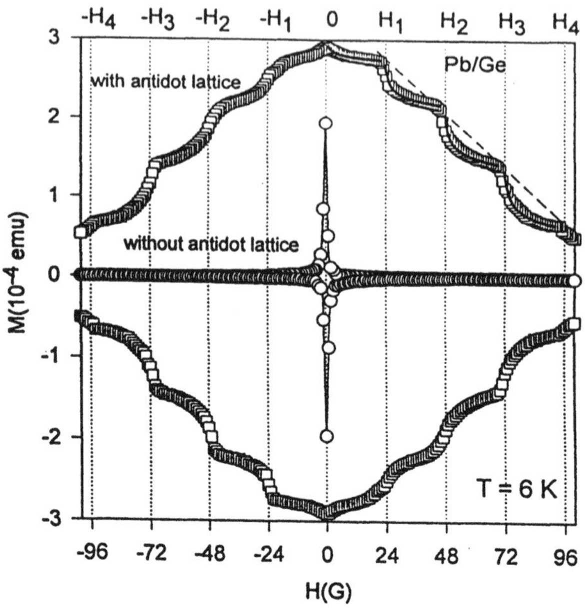

Figure 2. Magnetization loop $M(H)$ at $T = 6$K of a $(Pb(100Å)/Ge(50Å))_2$ multilayers with and without the triangular antidot lattice ($d = 1\mu m$). The solid line is a fit following the logarithmic dependence of the magnetization between the matching fields H_m and H_{m+1}. The dashed line is demonstrating the validity of the linear behaviour of $M(H_m)$. The loops were measured for $M > 0$ and symmetrized for clarity for $M < 0$.

well known $M \propto -\ln(H)$ behaviour of type-II superconductors in the London limit in the case of films with multi-quanta vortex lattices stabilized by the antidot lattice should be rewritten as $M \propto -\ln(H - H_m)$. Therefore, the whole magnetization curve of superconducting films with an antidot lattice is now successfully described (see solid line in Fig. 2 by the simple $M(H)$ expression, derived for interacting multi-quanta vortices in the London limit. This consistent quantitative analysis of the magnetic behavior convincingly demonstrates the existence of multiple-Φ_0 FL lattices in superconducting films with antidot lattices and opens new possibilities to control the width of the $M(H)$ hysteresis loop and critical current in nanostructured superconducting films. In films with antidot lattices the strongly enhanced critical current is close to its theoretical limit - depairing current [10].

3. COMPOSITE FLUX LATTICES

The crossover between the multiquanta and composite [6] vortex lattices is realized in fields $H > H_{ns}$ when the antidots are saturated and Φ_0 -vortices are forced to occupy interstices. The characteristic crossover field follows the temperature dependence expected from the saturation number $n_S = r/2\xi(T)$. The composite flux lattices are observed when the radii of the antidots are sufficiently small or/and temperature is not too close to T_c and H exceeds the limiting field H_{ns}. Also the pinning potential at interstices should not be very shallow to provide a weaker, but still sufficient pinning to form a softer interstitial flux solid. The composite flux lattices are characterized by the coexistence of the two

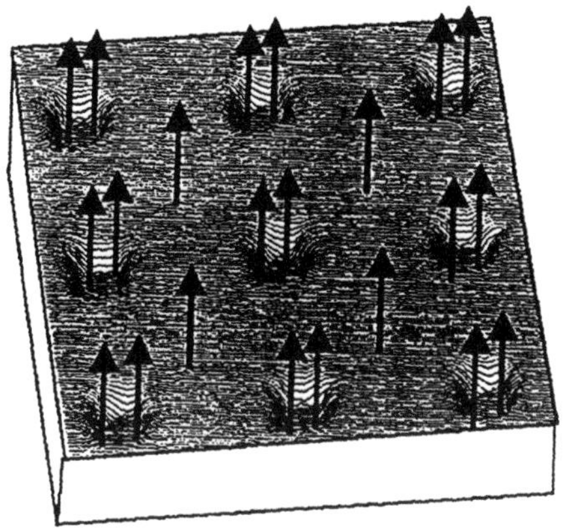

Figure 3. Schematic presentation, using an AFM picture, of the flux line lattice (arrows) in a (Pb(100Å)/Ge(50Å))$_2$ multilayer with the antidot lattice at the third ($H = H_3$) matching field. The composite flux lattice is characterized by the co-existence of the two interpenetrating flux lattices at interstices (Φ_0-vortices) and at antidots ($2\Phi_0$-vortices).

-weakly and strongly pinned - interpenetrating flux lattices at interstices (Φ_0-vortices) and antidots ($m\Phi_0$-vortices), respectively (Fig. 3) [6,9,10]. These coexisting lattices have recently been directly visualized in the Lorentz-microscopy [11] and magnetic decoration [12] experiments.

4. MULTI-QUANTA VORTEX LATTICE COEXISTING WITH THE INTERSTITIAL VORTEX FLUID

This flux phase is formed when $T \to T_c$ and the interstitial pinning potential becomes very shallow and thus can not prevent the melting of the caged interstitial pinning Φ_0-vortices. The melting transition as a function of magnetic field is then observed exactly at $H = H_{n_S}$ (see for example the transition at $H = H_2, T = 6.7$ K in Fig. 4, where for $H > H_2$ there is still a finite width of the hysteresis loop, or a very abrupt first order phase transition at $H = H_1$ and $T = 6.8$ K in Ref. 6). In other words, the homogeneous $m\Phi_0$-vortex solid with $m = n_S$ produces in this case only the $B = constant$ background while the melting of the interstitial Φ_0-vortex phase causes the very sharp magnetization drop at $H = H_{n_S}$. The interstitial vortex fluid reversibly responses to the field variation and this results in a zero width of the hysteresis loop in fields $H > H_{n_S}$. The novel flux phases listed above can exist at

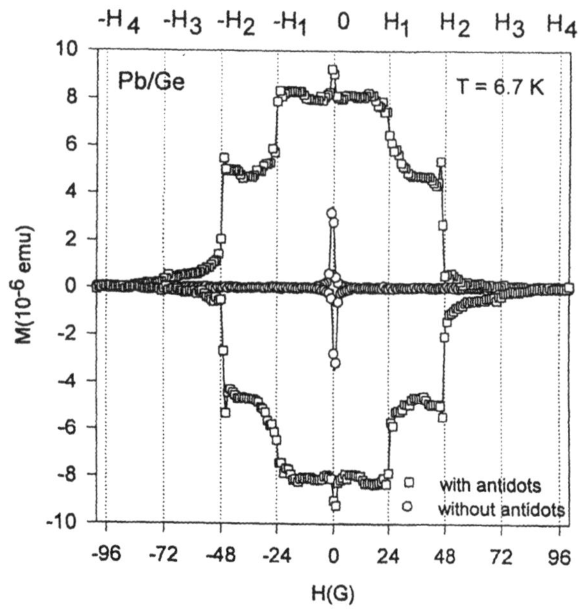

Figure 4. Magnetization loop $M(H)$ of a (Pb(100Å)/Ge(50Å))$_2$ multilayer with and without the antidot lattice at $T = 6.7$ K. At $H = H_2$ the interstitial flux solid melts, thus producing a sharp magnetization jump.

temperatures not too far from T_c, since at lower temperatures the tendency to form a conventional Bean profile starts to dominate and matching anomalies are suppressed, for example for Pb/Ge we barely see any $M(H)$ matching anomalies below 5 K.

5. CONCLUSIONS

In conclusion, the superconducting film with a lattice of relatively large antidots seems to demonstrate the novel single-terrace critical state [9] which appears at temperatures sufficiently close to T_c due to the multiple connectivity of the film and the stabilization of the $m\Phi_0$-flux lattices. The separation of the areas where flux penetrates from those where superconducting order parameter nucleates provides a kind of a "peaceful coexistence" of FLs pinned by antidots with the superconducting condensate in the space between them. Fabricating an antidot lattice to let flux go through, we are thus helping the order parameter between the antidots to sustain much higher currents and magnetic fields. Moreover, superconductors with an antidot lattice, having substantially reduced low critical fields H_{c1} and enhanced H_{c2} [13], still keep the product $H_{c1} \cdot H_{c2} = H_c$ constant. The presence of the antidot lattice also broadens the $H - T$ area where the London limit is valid. Using the two essential assumptions: B = constant (single-terrace critical state, especially at matching fields $H = H_m$) and $|\Psi|$ = constant, we have reached close to T_c a convincing quantitative description of the magnetization loops, including linear behavior of $M(H)$ at matching fields $M(H_m)$ and logarithmic behavior elsewhere: $M(H_m < H < H_{m+1}) - \ln(H - H_m)$. The existence of several novel flux phases - multiquanta vortex lattice, composite flux lattice and interstitial Φ_0-vortex fluid coexisting with the $m\Phi_0$-vortex lattices at antidots - has been made possible by fabricating regular arrays of antidots.

REFERENCES

1. P. G. de Gennes, *Superconductivity of metals and alloys* (Addison-Wesley, New York, 1966).
2. G. S. Mkrtchyan and V. V. Shmidt, Sov. Phys. JETP **34**, 195 (1972).
3. A. I. Buzdin, Phys. Rev. B **47**, 11416 (1993).
4. A. F. Hebard, A. T. Fiory, and S. Somekh, IEEE Transactions on Magnetics **1**, 589 (1977).
5. V.V. Moshchalkov, L. Gielen, M. Baert, V.V. Metlushko, G. Neuttiens, C. Strunk, V. Bruyndoncx, X. Qiu, M. Dhalle, K. Temst, C.D. Potter, R. Jonckheere, L. Stockman, M. Van Bael, C. Van Haesendonck, and Y. Bruynseraede, Physica Scripta **T55**, 168 (1994).
6. M. Baert, V.V. Metlushko, R. Jonckheere, V.V. Moshchalkov, and Y. Bruynseraede, Phys. Rev. Lett. **74**, 3269 (1995).
7. M. Baert, V.V. Metlushko, R. Jonckheere, V.V. Moshchalkov, and Y. Bruynseraede, Europhys. Lett. **29(2)**, 157 (1995).
8. H. Fang, R. Zeller, and P.J. Stiles, Appl. Phys. Lett. **55**, 1433 (1989); K. Ensslin and P.M. Petroff, Phys. Rev. B **41**, 12307 (1990).
9. V.V. Moshchalkov, M. Baert, V.V. Metlushko, E. Rosseel, M.J. Van Bael, K. Temst, R. Jonckheere, and Y. Bruynseraede, Phys. Rev. B **54**, 7385 (1996).
10. E. Rosseel, M. Van Bael, M. Baert, R. Jonckheere, V.V. Moshchalkov, and Y. Bruynseraede, Phys. Rev. B **53**, R2983 (1996).
11. K. Harada, O. Kamimura, H. Kasai, T. Matsuda, A. Tonomura, V.V. Moshchalkov, Science **274**, 1167 (1996).
12. A. Bezryadin and B. Pannetier, Journal of Low Temp. Phys. **102**, 73 (1996).
13. E. Rosseel, T. Puig, M. Baert, M.J. Van Bael, V.V. Moshchalkov and Y. Bruynseraede, *these proceedings*.

ELSEVIER

Physica C 282–287 (1997) 383–386

PHYSICA C

Josephson Plasma in Cuprate High T_c Superconductors

Masashi Tachiki

National Research Institute for Metals, 1-2-1, Sengen, Tsukuba, Ibaraki 305, Japan

The nature of the *I-V* characteristics along the c axis of $Bi_2Sr_2CaCu_2O_{8+\delta}$ is explained on a model in which the superconducting CuO_2 layers are coupled by the Josephson effect through the block layers. In this model the Josephson plasma is generated by the coupling between the Josephson current flowing along the c axis and the electromagnetic field. The longitudinal and transverse plasma waves propagating along the c axis and in the ab-plane respectively have quite different nature. The Josephson plasma waves in high T_c superconductors have been observed by reflectivity measurements of far infrared light and experiments of microwave absorption. The Josephson plasma strongly interacts with the vortices introducted by an external magnetic field. Therefore, the plasma frequency strongly depends on the vortex states such as the vortex liquid and solid states and on the vortex species like pancake and Josephson vortices.

1. INTRODUCTION

The superconducting properties of the cuprate superconductors are well described by a model in which the superconducting CuO_2 layers are coupled by the Josephson effect through the block layers. For example a peculiar nature of the I-V characteristics along the c axis of $Bi_2Sr_2CaCu_2O_{8+\delta}$ in the superconducting state is explained on the basis of this model. From this model we expect a unique excitation in high T_c superconductors. In the excitation, the Josephson current flowing along the c axis couples with the electromagnetic field and oscillates with a fairly low frequency in a range of Giga Hz and Tera Hz. We call the excitation mode the Josephson plasma. The longitudinal plasma propagating along the c axis has a different nature from the transverse plasma propagating in the a-b plane. These plasmas have been observed by the experiments of the reflectivity measurements using far infrared light and the microwave absorption measurements. The plasmas strongly interact with vortices and thus the plasma frequencies of the plasmas drastically change according to the various vortex phases in the high T_c superconductors.

2. I-V CHARACTERISTICS

Figure 1 shows the alternate stack of the superconducting CuO_2 layers and the insulating block layers in high Tc superconductors. The Josephson current flowing between the lth and $(l+1)$th layers is expressed as

$$j_c \sin P_{l+1,l}, \tag{1}$$

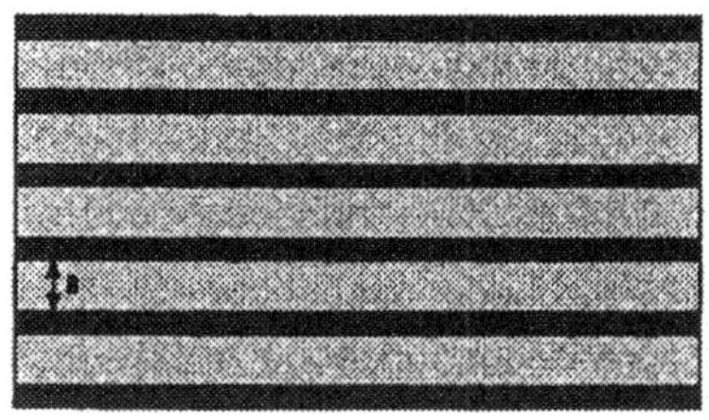

Figure. 1 Stack of the CuO_2 and block layers in high T_c superconductors

where $P_{l+1,l}$ is the gauge invariant phase difference between the lth and $(l+1)$th layers

$$P_{l+1,l} = \varphi_{l+1} - \varphi_l - \frac{2\pi}{\Phi_0}\int_l^{l+1} A_z dz, \tag{2}$$

φ_l being the phase at the lth layer, Φ_0 the unit flux, and A_z the vector potential. The time derivative of Eq. (2) is calculated as [1]

$$\frac{\hbar}{2e}\frac{\partial}{\partial t}P_{l+1,l} = sE_{l+1,l} - \frac{\mu^2\varepsilon}{s}\left(E_{l+2,l+1} - 2E_{l+1,l} + E_{l,l-1}\right), \tag{3}$$

with use of the Maxwell relations

$$E_z = -(1/c)(\partial/\partial t)A_z - (\partial/\partial z)A_o, \tag{4}$$

$$\rho_l = \frac{1}{4\pi\mu^2}\left(A_o(Z_l) + \frac{\phi_0}{2\pi c}\frac{\partial}{\partial t}\varphi_l\right), \tag{5}$$

$$E_{l+1,l} - E_{l,l-1} = (4\pi s/\varepsilon)\rho_l, \tag{6}$$

0921-4534/97/$17.00
PII S0921-4534(97)00281-5

where $E_{l+1,l}$ is the electric field between the lth and $(l+1)$th layers, A_0 the scalar potential, ρ_l the charge density of the lth layer, μ the charge screening length, and s the distance between the neighboring CuO_2 layers. The last term in Eq. (3) comes from the charging effect which couples among the layers.

For calculating the I-V characteristics we use a relation that the total current is equal to the current I supplied from an external source [1]

$$j_c \sin P_{l+1,l} + \sigma E_{l+1,l} + \frac{\varepsilon}{4\pi}\frac{\partial}{\partial t} E_{l+1,l} = I, \tag{7}$$

Machida solved the coupled equations of Eqs. (3) and (7) numerically and obtained the I-V characteristics shown in Fig. 2. [2] His calculation was done using a 50 layers sample. The result can explain the behavior of the I-V characteristics measured by Kleiner and Müller. [3,4]

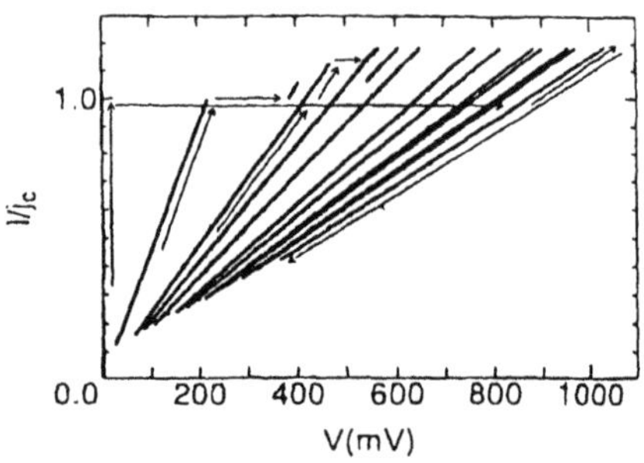

Figure. 2 Calculated I-V characteristics for the 50 superconducting layers coupled by the Josephson effect.[2]

3. LONGITUDINAL JOSEPHSON PLASMA

The equation for the longitudinal plasma is obtained from Eq. (3) and Eq. (7) where σ and I vanish as [5]

$$\frac{\partial^2}{\partial t^2} P = v_B^2 \frac{\partial^2}{\partial t^2} P - \omega_p^2 P, \tag{8}$$

with

$$\omega_p = \frac{c}{\lambda_c \sqrt{\varepsilon}}, \lambda_c = \sqrt{\frac{\hbar c^2}{4 j_c e s}}, v_B = c\frac{\mu}{\lambda_c}.$$

The order of magnitude of v_B is the Fermi velocity. In deriving Eq. (8) we used a continuous approximation assuming that the wave number k is much smaller than $2\pi/s$, s being the distance between the CuO_2 layers. Solving Eq. (8) we obtain the angular frequency for the longitudinal plasma wave as

$$\omega = \left(\omega_P^2 + v_B^2 k^2\right)^{1/2}. \tag{9}$$

The frequency dispersion for a neutral superconductor is $v_B k$, since the electronic charge e and thus ω_p vanish. This excitation mode with gapless dispersion is the Nambu-Goldstone mode which arises from the gauge broken symmetry in the superconducting state. The finite gap ω_p comes from the Anderson-Higgs mechanism, which takes account of the Coulomb interaction. Therefore, the longitudinal plasma wave is the phase wave dressed by the longitudinal electric wave.

The plasma frequency is much lower than the large energy gap of the high T_c superconductor. Therefore, the plasma damping is very weak and thus the plasma is quite stable. This is the reason that the longitudinal plasma mode was first observed in the high T_c superconductors as will be mentioned in section 4.

4. TRANSVERSE JOSEPHSON PLASMA

The equation for the transverse Josephson plasma is derived from Eq. (2) and the Maxwell equations [5]

$$(\partial/\partial x) H_{l+1,l} = (\varepsilon/c)(\partial/\partial t) E_{l+1,l} + (4\pi/c) j_c \sin P_{l+1,l}, \tag{10}$$

$$(\partial/\partial x) E_{l+1,l} = (1/c)(\partial/\partial t) H_{l+1,l}, \tag{11}$$

as

$$(\partial^2/\partial x^2) P_{l+1,l} = (\varepsilon/c^2)(\partial^2/\partial t^2) P_{l+1,l} + (1/\lambda_c^2) P_{l+1,l}. \tag{12}$$

Solving Eq. (12), we obtain the angular frequency of the longitudinal plasma as

$$\omega = \left(\omega_P^2 + c^2 k^2/\varepsilon\right)^{1/2}. \tag{13}$$

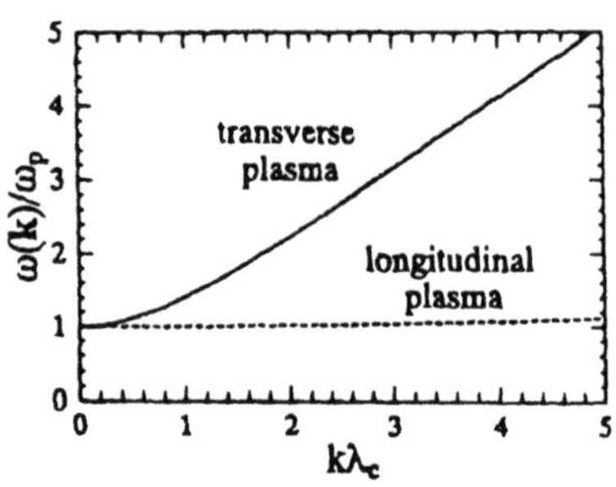

Figure. 3 Dispersion relations of the Josephson plasma waves.

The longitudinal plasma is a composite of the wave of

the Josephson current and the electromagnetic wave. This plasma is similar to the plasma polariton in semiconductors. The gap of the plasma is the same as that of the transverse plasma. However, the dispersion is much stronger than that of the transverse plasma as shown in Fig. 3.

5. OBSERVATION OF THE PLASMA WAVES

The first experiment to observe the Josephson plasma was done by Tamasaku, Nakamura, and Uchida [6] by measuring the reflectivity for infrared light in bulk single crystals of $La_{2-x}Sr_xCuO_4$. They observed a plasma edge structure in the frequency range of 20 cm^{-1}-50cm^{-1}. Recently they measured the reflectivity for thick films with the surfaces parallel to the c axis of $La_{2-x}Sr_xCuO_4$ and observed an oscillatory behavior of the reflectivity as shown Fig. 4. [7]

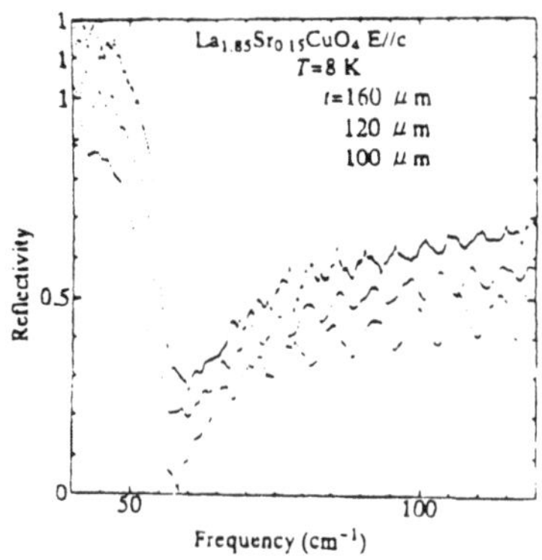

Figure. 4 Frequency dependence of the reflectivity for three films of $La_{1.85}Sr_{0.15}CuO_4$. [7]

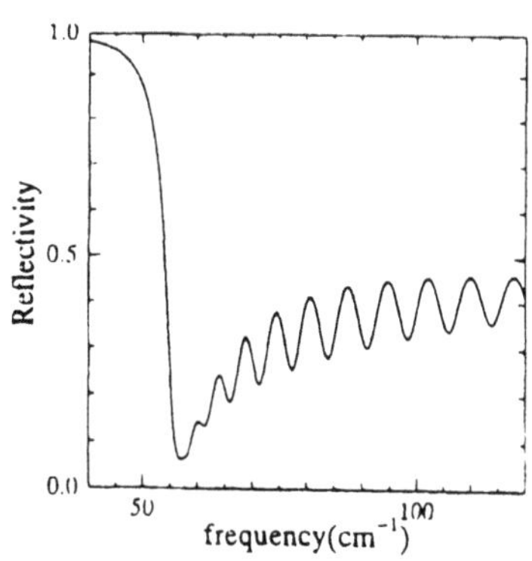

Figure. 5 Frequency dependence of the reflectivity for a firm with thickness of 100μm, calculated by assuming the d-wave pairing. [8]

The incident infrared light excites the plasma waves in the film, and the waves reflected at the surfaces of the film interfere with each other. The oscillatory behavior of the reflectivity comes from the interference effect of the excited plasma waves in the film. The frequency dispersion obtained from the interference pattern coincides with that of the transverse plasma. The frequency dependence of the reflectivity of a film with thickness of 100μm calculated by Takahashi assuming a d-wave pairing is shown in Fig. 5. [8] The calculated reflectivity is in good agreement with the experimental one, indicating that the superconducting pairing in $La_{2-x}Sr_xCuO_4$ is of a d-wave.

The longitudinal Josephson plasma was measured by absorption of a microwave. Tsui et al. [9] observed a sharp peak of the microwave absorption in $Bi_2Sr_2CaCu_2O_{8+\delta}$ and Matsuda et al. [10] confirmed that the peak originates from the plasma excitation by the polarization experiments. After that many experiments on the plasma resonance have been done. In the experiments a thick film of $Bi_2Sr_2CaCu_2O_{8+\delta}$ is placed at the position in a cavity resonator, where the microwave electric field has a maximum amplitude and the microwave magnetic field has a node. The sample size is small enough compared with the cavity dimension, ensuring that the microwave electric field is uniform and is exerted perpendicular to the ab-plane of the sample. In this setup the microwave electric field excites the plasma wave with the zero component of the wave number vector parallel to the ab-plane. The excited plasma waves are standing waves with amplitudes varying in the direction of the c axis. If the film is assumed to extend infinitely in the ab-plane, the components of the wave number vector parallel to the c axis are finite and the components in the other directions are zero. Therefore, the excited plasma waves in this case are the longitudinal plasma waves.

We calculate the absorption power of the signal due to the longitudinal plasma resonance as a function of the microwave frequency. In the microwave experiments the power absorption is usually measured by fixing the frequency and varying an external magnetic field. For inverting the frequency dependence of the absorption power to the magnetic field dependence, we use the experimental fact that the plasma frequency deceases in an external magnetic field B in proportion to $B^{-1/2}$. The decrease of the plasma frequency comes from the fact that the effective Josephson coupling between the layers is weakened by the vortices introduced by the external magnetic field. The calculated absorption power is shown as a function of the normalized magnetic field B/B_0 in Fig. 6 [8]. An example of the experimental data for the microwave absorption in $Bi_2Sr_2CaCu_2O_{8+\delta}$ is also shown by the dotted line in Fig. 6. [11] The shapes of the theoretical and experimental absorption peaks are in good agreement with each other. In the calculation we used the quasiparticle conductance for the d wave pairing state. If we assume the s wave pairing state, the quasiparticle conductance is zero for the frequencies smaller than the superconducting energy gap, and thus the absorption power vanishes at low temperatures, being inconsistent with the experiments.

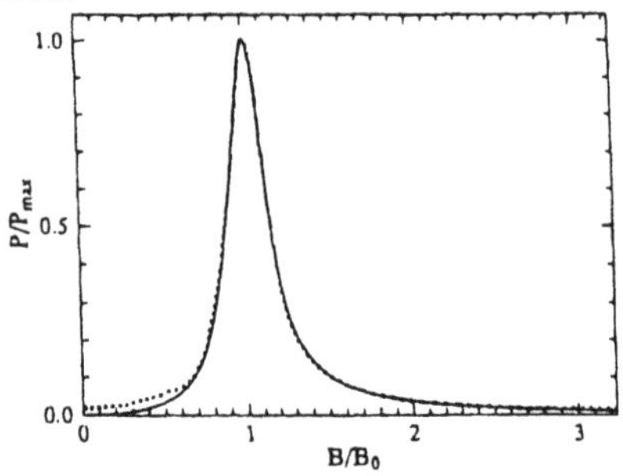

Figure. 6 Magnetic field dependence of the microwave absorption power due to the plasma resonance. The solid and dotted curves indicate the theoretical and experimental values.

6. INTERACTION BETWEEN THE PLASMA WAVES AND VORTICES

The vortices introduced by an external magnetic field modify the phase difference between the layers and change the effective Josephson coupling responsible for the plasma oscillation. The phase difference is divided into two parts; one is the contribution from the vortices $P^{v}_{l+1,l}$ and the other is the contribution from the plasma $P^{P}_{l+1,l}$:

$$P_{l+1,l} = P^{v}_{l+1,l} + P^{P}_{l+1,l}. \tag{14}$$

If we insert Eq. (14) into Eq. (1) and pick up the term linear in $P_{l+1,l}$, we have

$$j_c \cos P^{v}_{l+1,l} \cdot P^{P}_{l+1,l} \tag{15}$$

From Eq. (15), we see that the effective Josephson coupling responsible for the plasma oscillation is

$$J_{eff} = j_c \cos P^{v}_{l+1,l}$$

Since the plasma frequency is proportional to the square root of the effective Josephson coupling, the plasma frequency is obtained as

$$\omega_P(B) = \omega_P(o) \langle \cos^{v}_{l+1,l} \rangle^{1/2}. \tag{16}$$

where $\langle \cdots \rangle$ denotes the spatial average for the vortex distribution. The plasma frequency strongly depends on the vortex states such as the vortex liquid and solid states. The pancake and Josephson vortices change the plasma frequency in a different way. The plasma frequency change due to the vortices have intensively been investigated by Bulaevskii et al. [12] and Koshelev. [13] Here we show the result of the numerical simulation for $\langle \cos P_{l+1,l} \rangle$ by Machida. [2] He applied magnetic field parallel to the c axis and used the time-dependent Ginzburg-Landau equation. The results for three magnetic fields are shown by the circles, squares, and triangles in Fig. 7.

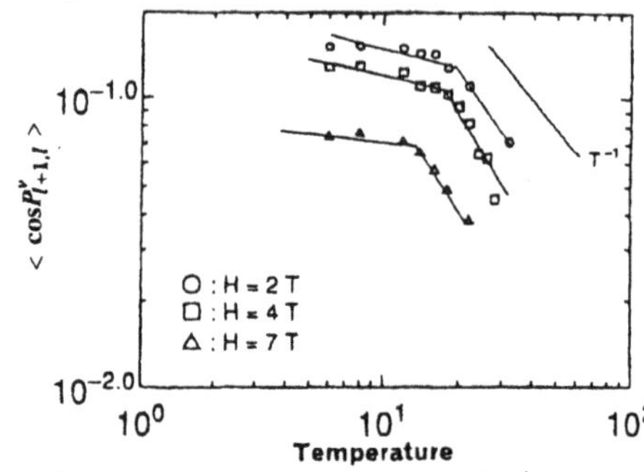

Figure. 7 Temperature dependence of $\langle \cos P^{v}_{l+1,l} \rangle$ for three magnetic fields. [2]

The kinks of the lines correspond to the vortex melting points. In the vortex liquid state above the melting point, $\langle \cos P^{v}_{l+1,l} \rangle$ is inversely proportional to temperature in agreement with the experiments.

The author would like to express sincere thanks to Dr. T. Koyama, Dr. M. Machida, Prof. S. E. Shafranjuk, and Dr. S. Takahashi for their kind collaborations.

REFERENCES

[1] T. Koyama and M. Tachiki, Phys. Rev. B, 54 (1996) 1618
[2] Machida, private communication
[3] R. Kleiner et al., Phys. Rev. Lett. 68 (1992) 2394; Phys. Rev. B49 (1994) 1327
[4] P. Müller, Physica C 235-240 (1994) 289
[5] M. Tachiki, T. Koyama and S. Takahashi, Coherence in High Temperature Superconductors, edited by G. Deutscher and A. Revolevschi (World Scientific, Singapore, 1996)
[6] K. Tamasaku et al., Phys. Rev. Lett. 74 (1995) 801
[7] S. Uchida, private communication
[8] S. Takahashi, private communication
[9] O. K. C. Tsui et al., Phys. Rev. Lett. 73 (1994) 724
[10] Y. Matsuda et al., Phys. Rev. Lett. 75 (1995) 4512
[11] K. Kadowaki, private communication
[12] L. N. Bulaevskii et al., Phys. Rev. Lett. 76 (1996) 1719; Phys. Rev. B 53 (1996) 6634
[13] A. E. Koshelev, Phys. Rev. Lett. 77 (1996) 3901

ELSEVIER

Physica C 282–287 (1997) 387–390

PHYSICA C

Intrinsic Josephson effects on BSCCO 2212 single crystal whiskers

Yu.I. Latyshev[a,b], P. Monceau[b], V.N. Pavlenko[a]

[a]Institute of Radio-Engineering and Electronics, 11 Mokhovaya St., Moscow 103907, Russia

[b]Centre de Recherches sur les Très Basses Températures, laboratoire associé à l'Université Joseph Fourier, C.N.R.S., BP 166, 38042 Grenoble-Cedex 9, France

Micron scale stacked junctions have been fabricated from $Bi_2Sr_2CaCu_2O_8$ whiskers. For in plane size L less than 20 μm measurements of the critical current I_c along the c-axis as a function of the parallel magnetic field H clearly demonstrate the intrinsic dc Josephson effect. For larger values of L a dimensional crossover to monotonic, size independent behaviour of $I_c(H)$ is observed. Flux flow steps on current voltage characteristics of overlap stacks have been found in high parallel magnetic fields, due to the collective motion of Josephson vortex lattice.

1. INTRODUCTION

Layered high T_c superconductors such as $Bi_2Sr_2CaCu_2O_8$ (BSCCO) are known to be considered as a stack of 2D-superconducting layers linked by Josephson coupling [1]. However it was shown [2] that the direct observation of intrinsic Josephson effects (IJE) when the current is driven across the layers can only be revealed for stacked structure with lateral size L_{ab} smaller than the Josephson penetration depth λ_j given by $s\ \lambda_c/\lambda_{ab}$, where s is the spacing between elementary superconducting CuO layers and λ_c, λ_{ab} are the anisotropic London penetration lengths. In BSCCO, λ_j is of the order of a few microns.

First evidence of IJE was obtained on rather large BSCCO single crystals with $L \sim$30-100 μm [1]. A great experimental effort is made for preparation of stack junctions with smaller sizes [3].

We have developed a method of fabrication of micron scale stacked junctions from single crystal $Bi_2Sr_2CaCu_2O_8$ whiskers [4,5]. Hereafter we describe our measurements which clearly demonstrate the existence of the intrinsic Josephson effects.

2. EXPERIMENTAL

Junctions have rectangular geometry in the ab-plane, the edges being parallel to the a- and b-axis (Fig. 1). Different junctions have been prepared with dimensions L_a, L_b between 200 μm and 5 μm. Along the c-axis typically they contain 20-100 elementary junctions. The critical current density across the layers, I_c, measured at 4.2 K with a voltage criterion of 1 μV was $5 \times 10^2 - 2 \times 10^3$ A/cm^2. This value is about three orders smaller than the longitudinal value of 5×10^5 A/cm^2 in samples from the same batch [4].

3. dc INTRINSIC JOSEPHSON EFFECT

For small junctions with in-plane size L_{ab} smaller than λ_j, I_c across the layers as a function of the magnetic field H parallel to the layers exhibits [2] a Fraunhofer behaviour such as:

$$I_c(H) = I_c(0)\sin(\pi sLH/\phi_0)/(\pi sLH/\phi_0) \qquad (1)$$

with ϕ_0 the flux quantum, L the junction size perpendicular to H. $I_c(0)$ is the maximum Josephson current across the layers which is defined by current density $J_c(0)$: $J_c(0) = c\phi_0/(8\pi^2 s\lambda_c^2)$. The first minimum of $I_c(H)$ appears at $H_1 = \phi_0/sL$.

For junctions with larger size, the Josephson behavior is disturbed by Josephson vortices entering the junction at fields $H > H_{c_1}$. In the case of $L > \lambda_j$, a recent calculation [6] predicts an universal size independent decrease of $I_c(H)$:

$$I_c(0) - I_c(H)/I_c(0) \approx \sqrt{H/H_0} \qquad (2)$$

with $H_0 = \phi_0\lambda_{ab}/\pi^2 s^2\lambda_c$ and $I_c(0)$ the same as in (1): $I_c(0) = s\ J_c(0)$.

PII S0921-4534(97)00282-7

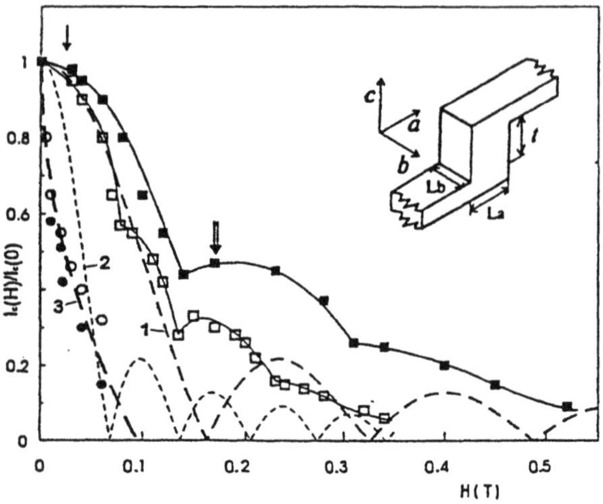

Figure 1. Normalized dependences of critical current of $Bi_2Sr_2CaCu_2O_8$ junctions across the layers $I_c(H)/I_c(0)$ at T =4.2 K with H parallel to the layers for samples of different sizes L: □ 8 μm, □ 20 μm, • 40 μm, ∘ 200 μm. Solid lines are guides for the eyes. Dashed curves 1 and 2 correspond to Eq. 1 for $L = 8$ and 20 μm, respectively. Curve 3 corresponds to Eq. 2 for $H_0 = 950$ Oe. Insert shows the geometry of the junction.

We have studied the $I_c(H)$ behaviour for a number of stacks with different lateral sizes. For large junctions with 40 μm $< L <$ 200 μm, we have observed a rapid monotonous drop of $I_c(H)$, independent of the junction size. The data follow the expected $\sqrt{H/H_0}$ law. Using the theoretical expression for H_0, we deduce $\gamma = \lambda_c/\lambda_{ab} \sim 1300$. Using this value for γ we estimated the value of maximum Josephson current density for our junctions. It gives $J_c(0) \sim 1.5 \times 10^3$ A/cm^2 which is quite consistent with our experimental values.

With $L <$ 30 μm, oscillations in $I_c(H)$ begin to appear. For the junction with $L = 8$ μm, we observed three oscillations with a period of 0.15 T. The period of oscillations decreases with an increase of the sample size as directly demonstrated on a junction with $L_a = 8$ μm and $L_b = 20$ μm by rotating H in the ab-plane (Fig. 1). Our data for junctions with a size smaller than 20 μm prove the predicted Josephson behaviour of $I_c(H)$ [2]. When L is increased above 20 μm, a crossover occurs to the behaviour predicted in [6]. This crossover is illustrated in Fig. 2 where the dependence of the magnetic field, H^*, which suppresses I_c to half its value at zero field $I_c(0)$ is plotted as a function of the junction size [7].

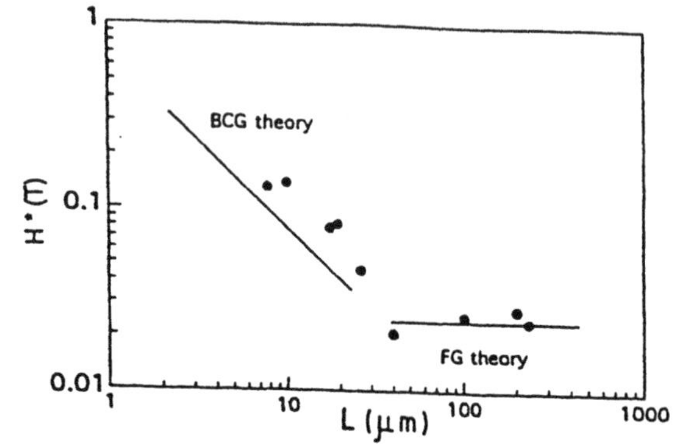

Figure 2. Dependence of the characteristic magnetic field H^*, as a function of the junction size L. Straight lines correspond to theories [refs. 2 and 6] calculated from Eq. 1 and Eq. 2.

The current voltage characteristics (IVC) of stacked junctions at low temperatures are known to have hysteresis and a multiple branch structure [1]. With current increase the voltage jumps to higher and higher branches until all the elementary junctions become resistive. This limiting quasiparticle branch should correspond to the voltage $2\Delta \cdot N$, with 2Δ the energy gap and N the total number of elementary junctions. Usually this value is 30-50% smaller [1] due to the gap suppression by heating effects [8] or quasiparticle injection [9]. Both effects often manifest themselves as a S-shaped form of the quasiparticle branch. Our experiments showed that S-shaping of the IVC can be essentially eliminated with a decrease of the overlap length L_a of the junction to 6-4 μm (overlap type junction). The IVC of the overlap junction with $L_a = 4$ μm is shown in Fig. 3. It has no S-shaped branch at high dissipation levels. We consider that a decrease of the overlap length leads to a better junction thermocoupling with the rest of the crystal.

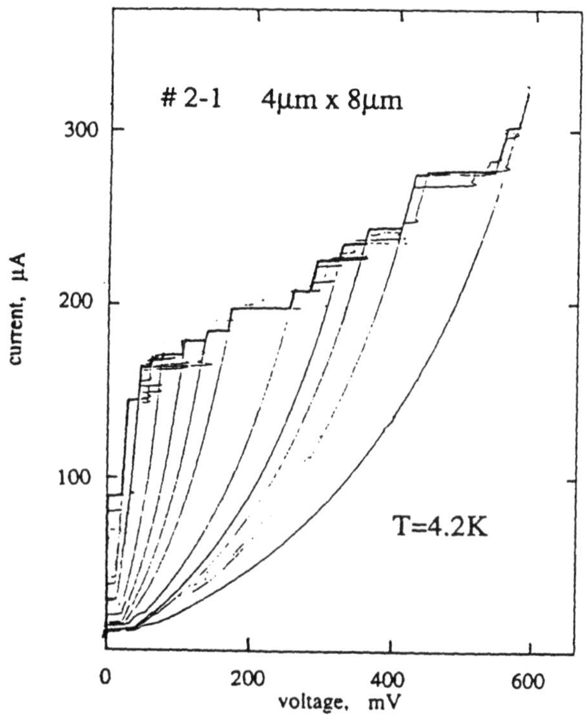

Figure 3. I-V curves of a 4 μm × 8 μm BSCCO stack junction at 4.2 K.

4. FLUX-FLOW MODE IN PARALLEL MAGNETIC FIELD

For long junctions with $L > \lambda_j$ the ac Josephson effect can be associated with the fluxon motion when H is applied parallel to the junction. Fluxons can be moved by a current flowing through the junction and their motion leads to electromagnetic radiation. The frequency of radiation ν is given by Josephson relation: $\nu = (2e/h)V = cV/\Phi_0$, where V is a dc voltage induced by the fluxon motion. In flux-flow regime, fluxons are created at one boundary of the junction and annihilate at the other boundary. The radiation frequency ν is determined by the fluxon velocity v and the spacing between moving fluxons Λ as $\nu = v/\Lambda$. The voltage corresponding to the flux-flow mode is then:

$$V = (v/c)(\Phi_0/\Lambda). \quad (3)$$

For layered superconductors it was recently shown [10] that in high enough parallel field a triangle lattice of Josephson vortices is formed. For $H > \phi_0/\gamma s^2$ the non-linear regions of vortices overlap strongly because Λ along the ab-plane becomes smaller than λ_j. Such a dense lattice was shown to be rigid enough to move as a whole [11] when the current is driven across the layers. A resonance can appear when the lattice velocity $\bar{v}$ is equal to $\bar{c}/2$, where $\bar{c}$ is the Swihart velocity given in [11] as:

$$\bar{c} = cs/(\lambda_{ab}\sqrt{\epsilon_c}) \quad (4)$$

where ϵ_c is the dielectric constant between superconducting layers.

We have searched such an effect in our overlap structures with $L_a = 8$ μm, $L_b = 20$ μm and with $H \parallel a$. H was varied up to 1.5 T. That is about twice the characteristic field $\phi_0/\gamma s^2$. For $H \approx 0.5$ T, $I_c(H)$ becomes negligibly small, $I_c(H)/I_c(0) \sim 0.5\%$, and a step of low differential resistance appears on the IVC. With the increase of H the step moves to higher voltages (Fig. 4). With current increase the step changes its curvature. The step is hysteresisless. It means that all the elementary junctions are synchronized at this state. With further current increase, IVC instabilities appear and then the stack comes to the multiple branch, hysteretic state, which is similar to the one observed above I_c at zero magnetic field. We associate the observed step as the flux-flow step for collective motion of the Josephson vortex lattice [11]. The maximum voltage of the step position varies approximately proportionally to H (see inset Fig. 4). Using Eq. 3 and taking into account that $v = \bar{c}/2$, $\Lambda = \phi_0/Hs$, the resonance peak voltage V_r is obtained as:

$$V_r = \frac{1}{2}N\frac{\bar{c}}{c}sH \quad (5)$$

with N the number of the layers of the moving lattice. Exactly the same expression has been deduced in [11] where the resonance electric field across the layers E_0 was found as $E_0 = H\bar{c}/2c$. The ratio $\bar{c}/c$ can be estimated as 2.8×10^{-3} for $\lambda_{ab} = 1700$ Å, $s = 15$ Å, $\epsilon_c = 10$. Using that value and $N = 30$, $H = 1.14$ T, we get from Eq. 5 $V_r = 20$ mV which is in agreement with the experimental value, $V_0 = 24$ mV.

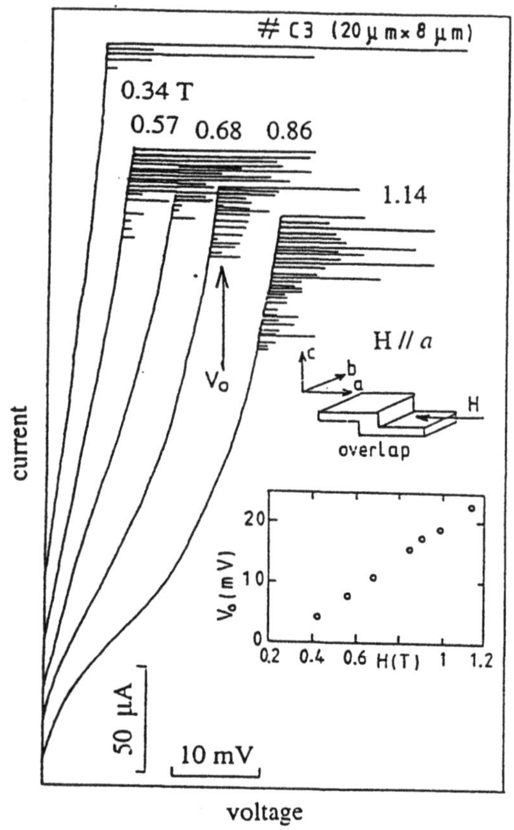

Figure 4. Series of I-V curves of BSCCO 8 μm $\times$ 20 μm overlap stacked structure in various magnetic fields $H \parallel a$ at 4.2 K. Inset shows the dependence of the maximum step voltage V_0 as a function of H.

The value of the Swihart velocity $\bar{c}$ in BSCCO stacks can be directly extracted from the experiment using Eq. 5. We found 9×10^7 cm/s at 4.2 K. Because the Swihart velocity is rather high, we can expect generation of the high frequency electromagnetic field near $V = V_0$. The generation frequency can be estimated as: $\nu \approx \bar{c}sH/2\phi_0$ which gives $\nu = 0.3$ TGz for $H = 1$ T. It points out the ability of the overlap stacks to be used as high frequency local oscillators in various devices of superconducting electronics.

5. CONCLUSIONS

We have developed micron scale stacked structures from BSCCO 2212 single crystal whiskers. We have shown a dimensional crossover for intrinsic dc Josephson effect for a junction size of $\sim$ 20 – 30 μm. Steps in the current voltage characteristics under a parallel magnetic field on overlap stacked analogues of long Josephson junctions have been found to be due to a collective motion of a Josephson vortex lattice.

Acknowledgements — We acknowledge fruitful discussions with L.N. Bulaevskii, A.F. Volkov, A. Ustinov. The work has been supported by Russian State Program on HTSC under the project N95028.

REFERENCES

1. For a review P. Müller, *"Intrinsic Josephson Effects in Layered Superconductors"*, ed. R. Helbig, Festkörperproblem, Advances in Solid State Physics, Vol. 34 (Vieuweg, Braunschweig/Wiesbaden 1994) p. 1.
2. L.N. Bulaevskii, J.R. Clem, and L.I. Glazman, Phys. Rev. B 46 (1992) 350.
3. P. Seidel *et al.*, p. 63; K. Schlenga *et al.*, p. 60; R. Dömel *et al.*, p. 147 in Proc. of the ICEC95 Conf., Nagoya (Japan) 1995 (Inst. of Physics Publ., Bristol 1996).
4. Yu.I. Latyshev, I.G. Gorlova, A.M. Nikitina *et al.*, Physica C 216 (1993) 471.
5. Yu.I. Latyshev and J.E. Nevelskaya, Physica C 235-240 (1994) 2991.
6. M.V. Fistul and G.F. Giuliani, Physica C 230 (1994) 9.
7. Yu.I. Latyshev, J.E. Nevelskaya, and P. Monceau, Phys. Rev. Lett. 77 (1996) 932.
8. T. Yasuda *et al*, Proc. of LT-21, Czech. J. Phys. 46 suppl. S3 (1996) 1265.
9. A. Yurgens, D. Winkler, Y.M. Zhang, N. Zavaritsky and T. Claeson, Physica C 235-240 (1994) 3269.
10. L.N. Bulaevskii and J.R. Clem, Phys. Rev. B 44 (1991) 10234.
11. L.N. Bulaevskii, D. Dominguez, M.P. Maley, A.R. Bishop, and B.I. Ivlev, Phys. Rev. B 53 (1996) 14601.

ELSEVIER

Physica C 282–287 (1997) 391–394

PHYSICA C

Interlayer Phase Coherence in the Vortex State of $Bi_2Sr_2CaCu_2O_{8+\delta}$ Proved by Josephson Plasma Resonance

Y. Matsuda[a], M.Kosugi[a,b], M.B.Gaifullin[a], K.Kumagai[a], K.Hirata[c], N.Chikumoto[d], M. Konczykowski[e], S.Watauchi[f], J.Shimoyama[f], and K.Kishio[f]

[a]Department of Physics, Hokkaido University, Sapporo 060, Japan

[b]Department of Physics, Aoyama-Gakuin University, Setagaya, Tokyo 157, Japan

[c]National Research Institute for Metals, 1-2-1 Sengen, Tsukuba, Ibaraki 305, Japan

[d]Superconductivity Research Laboratory, ISTEC, Minato-ku, Tokyo 105, Japan

[e]Ecole Polytechnique, Laboratoire des Solides Irradies, 91128 Palaiseau Cedex, France

[f]Department of Superconductivity, University of Tokyo, Bunkyo-ku, Tokyo 113, Japan

The interlayer phase coherence in the vortex state of $Bi_2Sr_2CaCu_2O_{8+\delta}$ is studied by Josephson plasma resonance in a wide range of temperatures and magnetic fields. The plasmon frequency changes drastically at the first order vortex liquid-solid melting transition line at low field. The vortex lines are shown to be almost decoupled just above the melting line. The effect of columnar defects on the plasma frequency is also investigated in a heavy ion irradiated $Bi_2Sr_2CaCu_2O_{8+\delta}$. We show the direct evidence that the strongly correlated vortex-line-liquid is formed by the columnar defects.

1. INTRODUCTION

It is well established that high-T_c superconductors (HTSC) behave as stacks of superconductor-insulator-superconductor Josephson junctions [1]. Recently, collective plasma oscillation of Cooper pair through the insulating barriers has attracted much attention. When the Josephson current couples to the electromagnetic field, the Josephson plasma mode in which the supercurrent and the electric field oscillate normal to the CuO_2 planes in the crystal is generated [2][3]. Since this plasmon mode lie well below the superconducting gap, the plasma damping processes such as Landau damping and optical phonon damping are prohibited from occurring and a very stable mode appears in the superconducting state. Recently, a resonance arising from the Josephson plasmon has been discovered in the vortex state of $Bi_2Sr_2CaCu_2O_{8+\delta}$ [4, 5]. Similar resonance has also been reported in $Bi_2Sr_2CuO_{6+\delta}$ [6] and κ-(BEDT-TTF)$_2$Cu(NCS)$_2$ [7] with extremely large anisotropies.

One of the most important properties of the Josephson plasmon is that it strongly interacts with the vortices created by the external fields. Therefore Josephson plasma resonance (JPR) is a powerful new means of investigating the properties of the vortex state, enabling us to determine the magnitude of the interlayer phase coherence. The vortex phase diagram of HTSC has been a subject of intense study. Especially in $Bi_2Sr_2CaCu_2O_{8+\delta}$ with extremely large anisotropy, the vortex liquid phase occupies a significant part of the $B-T$ phase diagram. However, it remains uncertain how the interlayer phase coherence changes when the vortices freeze in the vortex solid phase, and how the loss of the interlayer phase coherence occurs with change in temperature and field in the liquid phase.

In this paper we have performed the JPR measurement in a wide range of magnetic field in $Bi_2Sr_2CaCu_2O_{8+\delta}$, ranging from well above to below the first order melting transition line at low field [8]. In almost the entire region of the liquid phase, the plasmon frequency is well described

PII S0921-4534(97)00283-9

by the high temperature expansion theory. We show that the vortex lines are almost decoupled just above the melting line. On the other hand, introducing the columnar defect changes the nature of the vortex state drastically. We provide a strong evidence that the strongly correlated vortex-line-liquid is formed due to the columnar defects.

2. EXPERIMENTAL

The single crystals of $Bi_2Sr_2CaCu_2O_{8+\delta}$ have been grown by the traveling-solvent-floating-zone technique. The samples with different oxygen stoichiometry were obtained by annealing in the vacuum. The crystal of typically 0.5x0.5x0.05mm^3 was placed in a rectangular cavity with TE_{102} or TE_{101} mode made of copper. The magnetic field was applied parallel to the c-axis. The sample was placed at the maximum microwave electric field $\mathbf{E}_{ac}$ (minimum $\mathbf{H}_{ac}$) and oriented such that $\mathbf{E}_{ac}$ is parallel to the c-axis. This configuration generates the JPR mode in which supercurrent oscillates normal to the CuO_2 planes. Figure1 shows the sharp resonance mode arising from the Josephson plasmon excitation in $\mathbf{H}||\mathbf{E}_{ac}||$ c configuration for as-grown sample [4]. Subsequently the samples were irradiated by 5.8GeV Pb-ions at GANIL (Caen, France) in order to introduce the columnar defects parallel to the c-axis. The irradiation dose expressed in terms of a matching field B_Φ were B_Φ=0.3 and 1T.

3. PLASMA FREQUENCY IN THE VORTEX STATE

The Josephson plasma frequency ω_{pl} is related to the c-axis critical current j_c^c by $\omega_{pl} = \sqrt{8\pi^2 cs/\epsilon_0\Phi_0 j_c^c}$, where ϵ_0 is the dielectric constant, s the layer spacing, and Φ_0 the flux quantum. In $Bi_2Sr_2CaCu_2O_{8+\delta}$, ω_{pl} in zero field is above or within the microwave window. In a Josephson coupled superconductor, external field suppress ω_{pl} strongly [3]. This suppression is caused by the Josephson strings that are created by the deviation from the straight alignment of the pancake vortices along the c-axis. When ω_{pl} coincides with the microwave frequency in fields, the resonant absorption of the microwave occurs. The field dependence of the plasmon frequency is written as [3, 9]

$$\omega_{pl}^2(B,T) = \omega_{pl}^2(0,T) < \cos\varphi_{n,n+1} > . \quad (1)$$

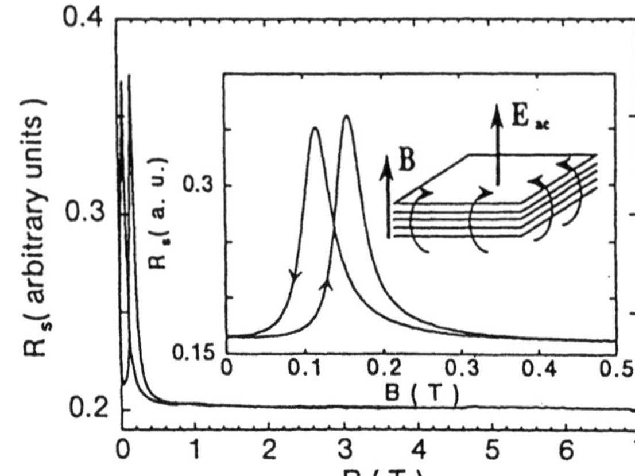

FIG. 1: Microwave absorption P_{abs} as a function of magnetic fields for $\mathbf{E}_{ac}$ || $\mathbf{B}$ || c at 45GHz (as-grown sample, T_c=87K). The inset shows P_{abs} at low fields.

Here $< \cos\varphi_{n,n+1} >$ represents the thermal and disorder average of the cosine of the gauge invariant phase difference between layer n and n+1. The field affects ω_{pl} through the phase factor. We note that $< \cos\varphi_{n,n+1} >$ remains finite in value even in the liquid phase where the d.c. critical current vanishes because the short range phase coherence survives in the liquid state [9].

4. PLASMA RESONANCE IN THE VORTEX LIQUID PHASE

Figure 2 displays the temperature dependence of the resonance field B_0 for underdoped sample (T_c=85K) at 45, 55 and 80GHz [10]. With decreasing T, B_0 increases, peaks, then decreases rapidly with a distinct cusp. The cusp temperature T_{cusp} locates very close to the irreversibility line IL which separates the states with pinned vortices from those with mobile unpinned vortices [4]. It has been pointed out that the origin of the rapid decreases of B_0 below T_{cusp} is attributed to the unusual two dimensional critical state of $Bi_2Sr_2CaCu_2O_{8+\delta}$ [10]. Here we focus on the resonance above T_{cusp}. A first order melting transition determined by the magnetization jump in SQUID magnetometer is also depicted in Fig.2 [11]. The resonance occurs just above (below) the melting line for 55GHz (80GHz). When going across the melting line, T-dependence of B_0 changes drastically from concave to convex [10].

Koshelev has recently shown that plasmon frequency in the liquid phase can be expressed as,

$$\omega_{pl}^2 = \frac{2\pi s\Phi_0}{\varepsilon_0 k_B T B} j_J^2, \qquad (2)$$

from the high temperature expansion theory, where j_J is the Josephson critical current in zero field which is related to the anisotropy factor $\gamma(\equiv \lambda_c/\lambda_{ab})$ by $j_J = c\Phi_0/8\pi^2\gamma^2\lambda_{ab}^2 s$ (λ_{ab} and λ_c are the in-plane and out-of-plane penetration lengths, respectively) [9]. This expression is valid if the layers are almost decoupled. We analyze our data in accordance with Eq.(2).

The frequency dependence of the resonance field $\omega_{pl}^2 \propto 1/B^{\mu}$ with μ=0.7-0.9 is obtained from 45-55GHz data. The inset of Fig. 2 shows $< \cos\varphi_{n,n+1} >$ at a constant field for 45, 55 and 80GHz calculated from Eq. 1 (We take μ=0.8). To obtain $< \cos\varphi_{n,n+1} >$, we assumed the Ambegaokar-Baratoff model for $j_J(T)$, $j_J(T) \propto \Delta(T)\tanh(\Delta(T)/2k_BT)$. For both 45GHz and 55 GHz, $< \cos\varphi_{n,n+1} >$ is inversely proportional to T in almost the entire region above T_{cusp}. The relation, $< \cos\varphi_{n,n+1} > \propto B^{-0.8}T^{-1}$, for 55GHz is very close to that resulting from to the high temperature expansion theory (Eq.2), suggesting that *the vortex lines are almost decoupled just above the melting line.* This fact implies that the solid vortex line lattice undergoes a *transition to the decoupled vortex gas* rather than to the correlated vortex line liquid at the melting line, i.e. the first order transition is "sublimation".

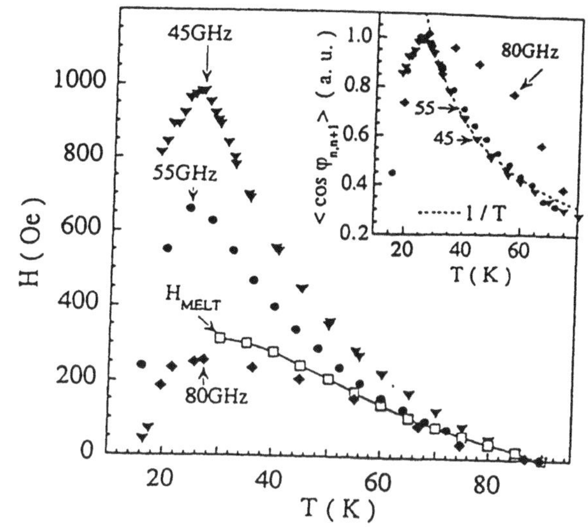

FIG. 2: The resonance fields vs T for optimally doped crystal (T_c=89.5K) at different frequencies in $\mathbf{H}\| c$; 45GHz (filled triangles), 55GHz (filled circles) and 80GHz (filled diamonds). The open squares represent the first-order melting transition fields from magnetization (H_{MELT}). Inset: Temperature dependence of $< \cos\varphi_{n,n+1} >$ at a constant field for 45GHz (filled triangles), 55GHz (filled circles) and 80GHz (filled diamonds). Data are normalized at the cusp temperatures. The dashed line represents the $1/T$ curve.

5. EFFECT OF COLUMNAR DEFECTS

Recently, heavy ion irradiation have been used to introduce well controlled defects into the material to enhance the pinning. The irradiation turned out to enhance the critical current density while moving the IL to higher field and temperature regime. However it remains uncertain how the interlayer phase coherence changes by the introduction of the columnar defects. Figure 3 shows T-dependence of B_0 for pristine and irradiated samples together with the IL. Due to the presence of columnar defects the IL is shifted toward higher temperatures and fields [12]. The irradiation also causes the drastic increase of the resonance fields, indicating the revival of the c-axis phase coherence [13]. The resonance fields increases with dose. We note that temperature and frequency dependence of the resonance for pristine sample is well described by Eq.(2).

In both irradiated samples three characteristic features are observed. The first feature occurs at the characteristic temperature T_{dec} ~70K. Above T_{dec}, the magnitude and temperature dependence of the resonance fields of irradiated samples are almost same as those of pristine sample. This indicates that the vortices are decoupled to the in-

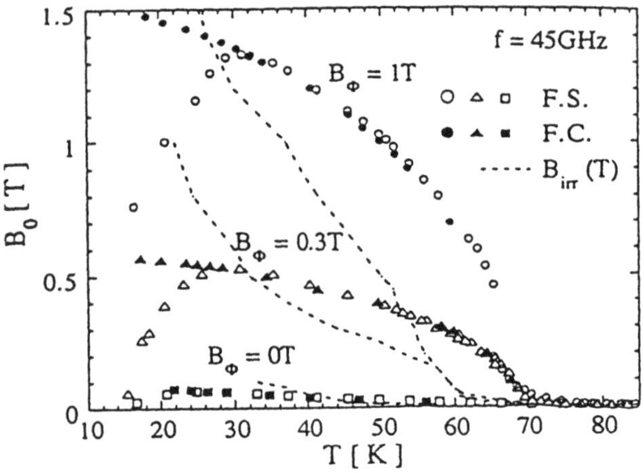

FIG. 3: Temperature dependence of the resonance fields B_0 for pristine(diamonds) and irradiated samples with B_Φ=0.3 (circles) and 1T (triangles) at 45GHz. Open and filled symbols represent the data taken at field sweep (FS) and field cooling (FC) conditions, respectively. The dashed lines represent the irreversibility lines determined by magnetization measurement.

dependent pancake gas due to the strong thermal fluctuation even in the presence of the columnar defects. Below T_{dec} B_0 increases rapidly, suggesting that the pancakes start to line up. Because the increase of B_0 occurs well above the IL, this result provide a strong evidence that the vortices behave as the strongly correlated *line-liquid*. The second feature occurs at a characteristic field B_Φ. The slope of T-dependence of the resonance field shows a slight change at $B \sim B_\Phi$. Above $\sim B_\Phi$, B_0 increases gradually with decreasing T. The third feature occurs at the IL. Below the IL, B_0 taken by the field sweep condition decreases rapidly with a distinct cusp with decreasing T, while B_0 for field cooling increases monotonically without showing a cusp behavior. This behavior of the resonance in the irradiated samples is same as that of the pristine sample [10]. Thus the introduction of the columnar defects alters the nature of the vortex state drastically by yielding the large enhancement of the c-axis coherence.

ACKNOWLEDGEMENTS

We gratefully acknowledge helpful discussions with L.N.Bulaevskii, T.Hanaguri, A.E.Koshelev, A.Maeda, T.Shibauchi and T.Tamegai for the helpful and stimulating discussions. We especially thank M.Tachiki for numerous discussion and continuing interest in this experiment. We are grateful to T.Mochiku and K.Kadowaki for providing as-grown $Bi_2Sr_2CaCu_2O_{8+\delta}$.

REFERENCES

[1] R.Kleiner and P.Müller, Phys. Rev. B **49**, 1327 (1994).
[2] M.Tachiki, S.Koyama, and S.Takahashi, Phys. Rev. B **50**, 7065 (1994).
[3] L.N.Bulaevskii, M.P.Maley and M.Tachiki, Phys. Rev. Lett. **74**, 801(1995).
[4] Y.Matsuda, M.B.Gaifullin, K.Kumagai, K.Kadowaki, T.Mochiku, Phys. Rev. Lett. **75**, 4512 (1995).
[5] O.K.C.Tsui, N.P.Ong, and J.B.Peterson, Phys. Rev. Lett. **76**, 819 (1996).
[6] S.Sakamoto *et al.* Phys. Rev. B **53**, R14749 (1996).
[7] T.Shibauchi, A.Mashio, M.Sato, T.Tamegai, H.Mori, S.TAjima and S.Tanaka, a preprint.
[8] E.Zeldov *et al.*, Nature (London) **375**, 373 (1995).
[9] A.E.Koshelev, Phys. Rev. Lett. **77**, 3901 (1996).
[10] Y.Matsuda, M.B.Gaifullin, M.Kosugi, K.Kumagai and K.Hirata, Phys. Rev. Lett. in press (1997).
[11] K.Hirata, T.Mochiku, and K.Kadowaki, Proceedings of the 8th International Workshop on Critical Currents in Superconductors, edts. T.Matsushita and K.Yamafuji, World Scientific Publishing Co. Pte. Ltd., 1996, p145.
[12] For example, D.Zech, S.L.Lee, H.Keller, G.Blatter, B.Janossy, P.H.Kes, T.W.Li, and A.A.Menovsky, Phys. Rev. B **52**, 6913 (1995).
[13] M.Kosugi, Y.Matsuda, M.B.Gaifullin K.Kumagai, N.Chikumoto, J.SHimoyama K.Kishio, K.Hirata, and M.Konczykowski, a preprint.

ELSEVIER

Physica C 282–287 (1997) 395–398

PHYSICA C

High Tc coplanar resonators for microwave applications and scientific studies

C. E. Gough, A. Porch, M.J. Lancaster, J.R. Powell, B. Avenhaus, J.J. Wingfield, D. Hung and R.G. Humphreys[a]

Superconductivity Research Group, University of Birmingham, Birmingham B15 2TT, UK.

[a] Defence Research Agency, Malvern, Worcestershire, WR14 3PS, UK.

We briefly introduce the use of microwave HTS coplanar resonators for measurements of absolute values of λ(T,B) and microwave losses with examples of temperature, static-field and power dependences. A number of applications are described including slow-wave resonators, current-switchable band-stop filters, and an ultra-sensitive ESR cavity. All these structures are easy to fabricate and offer substantial advantages over conventional technologies.

1. The coplanar resonator

Coplanar transmission line structures can be patterned very conveniently from high quality HTS films. A particularly simple in-line coplanar resonator structure is shown in Fig.1. The inductance per unit length of the transmission line is very approximately ~ $\mu_o [s + (\lambda_L^2/t) f(\lambda_L/t,s,w)]/w$, where s is the gap between the central conductor and ground planes, w is the width of the centre conductor, t is the film thickness and f is a geometric factor of order unity. The second term is the kinetic inductance associated with the supercurrents flowing near the planar edges. Porch et al. [1] have used numerical techniques to determine the current profile for a number of coplanar geometries, as illustrated in Fig.2, and hence obtain values for the kinetic inductance as a function of λ_L/t.

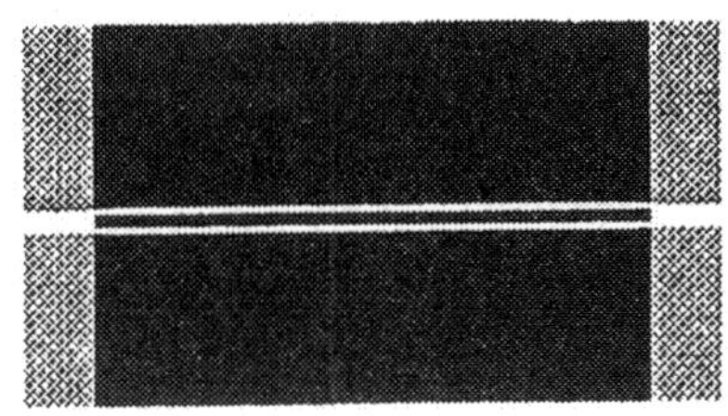

Fig.1 A typical in-line coplanar resonator, line length 8mm, width w=200μm and spacing s=73μm.

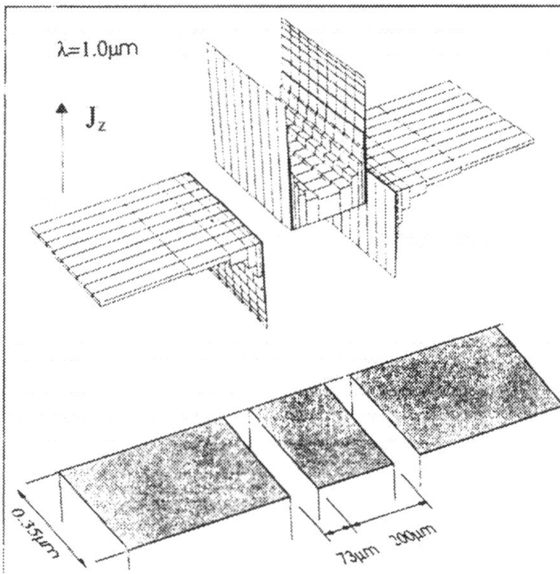

Fig.2. The current distribution on the resonator cross section [1].

2. Resonator measurements

By fabricating resonators with different spacings from the same HTS film, and assuming the same temperature dependence of $\lambda_L(T)$ for both resonators, it is relatively straightforward to extract absolute values for $\lambda_L(T)/t$ using the calculated current distributions, and hence the temperature dependent penetration depth [2], provided the film thickness is known.

Measurements of the temperature dependent resonant frequencies and Q-values of two resonators with different spacings, therefore, enables absolute determination of $\lambda_L(T)$ and σ_1 (and hence R_S).

0921-4534/97/$17.00
PII S0921-4534(97)00284-0

Figure 3 below shows the frequency shift as a function of temperature for two resonators with different line spacings (10 and 73 μm), from which we evaluate absolute values of $\lambda_L(T)$, as plotted at low temperatures in Fig.4.

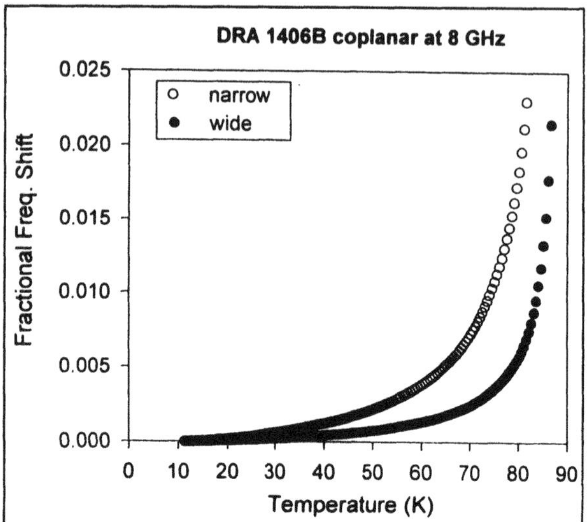

Fig 3. Frequency shifts for two resonators on the same film with different line-spacing

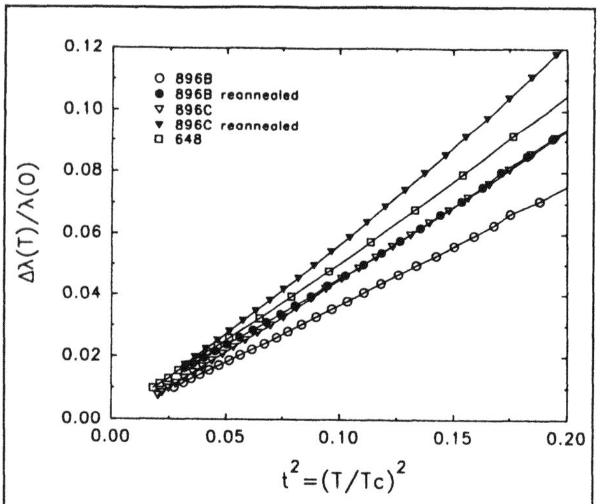

Fig.4. Fractional changes in derived $\lambda_L(T)$ for a series of samples.

We find that $\lambda_L(T) \sim T^2$ over quite a wide temperature range, in contrast to measurements on single crystals, which show a T-dependence, believed to originate from the d-wave nature of the superconducting wave-function[3]. The T^2 dependence for thin films is probably associated with the high density of defects in thin film samples, smearing out the nodes in the energy gap. However, in a detailed study of ~20 thin film resonators, no correlation was observed between residual losses and the magnitude of the T^2 dependence [4].

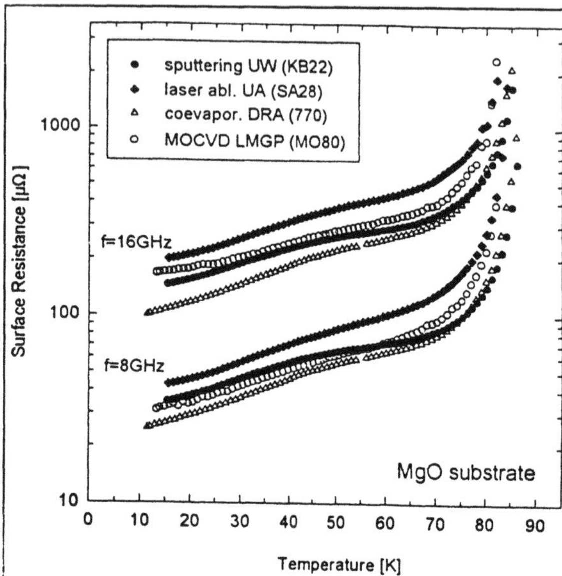

Fig 5 The temperature dependence of the surface resistance for a number of samples grown by different techniques at 8 and 16 GHz.

Fig. 5 illustrates the expected ω^2-dependence of R_s derived from the Q-factors and $\lambda_L(T)$. There is often considerable variation in performance of films from a single source; however, the best films have comparable performances irrespective of growth route[5] with values identical to those obtained by dielectric resonator measurements of the surface impedance on unpatterned films. The superconducting properties are therefore not significantly degraded by the currents flowing close to the lithographically patterned edges.

Coplanar resonators with their high Q-values are ideally suited to mixed state measurements of microwave properties, providing useful information on vortex pinning and viscous losses [6]. As an example, Fig. 6, shows the dependence of surface reactance X_s on static magnetic field perpendicular to the film, derived using the double coplanar resonator method7: a similar field dependence is observed for R_s. The linear field dependence implies strong pinning of flux lines with little evidence of magnetic interactions between flux lines up to ~4T. For fields up to 4T parallel to the planes, there is very little additional loss below ~50K, and a relatively small increase in losses even at liquid nitrogen temperatures.

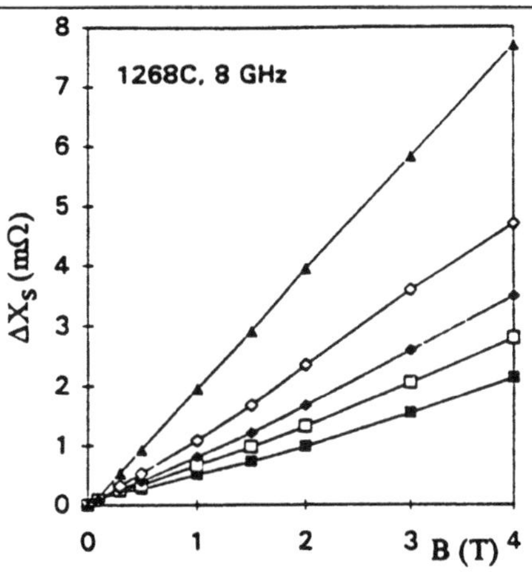

Fig.6 The field dependence of ΔX_s derived from coplanar measurements.

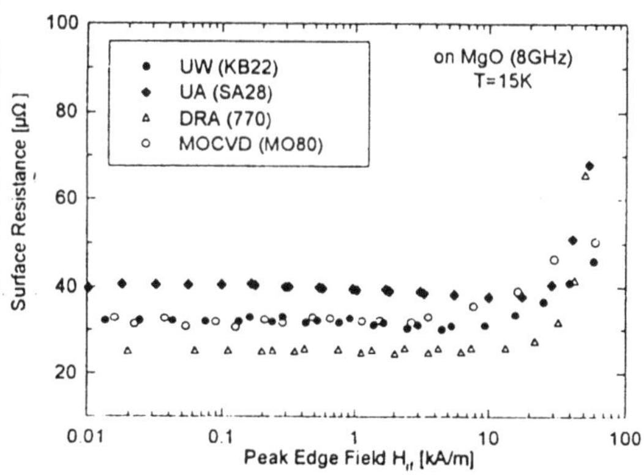

Fig.7 Derived values of R_s as a function of rf-field for a number of films deposited by various techniques at 8GHz and 15K.

The high Q of the coplanar resonators and the concentration of currents near the film edges make them ideal for investigating non-linear microwave power effects using relatively modest input powers. Fig. 7 shows the rf field dependence of microwave losses for a number of thin films from various sources. At low temperatures, non-linearity sets in for these high quality films at microwave fields corresponding to current densities close to the dc critical current of ~ 10^7 Acm^{-2}. Unlike conventional superconductors, microwave flux enters the film edges when the edge-current density exceeds J_c because of the very small coherence length and relative ease of vortex nucleation. In poorer quality films, non-linearity can set in at much lower current densities [5], but there is no correlation with the quality of the film measured by residual surface resistance.

3. Applications

Coplanar structures are used extensively as the basis of a number of planar microwave circuits. We describe three examples: a switchable band-stop filter, a fractal slow-wave resonator, and a coplanar resonator for ultra-sensitive ESR measurements.

3.1 Switchable filter

Small, low frequency, coplanar, interdigital resonators can readily be patterned using HTS films [8]. For example, a LC lumped-element resonator can be made by a number of interdigital capacitors of capacitance C between two conducting plates with a parallel inductive strip, which can be driven normal by a dc current. A switchable band-stop filter can then be realised by adjusting the current through a cascaded series of such resonators with

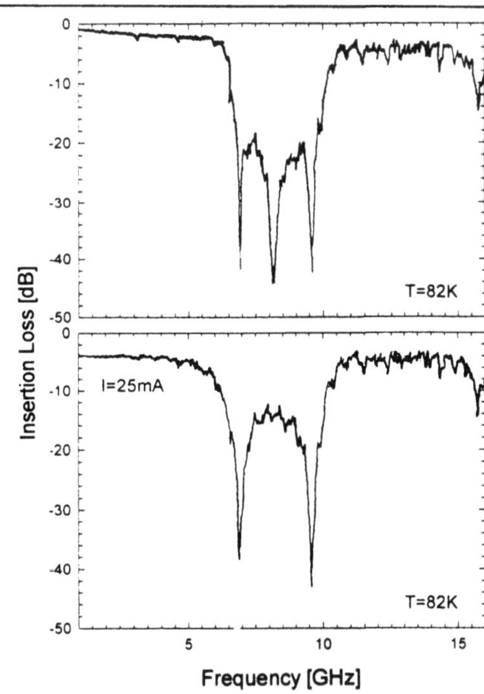

Fig.8 Switchable band-stop filter response.

different widths (10-16 μm) of central conductors: the response of one such filter is shown in Fig 8.

3.2 Slow-wave resonator

A natural extension of the above idea is to use the high interdigital capacitance to fabricate a slow-wave resonator, in which the space between the central conductor and ground plane is filled by a matrix of interdigital capacitors to increase the effective capacitance per unit length C above the TEM value $(c^2L)^{-1}$. In this way, the phase velocity along the line can be reduced by a factor as large as 10, enabling a ten-fold reduction in fundamental frequency for a given resonator length. A four-fold reduction is shown for a fractal interdigital geometry in Fig.9, where the fundamental resonant frequency has been lowered from ~ 8 to 2 GHz. HTS allows miniaturisation without the associated loss encountered using conventional material[9].

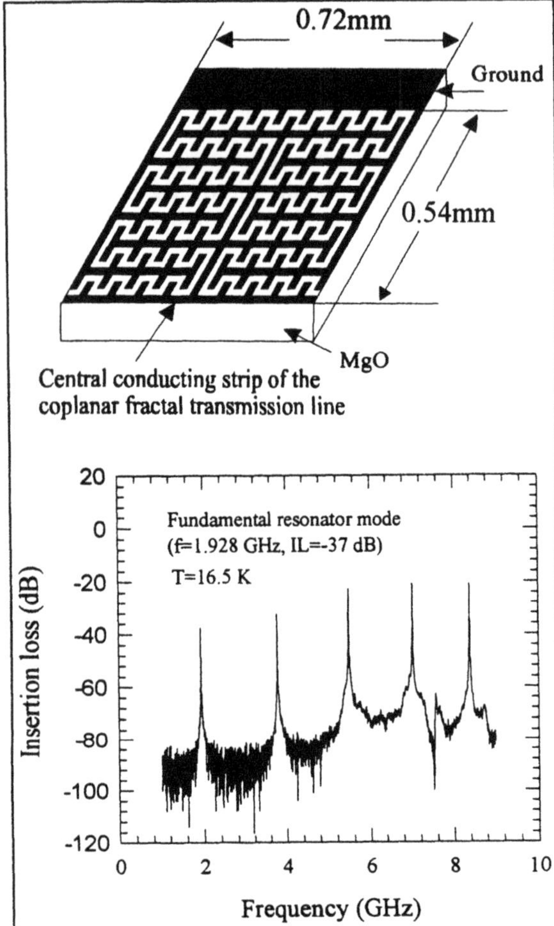

Fig.9 Fractal coplanar resonator and its frequency response; at 2 GHz and 16K the Q is ~10,000.

3.3 ESR resonator.

Our final example makes use of the very high concentration of electromagnetic field in the small spacing s between the central conductor and ground planes in a simple, straight coplanar resonator. The effective resonant volume of the resonant "cavity" is therefore ~ s^2l, compared to l^3 for a conventional ESR cavity, where $l \sim \lambda_{\text{free space}}/2$. For a small sample of size < s, the effective filling factor (which determines the size of the ESR signal) is therefore increased by a factor $(l/s)^2$, which could be as large as $(1\text{cm}/10\mu\text{m})^2 \sim 10^6$. At the field required for 10GHz ESR measurements (~0.3T), there is relatively little increase in microwave losses for fields parallel to the film, so that Q > 1000 can still be achieved. The cavity can then be integrated in a conventional ESR spectrometer configuration to provide a very sensitive system for small samples. Initial results, in a far from optimised system, are shown in Fig.10 for a small sample of DPPH (~10^{14} spins) - already comparable with the best commercial systems [10].

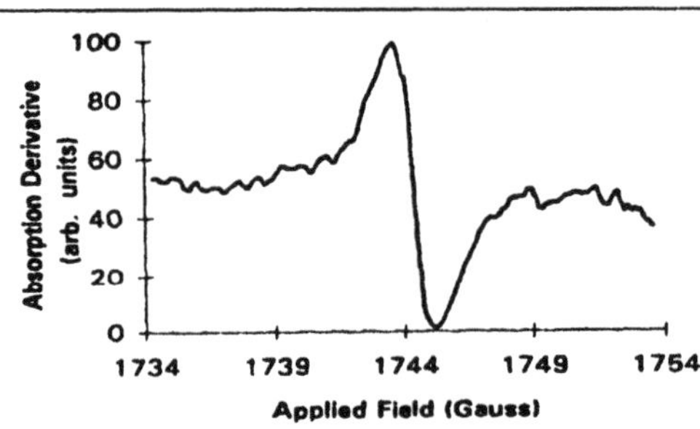

Fig.10 ESR signal for a small DPPH sample at 77K using an HTS coplanar resonator.

ACKNOWLEDGMENTS

This research has been support by the EPSRC and an ESPRIT - Supermica programme.

REFERENCES

1. A. Porch et al., IEEE-MTT **43** (1995) 306
2. A. Porch et al., IEEE-ASC **3** (1993) 1719
3. D.A. Bonn et al., Phys.Rev.B **50** (1994) 4051
4. A. Porch et al., Physica B **194-196** (1994) 1605
5. B. Avenhaus et al., IEEE-ASC **5** (1995) 1737
6. M.W. Coffey et al., Phys.Rev.Lett. **67** (1991) 2219
7. J.R. Powell et al., Proc. EUCAS-95, p1119
8. B. Avenhaus et al., Electronics Lett. **31** (1995) 985
9. M.J.Lancaster et al., IEEE-MTT **44** (1996) 1339
10. J.J. Wingfield et al., Proc. EUCAS-95, p1087

ELSEVIER

Physica C 282–287 (1997) 399–402

PHYSICA C

Harmonic Frequency Mixing at Millimeter Waveband Using High Tc Josephson Junction and Its Applications

P.H.Wu, H.B.Wang, Q.H.Cheng, and W.W.Xu

Department of Electronic Science & Engineering , University of Nanjing, Nanjing 210093, P. R. China
Center for Advanced Studies in Science and Technology of Microstructures, Nanjing 210093, P. R. China

When a signal at the frequency f_S and a local oscillator at the frequency f_L are both applied, a high T_C Josephson junction can be operated as a harmonic mixer to give rise to output at the intermediate frequency $f_{IF}=|f_S-nf_L|$ where n is an integer called harmonic number. Good results have so far been demonstrated for f_S up to about 100 GHz and n up to about 120. The laboratory application of such mixing to the phase locking of signal source at millimeter waveband has also been explored. Further attempts to push the operating frequency to submillimeter and far infrared wavebands are being made.

1. INTRODUCTION

In harmonic mixer, a signal at the frequency f_S and a local oscillator (LO) at the frequency f_L are both applied to a nonlinear element (e.g., Schottky diode, Josephson junction, etc.) and the output is at $f_{IF} = |f_S-nf_L|$, where n is an integer called harmonic number. One of the possible applications of harmonic mixing is that by measuring accurately f_L, f_{IF}, and n we can know f_S also accurately, thus transferring the measurement accuracy to much higher frequencies so long as n is large enough.

With conventional, or low T_C, superconducting junctions much work has been done on harmonic mixing; f_S up to a few terahertz and n as large as a few hundreds are possible.[1-9] There is little doubt that harmonic mixers based on low T_C Josephson junctions are superior to those based on Schottky diodes.

Since the discovery of high T_C superconductivity, attempts to do harmonic mixing in liquid nitrogen have become attractive because for harmonic mixing much power is usually available at both the signal and the LO frequencies, thus the deterioration in the mixer's noise performance caused by switching from liquid helium to liquid nitrogen does not seem a major obstacle. Furthermore, the large energy gaps in oxide superconductors have raised some hope that one can push the frequency limit of harmonic mixing to a value at least ten times higher than what is possible with low T_C superconductors, resulting in fewer links in the chains in applications such as frequency metrology. Therefore, recent years have witnessed quite some work on harmonic mixing based on high T_C superconducting junctions.[10-15] So far very nicely demonstrated have been the capability of high T_C superconducting junction to mix harmonically (up to about 100 GHz with a largest harmonic number of 120) , as well as the laboratory applications of such mixing to the phase locking of signal source at millimeter waveband used in quantum voltage standard. In this paper we shall review the recent progress in the field and discuss the possible developments we are expecting.

2. HARMONIC MIXING

An early version of our harmonic mixer is shown[11,15] where no attempt was made to optimize the coupling between the junction and the local oscillator, nor was the coupling between the junction and the signal optimized except that every effort was made so that Shapiro steps were as clearly observed as possible. A typical IF signal is shown in Fig.1 with the signal from a Gunn diode at the nominal frequency f_S of 104 GHz and the local oscillation from HP83621A at the frequency f_L of 2248.21 MHz showing a 46th harmonic mixing. The IF power was -106 dBm and IF signal-to- noise ratio (S/N) was 14 dB. A slight change in f_L shifted the intermediate frequency accordingly, and the signal frequencies

0921-4534/97/$17.00
PII S0921-4534(97)00286-4

determined from nf_L+f_{IF} agreed with each other very well within the accuracy of f_L and f_{IF}.

The IF output depended strongly on the dc bias current I_b fed into the junction and the LO power level. When the LO power was kept constant and the dc bias was increased monotonically, the IF output alternately decreased and increased. This oscillatory behavior was characteristic of harmonic mixing.[4,6] The LO power level had an even stronger effect on IF output when I_b was kept constant.[11] The IF output power (P_{IF}) depended on the local oscillator frequency as well.[14] We found that the IF power P_{IF} vs. harmonic number n roughly satisfied the relation $P_{IF} \propto n^{-2}$.[16]

In recent version of harmonic mixer,[18,19] the high Tc Josephson junction was integrated with an planar antenna so that the coupling between radiation and junction could be improved greatly.

3. PHASE LOCKING A GUNN OSCILLATOR AND ITS APPLICATION TO QUANTUM VOLTAGE STANDARD

As a possible application of harmonic mixer, we discuss here the phase locking of a Gunn oscillator at 3 mm waveband using a high T_C Josephson junction.[16, 20]

In a quantum voltage standard a high quality microwave source is used to irradiate the Josephson junction array to induce Shapiro steps.[21] In previous versions of Josephson junction voltage standard, microwave source such as commercial source locking counter EIP 578B was widely used. In this instrument, the output of a 10 MHz oscillator goes through a number of frequency multipliers to reach a frequency of 5.9 GHz, which is applied to a semiconductor harmonic mixer together with the output of a Gunn oscillator in the 3mm waveband. The output from this harmonic mixer is used to drive a digital phase discriminator. The typical phase noise of this source locking counter was measured to be about -25 dbc/Hz (±500 kHz),[22] which was close to its performance limit in the band. In our setup involving a superconducting junction (shown in Fig. 2), a local oscillator (Rohde & Schwartz signal generator) and a Gunn oscillator were applied to a high T_C Josephson harmonic mixer at 77 K. The output of the mixer was fed to an IF preamplifier (at ~ 40 MHz), a 40 MHz analog discriminator, and a driver, thus forming a phase-locking loop. The IF spectrum could be monitored on a HP 8561E spectrum analyzer. To achieve high frequency stability, the phase discriminator, the spectrum analyzer and the LO were connected together to a 10

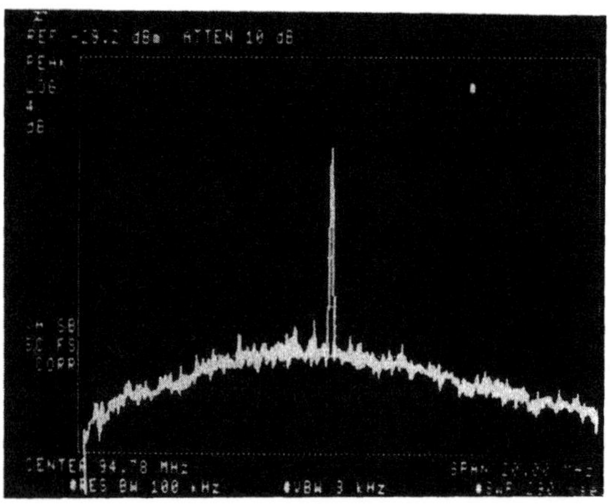

Figure 1 IF output from the mixing between a signal at the nominal frequency of 104 GHz and the 46th harmonic of LO at the frequency of 2.248 21 GHz.

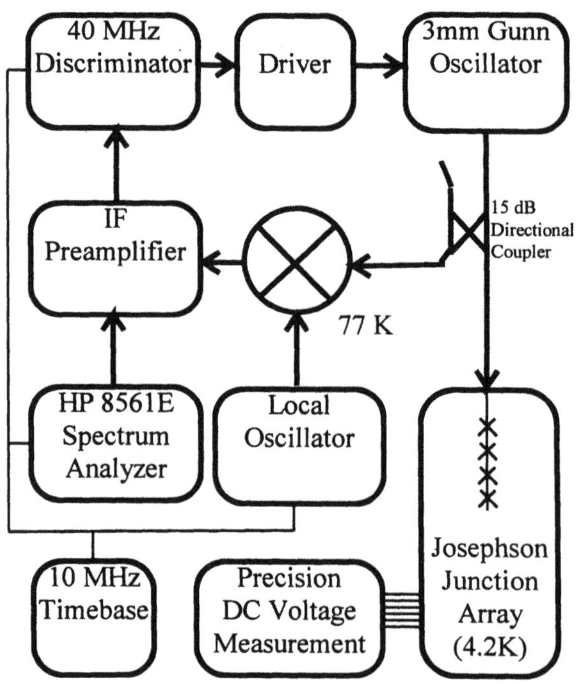

Figure 2 Experimental setup of Josephson junction array voltage system using a high-Tc Josephson junction mixer to phase-lock a Gunn oscillator.

MHz quartz oscillator time base with high stability which was calibrated against a subcarrier (4.433 MHz) of the color television signal (derived from Cs atomic clock) of the China Central Television (CCTV) with an uncertainty of 2×10^{-11}. The frequency instability of the phase-locked Gunn oscillator was directly measured to be less than ±2 Hz, with one second counting time.

The phase-locked Gunn oscillator was connected, via the main arm of a 15 db directional coupler, to an array of 3660 $Nb/NbO_x/Nb$ junctions fabricated by NIST (National Institute of Standards and Technology, Boulder, U.S.A.) and mounted on the bottom of a cryoprobe at 4.2 K.

The spectrum of the phase-locked IF signal is shown in Fig. 3 (a) and may be compared with that of the unlocked IF in Fig. 3 (b) for 99th harmonic mixing where f_{LO} was 901 MHz. The purity of spectrum was improved greatly when the phase-locking loop was working. Quantitatively, shown in Fig. 4 is the relationship between harmonic number and phase-noise at an offset frequency of ±500 kHz when the Gunn oscillator was phase-locked. The phase-noise of this 3mm oscillator ranged from -80 dbc/Hz to -59 dbc/Hz for harmonic numbers from 42 to 120. The improvement over that of EIP 578B source locking counter was remarkable.

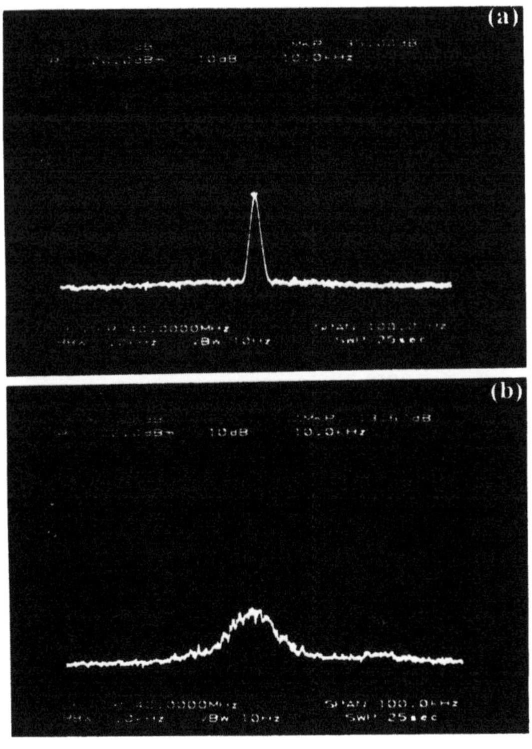

Figure 3 Typical spectrum of the IF signal for 99th harmonic number with the local oscillator frequency at 901 MHz when the phase-locking loop was (a) closed, (b) open.

Using the phase-locked microwave source, even with harmonic number up to 120 and a LO at 686 MHz, a 1.018V Zener reference was calibrated to check the property of the whole system. The nominal 1.018 V Zener reference was measured for about two hours using a procedure same as that used in other references.[21,22] As the result was similar to previous results obtained using other types of microwave sources[22] (e.g., source phase-locked by semiconductor harmonic mixer , or commercial source locking counter EIP 578B), we were sure that the whole system based on high Tc Josephson junction worked at least as effectively as other microwave sources did. The advantages of the present technology lie in its large harmonic number (compared with semiconductor mixers) which enables direct phase-locking between ~85 GHz and ~680 MHz, its better spectrum purity over narrow bandwidths (compared with EIP 578B) , and its cheapness (compared with EIP 578B).

4. DISCUSSIONS AND CONCLUSIONS

The highest frequency so far reported for high T_C Josephson junctions to work as harmonic mixers is around 100 GHz,[15, 16] far below the theoretical value expected from the energy gaps in oxide

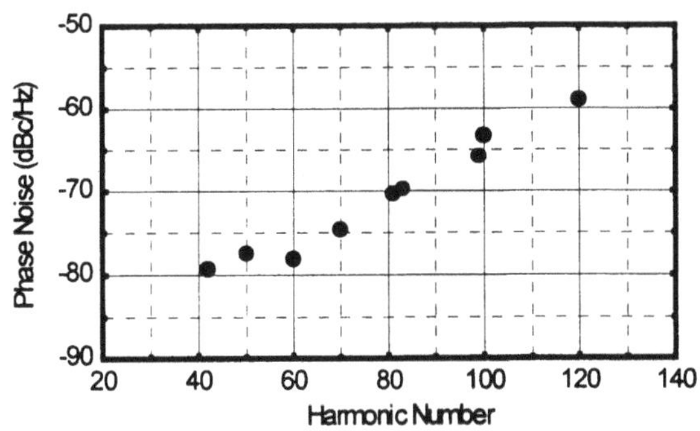

Figure 4 Dependence of phase noise at an offset of ±500 kHz on harmonic number of the high-T_C Josephson mixer when the 3mm Gunn oscillator was phase-locked.

superconductors. Presumably this is due to the lack of high T_C superconducting junctions with excellent high frequency performance. For superconducting junctions to work at high frequencies (terahertz and above), it is extremely important to have IV curves with voltage onset as steep as possible, to have distinctive current rise near the gap voltage, and to have a product of the critical current I_C by the normal resistance R_N as high as possible.[11, 17] In low T_C junctions, e.g. Nb point contacts, we know very well how to control the fabrication process to make such good junctions;[9, 17] however, for high T_C junctions it seems difficult at present to do the same. Nevertheless, recent progress in this respect is very promising and a product I_CR_N of 10 mV (at 4 K) has been reported.[18] We believe that very soon people will be able to operate high T_C superconducting harmonic mixers at terahertz region.

From a practical point of view, it is also desirable to develop a portable block of harmonic mixer with Josephson junction , reference quartz oscillator, ordinary voltage controllable Gunn oscillator, and phase locking loop all included in it. If this can be done, then anyone who has a high quality synthesizer can purchase such a block and easily get a high quality oscillator at much higher frequency.

To conclude, we demonstrated that nice harmonic mixing results could be obtained up to about 100 GHz . The IF output and S/N ratio could be made large enough to drive a phase locking loop. When the signal source was phase locked, its spectral purity was greatly improved and its phase noise in the vicinity of the carrier frequency appreciably reduced. Such a phase-locked source could find applications in , say, Josephson voltage standard, superconducting receiver, etc. More work is needed so that the highest frequency, at which a high T_C superconducting harmonic mixer is able to work, can be pushed to submillimeter and far infrared wavebands.

REFERENCES

1 D. G. McDonald, V. E. Kose, K. M. Evenson, J. S. Wells, and J. D. Cupp, Appl. Phys. Lett., 15 (1996) 121.

2 D. G. McDonald, A. S. Risley, J. D. Cupp and K. M. Evenson, Appl. Phys. Lett., 18 (1971) 162.

3 D. G. McDonald, A. S. Risley, J. D. Cupp, K. M. Evenson, and J. R. Ashley, Appl. Phys. Lett., 20 (1972) 296.

4 T. G. Blaney and D. J. E. Knight, J. Phys. D, 6 (1973) 936.

5 T. G. Blaney and D. J. E.Knight, J. Phys. D, 7 (1974) 1882.

6 T. G. Blaney and D. J. E.Knight, J. Phys. D, 7 (1974) 1887.

7 T. G. Blaney, N. R. Cross and D. J. E.Knight, J. Phys. D, 9 (1976) 2175.

8 T. G. Blaney, N. R. Cross, D. J. E. Knight, G. J. Edwards, and P. R. Pearce, J. Phys. D, 13 (1980) 1365.

9 P. H. Wu, N. R. Cross, and T. G. Blaney, J. Phys. E, 16 (1983) 331.

10 J. M. Song, P. H. Wu, Q. H. Cheng, S. Z. Yang, J. Chen, and D. Jin, IEEE Trans. Magn. 25 (1989) 911.

11 P. H. Wu, Y. H. Xu, and C. Heiden, Appl. Phys. Lett., 57 (1990) 1265.

12 T. Nozue, Y. Yasuoka, J. Chen, H. Suzuki, and T. Yamashita, IEICE Trans. Electron., E75-C (1992) 929.

13 T. Matsui and H. Ohta, IEEE Trans. Appl. Supercond., 3 (1993) 2421.

14 H. B. Wang, W. W. Xu, Q. H. Cheng, and P. H. Wu, J. Appl. Phys., 75 (1994) 2286.

15 H. B. Wang, W. W. Xu, Q. H. Cheng, and P. H. Wu, to be published.

16 H. B. Wang, W. W. Xu, Q. H. Cheng, Z. M. Ji, R. Z. Gao, P. H. Wu, and Z. C. Xu, Appl. Phys. Lett., 66 (1995) 370.

17 D. A. Weitz, W. J. Skocpol, and M. Tinkham, J. Appl. Phys., 49 (1978) 4873.

18 P. A. Rosenthal and Erich N. Grossman, IEEE Trans. MTT, 42 (1994) 707.

19 W. W. Xu, H. B. Wang, M. F. Xu, Q. H. Cheng, and P. H. Wu, Chinese J. Low Temp. Phys., 17 (1995) 1.

20 H. B. Wang, W. W. Xu, Q. H. Cheng, P. H. Wu, S. Q. Xue, H. H. Li, and X. C. Shen, J. Phys. D, 30 (1997) 274.

21 Y. Sakamoto, H. Yoshida, T. Sakuraba, A. Odawara, Y. Murayama, and T.Endo, IEEE Trans. Instru. and Meas., 40 (1991) 312.

22 S.Q. Xue, H.H. Li, R.M. Liu, X.C. Shen, and P.Deng, Modern Metrology and Measurement (in Chinese), 2 (1993) 23.

ELSEVIER

Physica C 282-287 (1997) 403-406

PHYSICA C

Single Flux Quantum Logic and its Non-Latching Operation

Yoichi Okabe and Hiroki Kodaka

Research Center for Advanced Science and Technology, University of Tokyo
4-6-1 Komaba, Meguro-ku, Tokyo 153, Japan

We review a few type of single flux quantum logic (SFQL) and discuss merits and defects of each circuit. At the same time, a new type of SFQL is gave an outline. The new circuit is a level logic system and operates in a non-latching mode like a semiconductor FET circuit. We also mention problems to be solved when SFQL is fabricated by using a high temperature superconductive material.

1. Introduction

Semiconductor circuit is now approaching to a limitation of integration and speed. The limitation was once considered to be solved by a huge parallel computation, but now it was revealed to be very difficult to manage so many processors efficiently. Therefore, essential solutions are necessary now, that is to say, more high speed operation of device. As is well known, superconductor has the highest efficiency in all materials and a superconducting device, that is, Josephson junction, is the fastest in all thinkable electronic devices. In order to apply a Josephson device to a logic circuit, many efforts have been made. Especially, single flux quantum logic (SFQL) that utilize a unique phenomena of superconductivity, quantization of magnetic flux, is the most promising in the point of its high speed operation and low power consumption. The operation frequency of SFQL achieves a few hundred giga hertz and the power consumption is almost atto joule (10^{-18} [J]). These values are at least hundred times superior than those of CMOS circuits.

In this paper, we review a few type of SFQL that is already proposed and these principles are mentioned. After that, problems that must be solved when a high Tc material is used are mentioned. Finally, we explain our new SFQL circuit named non-latch single flux quantum (NSFQ) and show results of a computer simulation.

2. Working Principle of SFQL's

2.1. History of SFQL

Josephson logic is one of the fastest digital circuits. Delay time of a conventional Josephson voltage logic is about ten pico seconds and it was demonstrated by IBM group[1]. However, since a lot of fluxons are managed in the conventional logic system, a performance of a Josephson junction is not exhibited fully. On the other hand, to control a single fluxon is much faster and generally, the delay time is less than one pico second in a SFQL system.

A logic status of SFQL is essentially represented by one fluxon is trapped or not in a superconducting loop. Since a Josephson junction is used in order to control a fluxon, most circuits are based on one junction SQUID's chain (rf-SQUID) as shown in Fig. 1(a)[2], or two junction SQUID's chain (dc-SQUID) as shown in Fig. 1(b)[3]. Both cases can propagate logic status, because a logic transition of one SQUID triggers a transition of the next one, and so on. In these SQUID gate circuits, logic of "AND" is normally realized by a single loop with high threshold, and it transits only when two previous inputs transit. In the same way, "OR" can be realized by a single loop with low threshold, and it transits when one of the previous inputs transits. "NOT" is very difficult and normally it is realized with so-called dual rail logic, which doubles the size of circuits.

0921-4534/97/$17.00
PII S0921-4534(97)00287-6

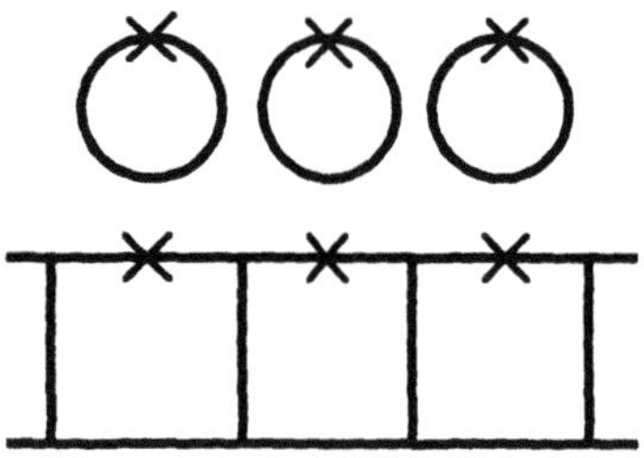

(a)

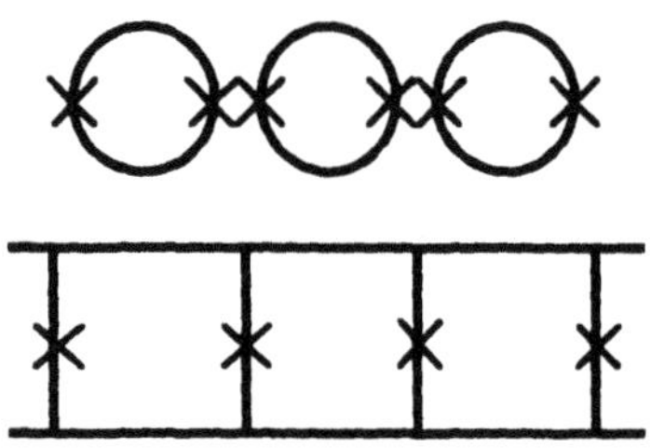

(b)

Fig. 1:
(a) The principle of rf-SQUID chain and its equivalent circuit.
(b) The principle of dc-SQUID chain and its equivalent circuit.

Josephson transmission line logic or parametric quantron family were also proposed[4]~[9]. However, each of these circuits has its merit and demerit. A latching logic prevents high speed operations and a pulse logic system is weak in negative logic. Circuit, that operation is very similar to FET, that is to say, non-latching operation and level logic system, is desirable. In third section, the principle of such a circuit is mentioned.

2.2. Application to High Tc Materials

In principle, a gap voltage of superconductor must be large to decrease the delay time, and in order to dump an oscillation, a junction capacitance must be reduced and size of the order of 0.1[μm] square junction is demanded. Moreover, since a critical current of a junction decides a threshold of SFQL, a fluctuation of the critical current must be controlled carefully. Unfortunately, in high Tc materials, the apparent gap structure could not be realized yet. Though the ideal gap voltage is of the order of 10[mV], the effective gap is about 1[mV] or less. In addition, micro- fabrication technology is still the order of 1[μm].

SFQL circuits of more than a hundred junctions have been demonstrated by using niobium tunnel junctions and showed very high speed and low power operations. For high temperature superconductive junctions, though SFQL circuits of several ten junctions were demonstrated[10], the speed is limited by an imperfection of fabrication technology. However, process technologies for oxide materials is still growing up and we believe that further development of micro-fabrication technology will make superconductive electronics promising.

3. Non-latching Operation of SFQL

As a variation of rf-SQUID chain[1], the non-latching operation is possible. It is well known that a characteristic of rf-SQUID only depend on its LIc product. (Fig. 2). Here, L is a loop inductance and Ic is a critical current of the Josephson junction. Therefore, if the LIc product is small, it is possible for this SQUID to operate in the non-latching mode shown in Fig. 3 even if an input signal is not so large. This is the principle of non-latch single flux quantum logic (NSFQ). Certainly, a stability of this circuit is spoiled and a margin decreases more or less compared with a latching circuit. However, an ultra fast operation, the most attractive performance of a Josephson device, has no equal and a few hundreds of giga hertz operation is fully possible. In following section, we show results of NSFQ operation.

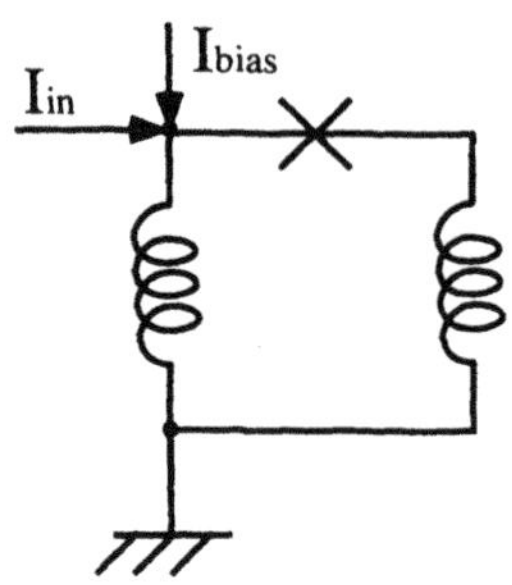

(a)

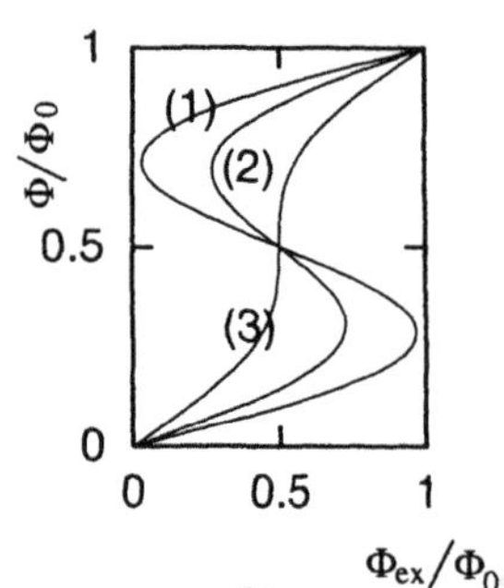

(b)

Fig. 2:
(a) A circuit of rf-SQUID.
(b) Its characteristic between an external flux and an internal flux. $LIc/\Phi_0 = 0.7, 0.45, 1/2\pi$ respectively.

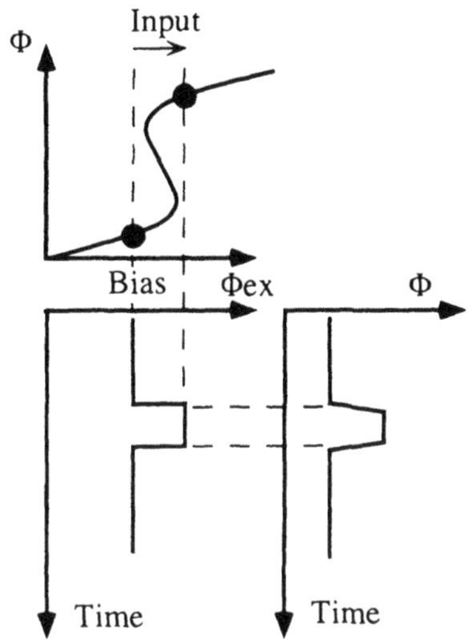

Fig. 3: The general concept of a non-latching operation of SFQL.

3.1. AND, OR logic

"AND" logic and "OR" logic are the same in the point that both of them have two inputs and one output. Such logics are realized by a circuit shown in Fig. 4(a). As is mentioned above, "AND" logic is achieved by a high threshold SQUID, and "OR" logic is achieved by a low threshold SQUID. Therefore, general ideas of these two logics are as in Fig. 4(b). As it is clear from this figure, the simplest parameter to classify two logics is the bias flux. Also the coupling constant is a parameter to classify two logics. When the coupling is strong, "AND" logic is realized and "OR" logic is realized by a weak coupling. Simulation results are shown in Fig. 4(c) and (d). In these simulations, a bias flux of input SQUIDs is $0.4\Phi_0$ and an input signal is 0.3 Φ_0.

In the case of AND logic (Fig. 4(c)), an output bias is 0.54 Φ_0, and in the case of OR logic, the bias is 0.46 Φ_0. Values of other parameters are as written in the caption.

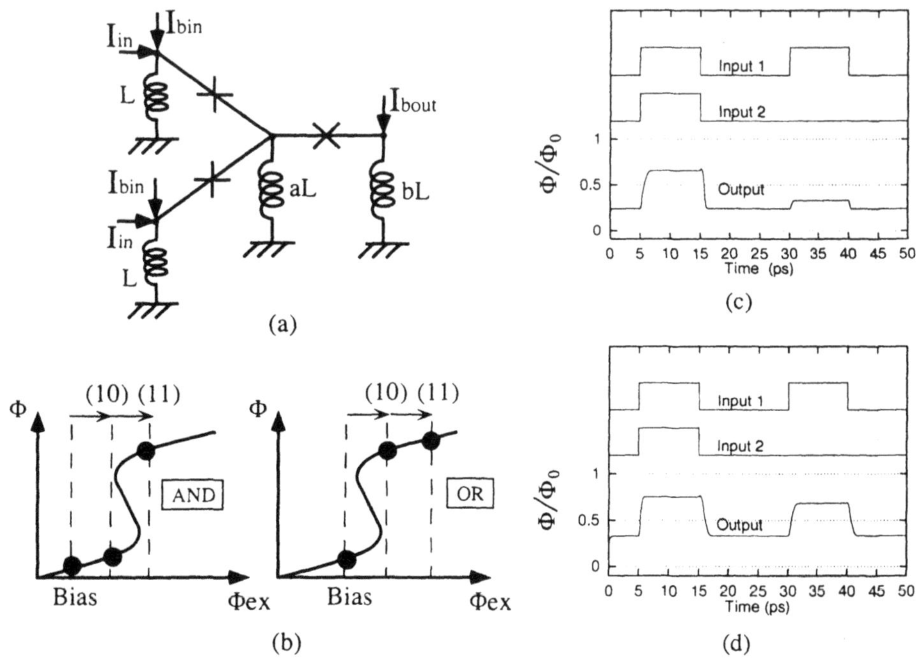

Fig. 4:
(a) A two inputs and one output circuit. Inputs and an output are coupled magnetically.
(b) The general idea of AND, OR logics. Each logic is accomplished by adjusting bias fluxes.
(c)(d) AND and OR operations (a=0.32, b=1.0, LIc=$0.12\Phi_0$. Only output bias is different between two logics).

3.2. NOT logic

As is already mentioned, negative logics like "NOT" or "NAND" are very difficult to be realized. However, for NSFQ system, such logics are not so difficult. Fig. 5(a) is a general idea of NOT logic. In order to realize this idea, an input SQUID and an output SQUID must be coupled negatively shown in Fig. 5(b).

Fig. 5(c) is a simulation result. As it is clear from this result, an output fluxoid is inverted by an input flux. Values of other parameters are as written in the caption.

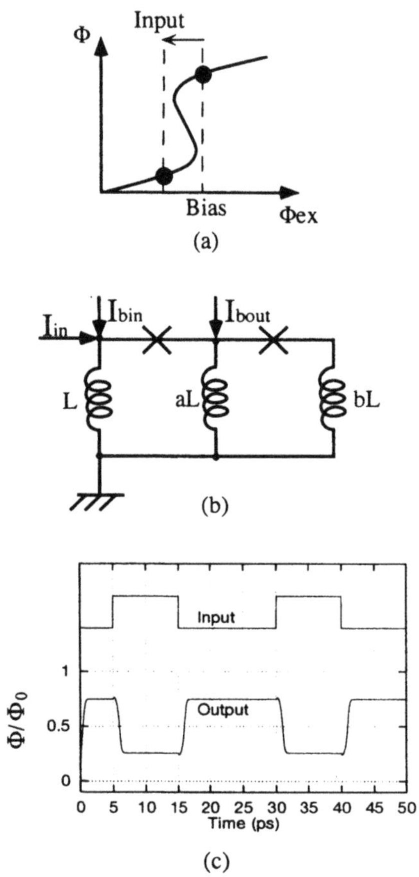

Fig. 5:
(a) The general idea of NOT logic.
(b) The circuit that couples negatively.
(c) Simulation result of NOT logic. (a=b=1.0, LIc= $0.09\Phi_0$. The input and output biases are 0.85 Φ_0 and 0.5 Φ_0 respectively.)

4. Conclusion

We reviewed single flux quantum logic circuit. We mentioned its application to high Temperature superconductors and pointed out that further progresses of micro-fabrication technology are indispensable. Finally, we proposed a non-latching operation of SFQL in order to solve defects of other SFQL circuits. As a result of computer simulation, it was revealed that all fundamental logic operations are realized.

Reference

1. "Josephson computer technology: An IBM research project," IBM J. Res. Develop., vol.24, Mar. 1980
2. H. Tamura, Y. Okabe and T. Sugano, IEEE Trans. ED., vol.27, pp.2035-2036, 1980
3. K. K. Likharev, O. A. Mukhanov and V. K. Semenov, IEEE Trans. Magn., vol.23, pp.759-762, 1987
4. K. Nakajima, Y. Onodera and Y. Ogawa, J. Appl. Phys., vol.47, pp.1620-1627, 1976
5. K. Nakajima, H. Sugahara, A. Fujimaki and Y. Sawada, J. Appl. Phys. vol.66, pp.949-955, 1989
6. S. Sakai, H. Akoh and H. Hayakawa, Jap. J. Appl. Phys., vol.22, pp.L479-481, 1983
7. K. K. Likharev, IEEE Trans. Magn., vol.13, pp.242-244, 1977
8. Y. Okabe, A. Inoue and T. Sugano, Proc. 17th Int.Conf. on Low Temp.Phys., pp.445-446, 1984
9. E. Goto, " Proc. 1st RIKEN Symp. Josephson Electronics, pp.48-51, 1984
10. S. Shokhor, B. Nadgorny, M. Gurvitch, V. K. Semenov, Yu. Polyakov, K. K. Likharev, S. Y. Hou and Julia M. Phillips, Appl. Phys. Let. , vol. 67, pp. 2869-2871

ELSEVIER

Physica C 282–287 (1997) 407–410

PHYSICA C

Eddy Current Nondestructive Material Evaluation based on HTS SQUIDs*

M.Mück, M.v.Kreutzbruck, U.Baby, J.Tröll and C.Heiden

Institut für Angewandte Physik der JLU Giessen, Heinrich-Buff-Ring 16,
35392 Giessen, Germany

High Temperature Superconductor (HTS) Superconducting Quantum Interference Devices (SQUIDs) are promising sensors for applications in eddy current nondestructive evaluation (NDE). Due to their high field sensitivity at low frequencies, they are especially suitable for applications, where a large penetration depth is required. We have investigated two different SQUID-based NDE systems, one of which is optimised for testing felloes of aircraft wheels. The second system allows for testing planar structures using a motorised x-y-stage, which moves the cryostat above the planar samples. As sensors 3 GHz rf SQUIDs made from YBCO were used, having a field noise of about 1 pT/√Hz. This results in a dynamic range of our SQUID system of about 155 dB/√Hz. In most cases, the SQUIDs have been cooled by immersing them in liquid nitrogen. We have however also developed a cryosystem, which allows for cooling the sensors by a Ne-gas flow. In planar test structures we could detect flaws with lengths of 10 mm, having a height of 0.6 mm in a depth of 13 mm. In aircraft felloes, flaws located at the inner surface of the felloe (thickness 8 mm) were easily detectable despite a high static background field of up to 0.5 G caused by ferromagnetic structures inside the felloe. For flaws in a depth of 5 mm, the spatial resolution of both systems was about 8 mm without applying image postprocessing.

1. INTRODUCTION

The high field sensitivity of SQUIDs at low frequencies makes them ideally suitable for nondestructive evaluation [1-6] of relatively thick conductive objects, like layered stacks of aluminium sheets or felloes. In contrast to a direct injection of dc or ac currents into the samples to be tested, the induction of eddy currents by an external excitation coil generates a locally limited current distribution. Since no electrical connection of the sample is required, eddy current NDE is easier to use from a practical point of view, however, the choice of the optimum measurement parameters like e.g. the excitation frequency is more critical. Furthermore, the calculation of the current flow in the sample from the measured field distribution tends to be more difficult than in case of a direct current injection.

NDE measurements of real samples, e.g. along a curved aircraft wing, require a three dimensional moving or tilting of the SQUIDs. This is not possible when using a conventional dewar for liquid coolants. Although special dewars are available for this purpose, a cryocooler is a promising competitor to liquid coolants in this case.

2. EXPERIMENTAL CONFIGURATION

We have used HTS rf SQUIDs [7] in our experiments, which have been operated at 3 GHz. The field noise of the sensors was about 1 pT/√Hz at signal frequencies above 1 Hz, and the field to flux transfer coefficient was about 50 nT/Φ_0. The SQUID readout system had a slewrate in excess of $5\cdot10^4\,\Phi_0$/s and a dynamic range of $\pm$ 500 Φ_0. This allowed for the operation of the system in an unshielded environment. To minimise the excitation field at the location of the SQUID, we used either a gradiometric excitation coil shaped like a double D, or a circular field coil with local compensation of the excitation field at the SQUID by a second coil. In

* This work is supported by the German Federal Ministry of Education and Research (BMBF) under Contract No. 13N6677/8 and in part by the Prize of the Justus-Liebig-Universität Giessen.

PII S0921-4534(97)00288-8

both cases, the excitation field at the sample can be as high as 5 mT. With the gradiometric excitation coil, the excitation field at the location of the SQUID could be reduced by a factor of about 200 (less than 20 Φ_0 at the SQUID). With a local compensation of the field at the SQUID by a second coil, the reduction could be as high as 5000.

We have set up two prototypes for NDE testing. The first system makes use of a motorised x-y-stage (scanning range 1 m × 1 m) which is used for two dimensional scanning of flat samples. A simple glass dewar is used for cooling the SQUIDs. The distance between SQUID and room temperature sample is about 7 mm. The excitation coil is attached to the bottom of the dewar and can be shifted horizontally by means of two set screws to adjust for minimum excitation field at the SQUID. The dewar is mounted to the moveable part of the x-y-stage and is thus scanned above the fixed test object. The stage is controlled by a PC, which moves the stage with an accuracy of better than 100 μm in both directions.

The z-component of the magnetic field above the test specimen is sensed by the SQUID and further processed by a lock-in amplifier. A 16-bit analogue to digital converter is used to record the position of the SQUID and the output signal of the lock-in amplifier with a PC.

In addition to the liquid nitrogen dewar, a cryocooler was used in some cases to cool the SQUIDs. In order to have a larger separation between SQUIDs and cooler, we have developed a closed cycle gas flow cryo system. This system uses a miniature Stirling cooler, which cools neon gas down to about 50 K. A transfer system consisting of two flexible coaxial tubes with a length of 2 m transports the cold Ne gas to the SQUID mount and cools the SQUIDs to about 65K. The flexible transfer lines allow for a 3D movement of the SQUIDs.

For testing aircraft felloes we used a different setup, where the felloe is rotated by an electric motor with a speed of 6-20 rpm. The SQUID is placed inside a conventional dewar, which is mounted close to the outer surface of the rotating felloe.

3.RESULTS

Most of the investigated test objects consisted of aluminium with an electrical conductivity of 15 to 30 MS/m. With a thickness of the samples of up to 25 mm, an optimum excitation frequency of the order of a few hundred Hz is calculated. When measuring the external noise in our laboratory, it turned out that at these frequencies the external noise was somewhat lower than the intrinsic noise of our SQUIDs, which is on the order of 1 pT/√Hz. At least in case of these high excitation freuencies, the use of an electronic gradiometer is not necessary. We think that in case of large test samples, a field noise of 1 pT/√Hz is sufficient for all measurements, since the thermal magnetic noise of the test samples at room temperature can easily be as high as a few pT [8].

Without any mathematical postprocessing, the spatial resolution of the system was found to be about 8 mm, which corresponds nicely to the separation between sample and SQUID. The spatial resolution can be improved further by using suitable differentiating filters (calculating dB_x/dx^2), however on the expense of a slight reduction in the signal to noise ratio.

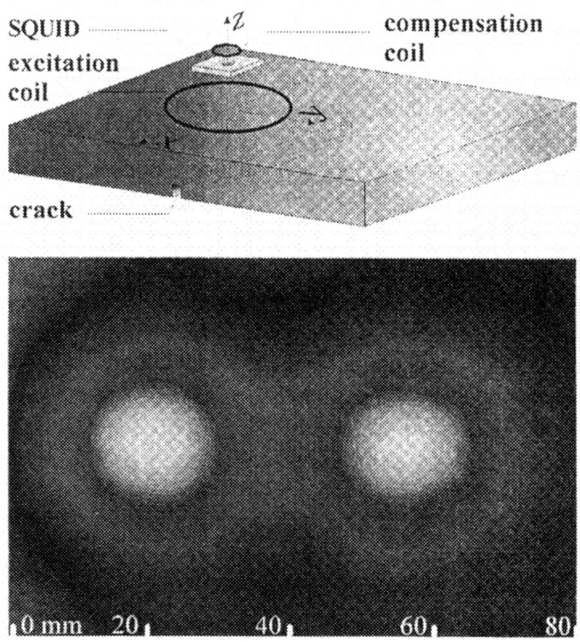

Fig.1 A small slot (40x1.2x0.15mm) below 13mm plate of aluminium. circular exitation coil: ∅=20mm, 90Hz, 50mA

Fig. 1 shows the result of a measurement of an aluminium sheet with an artificial slot (40 mm long, 0.15 mm wide and 1.2 mm deep) covered by a 12.7 mm thick aluminium plate. This measurement was performed using a circular field coil and an excitation frequency of 90 Hz. Since we measured the z-component of the eddy current field, we mainly detect effects from currents flowing in the x-y plane of the sample. These currents are maximum at the outer edges of the slot, leading to a corresponding high field strength at these points (see Fig. 1).

We then measured various test objects representing original aircraft parts. Fig. 2 shows the two dimensional field distribution above a test object consisting of three aluminium sheets bolted together by aluminium rivets. Close to some rivets, artificial cracks of 3-5 mm length had been introduced in the bottom aluminium sheet. A gradiometric coil with a diameter of 9 mm was used here. Due to the depths of the cracks of about 7 mm, the optimum excitation frequency was about 1-2 kHz. A periodic pattern can be observed in Fig. 2, which is due to the distortion of the eddy current distribution by the aluminium rivets. The cracks lead to a further distortion of the eddy current flow, which results in an increase in the magnetic field amplitude at positions close to the cracks (white spots in Fig. 2). As a guide, the position of the rivets (black circles) and the cracks (black lines adjacent to circles) are also given in Fig. 2. The figure shows the raw data, and no further image processing was used.

The measurements discussed above could be carried out with both, SQUIDs cooled by liquid nitrogen or cooled by the gas cooling system described in chapter 2.

Additional measurements have been carried out on aircraft felloes, which contain a number of ferromagnetic steel bars used as heat shields and venting holes, and have a varying thickness. Since for rapid testing it is undesirable to remove the steel parts, the signals caused by possible cracks should be separated from the signals of the steel parts by suitable filtering techniques. Again a gradiometric excitation coil was used, with a distance between the SQUID and the felloe of about 15 mm. Fig. 3 shows measured raw data of an aircraft felloe which contains six steel bars (with a separation of 60°) and

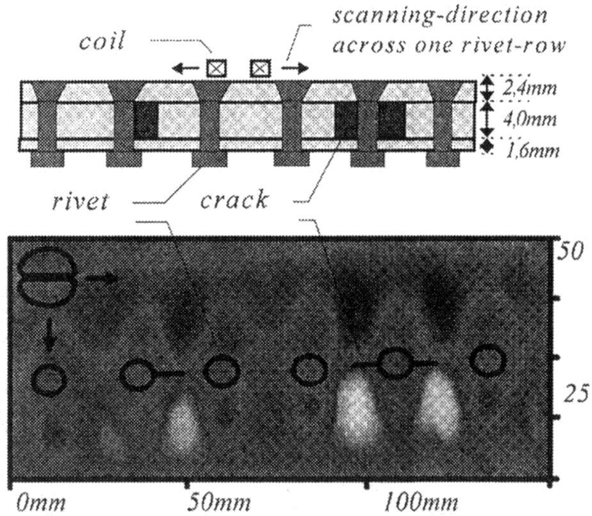

Fig. 2 Response of a rivet plate of a A-320 Double D: ∅=9mm, 1kHz, 150mA

three tapped holes (120°) with a diameter of 10 mm. A crack which was only about 4 mm long and close to one of the holes could easily be seen in the amplitude (Fig. 3a) and the phase (Fig. 3b) data.

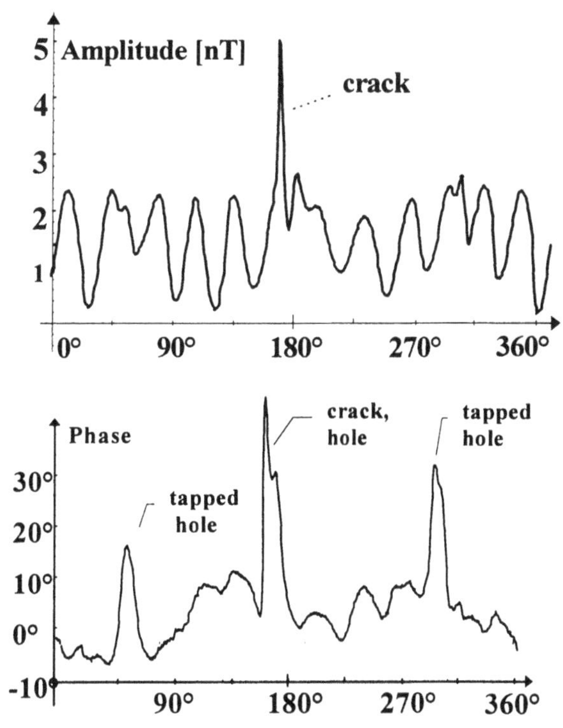

Fig. 3a /3b Measurement of one rotation of an aircraft fellow. Periodic amplitude and phase signals stem from the ferromagnetic steelbars and venting holes. Double-D coil: ∅=25mm, 220Hz, 750mA

To distinguish the crack from steel bars and holes, we developed a simple algorithm, which combines amplitude and phase data in a way that the signals from steel bars and venting holes vanish, but the crack is still visible (see Fig. 4).

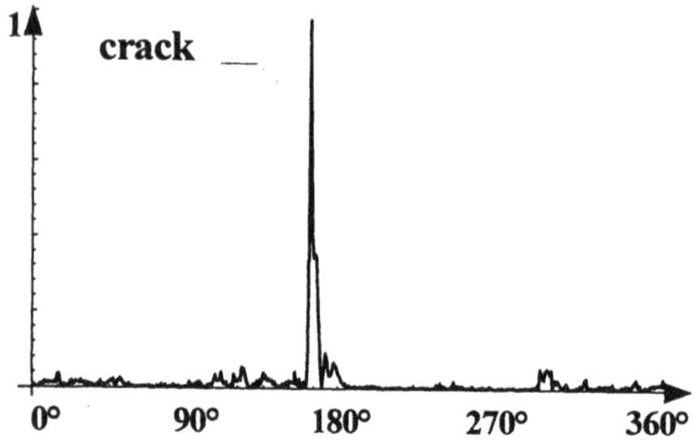

Fig. 4 Results of our correlation procedure for combining the amplitude and phase signals to supress undesired signals caused by different geometrical structures and varying conductivities.

4. NUMERICAL SIMULATIONS

We have performed some numerical simulations of the eddy current distribution in certain test samples, using the finite element method. Depending on the size and the depths of the cracks the eddy current perturbation acts approximately like a magnetic dipole. A small volume in close vicinity of the defect is defined as active volume, where defect induced perturbations occur. The integral of the eddy current density in this volume leads to the current flowing which in turn is proportional to the magnetic field measured by the SQUID. Fig. 5 shows results of these simulations to determine the optimum excitation frequency. The two upper curves show the integral eddy current flowing in the active volume with and without a crack in this volume. The difference of both curves gives the change in the eddy current distribution caused by the crack. The results of these simulations are in good agreement with our measurements.

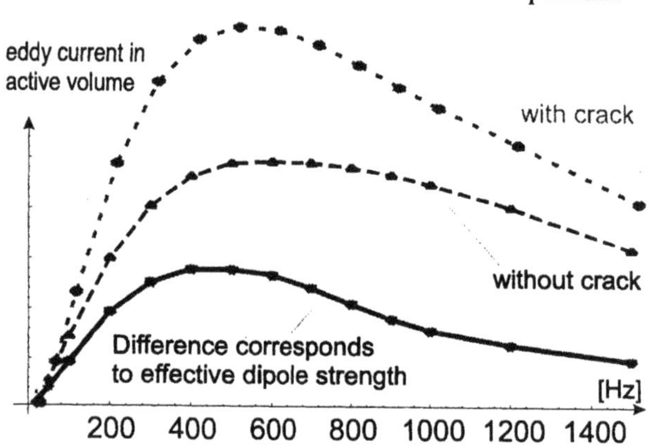

Fig. 5 most effective excitation frequency for a crack (20 x 5 x 0,1mm) in a depth of 5mm.

5. CONCLUSION

We have demonstrated a SQUID based prototype for eddy current NDE, which can be operated in an unshielded environment. A cryocooler operated version could also be realised which allows for a three dimensional moving and tilting of the SQUIDs. Various measurements on original aircraft parts could be carried out, and technically relevant cracks could easily be detected.

ACKNOWLEDGMENTS

The authors are grateful to W.B.Klemmt of DASA, F.Schur of Lufthansa, M.Junger and J.Rohmann of Rohmann GmbH and Y.Zhang, H.Bousack and H.J.Krause of Research centre Jülich for supplying the various test samples and for helpful discussions.

REFERENCES

[1] H.Weinstock, IEEE Trans.Mag. MAG-27, 3231 (1991)
[2] RR.Gans and R.M.Rose, Journal of Nondestructive Evaluation 12, 199 (1993)
[3] A.D.Hibbs, Review of Progress in Quantitative NDE 12, 1129 (1993)
[4] R.Hohmann, H. -J.Krause, H. Soltner, Y. Zhang, C.A.Copetti, H.Bousack and A.I.Braginski, M.I.Faley, accepted for publication in IEEE Trans. Appl.Supercond. AS-7 (1997)
[5] A.Cochran, J.C.Macfarlane, L.N.C.Morgan, J.Kuznik, R.Weston, L.Hao, R.M.Bowman and G.B.Donaldson, IEEE Trans.Appl.Supercond. AS-4, 128 (1994)
[6] Y.P.Ma and J.P.Wickswo, Review of Progress in Quantitative NDE 13, 303 (1994)
[7] The rf SQUIDs were supplied by Y. Zhang of Research centre Jülich
[8] L.Vant-Hull, R.A.Simpkins and J.T.Harding, Phys.Lett. A 24, 736 (1967)

ELSEVIER

Physica C 282–287 (1997) 411–414

PHYSICA C

Application of high temperature superconducting quantum interference devices in transient electromagnetic method for magnetotelluric soundings*

S.G. Wang[a], L.H. Zhang[b], C.J. Wang[c], and Y.D. Dai[a],

[a]Department of Physics, Peking University, and State Key Laboratory for Mesoscopic Physics, Beijing 100871, China;

[b]Institute of Physics, Chinese Academy of Sciences, and National Laboratory for Superconductivity, Beijing 100080, China;

[c]Institute of Geophysical and Geochemical Exploration, Ministry of Geology and Mineral Resources, Langfang, Hebei 102849, China

We have explored the application of high temperature superconducting interference devices (SQUIDs) in the artificial source magnetotelluric sounding. Field experiments were performed to test the high T_c SQUID in transient electromagnetic source. It showed that the SQUID sensor has some advantages over the conventional inductive sensor, and also revealed some problems. Theoretical studies have been carried out to investigate the prospects and limitations of the SQUID sensor.

1. INTRODUCTION

SQUIDs have been employed as the magnetic sensor in magnetotelluric (MT) soundings [1]. Not only the simplification of cryogenics for high temperature SQUID reduces the cost of operation, but also eases handling and maintenance, especially in the field experiment. Measurements on natural source MT have demonstrated that the SQUID sensor is superior to the loop sensors in the frequency band below 0.01 Hz [2], as the sensitivity of the former decreases proportional to the frequency f, while the latter as f^2. It is interesting to explore the application of SQUID sensor in the artificial source MT.

We will give a brief description about the artificial source MT in the next section. Some theoretical analysis to compare the SQUID sensor with inductive one are given in Section 3. The results of field experiment are shown in Section 4, and consequently problems and prospects are discussed and a few remarks are presented.

2. TRANSIENT ELECTROMAGNETICS

MT technique with an artificial source measures the electromagnetic response of the earth to an excitation generated by a transmitter. The transmitter is coupled to the earth electrically via a ground electric dipole, or magnetically via a loop, as shown in Figure 1. According to the data analysis it can be classified into two regimes: the time domain and the frequency domain. The former is known as transient electromagnetics, which is acronymized as TEM in the area of geological exploration, while the latter is the controllable source audio magnetotellurics (CSAMT).

In the CSAMT, three magnetic components and two horizontal electrical components are measured, and the appearance resistivity of the earth is

* Work supported by National Center for Research and Development on Superconductivity of China

PII S0921-4534(97)00289-X

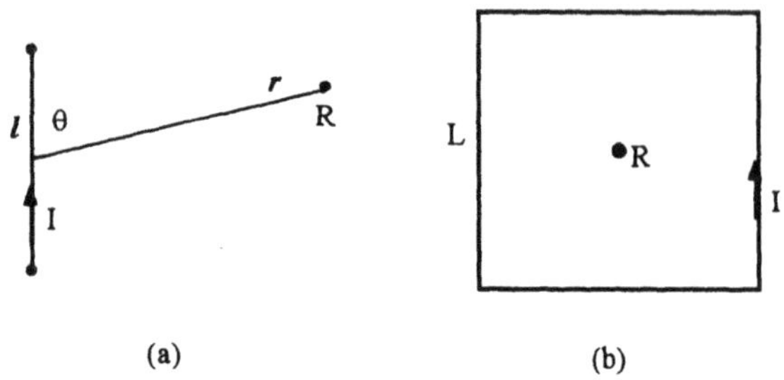

Fig. 1. Sketch for TEM system with ground dipole (a) and loop (b) transmitters. "R" indicates the receiver site.

derived as a function of frequency, which is related to the depth of the ground.
The TEM method is to register the decay of the secondary magnetic field induced by the transient current through a transmitter. The conventional magnetic sensor is an inductive coil, in which a magnetic core may be inserted to increase the effective area. It measures the induced voltage. The data are then converted into magnetic induction through integration. The sensor can be placed in the center of the loop transmitter, or a distance larger than the length of the ground dipole. The latter is referred to long off-set transient electromagnetics (LOTEM). SQUID is a magnetic sensor of high sensitivity, and broad band frequency response. It provides a direct detection of magnetic induction intensity.

3. THEORETICAL ANALYSIS

Since the SQUID and the inductive sensors detect the magnetic induction intensity and the induced voltage respectively, we shall compare the behaviour of the electric and magnetic properties, in the frame of a homogeneous semi-space with resistivity ρ. When a sinusoidal current $Ie^{i\omega t}$ is applied to a ground dipole of length l, the vertical component of the magnetic field at r is [3]

$$B_z = \frac{3\mu_0 P_E}{2\pi k^2 r^4}\left[1-\left(1+kr+\frac{1}{3}k^2r^2\right)e^{-kr}\right]\sin\theta, \quad (1)$$

where $k=\sqrt{-i\mu_0\omega/\rho}$, and $P_E=Il$ is the electric dipole moment. The induced voltage in the coil of an effective area q is

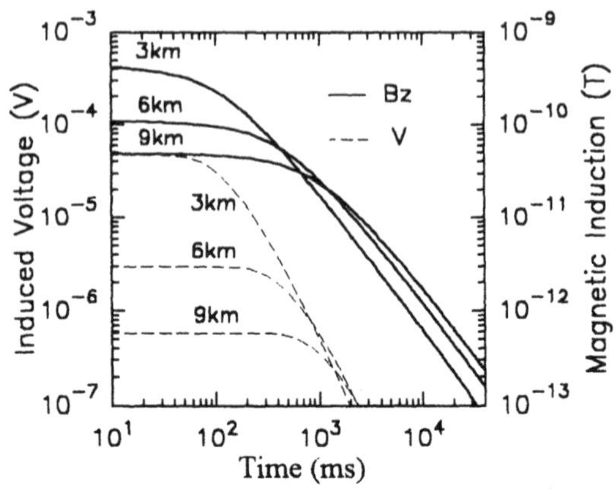

Figure 2. Magnetic induction (solid line) and induced voltage (dashed line) calculated for LOTEM at different distances.

$$V = -q\frac{dB_z}{dt} = -i\omega B_z$$

$$= \frac{3q\rho P_E}{2\pi r^4}\left[1-\left(1+kr+\frac{1}{3}k^2r^2\right)e^{-kr}\right]\sin\theta, \quad (2)$$

If the distance r is so large that Re(kr)>>1, the induction intensity and induced voltage become

$$B_z = \frac{3\mu_0 P_E}{2\pi k^2 r^4}\sin\theta = \frac{3i\rho P_E}{2\pi\omega r^4}\sin\theta, \quad (3)$$

$$V = \frac{3\rho P_E}{2\pi r^4}\sin\theta. \quad (4)$$

As one sees that the induced voltage is independent of excitation frequency, while the induction intensity is inversely proportional to the frequency. The latter is difficult to detect at high frequency. Therefore, it may be less advantage to use SQUID sensor for CSAMT method (frequency domain).

For the TEM method, a transient excitation, which is of a broad band spectrum $1/i\omega$, is applied to a transmitter. The signal at distance r on the axis (θ=π) is the integration over frequency to infinity. Dai and Wang [4] made a comparison between the magnetic induction and induced voltage in LOTEM. They showed that the former is much easier to detect, especially in a low resistivity area, *e.g.* a sea shore. Some curves calculated from the analytic expressions for B_z and V, with l=2000 m, I=20 A, ρ=20 Ω, q=20000 m^2, are shown in Fig. 2. If we takes the sensitivity criteria of 1 pT and 1μV for SQUID and inductive sensors respectively, one can see that the signal at 3 km could be detected up to 100s for inductive sensor, and up to 1000s for SQUID. The voltage is not readable at 9 km, but the late time signal of magnetic induction is even higher compared to that at 3 km.

TEM is employed to explore the metallic mines underground. They have demonstrated that a much stronger anomaly, compared to the homogeneous semi-space, can be observed in magnetic induction than in the induced voltage, for a mine deep underground.

4. RESULTS OF FIELD EXPERIMENT

The SQUID sensors were tested incorporating with TEM and LOTEM systems in field experiments. The transmitters were a loop of edge length of 50m to 200m in TEM, and a ground dipole of length of about 1.6 km. An excitation of 50% mode was adopted, in which the current is turned on, then switched off to zero, then turned to negative, and off again, with frequencies of 0.3125 ~ 16 Hz, and falling edges of 10 ~ 100 μs. The secondary magnetic field were measured by high T_c SQUID sensors. The data were measured, and averaged in synchronization with the falling edge. Figure 3 shows the induced voltage and secondary magnetic field, averaged over about 100 periods, as a function of time. The time origin is synchronized to the current switching off. It demonstrates that one can detect magnetic field in a much later time.

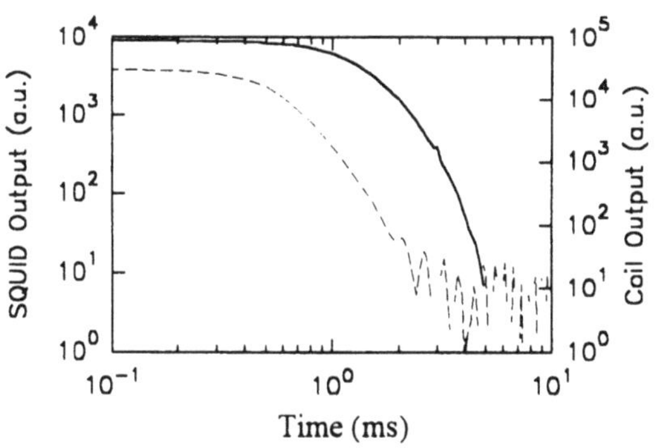

Figure 3. Outputs of SQUID (solid line) and inductive coil (dashed line) measured in TEM.

Two main problems were encountered in the field experiment:

- SQUIDs unlocking occurred, especially when they were mounted at the center of the loop transmitter.
- The slow drift was observed. It led to the saturation of the preamplifier in the data acquisition system. Then measurement had to be restarted.

5. PROBLEMS AND PROSPECTS

Both theoretical analysis and field experiments revealed some problems in the application of the SQUID sensors. Improvements must be made to the SQUID system, especially the electronics, in order to meet the requirements.

5.1. Dynamic range

At a site far from the urban area, the magnetic fluctuation of the natural field is typically a few nT. The initial magnetic induction intensity excited by a ground dipole transmitter can be calculated from Eqn. (1) with $\omega \to 0$, *i.e.*

$$B_z = \mu_0 Il/4\pi r^2 \tag{5}$$

For LOTEM, the initial field is weak as the distance is large. The field at the center of a square loop transmitter is 4 times of that generated by a dipole with $r=l/2$. If l=100 m, and I=1 A, the intensity is 40 nT. Thus, a dynamic range of a few tens nanotelsas is required for TEM.

5.2. Slew rate

SQUID operates in the flux locking mode, in which a feed-back flux is supplied to compensate the external variation of magnetic field. If a transient signal is so large as to exceed its slew rate, the system is unlocked. The auto-reset could bring it to an locking state, which may differ from the previous one by a number of flux quanta. Besides an external disturbance, unlocking occurs at the falling edges of TEM. A slew rate of a few mT/s is required for a 10 μs falling edge. For, the LOTEM application, a slew rate higher than 100 μT/s would be enough at a site far from the urban area.

5.3. Frequency response

Although SQUID is a broad band sensor, the system, such as electronics, cryogenic container, *etc.* will subject to some frequency limit. Dai *et.al.* has made a numerical integration based on Eqn. (1), but to a cutoff frequency, ω_c [5]. They have demonstrated that the high frequency response is crucial in TEM measurement. A deviation from the decaying trace calculated analytically is found as the cut-off frequency is lower than 100 kHz. For LOTEM, however, the trace follows the rigorous one even at a cutoff of 500 Hz.

It should be noticed that a fiber glass dewar is of a cutoff frequency of the order of kilohertz. And the metallic part around the SQUID sensor also distort field distribution depending on frequency.

5.4. 1/*f* noise

The average through stacking the data detected of many periods and folding them from positive and negative parts reduces the noise in the high and middle frequency bands. However, the low frequency noise with period longer than the duration in which a set of data is measured will entangle with the signal. These noises can be the slow evolution of the natural field or the intrinsic low frequency noise of the equipment. The former could be up to a few $nT/\sqrt{Hz}$. This slow drift can severely distort the late time signal, which is of great interest, as previously mentioned. The elimination of $1/f$ noise is crucial for the SQUID applications in the transient measurement.

5.5. Prospective improvements

Many efforts have been made to surmount the hurdles mentioned above. A specially designed high temperature SQUID with a dynamic range of 20 nT, a slew rate of 5 mT/s, and a frequency response to 400 kHz has been achieved. This can well satisfy the application of TEM. To eliminate low frequency drift, some techniques, such as synchronized reset, simultaneous compensation, *etc.* are also under investigation.

4. REMARKS

We have investigated the application of hig T_c SQUID in transient electriomagnetics. The theoretical analyses showed some merits, especially in the late time signal detection, in replacing SQUID for conventional inductive sensor. It also showed that the requirements of dynamic range, slew rate, frequency response are high, especially for TEM with loop transmitter. These requirements can be met with specially design. Some improvements are invoked to eliminate the low frequency drift. The field experiment was carried to test the SQUID sensor. It demonstrated that the SQUID sensor would be of benefit to obtain the information of deeper layer. The high sensitivity and small volume also enables simultaneous measurements at different sites. Although some problems remains to be solved, the application of SQUID in magnetotelluric sounding is prospective.

The authors are grateful to the important contributions in the field experiment from Y. Zhang, U. Clemens, U. Kalberkamp, U. Matzander, A. Chwala, C.S. Yang, L. Chen, S.Z. Wang, J.Z. Li, X.D. Chen, Y. Zhao, and D.F. He, and the computerized data acquisition system and simulation from Z. Tao, X.M. Huang and K. Xiu. They are benefited from the stimulating discussions with B.Y. Jiang, H. Burkhardt, A.I. Braginski, and Z.Z. Gan. Two of them (SGW and LHZ) also wish to thank the members of ISI and ZEL groups for the hospitality and collaboration during their visit in KFA, Jülich.

REFERENCES

1. John Clarke, in *New Superconducting Electronics*, Wainscot and Ralston (eds.), Kluwer Academic, Dordrecht (1993) p123-180.
2. C.J. Wang, *et al. Proc. Nat'l Conf. on Supercond. Thin Films and Devices*, (1994) 106.
3. K.-M. Strack, *Exploration with Deep Transient Electromagnetics*, Elsevier, Amsterdam, (1992)
4. Y.D. Dai, and S.G. Wang, Chin. J. Low Temp. Phys. 1 (1995) 11.
5. Y.D. Dai, X.L. Cai, S.Z. Mei, S.Z. Wang, S.G. Wang, (1997), these volumes.

ELSEVIER

Physica C 282–287 (1997) 415–418

PHYSICA C

Novel High-T_C Superconducting Three Terminal Devices

Zdravko G. Ivanov

Department of Physics
Chalmers University of Technology and University of Göteborg
S-412 96 Göteborg, Sweden

Working principles and characteristics of three terminal devices, with electrodes of high-T_C superconductors (HTS) and a transport channel controlled by a gate electrode, will be presented.

1. INTRODUCTION

The current transport in a number of superconducting devices can be controlled by a third electrode offering a transistor-like performance with a nondissipative on-state transport channel. A controlled Josephson junction with its ultra-low switching energy and picosecond switching time looks superior to semiconductor transistors. Several new devices with conventional superconductors were suggested and tested, however, a real transistor able to compete with Si and GaAs field effect transistors was not developed. High-T_C superconductors offer some new features that may give advantages in device performance: anisotropic crystal structure, higher T_C, possibility to modify their properties by doping, possibility to epitaxially grow multilayers of superconductors, semiconductors, insulators and metals. Several recent reviews thoroughly analyze the physics of the high-T_C three terminal devices[1-3]. Here, we report on our latest approaches in developing three terminal devices, including an YBCO SET transistor.

2. GATE CONTROLLED JOSEPHSON TUNNELING

A gate electrode, capacitively or directly coupled to an HTS Josephson junction may control the Josephson current. Two types of structures based on bicrystal Josephson junctions coupled to a metallic gate electrode are presented in Fig. 1.

A high electric field applied to the gate will modulate the carrier density in the Josephson channel and influence the Josephson current in the Josephson Field Effect Transistor (JFET) (Fig. 1, Scenario 1).

Injection of quasiparticles with energy $eV \geq 2\Delta$ causes a nonequilibrium population in the junction and depresses the supercurrent in the Josephson Injection Transistor, JIT (Fig. 1, Scenario 2).

PII S0921-4534(97)00292-X

Our approach to the device design was to introduce in situ a weak link of a material with low carrier concentration into an HTS microbridge. We choose bicrystal weak links since in this case the artificial grain boundary between two superconductors is formed in situ and its transport properties can be varied by changing the misorientation angle between the two halves of the bicrystal or by doping the grain boundary with different elements.

The insulator plays an important role in both devices and we shall discuss also our progress in epitaxial growth of HTS/ferroelectric double layers.

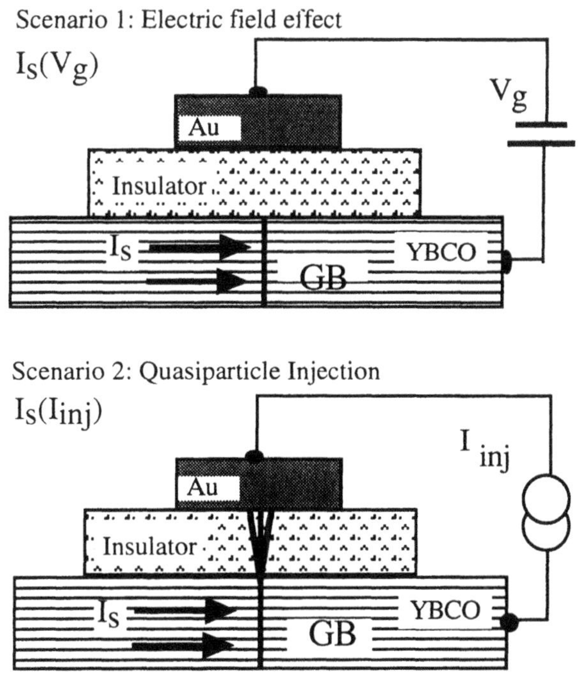

Figure 1. A schematic cross-section of a three terminal superconducting device based on a bicrystal grain boundary junction.

2.1 Voltage controlled Josephson junction

Several approaches to device design, all based on bicrystal grain boundary junctions, were developed [4–6].

We have developed a planar field effect device consisting of a 0-45° bicrystal grain boundary (GB) junction in an $Y_1Ba_2Cu_3O_x$ (YBCO) microbridge,: which was covered by a 300 nm amorphous $SrTiO_3$ (STO) layer and a 4 μm wide gate electrode [4]. The current transport through the GB was regulated by the voltage of the insulated gate. We have observed a strong (more than 50%) enhancement of the supercurrent at positive gate voltage and almost no change, or a slight decrease, at negative voltage, Fig. 2.

The Delft group has studied the field effect in a device with an overdoped $(Ca_xSm_{1-x})Ba_2Cu_3O_y$ channel and with a design similar to the one in Fig. 1a [5]. They observed an I_c enhancement at negative V_g and a depression at positive V_g. This $I_c(V_g)$ dependence was explained by the carrier-depleted grain boundary region.

These two approaches used an amorphous insulator (giving not free of defects interface between the insulator and transport channel). Mannhart suggested and tested a novel inverted structure with a Nb doped STO substrate as a gate electrode [6]. The insulator and YBCO were grown on top giving a high quality interface.

All JFET devices demonstrate a gate control of I_c but suffer from absence of a voltage gain.

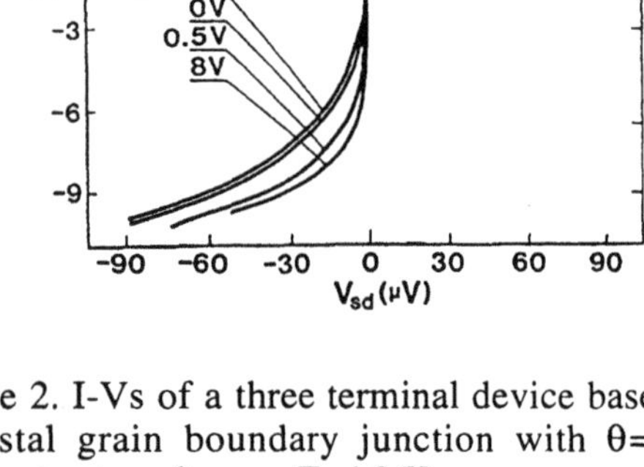

Figure 2. I-Vs of a three terminal device based on a bicrystal grain boundary junction with θ=45° at different gate voltages. T=4.2 K.

2.2 Current injection controlled Josephson junction

As it has been demonstrated by the Twente group, the injection of quasiparticles in an YBCO microbridge can be an effective tool to depress the supercurrent [7]. We shall discuss the case when the quasiparticles are injected in the region of the weak link or in the surrounding region with dimensions less than the quasiparticle recombination length.

We have studied a Josephson junction prepared on a bicrystal substrate with a grain boundary doped with Fe [8,9]. The device geometry is similar to the one presented in Fig. 1. It consists of an YBCO GB junction in situ covered with an epitaxial STO layer and, at a later stage, with a thin amorphous STO layer. The dependence of the current between the gate and superconducting electrodes, I_g, on gate voltage is highly nonlinear. At V_g below ±3V the I_g is less than 200 nA. As Vg increases, the I_g starts to increase rapidly.

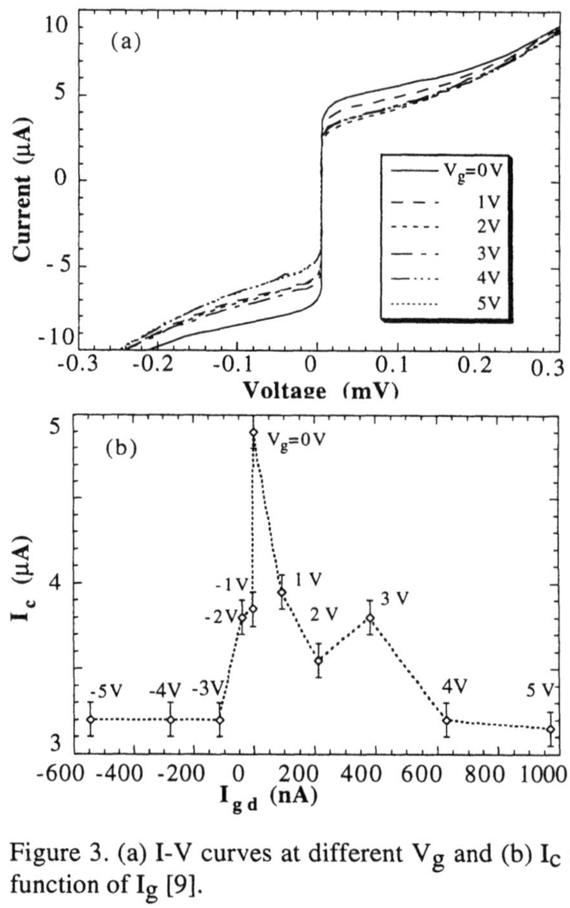

Figure 3. (a) I-V curves at different V_g and (b) I_c as function of I_g [9].

Figure 3 presents I-V curves and junction critical currents at different V_g. I_c decreases symmetrically for $Vg=\pm 1$ and $\pm 2V$. At gate voltages of $\pm 3V$, $\pm 4V$ and $\pm 5V$, the I_g is comparable with I_c but the latter is constant. These results can be interpreted in the following way: at low V_g the I_g is injected directly into the grain boundary region which has dimensions less than the quasiparticle recombination length and for $\pm V_g$ causes a symmetric depression of I_c. When the $|V_g|>3V$, I_g increases rapidly, probably due to an avalanche breakdown of some weak channels in the insulator and I_g is injected in the electrodes. This will not cause any change in I_c.

These first results show that a quasiparticle injection in the weak link can give current gain of 10-12 at low voltages even in an uncontrolled structure. A junction coupling to the gate electrode in a controlled way is needed to improve the device performance.

2.3 Insulator properties

Insulators with high permittivity, such as in $SrTiO_3$ (STO) ferroelectric films, are desirable in order to decrease the gate voltage to the level of a few mV. The maximum permittivity, ε, is observed at the Curie temperature, T_{Cur}, which for STO thin films is about 50K and below. By doping STO films with Ba, the T_{Cur} can be shifted to higher temperatures.

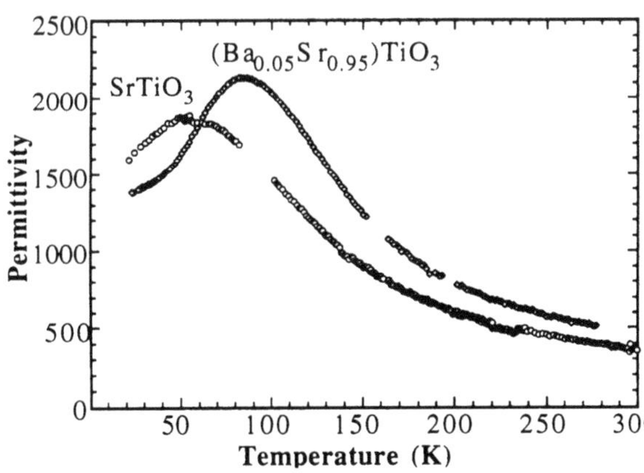

Figure 4. Permittivity vs temperature of $SrTiO_3$ (400nm) and $(Ba_{0.05}Sr_{0.95})TiO_3$ (480nm) films measured at 1 MHz. Note the shift in Curie temperature for the Ba doped film [10].

Recently we succeeded to grow high quality $SrTiO_3$ and $(Ba_{0.05}Sr_{0.95})TiO_3$ thin films by laser ablation [10]. Permittivities of 2000 and higher were measured in planar YBCO/STO and YBCO/BSTO capacitors, Fig. 4 [10]. Microwave measurements of the same structures indicated that these numbers are valid in a frequency range 0.1-26 GHz. It is clearly demonstrated that a 5% Ba doping shifts T_{Cur} with 20 K. These very high values for ε and the possibility of controlling them by varying the doping level give the promise that thin BSTO films, about 50 nm and less, will decrease the V_g values but at the expense of an increase in the device input capacitance, $C_{g\text{-}d}$.

For a 50 nm thick insulator and a $1\mathrm{x}1\mu m^2$ gate electrode, that will give about 0.3 pF for $C_{g\text{-}d}$ and the charging time, RC, will be 0.3 ps for a 1Ω gate resistance. Note that the permittivity is decreased under external electric field. This can be a source of nonlinearity in the input circuit but if the V_g variations are of the mV scale, the nonlinearity can be neglected.

3. YBCO SET TRANSISTOR

Another type of voltage controlled device is a Single Electron Tunneling (SET) transistor. The device consists of two small tunnel junctions connected in series by a small island and capacitively coupled to a gate electrode. The island may change its charge only by tunneling and therefore the charge is quantized in units of the elementary charge. The electrons may tunnel one by one when the applied voltage is above the Coulomb blockade voltage. By applying a voltage to a gate electrode, one can lift the Coulomb blockade and modulate the output voltage. SET transistors are, in general, realized in metals or conventional superconductors.

An YBCO SET transistor was recently fabricated and tested (Fig. 5a) [11]. An ultra small YBCO island was coupled to two 0.2μm wide Au electrodes through native tunnel barriers. A third electrode placed near the island was used as a gate electrode. The YBCO island, 1μm long, 0.2μm wide and 0.15μm thick, was patterned by e-beam lithography and Ar^+ milling. At temperatures below 0.5 K the blockade can be periodically lifted by varying the gate potential. In case of a current bias device the voltage across the device oscillates with applied V_g showing a transistor behavior. The latter confirms that the charge transfer mechanism through the surface barrier is the tunneling one.

(a)

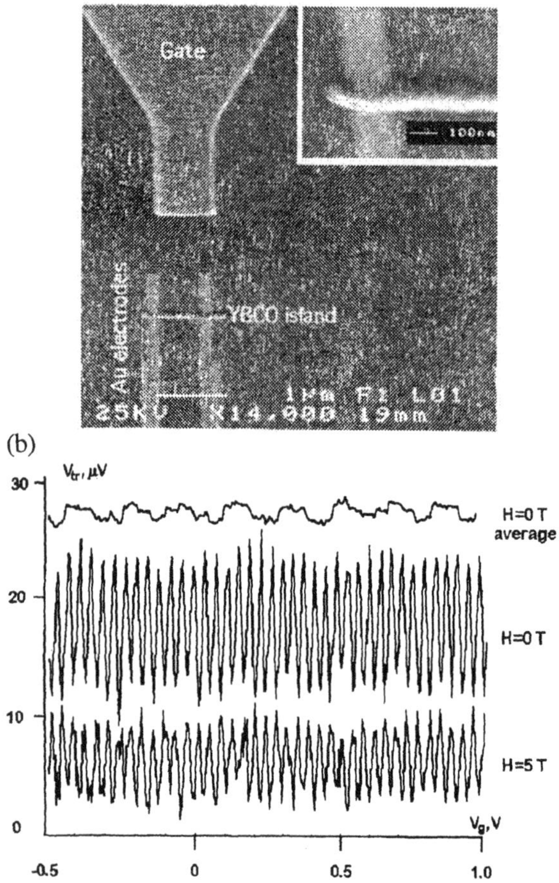

Figure 5. (a) A SEM image of an YBCO SET transistor and (b) voltage oscillations of a current biased transistor under gate voltage measured at 0 and 5T magnetic field. T=30mK [11].

The high-T_c SET transistor does not give any advantages compared to its metal counterpart however it can be used as an effective tool to study the energy spectrum of high-T_c superconductors. In particular, it was predicted that the parity effect in a SET transistor of conventional superconductors will be violated in a transistor with a superconducting island with nodes in the superconducting wave function.

Fig. 5a shows that the period of the voltage oscillations under 0 and 5 T magnetic field is constant. This, in addition to the fact that the parity effect was not observed at temperatures down to 20 mK, may be considered as an evidence that the YBCO superconductor can be described by a superconducting wave function with nodes.

4. SUMMARY

Several devices that allow the control of static and dynamic properties of Josephson junctions have been tested. These are interesting devices but not enough mature to be applied. They can be employed to control or to tune the supercurrent and to be used in fundamental experiments, as for example SET transistors. The use of JFETs and especially JITs with their high current gain, looks promising but bicrystal junctions should be replaced by artificial junctions.

The cooperation and discussions with P. Delsing, S. Kubatkin, R. Shekhter, E.A. Stepantsov, and A. Tzalenchuk during the experimental studies is acknowledged. We acknowledge the financial support from the Swedish Board for Technical Development (NUTEK) and Consortium on Superconductivity.

REFERENCES

1. J. Mannhart, Supercond. Sci. Technol., 9 (1996) 49.
2. J.E. Nordman, Supercond. Sci. Technol., 8 (1995) 681.
3. T. Claeson, Report at Future Electron Devices Symposium, Tokio, March 4-5, 1996.
4. Z.G. Ivanov, E.A.Stepantsov, A.Y. Tzalenchuk, R.I. Shekhter, and T. Claeson, IEEE Trans. Appl. Superconductivity, 3 (1993) 2925.
5. Z.W. Dong, V.C. Matijasevic, P. Hadley, S.M. Shao, and H. Mooij, IEEE Trans. Appl. Superconductivity, 5 (1995) 2879.
6. B. Mayer, J. Mannhart, and H. Hilgenkamp, Appl. Phys. Lett., 68 (1996) 3031.
7. K. Joosse, Yu. Boguslavskij, L. Vargas, G. Gerritsma, and H. Rogalla, IEEE Trans. Appl. Superconductivity, 5 (1995) 2883.
8. Z.G. Ivanov, E.A. Stepantsov, A.Y. Tzalenchuk, and T. Claeson, Physica B, 194-196 (1994) 2187.
9. Z.G. Ivanov, E.A. Stepantsov, and A.Y. Tzalenchuk (unpublished).
10. R.A. Chakalov, Z.G. Ivanov, Yu.A. Boikov, P. Larsson, E. Carlsson, and S. Gevorgian, to be submitted to J. Appl. Phys.
11. S.E. Kubatkin, A.Ya. Tzalenchuk, Z.G Ivanov, P. Delsing, R.I. Shekhter, and T Claeson, Czech. J. Phys., 46 (1996) 2305.

ELSEVIER

Physica C 282–287 (1997) 419–422

PHYSICA C

HTS SNS Josephson junctions: interfaces and mechanisms

K. Char

Conductus, Inc.
969 W. Maude Avenue, Sunnyvale, CA 94086, U. S. A.

$YBa_2Cu_{3-x}Co_xO_{7-\delta}$ goes through a metal-insulator transition near x = 0.4. When these doped YBCO materials are used as barriers in SNS junctions, a clean interface can be formed due to their excellent match of lattice constants and thermal expansion coefficients. By using three doping levels, x = 0.12, 0.21, and 0.42, we have estimated the effects of the dopant from the temperature and thickness dependences of I_c and R_n. In the framework of conventional de Gennes proximity effect theory, we have observed a crossover from the clean limit to the dirty limit as the doping level approaches the metal-insulator transition. In the case of x = 0.42, which no longer superconducts on its own, we had to invoke a pair-breaking mechanism by a cutoff scattering rate in order to explain the slower increase of the critical currents at low temperatures than the standard proximity effect would predict. Several ideas on the possible origins of such a cutoff scattering rate will be discussed. By adding Pr dopants in the Co-doped barrier, we were able to deduce the elastic scattering effect of Pr dopants, which results in smaller diffusion constants. The interfaces YBCO forms with other barriers such as $(CaSr)RuO_3$ and $(LaSr)CoO_3$ have been investigated in detail. The defect states on the surface of YBCO seem responsible for the observed zero bias conductance anomaly.

1. OVERVIEW

SNS or SIS junctions have been important tools both for studying the nature of superconductivity on a microscopic scale and for developing superconductive electronics such as SQUIDs and SFQ circuits. In the case when there is no significant interface problem between the barriers and the superconductors, the possible I_cR_n products as a function of bulk barrier resistivity is schematically described in Fig. 1.

When the barrier is in the metallic regime, the proximity effect in the barrier is responsible for a Josephson current, and the I_cR_n product can be as high as the gap as long as the barrier is thin and the junction is made very small because of the high J_c in these junctions. When the barrier is insulating and the thickness of the barrier is very thin, the dominant transport mechanism is tunneling. In this case the I_cR_n products are also as high as the gap. However, when the barrier resistivity is between the metallic and insulating regimes, the transport mechanisms are not well understood, even though there is some evidence for strong inelastic effects.

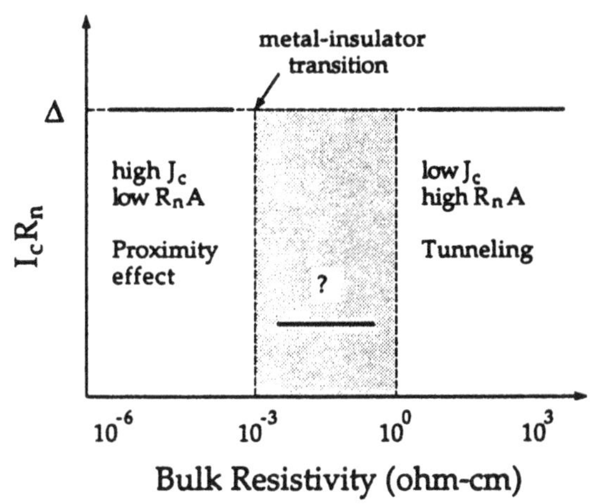

Fig. 1. Schematic description of possible I_cR_n as a function of bulk barrier resistivity.

0921-4534/97/$17.00
PII S0921-4534(97)00293-1

In this paper we will explore the barriers in the metallic regime as well as the barriers close to the metal-insulator transitions by using Co-doped YBCO. Since the lattice constants and the thermal expansion coefficients of doped YBCO materials can match well with those of YBCO, a clean interface can be formed in an edge junction geometry if a care is taken in forming the edges. In addition, if a dopant can be added in a microscopically uniform way, a barrier such as $YBa_2Cu_{3-x}Co_xO_{7-\delta}$ can be used as a "normal" barrier to realize SNS Josephson junctions without the complication of uncontrolled interfaces.

2. BARRIERS IN THE METALLIC REGIME

The temperature dependences of I_c and R_n of 4 % (x = 0.12) Co-YBCO barrier junctions [2] for three different barrier thicknesses are shown in Fig. 2. The conventional proximity effect given by de Gennes [1] predicts

$$I_c = \frac{18.5\text{mV}}{R_n}(1-t)^2 \frac{\kappa d}{\sinh(\kappa d)}$$

where $t = T/T_c$, $T_c = 88$ K, $I_0 = \pi\Delta_0^2/4R_nk_BT_c = 18.5$ mV/R_n in the case of BCS gap $2\Delta_0 = 3.52$ k_BT_c, d is the thickness of the barrier, and κ^{-1} is the decay length. In the case when the barrier has its own superconducting transition, the decay length has the form of $\kappa^{-1}(T) \propto (T/T_n - 1)^{-1/2}$, whether the barrier is in the clean limit or in the dirty limit. However, the physical constants one gets from the fitting are the diffusion constant in the dirty limit case and the BCS coherence length in the clean limit case. The diffusion constant we obtained was smaller than that obtained for the 14 % Co-doped barrier case [2], which indicates the 4 % Co-doped barrier is in the clean limit. From the fitting, the BCS coherence length of the 4 % Co-doped barrier was measured to be 28 Å.

Similar types of analyses were performed on the 7 % Co-doped barrier ($T_n = 40$ K), and the BCS coherence length has been found to be 35 Å, consistent with the previous analysis of the barrier being in the clean limit.

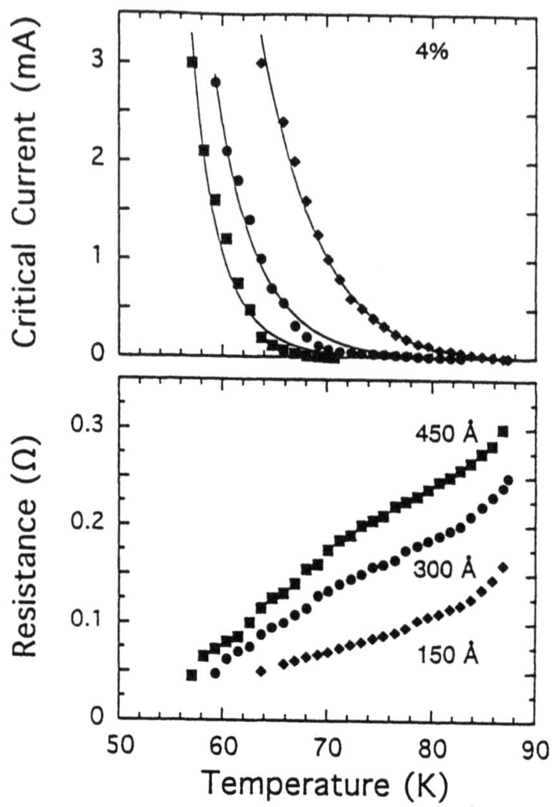

Fig. 2. Temperature and thickness dependence of I_c and R_n of 4 % Co-YBCO barrier junctions.

3. BARRIERS NEAR THE METAL-INSULATOR TRANSITION

When the Co doping is increased to 14 %, the barrier no longer superconducts, very close to the metal-insulator transition. Fig. 3 shows the temperature dependence of I_c and R_n of 14 % Co-YBCO barrier junctions with three different barrier thicknesses. [2] The dotted lines are fits to the de Gennes relation. The fits were very poor below 20 K. The divergence of the critical current was much slower than what the proximity effect would predict. When there is pair breaking scattering in the barrier material, the lifetime of superconducting pairs inside the barrier is modified. The effective lifetime will be given by

$$\frac{1}{\tau_{eff}} = \frac{1}{\tau} + \frac{1}{\tau_{pb}} = \frac{2\pi k_B T}{\hbar} + \frac{2\pi k_B T_{pb}}{\hbar}$$

The net effect is that T is replaced by T + T_{pb}. The solid lines are fits with a pair breaking scattering rate of τ_{pb} = 6 x 10^{-14} sec and a diffusion constant D = 6.6 cm/sec^2.

The pair breaking effect can be attributed to an inelastic process such as magnetic scattering in conventional s-wave superconductors. However, in a d-wave superconductor, a strong elastic scattering can also result in a pair breaking effect.

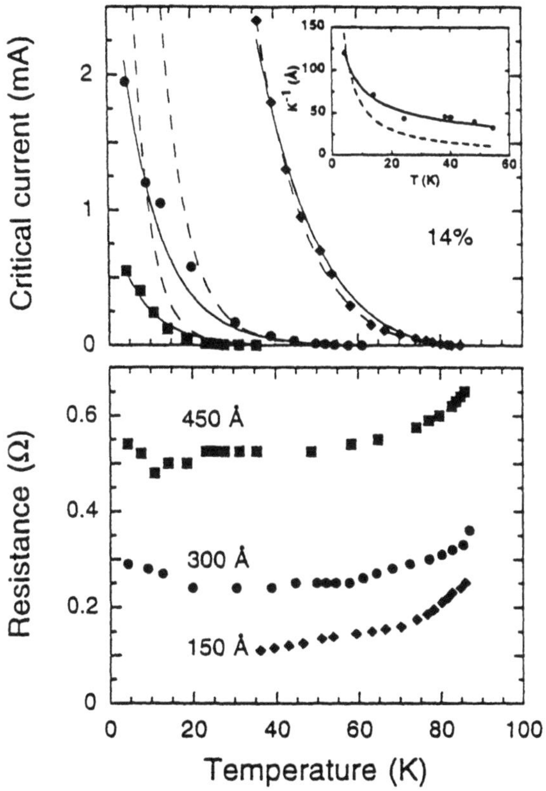

Fig. 3. Temperature and thickness dependence of I_c and R_n of 14 % Co-YBCO barrier junctions.

4. BARRIERS BEYOND THE METAL-INSULATOR TRANSITION

When Pr is doped into $YBa_2Cu_{2.7}Co_{0.3}O_{7-\delta}$ (T_n = 20 K), the system goes through a metal-insulator transition, as shown in Fig. 4. $Y_{0.8}Pr_{0.2}Ba_2Cu_{2.7}Co_{0.3}O_{7-\delta}$ [3] is slightly more insulating than 14 % Co-doped YBCO and has been used as a barrier material to investigate the effect of the Pr dopants.

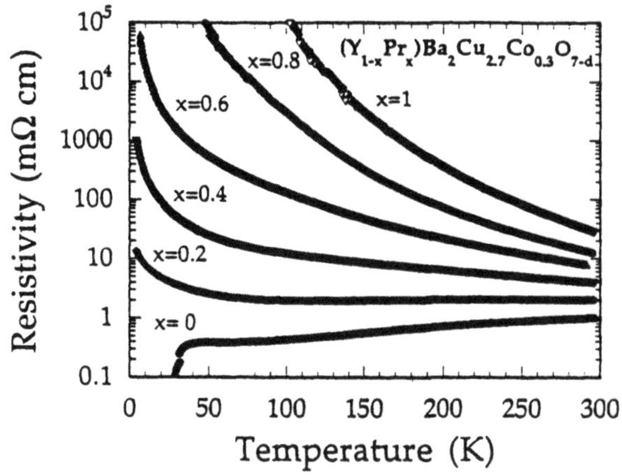

Fig. 4. Metal-insulator transition in $Y_{1-x}Pr_xBa_2Cu_{2.7}Co_{0.3}O_{7-\delta}$ thin films.

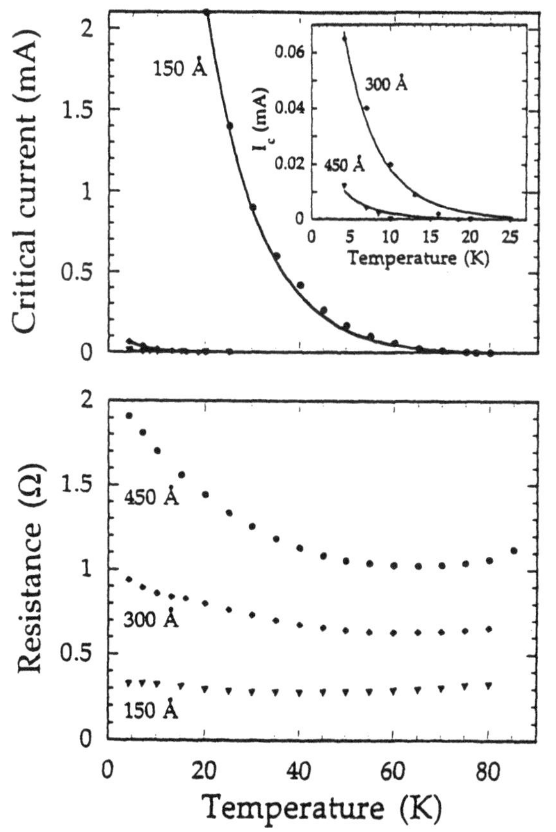

Fig. 5. Temperature and thickness dependence of I_c and R_n of $Y_{0.8}Pr_{0.2}Ba_2Cu_{2.7}Co_{0.3}O_{7-\delta}$ barrier junctions.

The temperature dependence of I_c and R_n, shown in Fig. 5, has been analyzed in a manner similar to that described earlier for 14 % Co-YBCO. The fittings provided a diffusion constant D = 2.3 cm^2/sec and a pair-breaking scattering time τ_{pb} = 8.6 x 10^{-14} sec. [3] The reduced diffusion constant in the barriers is evidence of an elastic scattering effect of the Pr dopants.

4. I-V CHARACTERISTICS OF INTERFACE RESISTANCES

When edge contacts are made between YBCO and other epitaxial conducting oxides such as (CaSr)RuO_3 and (LaSr)CoO_3, a consistent interface resistance of the order of 10^{-8} Ωcm^2 has been measured. [4] In the case of (CaSr)RuO_3, very reproducible zero bias conductance peaks, as shown in Fig. 6, have always been observed, as in the case of YBCO interfaces with metals such as Pb, Ag, and Au.

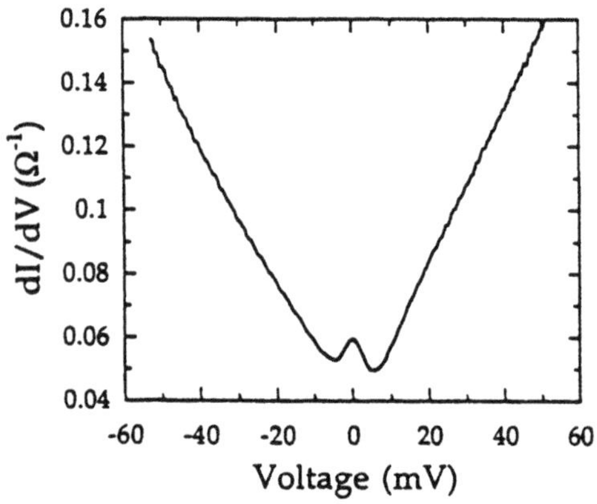

Fig. 6. Conductance-voltage characteristic of a YBCO/$SrRuO_3$ edge interface at 4.2 K.

On the other hand, YBCO/(LaSr)CoO_3 contacts did not exhibit any zero bias anomaly, as shown in Fig. 7(a). However, when annealed in vacuum (200 C, 30 min, Fig. 7(b)), the resistance went up and a zero bias conductance peak appeared. When it was annealed again in oxygen (350 C, 12 hr, Fig. 7(c)), the resistance was reduced to the original value, but the zero bias anomaly survived, suggesting that the zero bias anomalies may result from a very stable oxygen defect state at the interface. [4]

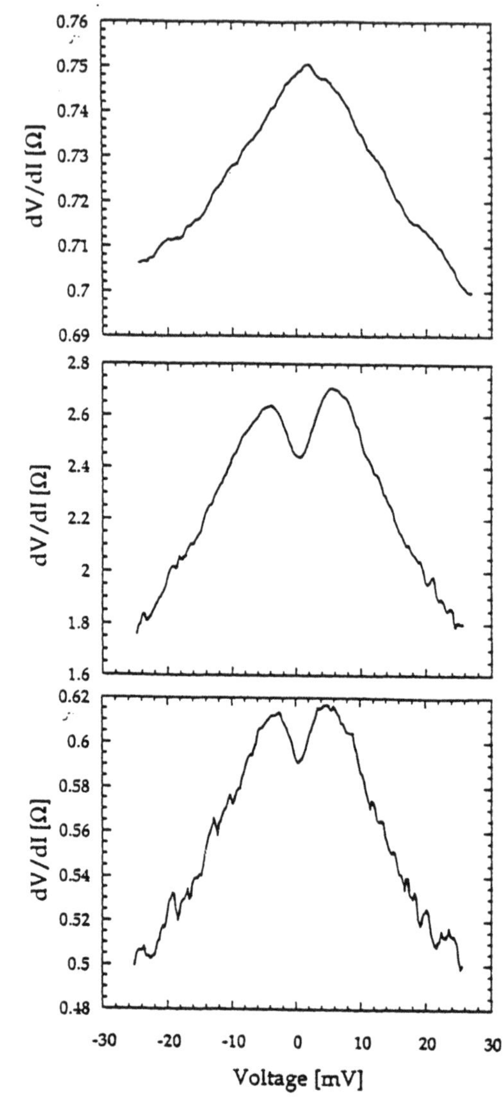

Fig. 7. Resistance-voltage characteristics of a YBCO/$La_{0.5}Sr_{0.5}CoO_3$ edge interface at 4.2 K after a series of heat treatments.

REFERENCES

1. P. G. De Gennes, Rev. Mod. Phys. **36**, 225 (1964).
2. L. Antognazza, S. J. Berkowitz, T. H. Geballe, and K. Char, Phys. Rev. B **51**, 8560 (1995).
3. L. Antognazza, K. Char, and T. H. Geballe, preprint.
4. L. Antognazza, K. Char, and T. H. Geballe, Appl. Phys. Lett. **68**, 1009 (1996).

ELSEVIER

Physica C 282–287 (1997) 423

PHYSICA C

Extended Abstract

Dynamics of Stacked Josephson Junctions

N.F. Pedersen

Physics Department , Technical University of Denmark, DK-2800 Lyngby, Denmark

Some high Tc superconductors such as for example the layered BSCCO type of materials have been shown to consist of a stack of intrinsic Josephson junctions. Electronic components fabricated from such materials have the potential for interesting applications, since the stacked structure gives the possibility of utilizing such multi-junction phenomena as fluxon phase locking, impedance matching etc. A disadvantage from the applications point of view is that such coupled long Josephson structures have an extremely complicated and not fully understood dynamical behavior. We have undertaken a study of the fundamental dynamic behavior of such systems by detailed numerical simulations of the coupled sine-Gordon equations describing such Josephson stacks. These equations, derived by Sakai et al., (ref. 1) are intended to describe superconductor-insulator-superconductor multilayer systems fabricated for example by alternating layers of Niobium and Al-oxide. In the limit of ultra thin superconductor films, these equations give a model for the description of fluxon motion in the intrinsic stacks of the BSCCO type, as elaborated in ref. 2. To keep the calculations relatively simple, we started by investigating numerically a system consisting of two vertically stacked long Josephson junctions described by two strongly coupled sine-Gordon equations (ref. 3,4). For that system we find two different Swihart velocities due to the coupling between the two long Josephson junctions. For an annular geometry, we find 'clean' fluxon shapes with identical fluxons in the two junctions corresponding to in-phase fluxon locking in the two junctions. In addition we also find out of phase fluxon locking due to fluxon-fluxon repulsion, as well as other modes - depending on the initial conditions and the biasing of the two junctions. For the overlap geometry, particularly with an applied magnetic field, we find the locked fluxon steps (flux flow steps) overlapped with Fiske modes of the two different velocities in a complicated pattern. As mentioned we may consider the Niobium systems as an experimental model to study the dynamical behavior of BSCCO intrinsic junctions, and we undertook experiments on two-stacks of Niobium. (ref. 5,6). Mutually phase locked fluxon-antifluxon states were observed, and the observed propagation velocities were in agreement with the theoretical predictions. The regions of stable phase-locking were measured in addition to other phenomena - particularly in a magnetic field. The samples could be considered three terminal devices , and the three terminal properties were investigated experimentally as well as numerically. Conclusions for the BSCCO systems are discussed.

REFERENCES

1. S. Sakai, P. Bodin, N.F. Pedersen, Fluxons in thin film superconductor-insulator superlattices J. Appl. Phys. 73, 2411 (1993)
2. R. Kleiner, P. Müller, H. Kohlstedt, N.F. Pedersen and S. Sakai, Dynamic behavior of Josephson-coupled layered structures, Phys. Rev. B 50, 3942 (1994)
3. A. Petraglia, A.V. Ustinov, N.F. Pedersen and S. Sakai, Numerical study of fluxon dynamics in a system of two stacked Josephson junctions, Journal of Appl. Physics 77, 1171 (1995).
4. S. Sakai, A.V. Ustinov, H Kohlstedt, A. Petraglia and N.F. Pedersen Theory and experiment on electromagnetic wave propagation velocities in stacked superconducting tunnel structures. Phys. Rev. B50, 12905 (1994)
5. G. Carapella, G. Costabile , A. Petraglia, N.F. Pedersen, and J. Mygind Fluxon Dynamics in stacked Josephson junctions. Appl. Phys. Lett. **69**, 1300 (1996)
6. G. Carapella, G. Costabile, G. Filatrella, M.H. Manscher, J. Mygind, M.A. Nordahn, N.F. Pedersen, and A. Petraglia, Mutual phase-locking of fluxons in stacked long Josephson junctions: simulations and experiments. Proc. of the Appl. Sup. Conf. 1996.

PII S0921-4534(97)00295-5

ELSEVIER

Physica C 282–287 (1997) 424–427

PHYSICA C

PROGRESS IN BSCCO-2223 TAPE TECHNOLOGY

A. P. Malozemoff, Q. Li and S. Fleshler

American Superconductor Corp., 2 Technology Dr., Westborough MA 01581 USA
ASC multifilamentary composite tape based on BSCCO-2223 high temperature superconductor (HTS) and a scalable rolling process has reached a record superconductor critical current density J_c (77 K, 0 T) of up to 58,000 A/cm^2, and engineering (full cross-section including metal sheath) J_e (77 K, 0 T) up to 15,000 A/cm^2. Pilot manufacturing of long lengths (100-400 m) has permitted a variety of increasingly significant prototypes, including a 200 hp ac synchronous motor, a 3300 A (77 K) 50 m machine-stranded cable conductor, a 5 kJ SMES system, and an operational ion-beam switching magnet installed in a tandem accelerator for carbon mass spectrometry dating. Recent ASC progress in manufacturing with Je over 8000 A/cm^2 creates new opportunities for yet more advanced prototypes and commercial development.

1. WIRE PROGRESS

This article reviews recent progress at American Superconductor Corporation (ASC) in BSCCO-2223 tape technology and some of the recent prototypes which this technology has facilitated. Following on recent reports[1,2], ASC's BSCCO-2223 multifilamentary composite wire technology continues to advance. Short lengths of rolled 85 filament BSCCO/Ag tapes have been synthesized with J_c (77 K, 0 T) up to 58,000 A/cm^2, and, in a separate sample, engineering (full cross-section) J_e (77 K, 0 T) up to 15,000 A/cm^2. The J_c standard deviation σ of 12 samples is less than 2%, demonstrating reproducibility. These results represent the first time that the electrical performance of commercially interesting multifilamentary wires has approached parity with that of pressed monofilamentary samples[3,4]. It has also been recognized[5] that at high J_e levels, self field effects become increasingly important, suppressing the current density from its true zero-field value. This effect is believed to be in the range of 15% for the above wires, implying that for the J_c 58,000 sample mentioned above, the true zero-field Jc approaches 66,000 A/cm^2. In addition, 313 filament tapes have achieved I_c values of 125 A (77 K, 0 T, σ = 1.7 A)[2].

Much of the progress has come from careful control of the deformation to avoid sausaging instabilities and improvement of grain connectivity by heat treatment development[2]. Furthermore, techniques have been found to improve the field dependence of these wires. For example, Li et al.[2] and Parrell et al.[6] recently reported increasing the "retention", that is, the relative $J_c(B,T)$ normalized to J_c (0,T), by a factor of 3 at 0.2 T field perpendicular to the plane, by means of a slow cooling procedure at the end of the final heat treatment. There is some evidence that this process alters Tc of the 2223 phase and also leads to a slight decomposition of the BSCCO-2223 phase into a Pb-rich 3221 phase. This suggests that the improvement in $J_c(B,T)$ retention stems from a change in the oxygen content and/or cation stoichiometry of the 2223 phase. We presume that either the CuO interlayer coupling has become stronger, or an effective flux pinning center has been introduced into the 2223 grains.

The ac losses in these tapes are also of interest. Results of an ac susceptibility measurement at 66 K and 50 Hz on a 150 µm-thick tape with an 85 filament core 100 µm thick (25 µm thick silver sheath on both sides) are shown in Fig. 1 with ac field applied parallel to the tape plane. The losses can be well described by a conventional Bean critical state model[7] for hysteretic losses in infinite slab as shown by the solid line, with the αH^3 region representing partial flux penetration through the thickness of the core and the approach to a linear H region

PII S0921-4534(97)00296-7

representing full penetration. This interpretation assumes that the individual filaments are fully coupled together by induced currents in the matrix, but that the eddy losses are negligible in size compared to the hysteretic losses. Thus the effect of the coupling is to increase the effective dimension for the critical state analysis from the thickness of the individual filaments to the thickness of the 85-filament core as a whole. The loss level at 0.1 T rms works out to be 1.1 mW/Am, where the loss has been normalized to the dc current level at 66 K and 0.1 T dc at 1 μV/cm. This level is of interest for applications such as transformers because, coupled with a typical 77 K GM cryocooler efficiency of 0.1, it represents half the net room temperature dissipation Jρ (per Am) of copper (resistivity 2 μΩcm) operated at 100 A/cm^2. To push below this level and further increase the efficiency advantage of HTS at this applied field, it will be necessary to either increase the resistivity between the filaments or reduce the total thickness of the multifilamentary core.

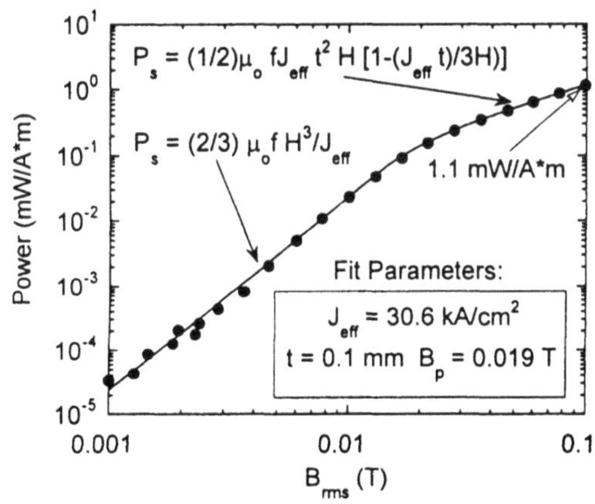

Figure 1. Ac (50 Hz) susceptibility measurement of loss, normalized to total current, vs. rms field applied parallel to the tape plane, of a 150 μm thick 85 filament BSCCO-2223 conductor with a measured J_e = 12500 A/cm^2 in 0.1 T parallel field dc, at 66 K. J_{eff} is fit to simple Bean model for net current density of multifilamentary core, t is core thickness and B_p represents the peak field for full penetration.

The short length wire technology is rapidly being scaled up into pilot manufacturing at ASC. J_e results on long lengths (mostly 100-400 m, with some work up to 1 km) are recently exceeding 8000 A/cm^2 (77 K, 0 T, 1 μV/cm) on a regular basis[10]. Since all the prototype work described below was done on earlier wire with J_e performance of 5000 A/cm^2 or less, one can expect a new and more powerful generation of prototypes and commercial product in the near future.

2. PROTOTYPES

A major prototype developed on the basis of the earlier generation of ASC BSCCO-2223 composite tape is the 125 hp ac synchronous motor built and tested up to a level of 285 hp by the combined effort of ASC and Rockwell/Reliance, supported by the USDOE through the Superconductivity Partnership Initiative (SPI) program. Four sets of 6 pancake coils of racetrack shape and differing lateral dimensions were wound from a total of 5200 m of 85-filament BSCCO-2223 wire of 0.25x2.5 mm^2 cross-section[11]. These were assembled into four racetrack coils with an overall beveled shape to fit snugly into the round cross-section of the rotor. At the design-point 80 A current level and 710 turns, the coils generated 56,500 A-turns. At 100 A, the coils had losses below 0.18 W/coil, well below the design limit of 1.25 W/coil. Because of the somewhat rounded current-voltage characteristic, these coils could be run at a significant overcurrent. The coils were mounted in a modified ac synchronous 4-pole motor[12] which was operated at 1800 rpm. The rotor was cooled to 27 K by a helium gas stream. Optimum efficiency was achieved at the 125 hp design point, but the motor was pushed up to 285 hp in subsequent tests.

This success, achieved in early 1996, is a major benchmark for HTS technology in rotating machinery. The SPI program is going forward to build 1000 hp and 5000 hp prototypes in the next several years, with coils for the 1000 hp motor already being fabricated at ASC.

Another major demonstration[9,10] in mid' 1996 was the machine-stranding of a 50 m multistrand conductor assembly by Pirelli Cable Corporation (PCC) of Lexington SC, based on the delivery of 6000 m of ASC multifilamentary wire with current levels on average above 30 A. This program is supported by the Electric Power Research Institute (EPRI) and the USDOE SPI program and targets a retrofit underground ac transmission cable application. A quality control program at ASC has confirmed tape performance within specification, including the first significant environmental testing (7 days at 70 C and 95% relative humidity, after 10 thermal cycles from room temperature to 77 K and a several day thermal stability test at 125 C). The wire was stranded using industrial-scale stranding equipment at the Lexington PCC facility into a 50 m multilayer multistrand conductor assembly with helically counterwound layers, which demonstrated 3300 A dc current at 1 μV/cm. Ac testing is in progress. The significance of this result is to confirm the robustness of the HTS technology in an industrial cabling environment. The full engineering and construction of a 3-phase cable with liquid nitrogen cooling, room-temperature dielectric and 115 kV operation is underway, and commercial power transmission cables are envisioned in a three year timeframe.

Another significant prototype[13] has been delivered by ASC to the Gesellschaft fuer Innovative Energieumwandlung and Speicherung (EUS) GmbH in Gelsenkirchen, Germany. This is a 5 kJ HTS SMES magnet and cryogenic system, based on a conduction-cooled and cryocooled HTS magnet operating at 25 K and 100 A. The substantial solenoidal coil, built out of pancakes and epoxy-vacuum-impregnated, measures 37 cm i. d., 44 cm o. d., and 9.8 cm high, and generates 140,000 A-turns. The coil is suspended in a single-wall stainless steel vacuum vessel lined with multilayer insulation. Conduction-cooled leads are cooled by a first single-stage GM cryocooler, and a second single-stage cryocooler cools the magnet. Both cryocoolers run from the same compressor and are supplied by Leybold. The peak radial field is 0.78 T. The SMES will be used as part of a model laboratory electric system with a power converter structure allowing the study of different application scenarios. The system was installed at EUS in mid' 1996 and is under test.

A more recent operational HTS magnet[14], shown in Fig. 2, has been developed in a collaboration of ASC with Alphatech International of Auckland NZ, Industrial Research Ltd. of Lower Hutt NZ, and Isys of Sunnyvale CA USA, and supported by the NZ Technology for Business Growth Programme. This magnet is designed to provide ion-beam-switching for a beam line of a 6 MeV tandem van de Graaff accelerator in the Accelerator Mass Spectrometry laboratory of the Institute of Geological and Nuclear Science (IGNS) in Lower Hutt NZ. Applications include carbon dating and studies of beryllium isotopes for archeological and geophysical studies. The current is set + or - to deflect the ion beam +/- 25° or 0° into any of three target beam lines.

The HTS coils built at ASC consist of a pair of racetrack coils with clear-bore dimensions of 35x62 cm^2, constructed from pancakes and epoxy-vacuum-impregnated. The operational current of 100 A generates a total of 9200 A-turns, which is coupled to an iron yoke and pole faces generating 0.72 T in the airgap. The design limited the local field on the HTS coil to 0.14 T. The coils were conduction-cooled and coupled to a single-stage Leybold GM cryocooler. The operational temperature is 25 K, but the system has been operated at full current at temperatures up to 50 K.

This system sets a number of benchmarks. Most importantly, it is the first fully operational HTS magnet system; no longer just a prototype, this magnet is expected to be in operation over years to come. Secondly, it is the first "distributed" conduction-cooled superconducting magnet in the 25-50 K range: the cryogenic envelope surrounds only the two large racetrack coils and leaves room for the iron yoke and beam line. Thirdly, it is the first significant superconducting magnet coupled to an iron yoke; this applications area is an interesting one for HTS BSCCO-based coils since local fields can be kept low so that the system can be operated at relatively high temperatures. Finally, it is the first HTS application for accelerators; this is promising area for early entry of HTS technology.

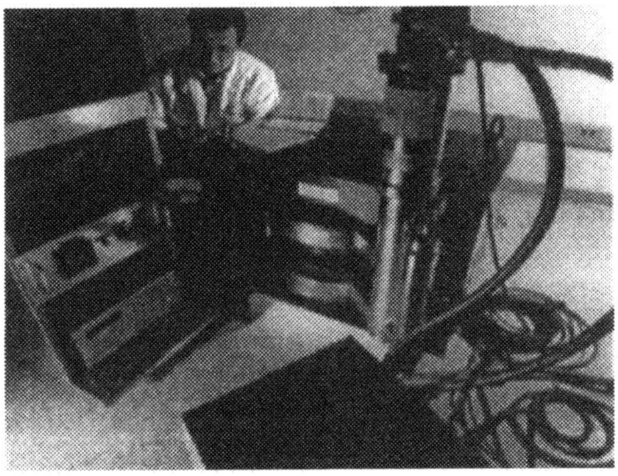

Figure 2. Ion-beam-switching magnet system, with iron yoke, HTS coils, GM cryocooler and control panel, developed for an accelerator mass spectrometry laboratory in New Zealand.

Finally, we mention here the delivery of over 10,000 m of ASC multifilamentary HTS wire to ABB Secheron, in Geneva, Switzerland, for the construction of a 630 kVA 3-phase transformer[15] operating at 50 Hz and 18.720/420 V. The system has a conventional room-temperature iron magnetic circuit. The HTS coils operate at liquid nitrogen, in a superinsulated cryostat. This project is also supported by the Swiss Federal Office of Energy, the Swiss Utilities Fund and Electricite de France. The manufacturing and factory testing of this unit has been completed, and installation in a Swiss utility in Geneva is underway.

3. CONCLUSIONS

The significant array of major prototypes and magnet products confirms the rapid ongoing progress of HTS BSCCO-2223 tape technology. HTS Current leads have already been commercialized. One can expect HTS magnet applications to be next. The major HTS applications in the energy and power sector, including motors, power transmission lines and transformers, will take several more years as engineering and beta tests are completed, but they too are advancing rapidly in the context of well defined programs, both at ASC and in other organizations around the world.

The authors thank their colleagues at ASC, in the Materials R+D and Engineering Products departments, for discussions. They also thank their many colleagues at the organizations mentioned in the text, coworking with ASC to achieve new state-of-the-art HTS applications.

REFERENCES

1. A. P. Malozemoff, in *Proc. 10th Anniversary HTS Workshop*, ed. B. Batlogg et al. (World Scientific, Singapore, 1997), p. 47.
2. Q. Li et al., Appl. Sup., to be published.
3. Q. Li, K. Brodersen, H. A. Hjuler and T. Freltoft, Physica C **217**, 360 (1993).
4. Y. Yamada et al., Proc. Fifth International Symp. on Supercond., p. 717 (1993).
5. L. Gherardi et al., Mat. Sci. and Eng. B**39**, 66 (1966).
6. J. A. Parrell et al., Appl. Phys. Lett. **69**, 2915 (1996); also to be published.
7. M. Wilson, *Superconducting Magnets* (Clarendon Press, Oxford, England, 1983), p. 163.
8. J. M. Seuntjens et al., Appl. Sup., to be published.
9. A. Bolza, P. Metra and M. M. Rahman, in *Proc. 10th Anniversary HTS Workshop*, ed. B. Batlogg et al. (World Scientific, Singapore, 1997), p. 595.
10. D. Buczek et al., Appl. Sup., to be published.
11 J. P. Voccio, B. B. Gamble, C. B. Prum and H. J. Picard, ibid.
12. R. F. Schiferl et al., in Advances in Cryogenic Engineering, vol. 42 (Plenum Press, New York 1977), p. 977.
13. S. Kalsi et al., Appl. Sup. to be published.
14. S. Kalsi et al., Pooke et al. to be published.
15. S. Dale, private communication

ELSEVIER

Physica C 282–287 (1997) 428–430

PHYSICA C

The Combinatorial Synthesis and Evaluation of Functional Materials

X.-D. Xiang and P. G. Schultz

Materials Sciences Division, Lawrence Berkeley National Laboratory, Berkeley, CA 94720

In order to increase the rate of materials discovery and optimization, as well as increase our understanding of increasingly complex modern materials, we have developed the strategyies and technologies for the combinatorial synthesis and evaluation of functional materials. In this approach large collections (libraries) of different materials are rapidly synthesized, processed and screened for specific properties of interest.

1. Introduction

Materials play a key role in the advancement of science and technology, the development of semiconductors being a good example. Many new or improved materials are needed including superconductors for the electronics and energy industries, phosphors for the display industry, dielectrics and electro-optical materials for the electronics and communications industries and heterogeneous catalysts for the chemical industry. Even though our understanding of the electronic, magnetic, optical and chemical properties associated with these materials has increased considerably in recent years, the ability to predict structure-property relationships is only possible in a limited number of cases, and involves considerable theoretical effort. This lack of predictive ability is not surprising in light of the complexity of modern materials and the importance of defects/dopants in controlling properties. As a result materials are typically discovered and optimized by an empirical trial and error process that is both time consuming and costly. We have recently shown[1] that combinatorial methods, an approach in which large collections (libraries) of materials are rapidly synthesized, processed and screened for specific properties of interest, can significantly facilitate the materials discovery/ optimization process. Methods for generating libraries of materials (hundreds to tens of thousands of different compositions at a time) have been developed that include thin film and solution based synthetic methods. Rapid scanning or parallel detection systems have also been developed to characterize the electronic, optical, and magnetic properties of these materials, as well as methods to subject libraries to different processing conditions (temperature, atmosphere, pressure).

2. Materials Library Fabrication

We have developed two technologies, combinatorial thin film synthesis and solution based synthesis, to fabricate materials libraries.

a) combinatorial thin film synthesis In this approach, different thin film deposition methods (laser ablation, sputtering, evaporation) and masking techniques (physical masks and photolithography) are combined to generate libraries of diverse materials with different compositions through combinatorial masking strategies.

A number of masking strategies can be used to address different types of discovery/ optimization problems. These include: i) binary masking strategy, in which only half of the substrate area is exposed to precursor thin film deposition with different patterns (1, 2, 4,... strips oriented in two possible directions) in each step. After N steps, the number of different compositions is 2^N, which includes all possible combinations of N elements (or precursors). This strategy is simple and effective for a broad search. However, sites having less than two elements and more than eight elements may not be very interesting or too complex for certain applications. ii) a quaternary masking strategy. In many cases, it is desirable to study compositions of N (or less) components, ABCD..., where each components

PII S0921-4534(97)00297-9

(A or B or...) is chosen from a group of four or more precurcsors. A quaternary masking strategy can be used to accomplish this. In each step, one quadrant of the substrate area is exposed to thin film precursor deposition. For component A, a mask with a window quarter of the substrate area can be rotated by 90° in each step to deposit 4 different choices of precursors; For component B, a mask with 4 windows, each exposing only a quarter area of the A layer window, can be rotated by 90° in each step to deposit 4 different choices of precursors for B layer; For component N, a mask with 4^{N-1} windows, each exposing only a quarter area of the window area in the previous layer is used. After 4N steps, one ends up with 4^N different combinations of N elements. This strategy can be extended to higher order libraries, such as 9 based (with a choice of 9 precursors for each component) there the similar symmetry reduces the required number of masks. iii) a x-y shutter masking system. An x-y shutter is flexible and can be used *in situ* to perform simple tasks, such as the gradient optimization or AB cross strip combinations (where A is chosen from 16 different precursors and B is chosen from 16 different precursors). Currently, 1 inch x 1 inch materials libraries containing 1024 different compositions are routinely generated using both binary and quaternary masking strategies in combination with physical masks or photolithography. After lead compounds are identified in these libraries, careful optimizations are performed using conventional gradient phase diagram approach with a x-y shutter equipped laser ablation deposition system.

b) Solution based combinatorial synthesis using muti-component inkjet delivery system. For many materials, e.g., polymers, heterogeneous catalysts and powder phosphors, synthesis is best accomplished by solution phase methods. In addition, solution based methods allow mixing at the molecular level, thereby reducing the need for high temperature interdiffusion in solid state synthesis and facilitating the isolation of metastable phases. In contrast to most current solution phase deposition systems used in the synthesis of libraries for pharmaceuticals or genomics, in which excess reagents are used, the ability to accurately control deposition volume is critical in determining the final stoichiometry of solid state materials. One approach to this problem involves the use of robotics to accurately pipet solutions; another involves the use of pneumatic valves. Neither has been implemented on a scale that allows submicroliter control over reagents, a requirement for library densities on the order of hundreds of members per $inch^2$.

We have developed a microscale scanning multi-inkjet delivery system with nanoliter delivery capabilities. This system consists of an array of single nozzle, "drop on demand" inkjets, that use a PZT piezoelectric disc as the electrical pulse to pressure transducing element. Droplets are ejected through a 50 μm sapphire nozzle from the pressure chamber of each inkjet which is connected with Teflon tubing to chemical reservoirs under positive pressure. The droplet size is on the order of 500 picoliters with reproducibility better than 99% and a maximum delivery speed of 2,000 droplets/s. This system has been used successfully to generate libraries of 100 members/ $inch^2$ in order to optimize phosphors. A system that would generate libraries with 1,000 sites/$inch^2$ is currently under development. These systems have added considerable versatility to our combinatorial synthesis capability and potentially to that of the pharmaceutical and biotech communities.

3. Materials library processing

Materials processing is a critical step in combinatorial synthesis. Since each site in a thin film materials library is composed of multi-layered thin film precursors, low temperature diffusion is necessary to uniformly mix precursors before high temperature phase formation. Multiple processing conditions (temperatures, atmospheres, pressures) are applied to identical libraries to identify optimized processing conditions. In addition, a temperature gradient oven has been used to identify the best processing conditions. Considerable effort is currently being made to further combinatorialize the processing procedure.

4. Materials Library Screening

Electrical impedance measurements are crucial in many applications, including measurements of superconductivity, magnetoresistivity, and ferroelectric properties. We have developed several systems for a range of different applications. For many measurements (e.g., superconductivity and magnetoresistivity) variable temperature contact measurements are necessary. A contact probe consisting of a matrix of spring loaded contact pins has been fabricated that allows multi-sample library measurements from 4.2K to room temperature. This system is capable of measuring 4-probe conductivity of 64 samples in one experiment. The probe can be fit into a superconducting magnet of 12 Tesla for conductivity measurements at various temperatures. The multi-sample measurement is accomplished by a computer controlled multi-channel switching and data acquisition system. Temperature control and data acquisition are fully automated by computer software. The areal density of the libraries that have been successfully measured with this system are 64-128/square inch.

Non-destructive scanning measurement systems with high resolution are most desirable for high density library screening. We have developed a scanning tip microwave near-field microscope (STMNM) to perform high resolution high sensitivity electrical surface impedance imaging.[2] This microscope is based on a novel structure which uses a sharpened tip as a part of a high Q microwave reasonator. As the tip scans over the sample surface, the changes in resonance frequency (f_0) provide information on the sample's dielectric constant/conductivity, while changes in Q factor give information on the sample's microwave energy loss. Since the resolution is limited by the tip geometry, the intrinic resolution of STMNM is comparable to STM and AFM. More importantly, since the sensing device is a high Q resonator (with a Q of several thousand), the sensitivity to differences in electrical impedance (dielectric constant and tangent loss) are extremely high. Using this system we have obtained the first sub-micron (~0.1 μm = $\lambda/1,000,000$) resolution profiles of dielectric constant and tangent loss in dielectric materials and have used this microscope to image libraries of dielectric/ferroelectric materials. A low temperature version of the microscope has been also developed and scanning images of superconductors (YBCO) has been successfully obtained.

An optical spectromicroscope and an optical polarization microscope have also been developed to screen optical and magnetic properties, respectively.

5. Applications to different classes of materials

With this infrastructure developed, we have begun to explore a wide range of interesting materials. New magnetoresistive materials[3] and phosphors[4] have been discovered and optimized using this approach. Currently, we are applying this approach to the discovery and optimization of superconductors, ferroelectric materials, and optical materials.

Acknowledgments
We would like to acknowledge the important contributions from Gabriel Briceno, Hauyee Chang, Sung-Wei Chen, Chen Gao, Yulin Lou, Ichiro Takeuchi, Kai-An Wang, Tao Wei, W. G. Wallace-Fredman and Young Yoo to this project. This work was supported by the director of Advanced Engergy Projects Division, Office of Computational and Technology Research, U.S. Department of Energy under Contract No.DE-AC03-76SF00098 and Office of Naval Research, Order No. N00014-95-F-0099.

6. References:

1. X.-D. Xiang et al. *Science* **268**, 1738-1740 (1995).
2. T. Wei, X.-D. Xiang, W.G. Wallace-Freedman and P.G. Schultz, Appl. Phy. Lett. **68**, 3506-3508, (1996).
3. Briceno, G., Chang, H., Sun, X., Schultz, P.G.; Xiang, X.-D. *Science* **270** , 273-275 (1995).
4. Xiao-Dong Sun, P. G. Schultz, Chen Gao, Jingsong Wang and X.-D. Xiang, Appl. Phy. Lett., to be published.

ELSEVIER

Physica C 282–287 (1997) 431–434

PHYSICA C

Preparation and characterization of melt-textured Nd_{1+x} Ba_{2-x} Cu_3O_y made in air

A. M. Hu[a,b], Z. X. Zhao[a], W. Gawalek[b], M. Wu[b], T. Strasser[b], P. Goernert[c]

[a]National Lab for Superconductivity, Institute of Physics, The Chinese Academy of Sciences, Beijing 100080, P. R. China

[b]Institut fuer Physikalische Hochtechnologie e. V., Postbox:100239, D-07702, Jena, Germany

[c]SurATech GmbH, Goeschwitzer Str. 22, D-07745, Jena, Germany

High quality melt-textured Nd_{1+x} Ba_{2-x} Cu_3O_y bulks have been prepared through a novel technology carried out in air. In present study, a modified melt processing technology is employed to prepare the melt-textured Nd_{1+x} Ba_{2-x} Cu_3O_y samples, and a post-annealing in flowing Ar is used to obtain the high Tc phase. AC susceptibility measurement shows that the critical temperature has reached 94.5K. The maximum levitation force is 7.4N for a sample with a diameter of 18mm. The Jc values are 2.4×10^4 A/cm^2 (0T,77K), 1.7×10^4 A/cm^2 (1T, 77K) and 3.3×10^5 A/cm^2 (8T, 21K). The fishtail effect is relatively obvious in 40-60K.

1. INTRODUCTION

Many properties of melt-textured Nd_{1+x} Ba_{2-x} Cu_3O_y become of worldwide increasing interests. Its Tc and Jc are higher than those of YBCO. Especially in high field regime, the solid solution Nd_{1+x} Ba_{2-x} Cu_3O_y leads to an enhancement of Jc in 1-3T [1,2]. Such a highly field-induced maximum Jc is not observed in the comparable melt-textured YBCO materials. This indicates that it is possible for NdBCO bulks to trap higher magnetic field than YBCO in the same size or to obtain stronger levitation force working at the same temperature. In addition, NdBCO has a higher decomposition temperature and much faster rate in phase formation than YBCO [3,4]. These characteristics can bring one some advantages in material processing.

However, it is difficult to prepare NdBCO through the conventional melt process. Many research groups report that the preparation of high Tc melt-textured NdBCO must be carried out in a reduced oxygen partial pressure or flowing Ar atmosphere [1,2,5]. Moreover, it is discovered that a large single domain is difficult to be obtained in this atmosphere [6]. We have developed a novel melt process to obtain high Tc melt-textured Nd_{1+x} Ba_{2-x} Cu_3O_y bulk samples in air [7]. In this process, a modified melt-textured process is employed to prepare high textured bulks, and a following Ar heat treatment is used to obtain high Tc phase. The characterization of the obtained samples is comparable with those reported results [8].

In this paper, we report some recent developments of this novel technology. Tc has reached 94.5 K. Jc is 3.3×10^5A/cm^2 in 8T at 21K. The fishtail effect is relatively obvious in the temperature regime of 40-60K.

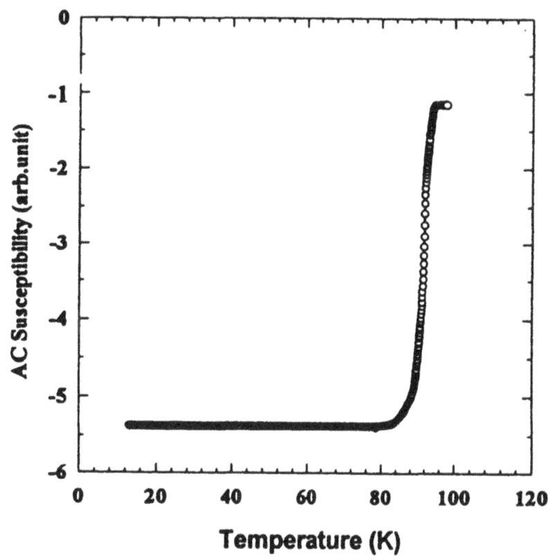

Fig 1. AC susceptibility transition versus temperature for melt -textured NdBCO samples

PII S0921-4534(97)00303-1

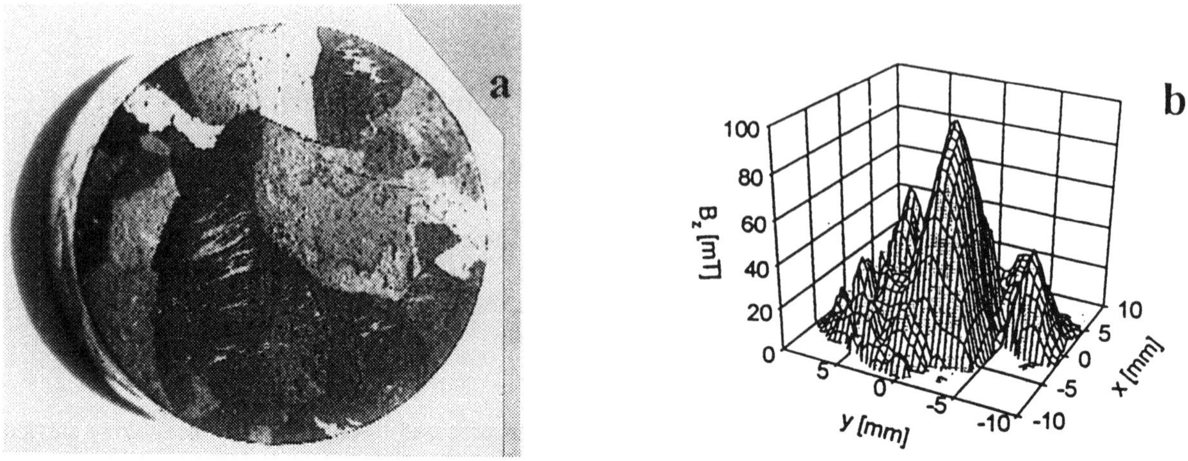

Fig 2. a. An illustration of the melt-textured NdBCO sample under the polarized light. b. Trapped magnetic field profile for the same sample. The diameter of the sample is 18.5 mm.

2.EXPERIMENTAL

The new melt process composes of four steps: the synthesis of precursors through a solid state reaction, a modified melt-textured process in air; an intermediate thermal treatment in flowing Ar and oxygenation. The precursor has a nominal composition of $NdBa_2Cu_3O_y$ + $0.4Nd_4Ba_2Cu_2O_Z$. The detailed process for each step was described in Ref. [7,8].

The critical temperature of the sample was measured by the AC susceptibility transition. Magnetic levitation force and trapped magnetic field were measured with a three-dimensional measurement system [8]. For the trapped magnetic field mapping, a 2T field is employed to magnetize the sample. The magnetization measurement was performed using a vibrating sample magnetometer(VMS, Oxford Instruments) with a 14 T superconducting magnet. Microstructure was examined by SEM. The phase composition was measured by energy dispersive x-ray spectroscopy(EDX).

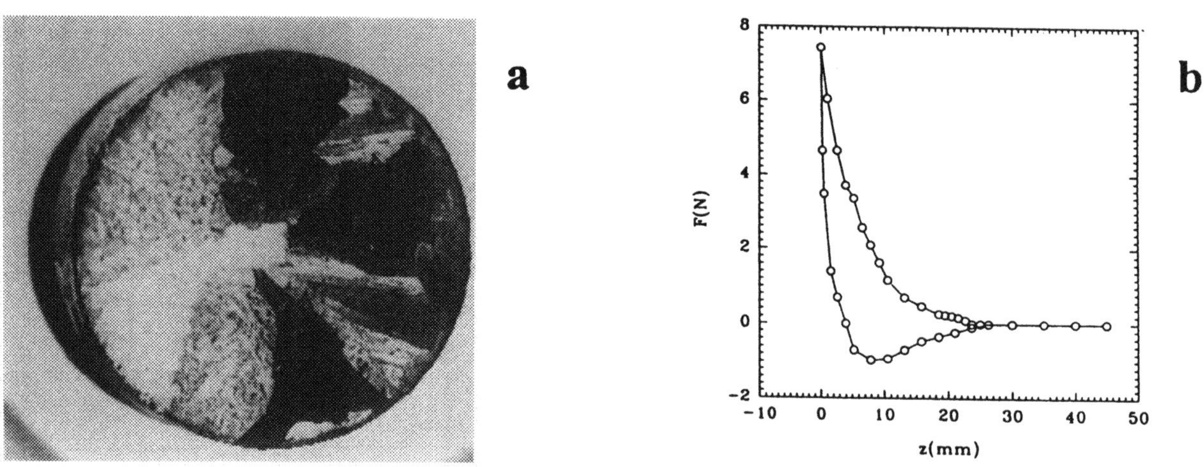

Fig 3. a. An illustration of the melt-textured NdBCO sample under the polarized light. b. Magnetic levitation force versus the distance between the NdFeB permanent magnet and the superconductor. The diameter of the sample is 18 mm.

3. RESULTS AND DISCUSSION

Fig.1 shows the AC susceptibility transition versus temperature for melt-textured NaBCO samples undergoing the presently thermal treatments in flowing Ar and O_2. Tc is 94.5K . The transition width is larger than that of the sample obtained through OCMG [1,2]. This suggests that the optimal oxygenation has to be investigated yet. Fig.2a illustrates a polarized optical micrograph of the polished surface of a cylindrically melt-textured $Nd_{1-x}Ba_{2-x}Cu_3O_y$ sample. Two big grains can be seen in the center of the sample. The lengths of the grains are more than 1 cm. Fig. 2b shows the trapped field profile at 77K for the same sample whose crystal structures shown in Fig.2a. The maximum peak has a value of 101mT at the position (x,y,z) = (1.5,1,0.25) mm. Fig.2(a, b) show that two grains in the center of the sample form only one magnetic domain (a domain is a volume in which the induced critical current can form a circuit), which is corresponding to the biggest trapped field peak in Fig.2b.

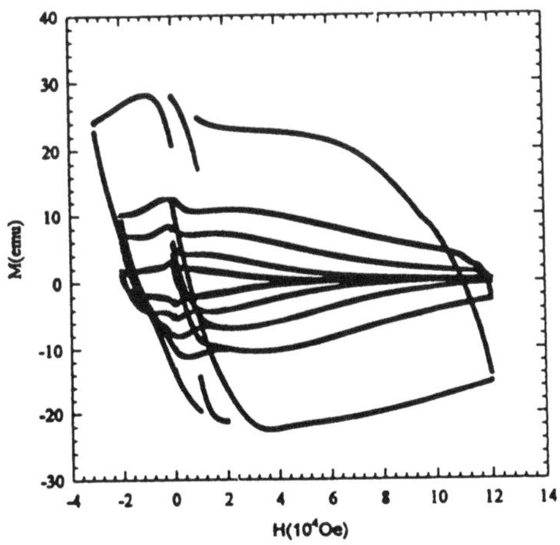

Fig 4. Magnetization hysteresis loops for melt-textured NdBCO under different temperatures. From inside to outside: 70K, 60K, 50K, 40K and 21K.

Fig.3(b) shows the levitation force versus the distance between the sample and the permanent magnet at 77K. A NdFeB permanent magnet with a diameter of 17 mm, a height of 12 mm and a surface magnetic field of 0.36T is used to measure

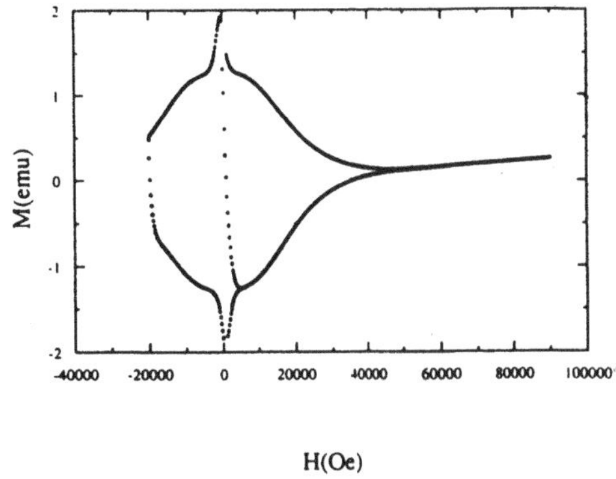

Fig 5. Magnetization hysteresis loop for melt-textured NdBCO at 77K.

the levitation force. The velocity of the permanent magnet towards or away from the superconductor is 3 mm/s. The minimum distance between the permanent magnet and the MTG bulks is 0.5 mm. The illustration of the corresponding sample under the polarized light is shown in Fig.3(a).

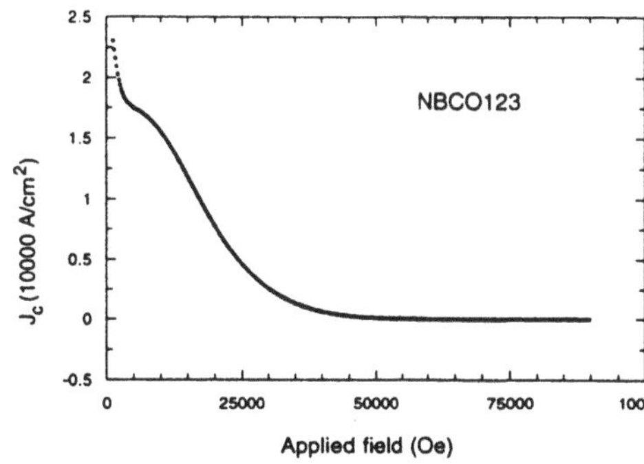

Fig 6. The magnetic field dependence of the critical current density for melt-textured NdBCO at 77K.

Figs. 4,5 show the temperature dependence of magnetization hysteresis loops for the melt-textured NdBCO sample. No pronounced fishtail effect can be seen at 77K (in Fig.5). This result is different from the report in Ref. [1,2].The fishtail effect is relatively obvious in the temperature regime of 40-60K. These similar field-induced magnetization enhancements are observed in melt-textured YBCO. In YBCO, the flux-pinning is due to the oxygen deficiency [9].In our sample, a small chemical variation between Nd and Ba has been detected. EDX shows that x=0.06 for the Nd1+xBa2-xCu3Oy solid solution[8]. But Fig.4,5 don't show the similar flux pinning effect reported in[1,2]. Further investigations on the pinning mechanism in our sample obtained in the present process will be given later[11]. The temperature dependence of the critical current density at 77K is calculated in the extended Bean model[10] and shown in Fig.6.

In summary, high quality melt-textured NdBCO bulks have been successfully prepared through a novel melt process in air. The maximum levitation force has reached 7.4N. The flux pinning mechanism for our samples is different from the previous reports.

ACKNOWLEDGMENTS

This work is supported by Chinese National Joint R/D Center on Superconductivity and National Natural Foundation Committee.

REFERENCES

1. S. I. Yoo, N. Sakai, H. Takaichi, H. Higuchi, and M. Murakami, Appl. Phys. Lett. 65, 633(1994)
2. M. Murakami, S. I. Yoo, T. Higuchi, N. Sakai, J. Weltz, N. Koshizuka and S. Tanaka, Jpn. J. Appl. Phys. 33, L715(1994)
3. M. Morita, M. Sawamura and S. Takabayashi, K. Kimura, H. Teshima, M. Tanaka, K. Miyamoto, and M. Hashimoto, Physica C 235-240 (1994) 209
4. K. Salama, A. S. Parikh and L. Woolf, Appl. Phys. Lett. 68(1996)1993
5. W. Bieger, G. Krabbes, P. Schatzle, L. Zelenina, U. Wiesner, P. Verges, J. Klosowki, Physica C 257(1996)46
6. G. Krabbes, P. Schatzle, W. Bieger, G. Fuchs, U. Wiesner and G. stover, IEEE Trans. Appl. Super. March, 1997.
7. A. M. Hu, S. L. Jia, H. Chen, Z. X. Zhao, Physica C 272(1996)297
8. A. M. Hu, Z. X. Zhao, M. Wu, C, Wende, T. Strasser, B. Jung, G.Bruchlos, W. Gawalek, P. Goernert, Submitted to Physica C.
9. M. Daeumling, J. M. Seuntijens and D. C. Larbalestier, Nature 346(1990)332, J. L. Varges and D. C. Larbalestier, Appl. Phys. Lett., 60(1992) 6893
10. E. M. Gyorgy, et al., Appl. Phys. Lett., 55(1989)283
11. A. M. Hu, Z. X. Zhao, et al. to be published.

Physica C 282–287 (1997) xcv–cxiv
North-Holland

List of Contributors

Aarts, J. 2083, 2085
Abakumov, A.M. 61
Abbas, F. 1579, 2535
Abell, J.S. 1093, 2007
Abrego Castillo, I. 2181
Abulafia, Y. 2209, 2225
Acha, C. 817, 857, 1167, 1819
Achsaf, N. 140
Adachi, S. 45, 643, 887, 889, 971, 977
Adachi, T. 1233
Adelmann, P. 1059
Adrian, H. 583, 655, 669, 1883
Aeppli, G. 231
Afanassyev, D.N. 979
Afonso, S. 613, 641, 683, 685
Aftab, M. 1225
Ageev, N.V. 1279
Agterberg, D. 214
Aguila, E. 819
Ahmed, F. 629
Ahn, D. 2523
Ahn, K. 615
Ahn, S.-A. 1707
Aigner, M. 719
Akimitsu, J. 153, 997, 1115
Akinaga, M. 635, 1139
Akita, S. 2439
Akoshima, M. 1265
Aksay, I.A. 232
Al-Harthi, A.S. 2273
Alario-Franco, M. 857
Alario-Franco, M.A. 781, 801
Albino Aguiar, J. 779, 807
Albrecht, M. 1117
Aleksandrov, K.S. 2447, 2449, 2453
Alexandrakis, G.C. 1901
Alexandrov, A.S. 269
Alff, L. 1477, 1485
Allen, J.L. 647
Alleno, E. 1317
Allenspach, P. 1327
Alloul, H. 226, 1389
Almasan, C.C. 1127, 1129, 1147, 2327
Almeida, M. 955
Altenburg, H. 855
Alvarez, G.A. 1483
Alyoshin, V.A. 61
Amador, U. 781, 1001
Amato, A. 1353
Ambrosch-Draxl, C. 1635, 1637
Ammor, L. 1555, 1983
Ananyev, A. 1337
Andersen, N.H. 781, 1089, 1289
Andersson, M. 1959
Ando, Y. 240, 1217
Andreev, D.V. 1609
Andrikidis, C. 2123
Anezaki, S. 2399
Ang, S. 685
Ang, S.S. 683
Anikeeva, O.B. 1157
Anshukova, N.V. 1065, 1067
Antipov, E.V. 61, 65, 857, 899, 1081, 1381
Antonenko, S.V. 1291
Anzai, H. 1895
Ao, P. 367
Aoki, R. 535, 629, 979
Aomine, T. 1989
Apakina, V.N. 1583, 1585
Arai, T. 1895
Aranson, I. 355
Arao, M. 1111
Arao, T. 1489
Arenas, B. 671
Artemenko, S.N. 1845
Asab, A. 909
Asayama, K. 210, 1115, 1351, 1365, 1367, 1369, 1373, 1385
Ashihara, A. 997
Assmus, W. 719, 879, 1437, 2589
Atou, T. 827
Attanasio, C. 2019
Aubin, H. 1505
Autric, M. 1145
Avella, A. 1757, 1759
Avenhaus, B. 395, 2527
Avila, M.A. 2201
Awaji, S. 2287
Awana, V.P.S. 779, 807
Awaya, Y. 2105
Axnäs, J. 1221, 1535
Azoulay, J. 2165
Azuma, M. 149

Babic, E. 2133
Babić, E. 2297, 2299
Babonas, G.-J. 749, 773
Babushkina, N.A. 1149, 1163
Baby, U. 407
Baca, E. 709
Badeche, T. 1161
Bae, M.-K. 1293
Baenitz, M. 1381
Baert, M. 379, 1567
Baggio-Saitovitch, E. 1941
Baggio-Saitovitch, E.M. 489, 1633, 1939
Bai, W.P. 2381
Bairu, Z. 681
Bakharev, O. 227
Balaev, D.A. 2447, 2449, 2453
Balagurov, A.M. 61, 1353
Balakirev, V.F. 1085
Balakrishnan, G. 2027, 2187
Bandyopadhyay, B. 1039, 1049, 1331, 1339, 2001
Banerjee, S.S. 2027, 2043
Bang, Y. 1851
Baños, L. 819
Bao, X. 1255
Baran, M. 1607, 2161
Barbut, J.M. 2623, 2625
Barca, O. 2583
Barentzen, H. 1777
Barholz, K.-U. 1481
Baril, L. 2207
Bark, K. 2371
Barrault, M. 2623, 2625
Basco, F. 1683
Baskaran, G. 1683
Bäuerle, D. 2147, 2321
Baumann, T. 2639
Baumgarten, C. 2589
Bäuml, B. 1829
Bazarov, V.V. 589
Béal-Monod, M.T. 1841
Beaugnon, E. 461, 525, 539
Becker, B. 331
Bedard, B. 1153, 1213
Bednorz, J.G. 37
Beduz, C. 2555, 2557, 2559, 2593
Behnia, K. 1505
Bei, Y. 1439
Beilin, V. 2213
Belevtsev, B.I. 1223
Bellot, P.-V. 1357, 1359
Bellot, P.V. 1361
Belmont, O. 2623, 2625
Belogolovskii, M. 1481
Belova, L.M. 1149, 1163
Ben Ali, A. 1887
Ben Ali, A. 1801
Ben-Dor, L. 1339
Beňačka, Š. 639
Benkraouda, M. 311
Bennaceur, R. 1801, 1887
Bennemann, K.H. 1681, 1775
Benoit, W. 1953, 2021
Berastegui, P. 1377

Berberich, P. 107
Bernal, O.O. 1393
Bernhard, C. 236
Berry, S. 2259
Berseth, V. 1953, 2021
Berthier, C. 227
Bertinotti, A. 1073
Beskrovny, A.I. 1353
Beuran, C.F. 162
Beuran, F. 929
Beuran, F.C. 631
Bhasale, R. 2587
Bhat, S.V. 1975
Bhattacharya, S. 2027, 2043
Bian, W.M. 1607
Bianconi, A. 1081
Bierleutgeb, K. 2321
Birgeneau, R.J. 85, 170
Bishop, A.R. 1645, 1827
Bishop, D.J. 355
Blackstead, H.A. 1005, 1513, 1551
Blanchard, N. 1389
Blatter, G. 19, 319, 1301
Blumberg, G. 1009, 1017, 1027
Bobba, F. 709
Bobroff, J. 226, 1389
Bobrovskii, V. 1335
Boccara, A.C. 1025
Boebinger, G.S. 240, 1217
Boekema, C. 2069
Boffa, M. 487
Böhmer, C. 1309
Boikov, Yu.A. 111
Bok, J. 294, 1025, 1665
Bokor, J. 252
Bolaños, G. 711, 2419
Bommeli, F. 1475
Bonfait, G. 955
Bontemps, N. 1013
Bonvalot, M. 539
Bordet, P. 65, 817, 857, 951, 1081
Borsa, F. 359
Bose, M. 799
Böttger, G. 1377
Boudissa, M. 1979, 2253, 2263
Bouffard, S. 2189
Boulesteix, C. 1161
Bouquet, F. 1425
Bourgault, D. 539, 2313, 2623, 2625
Boutet, M. 487
Bouvier, J. 294, 1025, 1665
Boyd, I.W. 1231
Bozovic, I. 252
Bracanovic, D. 2035, 2037
Braden, M. 1037, 1059
Brandstätter, G. 1309, 2089
Brandt, E.H. 343
Bratsberg, H. 1469
Bravi, C. 2403
Brinkmann, D. 1345, 1371
Brinkmann, M. 2265
Brom, H.B. 1393, 2121
Bromley, S.T. 1815
Broun, D.M. 1467
Broussard, P.R. 93
Brown, W.D. 683, 685
Browning, V.M. 93
Brunen, J. 599
Brusov, Paul 1833, 1881
Brusov, Peter 1833, 1881
Brusova, N. 1833, 1881
Bruynseraede, Y. 379, 1567
Budiarto, E. 252
Budnick, J.I. 2191
Bugoslavsky, Y.V. 2223
Bugoslavsky, Yu. 2151, 2155
Bugoslavsky, Yu.V. 1957, 2149, 2235
Buisson, O. 1595
Bukowski, Z. 541
Büttner, H. 1827
Button, T.W. 2511, 2527

Cabañas Moreno, J.G. 637
Cabeza, O. 2583
Cai, C. 469, 2127
Cai, C.-B. 499, 771
Cai, C.B. 2047, 2333, 2389, 2395, 2649
Cai, J. 2527
Cai, Q.-Y. 2507
Cai, Q.Y. 2499, 2505
Cai, X.L. 2483, 2487
Cai, X.Y. 2463
Cai, Y.M. 497
Calemczuk, R. 1425, 2059
Calles, A. 1623, 1913
Camerlingo, C. 2011
Campbell, D.K. 1875
Campuzano, J.C. 218, 999
Canals, B. 1885
Canepa, F. 573
Canfield, P.C. 355, 1475
Cang, H. 941
Cannelli, G. 1429, 1453, 1457
Cantelli, R. 1429, 1453, 1457
Cao, B.S. 777, 783, 2513, 2515, 2517, 2609
Cao, G. 757, 759, 761, 973, 1057, 1087, 1097, 1177, 1935, 2093
Cao, G.H. 897
Cao, L. 1593
Cao, L.X. 713
Cao, L.Z. 545, 843, 861, 863, 919, 1153, 1165, 1213, 1239, 1415, 1417, 2615
Cao, S. 1277
Cao, X.W. 1993, 2131, 2217
Cao, Y. 57, 653, 883, 885, 1243
Cao, Y.-M. 621
Cao, Z. 1691
Cao, Z.-L. 1791
Caplin, A.D. 1957, 2149, 2151, 2155, 2223, 2227, 2235
Capponi, J.J. 65, 857, 895, 899, 951, 1167
Carballeira, C. 1181
Cardona, M. 222, 557, 599, 1007, 1011, 1047
Cardoso, C.A. 2201
Carretta, P. 359
Carter III, C.F. 2541
Castellani, C. 260, 1821
Castro, H. 2343
Castro, J.J. 1623, 1913
Catherinot, A. 1035
Cava, R.J. 27
Celani, F. 2157, 2329
Chacón, M. 711, 2419
Chaillout, C. 65, 817, 857, 895, 899, 951, 1167
Chainani, A. 218
Chakraverty, B.K. 290
Champeaux, C. 1035
Chan, F.T. 613, 641, 683, 685
Chan, P.W. 619, 701
Chan, W.C. 1455
Chandrasekhar Rao, T.V. 799, 2027, 2043
Chang, B.-Y. 1039
Chang, H.-S. 1281
Chang, H.S. 2543
Chang, S. 2651
Chang, S.A. 485, 2413
Char, K. 419
Charfi-Kaddour, S. 1801, 1887
Chattopadhyay, T. 2089
Chau, R. 1475
Chaud, X. 461, 525, 1951, 2623
Chaudhury, R. 1797
Chauhan, H.S. 447, 503
Chavira, E. 637, 639, 819, 1825
Che, G. 747
Che, G.C. 755, 933, 1877, 1879
Chen, B. 2099, 2417
Chen, B.M. 721
Chen, C. 445, 537
Chen, D.B. 1435
Chen, D.F. 833
Chen, D.Y. 73
Chen, F. 1243, 1245
Chen, G. 1097
Chen, G.-H. 2471
Chen, G.-L. 649
Chen, G.H. 2445
Chen, H. 533, 605, 633, 933, 1537, 1877, 1879
Chen, H.Q. 1075

Chen, J. 753, 1435, 2267, 2325, 2407, 2467
Chen, J.-y. 2641
Chen, J.L. 2113, 2217
Chen, J.M. 961, 981
Chen, K.-X. 2325
Chen, K.X. 2381, 2407
Chen, K.Y. 613, 641, 685
Chen, L. 605, 657, 1451, 2413, 2461, 2477, 2489, 2545
Chen, L.-X. 1205
Chen, L.F. 1603, 2413
Chen, L.L. 497
Chen, M. 2639
Chen, N.L. 753, 1435
Chen, Q. 1159, 2097, 2127, 2129
Chen, Q.S. 2547
Chen, Q.Y. 653
Chen, S.P. 575, 1531
Chen, T.-M. 789, 915
Chen, T.G. 1091
Chen, T.P. 2233, 2383
Chen, W.M. 517, 753, 787, 833, 1435, 1611, 2347
Chen, X. 797
Chen, X.-j. 903, 905
Chen, X.A. 721
Chen, X.H. 1165, 1213, 1239
Chen, Y. 2513
Chen, Y.C. 1871
Chen, Y.Y. 1403
Chen, Z. 705, 2641
Chen, Z.H. 697, 699, 2513
Chen, Z.J. 545, 843, 2267
Chen, Z.Y. 795, 805, 871, 917, 939, 943, 2131
Chen Changkang 1287
Chen Liquan 931
Chenavas, J. 65
Cheng, Q.H. 399
Cheng Guosheng 969, 1283
Chern, J.D. 1433, 2451
Chernyaev, S. 2613
Cherrey, K. 1917
Cherubini, R. 2095, 2179
Cheung, J.T. 619, 701
Chien, F.Z. 73
Chikumoto, N. 371, 391, 2073, 2143, 2295
Chmaissem, O. 97
Cho, B.K. 1475
Cho, H.S. 1313
Choi, C.H. 2123, 2359
Choi, C.K. 1249
Choi, E.S. 1249
Choi, H.-Y. 1923
Choi, J.-H. 1039, 1293, 2001, 2003
Choi, M.K. 663
Choi, S.S. 1461
Chong, I. 999
Choudhary, M.F. 2273
Choudhury, P. 1049
Christensen, N.E. 1619, 1621
Chromik, S. 639
Chu, C.W. 57, 375, 653, 883, 885, 1243, 1245, 2553
Chu, W.K. 653, 1515
Chuang, Y.C. 2601
Chun, S.H. 2335
Chung, B.-D. 663
Chung, H.S. 2601
Chung, J.M. 2543
Chung, M. 1281
Chval, J. 2585
Chyan, Y.K. 1031
Cimberle, M.R. 573, 679, 1183
Citro, R. 1695, 1789
Ciurchea, D. 1257, 2365
Claeson, T. 111, 623, 1075, 2293
Clem, J.R. 311
Cohen, L.F. 1577, 2149, 2155, 2187, 2227, 2235
Collin, G. 226, 1389
Collings, E.W. 2597
Colson, D. 1007, 1013, 1073, 1199
Combescot, R. 1013
Conde-Gallardo, A. 637, 639
Conder, K. 77, 182, 202, 483, 527, 901, 1051, 1371
Continentino, M. 1939
Contour, J.-P. 689
Contour, J.P. 1025
Cooke, D.W. 2069
Cooksey, J.W. 683, 685
Cooper, J.R. 145, 1229, 1287, 1405, 1411
Cordero, F. 1429, 1453, 1457
Corega, C. 1257
Corson, J. 252
Corti, M. 359
Costa-Quintana, J. 1631
Cowan, D.L. 2203
Cowie, A. 1577
Crabtree, G.W. 327, 355, 2231, 2359
Crooks, R.J. 449
Cubitt, R. 2089
Cucolo, A.M. 709
Cui, D.-f. 2531
Cui, D.F. 699, 2515
Cui, X.T. 653
Cui Yujian 969
Cummins, T.R. 995
Cunha, A.G. 489
Cuoco, M. 1705
Currie, D.B. 449
Czajka, R. 1503
Dabrowski, B. 97, 541, 959
Dagotto, E. 1729, 1867
Dai, X. 1629, 1669, 2509
Dai, Y. 1537, 2409
Dai, Y.-D. 2473
Dai, Y.D. 411, 2087, 2381, 2407, 2479, 2483, 2487
Dai Xianqi 969, 1283
Dalakova, N.V. 1223
Dam, B. 559, 665, 2303
Dan, N.H. 785
D'Anna, G. 2021
Das, M.P. 1709
Dat, D.T. 785
Davidson, A. 2503
Davis, L.E. 2511
Davydov, D.N. 1487
Dawson, W.K. 1383
de Andrade, M. 1465
de Andrade Jr, R. 2051
de Boer, F.R. 1295, 1447, 2111
de Groot, D.G. 665
de Groot, P.A.J. 1969, 1971, 2017, 2031, 2033, 2035, 2037, 2039
de Kok, M.L. 1393, 2121
de Lima, O.F. 2201, 2251
De Luca, R. 2317
de Marneffe, J.-F. 2357
de Mello, E.V.L. 1819
De Wilde, Y. 355
Degiorgi, L. 1475
Delamare, M.P. 501, 507
Deligiannis, K. 1969, 1971
Deltour, R. 1257, 1267, 1487, 2357, 2365
Demsar, J. 186
den Hertog, B.C. 1709
den Ouden, G. 2419
Deng, J.-N. 2603, 2647
Desgardin, G. 501, 507, 511, 2573
Desmicht, R. 471
Deutscher, G. 140
Deville Cavellin, C. 162, 929
Deville-Cavellin, C. 631
Dew-Hughes, D. 2617
Dhalenne, G. 953
Dhalle, M. 2145
Dhallé, M. 1173, 1305
Di Castro, C. 260, 1821
Di Gioacchino, D. 2157, 2329
Di Leo, R. 709
Di Matteo, T. 2317
Di Trolio, A. 487
Diaspro, A. 679
Díaz, A. 2315
Digor, D.F. 1717
Dikstein, I.E. 1707
Ding, D.S. 591
Ding, H. 218, 999

Ding, Q. 2389
Ding, S.Y. 517, 845, 2047, 2075, 2139, 2281, 2333, 2389, 2395, 2397, 2615
Dogotaru, L.A. 1717, 1719
Doh, H. 1697
Doh, Y.J. 1281
Doidge-Harrison, S. 1889
Domange, J-L. 631
Dong, C. 533, 747, 931, 933
Dong, D.J. 777, 783, 2513, 2515, 2517, 2609
Dong, H. 1241, 1431, 1439, 2247
Dong, J. 1853
Dong, J.M. 1649
Dong, X.L. 1877, 1879
Dong, Z.F. 763, 1879
Doornbos, G. 347, 2303
Doria, M.M. 1595
Dou, S.X. 437, 467, 505, 877, 1991, 1995, 2125, 2133, 2283, 2297, 2299, 2587, 2591, 2595, 2597, 2621, 2627, 2629, 2643
Dow, J.D. 1005, 1513, 1551
Downes, J. 1237
Doyle, R.A. 323, 2023, 2067
Dravin, V.A. 1583, 1585
Drossos, G. 2511
Drost, R.J. 2023, 2241
Drouet, M. 689
Du, H. 2445
Du, J. 1171, 1207
Du, J.H. 497
Du, L. 2609
Du, X. 859
Du, Y.L. 659
Du, Z.L. 57, 883, 1243
Duan, X.F. 1879
Duan, Z.-Z. 2603
Duan, Z.Z. 2605, 2615
Dubey, P. 1831
Duc, F. 1081
Duffy, D. 1729, 1867
Duginov, V.N. 1353
Duhalde, S. 569
Dumas, J. 955
Dunne, L.J. 1787
Durand, O. 689
Durner, R. 1191
Durrell, J. 1025
Dutoit, B. 2343
Dutyshev, V.N. 1949

Ebbesen, T.W. 2279
Eccleston, R. 1335
Eccleston, R.S. 149
Eckstein, J.N. 252
Eder, R. 1731, 1733
Edwards, P.P. 145, 909
Efimov, V.B. 1285, 1903, 1921
Egami, M. 463
Eisaki, H. 153, 1119, 1121, 1133, 1323, 1365
Elliott, M. 1231
Ellis, D.E. 1633
Eltsev, Yu. 1221, 1535
Eltsev, Yu.F. 1583, 1585
Emel'chenko, G.A. 2089
Emelyanov, D. 881
Emery, V.J. 174
Emmerich, F.G. 489
Endo, A. 447
Endo, T. 1479, 1591
Endoh, Y. 85, 170
Enomoto, Y. 597, 1483, 2519
Erb, A. 89, 256, 315, 457, 459, 1399, 1401, 1425, 1601, 1953, 2059, 2145
Eremenko, V.V. 1427
Eremin, M. 1387
Eremin, M.V. 1659
Erts, D. 1075
Esashi, M. 2467
Escribe Filippini, C. 2207
Escribe-Filippini, C. 2049
Esquinazi, P. 1523
Eustache, B. 162, 929

Fabricius, G. 1619
Face, D.W. 2541
Faizrakhmanov, I.A. 589
Falcony, C. 637, 639
Fan, C. 1935
Fan, C.-G. 1261
Fan, H.F. 101
Fan, J.D. 1703, 1721, 1725, 1727
Fan, X. 2105
Fan, Y.-C. 601, 603
Fan, Z.-G. 2647
Fan, Z.G. 495, 497, 2393
Fang, J. 1993, 2131, 2217
Fang, J.-R. 2607
Fang, J.W. 659
Fang, L. 2433
Fang, M. 2097, 2127, 2129
Farmer, J.W. 2203
Farner, S. 1413
Fastampa, R. 1549, 1985, 2345
Faulques, E. 1937
Fedoroff, M. 953
Fedorov, V.E. 1099, 1137
Feenstra, R. 105
Fehske, H. 1779, 1827, 1829
Feigel'man, M.V. 2053
Felner, I. 1317, 2213
Fendrich, J. 2291
Feng, J. 983, 1611, 2171, 2175, 2307
Feng, Q.R. 465, 543, 645, 831, 859, 879, 1437, 2381
Feng, S. 1687
Feng, S.Q. 859, 879, 923
Feng, Y. 529, 1153, 1165, 1213, 1415, 1417, 2177, 2249
Feng, Y.J. 2459
Ferdeghini, C. 573, 679, 1183
Ferreira, J.M. 779, 807
Ferretti, M. 1429, 1457
Feucht, A. 1191
Fil, D.V. 1045
Fischer, H. 2089
Fischer, K. 2645
Fischer, Ø. 315
Fisher, R.A. 327
Fisher, S. 2593
Fisher, S.J. 2557
Fitzgerald, A.G. 601, 603
Fleshler, S. 424
Flower, N.E. 236
Flükiger, R. 89, 256, 459, 1173, 1305, 1601, 1953, 2145, 2571
Fong, H.F. 232
Fontes, M.B. 1939, 1941
Foong, F. 919, 1153, 1213
Forkl, A. 675, 2311
Forsthofer, F. 2431
Fossheim, K. 1469, 2279, 2617
Fournier, Th. 2313
Franck, J.P. 198
Frassanito, R. 1301
Freitag, B. 571
Frello, T. 1089
Frenkel Ben-Yakar, L. 1331, 1339
Frey, U. 583, 655
Freye, D.M. 303
Frikach, K. 1977, 1979, 2253, 2263
Fruchter, L. 1073
Fu, C.-C. 1947
Fu, C.C. 1943
Fu, L. 1557, 1945, 2341
Fu, X. 2635, 2651
Fu, X.K. 2309
Fu, Y. 469
Fu, Y.-X. 499, 771
Fu, Y.X. 2047, 2333, 2389, 2395, 2649
Fuchs, D.T. 2023
Fudamoto, Y. 1355, 1465
Fuhrer, M. 1547
Fuhrer, M.S. 1917, 2041
Fujii, H. 2567
Fujii, T. 1169
Fujikami, J. 2339
Fujimoto, H. 2137, 2237
Fujimura, K. 477
Fujino, Y. 1519
Fujisawa, H. 997

Fujisawa, M. 963
Fujishiro, Y. 149
Fujita, T. 210, 1083, 1151, 1229, 1255, 1369, 1413
Fujiwara, K. 1349
Fujiwara, M. 149, 691
Fukami, T. 1989
Fukase, T. 1103
Fukuda, Y. 993
Fukushima, K. 2105
Fukuyama, H. 124, 1683, 1685
Fulde, P. 1749
Funahashi, R. 945, 2561, 2579
Fung, P.C.W. 1247, 1523
Furrer, A. 1327, 1335, 1377
Furukawa, Y. 1103, 1105
Furusawa, T. 1211

Gagliano, E. 1875
Gagnon, R. 1505, 1969, 1971, 2017, 2031, 2033, 2035, 2037, 2039, 2183
Gaifullin, M.B. 391, 2073, 2221, 2429
Gajewski, D.A. 1127, 1129, 1147, 2327
Gallop, J. 1579, 2415, 2485
Gallop, J.C. 1577, 2535
Gameson, I. 145, 909
Gammel, P.L. 355
Gan, Z. 2409
Gan, Z.-Z. 1783
Gan, Z.Z. 551, 693, 1015, 2087
Gao, L. 57, 883, 885, 2553
Gao, Z.X. 2087
Garcia, C. 2553
Garif'yanov, N.N. 1321
Gasser, C. 643
Gasser, U. 1327
Gasumyants, V.E. 1267, 1279
Gautier, E. 817, 857, 895, 899, 951
Gautier Picard, P. 2313
Gautier-Picard, P. 461
Gauzzi, A. 689
Gavalek, W. 1247
Gawalek, W. 431, 2645
Gazda, M. 1575
Ge, Y. 2047, 2075, 2333, 2389, 2395
Geerk, J. 625
Gehringer, P. 1545, 2321
Geibel, C. 240
Geng, J.F. 2347
Geng, R.L. 1273
Genoud, J.-Y. 89, 315, 457, 459, 1399, 1425, 2145
Geraschenko, A. 1361
Gerashenko, A. 1337, 1347
Gerbaldo, R. 2095, 2179
Gerber, Ch. 132
Gerhold, S. 719
Germi, P. 1951
Gernoth, K.A. 1617
Gervais, F. 1035, 1141
Geshkenbein, V.B. 319, 1215
Ghigo, G. 2095, 2179
Ghinovker, M. 1299
Ghivelder, L. 2181
Ghosh, B. 1039, 1049, 2001
Ghosh, K. 2043
Giamarchi, T. 363
Giannini, E. 679
Giller, D. 2209, 2225
Gimm, T.-H. 1699
Giordanengo, B. 1939, 1941
Gippius, A.A. 1381
Giraldo, J. 1253
Giratá, D. 671
Giura, M. 1549, 1985, 2345
Glover III, R.E. 2081
Gmelin, E. 807
Göb, W. 1227, 2321
Godart, C. 1317
Goernert, P. 431
Goko, T. 1151
Goldberg, I.E. 1279
Goldgirsh, A. 2213
Goldschmidt, D. 140
Golovashkin, A. 1929
Golovashkin, A.I. 1065, 1067, 1291
Golub, A. 1963
Gómez, M.E. 671, 709, 711, 2419
Gomina, M. 2573
Gömöry, F. 2379
Gondaira, K.-I. 963
Gong, C. 1651
Gong, C.-D. 1641, 1671, 1689
Gong, Q.L. 497
Gong, S. 469
Gong, S.-M. 499, 771
Gong, S.M. 617, 1091, 2649
Gong, Z. 2527
Gong, Z.H. 2521
Gonnelli, R.S. 1473, 1501, 2403
González, M.T. 2315
Goo, D. 609, 615, 2337
Gooding, R.J. 1793
Goodwin, T.J. 745
Gopinath, C.S. 995
Gordeev, S. 2035, 2037
Gordeev, S.N. 2017, 2031, 2033, 2039
Gorelov, Y.A. 2491
Goren, S.D. 1331, 1339
Goretta, K.C. 2359
Gorius, M.F. 951
Gorlova, I.G. 875, 1533, 2077
Gormezano, A. 849
Görnert, P. 2645
Goshchitskii, B. 1335, 1337
Goto, J. 153
Goto, M. 1133, 1355
Goto, T. 891, 1103, 2287, 2463
Gou, C. 833
Gou, Y.S. 575, 1493, 1531, 1573
Gough, C. 537
Gough, C.E. 395, 1093, 1597, 2007, 2437, 2527
Goupil, Ch. 2319
Gozzelino, L. 2095, 2179
Grabowski, S. 1681, 1775
Grachev, V. 1161
Gradys, E. 2305
Grajcar, M. 1481
Grandini, C.R. 1457
Grantham, C. 2627, 2629, 2643
Grassano, G. 679, 1183
Grasso, G. 1173, 1305, 2571
Greco, A. 1741
Greenberg, J.H. 1339
Greenblatt, M. 955
Greene, R.L. 2209, 2261
Grepstad, J.K. 2521
Griessen, R. 347, 559, 665, 2219, 2265, 2291, 2303
Grigorashev, D.I. 867
Grilli, M. 260
Grindatto, D.P. 2571
Gritzner, G. 1423
Grivel, J.C. 1173
Grosche, F.M. 303
Grover, A.K. 1975, 2027, 2043
Grushko, Yu.S. 1915
Gu, B.L. 777, 783, 2513, 2609
Gu, G.D. 865, 1495, 1955, 2107, 2331
Gu, H.W. 2413
Gu, J.H. 591
Gu, S.W. 1657
Guasconi, P. 573
Gudenko, S. 1929
Guenzburger, D. 1633
Gumarov, G.G. 589
Guo, J. 1241, 1431
Guo, J.D. 543, 645, 923, 1449
Guo, L.P. 115, 553
Guo, S.-q. 351
Guo, S.Q. 777, 783, 933, 1325, 1877, 1879, 2269, 2609, 2615
Guo, T. 2637
Guo, Y.C. 2283, 2595, 2597
Gurov, A.F. 733
Gusakov, V.E. 803
Gustafsson, M. 623
Gwak, J.-H. 1039
Gygax, F.N. 1353

Ha, D.H. 1605, 1981, 2025
Ha, Y.S. 1235, 1249, 2523
Haage, T. 557, 661, 675, 2311

Haas, S. 1729, 1867
Habermeier, H.-U. 557, 661, 675, 1047, 2311, 2551
Haga, Y. 2043
Hahn, T.-S. 1461
Haibach, P. 655
Halimi, R. 1979, 2263
Hama, Y. 1233
Hamada, K. 1385
Hamasaki, K. 2399
Hamdan, N.M. 1445, 2273
Hamet, J.F. 647
Hammann, J. 1073, 2267
Han, C.Y. 1189
Han, G.C. 1603
Han, R. 1691, 1811
Han, R.Q. 2529
Han, R.-S. 1791
Han, S. 2633
Han, S.-H. 1571
Han, S.H. 1127, 1129, 1147, 2123, 2327, 2331
Han, S.K. 2523
Han, W. 567
Han, Y. 795
Hanaguri, T. 1125, 1303, 2375
Hancotte, H. 1487
Haneda, H. 2323
Hanic, F. 639
Hao, J.M. 703, 2103, 2433
Hao, L. 1579, 2415, 2485, 2535
Harabaev, N. 1833, 1881
Harabor, A. 2365
Harada, K. 1753
Hardy, V. 927, 2319
Hargreave, T. 445
Harneit, W. 2207
Harrington, C.M. 2541
Harus, G.I. 1149
Haruyama, Y. 765
Hasanain, S.K. 1225
Hasegawa, K. 1563
Hasegawa, T. 1479
Hash, M. 441
Hatada, K. 793, 1349, 1379
Hatterer, C.J. 929
Hauchecorne, G. 1025
Hauck, J. 453, 2613
Hayakawa, H. 811
Hayashi, S. 607
Hayden, S.M. 231
Häyrynen, T. 1617
He, A. 2527, 2635
He, A.S. 2517, 2609
He, D. 2481
He, D.F. 2487
He, G.H. 1449
He, J. 1177
He, M. 2515
He, Y. 1511, 2437, 2527
He, Y.-F. 2603, 2647
He, Y.S. 2407
He, Y.Y. 831, 859, 2331
He, Z.H. 547, 1247
Hechtfischer, G. 2431, 2435
Hector, A.L. 449
Hedo, M. 2043
Heiden, C. 407
Heine, G. 1561
Heinze, M. 1381
Heiras, J. 671, 709, 711, 2419
Hejtmánek, J. 2577, 2585
Helm, Ch. 2431
Henggeler, W. 1335
Hennion, B. 1375
Hensel, B. 1305, 2571
Héritier, M. 1801, 1887
Herlach, D. 1353
Herrenden-Harkerand, W. 1231
Hervieu, M. 41, 507, 647
Herzog, R. 1569
Hess, C. 955
Hewat, A.W. 911
Higemoto, W. 1383
Higgins, M.J. 2027
Higo, S. 707
Higuchi, T. 371, 2137, 2237
Hikosaka, H. 521
Hilgenkamp, H. 132
Himeda, A. 1521
Hinks, D.G. 97, 2359
Hirabayashi, I. 585, 595, 809
Hiraga, K. 935
Hirakawa, K. 1095
Hirata, K. 391, 475, 2073, 2221
Hirate, N. 1591
Hisaki, T. 1265
Hissa, J. 1617
Ho, J.C. 1403
Ho, J.S. 915
Hodby, J.W. 145, 445, 537, 1287
Hodges, J.P. 909
Hoekstra, A.F.Th. 347, 2265
Hofer, J. 136
Hoffmann, W. 1381
Hojczyk, R. 731
Holiastou, M. 583
Hollauer, G. 1595
Holm, W. 1221, 1535
Holmes, J.E. 2527
Holtzberg, F. 1943, 1947, 2053
Homma, N. 2631
Hong, I.V. 785
Hong, J. 1743
Hong, S.-P. 1697, 1699
Hong, S.M. 1325
Honma, T. 791
Hor, P. 1629
Horibe, Y. 1071
Horie, M. 1591
Horii, S. 809, 1443
Horng, H.E. 1031, 1143, 1433, 1997, 2443, 2451
Horng, L. 2113, 2153, 2361
Horovitz, B. 1963
Horsch, P. 1739, 1751, 1805
Horvat, J. 505, 877, 1991, 2125, 2133, 2283, 2595, 2621
Horvatić, M. 227
Hoshino, T. 2565
Hosokawa, E. 477
Hosokawa, T. 1489
Hou, J. 735
Hou, J.G. 565, 651, 737, 739
Hoya, K. 1499
Hoyos, R. 671
Hsieh, M.C. 1573
Hsu, T.I. 717, 1329
Hsu, Y.Y. 717, 1329
Hu, A.M. 431
Hu, J.Z. 93
Hu, K. 1033
Hu, L. 2525
Hu, Q.-H. 1075
Hu, Q.Y. 1995, 2591
Hu, S.F. 961
Hu, W.Y. 1625, 1627
Hu, W.-Z. 971, 977
Hu, X. 1439, 1537, 2057
Hu, X.-D. 2473
Hu, X.D. 2479
Hu, Y.F. 693
Hu, Z.H. 645
Hu, Z.Q. 843
Hua, C.Y. 2615
Hua, L. 1447, 1449, 1615, 2111, 2601
Hua, P. 2635, 2651
Hua, Z.Q. 703, 2103
Huang, H. 1275
Huang, H.S. 1273
Huang, J.C. 1403, 1407
Huang, M.Q. 605, 657, 2461, 2477, 2489, 2545
Huang, S.-L. 1469, 2279, 2617
Huang, W.N. 767, 1069
Huang, W.S. 1403
Huang, W.W. 1877
Huang, X. 1451
Huang, Y. 931, 1021
Huang, Y.J. 2153
Huang, Y.N. 1053, 1061, 1523
Huang, Y.Z. 1877
Huang, Z. 1687
Hubbard, F.H. 601

Hudáková, N. 2585
Hughes, T. 2559, 2593
Huh, Y. 1491, 2469, 2543
Huijbregtse, J. 559
Hults, W.L. 2065
Humphreys, R.G. 395
Hung, D. 395
Hung, K.C. 907, 983, 1559, 1611, 2171, 2173, 2175, 2243, 2245, 2307, 2367
Hunt, M.B. 202
Hüsser, P. 1639
Hussey, N.E. 1229, 1287
Huth, M. 1883
Hwang, Y.S. 1491

Iavarone, M. 355
Ichikawa, F. 1989, 2277
Ichikawa, N. 240, 1217, 1355
Ichimura, K. 1895
Ida, T. 971, 977
Ido, M. 1499
Ignatenkov, A.N. 1149
Ihara, H. 509, 957, 1973
Ihle, D. 1779, 1781
Ikeda, H. 1205, 1219
Ikeda, S. 210, 1105, 1369, 1413
Ikeda, S.-i. 729
Ikegawa, S. 673
Ikuta, H. 335, 1269, 2015, 2055, 2239
Ilonca, G. 1257
Ilonca, Gh. 2365
Ilonca, M. 1257
Imao, H. 473
Inada, Y. 2043
Indenbom, M.V. 1953, 2021
Ino, K. 2041
Inoue, Y. 1071
Ioffe, L.B. 1215
Ionescu, M. 437, 877, 1991, 2133, 2621
Iri, T. 963
Irvine, J.T.S. 727, 815, 1085, 1889
Isa, M. 691
Ischenko, V.V. 855
Ishida, K. 210, 1115, 1351, 1365, 1367, 1369, 1973
Ishiguro, T. 2323
Ishii, O. 937
Ishikawa, H. 945, 2561, 2579
Ishimaru, Y. 597
Isobe, M. 811, 813, 835, 949
Isoda, S. 2289
Ito, T. 479, 481, 1079, 1123, 1465, 1477, 1485
Ito, Y. 585, 595, 2285
Itskevich, E.S. 1987
Ivanenko, O.M. 1101, 1179
Ivanov, V.A. 1723, 1905, 1907, 1925
Ivanov, Z.G. 111, 415, 1075
Ivanova, L.I. 1065, 1067
Ivkov, J. 2297
Iwasa, Y. 307
Iwata, M. 585
Iyo, A. 509, 957, 1973
Izumi, F. 811, 813, 911
Izumi, M. 971, 977

Jacobo, S.E. 569
Jacobs, T. 256
Jaeger, H.M. 2359
Jaklič, J. 1803
Jakob, G. 669
Janeček, I. 2373
Jansen, A.G.M. 1487, 2049, 2357
Jansen, L. 2017
Jegoudez, J. 471
Jei, L. 2175
Jensen, W.R. 97
Jeong, Y.S. 663
Jergel, M. 637, 639
Ji, P. 1607
Ji, W. 833
Ji, X. 1041
Ji, Z.M. 2457
Jia, C.-L. 731
Jia, S.L. 533, 633, 933
Jia, X.L. 1325
Jia, Y.B. 1239
Jia, Y.X. 1547
Jiang, C.N. 1127, 1129, 1147, 2327
Jiang, H. 2325
Jiang, J.-F. 2507
Jiang, J.F. 2499, 2505
Jiang, J.H. 713
Jiang, L. 1171, 1207
Jiang, Q. 2267
Jiang, S.S. 753, 787, 833
Jiang, W. 1943, 1947
Jiang, X. 2081
Jiang, X.H. 2377
Jiang, Y.S. 2463
Jiao, Y. 2247, 2633
Jiao, Y.L. 485, 2087
Jiao, Z. 757, 759, 761, 775, 797, 973, 1177, 1935, 2097, 2127, 2129, 2417
Jiao, Z.-k. 903, 1197
Jijun, S. 681
Jiménez-Sandoval, S. 637
Jin, B.B. 591, 2539
Jin, C.Z. 2609
Jin, D. 49, 769, 1033, 1325, 1341, 1395, 1799, 1933, 2267, 2599
Jin, H. 545, 843, 861, 863, 1259, 1395, 2341
Jin, J.R. 2175
Jin, J.X. 2627, 2629, 2643
Jin, K.-j. 1869
Jin, O. 115, 697
Jin, R. 1209
Jin, X. 517, 593, 753, 787, 833, 845, 983, 1435, 1611, 2171, 2173, 2175, 2243, 2245, 2281, 2307, 2347, 2367, 2393
Jin, X.-J. 621
Jing, X.N. 1553
Jirák, Z. 2585
Johansen, T.H. 1469, 2279
Johansson, L.-G. 1075
Johnson, H.L. 2431
Jones, I.P. 1093
Joo, E.J. 1313
Jooss, Ch. 557, 2311
Jorgensen, J.D. 97
Jorrensen, J.D. 959
Jourdan, M. 1883
Ju, X. 725, 913, 975, 1087
Juang, J.Y. 575, 1493, 1531, 1573
Julien, M.-H. 227
Jumas, J.C. 741
Jun, J. 483
Junod, A. 457, 1399, 1401, 1425, 2059

Kabanov, V.V. 2231
Kadowaki, K. 475, 715, 1055, 1315, 1409, 1599, 2187, 2221, 2421, 2423, 2429
Kaiser, A.B. 1251, 1271
Kaiser, G. 2527
Kajimura, K. 1477, 2257, 2405
Kakeya, I. 1599, 2423
Kaldis, E. 182, 1051
Käll, M. 1089
Kallio, A. 1617
Kamada, K. 503
Kamada, S. 853
Kambara, M. 81, 447
Kambara, T. 2105
Kambe, S. 935, 937, 1211
Kanazawa, I. 1763, 1891
Kanda, N. 1297
Kanehashi, K. 821
Kaneko, T. 1519
Kang, J. 2493
Kang, J.F. 693, 2529
Kang, K.Y. 2523
Kang, L. 2457, 2539
Kang, M. 1009, 1017, 1027
Kanoda, K. 299, 1897, 2013, 2063
Kao, F.S. 789
Kao, H.-C.I. 767, 1069
Kao, R.T. 1493, 1531
Kapelnitsky, S. 773
Karamov, S.A. 1667
Karpinski, J. 77, 483, 527, 901, 1345
Karpiński, J. 2161
Karuzskii, A.L. 1581, 1583, 1585
Kasai, M. 218

Kashiwaya, S. 1477, 1485, 1521, 1855, 1857, 2405
Kastner, M.A. 85, 170
Kato, H. 1441
Kato, M. 1233, 1265, 1343
Katsoura, Ch. 1569
Kawabata, C. 1999
Kawaguchi, K. 2501
Kawai, T. 105
Kawakami, A. 2465, 2533, 2549
Kawakami, M. 629
Kawamoto, A. 2063
Kawano, H. 1055, 1315
Kawano, I. 707
Kawano, K. 2237, 2289
Kawasaki, M. 1297
Kawashima, T. 513, 949
Kayani, A.N. 869
Kaydanov, V.I. 1279
Kazumata, Y. 2135
Kee, H.-Y. 1743
Keimer, B. 232
Keller, H. 136, 190, 202
Keller, J. 2431
Kelly, M.E. 1901
Keren, A. 1355
Kes, P.H. 323, 331, 1543, 2023, 2067, 2083, 2085, 2101, 2241, 2259
Keurentjes, A. 2085
Kezuka, H. 521, 523
Khaibullin, I.B. 589
Khaliullin, G. 1749, 1751
Khan, M.N. 869
Kharlamov, V.S. 1333
Khasanov, R. 2159
Khasanova, N.R. 813
Khaykovich, B. 323, 2067
Khim, Z.G. 2335
Khrustalev, B.P. 2447, 2449, 2453
Khrustov, V. 2569
Kida, N. 945
Kikuchi, M. 521, 821, 827
Kikugawa, N. 1151
Kilian, R. 1749
Kim, D.K. 695
Kim, H.B. 1313
Kim, H.C. 2335
Kim, H.-D. 997
Kim, H.D. 2601
Kim, I. 1491, 2543
Kim, I.-S. 1605
Kim, J. 609
Kim, J.H. 1023
Kim, J.Y. 663
Kim, M.-S. 1293, 1987, 2001, 2003, 2025
Kim, S.F. 1203
Kim, S.-J. 1039, 2467
Kim, W.-S. 1987, 2003
Kim, Y.-I. 1605
Kim, Y.W. 707
Kimura, T. 240, 244
Kinase, W. 1753
Kinder, H. 107
Kindo, K. 1599, 2423
Kirtley, J.R. 4
Kiryakov, N. 881
Kishida, S. 473, 477
Kishine, J. 1771
Kishio, K. 240, 335, 391, 515, 1269, 1479, 2009, 2015, 2055, 2073, 2135, 2143, 2239, 2285, 2323, 2439
Kitaev, Yu.E. 1915
Kitaguchi, H. 2567
Kitano, H. 1125
Kitaoka, Y. 149, 210, 1115, 1351, 1365, 1367, 1369, 1373, 1385
Kitazawa, K. 335, 1479, 2015, 2055, 2239
Kiuchi, M. 2339
Kivelson, S.A. 174
Klamut, J. 541, 1463
Klamut, P.W. 541
Kläser, M. 2149
Klein, M.V. 1009, 1017, 1027
Klein, T. 2049, 2207
Kleiner, R. 2431, 2435
Klimczuk, T. 1503
Klinkova, L.A. 443
Klosowski, J. 505
Kluge, T. 669
Knee, C.S. 849
Knížek, K. 2585
Koba, S. 707
Kobayashi, H. 2015
Kobayashi, N. 691, 2009, 2117
Kobayashi, T. 2135, 2285, 2463
Kobelkov, A.G. 1845
Koblischka, M.R. 1469, 2193, 2279
Kodaka, H. 403
Kodama, Y. 2257
Koh, G.H. 2335
Kohara, T. 2105
Köhler, A. 571
Kohn, K. 963
Kohno, H. 124, 1683, 1685
Koike, K. 1211
Koike, S. 1765
Koike, Y. 1233, 1265, 1343
Koinuma, H. 1297
Kojima, H. 1957
Kojima, K. 1465
Kojima, K.M. 1355
Kojo, H. 371
Koleśnik, S. 1961
Kolesnik, S.G. 1915
Kolesnik, S.N. 1915
Kolesnikov, N.N. 1421, 1543
Kolodziejczyk, A. 1423
Komiya, N. 153
Komori, K. 1269
Konczykowski, M. 323, 391, 1943, 1947, 2023, 2045, 2053, 2067, 2073, 2101, 2143, 2189, 2241, 2259, 2427
König, W. 1011, 1423
Konishi, M. 2561
Kopnin, E. 77, 901, 2161
Kopnin, E.M. 483
Kopylov, V.N. 733, 1421, 1543
Korn, C. 1331, 1339
Kornecki, M. 2203
Korotaev, V.A. 1735
Kosaka, S. 607, 2071
Kosaka, T. 1443
Koshelev, A.E. 355
Koshizuka, N. 865, 1495, 1955, 2099, 2107, 2177, 2331, 2455
Kosseva, I. 445
Koster, G. 665
Kosugi, M. 391, 2073, 2221
Kottman, P. 2199
Kouba, R. 1635
Kovalev, O.S. 1391
Koyama, S. 587, 597
Koyama, T. 1861
Koyama, Y. 1071, 1111
Koyanagi, M. 1477, 1485, 2405
Kozeeva, L.P. 1135, 1157
Kozlov, M.E. 1041
Kozłowski, A. 1423
Krabbes, G. 505
Kraiskaya, K.V. 1065
Krasnosvobodtsev, S.I. 1291
Kräußlich, J. 571
Krinetskii, I.B. 1065
Krishana, K. 244
Krivenko, S. 1749
Krivov, V.V. 1145
Kronmüller, H. 2311
Kruchinin, S. 1569
Kruchinin, S.P. 1397, 1509
Krüger, Ch. 527
Krusin-Elbaum, L. 375
Ku, H.C. 717, 1329
Kubo, Y. 1441
Kubota, R. 1499
Kudriavtsev, E.M. 1145
Kulikov, D.V. 1333
Kumagai, K. 391, 1103, 1105, 1107, 1343, 1363, 2073
Kumagai, K.-i. 2221, 2429
Kumakura, H. 2567
Kumigashira, H. 997
Kuo, C.C. 2443
Kupfer, H. 2227
Küpfer, H. 2151, 2155

Kuroda, K. 2099
Kusano, K. 733
Kusevic, I. 2133
Kušević, I. 2297, 2299
Kushida, T. 707
Kusumori, T. 577, 1589
Kusz, B. 1575
Kuwata, S. 1753
Kuzemskaya, I. 1987
Kuzmichev, N.D. 1131, 2091
Kuznetsov, M.S. 867
Kvitkovič, J. 2109
Kwok, W.-K. 2291
Kwok, W.K. 327
Kwon, C. 2081

Lacey, D. 2511
Lacroix, C. 1885
Lagues, M. 631
Laguës, M. 162, 929
Lai, Y.-L. 789
Lai, Y.S. 547
Laisheva, L.V. 1915
Lam, C.C. 833, 907, 925, 967, 983, 1435, 1559, 1611, 1657, 2171, 2173, 2175, 2243, 2245, 2307, 2367
Lampakis, D. 1051
Lancaster, M.J. 395, 2527
Landinez Tellez, D.A. 779, 807
Lanfredi, A.J.C. 2051
Lang, W. 1227, 1545, 1561, 2321
Langan, R. 2037
Langan, R.M. 2017, 2031, 2033, 2039
Lankes, F. 655
Lanucara, M. 1587
Lanzara, A. 1081
Lapinskas, S. 847
Lapsker, I. 2165
Larica, C. 489
Larionov, I.A. 1659
Larkin, A.I. 1215
Larkin, M. 1355, 1465
Latyshev, Yu.I. 387
Lauriat, J.-P. 929
Laval, J.Y. 162, 2305
Law, L. 2555
Law, L.N. 2559
Le Bras, G. 1073
Le Doussal, P. 363
Le Floch, S. 817, 857, 951
Lebed, A.G. 1859
Lebedev, O. 61
Leblanc-Soreau, A. 1937
Lecomte, J. 1555, 1983
Ledbetter, H. 1077
Lee, C.H. 85
Lee, C.M. 1657, 2243, 2245
Lee, C.Y. 2543
Lee, E. 743
Lee, E.T.C. 909
Lee, H. 1281
Lee, H.S. 1313
Lee, J.H. 1235, 1249
Lee, J.J. 1455
Lee, J.Y. 1461
Lee, K.F. 1069
Lee, K.W. 1981, 2025
Lee, M.S. 2543
Lee, P.A. 279
Lee, S. 881, 2003
Lee, S.F. 1467
Lee, S.-G. 2469, 2475
Lee, S.G. 695, 1491
Lee, S.H. 232
Lee, S.-I. 1039, 1293, 1987, 2001, 2003, 2025, 2061
Lee, S.S. 2523
Lee, S.-Y. 1039, 2001
Lee, S.Y. 663
Lee, W.C. 1313, 2061
Lee, Y.S. 170
LeFloch, S. 65
Lei, M. 1077
Leibold, B. 1047, 2551
Leite, E.R. 2051
Leitner, J. 439
Lejay, P. 1199
Leonjuk, L.I. 1065
Leonyuk, L. 749, 773
Lerinman, N.K. 1149
Leung, P.W. 1793
Levin, G.A. 1127, 1129, 1147, 2327
Lewis, K. 2553
Li, C. 705
Li, C.S. 2611
Li, C.-Y. 2325
Li, F.H. 101, 829
Li, F.Q. 863
Li, H. 2527
Li, H.-C. 649
Li, H.-c. 351
Li, H.C. 1993, 2377
Li, J. 705, 835, 2249, 2387
Li, J.Q. 557, 661, 675, 811, 907, 1611
Li, J.W. 49, 633, 721, 723, 755, 777, 933, 1311, 1877, 2309
Li, J.-X. 1641
Li, J.Z. 2479
Li, K.B. 2131, 2217
Li, L. 115, 549, 553, 561, 563, 605, 633, 697, 699, 705, 713, 1989, 2413, 2461, 2477
Li, L.F. 2599
Li, M. 593, 621
Li, M.-R. 1671
Li, N. 2513, 2517
Li, Q. 424, 1495, 2081
Li, R.K. 917
Li, S. 1553
Li, T. 1783
Li, T.F. 1239
Li, T.W. 1543, 2067, 2101, 2241
Li, W.H. 2515
Li, X. 1523, 1553, 1663
Li, X.-G. 619, 701, 839, 841, 989, 1021, 1033, 1155, 1159, 1341, 1945
Li, X.J. 713
Li, Y. 625, 653, 721, 735, 737, 739, 975, 1057, 1087, 1097, 2093
Li, Y.J. 551
Li, Y.P. 777
Li, Y.Y. 2599
Li, Z. 1673
Li, Z.-R. 1447
Li Xigui 969, 1283
Lian, G. 1537, 2409
Lian, G.J. 551, 645, 693, 1015
Liang, J.K. 763, 769
Liang, K.F. 1247
Liang, R. 1341
Liang, W.Y. 981, 1565
Liarokapis, E. 583, 1051
Licci, F. 53
Licheron, M. 1035
Lichtenberg, F. 995
Liebich, W. 2321
Liechtenstein, A. 1739
Limonov, M.F. 1029, 1915
Lin, C.L. 2233, 2383
Lin, C.T. 491
Lin, C.W. 1407
Lin, F.F. 1143
Lin, F.Y. 1243, 2075, 2333
Lin, H.Q. 1871, 1875
Lin, J. 787
Lin, J.U. 1329
Lin, J.-Y. 1407
Lin, L. 533, 681, 2633, 2651
Lin, Q.M. 375, 1245
Lin, W. 1189, 2651
Ling, D.C. 73
Ling, X.S. 2191
Linzen, S. 571
Liou, S.H. 919, 1153, 1213
Litzkendorf, D. 2645
Liu, C.H. 547
Liu, C.-J. 1237, 1271
Liu, F.-S. 1643
Liu, G. 1241, 2247, 2649
Liu, G.D. 755, 933
Liu, G.R. 605, 657, 2461, 2477, 2545
Liu, H.K. 467, 505, 1995, 2125, 2133, 2283, 2297, 2299, 2587, 2591, 2595, 2597, 2627, 2629, 2643

Liu, J. 931, 1515, 2301
Liu, J.R. 653
Liu, J.Z. 549, 963, 2115
Liu, M. 1275, 1649
Liu, M.L. 1273, 2515
Liu, Q.M. 843
Liu, R.S. 961, 981
Liu, S.H. 721, 723
Liu, T.J. 2513, 2515, 2517
Liu, W. 563, 605, 633, 897, 1593, 1879
Liu, W.J. 833
Liu, X. 611, 625, 973, 1021, 1263
Liu, X.H. 617, 2547
Liu, X.T. 2547
Liu, X.Y. 2529
Liu, Y.W. 699
Liu, Z. 1537
Liu, Z.-Q. 2473
Liu, Z.-X. 2119, 2185
Lobb, C.J. 2081
Lobo, R.P.S.M. 1141
Loeffen, P.W. 182
Lonzarich, G.G. 303
Lopera, W. 709, 711, 2419
López, D. 375
López-Aguilar, F. 1631
Loram, J.W. 1405, 1411
Lou, F.M. 843, 861
Lou, Z.S. 917
Louati, R. 1801, 1887
Louchet, C. 2575
Loureiro, S. 65
Loureiro, S.M. 857, 895, 899, 1167
Lu, B. 829
Lu, C.C. 1019
Lu, D.H. 995
Lu, F. 1507
Lu, G. 759, 2381
Lu, H.B. 697, 699, 2513
Lu, H.X. 2547
Lu, J. 2531, 2537, 2539
Lu, J.H. 2451
Lu, L. 1547, 1553
Lu, Q. 2409, 2641
Lu, X.Y. 853
Lüders, K. 1381
Luiz, A.M. 1529, 2491
Luke, G.M. 1355, 1465
Lundqvist, B. 1221, 1535, 1959
Lundqvist, P. 1571
Luo, H.M. 795, 805, 871, 943
Luo, J. 1395, 1933
Luo, J.L. 49
Luo, J.S. 441
Luo, P. 1033, 1341
Luo, S. 2407, 2437
Luo, X. 1593
Luo, Y.Y. 547, 1247
Luo, Z.-x. 2531, 2537
Lykov, A.N. 2019

Ma, B.K. 627
Ma, G. 2509
Ma, K. 699, 705, 713
Ma, P. 1643, 2413, 2479
Ma, Q. 2093
Ma, R. 1057, 1097
Ma, W.J. 897
Ma, Y. 2619
Macfarlane, J. 2415
Mackenzie, A.P. 1185, 1229, 1467
MacManus-Driscoll, J.L. 2613
Mader, W. 571
Madsen, J. 1089
Maeda, A. 1125, 1303, 2375
Maekawa, S. 286, 991, 1785
Maeno, Y. 206, 210, 729, 1229, 1255, 1369, 1413
Maggio-Aprile, I. 315
Magishi, K. 1115
Magnusson, J. 2369
Mahajan, A. 226
Mai, Z. 705
Maignan, A. 927, 1539
Mair, M. 1423
Mairet, V. 631, 929
Majer, D. 323, 2023, 2067, 2189, 2209
Majoroš, M. 2109, 2199, 2205
Makarshin, L.L. 1609
Maki, K. 1343, 1837, 1839, 1841
Makova, M.K. 1903
Maksimov, I.L. 1949, 2197
Maksimova, G.M. 2197
Mali, M. 1345, 1371
Malik, S.K. 807
Mallozzi, R. 252
Malozemoff, A.P. 424
Malozovsky, Y.M. 1703, 1721, 1725, 1727
Maltsev, V. 749, 773
Manabe, C. 1499
Mancini, F. 1755, 1757, 1759, 1773
Mandal, J.B. 1039, 1049, 2001
Mandal, P. 1049
Manfrinetti, P. 573
Mannhart, J. 132
Manzoor, S. 1225
Mao, Y. 611
Mao, Y.J. 617
Mao, Z. 1263, 1307
Mao, Z.-Q. 1003, 1261
Mao, Z.Q. 787, 1415
Mao, Z.W. 795
Maple, M.B. 1127, 1129, 1147, 1465, 1475, 2327
Marcenat, C. 1425, 2059
Marchet, P. 1035
Marchevsky, M. 2083, 2085
Marcon, R. 1587, 1985
Marcus, J. 955, 2049, 2207
Marezio, M. 53, 65, 857, 895, 899, 1167
Marietti, Y. 1161
Marinaro, M. 1695, 1789
Marinel, S. 501, 507
Marini, M. 1817
Mariño, A. 1253, 2277
Maritato, L. 2019
Markert, J.T. 1023
Marks, L.D. 837, 959
Marohnić, Ž. 2299
Maroni, V.A. 441
Marrè, D. 573, 679, 1183
Marti, F. 1173, 1305
Martindale, J.A. 2065
Martínez, J.C. 2303
Martini, L. 2109, 2205
Martinie, B. 1555, 1983
Marucco, J.-F. 1073, 1389
Marucco, J.F. 226, 1199
Mason, T.E. 231
Mason, T.O. 837, 959
Massiot, P. 1331
Matl, P. 2183
Matsubara, I. 945, 2561, 2579
Matsuda, A. 1169
Matsuda, M. 2105, 2631
Matsuda, N. 1489
Matsuda, Y. 391, 2073, 2221, 2429
Matsuhata, H. 1079
Matsui, H. 2565, 2581
Matsui, Y. 513, 811, 813, 835, 949
Matsumoto, H. 1755, 1757, 1759, 1773
Matsumoto, N. 1107
Matsumoto, S. 1115
Matsumoto, T. 105, 811, 1443
Matsumura, M. 1371
Matsuo, S. 1497
Matsuoka, H. 2339
Matsuoka, S. 531
Matsushita, A. 1443
Matsushita, T. 2339
Matsuura, K. 2631
Matsuura, S. 1479
Matthes, A. 1481
Matveev, A.T. 949
Maurice, J.-L. 689
Maury, F. 451
Mawatari, Y. 2071, 2255, 2355
Mayama, H. 1419
Maza, J. 2315
Mazaki, H. 2211
Medvedev, Eu. 1337
Meen, T.H. 1407
Mei, S.Z. 2483
Meier, P.F. 1639, 1795

Meijer, G.I. 77, 483, 901
Mel'nikov, A.S. 1835
Mendels, P. 226, 1389
Menéndez, N. 1001
Meng, R.L. 2553
Menovsky, A.A. 1393, 2291
Mercey, B. 647
Meriani, S. 2599
Merrin, J. 1355, 1465
Mertens, M. 867
Merton, M. 2069
Mesot, J. 1327, 1335, 1377
Metlushko, V. 355
Metlushko, V.V. 379
Mezhov-Deglin, L.P. 1285, 1903, 1921
Mezzetti, E. 2095, 2179
Michel, C. 41
Michishita, K. 1441
Micklitz, H. 1941
Migliori, A. 1077
Miguélez, F. 2583
Mihailovic, D. 186
Mihalisin, T. 2233
Mika, K. 453, 2613
Mikhailova, D.A. 61
Mikhailova, Ju.V. 1677, 1715, 1761
Mikhalev, K. 1337, 1347, 1361
Mikheenko, P.N. 877, 1991, 2621
Mila, F. 1117
Milius, D.L. 232
Miller Jr, J.H. 1515
Minagawa, Y. 581
Minami, H. 1041, 1193, 1931
Minetti, B. 2095, 2179
Mirmelstein, A. 1335, 1399
Mirza, K.A. 1405, 1411
Mishima, A. 1769
Mishra, P.K. 2027, 2043
Mitamura, T. 2105
Mitberg, E. 1335
Mitchell, B.J. 815, 1085
Mitchell, J.F. 97
Mito, T. 1365, 1367
Mitsen, K.V. 1101, 1179
Miyagawa, K. 2063
Miyakawa, N. 1517, 1519
Miyamoto, S. 715, 1409
Miyashita, S. 2057
Mizuno, Y. 991, 1785
Mizutani, U. 809
Mochiku, T. 475, 1599, 2221, 2423, 2429
Mogilko, E. 2349
Mößle, M. 2435
Molinié, P. 741, 1937
Molinski, R. 77, 483, 901
Moliński, R. 2161
Momono, N. 1499
Monceau, P. 387
Monnereau, O. 1161
Monod, P. 1013, 1357
Monot, I. 501, 507
Mook, H.A. 231
Moon, S.-H. 663, 677
Morales, A. 637
Morán, E. 781, 1001
Moreo, A. 1729, 1867
Morgan, D.C. 1467
Mori, K. 1105, 1277
Mori, N. 2029
Môri, N. 153, 791
Moriga, T. 765
Morikawa, K. 1863
Morishita, T. 555, 587
Moriwaki, Y. 643
Morozov, N. 323, 2045
Moshchalkov, V.V. 379, 1567
Moskalenko, V.A. 1717, 1719
Moskvin, A.S. 1391, 1735, 1807, 1813
Mosqueira, J. 1181, 1539
Mota, A.C. 1301
Motida, K. 1063
Motohashi, T. 515
Motoi, Y. 673
Motoyama, N. 153, 1119, 1121, 1323
Moudden, A.H. 1375
Moudden, H. 493
Mouloud, S. 1883
Moussy, J.B. 162
Mück, M. 407
Müller, K.A. 3, 190, 202
Müller, P. 2431, 2435
Mun, M.-O. 2061
Munakata, F. 1111
Murakami, H. 979
Murakami, M. 371, 503, 1043, 2137, 2141, 2143, 2237, 2295
Muralidhar, M. 503
Murashov, V. 437
Murashov, V.A. 505
Murayama, C. 2029
Murayama, N. 1589
Murzin, V.N. 1583, 1585
Mutka, H. 1377
Muto, H. 577, 1589
Muzychka, A. 1335
Mygind, J. 2459
Myoren, H. 2467

Nachumi, B. 1355, 1465
Nagai, H. 595
Nagaosa, N. 283
Nagarajan, R. 69
Nagase, S. 1591
Nagashima, K. 2237
Nagata, A. 853
Nagata, R. 707
Nagata, S. 1107
Nagata, T. 153, 997, 1115
Naito, M. 965
Naito, T. 691, 2117
Nakabayashi, I. 765
Nakajima, K. 2467
Nakamura, A. 1519
Nakamura, F. 1083, 1151
Nakamura, M. 81
Nakanishi, K. 643, 971, 977
Nakanishi, T. 153, 1767
Nakano, S. 811
Nakano, T. 1499
Nakao, K. 1955
Nakatsuji, S. 729, 1895
Nakayama, Y. 2015
Nakazawa, Y. 1897, 2013
Nameki, H. 993
Nan, H.-Y. 1043
Nappi, C. 2011
Nara, K. 1441
Narlikar, A.V. 779, 807
Naumov, N.G. 1099, 1137
Navarro, O. 819, 1745, 1825
Navau, C. 2653
Nazarenko, A. 1729, 1867
Nechiporenko, I.N. 1427
Nestler, B. 519
Nevriva, M. 439
Newman, D.J. 1815
Newrock, R.S. 2081
Nguyen van Huong, C. 2305
Nhien, S. 511
Ni, Y. 931
Niarchos, D. 583
Nicolas-Francillon, M. 451
Nicolsky, R. 1529, 2411, 2491
Nideröst, M. 1301
Nie, J.C. 605, 657, 2461, 2477, 2489, 2545
Nie, R. 1439, 1537, 2409
Nie, R.J. 2479
Nigro, A. 709
Nikitenko, V.I. 2231
Nikolaev, E.G. 2121
Nikolaev, R.K. 1921
Nikolaichik, V.I. 443
Nishi, Y. 1753
Nishihara, Y. 481, 1605, 1975, 1981, 2041, 2257
Nishimoto, S. 1731, 1911
Nishimura, K. 1277
Nishino, J. 1297
Nishiyama, K. 1383
Nishizaki, S. 210, 1229, 1369, 1413
Nishizaki, T. 691, 2117
Nissen, H.-U. 2571
Niu, J.S. 627

Nizhankovskii, I. 1203
Noce, C. 1705, 1713
Noguchi, T. 2549
Nohara, M. 149
Noji, T. 1265
Nomura, K. 1895
Nonose, N. 715
Nordblad, P. 2369
Normand, B. 1113, 1117, 1685
Nose, M. 687
Noudem, J.G. 2623, 2625
Nováková, K. 851
Nozawa, K. 1123
Nozdrin, V.S. 1291
Nu, H.-b. 2531
Nücker, N. 995
Nunez-Regueiro, M. 817, 857
Núñez-Regueiro, M. 539, 1167
Núñez Regueiro, M.D. 987
Núñez-Regueiro, M.D. 1885
Nyhus, J. 1255

O, B.-h. 1023
Obara, E. 2071
Obara, H. 607, 2255, 2355
Oda, M. 1499
Oda, S. 2495
Odashima, S. 1861
Ogata, M. 1521
Ogawa, K. 1563
Ogborne, D.M. 849
Ogushi, T. 707
Oh, B. 663, 677
Oh, G.-Y. 1923
Oh, S. 2337
Ohira, S. 1383
Ohm, T. 955
Ohnishi, N. 935
Ohshima, E. 521, 821, 827
Ohshima, S. 935, 937, 1211
Ohsugi, S. 149, 1367, 1373
Ohta, T. 811, 911
Ohta, Y. 1731, 1733, 1911
Ohtsuka, H. 555
Oka, K. 479, 481, 1079, 1123, 1465, 1477, 1485, 1975, 1981, 2041, 2257
Okabe, Y. 403
Okada, Y. 1443
Okaji, M. 1441
Okajima, Y. 1319, 1419
Okamura, I. 1103
Okayasu, S. 2135
Okomel'kov, A.V. 2497
Okuya, M. 240, 2055, 2239
Oleinikov, N.N. 855
Oleynikov, N.N. 867
Olin, H. 1075
Ollitrault, R. 451
Olsson, E. 111, 623
Ong, C.K. 1603
Ong, N.P. 244, 2183
Onishi, M. 707
Onoda, M. 811
Onogi, T. 2169
Onuki, Y. 2043
Ooi, S. 1965, 2023, 2029, 2427
Oota, A. 2289
Orenstein, J. 252
Orlando, M.T.D. 489
Ormeno, R.J. 1467
Ortiz, W.A. 2051
Osafune, T. 1119, 1121
Osofsky, M.S. 93
Osorio, J. 671
Ospishchev, S.V. 2447
Ota, S. 1865
Ota, S.B. 799, 1865
Otabe, E.S. 2339
Otsuka, M. 2295
Ott, H.R. 1209
Otzschi, K.D. 837, 959
Oudovenko, V.S. 1739
Oussena, M. 1969, 1971, 2017, 2031, 2033, 2035, 2037, 2039
Ovchinnikov, A.S. 1391, 1813
Oviedo-Roa, R. 1745
Ozono, Y. 707

Paasi, J. 2199
Pace, S. 487, 2157, 2317
Pack, M.J. 751, 849
Paek, U. 1099, 1137, 1157
Paissan, G. 569
Pal, A.J. 1049
Palenzona, A. 573
Pallecchi, I. 679
Palles, D. 583, 1051
Palnichenko, A.V. 733, 1421
Pan, F.M. 787
Pan, S. 1097
Pan, S.-h. 1869
Pan, S.M. 2547
Panagopoulos, C. 145, 909
Panfilov, A.S. 1223
Pang, P. 2541
Pang, W. 1021
Panov, Yu.D. 1735, 1813
Paracchini, C. 1175, 2079
Parage, F. 1595
Park, J.-C. 1605
Park, J.-N. 1293
Park, J.C. 2025
Park, Y.K. 1491, 1605, 2025, 2543
Park, Y.W. 1235, 1249, 2523
Parmon, V.N. 1609
Partiot, C. 162
Paschoud, E. 1305
Pashitskii, E.A. 1195, 1843
Passner, A. 240
Patapis, S.K. 1397, 1509, 1569
Patil, N.G. 2027, 2043
Paul, D.McK. 2027, 2187
Paul, W. 2639
Paulsen, C. 955
Pavlenko, V.N. 387
Pe, T. 311
Peacock, G.B. 145, 909
Pechen, E.V. 1291
Pedersen, N.F. 423, 2459, 2503
Peerdeman, F. 665
Pelloquin, D. 927
Penc, K. 1117
Peng, H.T. 2111
Peng, J.L. 2209, 2261
Peng, N. 727
Peng, Z.Q. 697, 699
Peng, Z.S. 2103
Pennington, C.H. 2065
Penny, M. 2557
Pentegov, V.I. 1843
Pereira, L.A.A. 1529, 2411, 2491
Perestoronin, A.V. 1583, 1585
Pérez, L.A. 1745
Perin, A. 1173
Perkins, G.K. 1957, 2149, 2151, 2155, 2187, 2223, 2227, 2235
Perrin, C. 1331
Perroux, M. 951
Pessaud, S. 1035
Peters, K. 491
Petit, S. 1375
Petrov, M.I. 2447, 2449, 2453
Petrýdes, K. 893
Petukhov, V.Yu. 589
Phillips, N.E. 327
Pines, D. 273
Pinfold, S. 2017, 2035, 2037, 2039
Piskunov, Y. 1363
Piskunov, Yu. 1347, 1361
Pistolesi, F. 1817, 1821
Plakida, N.M. 1737, 1739
Plecenik, A. 1481
Plecháček, V. 2199, 2373, 2577, 2585
Plesenkova, O. 881
Poddar, A. 1299
Podlesnyak, A. 1335
Podobnik, B. 186
Poeppelmeier, K.R. 837, 959
Poissonnet, S. 2575
Pokrovskii, V.Ya. 1533, 2077
Polák, M. 2109
Polichetti, M. 2317, 2379
Pollert, E. 2585
Pomjakushin, V.Yu. 1353

Ponomarev, A.I. 1149
Ponomarev, A.N. 1353
Pop, A.V. 1257, 2365
Poppe, U. 731, 2303
Porcar, L. 2623, 2625
Porcescu, I.G. 1717, 1719
Porch, A. 395, 1597, 2527
Posazhennikova, A.I. 1847, 1849
Possanzini, C. 1985
Postrekhin, E.V. 2649
Poulakis, N. 583, 1051
Poulsen, H.F. 1089
Powell, J.R. 395, 1597
Pradhan, A.K. 2099
Pramana Dn., A.A. 1445
Prassides, K. 307
Preis, Ch. 2431
Prellier, W. 647
Prelovšek, P. 1803
Prieto, P. 671, 709, 711, 2419
Prischepa, S.L. 2019
Provost, J. 501, 507, 2319
Proyer, S. 1545, 2147
Prozorov, R. 1299, 2209, 2215, 2225
Prusseit, W. 93, 107
Puig, T. 1567
Pulzara, A. 711, 2419
Putilin, S.N. 61, 857
Putti, M. 573, 679, 1183
Puzniak, R. 1607, 1955
Puźniak, R. 1459, 2161

Qadri, S.B. 93
Qi, S.R. 1611
Qi, Y. 629
Qian, M. 591, 2539
Qian, M.C. 1625
Qian, Y. 823, 921, 973
Qian, Y.T. 795, 805, 871, 917, 939, 941, 943, 947
Qiao, G.-W. 1447, 1613, 2607
Qiao, G.W. 1295, 1449, 1615, 2111, 2601
Qiao, Z. 747
Qin, M.J. 2195, 2229, 2281, 2385
Qin, R.S. 1449
Qin, X.C. 465, 831
Qin, Y.L. 115, 563, 633
Qiu, J.-W. 649
Qiu, L. 2075, 2333, 2389, 2395
Qiu, M. 2633
Qiu, X.D. 885
Qiu, X.G. 1297
Qiu, Y.M. 1239
Qu, Z.J. 653
Quintana, G. 569

Radaelli, P.G. 65, 1081
Radousky, H.B. 745
Raffa, F. 1371
Ramakrishnan, S. 2027, 2043
Ramakrishnan, T.V. 290
Ramakumar, R. 1909
Ramallo, M.V. 1181, 1541
Ramírez-Castellanos, J. 813
Ramšak, A. 1711
Ramšak, A. 1805
Ran, Q. 1933
Rao, G. 1395
Rao, G.H. 763, 769
Rapp, Ö. 1221, 1535, 1571, 1959
Rassau, A.P. 2031, 2033
Rastogi, A. 1975
Raveau, B. 41, 647, 927
Ravikumar, G. 2027, 2043
Razzitte, A.C. 569
Recchia, C.H. 2065
Rector, J. 665
Rector, J.H. 559
Reed, R. 2415, 2485
Reeves, M.E. 93
Régnier, P. 2575
Reichardt, W. 1037
Ren, C. 611, 2047, 2075, 2333, 2389, 2395
Ren, C.-X. 649
Ren, H. 2247, 2633
Ren, H.-T. 2119, 2185
Ren, H.T. 485, 2087, 2309, 2381
Ren, X.Y. 2539
Ren, Y. 775, 2275, 2301, 2417
Ren, Y.-h. 1197
Ren, Z.F. 136, 2219
Renk, K.F. 655
Renner, Ch. 315
Repaci, J.M. 2081
Revaz, B. 315, 457, 1399, 1425
Revcolevschi, A. 140, 493, 953, 1375
Rèza, A. 773
Rial, C. 781
Ribault, M. 1505
Ribeiro, R.A. 2251
Rice, T.M. 214, 1113
Rieder-Zecha, S. 107
Riera, J. 1729, 1867
Rigamonti, A. 359
Rinderer, L. 635, 1989, 2277, 2343
Ritter, F. 879, 1437
Rives, V. 743
Rodriguez, C.O. 1619, 1621
Rodriguez, H. 2277
Rodríguez, J.E. 1253
Roger, J.P. 1025
Röhler, J. 182
Rolandi, R. 679
Romanenko, A.I. 1135, 1137, 1157
Romano, A. 1705
Romanò, L. 1175, 2079
Romano, P. 709
Romans, E. 2415
Rønnekleiv, A. 2521
Roos, J. 1345, 1371
Rosengren, A. 847
Ross, K. 57, 883
Rosseel, E. 379, 1567
Rossel, C. 136, 2219
Rouchaud, J.C. 953
Rouessac, V. 511, 2573
Roulin, M. 1399, 1401, 1425
Rozova, M.G. 61
Ruan, K.Q. 545, 843, 861, 863, 1153, 1165, 1213, 1239
Ruan, Y. 1935
Ruiz-Aragón, M.J. 1001
Rullier-Albenque, F. 1199
Rusakov, A. 1929
Rusakov, A.P. 1065, 1067
Rusakova, I. 57, 883
Russell, G.J. 865, 1955, 2107, 2331
Russo, M. 2011
Ruyter, A. 1555, 1983, 2319
Ryan, J.F. 186
Rydh, A. 1959
Rykov, A.I. 1029

Sabirzyanova, L.D. 1149
Sacuto, A. 471, 1013
Sadovskii, M.V. 1847, 1849
Sadowski, W. 1503, 1575, 1961
Saha, S. 2027, 2043
Sahni, V.C. 2027, 2043
Sahoo, B. 799
Saikawa, T. 1773
Saini, N. 1081
Saito, A. 2399
Saitoh, T. 503
Sakai, K. 535, 629
Sakai, N. 371
Sakamoto, S. 2375
Sakita, S. 1083, 1151
Salamo, G. 641
Salamo, G.J. 613, 683, 685
Salem-Sugui Jr, S. 2181
Salk, S.-H.S. 1697, 1699, 2371
Salk, S.H.S. 1707
Salvador, P.A. 837, 959
Samadi Hosseinali, G. 2147
Samoilov, A.V. 2053
Samokhvalov, A.V. 2163
Samuely, P. 2049
Samukawa, G. 937
Sanada, N. 993
Sanchez, A. 2653
Sánchez, D.R. 1939, 1941
Sánchez-López, M.M. 1631

Sandalov, I.S. 1631
Sanfilippo, S. 1951, 2313
Sarnelli, E. 2011
Sarrhini, O. 1979
Sarti, S. 1549, 1985, 2345
Sasagawa, T. 335, 2239
Sasago, Y. 977
Sasaki, T. 2009
Sashin, I. 1335
Sato, H. 965, 1233
Sato, K. 2339
Sato, M. 555, 963, 2427
Sauerzopf, F.M. 1333
Sawa, A. 607, 2071, 2255
Sawh, R.-P. 2275, 2301
Sawicki, M. 1961
Saya, T. 509
Scalapino, D.J. 157
Schäfer, E. 1191
Schafroth, S. 1639
Schaper, L.W. 683, 685
Schätzle, P. 505
Schenck, A. 1353
Schieber, M. 2213
Schilling, A. 327
Schlenga, K. 2431, 2435
Schlenker, C. 955
Schlesinger, Y. 2349
Schlueter, J.A. 1901
Schmalian, J. 1681, 1775
Schmidbauer, W. 145
Schmidt, B. 2045
Schmidt, M. 995
Schmidt, O. 1089
Schmitz, G.J. 519
Schnack, H.G. 347
Schneider, J.R. 1089
Schönherr, E. 491
Schroeder, A. 231
Schroeder, J. 2357
Schultz, P.G. 428
Schuppler, S. 995
Schweiss, P. 1059
Schwer, H. 77, 483, 901, 2161
Scott, S. 685
Sedmidubský, D. 439
Seeßelberg, M. 519
Segawa, K. 503
Ségransan, P. 227
Seidel, P. 571, 1481, 2527
Sekine, H. 1259
Sekinger, T. 1047
Sekita, Y. 1973
Sekizawa, K. 1105
Selig, H. 1339
Semenov, A.V. 1195, 1843
Semerad, R. 107
Señaris, M.A. 801
Senoussi, S. 1977, 1979, 2253, 2263
Seow, W.S. 2023
Sha, J. 797, 2417
Shabanova, N.P. 1291
Shaihutdinov, K.A. 2449, 2453
Shaked, H. 1331, 1339
Shames, A. 1331, 1339
Shamoto, S.-I. 963
Shan, Y.-Q. 2647
Shan, Y.Q. 495, 497
Shang, P. 1093
Shang Jiaxiang 1283
Shao, H.-M. 2387, 2391
Shao, H.M. 907, 913, 925, 967, 983, 1559, 2173, 2175, 2243, 2245, 2307, 2367
Shao, L.W. 967, 983, 1611
Shapiro, B.Ya. 1299
Shaposhnikova, T. 2159
Shapoval, E.A. 1655
Shaulov, A. 1299, 2209, 2215, 2225
Sheen, S.R. 73, 1403, 1407
Shelton, R.N. 745, 963, 2115
Shelushinina, N.G. 1149
Shen, B. 2499, 2505, 2507
Shen, B.G. 1295
Shen, H.M. 1053, 1061
Shen, J.-L. 1653, 1873
Shen, K. 2347
Shen, Y.Q. 2459
Shen, Z.-X. 123
Shen, Z.-Y. 2541
Sheng, S. 659
Shengelaya, A. 190
Shenoy, S.R. 1645
Sheriff, E. 2215
Shi, D. 469, 1673, 2329
Shi, D.Q. 703
Shi, J. 1853
Shi, L. 805, 1021, 1159, 1239, 1261
Shi, S.C. 2549
Shi, Y. 763, 769
Shibata, K. 707
Shibauchi, T. 1965, 2029, 2427
Shiina, Y. 1489, 1519
Shimada, D. 1489, 1517, 1519
Shimada, T. 935
Shimakage, H. 2401
Shimizu, H. 793, 1349, 1379
Shimomura, M. 993
Shimoyama, J. 240, 335, 391, 1269, 1479, 2015, 2073, 2135, 2285, 2323, 2439
Shimoyama, J.-i. 515, 2055, 2239
Shin, S. 963
Shiohara, Y. 81, 447, 463, 531, 581
Shiomi, T. 1211
Shirai, M. 979
Shirane, G. 85, 170
Shlopak, N. 1613
Shlyakhtin, O.A. 855
Shpanchenko, R.V. 61
Shustov, L. 1357, 1359
Shustov, V.A. 589
Si, M.S. 2433
Si, W. 1187
Sidorov, M.A. 1735, 1813
Siebold, Th. 1181
Sigrist, M. 214
Sikora, A. 541
Silva, E. 1549, 1587, 1985, 2345
Silver, R.N. 1829
Šíma, V. 2585
Simkin, V.G. 1353
Simon, Ch. 1539, 1555, 1983, 2319
Šimundić, P. 2297
Singer, J.M. 1795
Singh, R. 779, 807
Singh, R.K. 1919
Sinh, N.H. 873
Sinkler, W. 837
Sirenko, V.A. 1427
Siri, A.S. 573, 679, 1183
Skelton, E.F. 93
Skośkiewicz, T. 1961
Slaski, M. 537
Slavkin, V.V. 2091
Sleight, A.W. 565, 651
Smina, A. 1983
Smirnov, V.P. 1915
Smith, D. 186
Smith, J.L. 2065
Smith, P. 2511
Smith, P.A. 2527
Smithy, R. 625
Smrčková, O. 851, 893
Sobolewski, R. 1227
Soh, D.W. 495
Solovjanov, S. 1387
Somekh, R.E. 1569
Song, H. 757, 759, 761, 1177
Song, N. 2651
Song, P.C. 783, 2609
Song, Q.X. 2433
Song, S.Y. 2543
Song, W. 2335
Song, Y. 1675
Song, Y.S. 1249
Sonier, J.E. 1465
Soreau-Leblanc, A. 741
Soret, J.C. 1555, 1983
Sorrell, C.C. 2563
Soto, J.R. 1623, 1913
Spallone, A. 2329
Sparvieri, N. 487
Spears, M. 537
Speckmann, M. 669

Spiller, D.M. 2593
Sridhar, S. 256, 1601
Srikanth, H. 256, 1601
Stachiotti, M.G. 1619
Stacy, A.M. 801
Staines, M.P. 1237
Stangl, E. 1545, 2147
Steel, D.G. 2375
Steglich, F. 240
Stein, T. 1127, 1129, 1147, 2327
Stepanov, V.A. 1473, 1501, 2403
Stepantsov, E.A. 1075
Stevens, C.J. 186
Sticher, U. 1047, 2551
Stockham, G.J. 1889
Stoll, E.P. 1639
Stoll, S. 801
Stolypina, L.Yu. 1267
Storey, B.E. 601, 603
Strach, T. 599
Strasser, T. 431, 2645
Štrbík, V. 639
Strinati, G.C. 1817, 1821
Strohm, T. 222
Su, J. 1593
Su, X. 1901
Su, X.-D. 1447, 1613, 2607
Su, X.D. 1295, 1615
Su, Y. 1935
Su, Y.J. 897
Su, Y.X. 725, 913
Su, Z.-B. 1647, 1927
Su, Z.B. 1669
Subrahmanyam, V. 1645
Subramaniam, C.K. 1271
Sudershan, Y.S. 1975
Suenaga, M. 1495
Sugano, R. 2169
Sugano, T. 643
Sugawara, H. 1479
Sugawara, K. 853
Sugiura, S. 1591
Suh, B.J. 359
Suh, J.D. 579, 667, 695, 2469, 2475
Sulpice, A. 857, 1951, 2313
Sumida, M. 531
Sumioka, H. 1269
Sun, A.M. 517, 2139, 2281, 2347, 2385, 2393, 2397
Sun, J.-J. 1989
Sun, K. 833
Sun, P. 627
Sun, X. 839, 841, 1159, 1341
Sun, X.-F. 825, 989
Sun, Y. 627, 771, 1171, 1207, 2271, 2615
Sun, Y.G. 627
Sun, Y.Y. 57, 883
Sun, Z. 567
Sung, G.Y. 579, 667, 695, 1491, 2469, 2475
Suo, H.-L. 567
Surdeanu, R. 665, 2219, 2241, 2291, 2303
Suris, R.A. 1333
Susła, B. 1503
Sutanto, D. 2643
Suter, A. 1345
Suter, H.U. 1639
Suzuki, H. 2501
Suzuki, K. 2519
Suzuki, M. 985, 1095, 1497, 2565, 2581
Suzuki, S. 1379, 2495, 2501
Suzuki, T. 1083
Suzuki, Y. 993
Svedlindh, P. 2369
Svistunov, V.M. 979
Sýkorová, D. 851, 893
Syono, Y. 521, 821, 827, 993
Szabó, P. 2049
Szymczak, H. 1607
Szymczak, R. 773

Ta Phuoc, V. 1555, 1983
Tachiki, M. 383, 2057, 2421, 2425
Taguchi, Y. 1103
Taillefer, L. 1505, 1969, 1971, 2017, 2031, 2033, 2035, 2037, 2039, 2183
Taira, M. 985, 1095
Tajima, S. 1029, 1217
Takagi, H. 149, 1123
Takahama, M. 2287
Takahashi, H. 153
Takahashi, K. 2501
Takahashi, S. 2421, 2425
Takahashi, T. 218, 765, 997, 999
Takai, Y. 585, 595
Takane, N. 963
Takano, M. 149, 999
Takano, Y. 1105
Takasaki, S. 1895
Takase, J. 1083
Takashima, H. 1477, 1485
Takasugi, Y. 1211
Takayama-Muromachi, E. 513, 811, 813, 835, 949
Takemoto, S. 765
Takenaka, K. 1133
Takeuchi, S. 1233
Takeuchi, T. 765, 999
Takeya, H. 715, 1055, 1315, 1409
Takeya, J. 2439
Takezawa, N. 2105
Takizawa, T. 2141
Talanov, Yu. 2159
Tallon, J.L. 236, 1405
Tamasaku, K. 240
Tamegai, T. 1965, 2023, 2029, 2427
Tamura, T. 45
Tan, B.T.G. 1603
Tanabe, K. 45, 643, 887, 889, 911, 971, 977
Tanaka, I. 1957
Tanaka, K. 957, 1973
Tanaka, N. 587
Tanaka, S. 977, 985, 1043, 1385, 1497, 2561, 2565, 2581
Tanaka, T. 1367
Tanaka, Y. 509, 1477, 1485, 1521, 1855, 1857, 2211, 2405
Taneda, T. 1479
Tang, B. 1273, 1275
Tang, C. 2093
Tang, D.S. 1189
Tang, H. 1295, 1447, 1449, 1613, 1615, 2111, 2607
Tang, H.X. 2441
Tang, H.Y. 1271
Tang, J. 1443
Tang, K. 921
Tang, K.B. 939, 941, 947
Tang, N. 1933
Tang, W.H. 763, 769
Tang, Z.-M. 649
Tang, Z.-x. 2531, 2537
Tani, B.S. 441
Tanigaki, K. 307
Taniguchi, H. 2013
Taniguchi, S. 585
Tanikawa, T. 477
Tanimoto, M. 791
Tanuma, S. 733
Tanuma, Y. 1857
Tao, H. 561
Tao, H.J. 563, 633, 1507
Tao, Y. 725, 913
Taoufik, A. 1977, 1979, 2253, 2263
Tatara, G. 2105
Tatsuki, T. 45, 887, 889
Tchernyshöv, O. 1355
Tchernyshyov, O. 1809
Tebano, A. 647
Teng, H.Y. 1649
Teng, X.K. 2605
Terada, N. 509, 1477, 1485, 1973
Terai, T. 2135, 2285
Terasawa, M. 2105
Terashima, T. 999
Terborg, R. 519
Testa, A.M. 487, 2157
Tetenbaum, M. 441
Theilig, Th. 1191
Theunissen, M.H. 331
Tholence, J.-L. 955
Tholence, J.L. 65, 817, 857, 951, 1167
Thomas, J.V. 2151

Thompson, J.R. 375
Thürk, M. 2527
Tian, L. 1927
Tian, M.-L. 1003
Tian, M.L. 861, 863
Tian, W. 1053, 2387, 2391
Tian, Y.C. 861
Timofeev, V.N. 875, 1533
Tirbiyin, A. 1977
Titova, S.G. 1085
Tixador, P. 2623, 2625
Tochihara, S. 2211
Togano, K. 2567
Togonidze, T.G. 1543
Toh, Y.H. 1329
Tohyama, T. 286, 991, 1785
Tokiwa, K. 509, 957, 1973
Tokiwa-Yamamoto, A. 45, 887, 889, 911
Tokumoto, M. 509, 957, 1973
Tokunaga, S. 2519
Tokunaga, Y. 1351, 1367
Tokura, Y. 218, 1103
Tokutaka, H. 473, 477
Tomitsuka, I. 1219
Tomy, C.V. 2027
Tong, M.Z. 2499, 2505
Tong, X. 1933
Tonomura, A. 339
Tornau, E.E. 847
Tornero, J.D. 1001
Torrón, C. 1539
Totty, J. 2155, 2227
Totty, J.T. 2187
Toulemonde, P. 857, 895, 899, 951
Tournier, R. 461, 525, 539, 1951, 2313, 2623, 2625
Toyoda, K. 473
Tranquada, J.M. 166
Trapper, U. 1779
Trequattrini, F. 1429, 1453, 1457
Tretyakov, Y. 881
Tretyakov, Yu.D. 867
Tripodi, P. 2157, 2329
Tróchez, J.C. 1939, 1941
Trodahl, H.J. 1237
Trokiner, A. 1357, 1359, 1361
Tröll, J. 407
Trushin, Yu.V. 1333
Tsay, H.L. 1407
Tsay, Y.N. 1495
Tseng, J.L. 73
Tseng, T.Y. 1573
Tsubakino, H. 2105
Tsuboi, T. 1303
Tsuchimoto, M. 1863, 2631
Tsuchiya, Y. 2375
Tsuda, N. 1489, 1517, 1519
Tsuei, C.C. 4, 375
Tsuji, S. 1107, 1343
Tsukada, I. 687
Tsukamoto, A. 971
Tsukamoto, T. 957, 1973
Tunstall, D.P. 1889
Tyler, A.W. 1185, 1467

Uchida, S. 12, 153, 240, 1119, 1121, 1133, 1217, 1323, 1355, 1365, 1465
Uchinokura, K. 687, 977
Ueda, K. 2105
Uehara, M. 153, 997, 1115
Uemura, Y.J. 194, 1355, 1465, 1809
Uen, T.M. 575, 1493, 1531, 1573
Ueno, K. 945, 2561, 2579
Ueno, S. 1477, 2405
Ugolkova, E.A. 1723, 1905, 1925
Uimin, G. 1289
ul Haq, A. 869
Ullmann, J. 375
Ulrich, R.K. 683
Ulyashin, A. 1613
Uma, S. 807
Umeda, M. 957, 1973, 2071, 2255
Umeda, T. 81, 447, 531
Ummarino, G.A. 1473, 1501, 2403
Uno, Y. 1151
Unternährer, P. 2639
Urban, K. 731
Usagawa, T. 597
Usami, K. 2463
Utagawa, T. 1483
Utz, B. 107
Uwe, H. 1041, 1193
Uzawa, Y. 2401, 2465, 2549

v. Kreutzbruck, M. 407
Van Bael, M.J. 379, 1567
van Berkum, J. 665
van der Beek, C.J. 1953, 2021, 2101
van der Marel, D. 1023
Van Harlingen, D.J. 128
van Saarloos, W. 178
Van Tendeloo, G. 61
Varlamov, A.A. 248
Varma, C.M. 264
Varshney, D. 1831, 1919
Vašek, P. 851
Vašek, P. 2373
Vashakidze, Yu. 2159
Vasyutin, M.A. 1131
Vavilova, E.L. 1321
Veal, B.W. 1309, 2191
Vecchione, A. 487
Venkatesan, T. 2081
Vera, C.M.C. 569
Vera, O. 2069
Verdyan, A. 2165
Verkhovskii, S. 1337, 1347, 1357, 1361, 1929
Viallet, V. 1007, 1013, 1073, 1199
Victoria, M. 1305
Vidal, F. 1181, 1539, 1541, 2315
Vieillefond, P.A. 1199
Vietkin, A. 1375
Vietkine, A. 140, 493
Vikhreva, O. 2569
Villani, D. 1755, 1757, 1759, 1773
Villard, C. 2313
Villard, G. 927, 2319
Villars, G. 2575
Vinokur, V. 323
Vinokurov, A.L. 855
Visani, P. 1301
Vladimir, M.I. 1717
Vladimirskaya, E.V. 1267
Vlasko-Vlasov, V.K. 2231
Voit, J. 1747
Volchkov, N.A. 1585
Volkozub, A.V. 2031
von Zimmermann, M. 1089
Vostner, A. 2089
Vulcanescu, V. 1073

Wachter, P. 1475
Wagner, G. 1569
Wagner, J.L. 97
Wahl, A. 1539, 1555, 1983
Waki, K. 2137
Wakimoto, S. 170
Waldram, J.R. 1467
Walkenhorst, W. 2431, 2435
Walker, E. 89, 256, 315, 459, 1173, 1399, 1401, 1425, 1601, 2059, 2145
Walker, I.R. 303
Wan, L. 1261
Wang, B. 735
Wang, C. 705, 747, 1745
Wang, C.-L. 1927
Wang, C.J. 411
Wang, C.M. 767, 1069
Wang, C.Y. 843, 861, 863, 1153, 1165, 1213, 1455, 2615
Wang, F. 555
Wang, H. 735, 2525
Wang, H.B. 399, 2457
Wang, H.S. 697, 699
Wang, H.X. 871
Wang, J. 501, 507, 511, 703, 1537, 1935, 2433, 2471
Wang, J.-X. 2603, 2647
Wang, J.H. 136, 2219
Wang, J.R. 529
Wang, J.T. 2233, 2383
Wang, K.-Q. 1003
Wang, K.G. 1607

Wang, K.Q. 1153, 1415, 1417
Wang, L. 839, 1021, 1159
Wang, L.-B. 989
Wang, L.B. 863
Wang, M.J. 73
Wang, N.L. 240
Wang, Q.-H. 1967, 2353
Wang, R.-F. 2471
Wang, R.-L. 649
Wang, R.-I. 351
Wang, R.-P. 1003, 1261
Wang, R.F. 2445, 2479
Wang, R.K. 2587
Wang, R.L. 1993, 2377
Wang, R.P. 1239, 1259
Wang, S. 823, 1537, 1593, 2409
Wang, S.-G. 2473
Wang, S.D. 925
Wang, S.G. 411, 2087, 2407, 2479, 2483, 2487
Wang, S.J. 1493, 1531
Wang, S.Z. 2407, 2479, 2483
Wang, W. 2093
Wang, W.-H. 975
Wang, W.G. 2297
Wang, W.H. 495, 497
Wang, X. 2525, 2651
Wang, X.-p. 2531
Wang, X.B. 2457
Wang, X.H. 485, 2587
Wang, X.L. 467, 1561, 2125
Wang, X.P. 659, 703, 2513, 2517
Wang, X.Y. 495
Wang, X.Z. 1561
Wang, Y. 737, 739, 1057, 1097, 1395, 1799, 1933, 2093
Wang, Y.-B. 975, 1087
Wang, Y.-J. 1671
Wang, Y.-N. 2387, 2391
Wang, Y.-p. 1201
Wang, Y.-Z. 1447, 1613, 2473, 2607
Wang, Y.B. 725, 913
Wang, Y.J. 2513, 2515, 2517
Wang, Y.L. 485
Wang, Y.N. 1053, 1061, 1523
Wang, Y.Q. 2553
Wang, Y.Y. 2529
Wang, Y.Z. 1295, 1449, 1615, 2111, 2601
Wang, Z. 705, 1451, 1651, 1661, 2399, 2401, 2465, 2533, 2549, 2619
Wang, Z.D. 1525, 1527, 1967, 2353, 2441
Wang, Z.H. 1993, 2131, 2217
Wang, Z.Y. 1091
Wang, Z.Z. 1553
Wang Xizhong 969, 1283
Wanklyn, B.M. 445, 537
Warmont, F. 2319
Warthmann, R. 557, 2311
Watanabe, H. 1105
Watanabe, K. 2009, 2287
Watanabe, M. 963
Watanabe, T. 957, 1169, 1973, 2141
Watauchi, S. 391, 2009, 2015, 2055, 2323, 2439
Weber, H.W. 1309, 1333, 1961, 1995, 2089, 2147, 2591
Weht, R. 1621
Wei, C.-D. 2119, 2185
Wei, C.D. 2381
Wei, J.C. 2113, 2361
Wei, L. 681, 725, 913, 975, 1087
Wei, T. 2093
Wei Shoutao 969
Weinstein, R. 2275, 2301
Wellein, G. 1827, 1829
Weller, M.T. 449, 751, 849
Wellhöfer, F. 2527
Wells, B.O. 170, 1793
Welp, U. 327, 355, 2231
Wen, H. 2097, 2127, 2129
Wen, H.-h. 351
Wen, H.H. 347, 2265, 2281
Wen, J. 597, 2177
Wen, J.G. 2455
Werner, M. 1333
Westerholt, K. 2265
Wheeler, R. 375
Wijngaarden, R.J. 665, 2219, 2241, 2291
Wilke, T. 2589
Wilker, C. 2541
Willemin, M. 136
Willemsen, B.A. 256
Williams, G.V.M. 236
Williams, J.M. 1901
Wingfield, J.J. 395, 1597
Winkler, D. 623, 2293
Winterfeldt, S. 1781
Wirth, G. 2155, 2187, 2227
Wisniewski, A. 1607
Wiśniewski, A. 2161
Wissing, K. 867
Witteveen, J. 1393, 2121
Wittorff, V.W. 1287
Wolf, E.L. 1507, 1985
Wolf, T. 505, 2149, 2151, 2155, 2235
Wolf, T.H. 2227
Wolf, Th. 1089
Wolfus, Y. 1299, 2209, 2225
Won, H. 1837, 1839
Wondre, F. 445, 537
Wong, K.H. 619, 701
Woodall, P. 2527
Woodwardand, P.M. 651
Wu, B. 649
Wu, B.M. 659
Wu, D. 2525
Wu, F. 49, 533, 633, 755, 933, 1877, 1879
Wu, H. 2183
Wu, J.M. 1433
Wu, K. 2267, 2381
Wu, K.H. 575, 1493, 1531, 1573
Wu, L. 705
Wu, M. 431, 1247, 2645
Wu, M.K. 73, 1403, 1407
Wu, N.L. 57, 883
Wu, P.H. 399, 591, 2457, 2459, 2539
Wu, P.J. 605, 657, 2461, 2477, 2489, 2545
Wu, R.X. 591, 2539
Wu, S.H. 2605
Wu, W. 619, 701, 705, 839, 841, 1087, 1159, 1307, 1341
Wu, W.-B. 825, 989
Wu, W.-F. 975
Wu, X. 2249
Wu, X.-J. 45, 887, 889
Wu, X.S. 753, 787, 833, 967, 1435, 1559
Wu, X.-Z. 2603
Wu, X.Z. 529, 1607, 2605, 2611
Wu, Y. 115
Wu, Y.S. 1189
Wu, Z. 2511
Wyder, P. 1487, 2049, 2357
Wysokiński, K.I. 1701

Xi, Z. 523
Xi, Z.P. 2611
Xia, Z.J. 1015
Xiang, T. 145, 1713
Xiang, X.-D. 428
Xianyu, Z. 1293
Xiao, L. 485, 2087, 2119, 2185, 2247, 2309, 2381, 2633
Xiao, M. 1673
Xiao, Y.G. 2309
Xiao, Z.L. 2363
Xie, B. 735
Xie, M. 721, 723
Xie, Y. 1691
Xie, Y.-N. 1791
Xin, H. 823
Xing, D.Y. 1649, 1853
Xing Huaimin 969
Xiong, G. 1537, 2409
Xiong, G.C. 551, 645, 693, 1015, 2267, 2407, 2529
Xiong, H.Q. 1611
Xiong, J. 1187
Xiong, J.W. 777
Xiong, Q. 613, 641, 685, 885
Xiong, S.G. 2605
Xiong, S.Y. 777, 783
Xiong, X. 705
Xiong, X.M. 547
Xiong, Y.F. 49, 897, 1325

Xiu, K. 645, 1537, 2409, 2487
Xu, B. 713, 1877, 1879, 1989
Xu, C. 1033, 1095, 1263, 2581
Xu, C.Y. 721, 723, 919, 1311
Xu, F.-Z. 2471
Xu, F.Z. 2445
Xu, G. 1171, 1207, 1263, 1307
Xu, G.-J. 1003, 1261
Xu, G.J. 787
Xu, H.-C. 601, 603
Xu, J. 593, 625
Xu, L. 621
Xu, M. 2359
Xu, Q.L. 919, 1153, 1213, 1417
Xu, S.Y. 1603
Xu, W. 737
Xu, W.W. 399, 2457
Xu, X. 921, 1241, 1307, 1431
Xu, X.-F. 567
Xu, X.-J. 1557, 2341
Xu, X.J. 1993
Xu, X.L. 645, 923
Xu, X.N. 517, 845, 2139, 2281, 2385, 2393, 2397
Xu, X.S. 1053
Xu, X.Z. 162, 631, 929
Xu, Y. 713, 1341
Xu, Z. 775, 797
Xu, Z.-a. 903, 1197
Xu, Z.Y. 1877, 1879
Xue, L.X. 2381
Xue, Y.Y. 57, 883, 885, 1243, 1245, 2553

Yakhmi, J.V. 1039, 2001
Yakubovskii, A. 1347, 1357, 1359, 1361, 1929
Yakubovsky, A. 1363
Yakushi, K. 1907
Yamada, J. 1895
Yamada, K. 85, 170, 715, 2009
Yamada, M. 1319
Yamada, N. 809, 963, 1441
Yamada, R. 2029
Yamada, S. 1591
Yamada, Y. 809, 1443
Yamaguchi, H. 1079, 1123
Yamaguchi, M. 1733
Yamaguchi, Y. 1975, 2257
Yamaji, K. 1765, 1767, 1859, 1909
Yamamoto, E. 2043
Yamamoto, H. 957, 965
Yamamoto, T. 2501
Yamanaka, A. 1363
Yamanaka, H. 1367
Yamasaki, H. 607, 2071, 2255
Yamashiro, M. 1857
Yamashita, T. 2467
Yamauchi, H. 1385
Yamaura, N. 2141
Yamaya, K. 1319, 1419
Yamazaki, H. 963
Yamazaki, M. 2141
Yamazoe, K. 1367
Yan, D.-D. 1647
Yan, S. 1241, 1395, 1431, 1439, 2247
Yan, S.L. 2281, 2433
Yanagisawa, T. 1765, 1767
Yang, B. 2525
Yang, B.C. 2103
Yang, C.-W. 2603, 2647
Yang, D. 561
Yang, D.G. 563, 633
Yang, F. 1933
Yang, G. 1093, 2007
Yang, G.-z. 1869
Yang, G.Q. 617
Yang, G.Z. 697, 699, 1273
Yang, H. 561
Yang, H.C. 1019, 1031, 1143, 1433, 1997, 2443, 2451
Yang, H.D. 1407
Yang, H.T. 563, 633
Yang, I.-S. 663, 1039, 2001
Yang, J. 1873, 2413
Yang, J.L. 763, 769
Yang, K. 2531, 2537
Yang, L. 805, 825, 863, 921, 943, 947
Yang, Q.-S. 2471
Yang, Q.S. 2115, 2445, 2615
Yang, S.Y. 1433, 1997, 2443
Yang, S.Z. 591, 2385, 2457
Yang, T. 605, 657, 713, 2413, 2461, 2477, 2489, 2545
Yang, T.J. 2113, 2361
Yang, T.R. 1019, 1031
Yang, W. 2249
Yang, W.M. 529
Yang, X. 1783, 2087
Yang, X.-b. 649
Yang, Y. 115, 533, 2555, 2557, 2559, 2593
Yang, Z.-Q. 1447, 1613, 2607
Yang, Z.Q. 1295, 1615, 2111, 2601
Yao, X. 81, 1043
Yao, X.-x. 351
Yao, X.X. 517, 845, 925, 967, 1559, 2047, 2075, 2139, 2195, 2229, 2281, 2333, 2385, 2389, 2393, 2395, 2397
Yao, Y.D. 1403
Yao, Y.S. 49, 897, 1325
Yarmis, D. 1569
Yarunin, V.S. 1925
Yashchin, E. 2213
Yasuda, M. 853
Yasuoka, H. 119, 149, 2211
Yau, J.K.F. 2563
Yau, S.-T. 467
Ye, L. 1811
Ye, M. 1267
Ye, P. 2651
Yeh, N.-C. 1943, 1947
Yeh, W.J. 2005, 2351
Yelon, W.B. 807
Yépez, E. 1623
Yeshurun, Y. 1299, 2209, 2215, 2225
Yi, B. 2249
Yi, H.R. 623
Yi, Y.-S. 1647, 1927
Yin, B. 721, 723, 777, 1189, 1311, 1877, 1879, 2103, 2309
Yin, D. 2407
Yin, D.-L. 2325
Yin, D.L. 2267, 2381
Yin, W.-G. 1689
Yokoya, T. 218, 997, 999
Yonemitsu, K. 1109
Yoo, J. 615
Yoo, J.M. 2601
Yoo, S.-I. 2137
Yoo, S.I. 371, 2295
Yoon, J. 615
Yoshida, A. 1119
Yoshida, C. 937
Yoshida, K. 210, 1367, 1369
Yoshida, Y. 585, 595
Yoshikawa, A. 515, 1269
Yoshinari, Y. 226, 1389
Yoshioka, J. 2295
Yoshisato, Y. 2501
Yoshizaki, R. 1205, 1219
Yoshizawa, H. 1055, 1315
Yotsui, K. 1119
You, L.-P. 2473
Youm, D. 609, 615, 2337
Yu, D.-P. 2473
Yu, D.P. 543, 645, 923
Yu, H. 1893
Yu, H.W. 1143, 2443
Yu, L. 1647, 1669, 1899, 1927
Yu, L.K. 2351
Yu, S.-C. 1987, 2025
Yu, W. 1177
Yu, W.-h. 1197
Yu, W.-S. 1447
Yu, Y. 517, 757, 759, 845, 2139, 2195, 2393
Yu, Z.Q. 2005
Yuan, G. 2651
Yuan, G.S. 703, 2413
Yuan, H.J. 2513, 2515
Yuan, J. 611, 617
Yuan, M.P. 617
Yuan, P. 1015

Yun, Z. 681
Yurgens, A. 2293

Zaanen, J. 178
Zachar, O. 174
Zagoulaev, S. 1357
Zaitsev, R.O. 1667, 1677, 1679, 1761
Zakharchuk, N.F. 1099, 1137
Zakharov, A.A. 1353
Zalecki, R. 1423
Zaleski, A.J. 1463
Zama, H. 555, 587
Zangger, B. 1637
Zannella, S. 2109, 2205
Zavaritsky, N.V. 2293
Zavaritsky, V.N. 1565, 2167
Zavrtanik, M. 186
Zazunov, A.L. 1045
Zegenhagen, J. 557, 599, 675, 1011, 2311
Zeldov, E. 323, 2023, 2045, 2067, 2189, 2209, 2259
Zen, R. 2269
Zeng, D.-C. 1447
Zeng, J. 739
Zeng, R. 2651
Zeng, Z. 1633
Zeng, Z.Y. 2139, 2393, 2397
Zettl, A. 1547, 1917, 2041
Zeyher, R. 1741
Zhabrev, G.I. 1291
Zhai, H.-Y. 2471
Zhai, L.-H. 567
Zhai, Z. 256, 1601
Zhang, A.W. 939
Zhang, B. 777
Zhang, C. 1295, 1447, 1613, 1615, 2111, 2525, 2601, 2607
Zhang, C.P. 529
Zhang, D.-l. 1201, 1553
Zhang, D.F. 713
Zhang, F. 611
Zhang, H. 465, 469, 499, 543, 611, 617, 771, 831, 837, 859, 879, 959, 1275, 1437, 2271
Zhang, H.Y. 1273
Zhang, J. 2635
Zhang, J.-L. 2325
Zhang, J.B. 721, 723, 1311
Zhang, J.X. 547
Zhang, J.Y. 2457
Zhang, K. 843, 861, 863, 2359
Zhang, L. 1275, 2115
Zhang, L.F. 1273
Zhang, L.H. 411
Zhang, L.W. 783, 1091, 2515, 2517, 2609
Zhang, M.-L. 1693
Zhang, P. 2249, 2527
Zhang, P.-X. 2603
Zhang, P.X. 529, 661, 1047, 1607, 2551, 2605, 2611
Zhang, Q. 523, 757, 759, 761, 775, 797, 973, 1177, 1935, 2097, 2127, 2129, 2417
Zhang, Q.-M. 2391
Zhang, Q.-r. 903, 905, 1197
Zhang, Q.-s. 2531, 2537
Zhang, Q.M. 1053, 1061, 1523
Zhang, Q.S. 2539
Zhang, S.-C. 265
Zhang, S.Y. 659
Zhang, T. 1241, 1431, 1439, 2093, 2247
Zhang, W. 1231
Zhang, W.H. 939
Zhang, W.Q. 871
Zhang, X. 753, 775, 1177, 1435, 1935
Zhang, X.-j. 1197
Zhang, X.X. 2513, 2515, 2517
Zhang, Y. 561, 693, 839, 841, 921, 973, 1021, 1033, 1155, 1159, 1171, 1263, 1307, 1341, 1663, 2481
Zhang, Y.-H. 825, 989, 1003, 1261, 1557, 1945, 2341
Zhang, Y.G. 941
Zhang, Y.H. 619, 701, 737, 787, 843, 861, 863, 919, 941, 947, 1239, 2271
Zhang, Y.Z. 563, 633, 713, 2351
Zhang, Z. 923, 1275, 2473, 2599
Zhang Jincang 969, 1283
Zhang Ying 969
Zhao, B. 1989
Zhao, B.R. 563, 633, 713, 1201, 1877, 1879
Zhao, G.-m. 190, 202
Zhao, H. 1653
Zhao, J.H. 713
Zhao, L.H. 1877, 1879
Zhao, L.Z. 721, 723, 1311
Zhao, S.P. 2445
Zhao, W.C. 843, 861, 863
Zhao, W.L. 1449
Zhao, X. 841
Zhao, X.-R. 989
Zhao, Y. 465, 831, 859, 879, 1437, 2123, 2331, 2359
Zhao, Y.D. 947
Zhao, Y.G. 777, 783, 2513, 2517, 2609
Zhao, Z. 561, 747, 931, 1057, 1187, 2097, 2127, 2129, 2249
Zhao, Z.-x. 351
Zhao, Z.X. 49, 101, 431, 495, 497, 533, 563, 633, 657, 713, 755, 897, 933, 1325, 1877, 1879, 2103, 2309, 2377, 2461, 2489, 2545
Zhao, Z.Y. 1011
Zhdanok, O. 2569
Zhdanov, Y. 1357
Zhdanov, Yu. 1337, 1361
Zheglov, E.P. 589
Zhen, G.-q. 1385
Zheng, H.L. 2605
Zheng, L. 1155
Zheng, P. 545, 2267
Zheng, Q.Q. 1625, 1627
Zheng, X.G. 985, 1095, 1497, 2565, 2581
Zheng, Z.-S. 1515
Zheng, Z.Y. 845
Zhernov, A.P. 1163
Zhikharev, V.A. 589
Zhitomirsky, M.E. 214
Zhong, F.Q. 1247
Zhou, B.L. 1449
Zhou, D. 1471
Zhou, E. 611
Zhou, E.Y. 617
Zhou, G. 469, 839, 841, 1021, 1263
Zhou, G.-E. 825, 989, 1261
Zhou, G.E. 795, 805, 1239
Zhou, J. 2457
Zhou, L. 529, 1607, 1879, 2177, 2249, 2603, 2605, 2611
Zhou, L.-W. 649
Zhou, L.W. 2649
Zhou, M.-L. 567
Zhou, Q. 2615
Zhou, S.H. 2111
Zhou, W. 743, 909
Zhou, X. 222, 1007, 1011
Zhou, Y. 2527, 2635, 2637, 2651
Zhou, Y.-l. 2531
Zhou, Y.L. 591, 697, 699, 1273, 2377, 2513, 2515, 2539
Zhou, Y.Q. 115, 545, 1415
Zhu, A. 705
Zhu, C. 1593
Zhu, D. 1893, 1899
Zhu, J. 1663
Zhu, J.-S. 989, 1945, 2387, 2391
Zhu, J.-X. 1525, 1527
Zhu, M.H. 783, 2513, 2515, 2517
Zhu, T. 2637
Zhu, W. 2527
Zhu, X.-M. 367
Zhu, Y. 713
Zhukov, A.A. 1957, 2149, 2151, 2155, 2223, 2227, 2235
Zhuo, Y. 1293, 2003
Zhuravlev, M.Ye. 1723, 1905, 1925
Zhurkin, B.G. 1583, 1585
Ziemann, P. 1191, 2363
Ziq, Kh.A. 1445
Zoli, M. 1823
Zolliker, M. 1335
Zotov, S.D. 1145
Zou, G.T. 897
Zou, S. 611

Zou, S.C. 617
Zou, Y.H. 1015
Zou, Z. 479, 481
Zou, Z.G. 1515
Zuo, F. 1901, 2261
Zuo, J. 1033
Zuo, J.P. 2385
Zuo, T.-Y. 567
Zybtsev, S.G. 1203, 1533, 2077

Physica C 282–287 (1997) cxv–cxxix
North-Holland

Analytic Subject Index

$A_{14-x}Ca_xCu_{24}O_{41}$ (A = Sr, La) 991
$A_{14}Cu_{24}O_{41}$ 773
AF correlation 1811
AFM 303, 561, 563, 679
AFM order 1783
ARPES 999
ARPES-theory 987
a-YBCO/LSCO/a-YBCO trilayer 701
a-axis orientation 607
a-axis oriented films 605
ab initio 1621, 1639
ab-plane 1285
Abrikosov vortices 343, 2503
ac linear response 2347
ac losses 2035, 2559, 2593, 2603, 2647
ac magnet 2637
ac magnetic susceptibility 2329
ac magnetization 2383
ac measurement 2379
ac method 2045
ac penetration depth 2339
ac resistivity 1161
ac response 2173, 2175, 2229, 2243, 2367, 2397
ac susceptibility 843, 867, 1553, 2035, 2037, 2047, 2099, 2181, 2245, 2271, 2307, 2333, 2347, 2365, 2381, 2389, 2395, 2567
ac susceptibility measurement 825
achievements 3
acoustic attenuation 1651
acoustic properties 1045
activated hopping 1155
activation energy 2075, 2195, 2217, 2365, 2373
$(acuO_{2.5})_2(AlTiO_3)_m$ 959
adsorption 1609
aerosol deposition 637
Ag addition 2581
Ag doped 495
Ag-doped YBCO 2419, 2647
Ag-doping 653
Ag-plating 497
Ag substrate 567
(Al,C)-1223 509
Al-doping 517
$AlSr_2YCu_2O_y$ 813
alkaline earth niobates 37
alkaline solution 721
alloy sheath 2557
α value 1431, 1439
amorphous 681, 861
amplifier 2507
analysis 679
analytic theory 1777
analytical solution 1945
Anderson model 1717
Andreev reflection 140, 1691, 1853, 2399, 2411, 2509
angle-resolved photoemission 1673
angle-resolved photoemission spectroscopy 218, 995, 997
angular dependence 2045
anisotropic 1595
anisotropic dc moment 2271
anisotropic effects 2011
anisotropic Heisenberg model 1675
anisotropic pairing 1847, 1849
anisotropic resistivity 240, 1177
anisotropic superconductor 1073, 1905
anisotropic transport 675
anisotropy 146, 319, 665, 957, 1133, 1207, 1221, 1377, 1943, 1955, 1959, 1961, 1973, 1981, 1985, 1999, 2009, 2029, 2119, 2219, 2223, 2291, 2333, 2337, 2345, 2357, 2423
anisotropy of resistivity 1753
annealing 1243, 1333, 2609
annealing condition 859
annealing effect 887
anomalous behavior 1437
anomalous magnetization 2261
antennas 2511
antiferromagnet 1113
antiferromagnetic 273
antiferromagnetic correlation 1707
antiferromagnetic domains 166
antiferromagnetic fluctuation 1341, 1689
antiferromagnetic order 303, 1371, 1403, 1697
antiferromagnetic phase 1879
antiferromagnetism 1363, 1383
apical ions 1067
apical oxygen 1039
applications 3, 399, 403, 407, 411, 415, 419, 423, 424, 428
Ar atmosphere 431
arrowhead 2007
Aslamazov–Larkin model 1537
atomic layer epitaxy 105
atoms charge states 803
attempt time 2243
attenuation 1077
attractive and repulsive Hubbard models 1703
Auger electron spectroscopy 983
B-2212 tape 2579
BBO 1041
BCS condensation 194
BCS model 1643, 1849
BCS superconductivity 1817
BCS theory 367, 1077, 1665, 1819, 1909
BEDT-TTF 1911
BKBO 985, 1271, 1929, 2457, 2501
BPBO 1037, 1929
BSCCO 252, 477, 545, 1093, 1169, 1253, 1339, 1497, 1533, 1985, 2067, 2077, 2259, 2607, 2625
BSCCO-2212 979
BSCCO/Ag tapes 2297, 2299
2212-BSCCO single crystals 1561
BSCCO small particles 1511
BSCCO thin films 2277
BSCO 1187
BSSCO/Ag composites 2349
B-site cation 687
Ba–Bi–O system 443
Ba–Ca–Cu oxides 883
Ba–K–Bi–O oxide 721
Ba–K–Bi–O superconductor 723
Ba–K–Pb–O oxide 1311
$BaBiO_3$ 1931
BaCaCuO 57
$Ba_2(Cu_{1-2z}Ni_nZn_z)_3O_{7-\delta}$ 1563
Ba-site 1605
Ba-site Pr substitution 1197
Ba substitution 775
$BaTiO_3$ 699
$BaZrO_3$ 459
band structure 995, 997, 1621, 1713
Bardeen–Stephen model 2371
barrier 2021
Bean's model 1455, 2245
Bethe ansatz 1723
Bi 583
Bi-12212 867
Bi-2201 629, 999, 1211, 1261
Bi-2212 323, 383, 387, 467, 839, 841, 843, 845, 847, 861, 863, 865, 875, 877, 893, 1211, 1301, 1441, 1495, 1955, 1965, 1975, 1981, 2007, 2009, 2025, 2029, 2041, 2055, 2073, 2125, 2135, 2143, 2159, 2221, 2279, 2285, 2293, 2319, 2331, 2375, 2421, 2423, 2427, 2429, 2561, 2583, 2639
Bi-22212 2101
Bi-2222 1265

Bi-2223 511, 851, 853, 855, 1211, 2339, 2365, 2569, 2573, 2577, 2585, 2595, 2635, 2651
(Bi,Pb)-2223 869, 879, 2199
$Bi_4(Sr_{0.75}La_{0.25})_8Cu_5O_y$ 829
Bi-2212/Ag tape 2567, 2617
Bi-2223/Ag tape 1171, 1207, 2109, 2205, 2213, 2269, 2271, 2283, 2289, 2559, 2571, 2591, 2593, 2601, 2603, 2605, 2611, 2615, 2619
Bi-2212/Ag thick film 2609
Bi-2223/Ag wire 2597, 2629
Bi-2212 crystals 437, 1161, 2167, 2187, 2233, 2241, 2431, 2437
Bi-2223O/Ag tape 2587
Bi-2212 precursor powders 2617
Bi-2201 single crystals 1307
Bi-2212 single crystals 825, 1205
Bi-2212 tape 2621
Bi-2223 tape 424, 1995, 2555, 2563
Bi-2223 wire 2627, 2643
Bi–Sr–Cu–O system 439
Bi–Sr–Cu oxides 993
Bi-based cuprate 941
Bi-based high-T_c cuprates 1499
Bi-based superconducting single crystals 1209
Bi-based superconductors 473, 1003
Bi-based system 859
Bi-based tape 2153
Bi2233 columnar defects 1305
Bi-containing cuprates 873
Bi cuprate 2613
Bi(Pb)2201 1419
Bi2223 pinning 1173
Bi″2212 single crystals 476
BiSrBaCaCuO 1303
BiSrCaCeCuO 1259
BiSrCaCuO 1155, 1159, 1281, 1341
BiSrCaCuO crystals 989
$Bi_2Sr_2Ca_1Cu_2O_y$ single crystals 1589
BiSrCaLuO 1125
BiSrCaYCuO single crystals 1165
$Bi_2Sr_2CaCu_2O_8$ 1489, 1517, 1519
$Bi_2Sr_2CaCu_2O_{8+\delta}$ 1507
$Bi_2Sr_2CaCu_2O_{8+\delta}$ single crystals 1565
$Bi_2Sr_2CaCu_2O_{8+\delta}$ superconductor 1613
$Bi_2Sr_{2-x}La_xCuO_6$ 561, 563
$Bi_2Sr_{2-x}La_xCuO_{6+\delta}$ 633
Bi2201 system 1263
biaxial 611
biaxially oriented YBCO film 617
bicrystal 2473, 2545
bilayer coupling 1699
bilayers 1681, 1689
bipolarons 269, 1041
birefringence 1025
bismuth-containing oxides 1099
Bloch wave oscillation 2417
Bogoliubov equation 1861
Bogoliubov quasiparticles 1733
bond length 913
bond-order wave 1741
borocarbide superconductor 1055, 1315
borocarbides 69, 355, 573, 715, 717, 1317, 1327, 1409, 1633, 1939, 1941
boron–nickel compound 1667
Bose–Einstein condensation 194, 1629, 1817, 1821
Bose-glass transition 2053
bosonization 1109
bound states 1769
boundary condition 1761
buckling angle 1645
buffer layer 607, 641, 681, 695
bulk ceramic 2585
bulk high T_c superconducting magnet 1863
bulk materials 819
bulk pinning 2241
bulks combination 2633

C_{60} 737, 739, 1921
c-axis magnetoresistance 1561
$C_yCu_{1-y}Ba_2Ca_{n-1}Cu_nO_x$ 857
CDW&SDW 1233
C+implantation 2547
$(CU,X)Sr_2Ca_{n-1}Cu_nO_y$ 949
$Ca_{1-x}A_xCuO_2$ 77
$Ca_{1-x}Cd_xLaBaCu_3O_{7-y}$ 2129
Ca doping 49
CaF_2 and Cu 639
$CaLa_{1-x}Pr_xBaCu_3O_7$ 807
Ca_2PbO_4 2555, 2595
Ca_2RuO_4 206, 729
Ca-substitution 793
calorimetry 733
capping layer thickness 695
carbon 733
carbon-nanotube embedded 2279
carbonate 66, 895
carrier–phonon coupling 1061
catastrophic depinning 2257
cation 851
cation substitutions 53, 945
cavity 2527
c-axis 2041
c-axis coupling 1495
c-axis direction 2439
c-axis property 1023
c-axis resistivity 1129, 1155, 2293
c-axis tunneling 2431
c-axis V–I characteristics 2437
Cd-doping 939
c-direction 1285
Ce-based cuprates 943
Ce-doping 1933
$CeIn_3$ 303
CeO_2 buffer layer 613
$CeRu_2$ 2043
ceramic components 2639
ceramics 1593, 2091, 2569
chain cuprate 493
channeling 555
charge carrier density 747
charge correlation function 1871
charge-density distribution 971
charge-density waves 260, 955, 1345, 1747, 1827
charge distribution 981
charge dynamics 12
charge fluctuation 1391
charge order 174, 1393
charge reservoir 1243
charge–spin separation 1675, 1747
charge transfer 969, 1245
charge transfer complexes 1899
charge transfer mechanism 198
charger transport 1641
chemical control 915
chemical diffusion 621
chemical etching 629
chemistry 27
circular defect 2361
circularly polarized microstrip antenna 2513, 2515
classification 453
closed loop resonator 2523
Co-doping 833
co-evaporation 107
coating 497
coefficient 831
coexistence of superconductivity and magnetic orders 73
coherence length 1459, 1463, 1973
cold deformation 2213
cold-rolling 567
collapse of shear modulus 2183
collective creep 2251, 2253, 2257
collective electromagnetic wave excitation 2429
collective modes 387, 1685, 1833, 1881
colony connection 2591
columnar defects 375, 391, 2073, 2101, 2105, 2143, 2187, 2189, 2227, 2319, 2375, 2427
columnar pinning 2169
columns 2155
combinatorial synthesis 428
compact bosons 1849
competing interaction 1811
complex polymerization method 765
complex resistivity 2183

composite cooperators method 1773
composite crystal 811
composites 2449, 2453
composition 853
composition coefficient of T_c 789
compressibility 903
conductance 140
conductance oscillations 2441
conductivity 1193, 1533, 1647
conductivity tensor 2355
content of oxygen 833
controlled melt process 2601
Cooper instabiity 1677
Cooper-pairing 979
Cooper pairs 290, 1809
cooperative pinning 2203
coplanar 2495
coplanar resonator 395
coplanar waveguide 2525
copper germanate 953
copper layers 1357
copper oxycarbonates 647
corbino disk 1227
correlated systems 1701
correlation 1437
correlation function 1675, 2409
corrosion 719
corrosion dynamics 497
Coulomb interactions 1723, 1727, 1923
Coulomb repulsion 1811
Coulomb staircase 1493, 1531
coupled bilayer cuprates 1771
coupled grain model 1577
coupled microstrip lines 2529
coupled spin chains 1661
coupling 2205
coupling of Cooper pairs 1997
coupling strength 1831
covalency 1931
crack formation 575
creep 2019
critical current 375, 424, 609, 649, 665, 1173, 1203, 1305, 1995, 2095, 2109, 2133, 2145, 2179, 2199, 2215, 2269, 2275, 2277, 2283, 2291, 2297, 2301, 2303, 2305, 2315, 2343, 2449, 2563, 2569, 2595, 2605, 2615, 2619, 2623
critical current density 495, 501, 503, 507, 649, 675, 869, 1043, 1607, 2103, 2115, 2123, 2129, 2135, 2137, 2153, 2237, 2271, 2273, 2285, 2307, 2309, 2377, 2571, 2577, 2581, 2583, 2591, 2601, 2617, 2647
critical current distribution 2299, 2559
critical field 1543, 1567, 1945
critical fluctuations 1545
critical point 260
critical scaling collapse 2071
critical state 2083, 2211, 2223, 2229, 2385
critical state model 2307, 2315, 2653
critical temperature 1189, 1199, 1333, 1693
crossover 1817
crossover picture 194
crossover temperature 227
cryocoolers 407
cryomagnetic applications 2645
crystal field 1005, 1327, 1377
crystal growth 77, 81, 89, 446, 459, 463, 473, 477, 479, 493, 537, 719, 865, 953
crystal morphology 865
crystal structure 101, 541, 565, 651, 749, 829, 889, 911, 947, 949, 1059
crystalline axes 471
crystalline electric field 1335
crystalline structure 879
crystallinity 653
crystallization 437, 863
crystals 93, 491, 927, 1979
Cu-1234 1973
$(Cu,C,B)Ba_2Ca_{n-1}Cu_nO_{2n+2+d}$ 817
Cu-NMR 1373
Cu–O chain 1785
Cu–O chain structure 791
Cu–O linkage 977
Cu–O network 829
Cu–O plane 1639
$Cu–O_2$ plane dimpling 182
$CuGeO_3$ 953
Cu K-edge 447
63 Cu NMR 1115
CuO 1383
CuO_6 octahedra 787
CuO_2 plane 937, 1643, 1783, 1833
CuO planes 2151
Cu-substituted YNiBC 1249
$(Cu_{1-x}Tl_x)$-1234 957
Cu valence 1211
cube and rotated cube texture 567
cubic phase 579
cuprates 178, 269, 283, 751, 837, 1001, 1127, 1129, 1147, 1185, 1251
1212 cuprates 1085
current density distribution 2225
current-dependence 799
current leads 2577, 2589, 2651
current limitation 2625
current relaxation 2243
cylindrical texture 609

DLC films 2457
DOS peak 1935
d-pairing 1833, 1881
3d-substitution 779
d-wave 128, 146, 1301, 1385, 1399, 1411, 1521, 1767, 1855, 1967
d-wave pairing 1645, 1659, 1845, 2353, 2405
d-wave pairing symmetry 1525, 1527
d-wave superconductivity 1505, 1741, 1837
d-wave superconductors 1419, 1673, 1857, 1861
d-wave symmetry 1649
d–I–d junction 2405
dc SQUIDs 2451, 2477, 2489
decomposition 2007
decoration 2085
decoupling 1565, 2023
decoupling parameter 1675
defect and deformation 2133
defect structure 2303
defects 97
degradation 825
delayline 2525
delocatization effect 1057
demagnetization 2385
densing of states 1291
density functional theory 1635, 1637, 1639, 1791
density of Andreev states 2411
density of states 1759, 1835, 1839, 1913
density response 1751
depairing limit 675
dependence of magnetic field 985
deposition rate 579
destruction 1949
detector 2491
detwinned crystal 2035, 2037, 2039
detwinning 1177
device 415, 2495, 2499
device applications 731
diagram technique 1719
diamagnetic peaks 2251
diamagnetic signal 483
diamagnetic transition width 841
diamond film 2547
diaorder 1849
dielectric 705
dielectric permittivity 1585
dielectric resonance 1583
dielectric resonator 1579
diffraction 1081
diffussion 1091
dilute substitution 69
dilute vortex system 319
dimagnetic dip effect 1295
dimensional crossover 347, 1299, 1661, 1983, 2005, 2261, 2265
dimensionality 1509, 1569, 1989, 1993, 1997
dip effect 2251

discrete compounds 443
dislocation density 2587
disorder 323, 1137, 1385, 1745, 1847, 2017, 2061, 2067
disorder effects 1999, 2027
disordered materials 443
dispersion relation 1707
distortion effect 1649
domain wall 1523
dopant effect 1257
dopants 869
doped 1329
doped Hubbard model 1743
doped spin ladder 1109
doped YBCO 667, 1017, 1021, 2475
doping 162, 507, 633, 783, 927, 1093, 1101, 1361, 1513
doping effect 1289, 1499, 1665
doping hole concentration 873
doping induced in-gap states 105
double Hg layer 895
double ring resonator 2543
double sided film 1461
dry-type jet mill 523
d+s superconductivity 1841
Dy-doped 2305
dynamic 1449
dynamic behavior 423
dynamic losses 2157, 2329
dynamic magnetization process 1863
dynamic study 859
dynamical scattering 829

E–J characteristics 2255, 2341
EBCO crystal 2119
EPR 1929
ESR 1889
ET,Fe(CN)SNO 1893
EXAFS 182, 725, 913, 1081
Ea–Eag chain 1117
eddy current 407
eddy current loss 2603
edge barrier 2343
effective charge 1931
effective Hamiltonian 1733
effective mass 1459, 1891
effective misalignment angle 1995
effective pinning potential 2171, 2173
elasticity 2085
electric field gradient 1363, 1619
electric resistivity 1181, 1265, 1443
electrical conductivity 735
electrical power applications 2641, 2643
electrical properties 795, 871, 2415, 2501, 2597
electrochemical oxidation 1877
electrochemical synthesis 721, 1311
electrodeposition 2575
electromagnetic absorption 1975
electromagnetic resonant absorption 2423
electron and X-ray diffraction 1089
electron beam irradiation 2467
electron correlations 299, 1735
electron density distribution 977
electron diffraction 801, 923, 1111
electron interactions 1925
electron irradiation 1199
electron microscopy 61, 101, 743, 801, 895, 909
electron-pair symmetry 1077
electron photoemission 1423
electron spin resonance 1159, 1341
electron superconductors 1513
electron tunneling 1493, 1531, 1895
electron wave 339
electron–lattice interaction 1769
electron–phonon coupling 1037, 1719, 1723, 1825
electron–phonon interaction 190, 1109, 1253, 1473, 1489, 1501, 1517
electron–phonon spectral function 1519
electronic correlation 1103
electronic density of states 1895
electronic excitations 222, 286
electronic Raman scattering 1009, 1013, 1017
electronic resonant circuit 2627
electronic spectrum 1101
electronic states 206, 1083, 1099, 1497, 1623, 1773, 1931
electronic structure 869, 961, 963, 993, 995, 1007, 1283, 1625, 1627, 1631, 1783, 1905
electronic theory 1681, 1775
electronics high T_c material 1503
electrophoresis 2563
electroresistance 1163
element doping 2177
elemental substitution effect 757
eletromagnetic anisotropy 2055
elevated oxygen pressure 541
Eliashberg equations 1501, 1701, 1923
Eliashberg function 1473
Eliashberg theory 1825
energy dispersion 999
energy gap 222, 1077, 1643, 1757, 2463
energy loss 1445
energy spectrum 1715
enhanced cooling 2637
epitaxial films 551, 619, 1291, 2217
epitaxial growth 625, 643, 2413, 2465
epitaxial multilayers 111
epitaxial thin films 2131
equilibrium magnetization 2101
Er^{31}-EPR 1349
Er-doping 1349
η-pairing 1787
Ettingshausen effect 1547
Eu_2CuO_4 1325
$(Eu_{1.5-x}Pr_xCe_{0.5})Sr_2Cu_2NbO_{10-\delta}$ 745
evaporation of BaF_2 639
evolution 1093
exact diagonalization 991, 1705, 1731, 1733, 1793, 1827, 1829, 1875, 1911
excess conductivity 1571
excess oxygen 673, 1097, 1319
excitons 1041
extended Hubbard model 1875
extended *t–J* model 124, 1685
extended van-Hove singularity 218

F doping 763, 769, 891
f-electrons 1717
FLL melting 2055
1/*f* noise 1431, 1439, 2399
FT-IR 1049
fabrication 2553
far infrared spectroscopy 1023
fault current limiter 2639, 2641, 2643
Fe doping 753, 1097
Fe-doping effect 975
Fermi level 1015
Fermi liquid 273, 1797
Fermi surface 218, 999, 1621, 1637, 1669, 1759
Fermi surface nesting 1697
fermion pairs 1797
fermion-spin theory 1687
ferroelectric 713
ferroelectric/superconductive thin films 705
ferromagnetic phase transition 1727
ferromagnetic scattering 1527
ferromagnetic spin fluctuation 214
ferromagnetism 693, 1853
field-cooled magnetization 2233
field dependence 1597
field effect 2495
field-effect transistor 2475, 2497, 2499
field-induced pinning 2137
field-sweep rate 2255, 2393
field-sweep-rate dependence 2263
filament 891
film growth 557, 559, 599
film orientation 579
film stack 2631
films 252, 583, 2215, 2335, 2337, 2361, 2377
filter 2521
finite-element method 1967
first-order supercurrent 2435
first-order transition 2045, 2057
first principles 1635, 1637
fishtail 1971

fishtail effects 2119, 2125, 2145, 2161, 2177, 2185, 2207, 2209
Fiske resonance 2459
fission 375
flat band 1773
flat surface 597
floating zone method 865, 1409
fluctuation conductivity 1509, 1569
fluctuation effects 1209, 1539, 1541, 2163
fluctuation phenomena 1543
fluctuations 248, 1221, 1299, 1399, 1533, 1547, 1571, 2119
fluctuations magnetoresistivity 1535
fluctuations resistivity 1549
fluorinated 123 61
fluorite-type block 673
flux 2019
flux creep 331, 351, 2087, 2141, 2189, 2203, 2207, 2229, 2235, 2263, 2297, 2317, 2333, 2389, 2397
flux-creep velocity 2367
flux diffusion 2593
flux dynamics 351, 1175, 1953, 2079, 2345, 2379, 2381
flux flow 2141, 2335
flux-grown crystal 481
flux imaging 2231
flux jumps 2359
flux lattice 1949, 1963
flux-lattice melting 2051
flux-line cutting 1953
flux-line lattice 2089
flux-line-lattice dynamics 1401
flux-line localization 2095, 2179
flux-line orientation 1961
flux lines 2341
flux method 469
flux noise 2369
flux penetration 1469, 2219
flux phases 379
flux pinning 339, 371, 557, 1447, 1567, 1615, 1961, 1997, 2097, 2099, 2107, 2109, 2111, 2123, 2127, 2129, 2133, 2137, 2143, 2147, 2153, 2161, 2163, 2177, 2191, 2207, 2249, 2265, 2269, 2275, 2279, 2283, 2291, 2295, 2297, 2303, 2311, 2327, 2357, 2601
flux-pinning center 547
flux quantization 4, 2485
flux quantum 2493
flux structures 2343
flux transformer 1147, 2023
flux-transformer configuration 1281
flux trapping 2201
flux visualization 1469
Fokker–Planck equation 2355
formation kinetics 511
fourth harmonics of Nd:YAG laser 577
fracture toughness 2599
fragment symmetry carrier reservoir 1693
free energy 1699
frequency dependence 2157
frustrated Kondo lattice 1885
frustration 1117
fullerene superconductor 1923
fullerenes 307, 735, 737, 739, 1915, 1917, 1919, 1925, 1927
functional Schrödinger picture 1707
fundamental excitations 1595
fundamental science 3

GBCO 1993
g-on fractional exclusion statistics 1683
(Ga,C)-1223 509
γ-irradiation 989, 2273
γ-term 1897
gap 140, 1029, 1843
gap anisotropy 294, 1895
gap states 1709
gap symmetry 1337, 1505
gapless 1367
gapless S.C. 1373
gauge field 1763
gauge field fluctuation 1683, 1699
Gd^{3+}-EPR 793, 1379
Gd–Gd exchange 1379
GdBaCuO 625
$GdBa_2Cu_4O_8$ 1403
generic relationships 749
geometrical effect 2343
Ginzburg–Landau equation 1663, 1679, 1841, 1945
glass behavior 1999
glass phase 1963
glass transition 327, 2437
glassy exponent 2393
gold cuprate 951
Gor'kov equation 1653
gradiometer 2481
grain aligned sample 2253
grain alignment 487, 2619
grain boundaries 132, 1374, 1495, 2069, 2349, 2457, 2473, 2475, 2567, 2571, 2587
grain-boundary junction 2409
grain conductivity 2283
grain coupling 2579
grain size 843, 1455
granular superconductors 1555
granular $YBa_2Cu_3O_x$ 1591
granularity 1469, 2199, 2205, 2315
great magnetic resistant 1627
Green's function 1675, 1845, 2517
growth 551
growth kinetics 461
growth mechanism 549
growth mode 561
growth rate 491
growth simulation 631
gyro system 2649

2H-$NbSe_2$ 2027
H–T phase diagram 2013
HBCCO 913, 1013
HFSC 1881
HRTEM 571, 813, 2567
HTSC 967, 1611, 1881
HTSC ceramics 1145
HTSC + CuO 2453
HTSC + metal composite 2447
HTS coil 2629
HTSC + semiconductor 2449
HTS cylinder 2635
HTS inductor 2627
HTS interconnects 683
HV TEM 875
hairpin resonator 2487
Hall coefficient 791, 1143, 1617
Hall conductivity 1209
Hall effect 240, 1149, 1179, 1183, 1185, 1195, 1197, 1201, 1203, 1205, 1213, 1215, 1227, 2321
Hall mobility 1193
Hall probe 2247, 2289
Hall properties 1191
Hall resistivity 1207
Hall voltage 2373
Hao–Clem model 2003
harmonic 2389
harmonic mixing 399
heat capacity 715, 1409, 1421
heat-capacity jump 1655
heat treatment 2295, 2575
heating program 2605
heating rates 2555
heavy-fermion theory 1885
heavy fermions 303, 1883
heavy ion 2135
Heisenberg antiferromagnet 1671
Heisenberg model 811, 1781
helium boiling 2637
heterodyne mixer 2491
heterostructures 699, 709, 711
Hg-1201 61, 97, 1081
Hg-1212 1431, 2051
Hg-1222 921
Hg-1223 1241, 1431, 2003
Hg-1234 907, 1987, 2173
Hg-2212 899
$(Hg,Cr)Sr_2CuO_{4+\delta}$ 1049
$(Hg,Pb)Ba_2Ca_2Cu_3O_{8+y}$ 967
(Hg,Tl)-2212 911
(Hg,Tl)–Ba–(Ca,Y)–Cu–O system 889
(Hg,Tl)-22($n-1$)n 46, 887

$(Hg,V)Ba_2CuO_{4+\delta}$ 1049
$(Hg,V)(La,Sr)_2CuO_{4+\delta}$ 1049
Hg-1223 cuprate 925
Hg-1223 film 643
Hg-1223 phase 891
Hg-1223 superconductor 923
Hg-1223 tape 2553
Hg–Ba–Ca–Cu–O oxide 897
$HgBa_2Ca_{n-1}Cu_nO_{8+d}$ 857
$HgBa_2Ca_{n-1}Cu_nO_{2n+2+\delta}$ 77
$HgBa_2CuO_4$ 909
Hg-based compounds 1459, 2161
Hg-based cuprates 53, 881, 885, 893, 905, 919, 927, 2243, 2245
Hg-based films 645
Hg-based superconductors 881, 903, 1007, 1167, 1819
Hg-ceramics 1381
$Hg_{0.7}Cr_{0.3}Sr_2CuO_{4+\delta}$ 2001
$Hg_{1-x}Pb_xBa_2Ca_{n-1}Cu_nO_{2n+2+\delta}$ 901
$(Hg_{0.5}Pb_{0.5})Sr_2(Ca_{1-x}Y_x)Cu_2O_7$ 961
HgTlBaCaCuO 1153
HgTlBaCaCuO films 1213
high frequency 252
high frequency cut-off 2483
high oxygen pressure 457
high oxygen pressure synthesis 809
high parallel field 1565
high power 2541
high pressure 46, 49, 53, 57, 66, 77, 153, 811, 817, 911, 957, 1189, 1443
high-pressure effects 1167, 1819
high-pressure synthesis 509, 513, 789, 827, 887, 897, 949, 1325
high pulsed current 2625
high-resolution electron micrograph 897
high T_c 1497
high-T_c junctions 1491
high-T_c superconductivity 1065, 1805
high-T_c superconductors 773, 963, 1581
high temperature 555
high-temperature superconductors 1067
high voltage 2625
high voltage generation 2627
higher harmonics 2157, 2329
$HoNi_2B_2C$ 1409
hole concentration 767, 785, 797, 859, 903, 905, 1361, 1559
hole density 1211
hole doped cuprate 4
hole doping 945
hole-pairing 1745
hole pairs 1929
hole superconductivity 1513
Holstein t–J model 1829
homoepitaxial films 587
homoepitaxy 597
homogeneity 89, 649, 2377
homogeneous 437
homologous series 46
honeycomb 1117
hopping 2341
hopping conductivity 1223
Hubbard Hamiltonian 1819
Hubbard model 1387, 1697, 1705, 1711, 1715, 1719, 1745, 1755, 1757, 1779, 1787, 1795, 1799, 1867, 1907
hydrogen loading 2329
hydrogenation 1611
hyperfine field 1363
hysteresis 1583, 2023, 2031, 2469
hysteresis loop 1979, 2245

I–V characteristics 383, 1131, 1529, 1989, 2011, 2019, 2081, 2165, 2167, 2169, 2267, 2363, 2407, 2411, 2439, 2459
I–V curves 2299
IBAD 611, 617
IBAD YSZ layers 685
IF detector 2545
IPES 993
ice rule 847
impedance match 2487
impurities 283, 459, 669, 1205, 1551, 1709, 1749, 2447
impurity effects 149, 1351, 1367, 1851, 1861
impurity scattering 236, 1837
in-plane alignment 607
in-plane anisotropy 136, 1971
in-plane conduction 1535
in-plane defects 1199
inelastic neutron scattering 166, 1055
inelastic relaxation 1429, 1453, 1457
infinite layer 483, 627, 697, 971, 1011
infinite-layer compounds 105, 1335
infinite-layer cuprates 1035
infinite-layer structure 565, 651
infrared 1019, 1031, 1141
infrared rays 635
infrared reflectance 1011
infrared reflectivity 2425
infrared spectra 1041
inhomogeneities 93
insertion loss 2525
insulator–conductor transition 1101
integrated films 713
interatomic charge distribution 977
intercalation 1265
interface 419, 693, 2587
interferometer 128
intergrain resistivity 2315
intergranular pinning 2171
intergranular stacks 1991
intergrowth 877, 897, 2621
intergrowth structure 907
interlayer coupling 1698, 2221
interlayer spacing 1987
interlayer transport 1169, 1901
internal friction 1053, 1523, 1575
interplanar coupling 797, 905
interstitial oxygen 1453, 1457, 1575
intrinsic Josephson effect 387, 1565, 2293, 2439
intrinsic Josephson junction 2431
intrinsic model 2499
intrinsic pinning 1603, 1969, 2217, 2323
iodine intercalation 1981, 2025
iodometric titration 1435
ion beam 641
ion beam assisted deposition 613
ion bombardment 1191
ion implantation 589, 1043
ion irradiation 2095, 2105, 2179
ion milling 657, 2471
ionic size effect 761
irradiation 2135, 2143
irreversibility 2099, 2131, 2211, 2223
irreversibility line 375, 1381, 1951, 1983, 2051, 2095, 2111, 2117, 2135, 2159, 2167, 2179, 2181, 2201, 2281, 2285, 2345, 2381
Ising model 453
isothermal annealing 1091
isothermal partial melting method 2579
isotope effect 198, 202, 1753, 1813, 1917

Jahn–Teller effect 190, 1623, 1807, 1813, 1927
joint doping 755
Josephson coupling loops 1295
Josephson effect 1855, 2451
Josephson junction 399, 419, 423, 731, 1589, 2401, 2403, 2407, 2409, 2417, 2433, 2443, 2455, 2459, 2461, 2467, 2469, 2471, 2473, 2475, 2477, 2479, 2485, 2489, 2491, 2493, 2503, 2533, 2545
Josephson junction network 2437
Josephson plasma 383, 391, 1599, 2421, 2425, 2427
Josephson plasma resonance 1125, 2073, 2221, 2375, 2429
Josephson plasmon 1023
Josephson vortex 1565, 2323
Josephson vortex lattice 387, 2041
junction 2413, 2415, 2419, 2453, 2509, 2519
junction array 2545
junction fabrication 2445

$(K,Ba)BiO_3$ 2207
K–T transition 1175, 1555
60 K plateau 793, 1349

2Kf and 4Kf phonons 1109
kinetic inductance 2529
Knight shift 1373, 1795
Kondo effect 1395, 1885
Kondo magnetic impurity 1525
Kosterlitz–Thouless transition 2041, 2077, 2079, 2081

LBCO 753, 755, 2257
LNCO 781
LSCO 86, 140, 231, 244, 1151, 1373, 1411, 2257, 2425
LTT phase 1079
La-214 2143, 2359
$(La,Sr)_2CuO_4$ 537
$(La_{1-x}Sr_x)_2CuO_{4-\delta}$ 2239
La–Ca–Mn–O 1231
$La_{1.8}Ba_{1.2}Cu_2O_{6.1}$ 743
$La_{2-x}Ba_xCuO_{4-y}$ 1593
LaCaBaCuO 1069
$La_{1.5}Ca_xBa_{1.5-x}Cu_3O_y$ 767
La_2CuO_4 446, 535
$La_2CuO_{4+\delta}$ 1429, 1453, 1457, 1575, 1877, 1879
La_2CuO_{4+y} 170
La deficiency 535
$La_{2-x}K_xCuO_4$ 801
LaNdBaCuO 1255
$La_{0.5}Sr_{0.5}CoO_3$ 619
LaSrCuO 1079
$La_xSr_{1-x}CuO_{2-\delta}$ 565
$La_{1.85}Sr_{0.15}CuO_4$ 1465
$La_{2-x}Sr_xCuO_4$ 533
La system 1367
ladder compound 149, 153, 157, 162, 929, 1121
ladder system 1119
ladders 1113, 1117, 1343
λ_{ab} 1301
$\lambda/2$ transmission 2487
lanthanide 849
large area films 107
large area thin films 2543
large bulk sample 2645
large dimensions 1743
large Fermi surface 1773
laser ablation 577, 601, 645, 663, 1035
laser beam profile 663
laser deposition 573, 599, 679
laser induced 1275
laser-induced voltage 1247
laser molecular beam epitaxy 697, 699
laser-zone melting 2589
latent heat 327
lateral distribution of critical current 2289
lattice constant 937
lattice deformation 2103
lattice dynamics 1037, 1457
lattice effect 1057
lattice expansion 769
lattice instability 1061, 1065, 1261
lattice misdirection 593
lattice-parameter change 465
lattice parameters 747, 935, 941
lattice softening 1087
layer-by-layer growth 603
layered cuprates 41, 959
layered 2D system 1721
layered materials 1587, 1753
layered perovskite 37, 729
layered structure 945, 1539
layered superconductors 311, 319, 1653, 1859, 1957
lead cuprate 543
levitation 461, 2649, 2653
levitation force 485, 2631
Li intercalation 1451
lift-off 2461
linear response 1845
liquid-MO source 585
liquid source 595
$Ln_{2-x}R_xCuO_4$ 491
$L$$Ni_2B_2C$ (L = Lu and Y) 1475
local charge equilibrium 1693
local electrodynamics 1281
local electronic structure 969, 1639
local fields 1581
local gap 1643
local heat treatment 635
local Jahn–Teller distortions 202
local magnetic relaxation 2209, 2225, 2247
local moment model 1511
local moments 1709, 1749
local pairing 1735, 1807, 1813
local pairs 1815
local structure 725
localized magnetic moments 1359
lock-in 2151, 2155
lock-in transition 1957
logarithmic form 1797
logic circuit 2493
longitudinal and transverse plasma modes 1599
longitudinal optical phonon 1657
Lorentz microscopy 339
losses 2623
lossless resistor 2641
low-field limit 1201
low-T specific heat 1403
low temperature 1051
low temperature neutron diffraction 1085
low-temperature structural phase transition microstructure 1071
lower critical field H_{c1} 1301, 1309, 2205
$LuBa_2Cu_3O_7$ 963
$LuNi_2B_2C$ 355
Luttinger liquid 1747

MBE 162
MC simulation 2057
MOCVD 585, 587, 595, 617, 2495
MSR 69
MTG 485
MTG ring 2649
Madelung energy 991
magnetic 1893
magnetic Bi-2212 1329
magnetic coherence 231
magnetic field 252, 343, 2619
magnetic field distribution 2629
magnetic force 2113, 2653
magnetic hysteresis loop 2309, 2393
magnetic impurity 1489
magnetic levitation 501, 1455
magnetic measurements 1311
magnetic moment 1605
magnetic ordering 1289
magnetic orientation 525
magnetic oxides 1383
magnetic pair breaking 1933
magnetic phase diagram 2043
magnetic phase transitions 1941
magnetic properties 777, 783, 1651, 1779, 1955, 2181, 2213, 2379, 2653
magnetic pulsed compaction 2569
magnetic relaxation 2195, 2215, 2239, 2249, 2251, 2253, 2259, 2263, 2267
magnetic resonance peak 232
magnetic response 2211
magnetic scattering 1163
magnetic structure 1315
magnetic superconductors 69, 1907
magnetic susceptibility 811, 877, 899, 909, 1319, 1323, 2369, 2621
magnetic transition 1319
magnetism 307, 451, 745, 779, 1633, 1939, 2107
magnetization 457, 1073, 1225, 1313, 1539, 1877, 1951, 2091, 2117, 2139, 2151, 2155, 2199, 2223, 2255, 2359
magnetization anomaly 1981
magnetization behavior 1295
magnetization current density 2281
magnetization relaxation 2197
magneto-optical 1469
magneto-optical image 2087
magneto-optical visualization 2279
magneto-optics 2219, 2291, 2311
magnetocardiogram 2481
magnetometer 2477
magnetoresistance 1219, 1223, 1227, 1231, 1537, 1647, 1901, 2017, 2031, 2317, 2337

magnetoresistivity 1183, 1221, 2365
magnetostriction 1427
magnetotelluric sounding 411
magnetotransport 1547, 2049
magnetron sputtering 549, 2413
Maki–Thompson model 1537
manganocuprate 945
many-body effects 1631
many-body theory 174
marginal Fermi liquid 264
materials equation 2325
materials evaluation 428
maximum entropy method 971
McCumber parameter 2469
measurements 1147
mechanical measurement 2391
mechanical properties 2573
mechanism 174, 467, 1629, 1729, 1775
mechanism of high-T_c superconductivity 1737, 1739
Meissner effect 915
Meissner holes 2231
melt-growth 499, 501
melt-magnetization 2237
melt-powder 499
melt process 371, 2237
melt-processed $NdBa_2Cu_3O_x$ 531
melt-processed YBCO 2211
melt-texture growth 487, 507
melt-texture process 2613
melt-textured 529, 547, 2295, 2589
melt-textured growth 2599
melt-textured materials 431
melt-textured YBCO 2123, 2127
melting 1963, 2021, 2041, 2045, 2191
melting line 2049, 2395
melting of vortex lattice 1425, 1953
melting transition 2017, 2183
mercury-based superconductors 1039
mercury cuprates 66
mercury-free precursor 645
mesoscopic 2417
mesoscopic device 2399
mesoscopic superconductivity 2441
mesoscopic superconductor 1651
metal doping 915
metal–insulator transition 1105, 1107, 1111, 1161
metallic loop 1691
metallic tape 615, 703
metastability 2027
metastable states 1157
Mg-doped LSCO 787
MgO 629
MgO particles 2097
micro-Raman scattering 1043
microbridge 2321
microcurrent path 1171
micrograph 681
microscopic phase separation 194
microscopic theory 1471
microscopy 837
microstrip 639, 2511, 2525
microstrip antenna array 2517
microstrip resonator 2539
microstructure 115, 551, 801, 853, 1095, 2295, 2455, 2473, 2553, 2579, 2583, 2587, 2597, 2599, 2617
microwave 1467, 1573, 1583, 1585, 1601, 2519, 2537, 2543
microwave absorption 69, 1589, 1599, 2159, 2421
microwave applications 395
microwave bandpass filter 2523
microwave circuits 2541
microwave conductivity 1125
microwave devices 2521
microwave electrodynamics 1581
microwave exitation 2423
microwave losses 395, 1577
microwave measurement 256
microwave performance 2515
microwave properties 1597
microwave sub-systems 2527
microwave surface impedance 1579
mid-gap states 1521
Migdal's theorem 1473
miniaturise 2511
mirage technique 659
mismatch factor 831
mixed state 1149, 1213, 1399, 1545, 2015, 2375
Mo-substitution 827
model calculation 1289
model simulation 631
modulated microwave absorption 1591
modulated structure 901, 1753
modulations 801, 847
moiré fringe 925
molecular beam epitaxy (MBE) 631, 673, 929
Möller–Plesset theory 1639
molten salt precipitation method 723
moment method 2517
momentum distribution 1731
monoclinic 1093
Monte Carlo 1865
morphology 529, 563
morphology evolution 575
Mössbauer spectroscopy 975, 1001, 1087, 1941
Mott insulator 729
Mott transition 299, 1103, 1907
moving dissipation 2269
multi-critical point 2037, 2039
multi-shielding technique 2635
multichip modules 683, 685
multifilament wire 2339, 2629
multilayers 613, 681, 683, 685, 693, 731, 849, 1143, 1627
multilayers and superlattice 741
muon localization 1383
muon-spin relaxation 307
muon-spin resonance 1353, 2069

NBCO 81, 581, 607, 2099, 2193
NBCO/PBCO/NBCO multilayers 1483
NCCO 965, 1477, 1485, 2261
NCCO crystals 2209, 2225
NL-mode 1259
NMR 119, 226, 227, 236, 299, 1107, 1331, 1337, 1339, 1343, 1345, 1347, 1361, 1363, 1365, 1381, 1385, 1389, 1393, 1671, 1771, 1889, 2063, 2065, 2121
NMR/NQR 119
NQR 1345, 1351, 1363, 1367, 1369
NS junction 1691
n-type 1379
n-type superconductor 763
nanocrystalline Cu 1615
nanostructure 523, 861, 863
nanotubes 1925
Nb 2335
Nb-doped $SrTiO_3$ 2501
Nb film 2533
NbN junctions 2549
$NbSe_2$ 2085
Nd-123 2143, 2295
NdBaCuO 469, 677, 917
$NdBa_2Cu_3O_y$ 771
$Nd_{1+x}Ba_{2-x}Cu_3O_y$ 431
$Nd_4Ba_2Cu_2O_{10}$ particle size 531
NdCeCuO film 1163
$Nd_{1.85}Ce_{0.15}CuO_4$ 1465
$Nd_{2-z}Ce_zCuO_4$ 1513
$Nd_{2-z}Ce_2CuO_4$ 1005
NdLaCeCuO 747
necking 476
negative differential region 1529
negative magnetoresistance 691
Nernst effect 1279, 2009
neutron and X-ray diffraction 833
neutron diffraction 61, 97, 781, 911, 1353, 2089
neutron irradiation 1333, 1607, 2285
neutron scattering 86, 178, 231, 232, 1045, 1327, 1335, 1375, 1685
new materials 57, 543, 647, 741, 951, 1937
new superconducting cuprate 509
next nearest neighbor transfer energy 1767
Ni 615, 1389

nickel-substituted cuprates 1271
nitrogen annealing 533
noise 2335, 2415
noise power spectrum 1433
noise temperature 2401
non-copper oxides 1937
non-copper-oxides superconductor 741
non-cuprate superconductor 218
non-destructive method 2559
non-equilibrium electrons 1015
non-equilibrium superconductivity 2411
non-Fermi liquid 260, 1711, 1799
non-Fermi liquid theory 1801
non-linear 1601
non-linear conductivity 2077
non-linear electrodynamics 252
non-linear resistivity 1303
non-linear response 2407
non-linear vortex response 1947
non-logarithmic regime 2197
non-phonon superconductivity 1667
nondestructive evaluation 407
nonlinearity 2389
normal state 231, 240, 248, 273, 279, 303, 1057, 1195, 1219, 1235, 1241, 1267, 1273, 1437, 1641, 1681, 1763, 1789, 1801
normal state energy gap 123
normal state magnetism 226
normal state properties 1887
novel crucible 89
Nozière–Vinen model 2371
nuclear magnetic resonance 359, 1103, 1105
nuclear spin lattice 1343
nuclear spin-lattice relaxation time 1105
nucleation rate 623
number of CuO_2 layers 905
number parity effect 1651
numerical analysis 2157
numerical calculation 1803
numerical methods 1705
numerical simulation 519, 1863, 2353, 2507, 2631

O and F content 513
OCMG 503
O displacement 1025
^{17}O NMR 1357, 1359
oblique field 1965
on-site Coulomb correction 1625
one-dimensional 1323
one-dimensional conduction 1133
one-hole spectral function 1785
optial response 186
optical conductivity 991, 1121, 1141, 1179, 1651, 1743, 2425
optical dilatometer 1441
optical phonons breathing mode 979
optical properties 1475
optical transitions 1723
optimization 1559, 2585
optimum doping 885
orbital occupancy 803
order/disorder 813
order parameter 132, 2403
order parameter symmetry 1655
order–disorder phase transitions 847
ordered localized holes 1067
ordering 807
ordering kinetics 1089
organic analogues 519
organic conductors 1899, 1901, 1909
organic metals 1899
organic superconductors 299, 1859, 1887, 1889, 1891, 1895, 1897, 1903, 2013, 2063
orientational order 307
oriented growth 655
orthorhombic 1617
oscillator 2531
out-of-plane conduction 1535
out-of-plane conductivity 1229
out-of-plane resistivity 1219
out-of-plane transport 1565, 2167
outgrowths 553
overdoped regime 1357, 1359
oxidation states 967
oxides 1391, 1807
oxidized system 755
oxidized treatment 765
oxy-chloride superconductor 933
oxycarbonate 41, 817, 817
oxyfluoride 449
oxygen 1609
oxygen content 437, 517, 831, 1135, 1193, 1435, 1445
oxygen deficiency 671, 1959, 2147, 2239
oxygen-deficient perovskite LaSrCuO 1123
oxygen-deficient YBCO 1613
oxygen diffusion 527
oxygen doping 1321
oxygen interaction 37
oxygen nonstoichiometry 441
oxygen ordering 1089
oxygen partial pressure 535, 771
oxygen pressure 591, 821
oxygen redistribution 657
oxygen reduced 477
oxygen stoichiometry 527, 781, 1073, 1181
oxygen treatment 771
oxygen vacancy 747, 935
oxyhaloids 1067
PBCO 49, 479, 481, 681, 809
PBSCO 777
PGO barrier 2463
PIT method 2607
PLD (pulsed laser deposition) 549, 559, 627, 1231
PMP sample 1607
$P_4W_{14}O_{50}$ 955
p–d model 1695
pair breaking 985, 1551
pair-breaking parameter 1679
pair symmetry 1477, 1897, 2403
pair transfer 1909
pair tunneling 4
pair tunneling mechanism 1851
pairing 174, 1497
pairing instability 264
pairing interaction 123
pairing mechanism 157, 2089
pairing-state symmetry 136, 1485
pairing symmetry 4, 132, 146, 1413, 1515, 1853
palladium 449
pancake 2007
paraconductivity 1569
paraelectric mechanism 1645
paramagnetic current 2371
paramagnetic Meissner effect (PME) 1295, 1297, 1511, 2201, 2369
paramagnetism 843
partial-melt process 2287, 2611
partial melting 515, 853, 891
particle size 893
particle–hole excitations 1721, 1725
partition coefficient 953
passive devices 2535
pattern formation 1095
patterned layers 685
patterned substrates 623
Pb-1212 821
Pb-3212 939
(Pb,Cu)1212 815
$(Pb,Cu,Cd)_3Sr_2(Nd,Ce)_2Cu_2O_z$ 871
(Pb,Bi)-2223/Ag tapes 2557
Pb and La-doped 1261
Pb/BSCCO tunneling junction 2435
Pb-doped Bi-2223 1091
Pb-doped Hg-1223 2175
Pb doping 901, 907
Pb-12$(n-1)n$ 46
peak effect 371, 503, 1181, 1947, 1965, 2029, 2043, 2137, 2155, 2191
Peierls *d–p* model 1769
Peierls phase transition 1061
Peierls transition 955
pellet pressure dependence 2383
Peltier effect 1247

penetration depth 146, 395, 1459, 1461, 1463, 1465, 1467, 1663, 2025
penetration of magnetic fields 3
percolation 2299
periodic superlattice 37
peritectic growth 519
peritectic recombination 525
peritectic transition 767
permanent magnets 2633
permittivity 111
perovskite 727, 731, 751, 837, 849
perovskite oxide 687
perovskite structure 1111
persistent current 1691
perturbation 1875
perturbative approach 1721, 1725
phase 1971
1212 phase 805, 943
1222 phase 805, 943
123 phase 795, 917
124 phase 795
phase anisotropy 128
phase coherence transition 1555
1212 phase cuprate 775
phase diagram 319, 335, 439, 446, 727, 1101, 1667, 1677, 1775, 1819, 2015, 2039, 2387
phase distribution 2557
phase fluctuation 290
phase noise 2531
phase order 174
phase sensitive tests 4
phase separation 170, 236, 661, 1223, 1321, 1353, 1871, 1875, 1877, 1879
phase separation unrestricted Hartree–Fock 1873
phase shifter 2519
phase slippage 2453
phase stability 441, 515, 717, 815
phase transition 733, 923, 1059, 1157, 1313, 1415, 1417, 1523, 1541, 1655, 2351, 2391
phenomenology of type II superconductors 19
phonon and electron states 1915
phonon anomaly 1055, 1063
phonon-like harmonic fluctuations 2185
phonon mechanism 1825
phonon spectroscopy 1937
phonon vibration mode 1033
phonons 1007, 1029, 1277, 1635, 1751, 1831, 1913, 1919
photoconductivity 671
photoemission 123, 965, 1867
photoemission spectrum 973, 1935
photoirradiation 1283
photon-assisted tunneling steps 2549
photoresponse 671
physical properties 713
pinning 323, 363, 1305, 1427, 1951, 2007, 2067, 2083, 2141, 2165, 2313, 2361
pinning ability 2591
pinning array 379
pinning center 1139, 2173, 2301, 2373
pinning force 2193
pinning potential 1993, 2203, 2249
planar defects 2241
planar junctions 711
planar pinning centers 2355
plasma evaporation 581
plasma frequency 1595
plasma plume collision 601
plasma spray 489
plasma treatment 1613
plasmons 1831
plastic creep 347, 2209
plastic flow 331
plastic motion 2075
point contacts 1481
point disorder 1969
polarizing light 471
polaron formation 1827, 1829
polaron hopping 1429, 1453
polaronic charge carriers 202
polarons 186, 190, 1823
polarons pairs 1429
polymerisation 307
positron annihilation 1087, 1611, 2093
positron annihilation technique 975
positron lifetime 969, 1283
powder diffraction 449, 751
powder microstructure 855
powders 855
power 2513
power amplification 2491
power dependence 1577
power handling capability 2539
Pr-1248 1407
PrBaCuO 1137, 1177
$PrBa_2Cu_3O_7$ 963
$Pr_{1-x}Ce_xBa_2Cu_3O_{7-d}$ 783
PrCeCuO 2265
Pr_2CuO_4 541
Pr doping 789, 2309
Pr ion 1293
Pr ordering 807
$Pr_{0.7}Sr_{0.3}MnO_3$ 693
praseodymium-based cuprates 805
precursor 2609
precursor powder 2595
preformed pairing 1017
preformed pairs 1215
preparation parameter 2579
pressure 817, 857, 1245, 2569
pressure dependence 1151, 1619
pressure-dependent T_c 198, 1553
pressure effect 791, 885, 903, 2029
Pr4f–O2p hybridization 757, 973
Pr4f hybridization 1133
1/8 problem 1083, 1217, 1255
processing 2597
processing condition 529
processing parameter 2651
properties 1075, 1329
protective layers 589
proximity effect 1761, 2445
pseudobinary phase diagram 2613
pseudogap 124, 226, 236, 260, 987, 1053, 1169, 1377, 1405, 1507, 1561, 1809, 1815, 1839
Pt 615
Pt addition 499
Pt or CeO_2 addition 531
pulling method 463
pulse tube cryocooler 2527
pulsed field magnetization 1863
pulsed laser deposition 575, 619, 643, 647, 667, 677, 701, 1563

quality factor 2399, 2535
quantum antiferromagnets 1661
quantum creep 2265
quantum critical point 1113
quantum fluctuation 331
quantum flux 1589
quantum hydrodynamics 1821
quantum localization 1727
quantum logic 403
quantum mixing 2549
quantum Monte Carlo 1795, 1875
quantum transport 1527
quantum tunneling 1927, 2267
quantum well 2509
quasi-optical coupling 2549
quasi-particle coherence 244
quasihomostructures 1483
quasiparticle bands 1689
quasiparticle dynamics 186
quasiparticle tunneling 1473, 1501, 2405

$(R,Pr)_2Ba_4Cu_7O_{14+d}$ 789
RBCO 371, 599, 797, 803, 2103
RBaCuNbO 1363
(RE,Y)-123 phase 2093
RE-123 superconductors 81
RE–Ba–Cu–O 447
$RE_{2-x}Pr_xCuO_{4+\delta}$ 541
RF Flash Plasma evaporation 581
RHEED 563, 597
RHEED analysis 679
RLCSJ model 2533
RSJ behavior 2475
RSrCuCaO 1363

R124 superconductor 1443
RTEM 835
RVB 283
RVB state 1669
radiation effect 2301
Raman scattering 222, 1007, 1011, 1027, 1029, 1915
Raman spectra 737, 1021, 1263
Raman spectroscopy 583, 599, 1039, 1047, 1049, 1051
random potential 2191
rare earth cuprates:123 2379
rare earth titanates 37
Rb_3C_{60} 1917
Re–Ba–Cu–O 519
Re-doped Hg1223 superconductors 1269
Re substitution 827
realing 1797
recrystallization texture 567
reentrance 1697
reentrant melting 1975
reentrant peak effect 2027, 2117
reentrant superconductivity 1315
reentrant transition 1139
regular pinning array 1567
relaxation 2115, 2139
relaxation effect 2193
relaxation function theory 1671
relaxation rate 347, 2265, 2291
relaxation time 1343
remanent magnetisation 1957
remnant field 2087, 2153
renormalization group 1109
repulsion 1609
resistance 543, 799, 2023, 2453
resistance fluctuation 1433
resistive transition broadening fluctuation 1557
resistivity 451, 619, 665, 787, 823, 1035, 1137, 1147, 1149, 1151, 1159, 1175, 1179, 1185, 1249, 1277, 1313, 1415, 1647, 1687, 1823, 1919, 2021, 2033, 2039, 2079, 2319, 2447
resistivity anisotropy 1129, 1135
resistivity measurement 2015
resonant cavity 1603
resonant tunnelling 2509
resonators 2531, 2533, 2535
reversible fluxoid motion 2339
reversible magnetization 1293
review 3
rf SQUIDs 2479, 2481
Rietveld 947
Rietveld analysis 1069
Rietveld refinement 765, 911
Rietveld structural refinements 833
rocking curve 476
rotation effect 1225
Round Robin test 2615
Ru-1212 947
Ru-based double perovskite 73
ruthenate 729
ruthenium 1677

S–I–S junction 1483, 2171, 2463
S–N–S junction 1529, 2411, 2441, 2447
S–N transition 2623
SO(5) theory 265
SQUID magnetometry 1309
SQUIDs 128, 403, 407, 411, 1043, 2471, 2483, 2485, 2487
STM 355, 557, 1479, 1503
STM/STS 1499, 1895
STS observation 315
SUEFET 695
s-wave component 2435
s-wave symmetry 140
$(Sc_{1-x}In_x)Ba_2CuO_y$ 935
scaling 1183, 2019, 2193
scaling behavior 2003, 2137, 2185, 2309
scanning 575
scanning tunneling spectroscopy 105, 2443
scattering 669, 1391, 1939
Schottky junction 2501
screened Coulomb interaction 1725
screening current density 1461
screw dislocations 907
second magnetisation peak 2259
second peak 2257
second peak effect 2239
secondary ion mass spectroscopy 621
Seebeck effect 1247
seeded growth 487
self-assembly 1095
self-consistent Born approximation 1689
self-energy 1631
self-field effects 2281
self-flux method 477, 927
self-organized criticality (SOC) 2363, 2367
self-pumping 2401
series arrays 2489
Shapiro steps 2435, 2451
sheet resistance 1187
shell model 1735
shielding effect 2635
shock compaction 521
shock-induced defect 521
short interlayer distance 2001
short-range order 835, 1781
shunted resistor 2415
Si membranes 2467
SiO_2 interlayers 685
sign reversal 1201, 1209
silver addition 2565
silver doping 845
simulation 1815, 1869
single-crystal growth 457
single crystal YBCO 1043
single crystals 153, 463, 467, 469, 471, 473, 483, 493, 555, 715, 773, 841, 901, 953, 1135, 1145, 1157, 1309, 1421, 1955, 1981, 2055, 2091, 2107, 2125, 2145, 2465
single domain 485, 505
single electron transistor 2507
single electron tunneling transistor 2505
single field 1891
single flux 403
single-particle spectrum 1911
singlet band 1387, 1659
sinter-forging process 2573
sintering 853
size dependence 1765
size effects 1005, 2077, 2081
skull-melting 719
slave boson 1683
slave-boson mean-field theory 124
slave-boson technique 1779
Sm-123 seed 505
$Sm_{2-x}Ce_xCuO_{4-y}$ 1023
small antennas 2511
Sn doped YBCO 2249
sol–gel method 511
sol–gel process 921
solid-state method 511
solitary-type 1131, 1145
soliton 1763
space charge 705
spatial dispersion 1583, 1585
specific heat 327, 1153, 1397, 1399, 1401, 1405, 1407, 1411, 1413, 1415, 1417, 1419, 1423, 1425, 1755, 1789, 1897, 1913, 2057, 2059
spectra anomalies 1001
spectral 999
spectral function 1793
spectral properties 1803
spectroscopy 1503
spin bipolaron 1777
spin-charge stripe order 1217
spin density wave 1123
spin dynamics 1375
spin excitations 124, 232, 1685
spin fluctuations 86, 170, 178, 210, 1159, 1347, 1397, 1551, 1673, 1737, 1739, 1763, 1771
spin gap 12, 119, 124, 227, 279, 283, 1003, 1115, 1117, 1153, 1165, 1307, 1341, 1343, 1365, 1749
spin ladder 493
spin-ladder system 811, 1115

spin-lattice relaxation rate 1345
spin-lattice relaxation time 1371
spin magnetic susceptibility 1695
spin order 1393
spin pseudogap 1641
spin susceptibility 273, 1337, 1781, 1801
spin triplet pairing 214
spinel compound 1107
spiral 559
spiral growth 467, 603, 2125
splayed-columnar pin 2169
splayed configuration 2105
splayed glass 1947
spray drying 881
spray pyrolysis 569, 643
sputtering 633, 703
sputtering method 597
$(Sr,Ca)_nCuO_{n+1+d}$ 627
Sr–Cu–O film 707
Sr–Nb–O 727
Sr-based 795
SrCaCuO 1115
$Sr_2Ca_{n-1}Cu_nO_y$ 513
$Sr_{2.5}Ca_{11.5}Cu_{24}O_{41}$ 153
$Sr_{14-x}Ca_xCu_{24}O_{41}$ 997
$Sr_{14-x}Ca_xCu_{24}O_{41+\delta}$ 811
$Sr_2Ca_{n-1}Cu_nO_{2n+\delta}F_{2+y}$ 513
SrCaO 1343
SrCaTiO 1111
SrCuO 1119, 1121
$SrCuO_2$ 1011, 1323
$SrCu_2O_3$ 149
$Sr_{0.73}CuO_2$ 901
$Sr_{1-x}CuO_{2-\delta}$ 651
Sr_2CuO_3 1323
Sr_2CuO_{3+x} 539
SrF_2 additive 537
$Sr_2Nd_{0.5}Ca_{0.5}Cu_2O_{5+y}F_{1+d}$ 835
SrRuO 1229, 1369
Sr_2RuO_4 206, 210, 214, 218, 729, 1413, 1713
Sr substitution 895
$SrTiO_3$ 557, 1019
stability 2649
stacked junction 387
stacked structure 423
staging 170
statistical mechanics of vortex matter 19
statistical probability 2505
step edge 2459, 2461, 2471, 2477, 2479, 2489, 2545
step-edge junction 2457
sticking coefficient 631
stoichiometry 815
strain relaxation 689
stress-field pinning 2093
stripe pattern 1233
stripe phase 1811
stripes 166, 178
stripline edge 2539
strong and weak link 1591
strong correlations 1625, 1747, 1787
strong electron correlations 1717, 1737, 1739, 1741
strong electron–phonon coupling 1063
strong links 2595
strong pinning 2101, 2189, 2203
strongly correlated electron–phonon systems 1827
strongly correlated electrons 1711, 1805
strongly correlated system 286, 1791
structural anomalies 1085
structural disorder 1337
structural distortion 839
structural homogeneity 841
structural instability 1081
structural minals 749
structural model 571
structural properties 745, 1635
structural transition 753, 1083, 1217
structure 27, 66, 97, 453, 483, 517, 533, 759, 763, 781, 807, 817, 867, 883, 899, 933, 1073, 1075
1212 structure cuprates 823
structure distortion 1263
structure model 2505
structure of vortex lattice 1953
structure refinement 917, 943
structured anomaly 1239
subgap 2431
substituted cuprates 451
substitution effect 765, 809, 1233, 1605, 2107
substitution for Cu 1267
substitutions 66, 73, 226, 447, 509, 785, 857, 873, 881, 1389, 1447, 1487, 1981
substrate effects 2277
substrate miscut angle 593
super fluidity 1821
superconducting adsorbent 1609
superconducting carrier density 1459, 1463
superconducting coil 2641
superconducting condensate 1523
superconducting fluctuations 1537, 1561, 1839
superconducting gap 1027, 1141, 1405, 1475, 1479, 1487, 1499, 1507, 1531
superconducting gap parameter 1765
superconducting materials 27
superconducting order parameter 256
superconducting properties 1447
superconducting rings 2639
superconducting transition 1545
superconducting transition temperature 1563
superconductivity 49, 769, 1069, 1185, 1325, 1451, 1515, 1633, 1909
superconductivity and magnetism 1317
superconductivity suppression 757, 1395
superconductivity theory 1701
superconductor defects 875
superconductor/ferroelectric heterostructure 111
superconductor–insulator transition 1127, 1187
superconductors 869, 1347, 1435
supercurrent distribution 2409
superexchange 1659
superfluid density suppression 1355
superlattices 687, 689, 697, 1143, 2443
supersaturation 579
(110) surface 597
surface acoustic waves (SAW) 2521
surface barriers 1301, 2259, 2379
surface impedance 1467, 1587, 1597
surface impedance Josephson 1601
surface microwave resistance 553
surface morphology 629
surface optical phonon bipolaron 1657
surface particles 577
surface pinning 2139
surface polarization 1493
surface resistance 591, 1603, 1985, 2533, 2537, 2539
surface state effects 1653
surface structure 555, 909
surface superconductivity 2201
susceptibility 525, 931, 1387, 1991, 2157
switched band-stop filter 2527
symmetry 128, 1843
symmetry of superconducting 1013
synthesis 46, 49, 57, 66, 539, 723, 743, 775, 779, 783, 805, 819, 823, 867, 871, 893, 899, 919, 921, 933, 939, 941, 951
synthesis of trilayered system 707
synthesis strategy 41

T_1 1371
T_c 1559
T_c-curve 1617
T_c-depression 669
T_c-increase 1167
T_c-suppression 761, 1071, 1851
T_c-variation 717
TBCCO 637, 639, 641
TEM 115, 623, 689, 889, 925, 2455, 2571
TEP 1259
TGA 1407
3D glass state 2005
3D-*XY*/JJA model 1999
T-linear resistivity 1307
T′ phase 1059, 1325
TSFZ method 479, 481

TSSG method 81
T*-type structure 765
2D electron system 1753
2D Fermi liquid 206
2D glass state 2005
2D melting 2167
2D nature 1987
2D pancake 311
t–J model 279, 286, 1521, 1683, 1687, 1699, 1707, 1729, 1731, 1741, 1751, 1773, 1777, 1785, 1793, 1803, 1805
t–V model 1871
tapes 877
target surface modification 663
temperature dependence 913
temperature fluctuation 1433
temperature gradient 529, 545
template layer 605
tensile test 1053
terahertz 2549
ternary RE123 503
terrace effect 1869
tetra–ortho phase transition 1613
tetragonal structure 543, 1617
tetragonal superconductor 1139
texture 507, 611, 615, 2609, 2611, 2623
texture hardening 2607
textured ceramics 2573
textured YBCO 495
textured YBaCuO 1247
thallium 849
thallium superconductors 1423
theoretical model 264, 265, 2497
theory 290, 294, 1629, 1729, 1867, 1887
thermal analysis 919
thermal annealing 2285
thermal assisted flux flow 2175
thermal conductivity 244, 1277, 1285, 1287, 1505, 1903, 1921
thermal diffusivity 659
thermal dynamic behavior 465
thermal expansion 879, 1065, 1441, 1449
thermal fluctuations 359, 2051, 2493
thermal parameters 1913
thermal processing 2555
thermal stability 869
thermal treatment 629
thermal vortex 1281
thermally activated flux creep 2115
thermally activated model 2341
thermo-mechanical treatments 2271
thermochemical data 439
thermodynamic parameter 1293
thermodynamical quantities 1705
thermodynamics 441
thermodynamics and kinetics 527
thermodynamics of d-pairing 1397
thermoelectric effects 655, 1275
thermoelectric power 1003, 1233, 1235, 1239, 1241, 1243, 1245, 1249, 1261, 1265, 1269
thermoelectric response 1273
thermoelectric voltage 2551
thermogravimetric analysis (TGA) 446, 539, 745
thermomagnetic 1547
thermopower 673, 959, 1197, 1237, 1245, 1251, 1253, 1255, 1267, 1271
thick films 489, 2561
thickness dependence 2071, 2113, 2215, 2551
thin film growth 739
thin film material 929
thin films 115, 347, 549, 553, 561, 563, 565, 569, 571, 573, 575, 577, 581, 585, 589, 591, 593, 595, 601, 603, 607, 609, 611, 613, 615, 621, 623, 625, 627, 633, 635, 637, 639, 641, 647, 649, 651, 653, 655, 657, 659, 663, 665, 669, 671, 677, 679, 697, 707, 709, 711, 731, 735, 737, 929, 965, 1025, 1047, 1075, 1203, 1227, 1297, 1509, 1545, 1563, 1569, 1577, 1835, 1869, 1883, 1945, 1949, 1977, 1983, 1993, 2011, 2071, 2103, 2113, 2165, 2197, 2219, 2311, 2321, 2363, 2385, 2413, 2419, 2425, 2433, 2521, 2523, 2537, 2539, 2541, 2631
thin superconducting films 1587
three-band Hubbard model 1789, 1873
three terminal 415
Ti–Mo Co-doping 917
Ti-doping 2241
tight-binding model 1713
tilted substrate 2551
tilting 937
time decay 2317
time dependence 2233
tip growth mechanism 545
titanate 837
Tl 1369
Tl-2201 97, 1075, 1421
Tl-2212 1983, 2121, 2433
Tl-2223 1189
$(Tl,M)(Ba,Sr)_2Ca_2Cu_3O_z$ 915
Tl-1223/Ag tape 2111
Tl-2223 ceramic 2005
Tl-1234 film 2115
Tl-1223 system 827
TlBaCaCuO 1027, 1385
TlBaCaCuO films 1275
TlBaCuO 1185
Tl-based cuprates 2273
Tl-based superconductor 1097
Tl-film 2431
$(Tl_{0.5}Pb_{0.5})Sr_2(Ca_{1-x}Y_x)Cu_2O_7$ 981
Tl-system superconductor 975, 1087
top-seeding 501
topological defects 2085
torque magnetometer 136
transformed products 883
transient electromagnetic method 2483
transient electromagnetics 411
transient heating 1449
transient spectra 1015
transition temperature 2377
translational boundary 675
transmission lines 2529
transport 269, 1219, 1251
transport and magnetic properties 1889
transport critical current density 2313
transport current 1691
transport current density 2287
transport entropy of a flux line 2009
transport measurement 1171
transport properties 132, 162, 240, 248, 557, 655, 669, 673, 707, 777, 783, 1033, 1119, 1123, 1139, 1199, 1229, 1237, 1257, 1279, 2449
trapped field 2237
trapped flux 2633
tricrystal 1515
trilayers 2465
tunability 111
tunnel junction 2465
tunneling 709, 711, 965, 1471, 1481, 1483, 1489, 1491, 1497, 1507, 1517, 1519, 1799, 2049, 2445
tunneling conductance 1857
tunneling gap measurement 1487
tunneling spectroscopy 315, 979, 985, 1477, 1479, 1485, 1855, 1883
tunneling states 1823
turbulent flux structures 2231
twin planes 1951, 2149, 2313
twinning 1093
two-chain Hubbard model 1765
two-coil method 1461
two-component superconductivity 186
two-dimensional 1797, 1891, 2001
two-dimensional Hubbard model 1767
two-dimensional *t–t′–U* model 1759
two-dimensional vortex motion 691
two-dimensionality 877, 1151, 1991, 2147, 2621
two-fluid model 1581
two-leg Hubbard model 157
two phase co-existence 833
two-stage image processing 101
type-II superconductors 319, 343, 363, 367, 1401, 2325, 2397

U(*J*) relation 2195
ultra thin films 661, 667
ultrafine particle 523

ultrafine powder 547
ultrasonic 855
ultrasonic attenuation 1593
ultrasonic velocity 1593
unconventional oxidation states 1099
unconventional pairing 210, 1703
underdoped 1377
underdoped cuprates 12, 279, 987, 1063, 1641, 1669, 1809, 2327
underdoped–overdoped 182
universal scaling law 335
unloaded quality factor 1573
untilted and tilted substrates 563
untwinned crystal 2117
untwinned single crystals 1221
upper critical fields H_{c2} 1009, 1291, 1400, 1557, 1761, 1837, 1847, 1859, 1993, 2049
uranium 2275, 2301

valence analysis 821
valence bond structure 977
valence states 1137
van Hove scenario 1649, 1665
van Hove singularity 97, 294, 1641
vanadium effect 785
variable range hopping 1127
variable-temperature X-ray diffraction 1239
variational Monte Carlo method 1765, 1767
vector magnetisation measurements 1957
velocity 1131, 1145
very high magnetic field 2059
vias 683
vibrating-reed technique 2387
vibration modes 1019
vibrational states 1623
voltage noise 1977
vortex 339, 1521, 1835, 1971, 2113, 2155, 2361, 2387
vortex chain state 1965
vortex cutting 2365
vortex-density fluctuation 1303
vortex depinning 2373
vortex dynamics 19, 335, 367, 1943, 1959, 2013, 2031, 2033, 2063, 2065, 2149, 2151, 2169, 2187, 2227, 2235, 2323, 2325, 2327, 2351, 2353, 2355, 2371, 2391
vortex entanglement 2053
vortex flow structure 2503
vortex fluctuations 1983, 1987, 2003, 2011, 2025
vortex glass 347, 351, 1425, 1989, 2047, 2061, 2395
vortex-glass transition 2015, 2071, 2131
vortex imaging 2083, 2085
vortex instability 2363
vortex interaction 319
vortex lattice 363, 379, 1977, 2183
vortex-lattice melting 323, 327, 1399, 1965, 2015, 2023, 2029, 2037, 2057, 2063, 2067
vortex-lattice melting dimensionality crossover 2331
vortex-lattice phase transition 2021, 2259
vortex-lattice state 2159
vortex liquids 331, 391, 1943, 2073, 2347
vortex loop 2163
vortex melting 2027, 2043, 2059, 2061, 2351
vortex mobility 2121
vortex motion 2105
vortex pairs 2079
vortex phase diagram 2053
vortex phase transition 1303, 2075
vortex pinning 19, 375, 1979, 2321, 2349, 2497
vortex solids 331, 1943, 2221
vortex state 1837, 1841, 2069
vortex structure 1967, 1975
vortices 311, 315, 355, 359, 1297, 1331, 1339, 1865, 2165

wave 1131
weak ferromagnetism 1321, 1325
weak inversion region 2499
weak links 653, 2213, 2383, 2561
wet-type jet mill 523
whiskers 387, 545, 875, 1533, 2077, 2561
wire fabrication 2565

XANES 447, 961, 981
XAS 963
XPS 517, 845, 989, 993, 1021, 1451
XRD 627, 689, 889, 931, 971
X-ray diffraction 901, 1079
X-ray diffraction light scattering 1045
X-ray photoelectron spectroscopy 967, 983
XY-like 1943
XY-model 1865
X-ray characterization 93
X-ray diffraction 619, 701, 745, 753, 787, 819, 825, 839
X-ray Rietveld analysis 759
X-ray scattering 655

Y-123 527, 1279, 2359
Y-124 1619, 2565, 2581
Y-247 527
Y-211 substrate 505
Y–Cu–Se 931
Y–Pr–Ba–Sr–Cu–O 759
YBCO 93, 107, 182, 256, 327, 419, 461, 465, 471, 485, 489, 497, 499, 505, 507, 525, 529, 547, 549, 551, 553, 555, 557, 559, 567, 569, 571, 575, 577, 579, 585, 587, 591, 593, 595, 603, 607, 609, 611, 613, 615, 621, 623, 653, 657, 659, 661, 665, 669, 671, 679, 681, 683, 685, 689, 693, 699, 703, 719, 785, 793, 799, 833, 1035, 1047, 1051, 1089, 1141, 1225, 1287, 1331, 1349, 1351, 1425, 1439, 1445, 1477, 1485, 1491, 1549, 1573, 1605, 1621, 1969, 2009, 2011, 2017, 2031, 2033, 2035, 2037, 2039, 2047, 2059, 2065, 2071, 2089, 2091, 2097, 2117, 2139, 2141, 2145, 2159, 2165, 2177, 2237, 2247, 2253, 2275, 2301, 2305, 2309, 2311, 2321, 2333, 2337, 2363, 2383, 2389, 2413, 2457, 2459, 2463, 2469, 2495, 2513, 2515, 2519, 2523, 2547, 2589, 2633
YBCO 20% Ca 1405
YBCO crystals 1299, 2149, 2227, 2235
YBCO disk 2255
YBCO fiber 2287
YBCO films 1191, 1273, 2479
YBCO/LCMO multilayer film 691
YBCO/PBCO multilayer 1989
YBCO/PBCO/YBCO trilayer structure 709
YBCO/PYBCO superlattice 1997
YBCO particle 521, 523
YBCO single crystals 1597
YBCO tapes 2575
YBCO thin films 1537
YBaCuO 1237, 1289, 1295, 1333, 1389
$YBa_2Cu_3O_{7-\delta}$ 1615
$YBa_2Cu_3O_7$ 1025
$YBa_2Cu_4O_8$ 457
$YBa_2Cu_3O_{7-\delta}/Pr_{0.5}Y_{0.5}Ba_2Cu_3O_{7-\delta}$ superlattice 1031
$YBa_2Cu_3O_{7-x}/SrTiO_3$ bilayers 695
$YBa_2Cu_3O_{7-x}$ superconducting films 605
YBa_2Cu_3O thin films 1603
Y-based compound 1401
Y-based cuprates 2357
Y-based superconductors 761, 973
Y(Ho)BCO 2645
YNSCMO 1033
$YNiB_2C$ 1451, 1935
YNi_2B_2C 725, 1475, 2061
$Y(Pd_{1-x}M_x)_2B_2C$ 717
Y211 phase 499
Y211 phase particles 2127
$Y_{1-x}Pr_xBa_2Cu_3O_{7-\delta}$ 1395
$Y_{1-x-y}Pr_xCa_yBa_2Cu_3O_7$ films 1015
$(Y_{1-x}R_x)Ni_2B_2C$ 1933
YSZ buffer 617
YSZ buffer layer 611, 613, 703

YSZ single crystals 609
$YSr_2Cu_3O_y$ 515
Y substitution 851
yttrium cuprate 1077
zero-bias anomaly 1481, 1493, 1525
zero-bias conductance peaks 1857
zero energy states 1857
zinc 1389
zinc doped 1287
zinc subsititution 1355
Zn doping 12, 1411, 1553

www.ingramcontent.com/pod-product-compliance
Ingram Content Group UK Ltd.
Pitfield, Milton Keynes, MK11 3LW, UK
UKHW050418051026
15470UKWH00022B/1729

9 780444 828354